Machining and CNC Technology

Machining and CNC Technology

Michael Fitzpatrick

Boston   Burr Ridge, IL   Dubuque, IA   Madison, WI   New York   San Francisco   St. Louis
Bangkok   Bogotá   Caracas   Kuala Lumpur   Lisbon   London   Madrid   Mexico City
Milan   Montreal   New Delhi   Santiago   Seoul   Singapore   Sydney   Taipei   Toronto

Higher Education

MACHINING AND CNC TECHNOLOGY

Some ancillaries, including electronic and print components, may not be available to customers outside the United States.

This book is printed on acid-free paper.

1 2 3 4 5 6 7 8 9 0 VNH/VNH 0 9 8 7 6 5 4

ISBN 0–07–825090–0

Publisher, Career Education: *David T. Culverwell*
Publisher, Trades and Engineering Technology: *Thomas E. Casson*
Managing Developmental Editor: *Jonathan Plant*
Developmental Editor: *Patricia Forrest*
Marketing Manager: *Lynn M. Kalb*
Senior Project Manager: *Kay J. Brimeyer*
Production Supervisor: *Kara Kudronowicz*
Senior Media Project Manager: *Sandra M. Schnee*
Media Technology Producer: *Janna Martin*
Cover Designer: *Rick D. Noel*
Cover Photography Courtesy of: *Haas Automation, Oxnard, California*
Senior Photo Research Coordinator: *John C. Leland*
Supplement Producer: *Brenda A. Ernzen*
Compositor: *The GTS Companies/York, PA Campus*
Typeface: *10/12 Stone Serif*
Printer: *Von Hoffmann Corporation*

Cover Dedication:
Haas Automation stands out in their efforts to support and invest in technical education. With several levels of involvement, Haas supports better manufacturing through full and partial partnerships with progressiveschools. Through their generosity they make it possible to upgrade machining technologies here in North America and worldwide. My fellow instructors and I salute Haas Automation and Scott Rathburn—Marketing Manager and Editor of their own publications, "CNC Machining."

The credits section for this book begins on page 1050 and is considered an extension of the copyright page.

Library of Congress Cataloging-in-Publication Data

Fitzpatrick, Michael, 1945–
 Machining and CNC technology / Michael Fitzpatrick. — 1st ed.
  p. cm.
 Includes index.
 ISBN 0–07–825090–0 (alk. paper)
 1. Machining—Handbooks, manuals, etc. 2. Machine-tools—Numerical control—Handbooks, manuals, etc. I. Title: Machining and computer numerical control technology. II. Title.

TJ1165.F54   2005
671.3'5—dc21   2001040125
                               CIP

www.mhhe.com

While there were countless others along the way, these four made all the difference in my career and life. Without them I question whether this book would have been.

To Linda, my wife
for never complaining about the time taken from us to do this, for believing, giving, and forgiving.

To Jan Carlson
for demonstrating with acts, what a caring professional should be, and especially for the encouraging space to grow.

To Bill Simmons
for trusting me with more than just your tool box, for your gentle guidance. We all miss you, Uncle Bill.

To Bill Coberley
for marketing me in the beginning and for being a life-long friend.

Table of Contents

About the Author

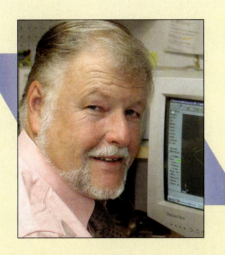

As if it was yesterday, I remember carrying my new toolbox down the isle at Kenworth Trucks of Seattle. Scotty, the crusty drill press operator, stepped away from his machine and planted himself right in front of me. Without a welcome, he raised his bushy eyebrows, poked two fingers into my chest, and said "You see all these men here?" He waited. At eighteen, I recall only nodding, unable to speak. He went on, "Each one of us will show you everything we know if you pay attention. We'll give you lifetimes of experience, but know this, lad, it comes with an obligation. Someday you'll pass it on."

Hello, I'm Mike Fitzpatrick, your machining instructor in print. Since you've honored me by studying my book, I thought it might be a confidence builder to tell a little about why I'm qualified to pay forward to you what Scotty and countless other fine craftsmen taught me.

I began that apprenticeship on the first Monday after high school graduation, in 1964. A year or so later, I was given the life-altering opportunity to be their first employee to run the first Numerical Control machine brought to the Seattle area, other than the ones at the Boeing Aircraft Company. Nothing like the computerized machines you're about to learn, that NC machine was a turret head drill press, run by paper tapes. Not far from a music box in its technology, it was primitive compared to the machines in your training lab. Still, it was enough to hook me for life. So, with a year of applications and interviews, I transferred to Boeing where I completed my machining certificate. There I learned to run programmed machines that had basements, and ladders to get up to the cutter head!

Passing the tough final with a 100% score, I qualified to take the even tougher test to become a tool and die apprentice. I made it and finished my training in 1971. That totaled 12,000 hours of rigorous on-the-job training under a whole army of skilled people. It also came with many hours of technical classes. Since then I've either been a machinist/tool maker or taught others for my entire adult life. For the last 25 years I've taught manufacturing in technical schools, private industry, a high-school skills center, a junior high school, and in two foreign countries.

Today I can stand in front of anyone and say with pride, "I'm a journeyman tool and die maker and a master of my trade." Nearing the end of my journey, Scotty's imprint calls me to pass it forward. But don't forget, what we instructors and machinists give you comes with the same obligation.

One trait we clearly see you'll need far more than we did is adaptability. Beyond imparting skills and competencies, this book has a mission: to start its readers down the long, ever-accelerating technology path. Clearly, the machinist of the future is one who can see and adapt to a changing future. When you do pass the baton forward, the trade won't be anything like that found in this book. But I'm confident it will be passed, because machinists have a long history of adaptation.

Preface

Proudly we look back to see we've been right at the cutting edge of the computer revolution. We began using programmed machine tools over fifty years ago. That predates designers using computer-aided drawing or scientists doing research on giant mainframes. Although the lines I've drawn below to define eras are fuzzy, programmed machine tool evolution can be divided into three generations, based on the way they were used in industry and how they were taught in the tech schools:

First Generation: 1940 to 1965

They began as lab experiments, then for twenty-five years slowly appeared in progressive shops. Like my tape drill at Kenworth, at first, only a few appeared within a manufacturing region. But nearing this era's end, about half of the big shops had one or more tape-driven machines. However, during this entire era, *NC was always considered a specialty.* Most machining was still completed on manually operated or automatic equipment. Programming was a labor-intense, time-consuming task. Purchasing an NC machine (tape driven with no computer) could only be justified if the shop made thousands of similar parts or if the work they wanted to produce was impossible by any other means. Since numerical control skills were a specialty, they were taught in few schools, and only *near the end of the era. NC jobs were never given to beginners.*

Second Generation: 1965 to 1990

This era might be called the big-bang. It begins with an estimated 20/80 ratio of programmed equipment compared to manually operated, but it ends somewhere around 90/10! During the middle, PCs become affordable, and software springs forth that anyone can use. Programming becomes a desktop task. With processor chips speeding up, programmed machines becoming ever-more affordable and capable, work is designed specifically for CNC manufacturing. Nearing the end of this era, *all mainstream manufacturing is done on programmed machine tools. Schools teach the subject as an advanced study* near the end of the machining course.

Third Generation: 1990 to the Present

Programmed machine tools now represent nearly 100% of manufacturing and of greater impact to you, of new jobs. *Entry level people usually start in the shop as CNC operators.* Flexible and friendly, the machines and programming systems are so quick and easy to learn that they are now practical even for one-of-a-kind work such as mold making and die work, as well as production. *Schools integrate and teach CNC as an entry level subject—starting from the first lesson on the first day.*

This book was specifically written to serve the third generation student. To do so, subjects have been grouped into four large career partitions:

Part 1 Introduction to Manufacturing

Manufacturing is a world of its own. Sections 1 through 8 are designed to open the door. It provides the background needed to fit into the shop, understand the rules, read and interpret the drawings, to be comfortable with extreme accuracy, and especially, to be safe.

Part 2 Introduction to Machining

Sections, 9 though 16 teach how to cut metal the right way. These lessons assume you'll eventually perform them on CNC equipment, but probably practice first on manually operated machines because they are simple, safe places to learn setups and operations.

Part 3 Introduction to CNC

Now we get to the text core, how to apply Parts 1 and 2 to setting up, programming, and running CNC machine tools. In Sections 17 through 24 we will learn how to professionally manage a CNC world. Because they move at lightning speeds with lots of power behind them, safety must be integrated into everything we study.

Part 4 Advanced and Advancing Technology

The evolution isn't over—not even close! These four sections, 25 though 28, set the tone for your career after graduating. The best is yet to come so let's get started!

So, many thanks for using my book to start your manufacturing career. It's an honor to be your instructor. Here's what I can pass on about our trade.

Mike Fitzpatrick

Acknowledgements

My deepest gratitude goes to these major contributors, without whom this book would not have come to be.

Bates Technical College (*Tacoma, Washington*) *Bob Storrar, Lead Instructor*
A complete program of machine instruction focused on the student's future. Bob brought his skill and knowledge as a graduate tool and die maker with 16 years industry experience and 15 years teaching at the college level to this book. Thank you for believing in this project, for your conscientious editing and encouragement, and especially for the fellowship.

Lake Washington Technical College (*Kirkland, Washington*) *Mike Clifton, Head Instructor*
A growing program of machine instruction dedicated both to career professionals and those interested in upgrading their skills. Thanks Mike, for contributing 25 years experience in advanced research and manufacturing, plus apprenticeship training, and for taking great interest in the photo manuscript.

CNC Software, Inc. Mastercam—*Mark Summers, President; Dan Newby, Training Director*
Thank you Mark, for believing in education and Dan, for your editing and guidance; and many thanks to your entire team for improving our trade and supporting education worldwide.

Auburn Comprehensive High School (*Auburn, Washington*) *Ron Cughan, Metal Trades Instructor*
A quality metal manufacturing program at the high school level, serving the Auburn Valley. Thanks for taking the extra time to edit this book so well, for believing in it, and being a friend!

Milwaukee Area Technical College (*Milwaukee, Wisconsin*) *Patrick Yunke & Dale Howser, lead instructors.*
Offering a nationally recognized, 2-year, tool-and-die-making diploma. MATC graduates learn die and mold making and qualify for Wisconsin's apprenticeship certificate. Thus they often serve full apprenticeships in the highly paid tooling area of manufacturing.

Dale Howser Sr.: 28 years journeyman tool-making experience with 15 years teaching these subjects. Dale holds degrees in tool and die making from Milwaukee Area Technical College and Voc. Ed. from Stout University. He also develops and works on educational materials for the Precision Metalforming Association and Wisconsin's Apprenticeship programs.

Patrick Yunke: A graduate of Wisconsin's Madison Area Technical College die making program and Stout University for Vocational Education, Patrick brings many years experience in all aspects of precision die and metal and plastic mold-making to MATC, where he has taught for 15 years. He has also been a consultant to industry for manufacturing and custom educational programs.
Many thanks to you both, for your expertise and for supplying great photos from your beautifully organized shop.

NTMA—National Tooling and Machining Association—*Dick Walker, President*
Many thanks for being at the root of this new book in the beginning, for investing

time and energy in it, and for the 45 drawings donated from your training materials.

NTMA Training Centers, California— *Max Hughes, Dean of Instruction*
Thanks Max, for the assistance with the CNC portion of this book.

Mr. Keith Ellis—*Northwest Metalworker* Magazine
Thanks for the wonderful cartoons to emphasize safety

Haas Automation, Inc. *(Oxnard, California) Scott Rathburn, Marketing Manager, Sr. Editor CNC Machining*
Thanks, Scott, for your commitment to machine tool education. (See cover dedication—copyright page), and for contributing support, time, energy, and many photos to this book.

Boeing Commercial Airplane Co.— *Tim Wilson, Apprentice Instructor*
Thank you, Boeing Apprenticeship, for giving me the best education possible when I began my career, and Tim for your ongoing support of quality apprenticeship, for help in planning and executing this book, and for being a lifelong friend!

Brian Mackin—McGraw-Hill Publishing
You were the one that saw a paradigm shift in manufacturing and that a new book had to be written. For that vision, I thank you. I'm not sure this would have happened without your guidance and support.

The Entire Career Education and Technical Team at the Higher Education Division—McGraw-Hill Publishing
Kay Brimeyer, Pat Forrest, Roxan Kinsey, Tom Casson, John Leland, Rick Noel, David Culverwell, Jill Peter, Michael Whitaker, and the entire Dubuque McGraw-Hill Group
No kidding, until I crossed paths with you all, I had decided this was my last book—but I've changed my mind! This

has been a totally positive experience, even through the 'hard spots.' Thanks folks. You are more than a group; you are a team composed of very likable, positive, and refreshingly real people. I hope to be working with you again.

Pat Steele—Manuscript Copyeditor
What often is the worst phase in book-writing became a wonderful experience. With trust in your editing, we developed one voice, then wrote this book together and became friends. Thanks Pat.

Northwest Technical Products, Inc.— *Vic Gallienne, President*
Serving the needs of the scientific, career, and technical educational community of the Pacific Northwest; thank you, Vic, for getting this project on the right track with Mastercam.

Brown and Sharpe Corp. *(Rhode Island) Metrology equipment*
For their commitment to education in metrology in technical schools and colleges.

Kennametal Inc.—*Kennametal University, dedicated to finding better methods and to educate anywhere machining is taught or applied.*
Thanks for the data, advanced tooling photos, text, and charts.

Iscar Metals—*Bill Christensen, Advanced tooling photos and text*
Advancing knowledge through research and education; thank you, Bill, for the HSM article.

Coastal Manufacturing—*Joel Bisset, Quality Assurance Manager*
Thank you, Joel, for editing the SPC articles, and for your lifelong commitment to quality in North American manufacturing.

Northwood Designs—MetaCut Utilities—*Bill Elliot, President/CEO, and Paul Elliot, Senior Software Engineer. Developing utility software for world manufacturing.*
Thank you for all your support and encouragement and for allowing us to

use your wonderful toolpath verification programs in this book and within Mastercam.

Sandvik Coromat—
Thank you for photos from *Modern Metal Cutting*.

SME—Society of Manufacturing Engineers—Westech Machine Tool and Productivity Exposition

Optomec—*Text and photos of LENS®* *process.*
Thanks for showing us a great new technology.

In addition, I want to thank the following reviewers of the final version of the manuscript:

Richard Granlund, *Hennepin Technical College*

Thomas E. Clark, *National Institute of Technology*

Martin Berger, *Blue Ridge Community College*

Totally Integrates Manual and CNC Instruction!

Machining and CNC Technology, provides **the most up-to-date approach** to Machine Tool technology available, with totally integrated coverage of manual and Computer Numerical Control-based equipment.

Motivational chapter features, such as *Key Points, Trade Tips,* and *Shop Talk* are included to show students the practical side of the subject.

◆ Key Point
◆ Trade Tip
◆ Shop Talk
◆ Critical Thinking

KEYPOINT

For now, understand that changing to a curved cut within a different primary plane requires a new code to be entered to signify which plane is wanted.

Axis ID on a CNC Machine

When facing a new machine for the first time, the **world orientation** of its axis set (relationship to the floor and to the operator) can often be identified this way, in this order (Fig. 17-3).

Z The axis parallel to and opposing the main spindle

X Usually the longest axis, *usually* parallel to the floor

Y The axis perpendicular to both *X* and *Z*

These are conventional guides, not a standard. For any given CNC machine, the set needn't be in any given relationship to the world. While the axis set remains orthogonal and the axes are in the same position to each other, the axis set can be rotated to any world position.

Try it yourself. Use the right-hand rule for a mill in your shop. Depending on the world perspective of the axis set, you may find your hand in any position, but it will be found to fit the rule. There are almost no violations to

World Axis Orientation

Figure 17-3 *The primary axes as they apply to three familiar machines.*

Trade Tip

When faced with an unfamiliar CNC machine, always look for the Z axis first as it will be the easiest to identify. The Z axis carries the tool to the work, as with lathes, or the work to the spindle or visa versa, on mills. Then with the Z axis under control, apply the **right-hand rule** to identify the other two axes (Fig. 17-4).

Figure 17-4 The right-hand rule helps identify the machine axes.

KEYPOINT

Right-Hand Rule

Pointing the thumb of your right hand along the positive *X* axis direction, your first finger points out the positive *Y* axis direction. Finally, the raised middle finger points out the positive *Z* axis.

the orthogonal set, but it can be found lying in some odd positions relative to the world, especially on advanced CNC equipment and robots.

The most common example of skewed world orientation, shown in Figs. 17-5 and 17-6, is a *slant bed lathe,* where the *X* axis has been tilted relative to the floor. This modification makes chip and coolant ejection more efficient and improves operator access for setting up tools.

ShopTalk

Oops! Some time ago, a few early programmed lathes were given reverse axis sign values for their Z axis only. The grand idea was to eliminate the negative sign on most Z axis coordinates, and thus create shorter programs. But this "improvement" led to so many serious crashes due to the nonstandard axis set that they were never produced again.

Visual Focus photos and line art make concepts easier for students to understand and apply the information presented.

Figure 15-15 *Learn to identify temperature by glow.*

Integrated Coverage of CNC Technology

UNIT 28-1

What Does a CMM Do? How Does It Work?

Introduction: Without experience, most assume any computer measuring equipment does what the manual instruments and processes do, but they do it faster and more accurately. Yes, that's true in many cases (but not all), but it's only a part of the reason they improve quality and profitability. In Section 28, we're focusing on computer-coordinate machines; however, similar arguments could be presented for other computer-directed measuring equipment as well.

KEYPOINT

CMMs verify geometric requirements in ways no other process can. Proof coming up.

KEY TERMS

Conformal axis system
Axes are assigned to the work independent of machine axes also called a floating axis set.

Digitizing
Programming by hand probe motion—see teach-learn.

Hit (CMM)
Shop lingo for one recorded touch of a data collecting probe.

Nontactile measurement
Using optical methods to measure surface elements.

Scanning CMMs
Two types both of which collect entire surface element data. The tactile type maintains its probe in constant contact with a surface. The nontactile type collects entire surface element without touching.

Shop hardened
An instrument built to withstand the temperature swings and dirt of a shop environment.

Tactile measurement
Measuring by touching the object with a probe.

Teach-learn (read-learn)
Programming by recording hand probe movements and menu picks.

Computer Inspection Advantages

Cures Choke Points

High-speed mills and lathes often machine parts faster than they can be measured with mics, electronic height gages, and indicators. Without a CMM, there are three options.

In the least desirable, the operator must measure the sample while the machine cranks out more questionable parts! Or, the halt button must be used until the sample part is proven right. Either way, time is lost.

The third choke point solution is to load the cutting tool drum with a touch trigger probe (information coming) and then write inspection routines into the machining program itself (Fig. 28-1). In previous sections, we've seen probes testing cutting tools to autoload offsets. But, not all feature inspection lends itself to *in-process inspection*, nor can all CNC machines perform it. There are many times when the part must still go to the stand-alone CMM.

Figure 28-1 *In-process inspection helps check key features as they are made and keeps real-time SPC up to date.*

Mastercam v9

.8061
.8029
6X–1.7952
28X–1.7892
28X–1.7868
N2330X–1.7782
332X–1.7752
334X–1.7711
36X–1.7617
38X–1.7578
2340X–1.7535
2342X–1.7434

Complete Education System

Hands-On Training:

For the Student:

◆ Textbook with Student CD-ROM
ISBN 0-07-829860-1

For the Instructor:

Classroom Performance System (CPS) available for in-class testing and quizzing, and class management (0-07-294206-1).

Instructor's Manual with Instructor's Productivity Center CD-ROM (0-07-825097-8) provides solutions, teaching resources, and a computerized test generating system (that also can be used directly with the CPS system). PowerPoint and ExamView (0-07-311223-2)

Common stainless alloys (300 series) will not hold on a magnet at all. The added nickel and chromium break up the metal's crystal structure to the point where there is no net attraction to a magnet. In theory, each crystal grain would adhere to a magnet but combined, their microscopic nature cancels this property. Some with a higher iron ratio will hold weakly (400 series). Still others hold nearly as well as regular steel (PH series).

Some stainless will spark similar to CI, with dull red sparks when ground. Other alloys will barely spark at all.

Stainless chips generally do not change color, however, there are a few special exceptions.

Figure 5-17 Composite materials are combinations of fibers, resins, and metals. They have an amazing strength-to-weight ratio.

KEYPOINT
A heavy, bright silver metal with no sign of rust that does not hold to the magnet is definitely stainless.

ShopTalk

Look it Up or Ask Before machining any new material, find out the machinability and hardness either from a journey level machinist or by looking it up in a machinist's reference book. There are lots of surprises within the subject of metallurgy. For example, there are two metals that actually catch on fire, given the right conditions. They are magnesium and titanium. Solid blocks of these metals are impossible to ignite but when their chips are surrounded with room oxygen and heated as during machining, they will burn! When machined some metals create carcinogenic particles that must not be breathed or exposed to bare skin!

Composite Materials

Composites are manufacturing materials made of combinations (Fig. 5-17) of a honeycomb core, an outer skin, and a resin to hold it all together. The cores can be metal or plastic fiber and the skin can be plastic or metal. These highly engineered materials are extremely strong compared to their weight. Because the cutting of these materials requires special cutters and processes, you probably won't see them in the training lab, but they are used in industry and gaining popularity.

In machining composites (Fig. 5-18), you must wear breathing protection from fumes or dust. Another precaution is fire prevention, as many composites are part metal and part plastic. Special cutters must be used that look more like a meat slicer than regular metal cutters. Holding composites during machining is a challenge as they are extremely strong for their weight

Figure 5-18 A composite being machined.

along a given axis but they can be crushed if not protected.

Physical Characteristics of Metals

Beyond selecting the right alloy for a job, there are more work order requirements that must be certified—the grain direction and the heat-treat condition, how hard it is.

Machining and CNC Technology

PART 1

Introduction to Manufacturing

This entire book is about getting a job in a machine shop. But more important, it's about keeping that job and advancing career responsibility and pay. To do so successfully, you need to know what is expected of you from the very first day. Like any workplace, there are tasks, procedures, and rules to be followed. Some are based on formal skills or rules, while some are informal and generally accepted by your fellow workers.

Part 1 of this book is designed to impart these skills and clarify these expectations to help you

Section 1   Know how to look, act, and be professional

Section 2   Understand, use, and speak the precision numbers of the trade

Section 3   Read early lab drawings and be ready for more engineering drawing instruction

Section 4   Identify and use geometric dimensioning and tolerancing (GDT) found on drawings

Section 5   Work with the materials, documents, and processes before and after the machining occurs

Section 6   Use five basic measuring tools, the right way always with the best repeatability

Section 7   Use specialized measuring tools when the five basic tools aren't the right choice

Section 8   Perform simple part layout and know when it's right or wrong for the job

Section 1

Professionalism in Manufacturing

Introduction

This book represents, a manufacturing world where computer-aided design/computer-aided manufacturing (CAD/CAM) and computer numerical control (CNC) have changed everything. Planned by a group of industry leaders and instructors, we use today's job market and tomorrow's career needs as our guide. To make room for the new subjects needed for career success, every effort is made to eliminate old technologies and skills no longer relevant to mainstream employment. The constant goal is to equip you, the beginning machinist, with the right competencies to get and keep that vital first job.

During those planning meetings, we also discussed the fact that students often breeze past the usual opening chapters to get to the "real training." So why then are we starting with a section on professionalism and it's companion subject, safety? Because a critical part, perhaps the most critical part, of your training has nothing to do with measuring, reading prints, and running programs on machines, yet getting this part right will have everything to do with your success! It might be called work ethic, team spirit, or job readiness. It's often called attitude on an employee evaluation or a grade report. No matter what you call it, it adds up to how you walk the walk of a skilled craftsperson. A large part of the separation between ordinary and skilled workers is a professional attitude.

While a whole lot more could be said on the subject, these units are enough to get you started fitting into a machine shop environment and starting the life-long process of being a professional. Taken to heart, the overall message of this section will make a real difference in your career.

Objectives

Dressing for Career Success

Introduction: In Fig. 1-1, which person would you want making precision parts for your new car or outboard motor? In truth, they might all be good machinists but—well, you get the picture. The concern here isn't styles of clothing or grooming, it's about being right for a precision shop environment.

KEY TERMS

Chips
Metal particles of waste removed from the workpiece by machining.

Natural fibers
Cotton or wool cloth, which tends to resist hot chip damage and melting, thus protecting the wearer from burns.

Synthetic fibers
Plastic cloth such as nylon and polyester, which tends to melt when hot chips touch it.

Z87 or Z87.1
The mark found on safety glasses approved for shop work, which means they will protect your eyes from the front and side in a dangerous environment.

Getting Ready for the Work Environment—Eye Protection Always

Figure 1-2 shows several types of acceptable eye protection. Every shop supplies one or more varieties. The best choice is the one you find comfortable and tend to leave on 100 percent of the time! Many machinists who do not otherwise wear glasses, prefer full-vision, wraparound lenses because they are all clear material and the wearer doesn't see a frame.

KEYPOINT

Safety glasses will bear the mark **Z87** or **Z87.1** on the ear piece if they have passed strict testing and are acceptable for shop work.

Clear or Yellow

Either lens is acceptable. Yellow lenses offset the blue of common fluorescent lighting and many feel that the correction boosts their ability to read precision tools. Never select dark glasses unless you must work near electric welding flashes because they dull your ability to see details.

Prescription Eyewear

Most wraparound safety glasses can be worn over prescription glasses. But the law requires

Figure 1-1 *Which machinist looks right for the job—and more importantly, for a career?*

Figure 1-2 *The best choice for eye protection is the one you find most comfortable and tend to leave on.*

prescription glass lenses to be made from tempered glass or high-impact plastic, so it is acceptable to wear them alone as long as side shields are added. It's a fact that many eye injuries occur from the side rather than straight on, so shatterproof lenses aren't enough protection by themselves.

Extreme Danger Areas

When performing tasks such as disk grinding with lots of flying debris, you'd be wise to double protect your eyes by adding a full-face shield over safety glasses.

Just Do It!

Be a trendsetter—wear safety glasses even when others don't. Modern CNC equipment usually features containment shielding so the operator feels safe from flying metal particles, but machinists forget they must occasionally walk though the shop past other unprotected machines. Make safety glasses a habit by putting them on when first entering the shop. Here's an attitude check: you've got it right when you feel strange without your safety glasses. No kidding, I've gone home still wearing them! If those provided aren't comfortable, then find an industrial safety supplier and buy a pair just right for you—that's the kind of pro we're talking about.

Hearing Protection

Machine shops can be noisy places. Some cutting operations are loud enough to cause hearing loss over time. Prevent permanent damage right now while your hearing is good. Get in the habit of wearing ear protection (Fig. 1-3) where noise gets above moderate—see the chart showing the range of common sounds. The two common types of hearing protection are expandable foam inserts that fit every ear shape and the muff type that fit over your ears.

Online Safety Equipment Suppliers Search for *industrial safety supply* + *eyeware* or *personal safety equipment.*

Figure 1-3 *Machinists protect their hearing when shop noise is loud or high pitched.*

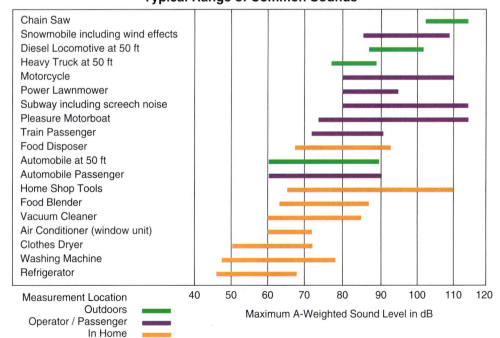

Typical Range of Common Sounds

Chain Saw
Snowmobile including wind effects
Diesel Locomotive at 50 ft
Heavy Truck at 50 ft
Motorcycle
Power Lawnmower
Subway including screech noise
Pleasure Motorboat
Train Passenger
Food Disposer
Automobile at 50 ft
Automobile Passenger
Home Shop Tools
Food Blender
Vacuum Cleaner
Air Conditioner (window unit)
Clothes Dryer
Washing Machine
Refrigerator

Measurement Location
Outdoors
Operator / Passenger
In Home

40 50 60 70 80 90 100 110 120
Maximum A-Weighted Sound Level in dB

Figure 1-5 *Two hot chips melted onto this nylon windbreaker while the center one went completely through to possibly burn the person within!*

Earplugs remove the loud spikes but allow controlled hearing. For extreme situations, earmuffs are supplied by your employer. Either way, don't give away your most immediate control sense by not protecting it. And by the way, to control your machine it is important to not use personal music earphones—sorry, but they're out when running a machine.

Shop Clothing

The main danger of loose clothing is that it can be caught by moving machinery—but you already knew that. While Fig. 1-4 was set up to be entertaining, the real act of being caught in a machine is no fun—think about it! Nothing loose—sleeves, necklaces, untucked

Figure 1-4 *Seriously, loose clothing does get caught in moving machinery!*

shirts—nothing! Well-fit, **natural fibers** such as cotton or wool, without pockets or tie cords hanging out, are essential.

Why Natural Fibers?

The by-product shavings made when cutting metal, are called **chips.** They're hot, as much as 1,000 degrees at times! While today's CNC machines won't operate with the safety guard open, thus lessening the danger, many older machines will. Flying hot chips must be dealt with. As we study machine operation we'll see several ways to control them, but one action the pro takes concerns the kind of fabrics worn on the job.

When hot chips contact **synthetic fibers** such as polyester, rayon, or nylon, they stick then melt through (Fig. 1-5). So your shirt or pants are ruined and the hot metal is held against your skin, which can singe. Ouch! Not only that, but getting singed makes it hard to concentrate on the task at hand, decreasing productivity and increasing safety risks.

KEYPOINT

Where hot chips cannot be contained by any other method, *wearing* natural fibers such as cotton denim or wool can work. They will just bounce off your clothing.

Aprons and Shop Coats

An apron or shop coat can be a good choice. But be aware that not all are designed for machining. Some are better suited for lab work, where machinery won't grab loose ties or pockets. Find one with internal pockets and no loose belts tied in front. Long sleeves are not smart for obvious reasons. As a last resort, roll long sleeves up tightly, don't just push them up so they can fall back down at the worst possible time. A professional approach is to keep a short-sleeved work shirt in your locker.

How About Shoes?

Your work shoes have three safety aspects. The first two are easy, but I bet the third will surprise you. Work shoes provide

◆ Protection for your feet from falling objects.

◆ Nonslip soles designed for a shop where chips, coolants, and oils are often on the floor.

◆ Protection from fatigue.

Athletic shoes are a poor choice. Even though they're comfortable at first, they aren't designed to stand up to a shop environment. Many companies ban them from the shop floor.

Steel Toes Shoes or boots with steel toe caps are better than shoes without and may be required in your school lab or on the job. No matter how careful you are, there's always the inevitable falling heavy object. Don't fall for the old tale that someone knows someone who had their toes severed by the steel insert collapsing when something really heavy fell on it. That story is always being passed around. Think about it: if the thing was that heavy, it would have done the same damage with or without the steel protection!

No Accessories

Jewelry catches on moving machinery (Fig. 1-6). Here's another aspect you may not know: it also conducts electricity and heat. In addition to the safety aspect, jewelry should be left in your locker because the machine environment ruins it. Look smart—leave accessories behind.

Hair Up

Think about this: would you reach up and pull a hand full of hair out of your scalp? Painful but that would be just a few hairs! So now, imagine a machine wrapping up most of your hair along with a patch of scalp. No kidding! I've seen it happen twice with my own eyes, to a careless machinist. Long hair can be dangerous (Fig. 1-7). It gets caught up by static electricity produced during machining and is blown by the air currents swirling around moving tools. A hair band, a bandanna, or a hat are right when your hair is longer than about two inches.

That's it. With these dress-for-success guidelines, you're ready to step into the shop and, if you see the bigger picture, into a career.

Quality Could be Linked to Good Work Shoes or Boots
As a machinist, you're going to be on your feet all day, usually on concrete. Guess when most folks make the most mistakes? That's right, at the end of the day when they're tired. Good-quality work shoes offset some of the problem and help keep your mind sharp.

Figure 1-7 *Hair does catch on moving machinery. Keep it out of harm's way.*

Figure 1-6 *Jewelry is dangerous. It catches on machines and chips and conducts heat and electricity.*

UNIT 1-1 REVIEW

Review the Key Terms

Chips
Waste metal particles removed from the workpiece by machining having five possible dangers: heat, sharpness, projectiles, slipperiness on the floor, and strength enough to pull hands into machinery.

Natural fibers
Cotton and wool clothing, which tends to reject hot and sharp chip damage.

Synthetic fibers
Plastic cloth such as nylon and polyester, which tends to melt when hot chips touch it.

Z87 or Z87.1
The mark found on safety glasses approved for shop work, which means they will protect your eyes from the front and side in a dangerous environment.

Replay the Key Points

- Eye protection is best if it's comfortable and is thus left on all the time.
- Well fitting coats or aprons made of natural fibers are best and must feature no loose pockets, sleeves, or outside ties.
- Ear protection is necessary in many work areas.
- Footwear must be designed for shop use and comfort.
- Any loose item is a danger, including jewelry and hair.
- Safety glasses will bear the mark Z87 or Z87.1 on the earpiece if they have passed strict testing and are acceptable for shop work.
- Natural fibers are safest and last longer in a shop environment.
- Athletic shoes, although comfortable at first, do not stand up to the shop environment.

Respond (Answers found at the end of Section 1)

1. Are *yellow*-tinted safety glasses acceptable in the shop? Are *brown*- or *green*-tinted glasses?
2. Describe the two reasons to avoid synthetic fibers and wear natural fiber clothing when running machines.
3. What are three aspects of footwear with regard to safety?
4. Why is hearing such a vital issue for a machinist?
5. For what two reasons does long hair become tangled in moving machines?

UNIT 1-2

Handling Materials

Introduction: Working in a machine shop requires the use of many materials. Some are technical chemicals with very specific precautions and earth issues. Some are consumable supplies meant to be used up but not wasted. Others are heavy and expensive. This section lays out the ways to handle them all—like a pro. Doing so is an excellent way to demonstrate a well-tuned attitude.

KEY TERMS

Choker strap
A nylon strap that cinches tightly around heavy objects with a strong eye loop on each end to be attached to the crane.

Heat lot number
The original quality control number for a specific metal bar.

Intervertebral disk (disk)
The flexible cushion between spinal vertebrae that can be damaged by the wrong lifting techniques.

Jib crane
A heavy lifting devise that may be fastened to a wall or column to swing in an arc, or on portable rollers and often called a cherry picker.

Traceability
The ability to link a given bar of metal from "birth" all the way to its final shape and individual serial number.

Lifting Heavy Materials

Always Use a Machine if You Can!

After all, we are *machinists*. The very last thing we use to get a job done is muscle power (Fig. 1-8). In most facilities, you will find one or more of these devises:

- Overhead cranes moving on rails
- Rolling lift tables
- Floor jacks for flat pallets of workpieces and boxes
- Fork lift trucks
- Portable **jib cranes,** often called "cherry pickers"
- Jib cranes fastened to a column to pivot around a circular area

a.   b.   c.

Figure 1-8 *Smart machinists use equipment, not muscle, to lift heavy objects.*

Use Your Legs (and Your Head) Not Your Back

All that said, there comes a time when lifting by muscle power is the only way to get the job done. At these times, the wise machinist asks for help from others. Two lifters are better than one. In North America, it's estimated that two out of five adults suffer from back pain that could have been prevented.

Here's the real point: It doesn't just happen later in life, it starts right now—today! Unlike the way most people see it, back injuries don't actually happen once when you're 40 or so, they are the product of lifting wrong over a long time. Everyone has heard about the straight back, bent legs method of lifting. It's best, but why?

To understand, let's look at your spine as a piece of machinery: a human lifter is a crane of sorts. In Fig. 1-9, the person on the left has turned her back into a long lever with the pivot point right at her lower back. *This wrong action causes pressure in the range of hundreds of pounds per square inch right on the lower back.* In contrast, the person on the right, bent at the knees, is focusing the pressure on his leg muscles not his back.

Disks in Your Back

Now focus on that pressure point, the lower back. Your back is well engineered with cushions between the vertebrae, called **intervertebral disks** (or just **disks**). They provide flexibility and they absorb shock. They also keep the vertebrae apart so nerves are not pinched as you move. Disks are critical and can be damaged (Fig. 1-10). When they are it can really hurt and even immobilize you!

The disks are tough, fluid-filled sacks with several layers holding the fluid center in. The wrong lifting technique overpressurizes them, resulting in a blowout. Here's the part most people do not know: When the disks first

Your Strongest Point Is Your Smarts Not Your Muscles Demonstrating superstrength doesn't impress new employers or fellow workers. It may show enthusiasm but it also shows immaturity.

Longer lever

More disk
compression

Back muscles,
more force

Spine

(a) Wrong way to lift.

Shorter lever

Less disk
compression

Less use of
back muscles,
less force

Spine

(b) Right way to lift.

Figure 1-9 *Lifting the wrong way and the right way. The worker on the left is setting up a painful future! Note that the spinal disk is under a lot more pressure because she is bending her back, not her knees.*

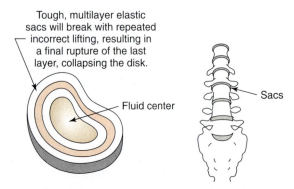

Tough, multilayer elastic sacs will break with repeated incorrect lifting, resulting in a final rupture of the last layer, collapsing the disk.

Fluid center

Sacs

Figure 1-10 *A spinal disk is a tough, multi-layered shock absorber. But lifting incorrectly can cause progressive damage, leading to a disaster in the future.*

stretch and break, it's an inner layer or two that goes, and usually the injured person doesn't even know. The layers do not often break all at once, as is commonly thought. But, with time and continued bad technique, the last one finally ruptures, resulting in enough pain that the person cannot stand up straight!

KEYPOINT

Lifting wrong is not only the start of a painful future, it creates a poor professional image.

Carrying Materials

Another aspect of handling heavy metals, is how to carry long bars of metal through the shop. Be cautious of the forward end—it's a ram! If the bar is too heavy to safely hand carry (beyond 40 pounds) or if it's over 8 feet long, there are two acceptable methods.

1. Crane Carry—One Person

In the unbalanced carry, hold the bar with a nylon **choker strap** near the leading edge and pull the remaining bar along on the floor using the overhead crane. One person can perform this carry by being at the forward end. The process can leave marks on the shop floor, depending on the weight of the bar, but it's safe. A choker strap is a strong nylon flat strap with loop ends that slip closed with pressure. Your instructor or shop lead will demonstrate this aspect of lifting.

2. Two Person

For the balanced carry, choke the bar near the middle and carry it with both ends off the floor. This requires a second person to walk at the leading edge, to prevent hitting people or objects (Fig. 1-11).

Crane Overswing

Remember, while it moves smoothly using the crane, the heavy bar is a mass in motion that can't "stop on a dime." It will swing forward when you stop pushing or driving the crane forward. The key point is control of the leading edge (Fig. 1-12).

Heavy lifting is a skill in its own right, requiring knowledge of rigging (the lifting devices that contact the work), lift machines, and the physics of lifting. Due to the specialized skills and knowledge required, many shops employ a lifting crew. But this subject is also considered part of the machinist's duty. If you are the slightest bit unsure about how to lift or move an object, get help.

Figure 1-11 *Use a crane and choker to haul heavy bars, while an assistant helps control the metal bar!*

Figure 1-12 *Two people hand-carry long bars by preventing the leading edge from harming people and objects.*

TOOL & DIE STEEL

A 2 Annealed Air Hardening 5% Chrome		D 2 Annealed Air Hardening High Carbon/Chrome		O 6 Annealed Oil Hardening Graph-Mo®	
A 6 Annealed Air Hardening Low Temperature		H 13 Annealed Air Hardening Hot Work		S 5 Annealed Oil Hardening Shock Resisting	
A 10 Annealed Air Hardening Graph-Air®		O 1 Annealed Oil Hardening		S 7 Annealed Air Hardening Shock Resisting	

Figure 1-13 *Correctly stored and color code identified raw materials.*

Never reach under, or worse, walk under a heavy object during a lift.

Storing Metals

Many times our school has received a big donation of expensive metal from a local shop because careless workers cut the color-coded identification off the bar end or lost the identification tag. Once that is gone, the bar is useless in most cases, especially when the shop makes people-moving products. In such situations, where lives are at stake if the final product fails, once the **heat lot number** (the original manufacturing quality control number) is lost, the metal is not only worthless, it's illegal to use!

When the design specs must be followed exactly, as in making airplane parts, for example, a continuous history must be traceable from the foundry to final part number. This is known as **traceability.** A specific part will have a specific number assigned to it, called the serial number (S/N). The manufacturer must be able to provide documents tracing the S/N to the heat lot of the metal's original manufacture. Material can be tested in an independent lab then recertified, but that's usually more costly than just buying more metal!

One of several good methods of storing metals is a rack. Note that the bars cannot fall out of the safety storage rack and that short bars are not kept here. "Shorts" are kept on smaller shelves where they won't fall through.

Note that they are sorted by type and alloy such that the color-coded identification is facing out (Fig. 1-13). Later, we'll take a look at material identifications and alloys.

These are not small details. You must have exactly the right material for the design and it often must be kept in storage for a specific job.

UNIT 1-2 REVIEW

Review the Key Terms

Choker strap
A nylon strap that cinches tightly around heavy objects with a strong eye loop on each end to be attached to the crane.

Heat lot number
The original quality control number for a specific metal bar.

Intervertebral disk (disk)
The flexible cushion between spinal vertebrae that can be damaged by the wrong lifting techniques.

Jib crane
A heavy lifting devise that may be fastened to a wall or column to swing in an arc, or on portable rollers and often called a cherry picker.

Traceability
The ability to link a given bar of metal from "birth" all the way to its final shape and individual serial number.

Replay the Key Points

◆ Always use a lifting devise for your first choice.

◆ If physical lifting is absolutely necessary, get help—all should bend their knees.

◆ Carry long bars safely by protecting the leading edge.

◆ Get help with any lift.

◆ Store long bars in a safety rack and short pieces away from the rack on shelves designed for this purpose.

◆ Mind the color coding and/or stamps—*never cut it off the bar.*

Respond (Answers found at the end of Section 1)

1. Name at least three shop lifting aids other than human muscle?

2. A choker strap can be used to carry long material through the shop as long as the forward end of the bar is high in the air. Is this statement true or false? If it's false, what could make it true?

3. Why is material identification important for many types of work?

4. Describe a disk in your spine?

5. Where must short bars be stored?

UNIT 1-3

Handling Shop Supplies

Introduction: Today's professional must know the rules when handling materials. Besides this being the law, ever-tighter budgets require it. Beyond the critical issues of waste and a ruined planet, the fines for ignoring the rules can cost both dollars and company image. Machinists must use chemicals and consume materials with an awareness of the issues concerning each. That means conservation, recycling, and lawful disposal.

KEY TERMS

Hydrostatic bearings
Found on modern CNC machines, these require an exact oil viscosity to separate two sliding components apart as they move.

Material safety data system or sheet (MSDS)
The product of the federal "Right to Know" act that ensures that workers must be provided the information needed to handle materials safely—found on MSDS sheets.

Spindle oil
Lubricant designed to prevent rolling friction.

Viscosity
The rated property of lubricants to resist thinning out and flow rate.

Way oil
A lubricant used to control sliding friction in machines (the part that slides is called the machine's ways).

Your Right (and Obligation) to Know About Shop Chemicals

Promanagement begins by finding out what hazards a material might have and taking the right actions based on those potential hazards. Those issues and the correct use of each chemical are found in documents called **MSDS sheets**—the **material safety data system** (Fig. 1-14). Mandated by federal law, manufacturers of chemicals must supply these sheets and employers must have them on file for all to read.

In general, machine shop chemicals are not overly dangerous. Still, some will ignite, some people's skin can react, and some produce fumes that must not be inhaled. Nearly every one involves some special precaution. Instructors and employers need your cooperation in handling these chemicals and that means being informed.

The six precautions are

◆ Fire hazard

◆ Chemical stripping of skin oils—direct contact

◆ Fume *toxicity* and oxygen exclusion (displaces breathing oxygen)

◆ Eye irritation

◆ Allergic reactions

◆ Contamination of other fluids or chemicals, with possible side reactions

For example, we sometimes use acetone to clean and remove dyes from parts. Using the MSDS in Fig. 1-14, we find that

◆ Acetone must be stored in a fire-resistant container.

◆ It should not be in contact with skin or eyes.

◆ You must avoid breathing excessive or concentrated acetone fumes.

MATERIAL SAFETY DATA SHEET

CRC.   **SILOO**

CRC Industries, Inc. • 885 Louis Drive • Warminster, PA 18974 • (215) 674-4300

PRODUCT NAME CLEAN-R-CARB (AEROSOL)   #- MSDS05079
PRODUCT- 5079, 5079T, 5081, 5081T

(Page 1 of 2)

1. INGREDIENTS

	CAS #	ACGIH TLV	OSHA PEL	OTHER LIMITS	%
Acetone	67–64–1	750 ppm	750 ppm		2–5
Xylene	1330–20–7	100 ppm	100 ppm		68–75
2-Butoxy Ethanol	111–76–2	25 ppm	25 ppm	(skin)	3–5
Methanol	67–56–1	200 ppm	200 ppm		3–5
Detergent	—	NA	NA		0–1
Propane	74–98–6	NA	1000 ppm		10–20
Isobutane	75–28–5	NA	NA	1000 ppm	10–20

2. PHYSICAL DATA : (without propellent)
Specific Gravity   : 0.865   Vapor Pressure   : ND
                                      % Volatile   : > 99
Boiling Point   : 176 F initial   Evaporation Rate : Moderately fast
Freezing Point   : ND   Vapor Density   : ND
Appearance and Odor: pH : NA
      A clear colorless liquid, aromatic odor

Solubility   : Partially soluble in water.

3. FIRE AND EXPLOSION DATA
Flashpoint   : −40 F Method : TCC
Flammable Limits   : propellent   LEL : 1.8   UEL : 9.5
Extinguishing Media   : CO_2, dry chemical, foam
Unusual Hazards   : Aerosol cans may explode when heated above 120 F.

4. REACTIVITY AND STABILITY
Stability   : Stable
Hazardous decomposition products
                   : CO_2, carbon monoxide (thermal)

Materials to avoid   : Strong oxidizing agents and sources of ignition.

5. PROTECTION INFORMATION
Ventilation   :   Use mechanical means to insure vapor conc. is
               :   below TLV.

Respiratory   :   Use self-contained breathing apparatus above TLV.

     Gloves : Solvent resistant   Eye & Face : Safety glasses
Other Protective Equipment : Not normally required for aerosol product usage.

Figure 1-14 *An MSDS sheet for acetone. When in doubt, read these bulletins. It's your right to know.*

So, using acetone right requires several actions on your part. What would they be given the precautions just listed? Use a steel, fire-proof container. Wear chemical-resistant gloves and work in a place where air circulates—never a small room with closed doors and windows.

KEYPOINT

Although shop chemicals are generally not extremely dangerous, each has its own set of precautions and procedures found in its own MSDS. Each also has a correct disposal procedure that must be followed.

The Five Kinds of Shop Chemicals

1. Lubricants
2. Coolants
3. Solvents/coatings
4. Cleaning products
5. Gasses (compressed and liquefied)

After being in this trade for a while, you might be called on to use compressed gasses for welding and brazing. We also use liquefied nitrogen for deep chilling when shrinking metal parts for special assemblies. But these tasks are beyond beginners, so we do not study gas safety here. However, be aware there are some highly specialized skills needed to use them. That leaves the other four categories, however, we concern ourselves here primarily with categories 1 and 2.

Using Lubricants the Right Way

Other than fire hazard, today's machine lubricants are safe and *benign* (not toxic).

Figure 1-15 *Specialized lubricants must be stored in fireproof surroundings and kept free from contamination.*

double problem: it creates inaccurate machine movements, but even worse, a thinner film than that required can let metal touch and rub with sure damage to follow!

KEYPOINT

Keep oilcans labeled and stored in a marked, fire-protected area with their tops closed to avoid contamination (Fig. 1-15).

Trade Tip

You Should Know! Fast, modern computer-controlled (CNC) machines are fussy about getting the exact oil for the duty and there are many kinds. An expensive machine can be damaged by adding the wrong lube nearly as fast as by not adding any at all? Of highest priority: oil thickness and resistance to flow, called **viscosity,** is absolutely critical—both from the standpoint of the precision action of the machine and to prevent wear.

Special Oils Must be Identified and Kept Clean

Several application-specific lubricants are necessary to keep most shops working smoothly, especially where there is a variety of CNC machines. Both sliding and rotating equipment must have lubrication, but especially faster machines equipped with high-tech **hydrostatic bearings.** Hydrostatic bearings depend on an exact thickness of oil between moving components. When they are used on a sliding machine, components actually "float" on a predictable thickness of oil much like an air hockey puck on the table. Introducing the wrong oil into the gap causes a

Sliding Lubricants and Rolling Action Lubes

Oils are used in two very different ways: to prevent sliding friction or rolling friction. The sliding type oil is called **way oil** because the machine parts it lubricates are also called "ways" (discussed in lathe and mill training coming up). Way oil is thicker than spindle oil. Even thicker yet, grease is occasionally used to prevent sliding friction in some older machines, but it's far less common today. However, if the machine needs grease, then using way oil would be a big mistake!

For bearings that rotate, **spindle oil** is the correct lubricant. There are more kinds of spindle oils than ways due to the wide differences in the kinds of spindle bearings. This is the area where you must be doubly sure to have exactly the right oil. Some bearings are made of ceramic composites, while others are ceramic/steel and still others are steel alone. Some spindles are refrigerated to maintain perfect accuracy, while others aren't. Spindle accuracy and tool life depend on lubrication.

Be aware of the lube needed, don't guess.

Here's a hint that will be noticed by others in the shop. Never leave open oilcans in the work area or even in the storage area. They become contaminated with chips and dirt. They could even catch on fire from heated flying chips.

How Do You Know Which Lube to Use?

Lubrication instructions are usually found on a metal plate fastened to the machine, right at the input point. If no plate or other tag is visible, look it up in the owner's manual (Fig. 1-16). Or just ask.

Using Coolants

For many reasons you will soon learn, coolants often make machining safer and faster. Today's coolants are water-based fluids that are either flooded or sprayed over the cut area. When they are used correctly, tools often last longer, productivity increases, and horsepower is reduced along with excess heat. Today most shops mix coolants such that one dilution can be used throughout the shop, but others prefer to custom mix for a specific application. For example, a thicker mixture is needed for milling,

FILL WITH SUNOCO #1180 WAY LUBE

Figure 1-16 *Be absolutely certain to add the right oil, especially on CNC equipment!*

but a thinner one works best for grinding. The important issues for coolants are

◆ If a vapor or spay is produced in the machine operation, it must not be inhaled. Containment barriers and a breathing mask are two good solutions.

◆ The undiluted syrup used to make coolant, is ultracostly. A barrel of the unmixed syrup can cost a week's wages for a journeyman!

◆ Coolants must be maintained by checking density and adding fresh water because the heat from machining evaporates the water and thickens the dilution ratio.

◆ Coolants become contaminated with lube oils from the equipment and dirt from machining, called "tramp oil." Tramp oil must be skimmed or filtered out.

◆ Coolants can be revitalized by chemical treatment and filtration, but at some point coolants must be disposed of. When the time comes, disposal must be done in the right, legal way!

Coolants come in three forms: synthetic, organic, and special threading compounds. *Synthetic coolant* is the most common, especially for CNC work. Synthetics are nonclinging, meaning that they run off the workpiece easily, thus parts come away clean from the machine, and the shop stays cleaner too. Synthetics last longer in the machine compared to older organic oils. *Organic coolants* can rot due to bacterial action, over time. Finally, even though their main ingredient is water, synthetic coolants prevent rust on machines and on steel or iron parts similar to antifreeze mixed with water in your car's engine block.

Figure 1-17 *Synthetic coolants are highly designed shop chemicals that are mixed with water to make machining more efficient and safe.*

Highly engineered synthetics (Fig. 1-17) are designed to be both nonallergenic to people and nonreactive to metals. They are supplied as a concentrate that is mixed at dilution ratios of one part concentrate with from 20 to over 50 parts water.

Cutting oils and cutting compounds (Fig. 1-18) are lubricants that tend to cling to the work, called the wetting property. Different from synthetics, these oils and compounds stay on the tool or work by design. Each has its use. They are either brushed or poured on during the operation. A few of these products are designed for thread machining and are exceptionally effective, but they're also expensive! They pay for themselves in results, but please don't waste them. If not used sparingly, they can contaminate the water-based coolants in the machine reservoir, which becomes tramp oil.

Figure 1-18 *Thread-cutting compound makes a huge difference in thread quality and tool life, but it is expensive.*

Recycling and Disposing of Waste

All responsible persons working with chemicals today must understand how to sort and dispose of them. That too can be found in the MSDS sheet. Very few chemicals can be released as-is back into the environment.

Most coolants, solvents, and oils must be turned over to a company equipped to either burn or chemically neutralize them based on the type. This is a very specialized operation that is regulated by state and federal agencies. Not surprisingly, it can be more expensive to dispose of chemicals than to buy them. Recycling within the shop whenever possible makes good sense. Your shop will undoubtedly sort the various kinds of scrap and have a well-defined waste program.

Try This Online Research
Have your instructor suggest one solvent, cleaner, or other shop chemical. Now search out how it must be handled. Words that might work are *Environmental Protection Agency* and *chemical disposal procedures*. You might also seek a home page from a local agency that accepts chemicals for disposal.

UNIT 1-3 REVIEW

Review the Key Terms

Hydrostatic bearings
Found on modern CNC machines, they require an exact oil viscosity to perform right.

Material safety data system or sheet (MSDS)
The federally mandated system of documents that spell out precautions and handling instructions for all chemicals used in the workplace.

Spindle oil
Lubricant designed to prevent rolling friction.

Viscosity
The property of lubricants to resist flow and thinning out.

Way oil
A lubricant used to control sliding friction in machines (the part that slides is called the machine's ways).

Replay the Key Points

◆ Know the correct use and potential hazards of all shop chemicals.

◆ Store all chemicals and oils in designated containers and areas.

◆ Be absolutely certain that you're using the right lube for the application.

- Two kinds of coolants are used in the shop: *synthetic coolants* are syrups mixed with water and *cutting oils* and compounds are wetting lubricants for cutting metal.
- Always ask for supervision or read the MSDS sheet if in doubt about how to handle a chemical.

Respond (Answers found at the end of Section 1)

1. We find information about shop chemicals on MSDS sheets. Is this statement true or false? If false, what will make it true?

2. Synthetic coolants are mixed at dilution ratios of water to syrup, of from
   A. 20 to 1 up to 50 to 1
   B. 10 to 1 up to 20 to 1
   C. 1 to 20 up to 1 to 50
   D. 1 to 10 up to 1 to 20

3. Machine lubrication falls into two general types (other than grease used on old machines). Name them.

4. Without looking back, name as many precautions as you can recall with regard to shop chemicals (there are six).

5. The sliding parts of machines that must have a specific kind of lubrication are called_____.

UNIT 1-4

Maintaining Equipment and the Work Environment

Introduction: Our last professionalism subject is not unlike taking care of your car. We all know a clean car gets better mileage but more to the pro point, a messy car says a lot about the driver! Organization equals efficiency. Need proof? When working on your car, how much time do you spend looking for that wrench you just used a few minutes ago?

While the mileage issue for your car might be false, long life and precision results on machine tools is true if they are maintained with care. Handing over the keys to a new CNC machine means management has trusted you with the value of a dozen or more high-dollar automobiles and more, with the profit that can be made with it if it keeps running.

Managing an Efficient Workspace

Along with the machines and tools of the trade, the work area is also a tool in itself, and like all tools, requires attention and proper care. A quickly recognized trait of well-trained pros is the pride they show in the way they organize and maintain their work area. But the subject involves more than pride—if managed right, tasks go faster and have more reliable results.

There are several names for the science of efficient work areas and environments, but the most common in modern manufacturing is **lean manufacturing.** The heart of lean manufacturing is the finding the most efficient, logical way to organize and manage a working space. Using the lean concepts we count the steps and minutes required to complete a task, then do whatever it takes to reduce them. After observing the task, or a process and the people doing it, we ask what tools and supplies are needed and whether they can be placed closer to the task. Once they are provided, far more importantly, a system is created to keep them there consistently. Need proof it works? When working on your car—how much time do you spend looking for that wrench you just used a few minutes ago but didn't put back?

So you can see that lean focuses on common sense applied to shop procedures. While

Figure 1-19 *A lean shadow board and machine.*

Figure 1-20 *Machine accessories must be stored and transported correctly. Make sure they are on shelves made for heavy storage.*

there's a lot more to the science, it starts with people having exactly what they need to do a job right where they need it and eliminating clutter. It also means returning tools to their place throughout the shift—not just at quitting time. Lean people constantly control the space in which they work. It's a habit.

A picture says it all, see Fig. 1-19. Your company may sponsor workshops where teams work together to get their area up to this level of efficiency. But there's nothing wrong with you becoming a committee of one! Start with your toolbox and the immediate area in which you work. Here are a few more ways to keep the workspace humming.

Safely Storing Machine Accessories

Most machine accessories are precise, heavy, and expensive. Use lifting equipment to move them to avoid back injury and to minimize the chance of dropping them. Store them on shelves made for the load they represent and set them back from the edge (Fig. 1-20). If the stored item is a machine accessory or finished metal part prone to rusting and if it will not be used for some time, a light coat of protective oil is in order before storage. Use a rag or a spray container to put on just enough.

Removing and Handling Chips

Chips are the ever-present challenge for machinists. They come off the work in two varieties: short and long. Short broken pieces are the ones we try to produce because they're

less dangerous and easier to clean up. The long ones can be nasty since they can snag and cut hands. But whatever the chip shape, here is a list of possible hazards to control (Fig.1-21):

◆ Chips are *hot* right after they are made, they may be up to 1,000 degrees Fahrenheit in standard machining and even higher when special ceramic cutters are used. Worse yet, after they are made, they actually get hotter as they lie on the floor or down your collar. Why? You'll get the clues needed to solve this mystery in chip-making theory studied in Section 7.

◆ Chips are *sharp*—as sharp as a razor at times—sharper actually since the edge may be only a few microns wide due to the way they are sheared off the workpiece!

Figure 1-21 *Long stringy chips are strong* razor wires *that catch clothing and cut skin. All chips, even the safer "C" shapes on the right can be dangerously hot.*

- Chips *fly like bullets*—as much as 150 miles per hour on manually operated machines but up to 250 or more for CNC high-speed machining!
- Chips are *strong*. The long string type can catch and can drag the unwary machinist into machines—this is the greatest danger. They can also cut deeply if drawn across your skin.
- Chips are *slippery* on bare concrete floors.

Thinking of signing up for some other class right now? Hold on, with some preparation and prevention, these nasty critters can be tamed and controlled. There are lots of ways that we'll learn as we study drilling, milling, and turning.

Two Different Ways to Cleanup Chips

For now, lets talk about getting them out of the area. There are two ways we remove chips from machinery, depending on what the machine is doing.

- When the machine is operating and making chips.
- When it's stopped, for example, when the chip bin is full at shift change or at the end of the job.

For safety, each requires a very different action on your part.

Machine Not Running Chips pile up fast (Fig. 1-22). With today's superfast spindles and axis drives, literal bales of them can be made in just seconds. Many CNC machines feature some form of automatic removal system, while others require the machinist or an unskilled helper to intervene by blowing, raking, or brushing them away. We dispose of chips as we go but sometimes they get ahead of us and even the automatic chip conveyors on CNC machines get backed up, halting progress if not dealt with. Whatever the reason, it's time to stop machining and get them out of the way. That means cleaning the chip bin, which almost always puts some part of your body at risk if the machine were to start.

Lock It Up or Lock It Out

Before reaching in to clear chips out be absolutely sure the machine is locked or blocked from accidental start-up, say if you bump it on, or if someone else might start it by accident (Fig. 1-23). Many CNC machines feature an **axis lock** function. Other machines such as manual lathes, require throwing the main power switch. But that might be

Figure 1-22 *Chips can pile up quickly and must be removed from the machine.*

a disaster on an older computer-driven machine because of the loss of data and/or setup positions.

KEYPOINT

Never reach into a machine to remove chips unless it is locked up or out. If you must go to the back of the machine, tell someone you are there so it can't be accidentally started.

Figure 1-23 *Lock it out or lock it up—never perform maintenance on a machine that can be started accidentally.*

If neither locking out or turning off is possible, then

A. Turn CNC axis overrides to zero whereby no movement will occur.
B. Place a sign over the panel *"Operator Cleaning Chip Bin."*
C. Tell fellow machinists you are cleaning in the back or out of sight.

Trade Tip

Chip Breaking There are two ways of breaking stringy chips into safe little packets. In the first, using heavier machining action, forms thicker, stiffer chips that cannot take the bending action of being removed from the parent metal. They then break into small C shapes. The other answer is using a cutting tool with a chip breaking dam (Fig. 1-24). The obstruction redirects the flowing chip such that it snaps off. We'll see more on both techniques in tool geometry later.

Broken "C" shape chip
Chip breaker
Cutter
Metal being cut

Figure 1-24 *A chip breaker bends the flowing chip so quickly that it snaps off into small "C" shapes rather than a long string.*

Now with the machine safely locked up or locked out, put on gloves and rake the chips out safely. If you are assisting other machinists (you aren't the machine operator), be absolutely sure that the person who is the operator knows you are working in the machine chip bin.

Clearing Chips from Operating Machines

This is a very different procedure from the standpoint of safety. It's often necessary to clear away chips from machines as they are being made. This is a machinist's prime duty. Here are the details

◆ If your machine is moving *do not wear gloves.* Lacking sensitivity, gloves put your hands at increased risk of being caught in the machine.

◆ Never reach in the machine with your hand or a gripping tool such as a pliers or vise grip. Can you guess why? Because you can't let go fast enough if the chips or machine catch the gripping tool.

Here are four safe ways to remove chips from moving machines:

◆ A sturdy brush
◆ Coolant stream
◆ Compressed air
◆ A chip rake or chip hook

The brush is a safe, self-explanatory tool used on drills and mills but not lathes and CNC operations since your hand is too close to the action. Although they sometimes get "eaten" by the machine, you can't hold on tightly enough to a brush to get hurt. A brush is the first choice of beginning machinists in training where chip volume is low.

Using a Chip Hook or Chip Rake

The longer chip hook pulls long stringy chips out of the machine, while a rake removes broken chips. Both tools are made with handles that can't grab your hand should the business end accidentally get caught in the machine. The handle should not turn back

500 #   8 #

Figure 1-25 *Sorting chips from scrap metal then recycling each improves profitability plus it saves 80 percent of the energy required to mine then refine raw metal.*

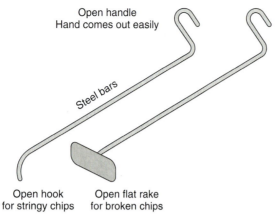

Open handle
Hand comes out easily

Steel bars

Open hook
for stringy chips

Open flat rake
for broken chips

Figure 1-26 *Chip hooks and chip rakes feature open handles designed to not get caught in moving machinery.*

Figure 1-27 *Compressed air is a good chip removal and cleanup tool when correctly used.*

into a closed loop. Figure 1-26 shows the right shape. Notice too that the hook end is not a full 90 degrees either. The open hook tends not to catch on the machine and also to release the chips easily.

Compressed Air

Nearly every shop uses compressed air to power hand tools, to actuate machine actions such as pneumatic clamps, and also for cleanup work. Air is a safe and efficient tool used for chip removal in modern machining. But it can be misused against both people and machines. Use these two pro guidelines to stay out of trouble (Fig. 1-27).

Guide 1. Personal Danger—Never blow off chips from your skin or clothing or point the air at fellow workers! You can embed chips in skin or even inject air under the skin. It is possible to introduce air into veins and arteries, where serious consequences can occur. Similar to the embolism experienced by scuba divers, air in your blood can stop the flow to your heart or prevent circulation to your brain. Any body opening and the organs within can be ruptured with compressed air (even from a distance). It should be obvious: never point the airstream, with or without ejected chips, at someone or at other machines. At the *very least*, damage to computers, delicate measuring tools, and your reputation can occur. When using air on a manually operated machine it's best to place a guard wall behind or beside it to protect others. CNC machines often feature full containment shields that do the job nicely.

Guide 2. Machine Safety—Never point compressed air at seals on the machine. All machine tools are equipped with seals designed to keep dirt and metal chips away from bearings and precision flat sliding surfaces. However, these tough seals are not able to stop pressurized air. Air can force chips under the seals, defeating their ability to stop further invasion.

KEYPOINT

A carelessly air-injected chip in a machine seal, lifts the seal, thus defeating it's usefulness. The embedded chips can also scratch precision surfaces.

Used correctly, compressed air works well to keep chips out of the immediate machining area, especially when the machine is moving. An airstream enables chip ejection without holding any solid object between you and the machine. It can't get caught and pull you into the machine. However, air cleaning and chip ejection is controversial. Many supervisors, instructors, and textbooks rule that it can't be used. But in the real world it's used pretty much everyday.

KEYPOINT

Ask First
Be aware that due to the potential for machine damage some shops have a strict rule against using air as a chip cleanup tool.

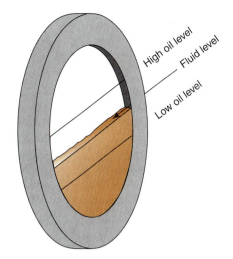

Figure 1-28 *Sight glasses are small windows in the machine's oil sump that show the fluid level.*

High oil level
Fluid level
Low oil level

Maintaining the Work Area

As boring as a list might be, it can't be avoided here if you are to know what the other machinists will expect of you when you enter the craft. Here are a few unwritten rules on how to keep your shop lean and mean:

- As much as possible, keep measuring tools away from the machine and in their case.
- When laying measuring tools or wrenches down, never lay them on sliding machine parts such as lathe ways. Do put a shop towel under them.
- Keep clutter such as rags, metal scraps, and especially old paperwork picked up. Keep personal items to a minimum too.
- Keep chips swept and picked up.
- If you see something that needs maintaining, fix it or report it, but don't ignore it.

Checking Machine Fluid Levels

Oils are consumed and coolant levels fall due to splashing, evaporation, and small amounts leaving on the work itself, called *clinging.* Nearly all newer CNC machines feature sensors and low-fluid alarms, but not all are so equipped and no manually operated machines have this protection. To determine fluid levels in these machines, there might be a dip stick or a **sight glass,** such as the one in the drawing. This is simply a little glass window into the side of the reservoir. Add the lubricant or coolant to the indicated full line (Fig. 1-28).

Lubrication Flow Glasses

When maintaining older machine tools, you must know the difference between a sight glass for checking fluid levels and a **flow glass** for coolants and lubrication. A *sight glass* has a line on the clear window, its purpose is to see critical fluid levels. It's like a dip stick, telling you how much oil or coolant to add.

No line on the glass or around the window's outside rim, means it is a *flow glass.* It's a foolproof way of seeing that the fluid pump is working. A flow glass shows when oil is moving up to the high point, the point where the glass window is located. Found on manual machines not CNC, when the pump is working, the oil flows down over this glass.

Caution, machines may have more than one sump with different oils in each.

On many smaller machines like those used first in a typical machining course, there is probably a central, manual lubricating oil pump for sliding surfaces. Its handle must be lifted or pumped each time you step up to the machine and once each hour of operation. This is sometimes referred to as a **one-shot lubricator** (Fig. 1-29).

Danger!
wide gap

Correct gap

Adjustments to Equipment

Naturally, there are many routine adjustments that must be made on machining equipment—to walk the walk, don't walk away from them. Make adjustments when

Figure 1-30 *Due to grinding wheel wear, this tool rest gap has become dangerously wide and must be adjusted closer to the wheel to avoid pulling in small parts and fingers!*

they need to be done and on a schedule too. Safety guards, for example, or eye shields, and doors that close for containment, chuck guards, limit switches and belts—they're everywhere in a machine shop and they are part of your job.

Tool rests on bench grinders (Fig. 1-30) are a perfect example. As the grinding wheel is used, the gap widens. To minimize the pinch point created between the grinding wheel and rest, keep adjusting it to a minimum of 1/16 in. or less. Zero clearance is acceptable because items cannot be pulled into the gap, but it could result in the tool rest being eventually worn away.

PULL HANDLE ONCE EVERY
4 HOURS OF OPERATION.
FILL WITH SUNOCO #1180
WAYLUBE

Figure 1-29 *A one-shot central lubricator. Use at least once per half day or on the schedule set by your instructor.*

UNIT 1-4 REVIEW

Review the Key Terms

Axis lock
One of the options a machinist might have to prevent any possibility of a machine moving while they are cleaning chips from the bin.

Chip breaking
The action of breaking chips into easier to handle, short segments.

Flow glass
A glass window in an older machine that indicates the pump is working as oil flows over the window.

Lean manufacturing
The study of organizing and managing an efficient work environment.

One-shot lubricator
A manual lubrication pump found on smaller machine tools such as school machines.

Sight glass
Functions like a dip stick by allowing the fluid level to be seen and compared to a line.

Replay the Key Points

◆ Depending on whether the machine is working or stopped, safe chip removal actions are very different. Identify or describe them.

◆ Air is a safe chip removal and cleanup tool, but some shops ban its use in this capacity.

◆ Lubrication is technical and often the machine operator's responsibility. Using the wrong oil can damage technical equipment almost as soon as using none at all!

◆ Savvy machinists keep their eyes open for conditions that need attention and they fix problems themselves.

Respond (Answers found at the end of Section 1)

1. Why is air not always permitted as a cleanup tool?

2. Gloves are acceptable when removing long stringy chips. Is this statement true or false? If false, what makes it true?

3. Why are flow glasses not found on modern CNC machines?

4. Name the four safe methods of removing chips from moving machinery.

5. Complete this sentence: Shop-made chip hooks and rakes are made such that: _____.

Section 1 Review

Unit 1-1

Not one part of Unit 1 or all of Section 1, for that matter, has been filler material. It was very deliberately placed here. Demonstrating a professional attitude, in the way you dress for the shop environment is one of the best tools you can develop to show your instructor and employer early on that you are on your way to becoming a top gun.

Unit 1-2
Unit 1-3
Unit 1-4

I once took my CNC class on a tour of a local shop. As we walked in I saw an apprentice I had previously trained, putting away a five gallon oil-can. Not realizing we were watching her, she carefully wiped the top, then rotated it so the label could be read. Only then did she close the door to the fireproof cabinet. I was truly proud but also noticed her supervisor nod toward me with approval!

Believe me, it's details like that, taking care of shop supplies, knowing their value and using them responsibly along with a good respect for machine accessories, and tools that give the beginner the walk of a journeyman. They will be noticed, or more importantly, when they are ignored, they get noticed even faster.

Questions and Problems

1. Describe a well-dressed machinist. Use at least five dos and two don'ts. (Obj. 1-1)

2. True or false? It's OK to listen to music with a personal player as long as you leave one earphone out to be able to hear your machine. If it is false, why? (Obj. 1-1)

3. List the best to worst ways to lift heavy objects. (Obj. 1-2)

4. Name two professional precautions (of three) for storing metal bars. (Obj. 1-2)

5. Explain why we use the bent knees lifting technique in ten words or less. (Obj. 1-2)

6. What would you expect to find on an MSDS sheet? (Obj. 1-3)

7. Lubricants fall into two general groups to prevent _____ and _____ friction. (Obj. 1-3)

8. True or false? It's OK to substitute lubricants as long as the new one has a higher viscosity than the required oil so it prevents friction better than the original. (Obj. 1-3)

9. Why do we not wear gloves around moving machinery? (Obj. 1-4)

10. True or false? Pressurized air is permitted as a cleanup and chip removal tool in all modern shops. If it is false, why? (Obj. 1-4)

Critical Thinking

11. A bar of spring steel has been misplaced, but you have found another that you are sure is the same material. Can it be substituted for the job? Is there a way to maintain the traceability? (Obj. 1-2)

12. True or false? Dark glasses impair vision and they are never worn in the machine shop. If it's false, why? (Obj. 1-1)

CNC Question

13. Your CNC mill has stopped after completing a part cycle and a red light flashes while your operator's panel tells you that the way lubricant is low. Name three or more professional steps needed to get the machine up and running again? (Obj. 1-2)

14. Why do you suppose the machine completed the cycle in Question 13 before it shut down due to low supply of a vital fluid? (Obj. 1-2)

15. Why must a specific lubricant be used in most modern CNC bearings? (Obj. 1-3)

ANSWERS 1-1

1. Yes for yellow lenses, no for dark unless working near incidental welding light. Note: dark lenses will not protect your eyes from looking directly at a welding flash, they only protect from incidental light reflected off walls and ceiling.
2. Natural fibers stand up to hot chips so they don't melt or burn the wearers—breaking their concentration.
3. Traction, foot protection, and fatigue
4. Besides the joy of hearing, it's your prime control of the process.
5. Air currents and static electricity

ANSWERS 1-2

1. Three or more: overhead cranes moving on rails, rolling lift tables, floor jacks for pallets and tub skids, fork lift trucks, portable jib cranes often called "cherry pickers," jib cranes fastened to a column to pivot around a circular area
2. False The forward end high makes it out of your reach and out of your control.
3. For two reasons: Once the ID is lost the metal cannot be proven without a lab test, as to what alloy it is. Traceability demands that there be a trail from original manufacture all the way to a specific part. Loosing the ID means that it is gone.
4. A fluid center surrounded by several layers of tough skin
5. On a shelf where they cannot fall through

ANSWERS 1-3

1. True
2. A
3. Sliding (way oil) or rolling (spindle oil)
4. The precautions are fire hazard, chemical stripping of skin oils—direct contact, fume toxicity and oxygen exclusion (displacing oxygen), eye irritation, allergic reactions, contamination of other fluids or chemicals with possible side reactions.
5. Ways

ANSWERS 1-4

1. It may be banned due to potential damage to machine seals.
2. False Use gloves only when the machine is not moving.
3. They are equipped with oil level sensors and alarms.
4. Chip hooks/rakes, airstream, coolant stream, brushes
5. Neither end can catch on machines or hands

Answers to Section Review Questions

1. Do wear: eye protection, shoes designed for the shop, hearing protection, tight-fitting clothing of natural fibers, a shop apron or coat. Don't wear jewelry or gloves or leave long hair un-controlled. Any shop coat or apron should not have loose ties or pockets.
2. False Any disturbance to hearing impairs ma-chine control.
3. Mechanical devises (crane, shop lift, fork lift, etc.), two or more people working together, all using correct technique, one person using the bent knee method
4. Never store short bars in the long bar rack where they could fall through. Store them with their ID visible and never cut the color code or stamp from the bar.
5. To prevent excessive pressurization and dam-age to disks in the spine.
6. Instructions for the safe use and disposal of specific chemicals plus storage and special precautions
7. Sliding and rolling friction
8. False

9. Gloves desensitize your hands, which tends to help get fingers caught in machinery.
10. While air is used by many as a cleanup tool, it can also cause damage to machine seals and other items in the area. It may be banned for this purpose in many shops.
11. The short answer is *no,* at least not by you. However, if the heat lot for the new bar is on record and its specification matches the lost bar, then the paperwork can be cut to make the substitution—the traceability thread is not lost this way.
12. It's mostly true. But we sometimes must work around electric welding. Then they are OK to prevent incidental light damage to eyes.
13. A. Determine the specific way lube that's right for that machine.
    B. After getting the right oil, make sure the funnel and can are clean before opening the oil port on the machine (to avoid con-tamination of the reservoir).
    C. Fill until the sight glass line shows the reservoir is full.
    D. Close the can and return it to the storage area (to prevent fire hazard and to keep the can from being contaminated).
14. Modern CNC machines tell the operator when necessary fluids are low long before they reach the critical point. Therefore, the control allows finishing a cycle before stopping.
15. They depend on an exact viscosity for accuracy as well as long life.

Section 2

Math Skills Self-Review

Introduction

Working in manufacturing means math—lots of it. It doesn't matter whether you're a top toolmaker, a programmer, or a beginner, nearly all your actions in the shop will be based on numbers and most must be calculated in some way. There's no escaping it. Every step from drawing the design to final inspection requires math.

Today, calculators are as common in the machinist's toolbox as micrometers. This need for number skills is only going to increase as we commit further to technology. Are you ready or would a little review be in order? Section 2 helps you answer that question and offers some hints about how to approach shop math. In addition, Section 2 is provided as a self-appraisal and a warm-up. If after trying the problems your score is below expectation, then a refresher math course or some self-study is indicated. Here are the goals of Section 2 to get those gray cells humming.

Objectives

UNIT 2-1

Understanding Precision

Introduction: What does precision mean? The answer varies depending on the science or profession in which the word is used. In the machining trade, it means cutting and measuring part features to within a few thousandths of an inch or decimal parts of a millimeter. Miss the target by too much or too little and the work is scrap!

Working all day long with small decimal numbers, machinists have developed their own way of pronouncing them. Besides sounding in-the-know, learning this trade lingo helps eliminate misunderstandings and more importantly for the student, it greatly speeds up mastery of measuring tools, print reading, and machine operations.

KEY TERMS

Imperial
The units based on imperial England—the ruler that includes feet and inches.

SI—International System of Units
The system of measurements based on metric values.

Tenth (of a thousandth)
0.0001 in.—to a machinist one-tenth of our basic spoken unit, the thousandth of an inch, thus it's a tenth in shop lingo.

"Talking Precision"— Thousandths of an Inch

When working in inches we machinists refer to the inch as our working unit, but we speak as though it is the thousandth of an inch, and that's different from the regular non-precision world. How do you pronounce: *0.01 in.?*

If you said either "one hundredth inch" or "point zero one inch," you were correct for general math or when using metric tools and dimensions "point zero one" is correct. But when machining to inch dimensions, you will say "ten thousandths of an inch"! Or it

might be shortened to ten thousandths. Or even "ten thou" in slang.

KEYPOINT

In the shop, we pronounce inch decimals as though they were all extended to the third column and 0.01 becomes *0.010 in.*

Trade Tip

When a decimal inch number isn't extended to the third column, add the missing zeros to make it into thousandths of an inch.

TRY IT
A. 0.13 B. 0.013 C. 1.31
D. 0.2 E. 0.25 F. 0.303

ANSWERS
A. One hundred thirty thousandths of an inch—or one hundred thirty thousandths
B. Thirteen thousandths
C. One inch, three hundred, ten thousandths
D. Two hundred thousandths
E. Two hundred, fifty thousandths
F. Three hundred, and three thousandths

How Small is a Thousandth of an Inch?

It's difficult to visualize 1 inch divided into one thousand parts (0.001 in.), however, here are some everyday examples:

A human hair is from 0.002 to 0.005 in. thick.

The paper this book is printed on will be from 0.003 to 0.004 in. thick.

A dime is about 0.04 in. (forty thousandths) and a stick pen around 0.390 in. in diameter—three hundred ninety thousandths of an inch or shop-shortened to: three hundred ninety thousandths.

Pronounce the horizontal and vertical dimension on the part print in Fig. 2-1 and the hole diameter?

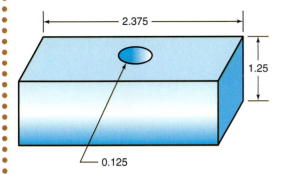

Figure 2-1 *Say these dimensions to the thousandth of an inch.*

ANSWERS
Two inches, three hundred seventy-five thousandths
One inch, two hundred fifty thousandths
One hundred, twenty-five thousandths inch diameter

Smaller Yet—Tenths of a Thousandth

Thousandths of an inch aren't precise enough for some designs so we further divide them into 10 smaller parts. As a machinist, you will pronounce these as **"tenths."** Again, quite different from the outside world, not tenths of an inch (0.1) but tenths of our spoken base unit, the thousandth.

For example, *0.0001* in. is one tenth of a thousandth.

A human hair might measure 0.0037 in. thick; larger than three thousandths by seven tenths. That's pronounced three and seven tenth thousandths or shop-shortened to three and seven tenths. As odd as that sounds to an outsider, any person in the trade would understand exactly. Say to an inspector or engineer, "The diameter is five tenths too small" and they would know that 0.0005 in. more must be taken out of the hole to be the right size (Fig. 2-2).

TRY IT
0.1259

ANSWER
One hundred, twenty-five thousandths *and nine tenths.*

Figure 2-2 *This machining inspector needs math skills to read the drawing and then program this computer coordinate-measuring machine (CMM).*

Trade Tip

First, pronounce the thousandths then add on the tenths.

TRY IT
Practice pronouncing these with another student.
A. 0.0067   B. 1.5678
C. 0.9999   D. 0.0878
E. 0.0087

ANSWERS
A. Six thousandths and seven tenths or six and seven tenths thousandths (Either expression would be OK even though it isn't correct outside the machining field.)
B. One inch, five hundred, sixty-seven, and eight tenths thousandths
C. Nine hundred, ninety-nine and nine tenths thousandths or nine hundred, ninety-nine, and nine tenths
D. Eighty-seven thousandths and eight tenths
E. Eight and seven tenths thousandths

Getting the Most Accuracy Using Your Calculator

When the calculations are to be made into solid metal, there simply can be no answering less than 100 percent right. In the real world, anything less results in scrap and lost profit! So before moving on to the review problems, here are a few suggestions on how to get the best results when using a calculator in applied math.

Rounding

Never round any number until the final result.

Then, *always round to the nearest tenth (0.0001 inch).*

Use the whole decimal number during calculations.

Keep all numbers and results on the calculator using the memory keys.

Round up if the trailing digits are 5 or greater. Round down if they are less than 5. Example: rounding to the nearest tenth (of a thousandth of an inch):

1.346*652 in.* = 1.3467   (Round up)

1.346*648*   = 1.3466   (Rounded down)

Take Advantage of "The Mystery Digits"

The display isn't the entire amount the calculator is using. In some calculations where decimals extend very far from the point, the processor will be using a number that extends two or three places beyond the display, depending on the calculator sophistication.

If the screen number is written on paper to be used later, the unseen decimals are lost. However, by using the calculator's memory to record intermediate results, the entire number is stored, including the unseen digits. Usually we do not need this exceptional accuracy, but it can make a difference when doing several multiplications of the number in a series. While the difference might be small, keep in mind we deal in small. Missing this hint might result in a loss of one or two tenths—just enough to cause a problem.

Calculator Rounding

Your calculator may have a decimal fix function (rounding). Use it to round results to the job tolerance of thousandths or tenths. But doesn't that diminish the accuracy? *No.* Unlike pencil and paper rounding, nothing is lost in the calculator. Only the display is rounded. The internal number continues to the full capacity of the processor.

Little Hints

Use Memory to Record Intermediate Results By all means, write them on paper too, but *do not depend on written numbers,* they create a chance to introduce errors.

Use Fresh Data When given the choice between two numbers to solve a problem, always choose the number that is closest to the original information on the print. Do not use calculated results if they can be avoided. They may be rounded, transposed, or wrong. Next is a hint to avoid that.

Talk to Yourself During calculator entry avoid accidental transposition and "fat finger" errors by saying them aloud slowly as you touch the keys. While doing so, look at the display. No kidding, this works when writing numbers on paper too. It's OK to talk to yourself in shop math!

Simplifying Fractions To convert a fraction into a decimal number divide the numerator (top) by the denominator (bottom); that's the rule. But modern calculators often feature a direct fraction key. It may be an *A/B* on the keypad. There may be several fractional keys involving *A/B* and *C* as variables. These keys allow duel inputs in a single problem, both decimal and fractional numbers can be used at the same time. If an older print is dimensioned with fractions, using this key can speed up and simplify the math. Find it and teach yourself to use it. It will be very helpful in the upcoming problems.

KEYPOINT

Use your calculator for rounding numbers with no loss of accuracy, for simplifying fractions and for storing intermediate numbers carried out to the capability of the calculator.

Conversion Review: Metric and Imperial Units

Occasionally, it becomes necessary to convert between **Imperial** (foot-inch) and metric **(SI)** units, especially when making products for the world market. As we enter the world competition, expect more work to be dimensioned in SI units. Here's a brief refresher on conversion. Most calculators feature direct conversion functions—read the manual or experiment with these examples:

Converting to Metric from Imperial Units

Multiply by To obtain
inches 25.4 millimeters

Examples:
How many millimeters are in 2.125 in.?
$2.125 \times 25.4 = 53.975$ mm

Convert 0.0935 in. to millimeters
$0.0935 \times 25.4 = 2.3749$ mm

Converting from Metric to Imperial Units

Multiply by To obtain
millimeters 0.03937 inches
in thousandths

Examples:
How many thousandths are in 23 mm?
$23 \times 0.03937 = 0.9055$ in.

Convert 245 mm to inches
$245 \times 0.03937 = 9.6457$ in.

TRY IT

A. On a separate sheet of paper or on keys, using Fig. 2-3, convert the imperial dimensions to millimeters.
B. From Fig. 2-4, convert the metric print to imperial units.

ANSWERS

A. Converting to metric units:
   Width 4.500 in. = 114.3 mm
   ($4.5 \times 25.4 = 114.3$ mm)
   Height 2.875 in. = 73.025 mm
   Slot 0.75 in. = 19.05 mm
B. A metric block $80 \times 40 \times 14$ mm is 3.1496 by 1.5748 by 0.5512 in.

Figure 2-4 *Convert this metric print to imperial (inch) units.*

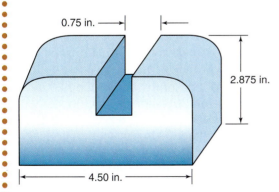

Figure 2-3 *Convert these dimensions to metric units.*

UNIT 2-1 REVIEW

Review the Key Terms

Imperial
The units based on imperial England—the ruler includes feet and inches.

SI—International System of Units
The system of measurements based on metric values.

Tenth (of a thousandth)
0.0001 in.—to a machinist this is one-tenth of our basic spoken unit, the thousandth of an inch, thus it's a tenth in shop lingo.

Replay the Key Points

◆ Machinists pronounce decimals as though the thousandth of an inch was their base.

◆ Using a calculator to its fullest requires study of its various functions.

Respond

Move on to Unit 2-2 to solve the first set of self-evaluation problems.

UNIT 2-2

Shop Math Self-Evaluation

Introduction: These problems are typical of those encountered in early training. A score below 100 percent indicates some math review might be in order. But not in every case, see Instruction B.

R 3/8" ± 0.010

Figure 2-5 *Problems 1 and 2, hinge bracket.*

Kerf
The narrow band of material removed by the saw blade's thickness.

Predictable error point (PEP)
Errors that often occur; machinist should be aware of PEPs to avoid them.

Unit 2-2 Instructions

A. Answers are found at the end of this section. Look up each on completion of the problem. If a detailed explanation is needed, turn to Unit 2-3.

B. Some problems have built-in **PEPs (predictable error points).** They are traps we all stumble into occasionally. The full explanation will clarify how to avoid them in the future. If you fell for one or more, *they are not indicators of need for more review.*

C. Engineers, programmers, and machinists often draw sketches to organize their thinking. This is a good time to get started. Record calculations on a sketch drawn close to a scale model of the problem, but don't forget to use calculator memory too.

D. Since these are real shop problems, there may be terms or symbols you do not understand. If so, ask your instructor for an explanation.

E. The goal is 100 percent—no less. Take enough time to be positive about answers.

F. On completing and correcting these problems go to the answers to Unit 2-2, to see if the warm-up was enough or if further review is in order. You are the judge.

Basic Shop Problems

1. A hinge bracket requires milled corner radii (plural of radius), a rounded corner. The print calls for a $\frac{3}{8}$-in. radius with an allowable tolerance of plus/minus 0.010 in. on the dimension. You have found a cutter that will cut a 0.275-in. radius. Can you use it for this job in Fig. 2-5?

2. What size range of radius cutters could be used for hinge bracket of Fig. 2-5? In other words, what is the biggest and smallest radius that would be acceptable for this job?

3. Using Fig. 2-6, the drill gage, what is the vertical distance between the top left, $\frac{3}{16}$-in. and

The wording in Problem 7 is typical of a shop work order—abbreviated without conjunctions.

the $\frac{9}{16}$-in. hole? (*Hint:* This is the center-to-center distance of two holes in the same vertical column.)

4. You need to drill and ream a series of holes in a metal bar per the drawing in Fig. 2-7. There is an initial 1.0 in. from the edge to the center of the first hole (called *edge* or *end spacing*) on both ends. Then the five holes are equally spaced within the remaining distance. All holes are $\frac{5}{8}$ in. in diameter.
   A. What will be the equal spaced, center-to-center distance?
   B. What is the edge-to-edge distance if they are drilled per the drawing?

5. The print shown in Fig. 2-8 calls for a milled step that is $1\frac{5}{16}$ in. The tolerance is plus 0.015 in. and minus 0.000 in.
   A. The actual dimension measures 1.318 in. after machining. Is it acceptable or not?
   B. How much more must be milled off for the part to be $1\frac{5}{16}$ in.?

6. You have been given a job that requires 25 parts made from a bar that is $21\frac{5}{8}$ in. long. Per Fig. 2-9, each part is to be $\frac{3}{4}$ in. long and each saw cut requires an additional 0.10 in. of material (called **kerf**). Can you make all 25 parts from the bar. If not, how many can you make from it?

7. The work order instructs: "Opp 20 (operation 20)—Rough machine rectangular hole w $\frac{1}{16}$" excess." It is to be finished later. For now, you are to leave $\frac{1}{16}$ above the finished sizes, material to be finished later, on all four inner sides of the hole (see Fig. 2-10). What will be the rough, inner width and length when you are done?

Figure 2-6

0.1875
0.2500
0.3125
0.3750
0.4375
0.5000
0.5625
0.6250 Hole diameters

2.00
1.50
0.38
0.50

0.12 R
All corners

No radius

5.00
3.75

3.25
2.88
2.50
2.13
1.75
1.25

59.00°

1.00
0.63

0.63
1.25
2.00
3.00

A
B

Drill gage project

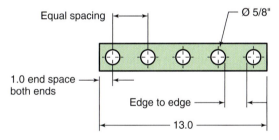

Equal spacing
Ø 5/8"
1.0 end space both ends
Edge to edge
13.0

Figure 2-7 *Problem 4.*

3/4"
Saw kerf
21 5/8"

Figure 2-9 *Problem 6.*

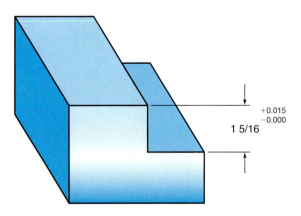

+0.015
−0.000
1 5/16

Figure 2-8 *Problem 5.*

2.625
1.00
1/16" Machining excess all sides

Figure 2-10 *Problem 7.*

UNIT 2-3

Problem Solutions

Introduction: Here are explanation details for the basic shop problems in Unit 2-2. Most problems required fractions converted to decimals, then combined by adding, subtracting, multiplying, and dividing. Problem 7 could be solved directly in fraction or in decimal form. Remember if you fell into one of the PEPs (common traps) as in Problem 4, that is not an indicator of a need for more training, it's a lesson in avoidance.

KEY TERM

Pi (π)

The ratio comparison of a circle's diameter compared to its circumference equal to 3.1415926; Any circle is just over three times bigger around than it is across.

Solving Problems 1 and 2

These problems involve changing a fraction to a decimal then adding and subtracting the tolerance to see what range is acceptable.

1. $\frac{3}{8}$ in. = 0.375
   Solution—No. 0.275 in. is far too small for this job. Did you determine that 0.275 was an acceptable size? A common PEP in machining and measuring is the gross 0.100-in. error. See the answer for Problem 2, watch out for this PEP—it does happen!

2. The decimal tolerance was added or subtracted to the 0.375 yielding *0.365* minimum to *0.385* in. maximum, for possible cutter size.

Solving Problem 3

Find the right figures on the print then subtract decimal numbers. The real trick wa determining which hole was the $\frac{9}{16}$ in., the bottom left hole. See Fig. 2-11: 3.25 in. − 0.63 in. = *2.62* in.

Solving Problem 4

Did you answer 2.20 in.? The PEP here is that five holes enclose *four spaces! Solution:* The total distance was 11 in. divided by 4 = 2.75 in. center-to-center distance (see Fig. 2-12). To find the edge-to-edge distance, subtract the hole radius two times from 2.75. The diameter is $\frac{5}{8}$ in. converted to decimal = 0.625 in. 2.75 − 0.625 = 2.125 in. edge to edge.

Center to center is 2.750 in.

Edge to edge is 2.125 in.

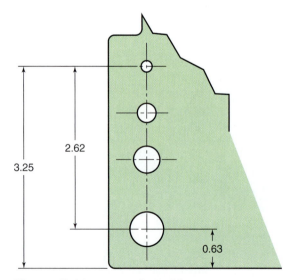

Figure 2-11 *Answer for problem 3.*

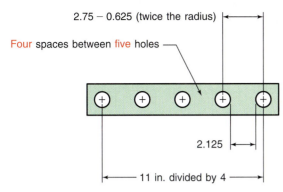

2.75 − 0.625 (twice the radius)

Four spaces between five holes

2.125

11 in. divided by 4

Figure 2-12 *Answer for problem 4.*

smaller. There are three different ways to solve this:

1. Use a decimal chart and find $\frac{1}{8}$ in. less than each dimension.
2. Convert the finished pocket to decimal form then subtract 0.125 from each dimension.
3. Solve the problem in fractional form, subtracting $\frac{1}{8}$ in. from each dimension. The width would then be $2\frac{5}{8}$ in.

$2\frac{5}{8}$ minus $\frac{1}{8}$ = $2\frac{4}{8}$ reduces to $2\frac{1}{2}$ in.

1 inch is equal to $\frac{8}{8}$ minus $\frac{1}{8}$ = $\frac{7}{8}$ in.

Solving Problem 5

This was a straight conversion and subtraction of decimal numbers. The $1\frac{5}{16}$-in. size becomes 1.3125 in. The milled size is 1.318 in., within the tolerance range by 0.0055. That is the material to be removed for perfection.

A. Yes—it is within the tolerance.
B. 1.318 − 1.3125 = 0.0055 in. can be removed.

Solving Problem 6

This problem involved a division of decimal numbers. The material consumed for one part was 0.85 in. long with the 0.10 *kerf* added to the 0.75 width.

$21\frac{5}{8}$ = 21.625 divided by 0.85 = 25.44 in.
Yes, you can make 25 parts.

Solving Problem 7

Since $\frac{1}{16}$ in. is to be left on both sides of each dimension, the rough pocket will be $\frac{1}{8}$ in.

Working within Fraction Form

Before subtracting one fraction from another, they must be in terms of a *common denominator*. Do you remember how to convert a given fraction to one with a new denominator? Recall that the denominator is the number on the bottom of the fraction. Follow this example.

Example Problem: What is $\frac{1}{32}$ in. less than $\frac{7}{8}$ in.?

Divide the old denominator (8) into the new denominator (32) and find that there are 4 (32nds per numerator). Now, since there are 7 numerators, multiply $7 \times 4 = 28$. $\frac{7}{8}$ in. = $\frac{28}{32}$ in.

Result, there are $\frac{28}{32}$ in $\frac{7}{8}$ in.

First convert $\frac{7}{8}$ in. into 32nds to subtract like units:

$$\frac{7}{8} \text{ in.} = \frac{28}{32} \text{ in.}$$

Now subtract

$$\frac{28}{32} - \frac{1}{32} = \frac{27}{32} \text{ in.}$$

Section 2 Review

Introduction: Now, with a review of the previous problems, see how much improvement has been achieved. Remember, ask for help with unfamiliar terms or symbols.

Instructions

A. This is a self-test—not a warm-up.

B. Do not look at each answer—be certain you are right on all problems.

C. Correct your own work and conclude readiness for lab experiences.

Questions and Problems

Write these decimals in words (Objs. 2-1 and 2-2)

1. 0.809 in.
2. 0.056 in.
3. 2.345
4. 6.09
5. 0.12 in.
6. 0.0089 in.
7. 0.0324 in.
8. 3.0506 in.
9. 0.5427 in.
10. 5.3387 in.

Convert imperial numbers to metric (Obj. 2-1)

11. 4 in. = _____mm
12. 2.5 in. = _____mm

13. 4.75 in. = _____mm
14. 20.0 in. = _____mm

Convert metric numbers to imperial inches—to the nearest tenth of a thousandth

15. 25 mm = _____in.
16. 120.5 mm = _____in.
17. 358 mm = _____in.
18. 225 mm = _____in.
19. 4.75 mm = _____in.
20. 2.5 mm = _____in.

Critical Thinking

21. A print calls for one rectangular part made from sheet steel. It is to be machined to $3\frac{3}{4}$ by $5\frac{3}{8}$ in. when finished. (See Fig. 2-13.) Operation 30 requests 0.150 in. extra machining metal added to every edge of the rough sawed material. The saw kerf consumes 0.060 in. for each cut. How much material will be consumed from the sheet for the one rectangle? (Objs. 2-1 and 2-2)

22. A customer asks for 50 tool hooks. This job is made from $\frac{3}{8}$ in., round steel rod. Each hook requires a length of $4\frac{3}{8}$ in. material plus 0.06 lathe, parting tool, kerf where it is cut away from the bar. How much material in inches, must you bring to your lathe to make all 50 parts—rounded to the nearest inch? Answer in feet and inches, rounded to nearest inch. (Obj. 2-2)

Figure 2-13 *Problem 21.*

23. The print requires a custom bored hole of 0.875 in. diameter. You need to select a drill bit that is $\frac{1}{32}$ in. smaller than the finished bore size, for machining excess (metal left to machine after drilling). What size drill will be used to predrill this operation? (Obj. 2-2)

24. Using the decimal chart in the *appendix,* what is the next nearest whole millimeter, metric drill diameter, that is *smaller* than the inch drill of Question 23?

CNC Question

25. In writing a drill routine for a CNC mill, you need to complete the **S** command word to tell the controller to revolve the given spindle **S**peed. The numbers following the **S** prefix will tell the controller the RPM speed. Calculate the RPM then fill in the command statement **S_____**.

For this hardened steel workpiece, the recommended surface speed is 80 ft per minute. The drill diameter is $\frac{7}{16}$ in. The formula to find the correct drill RPM for drilling a steel workpiece is

$$\text{RPM} = \frac{\text{surface speed} \times 12}{\text{Pi} \ (3.1416) \times \text{drill diameter}}$$

$$\text{RPM} = \frac{\text{surface speed} \times 12}{\pi \times \text{drill diameter}}$$

$$\pi = 3.1416$$

(Objs. 2-1 and 2-2)

ANSWERS 2-2

1 and 2. *Absolutely not.* The range of acceptable radii is 0.365 to 0.385. The 0.275 cutter is far too small. If you said yes, you fell into a common machinist trap—"the 0.1 error."

3. 2.6

4. Center to center is 2.750 in.; edge to edge is 2.125 in.

5. Yes—it is within the tolerance. You can remove another 0.0055 in.

6. Twenty-five full parts, with just a bit left over.

7. The rough pocket will be $2\frac{1}{2} \times \frac{7}{8}$ or 2.500×0.875 in decimal form.

Could you do these problems with 100 percent accuracy? If so, then there is no need for further review. Skip to the Critical Thinking review problems to solve the challenge problems and complete Section 2. If you did not achieve 100 percent accuracy, then go to Unit 2-3 for complete solution explanations. Remember, missing the predictable error points was not a math failure. Then, if the solutions make sense and you feel they helped, go to the challenge problems in the review section. If you then find that you can achieve 100 percent on those problems, which are slightly harder, it was just a case of needing to warm up to shop math. If not, ask your instructor for a math review suggestion.

Answers to Section Review Questions

Note that "of an inch" could be said where the parentheses appear but would probably not be added in real shop life.

1. Eight hundred, nine thousandths (of an inch)

2. Fifty-six thousandths (of an inch)

3. Two inches, three hundred, forty-five thousandths ( )

4. Six inches, ninety thousandths ( )

5. One hundred, twenty thousandths ( )

6. Eight thousandths and nine tenths *or* eight and nine tenth thousandths ( )

7. Thirty-two and four tenth thousandths ( ) *or* thirty-two thousandths and four tenths
8. Three inches, fifty thousandths and six tenths *or* three inches, fifty and six tenths thousandths ( )
9. Five hundred, forty-two and seven tenths
10. Five inches, three hundred, thirty-eight and seven tenths
11. 101.6 mm
12. 63.5 mm
13. 120.65 mm
14. 508 mm
15. 0.9843 in.
16. 4.7441 in.
17. 14.0945 in.
18. 8.8583 in.
19. 0.1870 in.

20. 0.0984 in.
21. The finished rectangle will be 3.75 × 5.375 in. The larger rough rectangle as sawed will be 4.05 × 5.675 in. Adding 0.06 kerf to each edge sawed yields the final consumed rectangle taken from the sheet

4.11 × 5.735 in.

(*Hint:* There were only two saw cuts to remove this product from the sheet!)
22. You need 221.75 rounded to 222 in. of steel rod. In feet and inches, 18 ft − 6 in.
23. The drill will be $\frac{27}{32}$ in.
24. A 21 millimeter drill is the next smaller, whole millimeter drill down from the $\frac{27}{32}$ in. drill.
25. The correct RPM is 698.4.

Section 3

Reading Technical Drawings

Introduction

Technical drawings and their companion documents, work orders (WOs), are the means by which machinists receive formal instructions. They also tie the whole industry together, linking management, customers, engineers, planners, programmers, and quality assurance people. These two very different documents work together to create a foolproof way to deliver the goods. That is, they're foolproof if all involved understand how to read them!

The drawing is the master engineering document. In pictures, symbols, and words, it shows exactly how the parts are to be shaped and/or assembled, but it doesn't tell the machinist how to make them. The work order serves that purpose. The main body of a work order is a carefully planned sequence of operations. They are written, step-by-step instructions of how to follow the drawing and produce the part. Learning to use WOs is so important that we'll explore them twice in later sections. Section 3 is about reading the drawings, to equip new machining students with just enough baseline information to get started with lab assignments. A formal course in print reading will be necessary in addition to this start-up information.

Online Electronic Documents or Paper

Today nearly all drawings are created with computer software called CAD (computer-aided drawing/design). That's the way the illustrations in this section were drawn. While CAD drawings can be printed on paper for shop use, they are also computer files.

Drawings printed on paper are useful in many ways in the shop. They probably won't ever be completely eliminated. But more and more, machinists are receiving their drawings and work orders in digital form, at their CNC workstations. Not too long ago, this made it necessary for the machinist to have access to a PC, but not today, if they work in a progressive shop.

Many CNC controls have evolved from single-purpose program managers to multitask computers, called PC-based machines. These are accessed with standard keyboards and run by the same operating systems as those found on desktop PCs, shop data is integrated (networked), and machinists can access programs, drawings, and work orders right at their CNC stations. They can communicate with programmers and with other support people who bring cutters or lubricants, for example, or with other machinists in other locations. Using these stations, the machinist becomes a data manager and can input job tracking, information, quality assurance data, and program requests and can return suggested edits to the programmer and perform many other tasks—all from their CNC control. The shop foreman can instantly see the number of parts completed, the number that were not acceptable (hopefully, none), and generally do a much better job of managing the shop.

This paperless trend not only saves storing thousands of hard copy drawings and the people required to manage them, it also makes it possible to control and instantly distribute design updates to everyone working on the same job, worldwide. It also saves forests.

ShopTalk

Technical drawings are also called several different names in the shop—prints, drawings, engineering drawings, and blueprints. But why "blue" when they are black and white? Read on.

Objectives

Unit 3-1 Orthographic Projection (Pages 48–53)
Identify the six orthographic views
Visualize objects using third-angle projections
Solve simple visualization problems typical of entry-level training
Identify auxiliary, detail, and cross section views

Unit 3-2 The Alphabet of Lines (Pages 54–56)
Visualize object, hidden, and phantom lines on drawings
Recognize leader, break, center, extension, and dimension lines on drawings
Define the correct weight and form of a given line

Unit 3-3 Putting It All Together—Challenge Problems (Pages 56–59)
Select the correct objects given limited views
Visualize three-dimensional objects given orthographic views

Orthographic Projection

Introduction: In Unit 3-1, we'll look at the system used by the designer to lay out the various views you see on paper—**orthographic projection.** Projecting means showing something on a flat surface, a computer screen or paper, for example. The Latin root, *ortho* means lying at 90 degrees and, of course, *graphic* implies pictures. So, orthographic drawings are various views of the same object taken from viewer perspectives 90 degrees to each other.

When assigned a job with a set of instructions, the machinist must be able to get to work as soon as possible. But it's critical that no metal is removed until the true nature of the part is fully understood. That means efficiently visualizing the part shown on the drawing.

KEY TERMS

Auxiliary view
A view taken as though an extra sheet of glass was added to the box that shows details not seen from any other perspective.

Detail view
A view that is magnified larger than the general drawing to show fine details not clearly seen otherwise.

First-angle projection
The second most popular method of projecting orthographic drawings—one that does not change the image but rearranges its location on the paper.

Fold lines
Imaginary lines where a paper drawing could be folded to re-create a model of the glass box; fold lines simulate hinges between the glass surfaces.

Orthographic projection
Views taken at various 90-degree perspectives.

Projection lines
Lines that are not on the drawing but when added connect details in one view to the next; when drawn a projection line runs 90 degrees to the fold lines.

Section view
A view taken internally to show details not clearly seen from the outside of a part.

Third-angle projection
The more common system of arranging the views in North America: the imaginary glass box in which the part is envisioned is the third of a possible four.

Print-Reading Skills Needed

No matter in what form the drawing comes to you, on a computer screen or on paper, the skill of understanding and applying it to your work remains the same. Print reading divides into two categories.

1. *Visualization* is the ability to see the image and compile an idea of what the designer saw when creating the drawing. It is the ability to look at a flat, two-dimensional image on screen or paper and then create a three-dimensional (3-D) object in your mind. This skill is where we'll concentrate our efforts in Section 3.

2. *Interpretation* is the second and far more complex set of print-reading skills. It comprises what might be called the "legal" aspects of taking a design from drawing to reality. Interpreting a drawing requires an understanding of the rules, symbols, nomenclature, and procedures of manufacturing.

Here are just three examples of the many things machinists must understand to correctly interpret a drawing that we'll be learning in other sections of this book, when they are needed:

1. Drawing and part-numbering systems used in industry to make the exact detail being requested by the work order

2. The revision system for updating prints when designs change

3. Reading the dimensions and tolerances on the drawing to produce the right size and shape object

The Six Standard Orthographic Views

There are six possible standard views but it's rare that more than three or four will be needed to convey the image for any one part. They can be envisioned as projections of the part out to one of six possible sides of an imaginary glass box surrounding it: *top, front, bottom, rear, right,* and *left* (Fig. 3-1). (CAD drawings might also include a pictorial view to speed up visualizing.)

The imaginary glass box provides a model of how each view relates to the overall flat drawing. It can be envisioned as the surfaces upon which the views were projected, from the object out to the glass—then folded out

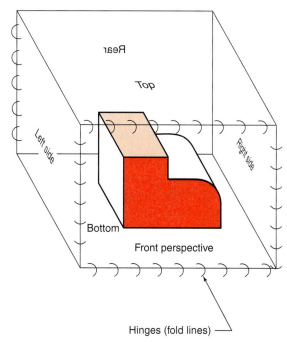

Figure 3-1 *The* imaginary *glass box surrounding the object.*

to become a drawing. While drawings aren't actually made in this way, Fig. 3-2 illustrates the best way to start learning visualization. Each view on the paper is as though the box had hinges and each surface was unfolded.

All views that show the height of the object might be called *elevation* views: front elevation, right side elevation, back elevation, and so on. But they are usually just called front, right-side, rear, left-side (or left/right-end) views. The term, elevation is more correctly used in architectural (building) drawings.

Third-Angle Projection

While there are other systems of projecting objects to flat surfaces, the most common in North America and elsewhere in the world, is

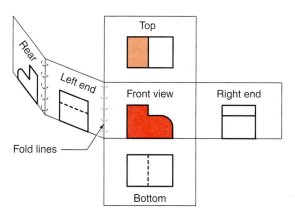

Figure 3-2 *Each drawing view is rendered as though the viewer is directly facing the various surfaces of the glass box.*

Figure 3-3 *The symbol for a third-angle* orthographic *projection drawing is a cone from the top and the right side.*

third-angle projection. In this system, the print readers envision each view as though they were standing outside the box and the view they see in any given position (top, front, and so on) is projected to the glass surface *between the object and viewer.* In other words, each view is as though you were walking around the box and looking in from each side.

The symbol used to notify the reader that the drawing was produced with third-angle projection is a cone as seen from the front and side views as if it were in the box. Where there is a chance that there might be more than one method used, such as for an international corporation or for one that has international customers, you will see this symbol in the drawing title area to indicate that the drawing is a *third-angle* projection (Fig. 3-3). With third-angle projection, the top front edge of the box is considered the prime edge (or hinge) out of which all other views unfold.

KEYPOINT

Third-angle projected images are created as though the projection surface (glass) is between the viewer and the object. These are the drawings you will use in this book and probably in your lab training too.

White Lines on a Blue Background

Figure 3-4 *A page out of history, a blueprint.*

First-Angle Projection

The other method used in some places in the world is similar and works equally well compared to third-angle drawings. Here the views are arranged differently as though the image was projected out of the object to the far glass surface—the surface away from the viewer on the other side of the box. It does not change the appearance of the six standard views. Drawings produced with **first-angle projection** rearrange the image placement on the paper.

For first-angle projection, the glass box is still used but the prime hinge line is elsewhere on the box. To see the difference in first- and third-angle drawings, we examine the four possible quadrants created by intersecting a pair of glass planes, as seen in Fig. 3-5. This is the universe of four boxes. Much of the industrialized world use the third box but fewer use the first.

The Glass Box Universe

If two imaginary glass planes are intersected, they create four possible glass box areas. Their mutual intersection line becomes the prime edge or hinge to each. In third-angle drawings, the object is placed in the third possible box of the four. First-angle drawings are then taken from the first possible box. Due to common usage in North America, from this point forward in Section 3, we will study only third-angle drawings. We'll leave

first-angle drawings for your formal print-reading course. Neither system is better. Once understood, they are used with equal ease. In today's world of international competition, it's very likely you will encounter first-angle drawings.

Fold and Projection Lines

To speed up visualization of the object, mentally place imaginary *fold lines* between each view. They represent where the hinges would have been between the glass surfaces if the box was real. As long as the views are correctly arranged on the paper, it is possible to cut, then fold the print along these lines to make a paper model of the box, complete with the projected views.

The drafter (person creating the drawing) uses the concept of fold and **projection lines** when generating the image on paper or computer screen. You can use them to visualize the object. For example, notice in Fig. 3-6, in both front and end views, the bottom of the object is touching the bottom of the box and the top is the same height in adjoining elevation views. The bottom of the box is a projection line. Details are found directly across the fold line from each other. They project from view to view. Again in Fig. 3-6, to ensure that the step was the same height in both views, a temporary projection line was extended from the front view to the right side. Then the vertical lines were trimmed to it. The last step would be to delete the projection lines between the views.

Ruler Check

To see how this works, check out the hole in Fig. 3-6 with a straight edge or ruler. Extend the dotted lines representing the hole, on the right, over to the front view.

KEYPOINT

The third-angle box uses the top, front hinge as its prime or parent hinge. All other views radiate out from this front view and the prime fold line (studied next).

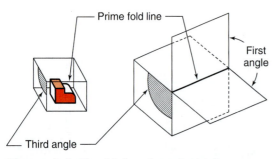

Figure 3-5 *The third-angle projection box comes from the third quadrant of a possible four.*

Figure 3-6 *Details project to the same height above the bottom of the glass box in both views.*

Auxiliary, Section, and Detail Views

There are several other views that can be added to clarify details not shown well in any of the six standard perspectives. We will examine the three most common: **auxiliary views, section views,** and **detail views.**

Auxiliary Views

Sometimes, the six-sided glass box doesn't provide a view perspective that reveals the true size or shape of the object, or some detail on the object, in the right perspective. For example, in Fig. 3-7, in every standard view, the hole in the angular surface appears as an ellipse or it is hidden from view. To show this hole in true perspective, an extra (*auxiliary*) sheet of glass is added to provide a straight-on view of the detail. In other words, to look directly into the hole where it appears as a circle. Notice that the hinge/fold line is on the top of the box where the angular surface is shown as a straight line. This is the only view from which the new auxiliary view can be projected as a true surface.

Removed or Rotated for Convenience
Due to the auxiliary projection rule, we often

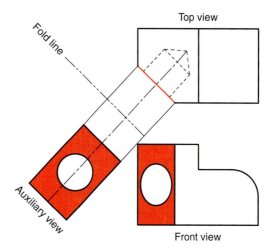

Figure 3-8 *This auxiliary view is in its correct position, across the fold line.*

KEYPOINT

Auxiliary Rule
An auxiliary view must be projected out of a previous view where the face to be shown appears as a straight line.

find auxiliary views projecting or folding out to odd positions on a standard orthographic drawing, as seen in Fig. 3-8. To solve this problem, they are usually transferred to a better location on the drawing and a note is *usually* provided next to the view to let the reader know this has been done—that the view is not found directly across the imaginary fold line but put elsewhere to make the drawing more compact. The view will be found in a more logical place on the paper but not in the place it would have been projected (Fig. 3-9).

KEYPOINT

When the auxiliary view is no longer in its correct ortho position, the projection line ruler trick will not work. Remember, auxiliary views are drawn to show some detail(s) of the object not shown clearly in the standard view perspective.

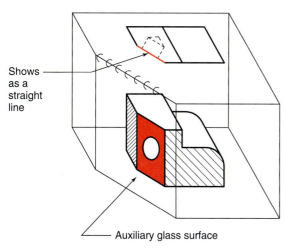

Shows as a straight line

Auxiliary glass surface

Figure 3-7 *An extra (auxiliary) sheet of glass is added to show the hole in true size and shape.*

Section Views

Another secondary view arises when the designer needs to show some internal feature not shown clearly from the outside of the

Cutting plane line

Figure 3-11 *The cutting plane line shows where the slice has been taken. It's wider than all other lines on the drawing.*

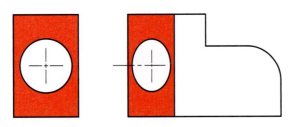

Auxiliary view (rotated)

Figure 3-9 *The auxiliary view has been moved to a better location on the paper but no longer projects back to the top view.*

object. There are several types of section views, but they all share the concept that they are drawn as though material has been sliced away from the object and you are seeing the inside of the part. The section view is depicted as it would appear if the material between the viewer and the remaining object had been removed.

Here are two common section views, the *full section* and the *broken out section*. Each clarifies inside details of this machined brass, hose fitting (Fig. 3-10).

Two rules are used as guidelines for creating section views:

1. Cutting Plane Rule

The theoretical surface along which the material has been sliced open is called

Broken out section

Full section

Figure 3-10 *Two different kinds of section views show internal detail.*

the cutting plane. In Fig. 3-11, note that the extra heavy, dashed line with arrowheads at each end is the cutting plane line. It represents the edge of the slice. The section view created by the cutting plane will be seen as though the viewer was looking in the direction of the arrowheads. The material that will be removed in the section view is behind the arrowheads, between the viewer and remaining object. If at all possible, the section view is placed in its correct orthographic position on the paper. But as with auxiliary views, it may be necessary to locate it elsewhere on the drawing (Fig. 3-11). The person making the drawing may or may not choose to draw the cutting plane line, as was the case in Fig. 3-10. Also, there may or may not be arrowheads, depending on the simplicity of the section view.

2. Sectioning Rule

After the section view is drawn, light lines will probably be added to clarify where the internal cut has been made. Section lines are not required but are added only when the designer feels they clarify a view or detail. Sectioning lines are also called *crosshatch* lines, as seen in Figs. 3-10 and 3-11.

Detail Views

Detail views are magnifications of small features not seen clearly at the scale of the entire drawing. They might be labeled as to how much they have been scaled up (blown up) and a circle might be shown to indicate from where the view has been taken—again, these are options the person drawing may or may not use. For example, **"3X—Size"** in Fig. 3-12 means the view is three times the normal drawing size.

Front view

Detail view (3X)

2.5 mm

Figure 3-12 *A typical detail view—three times the size of the actual drawing.*

UNIT 3-1 REVIEW

Review the Key Terms

Auxiliary view
A view taken as though an extra sheet of glass was added to the box that shows details not seen from any other perspective.

Detail view
A view that is magnified larger than the general drawing to show fine details not clearly seen otherwise.

First-angle projection
A second method for orthographic projection that places the image on the far side of the box from the viewer (first-angle projection is set aside for a formal print-reading course).

Fold lines
Imaginary lines where a paper drawing could be folded to re-create a model of the glass box; fold lines simulate hinges between the glass surfaces.

Orthographic projection
Views taken at 90-degree perspectives.

Projection lines
Lines that are not on the drawing but when added, connect details in one view to the next; when drawn a projection line runs 90 degrees to the fold lines.

Section view
A view taken internally to show details not clearly seen from the outside of a part.

Third-angle projection
The more common system of arranging the views here in North America: the imaginary glass box in which the part is envisioned is the third of a possible four.

Replay the Key Points

◆ There are six standard orthographic views: front, top, right and left end, bottom, and rear.

◆ All the standard views are pictures at 90 degrees to each other.

◆ The *front* view should depict the object best. Study it first when starting a new job. It is the parent view out of which all other views originate.

◆ Auxiliary, detail, and section views clarify a detail not clearly shown in the standard views.

◆ Section views show internal detail obscured by material. These views shows the details as though the intervening material has been removed.

Respond

1. In your own words describe an orthographic projection.

2. To interpret a drawing means to collect all the views together in your head then create a 3-D image in your mind. Is this statement true or false? If it is false, what will make it true?

3. In North America, do we use first- or third-angle projections most commonly?

4. While they aren't on real drawings, name two imaginary lines that help in understanding orthographic projections.

5. Match the letter to the number(s).

   A. Section views
   B. Auxiliary views
   C. Detail views

   1. A view that expands some small feature that can't be seen clearly otherwise.

   2. A view that is projected to an imaginary glass surface that is not one of the six standard views.

   3. A view that shows some internal aspect of the object.

   4. A view that has some material removed from the object.

   5. A view that is at a larger scale than the main drawing (magnified).

UNIT 3-2

The Alphabet of Lines

Introduction: A second family of visualization clues helps readers build the 3-D image in their mind by making various lines trigger different images. We will look at seven basic types. We've seen the eighth and ninth as cutting plane and section lines in Unit 3-1. Once you've become familiar with these different lines, it will be much easier to *see* the image. With some practice these lines almost make details jump out of the paper into your mind!

KEY TERMS

Break lines
Depict material that has been removed from the drawing *but not from the real part* to simplify or compact the drawing only.

Dimension line
A thin, continuous line that indicates size and shape, usually terminates in arrowheads.

Extension line
A thin, continuous line that leads the eye off the object to provide a dimension location.

Hidden line
A lightweight, dashed line that depicts surfaces and edges not directly seen in that perspective.

Leader line
A thin, continuous line that conveys special information or directs a note to a specific location.

Object line
A solid line of heavy weight that depicts surfaces and edges that can be seen in that perspective.

Phantom line
A dashed line of light weight that shows an auxiliary position or a related part not actually on this drawing.

Symmetrical
The same on both sides of a central axis.

Visualization
The process of mentally assembling an orthographic drawing.

Form and Weight of Lines

The clue as to what a line conveys is its form. For example, whether the line is composed of dashes or solid lines or combinations of both changes its meaning. Weight is the relative width compared to other lines on the drawing. Weights range from light section lines to the extra-heavy cutting plane lines (the widest line on any drawing).

1. **Object Lines**
   When visualizing a drawing, the foremost lines you should see are the **object lines.** These solid, heavy lines represent edges and surfaces that are visible. Thus these lines are also called *visible lines.* Other than cutting plane lines, they are the heaviest line on the drawing.

2. **Hidden Lines**
   These dashed lines depict details unseen to the eye; as though an X ray has revealed further information inside or on the opposite side of the object. **Hidden lines** are lightweight lines and are composed of equally spaced dashes. Hidden lines represent real surfaces, edges, and features obscured by other material. They are lighter in weight than object lines but thicker than section lines.

3. **Phantom Lines**
   Phantom lines are repeating double dashes and slightly longer lines. They are used in two different ways. They represent alternative positions of an object. For example, they might show the remaining part of the hinge in the full open position. Second, phantom lines depict an object that relates to the object being drawn but is actually not in the drawing. As an example, see the second part of the hinge in Fig. 3-13.

4. **Centerlines**
   Centerlines do not depict material, they show the central axis of an object or feature. They are medium-weight lines with a single dash and a longer line. Often, but not always, the object may be **symmetrical** (the same) on both sides of the line. If this is so, there may be a note or symbol stating this. There are several centerlines shown in the common usage in Figs. 3-13 and 3-14.

5. **Dimension Lines**
   Dimension lines and the next two types of lines are of the same weight and

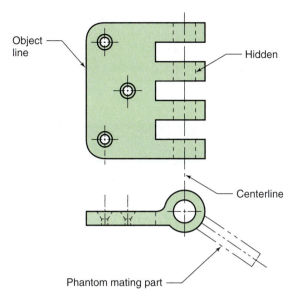

Object line

Hidden

Centerline

Phantom mating part

Figure 3-13 *Phantom lines are depicting alternative positions or related parts that aren't on this drawing.*

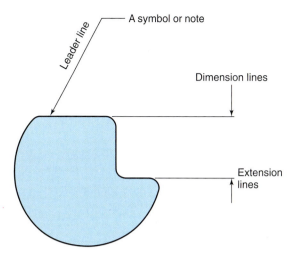

Leader line

A symbol or note

Dimension lines

Extension lines

Figure 3-15 *Leader, extension, and dimension lines all have the same form and weight.*

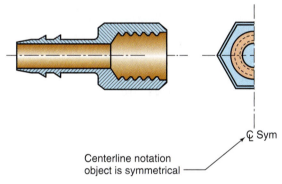

Centerline notation object is symmetrical

℄ Sym

Figure 3-14 *This fitting is the same on both sides of the centerline axis.*

form. They do not depict material but are used to clarify the meaning of dimensions and notes. They are lightweight indicators of a distance or size on the object. They almost always terminate in arrowheads. They may be broken by a dimension, but their form is continuous. They often are used in conjunction with extension lines, as seen in Fig. 3-15.

6. **Extension Lines**
   Extension lines are detailing lines that extend a feature out to where a dimension line can be used to show the size without interfering with the view. Because they

don't depict material, extension lines do not connect to the object by a small gap. The gap helps to avoid leading your eye off the object when **visualizing** it. They are the same weight and form as dimension lines, continuous and light.

6. **Leader Lines**
   Leader lines are utility lines and are similar to both dimension and extension lines in that they are not part of the object but clarify details about the object. They point to a particular zone or detail with an arrowhead at the terminus. They point out where a symbol or note applies to the part. Or, because the note is too large to place at that location, they may include a letter or number that is keyed to a note found elsewhere on the drawing. Leader lines might be curved but more commonly they are slanted, straight lines. The curve or slant is an attempt to avoid confusion with object lines.

7. **Break Lines**
   Break lines show where unneeded material has been removed to simplify the drawing. For example, a flag pole drawing needn't be 50 feet long if the pole is the same from top to bottom, or as seen in Fig. 3-16, this roller is the same throughout its length and can be broken to save space. There are several types of break lines seen on drawings, but they are all easy to recognize.

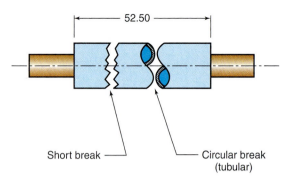

Short break ———⌐   ⌐——— Circular break
                                    (tubular)

Figure 3-16 *Two kinds of break lines show removed material to save space.*

3. Why do designers use break lines?
4. A centerline depicts the _____ of a feature.
5. What is another name for object lines?

UNIT 3-2 REVIEW

Review the Key Terms

Break lines
Depicts material that has been removed from a drawing.

Dimension line
A thin, continuous line that indicates size and shape, usually terminates in arrowheads.

Extension line
A thin, continuous line that leads the eye off the object to provide a dimension location.

Hidden line
A lightweight, dashed line that depicts surfaces and edges not directly seen in that perspective.

Leader line
A thin, continuous line that conveys special information or directs a note to a specific location.

Object line
A solid line of heavy weight that depicts surfaces and edges that can be seen in that perspective.

Phantom line
A dashed line of light weight that shows an auxiliary position or a related part not actually on this drawing.

Symmetrical
The same on both sides of a central axis.

Visualization
The process of mentally assembling an orthographic drawing.

Respond

1. Why does the designer place arrowheads on a cutting plane line?
2. True or false? A phantom line shows material that is hidden behind other features.

UNIT 3-3

Putting It All Together—Challenge Problems

Introduction: Now, let's investigate combining the clues to begin the visualization process. After a few hints on how to assemble views in your head, we'll tackle a few puzzles. Remember, your goal is to gain enough baseline understanding to read project guides and drawings in training.

KEY TERMS

Nicknaming
The mental trade trick that helps put a visual handle on a drawn object.

View-to-View (V-T-V)
The process of mentally assembling an orthographic drawing.

A Trade Tip for Forming the 3-D Picture

I've found that this tip cuts the time required to visualize a new drawing. It's absolutely critical that the machinist have a clear picture of the object before putting the cutter to the metal. Costly mistakes are made when we do otherwise! But time is also limited; the job needs to move forward. Listed next are a set of mental tools to aid in assembling the image ASAP. Follow them in the order presented.

1. **Study the Front View First**
   It will depict the object best. But don't stay there too long; just get the rough outer shape at first, a rectangle, "L" shape, or oval, for example. Don't try to see every detail at one time.

2. **Choose One Major/Obvious Front View Feature**
   This is a single feature that shows clearly in the front view; a hole, a thread, or a large

surface for example. Then relate it to other views using the ruler projection method and the alphabet of lines. This is called the **V-T-V (view-to-view)** process.

3. **Repeat the V-T-V Process**
Continue using features you *do understand*, V-T-V, several times more, choosing finer details each time. This solidifies general shape and gets the image ready for the next phase.

4. **Now, Choose Details That You Don't See Clearly in the Front View**
Start to investigate lines, and features V-T-V that you didn't see clearly. Use the projection line trick between adjacent views until they begin to clarify.

5. **Mentally Sum up a More Detailed Overall Image of the Object**
At this point, you should clearly see a refined definition of the overall shape of the object. For example, a cylindrical shape with a tab on the top, a cubical shape with two big holes, an elbow with a slot through one end, and so on. If this does not work, start over with the V-T-V process.

6. **Nickname the Object if You Can**
As soon as possible at any point in this entire process, try **nicknaming** the object with a handle of some everyday thing, if you can see a relationship. This tip within a tip, does wonders to complete the imagery and it really helps in talking about the part with others in the shop. For example, I have worked on parts we all called hockey sticks, frying pans, pork chops, paper hats, and Tootsie pops! Can you see them even without a print (Fig. 3-17)?

This little trade secret not only locks in what you see, the nickname provides a real grip on the image when discussing it with others. Using this communication tool, I might say, "we need to drill the hole in the handle of the fry pan." Instantly you know approximately where the hole is to be drilled and where to look on the drawing for the dimensions! It works!

Special Situations

There are times in industry when the product is complex but there isn't adequate time to visualize it completely before it's time to cut metal. This is an undesirable situation we all must face at some time. There are three ways to help get over the queasies of making a part you truly haven't visualized.

Figure 3-17 *Can you guess what this part was nicknamed in our shop?*

Sample Parts

If these parts were made previously, the shop may keep an old one around that cannot be passed on to the customer due to imperfections. Called an SPT (sample part template), it might be painted a bright color, usually red, and it may be deliberately cut or stamped with the words scrap to guarantee that this one is never returned to the production line.

Follow the Work Order and Ask Your Foreman

During times when you must "shoot without seeing the target," rely on the step-by-step instructions in the work order. Ask others who may have made these parts previously when you are in doubt. Primarily, your shop lead's job is to be there to assist in such tough situations too.

KEYPOINT

If you don't have a complete visualization of the work but need to get going on it, then don't guess, ask!

Digital Images

In many progressive shops, there may be a pictorial image of either the CAD-drawn part or of the CNC tool path or both, as shown in

Fig. 3-18. The tool path is a likely source of information since it is almost always checked by the programmer before releasing the new program to the shop. These files can put a turbo-boost to the visualization process. By the way, we called the parts shown in Fig. 3-17 boomerangs.

Figure 3-18 *Screen images of the program tool path can often help visualize objects. Courtesy of Mastercam (tool path) and Metacut Utility Software (solid model).*

UNIT 3-3 REVIEW

Review the Key Terms

Nicknaming
The mental trade trick that helps put a visual handle on a drawn object.

V-T-V
The view-to-view process of mentally assembling an orthographic drawing.

Replay the Key Points

◆ Never guess at the shape to be machined, always ask shop leads.

◆ Use the view-to-view process to speed up the visualization process.

◆ Nicknaming parts is a healthy way to increase visualization and communication.

Respond

1. Why would you study the front view first when presented a new drawing?

2. How do the work order and drawings work together to produce a finished product? What are their purposes in the manufacturing process?

3. When performing the view-to-view process of visualizing a drawing, what three clue sets are used? (*Hint:* They are all lines.)

4. Other than visualizing the part using orthographic projection, name some other ways to "see" the part before machining it. In other words, what are some other sources of information as to how the three-dimensional part looks in reality?

5. Four kinds of special views can be added to the drawing to clarify details not shown well on the standard six orthographic views. What are they and what information do they add to the visualization?

Section 3 Review

Unit 3-1

Being a machinist also means you'll be an information manager. Working hand in hand with work orders and programs, prints are *the* prime directive in our trade. They show us what but not how. While there are other ways to visualize a part shape using sketches, models, and high-tech screen images, projecting views outward to flat planes, from several sides of the object, is the undisputed best method not only to convey part shape but to record specifications and dimensions. The views fall into the following categories:

Six standard views taken from mutual 90-degree perspectives

Details blown up beyond the drawing scale to magnify small features

Auxiliary views taken from nonorthographic perspectives of the part section views

Internal views that reveal details not seen clearly from outside the object

Unit 3-2

Visualizing an object is a matter of adapting your mind to using a universal set of clues—or as Unit 3-3 puts it, putting it all together, to reassemble the flat views into a 3-D concept. One of the strongest sets of clues is the alphabet of lines. Using relative weight and form they show

Visible lines and surfaces—directly seen from that perspective

Hidden lines and surfaces—obscured by other parts of the object

Phantom lines and surfaces—represented but not actually on this object

Dimension and centerline details—to add clarity to specifications

Unit 3-3

This section was provided not only to help you understand the first few lab assignment sheets but also to demonstrate the need for further training in print reading. Time and again, industry studies show that frustrated trainees say that the greatest blockage to career success is a lack of print-reading skills. Be sure to take a formal course in this skill. But far beyond that upcoming course, the learning process for print reading continues as lifelong self-education. Thanks to rapidly changing technology, the long-term practicum will never end!

Questions and Problems

1. Print reading falls into two general categories. They are
   A. Visualization and tolerancing
   B. Interpretation and dimensioning
   C. Visualization and interpretation
   D. Projection and orthographic
   (Obj. 3-1)

2. Name the orthographic view that should convey the most information about the general shape of the object. (Obj. 3-1)

3. True or false? In North America we use the first-angle system but a few countries use the third. If false, what makes it true? (Obj. 3-1)

4. When rotating between views, what two imaginary lines, not found on the drawn object, help relate details? (Obj. 3-1)

5. A centerline has what form and weight? (Obj. 3-2)

6. List at least four lines used on orthographic drawings that are continuous (unbroken) and their relative weights. (Obj. 3-2)

7. A view must be taken internally—with material removed to see it clearly. What kind of view is that? (Obj. 3-1)

8. Referring to Question 7, what line is used to show exactly where the internal view is taken on the object? What form and weight are used for this line? (Objs. 3-1 and 3-2)

9. Referring to Question 8, what do arrowheads show when attached to the line that shows where the section view has been taken? (Two facts) (Obj. 3-2)

10. On a separate piece of paper, sketch the section view as it would appear in the indicated space on the drawing in Fig. 3-19. (*Hint:* See Problem 11 for more information.) (Obj. 3-3)

11. The arrowheads have been omitted from the cutting plane line in Fig. 3-19. If you were to add them, would they correctly point left on the page or to the right? Explain. (Objs. 3-2 and 3-3)

12. Given the front view in Fig. 3-20, identify which object(s) is/are depicted as pictorial views? Note, object A, the center hole, extends through the part. (Obj. 3-3)

13. On a separate piece of paper, assuming Fig. 3-20 is object A, sketch the correct top view, and assuming Fig. 3-20 is object C, sketch the correct top view. (Obj. 3-3)

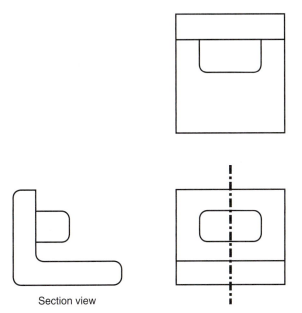

Section view

Figure 3-19 *Sketch the section view on a piece of paper.*

Front view

Figure 3-20 *Which object(s) is(are) shown with this front view, A, B, or C?*

Critical Thinking

14. Why must a rear view always be hinged to either the right- or left-side view? See Unit 3-1 Figs. 3-2 and 3-3. (Obj. 3-1)

15. In your own words, describe why a designer might choose to use an auxiliary view. (Obj. 3-2)

CNC Questions

16. A work order and print have been given to you to make a rush order of parts. The object is complex and the foreman tells you the job is "hot." "Load the program and make the setup, then start machining the parts right away," he says. You are uncomfortable with your situation because there's no time to visualize the object. What can you do to get a quick visualization of the object ASAP?

17. A. Which object in Fig. 3-21 is depicted in Fig. 3-22?

B. Are the two orthographic views in Fig. 3-21 complete without the pictorial view? Explain. (Obj. 3-3)

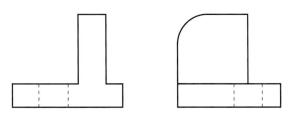

Problem 17—Missing top view

Figure 3-21 *Using Fig. 3-22, identify this object.*

Figure 3-22 *Which object is depicted in Fig. 3-21?*

18. On a separate piece of paper, sketch three views of Fig. 3-23, in their correct orthographic relationship. No size is necessary, fit your drawing to your paper. Note the view that is to be the front perspective in your drawing is indicated by an arrow.

19. Regarding the answer to Problem 18 (Fig. 3-27), in your opinion, did the engineer select the best front perspective view of the saddle

block to use for the front view (Fig. 3-23)? Explain.

20. Given these views of the slotted bushing (Fig. 3-24), find the error in the orthographic drawing (not the dimensions). Explain.

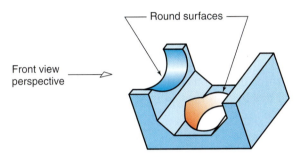

Problem 18—Saddle block

Figure 3-23 *Using this pictorial view, draw three orthographic views of this saddle block.*

Problem 20—Slotted bushing

Figure 3-24 *What's wrong in the orthographic views?*

Section 3 Answers

ANSWERS 3-1

1. Orthographic drawings are various views taken from 90-degree perspectives of the same object. In other words, looking at it from several sides.
2. False. That's the definition of visualizing, not interpreting.
3. Third-angle projections
4. Fold and projection lines
5. A = 3 and 4; B = 2; C = 1 and 5

ANSWERS 3-2

1. To show the direction in which the view has been taken. To show on which side of the cutting plane line the material has been removed, behind the arrowheads.
2. False. A phantom line shows alternate positions or objects that relate to the drawing but aren't on this drawing.
3. To depict material that is actually on the object but not needed to define the object. The material has been removed in the drawing, to save space.
4. Axis
5. Visible lines

ANSWERS 3-3

1. The front view is chosen by the person creating the drawing to be the best representation of the object.
2. The work order tells how to make the parts to the drawing specifications. The drawing provides the shape, size, and other specifications.
3. Projection lines, fold lines, and the alphabet of lines
4. Screen images of a CAD drawing or program (Fig. 3-18); pictorial views; sample parts;* find out if the object has been given a "nickname" by other machinists.*
5. Detail view, depicts small features not shown well in the scale of the drawing (blown-up details); section views, show internal details not seen clearly from the outside (as if material is sliced away to see inside the object); auxiliary view, a view taken from another perspective to clear up size and shape not shown clearly from the standard six perspectives (as though another sheet of glass has been added to the glass box); and pictorial (isometric), a graphic representation of the overall part—adds a picture of the object as well as flat views.

Answers to Section Review Questions

1. C. Interpretation and visualization
2. The front view usually is chosen to most clearly depict the shape of the object.
3. False. We use third-angle projection, a few others use first.
4. Fold and projection lines
5. Long dash and short, using thin weight
6. Object lines, heavy; extension, dimension, and leader lines, light; section lines, lightweight
7. Section view
8. Cutting plane line, extra heavy dash and lines
9. That material has been removed from behind the arrowheads, between the viewer and the cutting plane line; the view will be in the direction of the arrowheads in the section view.

*Assuming the parts have been made before.

10. The 45-degree section lines are optional but should be there to clarify the material at that section.

11. They should point to the left, as shown in Fig. 3-25. Either or both of these answers are correct. (A) Because the placement of the section view means that the view is seen as though material is removed to the right (behind the left-facing arrowheads) of the cutting plane line. (B) Because the section view's perspective is from the right side of the part as though you were looking toward the opposite side (left in the front view).

12. It could be either object A or C. (Because the center circle is unbroken, the answer cannot be object B.)

13. The two top views for objects A and C (Fig. 3-26).

14. To hinge the view otherwise would mean that its elevation would not be vertical. In other words, the part would not be right side up if it weren't hinged out of the two end views. Try it, return to Fig. 3-2 and draw it as it would hinge from the top view for example. It comes out upside down on the paper!

15. Your answer should be similar to or a combination of the following:

All other views do not show the object in true size and shape.

None of the standard six views are looking directly at the object or detail.

None of the other views clear up some detail or feature without looking directly at it.

The six glass planes do not show the detail in true perspective.

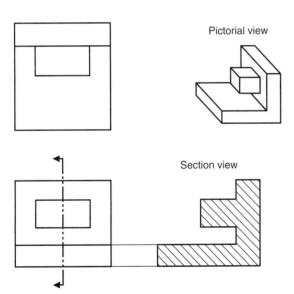

Figure 3-25 *The section view as it would be drawn, showing the rectangular center tab to be solid metal.*

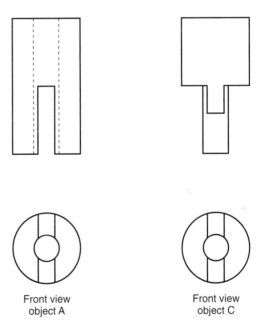

Front view object A

Front view object C

Figure 3-26 *As the top views appear. Note that the views must lie in this orientation to be correct.*

16. A. Ask if a sample part exists (or one very similar to this part).

B. See if the shop documents might include a pictorial view (remember this could be an electronic document, there may be a digital view of the object).

C. Seek someone familiar with these parts and ask questions about the setup and part's nickname and general shape.

17. A. It must be showing object B.

B. The ortho views must include a *top view* to be complete. The two views do not show whether the corner on the base is round or not or if the hole is square or round! You only knew those things from the pictorial views. If they were missing you could not decide the shape and size of the object.

18. The three views would appear as shown in Fig. 3-27.

19. Probably not. (This is arguable for this object.) I probably would have chosen the front view to be the one facing you in the pictorial view because it gives a better initial impression of the overall shape of the object. (The right-side view in the answer, see Fig. 3-26 for front perspective.)

20. The 0.625 diameter hole does go all the way through both walls of the bushing so the hole would appear unbroken in the side view, as shown in Fig. 3-28.

Figure 3-27 *The three views as they must appear given the indicated front view.*

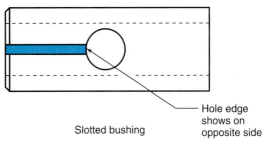

Slotted bushing

Hole edge shows on opposite side

Figure 3-28 *A very small error—the hole would appear to be unbroken.*

Section 4

Introduction to Geometrics

Introduction

Over 40 years ago at the leading edge of the computer revolution, as the jet age gained altitude and the ancestors of today's CNC machines operated without computers, like music boxes, it became obvious that the system of dimensioning and tolerancing engineering drawings needed upgrading. The methods that had brought technology that far no longer provided the flexibility or the feature control required to serve the developing products of the time, or of the foreseeable future.

The group of military leaders, engineers, and manufacturers that came together to solve that need could see that the manufacturing world would continue to evolve. Wisely, they set out to create a new dimensioning and tolerancing system to serve not only the technology of the time but one that would have the ability to adapt as new needs arose. Their new system eventually became known as geometric dimensioning and tolerancing (GDT).

As visionary as they were, they could not have foreseen the exploding technical age we live in today. Yet the system they compiled still serves well in an age where everything has changed. GDT was to be a living system with its users meeting regularly to oversee updates.

Here in the United States, the American Society of Mechanical Engineers in their Y14 committee assume the duty of reviewing the system and writing and publishing needed changes. Similar groups do the same in every industrialized nation. They then regularly bring their ideas to the International Standards Organization (ISO) committee, the same global group that monitors the metric system.

As in Section 3, our overall goal in Section 4 is to support early tech school lab assignments; to familiarize machining students with enough information to get started in training. Section 4 will also

> **KEYPOINT**
>
> There will be a difference in setups and measuring methods when the drawing uses geometric dimensioning and tolerancing, compared to working from a drawing using the older dimensioning and tolerancing system.

demonstrate the different way you must proceed when the drawing uses geometric dimensioning and tolerancing.

Our third goal is to recognize the geometric symbols and gain an idea of what they mean. We provide enough background to demystify these symbols, and to prepare you for a full study of the subject later in your career training.

Objectives

What Is Geometric Dimensioning and Tolerancing and Why?

Introduction: The GDT system is designed to better serve *function* compared to designs using previous methods (sometimes called coordinate dimensioning, but we'll refer to it as *standard dimensioning* here). Function is the way the part features we machine must work or assemble and operate. In order to understand this improved focus, we'll examine a few baseline definitions and look briefly at the meaning behind the symbols used in the system. This lesson is designed to provide proof of why we use GDT today.

KEYPOINT

GDT was invented to better serve *function*.

Symbolic Language Let's make a comparison between the two different systems: geometric and standard dimensioning and tolerancing. In Fig. 4-1, the location and tolerance for a hole is shown using examples of both methods. A second specification controls the roundness of the hole to 0.5 mm (the circle in the second box). At first glance, it seems that the GDT callouts simply restate in symbols that which was always there. Not so. Each and every geometric symbol was created to solve an unmet need or to expand some capability not found in standard dimensioning and tolerancing.

We will study what those symbols mean in Unit 4-3, but we've now come to the first advantage to GDT: It's a symbolic language that eliminates words on drawings. Words can be ambiguous, but with a bit of study, everyone knows the exact meaning of a symbol.

Symbols also save drawing clutter and shorten the CAD file compared to lettering and notes. But of greatest import, they are clearly understood worldwide. Nations can share drawings using geometric controls regardless of the language spoken.

However, these symbols only clear up drawing information when the print reader studies

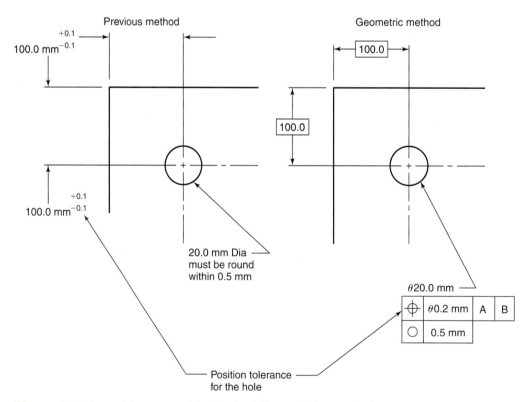

Figure 4-1 *Geometric compared to standard dimensioning methods.*

their meaning. For example, the boxes around the 100-mm dimensions in Fig. 4-1, have a definite meaning and purpose—we'll discover it later.

KEY TERMS

Datum
A starting point reference that many controls are based on.

Feature
Any single aspect of a part that can be dimensioned and toleranced, for example, a thread, a slot, and a hole.

Form
A control of the outside surface shape of regular objects (one exception—an axis, which can be controlled with straightness).

Function
The way things work or assemble.

Location
Controls the position of a feature's axis with respect to a datum; the controls are position, symmetry, and concentricity.

Orientation
Controls the angle of a feature's axis or surface with respect to a datum; angularity, parallelism, and perpendicularity are the controls of orientation.

Profile
Controls the outside surface shape of objects where the shape cannot be defined with form controls.

Runout
Controls surface wobble on objects that rotate around a datum axis.

Geometric Principles for Machinists?

Without background study, geometric concepts appear to be focused solely on dimensioning and tolerancing of prints. They seem like a new way to convey the same old information, and the underlying information itself, seems to remain the same. One might mistakenly think that geometric or not, the machinist's actions remain the same—this is definitely not so! The concepts behind the symbols affect much that you do in setting up machinery,

planning cut sequences, writing programs, measuring results, and making holding tools. These are all based on geometric principles.

Two Important Definitions— Characteristics and Features

Take a moment to scan Fig. 4-2. The 14 symbols in the right-hand column are the keys to the system. We'll return to this chart again. You might want to put a marker here.

Characteristics Are the Target Each symbol in Fig. 4-2 represents a desired condition or relationship (the target) for one aspect of the workpiece—it must be round or straight, for example, which is known as the geometric *characteristic*. Recognizing each symbol and what it stands for is a main goal of this entire section. Once the characteristic is defined, then a tolerance for acceptability will be placed on it.

A geometric dimension and tolerance does this

1. **Define the Target Characteristic**
   A desired *characteristic* is called out. For example, the hole in Fig. 4-1 is to be round.

	Type of tolerance	Characteristic	Symbol
For individual features	Form	Straightness	—
		Flatness	▱
		Circularity (roundness)	○
		Cylindricity	⌀
For individual or related features	Profile	Profile of a line	⌒
		Profile of a surface	⌓
For related features	Orientation	Angularity	∠
		Perpendicularity	⊥
		Parallelism	//
	Location	Position	⌖
		Concentricity	◎
		Symmetry	⩵
	Runout	Circular runout	↗ *
		Total runout	↗↗ *

*Arrowheads may be filled or not filled

Figure 4-2 *The five groups of geometric controls.*
Source: American Society of Mechanical Engineers.

2. Set Allowable Variation—The Tolerance

This is the allowable variation from the perfect target (the tolerance). In this example, it defines how far the hole may depart from perfect roundness 0.5 mm, found in the lower frame.

Features Are Anything That Can Be Dimensioned and Toleranced A characteristic applies to a single **feature** on the workpiece; a slot or a thread or a hole, for example. The characteristic then is the target for that individual feature.

The GDT Definition

Let's put this all together: GDT is a system of 14 controls that define a desired condition (characteristic) for some feature on the part. They also show how much maximum variation is permissible for the feature to pass inspection. All 14 characteristics solve some functional need that was unmet before GDT was introduced.

Five Characteristic Groups

GDT controls fall into five families, sorted by purpose. Returning to Fig. 4-2, notice that the symbols are grouped as

Form	Controls the outside surface shape of regular features—round, straight, flat, and cylindrical shapes. (There is one exception to this where straightness can be applied to a feature's centerline.)
Profile	Controls the outside surface shape of contoured features. Items that aren't straight, flat, or round. Profile is very similar to form, except for the nature of the outside shape.
Orientation	Controls the relationship of one feature with respect to a datum. For example, the angle of a surface or the perpendicularity (squareness) to a datum.
Location	Controls the position of a feature's center with respect to a datum.
Runout	Controls the wobble on the outside of a spinning object.

KEYPOINT

A **datum** is a starting point reference that many controls are based on. We'll investigate datums further in Unit 4-2.

Geometric Advantages

Geometric Principles Underlie All That We Do The prints you use may or may not use GDT, but the geometric design is by far the more thought out version. Even more to the point, *the challenge of the work itself is, and always has been, geometric.* That's the truth on which the Y14 committee based their work.

KEYPOINT

Manufacturing is geometric whether or not the design print is GDT.

Reasons to Create Geometric Controls: In addition to the universal language already discussed, a geometric design is more in tune with the function of the product, to provide

- More *control and flexibility* for the designer and the shop compared to older methods.
- All the *full, natural tolerance* possible within the function of the object, yet make the best quality product. (GDT doesn't lead to large tolerances. What GDT does do is remove the need for excessively tight tolerances "just to be sure".)
- An instant understanding of the *functional priorities* of the features to be machined.

Understanding design priorities tells the users where to begin machining. Functional priorities are the order of importance of features on the part. Understanding these priorities is important to the machinist because they indicate

How to hold the part for machining.

Which cuts to take first.

How to measure the results.

As a machinist, you do not need to understand how the object being made actually works, but you do need to know what part features are the most critical—that is, which has the highest priority. From a geometric design, you will quickly identify the critical functions and features. Then you will base setups and the sequence of operations on those features and functions. We'll see examples of this in Unit 4-2.

If you had to sum up the reason the GDT system was created in one word, you would say **function.**

The GDT system better serves the way we make objects, and the way they work. One final point: GDT is a true system. As you learn more, you will begin to see many relationships and interconnections between controls and principles. Looking for the connections between concepts and controls is called a systems approach to learning. We'll explore it a bit in upcoming units.

UNIT 4-1 REVIEW

Review the Key Terms

Datum
A starting point reference that many controls are based on.

Feature
Any single aspect of a part that can be dimensioned and toleranced, for example, a thread, a slot, and a hole.

Form
A control of the outside surface shape of regular objects (one exception—an axis, which can be controlled with straightness).

Function
The way things work or assemble.

Location
Controls the position of a feature's axis with respect to a datum; the controls are position, symmetry, and concentricity.

Orientation
Controls the angle of a feature's axis with respect to a datum; angularity, parallelism, and perpendicularity are the controls of orientation.

Profile
Controls the outside surface shape of objects where the shape cannot be defined with form controls.

Runout
Controls surface wobble on objects that rotate around a datum axis.

Replay the Key Points

- Every box, circle, or symbol on the GDT print conveys a meaning that the machinist must understand. Not to know is to work with a handicap!
- The geometric system was invented to solve technical function problems arising in manufacturing.

- Every aspect of the system is built around function.
- GDT enables flexible control of the entire manufacturing process.
- GDT is a living standard. It is overseen by an international group that ensures that it evolves with the times and the technology. The machinist must stay up to date with revisions as they occur.
- Because it's symbolic, GDT transcends language barriers.

Respond

1. What group oversees GDT in the United States?
2. What is the basis of the GDT system? What does it serve better than the previous system?
3. When making a machine setup using a GDT drawing, what do the functional priorities show the machinist to do? How do you start a GDT job? List three items.
4. List and briefly describe the five characteristic control groups.
5. What is the one exception to the form control of straightness? What is it?

UNIT 4-2

Recognizing Datums in Designs and Using Them in the Shop

Introduction: The use of datums is one of the most important differences between a GDT design and one that is not GDT. Nine of the 14 characteristics originate from a datum reference. For example, the location of a hole must be positioned "from" something. That reference is always a datum in GDT. In Fig. 4-3, two datums are called out: datums A and B. They are symbolized by the placement of boxes around a capital letter.

KEY TERMS

Datum
A theoretically perfect surface or axis used for reference. The datum is established by features on the part—it is not on the part. It simulates the assembly world outside the part.

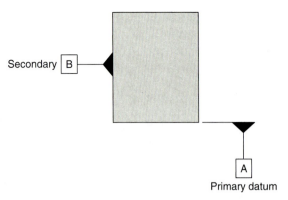

Figure 4-3 *Datums are indicated as capital letters in a square box, connected to the object.*

Datumization (author's term)

Accounting for the irregularities of a real object by placing it against a theoretically perfect reference surface for measuring or machining. Or an averaging of the irregularities on the feature to derive an axis datum. Figure 4-18 uses a precision ground pin in a hole to datumize its location.

Datum feature

Any regular part feature that is used to establish a datum on a drawing.

Feature

Any detail on the print that is dimensioned and toleranced.

Formal datum

A feature assigned a capital letter datum on the print.

Informal datum

Any feature used for reference for another feature. Does not have an assigned letter on the print.

In GDT the referencing of a feature "from something" is always based on a datum.

Datums Perform Three Vital Tasks

1. **Establish an Exact Basis for the Geometric Control**
2. **Eliminate Ambiguities (Uncertain Facts and Guesswork)**
   Without datums, craftpeople find themselves guessing at setups and measurements, as will be clearly demonstrated.

3. **Set Functional Priorities**
   A geometric print helps decide how to start the job. By the datum priorities, it shows what machine operations should be first, how to hold the work, and what features are critical. The job is easily organized. It's not always possible to go right to the first priority, but even if intermediate cuts must be made, the target is to establish Datum A (if not already established) as soon as possible, then B, C, and so on.

Make sure you understand all three of these concepts when we complete Unit 4-2, they are the main points. These concepts will appear every time you machine and measure a product.

Datums—The Exact Basis, the Starting Point

Without a **datum,** many measurements (and machine operations, too) are not scientifically exact. To illustrate this, answer the question posed in Fig. 4-4: Exactly how tall is the object with it's exaggerated irregularities? It's difficult to tell. Where do you place the measuring tool? Measuring in different places produces different answers. The problem is that the measurement has no formal basis; and the question of tallness cannot be answered. **Thickness can be measured but not tallness.** However, if this part is to be assembled with another, the functional question is how tall, not how thick it is. The problem lies in the way the drawing is dimensioned. It simply shows that a pair of surfaces must be 40 mm apart but does not clarify how to treat the measurement.

Figure 4-4 *How tall is this machined block?*

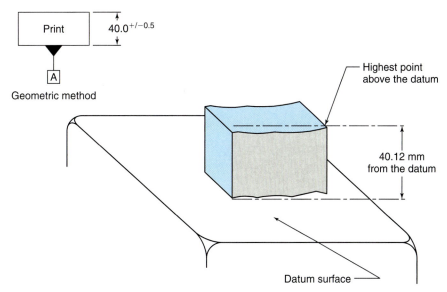

Print   40.0$^{+/-0.5}$

A

Highest point
above the datum

40.12 mm
from the datum

Datum surface

Figure 4-5 *The flat layout table is the datum. It answers the question.*

A datum solves the problem. In Fig. 4-5, laying the part on a precision flat table, called a layout or inspection table, the answer becomes clear. The *functional* height is the tallest point above the datum—above the table on which the part rests. That's how much vertical space the object occupies and how much assembly space it will need to fit into another part with two parallel surfaces—a slot, for example. So, reworded in terms of functional assembly; what is the smallest slot in which this block can fit?

So, the designer has assigned a datum to the bottom of the block as shown by the A in the box. It is the exact starting reference surface from which the functional height *must* be determined. To measure otherwise is wrong. It must be measured up off the datum.

The datum priority for the part of Fig. 4-3 is that Datum A holds the highest functional priority (order of importance to the function of the part) while Datum B is secondary. The alphabetical order shows their ranking in terms of importance to the part.

A datum might be indicated by an extension of the surface, as with Datum A, or connected to the surface itself, as with Datum B. On a drawing, when a box is drawn around a capital letter, it is called a *datum identifier* or *datum frame*. That symbol signals the machinist that the feature to which it is connected is significant and the datum will be used for reference for other features on the part.

KEYPOINT

The datum is not the part but rather is represented by the flat table upon which the part rested. The datum, which is considered perfect, accounted for irregularities on the real part. It simulates the assembly world outside the part.

KEYPOINT

The datum is needed not only to measure the part, but is also used as a foundation to machine it as well, as shown in upcoming examples.

Datums on Prints

The designer will specify the part feature that is to establish the datum. For example, datum feature A is the bottom of the part. It will be chosen in order of importance to the functionality of the product. Note that the crowfoot, triangle in Fig. 4-6, connects to

Part   Datum
            feature

Datum   A

Figure 4-6 *Datum A is established by the feature's surface, called datum feature A.*

the outside of the part surface—that symbol-izes the datum is touching the outside of the feature—*it is not the* **feature.** And that fact makes a big difference in how you are to pro-ceed with setups and machine cuts.

Two Kinds of Datums—Surfaces and Axis Centers

Datums Usually Contact Outside Sur-faces Any regular feature of the object can be assigned a datum. The designer chooses the order of importance based on the func-tion of the object or the assembly procedure.

Surface Datums When a datum is con-nected to a part surface, the feature it contacts is called a **datum feature.** So when estab-lishing datum contact during machining or measuring, the datum feature is held against a gauge-quality surface such as a pre-cision table, a ground vise jaw, or a milling machine table. In this way, the datum is established relative to the workpiece feature.

When the entire work surface isn't needed to establish the datum, specific points on the surface can be assigned that are called *datum tar-gets.* The drawing will specify where the specific points of contact are to be located. The machinists must then incorporate them into their holding setup or measurement.

Center Axis Datums For functional rea-sons, it's sometimes necessary to use the cen-ter of an object as a datum. An example is an automobile axle. Its most critical function is that all features are centered around the axis—so the axle's centerline is Datum A. Axis datums can also be flat planes at the center of a slot or a square tab, for example.

Axis Datums on Prints Figure 4-7 illus-trates two ways an axis datum might be

Figure 4-8 *The round test pin establishes the centerline datum.*

specified on a print; either by connecting the crowfoot to the centerline axis itself, as on the left, or by connecting it to the di-mension (the slot).

If the datum is not the center axis, then the crowfoot will contact the surface of the ob-ject or an extension of the surface.

KEYPOINT

If *any* GDT callout connects to the feature's dimension, then it applies to its center axis not its surface.

Axis Datums Are Mathematical Working with the center of a feature always requires a bit of detective work because the datum does not exist as an entity until the entire surface that establishes it, along with its irregularities, are summed up. For example, in Fig. 4-8, a hole's center has been found by inserting a precision round pin. The largest pin that will just slip into the hole determines the largest circular space available—the functional hole. Then the theoretically perfect pin's center be-comes a reliable datum.

Your Shop Actions— Machining and Measuring

Datums Define Priorities for the Job

While there can be many datums on the print, there is one master, called the *primary* datum. It is the one that should be used to start the job as much as possible. There are exceptions, but the goal is to make machine cuts that establish that reliable datum feature surface as soon as possible if it doesn't already exist. If the primary datum is already there,

Center axis datums

Figure 4-7 *Two ways an axis datum can be defined.*

then establishing Datum B using Datum A as the reference becomes the first machining objective.

Sometimes, it's necessary to make one or more intermediate cuts that lead to establishing the datum feature. As explained earlier, the machinist plans early operations around the datum priorities in their order on the drawing. The lowest letter of the alphabet found on the print (usually A) is the primary datum for that design. The designer created the design outward from it. Certain letters are not used to designate datums, O and I, for example, due to possible confusion with numbers or other letters.

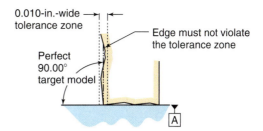
Figure 4-10 *The controlled edge must not violate a 0.010-in.-wide control zone built around a perfect 90-degree model extending from Datum A.*

![KEYPOINT]
> A well-written job plan or program sequences operations such that functional priorities are achieved in order, based on datum priorities.

Measuring a Perpendicular Edge—First Example

If datum priorities are ignored, several errors can occur, causing quality problems with the product and extra costs too. Consider measuring a square corner, defined on Fig. 4-9. The print specifies a 2 × 4 rectangle with the left edge at 90 degrees to Datum A.

We'll discuss the tolerance meaning in just a bit, for now note which edge is datum feature A—the bottom edge in the drawing. The information next to the part is enclosed in a special GDT symbol called a feature control frame. The information found in the frame states that

This surface shall be perpendicular, within 0.010 in., with respect to Datum A

In other words, for this part to pass inspection, the left edge must be proven to be square to Datum A when the lower edge (Datum feature A) is resting on the datum surface (Fig. 4-10). Once the datum reference is established, it must be proven that the left edge is within perpendicularity tolerance. The left edge can vary in any way as long as it does not violate the 0.010-in.-wide tolerance zone, that is perfectly square to Datum A. Figure 4-11 shows the right way to measure it. A precision square is used. Notice that in both methods, Feature A is against a flat datum surface, the square's base or the flat table, and either becomes Datum A. Only then is the variation correctly checked along the controlled edge. As long as no gap is found to exceed 0.010 in., the edge is within the perpendicularity tolerance. Both tests are looking for variation of the controlled edge with respect to Datum A.

Now, using the right method, suppose that the error found is beyond the allowance—that is, the corner fails the square test when a

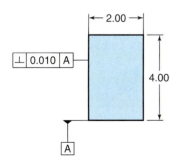
Figure 4-9 *This edge shall be perpendicular (90 degrees) within 0.010 in. with respect to Datum A.*

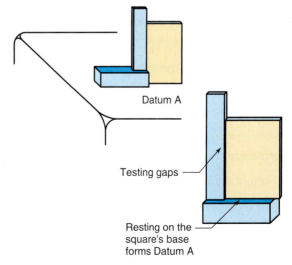
Figure 4-11 *Two correct ways of establishing Datum A then measuring the controlled edge.*

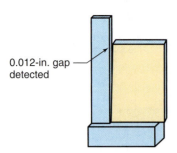

Figure 4-12 *This part fails the test by showing a 0.012-in. gap.*

Figure 4-14 *The first correct cut establishes datum feature A.*

0.012-in. gap is detected (Fig. 4-12). But suppose also that by not heeding the datum priority, the bad part is incorrectly measured in this way (Fig. 4-13).

In Fig. 4-13, the machinist has incorrectly placed the wrong feature against the datum, then tested for a gap on the wrong edge. Clearly under the GDT system, this is the wrong test! Since the incorrectly tested edge is shorter, it shows less error. While the perpendicular requirement is actually out of tolerance, it goes undetected due to a priority mix-up and a gap of only 0.006 in. results.

Guessing without a Datum From this example, can you see that without a geometric datum basis for the control, there could actually be two answers to the square corner? Recall that datums eliminate ambiguities and clarify functional priorities. Before GDT, the dimensioning and tolerancing methods defined only that two edges were to be square to each other not which edge was to be square to which.

Machining the Square Corner

There is a right way to machine this example job too. Again, it must be based on geometric principles. It must be started by establishing datum feature A first (Fig. 4-14). Then the rest of the procedure follows, referenced from it. Track datum feature A as it progresses through the next three drawings. In the first opera-

tion, datum feature A is machined. In Figs. 4-15 and 4-16, the second setup is made such that Datum A (the precision vise jaw) touches feature A. The 90-degree cut is made relative to that datum.

The precision (nonsliding) vise jaw becomes Datum A for cuts 2 and 3.

The vise jaw, which is considered perfect, averages out the part irregularities. Here in *Machining and CNC Technology,* we'll call the contacting of the real part surface against a theoretically perfect surface, **datumization.** (that's my definition).

Figure 4-15 *The second cut establishes the 90-degree relationship to the vise jaw Datum A.*

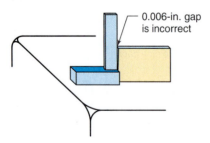

Figure 4-13 *The wrong way to make the measurement—a bad part might be accepted!*

Figure 4-16 *The third cut is parallel to the controlled edge—square to Datum A.*

Now, in Fig. 4-17, a second control has been added—the edge opposite Datum A must be parallel. So the part is set up vertically and the floor of the vise becomes Datum A. At this time, it would also be machined to the 4.00-in. dimension. Cuts 3 and 4 could be performed in different orders, but all these operations depend on a well-maintained, accurate mill vise touching Datum Feature A.

While this squaring up of edges is a simple example, it represents much more complex machining plans for larger objects. They must be sequenced with datum reference. Planning the operations by the datum priorities becomes even more critical when we use CNC machining, because doing otherwise introduces needless variation. Each and every dimension and/or control on the print that refers to Datum A must be set up in some way such that the holding tools contact feature A to establish a reference. It follows similarly for Datums B then C, if they exist.

We'll explore this squaring up process further in Section 12, Mills and Milling, but for now, do you see the functional use of datums in machining? The concept is absolutely vital to machining success. Even when the drawing is not geometric, these decisions must still be made. However, using nongeometric drawings, the choices are often clouded since the functional priorities are not always clear.

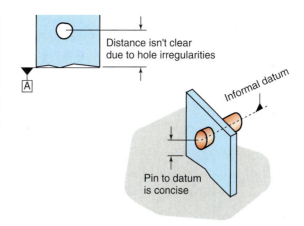

Figure 4-18 *How far is this drilled hole from Datum A? The pin solves the problem.*

Formal and Informal Datums

The datums shown thus far have been assigned a capital letter by the designer. They are **formal datums** on the print. But there are times in the shop when you'll see a need for the concept without formal assignment. Figure 4-18 poses a question to aid in seeing this concept. How far above the datum is the hole?

This hole has feature irregularities, as all do. It's unclear where the hole center might lie. To answer accurately, we must datumize the irregularities to find the functional round space available by inserting the largest round test pin that just fits. Considered perfect, the pin's surface establishes the **informal datum** at its center axis. An exact measurement of the functional distance above the datum to the hole is now possible.

Sometimes an informal datum is expressed on a drawing as shown in Fig. 4-19. The drawing makes it

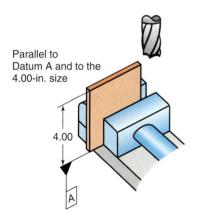

Figure 4-17 *Parallel to Datum A—the vise floor.*

Figure 4-19 *An informal datum.*

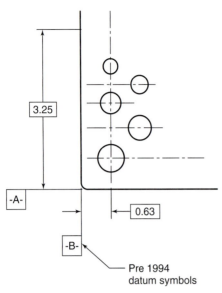

Figure 4-20 *Datums on older drawings.*

clear that an assembly condition exists that is dependent on the spacing between the two holes. Also shown is the fact that the left hole is datum to the right.

Datums on Drawings Previous to 1994

You will encounter older geometric drawings and texts that show datums slightly different than the way I've been illustrating them. Shown in Fig. 4-20, the datum will still be in a box but it will be between two dashes; -**A**- or -**B**- for example. The leader line to the datum feature will not terminate in a crowfoot triangle. Why, because a change was made to make all drawings similar, worldwide—the living system at work.

UNIT 4-2 REVIEW

Review the Key Terms

Datum
A theoretically perfect surface or center used for reference. The datum is established by features on the part.

Datumization (author's term)
Accounting for the irregularities of a real object by placing it against a theoretically perfect surface for measuring or machining. An averaging of the irregularities on the feature. Figure 4-18 uses a precision ground pin in a hole to datumize it.

Datum feature
Any regular feature that is used to establish a datum on a drawing.

Feature
Any detail on the print that is dimensioned and toleranced.

Formal datum
A feature assigned a capital letter datum on the print.

Informal datum
Any feature used for reference for another feature. Does not have an assigned letter on the print.

Replay the Key Points

◆ A datum can be assigned to any regular feature on a part: a surface, a hole, a centerline, or even a set of *designated points* (not covered in the text).

◆ A datum is not on the part but it is established by the part.

◆ A datum is considered perfect for the purpose of control of individual features.

◆ A datum simulates the assembly world outside the feature.

◆ A datum accounts for and averages part irregularities.

◆ A datum eliminates ambiguities and sets functional priorities.

Respond

1. On a drawing, a rectangular box around a capital letter is called a _____.

2. What single word sums up the reason or purpose behind the GDT system?

3. Explain why starting a geometric machining job from a GDT drawing is easier than from one that does not use GDT.

4. List the five types of tolerance groups for feature control symbols.

5. Many of the 14 feature controls require a starting point basis. What is it?

6. Describe a datum that is assigned to a feature surface on a drawing.

7. Besides surfaces, datums can be assigned to _____.

UNIT 4-3

Geometric Controls

Introduction: In controlling a feature's surface or its axis, all 14 geometric characteristics perform the same three tasks:

1. Define the perfect model of the desired characteristic

2. Denote the allowable variation from that perfect model—the tolerance

And, a new concept:

3. Imply how the characteristic must be measured

The characteristic doesn't tell the machinist which measuring tool or method to use, but it does specify datum reference (or not). That is the first factor—must you set up one or more datums to correctly measure the feature? There are other unique factors to verify that each control has been satisfied. If they aren't heeded, then mix-ups such as the square corner example in Fig. 4-10 can arise.

As we study individual control symbols, keep in mind that all 14 follow the three-stage process of defining the perfect target, then the tolerance. Then by knowing the nature of the control, you understand the method of proving that it's right—how to measure it.

KEY TERMS

Bonus
Additional tolerance possible based on function and actual measured feature size.

Contour
Controls that deal with irregular-outside-shaped features. Contour may or may not need a datum reference, depending on function.

Elements
Surface lines running in the direction of the control.

Entity
Either a surface element or a feature center that must be shown to lie within the tolerance zone to be within compliance.

FIM (full indicator movement)
The measured difference in high and low points on the surface of an object, rotated about its axis through 360°.

Form
Four GDT controls that define the outside shape of regularly shaped objects. Form has no datum reference.

Location
Three controls that define the tolerance for the center of a feature with respect to a datum.

Orientation
Three controls that define the angular relationship of a surface or feature center, with respect to a datum.

Perfect model
The goal of the desired characteristic.

Runout
Two controls that deal with surface wobble on rotating objects (axis datum).

Surface or Axis Control

The measurement method is determined by whether it applies to the feature's surface or its axis? We need to discuss both types.

KEYPOINT

In GDT, the control applies to either one or the other—to the feature's outside surface or to its center axis.

Elements on the Outside

If the control deals with a feature's surface, such as roundness in Fig. 4-21, then the control entity is called an *element*. Elements lie on the outside surface of a feature. **Elements** are lines that could be drawn on the surface with a pencil. They run in the direction of the control. Figure 4-21 illustrates two sets of elements—the roundness and straightness on a cylindrical pin. For the pin to be within the roundness tolerance, every circular element tested must be able to fit within a perfect pair of concentric circles—the tolerance zone. The distance between the two tolerance zone circles is the tolerance.

If the control is straightness, the elements run in a straight line, with the direction indicated by the control symbol line. We'll see pictorial examples shortly. Recall the measuring of the square edge in Unit 4-2, the machinist was testing straightness elements running perpendicular to the datum.

Surface elements

Round elements

Straight elements

Testing an element

Roundness tolerance

The element may vary in any way, but must not violate the tolerance zone.

Figure 4-21 *Elements are lines on outside surfaces. There are two kinds here—round and straight elements.*

There are an infinite number of elements on any given surface that could be tested to prove the feature is within the specified tolerance. The number chosen to measure is a matter of experience. For example, suppose the pin is 1 in. long. Then testing at two or three circular elements would probably be OK. But suppose that it was 3 in. long and perhaps had been made on a worn lathe. In such a case, testing more elements would be wise.

Control of Center Axes Is Mathematical

When the control deals with the axis of a feature, rather than a surface, the axis exists only after some inspection of the whole surface that creates it. For example, in controlling the location of a drilled hole (Fig. 4-22) its centerline must fall within the tolerance zone. But there is no centerline until the sum of the hole's roundness and straightness (cylindricity) is accounted for.

Figure 4-22 follows the three-step process:

Figure 4-22 *The feature control frame specifies. The position of this hole's centerline shall be within a round tolerance zone of 0.5 mm with respect to Datums A and B.*

In Fig. 4-22, the three-step process is defined within the *feature control frame* (FCF) (the rectangle containing the callout). We will study the characteristic of position further later, for now it's used to illustrate the use of the feature axis as a control entity. This drawing states that

The position of this hole's centerline shall fall within a round 0.5-mm tolerance zone located with respect to datums A and B.

Figure 4-22 follows the three-step process:

1. For a *perfect model* the target for the hole's axis location is a point that's 20 by 30 mm from datums A and B (from the datums, not the part edges). That's the quality goal.

2. A *tolerance zone* of 0.5-mm diameter is defined around the target. The hole's centerline must fall inside the round tolerance zone.

3. It *implies* how it is to be *measured*. This control applies to an axis positioned relative to two datums. Therefore, any test of its location must use the datum concept and determine where the axis of the controlled feature lies with respect to datums. This aspect is never shown in the control callout, but it must be known by the user.

How do you know the control applies to an axis? The control frame is attached to the hole's size dimension, the control applies to the center. We've seen this same rule used for indicating datums that are axes.

A Functional Test

Proving the hole axis location with the largest ground pin that will just slip into it with no movement shows where the largest round space lies. That's the function if, for example, a bolt is to be put into the hole. The ground pin occupies the largest cylindrical space available. We call that kind of measurement a *functional test*. It is functional because it does two things: It detects the biggest bolt that could be placed in the hole and it simulates the axis location of that bolt hole. The test pin's centerline then becomes the control element.

There are only the two possible control entities: outside elements or center axes. You must determine which by the way the feature control frame is applied. If the frame connects directly to the center of the feature or to its dimension, as on the two drawings on the left in Fig. 4-23, then the **entity** is an axis.

The Special Case—Straightness of an Axis

Figure 4-23 illustrates the special case for straightness whereby it is applied to the center of the pin. In this application, the axis must lie within a cylinder of 0.005 in. diameter. On the right hand drawing, the frame connects to the feature's surface, thus it is controlling straightness elements.

Basic Dimensions

Another fundamental geometric symbol is the box found around the two dimensions in Fig. 4-22. They indicate that the dimension defines the **perfect model.** When boxed in this fashion, they are called *basic* dimensions. Basic dimensions always originate from datums and they specify perfect targets (true location in this example). They show where the bull's-eye is located. They have no tolerance themselves, but tolerance will be built around their target location.

That is very different from standard dimensioning where the distance to the hole's center would have a tolerance. Functionally, in the nongeometric method, the target point can move. It has tolerance, *but not in the geometric system.* Here the bull's-eye remains stationery while the tolerance is built around it.

The shape of a geometric tolerance zone built around the target can vary, but the tolerance is always the distance across the zone. In Fig. 4-22 it's a circle 0.5 mm across. OK, it's time to see how the three-step process applies to all of the controls.

The axis shall be straight within 0.005-in. diameter

The surface shall be straight within 0.005 in.

Figure 4-23 *Different ways of specifying either the axis or a surface element—control entities.*

Summary

To correctly *measure* any geometric control, you must determine:

Does the control require datum reference or not?

Does it apply to surface elements or to an axis?

The Fourteen Geometric Characteristics

As we go through these controls, our objective is to achieve familiarity. Still, you might start to see deeper into the system to discover inter-relationships between the controls. Many individual controls contain further controls buried within as subsets. Measuring a feature for compliance to the stated control will also verify a second and even a third control. For example, testing cylindricity (to be a perfect cylinder) to be within 0.005-in. tolerance, also tests for roundness and straightness within 0.005 in. When one control is a subset of another, it's called *embedded*.

You might also look for ways controls could be combined as basic building blocks might be. When a specific control isn't available, two or more individual controls can be put together to form something new. For example, after a bit of study, how might we combine two or more controls to define a perfect cone (see Fig. 4-24)? Actually, there are several ways that it could be accomplished; some without a datum reference and some with. The composite control will need some way of defining the angle of the cone and its roundness. Keep this challenge in mind as you'll be asked to build the cone control in the final questions. To do so, you'll need to select controls from the 14 choices that follow.

Group 1—Controls of Form

All four controls in the **form** group, apply to the outside surface of a feature thus they prove *elements* are within tolerance. (Except straightness can be applied to an axis.)

◆ **Straightness** ———

◆ **Flatness**

◆ **Roundness**

◆ **Cylindricity**

Straightness ———

In Fig. 4-25, a straightness control has been called out for the top of the part. To pass inspection, it must be proven that each element tested lies within a tolerance 0.003 in. wide. This means that, as shown on the bottom of the drawing, each element can vary in any way, as long as it does not go outside the 0.003-in. sandwich.

The most convenient way to test straightness is to place a straight template against the surface, with the edge running in the direction of the control element (left to right in the drawing). A high-quality precision ruler could do the job, or an implement that has no graduations called a precision straight edge. Then using a 0.003-in.-thick gage (a thin metal feeler gage), check to see that no gap is greater than the drawing tolerance between the surface and the straight edge. See Fig. 4-26.

Figure 4-25 *Straightness per print, and its meaning.*

Figure 4-24 *In GDT, controls can be put together to define functions such as this cone.*

Figure 4-26 *Testing for straightness.*

element at a time, and that element has a defined direction.

To test the lunchbox, one would place the straight edge parallel to the control direction, in several places on the top, and then look for gaps exceeding the 0.010-in. tolerance. How many elements are tested is a matter of how reliable the feature in question might be. Because the lunchbox is a formed dome shape, front to back, many elements should be checked, while the cone might be tested in three or four locations.

KEYPOINT

Straightness is a single-element control. It is two dimensional.

Single-Element Test—Measuring Factor 3
Straightness elements are tested one at a time, which is known as a *single-element test.* Another way to word this is that a straightness test is two dimensional. It's important to understand this concept for it's a third factor that dictates how a given control is to be inspected. Several controls yet to be studied are also single-element controls.

When the control is a single-element one, each must pass its own test. This means each element must be shown to fit between two lines 0.003 in. apart in the example. None of the tests are connected. Figure 4-27 illustrates the concept. Both of these objects are straight in one direction only. A precision straight edge could be used to inspect them. Straightness does not control the surface except one

Exception—Straightness Applied to Axes
As mentioned previously, straightness can be applied to the center axis of a feature. Figure 4-28 is a good example of how much meaning can lie behind GDT symbols. Not the surface but the pin's axis has been toleranced to be straight within 0.003 in. We know the control applies to the centerline because the feature control frame connects to the dimension. Second, we now know that the tolerance zone is not two dimensional (2-D) because of the theta symbol just before the 0.003 in. This tells the reader that the tolerance zone is cylindrical.

As illustrated in Figure 4-28, the tolerance boundary becomes a 0.003-in.-diameter cylinder. The axis of the pin can vary in any way, as long as it doesn't violate that tolerance zone cylinder.

Figure 4-27 *A control of straightness applied to the top contour on a lunchbox and to a cone.*

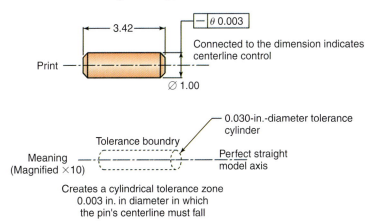

Figure 4-28 *Here straightness has been applied to the center axis of a feature, not to the surface.*

Flatness Print Application

0.2-mm tolerance

All elements on the controlled surface must fit within the tolerance zone

Figure 4-29 *Flatness as seen on the drawing and its interpretation.*

Flatness

Flatness is the three-dimensional extension of straightness (Fig. 4-29). Flatness is an *all-element* control, meaning that the entire surface must be tested at one time. When the control requires an all-element test, individual elements exist but they are connected into a surface. Here, the lunchbox bottom has been given a flatness tolerance of 0.2 mm. The feature control frame reads "this surface is to be flat within a tolerance of 0.2 mm."

Testing Flatness Different from straightness, *all elements* on the controlled surface must be shown to fit within the three-dimensional (3-D) sandwich of 0.2 mm width in the example. Theoretically, to measure flatness, a perfect, flat template is placed against the surface and gaps are measured similar to checking straightness with a ground edge. However, that is impossible if the surface is very big.

To test this flat surface with a straight edge template first using right-to-left elements then turning it 90 degrees and testing front-back elements **will not work**—that's a test of bidirectional straightness. Think this through, every single line element both left to right and front to back, could pass a 0.2-mm straightness test, yet their total variation would not fit into the 0.2-mm flat, three-dimensional sandwich.

So, how does one test for flatness? There is one practical shop method that uses a dial test indicator (see Section 6) to sweep the surface and look for the highest high point compared to the lowest low point.

However, while this test works OK, it is truly a test of parallelism—not flatness. It shows that the entire surface is parallel to the table datum.

See Fig. 4-36 later in this unit to see how this is done. While it tests parallelism, flatness is embedded within. Therefore, if the object passes the parallelism test within 0.2 mm, it must be flat within 0.2 mm as well. It might be tilted slightly yet be perfectly flat, but nevertheless, it passes the test.

KEYPOINT

Flatness is the 3-D version of straightness.

Keep this 2-D/3-D relationship in mind as we explore further controls. Understanding the difference aids in choosing the right way to measure a given control. Here's the next example, roundness is a two-dimensional control, while cylindricity extends roundness to a three-dimensional control.

Roundness

Continuing our systems approach to GDT, roundness can be envisioned as straightness rolled into a perfect circle. The tolerance zone then becomes a pair of concentric circles, whose distance apart forms the tolerance zone for each single element.

In Fig. 4-30, roundness is shown the way it would appear on a print. The second illustration is how it is evaluated. Each circular element must be round within 0.005-in.—that

Roundness

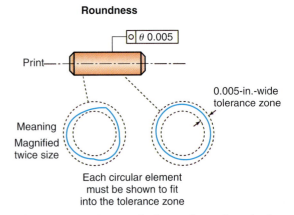

Print

θ 0.005

0.005-in.-wide tolerance zone

Meaning Magnified twice size

Each circular element must be shown to fit into the tolerance zone

Figure 4-30 *A control of roundness is a single-element control.*

is, each element must be able to fit into a pair of perfect circles that have 0.005 in. between them.

Each element could be a different diameter within the overall diameter tolerance, as with a tapered roller bearing, but each must be round within its own 0.005-in. tolerance. In other words, the roundness tolerance zone conforms to each element.

Trade Tip

Don't Fall into the Round Trap! Roundness can be tricky to evaluate at times. It is common to evaluate it using a distance-measuring tool called a micrometer (Section 6). However, the problem is that micrometers measure distance, not roundness. The shape shown in Fig. 4-31 is rarely produced if the machine is set up correctly. But it does happen every so often! Called a trichoid, it is not round with three lobes and rounded corners. Even though the trichoid is not round, when tested with a standard micrometer, every point measures the same distance across!

To detect the trichoid, a second type of micrometer is used that supports the object in a three-point contact, called a tri-anvil micrometer. We will study micrometers and other testing methods later, but for now, be aware there is much more to learn about geometric evaluation.

A Trichoid

Distance across is always equal but it is not round

Figure 4-31 *A trichoid measures the same distance across, everywhere, yet it is not round.*

Form Controls Use No Datum Have you noticed that none of the form controls were referenced from a datum? That's because they don't need one. Form control needs no outside reference. Think of the control zone as a custom-fitting sandwich that applies to each element or surface "as is." If the element can be shown to fit into its own custom sandwich zone, then that element passes the test. In other words, an object is simply straight without any outside reference.

Cylindricity

All circular elements must fit into the zone (the entire surface)

Figure 4-32 *A form control of cylindricity.*

Cylindricity

Cylindricity is a three-dimensional, all-element control extending roundness into a cylindrical surface. The tolerance zone then is a pair of perfect floating cylinders into which all circular elements on the controlled feature must lie (Fig. 4-32). In other words, it must be proven that the entire cylinder can fit into two heoretical tolerance cylinders that are the specified distance apart (0.015 in. in the example). Similar to other controls of form, the tolerance zone conforms to the surface as it is.

Group 2—Controls of Contour

The geometric **contour** controls were created to do two things: primarily to define and limit the surface of odd-shaped objects and to orient (position) the shape relative to datums. For example, they are used in making a die to form a rain gutter or the top of the lunchbox we've been looking at.

Both single-element and all-element controls are used to define the outside surface of features. Both are similar to form controls of straightness and flatness. If we were to take a straightness control and bend it into some special shape, then it would become the contour of a line (2-D). Then if we were to take flatness and bend it in one direction only, as with a sheet of paper, without wrinkling, its special shape would become contour of a surface (3-D).

◆ **Contour of a Line**

◆ **Contour of a Surface**

Contour of a Line

Print

Meaning
×3

0.03-in.-wide
tolerance zone
that is parallel to
Datum A
(in this example)

Each element must fit
within its own zone

Datum A

Figure 4-33 *Contour of a line is a 2-D control (single element). This contour is oriented to the assembly world with reference to datum A.*

There are two differences between contour and form: For contour, the perfect model must be defined. This can be by standard dimensioning if the shape is simple enough or by a master template. The master contour can also be defined with a table of *X* by *Y* data points, usually in a database, or the shape can be defined with equations.

The other difference from form is that these two controls may or may not require a datum, depending on function. While the form controls never refer to a datum, contour controls may need a reference to orient the shape to the assembly world. For example, in the lunchbox in Fig. 4-33, since the top dome shape must be pointed straight up, the bottom must be assigned a datum and the contour referenced from it.

KEYPOINT

Controls of contour may or may not require datum reference, depending on function.

2-D and 3-D Versions

The two contour controls exhibit the 2-D/3-D relationship. The single-element line contour extends into a 3-D surface contour. To do so, a straightness element is added as the surface is extended.

Contour of a Line

Line contour is similar to roundness in that a single element is tested against a template cut to the right shape for evaluation.

Figure 33 states

This surface shape shall be contoured within 0.03 in. with respect to Datum A.

You can see that the function is that the overall dome should be parallel to the bottom of the box or it would appear skewed. Therefore, this control refers to a datum. If the contour was all that mattered, similar to roundness, then there would be no datum reference in the control—stamping out a rain gutter, for example. What difference would it make if the die was tilted, as long as the finished gutter was the right shape?

Contour of a Surface

Following the 2-D/3-D relationship, line contour is extended out with a straightness element to form a surface control. All elements must now fit into a shaped sandwich as shown in Fig. 4-34. The elements in question are the form shape plus straightness, at 90° to them. Think of the control zone as a flatness control—but bent on one axis only. Within current ASME standards, there will always be a straightness element on the surface in one direction. The GDT system is not used to define surfaces that curve in two directions, called compound curves, other than spheres (next).

KEYPOINT

Since the top of the object in Fig. 4-34 is being formed only, there is no datum to control the orientation of the contour in this example. The shape is all that matters.

Contour of a Surface

Print

0.5 mm

Meaning
×3

A three-dimensional contoured sandwich 0.5 mm across. All elements must fall within

Figure 4-34 *Contour of a surface. This control has no datum reference. It is its own reference in this application.*

Contour Footnote—Spheres

There is one exception to 3-D shapes defined by the geometric system: the sphere. A sphere is curved in two directions and has no embedded straightness element. The special case of surface contour is *spherocity* (being spherical), which is the only geometric contour control without an embedded straightness (ASME Y14.5 M 1994 Standards).

Group 3—Controls of Orientation

Now we look at the three controls that establish the relationship of a feature with respect to a datum. They have two new properties. Each **orientation** control can be either a single-element 2-D control or a 3-D all-element control, depending on the function of the object. Second, controls of orientation can apply either to a surface or to an axis, depending on function.

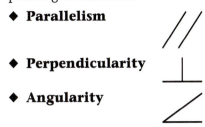

◆ **Parallelism**

◆ **Perpendicularity**

◆ **Angularity**

These three controls are nearly the same. Each defines a perfect angular relationship with respect to a datum. The difference concerns

The angle at which the model lies relative to the datum.

Whether the control applies to a surface or to an axis.

Whether it is a single- or an all-element control.

From this point on, including orientation, all controls refer to a datum.

Evaluating Orientation

The controls of orientation are much easier to evaluate than those of form due to the addition of the reference datum. Typically, a dial test indicator (a precision distance measuring tool) is used on the layout table. The table becomes the datum.

Control of Parallelism

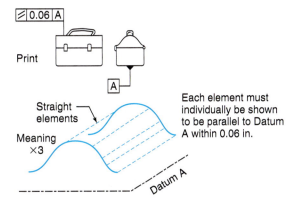

Figure 4-35 *A control of parallelism for the top of the lunchbox. In this example, it's a single-element control.*

In Fig. 4-35, the control of parallelism is applied to the top of the lunchbox. In this case, it is a single-element test. The parallelism does not apply to the entire surface but to each right-left element tested. You might envision that test as straightness that refers to a datum. In fact straightness is embedded in the parallel test.

Parallelism

The perfect target lies at 0 degrees with respect to the datum. Here, in Figs. 4-35 and 4-36, each element must be parallel to the datum but not to the entire surface. This is a single-element control. The feature control frame states that

Each element must be parallel within 0.060 in. of Datum A.

Figure 4-36 *Using a dial test indicator to evaluate the parallelism of a surface.*

Critical Thinking Question 1

OK, let's test your understanding of the system. Parallelism can be an all-element control as with the object shown in Fig. 4-36. That means the entire surface must be parallel to the datum table. Therefore, two controls are embedded within. Can you name them? This is easy, but the next question isn't. The answer is found just before the Unit 4-3 Review.

Perpendicularity

Manufacturing is an orthographic, *X-Y-Z* world. Ninety degree relationships are common. Perpendicularity is a convenience control for the often occurring, special case where the orientation is to be 90° with respect to the datum. While we could use angularity to do the same job, the GDT system includes special cases such as this.

In Fig. 4-37, the perpendicularity has been applied to the entire end of the lunchbox, therefore it's an all-element control. The whole surface must fit into the 0.15-mm vertical tolerance zone sandwich. Similar to the parallelism of a surface, this perpendicularity control embeds flatness. Flatness then contains straightness as a further subset.

Think Critical Again Question 2

Talk this over with a fellow student. In solving it, you will have gained insight into GDT. Can embedded controls be toleranced differently than the parent control? For example, in Fig. 4-37, with the perpendicularity tolerance of 0.15 mm on the lunchbox end, could its end flatness then be toleranced to 0.2 mm or to 0.1 mm—which or both? After making your decision, find the

Control of Angularity

Figure 4-38 *A control of angularity.*

answer. Part of the solution is coming up in a Key Point, with the full answer found just before the Unit 4-3 review.

Angularity

Notice in Fig. 4-38 that the 60° angle is placed in a basic frame. This signifies that the angle is the perfect model. Other than the basic frame, the concept is no different from perpendicularity or parallelism, where the target angle was understood to be a perfect 90° or 0°.

For perpendicularity, it's understood that the 90° has no tolerance; it is the target. So too, the 60°, also has no tolerance. The tolerance is built around it.

Measuring a Geometric Angle One simple way to measure a geometric angle is shown in Fig. 4-39 using a precision protractor (an angle measurement tool discussed in Section 7). The protractor is *locked* at the exact 60° angle. The user then checks several elements for gaps that exceed the tolerance when compared to the protractor blade. Feature A must rest on the datum that is the basis for the measurement. Another way to envision this test is that it's similar to testing perpendicularity except the *square* in this case is at 60°.

Figure 4-39 *Measuring a geometrically defined angle by testing for gaps that exceed the tolerance.*

Control of Perpendicularity

Figure 4-37 *A control of perpendicularity.*

Group 4—Controls of Location

The three **location** controls always position the axis of a feature with respect to a datum.

- **Position**

- **Concentricity**

- **Symmetry**

Position

Following the three-step process, this control defines the perfect model position with respect to one or more datums. Then a tolerance zone is built around that perfect target. The tolerance zone will take on different shapes, depending on the feature to be controlled. We will limit our discussion to the position of a machined hole, which produces a round tolerance zone. (It's actually a cylinder if extended through the part but we'll restrict this demo to 2-D for now.) In Fig. 4-40, the model position defines where the hole's centerline should be, then the tolerance defines a zone where it can be to be located within tolerance. To be in compliance, the hole's axis must fall within the circular tolerance zone built around the target location. We know it's a perfect position by the basic box around the two dimensions.

Controlling Position

Figure 4-40 *A control of position.*

The feature control frame of Fig. 4-40 states that:

The centerline position (of this 9.0-mm hole) shall fall within a round tolerance of 0.5 mm with respect to Datums A and B.

In Fig. 4-40, the hole has been drilled slightly off target to the upper left, but its center falls within the 0.5-mm target circle. It is acceptable.

Concentricity

Concentricity is designed to control the center of spinning objects. It's a similar control to position in that the center of a feature is being controlled relative to a datum. The difference is that the datum is always an axis too. Concentricity specifies that the center of one feature must lie within a tolerance zone

Trade Tip

Give Yourself a Bonus (Tolerance That Is)!

Depending on function, the position tolerance for a given feature isn't always the "bottom line." **Bonus** tolerances can be earned as long as you know the rules. When the Ⓜ symbol is present within the control frame, there is a possibility of additional tolerance (Fig. 4-41). The idea is that tolerances must always be based on the tightest possible case for assembly. For example, suppose you were drilling a 1.00-in. hole with a plus/minus tolerance of 0.020 in.

The tightest assembly case, within the size range, would be a hole at 0.980 in. That's the maximum material condition (size) and is abbreviated as MMC. It is symbolized as an Ⓜ within a circle, as seen in Fig. 4-41. Then as the feature departs from the tightest condition (a bigger hole), the target size can increase in many instances, which is called bonus tolerance.

The feature control frame of Fig. 4-41, states: "The centerline position of this hole shall fall within a 0.010-in.-diameter tolerance, when the feature is at MMC size, with respect to Datums A and B."

Position tolerance applies to MMC hole if not MMC diameter. Bonus tolerance applies.

Figure 4-41 *This feature control frame includes the MMC symbol.*

What if the hole wasn't 0.980 in.? Any diameter greater than the MMC could have more location tolerance due to the larger assembly space it creates. That hole could move around a bit more and whatever was to go in it would still fit. Thus, based on function, if the designer can permit the bonus, the MMC symbol will be added to the control frame.

You'll need to study this more in your formal geometrics training, but in a condensed form, here's the four-step rules if the Ⓜ symbol is shown:

- ◆ Determine actual machined size
- ◆ Determine MMC size
- ◆ Find the difference in the two numbers which is the bonus
- ◆ Add that bonus to the stated tolerance in the frame

Let's try it. Suppose the hole in Fig. 4-41 is machined to 1.003 inch? Compute the position tolerance for this hole accounting for bonus. Try it before looking.

The bonus tolerance answer is this: 1.003 is 0.023 in. greater than the MMC size of 0.980. Your bonus is 0.023 in. added to the location tolerance. The answer is 0.033 in. for that hole. Doing this is possible only if the Ⓜ is present, but there may be a bonus just waiting to be claimed by the informed machinist. This is an example of the extra control that the geometric system imparts to manufacturing.

---

built around the axis of another feature (the datum). It's the axis-to-axis control shown in Fig. 4-42. For something to be concentric means it will share centers with something else; recall that in GDT, the "something else" is always a datum.

In Fig. 4-42, the tolerance zone is specified to be a 0.5-mm cylinder. The diameter of the cylinder is the limit for the object's centerline.

Never a Bonus for Concentricity Because concentricity is used to control functions such as vibration or the center of mass on spinning objects, it cannot be given a bonus tolerance. The function doesn't permit it. This is true for the next control as well, no bonus tolerances apply to symmetry.

Control of Concentricity

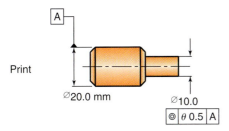

Print

⌀20.0 mm   ⌀10.0

◎ | θ 0.5 | A

Meaning

Datum A

The 10-mm feature's axis must fall within the 0.5-mm tolerance zone built around the axis datum

0.5-mm cylindrical tolerance zone

Figure 4-42 Concentricity *positions a feature centerline with respect to another feature centerline datum.*

Symmetry

Symmetry infers equal distribution of a feature about a central target (the same on both sides). Similar to position, symmetry requires that the controlled feature's centerline must lie within a tolerance zone built around a perfect target. (Fig. 4-43). The difference is that the target is usually a centerline too. Symmetry is a convenience control that locates objects that are not spinning. In truth, position can be used for the same purpose and has been in the past.

Control of Symmetry

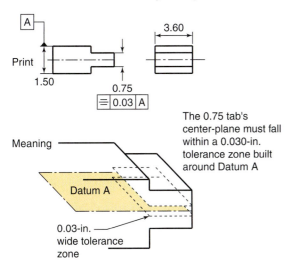

Print

1.50   0.75   3.60

≡ | 0.03 | A

Meaning

Datum A

0.03-in. wide tolerance zone

The 0.75 tab's center-plane must fall within a 0.030-in. tolerance zone built around Datum A

Figure 4-43 *Symmetry controls a feature that is equally distributed around a datum.*

Group 5—Controls of Runout

◆ Single-Element Runout

◆ Total Runout

Runout is designed to control the outside surface wobble of rotating objects with respect to a central axis datum. Different from concentricity, runout always applies to surface elements. Looking a little deeper, you might see form controls embedded in these controls too. In fact, runout can be described as a kind of form control with respect to a datum. Also similar to controls of form, the 2-D/3-D relationship is again involved.

Single-Element Runout

Two-dimensional runout, sometimes called circular runout, can be applied to round or tapered surfaces, as shown in Fig. 4-44, or to a flat end surface. Each element is an individual test. In Fig. 4-44, to be within tolerance no element may exceed 0.01-in. *FIM* (full indicator movement explained next).

Measuring Runout To measure runout, a dial indicator is placed against the surface in

Control of Runout (single element)

A   ↗ | ⌀ 0.01 | A   ↗ | ⌀ 0.01 | A

Print

⌀ 1.00   ⌀ 0.500   ↗ | ⌀ 0.01 | A

Meaning

Datum A

Rotating 360 degrees around Datum A, no single element may exceed 0.010-in. full indicator movement

Figure 4-44 *Runout as applied to three surfaces of the object.*

Figure 4-45 *Testing runout on a lathe, using a dial test indicator to find the FIM.*

Figure 4-46 *Total runout—an all-element, surface control is symbolized by the double arrows.*

question and the part is spun 360° about its datum axis. The indicator shows movement in thousandths of an inch or decimal parts of a millimeter. As the object spins, the highest point and lowest point are noted.

The difference between the high and low points is called the **FIM (full indicator movement)** or sometimes it's called the *total indicator reading (TIR)*. The FIM for one circle is the runout of that single element. To comply to a runout tolerance, the FIM for each element tested cannot exceed the tolerance (Fig. 4-45).

Each element requires an individual test and none may exceed the tolerance.

Total Runout—Control of a Surface

This is the all-element, 3-D version. It controls an entire rotating surface with respect to a datum axis. It is tested the same way as single-element runout, but this time the indicator must be moved over the entire surface to test all elements. The highest point is found and compared to the lowest point, and the difference is the total FIM for the surface. In Fig. 4-46, one might envision this control as cylindricity with respect to a datum.

Comparing Concentricity to Runout

At first it seems that concentricity and runout are the same controls, but while they both deal with rotational features and a central axis, they are different. Here are a couple of examples. Consider a camshaft lobe (Fig. 4-48). Its profile must rotate around an exact center—it must be concentric. But, to work right, the lobe has a designed amount runout. In the second example, the hex head is concentric to the bolt, yet it also has a lot of runout. From these do you see that

Concentricity controls the *axis* of a feature with respect to the axis of a second datum feature where both have the same center. *Runout* controls the surface of a feature, with respect to an axis datum.

Critical Thinking Answers

Answer 1

Straightness and flatness are embedded within the test of surface parallelism.

Answer 2

The flatness of the lunchbox in Fig. 4-35 could be toleranced tighter at 0.1 mm but to tolerance it to 0.2 mm is meaningless since it must already be flat within 0.15 mm to meet the perpendicularity requirement.

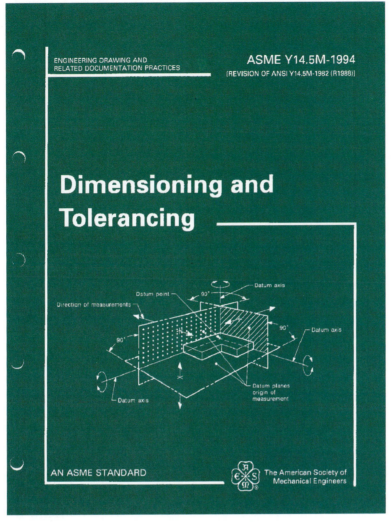

Figure 4-47 *ASME Y14.5, the living document that controls geometric application in the United States. Cover photo courtesy of American Society of Mechanical Engineers.*

Both features are concentric.
Both exhibit great amount of runout.

Figure 4-48 *Compare runout to concentricity.*

UNIT 4-3 REVIEW

Review the Key Terms

Bonus
Additional tolerance possible based on the function and the actual measured feature size.

Contour
Controls that deal with irregular-outside-shaped features. Contour may or may not need a datum reference depending on function.

Elements
Surface lines running in the direction of the control.

Entity
Either a surface element or a feature center that must be shown to lie within the tolerance zone to be within compliance.

FIM (full indicator movement)
The difference in high and low points in a 360 degrees rotation of a surface measured on a dial test indicator.

Form
Four controls that deal with the outside shape of regular-shaped objects. Form has no datum reference.

Location
Three controls that define the tolerance for the center of a feature with respect to a datum.

Orientation
Three controls that define the angular relationship of a surface or feature center with respect to a datum.

Perfect model
The goal of the desired characteristic.

Runout
Two controls that deal with surface wobble on rotating objects (axis datum).

Replay the Key Points

◆ A geometric control applies to either the feature's surface elements or its center axis.

◆ Elements are the outside surface control entity.

◆ For the feature to be within tolerance, each element tested must be shown to lie within the tolerance zone.

◆ There are an infinite number of elements on any surface.

◆ Each element runs in the direction of the control.

◆ Elements could be drawn as pencil lines upon the surface of the feature.

◆ There are interrelationships within the GDT system and many individual controls can be combined as building blocks, to create control definitions.

Respond

1. List the five groups of controls.

2. Sum up the reason for a GDT system compared to the standard dimensioning and tolerancing methods in 10 words or less.

3. Repeat Question 2 but respond using only one word.

4. A datum is always a reliable surface on the part. Is this statement true or false?

5. A drawing has datums A, B, and C assigned to surfaces. If it isn't already established on the part, which datum feature would probably be machined first? Why?

Section 4 Review

Ongoing Study

Section 4 was written to get you going in the lab with the right amount of GDT background. A second goal was to demonstrate that a lot of geometric knowledge remains unrevealed, but it would be too much information at this point in your training. After studying Section 4, as machining challenges come up in the shop, think about function and how the principles we've studied apply. You'll soon be ready for more formal study.

Unit 4-1

While technology was evolving slowly, the way part drawings were dimensioned and toleranced evolved more-or-less without central control, and that worked OK for the times. But the relatively recent rapid advances in technology showed that those old ways fell far short of meeting the functional needs of modern manufacturing. The new controls were carefully crafted by engineers, scientists, and manufacturing leaders not only to express drawing dimensions and tolerances, but also to

   Set functional priorities using datum
      references
   Specifically serve function
   Eliminate ambiguities

Unit 4-2

The datum reference is at the heart of the GDT system and is a major improvement over older methods. Datums tie the part to the real world by referencing its dimensions to theoretically perfect surfaces and lines outside the object. The design is built outward from a complete set of these references, starting with the most important function. The highest priority feature is assigned to the highest priority datum, and that makes all the difference in how you proceed as a machinist, inspector, or programmer. The older way created a design that left the part floating in space with no certain starting point for shaping it from raw metal to finished part. Using nongeometric prints, machinists find themselves guessing at how to get started. But datum reference clearly defines the functional targets and where to begin cutting. It doesn't show how to cut them. That aspect remains within the realm of skills. But using datum references, a geometric design clearly shows the functional goals.

Unit 4-3

The GDT system puts more responsibility on the machinist because so much information is implied by the symbols. Not knowing the rules of the game creates needlessly high costs in manufacturing. However, once the system is understood, the savvy craftsman has control that just wasn't available to earlier machinists using previous dimensioning and tolerancing methods. After datum symbols on the GDT drawing, nearly all the remaining information is conveyed in a rectangular data field called a feature control frame. Within the FCF, one of fourteen possible characteristics is signified with a symbol that you must be able to recognize. The allowable tolerance is also stated in the FCF along with other critical information both in symbolic and numeric form. All of which must be understood!

Questions and Problems

Answering the following questions about the GDT system will help prepare for the final test.

1. What characteristic does this symbol represent? (Obj. 4-1)

2. What two controls are embedded within the control of Question 1? (Obj. 4-3)

3. What does this symbol represent? (Obj. 4-1)

4. Describe the control of Question 3. (Obj. 4-3)

5. What is a datum? (Obj. 4-2)

6. True or false? A datum cannot be an axis since an axis is a mathematical entity. If it's false, what makes it true? (Obj. 4-2)

7. Name the document that standardizes GDT in the United States. (Obj. 4-3)

8. When a geometric control applies to a feature's surface, the control entity is an _____. (Obj. 4-3)

9. On a sheet of paper, sketch the GDT symbols for
   A. Profile of line
   B. The all-element (3-D) counterpart to straightness
   C. The three controls of orientation—in any order
   D. The three controls that define the angle of a feature's axis with respect to a datum

E. The location control(s) that *can have* a bonus tolerance

F. The controls that are *never* referenced from a datum

G. The four controls of form (Obj. 4-3)

10. Explain how a GDT drawing shows the machinist how to organize the first few cuts on the part. (Obj. 4-2)

Critical Thinking

These are tough—talk them over with a fellow student or two. Question 11 leads to 12 then to 13 with increasing difficulty. In Fig. 4-49, the larger diameter, datum feature A, is perfectly centered in the lathe chuck and being rotated about its axis. The smaller diameter is being tested with a dial indicator. The FIM is found to be 0.0035 in. over the entire surface.

11. Is this a test of concentricity or runout?

12. Which controls are embedded in this test?

13. What *range* of possibilities are there for the concentricity of the tested feature? In other words, how little and how much might its center be off from the axis of Datum A? Warning—this is a genuine brainteaser!

14. The 0.750-in. hole in Fig. 4-50 has a size tolerance of ± 0.005 in. It has a position tolerance of 0.010 in. The actual drilled hole is measured at 0.7467 in. What is its position tolerance after considering the drilled size?

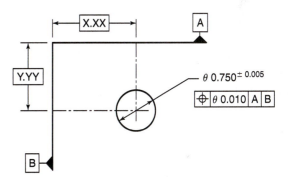

Figure 4-50 *What is the position tolerance if this hole is drilled to a 0.7467-in. diameter?*

Figure 4-49 *What is this test and what other tests are embedded within it?*

CNC Questions

15. Lathes machine round objects. On a CNC lathe *X* axis dimensions denote diameter positions of the cutting tool, while the *Z* axis dimensions denote length. If you command the tool to go to *X*3.500, it will be ready to cut a diameter of 3.500 in. You are planning out a program for a CNC lathe to machine an axle with many critical diameters. In terms of function, where would you expect the primary datum to be?

16. Referring to Question 15, at what value would the *X* axis be at when it was at that primary datum?

Section 4 Answers

ANSWERS 4-1

1. ASME Y14 (American Society of Mechanical Engineers)
2. Function
3. The functional priorities show you
   A. How to hold the part for machining.
   B. Which cuts to take first
   C. How to measure the results.
4. Form, controls outside surface shape of regular features—round, straight, flat, and cylindrical shapes; profile, controls outside surface shape of contoured features. Items that aren't straight, flat, or round; orientation, controls relationship of one feature with respect to a datum; location, controls the position of a feature's center with respect to a datum; runout, controls the wobble on the outside of a spinning object.
5. Straightness can be applied to an axis rather than the outside shape.

ANSWERS 4-2

1. Datum frame or datum identifier
2. Function
3. Datum priorities set the sequence of early cuts.
4. Form, profile, location, orientation, and runout
5. A datum
6. The datum is a theoretically perfect surface contacting the feature in question.
7. Axes (plural of axis)

ANSWERS 4-3

1. Form, contour, orientation, location, runout
2. As set of controls that offer flexible control
3. Function
4. It's false for two reasons. First, datums can be axes as well as surfaces and, second, datums are established by surfaces but they are not the surface.
5. Datum A because it holds the highest functional priority (primary datum) datums B and C are second and third.

Answers to Section Review Questions

1. Cylindricity
2. Straightness and roundness
3. Position
4. Controls a feature's axis, with respect to one or more datum(s)
5. A datum is a theoretically perfect surface or axis used for reference for orientation and location. It is not on the part but it is established by the part feature's surface and its irregularities.
6. False. An axis can be a datum as long as the irregularities of the feature's surface are summed up.
7. ASME Y14.5M
8. Element
9. A.
   B.
   C.
   D.
   E.

F. _____ , ▱ , ◯ , and ⌀

G. Same as F

10. The datum priorities indicate functional importance.

11. The test is touching and summing surface elements—it is a test of runout.

12. Straightness, parallel to the axis; roundness, of individual elements; cylindricity, of the entire surface since all elements have been tested together; concentricity, of the cylinder's axis

13. *Maximum Possible Error:* Given the FIM, it could be as far out as 0.0035 FIM in. assuming the small feature is perfectly cylindrical. Physically that would mean the two axes are 0.00175 in. apart (half the FIM). But it couldn't be off any more than that since concentricity is a subset of runout.

    Minimum Error: It could be 0.0000 since the surface might be perfectly centered but not cylindrical; for example, squashed but centered!

14. The true answer is 0.010 in.—since *there was no MMC permission to claim the bonus! No M symbol in the frame!* However, if you calculated 0.0117 in., you do understand the principal:

    MMC size = 0.745 in.   (tightest possible hole)
    Machined size = 0.7467 in.
    Possible bonus = 0.0017 in.
    (difference = bonus)
    Print tolerance = 0.010
    Result = 0.0117   (print plus bonus *if* it was permitted!)

15. Datum A would be the central axis.

16. The tool would be at $X = 0.0000$ (the center of the diameter).

Section 5

Before and After Machining

Introduction

Section 5 has two overall goals: to learn to read work orders and to learn to perform a collection of tasks found on them that aren't performed at the machines. These are the steps that get material ready for machining, or finish it afterward.

Some of these operations are actually machining since they shape the workpiece, but they are performed by hand at a workbench or on small machines that support the big CNCs. As such, they might be called offload, secondary, or benchwork. In most cases, they are less challenging than the CNC work, but not less important, for without them the job wouldn't be complete. Offload work is often assigned to the newly hired employee.

Objectives

Using Work Orders–Learning Job Planning

Introduction: Nearly all shops big and small, use some form of work order. They complement the drawing with formal, step-by-step directions that guide the job to the final conclusion. Work orders bring about order to the job, along with consistent quality and safety. There's usually a lot more information on them than the job planning sequence, although for the machinist, that's their central purpose. The dated work order lists:

Part Number   The specific product to be made, abbreviated *P/N*

Batch Quantity   Abbreviated *Qty*

Revision Level   Showing the design level, abbreviated *Rev*

Sequence of Operations   The logical step-by-step instructions, abbreviated *Opp*

The Material Type   The form and size of the raw metal provided

The Standards   to which the product is to be built

Instructions for Quality Assurance

Finishing & Packaging and other handling instructions

Special Requirements for tooling or other items needed

KEYPOINT

In addition to symbols, there are a lot of abbreviations used on manufacturing documents. They save time and space both on paper and in the database for the instructions.

Job Sequence Planning The heart of a work order is a set of carefully planned, specific instructions that guide the job from start to finish. They outline each step and operation that must be performed, in the required sequence. A lot of these instructions will be abbreviated, for example,

Operation	Instructions	Station
Opp 45	Drill 0.375 dia @ 3.0 ref Dat A × 4.375 ref Dat B	Bridgeport
Opp 50	Drill and tap 5/8-11 threads	Bench

operation 45 tells you to drill a 0.375-in. hole that's located 3.000 in. from datum A, its *reference*. To drill these parts the planner has scheduled them at a particular machine called a Bridgeport mill.

Finding the Best Sequence Each new design requires a puzzle-like solution to find the best way to go about making it. Often, that challenge can be solved several different ways. For example, the job just discussed could also be drilled at the drill press, perhaps back at operation 20. Each path solution creates a completed product, yet one or two paths will be faster, of higher quality, and of greatest impact to you, they will be safer. To find that best plan requires a full knowledge of all the machining processes available in the shop.

Because so much depends on good planning, work orders are written only by highly experienced and well-paid people. As a new employee your goal will be to follow the work order exactly without deviating unless given direct orders by your supervision. One way to learn planning when following a sequence is to think through why it was chosen as it was.

Here in Section 5 we will investigate how to best use work orders. Then, later in this text, after the study of lathes, mills, grinders, and drill presses, in Section 14, we'll look at the logic of creating a job plan. Then we'll study CNC program planning in Section 21. To get the ball rolling, at the end of this section, you'll be given a chance to solve a few planning puzzles yourself.

Followed—In Sequence The work order sometimes overrides the design print. Often we must deviate from the drawing by following special instructions on the work order. For example, leaving extra material on dimensions so the parts can be sent out to be heat treated, then returned to the shop for final machining. For example,

	Mill 3.00 to 3.060	
Opp 55	(grind excess)	Bridgeport

To miss this directive and make the parts exactly like the print, 3.000 in. instead of 3.060 in., would be a huge mistake! Sometime later, these parts will be moved to the precision grinding machine to finish. Here at operation 55, they are being rough machined, to be ground to size later. This is so critical, that in industries such as passenger aircraft, each operation in the plan must be certified complete by auditing inspectors before moving the job forward. Records must be kept to prove the certification of the steps.

Without proper records showing that the steps were followed in sequence, even though the parts are perfectly made, they are illegal to put on an airplane.

All journeymen are expected to keep their "paper work" straight. It seems complicated at first but not after using it for a while.

KEY TERMS

Detail/dash number
Individual identification of a part.

Job planning
A set of logical steps found on a work order, chosen to guide a job safely and efficiently through the shop maze.

Part number
Drawing and detail numbers combined to identify a single part.

Revision level
A letter showing how many times the drawing has been redrawn.

Secondary operations
Necessary tasks, usually not performed on the machines, needed to start or finish a job. Sometimes called offload work.

Work order
A master document initiating a job in the shop.

What You Must Know About Job Planning

Poor **job planning** leads to expensive or even dangerous traps in machining. It is possible to "machine yourself into a corner." That is, the poorly planned part is unfinished, yet there is no practical way to hold it or to machine the

Work Order S5-U1-1	McGraw Machining Company		Date: Jan 6, 20xx
Drawing FYM-101	Detail-1		Revision: New
Quantity 12	Customer: Student		Issued By: Fitz
Sequence	Operation		Complete
10	Saw blanks 0.125 aluminum 0.1 excess—all sides		
20	Square up—3.0 × 5.0 vertical mill		
30	Burr blanks		
40	Inspect operation 20		
50	Layout holes per print		
70	Locate, drill, and ream holes—drill press		
80	Burr holes		
90	Layout gage details		
100	Saw details with excess 0.1		
110	Profile mill shape		
120	Part finish—sharp edges and clean out cutter radii 1 place		
130	File 0.12 corner radii		
140	Part mark all hole sizes per print		
150	Final inspection		
160	Annodize—black		
170	Count wrap and package		
180	Ship to customer		
190			
200			

Figure 5-1

remaining operations. Or due to bad planning, it could be finished, but only at great cost or risk to the machine and its operator.

To simulate a real job, here in Section 5, we'll track the work order instructions for a batch of 12 parts. We'll make a dozen drill gages (QTY-12) using Work Order Number S5-U1-1 (Fig. 5-1) and the drawing in Fig. 5-2. Bookmark them, we'll be refer to them several times. That's typical of the way jobs proceed in the shop, going back to the work order and print often.

Part Numbers and Drawing Numbers

Often, a technical drawing shows a single detail part. However, it's also common for drawings to show several details—individual

Figure 5-2 *Drill gage drawing—sheet 1 of 1.*

parts but on the same drawing number. A detail is a single unique **part**—sometimes called a **dash number.**

Remember, different dash numbers can be found under the same drawing number.

Figure 5-3 depicts three dash numbers. To make any one of them, you would need the right drawing and the dimensions of the dash number called out on the **work order.**

203B-605-3 (Piston)
203B-605-5 (Rod)
203B-605-7 (Pin)

Part Number Breakdown

There are industry-to-industry differences, however, **part numbers** generally break down this way: In our example, the overall product is a 203B. For example, let's say it's a chain saw Model 203 and it's been engineered to level B. Then this drawing is a detail sheet

of the 605 section of the 203B saw—engine parts. Finally, there are three details on this drawing. For example, 203B-605-3 and the -5 and -7 (called the "dash five and dash seven" in shop lingo). The more complex the assembly on which the parts are to go, the longer will be the part number. This part number has three data fields and one dash number, others might have five or six depicting the complexity of the work they control.

When the drawing number (203B-605) is combined with the detail number (-3), it forms the part number: 203B-605-3.

Sometimes, there is little difference between one detail and the next. For example, a dash 5 part and the dash 6 might be mirror images of each other; the same but right- and left-hand versions. Or the -5 is 1.500 in. long while the -7 is 1.550 in. long; small things that matter a lot.

Figure 5-3 *Drawing number 203B-605.*

Multiple Sheets for One Drawing

Drawing number 203B605 might also have more than one sheet. Be on the watch for the note

"Sheet 1 of XX"

While it's correctly called "a" print or "a" drawing, it's not necessarily a single sheet of paper. This drawing might have several pages:

203B-605 Sheet 1 of 3

203B-605 Sheet 2 of 3

203B-605 Sht.3 of 3 (typical abbreviation)

If the drawing is a single sheet it might still say Sheet 1 of 1 to eliminate guessing.

Trade Tip

General Notes on Multiple Sheet Drawings *Caution!* Don't get caught without the needed information! The important, general notes for multisheet drawings are found on Sheet 1. Never work a job without having or knowing Sheet 1 even when the assigned detail is on another sheet.

Revisions—The Right Version of the Part Number

The final part number item to be coordinated by the machinist is the *version* of the part to be

made. In addition to ensuring that the drawing number and sheet number match the work order, you need to verify that the drawing you've been given is the right *version* of the part. That too will be recorded on the W/O (abbreviation for work order, also abbreviated just WO).

As original designs are improved and mistakes are corrected, drawings are updated to the latest version. These improvements are rigorously tracked and recorded, in what is called the **revision level** system. Each time the drawing is changed, it's given a new *REV* letter. Drawings start as *NEW* releases then, on the first design/drawing change, become *Rev A*. The second redraw is Rev B and so on.

Each progressive revision letter signifies that the drawing has been redrawn and any number of changes could have been incorporated into the design print at that time. Look in the upper right corner of Figs. 5-2 and 5-3, the sample prints. You must know the level of the drawing.

Trade Tip

Besides using letters to track revisions, in your shop, drawing revisions can also be tracked by number or by the date on which they were updated.

Check the Revision Box Look in the upper right corner of the drill gage drawing (Fig. 5-2). There have been no revisions, this is a NEW release drawing. It may be a very old design, but it has never been revised. The second print 203B-605, is at Rev A. It has been updated once, probably when the saw was improved. Brief notes in the Rev box give clues as to what those changes were. Normally, you needn't worry what the changes were, they are history as far as you are concerned. They have been incorporated onto the print. The important thing to ensure is that the next key points are followed.

KEYPOINT

When receiving a job, always verify that

A. The part number matches the WO.
B. The drawing revision level and work order revision level agree.

ShopTalk

Replacement Parts Keep in mind, shops also produce replacement parts for older products—*not the latest version!* In that case, a lot of detective work must be done. The changes made over time must be backtracked from the current Rev level back to the ones ordered. Each change must be investigated to see if it is compatible with the latest parts. Will they interchange or does the planner need to modify them to fit the older product? These can be incredibly complex research issues that must be done before releasing the job to the shop. The good news is that beginning machinists never have to do them! After the planner solves the issues, he or she writes a work order covering the steps needed to make the older version of the parts. All you must do is follow it. It will guide you to make the right parts for that job.

UNIT 5-1 REVIEW

Review the Key Terms

Detail/dash number
Individual identification of a part.

Job planning
A set of logical steps found on a work order, chosen to guide a job safely and efficiently through the shop maze.

Part number
Drawing and detail numbers combined to identify a single part.

Revision level
A letter showing how many times the drawing has been redrawn.

Secondary operations
Necessary tasks, usually not performed on the machines, needed to start or finish a job.

Work order
A master document initiating a job in the shop.

Replay the Key Points

◆ Learning the information presented here and in the next units won't change your tech school experience greatly, but

it will make you look very capable on the job.

◆ Even though the detail to be machined is found on a sheet other than number one, it would be wrong to proceed without the first drawing sheet without the notes and general tolerances.

◆ You could have the right drawing number and the right dash number and still make the wrong part. How? The wrong revision level!

◆ Often there are intermediate steps required before machining the final shape. They are found on the job planning within the work order.

Respond

Are you ready to use a work order? Answer the following to find out:

1. Name at least two reasons that sequence number 40 might be on work order S5-U1-1 (Fig. 5-1).
2. Why would it be a big mistake to drill all the holes to the sizes shown on drawing FYM 101 (Fig. 5-2)? Which WO sequence covers this?
3. According to WO S5-U1-1, what revision level drawing is to be used?
4. What does the work order instruction mean at sequence 120?
5. To what size must the 0.125-in.-thick blanks be sawn at sequence 10?

UNIT 5-2

Selecting the Right Material for the Job

Introduction: The first aspect of getting a job ready for the machines is to choose the right metal, in the right condition and form. While there are literally hundreds of different *alloys* (metal combinations), Unit 5-2 focuses on the five most common. As a new machinist you should be able to roughly identify them by color, relative weight, code marks, and other characteristics that we'll investigate.

It's not enough to get the right alloy, machinists must also see that the form and condition of the metal is right for the job too. It starts with knowing the necessary terms.

KEY TERMS

Alloy
Noun: The specific metal formula within a group. An exact combination of a base metal, secondary metals, and lesser amounts of minerals. Verb: To put one metal within another.

Alloy steels
A family of steels used for production parts. Various alloy steels range from plain mild steel through highly engineered high-carbon steels.

Aluminum
A lightweight, white/silver metal that does not adhere to a magnet. Has a very high machinability rating.

Brass/bronze
Two relatively heavy metal groups with a base of copper. Colors run from yellow-gold to reddish. Not all alloys adhere to magnets.

Cast iron
A dull gray heavy group of metals. They do adhere to magnets.

Castings
Molten metal is poured into a mold. Castings are weaker than forgings, but they can have many internal details.

Composite
A custom-made material composed of plastic resins, metals, and fibers. Composites provide strength-to-weight ratios not possible in standard metals.

Extrusion
A preform made by forcing heat-softened metal out through dies. Aluminum, copper, and brass are often supplied as extrusions.

Forging
A preform made by hammering and forcing metal into progressively more detailed shapes. Stronger than castings but less complex in shape. May be accomplished on heated or cold metal. Costly, but very strong.

Grain
The strength bias characteristic created by the way the metal was originally formed. Different metal forms (bars or extrusions, for example) have different amounts of grain structure.

Mild steel/low-carbon steel
A general utility steel with less than 0.3 percent carbon. Simple heat treatment to change its physical characteristics is not possible. Work hardening is not common or problematic.

Stainless steel
A metal grouping based on iron with added chromium and nickel. Some alloys will adhere to a magnet and others will not. Ranging in machinability from moderate to very low, corrosion resistant steel (CRES) will work harden.

Tool steels
A family of steels that are more complex in formula and more specific in application than alloy steels and more expensive. They are generally used to make cutting tools, dies, and molds that must endure extreme conditions.

Work hardening
A characteristic of some alloys that change hardness due to incorrect machining actions. Once work hardening has occurred, it is difficult or impossible to machine the object further until heat-treated back to a softer state.

Material Identification and Characteristics

Work Order S5-U2-1 Op 10 (Opp 10)

For your first trip into the world of metallurgy, we'll look at

Aluminum, Cast Iron, Brass, Steel, and Stainless Steel

These materials can be differentiated by color, weight, magnetic attraction, spark color, and pattern (when ground) and by their factory label. On completion of the reading, you will be asked to go into the shop and find examples of these metals then answer questions about them.

Machinability

One important characteristic of various metals is machinability. This is a rough guide to the comparative ease or difficulty with which a particular alloy can be cut. It's based on the hardness and toughness of the metal. A high rating indicates an easy metal to cut like aluminum, while a low rating means extra caution is needed. The low rated metals, such as hardened steel or stainless steel, require slower cutting speeds, extra sharp tools, and coolants. The higher rated metals, such as brass, can be cut much faster with far less concern for heat, cutting tool wear, and breakdown.

Metal hardness is the key. (Hardness is studied in Section 15.) As hardness increases, the machinability rating decreases. It becomes more difficult to machine. Cutters wear quickly, plus size and finish are more difficult to control. Some

Alloying and Alchemy For hundreds of years, creating a new alloy has been more a trial-and-error process with a bit of magic thrown in. At the dawn of the science, *alchemists* tried to make gold by melting and putting combinations of heavy and shiny metals together—but no luck. But from those early experiments, they discovered that many of their creations were greater than their parts, that they exhibit characteristics beyond those of their base elements! But today, given deep computing power and newly discovered math models of those mysterious properties, *metallurgists* (the new metal magicians) now can model the new alloy before ever melting and combining ingredients. So, using those scientific methods, do you suppose they might conjure up gold?

alloys exceed the limits of practical machinability due to their hardness. They cannot be cut with standard cutting tools. These must be machined by grinding or electrical spark erosion (all studied in this book). Or they must first be softened through heat treatment, then machined.

Metallurgy is a big subject but for now a background in the five groups is all that's required to get into the lab or start your new job as a machinist. Each group has very different machinability. It makes all the difference in the speed with which the work may be cut without destroying tools or the work itself.

Groupings and Alloys

Metal is classified with two headings—the *general family* and the *specific alloy*. The family name is based on a single, major ingredient. For example, in our basic five, steel is based on iron, while brass is based on copper. But, further, there are many different **alloys** within a group determined by their exact composition. Different alloying ingredients customize the metal for different purposes. So, starting with the base metal, each added ingredient provides some improvement for its use on a product.

For example, starting with iron, the mineral carbon is added for hardness and chromium and molybdenum (both metallic elements) are added for strength and toughness. With this unique composition of ingredients, it becomes *chrome-moly* steel, which is common in bicycle frames.

KEYPOINT

In most cases, the exact alloy called out on the work order and print must be used for that job. Your job during material prep, is to make it so—get and saw the right stuff.

Start-up Metal Facts

There's a lot to know about metals, even in this abbreviated first lesson. So much so that the study has a name: *metallurgy*. The scientist is a metallurgist. But, machinists need to know a lot about metals too. Collecting all the facts will be a career-long activity. For this lesson, associate only the highlights with the metal. Most of all, remember the facts that affect your machine setups like the

metal's machinability, the cutting tool's life expectancy when cutting it, and whether or not coolant is needed.

Physical Forms—Extrusions, Forgings, and Castings

We'll start by looking at three forms in which a given job might come to your machine other than in simple sawed blocks, sheets, or bars:

Castings   Forgings   Extrusions

All three save machining time because they are already roughly shaped toward the finished product. All five metals we'll be studying vary in terms of which of these forms may be supplied.

Forgings (Fig. 5-4) have been forcefully hammered from a solid block (called a billet) into a nearly finished shape. The forging hammers are massive machines that deform the metal by hitting it with great force, smashing it down in stages into precision dies. Forging might be accomplished on hot or cold metal. Forgings are exceptionally strong due to the working of the metal. Due to the initial tooling required to produce the dies and the individual handling of each part when forged, they are costly to produce yet they save much time in machining.

KEYPOINT

When the duty to which the part will be put requires the strongest version of the metal, then a forging is usually the answer.

Figure 5-4 *These parts will become drive joint sets, so they must be the strongest physical form of the metal, a forging.*

Figure 5-5 *An extrusion is forced out through shaping dies into long lengths.*

Figure 5-6 *Castings are made by pouring molten steel (clamp), iron (small anvil), or aluminum into a sand mold, which results in a rough surface as seen on the large coin.*

Extrusions are long complex shapes that are preshaped by forcing the metal through extrusion dies, much like toothpaste from a tube. To make an extrusion, the metal is heated to soften it to a plastic state, whereby it can be extruded through the die (Fig. 5-5). Not all alloys can practically be extruded. Aluminum, brass, and copper, all relatively soft metals, are the most common metals to be extruded.

An example of an extrusion might be the frame around a screen door or the radio antenna on your car. First, at the extrusion mill the metal is made into a continuous length of several hundred feet. It's then cut into 20-ft sections. The extrusion is further cut to the job length in the shop, then machined to final shape. Extrusions are also costly to create due to tooling costs to make the dies but they are supereconomical for production since they require far less handling than forgings.

Castings are the result of molten metal being poured or injected into molds or dies. Castings are made by two different processes, depending on the accuracy required: sand castings are less accurate and die castings are closer to the final shape and smoother.

Sand Casting Everyday castings are made by pouring the liquid metal into a hole in compacted sand. Beforehand, the cavity is created by tightly packing lightly moistened sand around a model of the casting. This is called a *pattern*. After packing, the pattern is carefully removed to leave behind a roughly accurate hole in the sand (Fig. 5-6).

After the liquid metal cools, the sand is broken away leaving the final shape. The surface of a sand casting is as rough as the sand from which it was molded. Typical examples might be the engine block in your car or an iron frying pan.

Die Casting When more precise castings are required, the liquid metal is poured into a precision die mold. These are the more costly version of a casting, since the dies are complex. But die cast metal is often accurate enough that many features require no further machining. Typical examples might be the metal frame in your car's radio or car's carburetor body.

Both casting types look similar to forgings but they are not as strong. Since they are molded not hammered into shape, castings can be made with complex internal features not possible with forgings, but they exhibit a large crude crystal structure so they haven't the strength of a forging or extrusion.

KEYPOINT

Castings are less strong than forgings but they may have intricate internal details possible only through casting.

The Five Basic Metals

Aluminum

Aluminum (abbreviated AL) is a comparatively lightweight metal that is a joy to machine. It has the highest machinability rating possible. It can be machined and sawed at very high removal rates with very little tool wear. (Caution, there are exceptions to that statement—called the super-high-silicone alloys.) Of our five metals, aluminum shows

Figure 5-7 *Aluminum is a light metal with the whitest hue of the metals we will look at.*

Figure 5-8 *Machining aluminum can result in long strings if not deliberately broken.*

the whitest hue of the silver colored metals we will examine (Fig. 5-7).

Some aluminum alloys can be heat-treated to exhibit differing physical characteristics (toughness and hardness), but none ever become hard enough to change their machinability. Aluminum is used extensively on aircraft and on other products where weight to strength is a factor. It is supplied in all forms: long bars, blocks, thin rolled sheets, castings, forgings, and extrusions. Cutting tools have a very long life when cutting most aluminum alloys—that is, they do not become dull for a long time.

Aluminum machines best with a coolant or oil on the cutter, but it can be machined dry. Coolant reduces heat, aids with chip removal, improves finishes, and lengthens tool life. Softer alloys of aluminum have a tendency to jam and clog cutters, known as loading the cutter. When this happens, the necessary space between cutter teeth is jammed with unwanted aluminum. All cutting comes to a stop and the cutter can even break if the situation is not corrected. Soft metal such as this is referred to as a *gummy* alloy and coolants are the main solution to the problem. See Fig. 5-8.

Aluminum Factoids

Aluminum will not adhere to a magnet.

Aluminum will not spark when ground.

There are a family of extremely high silicone aluminum alloys that wear on cutting tools faster than most steels. These superalloys are brittle but very wear resistant. They are used in automotive applications—piston cylinders, for example.

Hot aluminum chips do not change color during machining (many other metals will).

Aluminum's *malleability* (ability to be squished, stretched, or otherwise deformed without breaking) tends to cause long stringy chips on the lathe. As we have discussed previously, this is dangerous and should be prevented if possible.

Brass and Other Copper Based Metals

Brass is an alloy of copper with added tin. Adding zinc and various other ingredients creates a bronze. Bronze is tougher, harder, and stronger than brass but due to cost, is less common in school labs. However, you might see brass in training and perhaps the reddish colored, parent metal copper. Copper itself is very soft, thus little is machined from it. Due

Figure 5-9 *Typical brass products. Note the color range.*

to its great *malleability,* copper is extruded, rolled, and drawn (pulled) into pipes and electrical wire. Copper without other alloying ingredients is gummy when machined.

Brass The alloying ingredients of brass (Fig. 5-9) turn the soft red copper into yellow or red-yellow brass and make it tougher, harder, and durable under mechanical forces and friction. Copper is a heavy metal, thus brass and bronze are relatively heavy metals too, about twice the weight per volume of aluminum. There are many varieties of both brass and bronze differing in hardness and toughness. Brass and bronze can be cast and forged and are also commonly supplied as bars and plates. Brass is also extruded, but not as much as aluminum.

Many alloys of both brass and bronze have good antifriction bearing qualities and are used where metal-to-metal contact friction is encountered. They are also used for decorative items as well, since they can be polished to high luster (shine).

Most brasses are considered free machining (a term used for good machinability) just below aluminum. Coolants are used only to remove heat and flush out chips. Coolant does not improve the machine finish or tool life in most alloys other than the very hard bronzes. Since brass can be machined without coolant it is called a "dry" metal.

KEYPOINT
Brass can be machined without coolant but it's needed for the harder bronzes.

Brass and Bronze Factoids
Brass/bronze will not hold to a magnet.

Brass/bronze will not spark during grinding, however, it is considered poor technique to grind it. The soft metal tends to clog the pores in the wheel and render it useless until it is dressed to remove the contaminating metal. This is true for aluminum too—do not grind it on standard grinding wheels.

Due to the high copper content, both alloys are used where electricity is conducted as in contact springs.

Common yellow brass is the right stuff to make your small machinist hammer head.

Weird: Adding just 2 percent beryllium and 2 percent other metals (cobalt and/or vanadium and/or nickel) to plain copper, yields beryllium copper, which can then be hardened enough to machine soft steel even though it's made of 96 percent soft copper!

With some brass or bronze alloys, the chips turn fascinating rainbow colors during machining, while other alloys remain unchanged.

Cast Iron

Cast iron (Fig. 5-10) is a dull gray metal used extensively to make heavy equipment and ground transportation parts. Cast iron, often abbreviated to CI, is easy to machine with a relatively high machinability rating, however,

Figure 5-10 *A typical cast iron object, this machine table appears dull gray even after machining and grinding.*

a few sensitive alloys can work harden beyond practical machinability. Also, due to the casting process, CI can form a very hard, thin outer crust. Another unique challenge for CI: surprise pockets of unwanted trapped sand left over from the mold can cause catastrophic tool failure. These highly undesirable pockets always break the cutting tool. While CI is nearly always supplied in the form of castings, it is sometimes brought to the machine in cast blocks.

CI is considered a dry metal. It requires no coolant during machining except to wash out the chips and the free carbon dust produced and to cool the cutting tools.

Tools exhibit a medium-long life span when machining CI. The major reason for tool wear is due to work hardening and to the surface crust. If you do not have to cut through the crust, tools last very well.

Cast Iron Factoids

CI chips rarely change color but some alloys turn dull brown.

Cast iron will hold on a magnet very well.

When CI is ground, the sparks are a very dull red and compact (do not form star-like bursts). See Fig. 5-12.

Cast iron rusts very fast when exposed to moisture.

Cast iron is overloaded with free carbon, which must be removed by smelting (burned off by oxygen injection into the molten metal) before the iron can be turned into steel.

Due to the excess carbon, coolants become quickly contaminated when machining CI.

Figure 5-11 *Cast iron chips always break into small packets but they also create black carbon dust, which is a health issue.*

Figure 5-12 *When ground, cast iron creates a very dull red spark due to the excess carbon.*

When machined, CI produces small, broken chips along with carbon dust.

The carbon dust gets into your skin and is difficult to wash out.

Steel Alloys

When iron is refined, the excess carbon is first completely removed then very controlled amounts are reintroduced to produce steel. There are literally hundreds of steel alloys, subdivided into two broad categories by application:

1. Tool Steels

More expensive and generally the more technically complex alloys, tool steels are used to make cutters that machine other metals; to make dies and punches to shape or cut; or for any application demanding extreme hardness, friction resistance, durability, and strength. Due to their low machinability rating, tool steels can be very challenging to machine.

2. Alloy Steels

Alloy steels are commonly used for consumer parts; for example, automobile axles, fenders, and fence posts. However, some alloy steels are very highly designed for a given purpose and are nearly as challenging to machine as tool steel.

Both types are supplied as castings and forgings, bars, billets, and sheets but never as extrusions.

Steel Factoids

A little carbon goes a long way in steel! Similar to silicone in aluminum, the mineral carbon in steel is the most important ingredient after the base iron. Increasing the amount of carbon makes the difference in steel hardness, thus its machinability. Below 0.3 percent (3/10 percent) the steel is called a *mild steel*. Above the 0.3 percent level, the steel can be hardened to some degree, depending on the amount of carbon in the alloy. Carbon content rarely rises above 2.5 percent in most steels, with 4 percent the upper limit.

Mild steel is an alloy steel that's also called **low-carbon steel.** Mild steel has a single soft state. Later (in Section 13) you will learn of a special process called carburizing whereby the physical state of mild steel can be changed by adding carbon.

Steel machines best with a coolant.

All steels will hold to a magnet.

Steel sparks when ground. Mild steel sparks with bright white/orange stars, but as the alloy's carbon content rises, the sparks become globular and progressively redder (Fig. 5-13).

Mild steel rarely work hardens, nor can it be heat-treated to change its hardness. Other alloy steels do harden depending on carbon content.

Free cutting steel (FCS) is a steel alloy with an uniquely high machinability rating. It's made by adding lead and/or sulfur to the molten steel during the original pour in the *smelter* (a factory making raw metal from base elements). While the lead nearly disappears in the alloy, it modifies the steel structure to magically machine easily. *FCS* exhibits nearly all the characteristics of mild steel in application, yet it can be cut at nearly twice the speed.

KEYPOINT

Remember mild steel has less than 0.3 percent carbon present in its makeup.

Steel's machinability ranges from a bit more difficult than cast iron for mild steel to the very difficult, more technical and harder states of both tool and alloy steel, which must be cut with extreme care by slowing the cutter, adding flood coolants, and paying strict attention to cutter sharpness.

Mild steel is not prone to work hardening although it can occur. Usually, when it does happen, the hard spot can be machined away using a new sharp tool and reduced cutting speeds. Even so, it can burn up tools.

Steel chips commonly turn yellow-brown then blue when machined as frictional heat rises. The iron combines with room oxygen to form various oxides when heated above 400°F. In Fig. 5-14, the chips cut at a lower temperature are on the left and those cut at the highest temperature are on the right.

Two Steel-Forming Methods: Hot and Cold Bars and sheets are rolled into shape in

Figure 5-13 *Steel sparks range from a dull red for the higher carbon to bright orange stars for the lowest carbon steels.*

the mill, in one of two different ways: hot rolled (HRS) and cold rolled (CRS). The cold rolled steel is the stronger and finer finished material due to the cold working (stretching

Figure 5-14 *Steel chips change color with increasing machining heat.*

of the structure in the metal). The elevated temperature required to soften the steel during hot rolling steel (Fig. 5-15), forms a black scale of carbon on the outside. The shape of HRS is not as well formed compared to CRS, thus it requires more final machining.

While it's more expensive initially, cold rolled steel is smooth and shiny on the outside from the rolling action and often requires less machining to bring to size. Due to its clean appearance and better shape, CRS can often be used as is.

Stainless Steels

Stainless steels are also iron/steel alloys but with higher nickel and chromium contents than tool or alloy steels. **Stainless steel** can be cast and forged but not easily due to its tough nature, thus it's far more commonly supplied in bars and sheets. Stainless steel is also called corrosion resistant steel, abbreviated CRES.

(a) Hot rolled steel   (b) Cold rolled steel

Figure 5-15A-B *Compare the appearance of two steel forms: cold rolled steel CRS (shiny surface, right side) with the dark hot rolled steel HRS on the left.*

Within the CRES family, there are several subgroupings of stainless for different applications. The groupings could be characterized as soft but very resistant to corrosion (called the 300 series), the harder types used in jet engines and surgical instruments (the 400 series, which can be heat-treated as is), and the third series called *PH* (precipitation hardening), meaning that other elements can be added to the metal to make its surface harder.

Stainless steel looks slightly different from regular steel in color. It is difficult to describe in words, but it is easy to see the difference between regular steel and stainless (Fig. 5-16). I perceive stainless to be a bit more tan in hue—or perhaps just more bright. You'll get the chance to view samples of both coming up.

Rust is a good indicator that the metal *isn't stainless!*

Cutting tools don't last well when machining stainless. Various stainless steels exhibit machinability ratings from moderately difficult to downright nasty. Absolute control of cutting speed is essential. Maintaining a steady stream of coolant and monitoring cutter

sharpness are also critical. Even a hint of dullness and most alloys will work harden. When machining stainless, examine the cutters often and at the slightest hint of dullness, change them. Any slight change in sound or vibration or the finish on the work surface is a sure sign that disaster is just around the corner when machining stainless.

Stainless Steel Factoids

Magnetic attraction is an issue with the various stainless alloys.

Figure 5-16 *Compare the color of stainless steel to plain steel.*

Common stainless alloys (300 series) will not hold on a magnet at all. The added nickel and chromium break up the metal's crystal structure to the point where there is no net attraction to a magnet. In theory, each crystal grain would adhere to a magnet but combined, their microscopic nature cancels this property. Some with a higher iron ratio will hold weakly (400 series). Still others hold nearly as well as regular steel (PH series).

Some stainless will spark similar to CI, with dull red sparks when ground. Other alloys will barely spark at all.

Stainless chips generally do not change color, however, there are a few special exceptions.

Figure 5-17 *Composite materials are combinations of fibers, resins, and metals. They have an amazing strength-to-weight ratio.*

KEYPOINT

A heavy, bright silver metal with no sign of rust that does not hold to the magnet is definitely stainless.

Composite Materials

Composites are manufacturing materials made of combinations (Fig. 5-17) of a honeycomb core, an outer skin, and a resin to hold it all together. The cores can be metal or plastic fiber and the skin can be plastic or metal. These highly engineered materials are extremely strong compared to their weight. Because the cutting of these materials requires special cutters and processes, you probably won't see them in the training lab, but they are used in industry and gaining popularity.

In machining composites (Fig. 5-18), you must wear breathing protection from fumes or dust. Another precaution is fire prevention, as many composites are part metal and part plastic. Special cutters must be used that look more like a meat slicer than regular metal cutters. Holding composites during machining is a challenge as they are extremely strong for their weight

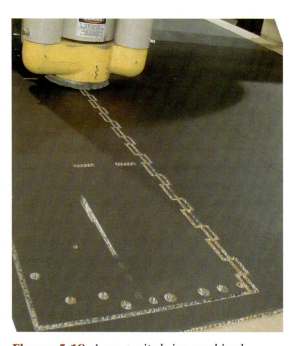

Figure 5-18 *A composite being machined.*

along a given axis but they can be crushed if not protected.

Physical Characteristics of Metals

Beyond selecting the right alloy for a job, there are more work order requirements that must be certified—the grain direction and the heat-treat condition, how hard it is.

We'll leave the study of hardness for a later section.

Grain Structure and Direction

When a metal is first cast after smelting, it becomes a solid ingot or billet. At that time, it is a block of connected crystals—not unlike plain ice. But when the block is worked into a consumer shape, bars, castings, extrusions, sheets, and forgings, for example, the working stretches the crystals into *stringers*. They create a **grain** direction in the metal. Each form exhibits varying amounts of grain structure. This grain direction gives the material a strength bias similar to a wooden board.

When flexed parallel to the grain direction, the metal breaks more easily while it resists breaking when bent across the grain. Depending on the function of the finished product, you can see that the grain direction in the raw material might be very important to identify before sawing raw stock for machining. When this is a critical factor (it often is), the need for grain identification and tracking will be noted in the work order.

Figure 5-19 is a computer simulation of how grain structure occurs in metal. On the left, a highly magnified section of cast metal shows large grains with no bias, but as the computer elongates the billet, simulating rolling into bars or sheets or hammering it by forging, the grains form the stringers—stretched out grains. It's easy to see that as the bar becomes longer so too do the stringers.

Cast billet   Rolled bar

Figure 5-19 *Grain structure is caused by elongating crystal grains, as illustrated in this computer-enhanced drawing.*

Determining Grain Direction There are three ways to know which way the grain bias lies within a piece of raw material:

1. **Parallel to Long Bars**
   In long bars, the grain stringers are elongated with length because that is the direction in which they were stretched during rolling.
2. **Parallel to Printed Identification**
   Lettering and numbers on sheets parallel the grain. When the metal is run through the final roll, the alloy and other information is printed on its surface. That's the grain direction because it's the direction the metal traveled through the rolling mill.
3. **Grain Stamp**
   A stamp mark that looks like an arrow (Fig. 5-20), sometimes with the word "GRAIN."

Material Identification and Traceability

Often, not only the exact alloy must be used for a job but it's heat-treat history and alloy certification must be linked all the way back to its original formation as raw metal. That certificate must stay with the job on certain people moving applications and for aerospace and military work. That critical paper trail is known as *traceability*. In military and commercial aircraft manufacturing, for example, each flight critical part (parts affecting safety of the aircraft) must show a serial number traceable all the way back to its birth by its *heat lot* number. The heat lot is the certificate number of the metal sample from which the individual part has been made. It is a pedigree that identifies the *crucible* (melting pot) from which it was born.

Suppose the material order falls three parts short of completing the job? In that case, more certified material can be substituted into the batch, *but not casually by the machinist!* The extra material can be substituted into the batch, but only when records prove it's the right stuff.

Caution–Long bars only! short bars can fall through

Figure 5-21 *Color coding keeps raw metal identified until the last piece is used (with the coding on it).*

Color Coding To quickly identify steel alloy bars, they are sometimes color coded on the end (Fig. 5-21). In a well-organized shop, they are placed in the rack with that ID toward one end.

KEYPOINT

When cutting sections from a color-coded bar, *never cut away the color coding unless you are taking the last piece of the metal.*

As discussed in Section 1, without the proper ID, metal becomes useless, even dangerous to keep around as it might accidentally be put to the wrong use!

UNIT 5-2 REVIEW

Review the Key Terms

Alloy
The specific metal formula within a group. An exact combination of a base metal, secondary metals, and lesser amounts of minerals.

Alloy steels
A family of steels used for production parts. Various alloy steels range from plain mild steel through highly engineered high-carbon steels.

Aluminum
A lightweight, white/silver metal that does not adhere to a magnet. Has a very high machinability rating.

Brass/bronze
Two relatively heavy metal groups with a base of copper. Colors run from yellow-gold to reddish. Not all alloys adhere to magnets.

Cast iron
A dull gray heavy group of metals. They do adhere to magnets.

Castings
Molten metal is poured into a mold. Castings are weaker than forgings, but they can have many internal details.

Composite
A custom-made material composed of plastic resins, metals, and fibers. Composites provide strength-to-weight ratios not possible in standard metals.

Extrusion
A preform made by forcing heat-softened metal out through dies. Aluminum, copper, and brass are often supplied as extrusions.

Forging
A preform made by hammering and forcing metal into progressively more detailed shapes. Stronger than castings but less complex in shape. May be accomplished on heated or cold metal. Costly, but very strong.

Grain
The strength bias characteristic created by the way the metal was originally formed. Different metal forms (bars and extrusions, for example) have different amounts of grain structure.

Mild steel/low-carbon steel
A general utility steel with less than 0.3 percent carbon. Simple heat treatment to change its physical characteristics is not possible. Work hardening is not common or problematic.

Stainless steel
A metal grouping based on iron with added chromium and nickel. Some alloys will adhere to a magnet and others will not. Ranging in machinability from moderate to very low, corrosion resistant steel (CRES) will work harden.

Tool steels
A family of steels that are more complex in formula and more specific in application than alloy steels and more expensive. They are generally used to make cutting tools, dies, and molds that must endure extreme conditions.

Work hardening
A characteristic of some alloys that change hardness due to incorrect machining actions. Once work hardening has occurred, it is difficult or impossible to machine the object further until heat-treated back to a softer state.

Replay the Key Points

◆ In many manufacturing situations, the exact alloy, heat-treat lot, grain direction, and history of the metal must be kept.

◆ The machinability of a metal is a relative rating that defines how fast the material can be cut and how long one can expect tools to last.

◆ There are a few metals that "can" be cut without coolant, however, it does help to remove heat and clear chips. Using coolant is rarely the wrong thing to do.

◆ Work hardening usually results from a machinist not heeding the correct cutting speed for a given metal or allowing a tool to dwell too long.

◆ Most forms of metal exhibit a grain bias that must be heeded when flexure of the product is a critical function.

Respond

1. A print calls for the work to be made from CI, describe the material you are seeking? In what form will it likely be found?

2. The work order requires aluminum extrusion to be cut for the job. Describe what you will be looking for.

3. Name the forms of metal that a work order might require to be sawed into blocks or lengths for a job.

4. Since forgings are more expensive than castings, when would a job require one?

5. Name the best way to identify the right CRES for a job that, once completed, is to be sold to the military.

UNIT 5-3

Sawing Material

Introduction: OK, you've selected the right alloy and the material is the form specified on the work order. The paperwork is in order, so now it's to be sawn into blocks or other shapes, for the upcoming machining. There is a chance to scrap the job right here, before the machinists even get it! The work order usually notes the amount of extra metal that must be left on each blank to allow the machining to occur, but not always. If not, deciding how much bigger the raw blanks must be before machining can be simple or complex depending on the part design. This unit will explain how to calculate excess material then discusses the sawing.

KEY TERMS

Abrasive sawing
Cutoff or *chop* sawing accomplished with a resinoid fiber reinforced grinding blade.

Annealing
Softening of a metal—annealing a band saw blade just next to the new weld area.

Cleanup cut
Machining raw material to create a reliable machined surface by removing a minimum amount of material from the workpiece.

Excess
A calculated amount of material left to finish dimensions also called machining allowance or cleanup material.

Friction sawing
Cutting metal by melting rather than cutting but accomplished on a band saw not an abrasive saw.

Kerf
The slot created when sawing. The kerf is wider than the blade due to the tooth set. Kerf must be calculated into excess when sawing material.

Pitch
The number of teeth per inch in a saw blade.

Scroll sawing
Sawing to a curved line only possible on a vertical saw.

Set (saw blade set)
Bending or warping saw teeth out from the base blade. Blade set creates a wider kerf than the blade to prevent drag and allow scroll sawing.

Silicone carbide
A hard abrasive used in resinoid cutoff saw blades.

Skip tooth
Band saw blades with extra space between their teeth for sawing soft material.

Stripping
A dangerous event where teeth are torn out of a blade, caused by selecting a blade pitch too course for the material.

Surface speed
The recommended speed of the cutting action expressed in feet or meters per minute of band saw movement when sawing.

Planning Excess Material

Go back and note operations 10 and 100 on the sample work order Fig. 5-1. At Opp 10, the raw material is sawn into blanks suitable for the upcoming machining. Then at Opp 100, an intermediate saw cut is taken to speed up the machining of the final profile. It's important to note that at each operation, the sawing includes enough *excess* to machine the final shape. **Excess** is a layer of extra metal to be machined away, it might also be called **cleanup cut** *material* or *machining allowance* material, which provides the amount needed to remove all rough material, this leaving a 100 percent cleaned (finished) surface.

Besides providing metal to finish, excess material might be planned for holding purposes during machining. For example, an extra amount is needed to hold the part in a lathe chuck while it is being turned into a shape, then it is removed at the last cut—this excess is called *chucking allowance*.

The Right Amount of Excess?

The answer can be as simple as reading the rough sawn size from the WO. Or in many job plans, the amount is noted, but the machinist must calculate the overall size including the excess. However, on a few, the machinist is left to make the decision of how much is right.

> How much excess? Usually, from 0.100 in. up to a maximum of 0.250 in. for normal machining (with many exceptions). For precision grinding, from 0.005 to 0.060 in. excess.

Planning too much excess robs profit due to extra machining time to remove it plus the cost of consuming too much metal. However, too little excess and the part will not *clean up*. That is, the surface will not be completely machined when it reaches the indicated size or shape. Saw marks will still remain. So, an error on the heavy side is better until your skills improve. But the goal is always to add a minimum amount.

Here are the factors to be considered:

1. Complexity of the part to be made—complex parts often require a bit more.
2. Previous experience with this part—proven production requires less excess.
3. Cost of the material itself—less costly material might have more excess.
4. How is the part to be machined? What machines are to be used? Are there

Figure 5-22 *Excess material must be left on the rough blank.*

fixtures to hold the material? What cutting tools are available?

5. Nature of the material—since aluminum machines so easily, more excess might be better (Fig. 5-22). However, leaving a large amount of excess on a stainless steel blank could be a real nightmare!

Trade Tip

While learning, or whenever in doubt, leave 0.25-in. excess on all surfaces to be machined (or as directed by your instructor). This is too much by industrial standards, but it's right for trainees.

Allowing for Saw Kerf Thickness

Here's another consideration: when sawing you must also allow for the thickness of the saw cut itself called the **kerf.** If a second part was to be sawn from the sheet in Fig. 5-22, approximately 0.100 in. would be needed for the kerf as well as the machining excess on both parts. Each type of saw tooth pattern has its own kerf thickness.

Sawing Metal

In today's shop, there are 10 very different possibilities for cutting blanks for machining. The choice can make a big difference in job success. This is a perfect example of the challenge of job planning. At the end of this section, you will be quizzed as to which might be best, given a specific material situation. The first few in the list would be expected in a school shop so we'll study them here, while the remaining are CNC—high-tech methods found in industry. We'll explore them in Section 26. Here are few clues toward making the right choice.

Common Methods

1. **Hand Sawing—Hacksaw**
   While too slow for production work, hacksawing is a needed machinist skill.

2. **Vertical and Horizontal Power Band Saws**
   A continuous cutting action is caused by a looped saw blade running on wheels. A complete shop must have both types.

3. **Abrasive Sawing**
   Cutting metal with a thin specially reinforced grinding wheel, this method is primarily dedicated to cutting long steel bars into smaller lengths.

4. **Oxygen-Acetylene Flame Cutting**
   Produces a very rough edge so this option isn't very suitable for most job plans. It cuts by rapid oxidation so only ferrous (iron-based) metals can be cut this way. It creates a HAZ (heat-affected zone), an undesirable band of modified metal next to the cut. Usually a last resort in machining, this method is used extensively in welding and fabrication work.

5. **Shearing**
   Shearing cuts with a slicing action and is used for relatively thin material such as sheet metal or for the drill gage blanks. Shears might be found in schools but are more likely in industry where shears cut up to $\frac{1}{2}$-in.-thick steel or more! This machine is banned for use by minors due to its extreme danger factor.

Industry Methods

1. **Plasma Cutting**
   A high-energy state, electrified gas cuts by injecting an electric current into a compressed airstream. The extreme heat allows a greater range of material to be cut compared to oxy-acetylene. But it creates an unwanted band of modified metal next to the cut called a HAZ.

2. **High-Energy Laser Cutting** (Section 26)
   This method uses highly focused heat to remove a narrow band of material. It can cut many kinds of metals but is usually applied to those not easily cut by other means. Using CNC drives can cut intricate shapes. Leaves a HAZ.

3. **High-Pressure Waterjet Cutting** (Section 26)
   This method uses a highly focused stream of water to cut an amazing range of materials from glass to chocolate and baby diapers without getting them wet! To cut metals and other hard materials, abrasive grains are injected into the stream. Using CNC drives, it can cut complex shapes. It does not leave a HAZ, but is relatively slow compared to sawing and shearing.

4. **Rigid Sawing**
   Similar to a very sturdy version of a table saw for wood, a rigid saw has a round disk cutter that spins while either the material or the saw blade is moved, depending on machine construction. It is used mostly to accurately cut very large plates to size and is found only in major industry.

5. **Reciprocal Sawing**
   This sturdy, power-driven saw is similar to a big hacksaw. Due to the blade cost and pinch point danger, these saws are fading from the industrial scene.

Hacksawing

Although unlikely as an industrial operation, hacksaws are basic to a machinist's tool box. With all sawing including hacksaws, there is a right blade for the material kind and thickness.

Selecting Blade Pitch Saw pitch is the number of teeth per inch (Fig. 5-23). In general, the coarser blade will cut faster. They are used for heavy cuts or for softer metal. Softer metal cuts better with bigger teeth but fewer per inch of blade. However, there is a limit for coarseness both for safety and for tool life.

Figure 5-23 *Select blade pitch such that three or more teeth are in contact with the work.*

Pitch is too coarse
for this thin material

Figure 5-24 Too coarse—*teeth can be stripped.*

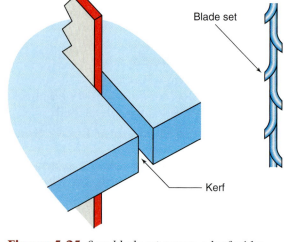

Blade set

Kerf

Figure 5-25 *Saw blade set causes a kerf wider than the blade's thickness.*

Three teeth touching, prevents thin metal from slipping between the teeth. The blade in Fig. 5-24, is too coarse—too few teeth per inch. When this happens, teeth will be broken away. Choosing a blade with this contact is critical for all kinds of saws, both hand or power. Teeth ripped out, as shown in Fig. 5-24, is called **stripping.** For hand sawing, stripping isn't dangerous but it will damage the blade.

ShopTalk

Stripping Hurts! When the teeth strip off power band saw blades, especially when sawing fast as with aluminum, they come out like machine gun bullets! No kidding, I have seen a student make this mistake and within a fraction of a second, he had a row of hot teeth embedded in his safety glasses! We hung those glasses on the saw as a reminder— *change the blade when sawing thin material or find another sawing process.*

Blade Set Causes a Wider Kerf
Tooth set is a saw blade feature necessary to make the cut wider than the base blade. Without tooth set, the blade would drag in the cut and be unable to cut curves. To produce a kerf wider than the blade, the teeth are either bent to alternate sides, as shown (Fig. 5-25), or for very fine teeth not easily bent, the entire blade is bent in a wavy pattern.

Correct Use of the Hacksaw
There are five guidelines beyond pitch selection when using a hacksaw.

1. **Mount the Blade so the Teeth Face Away from the Handle.**
   They cut on the forward stroke (Fig. 5-26). Often an arrow is

Trade Tip

New Saw Blades in an Old Kerf A problem is encountered when a worn blade is replaced with a new blade in the same cut both for hacksawing and for power sawing. When the old blade dulled, it lost its set. Its kerf became narrow. However, the new blade's set is wide again.

It's a disaster to begin sawing at the original place. The set on the new blade will be instantly flattened or the teeth will be stripped. To prevent this, start sawing some distance back, where the kerf was correct, then slowly saw back to the last position. When this reentry becomes time consuming, it might be better to just start a new cut from the opposite side of the work to meet the first.

painted on the blade indicating in which direction the teeth cut.

2. **Do not Force the Cut**
   From experience you will learn there is a "sweet" pressure beyond which any greater push results in broken or dull blades.

3. **Apply Pressure on the Forward Stroke**
   Depressurize the blade for the return stroke or lift it up instead of dragging it. This simple act will lengthen blade life.

4. **For Hard Materials, Slow Your Stroke Rate**
   Even when hand sawing, if the strokes are too fast, a blade quickly dulls. Saw from

Figure 5-26 *A hacksaw blade is correctly mounted when the teeth cut on the forward stroke.*

50 to 70 strokes per minute—around one per second. *Adjust your speed according to the machinability of the alloy being cut.*

5. Hacksaw Blades Can Shatter
Most hacksaw blades feature hard teeth with a soft back. They will not break all at once. However, some higher quality blades are hardened all the way across the blade (a better blade but brittle). These break instantly and if you are pushing too hard, a good knuckle skinning follows! Don't lean into the cut under any circumstance but especially when using full hard blades.

Trade Tip

Sawing Thin Material For material that's too thin, or where only coarse blades are available, rotate the material to cut along another dimension (side) or tilt the saw blade to make a longer cut, thus more teeth come in contact (Fig. 5-27). For thin tubing, insert a wooden plug to keep the teeth from stripping. For thin sheet material that can't be tilted, you can sandwich clamp it between plywood or sacrifice metal plates of a softer metal before sawing.

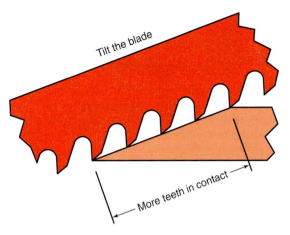

Figure 5-27 *Tilt the blade to achieve three or more teeth in contact.*

Power Sawing

Three types of saws are commonly found in training labs and industrial workshops:

		Abrasive Saws
Vertical	Horizontal	(Cutoff or
Bandsaws	Bandsaws	Chop Saws)

Saw Blades Selection To saw effectively using a vertical band saw, the machinist needs to understand saw blade selection. Blades need to be changed for the specific application.

Pitch is the main factor, with three-tooth contact even more important to avoid breaking teeth from the blade. Stripping of teeth on a vertical band saw set to a high blade speed for aluminum or brass can be a rapid fire, dangerous event. Three or more teeth contacting the work is more important on vertical saws where the user applies the forward movement of the work into the blade. On horizontal saws equipped with automatic down feed, two-tooth contact will work since the saw can't surge forward to strip teeth.

Tooth pattern and set Power saw blades differ from hacksaw blades in their cutting teeth patterns and in the types of **set** (Fig. 5-28). In general, **skip tooth,** *claw* or *hook,* and *buttress teeth* (Fig. 5-29) are chosen for cutting softer materials. The extra clearance is necessary to allow each tooth to temporarily contain the larger chip created due to the quicker cutting action. It's held in the hollows between teeth, until it exits the kerf to be ejected, before returning to the cut again.

Blade alloys Band saw blades are supplied in three varieties, high-carbon steel, high-speed steel, and bi-metal. Each has its correct application.

Raker set   Wavy set

Alternating set

Figure 5-28 *Blade set creates a kerf wider than the blade to prevent rubbing and allow curved cuts.*

Choosing the right tooth form

Standard tooth
General metals—best finish
Hand or power fed

Skip tooth
Agressive feeding in softer metals
Power feed is best to avoid stipping
General metals—rough finish
Useful in long cuts—space holds long chip

Hook tooth
Very agressive feeding—fast cutting
Limited to soft materials
Rough finish
Good for production sawing

Figure 5-29 *Choose the right tooth form based on the sawing method, the alloy hardness, and the size of cut. Choose coarser blades for longer cut surfaces or softer metals.*

Carbon-steel blades are the budget version. Their teeth are not as strong or as hard as those of the other two blades so they wear the fastest. But they have a great advantage in that they can be welded together. This is useful for several reasons.

Sawing Internal Features When cutting internal holes in the work (donut holes), a pilot hole is drilled near the cut line, then a blade is inserted through to be welded closed inside the hole. After sawing out the central *slug* (leftover unwanted metal—donut hole), the blade is cut open with shears, then welded back together for other work.

Using Bulk Blades Because carbon steel can be welded, it's economical to buy in 100-ft rolls then make blades as needed. That also means blades that break during use can be rewelded or even short sections can be welded in place to repair stripped sections.

High-speed steel blades feature the hardest teeth, thus they last the longest when sawing difficult metals such as stainless steel. They cannot be welded back together without special equipment and processes, thus they are considered a specialty blade in most shops. These are more costly than the other two blades.

Bi-metal blades fall between high-speed steel and carbon steel, in both price and performance, although in most applications they stand up almost as well as the higher priced high-speed steel blade. They are constructed with very hard teeth but softer backing material. You can recognize them by a dark band of heat-treated metal on the tooth side of the blade. Due to the dual advantage of price and performance, bi-metal blades are the most popular in industry, while carbon-steel blades are more popular in training labs.

Vertical Band Sawing

Often, an efficient job plan includes metal removal by sawing, as in operation 100 on our work order. Even in the modern world of high-speed CNC machining, sawing can be the fastest metal removal tool in the shop when used correctly. Sawing takes away whole "chunks" of metal rather than making chips. When planning a job, always consider the saws as an operation. In addition to efficient metal removal, those large scrap pieces are far more valuable for recycling. See Fig. 5-30.

Major Saw Types and Nomenclature Saws vary greatly as to their features and size. We'll concentrate on the two basic types most likely found in a tech school (Fig. 5-31).

Vertical Saw Safety is a Critical Issue Because the blade moves fast and the work is fed through by hand, you need to know about

Specific safe moves

Holding and pushing work through the saw

Calculating blade speed

Even with correct guards a saw blade must necessarily be exposed over some part of its length and those teeth face directly at you. Coarse saw blades, moving fast, will immediately cut anything—including your hand. To be safe follow the guidelines listed next.

Specific Safety
1. Adjust the Guard
Always have guards in place covering as much blade as possible.

Chunks Recycle Better
Compared to cutting away metal chips, sawing away chunks from the workpiece is an economical plan for another reason beyond fast removal, they command a better payback fee at the recycling plant. Clean large blocks can be recycled at the smelter more easily because they are closer to pure uncontaminated metal, thus they command a higher return yield to the shop. But to take advantage of this, the chunks must be stored separate from the chips.

Vertical saw operations

Contouring   Filing   Friction sawing

Fixture work   Square cutting   Angular cutting

Figure 5-30 *Standard and special band saw operations.*

**Major parts of
the vertical band saw**

Job selector dial

Blade tension indicator

Speed indicator

Blade cutter

Air jet

Guide post

Switch

Table

Welder

Blade guide

Motor cover

Speed control

Power feed handwheel

Blade tension handle

Power feed foot pedal

Figure 5-31 *The basic parts of a vertical saw.*

2. Brace Yourself

Brace your hip or knee against the saw base to avoid slipping forward if something moves unexpectedly (Fig. 5-32).

3. Hands Clear

Never put your hand directly in line with the saw blade. If pushing the work from directly in front of the blade is absolutely necessary, use a push stick (wood block or metal) to take the place of your hand. Always use a push stick or saw vise rather than your hand.

4. No Gloves

Gloves catch on things faster than bare skin.

5. Correct Guides

Make sure the blade guides are as low as practical for the material and that they are adjusted right. This keeps the blade tracking right, keeping if from wobbling as it enters the work.

6. Nonskid Flooring

If standing on a bare concrete floor, keep the small saw chips swept up, as they

Figure 5-32 *Always brace your body against the saw to avoid slipping forward.*

become slippery. (Note that bare floors in front of saws are banned under OSHA guidelines.) Better yet, make sure you are standing on nonskid surfacing (paint with added sand for traction). Even better yet, stand on a rubber skidproof mat if at all possible. That allows the tiny saw chips to fall through to the floor.

7. Brace Your Hands
To avoid sudden, unexpected movements of your hands if the work should accidentally shift, drop your wrists down onto the table (Fig. 5-33). Do not hold work from

Figure 5-33 *Wrists or palms in contact with the saw table, and not in front of the blade, create safe stability. If something moves unexpectedly, your hands won't move.*

the top, but rather put your hands on the saw's table then push the work. This action ensures that if the work slips, your hands won't go with it!

8. Danger Area to the Right
Do not allow others to stand to the right side of the saw table as that is where a broken blade will tend to eject.

9. Stable Work Only
Finally, never saw a workpiece that might tend to tip forward as the blade pulls downward. Sawing disks from round stock is the prime example of this potential danger. When the blade grabs it, it tends to rotate forward out of control, which often breaks the blade and jerks your hand with it. There are two ways to avoid unstable work. The best idea is to take it to the horizontal saw where the bar is firmly held in a vise and no hand movement is needed. If the work must be sawed on a vertical saw, use a special saw vise to firmly clamp the unstable work. Firmly clamped in the vise, it cannot shift or tip forward.

Set the Saw at the Right Blade Speed In setting up all machines, drills, lathes, mills, and saws, three things must be considered for choosing the right cutting speed:

Material type

Material hardness

Is coolant available

Cutting tool hardness is also a factor on most other setups, but for saws it's always a constant as the three blades are about the same hardness.

In general, the harder the work material, the slower the saw must cut to avoid burning or dulling the blade or work hardening the material. Vertical saws don't have coolant because there is no way to collect it to a reservoir, so your job will be to shift the saw's transmission to achieve the right blade speed based on the work material alone.

Correct blade speed can be obtained from a chart like the one shown in Fig. 5-34 as the job selector dial. To use a rotating selector such as this, place the material in the window then read the speed in the index. If there is no speed chart, refer to any comprehensive machinist reference book such as *Machinist's Ready Reference*©. It will list blade speed for sawing a given material in feet per minute for materials (generally under the topic cutting speeds).

Figure 5-34 *A metal plate attached to this vertical band saw shows the correct blade speed for various materials.*

KEYPOINT

Band saw cutting speeds are specified in feet per minute abbreviated as F/M or FPM.

Surface Speed For all machining where a cutting action takes place, there exists a magic number called *recommended* **surface speed,** sometimes it's also called *cutting speed*. This important number is needed to set up all equipment including saws.

We will discuss it again for drilling, then for turning on a lathe and for milling as well. For sawing, it's an easy number to envision. It's the velocity of the blade expressed in feet per minute (FPM) or it might also be in meters per minute (m/M). If your saw has no speed chart, remember these ballpark guides.

Material Being Sawed	Cutting Speed in FPM
Aluminum	300
Brass	200
Soft Steel	100

Although very low for industry, these numbers are easy to remember at the start. They are biased toward survival and safety. After some experience in sawing, they can be increased, especially when sawing aluminum.

Notice that these numbers reflect the metal's machinability. Steel, a hard metal, must be sawed slowly, while soft aluminum is sawed the fastest.

KEYPOINT

For any machine cut where chips are made, there is an optimum surface speed, in feet per minute.

If the selected surface speed is too slow, the cutting action is poor and production is low. Conversely, if it's too fast, the blade will dull quickly, then burn up if not removed from service, and worst of all, the material can be ruined if it work hardens.

Industrial surface speeds commonly range from a slow 30 feet per minute for sawing hard steel to 2,000 feet per minute for some alloys of aluminum.

Step by Step—Setting up a Vertical Saw
Here's a checkoff guide:

☐ Enter the blade speed table with the material type/hardness—find the correct cutting speed.

☐ Shift the saw's belts and/or transmission to obtain that recommended speed.

☐ Check the blade pitch for minimum tooth contact.

☐ Position and check all guards and blade guides.

☐ Be ready with a pusher or saw vise to force the work through the blade. Never have your hands in direct line with the blade!

Lab Verification *With correct shop attire, and with permission,* take your book to the vertical band saw in your training facility. Answer these questions:

1. Most saws have 10 or 12 blade speed selections. How many speeds does your saw have? What do you do when the chart recommends a blade speed between two of the speeds? (*Answer:* choose the lower speed and observe the results. From experience, you will come to know if you can shift up to the next speed.)

2. Does your saw have a speed selection plate or job selector dial, similar to the one in Fig. 5-34?

3. If so, what range of speeds does it show and which materials are shown?

4. What speeds could you select to saw steel? (*Note:* they will probably fall between 60 and 200 F/M.)

5. Observe—does your saw have a blade welding attachment? (Training is coming up on this.)

6. If your saw doesn't have a surface speed chart (blade velocity), where might you find this information in your shop?

Saw Vises A special vise made for vertical band sawing is a must (Fig. 5-35). It not only provides a safe way to push the work, the handle bars help steer through curved cuts as well. While keeping your hand out of the blade path, saw vises amplify the ability to direct the work using the extended handles.

Saw Accessories There are several vertical saw accessories used for safety and efficiency (see Fig. 5-36). For long work, the outboard support is used. For complex or specialty cuts, the chain jaw vise is used. It can be power fed if the saw has the capability. By stepping on a foot pedal, the chain pulls the work forward through the cut. Other accessories include work guides and stops.

Scroll and Friction Sawing

These two sawing operations are less common, but are within the beginner's skill level.

Scroll Sawing (Curved Cuts) is only possible on vertical saws. Cutting curves is a

Figure 5-35 *A saw vise provides a safe way to control scroll curve cuts on the vertical band saw.*

major advantage of a vertical band saws. The reason it can cut curves is that the kerf width allows the user to rotate the work relative to the blade within its own cut. Well-adjusted blade guides and sharp blades with a wide blade set are essential.

Workpiece fixture

Universal fence

Work support

Work-holding jaws with feed chain and cable

Cutoff gage or stop

Figure 5-36 *Saw accessories.*

Scroll sawing is possible
due to the kerf width

Figure 5-37 *Kerf width makes scroll sawing possible.*

Figure 5-38 *Here the safety guard has been removed to expose the upper* blade guides. *There is a second set below the table as well.*

From Fig. 5-37, you can see that if a tighter curve (smaller radius) is needed, then a wide blade must be replaced with a narrowor blade, to be able to turn within its kerf. When doing so, you must readjust the trust wheels on the saw guides such that the tooth set does not contact the side guides. If not, the blade set will be flattened on the wider blade.

KEYPOINT

When changing from a wide to narrow blade, always readjust the thrust rollers. Otherwise the narrow blade's set will be flattened if it touches the side rollers.

Adjusting blade guides When replacing one blade for another of the same thickness and width, no adjustment is needed. But, when a new blade size is mounted on the saw, the guides must be set to control it. This step is often overlooked by beginners, yet it will have everything to do with cut accuracy and blade life.

Notice in Fig. 5-38 that a power band saw features two sets of blade guides, one before the cut and one after. These guides stabilize the blade from wobbling and vibrating. In Fig. 5-39, a top view looking down on a guide set, notice that there are three rollers in a set. Two support the sides of the blade, the third is a thrust bearing. When you push the work against the saw, this bearing provides a rolling brace to keep the blade vertical.

Both upper and lower guide sets must be adjusted to fit the blade. Each set creates a

Band saw roller guides

Blade cross section

Figure 5-39 *Adjust the three-point suspension to fit the blade width and thickness, but never touching the tooth set.*

three-point suspension: the sides and the back of the blade. Set the thrust roller with as much blade contained within the side rollers as possible, *but not touching the tooth set.* The teeth must never touch the blade guides.

Some saws use solid guides rather than rollers for side support of the blade. These guides are usually made of brass/ bronze and due to wear must be constantly maintained see Fig. 5-40. When cutting curves they are consumed by the twisting force and friction of scroll work.

Solid guides work OK for light sawing but aren't practical in industry. Solid guides are adjusted close to but not touching the blade. They are common on saws meant to cut wood, rather than metal, due to the reduced side loads on the blade.

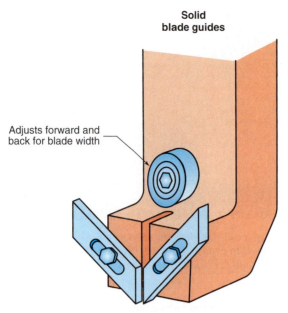

Solid blade guides

Adjusts forward and back for blade width

Figure 5-40 *Solid guides are found on some saws.*

Bringing the correctly adjusted upper blade guide down close to the work surface creates the best control over the saw path.

Friction Sawing This sawing technique requires a powerful saw capable of *surface speeds of 2,500 feet per minute or greater.* **Friction sawing** might not be considered a beginner skill in your shop due to the list of possible dangers and the problems it causes for both the saw and the workpiece. However, in sawing certain problem materials it works wonders when other methods won't do the job. For example, you can saw through another band saw blade using friction sawing. Ask for a demonstration and always check with your instructor before attempting friction sawing.

Friction sawing does not cut metal, it melts the work ahead of the blade. Why doesn't the blade melt too? Because any portion is in contact with the work for a fraction of a second, then it rotates through the entire saw loop to cool before once again contacting the cut. The workpiece cut area remains hot. Friction sawing has several advantages and four disadvantages.

Advantages You will be amazed at how fast you can saw once the metal starts to melt.

- Material that's too hard or too thin to be sawed by standard methods can be cut.

- Thin material that would strip teeth is easily friction sawed.
- A worn blade with no sharp teeth works best thus old, completely dull blades work best and extend their life span.

Disadvantages and safety precautions

- Friction cutting leaves a ragged burr.
- Heat may work harden a thin zone around the cut (the heat-affected zone).
- Fire!! See the upcoming key point.
- Freight train blade crashes.

The blade is traveling at such high speeds that should it break, it begins to fold upon itself as rear parts catch up to the crash. Thinner blades don't just stop when they break.

Caution!
When friction sawing, make sure all blade and safety guides and guards are as close to the work as possible and that nobody stands near and especially to the right of the saw frame where an escaping blade might go.

Heat and Fire
The work will heat rapidly near the cut line during friction sawing. Plus the outgoing swarf (tiny chips being removed) is molten. Before friction sawing, inspect the swarf bin and the saw's interior for wood or plastic waste left behind from previous cuts (see Fig. 5-41).

Remove all flammable material from the catch bin and around the saw.

Supersonic tooth stripping Be aware, tooth stripping can occur, except now at bullet speeds!! To avoid this problem, start your cut gently, then increase the pressure. Wear double eye protection—safety glasses plus a full-face shield are good ideas.

Horizontal Power Sawing

This technique is used to cut large bars and heavy work. There are four major differences and advantages of a horizontal saw over a vertical:

1. Self-Feeding
Built with a pivoting band saw frame on a horizontal bed, horizontal saws lower their frame and blade to automatically

Figure 5-41 *From above, nothing looks unusual while friction sawing this steel, but looking under the saw table, the stream of sparks and melted metal tells of the high heat in this process.*

control the cut rate. This means they can be left unattended to saw large work and that the three-tooth rule doesn't apply as long as the drop rate is adjusted such that tooth stripping cannot occur.

2. Any Length Workpiece

Between the saw guides, the saw blade is warped about 45 degrees relative to the plane of the saw wheels. This means that the frame sits back from the cut and therefore the work does not need to pass through the frame, as it does on vertical saws. On a horizontal saw, any length bar can be cut.

3. Vise Held Work

A long-range vise is part of the bed frame, which means no hands are needed to steady the work as with a vertical saw. The work remains stationary while the pivoting support arm and blade progress down into the work. This means that the machinist is safer without pushing the work through the saw by hand. These saws are used to cut off bars or blocks with little operator intervention. Straight and angular cutoff work is the only possibility on a horizontal saw (Fig. 5-42). Scroll and friction sawing are not done here.

4. Coolant Systems

Horizontal saws feature a pressure coolant system. Their construction allows coolant recovery into a sum pan at the base. Coolant means longer blade life and higher sawing speeds.

Figure 5-42 *A horizontal saw self-feeds down through the work at a controlled rate.*

**Major parts of
the horizontal band saw**

Saw frame

Guide posts

Frame weight

Motor and
step pulleys

Idler wheel

Blade tension
handle

Table

Saw
guides

Vise
handwheel

Switch

Vise

Coolant tray

Base

Band blade

Figure 5-43 *The main features of a horizontal saw. Note that the blade is warped 45 degrees to the plane of the wheels to allow sawing long stock.*

With guides adjusted and sharp blades, horizontal saws can cut stock to length with repeatabilities of 0.010 in.! Start to learn the various parts of the horizontal saw from Fig. 5-43. The saw is shown with the safety shield open to expose the idler wheel at the front but not the drive wheel at the back.

Special Features Where large quantities of material are sawn, several extra production features are added to horizontal saws (Fig. 5-44). The newer versions are also CNC controlled for quick setup and operation.

Figure 5-44 *Fitted with a bar feed, a horizontal saw can advance and cut raw stock blanks.*

Industrial versions of the horizontal saw, called cutoff machines, include

◆ Automatic or CNC stops, once the saw completes the cut.
◆ Automatic or CNC bar feeds. When a long bar is to be cut into short blanks, the saw can be set to automatically advance then saw then advance again.

Horizontal Saw Safety Horizontal saws are relatively safe compared to vertical types. But, this saw has its own particular precautions:

1. When lowering the moving blade to the work, go easy. New machinists often contact the work too fast, and then blade breakage and stripping can occur.

2. When setting the work in the vise to the correct cutoff mark, never have the saw blade moving. *It must be turned off until the work is clamped in the vise and your hands are out of the saw.* Incorrectly clamped work that suddenly shifts is a major reason that blades break on horizontal saws.

3. Do not try to catch the work with your hand as the part is cut off. Let it drop to the floor or set up a catching chute. Occasionally, at the moment of disconnection, the work will catch and pivot toward the blade. If you are holding it, your hand could be carried into the blade!

KEYPOINT

Blade guide adjustment is even more important on a horizontal saw because the guides warp the blade 45 degrees out of the plane of the drive and idler wheel to allow the sawing of long stock.

Abrasive Sawing

Abrasive sawing uses a thin abrasive cutoff wheel (Fig. 5-45), somewhat like a grinding wheel. It too is used to cut pieces off metal bars. Cutting steel is its primary purpose, but other metals can also be cut with this process as long as they don't clog the cutoff wheel. Compared to band sawing, abrasive cutting is very fast.

Figure 5-45 *An abrasive cutoff blade is strong on its rim but easily broken from side loads.*

Figure 5-46 *An abrasive saw in action.*

When using mineral grains to remove material, rather than cutting tools, we call the process *abrasion* or *abrading*, and it's also called grinding in many instances as well.

Abrasive Cutting Makes Chips Too The definition of abrasion is to scratch or rub away material. But, looking closely, we discover that abrasive processes remove metal by cutting with sharp-edged tools. There are chips being made, but they are nearly microscopic, and thousands are made per second. While a grinding wheel will abrade at slow speeds, it becomes efficient only if its speed is boosted up many times faster than that of cutting tools.

The cutting tools in this case, are the sharp edges on each individual grain within the grinding wheel. Those grains are much harder than cutting tools and are able to withstand extreme heat. So when abrasive sawing steel, for example, we see a spray of sparks exiting the cutting wheel (Fig. 5-46). These happen because the tiny hot chips react with oxygen in the room and literally burn up.

Safety Precautions for Abrasive Sawing
1. **Always Inspect the Wheel Visually Before Using It**
   Lift the guard and rotate it to be sure it is OK to use. Do this every time you use the saw!
2. **Double Eye Protection and Dust Masks**
   Both hot flying debris and the possibility of exploding wheels make it important to wear a full-face shield over your safety glasses, and a dust mask is advisable if you are doing a lot of this kind of metal cutting (Fig. 5-47).
3. **Stand to the Side**
   Avoid the hot swarf and possible flying wheel fragments. Stand to the side, never in line with the wheel when it's cutting!
4. **Check Guards**
   Be certain the wheel scatter guard is in place and drops down! These wheels can explode when used incorrectly!
5. **Fire Prevention**
   Be certain that nothing behind the saw is flammable! If walls or other items are in

Figure 5-47 *It's wise to wear double eye protection when using the "chop saw."*

the path, place a sheet metal spark catcher in the swarf path.

6. Look Before Operating
Let others in the area know when you intend to use the *"chop saw."*

7. Wear Ear Protection
Abrasive sawing is loud. If more than one quick cut is made, ear protection is a must.

Reinforced, Flexible Saw Wheels To keep the kerf narrow and concentrate the cutting action, abrasive saw wheels must be thin, but that makes them prone to shattering. To help them take the punishment, abrasive wheels are made differently from regular grinding wheels in three special ways:

The binder (the glue) holding the abrasive grains together, is a plastic resin, which after curing under low heat, can flex with the changing loads of this kind of abuse. The resulting flexible grinding wheels are called *resinoid* wheels.

In addition to the grains and binder, *cutoff wheels* contain fiber reinforcement similar to radial tires for your car.

The special grains, called **silicone carbide,** are even harder than normal grinding wheels. This is too technical for now, but we'll explore the various abrasive materials when we come to grinding operations.

Portability Advantage The compact size of abrasive saws is another advantage. They are small enough to be moved to the steel rack (Fig. 5-48). Instead of taking a heavy bar to the saw, the cutoff saw can be put on a rolling cart and taken to the steel rack. This increases safety as well as saving time. The bar is pulled out, the saw is cranked up to the bar's level, and it's cut off right in the rack!

Power Cutoff Saws Normal cutoff wheels are 12 or 14 in. in diameter. The industrial big brother is much larger in size, using wheels as big as 2 ft, they feature power-driven coolant. They are CNC controlled or semiautomatic, thus large material requiring lengthy cuts can be started and monitored by the machinist, who can do other tasks during the cut. It's unlikely you will encounter them in tech school.

Heat Factor High-carbon steels that are prone to work hardening can and will form a HAZ when abrasion sawed. They can create a glass-hard surface at the cut line (the heat-

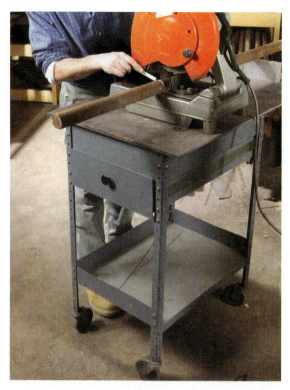

Figure 5-48 *Not a heavy machine, the abrasive saw can be taken to the steel rack rather than move heavy bars to the saw.*

affected zone). Additionally, a jagged burr forms at the exit point of the cut and the parts are often too hot to handle with bare hands.

Not for Soft Metals Aluminum and other gummy metals are not usually cut with manual abrasive saws. The problem is that the air spaces between the grains become clogged with removed material and they stop cutting. With no space between grains, there is no way for the edges to contact the work surface. They float over the work and only rub. This problem happens on all grinding wheels and is called *loading*. On small saws where no coolant is available, there are rub-on lubricants that do work, but most shops set a policy about cutting soft metal using abrasion. On the larger power saws, coolant solves the problem (Fig. 5-49). The dry version makes a real mess. Be sure to cut where the slurry of burned metal and spent abrasive won't contaminate other machines.

Band Saw Blade Welding

Our final sawing skill is making a blade from rolls of bulk material. To save time, most

Figure 5-49 *A grinding wheel loaded with soft aluminum cannot cut.*

Figure 5-50 *As the electric current passes through the blade joint, heat melts the joint together.*

Blade welders (Fig. 5-51) use electrical resistance to create the high heat needed. The trimmed blade ends are clamped into copper or brass jaws then forced together. An electric current is passed between the jaws and blade joint. The joint heats and as it melts, the jaws push the joint together (Fig. 5-52). When the electricity turns off automatically, the two halves are fused together.

shops buy prewelded blades ready to mount on the saw. However, the competent journeyman must know how to weld a blade for four reasons:

1. Broken blades that are still sharp can be repaired by welding.
2. Blades cost far less when bought in 100-ft rolls.
3. Without a blade welder, it is impossible to saw the inside of a part. Here the blade is inserted through a drilled hole then welded together once inside; for example, when sawing a circle shape out of the center of a plate.
4. Otherwise sharp blades that have a few stripped teeth can be repaired by cutting out the bad part and inserting a new section.

Note the blade welding attachment shown in Fig. 5-50.

Parts of the butt welder

Grinding wheel guard

Blade thickness gage

Etching clamp

Annealing button

Welding switch lever

Grinding wheel

Weld selector switch

Blade clamps

Figure 5-51 *The features of a band saw blade welder.*

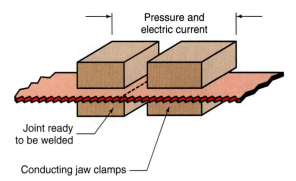

Figure 5-52 *The conducting jaws force the melting joint together.*

Annealing the Weld

As the blade weld rapidly cools, the metal just next to the weld becomes extra brittle and would snap instantly during use. Before putting the newly welded blade into service, it must be softened by **annealing.** The blade welder has provision for opening the jaws wider than the weld, to hold a longer portion of the blade than the weld area. Reheating the general area more slowly then cooling it even more slowly will soften the hard spots.

Annealing is a softening process that happens when metal is cooled slowly from an elevated temperature.

Final Shaping

After annealing, the bulging weld must be ground or filed thin and smooth to allow it to pass through the saw guides. When shaping the weld flat, make sure the joint is as thin as the original blade to allow passage through the saw guides. Many welding attachments feature an onboard grinder and a test slot to try the blade trim, however, a micrometer or caliper will also tell you if it is thin enough.

UNIT 5-3 REVIEW

Review the Key Terms

Abrasive sawing
Cutoff or *chop* sawing accomplished with a resinoid fiber reinforced grinding blade.

Annealing
Softening of a metal—annealing a band saw blade just next to the new weld area.

Cleanup cut
Machining raw material to create a reliable machined surface by removing a minimum amount of material from the workpiece.

Excess
A calculated amount of material left to finish dimensions also called machining allowance or cleanup material.

Friction sawing
Cutting metal by melting rather than cutting but accomplished on a band saw not an abrasive saw.

Kerf
The slot created when sawing. The kerf is wider than the blade due to the tooth set. Kerf must be calculated into excess when sawing material.

Pitch
The number of teeth per inch in a saw blade.

Scroll sawing
Sawing to a curved line only possible on a vertical saw.

Set (saw blade set)
Bending or warping saw teeth out from the base blade. Blade set creates a wider kerf than the blade to prevent drag and allow scroll sawing.

Silicone carbide
A hard abrasive used in resinoid cutoff saw blades.

Skip tooth
Band saw blades with extra space between their teeth for sawing soft material.

Stripping
A dangerous event where teeth are torn out of a blade, caused by selecting a blade pitch too course for the material.

Surface speed
The recommended speed of the cutting action expressed in feet or meters per minute of band saw movement when sawing.

Replay the Key Points

- ◆ To be safe a vise or push stick must be used on vertical band saws and your wrists should contact the table, never push on the part alone.

- ◆ For all sawing operations, blade pitch should be chosen such that three teeth are in contact with the work at all times.

- ◆ Horizontal sawing is for cutting large metal parts.

- Friction sawing is performed on vertical saws capable of 2,000- to 3,000-FPM blade speeds. It is used when the material is too hard to saw otherwise.
- Abrasive sawing is used primarily for quickly cutting long steel bars to length.

Respond

The questions here are based on these material cutting methods: vertical band saws, hacksaws, horizontal saws, abrasive saws, and shears.

1. You have a sheet of $\frac{1}{8}$-in.-thick aluminum from which approximately 2.25 by 5.25 in. rectangles must be cut for the drill gage. Of the five methods listed, which could be used?

2. Why would you not choose the abrasive saw for Question 1?

3. You have a sheet of $\frac{1}{8}$-in.-thick aluminum from which 400, 4-in. squares must be cut. Of the five methods listed here, which would be most practical?

4. The drill gage is to be made from $\frac{1}{4}$-in.-thick sheet steel. Would a horizontal saw be a good choice to cut the blanks?

5. Would the abrasive saw be a good choice for cutting the drill gage blanks? They are made from low-carbon steel.

6. You are going to first saw the blanks of Question 1, then the steel blanks of Question 4. What changes should be made to the saw?

UNIT 5-4

Part Finishing

Introduction: Work Order Operations 30, 80, 120, 130. Although we try to avoid it, machining often leaves raw, sharp edges, burrs, and occasionally incomplete transitions between surfaces. To correct these imperfections, hand work and part finishing is needed. Because this kind of work adds to manufacturing cost, programmers make every effort to eliminate the need for secondary finishing work. But clearly there are times when it cannot be avoided.

Apart from the danger of cutting the hand of the next person to lift a raw part, sharp edges look and feel wrong. As a craft master, you will be able to put your hand to a machined part and know it needs finishing.

A Time Proven Test Applying the skills discussed in this unit are often given as an undeclared test to new employees. The way you go about finishing parts will show your suitability for the craft. It's a chance to demonstrate patience, a steady hand, and most importantly a good attitude about quality and detail work.

Here in Unit 5-4 we'll cover a general background in part finishing—enough to get those lab assignments finished and ready to grade. There's a lot more to the subject than our coverage here, but because it becomes shop specific, the rest is be left for on-the-job experience. On completion of this unit, you should clearly understand how to handle machined products such that you improve their quality and finish while never marring or damaging them during handling.

KEY TERMS

Aluminum oxide
The common gray abrasive used in general grinding wheels.

Burrs
The unwanted jagged metal left after machining is completed; the tool used to remove burrs is also called a burr or a rotary file.

Draw filing
Holding the file nearly sideways. Produces a finer finish and better control of shape.

File card
The multifunction brush used to clean out loaded (pinned) files.

Lathe file
A flat file with teeth on two sides.

Mill file
A flat file with teeth on four sides.

Part-finishing wheels
A soft fiber and abrasive wheel that is very effective at finishing part imperfections. Often called a 3-M wheel.

Vitrified
The hard grinding wheel made by firing a clay binder and an abrasive until it forms a material similar to china or porcelain.

Putting the Final Touch on Work—Part Finishing

Work Order Operation 120, "File .12 inch Radii," and Operation 130, "Remove Cutter Radii 1 Place."

Three Kinds of Part Finishing

These part-finishing tasks fall into three categories:

1. *Surface finishing*
   A. Removing burrs and sharp edges
   B. Improving surface finish
   C. Final shaping
2. *Part marking* numbers and other information
3. *Secondary machining operations*
   A. Threading with taps and dies
   B. Assembly press fits
   C. Drilling, chamfering, reaming, and honing

These are mostly hand skills (Fig. 5-53), however, new technologies are providing some interesting methods of improving and marking parts. In this unit and Unit 5-5, we'll look at these new skills as well as the more traditional ones. Work imperfections can be corrected both with power equipment and by hand, using a variety of implements and tools.

Hand Tools

We're talking good old hand-eye coordination. It comes with practice with files, power tools called rotofiles and burrs, along with a variety of other finishing tools (Fig. 5-54).

Let's look at the various operations to finish the order of 12 drill gages. Operation 30, "Remove all burrs" and also the general note on the drawing: "Break all sharp edges 0.005 in." That means to remove the sharp edge by creating a

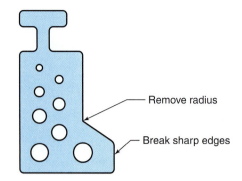

Figure 5-53 *Edges and a corner requiring hand operations.*

Figure 5-54 *A selection of hand work burr tools and an air motor tool.*

Figure 5-55 *Left: The burr is forming ahead of the bit. Right: The drill has burst through and bent it to where it cannot be cut off.*

small chamfer or corner radius no more than 0.005 in.

At this stage, the blank rectangle has been machined to the finished 2 × 5 in. size and the holes drilled. We will start by *breaking* the outside edge that is to put a small rounded corner on the sharp edge by hand. An edge break is from 0.005 to 0.015 in. There are five choices and any combination might be used.

1. Hand filing
2. Edge-breaking tools
3. Hand power tool
4. Pedestal burr wheel
5. Slurry vibration ("vibraburring")
6. Tumbling with burring stones

Hand Files

Files come in many shapes and with different tooth patterns and coarsenesses. For this job, you would select a flat file with fine teeth and a pattern of single- or double-cut rows. Let's look at each characteristic.

File Shape Little training is required for this aspect of files. Most of the shapes available are shown in Fig. 5-56. Many file shapes are supplied in two varieties: tapered and straight (Fig. 5-57).

Mill and Lathe Files Files with teeth on all four edges that are able to cut down and sideways, are called **mill files.** Those with teeth on the two larger surfaces are called lathe files. **Lathe files** can be used to file against an edge without metal removal from the edge itself (Fig. 5-57).

Tooth Pattern Machining files have four kinds of teeth, as shown in Fig. 5-58. Each has a purpose:

Single-Cut Files: Finishing Work With one row of teeth, they are used where metal removal is small and finish should be good. They would be a good choice to break the edges of the drill gage blanks. Due to the small space between teeth, single-cut files tend to clog or "load up," also called "pinning," especially when filing soft metal such as aluminum. The loaded metal then scratches the finished product.

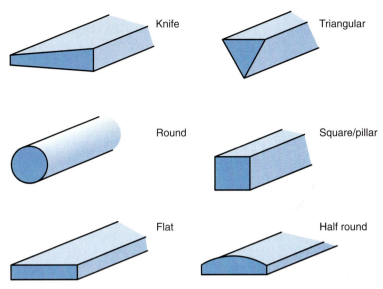

Figure 5-56 *Common file shapes.*

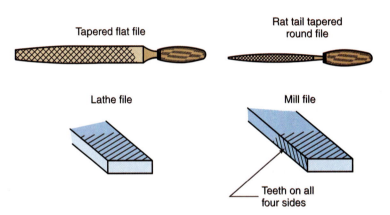

Figure 5-57 *Tapered files and mill files compared to lathe.*

KEYPOINT

Single-cut files are the slower metal remover but produce the best finish.

Double-Cut Files: Semifinishing and Roughing This file will shape metal well but leaves a rougher surface behind. Double-cut files remove more material faster than single-cut files. They wouldn't be the best choice to break the edges.

Bastard Cut Files These have teeth set at an angle to each other but not in a double-cut pattern. Notice the discontinuous teeth in the

Figure 5-58 *Four file tooth types. Each has a different purpose.*

pattern. Because of the way they lay to each other, this file cuts very fast. It is used for rough filing and heavy metal removal. It is a very poor choice to break the edges.

Rasp Cut Files Used for fast removal where surface finish is of no concern, this file works well on soft metals like aluminum, where tooth loading can be a problem with finer files. A rasp is made differently than a standard file in that individual teeth are formed by lifting metal up off the file body. They are sharp enough to cut flesh! Rasps are more commonly used in woodworking rather than metal work.

File Handles—Always Use Them!

File handles (Fig. 5-59) protect palms from a nasty stab from the sharp *tang* (the pointed piece that is forced into the handle). Never use a file without a handle. This is doubly true if filing on the lathe, where the machine action could catch and send the file back at high speed.

Filing the Drill Gages

To file the sharp edges of the drill gage requires a couple of skills:

1. Producing good straight edges without overfiling.
2. Avoiding marking the part in some other way while holding it.

There are two ways to remove metal using a file.

1. Reciprocal (back and forth) (Fig. 5-60)
2. **Draw filing** (angular shearing) (Fig. 5-61)

The draw file method might work better for the edges. To do it right, place your second hand on the end opposite the handle.

Use a file handle

Figure 5-59 *Always use a handle on files.*

Reciprocal filing

Figure 5-60 *Straight filing makes removal faster but can make it difficult to control shape.*

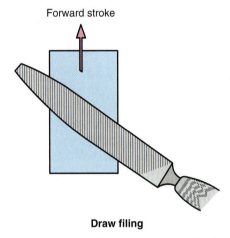

Draw filing

Figure 5-61 Draw filing *controls the shape better and produces a better finish.*

File Cleaning

Use a **file card** to remove particles of built-up metal in the file, a process called pinning. Brushing a light coat of cutting oil on the work (or plain chalk on the file) can often help, but there comes a time when you must stop because unwanted loaded metal is marring the surface as it is produced.

To clean a file, brace it against a bench and use the three parts of a file card/brush (Fig. 5-62). First, use the stiff wire "card" for loosening and removing big particles. Then use the bristle side to clean the entire surface. For difficult lumps of metal that remain, the file cards feature a fold-out metal pick.

Do No Harm to the Work— Holding It for Finishing

Many employers complain of tech school graduates causing more work than they solve

Figure 5-63 *Sanding tools can be held in an air motor.*

Figure 5-62 *Cleaning with the wire then bristle brush sides of a file card can remove built-up metal from the file to keep it cutting cleanly.*

by marring and scratching parts while working on them.

This is very important: always protect the work from marring. Don't cause mars by the way you handle or set parts down.

When holding parts in a vise, always use jaw protectors. These can be shop-made from a material softer than the work or they can be rubber or plastic protectors made specially for your vise. Here's another set of little things that do matter—the game rules of part handling:

Always set parts down gently.

Keep parts separated if possible by paper or in racks or boxes—don't allow them to bump together.

Never hammer or wrench against machined surfaces.

In machine setups, watch out for clamps and vise marks.

Never stack parts up where they could fall.

Using Rotoburrs

These are little files that cut by spinning at very high rates—from 10,000 to 20,000 RPM. The hand instruments that spin them, called roto-motors, die grinders, or air motors, are mostly air driven to keep them light yet powerful. Electric burr motors are also available for use where the work is not demanding. Sometimes called an air gun, this is a high-speed motor in which a variety of tools can be held (Fig. 5-63).

In addition to **burrs,** we also mount sanding tools in the spindle of a roto-motor (Fig. 5-64). The shaped sanding tools might be a good choice for the drill gage edges. There are disks, cones, and cylinders. With practice these tools can shape very well to a fine finish.

Figure 5-64 *Breaking the edge with a sanding cone (beehive) might be a good choice.*

Rotary Files (Sometimes Called "Burrs" or "Rotoburrs")

The files are made from carbide or high-speed steel. They are supplied in a wide range of shapes: cones, balls, cylinders, and many others. They are fast metal removers and can be used to form transitions and other part features. They are called rotofiles or rotoburrs (Fig. 5-65).

For breaking the drill gage, burrs are too aggressive. They would be good at removing large burrs from abrasive sawing or the rough edge of friction sawing, for example. Using them requires practice and a steady hand. They are used only for fast metal removal where further sanding or filing is expected. They are not appropriate for our drill gages.

Lubricating Air Tools

The turbine bearings require twice daily lubrication. The easy way is to put a few drops of air tool oil right into the air pressure inlet. In shops where they are used extensively, the air lines themselves are injected with the oil.

Power Machine Edge Burring and Breaking and Tumbling and Vibrating Burr Machines

There are several machine methods of removing burrs and breaking edges. Two are made

Figure 5-66 *Mechanical burring and edge-finishing machine.*

specifically for the purpose, the part finisher and the part tumbler (Fig. 5-66). These finishing machines both use loose abrasive in a slurry (water, abrasive, and sometimes a cleaner). This type of burring/finishing is automatic and very good for large qualities of parts. The parts are either vibrated rapidly or rotary tumbled against an abrasive medium. The tiny collisions remove sharp edges and burrs. The challenge is to find an abrasive medium that will get at the problem areas and not get stuck within details of the parts or remove too much material. There are hundreds of abrasive shapes and sizes from which to select (Fig. 5-67). These artificial stones are manufactured from abrasive grains and a resin binder.

The advantage of this type of burring and finishing is that it can finish hundreds of parts at one time while requiring no operator intervention. The disadvantage is that this process is slow and the entire surface is given the tumbled-matte surface. Many vibrating machines are noisy, so operators must wear ear protection when working around them or the machine must be placed in a sound-proof room.

Common Shop Machines for Part Finishing: Pedestal Burr Wheels and Disk and Belt Sanders

The final method by which you could finish the edge of the drill gage uses one of three common machines in the shop.

Pedestal/Bench Grinders

Used to sharpen tools and perform many utility removal operations, they are the same

Figure 5-65 *Rotary files (rotoburrs) are too aggressive to slightly break aluminum edges unless they are used with exceptional care.*

Figure 5-67 *A variety of abrasive shapes for mechanical finishing of parts.*

machine other than the addition of a pedestal (Fig. 5-68) to mount the grinder to the floor. Since our parts are aluminum, doing them on a regular grinding wheel would be banned due to loading the wheel, plus it would do a very bad job of breaking the edges of the drill gage. However, there are special wheels made for part finishing that can be mounted on pedestal grinders (Fig. 5-69). We will look at these special wheels after some introductory discussion of the more common bench-grinding operations and about grinding wheels in general.

Figure 5-68 *A pedestal grinder is used for tool sharpening but might be too rough for part finishing.*

Figure 5-69 *A variety of grinding and finishing wheels for a pedestal grinder. From left to right: Standard abrasive, high purity fine abrasive, brass wire wheel, steel wire wheel, and resinoid cutoff wheel.*

Grinding Wheel Construction

Grinding wheels are composites of three materials:

Abrasive grains

A *binder* to hold the grains together and provide strength to the wheel

Air spaces between the grains

Without the air spaces, the wheel wouldn't work very well. They create the necessary chip clearance, help coolants get to the cut (air in the case of a bench grinder), and provide temporary places for the removed chip to stay until thrown out by centrifugal force.

Vitrified and Resin Wheels The two kinds of grinding wheel binders used in machining are rigid (**vitrified**) and flexible (resinoid discussed previously under abrasive sawing) wheels.

Resinoid—The Flexible Binding These more shockproof wheels are pressed and heated using a plastic binder and abrasive grains. The abrasive is usually silicone carbide, but other abrasives are also used. These wheels can take more mechanical shock; for example, the wheels used on handheld right-angle grinders for cleaning up welds. But they can be broken and they do fly apart when misused.

Slitting Wheels On the pedestal grinder, when we need to cut a narrow slot in a part, a resinoid wheel would be used. An example might be cutting off a few coils of a spring. Or a slitting wheel (Fig. 5-70) might be used for any task where a regular grinding wheel would be too wide to do the job. They are special purpose wheels made for the pedestal grinder and are similar to but less strong than abrasive cutoff wheels.

Vitrified—The Rigid Binding The common, hard gray grinding wheels used for most bench and precision operations are held together with clay. This binding material is mixed with the abrasive grains, then fired similar to a ceramic cup or porcelain. The way the air space is created is unique, can you guess how it's done? Section 13 answers the question.

Aluminum Oxide—Everyday Abrasive In addition to the clay, the abrasive in most pedestal grinding wheels is made from **aluminum oxide.** Higher quality, pure aluminum oxide wheels are bright white. Not usually used for pedestal applications, the white wheels are used on precision grinding machines where long life and shape retention is needed and the expense is justified. Everyday aluminum oxide is gray.

Due to brittleness, vitrified wheels do break when misused. When struck or when work shifts rapidly under them, they can crack.

"Green Wheels on the Pedestal Grinder" Special vitrified grinding wheels are made from silicone carbide rather than aluminum oxide. They are for grinding extra-hard carbide tools only. The binder in these wheels is colored for easy recognition.

A Few Tips the Pros Use

1. As the wheel wears, the gap between the tool rest and the wheel increases. Keep this gap to less than 2 mm or $\frac{1}{16}$ in. Otherwise thin objects and the fingers holding them can be instantly sucked in! Loose clothing and hair go in through any gap, big or small.

2. Be certain there is a scatter shield on the upper portion of the wheel and that it is lowered close to the wheel. Its main purpose is to contain hot swarf, but in the rare incidence of a wheel explosion, it will also contain wheel fragments. (The upper guard is not shown in Fig. 5-71.)

3. Stand to one side when starting a grinding machine. If the wheel was damaged by the last user, it will blow up during acceleration.

4. Inspect mounted grinding wheels before starting the machine by rolling them while looking for cracks and chips.

Figure 5-70 *A thin and flexible slitting wheel cutting a slot in a steel part. (Safety guards removed for photo clarity only.)*

Figure 5-71 *The tool rest must be within 0.100 or less from the wheel.*

Mounting a New Wheel

Before the ingoing wheel is put on the grinder, perform

A. *The Visual Test*—Look them over for tiny cracks. However, cracks can be missed this way.

B. *The Ring Test*—Holding vitrified wheels on your finger or by a rod through the center hole, tap them lightly with a screwdriver handle or other plastic object at several locations. They will make a thud if they are cracked and will ring clear if unbroken. Think of a broken porcelain cup in the same test. It would sound similar. See Fig. 5-72.

5. Perform an acceleration test on new wheels. On initial turn on, after mounting a new wheel, touch the start button, then quickly press the stop button. Allow the wheel to coast for a second. Now press the start button again and let it accelerate a bit more before touching the stop. Do this three times, each time letting the wheel go a bit faster. Finally, start the wheel for 1 minute and just let it run with no load. This ensures that it is stable

Figure 5-72 *Perform a* ring test *of any vitrified wheel before mounting it on a machine.*

and safe. Of course, stand to the side when doing this.

6. When mounting any kind of wheel on any grinder, always use a paper gasket on both sides of the wheel. This ensures even pressure from the mounting flanges against the wheel and it helps to absorb shocks to the wheel. Do not overtighten the wheel nut.

7. Be cautious of the heat generated during grinding (Fig. 5-73). Handheld objects can heat rapidly when ground. Keep a water pot close by to cool the work.

8. When grinding, as much as possible, move your work over the entire surface of the wheel to avoid wearing a groove in the flat surface. And do not force it—you will find the right amount of minimum push that works best with practice.

Dressing Grinding Wheels

Grinding wheels self-sharpen to some degree. That is, the binder will break as dull grains drag on the work. The extra pressure causes them to either crack or shear away completely, thus exposing new cutting edges. However, with prolonged use, grinding wheels become dull and their surface becomes irregularly shaped. These wheels require a touch-up dressing. There are three choices (Figs. 5-74 and 5-75).

Wheel dressing tools

Diamond dressers

Carborundum sticks and blocks (a hard material)

Diamonds and carborundum are used more for technical precision grinding and will be studied later. For pedestal grinding, we use the wheel dresser.

Figure 5-73 *For safety and performance, always use paper gaskets when mounting grinding wheels on spindles.*

Figure 5-75 *The wheel dresser features a set of star-shaped cutters and guide lips to hold it straight on the tool rest.*

Caution!
Safety guards
in place

Guide lips

Tool rest

Your slip-on guide

Figure 5-76 *Dressing a grinding wheel using the guide lips.*

Figure 5-74 *The three methods of dressing a wheel: a wheel dressing tool, a diamond dresser, and a carborundum stick.*

Using a Wheel Dresser

Note in Figs. 5-75 and 5-76 that there are small guide tabs on the leading edge of the dresser. They are meant to hook over the front edge of the grinder's tool rest, to help move the dresser in a straight line from side to side. However, they work only as long as the tool rest is straight. If there is a groove ground in the rest from adjusting it too close to the wheel, then the tabs cannot be used as guides. To use the guide tabs, loosen then pull the tool rest back enough to allow them down past the wheel, then retighten the rest. The dresser is now tipped up until the dressing

wheels contact the spinning grinding wheel lightly. Moving it side to side while increasing the pressure, will cause the dresser wheels to spin rapidly and vibrate. That action plucks out grains. Have your instructor demonstrate how to do it the first time.

Part-Finishing Wheels

A very good option for breaking the edges of the drill gage, this soft part-finishing wheel is used on a pedestal grinder (Fig. 5-77). **Part-finishing wheels** come in several stiffness degrees and abrasive sizes, but they all have the ability to conform to the edge and remove just the right amount of material. It takes only a minute to learn to use one.

Wire Wheels

The last pedestal grinding wheel we might use to finish the edges of the aluminum drill

Figure 5-77 *A part-finishing wheel on a pedestal grinder.*

Trade Tips

Make a Straight Guide Most of the time, the rest is grooved on the front edge from incorrectly adjusting it too close to the wheel and from collisions during use. If damaged, it cannot be used to guide the dresser. So, most of the time, we lay the dresser on top of the rest and move it across the wheel, but that makes it difficult to get a perfectly straight wheel. Here's a tip to get a perfect dress every time: make a slip-on guide, as shown in Fig. 5-76. After pulling the rest back, slide it over the tool rest, so it's just past the worn edge. The guide must be made to fit the particular grinder, but most feature a similar rest. Make one in training, then bring it to your first job. Your coworkers will probably be dazzled unless they too studied in *Machining and CNC Technology*.

Don't Dull the Fresh Grains When dressing a grinding wheel, remove the minimum amount required to create a new true surface. Once the wheel is trued all the way across its face, do not go back over the surface lightly "just to be sure." That only dulls the grains you have just exposed.

ShopTalk

Robot Burring and Part Finishing The theme in this text is adapting to technology. It's the essential survival trait of the future-minded machinist. Here's the perfect example; a progressive French company, added a robot to their part-finishing system to automate the tedious job of burring and polishing work. Working faster and more consistently, it paces their high speed machining facility.

The point is, productivity goes up using about the same number of employees after the change as there were before, but their roles are drastically changed. The lower-skilled part finishers are replaced by programmers, and support tooling people. Are they the same people with new skills? Or are they displaced by the changing technology?

Photo courtesy: Stäubli Unimation®

gages is a wire wheel (Fig. 5-78). Wire wheels do take off burrs on work and can clean away rust or unwanted coatings. Their wires vary in stiffness and composition. The stiff variety is used to remove steel burrs and rust. However, only the most flexible version could possibly be used to break the edges of the drill gages. Wire wheels would be a wrong choice for aluminum parts, but not a bad one for finishing some steel products. Wire wheels come with two special precautions.

1. Safety

The wire bristles fly out like arrows! Be certain to wear adequate eye protection. Watch your eyes and the eyes of others working around you.

2. Contamination

Steel wire wheels (the typical type) will cause surface contamination in other metals. Stainless steel that has been brushed with a common steel wire wheel will rust from the microscopic particles of steel embedded in the material. There are stainless wire wheels and brass ones made to avoid surface contamination, but they are not common in most shops.

Disk Sanders

The disk might be chosen to finish the edges of aluminum. But since it is an aggressive metal remover, a fine grit disk would be used and still it might remove metal beyond the 0.005-in. target. The challenges would be not to over-round the edges and to create an even amount of break for the entire edge. Both disk and belt sanding machines are very aggressive metal removers. Although they are simple,

Scatter shield

Eye shield

Water pot for cooling work

Tool rest

Figure 5-78 *Wire wheels remove burrs and sharp edges but would be too crude for the drill gages. Be cautious of flying wires and the tendency to pull objects into wire wheels!*

Figure 5-79 *A disk sander can pull loose sleeves into its pinch point faster than you can react!*

sanders have two serious safety hazards that are often overlooked.

1. **Keep Loose Clothing Away from the Disk!**
   Both disk and belt sanders pull loose items into their pinch point at an amazing speed. Cloth instantly disappears! There's no time to react or pull back. If it's your shirt, you will be injured! See Fig. 5-79.

2. **Stay Out of the Disk Danger Zone**
   On disk sanders, the left side pulls the work down against the table—that's good. The right side lifts it up very fast and it usually flips your knuckles over to the left side of the disk—that's bad! See Fig. 5-80.

KEYPOINT

Never sand on the right half of the disk!

Hand Burring Tools

There are some new and innovative burr- and edge-dressing tools in tool catalogs. Shown in Figs. 5-81 and 5-82 are a couple of examples. These tools are pulled along the aluminum and both the top and bottom edge are given a chamfer at the same time!

Safe side

Danger zone

Figure 5-80 *There is a good side and bad side on a disk sander—know the difference!*

Getting a Feel for a Finished Edge—Hands On

A journey machinist is not likely to turn in a part for inspection or shipment without first testing for sharp edges. This test will tell whether or not a machined edge is finished.

Correctly dressed for the shop, find a piece of scrap steel and/or aluminum. For an experiment, lightly slide the unfinished edge across a piece of plastic such as a screwdriver handle. Using very little pressure, if it peels or slices a slight bit off the plastic or digs in and will not slide, the edge is too sharp to be a finished product.

Figure 5-81 *With a little practice, you could do a good job of breaking the drill gage edges with this edge finisher.*

Figure 5-82 *A hand burr knife made from a triangular file can be useful in removing sharp edges.*

Now select any method we've discussed, then break the edge just a bit, 0.005 to 0.010 in. Try it again, scrape the plastic and get the feel of a finished edge. Now touch it with your hand and compare it to the feel of the raw edge.

Finishing the Holes

So far, we have concentrated on the edges of the drill gage, but what about finishing the drilled holes? There are two methods beyond the vibrating and tumbling machines discussed. Hand tools or a drill press with a countersink.

Countersink by Hand or Power Countersinks are cone-shaped cutters used to machine small bevels at the entrance of holes—if the

Figure 5-83 *A countersink mounted in a portable drill motor could break the hole edges.*

cone-shaped hole produced is deep enough, it provides a nest for a screw head. They are also very efficient tools for breaking the sharp edges of the holes. We'll look at them more closely in Section 10 on drilling skills.

A countersink mounted in a portable drill motor could do a fairly good job of breaking the hole edges (Fig. 5-83). A countersink mounted in a file handle also works for this task as well (Fig. 5-84).

Figure 5-84 *A handheld, rotary hole finisher could break the edges quickly. (We call these swizzle sticks.)*

Burr Knifes and Other Hand Tools

Because of the way the are twirled around, these handy little tools are sometimes called swizzle tools in shop lingo (Fig. 5-84). They quickly break the sharp edge.

UNIT 5-4 REVIEW

Review the Key Terms

Aluminum oxide
The common gray abrasive used in general grinding wheels.

Burrs
The unwanted jagged metal left after machining is completed; the tool used to remove burrs is also called a burr or a rotary file.

Draw filing
Holding the file nearly sideways to the to the stroke. Produces a finer finish and better control of shape.

File card
The multifunction brush used to clean out loaded (pinned) files.

Lathe file
A flat file with teeth on two sides.

Mill file
A flat file with teeth on four sides.

Part-finishing wheels
A soft fiber and abrasive wheel that is very effective at finishing part imperfections. Often called a 3-M wheel.

Vitrified
The hard grinding wheel made by firing a clay binder and an abrasive until it forms a material similar to china or porcelain.

Replay the Key Points

◆ Burrs can be minimized during machining by keeping tools sharp.

◆ Taking just few minutes longer on setups and maintaining sharp tools can save hours of benchwork removing burrs.

◆ Air turbine rotary tools spin at 15,000 to 20,000 RPM—always protect your eyes when using them.

◆ When handling and finishing parts, do them no harm.

◆ There are many options for finishing work. Choose the one that delivers the quality required in the least amount of time.

◆ A true craftsman would never turn a part in for inspection with sharp edges—unless the situation simply wouldn't allow doing otherwise. For example, the CNC machine cycles too fast to permit burring the work.

Respond

1. Burrs are caused by
   A. Dull tools
   B. Metal's ability to stretch ahead of the cutter
   C. A cutting action that's too fast or hard
   D. Poor or no coolant
   E. All of the above

2. True or false? Files with teeth on their sides are classified as mill files.

3. Rasps are primarily used to file _____.

4. You need to break the edge of 500 parts with many edges and features. What is the best method? Why?

5. Name the methods that might be used if the order in Question 4 was for 20 parts.

UNIT 5-5

Part Marking and Identification

Introduction: Work Order Operation 140 Another secondary operation often given to beginners is marking parts for identification or other functions where size, model, serial number, and usage instructions must be made permanent on a machined part. On the drill gage drawing and WO Operation 140, the shop is to part mark the hole sizes next to the holes. There are several ways in which they might be marked:

Rubber stamping/stenciling/silk screening
Laser etching, which is computer driven
Engraving machines, which can be manual and computer driven
Steel stamp
Electrochemical etch

Destructive Marking—Use Caution!
With ink stenciling there is no physical change made to the parts. But there's nothing to learn about rubber stamping, stenciling, or silk screening. They are common nondestructive part-marking methods used in many industries.

Skipping them, we'll look at the next four. All these methods change the surface of the work in some way. Many parts have been sent to the scrap heap because a new worker identified them in the wrong place—check the print! Or ask your supervisor.

KEY TERMS

Ball-peen hammer
A steel hammer offering a sharp impact. It has one barrel and one hemispherical.

Deadblow hammer
A soft-faced setup hammer that does not bounce back due to sand or lead shot in the head.

Peen
To shape metal by small dents caused by the hemispherical head of a ball-peen hammer; head—the peen.

Part-Marking Methods

Electrical-Chemical Etching Part Marking

This is a process we should mention but not study. Modern programmable part-marking technologies have all but replaced this slow process that etches part information. Chemical etching harms the work surface by eroding metal through the combined action of acid and electrical current. It's not unlike stenciling. A template of the desired message is cut from acid-proof insulating material similar to stiff paper. This is taped or held tightly on the metal to be marked, which has been well cleaned beforehand. With the workpiece on a ground plate, the soft-faced electrode is dipped in acid then pressed down on the stencil. The metal below is exposed to erosion, while the stencil prevents it elsewhere. The result is a dark letter on a bright background once the stencil is removed.

Advanced Part-Marking Systems—Laser and CNC Engraving

These two methods are both specialties beyond the scope of Unit 5-5 but they are the most likely part markers encountered in modern industry. Both are controlled by programs written offline at PCs. Letters, numerals, symbols, and pictures are drawn onscreen then turned into a program via computer-assisted machining (CAM) software. The big difference lies in how the characters are cut into the part surface.

Laser Engraving

This method can mark almost anything, including diamonds. It can mark irregular surfaces and the marks made can be any size down to micron size. The laser engraves in a very different way from older tracing machines. First, the characters are cut into the work surface by a concentrated beam of coherent light (Section 26), which creates nearly any width line to a controlled depth.

Since heat etches the work surface, it's critical that the marks not be made on bearing surfaces or other locations where the heat-affected metal might cause a problem. The process does harm the surface, which might need to be sanded or buffed after engraving metal objects to remove roughness. Laser engraving (Fig. 5-85) is slower than mechanical engraving, but it has several advantages.

The laser part-marking machine works differently than other cutting machines in that neither the laser head nor the work moves as it does with engraving machines. Since they are light, laser emissions will reflect off mirrors. The laser beam is stationary but directed by a mirror capable of pivoting on two axes, controlled by a CNC computer. Pivoting mirrors is a relatively inexpensive system. Beyond their relatively low cost, laser part markers also have these characteristics.

◆ They can make any size character in any font within the computer driving them (mechanical engravers can do this too, but must have many font set templates).

Figure 5-85 *A programmable laser part engraver. Epilog laser—Golden Colorado.*

- Setting up the part marks is a matter of defining the part shape and typing in the message.
- Once set up, the program can be stored for later use and quick setup.
- They can mark over any surface, whether or not it is curved. The computer needs to know the shape and size of the surface upon which it is projecting its beam, and the surface composition, but once the computer has that definition, it can make accurate characters on variable surfaces.

CNC Engraving

Engraving is a cutting action using a tiny pointed cutter with limited depth capability. It's most commonly seen on trophies and awards, but is also used in manufacturing where many parts must be engraved per day.

Before CNC control, engravers (Fig. 5-86) traced letter and number templates set up in a track. A stylus was lowered into the template to guide the rotating cutter reproduced on the work surface. They were slow and their use is limited to flat surfaces. Today's computer-directed engraving workstation can do a whole lot more.

Steel Stamping

About as low-tech as one can get, this method of marking parts can result in a sore thumb! The stamps are aligned to a light pencil line then firmly hit just once, with a hammer. Steel stamping (Fig. 5-87) is common in tooling shops and schools, but is not practical in production. Stamping is harmful to the work and occasionally to workers. The new technologies are faster and neater and can mark

Thumb groove

Figure 5-87 *Stamps feature an alignment groove. Hold firmly and strike only once!*

Trade Tip

Groovy Stamps When using steel stamps, note that most popular brands have a groove or notch on the side toward you (the bottom of the letter). You need not look at the character, to be sure it is upright. Put your thumb on the groove as you hold them for striking.

KEYPOINT

Hit it Once
Striking a steel stamp twice usually makes a double image. When steel stamping, place the part on a firm, massive surface and avoid bouncing the part against it which mars the work on the side opposite the stamping.

Figure 5-86 *A computer-operated engraving machine.*

Trade Tip

Erasing Stamping It is possible to erase, or more to the point, eradicate a wrongly stamped letter! Remember, there was no metal removed — it was only displaced upward around the stamp. Using a ball-peen hammer, tap the letters lightly around their edges and they will close up. Now stamp right over the closed letters with the right information this time. Done carefully, it's difficult to see the original mistake. What is a ball-peen hammer? See the upcoming paragraphs.

surfaces that aren't flat. That said, it's the most likely method of stamping the hole sizes per Opp 140 on the 12 drill gages.

Machine Shop Hammers

Let's take a moment to discuss which hammer is used to hit the stamps. Hammers are also used to hit a big list of other shop implements such as punches and chisels. To keep this simple we'll divide hammers into three kinds, differentiated by the sharpness of the impact they create.

Ball-peen hammers have a sharp impact but are accurate.

Soft-faced setup hammers are less hard, but may be light or heavy.

Deadblow hammers are less marring, but do not bounce back; they have good impact.

Ball-Peen Hammers

Having the sharpest impact of the three, **ball-peen hammers** are used in many different ways in the shop. Although often called metalworkers' hammers, they are also used in other trades. Besides using the barrel end to tap punches and chisels, the hemispherical **peen** is used to deform metal—called peening. As seen in the Trade Tips, we closed the letters by moving the metal back over the stamp. Peening shapes, but does not remove, material. The classic example is found in forming a rivet head by hand (Fig. 5-88) to join two objects together. The rivet is put through a closely sized hole then held in place with a solid, heavy rivet setter to resist the impact of the hammer. Then, as shown in Fig. 5-88, the head is formed on the opposite side of the work by tapping the shaft until it creates the formed head.

Most machinists keep two or more ball-peen hammers in several different weights for light to heavy blows. This is the hammer you would use to steel stamp the drill gages while they rested upon a smooth iron or steel plate.

Typical ball-peen uses

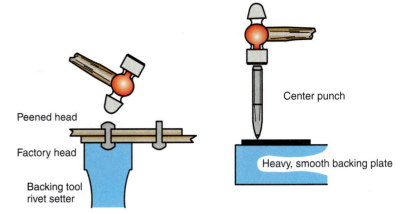

Peened head

Factory head

Backing tool rivet setter

Center punch

Heavy, smooth backing plate

Figure 5-88 *Two shop uses among many for the ball-peen hammer.*

Hammer Safety Precautions

For all hammers,

◆ Make sure the head is not loose. Loose heads have obvious consequences—perhaps the phrase to "lose your head" originated here?

◆ Wear safety glasses—hardened hammer faces as on the ball-peen and the punches they strike can splinter when struck hard.

◆ Use caution when striking a hard object with a hard hammer such as a ball-peen, because the hard steel being struck can dangerously shatter or splinter. Note that punches and other tools meant to be struck with a hard hammer are softened on the top to avoid splintering.

Soft-Faced Hammers

These impact tools are common in machine shops (Fig. 5-89). While there are several common types, we'll group them together to keep this brief. They are made of either moderately soft metal, usually bronze but never lead, or have soft faces of plastic or rubber. On some of these hammers, the faces are replaceable and there are a variety of hardnesses available to create different impacts. I prefer a very soft face on one end and a medium hard face on the opposite end. When they are replaceable, various faces are color coded for quick reference. Usually green is the softest while yellow is the hardest with

Why Not Lead? In the past, lead was a popular material for soft-face hammers. You might still find one in the back of a drawer but, although these hammers work well, *do not use them!* The heads tend to mushroom (bulge out at the sides) with use. Then fragments can shoot off. The lead also contaminates some metals if they are struck hard enough. Enough accidents occurred from lead hammers that the Occupational and Safety Hazard Act (OSHA) made them illegal for shop use.

Figure 5-89 *Soft-faced hammers come in several varieties. They are used extensively for setup tasks.*

Figure 5-90 *A deadblow hammer is used to drive this part down into the vise, without it bouncing back.*

other colors between. There is no industry standard for this coding.

In machining, soft hammers are constantly used

◆ To move machine accessories a bit at a time during a setup while never marring the tools or machine, for example, to tap a vise until it is parallel to a machine axis.

◆ To firmly seat parts into a vise or a chuck during a production load.

A soft-faced hammer would also be used to flatten a bent drill gage without damaging the surface. Their lack of sharp impact would make them useless for steel stamping.

Deadblow Hammers

A special version of the soft hammer, **deadblow hammers,** also called *reboundless* in tool catalogs, also have soft faces, which are usually replaceable. But they also feature a hollow head that's filled with lead shot or sand. When striking with them, the filler catches up with the head at impact and cancels the tendency to bounce back (Fig. 5-90). This greatly improves the effectiveness of the tool in many instances.

These hammers are used where it's important to set a workpiece firmly down into a vise or a chuck. Without them, using a standard soft hammer, driving a part against a surface can be frustrating—it comes back up from the bounce. With a reboundless hammer, the bounce is canceled and the part stays put. But even though they move objects very well, their soft face does no harm to the work surface.

Performing Secondary Offload Operations and Assembly

Introduction: In this unit we'll look at several crossover operations that could be done by hand or on the machines. The reason to perform them off the CNC machines is three-fold:

1. To free-up the high-production machines to do what only they can do.

2. To break up bottlenecks and speed the work forward when it can't all be done on the CNC machines.

3. To complete some operation not easily finished or even impossible to finish on a CNC.

Offload Machining Is Avoided if Possible Today every effort is made to avoid moving the workpiece to several stations if the planner can see ways to machine the part with one CNC setup (called *one-stop machining* in shop lingo). But there are times when it's more efficient to take the job off the big machines and do the work elsewhere.

If the secondary task involves chip making it is called "offload" machining. In following this kind of planning, some shops have small, less costly offload CNC machines that take work from their big brothers to remove burrs or finish operations when it speeds up work flow.

KEY TERMS

SECONDARY OPERATIONS

Arbor press
A machine used for pressing and depressing assemblies.

Blind hole
One that does not go all the way through a part.

Bottom tap
Used to finish more threads in a blind hole.

Die
A threading for making a bolt or stud—outside thread.

Honing
An abrasive process that removes small amounts of metal to create a close tolerance hole of good finish.

Plug tap
Cuts closer to the bottom of a blind hole than a starting tap but not as close as a bottom tap.

Press fit (P/F)
Permanent assembly by forced insertion.

Shrink fitting
Using heat and cooling to lighten the pressure of a press fit.

Tap
A threading tool used to make internal threads—nuts.

THREADS

Class
The fit of a thread—how tight or loose it is to be.

Clearance
The deliberate space between bolt and nut to allow movement.

Coarse threads
General purpose threads.

Fine threads
Sometimes called SAE threads. Fine threads have a shallower helix and thus create more translation force but less distance.

Form
The shape of a particular thread. The most common form is the triangular thread for both ISO metric and Imperial Unified threads.

Helix angle
The angle at which a thread's inclined plane works.

Nominal diameter
The basic design size of the bolt. Nominal is the drawer size for the bolt or nut, but not the real size.

Pitch
The term used to express either the pitch distance or the threads per inch.

Pitch distance
The repeating distance of a thread. On Imperial threads it's found by dividing pitch number into 1 in. For metric threads, it is specified in the callout.

Threads per inch
Abbreviated TPI and often called pitch.

Translation
The distance a nut moves on a bolt with one turn—the result of the wedge being driven by turning the nut on the bolt.

Truncation
Clipping the top and bottom points off a 60 triangular degree thread.

ShopTalk

TMI Sources At this stage, the body of information on threads and threading would be too much information, however, there are places to go when an in-depth reference is needed: *Machinery's Handbook®* CD, website, or hardbound book, or the Internet. Search for Standards + Screw Threads + Unified, or the American Society of Mechanical Engineers home site, then let your pointer find its way through their menus.

ShopTalk

SAE Threads Fine threads in the United States are also referred to as SAE threads because, in the distant past there was a lack of standardization so the Society of Automotive Engineers took on the task for use in their industry. We seldom use that term in machining because all threads were also organized by the ANSI some time ago, to become the Unified Screw Thread System. Then the entire jumble of thread types were collected together and standardized. That's where we get the UNF and UNC. So, UNF and SAE are the same thread—perhaps the first "double standard." Today, in an effort to streamline the entire gambit of manufacturing controls, the ANSI has become part of the Society of Mechanical Engineers (SME), the same group that deals with GDT standards.

Introduction to Threads and Threading by Hand

This is a first lesson in the technology of threads. There is enough to the subject that

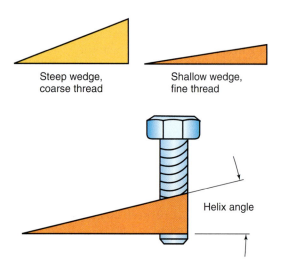

Steep wedge, coarse thread

Shallow wedge, fine thread

Helix angle

Figure 5-91 *A thread is a wedge wrapped around a cylinder. A fine thread is a shallow wedge, while a coarse thread is a steep one.*

Figure 5-92 *Coarse and fine bolts compared. The fine thread has a lower helix angle and smaller threads placed closer together.*

we'll discuss them twice more in *Machining and CNC Technology,* and yet we'll still be only scratching the surface! But for now, staying with our theme of W/O secondary operations, Unit 5-6 will provide what you need to know to make threads at the workbench.

Threads are inclined planes (wedges) but they are wrapped around a cylinder. In the drawing (Fig. 5-91), which wedge could move the most weight with the least forward push? The answer is the one on the right; the shallow-angled wedge. By contrast, the one on the left could move the weight the greater distance (known as the **translation** distance), but it would also require more force to do so.

Threads have shallow and steep counterparts—the shallow version being a fine thread, and steep version, coarse threads. The angle at which the thread lies with respect to the cylinder is called the **"helix angle,"** as shown in Fig. 5-91.

Standard Sizes

While there are many more kinds of threads, for every size of standard bolt (metric or Imperial) the American National Standards Institute (ANSI) and the ISO have designated a fine and coarse thread. For example: a half inch fine and a half inch coarse or the same for a 12 mm bolt (Fig. 5-92).

Fine threads, UNF (unified national fine), and ISO fine (metric threads) develop more closing force from a given amount of rotary torque, but they also require more turns to assemble compared to the coarse version. Designers tend to use fine threads for precision fastening and critical assemblies.

Coarse threads, UNC or ISO-C, are quicker to assemble. Since there are fewer thread grooves per inch, each individual thread is bigger than a fine thread. Due to the larger thread size, coarse threads have more lateral resistance to thread breaking called *"stripping."* They are common for everyday mechanical assemblies. The majority of your car is held together with coarse bolts (or duct tape, depending on student budget).

Main Thread Designations— Nominal Diameter and Pitch

To select the right cutting tool as called out on the print and work order, and to cut a specific thread, there are two numbers beginners should understand (Fig. 5-93).

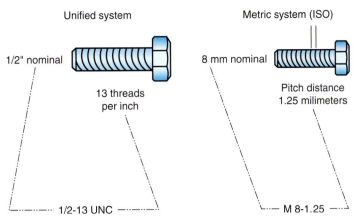

Unified system

1/2" nominal

13 threads per inch

1/2-13 UNC

Metric system (ISO)

8 mm nominal

Pitch distance 1.25 milimeters

M 8-1.25

Figure 5-93 *The thread terms a beginner should know.*

Bolt Nominal Diameter This designator is the theoretical outside diameter of the bolt; the designation size. In reality, the outer diameter is just a bit less than nominal. That is called the *major diameter.* The tiny difference (from 0.003 to 0.010 in.) between major diameter and nominal provides mechanical **clearance** so the nut turns on the bolt. In Fig. 5-93, the UNC bolt is a nominal half inch but would measure a bit less.

Pitch and Pitch Distance In the unified system, separated by a dash, the second designator tells the user if the thread is fine or coarse. For Imperial threads it is the *pitch number* abbreviated to **pitch.** It states how many **threads per inch** (TPI) are on the bolt—in this case, 13 TPI. This is an area where U.S. and ISO threads are slightly different.

The metric designator for the fine or coarse thread is a different number related to pitch. It's the **pitch distance,** the distance from thread crest to crest. It's often shortened to pitch as well which might be confusing at first. Pitch distance is the measurement between corresponding points on a thread. In Fig. 5-93, the metric bolt has a pitch distance of 1.25 mm.

Critical Question For an Imperial bolt then, how would you calculate the pitch distance for a 1/2-13 UNC thread? The answer is to Divide 1 in. by 13 to get 0.0769 in. pitch distance.

KEYPOINT

Unified threads are specified as nominal diameter - pitch number. Metric threads are specified as nominal diameter - pitch distance.

Thread Forms

There are several shapes or **forms** for threads, with each having a specific mechanical function. We'll look at them more closely in Section 14. However, the vast majority are the 60 degree triangle used both on metric and Imperial threads. It is built on an isometric triangle—one having all sides and all 60 degree angles equal.

Truncation (to cut off)

If left as a theoretical triangle with sharp points, the bolt would have a serious weakness. Shown shortly in Fig. 5-95, it would tend to break at the bottom of the vee. So, modifications are made to the points to improve the strength—at both the top and bottom of the thread shape. Figure 5-94 shows the formulas used. It's not critical that you understand them at this time, just be aware of their existence for later study.

The sharp point of the triangle is removed top and bottom—called **truncation.** This is done both to the internal (nut) and external (screw) threads for two reasons.

1. **Mechanical Strength**
   Look at the two threads in Fig. 5-95. Which would tend to break more readily and where? *Answer:* At the bottoms of the sharp triangles, while the clipped threads would be far less prone to breaking the shaft. The clipping is called *truncation.* Both the nut and bolt triangles must be truncated equally if they are to fit together.

Amazing Strength
Interestingly, with their tops and bottoms gone, triangular threads retain almost all of their ability to wedge—that is, to not strip after truncation. A truncated thread retains 96 percent of the original lateral translation strength. OK—maybe that's not "amazing," but it got you to look!

$$D(MAX.) = 0.7035 \times P \text{ or } \frac{0.7035}{N}$$

$$(MIN.) = 0.6855 \times P \text{ or } \frac{0.6855}{N}$$

$$C \& F = 0.125 \times P \text{ or } \frac{0.125}{N}$$

$$R(MAX.) = 0.0633 \times P \text{ or } \frac{0.0633}{N}$$

$$(MIN.) = 0.054 \times P \text{ or } \frac{0.054}{N}$$

International metric thread form

$$D = 0.6495 \times P \text{ or } \frac{0.6495}{N}$$

$$C \& F = 0.125 \times P \text{ or } \frac{0.125}{N}$$

American National thread form

Figure 5-94 *Formulas for 60 degree thread forms.*

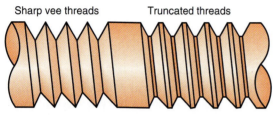

Sharp vee threads   Truncated threads

Figure 5-95 *A sharp thread would create a serious weak point in the shaft, while truncated threads reduce this problem.*

2. Less Jamming

Left on the thread, those pointed tops are prone to breaking off and jamming between the nut and bolt when heavily loaded by a wrench. But, with truncation, they resist breaking and jamming.

Thread Classes— Preciseness

Adding a third data field to the designator, the **class** shows the level of precision of the thread. This is only added to the designation when it's required by some special function. If not, the class is considered to be 2-everyday threads.

Class 1 designates loose threads where they must be constantly

turned. A good example is a "C" clamp thread.

1/2-13-**2**

Class 2 is for general purpose threads, such as the nuts and bolts you would buy at the hardware store.

1/2-13-**3**

Class 3 is a precision thread for critical assembly. It is the tighter tolerance of the three fits.

As the classes become tighter, the tolerance for acceptability becomes narrower. There are further refined classes beyond the three outlined here but they are beyond this discussion.

KEYPOINT

Class 2 threading tools are what you normally find in any tool room.

Introduction to Pipe Threads

When the threads are to go on pipes either for mechanical connections as on construction scaffolding, for example, or for carrying liquids or gasses, the thread designator will include a letter "P."

Pipe threads are similar to the fastener threads in that they also use the 60° triangle

1/2-inch bolt   1/2-inch tapered pipe thread

Figure 5-96 *Pipe threads are designated by the pipe's carrying capacity.*

form. They are different in three ways. For now, if you understand these differences that should suffice.

Tapered Threads There are two kinds: straight and tapered. The tapered version has its wedge wrapped around a gradual cone rather than a cylinder. The increasing size closes tighter as it is screwed together, to aid in sealing. The designator for tapered pipe threads is NPT (National Pipe Tapered).

Size is Based on Hole in pipe A second difference compared to fasteners is how the nominal size is specified. With bolts and nuts, it's the outside diameter of the bolt, but not with pipe threads (Fig. 5-96). Their size is based on the inside diameter of the pipe. That means that a $\frac{1}{2}$-in. pipe thread is a great deal bigger than a $\frac{1}{2}$-in. bolt, as shown in the drawing.

Different Pitches Pipe threads feature pitches different from those of everyday bolts and nuts. They cannot be interconnected, to avoid possible failures or misapplications. While there's a lot more to know about them, the chances are you will not be required to cut pipe threads during your early training, so we'll leave them until Section 14. But, when you see a designation of 1/2-14 NPT, you will know that it's a tapered pipe thread.

Pipe threads do not assemble with other kinds of threads for safety reasons.

Determining Thread Pitch When we need to check a given bolt for its pitch, we use a small template gage called a *screw pitch gage* (Fig. 5-97). The leaves are spread out, then each is held against the bolt in question until an exact fit is found, as shown in Fig. 5-98.

For now, that's all the background needed to select and cut standard threads with a tap or a die. A tap makes an internal thread, a

Figure 5-97 *A screw pitch gage features leaves, each with a different pitch number or distance if it's metric.*

13 TPI

Figure 5-98 *The pitch will become obvious when the correct leaf fits the thread.*

nut, and a **die** makes an external one. Now, here's how to cut the threads.

Threading with a Die

Threading dies are cutting tools used to make or to restore outside threads (bolts and studs). Studs are bolts with no head and they have threads on both ends. Threading dies are supplied in two general forms:

Cutting dies   To make a thread on a raw shaft

Chasing dies   To restore a damaged thread (Fig. 5-99)

Chasing Dies

These tools are normally easy to use. They are started then run down the thread just like a nut. In fact, their outside looks like an oversized nut and they are turned onto the bolt using a standard wrench.

Figure 5-99 *A chasing die nut is designed to re-cut a damaged thread.*

Skill Hints for Using a Chasing Die

Check the end of the shaft for damaged threads—if they are smashed, grind a small chamfer on the end before starting the chasing die. This helps the die start in the original threads.

Use cutting oil or thread compound and go slow even though they can be run on faster.

One side of the nut is designed to lead the cut—the side with the markings on it. Look for "start" on the die face.

While it's possible in an emergency to cut a new thread on a raw shaft using a chasing die, it's the wrong thing to do.

Figure 5-100 *A button die in a handle is called a die stock.*

They aren't designed to cut metal chips, only to remove dents and smooth out damage on existing threads.

Threading Dies

The affordable kind you'll likely encounter in schools are called button dies for their compactness. They are designed to cut new threads on round features. Button dies (Fig. 5-100) differ from chasing dies in that they provide an adjustment for thread size. Because they are usually round, they must be mounted in a die handle, as shown in Fig. 5-101, rather than being turned using a wrench.

Adjustable Quality button dies feature a relief slot through one side of their outer circle, as shown in Fig. 5-101. The spring slot and set screw allow adjustment to compensate for resharpening of a dull die. Each grind of the cutting edge makes it cut a slightly larger thread. Without the adjustment, the thread would become progressively larger until a nut would not screw on the bolt produced. The adjustable die is also used to create the different classes of threads. Finally, it can be used to set oversize for a first rough pass, then closed to the right diameter to make a final pass. This technique helps where the best finish and size are required on expensive work material.

In the drawing, notice the central adjustment screw wedges into a chamfer on the slot edges to force the die open. Holding screws on both sides (not shown), keep the die closed while the adjusting screw forces and holds it open against the spring pressure of the die.

Skill Hints for Threading Dies

Aligning the die to the shaft is important. Misaligned dies are the number one cause of failure when hand threading with a die.

Figure 5-101 *The adjusting screw opens the die slightly.*

To help, there may be an alignment guide as part of the die handle. It is composed of three movable pads and an adjustment ring that is used to ensure the die is going on straight. Use them, they help avoid "drunk" threads that wander off. If using handles that lack the alignment guides, use a square in two directions to be certain the die is on perpendicular to the shaft (Fig. 5-102). When threading, constantly watch the die handle to ensure it is not tipping relative to the shaft.

Make sure the intended shaft has a starting chamfer on the end.

A cutting oil or compound is essential.

A threading die can be used as a chasing die, but a chasing die should not be used as threading die.

Due to the lead-in angle, dies produce incomplete threads for the first two or three threads. To thread closer to the end of the shaft, finish the thread, then turn the die around and thread with its reverse side.

To set the die to the right size, open then screw it onto a good-quality screw or model part of the right size, pitch, and class. Rotate the die slightly while turning the adjustment screw out, and the clamping screws inward until the die drags a bit on the screw.

Figure 5-102 *When threading with a die, watch the die handle to ensure that it is perpendicular to the shaft.*

Figure 5-103 *An insert jaw die has the advantage of easy resharpening and quick size adjustment.*

KEYPOINT

When threading by hand, turn the die forward two or three turns then reverse it to break the chip.

Other Kinds of Dies

Fixed Size Dies You may encounter a button die lacking the size adjustment feature—a solid button die. They are tools purchased for their lower price only, since they can only be resharpened a few times before they must be thrown away due to progressively bigger threads. Solid dies of this type are usually supplied in Class 2 threads only, and are considered throw-away tools.

Insert Jaw Dies This is a very different type of threading tool (Fig. 5-103). Although more costly, they bring several features to the shop where a lot of hand threading is being performed. It's unlikely that you will encounter these dies in schools.

The inserts can be removed from the carrier jaw and easily resharpened on their sides using a standard tool grinder or surface grinder. This isn't so with button dies, which must be reground with shaped stones in the holes between cutting flutes. The jaw types can have their inserts replaced in just minutes where time is of the essence.

Taps for Internal Threads

Taps could be described as very hard bolts with cutting teeth. They cut threads inside a hole as they screw into it. Taps are supplied in several varieties based on the material to be cut, the method to be used, and the far end of the drilled hole to be threaded (Fig. 5-104). After determining the nominal size, pitch, and class of thread to be produced, in selecting a tap the questions are

♦ Does the hole go through or is it a blind hole?

♦ Must the threads go to the bottom of the blind hole or not?

♦ Can the chips go ahead of the tap or should they come back up the tap?

Figure 5-104 *Different kinds of taps for different purposes.*

Depending on the answers, there are three shapes of taps (Fig. 5-105) from which to choose: starting, plug, and bottom tap.

1. **Starting Tap**

   Sometimes called a taper tap, this is the general purpose tap that begins cutting best. Always use a starting tap first if you possibly can (even in a blind hole). Notice in the drawing, that there is a tapered portion on the first two taps that cuts the initial threads. This starting portion of the tap leaves imperfect threads that are not cut to full diameter. The taper tap leaves the most imperfect threads.

2. **Bottom Tap**

   Used only to finish a blind hole where perfect threads are required to the bottom of a blind hole. A bottom tap is used after the start, then plug taps are turned in as far as they will go. It is not used to start a thread

except in rare cases for extra shallow threads. These taps have no lead, thus they cut poorly compared to the other two.

3. **Plug Tap or Second Tap**

   Halfway between the other two forms, this intermediate tap cuts better than a bottom tap but it does not start the thread as well as a starting version.

Above the tapered portion, all taps produce a finished thread. There is no difference in the size of the thread produced by each tap except the lead part at the tip.

Tap Flutes Another difference in tap construction is the number of cutting edges with which they are supplied; two, three, and four *flute* (cutting edges) versions (Fig. 5-106). In general, taps with fewer flutes are used in softer material, while hard material is tapped using four-flute tools. For hand tapping, there is little performance difference in the three taps. However, when power tapping at a much faster RPM, the two- and three-flute taps tend to deliver chips out of the hole better while the four-flute tap has more cutting edges and a

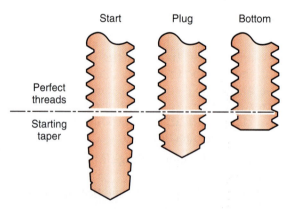

Figure 5-105 *A starting tap, a plug tap, and a bottom tap. The difference lies below the line, in the length of the tapered portion.*

Figure 5-106 *Compare two-, three-, and four-fluted taps—cross section.*

stronger center portion for threading harder materials but tends to choke on chips.

Spiral or Gun Pointed Taps The third difference in taps is the direction they drive chips as they cut the thread. By changing the angle of the cutting portion, the chip can either be sent ahead of the tap or back up the flutes, similar to a drill.

Notice in Fig. 5-104 that the spiral gun tap has an axial lead angle on its cutting edge. It sends the chip forward into the hole. This type works best in holes that are drilled through or are much deeper than the intended thread. The straight pointed tap lifts chips back out of the hole.

Taps break easily. Here's a group of hints how to avoid the embarrassment of breaking yours:

Sharp Tools This is critical. A dull tap breaks far easier than a sharp one. To recognize a dull tap, look at the cutting edge. If there is a tiny line of reflected light right at the cutting edge, caused by the rounding of what should be the clean intersection of two surfaces, it must not be used. A sharp edge will not reflect light. This visual inspection works for all cutting tools.

A sure symptom of dullness is if, during use, the tap makes any sound or drags even slightly more than it has been. When detecting even the slightest change, stop immediately and back it out! Any change in sound or feel usually signals that the next step is going to be a broken tap. Once broken in the hole, the tap is difficult to remove.

Removing Broken Taps Once broken inside the work, the tap cannot be drilled out as it is as hard as the drill or even harder! There are two viable options if the tap has snapped off leaving nothing to grab for backing it out, which is always the case! A broken tap remover (Fig. 5-107) is a set of strong, shaped wires that fit into the holes between the flutes. You'll need an experienced person to demonstrate these little tools but their success rate isn't wonderful. Most of the time, the broken tap is fragmented and tends to lock up inside the hole, in which case the part is thrown away or if it's costly enough, turn to the next option. Patience is the key to using tap removers.

Tap Burners These great little machines use electrical sparks to erode the broken tap out of the hole. They have a 100 percent

Figure 5-107 *Tap remover wires and guides are supplied in sets, to fit a specific tap.*

success rate if available. We'll explore this very different metal removing process called electrical discharge machining in Section 26. There's a lot that can be done beyond removing taps too hard to drill out.

Tap Alignment Next on the hint list for avoiding disaster, is getting the tap pointing directly into the drilled hole. In other words, make sure the tap is parallel to the hole and that it remains so during the cutting (Fig. 5-108). This is the number one reason taps break during hand tapping. A small square is useful (Fig. 5-109) or a guide called a tap block can be used to start parallel to the hole. With some practice, you will be able to get

Tap block 90 degrees starting guide

Clamp

Work

Figure 5-108 *A tap block helps us to align taps and keep them parallel with the hole.*

Figure 5-109 *A precision square can be used to check the tap.*

taps reasonably aligned by looking at the tap handle. Here are two more accurate methods used to get a tap aligned to the hole:

Tap Blocks An alignment tool with a series of holes slightly bigger than the nominal size for several different taps. The block is clamped over the intended hole as a guide for starting the tap straight. After several threads are in, the tap handle is removed and the tap block is slid up and off the tap.

Drill Press Guidance Here a technique called a "dummy center" (Fig. 5-110) is being used. The machinist has aligned the drill press spindle over the hole to be tapped, then bolted the part in this position, drawing it back to guide the tap.

KEYPOINT

When using a dummy center on a drill press, no power is used. The press guides hand tapping.

Figure 5-110 *Use a drill press and center to keep the tap aligned to the hole.*

Lubrication Is Essential Brush a light coat of cutting oil or, even better, a special threading fluid on the tap and in the hole. And keep replenishing it as the cut progresses. Lubrication makes a big difference in cutting action and chip delivery. If the thread is deep, don't go all the way in. Reverse the tap and bring it out of the hole, then blow the chips out of the hole and off the tap, relube and send it back.

The Right Moves—Tapping Actions A major cause of broken taps is chip loading. In certain situations the chips can't come out of the hole. You must stop every two turns or less and break the chip by backing up one half turn (Fig. 5-111). If the broken chips continue to clog the flutes, you will need to completely remove the tap after each succeeding two turns in, brush it off, and blow out the chips from the hole. Be careful in blowing chips out of a hole, hold a rag in the airstream to trap the oil and chips.

Listen and Feel If a tap squeaks while threading, it is from one of three causes:

A. Lack of lubrication—use cutting fluid.
B. It is dull and should not be used.
C. Some metals just squeak and nothing is wrong. Squeaking might occur in some brass/bronze and stainless. Also in harder types of steel.

Tap Handles

Taps must be mounted in handles to turn them by hand. The tee handle closes down

Two turns in
one-half back

Figure 5-111 *Breaking the chip is important when hand tapping.*

on the tap similar to a drill chuck. It is compact and is used where the tap is small and you wish to limit the force to avoid breaking it, or where space is limited to turn the handle (Fig. 5-112). The standard tap handle holds taps in a set of grooved jaws. It is used to turn taps where more force is required (Fig. 5-113).

Roll Form Tapping (Cold Formed Threads)

This kind of tapping isn't usually performed by hand, but it is worth mentioning here because it produces a stronger thread compared to cutting threads. It is performed with a tap somewhat similar to the others, except it has no cutting edges and removes no metal. See Fig. 5-104, right-hand column.

The tap deforms (ploughs) the top of the thread up by displacing metal from the grooves. The threads produced, due to the cold working of the metal, are much stronger than cut threads. They are rounded in form too which adds strength and long life. The roll form tap in Fig. 5-114 has gone partly through the work to show the gradual thread formation.

The hole drilled before tapping, is larger than the one for a cut thread. While getting the size right for regular tapping is critical, for roll threading, it's extra critical. Based on the 60 degree triangle, the shape of the threads formed is quite different (Fig. 5-114). Roll form tapping is not an entry-level skill and usually is not taught in schools, we'll leave further discussion for on-the-job training.

Tap Drill Holes

For any tap, whether it's for cutting or forming threads, the hole size drilled before tapping must be the right size. There is a correct drill size for every tap—it depends on nominal thread diameter and fine or coarse threads. In this section, we're assuming you are following the machines with offload work, so we won't cover selecting the drill or drilling the hole at this time; holes were drilled before offloading the work. However, remember the key point here.

KEYPOINT

There is a correct size drill for each tap size and very little tolerance for variation in that size.

Figure 5-112 *A tee tap handle is used for small taps.*

Figure 5-113 *A tap handle is used for large and small taps.*

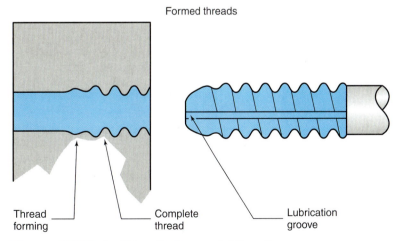

Formed threads

Thread forming

Complete thread

Lubrication groove

Figure 5-114 *A roll tap displaces metal from the grooves to form the ridges.*

Trying to tap a wrong-sized hole, will either break the tap if the diameter is too small or it will make weak or nonexistent threads if it's too large. For information on tap drill selection, turn to the Appendix decimal chart or go to Unit 10-5.

Finishing Holes—Work Order Operation 50

Two types of secondary operations that make drilled holes more smooth, accurate, and round:

Hand-Reaming   Machine Honing

Reaming

Reaming finishes a predrilled hole. There are two types of reamers: hand and machine. The hand-reamer is of interest here in Section 5. It removes very little metal but makes the hole round, straight, and smooth. Tolerances within 0.0005 in. (or closer) for diameter, straightness, and roundness can be met using a hand-reamer (Fig. 5-115). Reaming done right can produce a 32 microinch finish or better.

How much metal is removed during hand-reaming depends largely on the size of the hole, from 0.003 in. up to around 0.010 in. max is normal. With all reaming, by hand or by machine, larger holes usually call for greater amounts to be left for reaming, but there is a maximum reaming excess, at around 0.015 in. even for machine reaming.

Hand-reaming is often performed to resize a bushing after it is press fitted into a hole, or to remove paint or other coatings during final assembly or to put the final touch on

Figure 5-116 *Hand-reaming might be used to finish the holes in the drill gages after drilling undersize.*

holes, as shown in Fig. 5-116. The hand-reamer is mounted in a tap handle then pushed in while being rotated.

Skill Hints for Using Hand-Reamers

Always use cutting oil when reaming metals that are not easily machined, hard steel for example, a specific tapping fluid designed for that metal works well also.

Keep constant in-pressure on the reamer. If the reamer progresses about one quarter of its diameter into the hole per revolution, you are pushing about right. Reamers need steady progress into the hole to get good results.

The lead-in diameter (Fig. 5-115) will help ensure the reamer is parallel to the hole. Sight the tap handle from two directions to be sure before reaming very deep. Different from hand tapping, once the reamer starts right, it will tend to stay parallel and not wander.

Never turn it backward when in the hole—this will dull the cutting edges. When through, pull it out while turning it forward.

Square driver

Starting taper

Figure 5-115 *A hand-reamer features a square top to be used in a tap handle and a lead-in taper to improve the starting qualities.*

KEYPOINT

Reaming is a cutting process that removes from 0.003 in. diameter, up to a max of around 0.010 in. for hand-reamers and 0.015 in. for machine versions.

Honing

Honing is a secondary operation that finishes holes finer than reaming. It's an abrasive action used to improve diameter, roundness, and surface finish. A hole that's tapered or not straight can be corrected by honing, to bring it closer to a perfect cylinder. However, honing to achieve better straightness requires a lot of operator intervention, it's not simple to do. It's best if the hole is straight with no taper before starting to hone.

Although honing uses friction/abrasion rather than cutting, it's much slower than grinding, and it occurs at a very slow speed with no heat. In production machine shops, honing is performed on a machine specially designed for the task. In automotive machine shops, it's done using portable tools. Honing machines (Fig. 5-117) are common where extremely close hole tolerances and smooth finishes are a requirement, such as hydraulic cylinders, for example.

Honing is necessary when the metal must be hardened after drilling/reaming then hardened beyond machinability. The carbon scale on the hole's surface, resulting from heating the metal, must be removed by honing.

Honing uses a three-point suspension process: two floating pads that rub and oppose the abrasive stone and an abrasive stone on the opposite side that removes a tiny amount of metal. While rotating the honing head in the hole with cutting oil, the stone is forced out against the hole. It slowly abrades material to a very fine finish of around 8 to 16 microinches or better. The machine floods the cutting head and work with cutting oil to lubricate and flush metal particles.

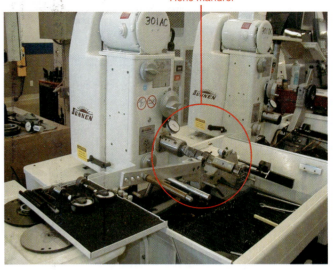

Hone mandrel

Figure 5-117 *Looking at the front of a hone head. Observe the two pads and the adjustable stone.*

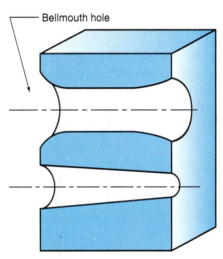

Bellmouth hole

Figure 5-118 *Preventing and correcting the bellmouth and taper are a constant challenge for a hone operator.*

Honing is a slow, hole finishing, abrasive process used to remove from 0.0002 up to a max of 0.003 in. of material from holes.

Honing removes far less metal than reaming: from 0.0002 up to a maximum of 0.003 in. with around 0.0015 in. being normal. Prior to honing, the hole would have been drilled and probably reamed. Honing is challenging in soft metal such as aluminum due to the loading of the stone, but it is possible. The constant challenge of operating a hone, is to correct taper and bellmouthing, as shown in Fig. 5-118.

Press Fit Assembly

Press fitting is a secondary operation used to permanently assemble one part into another. It uses friction to hold the part in place as a round bushing is pressed into a hole for example. While the object can be removed, press fitting isn't designed to be removed easily. Objects that are pressed in can be pressed out again, but it is considered a permanent assembly.

Amazingly, it is possible to force a pin or bushing that is slightly bigger than a hole into the hole! Why, because, metal stretches, but only a small amount.

A Typical Press Fit

For example in Fig. 5-119 we'll consider pushing a $\frac{3}{4}$ (0.750) nominal outside diameter bushing (Dash 2 part) into a reamed hole in the Dash 1. Either the bushing or the hole must have the press fit allowance built into its size. Usually that's not more than 0.001- to 0.003-in. difference in the two parts, depending on how tightly the press fit must be functionally.

Which Component Has the Allowance?

The **press fit** (abbreviated **P/F**) allowance is the amount of interference in size between components. This calculated amount of interference is usually built into the bushing because bushings are mass-produced and designed to fit into a common size hole. For example, a half-inch bushing might measure 0.5005 in., to be pressed into a 0.5000-in.-diameter hole. On the other hand, pins, due to their function, usually do not have the allowance built in and thus require the hole to be under the nominal by the interference amount.

Figure 5-119 *The press fit allowance (0.0015 in.) is built into the purchased bushing in this drawing. The mating hole then is produced by a standard 0.750-in. (3/4-in.) reamer.*

P/F bushings have the press allowance built into their outside diameter. P/F pins have the press allowance built into the receiving hole diameter.

Either way, the bored or reamed hole must be made for the correct press fit application. Select either a standard size reamer or a specially made undersize reamer for the job. Read the work order and print to know which part has the allowance and always measure components before press fit assembly.

How is the Press Fit Allowance Calculated?

There are three variables considered when computing a press fit allowance.

1. **Permanency of the Assembly Called the Class of Press Fit**
   This is the tightness based on the duty to which the component will be put.

2. **Nominal Size of the Objects**
   As the nominal size of the assembly grows larger, so does the P/F allowance required. For example, a 1.000-in.-diameter pin might use a 0.001-in. press fit allowance, while a 2.000-in. pin requires a 0.003-in. press. The amount can be found in machining references or other engineering references.

3. **Metal Alloy**
   Although a less important variable, different metals have different holding characteristics. For example, a bronze bushing might require a bit more interference in a steel assembly than a steel bushing pressed into the same hole.

Rule of Thumb—Press Fit Amounts

A permanent press fit is considered from 0.0015 to 0.0025 interference for the first inch in diameter of the objects to be put together. Then lesser amounts of interference are added

At the beginners level, the information you need to make a press fit assembly will be found in the work order and on the print.

per additional inch of diameter as the size of the object increases.

Heating and Cooling Metal for Press Fits

When interference amounts become large, the outer part is heated and the inner part is cooled to allow the parts to go together more easily. Heating one component while freezing the other temporarily reduces the force required. Then as the components' temperatures equalize, the fit tightens and becomes more permanent. The heat/cold method is called **shrink fitting** and is considered a nearly nonremovable press fit if both items are the same alloy because it would be difficult to reheat or cool individual pieces once they are in intimate contact.

Many objects made for a press fit will feature a short, *lead-in diameter* (Fig. 5-120). It will be smaller than the target hole size for a short distance so the object starts in with no resistance. This assists in alignment of the bushing or pin.

Machines That Perform Press Fits

Arbor Presses Whether hand or hydraulic, a press fit can be performed with vises and even hand driven tools, but press fitting is usually accomplished with an **arbor press** (Fig. 5-121). Several types are found in production, repair, and automotive machine shops:

1. Powerful electric-hydraulic
2. Manually pumped hydraulic

Figure 5-120 *This bushing has been specially made with a built-in press fit allowance and a lead-in diameter.*

ϕ 0.7515

ϕ 0.7480

Starting leader

Figure 5-121 *Arbor presses are used to force-assemble press fits.*

3. Hand-operated rack gear presses (Fig. 5-121)
4. Hand-operated screw presses

Skill Hints—Press Fit Assemblies

1. Although the parts are supposed to be the correct size, always do a quick check of both the hole and bushing before starting.

2. Make sure to place the lead end of the bushing or pin in the hole first!

3. A little lubricant such as grease or other high-pressure lube is recommended. Rub a thin coat on both the bushing and the hole. Check with your supervisor—some press fit assemblies cannot have a lubricant present.

4. Make sure the opening of the hole has a small chamfer, not a sharp edge that might scrape the bushing and possibly misalign it during assembly.

5. Make sure there are no burrs, dirt, or chips before attempting assembly.

6. After starting the bushing in, stop and check alignment similar to a tap. Don't attempt to push it all the way in without checking alignment. They can and do start in crooked by scraping material off the wall of the hole. This is difficult if not impossible to repair.

7. After pressing, measure the inside of the bushing. It will probably be slightly smaller due to the metal squeeze from the press. Bushings that close down depend on a relatively thin bushing wall and may require a final hand-ream.

UNIT 5-6 REVIEW

Review the Key Terms

SECONDARY OPERATIONS

Arbor press
A machine used for pressing and depressing assemblies.

Blind hole
One that does not go all the way through a part.

Bottom tap
Used to finish more threads in a blind hole.

Die
A threading for making a bolt or stud—outside thread.

Honing
An abrasive process that removes small amounts of metal to create a close tolerance hole of good finish.

Plug tap
Cuts closer to the bottom of a blind hole than a starting tap but not as close as a bottom tap.

Press fit (P/F)
Permanent assembly by forced insertion.

Shrink fitting
Using heat and cooling to lighten the pressure of a press fit.

Tap
A threading tool used to make internal threads—nuts.

THREADS

Class
The fit of a thread—how tight or loose it is to be.

Clearance
The deliberate space between bolt and nut to allow movement.

Coarse threads
General purpose threads.

Fine threads
Sometimes called SAE threads. Fine threads have a shallower helix and thus create more translation force but less distance.

Form
The shape of a particular thread. The most common form is the triangular thread for both ISO metric and Imperial Unified threads.

Helix angle
The angle at which a thread's inclined plane works.

Nominal diameter
The basic design size of the bolt. Nominal is the drawer size for the bolt or nut, but not the real size.

Pitch

The term used to express either the pitch distance or the threads per inch.

Pitch distance

The repeating distance of a thread. On Imperial threads it's found by dividing pitch number into 1 in. For metric threads, it is specified in the callout.

Threads per inch

Abbreviated TPI and often called pitch.

Translation

The distance a nut moves on a bolt with one turn—the result of the wedge being driven by turning the nut on the bolt.

Truncation

Clipping the top and bottom points off a 60 triangular degree thread.

Replay the Key Points

- Standard threads feature a coarse and fine series both for metric and Imperial sizes.
- Both metric and Imperial standard threads feature the 60 degree trianglar thread form.
- The thread designator tells the nominal size—the pitch or pitch distance—then the class (all separated with dashes).

- There is a slight difference between Imperial and metric designators in that the Imperial shows pitch number while the metric shows pitch distance.

Respond

1. Why do we truncate the standard thread form?
   - A. To save material—less metal.
   - B. To make the bolt smaller than the nut.
   - C. To remove the fragile tips that would break.
   - D. To lighten the bolt and nut.
   - E. None of the above.
2. Which of these threads would *not* use the 80 degree triangle form?
   - A. 3/8-16 UNC
   - B. M 12-1.25
   - C. 1/8-27 NPT
3. Describe this thread: 3/8-16 UNC-3.
4. True or false. This thread has a tighter (smaller) manufacturing tolerance than the one in Question 3, 3/8-16 UNC-2.
5. Between these two threads, which has the smaller individual threads: 1/2-13 UNC or 1/2-20 UNF?

Section 5 Review

Unit 5-1

Today's machinist is a manager as well as a craftsperson. There are many daily items that must be organized. And there are game rules—lots of them; to manage materials, to use and dispose of chemicals, and other systems for part numbers, dash numbers, and program revisions to track. While knowing the rules today in training won't make a great deal of difference, you're here to learn to safely and efficiently make chips, learning the rules and following them in industry will make all the difference in your career. In this unit, we've looked at the general system used in most factories and shops. But there are differences. Be sure to find out how your new employer processes

- Work orders
- Part numbers
- Revisions to the drawing and to related programs

Unit 5-2

One of the greatest obligations is to guarantee that the right metal is used to make the product. Metallurgy is a big subject, in fact it's an entire college degree. But we've started your knowledge bank using five selected metals:

- Steel
- Cast iron
- Brass/bronze
- Stainless steel
- Aluminum

Plus we've looked at the various physical forms in which metals are supplied beyond blocks, bars, and sheets:

- Castings
- Forgings
- Extrusions

But simply getting the right alloy in the right form isn't enough in some cases. The hardness and grain direction often must be included in the history of the material. In some situations such as transportation equipment, where lives are at stake, the metal's pedigree must be proven from smelter to the individual serialized part number! If not, the parts might be right physically, but they are illegal to use! Paperwork is the reality of our business but computers are making it easier.

Unit 5-3

Modern sawing equipment can shave hours from production. It's a skill well worth learning. When the saw is fine-tuned with the right blade pitch and cutting speed, and the blade guides are well adjusted, repeatability tolerances can be held within 0.020 in. or even less! Additionally, sawing removes large chunks of material from the raw stock, all at one time—an advantage that only saws have. All other machines make small chips from the removed material. That not only saves time but it makes recycling dollars and sense, too.

Unit 5-4

Part finishing is no fun! We all agree on that. But it must be done before the products can be delivered to the customer. Above all else, while removing burrs and smoothing rough spots in the machining, do no further damage to the parts. Be careful how you handle them and stack them. Protect parts from dropping or banging together. Do not mar them in any way.

Unit 5-5

Modern part marking has changed a great deal from the days when we stamped or stenciled part and serial numbers on the work. Today, we can use lasers to instantly cure powdered inks to the surface or the work or the high-intensity light can be used to modify molecular structure of the work surface without changing the material. The laser can remove colors from coatings to leave a bright, fine line of information. All this is possible with computer-directed part markers.

Unit 5-6

In previous machining textbooks, this subject was called benchwork. *The bench* was the place most of us began our apprenticeship. But today, the focus is more on supporting the CNC equipment to keep it producing work that only they can do. Work is only taken off the expensive machines when it removes a bottleneck in production. Still, the complete bag of machining skills requires the ability to perform:

- Hand threading with taps and dies
- Hand-reaming
- Press fit assembly

Questions and Problems

Are you ready to read work orders then perform secondary operations when they are specified?

1. List at least four vital kinds of information found on a work order and briefly describe each? (Obj. 5-1)
2. True or false? Part numbers are comprised of a detail number followed by a drawing number but separated by a dash. If it's false, what makes it true? (Obj. 5-1)

Questions 3 through 8, refer to the five basic metals: aluminum, cast iron, brass, steel, and stainless steel (Obj. 5-2),

3. Which has the highest machinability rating (most easily machined)?
4. Which is the dullest in color?
5. Which is the lightest in weight?
6. Which is often abbreviated to CRES?
7. Beyond bars and sheets, name three preshape forms that metal might be supplied in.
8. Of the two forms, which is the strongest and why, CRS or HRS?
9. Identify and describe the saw blade factor that creates a wider kerf than the basic thickness of the blade (the portion behind the teeth).

Identify two advantages imparted by this blade feature. (Obj. 5-3)

10. When power sawing on a vertical band saw, potentially dangerous tooth stripping occurs because the machinist failed to_____. (Obj. 5-3)
11. The common gray grinding wheel used on bench and pedestal grinders has its abrasive grains bonded together with fired clay. Identify the bonding process and abrasive type. (Obj. 5-4)
12. There are two methods of removing burrs and finishing edges on parts, that require no hand work by the machinist. Identify and briefly describe each. (Obj. 5-4)
13. List the five categories of part marking. Of these, which method(s) are not harmful to surfaces? (Obj. 5-5)
14. Explain the difference between the two numbers 20 and 1.5 in these two thread callouts.
   1/2-20   M12-1.5
   (Obj. 5-6)
15. In Question 14, what is the pitch distance for the $\frac{1}{2}$-in. thread? (Obj. 5-6)

Critical Thinking Questions

Your supervisor tells you "Make twenty 501B-3456-75 Rev J parts."

16. You have found a Rev K drawing of the correct number. Is it OK to use to make these parts?
17. In view of Question 16, what kind of parts are you most likely making?
18. There are three documents you'll need to do this job. Which are they and why these three?

CNC Questions

19. A program has been written to machine parts from stainless steel, however, you have been asked to make one prototype (experimental example) from aluminum to prove out the program.
   A. What editing can you do to the RPM?
   B. To the cut sequences?
20. Assuming the prototype part of Question 19 was inspected and found to be within tolerance for all dimensions, can you return to the original speeds and sequences and assume the machine will also make a good part from stainless?

Section 5 Answers

ANSWERS 5-1

1. To ensure that
   A. The right material has been cut for this job.
   B. The blanks are the right size to proceed.
2. Because the work is to be reamed to final size, as shown at operation 70.
3. Rev—New
4. Break the edges from 0.005 to 0.001 in. and file or machine out the rounded inside corner per the drawing note.
5. 3.200 in. by 5.200 in.

ANSWERS 5-2

1. A dull gray, heavy metal "casting"
2. White-silver, relatively light metal in a long thin complex shape
3. Bars, sheets, or extrusions but not forging or castings, they are prepped to go to the machines already.
4. Greater strength
5. You are looking for stainless steel—while stainless can be identified with tests like color, magnetic attraction, and so on, this is a controlled job requiring traceability—the only answer is by exact serial number/heat lot otherwise the material is illegal to use on this job! The exact stainless alloy was purchased specifically for this job—none other will do!

ANSWERS 5-3

1. The hacksaw, the vertical saw, and the shear
2. Aluminum would clog the wheel.
3. The shear
4. Yes
5. It might work but would leave a burr. The HAZ would exist but since this is low-carbon steel, it will not be hard. Also, if the excess is large it will remove the heated material.
6. Slow it to 100 ft per minute blade speed. You might be able to switch to slightly coarser pitch since the material is thicker.

ANSWERS 5-4

1. E—all can cause burrs
2. True, it is a mill file.
3. Soft metal
4. Either of the abrasive/slurry methods. They finish large batches. Send them out to a shop specializing in burr work if you don't have either method. You will save money over trying to do each part individually.
5. Files, air motor sanding, edge-breaking tools, pedestal grinder with soft burr wheel, both slurry methods but not economical in this case

ANSWERS 5-5

1. Any surface, line width, or font can be engraved.
2. To be certain the marks are in the prescribed location
3. Not bounce back thus they drive work into vises and chucks better
4. It is nondestructive, but not truly permanent.
5. Because, with use their head first mushrooms then dangerously splinters.

1. C—to remove the fragile points
2. None (semitrick question) These threads use a 60° triangle!
3. $\frac{3}{8}$-in. nominal size, 16 threads per inch, coarse thread series, and a class 3 thread
4. False. The Class 3 is more precise than the Class 2.
5. The 1/2-20 UNF

Answers to Section Review Questions

1. *Part numbers,* the specific product to be made and which print is used. *Revision level,* designs change—is the one you have the right dated drawing? Remember, you might not be making the latest design, these parts might be replacements for an older product!. *The sequence of operations* to completion of the job—the logical steps. *The material type,* what form and size is to be provided? *The standards* to which the product is to be built. *Instructions for quality assurance. Finishing packaging* and other handling instructions. *Special requirements* for tooling or other items needed.
2. False. A part number is a drawing number followed by a detail or dash number.
3. Aluminum
4. Cast iron
5. Aluminum
6. Stainless steel is abbreviated corrosion resistant steel (CRES).
7. Forgings, castings, extrusions
8. Cold rolled steel (CRS) is stronger due to the cold working.
9. *Blade set* cuts a wider kerf than the blade basic thickness. It is the bending of teeth outward to prevent rubbing and allow scroll sawing (curved) cuts.
10. Select the correct blade pitch—it is too coarse. At least three teeth must be in contact with the work at all times during a cut.
11. The vitrified, aluminum oxide grinding wheel
12. *Slurry vibration*—Parts are put into a rapidly vibrating vat containing specially shaped stones or abrasive grains in a liquid media with added cleaners; tumbler—parts are put into a rotating cylinder containing a similar medium as just described.
13. Stencil/rubber stamp/screen printing—All apply a nonharmful ink or paint; laser engraving; steel stamping; engraving; electrochemical etching.
14. The Imperial bolt has 20 TPI, while the metric thread has a pitch distance of 1.5 mm.
15. 0.050 in. (1/2-20 = 0.050 in.). Divide 1.0 in. by the pitch number.
16. Maybe. It depends on what changes were made to the Dash 75 when the drawing was updated to Rev K from J. The only way to tell for sure is for the planner to search backward.
17. You must be making customer replacement parts for existing products since there exists a later drawing revision. You were told to make them to an older design. Or there is a possibility of an error in the instructions—they simply told you to use the wrong revision level drawing.
18. 501B-3456 Sheet 1 at Rev J for the general notes and tolerances. 501B-3456-Sheet? at Rev J for the Dash 75 detail. Since there appears to be many parts under this drawing number, it's likely that it's a multisheet drawing. Work order with planning to machine the Rev J parts—not K. Again a planner's issue.
19. A. Speed up the RPM from three to ten times faster or more depending on the strength of the cutters.
    B. Stainless, being low on the machinability scale, will require more cuts, taking less metal per pass than aluminum. Although it's a challenge, you "could" edit the program to make bigger cuts in aluminum. *However, keep in mind your objective was to prove the program, it would be best to leave it in its original sequence!*
20. You have proven the dimensions in the program are good, which is a good sign that the program will run right in stainless. But the aluminum prototype did not prove the cutting action in a much more difficult material. We're talking machinability differences here. The stainless may not cut the same way as the aluminum.

Section 6

The Science and Skill of Measuring— Five Basic Tools

Introduction

In our past, delivering quality work meant a lot to the machinist's career. It was the pathway to advancement and to keeping happy customers. But in today's world of Internet commerce, rapid transoceanic shipping, and free trade, quality products, along with price, mean nothing less than survival! In machining, quality means accuracy; it's a do-or-die issue.

Here in North America, we remain strong in manufacturing not by working cheaper but by being more efficient and by delivering quality. To be competitive, today's machinist must be able to measure in five different ways using a variety of instruments, all of which you will learn about in Section 6 and in Sections 7 and 8.

Getting it right requires more than just studying the measuring tools. The fine-tuning happens after you learn about their purposes, the names of their parts, and how to read them. The best accuracy comes about with six kinds of skills.

Knowing the Measurement Task Understanding tolerances and which tool matches the exactness of the task.

Learning to Identify Inaccuracy Factors Understanding what might spoil a measurement using a given tool.

Controlling the Inaccuracy Factors Learning how to reduce inaccuracy factors to a minimum and how to choose the instrument and process that meets the need, and has the least inaccuracy factors.

Uncompromising Attitude Of the greatest import, mastery happens only when students develop an uncompromising attitude about accuracy from the start. You don't get that from a book.

Practice-Practice-Practice   Using the tools and processes in a true research fashion, with the goal to always refine measured results.

The Scientist's Approach   To develop the greatest ability in the least time, approach learning as a researcher might, studying the principles then participating in experiments to "zero in" on accuracy. Section 6 offers exercises that set that lifelong pattern of awareness and control.

Objectives

UNIT 6-1

Dimensions and Tolerances

Introduction: When machining a feature, the objective is to produce the specs found on the print, called the *nominal size* (meaning expected size). However, while perfection is the objective, variation is the reality. When a feature is measured and found to be different from nominal, the difference is called *variation*. Undesirable but the reality of machining, variation must lie within the specified tolerance, otherwise the part is scrap or needs to be remachined.

While the objective is to eliminate variation, it is inevitable that some will occur. This could be due to operator actions on the machines, tool wear, heat, mechanical movement of the machine, and hard to detect factors such as variation in the raw metal used. Also, there are times when we deliberately aim for the safe side of the tolerance—known as hedging. For example, in drilling a 1.000-in. hole with a size tolerance of plus or minus 0.005 in., a plan might aim at 0.997 in. That leaves 0.007 in. for unforeseen variation.

KEYPOINT

We can't eliminate variation but we can control it. The idea here in Sections 6, 7, and 8 is to not introduce needless variation *due to choosing the wrong measuring process.*

KEY TERMS

Bilateral tolerance
A variation range distributed equally around the nominal dimension.

Geometric angular tolerance
A linear variation built around a perfect model that lies at an angle with respect to a datum.

Limits
A tolerance method where the range is expressed as the maximum and minimum size usually with no nominal shown but it can be added.

Nominal
The target dimension on a print, for a given feature.

Protracted angular tolerance
An angle originating from a single point, creating a fan-shaped tolerance zone.

Tolerance
The range of variation that the design will tolerate and still work right.

Unilateral tolerance
A variation range extending in only one direction from the nominal.

Variation
The difference in the produced result and the nominal.

Five Kinds of Measuring in Machining

There are five aspects to measuring in a machine shop. Ability in all five will be immediately needed to complete your first few assignments in the lab. For example, consider drilling a hole in a part. You will need to measure (Fig. 6-1) the

◆ *Size*—We mean the diameter and/or depth of the hole.
◆ *Position*—Is the hole located correctly with respect to datums?
◆ *Form*—This concerns the shape limits. Form is controlled to some degree by size

Figure 6-1 *A source of pride: five basic measuring tools.*

tolerance, as a subset of size. However, the design could require a hole to be far closer to perfect roundness than its size tolerance range. Form is an embedded control within size.

◆ *Surface Finish*—Machined objects are given a max roughness in microinches. This specifies how deep the microscopic marks left behind by the machining action can be.

◆ *Orientation*—Is the hole perpendicular or parallel to a given reference datum? Orientation is also a subset of position. But it can be controlled separately as an embedded control.

Tolerances on Drawings

There are several ways a **nominal** dimension and **tolerance** might be called out on a print. The important issue with regard to measuring is to find the tolerances that apply to your task and know how "tight" they are. Then choose the right measuring tool and process to deliver the needed results.

Figure 6-2 shows several mixed examples of nominal dimensions and their given tolerances.

Four Kinds of Linear Tolerances

In Fig. 6-2, four different means of expressing linear tolerances are shown (straight-line distance). Examples are a thickness, a diameter, or a distance from an edge to the center of a hole.

1. **Bilateral Tolerance**
   On the right, the 5.00-in. dimension has a plus or minus 0.005-in. tolerance. The design will tolerate from 4.995 to 5.005 in. to either side of the target—a total range of 0.010-in. permissible **variation.**

2. **Unilateral Tolerance**
   At the top, the 1.50 nominal dimension is expressed as a minus 0.010-in. variation but no plus variation is permitted. Unilateral

Figure 6-2 *Different types of dimensioning and tolerancing.*

tolerancing means extending in one direction only from the nominal.

3. Tolerance Expressed As Limits

On the bottom we see a limit tolerance. It actually has a nominal dimension but it is imbedded in the middle of the range. The range remains 0.010 in. for variation and the nominal will usually be the middle but not always, depending on circumstances. If the nominal target is critical, then it can be inserted between the **limits.**

4. General Tolerances

A handy method is used when the object is very complex with lots of dimensions. The tolerances are compiled in a table near the bottom edge of the drawing. Then, the way the nominal dimensions are stated indicates what the tolerance is. Notice the box on the lower left of Fig. 6-2. It tabulates the code for exactness for any dimension not given an individual tolerance on the drawing.

ShopTalk

Statistical Process Control (SPC) and PC-Based CNC Controls Ask yourself why some automobile engines last for long periods while others fail within a short time. The answer is complex, of course, but a large part of the answer lies in how close to nominal the engine block was machined. Each engine was inspected to be within tolerance when shipped. But those that hold up for a long time most likely had near perfect parts. That's only possible when a machine shop knows exactly what it is capable of doing for accuracy. It takes more than smart machinists.

Section 27 presents an overview of a mathematical method used to determine the normal variation in any process then to track the process in real time to be certain the produced results are within the limits of normal—not within the print tolerance, but inside the capability of the process if everything is going right. You could apply SPC to serving hamburgers, to making clothing, or to any process that has an identifiable quality outcome target. Machining is a prime use of this mathematical tool.

Using a PC-based CNC control (or any PC), SPC highlights trends before any scrap is made. The CNC control can track all data input either by manual data entry (MDI or through automatic downloading of electronic tools). Today, machines even have the ability to make their own measurement. SPC also helps management make decisions about equipment and tool purchases and jobs to bid that will fall within their capabilities.

Don't miss this concept in Section 27. Be assured, the nations we compete with for customers use it to the max!

Dimensions that are noted on the drawing using two decimal places in the example print are *X.XX* (for example, 0.25 or 0.37) are given a tolerance of plus or minus 0.015 in. Where the dimension requires a tighter tolerance, the code is carried out to four places on this drawing: *X.XXXX* (for example, 0.2500 or 0.3700). The drawings show just the part to the right of the decimals because that is the tolerance code.

Can you find a dimension that has a tolerance of ±0.001 in. on the print?

Angular Dimensioning and Tolerancing

There are two very different methods of noting angles on prints, and these have different kinds of tolerancing: **protracted angles** and **geometric.** Each is measured differently but using the same tool.

Protracted Angles

Figure 6-3 expresses the angular tolerance to be plus or minus 2 degrees. Expressed this way, variation is measured with an angle-measuring tool called a protractor. Protracted angles start from a single point and radiate outward. They yield a fan-shaped tolerance zone that totals 4 degrees, as seen in the illustration.

Geometric Angles

The second and more useful way of dimensioning and tolerancing angles in terms of assembly function, is called the geometric method (see Fig. 6-4). A geometric angular tolerance range is a pair of parallel lines built around the perfect definition of the angle. In this case, 59 degrees. This yields a tolerance zone that is a parallel sandwich not a fan. A geometric angular tolerance is actually linear.

The real machined surface must lie within the sandwich lines, but it may vary in any way possible. It is measured by placing an exact angular gage against the surface and checking for

A protracted angle and tolerance

$59°\ {}^{\pm 2°}$

57° Minimum

61° Max

Figure 6-3 *Protracted angular tolerance yields a fan-shaped zone.*

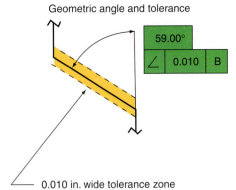

Geometric angle and tolerance

59.00°

∠ 0.010 B

0.010 in. wide tolerance zone

Figure 6-4 *A geometric tolerance for angularity is linear.*

gaps exceeding the tolerance. While this gage will be the protractor again, its use is different. This time, it's locked in position similar to a square, then gaps are checked between the blade and the object, always with datum reference as we learned in Section 4.

The geometric definition of angles works far better for the assembly of two parts where a maximum space between the two is of concern. When gap is the function, using a protracted, fan-shaped tolerance makes control difficult, while the geometric version is quite simple.

UNIT 6-1 REVIEW

Review the Key Terms

Bilateral tolerance
A variation range distributed equally around the nominal dimension.

Geometric angular tolerance
A linear variation built around a perfect model that lies at an angle with respect to a datum.

Limits
A tolerance method where the range is expressed as the maximum and minimum size usually with no nominal shown but it can be added.

Nominal
The target dimension on a print for a given feature.

Protracted angular tolerance
An angle origination from a single point, creating a fan-shaped tolerance zone.

Tolerance
The range of variation that the design will tolerate and still work right.

Unilateral Tolerance
A variation range extending in only one direction from the nominal.

Variation
The difference in the produced result and the nominal.

Replay the Key Points

◆ All dimensions must have an associated tolerance.
◆ The target for the dimension is called the nominal.
◆ Linear tolerances divide into four types:
   ◇ Bilateral, equal distribution around the nominal
   ◇ Unilateral, variation permitted in one direction from nominal
   ◇ Limits, the range of variation shown with no nominal
   ◇ General, tolerances not shown on the print but tabulated in a box
◆ Angular tolerances are either geometric or protracted.
   ◇ Protracted tolerances are stated in degrees. Protracted tolerances produce fan-shaped tolerance zones—measure with angular tools called protractors
   ◇ Geometric tolerances are stated as sandwich-like zones around a perfect model of the angle.
◆ Datums expressed on drawings are starting points for measuring and they must be heeded for correct results.
◆ Any feature can be used as a datum and must then be accounted for when measuring other features relating to it.

Respond

1. Variation stems from what obvious sources? (Name three.)
2. What are tolerances?
3. Nominal is the range of tolerance permitted. Is this statement true or false?
4. Linear tolerances can be bilateral, unilateral, and _____ (two more).
5. When there is no tolerance associated with a dimension, it will be found _____.

UNIT 6-2

Managing Accuracy

Introduction: Resolution and repeatability are two vital terms that define accuracy in our trade. It is important to know the difference between them and to understand how they affect results with any given measuring tool.

Resolution is the finest graduation on the tool or the finest possible controlled movement of any precision tool including machines. For example, one micrometer may be graduated with 0.001-in. divisions, while a second micrometer is graduated to 0.0001 in. The second has 10 times finer resolution.

Resolution is the result of the way the tool is made.

Often, a skilled user can measure finer than the resolution of the tool. For example, when using a 0.020-in. graduated ruler or the 0.001-in. graduated micrometer, experience will show that it's possible to "read between the lines." While that improves accuracy to some degree, these are skilled estimates. In terms of repeatability, it's far better to find a measuring tool or process with a resolution finer than the job tolerance. We machinists call that the *decimal point rule*—if possible, use a measuring tool or process whose resolution is 10 times finer than the job tolerance.

Repeatability is a combination of many factors. Repeatability is the overall trusted accuracy of the tool time after time. But most importantly, tools need users; the repeatability that finally counts is when it's in your hands. *Personal repeatability* is the objective of this unit.

There is no absolute answer as to the best repeatability. However, buying quality measuring tools, taking good care of them, understanding their inaccuracy factors (next), and practice combined with habits gained over time are the vital ingredients. When describing the tools in this Section and Sections 7 and 8, I will provide a target repeatability for each. Use it as a goal. It might be difficult at first, but it's easy to achieve with practice.

KEY TERMS

Calibration
Testing the tool against proven standards to adjust out or remove all error and to certify that it agrees with those standards.

Coefficient of expansion
The amount of linear growth with each degree of temperature change per unit of length.

Feel
The amount of pressure that must be applied to a measuring tool to be coordinated to the calibration.

Parallax
Not viewing a measuring tool from the correct perspective, which can cause errors in reading the tool.

Repeatability
Delivering the same results time after time.

Resolution
The way a tool is made; its smallest graduations or movement.

Standard (gage)
A master comparison tool that is certified to be a given dimension. A standard is used to test and correct error in shop measurement tools and it can be proven to be theoretically perfect within certain tolerances, depending on level of certification.

Ten Inaccuracy Factors

In all five kinds of measuring, there are spoilers that can creep in and degrade results. Each measuring tool we study has its own combination that determines the tool's repeatability. As an example, suppose a job tolerance of plus/minus 0.001 in. An electronic caliper reads to the nearest half thousandth (0.0005 in.) on the screen. But can you actually trust the measurements to this accuracy, time after time?

The answer is no; especially not at the beginner level. Why? This is partly due to inexperience, but it's also because of the tool's built-in inaccuracies. Calipers must be closed by hand pressure, that along with alignment of the tool to the surfaces to be measured are the two big ones. This tool has a "feel factor" and an "alignment problem" and probably isn't the best choice for this tolerance.

The Ten Spoilers

This list is arranged in order of common occurrences for all measuring tools but every process and tool has its own special combination. Note, tool quality is not listed as we assume professional tools are being used.

1. *Pressure* or **feel**
2. **Alignment** to the object
3. **Dirt and burrs**
4. **Calibration,** which involves testing the tool against proven standards to adjust out or remove all error and to certify that it agrees with those standards
5. **Parallax,** the error introduced by not sighting the tool from a correct perspective
6. **Tool wear**

7. **Heat**
8. **Damage** to the tool—bent, sprung and so forth
9. The **measuring environment** a place that is organized, clean, with good lighting, and a comfortable safe place in which to take the measurement
10. **Bias for a result**

The trick is to recognize which factors are greatest for a given tool and situation then control each. Here's how.

Getting a Handle on Pressure and Feel

Your sensitivity is the real challenge to many measuring tools and processes, even digital electronic instruments. Tools that are hand calibrated often require a feel. That is, they must be used with the same touch or pressure in measuring as that to which they were calibrated. The machinist must have a sense of that pressure. All craftspeople soon develop different sensitivities and preferences. Thus it's critical to either calibrate a particular tool yourself or that you test and "get the feel" of it if you did not calibrate it.

When selecting a measuring tool from a tool room, always test its feel against a standard. First measure an object that is an exact, known size—called a **standard** or sometimes a *gage*. This will check the calibration of the tool while showing its feel at the same time.

Some feel-type tools require no standard for testing such as the 0- to 1-in. micrometer or the 6-in. caliper. Both tools are closed to their zero position upon themselves, then are read to test their reading accuracy and feel. Tools of this sort are called *self-zeroing*. We'll see others in Sections 6, 7 and 8.

Gaining the All Important Sensitivity

Controlled experiments and practice are the keys. A good experiment to develop sensitivity is to measure an object that has a known size (but one that is unknown to you at the time to avoid biasing the results). Upcoming activities offer opportunities to practice this. Another method is to compare your results to those of a more skilled person(s). Third, measure any object using a given tool and process, then find a better process—one that has fewer inaccuracy factors, and make your own comparison. However, this third method won't be available until you learn several measuring tools.

Different footprints change the feedback

Full-contact round footprint

Straight line

Single-point footprint

Figure 6-5 *As part shape changes, pressure feedback challenges vary.*

Improving Tool Alignment

The second most common inaccuracy factor is getting the measurement along the axis intended. For example, when measuring the thickness of any object, if the measuring tool isn't aligned (perpendicular to the sides), it could be measuring an incorrect distance, as shown in Fig. 6-5.

Control for this factor is exactly the same as for pressure. In fact, you will be correcting both factors at the same time—measure and compare, measure and compare.

Different tools have different ratios of feel and alignment challenges. Differently shaped objects present different challenges, too. For example, when measuring the thickness of the three objects illustrated using a micrometer, the contact area changes, thus affecting your perception of the amount of pressure being applied.

The Easy One to Prevent: Dirt and Burrs

Although listed as third, for inexperienced machinists this might be first. The smallest burr can be from 0.003 to 0.005 in. Dust and atmospheric grime can account for another 0.001 error—this factor matters! The measuring faces of a tool become contaminated simply sitting on a workbench. They should be cleaned often using lint-free towels or clean paper.

KEYPOINT

Keeping tools in their case when not in use is one simple solution to preventing a buildup of grime.

Setting the Tool to Zero and Calibration

Zeroing index error (shortened to zeroing) and calibration are two different acts. Both terms must be understood by the machinist.

Zeroing Index Error Many but not all tools have a method of adjusting out constant error in the readings. Time, heavy use, wear, daily temperature changes, and shocks such as dropping the tool or allowing it to quickly heat up from contact with a hot metal part can all change the point where the tool reads zero (or its lowest value). Micrometers and calipers are such tools. When closed upon a standard (or upon themselves), they should read zero at their index line (training coming up). If not, the savvy machinist must stop and clean them, then adjust the error out of the tool. We'll see how in a few pages. Watching for tool error is a constant requirement, because they do get out of adjustment.

KEYPOINT

If you do use a tool that is wrong to make parts and they do not pass inspection, it is your fault no matter whose measuring tool it was that you used.

Calibrating Measuring Tools Knowing when and how to calibrate measuring tools is part of a journeyman's job. Calibration is a formal process. It becomes a supercritical requirement in some industries where people's lives depend on the exactness of the parts produced—so much so that the shop may employ a specialist to perform the calibration, or more commonly, they send their tools out to an independent lab to certify them. Calibration certifies that the tool reads perfectly all along its entire range of measurement. That certification is a legal document with traceability all the way back to national standards. A certified tool bears a sticker that shows the date of calibration, the lab that performed it, and when it must be recertified. Using the example of the micrometer, to calibrate it means to first zero the index error then compare its readings all along its range to various size standards.

To ensure constant and total calibration throughout the company, many quality conscious shops require master tools be tested on a schedule. Whether personal tools are certi-

Figure 6-6 *Be on the watch for a micrometer out of calibration.*

fied is a matter of shop management philosophy. Some shops do not allow personal tools in critical machining areas so that they can control the total calibration of their facility!

KEYPOINT

The mic in Fig. 6-6 was taken directly out of our lab's tool room—showing nearly 0.003-in. index error. The calibration of this tool is very suspect at this point. It might be fine, but after zeroing the index error, it should be tested against a few different size standards, 0.5000 and 1.0000 in. for example, to prove its calibration.

Trade Tip

Ask Before Adjusting! Before resetting the zero on any school or industry measuring tool, always ask! There may be a policy that bans doing so.

KEYPOINT

Detecting calibration error has two stages.

A. Check against a standard.

B. If wrong, stop and zero the error out right then—then test it against the standard again.

The purpose is not only to zero error but to refresh the feel/pressure at zero as well. While it's possible to add or subtract incorrect index errors until you can *get around to fixing the problem,* it's unprofessional to do so unless there is absolutely no time to do otherwise.

The Perspective Problem— Parallax

This is a visual problem that's easy to eliminate in some instances and difficult to avoid in others. It's caused from not sighting the tool straight on when measuring and reading it. As an example, think of yourself as a passenger in a car looking over at a classic sweep-hand speedometer needle. Would the speed appear at its true value or would you see it as slower or faster from your perspective? That perspective, sighting at an angle past the needle to the numbers, would make it appear as though the car were going slower than reality. That's parallax error (see Fig. 6-7).

Parallax affects either the numeric reading, as in this example, or it changes the alignment of the tool to an object such as a ruler to a part edge. Every now and then, due to a machine guard or the parts of the work-holding setup or deep bored hole, for example, we find ourselves in measuring situations where the viewing angle cannot be straight on. There are four possible control solutions:

1. Reach in "blind," get the measurement, and lock the tool in position. Pull it out to where it can be read without parallax. (The bad news is that locking often changes the reading.)

2. Read it in place and account for parallax. (In the car example, you might add 5 miles per hour to the reading—that's risky.)

3. Find another tool, process, or position so you can read at the correct viewing angle.

Figure 6-7 *This micrometer's edge is thinned to avoid parallax error.*

Thin edge helps avoid parallax errors

4. Anticipate the problem from the start. Avoid making setups that create these impossible-to-measure scenarios. This skill is usually learned by trial and error, over time.

Trade Tip

Electronic Measuring Tools A major advantage of electronic, digital tools is that they eliminate parallax factors with clearly read digital results.

Accounting for Tool Wear

The real problem here is undetected wear, especially when it is uneven. An example would be a micrometer used to measure round, machined pins, from say 0.250 to 0.450 in., repeatedly for a year. That area of the tool is worn in one area while the numbers out of this range have nearly no wear.

While there is a backlash adjuster to compensate for wear in most micrometers, for what is it set? If the worn part is corrected, the unworn sections will be too tight for use. Uneven wear usually means relegating the tool to a less important role within your toolbox. Many machinists keep two or more micrometers for this reason—one for daily use where tolerances aren't tight and one for extra precision tasks.

Another example of tool wear can be the faces of micrometers where they constantly rub against the work. They become dished or rounded. The most common example is rounded corners on rulers whereby the end zero becomes unclear. For rulers, I bet you already know the fix for this problem? Don't start measuring from the ruler's end but rather start from the 1-in. line.

Controlling Heat

Machining causes heat, which expands metals. If a part is significantly above room temperature, the measured size will be larger than when it returns to normal.

A predictable growth factor can be used, called the **coefficient of expansion.** Found in an engineering table, it is given in expansion units per degree, per inch of length. Every material has a different coefficient of expansion. Figure 6-8 shows two metal bars, one aluminum and one common steel. Notice they

Figure 6-8 _As metal heats, it expands, but various alloys expand at different rates._

Shrink Fits This expansion concept is useful when assembling parts with press fits too heavy to perform at room temperature (Section 5). For example, a large ring gear is fit to the rim of a flywheel. The only way to get the ring on without breaking it is to heat it while cooling the flywheel (or leave it at room temperature). When put together quickly before the heat equalizes, they slip together with little or no force. Then when temperature is equalized, the ring will tighten and not come off unless it is heated again. Using heat differentials to assemble work is called _shrink fitting._

both start at 15 in. long at room temperature but as their temperature rises, they grow at different rates. The aluminum expands faster!

Depending on tolerances, metal type, and overall size, expansion might be discounted as a problem since in the example we see only 0.033-in. growth in 250 degrees of a 15-in. object! However, where print tolerances are less than 0.002 in. and the machined object heats over 40 to 50 degrees beyond room temperature, you must do something about expansion and measuring. The control can take two different paths.

Solution 1: Compensate for Heat Measure the object and its temperature then mathematically compensate for the heat. Use the table of expansion values found in _Machinery's Handbook®_ by Industrial Press. Search the CD or look under the _coefficient of expansion_ for various materials within the printed index.

For example, in Fig. 6-8 (from _Machinery's Handbook®_), aluminum has a coefficient of expansion of 0.00001244 per inch unit per degree. A bar 10 in. long will lengthen 0.0124 in. when heated 100°F, while

a steel bar of the same length will expand 0.0063 in. over the same temperature change.

Multiply the change in temperature times the length, times the coefficient of expansion. The problem with this correction method is that an accurate thermometer must be present.

$$\text{Expansion} = \text{Coefficient of Expansion} \times \text{Temperature Rise} \times \text{Length}$$

Solution 2: Cool the Part or Reduce Heat in the Machining Process Skilled machinists reduce the creation of heat in the work by adding coolants, changing tool geometry, and adjusting machine speeds and force. All combine to reduce heat to the point where it has little affect on measurements. This is the more plausible solution within your control.

Heat in measuring tools themselves also has the potential to be a problem, but control is not difficult. Don't hold them in your hand for a long time and keep them out of the sunshine or away from other sources of heat. Heat in the machined object is usually a much greater problem.

Detecting Damaged Tools

Similar to wear, this inaccuracy factor is often hidden. For example, if the frame of a micrometer is closed using too much force, although stout, it can bend and the faces can become out of parallel (Fig. 6-9). The sprung tool will appear to be calibrated—that is, it will read zero when closed all the way, but, as seen in Fig. 6-10, it would be far from accurate on real measurements. With tools of this kind, check for damage by closing calipers

Sprung micrometer frame causes tip misalignment

Figure 6-9 _The frame of this micrometer is sprung, causing tip misalignment._

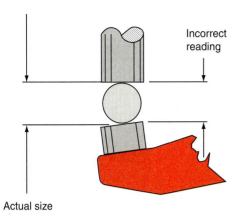

Incorrect reading

Actual size

Figure 6-10 *Results from the micrometer of Fig. 6-9 could not be trusted.*

and micrometers, then look for light gaps in the faces. But there are many other kinds of damage, too numerous to list, that must be watched for as well. The key point here is to not take tool accuracy for granted, inspect your tools.

Take Charge of the Measuring Environment

Accuracy is improved when performed in the right setting or perhaps better put, is degraded in the wrong setting. That means

Comfortable body positions

Clean parts with no burrs

Exceptionally good lighting, for example, if shadows cause distortion, bring in a portable light or better keep a work light as part of your tools

A good magnifier is useful no matter how good your vision might be. Clean up chips, burrs, and dirt before measuring and keep measuring tools safe and clean. It is convenient to lay calipers or micrometers on the machine but not professional! Keep them in their case away from the danger and dirt of the machine and from extra heat which will expand them. Develop an organized, patient approach to measuring.

Bias—The Built-In Error

This factor is psychological but real. If the machinist starts out to prove the part is right, rather than to measure it, errors happen. This isn't a matter of integrity, it's human nature to "make the reading" rather than "get the reading." In other words, without conscious thought, your hand

will either tighten the tool too much or too little, if that's what it takes to make the part the correct size.

The main control of bias is to be aware that it does exist, then check your actions to avoid over- or undermeasuring. But one can also look away as the final adjustment is made, then look back.

Options for Control and Accuracy

As you can see, other than damaged tools, machinists must develop control of all the various factors. Once detected, the damaged tool must be repaired or discarded. Feel, calibration, and alignment are the major challenges to accuracy but practice solves them. Once calibrated, a tool should remain so for a long period with normal use. The main control is simple awareness of the factors.

In today's integrated shops, the ultimate instrument is the CMM, the computer coordinate-measuring machine (Section 28). It's at the top of the evolutionary process. The good news is that CMMs eliminate all 10 errors studied, including heat in the work. If the computer knows the work temperature, it can automatically compensate readings. The bad news is they have their own list causing an uncertainty in the sub-ten-thousandth-in. range.

KEYPOINT

As we proceed to discuss the various measuring tools notice that there are processes with fewer factors. For example, a process that measures with less feel factor is going to be more reliable than one that requires more. Not only do tools vary in their factors, but ways of using them also varies. Sometimes there is a simple method and a more accurate method of measuring the same object and perhaps several other methods in between. The skill is to match the job tolerance to the tool, the process, and your personal repeatability.

Review the Key Terms

Calibration
Testing the tool against proven standards to adjust out or remove all error and to certify that it agrees with those standards.

Coefficient of expansion
The given amount of linear growth with each degree of temperature change per unit of length.

Feel
The amount of pressure that must be applied to a measuring tool to be coordinated to the calibration.

Parallax
Not viewing a measuring tool from the correct perspective, which can cause errors in reading the tool.

Repeatability
Delivering the same results time after time.

Resolution
The way a tool is made, its smallest graduations or movement.

Standard (gage)
A master comparison tool that is certified to be a given dimension. A standard is used to test and correct error in shop measurement tools and it can be proven to be theoretically perfect within certain tolerances, depending on level of certification.

Replay the Key Points

◆ Resolution of a measuring tool is the smallest readable graduation.
◆ Repeatability is your ability to deliver consistency.
◆ Every tool has its own list of inaccuracy factors.
◆ There are 10 possible factors that effect measuring:
  ◇ Pressure-feel
  ◇ Alignment to the object
  ◇ Dirt and burrs
  ◇ Calibration
  ◇ Parallax
  ◇ Tool wear
  ◇ Heat
  ◇ Damage to the tool (bent, sprung, and so forth)
  ◇ Control of the measuring environment
  ◇ Bias

Respond

1. Calibration error occurs when viewing a measuring tool from the wrong angle or perspective. Is this statement true or false?

2. Name two ways a conscientious machinist can control biasing results.

3. Name the two ways to deal with machining heat.

4. Describe three possible methods to help reduce pressure and feel-type errors. To zero in on accuracy. (There are four possible answers.)

5. What can you do to the measuring environment to improve accuracy?

UNIT 6-3

The Five Basic Measuring Instruments

Introduction: These are the tools you will need immediately in your training or on the job. Each has its own factor list. Start now to teach yourself right; after learning about the tool, dig into experiments with it.

Buying Your Own Tools There is a real advantage in knowing the feel of your own tool. In this unit, as we look at a new instrument, I will point out the necessary features and three levels of priority for tool purchase:

Priority 1—Buy as soon as you can. A P-1 tool is fundamental to your kit. Four of the five tools discussed here are priority 1 tools.

Priority 2—Buy after employment, perhaps within six months. Some companies have purchase plans set up with local suppliers for this.

Priority 3—Buy only after you are sure of the direction your career is going or perhaps not at all. Priority 3 tools are often supplied by the shop but are useful as personal tools on a case-by-case basis.

KEY TERMS

Cosine error
False readings caused by tilting the travel indicator's axis away from the direction to be measured—causes amplified results

Co-Linear
When two lines coincide (line up). The way a vernier scale is read.

Discrimination
In measuring, the ability to perceive a difference in two or more graduations on a given tool.

Nominal zeroing

The ability of electronic tools to set any point as zero—thus measuring a large quantity of parts by reading the variation only.

Progressive error

The error created by a test indicator's probe swinging in an arc yet reading in linear thousandths of an inch.

Vernier scale

A mathematical method of improving tool resolution by creating divisions one unit smaller than the real graduations.

Basic Tool 1—Rulers

Although they are always in the machinist's pocket, you may not have considered rulers to be precise instruments—think again! With practice, it's not only possible to measure to a repeatability as close as 0.005 in. using a ruler (even closer in some cases), it's expected of the journey machinist! Some tips for using a ruler well are shown in Figs. 6-11, 6-12, and 6-13.

> **KEYPOINT**
>
> Due to a difficult location or odd shape, once in a while, rulers are the only tool to do the job.

Rules are used as a fail-safe, backup check of any other process or tool. For example, after measuring a diameter with a micrometer, do a spot check with the rule. This double check avoids the "decimal place" errors such as 0.100 in. big or small.

Tech-Specs for All Rulers

Rule measurement is common for tolerances of 0.015 to 0.030 in.

Figure 6-11 *Alignment to the distance in question is a must.*

Figure 6-12 *"Split" the lines on the rule.*

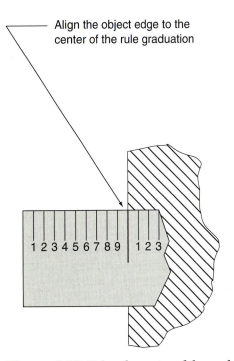

Figure 6-13 *Using the center of the graduation line makes measuring more accurate.*

Target Repeatability = 0.005 in. or 0.1 mm

This is the expected result of careful measuring with a rule after some practice and experience. You'll need an excellent rule, good lighting, practice, and perhaps a magnifier to achieve it.

Rule Inaccuracies

Be on the watch for

A. Alignment to the measurement

B. Parallax

C. Burrs and rounded part edges

Figure 6-14 *A pair of machinists' flex rules. Choose satin chrome or black finish, both excellent reading rules.*

D. Poor lighting
E. Damaged rules—especially the corners

Three Types of Machinist Rule

Type 1. Flex Rule These are 6, 12, or 18 in. long (150 or 300 mm). Flex rules are thin, universal tools (Fig. 6-14). The 6- or 12-in. rule is a priority 1 purchase. Longer versions are a priority 2.

Figure 6-15 *A spring or flat rule is useful when used with a hook attachment.*

Type 2. Spring and Hook Rules Sometimes called flat rules, these are about twice as thick and wider than a flex rule. These sturdy little rulers feature the same graduations as flex rules. Their advantage beyond rigid strength and long life is that some feature an attachment called a *hook*. It's useful in reaching through or across an object to measure its thickness. The upper rule in Fig. 6-15 has a drill point gage attachment used for resharpening bits.

Type 3. Ground or Rigid Rules Because they are used in several trades, several levels of accuracy and quality can be found for this rule in tool stores. Only the best are used in machining. Made of high-quality spring steel in 12- and 18-in. lengths, they are a priority 1 for purchase. The 24- and 36-in. lengths are priority 3 tools.

Rigid rules are often combined with a set of angle-measuring tools called a combination set. See Fig. 6-16. We will discuss the combination set later in Section 7. This set could be

Figure 6-16 *An 18-in. ground rule with combination set. Left to right, a square, a protractor, and the centering head.*

a priority 1 or 2 purchase, depending on the nature of local industry—check with your instructor.

Be aware that "bargain" rules are available stamped from plain sheet metal. These tools are OK for home shop work but not for machining. The ground rule intended for machinist uses features precision engraved divisions. Satin finish is a must. A high-quality ground rule is used as a template for testing straightness, too.

Basic Tool 2—Calipers

Calipers are the most versatile measuring tool in the machinist's kit. Modern calipers feature multiple measuring abilities using a dial face or an electronic digital readout. They are quick and easy to learn but require a bit of practice to master. They are priority 1. Buy an electronic digital type if your budget allows, or a dial type.

Calipers Are Not a Panacea

Since they are so quick and easy to use, there is a tendency among beginning machinists to overuse calipers. They have several accuracy factors affecting repeatability:

◆ Feel and alignment are the most prominent.
◆ Calibration can be a problem.
◆ Damaged or sprung jaws are not uncommon.

Four Functions

Modern calipers provide four different measuring abilities (Fig. 6-17):

1. *Inside measurement*—that between two opposing edges or surfaces as a slot or a hole
2. *Outside measurement*—a diameter or thickness
3. *Depth measurements*—from a surface to an opposing surface
4. *Step measurement*—from a surface to a surface with both facing the same direction

Note that step measuring is recommended only as an alternative when some other method isn't available. This function allows measurement from the head of the tool to the point of the farthest inside jaw. Step measuring is usually not the best process due to the estimated position of one or the other jaw. Using this function also requires knowing the *offset amount* of the step then

Outside

Step

Depth

Inside

Figure 6-17 *The four functions of a caliper: outside, inside, depth, and step measurement.*

adding it to the reading on the caliper (see Fig. 6-17).

Tech-Specs for Calipers

Correct Application Both electronic and dial types are used for tolerances above 0.001 in., even though the electronic readout features a 0.0005-in. resolution. Both are used in the same way and have the same accuracy challenges, except parallax is not a problem with electronic readouts. The major challenge to accuracy with all calipers is their feel and alignment factors.

Target Repeatability Due to different contact footprints and feel factors, various functions on the caliper produce slightly different repeatabilities.

Outside Measurement Dial types (depending on work shape and skill level) measure from 0.001 in. to 0.002 in., while electronic types (slightly better only because of the ease of reading the scale) measure from 0.001 in. to 0.0015 in.

Inside Jaws and Depth Probe Both types measure from 0.001 in. to 0.003 in.

Step Measurement Due to visual estimation of one jaw to the distance, in most cases, these measure 0.003 in. to 0.010 in.

It is possible to encounter older calipers with no dial or electronic readout, called vernier calipers. If your school has them, go to the unit 3, vernier height gages, here in this section for training on how to read them.

Electronic calipers (Fig. 6-18) have four distinct advantages. The first is their ability to switch between metric and Imperial dimensions, while most dial calipers are either metric or Imperial but won't serve both systems.

The second advantage involves what is called **nominal zeroing.** Electronic calipers can be zeroed anywhere along their range. This allows you to set the nominal size for a large group of parts as the zero. Thus set at zero, each new measurement reports the variation with no mental work to see if it's within tolerance. This feature is very useful. For example, if you are to measure 100 round pins of 0.375-in. diameter plus or minus a 0.003-in. tolerance. The electronic caliper is closed on a gage pin of exact size then zeroed with the touch of a button. This also sets the feel factor, too.

The third advantage is *direct reading of variation.* When each pin is measured, the variation will be displayed on the screen. This special ability makes inspection much faster for large lots of parts. You will see that other electronic tools share this nominal zeroing ability.

The final advantage is *data output.* Electronic calipers can send their measuring data to a host computer concerned with graphing the size variation of the samples. This enables statistical process control by directly entering results to the database. Electronic tools either plug into a data cord or they transmit this information without wires.

Older Vernier Calipers

You might encounter a third variety of caliper that has been displaced by technology; the vernier caliper (Fig. 6-19). A vernier caliper has no read-out or dial. It takes advantage of a math concept whereby two graduated scales are slid by each other as the caliper opens and closes. As calibrated lines on a vernier scale become coincidental (lined up) with graduations on a rule, the user can interpret measurements within about 0.001 to 0.002 in.

If you require vernier training, turn to height gages in this section then return here. Vernier calipers might be found in schools due to their simplicity and sturdy nature. In industry, longer lengths of 18 in. or more are still being used. Vernier calipers do have two advantages over the dial type and one over electronic types: They often feature dual metric/Imperial scales and they are physically compact and sturdy. They will also fit into tight spots not accessible to electronic versions.

Caliper Familiarization

Take a look at the calipers found in your training lab. It would be beneficial to locate the following features.

On dial types (Fig. 6-20) look for

A. Rotating bezel ring
B. Bezel lock screw

Figure 6-18 *An electronic caliper has several data functions.*

Figure 6-19 *Vernier calipers, an earlier measuring tool, are useful in tight spots, due to their compactness.*

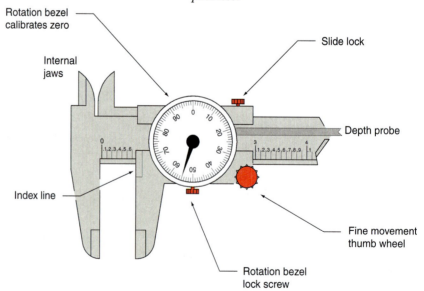

Figure 6-20 *The parts of a dial caliper.*

C. Slide lock screw

D. Depth probe

E. Index line

F. Fine movement thumb wheel

On electronic types, you'll see

A. On/off button

B. Slide lock

C. Depth probe

D. Metric/Imperial button

E. Data port (optional)

F. Zero reset button

Vernier types show

A. Vernier scale

B. Imperial scale, usually the lower scale in the United States

C. Metric scale, the upper scale

D. Index line (shows jaw position against ruler)

E. Slide lock

Zeroing Dial or Digital Calipers

After cleaning the jaws, use the fine movement wheel to close them to the pressure you feel is right. This is a light push of less than $\frac{1}{2}$ pound. Then be sure the readout or dial reads zero. If it doesn't, go back and check the cleaning of the jaws (Fig. 6-21). If they are clean then close them again and rotate the outer bezel ring or push the zero reset button on electronic tools, such that the caliper is set to zero at your pressure.

Dial Caliper Ranges Calipers using mechanical dials (Fig. 6-22) are supplied in one of two possible dial ranges: 0.100 in. per dial revolution or 0.200 in. per dial rotation. The

Figure 6-22 *Dial face calipers are supplied in two ranges.*

0.100-in. dial is slightly better due to the expanded numbers on the face. This is called the **discrimination** of the tool. The improved discrimination of the 0.100-in. dial aids accuracy.

Reading the Dial Caliper It's a snap to learn to interpret the numbers on these and even easier to read electronic types. See if you can read the caliper using the magnified dial face in Fig. 6-23.

On the caliper of Fig. 6-23, the index line is between 0.6 and 0.7 on the horizontal scale. The dial is at 0.055 (the index has not passed a full inch yet). The total is 0.655 in. Now take up your caliper and set it at 0.655 in.

Figure 6-23 *Read this dial caliper position.*

Figure 6-21 *To clean caliper jaws, close them on paper then slide it out.*

Figure 6-24 *Read this dial caliper position.*

Figure 6-25 *Read this caliper setting.*

Figure 6-26 *Read this caliper.*

KEYPOINT

Note that all three functions—outside, inside, and depth probe—are at the same reading at the same time. Modern calipers that feature this three-way action are known as *offset jaw* calipers.

On dial types, first the full inches and 0.100-in. increments are noted on the horizontal rule, at the index line. To that, the dial or vernier amount is added. Now read the three calipers shown in Figs. 6-24, 6-25, and 6-26.

Answers:
Figure 6-24 = 1.238 in.
Figure 6-25 = 3.192 in.
Figure 6-26 = 9.815

Basic Tool 3—Outside Micrometers

Micrometers are such basic trade tools that they are used as the logo of the craft. Sometimes they are formally called micrometer calipers but more often they're referred to as *mics* (sounds like Mikes) in trade lingo. The smallest, 0- to 1-in. mic, and the next size, the 1- to 2-in. mic are a priority 1 tool purchase with the 3-in. mic being a priority 1 or 2, depending on your budget. Beyond that, all others are a priority 2 or 3 for the beginner's tool box.

Micrometers are divided into three general types: outside, inside, and depth. Each is a separate function of the dial caliper but, due to improved resolution and fewer inaccuracies, they are from 10 to 20 times more accurate than calipers. Micrometers studied in this unit are outside types. Figure 6-27 shows basic mechanical micrometers, while Fig. 6-28 shows a direct reading electronic version. Inside and depth mics will be studied in Section 7.

Specialized Micrometers

There are many specialized kinds of micrometers that we will not study in this unit. For example, Fig. 6-29 shows an electronic thread mic. In industry, special micrometers are

Figure 6-27 *Micrometers from zero to three inches are normally owned by the machinist.*

Figure 6-28 *Both direct reading and electronic micrometers are easy to use and calibrate.*

Figure 6-29 *A thread micrometer set.*

used to measure deep grooves, or with very deep frames to measure sheet metal thickness. After learning the basics of micrometers here, adapting to the specialized mics will be easy when encountered on the job.

We will also not study reading electronic mics (Fig. 6-28) as there is virtually nothing required beyond what you will learn using mechanical types. Getting the right feel and alignment are the real skill challenges and they remain the same for all micrometers, no matter their construction.

Tech-Specs for Outside Micrometers

Correct Application Outside micrometers are used to measure thickness and diameters. Tolerances below 0.002 in. require micrometers. They are also used to measure and calibrate other measuring tools such as small hole gages or adjustable parallels (Section 7).

Target Repeatability

Inch Micrometers	
Vernier Scale	0.0001 to 0.0002 in.
Without Vernier	0.001 in. to ≈ 0.0005 in. by estimating
Metric Micrometers	
Vernier Scale	0.001 to 0.002 mm
Without Vernier	0.01 mm to ≈ 0.005 mm by estimating

Control of Inaccuracies The five factors for outside micrometers are feel, calibration, dirt, burrs, and alignment. Wear and damage can be problems, too. Parallax can also effect results. Use the joggle/null method to arrive at correct alignment. Here's how.

After cleaning the measuring faces as shown in Fig. 6-30, hold the micrometer as shown in Fig. 6-31 with one hand, then move it in smaller and smaller sweeps to obtain the lowest possible reading—the movements are described as joggles. The micrometer will stop changing at lowest possible value, called a *null*. We'll use the null concept again many times for measurement and machine setups.

To avoid overtightening, use the friction drive on the micrometer for developing your own personal feel. Parallax and tool heat can be factors but to lesser degrees than alignment and feel.

Figure 6-30 *Always clean the tips before making a critical measurement.*

Reading a Micrometer

Please obtain a micrometer and check its features as we proceed with explanations and activities.

Hand position is important to allow full control of all micrometer functions with the least disturbance to measurements and to achieve alignment. The position is different for small and large outside micrometers. For small micrometers up to about 4 in., it is held and aligned with one hand as in Figs. 6-31. This allows joggling the mic while reaching the friction/pressure control drive and spindle lock.

Before measuring, it's critical that calibration is checked and that the measuring faces (tips) be cleaned well. Always check for dirt on the work, too.

The friction/pressure drive is used to regulate the maximum amount of force brought to bear on the object being measured. Friction drives are supplied in two forms: a ratchet at the end of the micrometer or a friction drive on the thimble. The friction type is a better solution because it is easy to reach. See Fig. 6-34 for these micrometer terms.

Friction or ratchet drives reduce variation where several machinists must share the same tool. They are also useful for learning about feel. But most machinists do not use the

Friction drive

Lock

Figure 6-31 *Holding it right, all functions are within reach of one hand.*

Figure 6-32 *This is the correct method of holding a larger micrometer.*

pressure control after they become accustomed to their own micrometer.

Look at the study micrometers in your lab. What kind of pressure drives do they have? Notice the lock devise. There are two common types: lever or rotating ring.

Now hold the micrometer as in Fig. 6-31 or 6-32 if it's a large micrometer, and make sure you can rotate the thimble and reach the lock and the pressure control with your thumb. You may not be able to reach the ratchet type without moving your hand. If so, try it to get the feel of the tool.

Precision Features for Micrometers

When purchasing a micrometer, check for the following features (see Fig. 6-33). Please examine the student micrometer or your own as we proceed.

1. Carbide wear tips—look closely at the spindle and anvil tips. There will a slight color change to a darker metal. These are hard carbide tips for wear resistance.

2. Look on the sleeve. Is there a tenths scale? It will be visible as 10 numbered lines running parallel to the index line on the sleeve.

3. Hand heat insulation—both sides of the frame are insulated with a hard plastic pad where the large part of your palm contacts the frame. This prevents heat transfer to the frame.

4. In a clean environment, unscrew the thimble until the internal parts are visible. Notice the wear and backlash adjustment nut on the sleeve. This is not part of normal calibration and is adjusted only over a long a period of time. Caution! Be careful to not allow dirt into the precision threads when reclosing the micrometer!

5. Does it have a good protective case? It should. If not, be sure it is protected in a tool box drawer when not in use.

Figure 6-33 *Standard features on a micrometer.*

Carbide tips   Thumb lock

Rachet or friction drive

Insulation pads

Trade Tip

Caution! Never close your micrometer to zero when putting it away. If left in the tool box drawer as it is now, for any length of time, the micrometer in Fig. 6-34 will begin to change shape. That tends to spring the frame over time.

Figure 6-34 *Never close a 1-in. micrometer when storing it! Damage can result.*

Figure 6-35 *Read this micrometer.*

Index line

Figure 6-36 *Read this measurement.*

Developing Micrometer Skills

First, we'll learn to read the micrometer then we will do some practice measuring. Remember the best repeatability results from recognizing and controlling the inaccuracy factors.

Reading an Imperial (Inch) Micrometer

The micrometer shown in Fig. 6-35 reads 0.475 in. Notice the line parallel to the sleeve axis. This is the index line. The index line is where the measurement is to be read. Identify the index line on your micrometer then set it to the same reading. When doing so, notice that the gap opens between tips, by 0.025 in. for each full revolution, and that the thimble is divided into 25 parts. Each line on the thimble then represents 0.001 in.

Now read the micrometer in Fig. 6-36. This illustration shows 0.785 thousandths. First, the 0.700-in. graduation is uncovered along

the index line. Next, three 0.025-in. revolutions are visible, 0.775 in. Finally, the thimble is rotated an additional 0.010 in.

Set the student micrometer for the following numbers, then look at Figs. 6-37 and 6-38 to see examples.

A. 0.555

B. 0.840 (Notice that the number 15 is aligned with the index—to create the forty-thousandths.)

0.555 in.

Figure 6-37 *Reading, 0.555 in.*

Figure 6-38 *Reading, 0.840 in.*

0.840 in.

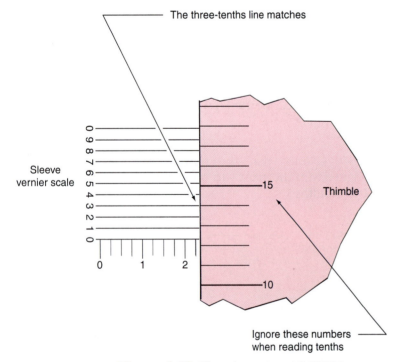

Figure 6-39 *The vernier scale is wrapped around the sleeve.*

Vernier lines

The three-tenths line matches

Sleeve vernier scale

Thimble

Ignore these numbers when reading tenths

Figure 6-40 *The micrometer at 0.2377 in.*

Reading the Vernier Tenths Scale

To accomplish this exercise, obtain a vernier micrometer as shown in Fig. 6-39. A vernier micrometer features 10 numbered lines running parallel to the index line on the sleeve. They subdivide each thousandth into 10 parts, 0.0001 in.

Set your micrometer to any number, say 0.637 in., for example. Now as you rotate the thimble past 0.637 note that it is possible to estimate how many tenths beyond. But, the **vernier scale** provides an exact reading between thousandths. Set the micrometer to an estimated four-tenths of the way between 0.637 and 0.638 in. (0.6374).

Tilt the micrometer up and observe the lines running parallel to the index line along the sleeve, starting at zero and ending on nine. These are the vernier lines. (Some scales start and end on a zero.) Find the vernier line that best matches a line on the micrometer's thimble. The number at the end of that vernier line indicates how many tenths (of a thousandth) the thimble has been rotated beyond a particular thousandth.

KEYPOINT

To obtain tenths, read the small numbers wrapped around the sleeve at the end of the vernier lines. Do not use the numbers on the thimble for this function.

Read the micrometer in Fig. 6-40. Note that the drawing is as though the vernier scale is laid flat. Notice that only the vernier line numbered 3, exactly matches a line on the thimble. Set your micrometer to this position.

Set your micrometer to the following sizes. Remember, first set the thimble to the correct

Figure 6-41 *The vernier scale will be on the side of the sleeve.*

thousandth and estimate the tenths—then tilt it up to see the vernier scale. Now turn the thimble just a bit more until the correct numbered vernier line on the sleeve coincides with a line on the thimble.

1. 0.4477 in.? See Fig. 6-41 for the first answer.
2. 0.3689 in.? Have your instructor check this one and the next.
3. 0.3333 in.?

A Pattern for Improvement

Paper micrometers like the ones we've been reading about here can only coach you to read the scale on the micrometer. Now it's time to use the micrometer in real measurement situations. Remember, the goal is to obtain consistent results that fall within the expected repeatability of the tool. That takes practice.

Improving ability with all measuring tools is a five-stage process:

1. *Measure* one or more objects (unknown size at the time, to avoid bias).
2. *Compare* your results to the best answer available of true size.

   The best are gage blocks (Section 7) or standards or prepared items with answers.

   The next best is to have a more skilled person test the measurement.

   The third method is to test the same measurement with a better tool or process.

   A last is to compare your results to those of two others of the same skill level, but you can see the potential problem with this.
3. *Determine tendency*. Did you tend to make a consistent error? Too much by say 0.003 in. or too small repeatedly.
4. *Correct* the problem (less pressure, more, better alignment, and so on). Have your

instructor show you ways of aligning the tool to the measurement.

5. *Try again.*

Your instructor may have an activity planned with objects and an answer sheet. If not, make one up and test each other. Make up an accuracy challenge between yourself and fellow students.

Zeroing out Index Error Before Measuring

Before measuring with a micrometer it's important to test the feel and the zero index. First, clean the two measuring tips with a clean lens tissue or other lint-free wiper. This act alone will prevent much misery in measurement. Next, close the mic to zero or upon a standard that's the lowest value of the micrometer. For example, to check the feel and zero of a 3- to 4-in. mic, use a 3.000-in. standard.

If the mic reads zero at a pressure you feel is OK, proceed with using it. If not, it's time to realign the mic. Realigning the zero index requires moving the thimble relative to the spindle—just a bit so that it reads zero, when closed to the right pressure, at its true zero position (or at its lowest value for mics larger than 1 in.). The task requires loosening but not disconnecting the spindle from the thimble. The numbered thimble is then adjusted to read zero when clean tips are closed at the right pressure. After adjustment, the thimble is relocked in place.

Note that it's often necessary to do this operation more than once as the locking stage disturbs the setting. Always double-check calibration after locking the mic (see Fig. 6-42).

If allowed, practice first on a 1-in. micrometer then on a larger mic

Caution! A reminder, zeroing may not be allowed at your level of training or even for beginners in industry where specialists perform the task. Ask about the policy.

Figure 6-42 *Calibrating against a standard.*

Figure 6-43 *Calibrating requires a special spanner wrench to unlock and relock the thimble.*

using a standard. There are two methods of loosening the thimble from the shaft:

Straight shaft—set screw
Tapered shaft—self-holding

The set screw type requires no instruction. If it is a self-holding taper, a special spanner wrench is required (Fig. 6-43). After loosening it one-quarter turn, tap the thimble with a screwdriver handle or some other firm, non-metal material; this breaks the taper loose. Ask your instructor or a machinist to demonstrate this action.

Reading a Metric Micrometer

Check with your instructor before studying the following information on metric micrometers. To stay focused on Imperial tools, you might skip forward then return here. Metric measuring is very important in a world economy. However, this skill may not be a required skill at this time. As with all metric implements, these measuring tools are easy to understand. The index line is graduated in full millimeters above and half millimeters below.

KEYPOINT

Each full rotation of the thimble opens one-half millimeter.

The thimble is divided into 50 parts thus each part equals 0.01 mm. To read the micrometer, add the uncovered full and half millimeters on the index line to the hundredths on the thimble.

Figure 6-44 reads 4.79 mm—4 full millimeters plus one 0.5 graduation plus 0.29 from the thimble. Now, read the metric micrometers shown in Figs. 6-45 through 6-48.

Figure 6-44 *Metric micrometer at 4.79 mm.*

Figure 6-45 *Reading 15.35 mm.*

Figure 6-46 *Reading 24.50 mm.*

Figure 6-47 *Metric mic reading 6.05 mm.*

Figure 6-48 *This metric micrometer reads 10.10 mm.*

Answers

1. Figure 6.45 reads 15.35 mm.
2. Figure 6.46 reads 24.50 mm.
3. Figure 6.47 reads 06.05 mm.
4. Figure 6.48 reads 10.10 mm.

Basic Tool 4—Height Gages

Height gages are dual-purpose tools used for both measuring and layout work. Layout work is a premachining activity taught in Section 8. Here we concentrate on their many measuring applications. They measure up from a flat datum surface such as a granite table or a milling machine's table.

Height gages are a priority 3 for ownership or perhaps a priority 2 in some rare cases where the job demands constant use of one. Due to their high cost, they are usually provided by the shop.

Figure 6-49 *A scribe can be used as a measuring probe in the height gage.*

Scribe and Indicator Attachments

When measuring, height gages are used in two ways:

◆ With a scribe as a touch probe. In Fig. 6-49 the machinist is measuring the height of a part above a vise bed. To do so, they have calibrated the height gage scribe to read zero on the vise bed before loading and machining the part. The machine table creates the flat datum but the vise bed is calibrated to be zero height above the datum. The part height is read directly on the height gage.

◆ Using an indicator (Fig. 6-50) as a more sensitive touch probe for improved accuracy—this superior process eliminates feel problems but can't reach into places

Figure 6-50 *An indicator is mounted on this height gage.*

Figure 6-51 *The indicator could be used to measure the height of the part.*

Ruler index line

Visually estimated height

Figure 6-52 *A height gage is an upright ruler with a pointer.*

the probe might. In Fig. 6-51, the indicator is zeroed on the workpiece of Fig. 6-49. We'll study indicators next.

Reading a Vernier Height Gage

It's entirely possible to find three kinds of height gages in industry: vernier, dial, and electronic. Unlike calipers, the high initial cost of vernier height gages means they will be used for some time. Since applying and reading electronic height gages requires little training beyond the skills of the dial or vernier type, we will concentrate our efforts on vernier types. The real technique is not in reading but in using them.

Electronic Height Gage Shop Policy

Your shop may or may not have a rule about keeping the electronic height gages away from the machine area. Mechanical versions are always allowed on the machines where hot chips and coolants are ever present. But in some shops, electronic versions are reserved for the layout/inspection tables. Today, however, shop-hardened versions of the electronic gages are able to withstand this attack. The electronic height gage adds flexibility in data manipulation, especially with nominal zeroing, as shown in Fig. 6-49.

A height gage (Fig. 6-52) measures distances above a datum—the flat surface on which it sits. In simplest terms, a height gage is a precision rule standing upright on a sturdy base. We could attach a sliding straight edge as a probe and read its intersection with the vertical rule. However, using this simplified

height gage, the best repeatability would only be slightly better than the rule graduations.

Understanding the Vernier Concept

Vernier scales are math-based devices that divide graduations on a rule into finer parts than can be seen or estimated with the eye. They improve the resolution of any tool to which they are added several times over. Once the vernier concept is mastered, transferring to a new tool using one, becomes far easier. We've seen one kind on micrometers, but a height gage is a good place to learn the principle.

The Vernier Divisions are Slightly Smaller When used to measure distances, vernier divisions allow us to create a 0.001-in.-resolution tool without actually dividing the rule into 1,000 parts. Here's how. The secret lies in a concept called colinearization, meaning that it is possible to tell when two lines coincide (line up to become **colinear**)—better than we can read graduations. You saw this on the micrometer vernier scale.

A linear vernier scale is divided such that each graduation is 0.001 in. shorter than the rule graduations. Each time a further vernier line coincides with a rule line, the vernier has moved 0.001 in. So, if the vernier is colinear with the rule at the sixth line, it has moved six-thousandths away from zero.

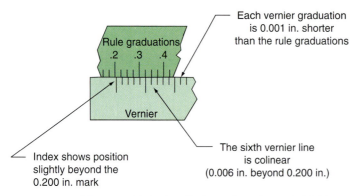

Rule graduations
.2 .3 .4

Each vernier graduation is 0.001 in. shorter than the rule graduations

Vernier

Index shows position slightly beyond the 0.200 in. mark

The sixth vernier line is colinear (0.006 in. beyond 0.200 in.)

Figure 6-53 *Detail view of vernier at 0.206 in.*

First, read the tool roughly (Fig. 6-53). The index line points slightly beyond the 0.200-in. line on the rule. Estimating the answer helps avoid big errors. Next, the true index position is determined by counting the vernier lines until the colinear pair are found. This vernier then reads 0.206 in. This reading could be on a height gage or on a vernier caliper.

KEYPOINT

On this linear scale, the vernier graduations are exactly 0.001 in. smaller than the rule graduations. But they could be divided into many other ratios on other tools.

There are two common Imperial height gage vernier sets:

The rule is graduated with 0.025-in. spaces.

The rule is graduated with 0.050-in. spaces.

It takes a moment to determine which is which by looking at the rule and vernier scale on your height gage. Try it yourself. Read the height gage in Fig. 6-54. Answer and instructions follow.

The rule of Fig. 6-54 is graduated with 0.050-in. lines. First, add up full inches and hundred-thousandths plus the graduations passed. In the illustration, 4 in. plus 6 full hundred-thousandths lines plus one, fifty-thousandths line = 4.650 in. (a large total).

To that add the vernier distance 0.006 thousandths = *4.656 in.*

Precision Features—Vernier Height Gages

Take a few moments to go to the lab and locate the following features on a real height gage. Compare the photographs in Figs. 6-51 or 6-55 to your height gage, if available. There will be differences depending on the kind: dial, vernier, or electronic.

1. *Fine movement screw.* This is usually on the back or side. It consists of a slide with a lock and a travel screw attached to the vernier slide (Fig. 6-55). All height gages have a method of moving the slide and probe finer than hand motion.

2. *Calibration lock and adjustment screw (vernier types).* Many but not all height gages have some form of calibrating the height gage to zero. On vernier height gages, this is accomplished by moving the actual rule relative to the upright body. This is where electronic height gages are clearly better since they can zero anywhere on their scale by pushing a button! Some height gages have a small

Figure 6-54 *Vernier scale reading 4.656 in.*

Figure 6-55 *The fine movement screw helps with final adjustment of the slide.*

Figure 6-56 *Always estimate first by reading the index line's position on the rule.*

range for calibration, adjustment while others allow a long range of calibration positions. Check to see if your height gage has this ability.

3. *Slide lock screw.* This locks the slide and attachments into position once they are set (all types).

4. Most importantly, *the index line.* Found on most height gages, the index points to the exact position of the tool on the rule. To avoid errors, always estimate the position using the index line on the rule, before reading a vernier (Fig. 6-56).

Zero Calibrating Height Gages with a Scribe

Most of the time, the height gage needs to be zeroed when the scribe touches the datum upon which it sits (Fig. 6-58). Usually a flat granite table or on a machine table. Zeroing the height gage is the same for all types, no matter how they are constructed. It's easy and quick but you must be on the watch for dirt and the feel factor discussed in the previous Trade Tip. After cleaning the table and height gage base, lower the scribe until it

Trade Tip

Paper Feelers for Improving the Feel Inaccuracy

Machinists are often faced with making height gage measurement for lack of a micrometer that has the size range. When using a scribe as a probe, the problem is always knowing when it has just touched the surface in question—it's not always easy to tell. We often press down too hard thus loosing accuracy.

To help control this factor, use a feeler gage or a piece of paper of known thickness as a sensitivity gage between the part and scribe (Fig. 6-57). Pull it in and out of the touch zone while slowly lowering the scribe toward the surface just until the paper drags. That's the light touch! But then, after locking the slide screw and reading the height gage, don't forget to deduct the thickness of the paper feeler from the total.

Figure 6-57 *A paper feeler improves sensitivity, but its thickness must be accounted for in the measurement.*

Set 0.000

Set 0.003

Figure 6-58 *Calibrating a gooseneck scribe at zero against the datum table.*

touches the table. Once the scribe is locked at the table surface, check the reading—is it calibrated? If not, reclean and try one more time. Still off zero? Then either unlock the vertical rule from the height gage and reset it to read zero against the vernier slide or turn the dial bezel to read zero.

Layout Tables

Height gages are used on special granite or older cast iron tables, also called surface plates (sometimes called a layout table or inspection plate). The granite table is an ultraprecision implement right at the heart of precision in a shop (Fig. 6-59). Layout tables form the master *X-Y* datum for the shop.

While layout tables are made from several kinds of granite, black is preferred because it

Figure 6-59 *The granite table is the prime datum and the basis for precision in a shop.*

is the most stable over a long period of time. In recent years, composite granite made from granite chips and plastic resins has gained popularity, too. Since granite is harder than the tools slid upon it, its surface should last indefinitely as long as it's treated right! There are some professional skills to be learned in the treatment of the table.

An understanding of these rules is of high priority in any machine shop:

1. **Avoid Contamination**
   Clean your work before touching the table. Never set dirty parts upon the table. Avoid leaning your hands or arms on the surface, too. Oil in your skin will make tools drag when moved. Wash your hands first to remove oils. When using the table always wipe the surface before and during use. Dust settles out of the shop atmosphere all the time. A final wipe with a clean, oil-free hand is the best. You will feel any speck of dirt or chip.

2. **They Chip and Break!!!**
   Never hammer, punch, or otherwise hit any object on the layout table. Move your work to the bench for that type of operation. Note that some shops permit light taps of automatic or even pilot punches that will be deepened later using a center punch but away from the table. Some shops will not even permit that risk! Don't guess, ask first!

3. **Precision Work Only**
   Use the table exclusively for layout, measuring, and machine accessory setup. Do not store or lay unrelated items on the table. Many will feature a cover that is placed over the table when not in use (Fig. 6-60).

4. **Care for the Table**
   Follow the shop recommendations for cleaning using residue-free solvents such as acetone (read the MSDS for acetone) and lint-free shop towels—not paper towels. Occasionally, a scrub with cleansing powder and a little water or a specialized cleaner, followed with an acetone wipe and a final dry wipe is necessary. Each shop will have their own preference. Then, cover the table when it is cleaned but not in use and set nothing on the cover.

A well-maintained layout table is a delight, to use. Precision tools slide as though they were on magic carpets. In fact, be sure to hold

Figure 6-60 *When not in use, cover the granite table to protect and keep it clean.*

Figure 6-61 *Two kinds of indicators—travel on the left and test or probe type right.*

onto precision tools, more than one has been given a slight push to see it go all the way across then crash to the floor if not held back!

Basic Tool 5—Indicators

Indicators are superuniversal measuring tools. As such, one or more, along with some type of indicator holder, are a priority 1 tool kit purchase. They are used throughout the shop in three different ways:

1. **Axis Control**
   On the machines to watch exact movements of axes. Usually on manual machines that lack a true micrometer hand crank or dial.

2. **Setup Alignment**
   To set up machine accessories or align accessories such as vises true to the machine or to align a workpiece in a chuck or vise.

3. **Measure**
   Improved measuring at the inspection table and on the machines.

It is this measuring aspect we will study. The other uses will be covered in the lathes and mill sections. The indicator will be mounted in a height gage.

Two Common Types

Dial test indicators (often abbreviated to DTIs) are supplied in two basic forms: probe (top) and travel types (Fig. 6-61). The differences in these two types are discussed later.

Both are precise and delicate and have a particular application. Both measure in thousandths of an inch or ten-thousandths of an inch depending on construction. The indicators graduated with 0.0001-in. divisions are referred to as *tenth indicators* in shop lingo.

Travel indicators are more sturdy and they measure greater ranges of movement. Probe indicators are smaller and versatile, thus they can be used in a greater number of applications. We will be using the probe type for all height gage measurements.

Here at last is a measuring instrument free of human feel problems. They are, however, subject to parallax errors and wear. That's the good news. But the bad is that both types of indicators have their own unique factor we will study in detail later. Once understood, it can be controlled.

Tech-Specs for Indicators

Correct Application Job tolerances below 0.001 in. require indicators. The tenth indicator (0.0001 in.) is one of the most accurate instruments in a machinists tool kit.

Target Repeatability Resolution varies with indicators supplied in 0.001-in., 0.0005-in., or 0.0001-in. graduations. Repeatability should be close to the resolution due to the lack of inaccuracy factors. Watch out for special progressive and cosine errors (coming up), which can degrade results.

Indicator Holders

All indicators require one of several kinds of mounting accessories that either clamp or magnetically hold the indicator in the setup or measuring situation. For the measuring discussion, the indicator is held in a height gage using an attachment bar supplied with the indicator, the same size as the scribe (Fig. 6-62).

To use your indicator while working in the machine area, one or two types of holders will be a priority 1 purchase also. Two useful kinds are shown here: the spindle clamp for milling machines (Fig. 6-63) and the magnetic mount used on all machines (Fig. 6-64). One or both are priority 1 tools.

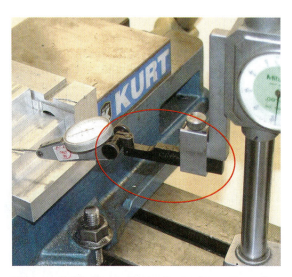

Figure 6-62 *The indicator bar replaces the scribe.*

Figure 6-63 *A spindle mount indicator on a vertical mill.*

Figure 6-64 *A magnetic base is useful for mounting indicators on machines.*

Three Methods of Measuring Distances

This is a perfect example of choices that must be made to match the job tolerance to a measuring process. Although they are performed with the same three implements, the flat datum table and height gage with a dial indicator attached, the three discussed next differ greatly in total distance that can be measured, expected repeatability due to different inaccuracy factors, and time required to perform the task.

◆ Dial Measuring
◆ Comparison Measuring
◆ Height Gage Scale Measuring

Dial measuring is accurate and quick but has a very limited range. In comparison, it is the most accurate method available, but it takes time to set up and perform. Height gage scale measuring expands the range that can be covered but lowers the resolution of the process. As we progress through further measuring lessons, more examples of using the same tools in different ways will appear.

Figure 6-65 *Reading parallelism FIM.*

the indicator on the height gage is first ze-roed on the master stack sitting on the table datum. Next the indicator is moved across the surface plate over to the part in question. There, the indicator falls short of the master by 0.0053 in.—the size is 2.4947 in. exactly.

The difference on the dial face is either added to or subtracted from the standard amount, depending on whether it shows taller or shorter. The result then becomes the measured height with nearly no inaccuracies! Free of alignment, feel, and calibration problems, this method is accurate to 0.0001 in. as long as the indicator is so graduated and the standard is clean and accurate.

KEYPOINT

Other than CMMs, this is the most accurate measuring method you will learn in this book.

1. Dial Measuring

This method directly detects the distance in question on the dial face. Therefore, the distance to be measured must fit within the indicator's range. Here the indicator is the calibrated measuring tool. In Fig. 6-65, parallelism is being measured. The indicator is mounted in the height gage then moved over the entire top surface. The total difference in the highest and lowest point is observed in thousandths of an inch. What was the term for that total difference?

2. Comparison Measuring

The second method allows measurement of larger distances beyond the range of the indicator, but its main advantage is near perfect accuracy. In Fig. 6-66, the indicator is used as a transfer tool to compare a part to a master standard.

The nominal dimension for the drill gage (2.5000 in.) is set up in precision gage blocks (hard steel blocks of the exact nominal size, studied in Section 7). Then

3. Height Gage Scale Measuring

Here the indicator is used solely as a sensitivity probe. Accuracy will be only as good as the resolution of the height gage since we will be reading distances from its scale. In Fig. 6-67 we are measuring the height of this transmission case from the lower shelf to the top. First the height gage is lowered until the indicator reads zero on its dial face at about half of its total travel. That sets how much pressure is brought to bear on the surface. Now the height gage is read at the first position location. Or it is reset to zero if it is electronic.

Gently pulling the probe off the surface to not disturb its position on the height

Indicated difference

2.5000–in. master gage

Figure 6-66 *Comparison measuring requires a master gage close to the size of the object.*

Read at second position

Read at position

Figure 6-67 *Measuring using the height gage as the instrument and the indicator as a pressure probe.*

Figure 6-68 *A set of precision gage blocks can be used for comparison measuring.*

gage, the slide is lifted up above the second surface then slowly lowered down until reaching zero on the dial face again (same pressure). The position is now read directly from the height gage's readout if it's electronic. Or if it's mechanical, the first position is subtracted from the present to yield the measured height.

Selecting the Gage Standard

To select the standard, an estimate of the distance to be measured is made by rule, calipers, etc., or the nominal size could be taken from the design print as in the example. The best master (Fig. 6-68) is chosen from precision gage blocks (studied in Section 7). It should be as close as possible to the expected part size within about 0.010 in. or less.

Machinists often shorten dial test indicator to DTI.

Comparing Indicators

Let's look more closely at the two common indicator types: test probe and travel indicators. Both do the same job but in a different way.

Travel or Plunger Type Indicators

Sometimes called long-range indicators, these DTIs are used at the machines for alignment and axis control, they can be used for measuring too, although they are a bit bulky for the task. The plunger moves in a straight line, while the dial rotates to indicate total distance traveled. They are supplied in many ranges of travel from 0.250 in. up to specialized versions detecting several inches. The 1-in. travel indicator is the most popular. Many but not all DTIs of this type feature one full revolution of the dial equal to 0.100-in. or 4-mm movement of the plunger. Some feature 0.200 per dial resolution.

Each single rotation of the main dial is called one *clock*. To track the total number of clocks on indicators featuring large spans, a second dial hand is often included. When using a travel indicator, read the label near the center of the face to discover the resolution and what one clock represents. Read the indicator shown in Fig. 6-69. This travel indicator reads 0.445 in. (Four clocks plus 0.045 in. on the dial face.) Notice the resolution note in the center of the dial face.

Control of Cosine Error

Travel indicators are subject to a particular kind of operator-induced problem called **cosine error** (Fig. 6-70). It is caused by not setting the plunger axis perpendicular to the direction to be measured. In the drawing, notice the distance to be measured A-B compared to the distance detected by the indicator B-C. The tilt of the probe axis *amplifies the result* to a larger false FIM (full indicator movement). Preventing cosine error is easy, if the probe axis is eyeball close to square to the measurement, results will be well within shop accuracy.

Figure 6-69 *A travel indicator with a clock counter.*

Correct position to measure step distance

Wrong

A–B = True distance
B–C = False distance

Figure 6-70 *Cosine error is caused by tilting the indicator.*

Test Indicators

Due to their particular accuracy challenge, these small indicators must be limited in total range to around 0.030 in. or 1 mm. Beyond that, they loose usefulness.

> **KEYPOINT**
> They measure not by a straight-line probe as with travel types, but rather by swinging the probe in an arc about a pivot point.

Due to their compact size, they are the more universal of the two types. If purchasing only one indicator—the test type should be the one for your kit since they can be used both on the machines for setup work and at the inspection bench for measuring. However, owning both kinds should remain a priority 1.

Body Position versus Probe Position

In using a test indicator, there are two setup positions that must be considered: the body and probe angles. The body position is noncritical. It's adjusted to keep the indicator out of the way of the setup and to position it such that it can be read (Fig. 6-71).

> **KEYPOINT**
> The test indicator's body can be set at any convenient angle with respect to the work being tested with no effect on accuracy.

Probe Position is Critical Setting the probe angle relative to the surface to be measured is very important. It should be a specific

All correct body positions

Figure 6-71 *Indicator body position has no effect on measurements.*

angle, usually 18 degrees. However, from 15 to 20 degrees will produce shop accurate results. This 18-degree angle is true for most (but not all) brands of test indicators—known as its base angle.

If not correctly positioned, a subtle inaccuracy is produced. The resultant reading will either be amplified or reduced relative to the true distance as the probe departs from the correct base probe angle. Look at Fig. 6-72 where the probe is measuring from the upper surface to the lower and see if you can reason out why the probe angle must be right. The secret is that as the probe moves in an arc, the gearing inside translates that into thousandths of an inch on the dial face. Truthfully, there is no arc that equals the true distance to be measured, however, the movement of these indicators is internally corrected to read within shop requirements when the probe is set approximately at 18 degrees relative to the surface being tested.

However, the extreme probe angle on the left, produces a much longer arc as it swings from the top to the lower surface. It shows a false distance bigger than the truth. Conversely, if the probe angle is set much less than 18 degrees, closer to parallel to the surfaces, the result will be reduced. The effect is known as **progressive error**—the farther

Extreme probe angle produces false result

Correct angle produces an accurate reading

True distance

Corrected for this arc

Figure 6-72 *Progressive error caused by missing the indicator's base angle.*

from the center of the base angle the probe swings the worse the error becomes.

Setting the Base Angle While any change in probe angle will affect results, the good news is that the 15- to 20-degree range is acceptable or even a bit more when tolerances aren't very tight. Now, can you see why these probe type indicators must have a limited travel range? Otherwise, the longer swing would produce too much progressive error. Keep in mind that the body angle in Fig. 6-73 has no effect on the result—only the probe angle matters.

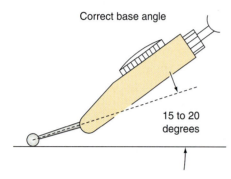

Correct base angle

15 to 20 degrees

Figure 6-73 *The correct base angle for accuracy.*

UNIT 6-3 REVIEW

Review the Key Terms

Colinear
When two lines coincide (line up). The way a vernier scale is read.

Cosine error
False readings caused by tilting the travel indicator's axis away from the direction to be measured—causes amplified results

Discrimination
In measuring, the ability to perceive a difference in two or more graduations on a given tool.

Nominal zeroing
The ability of electronic tools to set any point as zero—thus measuring a large quantity of parts by reading the variation only.

Progressive error
The error created by a test indicator's probe swinging in an arc yet reading in linear thousandths of an inch.

Vernier scale
A mathematical method of improving tool resolution by creating divisions one unit smaller than the real graduations.

Replay the Key Points

◆ There are two types: Plunge types, also called travel indicators, and test types.

◆ Both types have a possible error due to misaligning the indicator to the measurement to be taken.

◆ To avoid progressive error, the test probe must be set at 15 to 20 degrees to the direction to be measured.

◆ To avoid cosine error, the plunge type must have the plunger probe directly in line with the measurement to be taken.

Respond

1. How fine might a skilled machinist read a precision rule?

2. At what minimum (tightest) tolerance should the machinist forsake their calipers for outside micrometers?

3. Why are calipers not used as very accurate tools—what challenges their accuracy?

4. If a travel indicator is tilted relative to the direction to be measured, the result will be an amplified reading. Is this statement true or false? If false, what will make it true?

5. The error described in Question 4 is a progressive error. Is this statement true or false? If false, what will make it true?

Section 6 Review

Unit 6-1
One of the most basic mistakes made is to begin a job without knowing the exactness of the upcoming operations. Without that knowledge, the machinist might choose the wrong process both for shaping the part and for measuring the dimensions. Always look at the print to determine tolerances before beginning the job.

Unit 6-2
Although we've discussed many ways to control results and improve personal repeatability, the process has only just begun. Staying accurate is a lifelong process—not different from playing a musical instrument. It takes regular practice and awareness.

Unit 6-3
I taught machining for engineers for 1 year in New Zealand. On my first day in class, the veteran lab assistant handed me a micrometer and asked me to "check it out for feel." After I took it and measured a gage block, he said, "I can see you are going to work out just fine by the way you held that micrometer." No kidding, little things like correct hand position show you to be a well-trained pro.

Questions and Problems

1. Of the four kinds of tolerancing found on drawings, *bilateral, unilateral, general,* and *limits* which
   A. is stated as the minimum and maximum size.
   B. usually has no stated nominal size.
   C. is determined from a table on sheet 1 of the drawing.
   D. is determined by counting the number of places to the right of the decimal point.
   E. extends one direction from nominal. (Obj. 6-1)

2. A *protracted* angle is dimensioned and toleranced in what units? (Obj. 6-1)

3. True or false? Developing the best possible resolution for any given measuring tool, should be a personal goal. If it's false, what will make it true? (Obj. 6-2)

4. Define parallax error in two words. (Obj. 6-2)

5. List the 10 factors that might plague a given measuring tool? (Obj. 6-2)

6. What advantages are there to electronic calipers, compared to dial? (Objs. 6-2 and 6-3)

7. Complete this: Cosine error occurs when a _____ type dial test indicator is _____ relative to the measurement. (Obj. 6-3)

8. Of the three measuring techniques involving a test indicator and height gage,
   A. Which involves gage blocks?
   B. Which requires a datum table?
   C. Which does *NOT* use the DTI as a measuring tool but rather a pressure sensing probe? (Obj. 6-3)

9. The friction drive on a micrometer helps avoid _____. (Obj. 6-3)

Critical Thinking Problems

10. You have two micrometers in your tool box, one with a friction drive and one without.
    A. Which would be the most accurate for you alone?
    B. Which would be the most accurate when you and another machinist use them? Why? (Obj. 6-3)

11. Why would the repeatability of a dial caliper with graduations of 0.001 in. be similar to that of an electronic, digital caliper that reads to 0.0005-in. increments? (Objs. 6-2 and 6-3)

12. You wish to dimension and tolerance the diameter of a hole in which a precision pin must fit. The design calls for a slip fit between the pin and hole not closer than 0.001-in. clearance between the two. In other words, under no circumstance can the pin be any closer to the hole size than 0.001 in. so it will always

slide freely within the hole. The pins are commercially supplied at 0.875 in. with a guaranteed unilateral tolerance of plus 0.000 in. and minus 0.0005 in. (plus nothing, minus a half thousandth). You are to allow three-thousandths total variation on the hole size. Using a bilateral tolerance, what should the hole's nominal size be and the tolerance? (Obj. 6-1)

13. Explain how progressive error might be used to your advantage. This was not directly covered in the reading—discuss it with other students. (*Hint:* Review the uses of an indicator in the shop.) (Obj. 6-3)

14. Describe the principle on which all vernier scales are based. (Not how to read them but how they work.) (Obj. 6-3)

15. If a micrometer was heated above the temperature of the object being measured, would it measure larger or smaller than the true distance? (Obj. 6-2)

16. A. You are to machine a very accurate hole. What five controls might the engineer require for that hole? (Obj. 6-2)
    B. If the drawing used GDT, which of the five would require a datum reference? (Obj. 6-2)

17. Describe a geometric angle tolerance zone. (Obj. 6-1)

18. Describe one method of measuring a geometric angle using a protractor, use three facts. (Obj. 6-2)

19. The graduation lines on the sleeve of a metric micrometer represent how much distance each? (Obj. 6-2)

CNC Question

20. You are running a high-speed turning center, making 500 parts. Due to your sharp editing of the program, the part cycle time (one part completed) has been cut to less than 3 minutes! You are trying to burr the parts as they come off and stack them carefully in egg crates so there's not a lot of time to measure a critical diameter to a tolerance of five-thousandths inch. What is the fastest measuring method you can think of that will reproduce within that tolerance?

Section 6 Answers

ANSWERS 6-1

1. Heat, tool wear, operator actions
2. Undesirable but necessary ranges of acceptable variation for a dimension
3. False. Nominal means the target dimension
4. Limits and general tolerances
5. Near the bottom of the print in a table

ANSWERS 6-2

1. False (that's parallax)
2. Awareness and trying to not look as the final adjustment is made
3. Reducing heat in the process (better solution); calculating size for elevated temperature (requires thermometer)
4. A. Measure an object that has a known size but unknown to you. Then compare and adjust your tendency.
   B. Compare your results to another craftperson who has more skill.
   C. Compare your results with another method that has less inaccuracies.
   D. Measure a standard gage and retain the feel of the tool.
5. Be sure there is good lighting or use a portable light. Get into a comfortable position to make the measurement. Use a magnifying glass even when your vision is good. Prevent shadows. Move your perspective to avoid parallax. Find a different tool or method.

ANSWERS 6-3

1. It is generally held that 0.005 in. is possible. Note, a few feel that 0.003 in. is possible on a reliable basis with a magnifier. (I agree with this.)
2. As tolerances dip below 0.002 in., calipers aren't repeatable enough.
3. Alignment and feel problems
4. It is true. Readings appear larger than reality.
5. False. It is a cosine error.

Answers to Section Review Questions

1. A. Limits   B. Limits   C. General
   D. General   E. Unilateral
2. Degrees for both the dimension and tolerance
3. False. Developing repeatability is your goal
4. Wrong perspective
5. Pressure—feel; alignment to the object; dirt and burrs; calibration; parallax; tool wear; heat; damage to the tool—bent, sprung, and so forth; the measuring environment (organized, good lighting, comfort, etc.); bias for a result
6. Data output, metric/Imperial conversion, datum and nominal zeroing
7. Cosine error occurs when a travel (plunger) type dial test indicator is tilted relative to the measurement.
8. Of the three measuring techniques involving a test indicator and height gage:
   A. Comparison uses gage blocks.
   B. They all require a datum table.
   C. Height gage scale measurement doesn't use the DTI as a measuring tool.
9. Overforcing or cranking too hard
10. A. For you alone, either mic should be equally accurate unless you are a rank beginner, whereby the friction drive will help with developing a reliable feel.
    B. For two users, the friction device is the better choice as it helps remove differences in feel between multiple users.
11. Because both tools are challenged by the same two major inaccuracies: feel/pressure and alignment.
12. Since the largest possible pin could be 0.875-in. diameter (MMC size), then the smallest MMC hole must never be smaller than 0.876-in.

diameter. Adding the total permissible variation to the hole diameter, the largest hole size would then be 0.879 in. Half way between 0.876 in. and 0.879 in., the nominal size would then be 0.8775 in. to create a true bilateral tolerance.

13. When using the indicator to align machine axes or a vise, the object is to zero out error end-to-end, therefore the numbers matter less than both ends of the tested object being the same. By deliberately causing progressive error with the probe angle, the indicator is made deliberately more sensitive.

14. Actual (rule) graduations are slightly bigger than vernier. When the vernier slide is moved until a given line coincides with an actual line, the movement represents the difference between the two graduations, times the number of graduations passed. That is, if the seventh vernier line coincides and the difference between graduations is 0.001 in., the distance moved is 0.001 in. times 7 = 0.007 inch.

15. The reading becomes less than the true size.

16. A. Size, position, form, smoothness, and orientation
    B. Only position and orientation require a datum reference.

17. It is a distance between two lines or planes built around a perfect model of the angle with respect to a datum.

18. It is measured from the surface to a protractor blade *locked at the correct angle* with respect to a *datum.*

19. Above the horizontal line, each graduation represents 1.0 mm, while below the line each graduation equals 0.5 mm.

20. An electronic caliper (or micrometer) set to read zero at the nominal size of the object. It would then read in plus or minus thousandths and tenths. This is nominal zeroing and very useful when one doesn't have time to add and subtract numbers.

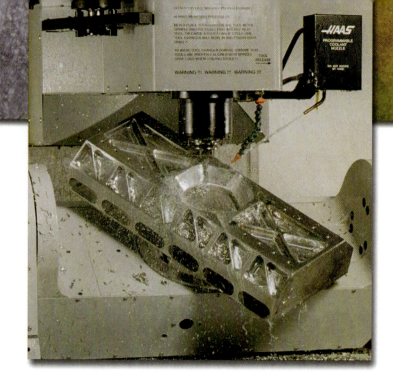

Single-Purpose Measuring Tools, Gages, and Surface Roughness

Introduction

The measuring tools and processes of Section 7 are more specialized than those we've just discussed. Within the five kinds of measuring, they fill in many gaps in the ability to perform close tolerance measuring in difficult situations. We'll also study measuring angles and surface roughness. On completion of Section 7, you'll have a fairly well rounded set of skills.

Similar to other technologies, measuring is a never-ending lesson that's evolving with advancing technology. You'll discover many new ways to use the tools we'll discuss here and in Sections 6 and 8. Plus, new instruments become available regularly. The best way to stay informed is through tool catalogs and flyers sent to your shop or home. Go to the Internet and search for precision + measuring tools, then sign up for their mailing lists both online and mailed.

Objectives

Measuring with Inside and Depth Micrometers

Introduction: These two micrometers measure linear distances using contact pressure applied by touch. In addition to touch variables, each has its own list of inaccuracy factors. Each requires practice to use correctly.

KEY TERMS

Base size
The smallest size possible to measure for inside or depth micrometers.

Drag
The relative tightness or looseness of a measuring tool, set by the wear adjustment ring within the head.

Jacking
The undesirable act of lifting the base of a depth micrometer up off the surface upon which it rests.

Joggling
Author's term for the act of lightly moving a measuring tool until the null point is detected.

Null—null point
A maximum (or minimum) value detected on a measuring tool when it is moved into alignment.

Inside Micrometers

Inside micrometers (Fig. 7-1) measure hole diameters within a range from around $1\frac{1}{2}$ in. up to 6 in. for the standard set, or expanded sets to 12 in. There are much larger sets owned by the shop, not the machinist, that measure up to several feet. A standard set would be a priority 2 or perhaps 3, for your purchase plan.

Tech-Specs for Inside Micrometers

Correct Application These measuring instruments are applicable where diameter tolerances are closer than 0.002 in. but not closer than around 0.0005 in. depending on the situation. Closer inspection of hole diameters requires bore gages, to be discussed in this section, and other hole measuring technologies beyond the scope of this section.

Figure 7-1 *The complete inside micrometer set includes the $\frac{1}{2}$-in. range head, extension rods, and the $\frac{1}{2}$-in. spacer.*

Inaccuracies In addition to the usual factors of parallax, wear, heat, and burrs, there are two new challenges to control for accuracy. Please obtain an inside micrometer now as we look at the features and factors (Fig. 7-2).

Double Calibration Note that the set is composed of a measuring head and several precision rods. On some sets, not only must the micrometer head be calibrated to remove zero error but each interchangeable rod must be adjusted, too. Adjusting out error in these tools is time consuming.

Inside micrometers can be checked for zero index error with outside micrometers but

Figure 7-2 *Inside micrometer components.*

Figure 7-3 *Inside micrometers and dial bore gages (studied next) are best calibrated against precision ring gages.*

To joggle the inside mic, hold the measuring head end against the hole wall then joggle the other end through.

introducing a tool with its own inaccuracies as a checking gage degrades results. The best solution is a precision ring gage (Fig. 7-3). A ring gage is a hardened steel, ground donut made to an exact inside diameter. Several must be owned by the shop if they are to calibrate inside mics and other inside measuring tools as well. Often, shops will purchase the major sizes such as 3.0-in., 4.0-in., 5.0-in. diameter, and so on.

Feel Challenge Differing from other kinds of micrometer measurement, the inside mic is not adjusted to fit the hole size when measuring smaller diameters—when the micrometer head is in the hole. That's because your fingers may not fit in. So, finding the **null** or **null point** when measuring (the largest micrometer size that will just pass through the hole) is a matter of trial and error. With the micrometer set to a trial size, insert it to see if it fits. Then take it out, adjust up or down, and try again until it slips through at the right touch for which it was calibrated.

The alignment method of lightly moving a measuring tool is called **joggling** in this book, you've already used it for other measuring tools in Section 6. It's a light wiggling side to side and back and forth, rocking it through the hole. If it falls through easily, remove it and turn the head out one- or two-thousandths and try again. If it won't go through reduce size until it will joggle through lightly.

As you'll soon see, developing a personal repeatability takes practice and patience. The only way you'll learn that "right feel" is to set up experiments following the investigation pattern from Section 6.

Set with More Drag The inside mic's head has no lock. Instead, it is set much tighter than all other micrometers. That's so it won't move when joggled and removed.

Target Personal Repeatability

Your target should be from around 0.0005 in., depending on hole diameter (larger holes are easier to measure) and the machined roughness in the hole.

There is no vernier scale on an inside micrometer to subdivide thousandths of an inch. It would be pointless, since their discrimination is hampered by the extreme feel factors and the way they must be used inside a hole.

Setting Up, Reading, and Using an Inside Micrometer

Set Up To ready the micrometer for use, you will need to insert the correct rod into the micrometer head. Make absolutely certain both the rod and mating head surfaces are clean before insertion then tighten the retaining cap.

Half Inch Range Now, notice that the inside micrometer head is smaller than other micrometers. It is limited to a 0.5-in. (or 13-mm) range. That keeps it compact enough to be used in smaller diameter holes.

The basic micrometer with the shortest rod measures from around $1\frac{1}{2}$ in., depending on the makeup, to $2\frac{1}{2}$ in. Each rod covers a range of 1 in., but in half-inch increments. The first half inch of the rod's range is achieved without the spacer. To set the micrometer up for the second half of the rod's range, the half-inch precision spacer must be added to the setup. Identify the spacer now. It is also shown in Fig. 7-1. Examples of this setup are seen in Fig. 7-4.

Inside Micrometer Ranges

Basic size — 1.500

Spacer — 1.500 + 0.500

New rod no spacer — 1.500 + 1.000

Figure 7-4 *Three ranges by the way the inside mic is set up.*

The ½-in. spacer in an inside micrometer set is a simple, small item that can be mistaken for a small bushing and misplaced easily. Without it, the set becomes useless for half of all its potential range.

Figure 7-5 *The extension handle helps reach into deep holes or tight places.*

Skill Tips—Reading Inside Micrometers Here are four little hints to help obtain the best accuracy from an inside mic:

1. **Ruler Double Check**
   After completion of the measurement, it's a good idea to double-check the results against a precision ruler. The ruler check will verify that you missed nothing in the math, which does happen since the total reading is a total of several elements (base head size + rod size + head reading + spacer if used).

2. **Extension Attachment**
   The set may include an extension handle similar to the one in Fig. 7-5. It is used to get the micrometer into places where you would not be able to reach with your hand alone.

3. **Check with an Outside Micrometer**
   When taking an inside mic out of your tool box always test the basic head against a calibrated outside micrometer unless a ring gage is readily available. It's also a good idea to double-check final results using an outside mic, just to be sure!

4. **Test Several Locations**
   It's good practice to test the hole in several locations to verify cylindricity. It might not be round or cylindrical! Three tests minimum at each circular element are recommended. In other words, measure each element three times, then test other circular elements. Vary the contact point with each test.

Reading the Inside Micrometer

When reading an inside mic, you will add three or four factors together for the final measurement, depending on whether the spacer is used or not.

Base Size First Determine the basic or **base size** of the inside micrometer—that is, determine the smallest measurement it can make with the first rod and no spacer. That size will often be engraved on the thimble or body of the tool. For example, in Fig. 7-6, if the thimble of this mic was at zero (0.000) it would be at 1.500 in. What does it read as set? *Answer* 1.502 in.—just slightly larger than its base size on the thimble.

Figure 7-6 *Read this inside micrometer.*

From that we see that any given measurement will be the total of:

Base head size + Head reading +

Rod + Spacer (if it's used)

Sound difficult? OK, maybe a little, but the double checks with a micrometer or ruler will help offset the added complexities. Try this one with a base size of 1.500 in.:

Answer 1.502

KEYPOINT

We'll assume a base size of 1.500 in. for all exercises here, but that's not always true with all inside mics! If it's not engraved on the tool, use your outside micrometer to find out for sure the smallest size with the base rod and no spacer.

Question? What would the reading be if the inside mic in Fig. 7-7 had the spacer?

Answers

A. 1.5 + 0.317 = 1.817 in. (no spacer)
B. 1.5 + 0.500 + 0.317 = 2.317 in. (spacer adds 0.500)

Rod Size Next Now, with the basic size solved for that micrometer, determine which rod is needed for your measurement. They usually feature a range of 1 in. to 2.500 to 3.500 in., for example, then 3.500 to 4.500 in. and so on. Try this measurement in Fig. 7-8.

The answer is 2.500 + 0.392 = 2.892 in.

Spacer Required? What would Fig. 7-8 read if we had the spacer installed? 3.392 in. So, you get the idea, add another 0.500 in. to the total reading. The challenge isn't in the tally of the reading, it's in using the mic with just the right feel. You'll need to set up

Figure 7-7 *Read this inside micrometer.*

Figure 7-8 *Read this micrometer with a longer extension rod.*

experiments and challenges that follow the same investigative path as those for outside micrometers.

Depth Micrometers

Like inside micrometers, depth micrometers are purchased in universal sets measuring from 1.5 in. up to 6 or 12 in. They are also a priority 2 in your purchase plan or a priority 3. Depth mics might be supplied by your employer, but they are handy to have in your tool box also. They too are more accurate than the depth probe function of calipers. Depth mics have their own inaccuracies that must be recognized and controlled with practice and experimentation.

Tech-Specs for Depth Micrometers

Correct Application Depth micrometers (Fig. 7-10) are used when measuring linear distance between two surfaces where the upper surface touching the mic's base, is flat. With practice, they deliver personal repeatabilities finer than 0.002 in. but not finer than 0.0002 or 0.0003 in. due to their own peculiar feel factor and head construction. Reading finer than 0.001 in. requires visual estimation as there is no vernier scale (Fig. 7-9).

Target Personal Repeatability This needs to be from 0.0003 to 0.0005 in., dependent on the situation.

Control of Inaccuracies The major factors for depth micrometers are calibration and feel.

Calibration Notice that these too have calibrating rods similar to the inside mic. Using the basic rod size (0 to 1 in.) with the head set to 0, test it against a flat precision surface to

ShopTalk

Inside Mics are Too Slow for CNC Production!

Although they're more accurate and more specific than calipers, inside mics are still a universal instrument not generally suitable for production work. For faster measurements with better repeatability, there are other methods of measuring inside diameters. The dial bore gage is the most prominent (Fig. 7-9). Bore gages are set up to measure a specific diameter with a repeatability of around 0.0002 to 0.0001 in. It's possible that your instructor has one or more in the lab but they aren't common in training. Using a ring gage, they are set to read zero at the nominal size for a specific measurement. With no math and being fast and repeatable within tenths, these tools are absolutely necessary for CNC production where messing around with inside mics would be way too slow and inaccurate. More to come on bore gages in Unit 7-3.

Figure 7-9 *A dial bore gage is far more repeatable and easier to use than an inside micrometer.*

Figure 7-11 *Calibrating the base rod (left), zero against the table. To check longer rods, use gage blocks.*

Figure 7-10 *This depth micrometer is being used to measure this precision ratchet part in the shop.*

Figure 7-12 *Lifting of the micrometer off its base can occur.*

KEYPOINT

To help avoid lifting the base, most machinists set the drag on their depth mic's head a bit loose to facilitate immediate feedback sensitivity.

check for zero error. Performing this test is a good way to get the feel of the zero pressure as well.

Calibrating all the measuring rods beyond 1 in. requires measuring down from a stack of precision gage blocks against the granite plate or other precision datum surface (Fig. 7-11).

Feel. The major bug with depth mics is the contact footprint of the tip of the small diameter probe rod. It's not easy to tell when it touches the surface being measured. One can accidentally lift the base up off the surface on which it rests—known as **jacking** the mic or lifting it (Fig. 7-12).

Both outside, inside, and depth micrometers feature a small drag adjustment ring inside the head, that allows changing the clearance between the spindle threads and the threaded thimble. It's used to set a given amount of **drag** between the two (the relative tightness or looseness of a measuring tool). This is the same adjustment to take up extra play in a micrometer when its threads are worn. To see this adjustment ring, the thimble must be backed out beyond normal measuring position.

You'll make the decision of how tight to adjust the drag on your micrometer head but only when buying your own. The drag on shop's depth micrometers will probably be set closer to the outside micrometer's head.

Reading a Depth Micrometer

It would be best if you had a depth micrometer set as we proceed. There's an oddity about these instruments.

Backward Readings In Fig. 7-13, note that the micrometer head reads opposite previous instruments. It is at zero when the thimble is backed all the way out (unscrewed). *You must read it by determining how much has not been covered up.* Look at Fig. 7-13. What is the reading? Now, read the magnified depth mic setting in Fig. 7-14. It's tricky.

Answer It's 0.012 in. beyond the 0.050 line but not yet to the 0.075 graduation line at 0.062 in. Add that to the 0.600-in. grads covered by the thimble and the 2.000-in. rod, for a total of 2.662 in.

Now, set your depth micrometer to the same readings as in Figs. 7-13 and 7-14.

Figure 7-13 *This depth micrometer reads 0.475 in. Can you see why?*

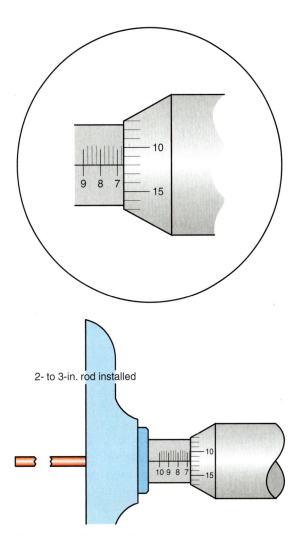

2- to 3-in. rod installed

Figure 7-14 *Read this depth micrometer.*

Skill Tips for Depth Mics

◆ Be certain the measuring tip is clean before using it.

◆ Use ultralight fingertip pressure on the thimble.

◆ Apply hard pressure to the topside of the base to avoid jacking it above the datum surface.

◆ After contact is made, back up and recontact three or four times to be certain.

◆ If the width of the base is too narrow for the measurement, it's possible to bridge the gap with a ground parallel bar, as shown in Fig. 7-15. Remember to deduct the parallel thickness from the actual measurement!

Precision Features for Depth Micrometers

◆ *Different Base Widths* There is no best size; however, a narrow base can fit where a wide one wouldn't.

Pressure ratchet

Ground parallel

Figure 7-15 *A ground parallel bar of known size can bridge larger gaps.*

◆ *Pressure Regulators* Depth mics may have a friction or ratchet pressure control device, which is good if more than one machinist is to share the tool (Fig. 7-15).

◆ *Carbide Tips* Less important than outside mics since they do not rub with as much pressure, some depth mics feature wear resistant carbide tips on the measuring rod faces.

UNIT 7-1 REVIEW

Review the Key Terms

Base size
The smallest measurement possible for inside or depth micrometer sets.

Drag
The relative tightness or looseness of a measuring tool, set by the wear adjustment ring within the head.

Jacking
The undesirable act of lifting the base of a depth micrometer up off the surface upon which it rests.

Joggling
Author's term for the act of lightly moving a measuring tool until the null point is detected.

Null—null point
A maximum (or minimum) value detected on a measuring tool when it is moved into alignment.

Replay the Key Points

◆ Every inside mic has a basic smallest diameter size that must be known to the user.

◆ Inside micrometer heads feature a range of $\frac{1}{2}$ in. to keep them compact.

◆ The inside micrometer kit includes a small spacer to extend the range of each measuring rod.

◆ Depth mics are read by determining the graduation lines covered—just backward from all other micrometer heads.

◆ Both depth and inside mics do not have a vernier scale as their feel inaccuracy makes their repeatability too coarse to measure to tenths.

Respond

1. True or false? Depth micrometers measure backward—the largest reading on the head is with the thimble backed all the way out (unscrewed to the end of their range).

2. The range of inside micrometers is _____ inches or _____ millimeters. Explain why the head of inside micrometers is different than all others.

3. The final reading of an inside micrometer is the sum of three or four elements. What are they?

4. What is the greatest accuracy challenge common to both inside and depth micrometers?

5. When tolerances dip below around 0.0005 in., the inside micrometer might not be the tool to use, especially where many measurements might be needed of the same dimension. What instrument would be better?

UNIT 7-2

Setup, Use, and Care of Precision Gage Blocks

Introduction: The word gage is used in many different ways within our trade and outside, too. The classic definition, when applied to shop measuring, means a tool that's used as a comparison template of the perfect shape and/or size. It is held against or next to the part in question and compared in one of several ways, which are discussed here.

In this unit, we'll look at a variety of gages, each of which is normally owned by the machinist. The exception to ownership are the gage block sets, which, due to their high cost, are shop supplied. We'll start with them because they are at the root of precision in most shops along with the granite table.

Functional gage
A gage that detects MMC condition. A functional gage simulates the worst case assembly to the mating component.

Gage (gauge)
In measuring, a gage is a template of the size or shape of an object. This definition is one of several as the word gage is used in other ways within measuring.

Jo blocks
A trade term for precision gage blocks, stemming from the original manufacturer the Johannson Tool Company.

Master gage
Any calibrating tool not used for shop or day-to-day inspection work.

Wring
Pronounced "ring," the act of placing two precision gage blocks together whereby they cling to each other. Wringing is proof of surface accuracy for gage blocks.

Precision Gage Blocks

Precision gage blocks, are a set of extremely accurate hard steel or ceramic blocks, supplied in graduated sizes. They are used in several ways within the overall measuring system. Many shops keep two or more sets, one for each duty listed below. They are manufactured in several different tolerance grades depending on the application starting with the closest to perfection:

1. **Master Calibration**
   The parent reference standard found in calibration labs but not in manufacturing, the **master gage** is used to calibrate tools farther down the pyramid. It will be used daily within calibration labs or rarely within inspection areas in manufacturing, but is not usually used for direct contact with work at the machines.

Figure 7-16 *A set of precision gage blocks.*

2. **Inspection Shop**
   They are used in the inspection area of the shop but not at the machines. Working gages are used for machine setups and in-work inspection.

Gauge or Gage? What's the difference? Answer found later.

KEYPOINT

No matter the tolerances to which they are made, even the shop grade precision gage block could be considered a perfect tool within a machinist's frame of reference. The only way to be more accurate is through scientific means!

Target Repeatability

Correctly used, gage blocks (Fig. 7-16) can be combined to reproduce a comparison size model to the closest 0.0001 in. with an overall repeatable tolerance of plus 0.00004 to minus 0.00002 in.* per block—inspection-quality blocks. Most sets can be combined to create a stack of from around 0.010 in. up to 8 or 10 in. The goal is always to choose the fewest number of blocks to achieve the target size—that's a trick we'll explore here.

Control of Inaccuracies

The whole point of gage blocks is that there are only three possible spoilers to repeatability—dirt, damage to the precision surface,

*That's not a misprint! These tools are made at subtenth tolerances—pronounced "plus forty to minus twenty millionths" in shop lingo.

How Are Gage Blocks Made? Many parts of this process are guarded trade secrets, but even the known basic steps are complex and labor intensive. First, they are machined from a highly refined tool steel without impurities. The alloy is custom designed to exhibit a very low coefficient of heat expansion and shape stability. The steel is progressively hardened through a series of steps that leave it without the internal stresses of normal hardened tool steel — it won't warp over time as other heat-treated steel will. Also, it will not warp with temperature change.

Next, after machining to within a few thousandths of the final size, they are precision ground in stages to within tenths of the target. Then the blocks are abrasion lapped (rubbed with loose abrasive grains) against other flat master lap blocks. This time-consuming process occurs in stages with finer and finer abrasive. The final step is to remove microscopic imperfections then inspect and grade them. They are sorted to become one of the four grades, then rust-proofed in a variety of ways and put in sets. The resulting gage blocks have pedigrees that are just one or two generations removed from international standard master-masters. Very few shops, worldwide, can produce gage blocks to these exacting standards and tolerances. That explains why seemingly simple little steel blocks cost so much!

Figure 7-17 *The right way to lift precision blocks from their case.*

and heat. All are easily overcome but the damage aspect requires some training.

KEYPOINT

Radical Care Required!
Because they are simple little blocks, untrained beginners might miss how carefully they must be handled and used. No kidding, one single thumbprint left behind will ruin them within just minutes! The surface is so fine that skin acid will quickly corrode it.

Here are the game rules for the savvy user:

Wash Your Hands First. Especially if you've been working hard enough to perspire.

Do Not Touch the Gage Surface. Even with clean hands. To help, the storage box is made such that the gages can be tilted up then grasped by the edge (see Fig. 7-17).

Clean Them Immediately if You Do Touch the Surface. Nearly every time I demonstrate how to not touch the surface, I accidentally do, with the whole class watching! It happens. But the pro-fix is to keep clean white paper or tissue handy to wipe off the oils and acids right away.

Clean the Surface Every Time They Are Stacked Together. Use clean white printer or notebook paper lying on a flat surface, pull the block across it three or four times. Lens cleaning tissue is ideal for this task, but not paper towels—they are not pure enough.

Keep the Box Closed. Even normal atmospheric contaminants can ruin gage blocks! But the dust that settles on them must also be wiped away before stacking together.

Preserving the Surface. To prevent corrosion, an oil spray or preservative wipe is required for storage in the box if they won't be used for a while.

Never Lay the Stack on a Machine Surface. Once the stack is set up, lay it on a clean rag on the workbench.

As extreme as that all sounds, one sure way to identify yourself as a beginner in your new job is to mistreat the **"Jo" blocks.**

Correct Application

In machining, you will use gage blocks in four different capacities:

1. Calibration of other tools such as outside micrometers.

Figure 7-18 *Gage blocks are inserted in the slot as a functional test of the slot's machined size.*

2. As a gage measuring tool. In Fig. 7-18, the machinist is using a stack of blocks as a limit test of the slot width. The test shown is called a **functional gage** test because it determines the true geometric width of the slot by accounting for ir-regularities. A functional gage tests a feature such as this slot (or a group of features) by simulating assembly to the mating component at its MMC size. In other words, it simulates the worst case for assembly. The largest gage block stack that will slip in is the true functional width. It is determined from the maximum material condition of the mating part. The stack can be adjusted up in size by one-tenth of a thousandth at a time.

3. As a comparison standard for indicator measurement, studied in Section 6.

4. To set up other tools or processes (see sine bars in this section).

Gage Block Accessories

Gage blocks are supplied in Imperial and metric sets with several levels of accuracy and overall size ranges. A shop might buy carbide end blocks to resist wear or clamp holders to keep them in a stack. Other accessories include special probes, jaws, and scribes that can be added to the blocks to use in special applications. These are items you will learn on the job as required. The important thing

here is to know how to handle gage blocks and how to set up a stack.

Skill Tip 1—Wringing Gage Blocks

The nature of gage blocks can be demonstrated. They are so flat and smooth that they will tightly hold together without any mechanical fastening. This is called **"wring-ing."** (As a bell would sound.) Ask your instructor to demonstrate this or see if you can wring the gage blocks yourself.

Back on earth, here's how we wring blocks (Fig. 7-19):

Put two clean blocks together at an angle to each other.

Push lightly and rotate into alignment.

At a certain point, they will adhere; a perfect test of their cleanliness and surface condition.

They will not come apart easily without rotation or sliding sideways, if correctly wrung! Several blocks may be wrung together to form an exact measuring stack.

Skill Tip 2—Making Up the Gage Block Stack Size

Often, a stack of blocks must be set up to a given, odd size (such as 3.7834 in., for example). There is a trick to getting the right blocks out in the correct order:

KEYPOINT

Always eliminate the smallest digit to the right of the decimal point first then the next smallest and so on.

Observe the gage block set in your lab. You will see three series of sizes. One with nine blocks having 0.0001-in. graduations (0.1001, 0.1002, 0.1003, and so on). There are two more with 0.001- and 0.010-in. increments. Then the lager blocks in quarter, half, and full inch increments. Metric blocks are far simpler due to the nature of the ISO system.

1. Place gage blocks together as shown.

2. Slide the upper block over the lower with a slight circular motion.

3. Slide the upper block half out of the engagement.

4. Slide the upper block back into full engagement.

Note: To separate, reverse order.

Figure 7-19

For this example (Fig. 7-20), you are to make up a stack of blocks that totals 3.7834 in. The first block to select is the 0.1004-in. block from the tenths series. Now subtract that from the total stack:

3.7834 − 0.1004 = 3.683 in. left

Now take out and clean the 0.103-in. block from the thousandth increment series and so on.

3.683 − 0.103 = 3.580 in. left

Next select the 0.180-in. block from the ten-thousandth increment series. Clean and wring it in, then subtract:

3.580 − 0.180 = 3.400 in. left

Finally, remove the 0.400-in. and 3.0-in. blocks to complete the stack. The final total is 0.1004 + 0.103 + 0.180 + 0.400 + 3.00 = 3.7834 in.

Select blocks in this order

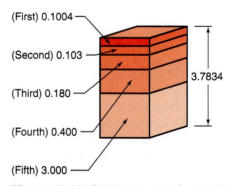

(First) 0.1004
(Second) 0.103
(Third) 0.180
(Fourth) 0.400
(Fifth) 3.000

3.7834

Figure 7-20 *Setting up a stack at 3.7834 in., eliminate digits from the right (smallest value) as you select and wring blocks.*

Review the Key Terms

Functional gage
A gage that detects MMC condition. A functional gage simulates the worst case assembly to the mating component.

Gage (gauge)
In measuring, a gage is a template of the size or shape of an object. This definition is one of several as the work gage is used in other ways within measuring.

Jo blocks
A trade term for precision gage blocks, stemming from the original manufacturer the Johannson Tool Company.

Master gage
Any calibrating tool not used for shop or day-to-day inspection work.

Wring
Pronounced "ring," the act of placing two precision gage blocks together whereby they cling to each other. Wringing is proof of surface accuracy for gage blocks.

Replay the Key Points

◆ Precision gage blocks are considered perfect in machine facilities.

◆ Avoid and immediately clean finger prints off the surfaces of the blocks.

◆ They are the standard for measuring in a shop.

◆ To set up a stack, remove the smallest digit first.

◆ Wringing is the test of perfection.

◆ Always store them in their special box, and if it is to be for a long term (1 day or more) wipe them with a preservative.

Respond

1. What two major challenges affect inside and depth mic accuracy?

2. True or false? An inside micrometer is read differently than an outside micrometer. If it's true, then how is it read differently?

3. True or false? A depth mic is read differently than an outside micrometer. If it's true, then how is it read differently?

4. List the blocks that should be taken out of the box, in order, to make up a stack of gage blocks to equal 4.4089 in.

5. Complete: Gages are used to _____.

UNIT 7-3

Gage Measuring

Introduction: In this unit, we'll look at a group of measuring tools that determine size and shape. We'll start with the hole gages, then look at several more tools that determine either size or shape elements. Each of the gages in this unit are very affordable and might all be considered a priority 1 for your measuring kit.

KEY TERMS

Go, no-go gage
A functional gage that tests the feature against upper and lower limits of its tolerance.

Planer gage
A type of adjustable parallel—might also be called a machinist's precision gage in tool catalogs.

Radius gage
A small flat template featuring several arcs all of which are the same radius curve.

Ring gage
A hardened steel donut with the inside diameter ground into a perfect circle of known size.

Snap gage
The same tool as a telescope gage.

Telescope gage
An expanding feeler used to determine inside diameter. After setting at best fit, they are removed and measured with a micrometer.

Small-Hole Gages

Ball Gages

These are inexpensive tools for measuring the diameter of moderately small holes—from around 0.7500 in. down to a 0.125-in. diameter. Found in sets of four (or more), each measures a range that overlaps with the next.

Correct Application Ball gages (Fig. 7-21) are used where tolerances are at or above 0.001 in. However, in a practiced hand they might deliver repeatabilities close to 0.0005 in. when no other method is available. Again, practice and awareness of their shortcomings is the key to repeatability. Due to the double feel (one time while the gage is in the hole and once when measured with a micrometer), one can never be sure of the result with these (and with telescope gages too!). As such, they must be considered utility, universal tools as

Figure 7-21 *A set of ball gages has a range from 0.1 in. up to around 0.5 in.*

Expansion nut

Tapered shaft

Split ball

Figure 7-22 *The ball gage in cross section.*

Figure 7-23 *A full set of telescoping gages.*

opposed to dead accurate methods. Ball and telescope gages are OK for many measuring situations where a dial bore gage isn't set up. But they are slow to use.

Gaging the Hole Ball gages are used as an adjustable, expanding feeler to determine the best-fit diameter. In the hole, while moving the gage in and out, the expansion nut is turned to expand the ball (Fig. 7-22). Once the ball drags in the hole, approximately the same as the micrometer that will be used to measure the gage, it is removed and measured.

Only practice will develop that sensitivity. Using them this way, ball gages are too slow for most modern CNC production.

KEYPOINT

Due to the way they are used, ball and telescope gages bring a double-feel challenge to the accuracy test. The feel within the hole must be the same as that applied by the micrometer to measure the ball. Both ball gages and telescope gages require experimentation and practice to get consistent results. When using them, seek out the smallest point within the hole's diameter. The gage set to that functional limit then detects the MMC size of a mating pin. Any larger pin will not pass the tight spot.

Telescoping Gages

Telescoping gages comprise another inexpensive, universal set for gaging holes bigger than those checked by the ball gages (Fig. 7-23). Starting at the smaller size of around 0.50-in. diameter, they cross over the inside micrometer's range at around 1.50 in.

Correct Application Similar to ball gages, telescope gages are used in an adjustable feeler test of the hole size that's removed and measured with a micrometer. They can deliver accuracy within 0.0005 in., but if there is an inside micrometer that can do the job, it's usually the better choice. Telescope gages require more practice to use correctly, than ball gages, due to their alignment/feel problem.

Different from ball gages, the control of the drag is preset with a telescope nut. Following the uncertainty principle, the variation between tightening that nut by two different users is often surprisingly great, as much as 0.001 in. or more!

Rock through *one time only*

Friction lock nut

Figure 7-24 *After expanding the telescope in the hole, snug the lock. Then rock it through, remove it, and measure it.*

- **TRY IT**

- **AN EXPERIMENT**
- Set up your own test comparing yourself
- with one or two other students to see how
- close you can come to consistency. You'll
- soon see what I mean. But, with good
- technique and patience, it's possible to
- reduce the variation between the test
- group. Here's how:

Using the Telescope Gage Right These gages are inserted in the hole then tilted at an angle, as shown in Fig. 7-24. The friction lock is released, which allows the telescope to expand larger than the hole's diameter. The friction lock nut is resnugged "just the right amount," then the tool is rocked through the axis of the hole—*one time only.* Not back. Once again, that right amount will be determined only with practice.

KEYPOINT

Do not rock the telescope gage back and forth—pass it through in one direction, one time only. The tightness of the friction lock determines the feel of the gage within the hole.

As the telescope rocks through, it is squeezed down to the hole size and does not change. At least that's the theory! The right amount to tighten the nut is light but firm—I can't describe it better than that.

Trade Tip

Due to the uncertainty of telescope gages, always make five or more tests of the same hole. Record results and look for a consistent average. This is not just a beginner's move—we all do it if there's no other instrument available to do the measurement.

As shown in Fig. 7-25, the micrometer is closed slowly and lightly on the telescope gage. Note the correct hand position. The problem is, using the fine-threaded micrometer, the contact feel can actually push the telescope farther closed to create a false reading. The solution would seem to be to tighten the lock after removal from the hole, but that often disturbs the position of the telescope. The only real control is to go easy on the measurement!

Dial Bore Gages

We've come to the best everyday solution for fast accurate hole measurement, but it is very expensive compared to the previous two. Dial bore gages incorporate a three-point contact with the hole, which produces a better picture of roundness compared to the two-point contact of telescope and ball gages.

ShopTalk

Snap Gages You might hear telescope gages called **snap gages** due to the way they spring open when the lock nut is released. Some feature a single sprung telescope section while others use a double expansion system to encompass a wider range of possible sizes per individual gage.

Figure 7-25 *The correct hand position for measuring a telescope gage with a micrometer.*

Figure 7-26 *Dial bore gages for measuring hole diameters.*

Arguably the most common tool in industry for hole measurement, they are expensive enough that most tech schools may not have more than one or two for demonstrating.

Not strictly following the definition of a gage, they report size on a dial (or electronic readout), as shown in Fig. 7-26.

Each dial bore gage covers a range of diameters by installing different probe rods (Fig. 7-27). But each is still limited by the curvature of the hole and the head size, thus shops must own several to cover a full working range of sizes.

Tech-Specs for Dial Bore Gages

Correct Application Different sized dial bore gages can measure small holes from around 0.100-in. diameter up to an unlimited size. They are suitable for tolerances finer than 0.0005 in. As mentioned in Section 6, they are one of the best tools for fast production where there is no time to use the telescope or ball gages and a reliable test is required. They're nearly foolproof to use with no inaccuracies other than the original setup to make them read zero at the nominal hole size, and dirt or burrs in the hole can destroy accuracy.

Because it's such a critical step, the bore gage may need to be set up by a lead person or the inspection department. If you will be using them repeatedly in CNC production, you may be supplied with a shop-made ring gage of the nominal size, to quickly check that the bore gage hasn't slipped from nominal.

Skill Tips for Use Similar to with a telescope gage, the head is placed in the hole at

Figure 7-27 *The right way to rock a dial bore gage.*

an angle, then rested on the contact points (Fig. 7-27). The probe is then rocked though the diameter, while you watch the dial. Holding the handle parallel to the hole's axis, in the side-to-side plane, it is rocked up and down. Doing so, the reading will swing from small to large then small again. The high point null is the correct diameter. This is not difficult to do and requires little practice to get near perfect results. It's a good idea to move the gage to several other locations within the hole, to test for taper.

Repeatability Quality dial bore gages deliver from 0.0001- to 0.0002-in. repeatability when correctly set up, maintained, and used. Because the probe spring always pushes against the hole's surface, the challenge of feel is eliminated. The major challenge to accuracy is the hole's roughness.

Setting Nominal Zero Although there are other more technical calibration stations used to set bore gages, the precision ring works well but adds further expense to owning them. Adjustment to their reading is made in one of two possible locations, depending on the gage construction: either the probe is screwed in or out, then locked with a jam nut or on the readout is adjusted to zero at the right size. (Both adjustments are common on many dial bore gages.)

No Math! Dial Bore Gages are Plus/Minus Reading Note in Fig. 7-27, that dial bore gages register variation in a plus and minus amount from zero similar to probe indicators. This test shows the diameter at 0.0007 in. larger than nominal zero. Most are equipped with small clip-on tolerance flags that can be set at the upper and lower limits, as depicted in Fig. 7-27.

Trade Tip

Note the Size In high-speed CNC work, where more than one dial bore gage might be required to inspect one part, it's a good idea to put a small piece of tape on the gage face or the handle, with the nominal size noted on it. Also, keep the **ring gage** handy and test the setup twice or more per shift; heat and other factors can cause them to slip as much as 0.00005 in. (fifty-millionths). Not much, but it could be a half to a third of your entire tolerance!

Ground Gage Pins

This gage can be used to check the size of very small holes. They are supplied in large sets similar to gage blocks, but not in series as with blocks. Pins simply start at the smallest diameter and increase in increments of 0.001, 0.0005, or 0.0001 in., depending on the set.

To gage a hole diameter, one starts testing with increasingly larger pins until the size that fits the best is found. Like gage blocks, they too are shop-owned tools due to extreme cost. Like gage blocks they must be treated with ultracare (Fig. 7-28).

Trade Tip

While they aren't as accurate as manufactured gage pins, machinists make their own utility pins in sizes they need often for a given job. While in tech school, if the opportunity arises, make as many test pins as you can find time to do. Don't forget to engrave the size on them if you have the ability!

Inaccuracies for Gage Pins

Pins might be envisioned as round gage blocks, and they too are nearly free of challenges. When testing holes, they do have the

Figure 7-28 *A set of gage quality pins must be treated with great care.*

inaccuracy of feel, but it's minor compared to that of ball and telescope gages. The possibility exists of a damaged pin, but that's rare as they are made of hardened tool steel.

Correct Application

Using best fit/feel, test pins can gage hole diameters of a few thousandths inch diameter up to around 1.000 in. In addition to diameter, they are also a geometric functional test of cylindricity. With the MMC pin inserted, you have found the largest cylindrical assembly space available. Left within the hole, that MMC pin can also be measured for position and orientation to a datum (Fig. 7-29).

Test pin

Largest cylindrical space available

Figure 7-29 *A gage pin tests for the available cylindrical space. This pin could now be measured for location and also orientation.*

Trade Tip

Pins as a Go, No-Go Test Often, two pins will be used; one at the lower tolerance diameter and one at the upper. If the "go" fits in and the "no-go" doesn't, the hole falls within the tolerance. When doing so in production, it's useful to tape them together in a pair, as shown in Fig. 7-30, then mark on the tape, red and green or go and no-go.

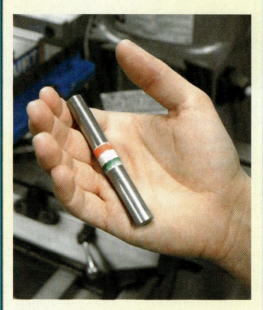

Figure 7-30 *With the lower hole size marked green and the upper limit in red, joining two gage pins forms a go, no-go tester.*

Choosing the Right Tool

Comparing Hole-Measuring Methods

The real skill is deciding which tool to use: calipers, inside micrometers, ball or telescope and dial bore gages, or test pins. You'll base your choice on

What's available.

How tight the tolerance is.

How quickly the test must be made.

Here Are a Few Deciding Factors

- Ball and telescope gages have a double-feel challenge, but they are usually readily available.
- Bore gages are the best, but they must be set up to a given nominal before they can be used.
- Unless already set up, a bore gage would be a slow choice for measuring only one hole!
- Gage pins determine not only diameter but also the geometric cylinder.
- Inside micrometers are universal, but they are slow and cannot measure small holes.
- Calipers are fast, but can't reach down into the hole, plus they have a low resolution compared to the other methods. Calipers make a good double check to ball and telescope gages though.
- There simply is no perfect answer! Every measuring task brings decisions—that's why we're called *skilled* machinists.

Adjustable Parallels

These are adjustable distance gages. They are useful tools for both measuring and setup work (Figs. 7-31 and 7-32). They can be used as a functional feeler gage for slots and grooves or can be preset as a **go, no-go gage** of thickness of an exact size (but are not as accurate as gage blocks).

Tech-Spec Comparisons for Hole Measuring			
Name	**± Repeatability (in.)**	**Skill Level**	**Speed**
Calipers	0.001 to 0.0015	Easy (overused by beginners)	Moderately fast
Micrometer	0.0001 to 0.0002	Practice required	Medium slow (math)
Ball gages	0.001	Practice	Slow (repeat test)
Telescope gage	0.001	Extra practice	Slow (must repeat)
Dial bore gage	0.0001	Easy, expensive	Fastest once set up
Ground pins	0.0001	Very easy, expensive	Moderate (try several)

Adjustable parallels are a priority 2 purchase. As you become more skilled, you will find many uses of these tools not only in measuring and layout but throughout the shop.

Figure 7-31 *A set of adjustable parallels are very useful in your measuring tools kit.*

Figure 7-32 *One use of an adjustable parallel is to measure the width with an outside micrometer or caliper.*

Trade Tip

Planer Gages Illustrated in Fig. 7-33 is a second type of adjustable parallel called a **planer gage** or *adjustable precision gage* in modern tool catalogs. They are a larger, more rugged version of the adjustable parallel, with potential uses all over the shop!

Planer gages feature two screw-on extension rods that increase their range by 1- and 2-in. increments. Also, there are several steps of exactly 1 in. built into the tool. They are also a 30/60/90 degree triangle, which is useful for checking common angles. The carriage can be removed to use the triangle alone. Some even feature a bubble level set in the frame!

Invented originally to set tool heights above the table on an uncommon machine in modern industry, the horizontal planer, these tools are being forgotten. But this is one item that shouldn't be made obsolete by changing technology. They are an excellent measuring and gaging tool. Here's just one example: As shown in Fig. 7-33, a slot width can be measured when the needed micrometer size is unavailable. These tools would be a priority 2 or even a 3 as compared to buying standard adjustable parallels, but don't forget them as your career develops.

Figure 7-33 *A planer gage has many shop applications.*

Radius Gages

These little flat templates are both a size and form evaluation tool (Fig. 7-34). To use them, hold them against the round feature in question to find the best fit (Fig. 7-35). The clue will be the visual gaps. Similar to evaluating straightness, you are looking for the widest gap between the gage and surface in question. **Radius gages** are a priority 2 purchase. They are supplied in metric or fractional or decimal inch sets.

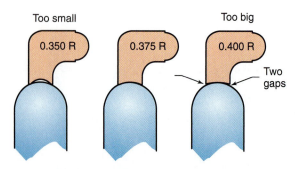

Figure 7-36 *Gaging a radius using three progressive sizes.*

Trade Tip

Use Light or a White Background When evaluating a radiused feature, a light background or a light source is very useful. To do so hold the part and gage up toward a light or place a piece of white paper in the line of sight beyond the object. This trade tip is useful in many ways throughout the shop, wherever visual estimating is done.

Reading the Gaps

It's not difficult to read the gaps to tell if the gage is too small or too big for the feature. In Fig. 7-36, three radius gages are shown testing the 0.750-in.-diameter round nose pin. One gap indicates the part's radius is bigger than that of the tool. Two gaps on either side tell the user that the gage is too big.

Figure 7-34 *A full set of radius gages in a protective pouch.*

Testing Machined Radii

Figure 7-35 *Using a radius gage to test the corners of a drill gage.*

UNIT 7-3 REVIEW

Review the Key Terms

Go, no-go gage
A functional gage that compares the feature to upper and lower limits of the drawing tolerance.

Planer gage
A type of adjustable parallel—might also be called a machinist's precision gage in tool catalogs.

Radius gage
A small flat template featuring several arcs all of which are the same radius curve.

Ring gage
A hardened steel donut with the inside diameter ground into a perfect circle of known size.

Snap gage
Shop lingo for a telescope gage.

Telescope gage
An expanding hole gage.

Replay the Key Points

- ◆ Both ball and telescope gages have serious feel challenges in terms of accuracy.
- ◆ Dial bore gages are the most accurate tool of the hole gages we have studied.
- ◆ Planer gages are a useful form of adjustable parallel.
- ◆ Radius gages can be fit to the feature by looking for gaps.

Respond

1. Name the possible measuring tools to determine a diameter of 1.500 in.

2. Same as Question 1 but within a tolerance of 0.0003 in. in one part? Rank your choices.

3. What tool must be used to calibrate a dial bore gage?

4. True or false? A planer gage is a versatile version of the adjustable parallel.

5. Radius gages evaluate two related items, they are: _____ and _____.

UNIT 7-4

Measuring Angles

Introduction: Unit 7-4 is all about angles and how to measure them. Before we tackle the measuring aspect, we need to take a look at how angles are expressed and toleranced. This background is necessary to understand how to apply the protractor—studied next.

KEY TERMS

Bracketing
Determining a colinear pair of lines by looking for the symmetrical mismatch to either side.

Complementary angles
Two angles that add to 90 degrees.

Decimal degrees (DD)
Angular expression in whole degrees and decimal parts.

Degrees, minutes, and seconds (DMS)
Angular expression in whole degrees, minutes (1/60 degree), and seconds (1/3,600 degree [60 × 60]).

Hypotenuse
The longest side in a right triangle—always opposite the right angle.

Protractor
An angle-measuring instrument.

Protracted angular tolerance
A tolerance zone that radiates outward from a vertex point.

Quadrant
Ninety degree segment of a circle. Four quadrants to one circle.

Sine bar
An angular setting and measuring tool based on the sine ratio of right triangles.

Sine ratio
The ratio comparison of the side opposite the angle in question to the hypotenuse side in a right triangle.

Vernier bevel protractor
An angular-measuring tool with a resolution of 5 minutes of arc.

Angular Expressions and Tolerances

Protracted and Geometric Angular Tolerances

For a quick review, recall that protracted angular tolerances start at a point and radiate outward. They produce a fan-shaped zone of plus or minus so many degrees. A geometric angle is a sandwich-shaped tolerance zone around a perfect definition. A geometric angle's permissible variation is a width. Knowing this, we'll apply the protractor to measuring the angle differently. See Fig. 7-37.

Decimal Degree (DD) Angles or Degrees, Minutes, and Seconds (DMS)

Using either expression, DD or DMS, a full circle is divided into 360 degrees. Then degrees are further divided into finer units. Either by extending the decimal or adding on minutes then seconds.

Decimal degree	60.22833°
DMS	60° 13′ 42″

When angles are expressed in DMS, degrees are divided into 60 minutes, and minutes are divided into 60 seconds.

Either expression might be seen on drawings. DD and DMS (or the symbol ° ′ ″) appear on your handheld calculator.

DMS While less useful from a mathematical and resolution perspective, DMS angles are

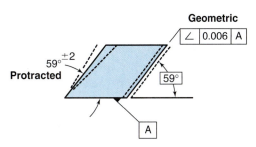

Figure 7-37 *A comparison of a protracted and a geometric angle.*

In engineering and mathematics, there are two more ways the circle is divided into major parts other than degrees. They are more useful for engineering and scientific calculations where extreme revolutions or long or fast cycles are part of the computation.

A full circle contains 2π radians or 400 gradients. While using gradients would simplify shop math, don't expect to see them showing up anytime soon. Changing tradition, redoing drawings, and buying new measuring tools would all need to be addressed. The DRG key (degrees, radians, and gradients) toggles from degrees to radians to gradients then back to degrees again. Make sure it's in degree units when we come to sine bars in this unit.

still found on most prints, especially older versions, and they are used on the vernier protractor coming up. Decimal and minutes are also found on many modern machine accessories. Degrees are signified by a circle superscript. Minutes receive a single slash (') and seconds are shown with a double slash (").

Example: 38° 30′ 45″

It's pronounced "38 degrees, thirty minutes, and forty-five seconds."

Decimal Degree Angles Simpler to use in every way, whole degrees are followed by decimal parts. Given the choice, always use decimal angles. Conversion is a good idea too—before starting a problem involving angles. For example, DMS angle of 38°30′45″ becomes *38.512°*. This decimal angle is pronounced as with all math; "thirty-eight point five, one, two degrees or thirty-eight and five-hundred, twelve-thousandths degrees"—either expression is OK. When pronouncing angles, we drop the thousandth convention used for distances in inches.

Plain Protractors—1° Resolution

These are the utility angle-measuring tools that have a resolution of 1 degree. Although they might be shaped differently, they are all called a flat or plain **protractor.** The second version within this group is the protractor attachment for a combination set—seen in Section 6. These protractors are used the same way to measure and scribe angles where tolerances are modest (Fig. 7-38). One tool from this group should be a priority 1 then the others a priority 2 for your kit.

Correct Application With a good magnifier, it's possible to use plain protractors slightly finer than 1 degree but not closer than 30 minutes. Because the graduated arc is larger, reading between the graduations is a bit easier on the combination set protractor. In addition to measuring angles, you might use these protractors for layout work, hand tool grinding, and also to set up a machine where angles are not closely toleranced. We will not study them further but defer to the next tool on the precision ladder.

Vernier Bevel Protractors—5-Minute Resolution

The **vernier bevel protractor** is shortened to the *vernier protractor* in this text. This

Measuring Angles

There are only three low-tech choices for measuring angles. Moving toward technical solutions, we find the computer coordinate-measuring machine and optical comparitor—both covered later in this text.

| Plain Protractors | Vernier Protractors | Sine Bars and Plates |

Figure 7-38 *A protractor for angular measurement comes in several forms.*

angle-measuring tool is a universal, high-quality instrument that might be purchased as a priority 3. They are often shop supplied. The vernier version is the most common, therefore we'll study reading them, however, they are also supplied in dial versions, a very good improvement but additional expense for an already costly tool. The dial type is far easier to read because, as you will soon see, the vernier on these instruments is challenging!

While the word bevel is another name for a chamfer, these tools (Fig. 7-39) are far more universal than just checking bevels, when measuring angular surfaces of machined parts. Figure 7-40 shows just a few of the ways these tools can be used with and without the 90-degree attachment. Note in Fig. 7-40 that two applications are referenced from a datum surface—the one in the height gage is ultimately parallel to the table on which the base rests. Also note in Fig. 7-40 that the long and short blades have different angles on their tips 30°, 60°, and 45°. These can be used as templates.

The vernier protractor is supplied in a kit in which the main blade, a short blade, a central measuring head with a built-in magnifier, and a right-angle blade are included.

Tech-Specs for Vernier Protractors

Correct Application Vernier protractors are the right tool for measuring or setting up angles where tolerances are finer than 1 degree but not finer than 5 minutes of arc. Besides measuring angles, vernier protractors are used for layout work and for machine setups where attachments or slides must be set at a precise angle to the vise or machine table.

Target Repeatability 5 Minutes It is not possible to read finer than the tool resolution even when using a magnifier. A magnifying lens is usually supplied as part of the kit. If not, it's a good idea to use one when viewing the scale.

Control of Inaccuracies

The major inaccuracy factors for vernier protractors are

Parallax—a real problem using the curved magnifier lens. Be certain to view it straight on.

Visual estimation of the blade alignment to the work surface.

Figure 7-39 *A vernier (bevel) protractor and measuring blades in a kit.*

Figure 7-40 *Three possibilities for measuring angles with a vernier protractor.*

Reading the small vernier lines is difficult and it's possible to read the wrong vernier scale (training is coming on this problem).

While the central measuring head seems big, it is necessary to allow the 5-minute discrimination. However, the large head can make the tool difficult to fit into tight places at times.

Reading a Vernier Protractor

It would be best to obtain a vernier protractor as we proceed. Notice on yours or the one in

Angle index line
twin vernier scales

90°
Quadrants

Figure 7-41 *The blade has been removed to show the four 0 to 90 degree quadrants.*

Degree graduations

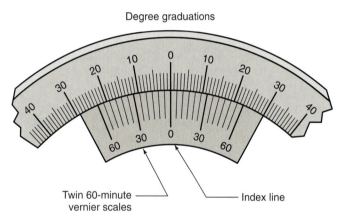

Twin 60-minute
vernier scales

Index line

Figure 7-42 *The vernier scale features an index line and two vernier scales. Use the "quadrant rule" to choose which vernier scale to use.*

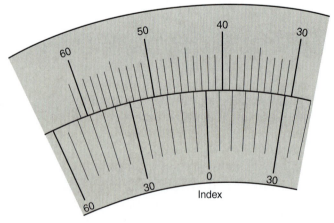

Index

Figure 7-43 *This vernier protractor reading is at 42 degrees and 25 minutes.*

Fig. 7-41, where it's shown with the blade removed, that the outer part of the measuring circle is divided into four 90-degree sections called **quadrants.** Two quadrants start from zero, then they are graduated up to 90 degrees. Then the adjacent quadrant descends back to zero. *For any given measurement, the index line will be one of the four. It's important to note that in which quadrant the index lies dictates which of the two vernier scales is to be read.*

Twin Vernier Scales

Notice in the magnified view of Fig. 7-42, that the vernier scale on the inner circle has a central index line that points at the actual angle on the outer circle. Then there are two vernier graduation sets that extend opposite directions from center. They are mirror image scales extending clockwise and counterclockwise. Use the quadrant rule to determine which vernier is used. Notice also that the vernier lines divide 1 degree into 12 parts $\frac{60}{12} = 5$ minutes of arc—that is the resolution of this tool. The vernier tells the user how many 5-minute divisions the index line has passed beyond a whole degree. Figure 7-43 illustrates a reading of 42 degrees and 25 minutes. It is 42 whole degrees and $\frac{25}{60}$ of the distance across to 43 degrees.

Trade Tip

Zeroing in on the Colinear Pair To help find the lined-up pair (Fig. 7-43) look for equally mismatched lines surrounding the colinear pair—the lines at 30 and 20 minutes. This trick, called **bracketing,** works for all vernier tools—the mismatch will be symmetrical in pairs around the centered lines.

Now read the protractor shown in Fig. 7-44 to be sure you have the concept, before moving on.

The answer is 23° 35′.

Figure 7-44 *Read this vernier protractor.*

Measuring Protracted Angles— Best Fit to Surface

If the angle is protracted as in Fig. 7-45, then place the protractor base on the edge of the object and use the fine movement screw until the blade fits best on the edge to be measured. The angle should fall within the stated **protracted angle tolerance.** In this example, plus or minus 2 degrees.

KEYPOINT

When measuring protracted angles with a plus/minus tolerance in degrees, you are moving the blade until it fits the object best. Sometimes that's hard to decide if it's bowed.

Measuring a Protracted Angle

Print callout

Rotate fine movement screw to measure best fit angle

Figure 7-45 *To measure a protracted angle, seek the best fit by rotating the fine movement screw.*

Measuring Geometric Angles— Gage Surface

If the angle is geometric, then instead of measuring by moving the protractor, you will be gauging the fit. First, set the protractor at the stated nominal (perfect model) angle—59 degrees—and lock it in position.

KEYPOINT

You are testing the surface with the blade locked in position at the target angle, looking for gaps. The protractor's base forms a datum of the reference surface. There is no guesswork.

Now hold the object in the protractor or place the set on a datum (Fig. 7-46) reference, then using a feeler gage or precision wires, check that no gap exceeds the tolerance along the blade. The geometric definition of angles is far more useful when fitting two machined parts together where the functional objective is to limit the maximum permitted gap between.

Gaging a Geometric Angle

Print callout

Testing for maximum gap 0.010 in.

Locked at 59 degrees

Datum B

Figure 7-46 *Gaging a geometric angle with a locked protractor.*

Trade Tip

Stay Out of the Complementary Trap! Avoid this predictable error when using any kind of protractor, be certain the angle measured is the target angle that you intend to measure. This problem arises from a geometry twist called **complementary** and supplementary **angles.** Angles are complementary when any two add to a total of 90 degrees and supplementary when they total 180 degrees. In some cases, it's easy to confuse the target angle with the complementary or supplementary angle on an object—always right next to the target!

This drill gage would be a prime example (Fig. 7-47). The angle needed must be tested from the supplementary side. Awareness of the trap is the real key. Want proof that it's easy to fall into this trap? I set you up. Return to Fig. 7-46 and see that truly, the protractor isn't measuring the 59 degree angle, but it's gaging the supplement. That's OK as long as you've done the math: what the protractor should read in Fig. 7-46 is 180° − 59° = 121°.

Variation is Reversed! Using the right-angle attachment can eliminate this problem in some cases, causing the reading to be direct. However, most of us just determine which angle we are actually measuring—the target angle, the supplement, or the complement. Then, when it is the complementary or supplementary angle that we have found, we know that the variation is reversed. In other words, if the supplement in Fig. 7-47 measures *big* at 123 degrees, the resulting target is actually *small* at 57 degrees.

The Final Confusion! But we aren't through yet. Notice in Fig. 7-48 that flat protractors have two scales running clockwise and counterclockwise, 0 degrees up to 180 degrees. These are supplementary values for the blade position. Setting the arm index at 59 degrees, as shown in the blown up view, also sets it at 121 degrees. When angular problems get complex, always use a flat protractor to check the vernier protractor to be sure. Given the next complexity of measuring angles, the flat protractor can solve many of the headaches.

Figure 7-47 *Be cautious when measuring that you do not confuse the target with the complementary or supplementary angles.*

Figure 7-48 *Use the double scales to verify the angle or its supplement.*

Remember the vernier protractor quadrant only goes up to 90 degrees then starts over. So the real angle to be evaluated on the vernier, in Fig. 7-47, is 31 degrees (121° − 90°).

Measuring and Setting Angles with Sine Bars—0.001-Degree Resolution

When angular tolerances are tighter than 5 minutes, we turn to a very different tool—the **sine bar** or its big brother the sine plate. If they are made accurately enough, these precision tools are able to measure or set up angle models within a thousandth of a degree. They are the most accurate angular tool in the machinist's box, or in the shop for that matter.

Sine bars function as parallel bars that tilt relative to a datum surface (Fig. 7-49). Although ultraprecise, they are also very sturdy so they can be used during machining setups—but they are more commonly used in measuring and making setups, where they would be removed once the angle is established.

Figure 7-49 *A sine bar is the ultimate angular measuring and gauging tool.*

Make Your Own Sine bars are often made by the trainee when in school, along with parallel bars. To make them right, it's necessary to cut and grind the pads that establish the distance between the circular rods to an exact number, usually 10 in. (but 5- and 1-in. sine bars are also used). The 10-in. bar is the more useful because of both its size and that it greatly simplifies the math. If purchasing your sine bar, it would be a priority 2.

Sine plates are wider tilting tables with their top surface hinged to a bottom plate that can be firmly held on the mill or grinder table. Sine plates usually feature threaded holes or magnets on their top surface to hold parts during light machining or angular grinding operations. We'll look at them again later as we explore machining techniques, but for now, we need to discuss the measuring capabilities of sine bars.

Both bars and plates are based on the trigonometry principle that any given angle will produce a specific ratio between the stack under the front of the bar as shown in the drawing and the distance between the round rods—called the **hypotenuse** (longest side of a right triangle). This ratio is called the sine ratio and abbreviated to *SIN* on your calculator (Fig. 7-49).

Figure 7-50 *What stack height produces 22.5° on this 10-in. sine bar?*

Depending on your calculator's logic, enter either

22.5 then touch SIN

or

SIN then touch 22.5 = **0.3826834**

There are two ways to look at that number:

Percentage of Hypotenuse It's a comparison of the opposite side (the stack) compared to the hypotenuse. In this case the opposite side of a 22.5° triangle is about 38 percent of the hypotenuse, 38.26834 percent to be more exact. So the last step is to calculate

0.3826834 × 10 in. = 3.8268-in. stack height

Compared to One Because a decimal ratio can be envisioned as compared to one unit (0.3826834/1) the ratio we've just obtained would be the stack height if the sine bar was 1 in. long but it's 10 times bigger, so multiply times 10.

0.3826834 × 10 in. = 3.8268-in. stack height

KEYPOINT

> The sine ratio is a comparison of the length of the side opposite a given angle to length of the hypotenuse side of a right triangle (the hypotenuse is always opposite the right angle).

KEYPOINT

> Either way you look at it, you arrive at this conclusion:
>
> Stack height = Sine Ratio × Bar length
>
> Multiply the sine bar's hypotenuse times the sine ratio to obtain the stack height.

Here are two examples. Please follow them on your calculator (Figs. 7-50 and 7-51).

Setting up an Angle In this case, we know the angle and want to set up a perfect model. For example, to hold a part up at 22.5° relative to the table datum, the floor of a milling machine vise. This set will cut the top off a part at 22.5°. Once the part is tilted at the right angle then clamped in the vise, the sine bar would probably be removed (Fig. 7-50).

We need to calculate the stack height for *a 10-in.* sine bar when it is elevated up to 22.5°. Be certain the calculator is working in degrees. Touch the DRG button until it does. Most calculators default to (start at) angles expressed in degrees when turned on. First we need to look up the SIN ratio.

Trade Tip

In calculations such as this, never round results until the last conclusion. Then round out to the nearest ten-thousandth inch for best accuracy.

TRY IT

A. Calculate the stack height for a 10-in. sine bar to create an angle of 27.35°.

B. Calculate the stack height for a 5-in. sine bar to set up an angle of 7°.

C. Calculate the stack height to create an angle of 18°27' for a 10-in. sine bar.

ANSWERS

Note that answers are rounded to nearest tenth of a thousandth inch.

A. A stack for a 10-in. sine bar must be *4.5942* in. to create an angle of 27.35°.

Sin 27.35 × 10 = 0.459424845 × 10.

B. The stack height for a 5-in. sine bar set up to 7° is *0.6093* in.

Sin 7° × 5 = 0.121869343 × 5 = 0.6093 in.

C. The stack height for an angle of 18°27' under the 10-in. sine bar will be *3.1648* in. 18°27' = 18.45°.

Sin 18.45° × 10 = 0.316476967 × 10 = 3.1648 in.

Measuring an Angle This time we need to measure to the best accuracy possible the angle on the undercut block of Fig. 7-51. Here the adjustable parallel comes in handy along with a ground parallel bar under the sine bar. It's placed under the front rod, then opened until the bar fits the angle surface perfectly. Now locked, it's removed and measured with a micrometer and found to be 2.9565 in. tall. To proceed we reverse the process of the last examples. We have the height of the opposite side and need the angle it produces on a given sine bar. In this example, it's a 5-in. sine bar:

First, divide 2.9565 by 5 to obtain the sine ratio for the mystery angle.

$$\frac{2.9565}{5} = 0.5913$$

Figure 7-51 *Measuring an angle with a sine bar.*

Then, depending on your calculator's logic, enter

0.5913 then touch 2nd + SIN 36.2493°

or

2nd + SIN 0.5913 = 36.2493°

TRY IT

Round your answer to the nearest thousandth of a decimal degree (three places, which is accurate enough for most but not all shop works).

A. A 5-in. sine bar is inserted into an angle on a part and the stack height measured to be 2.5566 in. What is the angle?

B. A stack height of 0.476 in. under a 1.0-in. sine bar indicates an angle of ____°?

C. An adjustable parallel is measured to be 0.7654-in. under a 10-in. sine bar. At what angle is the bar elevated?

ANSWERS

A. A stack of 2.5566 in. in a 5-in. sine bar indicates an angle of *30.752°*.

2.5566/5 = 0.51132 ratio;

2nd sin 0.551132 = 30.752°.

B. A stack height at 0.476 in. in a 1.0-in. sine bar indicates an angle of *28.424°*. 0.476 2nd Sin = 28.42447991 (Division by 1.0 was not needed, the sine bar is 1.0 inch long so the stack height is the ratio.)

C. 0.7654 in. under a 10-in. sine bar calculates to *4.390°*.

KEYPOINT

Here's what we just learned about the calculator: Touching the ratio button (SIN) with an angle on screen produces the corresponding ratio. Touching 2nd + ratio button (SIN) with a ratio on screen produces the corresponding angle.

UNIT 7-4 REVIEW

Review the Key Terms

Bracketing
Determining a colinear pair of lines by looking for the symmetrical mismatch to either side.

Complementary angles
Two angles that add to 90 degrees.

Decimal degrees (DD)
Angular expression in whole degrees and decimal parts.

Degrees, minutes, and seconds (DMS)
Angular expression in whole degrees, minutes (1/60 degree), and seconds (1/3,600 degree [60 × 60]).

Hypotenuse
The longest side in a right triangle—always opposite the right angle.

Protractor
An angle-measuring instrument.

Protracted angular tolerance
A tolerance zone that radiates outward from a vertex point.

Quadrant
Ninety degree segment of a circle. Four quadrants to one circle.

Sine bar
An angular setting and measuring tool based on the sine ratio of right triangles.

Sine ratio
The ratio comparison of the side opposite the angle in question to the hypotenuse side in a right triangle.

Vernier bevel protractor
An angular-measuring tool with a resolution of 5′ of arc.

Replay the Key Points

- Protractors are used to measure or gage angles depending on the drawing specification.
- Angles can be specified as DMS (degrees minutes, and seconds) or as decimals DD.
- Flat and combination protractors are semiprecision tools.
- Vernier protractors have a resolution of 5 minutes of arc.
- Use the rule of quadrants to select the correct vernier scale.
- Watch out for complementary and supplementary angle confusion when measuring angles.
- Sine bars are the most accurate measuring tool for angles.

Respond

1. Name the two big differences in measuring a protracted angle and a geometric angle.
2. True or false? Because DMS are the units found on many drawings and tools within the shop, they are obviously the more useful angular expression. If it's false, why?
3. State the quadrant rule.

4. What inaccuracy factors could spoil vernier protractor accuracy?
5. Which statement is true?
   A. The SIN ratio is a comparison of the opposite side of a right triangle to its hypotenuse.
   B. The SIN ratio is a comparison of the hypotenuse of a right triangle to its opposite side.
6. Supplementary or complementary angles add up to 90 degrees?

UNIT 7-5

Measuring Surface Roughness

Introduction: Now we turn to a different measuring skill—the fifth element, determining how rough or smooth the machining action has made a surface on a feature. When operating machine tools, you must control not only the size, shape, and position of features, but often the finish produced.

KEY TERMS

Cutoff
The lateral span from which the profilometer extracts the average.

Lay
The direction and nature of the surface roughness marks.

Microinch
One millionth of an inch, the base unit for Imperial surface finish callouts.

Micron
One millionth of a millimeter—the metric unit for surface finish callouts.

Profilometer
An electronic device that determines surface finish roughness.

Scratch plate
Used by scratching samples with fingernail then comparing to machined feature. Also called a surface finish comparitor.

Surface roughness
The smaller cyclic effect of the cutting action on machined surfaces.

Waviness
The larger cyclic effect in machining caused by vibration.

What You Need to Know About Roughness

Microinches (0.000001— one-millionth of an in.)

Roughness is expressed in **microinches** symbolized as μIn. There are other systems used elsewhere in the world, but they are similar enough that conversion is simple. Metric roughness is stated in **microns,** which are 0.001 of a millimeter.

The marks left behind from machining are the *result of working the metal.* If we look closely, they are like regular hills and valleys produced from the cutting, grinding, abrading, and eroding electrically and also from vibrations in the process. Machinists are challenged with controlling these irregularities to not rougher than a given microinch tolerance. Functional examples where surface finish control might be necessary are for

◆ Customer appearance.

◆ Smooth mechanical sliding action.

◆ Just the opposite, minimum roughness because bonding is going to occur to the surface (paint or glue).

◆ To create specific machined lines. This control specifies the how the surface is made. If the tiny surface lines were parallel to a flexing action, cracks could start along them. This becomes very important in some high-stress applications.

One microinch is 0.000001 in. and is symbolized as 1 μIn.

The Symbols Found on Prints

In Fig. 7-52 you see a closeup of a hole in the drill gage. The bracket symbol adjacent to it is the way the designer communicates a finish requirement. It's drawn a bit larger than normal in this illustration.

Maximum Roughness Notice the number in the bracket, *125.* That's the upper microinch limit for the finish of that hole—the roughest allowed. Any smoother finish is acceptable unless otherwise stated in the control bracket.

Drill and ream holes

Figure 7-52 *The bracket specifies a maximum of 125 microinches roughness (125 μIn).*

Specification of minimum roughness is rare but it can be used where a rough surface is desired. If both upper and lower were required, there would be second number below the 125, in the same data field within the bracket.

On engineering drawings, these bracket symbols might be found relating to individual features as with this hole. Or, you might also find a general finish requirement in the title block next to the general machining tolerances. It would specify the requirement for all machined surfaces not given a specific finish on the print. "Unless otherwise stated all surfaces to be: 125 μIn." Often the μIn will be left off, but is understood.

General Finish In the United States, if no finish is specified on the surface or in the general tolerances box, then a 125-μIn roughness is considered standard machine finish.

The drawing symbol used today for a finish requirement evolved from a simple check mark to indicate surfaces to be machined, to the one you see in the illustration. It has three data fields—the first and most commonly used is inside the bracket crook where the 125 appears, a second is above the horizontal line, and the third within the bracket.

Let's see what the 125 actually means. In Fig. 7-53, you see a magnified cross section of a machine cut. Points 1, 2, 3, and 4 would be averaged as would P5, P6, and P7. The sum of the two averages must not exceed the 125-μIn limit. It could be smaller, however—smoother. The 125 μIn then is the averaged distance from a median line up to the hills and down to the valleys. It's calculated over a horizontal span called the **cutoff** *distance.*

Machine Finish Cross Section

Figure 7-53 *Roughness peaks and valleys are specified relative to the median line about half way between.*

Cutoff Distance The cutoff distance is specified in the bracket, as shown in Fig. 7-58, it's stated at 0.010 in. wide. If this data field is blank, then a 0.030-in.-wide cutoff span is assumed. When performing everyday evaluation, the cutoff distance becomes a nonissue.

Industrial Surface Evaluation It only becomes important for you to know cutoff values when for functional reasons, it's in the symbol on the drawing. Then it becomes necessary to turn to an electronic roughness computer, such as the one in Fig. 7-54. This portable electronic machine, called a profilometer (pronounced with a short i), is common in inspection departments. The profilometer can go to the machine to check results without removing the work to the inspection department.

The big brother, surface evaluation computer (Fig. 7-55), can filter the data to see different aspects of the finish, as in only the **waviness,** for example, and print out microscopic graphs of the surface and so on.

Both types collect surface data with a tactile (touching) stylus. It works much as dial test indicator does, being pulled across the surface slowly as it senses and records the peaks and valleys in microinches. The contact stylus is a pointed diamond or similar material so it can fit between the peaks in order to get a clear picture of the surface. As its central processor collects data, it computes the roughness average over the specified cutoff distance, then reports on screen, dial, or on paper.

Figure 7-55 *A surface evaluation computer can determine and display any technical aspect of surface roughness.*

Figure 7-54 *This portable profilometer is checking for a 64-in. (microinch) surface finish on this aluminum part.*

Surface Roughness Comparison at Your Machine

Getting a Feel for the Surface (Literally)

Profilometers are necessary when surface finish specs are given with cutoff distances or they are finer than 32 μIn. But for the everyday application at your machine, many machinists use the surface finish comparitor plate (Fig. 7-56), or sometimes called a **scratch plate** because of the way it's used.

The plate features different sample patches of various kinds of machine finishes. Finding the closest representation of the finish you've made, the plate is scratched with a fingernail, then with the feel for the **surface roughness,** immediately scratch the surface in question! Go back and forth—it works amazingly well with a little experience. Be sure to scratch across the marks—not parallel to them. A scratch plate would be a priority 2 or 3 for purchase for your box, but is handy to have. Looking at the roughness is of some help too, but you can be fooled by how shiny the metal is due to surface conditions smaller than that of machining marks.

Lay Symbols

As the scratch plate photo shows, there are different ways a finish might be produced that make swirls, straight lines, or crosses on the feature's surface—called the **lay** of the finish. The various lay marks will become familiar as you operate equipment. If it's necessary to specify how the finish lays on the feature, then a symbol will be placed inside the bracket. Similar to geometric symbols each convey meaning to the machinist. Especially note Internet symbols. Figure 7-57 depicts a few of the more common ones and Fig. 7-58 shows a closeup of the bracket with the multidirectional lay symbol in the correct location.

For now, without an in-depth machine knowledge, it's not critical to understand how these lays are made,

Figure 7-56 *A scratch plate helps estimate finishes.*

but it is important to keep in mind that the roughness can look quite different between, for example, the straight lines made by a lathe and the swirl lines made by a mill cutter. Yet both would measure the same microinch roughness.

Waviness

Sometimes, a cutting action will produce two patterns on the work. The individual marks made by the cutting point or teeth, then a group pattern caused by vibration of the cutter, or by work vibration or even the machine itself. When this occurs a wave is formed. When it's important to control that, the waviness spec will be placed above the bracket. Again, this matters only for highly technical surface controls and cannot be measured except by electronic means.

KEYPOINT

If a lay is specified, it must be achieved as much as the other specs and tolerances on the drawing.

Symbol	Meaning	Illustrated

Figure 7-57 *Typical lay symbols and their meaning.*

Standard Finish Applications

Under normal conditions, specified finishes run from around 125 μIn down to a modestly smooth 32 μIn or perhaps 16 μIn. Notice that the numbers are each one-half the previous finish. For example, a moderate 125, 63, 32, and a fine 16.

You might encounter a rough 250 μIn or even rougher 500 μIn, depending on the nature of products made at your company. As you scan the examples notice that the various categories have gray areas where their function overlaps. Here are a few examples where these common finishes might be applicable.

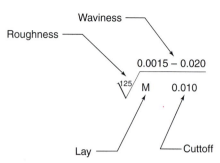

Figure 7-58 *Waviness, roughness value, and cutoff distance can all be found in the bracket.*

Don't Confuse Shiny with Smooth Light reflection is one of the inaccuracies of a scratch plate. Often, a finish will reflect light when it is not as smooth as it might seem. Don't be fooled by shiny surfaces—they may not be smooth.

For an example, look at the reflection of the sun in a lake. The flat calm lake would show you only one reflection, whereas the rippled lake would show you many reflections coming from many points—the rougher surface would actually reflect more light your way!

It is best to scratch and compare with your feel and use reflectivity as a subordinate evaluator. Reflectivity is caused by closing all pores in the work surface. It's a microscopic function below the resolution of normal machining. Reflectivity is a gage of the surface between the machine peaks and valleys until finishes below 8 μIn are encountered.

μIn	Function and Description
125	**General Machine Finish** Noncritical parts of an automobile engine or a lawn mower. Standard operations produce this finish such as drilling, turning, and milling. Fairly rough to the feel. Will leave a trail of fingernail if scratched.
63	**General Machine Finish—Better Quality** Gasket surfaces or a smooth nonbearing shaft for a sliding control rod in disk player, for example. Made in the same way as the last but taking one extra cut. Will also remove a bit of fingernail when scratched.
32	**Quality Finish** A shaft bearing or "mag" wheels before polishing. Generally requires special machining such as roughing and two finishing cuts. Or might require an abrasive (grinding) process to achieve. This finish costs more to produce. The last finish to remove fingernail. Any smoother will not.
16	**Precision Finish** Hydraulic cylinder wall for a jack or a plastic mold surface. Must be planned carefully with roughing and finishing operations then final abrasive work. This costly finish is only justified by special conditions.
8	**High-Quality Precision Finish** Surgical instruments or a contact lens mold, for example. In addition to standard machining and grinding, these finishes often require secondary abrasive processes to reach this level of smoothness.
4/2	**Lapped and Polished** Farther along in your career, you might encounter a 4 μIn and a remote possibility of a 2 μIn (might be on your scratch plate, too). These are produced with very technical methods requiring exceptionally fine abrasives and a well thought-out, three-level machining, grinding, and finishing plan.

UNIT 7-5 REVIEW

Review the Key Terms

Cutoff
The lateral span from which the profilometer extracts the average.

Lay
The direction and nature of the surface roughness marks.

Microinch
One millionth of an inch, the base unit for Imperial surface finish callouts.

Micron
One millionth of a millimeter—the metric unit for surface finish callouts.

Profilometer
An electronic device that determines surface finish roughness.

Scratch plate
Used by scratching samples with fingernail then comparing to machined feature. Also called a surface finish comparitor.

Surface roughness
The smaller cyclic effect of the cutting action on machined surfaces.

Waviness
The larger cyclic effect in machining caused by vibration.

Replay the Key Points

◆ Unless otherwise specified, surface finish is specified as a maximum value, any smoother finish is acceptable. 125 μIn is standard if not otherwise specified.

◆ Missing a surface callout can lead to scrap—these are tolerances.

◆ The cutoff distance is 0.030 in. unless otherwise specified.

◆ One of the main reasons for poor finish, is caused by a combination of tool vibration, work vibration, and machine vibration! The main control is reduced cutter speed.

◆ The symbol for "surface must be established by machining" is a horizontal line in the bracket.

◆ The circle in the finish bracket indicates no machining is to occur.

◆ The numbers specify the average height of the marks above and below the median.

Respond

1. True or false? When no surface roughness is called out in the general tolerances, then all surfaces are to be 125 μIn. If false, explain why?

2. Explain the two entries in the surface roughness symbol shown in Fig. 7-59.

 X

Figure 7-59 *Describe this roughness callout?*

3. What data are missing that might be in Fig. 7-59?

4. Which is the smoothest finish—125 μIn, 64 μIn, or 32 μIn?

5. At your machine there are two instruments that might test surface finish. They are:
   A. Micron meter
   B. Optical comparitor
   C. Surface finish comparitor
   D. Scratch plate
   E. Profilometer

Section 7 Review

Unit 7-1

Inside and outside mics are more specific than a caliper and they deliver better accuracy, but they must still be considered universal tools. In this unit we strove for an initial repeatability of 0.001 in. even though, in the hands of a skilled user, these two tools can deliver around 0.0005-in. repeatability (or slightly better—depending on situation and user). The two major challenges to using them right are feel and zero-error calibration. However, even though they can be challenging to use correctly, there are times when an inside or depth micrometer is the best tool for the job.

Unit 7-2

To an untrained observer it would be hard to see why a box of little steel blocks could cost a weeks wage for a machinist. But you now know the reason. These little gems are made with the greatest precision of any measuring instrument in the shop. In order to make them, the machine shop must own some of the most sophisticated equipment on the planet. Not only that, but they undergo many steps in their manufacture and each is performed by ultra-skilled people. Along with a couple of other gages and standards, gage blocks are the bottom line for precision in most machine shops. Even where CNC, coordinate machines are used, the gage blocks still rule as the basis for accuracy.

Unit 7-3

Here we lumped together a series of small measuring tools that each perform a specialized measuring function. They all have some inaccuracy. One final thought about them (and all others we've studied) is to be on the look-out for technical advancements that improve what they do. They leave plenty of room for improvement! We also tipped the argument about the use of the word gage. The classic meaning being a measuring template that compares some target characteristic against a perfect version. However, we saw that the word is used pretty loosely in our trade and that nearly any tool can be called a gage (gauge). For example, a dial bore gage or a surface gage—neither of which fit the definition.

Unit 7-4

There was a lot to learn about measuring angular surfaces. First, we needed to clear up a couple of terms: There are *two kinds of angle callouts on prints*. The protracted emanating from a single point outward and the geometric lying within a parallel sandwich tolerance zone with respect to a datum. And although each requires a very different measuring setup, they are measured with the same tools but used differently. There are also *two ways of expressing their value* degrees-minutes-seconds (DMS) versus decimal degrees (DD). Then, as with all other measuring tools, the final skill was to know the various choices available to measure angles depending on tolerance. When tolerances are above 1 degree, we use plain protractors but when they become tighter we have three options: vernier protractors, electronic protractors, and sine bars and sine plates. After that, we turn to technology to solve measuring problems: computer coordinate-measuring machines and optical comparitors.

Unit 7-5

While we use electronic profilometers to measure surface roughness, we also develop the ability to scratch a machined surface and come fairly close to its measured roughness. Similar to looking at a spinning cutter and knowing if it's right or wrong, working with surface finishes for a while will bring a sense of them with no scratch plate to use for comparison.

Final Review Quiz—Choosing the Right Measuring Tool

When faced with proving a part is within tolerance, nearly every dimension creates questions. Questions about the right tool and how it's used. Which will prove beyond doubt that the feature is of the best quality? This final review is designed to simulate that daily game. The upcoming problems are designed to stimulate discussion about ways others might get the same job done! As you'll see in the answers, the basic tools we've

ShopTalk

The Challenge of Computer Generated Parts Integrated machining has added a new twist to this game. Today we have the ability to quickly define and make part shapes that were only dreams a few years ago. We can draw a wild shape on CAD, then program it with CAM, and send it down to the controller to be made. At the end, since no errors came up in the process, we're pretty sure the final item is right, but stumped as to how to prove it! Computers can and do put us in the embarrassing position of producing a more accurate or more complex shape than our ability to measure with standard tools and processes! However, computers can be the solution, too. If it can be shown on screen, there is probably a CMM program that can be written to prove the accuracy of the shape. The technology of computer directed measuring will be found in Section 28.

learned thus far can be combined in many different ways!

Thus far, in Sections 6 and 7, we've been improving our understanding of the hardware and how to use it, but not how to evaluate parts. For nearly every measurement there will be several choices that *can* do the job. There's usually a fast method (the calipers or ruler—both good backups but seldom the best choice) and there will be a most accurate method or two, which will take time to set up and perform. Then there will probably be several others between with various mixes of speed and accuracy.

Different from a classic test, the upcoming problems may have several answers. If your choice delivers the accuracy required, then you've got it. It's more than possible that you will come up with a method the planning committee and I didn't think of, not on the answer sheet! That's great if you do.

Your instructor or another skilled machinist may have differing opinions or expanded ideas from those in the answers and that too is to be expected. Measuring is an art, there are going to be differences, journeyman to journeyman. Discuss answers with others to see how they made their choices. Your instructor may also have real parts similar to those shown in Fig. 7-60.

Guidelines for Choices

◆ First, before choosing the tool and process, know the tolerance(s) for the feature in question! The tolerance eliminates some tools and dictates which methods will deliver the accuracy required.

◆ Decide how quickly must you deliver results? This isn't normally required in training but it's a factor on the job. Time added to a part cycle due to a poor measuring choice cannot be justified. This is especially true on fast CNC equipment. For a one-time-only test, it's OK to use the slow method if it delivers. But when parts come off every few minutes, there's no time for it.

◆ Know your personal repeatability with each tool in the arsenal.

◆ With the preceding factors under consideration, determine which tools are available that will do the job. Keep the "rule of ten" in mind as you do so if at all possible.

The old masters in the shop gave two bits of advise about measuring: "Measure twice, then cut once!" and "It's a poor journeyman who blames his tools." The meaning is if you get it wrong because your tools are out of calibration, it's still your fault!

Figure 7-60 *All process measuring test.*

- *Never assume the tool you choose is in calibration!*
- Look for the spoilers and control them as you take the measurement.
- Always plan a backup, double check process especially if it's an expensive or critical job.

Problems 1 through 10

Beat the Author by earning over 100 points total on the first 10 problems. Challenge another student or your entire class in this game.

Game Rules

- **Legal Tools** Use only measuring tools we've discussed in Sections 6 and 7 but you may use them in any way possible that delivers the accuracy!
- You may also choose to use the granite table with right-angle plates to hold parts perpendicular to the table datum. Ground parallel bars are OK, in any way you choose. You may choose to lightly clamp the work on these utility items.

- **Base Points** The first bracket for each question sets the point value for finding a given number of solutions, the target for the question. *Finding fewer solutions than the target, no points are earned.*
- **Bonus Points** Each solution you find that is on the answer list, above the target number, earns the bonus value for that question. One bonus per answer above the target number.
- **Loosing Points** If your number of solutions falls short of the target value, then deduct the bonus value multiplied times the number below target. (You'll notice that it's possible to score well below zero in this contest!)
- **Double Bonus** Discovering a reasonable answer that measures within the resolution of the question, that isn't on the answer list, adds *twice* the bonus to your score, for each solution found beyond the list.

If you identify the *best accuracy solution*, add one double bonus to your total. There may be more than one best accuracy solution—but you get only one double bonus for best accuracy.

Questions and Problems

In addition to the objectives called upon here in Section 7, you will be using the tools and processes of Section 6 as well.

1. For Question 1, you may use any item of known thickness to test for gaps. Describe or sketch *three* ways the 55-degree angle might be measured within the print tolerance. (Obj. 7-4) (point value: 10; bonus value: 3)

2. Name *eight* methods that could measure the 0.375-in. dimension within ±0.005 in. (Objs. 7-1 and 6-3) (point value: 12; bonus value: 2)

3. Identify *three* methods to measure the 1.750-in.-diameter hole within 0.0005-in. tolerance. (Objs. 7-1, 7-3, and 6-3) (point value: 10; bonus value: 4)

4. Identify *two* methods to measure the 0.750-in. component of the 0.250-in. hole's position. Discrimination must be ±0.001 in. or better for your choices. (Objs. 7-2, 7-3, and 6-3) (point value: 8; bonus value: 5)

5. Find the *three fastest* ways to measure the 0.250-in. hole's diameter within ±0.005 in. In order of time consumed. Note, if your speed

order matches answers add one double bonus! (Objs. 7-3 and 6-3) (point value: 6; bonus value: 3)

6. Identify *three* methods of measuring the width of the 0.625-in. slot within ±0.0005-in. discrimination. (Objs. 7-2 and 6-3) (point value: 12; bonus value: 4)

7. Find at least *five* ways to measure the 0.125-in. depth within ±0.002 in.? (Objs. 7-1, 7-2, and 6-3) (point value: 10; bonus value: 3)

8. Using the tools and processes of Sections 6 and 7, find *one* way to measure the 2.500-in. dimension to the 0.625-in. slot's centerline within a tolerance of ±0.0003 in. (Objs. 7-2 and 6-3) (point value: 15; bonus value: 10)

9. How many ways can you list to measure the 0.625-in. thickness of the part within ±0.010 in.? Target is *nine* solutions. (Objs. 7-1, 7-2, and 6-3) (point value: 12; bonus value: 3)

Critical Thinking Problems

10. Is there any practical way to measure the size of the corner 0.75-in. radius other than radius gages? (point value: 5; no bonus value here)

End of contest, add up all points including deductions and check your score against your opponents (and me), then answer the remaining questions.

11. What does the box around the 55-degree angle mean? (Obj. 7-4)

12. Return to Question 2. What method would be the most practical and quickest? (Obj. 7-1)

13. If the surface of the hole in Fig. 7-60 was given a finish callout of 32 μIn, would it be possible to achieve using a standard drill? (Obj. 7-5)

14. The position tolerance for the 0.250-in.-diameter hole applies to its (A) surface, (B) centerline, (C) axis?

15. What is the distance of the 1.750 hole from datum B? How do you know?

CNC Questions

16. A program is running well. All key dimensions are within tolerance and the surface finish is better than 125 μIn. However, with each part cycle it seems to be getting a bit rougher. What might be the cause and fix?

17. CNC lathe is turning out gear shafts that have just one key dimension—they must be 4.500 in. long with a tolerance of ±0.003 in. The machine cycles (makes one part) every minute and 45 seconds. It has an automatic bar feeder and a finished part catcher so you are free to measure and record data with no duties between taking out the sample part. According to your job plan you are to measure every tenth part then enter the data into the SPC software on your controller to track real-time quality.
    A. What facts must be considered to make your choice of measuring process?
    B. Name at least three ways this dimension might be measured.

18. Explain what a 125 μIn finish means. Four facts minimum. (Obj. 7-5)

19. A surface finish bracket has only the 63 μIn with no other information. What then, is the cutoff distance? (Obj. 7-5)

20. What is the classic definition of a measuring gage? (Obj. 7-3)

Section 7 Answers

ANSWERS 7-1

1. False. Their lowest reading is obtained with the thimble backed all the way out.
2. $\frac{1}{2}$ in. or 13 mm; so the head can fit into small holes
3. Basic head size; the reading on the head; the rod range; and the spacer if used

4. Feel—both have challenges in this respect
5. The dial bore gage

ANSWERS 7-2

1. Calibration of both the head and individual extension rods. Feel is a problem for both.
2. False
3. True; by seeing which graduations are covered rather than uncovered

4. 0.1009; 0.108; 0.200; 4.000
5. Compare size or shape

ANSWERS 7-3

1. Calipers, telescope gages, dial bore gage, and inside micrometers
2. Best choice: inside micrometer; next best: dial bore gage (not first choice because it must be set up for one part); next: telescope gage

3. Ring gage
4. True
5. Size and form (roundness)

ANSWERS 7-4

1. Fan-shaped tolerances versus distance across zone; geometric angles are evaluated from a datum!
2. False. Decimal angles are easier to use due to simplified math. Gradients would also bring the same advantage!
3. The quadrant rule helps determine which of the two vernier scales to use. It is worded to

this affect: Each 90-degree segment of the protractor increases either clockwise or counterclockwise. The correct vernier also increases in the same direction.
4. Parallax from the curved magnifier lens; misreading the vernier scale
5. Statement A
6. Complementary angles

ANSWERS 7-5

1. False. It might be called out on individual part features.
2. 32-microinch maximum roughness with the lay running angular to the feature
3. Cutoff and waviness

4. 32 μIn, because it would have the smallest peaks and valleys
5. (E) plus either (C) or (D) (Surface finish comparitor and scratch plate are the same instrument!)

Review Problems

1. Remember, every solution should include datum reference for this measurement!
   Set and lock vernier protractor at 55° (or 35° depending on quadrant); protractor base resting on granite table; test for gaps exceeding 0.010 in. with feeler gage.
   Mount protractor in height gage—on granite table (Fig. 7-61).
   Hold part against right-angle plate on datum A; use sine bar set at 35° (complement of 55°) (Fig. 7-62); test for gaps with feeler gage.

2. Ruler with magnifier (very risky but could work with good technique)
   Caliper depth probe
   Caliper inside jaws with parallel bar on angular face (Fig. 7-63)
   Same as Fig. 7-63, but use adjustable parallel in space between parallel and part
   Depth mic (*Best Accuracy*)
   Lying on angular face, test with gage blocks in space between table and edge (*Best Accuracy*)
   Very slow! (Fig. 7-64)

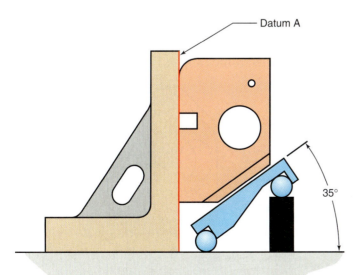

Datum A

Figure 7-61 *Measuring the angle to the table datum.*

Datum A

35°

Figure 7-62 *Measuring the angle with a sine bar.*

Measure between

Figure 7-63 *Measuring the 0.375-in. step with a parallel bar to outside edge.*

MMC gage block stack

Figure 7-64 *Measuring step with gage blocks.*

Lying on angular face with adjustable parallel in space

Lying on angular face using ground pins in space

Lying on angular face using upside-down comparison measurement with indicator probe and gage blocks. One extra block must be used as a top gage. (*Best Accuracy*) (Fig. 7-65)

Outside mic from 0.75-degree radius to each surface, difference is distance! (Fig. 7-66); only

Figure 7-65 *Upside-down comparison measuring of 0.375-in. step.*

Using Mics

Longer distance = 5.133

Shorter distance = 4.758

| 5.133 |
| − 4.758 |
| 0.375 |

4-5

Figure 7-66 *Measuring the step using a micrometer—the difference in both readings is the distance!*

works because there is a radius on the opposite corner, not a recommended method. There are alignment problems but it will work.

Caliper same as above, from radius corner.

3. Inside mics

Dial bore gage (*Best Accuracy*)

Calipers with inside jaws—but very risky, probably a poor choice

Telescope gages—may be beyond the range of some sets

4. MMC test pin in hole—part lying on datum B Indicator in height gage. Zero indicator on granite table. Move up to pin top. Subtract pin radius from result. Note, borderline accuracy, may not deliver 0.001-in. discrimination. Adjustable parallel between pin and table. Gage blocks at 0.875-in. using height gage, indicator comparison measurement (*Best Accuracy*) (Fig. 7-67). After finding distance off table, deduct half the MMC pin diameter.

Note, using the outside jaws of calipers to measure the distance from the hole's edge from datum B, might be a good backup check (0.625 in.) but this method isn't recommended for two reasons. First, it doesn't create datum B and it doesn't account for the hole's shape. The 0.750-in. dimension is to the hole's center not its edge. Using the largest ground pin (MMC) that just fits, establishes the round space center.

5. Dial bore gage, fastest after initial setup (*Best Accuracy*)

Go, no-go pin set

Dial or electronic calipers

Figure 7-67 *Gage blocks at 0.875 in. compared to top of MMC pin. Deduct half of pin diameter.*

6. Adjustable parallel with micrometer (*Best Accuracy*)

   Gage blocks finding MMC stack (*Best Accuracy*)(slow due to picking block until MMC fit is found)

   With part resting on datum B (Fig. 7-68). Electronic height gage with 0.0001-in. indicator. Ground block held on lower side. Zero on each surface. Read height gage scale to find difference is heights = slot width.

   Calipers—inside jaws. Calipers are questionable for 0.0005-in. discrimination even using electronic types.

7. Depth mic

   Depth probe on calipers

   Height gage with indicator—part lying flat, step side up; height gage scale measure step

   Same position step side up. The 0.125-in. gage block stack sitting on step. Comparison measurement (*Best Accuracy*)

   Using 1-in. micrometer measure full thickness 0.625-in. dimension, measure step dimension 0.500 in., subtract to find 0.125 in.

   Calipers are risky used in the same way as before as they could consume nearly all the discrimination in measuring twice—but they might work (borderline).

   Comparison measurement with two stacks—one at 0.625 in. and one at 0.500 in., sitting on table. Using indicator find top or bottom height relative to stack. Then determine other height and combine results (*Best Accuracy*).

8. Note, it's from a datum to a centerline. To make this kind of tolerance, exceptional cleanliness and care will be necessary! Part lying on datum B:

   Adjustable parallel in slot. (*Most Accurate*); gage block stack set at 2.8125 in.; find height of top of slot (Fig. 7-69) from granite plate by comparison; mic the parallel with vernier micrometer. Subtract half of slot width from height measurement.

   Alternative MMC gage block stack in slot. (*Most Accurate*)

   Parallel bar of known thickness held against datum B. Micrometer from parallel bar to closest side of slot (2.1875-in. plus parallel bar thickness). Note, the mic cannot be used directly as the 0.75-in. corner radius trims away flat surface in line with measurement.

9. Vernier micrometer (*Most Accurate*)

   Calipers outside jaws

   Ruler

   Height gage with scribe

   Depth mic to granite table

   Caliper probe to table

   Depth mic to table

   On granite table—parallel bar on top—adjustable parallel in space; mic adjustable parallel

   Height gage with indicator zeroed on table; measure on height gage scale

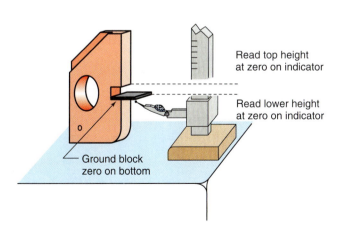

Figure 7-68 *Height gage scale measurement of a 0.625-in. slot width.*

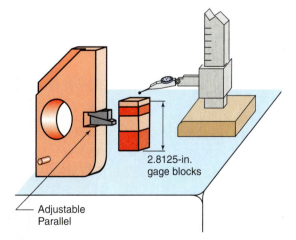

Figure 7-69 *Measuring the top of the slot finishes the inspection.*

Comparison measurement. Part on table, 0.625-in. gage block stack. (*Most Accurate*) Indicator in height gage.

10. No—not with the legal tools. There are two practical ways to determine the radius—one is an optical comparitor that casts a greatly enlarged projection of the object on a screen where templates blown up the same amount (big radius gages) are put against the screen for comparison. The second is the CMM (computer coordinate-measuring machine).

11. The box indicates it is a perfect model for the feature (basic dimension).
12. Depth micrometer
13. Almost certainly not
14. B. Centerline and C. axis because the position box is attached to the dimension not the surface—that indicates axis control.
15. It's "probably" 2.500 in. for two reasons: (A) the hole's centerline projects down to the slot's axis; (B) the hole has no horizontal distance given so it must be 2.500 in. or the engineer left off a dimension.

16. Since the run was going well, the problem is not cutting speed, so something is degrading. It is probably tool wear.

17. A. Tolerance is moderately tight at 0.003 in. Some processes will not be accurate enough. Total time to measure and record is over 17.5 minutes—adequate but includes time for entering data, too. Must also monitor machine during that time.

    B. (1) 4 to 5 vernier micrometer. (2) Comparison measuring with height gage. (3) Electronic caliper zeroed on nominal 4.500 in. (4) Electronic caliper. (5) Dial caliper. Note that all calipers are risky at ± 0.003-in. tolerance.

18. When no finish is specified, 125 μIn, it is considered standard roughness. It is the *averaged* distance in *microinches, above and below* a median line, over a specific distance.

19. The distance is 0.030 in., when no cutoff is specified in the bracket.

20. A tool that compares the perfect definition of the target characteristic to the real surface—a template.

Layout Skills

Introduction

Layout is a premachining operation whereby a template is *scribed* (by precision scratches) in straight lines and arcs, with points punched on the surface of the metal. Precisely laid out, they form a picture that indicates the limits of cuts, or the size and location of features such as drilled holes or slots.

A layout template is used in three interrelated ways, depending on how difficult it will be to measure the feature being machined.

A Fail-Safe Line   These marks are used as a yellow flag, warning line when making cuts. Using layout in this way, measurement of the final result is possible. The marks are there to prevent accidentally taking too much metal off before stopping to measure.

Last Word Definition   On rare occasion, when no practical measurement can be taken of the final feature shape, size, or location, the layout is used as the only stop line. Used in this way, layout is a checkered flag.

Eyeball—Quick Simple Method   When tolerances do not demand better resolution, or when a quick operation is called for with low tolerances, layout can be used for location or shape; for example, scribing then punching the location of drilled holes. They are then drilled by locating the drill directly over the punch mark, by visual alignment only. In machine lingo, that's called a "quick and dirty" method.

Target Repeatability

Generally layout is considered a semiprecision operation yielding approximately ±0.030-in. tolerances and repeatabilities. However, with good lighting and mastery of technique, one can deliver accuracy near 0.005 in.!

Previous to our CNC world, layout was often the only way to machine the work to final size or shape. Today however, CAD/CAM programmers can easily define nearly any part shape then CNC machine it to tolerances and repeatabilities 20 times finer than layout methods, no matter how skilled the machinist. Therefore, today lay-out is considered a non-value-adding, extra operation. The need for this skill in production manufacturing has all but evaporated. So this book's planning team debated eliminating its study, but decided some skill was necessary for these reasons:

KEYPOINT

Even though layout can deliver moderately accurate results, the reliability of this kind of machining depends solely on the user's eyesight, excellent lighting, and extreme skill!

Useful Tooling/Job-Shop Skill   Layout methods are used in the quick turnaround, job-shop environment where one or two parts must be made at a rock-bottom price in the least time possible. They are also used by the tool and mold makers to manufacture difficult shapes when no CNC equipment is available.

Enhances Granite Table Measuring Skills   The abilities gained in layout practicum complement and improve measuring ability, especially when reference to a flat datum is needed and height gages will be used. These are crossover tools that both measure features and lay them out using the scribe attachment.

Logical Sequence Training   The planning and execution of a layout is similar to planning cut sequences on a work order. The same kind of thinking is required to avoid traps in a set of operations and to not leave behind destructive scratches when the cuts are completed.

Objectives

The Purpose of Layout

Introduction: As a cost-adding operation, you should recognize when layout is needed and when it can be avoided. This brief unit demonstrates what it can do and the expected resolution. As your skills improve in machine setup and operation along with improving ways to measure odd- or difficult-shaped features, the need for layout should decrease.

KEY TERMS

Awl (scratch awl)
A tool borrowed from leather work, where it's used to punch holes. In the shop, it's a handy scribe with a large handle.

Fail-safe
A layout used to tell when the cut is close to finished size, to prevent cutting too much material away before measuring.

Last word
Used when no other method of defining and measuring the object is possible, a final template of shape.

Pilot punch (prick punch)
A 60 degree pointed punch used to precisely punch a small starter indentation at layout line intersections, normally followed by a sturdier, blunter center punch mark.

Splitting the line
The best accuracy occurs when the machinist cuts away half the layout line width on every cut referenced to the layout.

Witness mark
Small shiny marks lightly cut on the work surface to show when a cutting tool is touching the work in the correct location.

Applications

Do You Need a Fail-Safe Template?

This layout type indicates when to pause cutting to measure the feature for size and/or location. It is commonly used on extremely expensive work or when some other exact measurement is going to be used, but this extra step, or **fail-safe,** is warranted—*just to be sure.*

Used in this way, it's a *yellow flag* that the cut is getting close to the finished size—slow down. For example, in making the 12 drill gages, recall the work order called for rough sawing the blanks before machining. Layout was used to ensure the rectangle contained the right amount of excess for machining. You would saw near the lines but outside them.

Another example is when machining the rough blank to its final rectangular size, a finished size layout might be used to prevent cutting undersize, but measuring with a micrometer is the answer to precision.

Could one or both of these examples eliminate layout with the same results? In the case of the saw-out, it would be a good idea. In the case of the machine cut, the micrometer dials on the machine could do the job nicely. It might be a waste of time, but as a beginner, I might choose to use layout anyway.

Avoid Leftover Layout

Beyond the extra time used, a valid reason to not put layout on a workpiece is that the lines and points themselves are destructive to finished surfaces and all must be machined away at final size. Depending on the product being made, it can be a serious error if any of those scratches or center punch dents show after the machining is completed. In fact, in many manufacturing situations, leftover layout will cause the work to be scrap! A tiny scribe line extending beyond a bolt hole, for example, can be the start of a crack in the metal as it's flexed in day-to-day use.

KEYPOINT

No layout line or point should remain on the final product!

Do You Need a Last Word Template?

A **last word** template is the checkered flag—no other measurement will be or can be made. Here are a couple of examples.

When using the drill gage, layout might be used to mark and then machine the 59 degree angular edge relative to datum A, as seen in Fig. 8-1.

Before putting the blank on the tilting sine bar (Section 7), it would receive the *X-Y* lines that show where the 59 degree line begins.

Figure 8-1 *Laying out the angular line at the X-Y location.*

Figure 8-3 *An awl and a pocket scribe are used to make layout lines.*

Then with it tilted on the sine bar, one could scribe a guideline at the correct angle, intersecting the X-Y layout point on the blank.

Finally, the blank could be held in a precision vise such that the layout line is level with the table (parallel to the vise top) to cut it (Fig. 8-2). Although this method of locating the workpiece could not be used for tight tolerances, it works within 0.030 in. for roughing cuts.

KEYPOINT

In Fig. 8-2, the machinist would cut down until half the width of the layout line has been removed. We call that **splitting the line.**

Why Was Layout Needed?

Here, measuring the 59 degree angle relative to datum A isn't the challenge. It is in cutting

the angular line to the right X-Y intersection. In this situation, a layout is a good idea as long as tolerances permit it.

There are several ways to scribe the three lines in Figs. 8-1 and 8-2, and we'll look at others in a moment. A protractor with its blade set at 90 degree then 59 degree, a pocket scribe (a pointed tool similar to a pencil but with a carbide tip), or a scribe with a larger handle called a **scratch awl** can be used. These are a couple of ready methods (Fig. 8-3).

The height gage with the scribe attachment (Fig. 8-1) would be the most convenient way of scribing precision lines parallel to a datum. It would be elevated with the fine movement screw until the scribe point intersects the X-Y location, then the scribe point would be pulled across the work with just enough force to create a clean, well-defined line.

KEYPOINT

For the best accuracy, the scribe is pulled, not pushed!

Trade Tip

To quickly align the 59 degree layout line level to the vise jaw, lay a thin, ground-parallel bar on top of the jaw. Then level the layout line to it, tighten the vise, and remove the parallel. The line is now level with the vise top, and ready to cut. But it is safely above the hardened jaw.

Figure 8-2 *Cutting the angle to the layout.*

How Could a Quick Layout be Used?

An example of using a layout as the quick locator is found in drilling the holes in the gage. Assuming we are working on a drill press, but there is no drill jig to guide the drills to the right location on the work, and we have no micrometer dials on a drill press, layout becomes one answer to position the bit directly over the right location. It's possible, if one follows the upcoming procedure, to achieve ±0.030-in. location repeatability or slightly better.

Hole center locations can be laid out by crossing two *X-Y* lines relative to datums A and B. A fine punch mark is made at that point using a **prick** or **pilot punch,** with a light tap only. The tiny mark is then visually checked to see if it's on target, then tapped again. If still on location, a blunter *center punch* mark is driven deeper. A small pilot drill is centered over the punch and a test is made by just touching the rotating bit enough to make a shiny mark, called a **witness mark.** If the witness surrounds the punch mark, it shows that the alignment is probably OK. But if the shiny witness is off to one side, you know the part needs to be shifted more closely to location.

Several Other Hole Location Options

Whether or not to use layout to drill holes is an accuracy question. You will learn that there are two more drill press methods to align the drill over the *X-Y* marks in Section 10. They both yield results in the ±0.010- to 0.015-in. range. Then for finer location tolerances, there are two improved accuracy methods in Section 12, yielding ±0.005 in. or better.

Match the job tolerance to the location method.

If the pilot drill's *witness mark* shows good alignment, the pilot hole is drilled, followed by the full-size drill. This takes a lot of time compared to drilling on a CNC milling machine, but it does work if it's the best or only option and tolerances permit it.

UNIT 8-1 REVIEW

Review the Key Terms

Awl (scratch awl)
A tool borrowed from leather work, where it's used to punch holes. In the shop, it's a handy scribe with a large handle.

Fail-safe
A layout used to tell when the cut is close to finished size, to prevent cutting too much material away before measuring.

Last word
Used when no other method of defining and measuring the object is possible, a final template of shape.

Pilot punch (prick punch)
A 60 degree pointed punch used to precisely punch a small starter indentation at layout line intersections, normally followed by a sturdier, blunter center punch mark.

Splitting the line
The best accuracy occurs when the machinist cuts away half the layout line width on every cut referenced to the layout.

Witness mark
Small shiny marks made lightly cut on the work surface to show where a cutting tool is touching the work in the correct location.

Replay the Key Points

- Layout is an extra operation that adds time and cost but it sometimes is justified.
- Layout can be a useful fail-safe template.
- Layout can be a last word definition of shape, size, or location.
- Layout can be a quick and dirty method.
- Leftover layout lines and points cannot show on the final product!
- When cutting to a layout line, try to "split the line."

Respond

1. What are the two main functions of layout?
2. Beyond placing layout in the wrong location, how can the machinist ruin the workpiece with layout? Explain.

3. Although it was stated that one could achieve closer results, what is the expected repeatability of machining to a layout as the last word for size?

4. How would you describe layout to a person with no trade experience?

Critical Thinking Question

5. Although not covered directly, when does layout become a value-added operation? (*Hint:* There are three (or perhaps four) related reasons.)

UNIT 8-2

Layout Tools, Planning, and Practice

Introduction: Now, let's look briefly at the common tools used in layout work. Some you already know such as the height gage and the granite table. Others are new and quite useful both on the table and, as a key point, also throughout the shop for setup work and measuring.

Then we'll discuss how to get started doing your own layout. There is a useful final activity that should be enough to enter the training lab and make a simple layout on an assignment.

KEY TERMS

Dividers
Pointed layout tools used to scribe radii and divide distances into equal portions.

Hermaphrodite calipers
Used to scribe lines parallel to an edge. Not truly a caliper or divider, the dictionary meaning of "hermaphrodite" is having both male and female characteristics.

Layout dye
Surface stain to reduce glare and make layout lines more visible.

Master square
The rigid square used to test squareness and to lay out lines perpendicular to edges.

Right-angle plates
Used to test squareness and to hold objects perpendicular for layout.

Scribes/awls
Simple pointed styluses used to scribe lines.

Surface gage
A multipurpose tool used to scribe lines parallel to a surface, to transfer lines from a rule to a part, and to scribe lines parallel to an edge of a part.

Vee blocks
Ground blocks used to hold round work for scribing, measuring, and machining.

The Fifteen Basic Tools

Knowledge of the following tools is essential for layout work and for other aspects of set up and measuring.

The Granite Table

The same flat datum we've used for measuring, often called a *layout table*.

Layout Dye

Layout dye is an alcohol-based, water-thin, dark blue, purple, or red stain intended to dull the work surface (Fig. 8-4). Put on before scribing, the dye makes layout lines much easier to see by reducing reflected light. The newly scribed line shows as bright metal on a dark background. There are both brush-on and spray-on versions of layout die.

It's a stain or dye, not paint!

Use a thin layer and do not try to get even color coverage by painting on extra layers. A correct layout dye will just stain the surface and need not be consistent in color. Correctly applied, it dries quickly, yields crisp clean

Figure 8-4 *Correctly applied, the dye stains the surface. Do not go over it to create an even color.*

layout lines, and does not interfere with measuring. Put on too heavily, it can do just the opposite, obscuring lines. Plus it can actually become thick enough to affect precision dimensions when tolerances are less than thousandths of an inch.

The Height Gage with Scribe Attachment

Reading and setting the height gage was covered in Section 6. Here, we take a second look at the scribe for scratching the lines, not as a measuring probe (Fig. 8-5).

Scribe Points To be precise, the scribe points for height gages must be kept sharp. They are often carbide tipped to retain their edge longer. This carbide edge is brittle, be careful to not push the scribe too hard against the work nor bump it. Correct scribing pressure is about the pressure of a pencil on paper. And as with other layout tools, when scribing lines, pull the scribe do not push it.

Figure 8-6 *The scribe point is sharpened by grinding this surface only!*

Gooseneck Scribes The scribe shown in Fig. 8-6 features a dropped-down point such that it can touch the granite table at zero on the height gage. This offset point scribe is known as a *gooseneck.*

Sharpening scribes is not difficult but requires some background knowledge. Carbide must be ground using a special silicone-carbide (green) grinding wheel or a diamond composite wheel, because of the material's extra hardness.

Sharpen it Correctly! When sharpening the scribe, grind only the front face (Fig. 8-6). Grinding the wrong surface ruins its ability to be used as a measuring probe. Many shops prefer that only one person sharpen the scribes because of this potential problem.

KEYPOINT

Never grind any scribe surface other than the one shown in Fig. 8-6.

Right-Angle Plates (Angle Plates)

These plates are normally purchased in matched pairs of some useful dimension set such as 6 by 6 in. or 250 by 250 mm. When laying out objects such as the drill gage, **right-angle plates** are used to hold the work perpendicular to the table by clamping the blank to the plate. The plates used on the layout table are usually reserved for precision called class A.

Right-angle plates can also be used as supports for large work and for many other chores on the granite table and away from it. These are called class B plates, as shown in the lower half of Fig. 8-7. Thicker and stronger, they feature bolt-down slots and are correctly used on the machines for cutting and grinding duties.

Trade Tip

Most precision tool companies offer apprentice measuring sets. They are the most common measuring and layout tools bundled together into a well-fit case. To discover these bargains, look in any measurement tool catalog or ask your instructor or employer where they can be obtained.

Figure 8-5 *The height gage set up with a scribe attachment.*

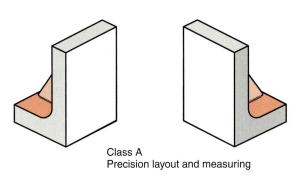

Class A
Precision layout and measuring

Class B
General machining and grinding

Figure 8-7 *Matched pairs of right-angle plates.*

Caring for Right-Angle Plates They require little care except cleaning and caution to not bump them. If they are bumped hard enough to cause a dent or nick, use a fine honing stone to remove the imperfection. Avoid using a file unless the nick is large.

Trade Tip

Air Hockey Accidents *Don't push, then let go, of accessories!* The granite table is extremely flat and so are the right-angle plates. If both are clean and in good condition, the plates (and many other table accessories) will glide over the table with near zero friction. Pushing a right-angle plate without holding on to it can result in an embarrassing continued slide right off the far edge of the table, damaging not only the expensive plate but your reputation, too!

Master Square

A **master square** is shown in Fig. 8-8. It's used to test perpendicularity of features, to align work at 90 degrees to the table, and to

Figure 8-8 *A master square is a common layout and inspection tool.*

lay out lines at 90 degree to surfaces. Smaller master squares (6 or 10 in.) are purchased by the machinist as a priority 1 or 2. Squares of 12 in. or larger are usually owned by the shop.

Depending on shop policy, it might be acceptable to take the master square to a machine when needed for alignment work or for testing a cut relative to the machine's table. But when doing so, exercise good, class A tool practices; keep it away from the machining action and keep it in a place safe from falling or from objects being laid upon it.

Master squares require the same care as the right-angle plates. Although they are only two pieces of joined steel, they are costly and precise. They should be kept in a case when not in use.

Testing a Master Square or Right-Angle Plate for Squareness Squares and angle plates can be damaged and therefore need to be tested from time to time (Fig. 8-9). This can be accomplished using any other master

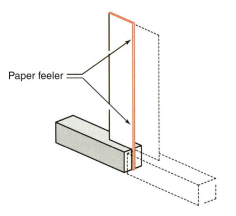

Paper feeler

Figure 8-9 *Testing the square for squareness.*

Step 1   Step 2

Testing squareness

Figure 8-10 *A second more reliable test of squareness.*

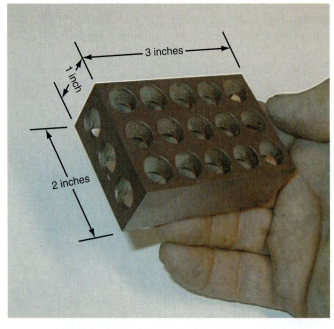

3 inches

1 inch

2 inches

Figure 8-11 *A 1-in. by 2-in. by 3-in. (one-two-three) parallel block is useful in layout and setups.*

square. Usually, a higher gage grade is used to perform the test. Two pieces of paper or two feeler gage leaves of the same thickness are used, one high and one low. When brought together on the two, tugging should produce the same amount of drag.

If an error is found, each square is then tested against a third, higher grade instrument. They can also be tested on a computer coordinate-measuring machine.

Trade Tip

With no explanation, can you see how this test works for squareness? It's a nifty little fixture you can make (Fig. 8-10).

Parallel Bars

We've referred to parallel bars previously, and will again several more times, since they are used all over the shop, for layout, measuring, and work holding during machining. Quality sets are made of hardened and ground tool steel.

They are usually ground to three exact dimensions, such as 0.500 by 0.750 by 6.000 in., in matched sets of two or four. Probably the most common parallel bar is the $1.0 \times 2.0 \times 3.0$ block (*one-two-three blocks*), as shown in Fig. 8-11.

Parallel bars are used constantly for setup work on milling and grinding machines and to support the work. They are also used in measuring and in layout work. It is quite common for student machinists to make their own parallel bars during training.

To care for them,

Protect them from rust.

Repair small nicks using a superfine honing stone.

Put them in a drawer or box when not in use.

Specialty Parallel Bars There are three specialty parallel bars that are quite useful. Although they aren't layout tools, this is a good time to get to know about them:

1. Slot Blocks

These parallel bars are ground to an exact size (usually 0.625 [$\frac{5}{8}$] in.) that taps into the table slots of milling and drilling machines and stays put until pried out with a

Figure 8-12 *Wavy parallel bars support thin work then flatten out as the vise is closed.*

Figure 8-13 *A vee block is useful in holding round stock and other regular shapes too. Notice the three different vees in these matched blocks.*

lever. They are used to align work to the machine's axis. Table slots are usually $\frac{5}{8}$ in. (0.625 in.) or $\frac{3}{4}$ in. We'll see examples of slot blocks in Section 12 on milling machine setups.

2. **Magnetic Parallel Bars**
   These parallel bars are laminated of alternating layers of magnetic and nonmagnetic metals (steel with brass or steel with aluminum), then they are ground to an exact size. When laid on a magnetic chuck (a work-holding device for grinders), they transfer the magnetic pull through themselves to the object being held.

3. **Wavy Parallel Bars**
   These bars are very thin and bent in a wave pattern. Their special purpose is to support thin work parallel to the vise floor until the vise is closed (Fig. 8-12). As the vise closes, the waves are flattened out. Wavy parallels support work like a wider parallel until the vise forces them to flatten.

Vee Blocks

Used for holding round objects for measuring, testing roundness, and for scribing on the work (Fig. 8-13), class B vee blocks are also used on the machines and on the layout table. Often a machinist will own two or more matched sets. Purchasing **vee blocks** would be a priority 2 for your tool box.

The shop will normally provide larger, class A versions for inspection and layout work. Class B vee blocks are used to hold work for drilling, grinding, and milling operations. There are magnetic versions of these tools, similar to parallels.

Example The task being performed in Fig. 8-14 is to scribe two lines at 90 degrees to each other that intersect at the exact center of the round bar.

Figure 8-14 *Scribing perpendicular lines on the center of a round bar using the vee block.*

Similar to parallel bars, vee blocks are made of hardened and ground tool steel. Hone nicks away and keep them in their box when not in use. A coat of oil prevents rust for long-term storage.

Sine Plates—Sine Bars for Angle Work

Discussed back in Section 7, sine tools are used to hold work at a very specific angle for layout work or machining.

Scribes and Awls

Also shown previously, hand **scribes** and scratch **awls** are used often in layout. The trade skill is how to sharpen one—see the trade tip.

Punches

There are four common punches used in layout work.

1. The Pilot Punch—60-Degree Point (Prick Punch)

This sharply pointed punch is almost a scribe. It is tapped lightly at an *X-Y* location. Its purpose is accuracy at picking up layout lines. A pilot punch provides a small spot for deeper punching or to make a pivot point for a divider when scribing a circle. Since it is meant to make a tiny dimple, a common name for this tool is a *prick* punch.

2. The Center Punch—90-Degree Point

The purpose of center punching is usually to mark and exact location, to provide a start for a drill. This punch has a blunt point for producing a wider mark without breaking the punch point. A center punch mark follows a pilot punch.

3. The Automatic Punch

One solution to the no-hammering rule on the layout table is the *autopunch*, sometimes called a *snap punch* due to the way it works. By pushing down, a spring is compressed, then at a given pressure, it triggers and the body of the punch snaps down on the point. The impact produces just the right amount to drive a neat pilot punch mark with no hammering (Fig. 8-17).

4. The Transfer Punch

This tool is used to transfer the center location of a drilled hole in one object, to a metal product below. These are often used in conjunction with a template, as shown in Fig. 8-18. The transfer punch will quickly and accurately mark the center location of each hole, guided by the template, into the metal below.

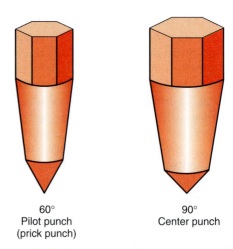

60°
Pilot punch
(prick punch)

90°
Center punch

Figure 8-16 *Comparing pilot to center punch points.*

Figure 8-17 *Pressing down on the red pad triggers this automatic punch to snap down, driving the point into the work.*

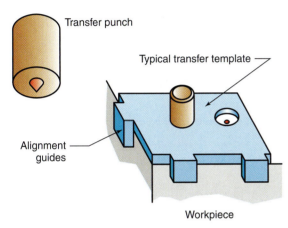

Transfer punch

Typical transfer template

Alignment guides

Workpiece

Figure 8-18 *Hole locations can be transferred from part above to metal below.*

Combination Set— Centering Head

You have seen this tool in Section 6, where we looked at the rule and the protractor head. In addition to the square and protractor heads, the centering head is the third attachment useful in finding and scribing centers of round objects (Fig. 8-19).

One edge of the rule is on the center of the vee. If it's rotated to two or three different positions on the bar, scribed lines will intersect on center.

Dividers

Dividers are used for two layout tasks: *scribing arcs and circles* and *stepping off divisions*, such as breaking a line into four equal parts—thus their name. You should own at least one pair of dividers. Relatively inexpensive, they would be a priority 2 purchase for your tool box.

Three scribed lines cross at center

Figure 8-19 *A center head attachment helps scribe centers of round objects.*

Dividers found in the machinist tool kit come in three forms:

1. Spring dividers
2. Center adjust drafting compass/dividers
3. Trammel and beam dividers

Spring Dividers These tools are specifically designed for layout work (Fig. 8-20). They can be purchased in a variety of sizes. The size refers to the largest radius circle they can produce—their maximum spread. Six- or ten-inch dividers are common for layout work. You might own a second pair for smaller circles and more precise work. The second divider could be the compass type.

Figure 8-20 *Spring dividers are useful for scribing arcs and stepping off divisions of equal length.*

Sharpen Your Spring Dividers Right First, close the points tightly, then lightly twirl them against a fine grinding wheel in the same fashion as a scribe. This results in a point of 30° to 40°, included angle.

If they are ground right, when opened as shown in Fig. 8-21, the conical point is split. The challenge is to grind the point such that the division between each leg is on center. Opened for use, the flat on each half forms the scribing edge. This inner flat is the place to hone them with a stone between grindings. Then for a quick resharpen between grinding, a honing stone can be stroked across the elliptical flats several times.

Sharpening dividers

Half cone point each side

Hone this flat area

Figure 8-21 *Sharpening dividers the right way.*

Figure 8-22 *An adjustable compass with a scribe instead of a pencil can be useful for scribing arcs and circles.*

Center-Bow Compass Dividers Designed for dual drafting and layout duty, these use either a drawing lead holder or steel point attachment. With their steel points in place, compasses work better than the larger spring divider for small radius layout. They are not as rugged as spring dividers, but they are very accurate especially for scribing small arcs (Fig. 8-22).

On these tools, the points are removed and sharpened individually. Due to the small size of the point attachments, they are difficult to sharpen and are often simply replaced as they become dull. When purchasing compass dividers, look for a quick release clutch and a sturdy center adjust. I would suggest they are priority 2 for your box.

Trammel Point Sets A large circle-scribing and dividing tool, this is generally supplied by the shop. Trammels clamp onto any convenient beam of wood or metal, as shown in Fig. 8-23. Either a pencil or a divider point can be held in each. The scribe points look very much like the hand scribe, except the center is eccentric—off center.

Off-Center Trammel Points for Fine Adjustment It's difficult to set a trammel at an exact position on the beam unless you know this trick. Grind the points off center just a bit, then twist them with the clamp snug but not tight. The cam action will provide fine movements, inward or out, along the beam!

Surface Gage

This is a universal tool used something like a height gage in layout work to scribe lines a given distance above the table datum. The scribe height is set against a height gage or rule by adjusting the tilt screw, as shown in Fig. 8-24. As the thumb wheel is rotated, the scribe mount pivots to adjust the point up or down. For layout work, surface gages offer one advantage over height gages—they feature drop-down guide pins in the base.

Scribing Parallel to an Edge The **surface gage** performs another task not possible with a height gage; by lowering a pair of edge

Figure 8-23 *Trammels scribe large arcs.*

Figure 8-24 *A surface gage is used for scribing lines and testing heights above a datum.*

Fine adjust
tilt screw

Scribing line
parallel to edge

Pin guides
lowered

Figure 8-25 *A surface gage with guide pins down scribes a line parallel to an edge.*

guide pins, it can also scribe lines parallel to edges, as shown in Fig. 8-25. These tools can also be used in many ways throughout the shop; as a nonmagnetic indicator holder and a vertical tool height pointer on the lathe or mill, for example. Purchasing a surface gage would be a priority 2 or maybe a 3.

Indicators on Surface Gages Many test indicators are equipped with a small stud attachment that will fit in place of the scribe in the holding clamp (Fig. 8-25). Moving the indicator by a surface gage rather than a height gage can be useful when the height gage is too big for the task or when the indicator should run parallel to a datum reference.

Hermaphrodite Calipers

Hermaphrodite calipers are single-purpose calipers used to scribe lines parallel to an edge

Hermaphrodite Calipers

Scribing parallel
to an edge

Figure 8-26 *A hermaphrodite caliper also scribes distances from edges.*

of a part, similar to the edge pins of the surface gage (Fig. 8-26). Buy these tools as a priority 2 or 3.

Radius Gages

Although their prime purpose is gage testing of size, radius gages can also be used as layout templates for small arcs. When doing so, be sure to tilt the scribe into the edge, as shown in Fig. 8-27, to avoid false distances.

Scribing arcs with a radius gage

Keep the point
against the curve

Figure 8-27 *Tilt the scribe to produce best accuracy.*

Tilting the scribe is necessary whenever your goal is to scribe accurately against a guide—radius gage, ruler, or square.

Planning a Layout

OK, here's the challenging part of layout, it can be simple and quick or it sometimes becomes a time-consuming nightmare of extra lines and mistakes! It all depends on preplanning. The very best lesson is practice so, Unit 8-3 is designed to challenge you to do a layout then compare your results to the pictorial answers found here.

Your course syllabus may include laying out the following activity on a metal blank or doing another shape of similar nature. If so, you may still want to do a practice run using the activity in Unit 8-3. There are several critical thinking traps even in the simple shape layout there. The goal is to learn to identify traps that lead to leftover layout or too many lines, then plan them out. In order to save time, frustration, and telltale lines left on the work, always start with some forethought—then a plan.

Here's a good example of the kind of errors one can encounter without planning: returning to the *X-Y* intersection point for the start of the 59 degree line (Fig. 8-1), note that the intersection borders on the finished part. The lines should not fall on metal that won't be removed (Fig. 8-28).

KEYPOINT

How can you avoid overscribing? By using a soft pencil and rule to rough in the *X* and *Y* lines before scribing them. It's that simple in this case, but is not always so. A few game rules are listed on page 284.

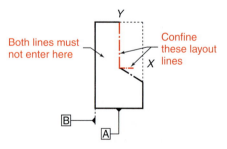

Both lines must
not enter here

Confine
these layout
lines

Figure 8-28 *To avoid leftover lines, plan out where they start and stop.*

Executing a Successful Layout

1. Think through the machining to be done. Which lines are truly required? Often, only a few will be needed. The least number and shortest lines are best. Extra lines only confuse the scene.

2. Now, take a "dry run" through the layout, think through the scribing of lines and points. As you do so, look for intersections where a rough pencil guideline is needed to place and limit scribe lines.

3. Toward efficiency, locate and pencil in all lines parallel to datum B first, then scribe the intersecting lines parallel to datum A.

4. Now, with the scribed lines as guides, scribe the lines parallel to datum B. Note that sometimes it's necessary to pencil in lines parallel to both datums before scribing.

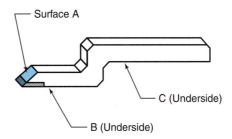

Figure 8-29 *Which surface(s) should be reground when sharpening a scribe point?*

- Never regrind a height gage scribe on any surface other than the front sloping face.
- Do not push accessories on the granite table, then let go.
- Never hammer in any way on the layout table!
- Use a soft pencil to establish boundaries for the upcoming scribe lines.
- *Plan the layout as you would a machining job.* Make a rough, soft pencil guide to avoid excess scribe lines.

Respond

1. Without going back in the reading, on a sheet of paper, list the 15 tools used in layout work and briefly describe each using a sketch if needed.

2. Layout is used in two different ways in machining, they are as a fail-safe template and sometimes as _____ _____.

3. True or false? When used as a last word definition, layout repeatability can be as close as 0.005 in., but the normal expectation is 0.060 in. If it's false, what makes it true?

4. A center punch features a point angle of _____°.

5. Which surface of a height gage, gooseneck scribe is ground when it's resharpened; A, B, C or A and C together (see Fig. 8-29)?

UNIT 8-2 REVIEW

Review the Key Terms

Dividers
Pointed layout tools used to scribe radii and divide distances into equal portions.

Hermaphrodite calipers
Used to scribe lines parallel to an edge. Not truly a caliper or divider, the dictionary meaning of "hermaphrodite" is having both male and female characteristics.

Layout dye
Surface stain to reduce glare and make layout lines more visible.

Master square
The rigid square used to test squareness and to lay out lines perpendicular to edges.

Right-angle plates
Used to test squareness and to hold objects perpendicular for layout.

Scribes/awls
Simple pointed styluses used to scribe lines.

Surface gage
A multipurpose tool used to scribe lines parallel to a surface, to transfer lines from a rule to a part, and to scribe lines parallel to an edge of a part.

Vee blocks
Ground blocks used to hold round work for scribing, measuring, and machining.

Replay the Key Points

- Layout dye is not put on thickly as would be paint.

UNIT 8-3

A Layout Challenge Game

Introduction: This unit is an opportunity to practice layout and it's a critical thinking exercise. There are no terms nor key points needed, it's all about practice. But there are answers.

Laying Out the Drill Gage (Fig. 8-30)

On a blank piece of paper, follow the rules here to create a *last word* layout to completely machine the shape and drill the holes using Fig. 8-30. *Hint:* Graph paper with engineering divisions (0.100 or 0.200 graduations) would make it easiest.

There are seven steps required to do this layout. Each is needed, but there are two or three valid sequences in which you might do them. Each step is discussed in the answers. But they are presented in the sequence in which I did the layout. Be sure you have covered them all in your plan and layout but not necessarily in the order presented.

The real test would be whether or not you could cut your part out from the paper using a knife or razor, and have the right shape and size, with no leftover layout lines. If so, you've got the idea.

Game Rules

◆ To simulate the real thing, use a soft pencil pressing lightly for the rough-out guide lines.

◆ Erase the pencil lines as often as needed as you would on real metal. Then employ a pen to represent the permanent scribe lines.

◆ If any pen lines show after the complete object is defined (machined to shape) your layout fails inspection.

◆ Use a compass to scribe each hole's diameter.

◆ The *X-Y* centerlines must not extend out beyond their final diameter. Note that smaller diameter holes require short intersection lines to stay inside their bounds.

◆ Assume that the lower left corner of the blank metal, the intersection of datum features A and B, is premachined to 90 degrees.

◆ The blank includes at least 0.100 in. for machining its height and width.

Challenge—Steps to Perform the Layout

1. *Pencil lines parallel to Datum B* (Fig. 8-31). Using a ruler, draw light pencil lines parallel to Datum B first. These will be used to guide the scribe lines parallel to Datum A.

2. *Scribe lines parallel to Datum A* (Figs. 8-32 and 8-33).

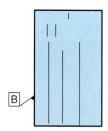

Rough pencil guidelines parallel to Datum B first

Figure 8-31 *All pencil guide lines parallel to Datum B.*

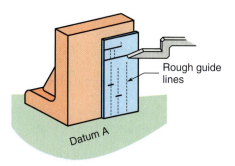

Rough guide lines

Figure 8-32 *Scribing lines parallel to Datum A.*

Horizontal scribe lines completed

Figure 8-33 *Lines scribed.*

0.1875
0.2500
0.3125
0.3750
0.4375
0.5000
0.5625
0.6250 Hole diameters

2.00
1.50
0.38
0.50
0.12 R All corners
No radius
59.00°
5.00
3.75
3.25
2.88
2.50
2.13
1.75
1.25
1.00
0.63
A
B
.63
1.25
2.00
3.00
Layout challenge

Figure 8-30 *Print for Unit 8-3 layout.*

Figure 8-34 *Do not scribe in the center of the tab.*

Critical Question 1

In Fig. 8-32, there is a deliberate error. Can you find it? What is the solution? The answer is found in Fig. 8-34.

Answer (Fig. 8-34) Notice that the horizontal line being scribed must be broken by the two guides. If not, that line would appear on the finished product!

Critical Question 2

The line 1.00 in. from datum A serves what purpose and where do you scribe it on the metal blank?

Answer It is the start point for the 59-degree line. It must be limited to the excess outside the finished metal (Fig. 8-35).

3. *Pencil in the 59 degree line.*

4. *Scribe lines parallel to Datum B* (Fig. 8-36).

5. *Scribe the 59 degree angle line.* There are three possible methods:

    The Protractor Using the intersection of the 1.00 in. by 3.00 in. lines as a locating point with the protractor base against datum feature B or A. (Watch out for complementary angle mix-ups.)

    Tilting the Blank on a Sine Bar

    Calculating the Far End of the Line
    The third method would be to calculate the vertical height of the intersection point above Datum A (Fig. 8-37).

Figure 8-35 *The start of the 59-degree line is 1.00 in. from Datum A.*

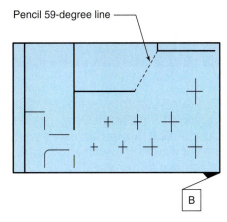

Figure 8-36 *Scribe lines parallel to Datum B limited by 59 degree pencil line.*

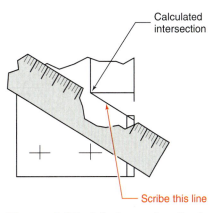

Figure 8-37 *Calculate and scribe the upper end of the line, then connect it with a straight edge line.*

KEYPOINT

The far end of the 59 degree line is found using trigonometry. Trig is a necessary competency for machinists.

6. *Scribe the corner radii.* Scribing the radii can be a challenge. This step might be skipped. Use radius gages as templates and hand scribe the lines (Fig. 8-38).

7. *Compass scribe the hole perimeters (might be skipped).*

Figure 8-38 *Complete the shape by scribing the corner radii using radius gages and holes using a compass.*

Section 8 Review

Unit 8-1
In some situations there's still no substitute for an accurate, well-thought-out layout. Even in the world of CNC, layout is a useful skill. But it can also be a waste of time and another way to risk scrapping the job! The skilled machinist knows the difference and uses layout only when needed.

Unit 8-2
The tools presented in Unit 8-2 add a lot of extra ability to granite table skills as well as providing the means to make a layout. But the most useful tool we discussed is *planning* the layout. I remember my first layout assignment when I was an apprentice at Kenworth Trucks. It took me half a shift and it was still wrong! Why? I didn't plan, I just started scribing lines all over the work!

Unit 8-3
So how did you do? By now you see why we on the planning committee weren't ready to send this skill to the technology graveyard. It's challenging and can be a great backup skill—once one practices it for a time.

Questions and Problems

1. What is the caliper that scribes lines parallel to edges? (Obj. 8-2)

2. What is another layout tool that can scribe lines parallel to edges? How? (Obj. 8-2)

3. What are the two main reasons to use layout? (Obj. 8-1)

4. Name the only layout tool that's allowable for making punch marks on the work while it rests on the granite table? (Obj. 8-2)

5. Name the punch with a sharp pointed 60-degree nose? What does it do? (Obj. 8-2)

6. Describe how a pencil is used in layout work? (Obj. 8-2)

7. Of the parallel bars we've discussed, which are
   A. Made to lightly press fit into the table slots?
   B. Used in vises to support very thin work?
   C. Made with three ascending dimensions—a very popular, utility size? (Obj. 8-2)

8. What general tolerance could one except to achieve when machining to layout? (Obj. 8-1)

9. True or false? Several coats of layout dye may be necessary to produce a uniform result in order to highlight scribe lines. (Obj. 8-2)

10. When *splitting the line* machining results can be as accurate as ____ inch? (Objs. 8-1 and 8-2)

11. Regarding Question 10, what repeatability results? Why? (Obj. 8-2)

12. Name the three attachments mounted on a combination set ruler? (Obj. 8-2)

CNC Questions

13. What has CAD/CAM done to the need for lay-out skills?

14. True or false? Layout would never be a good idea for a CNC program run. If it's false, what might make it true?

Critical Thinking Not Covered in the Reading

15. In your estimation, how accurate is drilling a hole by CNC methods as compared to drill press, layout location. Compare estimated repeatability.

Section 8 Answers

ANSWERS 8-1

1. As a fail-safe template that helps indicate when it's time to stop and measure with some other means: as a last word means of machining a size or shape that's not possible otherwise.
2. By extending layout lines onto surfaces where the layout will remain after machining. Left-over layout can scrap some objects!
3. It is generally accepted that ±0.030 in. is standard for machining to layout.
4. A visual template of lines and points scratched onto the surface of an object. It is used to guide the machinist as to where to cut the object.
5. Layout adds value if it's a last word or if it improves quality, reduces cutting time, or prevents scrap.

ANSWERS 8-2

1. The 15 layout tools: *layout table*, provides a perfect flat work datum surface for layout; *layout dye*, cuts out glare and makes layout lines more visible; *height gage/scribe*, to scribe very accurate parallel lines parallel above the layout table; *right-angle plates*, to test squareness and to hold objects perpendicular for layout; *master square*, to test squareness and to layout lines perpendicular to edges; *parallel bars*, used for measuring, scribing lines parallel to edges, holding work parallel to a surface, and a wide variety of uses throughout the shop; *vee blocks*, to hold round work for scribing, measuring, and machining; *sine bars/plates*, to hold work at a very accurate angle for scribing and for machining, measure very accurate angles; *scribes/awls*, to scribe lines; *punches*, to mark a location for a hole; *combination set*, the rule is used to measure and as a straightness template, the center head to find and scribe the center of round objects, the right-angle head to measure and scribe 90 degree lines, and the protractor head to measure and scribe angular lines; *dividers*, to divide out distances and to scribe circles; *surface gages*, to scribe lines parallel to a surface, transfer lines from a rule to a part, and scribe lines parallel to an edge of a part; *hermaphrodite calipers*, to scribe lines parallel to an edge; *radius gages*, to scribe parts of arcs and to measure arcs for size and roundness.
2. As the final word. The only definition of shape or location.
3. False. While it is true that we sometimes are able to pull off a tolerance of 0.005 in. using layout alone, 0.030 in. is considered standard.
4. 90 degrees. However, this isn't critical—a center punch need only be blunt to withstand the hard blows. From 80 degrees to 100 degrees work fine.
5. Surface A only.

Answers to Section Review Questions

1. Hermaphrodite calipers
2. The surface gage; by pushing guide pins down below the base
3. Fail-safe or last word
4. The automatic punch (snap punch)
5. The pilot punch or sometimes called the prick punch. It makes very light indentations to guide the upcoming center punch.
6. To make guide lines to prevent overscribing into forbidden areas.
7. A. Slot blocks
   B. Wavy
   C. 1-in. × 2-in. × 3-in. blocks (one-two-three blocks)
8. 0.030 in.
9. False. One coat. Uniform color is not required.
10. 0.010 in.
11. Remains at 0.030 in. as a best case, because eyeball estimation is required.
12. Square, centering head, and protractor
13. If the shape can be defined on the screen, it can be machined accurately with far better repeatabilities. Therefore, the need for layout skills is diminishing.
14. False. When trying a program for the first time on very expensive materials, it's wise to create a fail-safe layout.
15. Drill press with layout = ±0.030 in.; CNC = ±0.003 in. without a pilot drill and ±0.0015 in. with pilot drilling and perhaps boring the hole

PART 2

Introduction to Machining

Now we get started on the trade skills needed to shape metal using a cutter or grinding wheel. This subject is about the physics of making chips and how to contain the forces encountered on various kinds of machines. Today's CNC machines are so fast and powerful that we machinists can no longer estimate or guess how to write a program or set up a machine; everything must be right.

But the chances are your lab hasn't enough CNC machines for every student, so I'll present this material assuming you will be practicing basic setups, operations, and safety using manually operated machinery. Sometimes, however, there are big differences in using your hands to start an action or machine motion compared to the computer taking control. Those differences will be pointed outs so that when you get to Part III on CNC machining, you'll also understand the capabilities of CNC machines.

But don't get the idea that manually operated equipment is gone in industry. Not even! To prepare for this book I estimate I've visited 50 machine shops, large and small. In every one, there were some manually operated machines. Sure, they were far from the heart of the shop, but they were used for secondary operations when it was not practical to tie up the CNCs to do a job. Manually operated machines are also used for simple production when the tooling has been built and the cost of moving the job to the CNC doesn't make dollar sense. Manually operated machines are especially useful for building tooling for the shop or for repair work.

In Part 2, you will learn

Section 9   To gain complete control of the machining process by studying the physics of cutting metal and how to vary the cutters doing the work; to control longer tool life, better accuracy, finer finishes, and more efficient production, always with safety as the overall goal.

Section 10   To drill holes to the specified size, at the exact location—a baseline requirement for machinists. While your early training will be on manually operated drill presses, the skills gained will apply to all CNC programs that call on drill bits to remove metal.

Section 11   That lathes spin the workpiece then cut metal away using a stationary cutter; they create more potential danger and you need to completely understand setups and turning operations. Although this book assumes you'll be practicing early setups and operations using a manually driven lathe, the lessons are aimed at their safe, efficient performance on a CNC machine.

Section 12   That mills perform most metal shaping today. Using a computer to drive multiple axes of motion and to feed the work through any complex path, almost any shape can be milled.

Section 13   To turn to metal removal by abrasion (grinding it away) when ultra accuracy is required or when finish requirements or workpiece hardness exceed the ability of chip making—a skill that applies across the whole range of processes; every machine type has a counterpart that grinds.

Section 14   The range of threads to be machined and a bit about how to machine them. While this is our third investigation into it this deep subject, we have only just begun, but you will now have the background needed for the learning beyond Section 14—specific on-the-job training when specialized threads are required.

Section 15   More complete control through an in-depth understanding of metal; alloy identification and makeup; and how to change steel and aluminum's hardness and how to test the results. Beyond Section 15, metallurgy is one of the most fascinating frontiers with the potential to affect your future.

Section 16   To accept and succeed at the challenge of challenges! Planning out a logical sequence of operations that lead to the perfect part, in the least amount of time, with the safest possible path is an incredibly complex task that can make the difference in profit and loss—and survival as a company!

Section 9

Cutting Tool Geometry

Introduction

The upcoming sections begin the core instruction on drilling, turning, and milling operations. To get the most from that instruction we first take a close look at what happens when any tool cuts a chip away from a metal part, and why the tools that do it are shaped as they are. This is bedrock theory for all machining, even grinding, where the chips are microscopic, but chips just the same.

Surprisingly, with over 200 years of learning about the subject behind us, cutting tool technology is currently undergoing its greatest development other than during the period when tungsten-carbide became commercially affordable! Cutters are one of the few hardware issues evolving as fast as computers and the software that drives them. In fact, the two—cutting tools and CNC computers—are locked in a leapfrog growth, with each spurring the other to improve.

Each new generation of central processors crunches faster, enabling them to drive the machines at unheard-of velocities and RPMs. Then tools must be improved to take full advantage of the increased speed. Once the tools catch up, for a time, we cut with new cutter compositions, new coatings, or even gem crystals grown upon their surfaces, always aiming to improve performance. Tools are also shaped in radical new ways because computer simulations provide deeper insight into how a chip flows past the cutter at accelerated speeds. So, for a time, the tools can take all the abuse this generation of CNCs can dish out and the process goes on. But before we can discuss the basic operations and then the future, we need a starting point.

The first two units of this section explain how all cutting tools produce a chip. Then we'll look at variables that can be adjusted by a skilled machinist to improve finish, accuracy, tool life, and safety. Make sure you fully understand the material in the presented sequence. These units of learning are rungs on a critical ladder, one step leading to the next, with full understanding and control of machining at the top. Here are the goals of Section 9.

Objectives

Solve the Self-Heating Chip Mystery!

There's an enigma we will solve in Section 9: Metal chips become hot due to extreme friction during machining. OK, that's no surprise. But the mystery is that once they're cut away and lying on the floor, they seem to get hotter? Here's proof: due to that puzzling rising heat, steel chips begin to oxidize and turn rainbow colors right before your eyes, after the chip is fully formed. Why then when the friction is over? As we go forward, we'll pick up clues to solve this enigma.

UNIT 9-1

Four Universal Cutting Tool Features

Introduction: All cutting tools share four common shape features—rake angle, lead angle, corner radius, and clearance angle—each of which must be present to varying degrees for the cutter to work right. They are the basis of tool geometry.

The cutting bit shown in Fig. 9-1 is generic. It could be the tip of a drill bit as it rotates in the metal, moving to the left in the illustration. It could be the tip of a lathe bit as the work spins past to the right, or it could be the cutting edge of a milling machine cutter as it rotates and is fed sideways into the material. Representing all three, the figure shows two of the four features needed to cut metal.

For now, as a clear example, let's relate it to an operation you probably understand: drilling. Envision that this bit is spinning. We are looking at it from the side as it makes a spiral chip. While the sharp edge cuts, the action isn't a simple slice as one might assume. There are a few surprises here. As always, understanding the terms will be very useful in your career and to unravel the hot chip mystery.

ShopTalk

Cincinnati-Millacron (a famous CNC machine tool builder), in their instructional video "Cool Chips" illustrates the chip formation action to be similar to a snowplow moving snow.

KEY TERMS

Chatter
Unwanted vibration in the tool, work, or the machine.

Clearance angle
A relief behind the edge to enable cutting contact.

Corner radius (nose radius)
The rounding of the point or junction of the front and side of a tool bit's cutting edge.

Deformation
The permanent bending of any metal, in this case, of the chip as it is raked.

Lead angle/cutting angle
The angle of the bit compared to the angle of the cut. Lead spreads the chip width over a larger area.

Negative rake
The tool that bends the chip the most.

Figure 9-1 *A generic cutting tool showing the rake and clearance angles.*

Nose radius
Term used for corner radius on lathe tools.

Positive rake
The tool that bends the chip the least.

Rake angle/rake surface
The surface upon which the chip forcefully slides.

Two Distinctly Different Actions: Cut and Rake

Two different actions occur together when a chip is made. First, the cut occurs where the sharp edge slices the chip away from the parent metal. Then immediately after, the rake begins, and the now flowing chip is forced up, across the surface or face of the bit.

The Rake Surface

During this brief phase, the chip comes into forceful contact with the bit. As it slides up over the rake surface of the cutter, the chip is redirected, bent to flow in another direction.

There is a lot of friction at the contact point between chip and tool bit face, where the chip is raked. That pressurized rake zone is critical to the cutting action. (See the Shop Talk.)

Beyond the rake zone, the chip is fully formed. It continues to slide up the extension of the surface and out of the drilled hole in our example or off the bit in some other way in other operations. But the true rake occurs from a few thousandths to no more than one-eight of an inch beyond the cutting edge. The distance depends on how hard the bit is being pushed through the metal, a process called the *feed rate*.

If the feed rate is light, the rake can be a minor action compared to the cut. Or it can be the major part of the chip removal. In the drilling example, you might be exerting 5 or 10 pounds of pressure downward on the drill

The image labels: Rake angle, Rake surface, Cutting bit, Clearance angle, Parent metal (the work being machined)

press feed lever, producing a small chip. In that case, there is a very short rake.

The rake becomes ever more significant as feed rates climb. Pushing down with a force of 25 pounds, the chip pressure increases on the bit face and the rake zone widens away from the cutting edge. The rake contributes more to the formation of the chip.

In CNC production, we're always looking for ways to cut more metal per minute. As we feed ever harder, the rake phase becomes more important, as we'll see in just a while. We experiment with varying rake angles, improving the coolant formulas, and using better antifriction materials to help the chip slide up the rake face, all to achieve faster cutting and longer tool life.

KEYPOINT

As feed rates increase, the rake becomes more significant relative to the cut, and we use that fact to our advantage.

Chip Deformation

The **rake surface** of the bit redirects the chip as it flows. It bends it away from the direction it would take if the rake face wasn't in its path, somewhat like a wedge. This permanent bending action is called chip **deformation.** Remember that, it's one of the clues that solve the hot chip mystery! But here's another odd fact and perhaps another clue: of the heat produced in making the chip, only 25 percent comes from the friction between the bit and rake face during the deformation.

KEYPOINT

The bit does not split the chip away from the work as when wood is split, but rather, right after the cut, the chip is deformed to another direction by friction against the rake face.

Rake Angles from Positive to Negative

Changing the rake angle relative to the cut direction can have a dramatic effect on the cutting action and on the chip. We'll return to the various angles involved later, but notice in Fig. 9-2 that a **positive rake** deforms the chip the least, while a **negative rake**

Figure 9-2 *Positive and negative rake bits.*

bends it more. Changing rake angle is one of the control variables.

Investigate It!

Ask your tool room attendant for a relatively large drill bit of about a half-inch diameter or bigger. Look for one that has been used for some time. Now starting at one of the two cutting edges, trace a pencil up the spiral groove. The groove is called the *flute.*

Similar to the microphotograph in Fig. 9-3, there will probably be a small shiny portion next to the cutting edge where the dark or dull finish has been worn shiny. That's evidence of the rake zone friction.

Note that, beyond the rake around 0.050 in., the drill's surface returns to the normal color. Keep this drill close by, we'll look at it again.

Critical Ladder Check Point

Before moving on to clearance angles, be sure you can define the two different phases a chip undergoes during its formation. Can you describe them to another student? What happens to the chip during the rake phase?

Figure 9-3 *The small shiny spot is evidence of chip friction against this drill bit's rake face.*

Clearance (Relief) Angle

Return to Fig. 9-1 and note the **clearance angle.** It serves a very different purpose from rake—it enables the cutting edge to contact the work with no rubbing behind the edge. You might also hear it called a relief angle because that's what it does; it relieves the cutter behind the cutting edge, to prevent unwanted friction.

In Fig. 9-4, you see a correct bit compared with an incorrect bit with zero clearance. The bit on the right will not cut. Not only that, it will quickly heat up due to the rubbing between the clearance surface, sometimes called the heel of the bit, and the work.

Just Enough and No More!

As long as a minimum amount of clearance is present, the bit will work. Adding more has no effect on the cutting action. But with too much clearance, the bit will cut only for a while. Excessive clearance makes the bit weak due to the thinning of the cutting edge, as shown in Fig. 9-5. It usually breaks before it dulls. The correct angle is usually around 5 to 10 degrees.

ShopTalk

When a bit with inadequate clearance rubs the work without cutting or cuts poorly, it's sometimes called "heel drag."

KEYPOINT

Changing clearance has little effect on the actual chip formation. It is necessary to have only a minimum amount. We do not adjust clearance to achieve a result. It is not a control variable.

Excessive clearance not only weakens the bit, the thin edge also makes it a target for heat buildup as well. With less mass to conduct heat away, bits burn up quickly with excess clearance.

With clearance

Without clearance

Rubbing friction

Figure 9-4 *The bit on the left will cut while the right bit with* zero clearance *will only rub.*

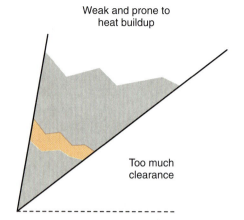

Weak and prone to heat buildup

Too much clearance

Figure 9-5 Too much clearance *creates a weak bit that is prone to heat buildup and breakage.*

There is one positive but very minor aspect to adding more clearance to a bit; the triangular, hollow space created by the relief angle face, enables coolant to collect, which is a good thing. However, when balanced against the potential of a bit breaking or burning up, coolant induction doesn't justify more clearance!

Identify the clearance angle on your example drill. In the shop, you'll need to identify the clearance angle on every cutting tool selected. Note that when resharpening a drill bit, you will be grinding the surface that makes the clearance, called the *heel* of the bit. (More on this in the Drilling Section.)

Critical Ladder Key Points

◆ Machinists change rake angles to achieve different results, but not the clearance angle.

◆ Clearance is a matter of "just enough" and no more.

Got the idea? If so, we step one rung closer to solving the chip mystery—which, as you may have guessed, has nothing to do with clearance.

The Lead or Cutting Angle

The third universal tool angle is called either the **cutting angle** or the **lead angle** on many cutting tools. (It's often shop-shortened to "lead," which rhymes with feed.) The cutting angle is an important chip control feature on all bits.

Lead is the angle of the cutting edge compared to the cut direction. It imparts many benefits to the action. To see the cutting angle on the example drill bit we rotate it 90 degrees. In Fig. 9-6, the cutting angle on both edges forms the point of the drill bit.

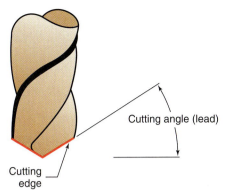

Figure 9-6 *The cutting angle, may also be called the* lead angle *on other cutting tools. The combined cutting angle of both flutes forms the* point angle *of this drill bit.*

What Is Lead?

To understand the action a lead angle imparts, envision yourself whittling wood with a knife. If you hold the knife square to the direction you wish to carve, the chip (shaving, in this example) will be the same width as the work (Fig. 9-7). The knife enters the wood all at once and exits the same way, abruptly—that's zero lead.

That action would be awkward and would not work well, so you would naturally rotate the knife to an angle of about 20 degrees relative to the direction of the whittle—that's a lead angle. Here's why.

Easy Entry and Exit

With lead, the knife would start into the work at a single contact point and the contact width would grow wider gradually to the full chip width. It would then slice cleanly as it moves forward, then exit the work gradually, too.

The reason a bit with lead slices better is that it imparts a diagonal sliding action as

Lead angle spreads the cutting force over a greater area

Figure 9-7 *A comparison of three lead angles.*

well as a straight forward motion. Look at Fig. 9-7. More cutting edge contacts the work with lead than without lead. Lead angles on metal cutting tools vary from 0 degrees up to a maximum of around 30 degrees, as on a drill bit.

Additionally, when lead is used, the chip produced is wider than the work. Because it spreads over a longer cutting edge, the tool also lasts longer in most cases. The chip is thinner and easier to shear away when lead is used.

KEYPOINT

Lead angle produces:

A. Smoother entry and exit.
B. Longer tool life due to more cutting edge doing the work.
C. Better finish on the work.

We change lead as a control variable.

Too Much Lead Can Cause Chatter

There is an upper limit to the benefits of increasing the cutting angle. Imparting too much lead produces a chip that's too wide. Beyond a certain point, when too much cutting edge comes into contact with the work, an unwanted vibration starts known as *chatter*. If left uncorrected, chatter can ruin the finish on the work and even break the tool.

That's one way lead is used as a control variable; to reduce chatter when it occurs, lessen the lead angle. This reduces the contact length between bit and work. Several other controls can also reduce chatter. We'll learn them under troubleshooting.

Too Little

Returning to the drill bit example, assume there is no lead angle. The bit would be flat on the tip, as shown in Fig. 9-8. On that flat-pointed drill, the chip width is the radius of the bit. That drill would be a nightmare to get started drilling. Trying to take the entire chip width the instant it touches the work, it would grab, try to walk off center, and then snap back repeatedly before it would start drilling. It might even break the drill.

However, if we grind it with a point angle (a cutting angle on both edges), the drill bit

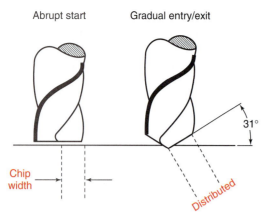

Abrupt start   Gradual entry/exit

31°

Chip
width

Distributed

Figure 9-8 *A drill ground without a point angle (lead) will be difficult to start into the work.*

would tend to start into the work more smoothly, at a tiny circle that gets wider as it goes down. The chip width increases to the full length of the angular cutting edge. So comparing this to the wood carving example, we see that the pointed bit exits out the far side of the hole the same way, with a tiny circle breaking through that gradually widens to the full diameter finish. That's because it now has lead angle.

Adding and Changing Corner Radius to the Cutting Edge

Corner radius is the fourth basic feature of tool shape. In Fig. 9-9, a drill is shown with a corner radius, which is a rounding of the cutting edge between the intersection of its front edge and side. The lines on the right represent the standard grind, unmodified, sharp-cornered drill bit. Corner radius is far more common on lathe and mill cutters, but it imparts the same benefits on drills or any cutting tool. (See the Trade Tips.)

Figure 9-9 *Corner radius is more common on other cutting tools but can be ground on a drill bit, too.*

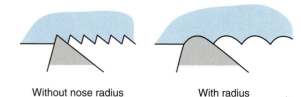

Without nose radius   With radius

Figure 9-10 *Finish is improved with corner radius.*

There are three benefits:

1. **Smoother Cutting Action and Chip Distribution**
Adding radius to a cutting edge has an effect similar to adding lead angle. It smoothes the cutting action and distributes the chip over a longer cutting edge. But corner radius adds two more positive benefits to the tool's geometry.

2. **Better Finish and 3. Stronger Tools**
With a corner radius, the tool is stronger and less prone to breaking at the point intersection. Sharp-cornered tools tend to chip or break at the point. The finish a radiused tool produces is also smoother. Figure 9-10 shows two lathe tools machining metal. With radius, the tool leaves tiny wavelike marks in the finished surface compared to sawtooth marks for the sharp-cornered tool. When that **corner radius** is put upon a lathe tool, as in Fig. 9-10, it's called **nose radius.**

Trade Tip

When no reamer is available have a journeyman demonstrate grinding a corner radius on your drill bit. The main improvement is better finish in the hole, and size control is also improved. Corner radius also prevents edge breakdown when drilling harder materials; the radiused bit lasts longer.

Too Much Radius

A large corner radius can cause chatter in the same way as too much lead angle; there is too much bit in contact with the work. If chatter arises, reduce radius to shorten the contact edge of the bit with the work. There is a time when chatter is such a problem that a sharp-cornered tool is the answer.

> **KEYPOINT**
>
> Corner radius is one of the control variable features.

Adding corner radius brings three benefits:

◆ Stronger tools
◆ Smoother machining action
◆ Improved machine finish

UNIT 9-1 REVIEW

Review the Key Terms

Chatter
Unwanted vibration in tool, work and the machine.

Clearance angle
A relief behind the edge to enable cutting contact.

Corner radius (nose radius)
The rounding of the point or junction of the front and side of a tool bit's cutting edge.

Deformation
The permanent bending of any metal, in this case, of the chip as it is raked.

Lead angle/cutting angle
The angle of the bit compared to the angle of the cut. Lead spreads the chip width over a larger area.

Negative rake
The tool that bends the chip the most.

Nose radius
Term used for corner radius on lathe tools.

Positive rake
The tool that bends the chip the least.

Rake angle/rake surface
The surface upon which the chip forcefully slides.

Replay the Key Points

◆ Rake angle is a control variable.
◆ As feed rates increase, the rake becomes more significant relative to the cut.
◆ During the rake phase, the flowing chip is deformed into another shape.
◆ Clearance is a matter of enough and no more.
◆ Clearance is not a control variable as long as the bit isn't rubbing on its heel.
◆ Lead is a control variable.
◆ Lead imparts gradual entry/exit in the cut, longer tool life, and better finish.

Respond

1. List and describe the three basic tool angles.
2. Of the three answers to Question 1, which are control variables?
3. True or false? The cutting angle is varied to achieve different results in machining.
4. Name the two phases of chip formation.
5. Of the two phases of Question 4, which becomes more significant as the feed force is increased?
6. What is the range of common rake zone widths?
7. What are two benefits of tool radius?

UNIT 9-2

The Physics and Forces of Chip Formation

Introduction: In Unit 9-1, we didn't explore all that happens inside the chip. After we complete this study, hard, unbending metal might seem more like clay or hot plastic as it yields to the cutter.

Machinists are expected to operate their machines at maximum efficiency yet not break or destroy tools. To do that, choices must be made based on the upcoming knowledge. Cutting just a little too fast can burn up or break the bit, requiring time to resharpen or replace it, and, often even more costly, lost production time to get the setup back up and running again. But running too slow to preserve the cutter brings reduced productivity. It's a decision to be balanced based on experience and knowledge of the facts presented here.

KEY TERMS

Boundary zone
The small space between interference point and cutting edge on a bit.

Built-up edge
The result of chip weld holding on too long, then allowing a small part to escape under the bit to reweld to the surface of the work.

Chip load
The amount of chip being taken as a combination of feed rate and depth of cut. See removal rate.

Cratering
The erosion of the rake surface of a bit due to welding.

Deformation
The bending and piling up of the chip during the rake phase.

Interference point
The concentrated center of contact pressure between chip and bit.

Negative rake
The bit deforms the chip through an acute angle, causing more bend than positive rake bit.

Plastic flow
The characteristic of a chip that causes it to deform without returning to its original shape.

Positive rake
The bit deforms the chip through a lesser angle, causing less deformation to the chip.

Removal rate
The combined result of depth or cut, feed rate, and cutting speed (RPM) rated in cubic inches per minute (or liters per minute). Equates directly to horsepower requirements for the cut.

Shear line
The line inside the chip along which the deformation occurs.

Welding
The instantaneous, heated adhesion of the chip to bit. Leads to cratering and built-up edge.

Force, Friction, and Heat

Choosing the right cutting speed and feed can prolong tool life in ways that might surprise you. To understand how, look closely at a chip being made. Notice that there is a pressure curve against the rake face. It starts out light near the cutting edge then increases to a focus point farther back from the edge, then lightens up again until the chip no longer contacts with the bit (Fig. 9-11).

Evidence of that gradient pressure can be demonstrated by looking at the rake face of a tool that's been used to cut steel chips for a long time with no coolant. When we do so, we see a worn spot that formed due to the extreme friction wearing away small particles of the bit. We'll look at ways to prevent **cratering** because it will eventually destroy the bit. But for now, we're using it to demonstrate how the pressure is distributed.

The center of the crater where the pressure was the greatest is a distance back from the cutting edge. That central focus point is called the **interference** or *contact* **point.**

KEYPOINT

Good cutting speed and feed rate choices push the interference point up the rake face, away from the cutting edge, such that the mechanical pressure, friction, and heat are distributed over a stronger part of the tool.

External Friction and Heat

Pressure at the interference point can be as high as 100,000 pounds per square inch between chip and bit. That's not a misprint, the total force is concentrated on microscopic contact points as the chip slides by. In a setup with no coolant or an extreme cutting speed, the resultant heat can actually cause **welding** of the chip to the bit. But each tiny weld must break as the chip pushes forward. This weld-break-weld cycle not only causes cratering, it can also affect the finish.

That high-pressure contact causes "external friction and external heat." Some of the external heat is conducted away in the bit and also in the coolant, if used, but most is conducted to the chip and the uncut workpiece as well. Everything heats up; work, bit, and chip can all become dangerously hot! Again, by knowing the forces and actions, a skilled machinist can control heat and even use it to good advantage as in high-speed machining (Section 26.)

So, here's a big clue to solving the "hot chip mystery." While the removed chip receives its share of the external heat, this clue doesn't quite explain why the chip continues to heat after removal. There is no more friction once it passes the rake surface.

Figure 9-11 *The pressure gradient on the rake face centers at the interference point.*

Center of crater

KEYPOINT

A chip can reach as much as 1,000°F or more.

Figure 9-12 *Uncontrolled friction and its crime partner, heat, can make any cutting tool fail too soon.*

Surprisingly, only 25 percent of the total heat created in machining is attributable to external friction.

Uncontrolled, heat burns up the bit and possibly work hardens the material. It's also personally dangerous as the heated chips fly off the work. Metal expansion due to heat can also cause havoc with measuring, as we've already seen (Fig. 9-12).

Deformation Heat Is Internal

The final clue to solving the mystery lies inside the chip as it is deformed up and away from the work. Notice in Fig. 9-13 that as the chip presses against the rake face, friction causes it to stack up. The chip actually gets thicker than the depth of cut.

As the chip is raked, there is a continuous, forced "flow" of metal along a real line called the **"shear line."** It occurs at the "shear angle," which is in the chip. Due to external friction against the bit, the chip is stacked thicker. As it does so, grains of metal are rubbed along the line that extends from point A, close to the bit tip in Fig. 9-13.

Point A

Shear line

Figure 9-13 *As the chip deforms it gets thicker than the depth of cut—along the shear line.*

Mystery Solved!

The greater part of the total heat in the chip-making process is from internal friction. The heat originates inside the chip and conducts outward.

It is desirable to increase the shear angle, thus shortening the shear line, which, in turn, reduces internal friction and heat. The shear angle can be changed with three elements of control:

◆ Coolants to lubricate the external friction
◆ Harder bit faces that resist friction so that chips slide past
◆ Varying rake angles on the bit

Investigate It

Machinists and programmers must understand these chip-making principles, especially when pushing the limit with fast CNC equipment. Here are four experiments that bring tool geometry concepts into clear focus.

Experiment 1—Modeling Clay

Using plastic clay and a wooden bit, we can simulate nearly all of the concepts we've discussed—rake, clearance, deformation, and shear line. Figure 9-14 shows a bit being fed

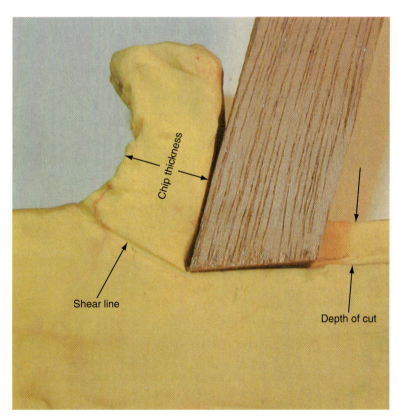

Chip thickness

Shear line

Depth of cut

Figure 9-14 *Chip deformation can be demonstrated with clay.*

through the material. Note the deformation and chip thickness relative to depth of cut. If you perform this experiment, watch closely, you will see the shear line (emphasized in the photo).

- Try varying the rake angle and observing the chip thickness.
- Try cutting with no clearance.
- Try cutting with an unsanded, rough wood tool then a smooth rake face and note the difference in the stacking. (Modern tooling is coated on its surface

with exceptionally hard materials and some are even micropolished on the rake face.)

Experiment 2—Look at Any Chip

To see firsthand evidence of this stacking effect, with safety glasses on, go to the shop and find any example chip made by any process and notice that it is smooth on one side and rough on the other. Do not take it from a moving machine. Is it obvious which side was toward the bit and which was away? With experience, you'll learn to read that difference. When cutting speed is too slow, the rough side shows signs of poor "lumpy" **deformation** (Fig. 9-15).

Experiment 3—Internal Heat

Hold a piece of thin sheet metal with the sharp edges removed to protect your hands, and bend it repeatedly. What happens at the bend? It heats up from the shifting metal grains. Eventually the metal's granular bond fatigues and it begins to break along the bend line.

Experiment 4—Blue Steel Chips

Your instructor may have examples of this phenomena. Evidence of chip heating can be seen when machining steel (Fig. 9-16). When it gets above 375°F, it begins to form surface oxides from the air in the room. Different temperatures produce different colors. The lower chip in the photo runs from near straw color at 430°F to brown at 450°F. The upper example starts at deep brown at 500°F and ends with deep blue at 550°F. These chips were made by programming constantly increasing downward force on the drill while increasing the RPM.

But the odd fact is that the chip comes off the work a shiny steel color, then lying on the floor, it begins to turn light yellow, then light blue, then dark blue as it heats from the inside out. A clue?

Machinability

Resistance to deformation differentiates various metals as to the relative ease or difficulty with which they can be machined. For example, soft aluminum will easily deform, while steel resists from 3 to 10 times more compared to aluminum. The softer metal will heat less, both externally and internally, and requires less horsepower to machine. That

Figure 9-15 *The smooth and rough side of typical chips indicate deformation.*

Figure 9-16 *Steel chips change color due to oxidation. Various colors indicate different temperatures.*

can be envisioned in the clay experiments. Warm clay versus cold, how would they each deform when the model bit is passed through?

KEYPOINT

There are two sources of chip heat: external from the rubbing of chip against tool and internal caused by the deforming or internal sliding of the chip against itself.

Now that you understand the heat source, let's see how to control them.

Changing Rake Angles

Figure 9-17 depicts exaggerations of three bits making a chip. Notice that the depth of cut remains the same, only the rake angle is changed, and that changes the thickness of the chip. As the rake becomes more negative, it's a thicker wedge, the chip must deform more.

KEYPOINT

Keeping all other variables the same, a chip made by a positive rake tool is thinner, closer to the depth of cut compared to one made with a negative rake tool. **Positive rake** bits require less energy to produce the chip.

As the rake surface becomes more negative, more deformation occurs, which results in more heat both externally from more pressure and internally from a longer shear line. A **negative rake** tool also consumes more energy (machine horsepower).

Positive rake— shortest shear line

Negative rake— longest shear line

Changing the rake angle

Figure 9-17 *Changing the rake angle affects the shear line length, altering chip thickness, which produces heat and changes the force required to machine.*

Coolants Shorten the Shear Line

Coolants also lubricate. They reduce external friction, which makes the shear angle go up, resulting in a shorter deformation and far less overall heat. So coolants play a double role in reduction of heat.

Harder Bits Shorten the Shear Line

If the bit material is harder, it offers less frictional resistance to the chip. In other words, the chip slips past easier. It has the same effect as adding a lubricant to the interference point.

Critical Question

Assuming no other variable changes, what differences would you expect to find in the machining if you changed to a more negative rake bit?

1. More external and internal friction
2. More force required to do the job
3. Hotter chips

So, from these traits, one might assume that a positive rake tool is the way to go—but it isn't always so. In fact, in production machining, negative rake tools are the more common. Why? The answer requires investigation.

ShopTalk

Neutral Rake In theory, in Fig. 9-17, the center tool is a neutral rake. However, in actual shop practice, a neutral rake bit performs much the same as a negative rake.

Why Not Always Use Positive Rake Tools?

That's a natural question. It seems that all factors are improved when we do. Less heat is produced, less force is required, and the chips are cooler. The simple answer is we would if we could, but the bit with positive rake is the weakest due to its shape. If we wish to push a positive rake bit harder—that is, to machine more metal per minute, the tip can snap off due to its thin shape. Also, positive rake tools don't conduct away heat as well due to their lack of mass. They are more prone to burning as well. But strength and mass aren't the sole reasons for using a negative rake tool. The longer answer is coming up—pushing the boundary.

Each rake angle has its correct application. Your objective is to understand them both and when to use each. Unit 9-3 will help, along with machine time.

Increasing the removal rate is the prime reason to select a negative rake tool.

Preventing Cratering and Built-Up Edges

As removal rates increase toward maximum, two problems can arise, both caused by friction at the interference point: cratering and edge buildup. As mentioned previously, cratering (Figs. 9-18 and 9-19) can destroy bits, while edge buildup can ruin the finish on the workpiece. Both are traced back to the high pressure at the interference point.

Edge Buildup

A second common problem associated with welding, is called **built-up edge.** Here, the weld occurs but this time holds on a little tighter. It doesn't instantly snap off. The flowing metal within the chip begins to back up. It disturbs the area upstream within the chip

Microscopic welds form, then break away, causing *cratering*

Figure 9-18 Cratering *will eventually break the tip off the bit due to the weak point it creates.*

Figure 9-19 Cratering *is seen on the tip of this carbide cutter after it has cut metal for some time.*

Figure 9-20 *A depiction of built-up edge escaping under the bit.*

behind the jam until the pressure becomes great enough to shear away the stubborn weld. Envision a large rock in a river. The water upstream piles up behind it, then escapes around it when the pressure is great enough.

The problem in machining is that if the buildup is not corrected, some of the disturbance escapes *under* the bit to become an irregular and permanent part of the machine finish. Figure 9-20 shows this disturbance in the clay model.

To simulate the weld a nail was driven into the rake face of the wood bit (Fig. 9-21). Notice the disturbed area has extended ahead

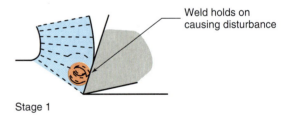

Weld holds on causing disturbance

Stage 1

Built-up edge escaping under the bit

Stage 2　　　　Escaped buildups

Figure 9-21 *The weld holds on a little longer, the disturbed area backs up behind the weld, and small amounts escape under the bit.*

of the cutting edge, ready to break away and become part of the work surface. Red clay was added to emphasize the disturbed area. But if you try the same experiment with a clay model, you'll see it with no marker.

The Solutions

The same control variables that prevent cratering also eliminate edge buildup.

Harder Bits or Electrocoated Rake Surfaces The harder the rake face, the less the chip will drag upon it. That's why most negative rake tools are made of tungsten-carbide due to its extreme hardness.

Lubricants in the Coolant Either cutting oil or coolant forms an antifriction barrier between the bit and chip. Reduced friction at the interference point shortens the shear line, thus lowering heat and pressure—the weld cannot form. However, there are several other advantages gained when coolants are introduced.

Coolants Cause Thinner Chips With less rake friction, the shear angle is reduced and the chip produced is thinner. Internal heat goes down along with the power required to make the cut. The overall result is that the bit performs like a more positive rake tool without changing its shape. The reduced heat means better tool life and better machine finishes (Fig. 9-22). And safety!

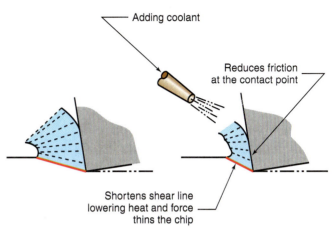

Figure 9-22 *Coolant reduces interface friction, thus reducing deformation force. The result is a thinner chip and a reduction in the heat and horsepower required.*

Pushing the Boundary— Why We Use Negative Rake Bits

Why does tool life often improve when we machine more metal per minute or switch to a negative rake bit that causes more friction and heat? The partial answer is that by pushing the point of contact away from the edge, we distribute the force over a stronger part of the tool. But there's more.

The deforming chip concentrates great pressure against the bit but not right at the cutting edge. A small area exists between the central force point and the cutting edge called the **boundary zone.** Return to Fig. 9-18. The boundary is illustrated as a small flat spot between the crater and the cutting edge and, in fact, it can appear that way on real tools too. The zone is subject to less force and friction than exists at the center of the crater farther up the bit.

The boundary zone, shown in Fig. 9-23 can be beneficially extended, farther back from the cutting edge by pushing the bit harder (more feed rate) or by taking a deeper cut (more bite) or by changing to a negative rake tool. Often, when the boundary zone is extended away from the vulnerable cutting edge, the tool lasts longer because the force, heat, and friction

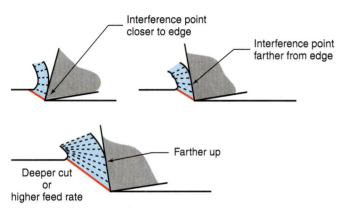

Interference point
closer to edge

Interference point
farther from edge

Farther up

Deeper cut
or
higher feed rate

Figure 9-23 *The boundary zone can be shifted away from the cutting edge by increasing the cutting force.*

have been shifted up to stronger parts of the bit. Can you see how this might prolong tool life?

However, the trade-off concerns how much force the tool and the setup can withstand. There is an upper limit to how hard one can push the setup. Knowing the limit is a matter of practical experience and calculating the right feed rates.

Plastic Flow

Plastic flow also concerns how speeding up the cut affects the way the chip deforms. If the deformation rate is increased—that is, faster machining—then the internal heating softens the chip and it flows more easily.

In each machine section, there is a unit on selecting speeds and feeds. This is the final information you need to make the right choices about correct removal rates.

KEYPOINT

Tool life can often be skillfully extended in one of two ways:

A. Increasing cutting *force* shifts or redirects the boundary zone away from the weak cutting edge, which increases depth of cut or feed rate.

B. Increasing cutting *speed* causes the chip to flow more easily.

However, both techniques have practical upper limits where the tool itself is compromised from either force or heat. You will balance these factors in your selection of speeds and feeds and depth of cut.

Trade Tip

Heavy Chip Load in Titanium and Stainless Steel
When cutting certain difficult metals, tools will dull quickly unless the cutting force is sufficient. Keeping the tool working hard is called a "heavy" **chip load.** Machining titanium, most stainless, and hard bronze are examples of metals that require a heavy chip load to prolong tool life.

When machining these metals, you must "bury" the cutter (a deeper cut depth) and keep it working hard (a heavy feed rate). Light cuts dull the cutter every time. A heavy chip load extends the boundary zone and relieves the cutting edge.

UNIT 9-2 REVIEW

Review the Key Terms

Boundary zone
The small space between interference point and cutting edge on a bit.

Built-up edge
The result of chip weld holding on too long, then allowing a small part to escape under the bit to reweld to the surface of the work.

Chip load
The amount of chip being taken as a combination of feed rate and depth of cut. See removal rate.

Cratering
The erosion of the rake surface of a bit due to welding.

Deformation
The bending and piling up of the chip during the rake phase.

Interference point
The concentrated center of contact pressure between chip and bit.

Negative rake
The bit deforms the chip through an acute angle, causing more bend than positive rake bit.

Plastic flow
The characteristic of a chip that causes it to deform without returning to its original shape.

Positive rake
The bit deforms the chip through a lesser angle, causing less deformation to the chip.

Removal rate
The combined result of depth or cut, feed rate and cutting speed (RPM) rated in cubic

inches per minute (or liters per minute). Equates directly to horsepower requirements for the cut.

Shear line
The line inside the chip along which the deformation occurs.

Welding
The instantaneous, heated adhesion of the chip to bit. Leads to cratering and built-up edge.

Replay the Key Points

- Chips deform as they contact the rake surface of the bit.
  - Chips get hot from external friction—25 percent of total heat.
  - Chips get hot from deforming internal friction—75 percent.
  - Bits are worn away by forced contact with the chip—welding occurs, causes cratering.
  - Several factors can affect the shear angle with differing results in the machining process.
- Changing the rake angle affects friction.
  - Positive rake causes less friction and less deformation in the chip.
  - Positive rake is the weakest, thinnest bit and thus heats faster.
  - Negative rake tools can push harder.
  - Negative rake tools are usually made of harder material to better resist the additional friction.
- There are several other control factors that affect the shear line and chip formation.
  - Coolants
  - Bit hardness
  - Speed and feed changes
  - Cutting angles
  - Depth of cut

Respond

Determine readiness to move up by answering these questions. This is a critical point beyond which you should not pass until you feel you have a solid understanding of shear line, rake angles, and chip formation.

1. Why do negative rake tools often last longer in a setup?
2. In what two ways do coolants reduce heat in the machining process.
3. Identify and explain two ways to control built-up edge problems in machine finish.

Critical Thinking Questions

4. A tool keeps breaking before it's dull. Identify at least two ways to prevent this.
5. From where does most of the heat originate in machining a chip?

UNIT 9-3

Taking Control–Adjusting Variables

Introduction: This unit is about putting what we've just learned to work for you. It's all about what happens when we change the tool geometry variables and the setup factors and why. At the end, we'll take a brief look at troubleshooting a few common problems.

Read through the following suggestions but don't try to retain it all. Then, when improving a setup or when things don't go as expected, return for suggestions. Experience on the machines will bring it into focus. Also ask your instructor and several journeyman machinist's, but don't be surprised if each offers a different fix.

Nearly every problem or improvement can be dealt with using several fix combinations. For example, if a tool is chattering, you would probably slow the cutting speed and possibly change the lead angle on the tool. Another control that would decrease chattering is to increase the feed rate and/or change the rake angle of the cutter. The setup may need to be strengthened too, since the vibrations may be emanating from the work-holding method not the workpiece. The corner radius might need to be lessened. It's a case-by-case challenge.

You'll soon develop a sense of what's wrong and your own ways of solving the problem. Doing this requires more than applying knowledge, it involves intuition based on experience—and that is the essence of being skilled!

KEY TERMS

Carbide grade
The relative hardness of the cutter running from very hard to tougher but softer at the opposite end of the scale.

Chatter

The unwanted vibration that sometimes occurs during machining. Chatter can destroy tools and ruin finishes.

Interrupted cut

A noncontinuous cutting action causing shock to the cutter.

Loading

Chips clog the cutter, destroying finishes and even breaking cutters.

Maximize

To make a setup more efficient with the goal of more parts per time unit.

Summing Up the Control Variables

Here are the possibilities in brief.

Changing Tool Shape

There are three shape variables:

Corner Radius   Cutting/Lead Angles
Rake Angles

Change corner radius to

Prevent tool failure

Improve finish

Reduce chatter

Form an inside rounded corner on the workpiece

Increase radius when

Finish is poor

Tools are breaking or burning up frequently

Decrease radius when

Chatter is a problem

Rake Angle Changes

Both positive and negative rake tools can deliver a fine finish when correctly applied.
Select positive rake when

The job requires moderate removal rates or for softer work such as aluminum or softer brass.

Very slow speed machining is performed, such as threading on a manually operated lathe.

Chip ejection is a problem with negative rake tools. It is more commonly encountered when machining softer alloys of aluminum. The chips are welding and clogging the cutter, called **loading.**

Change to negative rake tools when maximum metal removal is desired.

Harder work materials must be machined.

The tool breaks or burns up with positive rake.

Changing Cutting/ Lead Angles

Lead angles are changed for three reasons:

1. To smooth out the cutting action.
2. To increase the removal rate.
3. To form a shape as on the right in Fig. 9-24. The part is rotating while the lathe bit moves to the left through the work, leaving an angular surface and a radius.

Increase lead angle when

The zero lead tool (left side) is common for moderate metal removal setups.

The 20-degree bit would be used for a high volume cut.

Decrease lead angle when

The 20-degree lead tool might chatter, at which time a lead of 10-degree might be selected.

Forming

Sometimes, even on CNC lathes, the situation forces use of a cutter such as the 20-degree tool in Fig. 9-24 to create the shape of a part. The shape of the bit produces the shape of the work. A form tool has the effect of a very long lead angle.

With form tools, select a slow cutting speed and use lots of coolant or cutting oil. Also be sure the setup is very rigid. Due to the large

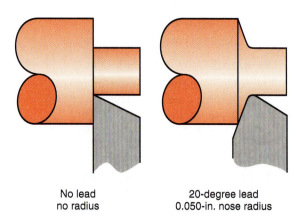

No lead   20-degree lead
no radius   0.050-in. nose radius

Figure 9-24 *Looking down on lathe bits, the bit on the left has zero lead while the bit on the right has 20 degrees, and a nose radius.*

amount of contact area between bit and work, chatter is the enemy when forming. The fixes are in this order:

Slow the RPM.

Improve setup rigidity.

Add cutting oil rather than coolant.

Sometimes fiddling with the feed rate (up or down case by case) can help, but it's the last item to try.

Changing Tool Hardness

Tool hardness change usually accompanies a change to negative rake to increase production, but not always. Every so often, we need to change to a less hard, but tougher cutting tool.

Hard Versus Tough Tools Carbide grade tools are supplied in a range of hardnesses that we'll discuss in detail later on. The range goes from very hard at one end of the scale to tough but softer at the other. High-speed steel tools are softer than the lowest grade of carbide, but they are tougher, too, and more able to withstand shocks and vibration.

Change to a Softer Tool When The carbide tool is breaking or chipping on the edge, before it is becoming dull from normal use.

HSS tools resist breaking on intermittent cuts too—that is, an in-out-in, hammering action such as a lathe turning a nonround object or one with a hole in its side. The bit must enter, exit, and enter the work for each revolution. This kind of abrupt action is called an **interrupted cut.**

Change to Harder Tool When

The job requires a large removal rate.

The tool is burning up before becoming dull from normal use.

The work is too hard and dulls the tool quickly.

Changing Cutting Speeds and Feed Rates

These are two vital controls that can be varied with the twist of dial for a CNC operator and a few shifts of levers or belts on manual machines. They are almost always the first variable to be adjusted for improvement or troubleshooting. In the drill, lathe, and mill sections of this book, we'll calculate the right speed for the job. Here, we will discuss this control in a general way.

A quick reminder—cutting speed is the speed with which the chip is flying off the cutter. It is the velocity of the cut. In drilling and milling operations, the cutter's rim is set to turn at the specified feet-per-minute rate. With lathe work, it's the work's rim that must revolve at the specified speed. In either case, the objective is to calculate and adjust the RPM to achieve the cutting speed in feet per minute (F/M) or meters per minute (M/m).

KEYPOINT

Cutting speed is the speed in F/M or M/m of the metal passing the tool or the tool passing the work. Changing cutting speed means adjusting RPM.

The goal is to find the sweet spot where the best finish is accomplished but tools also last well. Not calculating the correct RPM based on recommended cutting speed is one of the most common errors we instructors observe with new machinists.

Safety Machining too fast means that something is spinning very rapidly. Either the work or cutter might be exceeding safe limits. Lathe work is the more dangerous place to overspeed a machine. Here the heavy work and chucking device are rotating. Also, with increased speed, chips become hotter and fly faster.

Horsepower and chip ejection are also practical limits to adjusting speed upward too far. Often the chips can't get out of the cut area, and they load it regardless of whether a positive or negative rake is being used. The first fix is to bring coolant to the cut.

Decrease Cutting RPM When

Slightly below calculations, when setting up a machine for a first cut, it's smart to start out slightly below the calculated speed then work upward, observing the results to both the work and tool.

Tools burn up before dulling due to work hardness.

Chatter occurs. Reducing RPM is the most immediate fix for chatter.

The setup appears unsafe or unusual.

Increase Cutting RPM of the When

The action doesn't shear chips away correctly. Deformation chip is occurring in big chunks not flowing layers.

Tools break before dulling.

Here's an example of that. If a machine's power is turned off, but left to cut until it coasts to a stop, it will slow down while making a chip. Commonly, just before stopping completely, the bit will snap off due to poor chip deformation. This is more common with brittle carbide, but HSS will do the same in heavy cuts.

When examining a chip produced under this low-speed condition, you will see the back side is exceptionally rough due to the large chunks of deformed metal sliding then sticking (Fig. 9-25).

A cutting speed that is too slow results in tool breakage, poor finish, and slow production—inefficient removal rates.

Adjusting Feed Rates

Similar to cutting speed, feed rate changes have an effect on both tool life (Fig. 9-26) and work finish and accuracy. Calculating the rate

Extra rough chip is evidence of cutting speed that was too low

Figure 9-25 *An extra rough chip may be a sign of poor deformation.*

Figure 9-26 *Early tool failure from chipping can be caused by several factors.*

of cut is important but less critical than cutting speed because there is normally a range of acceptable feed rates.

The upper limit is due to the strength of the setup and the roughness of the finish produced. Fast feeds are the objective in terms of efficiency. Slow feeds are usually not a problem except that it's low-volume machining. They are commonly used for a final cut where finish and accuracy are of prime concern. Here are two exceptions to that.

Increase Feed Rates When

Work hardening is occurring due to rubbing in some sensitive metals such as stainless, high-carbon tool steel, or certain bronzes.

Chatter is occurring—sometimes a faster feed rate can stabilize chatter because a larger chip is forced against the rake face of the cutter.

Plan Heavy Then Fine Feeds A slow feed leaves finer machine marks and a smoother finish. Also, finer feeds exert less force on the tool, thus improving accuracy. A good plan usually includes one or more rough, heavy feed passes then a semifinish cut to stabilize the accuracy, then a final, fine-feed cut.

Decrease Feed Rates When

The work and machine are subject to questionable forces.

Machined finish is rougher than expectation.

Accuracy and repeatability are below the job tolerance.

Tools are breaking before dulling.

Problem Solving

Here are a few common problems and their associated causes. As you work toward more efficient operations keep in mind the first goal is safety, then accuracy. Speeding up is also desirable, known as **maximizing,** but never at the risk of the first two: safety and quality.

Problem	Possible Cause
Poor finish	Dull tool Excessive feed rate and/or depth of cut No coolant Inadequate tool clearance Low cutting speed Wrong rake angle No tool radius No lead angle on tool
Built-up edge	No coolant or poor injection Dull tool Tool too soft
Chatter/vibration	Excessive RPM speed Large tool radius Large lead angle on tool Poor holding method for workpiece Workpiece too thin or weak for operation Feed and/or depth of cut too light Tool too deep; depth of cut too great
Inaccurate/inconsistent repeatability	Dull tool Poor or loose setup/work-holding method No coolant or cutting fluid Excessive feed rate Excessive depth of cut Inadequate tool clearance Tool flex—too much overhang/excessive force for tool size
Excessive work heat	High RPM Dull tool Rake too negative Poor cooling action
Tool breakage	Excessive feed or depth of cut (number 1 cause) Weak tool Tool is too hard Setup not rigid or not tightened well Positive rake not strong enough Slow cutting rate causes poor deformation
Tool cratering	No coolant Coolant doesn't reach cut area—chips pushing it away Tool too soft—use harder grade
Chip load-up	RPM too high No coolant Dull cutter Negative rake tool
Parts too hot	RPM too high Too much cut depth Excessive feed Dull tool No clearance on tool

Problem	Possible Cause
Tool burn-up	Use carbide (harder) tool Feed rate too high RPM too high Depth of cut too great No cutting fluid No tool clearance Spindle in reverse (backward, it happens!) Hitting (or creating) a work hardened zone
High-pitch-squeal	Hard workpiece Excessive speed No clearance Excessive tool overhang—too long
Dangerous chips hot and fast	No coolant Excessive RPM No safety shield Not cleaning up chips as they buildup close to spindles and rotating work
Long, stringy chips	Feed or depth of cut too low Continuous feed No chip breaker on tool

Trade Tip

Squeals Sometimes a very high-pitched squeal will occur with no evidence of chatter marks in the work. This pesky irritant doesn't affect tool life or work accuracy, but it can set your teeth on edge and spoil your day. It occurs when machining very hard steels, some stainless steels and hard bronzes. More than irritating, prolonged exposure can result in hearing loss—wear external earmuffs for this kind of noise.

UNIT 9-3 REVIEW

Review the Key Terms

Carbide grade
The relative hardness of the cutter running from very hard to tougher but softer at the opposite end of the scale.

Chatter
The unwanted vibration that sometimes occurs during machining. Chatter can destroy tools and ruin finishes.

Interrupted cut
A noncontinuous cutting action causing shock to the cutter.

Loading

Chips clog the cutter, destroying finishes and even breaking cutters.

Maximize

To make a setup more efficient with the goal of more parts per time unit.

Replay the Key Points

◆ Surface speed is the velocity of the rim of the cutter or work (whichever is spinning).

◆ Achieving the right surface speed means calculating and adjusting RPM up or down.

◆ Choosing a surface speed too low results in poor finish, broken tools, and generally inefficient machining.

◆ Choosing a surface speed too high results in burned tools, work hardened material, extra hot fast flying chips, and dangerous conditions.

◆ Feed rates are important, but there is a wide range that produce acceptable results.

Respond

Check your comprehension of troubleshooting.

1. Identify at least three ways that chatter might be stopped.

2. True or false? Feed rates are based on surface speed. If false, what makes it true?

3. Which control variables lower chip heat in a setup? Briefly explain each.

4. What three tool bit shape variables can be adjusted?

Critical Thinking Question

5. When is it desirable to change to a softer tool bit?

Section 9 Review

Unit 9-1

We've only touched on the basics of tool geometry. As skills build with machine experience, we'll look at three more kinds of rake angles, for example. Look for compound rakes on lathe tool bits that create a bias for the tool's movement right or left. Look for milling cutters with various combinations of positive or negative radial (center outward) and axial rakes (parallel with the spindle) and for hooks in the rake faces. But they all start with the four basics: rake, clearance, lead, and corner radius. Every tool has these, even though in some cases, they might be zero, as with corner radii. Be certain that you can identify them on any tool you put your hand to. That's the bottom line—you must know cutting tool geometry.

Unit 9-2

Part of the machinist's paycheck comes from doing the research. Today, cutting tools are evolving at a fantastic rate, driven by the need for speed. Better tooling is one of the frontiers that I suspect (and hope) we'll never completely cross. As a journey machinist you will be bombarded with samples of new tools, with articles provided by the makers of them, and by end users like yourself, test driving them. There will be demos, brochures, and sales pitches to wade through. Although expensive, some new tools pay for themselves in days, yet some don't live up to their promise in your particular application. Part of the job includes doing smart research based on the knowledge in Units 9-1, 9-2, and 9-3.

It's easy for even a small shop to own hundreds of thousands of dollars in tooling! No exaggeration! In large companies, excess costs tied up in cutting tools are one of the most common physical challenges.

Unit 9-3

If ever there was a perfect test of logical thinking combined with guesswork, cutting tool troubleshooting has to be it. I can recall bets and near shouting matches between three or more machinists in our shop as to exactly what the problem might be and how to fix it. Each of us could and did make it work better, but the truth is the absolute answer doesn't exist. It's a science of variables. CNC machining makes it even more complex as we demand more parts per minute and per day. Learn this stuff well.

Questions and Problems

1. Identify the tool geometry (shape) features that can be used as control variables. (Objs. 9-1 and 9-2)

2. Briefly describe the purpose of each answer to in Question 1. (Obj. 9-1)

3. Of these four, which is not a control variable: rake, lead, clearance, and shear? Explain. (Objs. 9-2 and 9-3)

4. Of the four features in Question 3, which is not on the bit? Explain. (Obj. 9-2)

5. What happens to the chip if we change to a more negative rake angle? (Objs. 9-2 and 9-3)

6. In terms of machine requirements, what happens when we switch to a more negative rake tool? (Obj. 9-2)

7. What is the most likely reason to have a corner radius above 0.125 in.? (Obj. 9-1)

8. Name two corrections or changes to solve a built-up edge? (Objs. 9-2 and 9-3)

9. Is this statement true or false? The boundary zone is defined as the space between the point of interference and the cutting edge. (Obj. 9-2)

Critical Thinking Problems

10. In a given setup, the bit is prematurely breaking along the cutting edge before it becomes dull from normal use. Why and how might it be corrected? Identify at least four solutions. (Obj. 9-3)

11. A bit is burning up before becoming dull, identify at least four causes and their solution. (Obj. 9-3)

12. A bit has a gold coating on its surface. In terms of what we've just learned, why might that be?

13. The bit is changed from a noncoated to a coated version, as in Question 12. Identify at least two benefits assuming all other tool variables and setup factors remain the same.

14. What is the main reason to change to a carbide bit from a HSS?

15. The machine finish on a milled surface is too rough to pass inspection. Without looking at the Unit 9-3 troubleshooting chart, what changes might solve it?

16. In what order might you try your fixes for Problem 15?

17. A job must be taken to a manual milling machine that's limited to eight spindle speed selections. After doing the math (to be learned in Section 12) you determine that the calculated speed of 375 RPM, falls between two speeds at 250 and 400 RPM. What should you do?

18. Using the same machine as in Question 17, the needed RPM calculates to 300. Now what must be done?

CNC Questions

19. A standard carbide bit is producing 15 parts on a CNC lathe before it dulls and must be changed to a new, sharp cutter. However, you suspect a different cutter with artificial diamond coating can make 40 parts per cutter at a nearly 50 percent decrease in cycle time per part.

What questions should you ask yourself before making the change?

20. If your career is aimed at shops with CNC equipment, why must you stay informed of advancements in cutting tools?

Section 9 Answers

ANSWERS 9-1

1. Rake, the surface upon which the chip slides and is deformed; clearance, allows only the cutting edge to touch and prevents friction; lead, the angle of the cutter relative to the direction of cut, distributes the chip over a wider area.
2. Rake and lead
3. True
4. Cut and rake
5. As feed increases, rake becomes more important.
6. Nearly nothing to one-eighth inch more or less, depending on feed pressure.
7. Stronger tools, smoother cutting action, and smoother finish

ANSWERS 9-2

1. They are stronger and able to conduct heat away better than positive rake tools.
2. Lubricate contact, thus reducing external heat from friction; less friction results in less deformation, thus less internal heat.
3. Coolants lubricate thus welding doesn't occur; harder bits prevent friction against chip.
4. Less force = lower feed rate or less depth of cut; change to a stronger bit = negative rake
5. From deformation inside the chip

ANSWERS 9-3

1. Slow RPM, reduce lead angle, reduce corner radius, strengthen the setup, increase feed rate (or depth of cut)
2. False. Cutting speed, RPM is based on surface speed.
3. Coolants, lower heat by reducing both external and internal heat; positive rake, reduces shear angle with shorter deformation line; harder tool bit, reduces external friction which reduces internal friction
4. Rake, corner radius, lead
5. In terms of tool life and performance the simple answer is "never." Harder is better in every case other than chipping and breaking due to interrupted cuts and chatter. But softer equates to tougher tools.

Answers to Section Review Questions

1. Rake; lead; corner radius
2. Rake, the surface and angle of the bit that deforms the chip; lead, angling the cutting edge relative to the direction of cut to distribute the action over a longer cutting edge, also smoothes entry and exit into the work; corner radius, similar to lead, distributes the cut over a longer distance but also creates stronger tool and better finish.
3. Clearance—a minimum amount is all that's required.
4. Shear is within the chip.
5. More deformation, longer shear line, more friction and heat
6. More orce and horsepower required
7. It is a *form* tool.
8. Coolant and a harder bit
9. True
10. *Increase cutting speed,* it might be from slow deformation; *reduce feed rate or depth of cut,* lower the force; *tool is too hard,* change to a tougher tool bit; *weak tool,* add corner radius for strength, change to a negative rake, look for excessive clearance; *improve and tighten* setup vibrations, not rigid tightened well
11. Excessive cutting speed
Excessive feed and depth of cut

Too much negative rake angle—friction too high
Too positive rake and the heat is concentrating on edge
No coolant
Not enough clearance—rubbing
Tool too soft—change to harder grade

12. It must be a harder surface coating to reduce friction.

13. By reducing contact friction, the deformation heat goes down, leading to longer tool life. It also means the machine isn't working so hard to make the same cut.

14. They are harder so there is less friction at the contact area. Note, because tungsten-carbide tools are a great deal denser than steel ones, they are much better at conducting heat away too, but that is a side benefit compared to the hardness.

15. Dull tool, check and change tools if needed; chatter—slow the speed, reduce corner radius, lessen lead angle, change rake; improve setup rigidity; sometimes (not often) increasing feed will work

16. In the order presented in Answer 15 above.

17. Choose the 400 RPM since 400 RPM is only 6 percent faster than calculated. However, watch carefully for signs of tool burn-up and work hardening.

18. This occurs often on many manual machines. Initially, choose the 250 RPM—observe results. If you are seeing slow cut characteristics, try the 400 RPM, but use extra caution—you will be working 25 percent above calculated speeds. It might work, but tool life is sure to be shortened.

19. Does the increased production justify the extra cost of the tooling? (Do the math.) Do you have access to enough of these new tools to keep production going? Do you have a large inventory of the present tools, that should be used up? How many parts are left to machine, would stopping to retool be good or bad at this point? Would the higher cutting rates lead to lowered quality or safety?

20. This answer lies in efficient use of capacity. CNC machines are expensive and they must earn their keep. That means working them at the highest removal rates within their capacities. As machines evolve, they will be able to go faster. Not adopting tooling that paces this evolution, cancels much of the advantage.

Section 10

Drilling and Operations and Machines

Introduction

Learning to operate a drill press is less challenging than other kinds of machining, but don't mistake your experience there to be of lesser value. Beyond learning to use a manual drilling machine safely and efficiently, Section 10 is about an important family of operations where we spin a cutting tool then plunge it straight into the work to create a precise hole. Machining exact diameters in the right location to a specified depth is one of the most fundamental ways metal is shaped. The odds are that within any CNC program calling up three tools or more, at least one will perform some form of drilling to complete the work shape.

No matter whether the holes are produced by high- or low-tech means, a competent machinist must be able to make a drill bit start and stay in the right location, and must know how to drill deep holes while keeping the form straight and round. A third challenge is to avoid clogging the bit with chips and eventually breaking it. Last, drill setups must also include safety and long tool life. Today, the focus isn't on the minor cost of halting to replace or regrind an abused drill, due to poor setups, rather it falls on the higher cost of lost production due to stopping for any reason!

Beyond basics, there's another reason to learn drilling. Today in CNC mill programs, it's used for rough profile machining of extreme metals that otherwise would be difficult to cut. Due to the ability of CAM software to quickly create drill locations around the perimeter of the part, programmers create cut sequences that remove initial excess from complex shapes by drilling it away rather than cutting around the outside. The advantage is that a hard but brittle carbide drill will withstand an end push much better than a carbide mill cutter can take a sideways thrust. Drilling is an efficient CNC option compared to milling, and if done properly, the drilling cutter lasts longer than other kinds of cutting tools.

There are nine challenges to successful drilling:

- Choosing the right tool
- Surface finish per drawing specs
- Calculating speeds and feeds
- Removing chips from deep holes
- Achieving size, including roundness and straightness
- Drilling in the right location
- Drilling to the correct depth
- Making safe and strong setups
- Prolonging tool life

Virtually all the skills required to meet these challenges can be practiced on an everyday manual drill press like those found in your tech school lab. Once mastered, your ability will transfer to higher-level machine tools. In Section 10, we'll investigate these goals:

Objectives

Unit 10-1 Basic Drilling Tools (Pages 320–332)
Select the right drills for the job
Measure, name, and describe an industrial drill bit
Select a pilot drill for a reamer
Select the right shank type and adapter

Unit 10-2 Drill Presses and Setups (Pages 333–343)
Identify the various features on standard drill presses
Know the safe work habits and holding methods for vises, plates, fixtures, and clamps

Unit 10-3 Speeds and Feeds for Drilling (Pages 343–347)
Calculate or look up a starting RPM on any drilling or reaming operation
Access recommended surface speed charts in reference books

Unit 10-4 Drilling Per Print and Secondary Operations (Pages 347–359)
Setup and perform drill location by the layout-punch method to a 0.030-in. repeatability
Setup and perform drill location by the center find method to a 0.010-in. repeatability
Choose the right tap drill for a specific tap
Cut threads using taps on the drill press
Select and use a counter bore and counter sink/chamfer tool
Drilling setups for parallel and angle holes

Unit 10-5 Sharpening Drill Bits (Pages 359–365)
Safely and correctly sharpen a drill bit using a pedestal grinder
Recognize a dull drill needing regrinding

UNIT 10-1

Basic Drilling Tools

Introduction: Drills and reamers are supplied as small as a human hair and as large as 6 in. in diameter or larger in special cases. The depth of standard drilling operations is limited to around 5 to 10 times the drill's diameter, however, it's possible to drill far deeper using special techniques called pecking, or by switching to tools called spade drills and gun drills (all studied in Section 10).

We begin the conquest of hole making by looking at the basic tools: drills and reamers. And as usual, there are terms to learn.

KEY TERMS

Blind hole
A hole that does not go through—this leads to chip and coolant challenges, especially when tapping threads.

Dead center
The inefficient center of a drill bit that does not cut.

Drift pin (drill drift)
A small, flat wedge used to break the holding force of taper-shank tools.

Land
Similar to margins, a small surface on a cutting tool.

Margins
A narrow band where drills are full diameter.

Morse taper
A self-holding machine taper of $\frac{5}{8}$ in. per foot.

Morse taper adapter
Either a *sleeve* for upsizing or a *socket* for downsizing.

Reamer float (tool float for other kinds of cutting tools)
The tendency of cutting tools to pick up and follow previously machined shapes, such as a reamer following an incorrectly located drill hole even though the press is in the correct location.

Shank
The mounting end of the tool.

Tang
The torque drive tab on taper shank drills and reamers.

Turned-down shank
A drill larger than $\frac{1}{2}$ in. diameter, but featuring a straight shank. May be made from a damaged taper shank drill.

HSS Tough when Hot High-speed steel retains its full strength up to around 450°F and most of its strength up to an amazing 650°F!

Twist drill
A high-speed steel cutting tool made with spiral flutes.

Choosing the Right Drill and Reamer

The everyday **twist drill** (helical flutes) that you'll be using in lab assignments is commonly made from high-speed steel. HSS is also used for many other cutters because it has four great characteristics for this duty:

1. Can be resharpened using standard grinding wheels.
2. It flexes to withstand the unique forces of drilling.
3. HSS holds a sharp edge well, even when it heated to high working temperatures.
4. It resists softening from generated heat.

You might read an advertisement for drill sets made from high-carbon (HC) steel. Although the price will be tempting, they are inferior tools in terms of strength and edge life. Other than band saw blades, high-carbon steel is never brought into the professional shop. Additionally, the HC bandsaw blade is considered a throw-away grade. High-carbon drills and other cutting tools are just OK for home workshops if used in woodworking, but they are useless for work with metal.

When selecting a drill for a job here's what you need to look for:

◆ *Diameter*
◆ *Tool-holding method*
◆ *Drill geometry*

Four Different Drill Size Systems

Fractional drills (for example, $\frac{3}{16}$-in. diameter) are the more common. But using fractional increments doesn't create enough size differences for manufacturing. Therefore, three additional size systems are in use, and they fit between the fractions.

Number series	Sizes 1 to 80 (big to small)
Letter series	Sizes A to Z (small to big)
Metric size drills	Sizes 0.5 mm and up

It's confusing at first, because the system evolved without guidance. To organize the process, machinists use a chart such as that shown in Fig. 10-1, found on the wall of all shops, in the Appendix of this book, in the lid of their tool box, and in reference books such as *Machinist's Ready Reference*© by Prakken Publications (ISBN: 0-011168-90-7). Using a drill chart is a snap with just a bit of investigation.

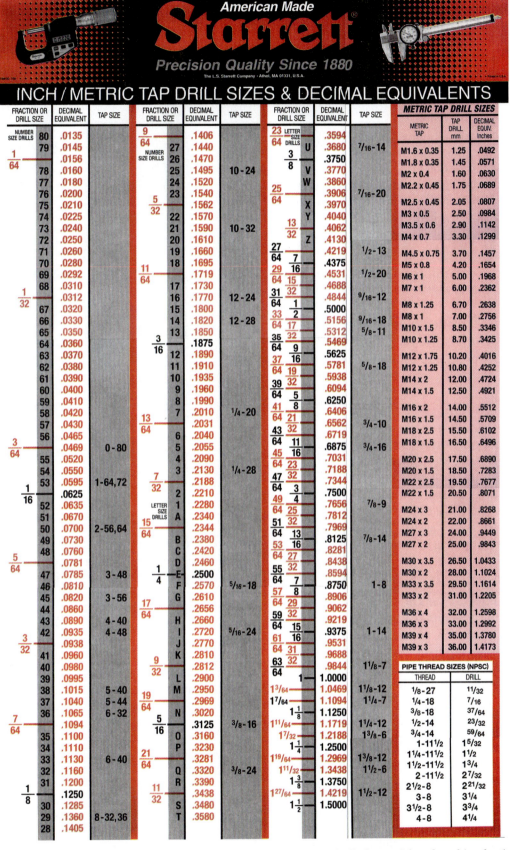

Figure 10-1 *A standard Drill decimal equivalent and tap drill chart. Often found in plastic "pocket charts" and on most shop walls. You will use this chart every day in this trade.*

For example, note on the chart of Fig. 10-1, that near the $\frac{3}{8}$-in. diameter drill (0.375 in.), a "U" size drill is found that measures 0.368 in. and a "V" drill at 0.377 in. That's not much difference in size, but it's enough to make a big difference on the final product.

Question? What drill size is recommended for tapping a $\frac{1}{4}$-20 thread?

Answer: Letter "F" drill, at 0.201 in. diameter.

Respond 10-1A

To be certain you have the idea, here are a few questions with answers found at the end of Section 10.

1. What are the two drill sizes surrounding the common fractional size of $\frac{3}{16}$ in.?
2. Looking at the drill chart, what is odd about the number size series? Why might that be?
3. What is the smallest number size drill and its decimal inch size on the chart?
4. What full millimeter drills are closest to $\frac{3}{8}$ in.? (Next metric size larger and smaller?)
5. What is the letter size drill closest to $\frac{1}{2}$ in.? What size is it?
6. Letter and fraction size drill bits get bigger in diameter as you scan down the chart. Is this statement true or false?
7. What is the decimal equivalent of the smallest letter size drill?
8. In general, which are the larger drills— letter or number size drills?

Selecting Drills and Reamers for Size

OK, from the chart you know the size you need. Now it's critical to know how to check the tool to be certain it's actually that size. There are three ways of determining the diameter of a drill bit or reamer:

◆ Size markings
◆ Measuring the drill
◆ Drill size gage

The drill size gage shown in Fig. 10-2 is similar to the one we've been using as an example in other units—a series of ascending hole sizes, used as a best-fit test of drill diameters. But, as we've seen, there can be such little difference between some of the smaller drills, that it must be considered a quick, rough sorting guide only. Like the drills, the size gages

Figure 10-2 *A drill size gage can quickly sort drills, but it must be taken as a rough measurement only.*

are supplied in each of the four series: metric, inch fractions, number, and letter sizes.

Diameter Markings

When they are made, drills and reamers (a hole-finishing cutter that follows a drill) are size marked near the **shank,** their holding end (Fig. 10-3). If the marks are readable, trust that the drill is the size shown *but never trust a reamer's size* (see *nonstandard sizes* next). However, on drills that are gripped by chucks, the size is often obliterated because the bit spun during use and the identification was wiped out. The final word on any tool's diameter is to measure it before use.

KEYPOINT

Make it a habit to measure a cutting tool before use. Like using the uncalibrated micrometer, if the cutting tool isn't the right size and you use it to make parts, the fault for the scrap or rework falls on the machinist, not the person handing the bad tool to them.

Figure 10-3 *Drill sizes stamped or engraved on the drill's shank may be obliterated.*

Figure 10-5 *The micrometer is placed on the margins.*

Measuring on the Margins

In Fig. 10-4 (looking at the tip of the drill) and in Fig. 10-5, notice the narrow band along the length of the drill on each side, the **margin.** Always measure the size of the bit on these margins, near the cutting edge, as this is the only full-diameter portion. The narrow margins provide side clearance in the hole. Behind them, the bit is relieved to reduce rubbing and allow coolant to penetrate down to the cut from above.

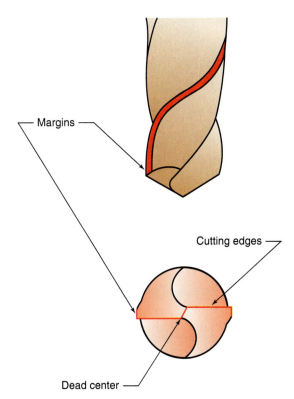

Figure 10-4 *Always measure on the margins of the drill bit.*

Standard and Nonstandard Diameters

While both drills and reamers are supplied in each of the four size series, reamers can be found in the shop in any diameter. There are two reasons that these odd-sized reamers lurk about the shop just waiting to trap the unsuspecting machinist:

Special Undersized Reamers Recall the press fit discussion of Section 5; to press a 0.500-in. bushing into a plate, one would ream the hole to 0.4985 in. That reamer is made by reducing a 0.500-in. standard size reamer by grinding the side cutting edges. When a reamer (or any cutting tool diameter) is modified, the tool grinder is *supposed* to re-mark the new size and they *should* be kept in a special drawer away from standard sizes.

Resharpening Makes Them Smaller When a drill bit is sharpened, only the clearance face is ground, thus it stays nearly the same diameter for it's entire service life. However, the side teeth of reamers eventually wear. They, as well as the end cutting edges, must be reground. At that time, the objective is to grind away a minimum amount of cutter diameter to restore the sharp edge, but no matter how little is ground away to sharpen it, the reamer ends up a nonstandard size. Again, they should be remarked when ground and stored away from standard sizes, but that doesn't always happen, or they don't get put away in the right place once they are used.

Oversize Reamers Or perhaps, a special-diameter reamer might be needed just slightly bigger than 0.500 in. to allow a precision slip fit to the hole. It is possible to buy a specially made reamer from 0.0001 in. up to 0.005 in. larger than a standard size.

KEYPOINT

Size markings may not be the true diameter with reamers. Always measure!

Selecting the Holding Method—Chuck or Taper

Straight- and Taper-Shank Tools

These are the two ways drills and reamers are held in drill presses. Each imparts some advantage.

Straight-shank drills (Fig. 10-6) must be held in a chuck that squeezes them between jaws. Under a heavy load, they can spin in the chuck. In contrast, taper-shank tools are mounted by tapping them into a receiving socket—they are more solidly held both for centering and for torque drive.

KEYPOINT

Small drills up to $\frac{1}{2}$ in. are usually straight shank, while sizes above are normally taper shank in industry for both manual and for CNC work.

There are exceptions to the size rule, where drills larger than $\frac{1}{2}$ in. feature straight shanks, called **turned-down shank** drills. They are used where the drill press hasn't the capacity to hold taper shanks, or they are made from damaged taper shank drills to continue using them. Using them isn't a recommended practice as they strain chucks beyond designed limits. There are also drills smaller than $\frac{1}{2}$ in. to be used in setups where a chuck isn't possible, but a small diameter is needed. Both types are not a standard practice.

Chucking Straight-Shank Tools

Straight-shank drills and reamers require a chuck to hold them in the machine (Fig. 10-7). It may be either the key type often called a "Jacobs" chuck due to a well-know manufacturer, or the more accurate and expensive, keyless chuck. Both feature three jaws that close simultaneously and remain on the center of the chuck as they do—called

Morse taper shank

Straight shank

0.375

0.625 MT 2

Figure 10-6 *Straight-shank and taper-shank drills.*

Figure 10-7 *Two varieties of drill chucks—with and without chuck keys. If using a key type chuck, never leave the key in the chuck!*

a universal jaw chuck. Due to the relatively quick changing of the bit, compared to taper tools, the chuck is the more universal drill holding method.

Chuck Limitations

While chucks are quick and easy to use, the grip force on the drill is limited with both types, key type or keyless. The keyless chuck has some advantage over the key type, in that as the drill torque increases, it tends to tighten its grip on the tool. However, that can lead to overtightening if the bit hangs up in the hole.

But both chuck varieties can allow the drill to slip. Even though the jaws are made harder than the drill shank, with time, this action degrades the centering accuracy of chucks. Their jaws can be reground but most often they are replaced.

Trade Tip

Tightening Drill Chucks When tightening either type of drill chuck, lightly tighten the tool then loosen it just a bit, two or three times, while joggling the tool, before the final squeeze. This helps eliminate pinching the bit, off center, between two of the three jaws (Fig. 10-8).

Safety Tip Never leave a chuck wrench in a chuck. If the press is accidentally turned on, the wrench suddenly becomes an unguided missile!

Figure 10-8 *Avoid pinching the bit between two of the three jaws.*

Holding Taper-Shank Tools

Taper-shank tools have three big advantages over straight-shank tools.

1. **Positive Drive**
   They are designed to carry greater torque loads without slipping. The flat **tang** Fig. 10-9, transfers torque from the spindle to the tool. The inner end of the tapered hole features a slot where the drive tang fits.

Inserting and removing taper shank tools

Figure 10-9 *Clean all tapers before assembly. Then wedge out with a drill drift pin.*

2. **Accurate Centering**
   Tapers stay more accurately on the center of the spindle, as long as neither the spindle nor the taper are damaged.

3. **Self-Holding Solid Design**
   If cleaned then taped lightly into a clean spindle, taper shanks stay put until driven out again. After wiping both taper and hole, insert, then tap them with a soft hammer. To remove them, use the **drill drift pin** (or just drift pin) shown in Fig. 10-9.

Morse Taper Sleeve Adapters

Made popular by the Morse Tool Company, tapers for drill tools, bear their name—the **Morse taper.** The cone shape of the taper shank is a universal, worldwide standard. (There are several other tapers used to hold tools in machines but only one for drilling tools.) The standard is $\frac{5}{8}$ *in. taper per foot.*

But there are six sizes of Morse taper: MT1 (smallest, near a half inch) up to MT6, the largest (Fig. 10-10). They are actually different segments of the same $\frac{5}{8}$-in.-per-foot cone. When mounting taper tools, (a **Morse taper adapter**) might be needed to increase or decrease the size of the tool to match the size of the drill press spindle.

Taper *sleeves* might be needed to increase the tool taper up to the spindle taper (Fig. 10-10). Taper *sockets* are used to reduce a taper down, as shown in Fig. 10-11. Both sleeves and sockets are marked as to their smaller and larger size for example MT 1–3, meaning that the it brings a number 1MT to a number 3.

Assemble Taper Tools Right—*Clean!!!*

When assembling taper shank tools or adapters, it's critical that both parts are perfectly clean. A single chip of even the softest

Figure 10-10 *A variety of Morse taper adapters.*

Increase size with a socket

1 Morse taper

3 MT

Figure 10-11 *A Morse taper socket increases taper size.*

metal caught between the taper and receiving tapered hole will dent the shank of the tool, which then destroys the centering and self-holding of the tool (until it's repaired).

The taper in the machine spindle is deliberately made with harder steel than the tools it accepts so it will tend to survive such a careless act. But they can be damaged too. A fast way to be seen as an amateur is to slam them together without taking the time to clean both components!

Drill Geometry

Varying the Point Angle

As we've discussed previously, the three basic tool angles are seen on any drill bit. The point angle is created by two lead or cutting angles on both edges relative to the axis of the drill. They combine to make the point angle.

KEYPOINT

A point angle of 118° is standard—for soft to moderate steel, aluminum, brass, and cast iron.

Most metal is drilled well with an included angle of 118° (Fig. 10-12). To check this angle we use a *drill point gage,* which tests that each cutting edge is 59° relative to the drill center-line—half the point angle.

When drilling materials such as hard steel, bronze, or stainless, we sometimes flatten the point angle to an included angle of 135° or more, not because it cuts better but because it makes the drill's outer corner stronger, and therefore less prone to chipping and to heat build up. The corner radius between the cutting edge and margin also helps the drill withstand drilling hard metals (Fig. 10-13).

Because they have less point angle, these flatter drills do not start well without a pilot hole. They tend to wander off location as they touch work material. To prevent *wandering,* the hole can be started with a standard drill point of the same diameter, but it is fed down only enough that the flattened drill will begin drilling fully within the hole. In other words, the starting drill creates a

Standard drill geometry

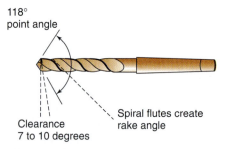

118° point angle

Clearance 7 to 10 degrees

Spiral flutes create rake angle

Figure 10-12 *Standard drill point geometry shows the three primary angles.*

90°
Not common for CNC work
Soft materials
Abrasive materials

118°
Standard point
General metals
Manual or CNC work

118° to flat°
Hard materials
Flat bottom holes
Must pre-drill-uncommon for CNC
Won't start a hole

Spur or pilot point
Thin materials-not for CNC work
Solves breakout problems

Step point
Drill and counterbore
Various shapes
Manual or CNC drilling

Spade drills
Deep holes
CNC drilling
Any size possible

Figure 10-13 *Modifications to standard points for various materials, some for CNC work some not.*

shallow hole that will not let the flat pointed drill wander.

For drilling plastics we sometimes use a more pointed drill than the standard 118° angle. But all soft or thin materials tend to grab and tear as the bit bursts through the opposite far side of the work. That leaves a jagged edge on the hole's rim. To prevent this problem, there is a different modification that can be ground on the point for plastics or thin sheet metal, called the *pilot point* or *spur-pointed drill*.

Pilot Point Drills

The pilot point stabilizes the drill from wandering as the vee-shaped cutting spurs cut the rim of the hole first, thus producing a good finish on the outer edge. These drills are sometimes called sheet metal points. They are also used for wood and plastic materials of any thickness (Fig. 10-14).

Spur
Pilot
90°

Figure 10-14 *A spur or pilot point drill is ground for thin cutting materials, plastics, and wood.*

Although drills can be purchased preground with this configuration, the spur point can be ground onto any twist drill with a large enough diameter (above $\frac{3}{16}$ in. or so). To create the tiny vee into the cutting edge on both sides of the point, the grinding wheel must be dressed with a 90° sharp corner, as shown in Fig. 10-15. The challenge is to compound elevate and rotate the drill to create the necessary side clearance on the point and the end clearance on the cutting edge spurs. Correctly held against the grinding wheel, the front and side faces of the grinding wheel produce all

Figure 10-15 *Grinding a spur point requires a well-dressed wheel. Note—safety guards are removed for clarity only!*

the necessary angles at the same time. Have your instructor or an experienced journeyman show you how.

Adding a Negative Rake Angle to Prevent Self-Feeding

Due to the helix angle and digging action of the rake face, drills sometimes self-feed into soft or thin materials. They literally screw themselves into the work as a wood screw might. In certain materials such as soft brass, self-feeding can occur even when no pilot hole has been drilled. This unwanted action can break the drill and pull work out of a setup. It's happened to me! An unmodified drill literally screwed its way through a brass door striker plate! There was no hole, the drill went right through but snapped off as it broke through the opposite side of the work. It fed down so fast that it pulled the quill feed lever out of my hand!

To avoid this happening to you, grind a tiny neutral or even negative rake **land** on the rake face of the drill at the cutting edge, as shown in Fig. 10-16. (Land means narrow flat spot similar to a margin.) No more land than 0.015 in. wide is needed to retard the self-feed tendency. Grinding more adds nothing to the action. However, the excessively large negative rake land is far more difficult to grind away, to restore the bit back to a normal rake when it's no longer needed on the drill.

Sharpness—Cutting Edge and/or Margin Wear

Drill damage is determined with your eye before mounting the tool in the machine. Look

for obvious nicks and chips in the cutting edge or rounded edges, especially at the outer corner (Fig. 10-17).

However, when the cutting edge has worn normally from use, it requires a closer look. The dull edge, which may be as small as a few thousandths of an inch, will look like a bright line right at the junction of the cutting edge with the rake surface. To detect it, rotate the drill until the room light reflects off it's worn radius. Cutting edge dullness is easily resharpened. Margin wear is not so easy to fix. The entire length of damage must be cut away using an abrasive saw then a new point reground from there.

.015 in.
flat spot

Figure 10-16 *Compare an unmodified drill to one with added negative rake to stop self-feeding.*

Figure 10-18 *Compare two drills—one with correct clearance (left) and one ground incorrectly without clearance.*

Clearance

When selecting a drill, verify that it has a correct clearance angle of around 6° to 10°. In industry it's unlikely that you will encounter a poorly ground bit—however, in training, experience has shown that clearance is often missed by beginners. Verifying the clearance angle (Fig. 10-18) and regrinding if needed, can prevent a burned drill or work hardened material.

The Dead Center

The flutes of a drill bit are held together by a solid core of metal in its center, called the *web*. In Fig. 10-18, looking at the tip of the bit, the web creates the **dead center,** so called because it has no rake angle, clearance, or cutting edge. Without modification, the dead center does not cut, it simply scrapes metal away. There are two choices as to how to proceed.

1. *Leave the dead center as-is.* While it's inefficient, it is acceptable point geometry in many cases, for example, when drilling aluminum, mild steel, or brass. Unmodified, the drill requires more pressure to force through the cut, but it will drill successfully. Predrilling with a smaller pilot drill relieves the pressure of forcing the dead center through the work. Of course, the pilot also has a dead center, but it's much smaller.

2. *Thin the web with some extra grinding.* There are two ways to do that. The simple method shown in Fig. 10-19, makes the web thinner but doesn't create very effi-

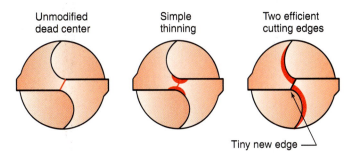

Figure 10-19 *Thinning the dead center helps the drill cut with less down feed pressure and heat.*

cient cutting edges. It's quick and easy to grind onto the point but less efficient. It is created by nicking the point on both sides using the corner of a grinding wheel.

The drill point on the right in Fig. 10-19 has been ground with two new cutting edges. This second method requires a sharp corner on the grinding wheel, a steady hand, and some practice, but it creates cutting edges at an angle to the original edges. Have your instructor demonstrate this skill after practicing grinding drill bits in Unit 5.

Trade Tip

Try an Experiment When hand-feeding a drill through work that has not been pilot drilled, it's obvious when the dead center is thinned and when it is not. Relieving a dead center can reduce the pushing force by as much as 50 percent in some cases. If your instructor can permit an experiment of this kind and has two spare drills of the same diameter to use for this test, thin the dead center of one but leave the second with its center unmodified. Use drills of $\frac{1}{2}$-in. diameter or bigger so that the feed resistance will be more pronounced. Chuck each in the press and compare the down feed resistance when drilling steel.

Reamers

Machine Reamers

These reamers are designed to be turned by a drill press or other machine (Figs. 10-20 and 10-21). They are supplied in both straight- and taper-shank versions. Machine reamers have multiple flutes of either straight or spiral design. The spiral fluted reamer tends to not chatter as readily as the straight flute. Straight flutes work OK but RPM must be kept to around 25 percent of the possible speed to prevent chattering. The spiral design not

Figure 10-20 *Spiral and straight fluted reamers.*

Reamer
Basic parts

Figure 10-21 *The main parts of a reamer.*

Specialty Reamers

There are several other choices of reamers beyond the standard machine reamer (Fig. 10-22).

Shell reamers are common in production where they have the cost advantage of interchangeable cutting heads on one shank. These compact tools are very easy to resharpen as well.

Expansion reamers can be adjusted to any size within a small range. Expansion reamers are supplied in both hand or machine versions.

Taper reamers are made to finish tapered holes. For example, the internal taper in a Morse taper sleeve. When using tapered reamers, use caution because pushing them just a little too deep can cause grabbing, whereby the reamer often breaks.

Reverse helix reamers are used in place of standard reamers, which are made with a right-hand helix. Although the right-hand helix helps to bring chips up and out of the hole, occasionally, a reamer with the standard helix will grab and tend to self-feed into a workpiece similar to a drill. This happens in soft material such as brass or lead. When self-feeding on a power feed machine, the reamer will tend to pull a taper shank out of the spindle. The correction is the left-hand spiral reamer, which cancels self-feeding but requires more force to push through the hole. They push chips forward into the drilled hole rather than removing them, so they work best when the hole goes all the way through the work.

Trade Tip

When Reaming First, use liberal amounts of coolant or cutting oil.

Second, push the reamer through the hole at a moderately high rate. (This is known as *crowding* the reamer in the trade.) Infeed should equal approximately $\frac{1}{4}$ the reamer's diameter per revolution. Often, instead of slowing the RPM, increasing the feed rate will produce a larger chip load and stop the chatter. When practicing on the manual drill press, experiment with various combinations to see this effect for yourself.

only tends to dampen chatter, it also delivers chips out of the hole better and it makes a finer finish due to the lead angle of the side cutting edges in the hole.

Finally, there's *backing the reamer out of the hole.* Normally it's OK to pull the spinning reamer out of the hole before turning off the spindle on drill presses or CNC machines. However, in softer metals such as aluminum or brass, that action can remove more material on the way out—making an oversized hole. To solve this problem, on manual presses, turn off the spindle at the bottom of the stroke, wait until it nearly stops, then quickly withdraw the reamer as it coasts to a stop. But don't let it stop turning completely, as that makes straight marks along the hole's wall, resulting from contact with the reamer's cutting edges. If it does stop rotate the spindle forward a half turn or so by hand as you withdraw the reamer. *Caution:* Do not rotate the reamer backward while in the hole, as that can destroy the cutting edge. If a CNC reaming operation has this double-cut problem, edit the program to slow the RPM below 150 before backing it out.

Reamers

Figure 10-22 *There are several other choices for reamers beyond the standard version.*

Drilling the Reamer Pilot Hole

Reamers require a pilot hole slightly smaller than their finished size. The size and location of the drilled pilot hole, sometimes called the *reamer drill*, is important for accurate, smooth results.

If the *pilot hole is too small*, the reamer will clog on the chips. While reamers are not as efficient at delivering chips up out of the hole as are drills, to help, always use a cutting fluid. Lots of it. Oil or flood coolant are essential for good finish, tool life, heat reduction, and chip ejection.

If the *pilot hole is too large*, there isn't enough metal remaining for the reamer to cut an effi-

cient chip. It may chatter and the drill marks may not be completely removed as well.

Accurate location of the pilot hole is also important. The reamer will tend to align to the drilled hole and follow it even if the spindle is misaligned to the drilled hole. The reamer will flex as it rotates and follow the drilled hole a bit like a plumber's snake. We refer to this as **reamer float.**

Selecting Drills for Reamers

As the reamer size increases, so can the excess left for it to cut, because the flutes become larger and can deliver more chip volume up out of the hole. There is some discretion here, based on material being reamed and the

Reamer Excess Chart

Inch Sizes Reamer Diameter Range	Excess for Reaming (Deduct from Finished Diameter)
$0-\frac{1}{4}$	$\frac{1}{64}$
$\frac{1}{4}-\frac{3}{4}$	$\frac{1}{64}-\frac{1}{32}$
$\frac{3}{4}-1$	$\frac{1}{32}-\frac{1}{16}$
1 and up	Can be greater than $\frac{1}{16}$ in. but serves no real purpose; $\frac{1}{16}$ in. is adequate.
Metric Sizes	
0–6	0.5 mm
6–12	0.5–0.75 mm
12–24	0.75–1.0 mm
24 and up	1.0–1.5 mm Can be greater but serves no purpose.

depth of the hole—the sizes are recommended. The chart is a rough guide to reamer drill excess to be left for reaming.

For an example, consider what fraction size bit would work well as a pilot for a $\frac{5}{8}$-in. reamer. Solve this from the drill chart or calculate the drill size.

Answer $\frac{5}{8}$ falls within the $\frac{1}{4}$- to $\frac{3}{4}$-in. range—subtract $\frac{1}{32}$ in. on the drill chart to derive a $\frac{19}{32}$-in. pilot drill.

KEYPOINT

This is a rough guide. A slight difference in pilot drill size, larger or smaller, will work well too. If the material is hard, choose less excess than suggested in the chart, if soft, slightly more machining excess is OK.

Trade Tip

Control by Sound During lab practice, make an effort to learn the normal sounds of sharp tools drilling, reaming, and tapping. Then, if the sound changes, that's a strong signal that the tool might be dulling or loading up with chips. Stop immediately if there is any doubt.

UNIT 10-1 REVIEW

Review the Key Terms

Blind hole
A hole that does not go through—this leads to chip and coolant challenges, especially when tapping threads.

Dead center
The inefficient center of a drill bit that does not cut.

Drift pin (drill drift)
A small, flat wedge used to break the holding force of taper-shank tools.

Land
Similar to margins, a small surface on a cutting tool.

Margins
A narrow band where drills are full diameter.

Morse taper
A self-holding machine taper of $\frac{5}{8}$ in. per foot.

Morse taper adapter
Either a *sleeve* for upsizing or a *socket* for downsizing.

Reamer float (tool float for other kinds of cutting tools)
The tendency of cutting tools to pick up and follow previously machined shapes, such as a reamer following an incorrectly located drill hole even though the press is in the correct location.

Shank
The mounting end of the tool.

Tang
The torque drive tab on taper shank drills and reamers.

Turned-down shank
A drill larger than $\frac{1}{2}$ in. diameter, but featuring a straight shank. May be made from a damaged taper shank drill.

Twist drill
A high-speed steel cutting tool made with spiral flutes.

Replay the Key Points

◆ The five factors that determine the right drill for the job are *size, shank type, point angle, sharpness,* and *dead center*.

◆ Drills and reamers can be specified either as *metric, inch fractions, letters,* and *numbers*.

◆ Always measure a reamer before use.

◆ Measure the drill or reamer diameter on the margin.

◆ The pilot hole needed for a reamer is a calculated amount smaller than the reamer.

Respond

1. Turn to the appendix drawing—drill gage, and using a drill chart, choose reamer drills for each hole size listed on the print up to $\frac{1}{2}$ *in*. Use the guide here in Unit 10-1 to find the correct excess. Note, choose the greater excess (smaller hole) of the two possible.

2. Using the drill chart, select the metric reamer drill to ream a 12-mm-diameter hole.

3. Of the holes in the drill gage, which will probably be taper-shank tools?

4. What is the standard drill point angle?

5. To drill hard bronze, what point angle might be needed to prevent edge breakdown 108° or 135°?

6. Why choose the point angle of Question 5?

7. What problem arises when we choose to modify the drill point geometry as in Question 5?

8. Describe the three possible options a machinist has in dealing with a drill's dead center.

Drill Presses and Setups

Introduction: Before investigating drill press setups, we need to take a brief look at the manual machines commonly found in labs and industry. There are four general types with many variations within each. They are classified by whether they have power feed and what type of drilling head they feature:

Standard drill presses

Power feed drill presses

Radial arm drill presses

Production drilling machines—multispindle/ multihead

CNC milling machines (often called machining centers) are rapidly replacing many of the production drills because they can also cut and shape the metal as well as drill it, all in one efficient setup—called a one-stop setup. So, we'll limit our investigation to the first three. However, the knowledge gained here will transfer to the production machines should they be encountered on the job.

KEY TERMS

Bushing (slip bushing)
A hard steel guide to ensure drill and reamer location.

Fixture (jig)
A work holding devise that accurately secures odd shapes and shortens part turnaround time.

Peck drilling
The in-out-in action used to break long chips into short segments and to clear chips during deep drilling.

Quill
A cylinder that contains the spindle bearings and spindle. The quill slides in and out to create the Z axis of most machines.

Radial (arm) drill press
A drill press for large work wherein the head is moved over the intended hole using the radial movement of an arm.

Sensitive drill press
A lightweight, high-speed drill press for small drills.

Standard Drill Press

At least one is found in every shop. These simple drilling machines are designed for light-duty drilling and reaming work up to around $\frac{1}{2}$-in. diameter. They subdivide into two general types.

Sensitive or Bench Top Presses

Because they break easily, drills smaller than 3 mm or $\frac{1}{8}$ in. require very high RPM for efficient cutting and a light in-feed touch by the operator. The **sensitive drill press** meets that need because it can achieve spindle speeds as high as 12,000 RPM or more. To increase operator feel, the sensitive press has a lighter **quill** and less powerful return spring compared to its big brother presses.

What is a Quill? Quills are used on many machines where the tool must spin and also slide in and out along an axis. In a CNC program, that axis is always identified as Z. On a drill press, the quill is the precision cylinder that contains the spindle bearings and rotating spindle (Fig. 10-23). The quill slides down by hand lever pressure and returns by spring action.

KEYPOINT

When a spinning cutter slides in and or out, it's always named the Z axis if it's the only quill on the machine.

Figure 10-23 *A sensitive press is best for small holes below $\frac{1}{16}$ inch (0.5 mm) as it can achieve 20,000 RPM and has a light quill return spring.*

Because the quill is relatively light on the sensitive drill press, the return spring can be weak, thus allowing good feedback of the cut resistance through the hand-feed lever. The operator can sense or feel how the drill is progressing. The sensitive press uses only straight-shank tools, tapered drives are not required here. Sensitive drills usually cannot reverse their rotation, which means a special attachment called a tapping head is required to perform the cutting of threads. We'll look at tapping heads in just a bit.

Standard Floor Mounted Presses

Figure 10-24 shows the larger, standard drill press, the everyday workhorse of all shops. Its major difference from the sensitive press is its greater size and power. These presses feature a lower RPM range from around 500 up to 3,000 or so, depending on make and motor, but always more horsepower compared to their little brother bench models.

Most of these presses feature a spindle taper with a number 2 or 3 MT. Taper-shank drills can be mounted directly in the spindle taper, or chucks can be inserted in the spindle taper. The chuck is normally left in the spindle since that makes exchanging tools faster. Similar to bench drill presses, most standard presses cannot reverse rotation.

Figure 10-25 *Step pulleys provide a limited number of spindle speeds.*

Both bench and floor mount presses feature

A. Variable speed for differing size drills. The speed selection may be by step pulley, which offers a limited number of RPM selections (Fig. 10-25). Or RPM may be changed with a variable cone pulley transmission, offering infinite speed adjustment (Figs. 10-26 and 10-27).

Variable Speed Drives This is a good time to understand variable speed cone transmissions as they are used on other small machines in the shop too. Figure 10-27 is a simplified illustration of a cone transmission. Rotating the speed selection dial (Fig. 10-26), causes one pair of facing cones to move apart, thus creating a smaller

Figure 10-24 *Standard drill press.*

Trade Tip

Changing RPM the Right Way If the spindle speed is selected by variable drive, *only change speeds when the spindle is running.* Incorrectly cranking the speed dial with the spindle turned off can dislodge the belt from the pulleys—similar to shifting a 10-speed bike while it's not moving.

Figure 10-26 *A variable speed dial.*

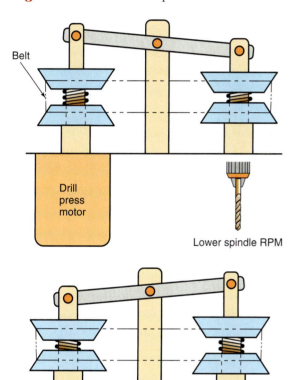

Figure 10-27 *The principal of a variable cone drive.*

Figure 10-28 *A drill press table swung out of the way to accommodate a part too tall to mount on the table.*

circumference for belt contact, while the other two cones move together at the same time, creating a larger circumference. This action changes the drive-to-driven ratio. As the cones shift diameters, belt tension remains the same always, but the motor drive to the spindle driven ratio changes continuously.

B. Vertically adjustable table—The work table can slide or be cranked up and down to accommodate large or small work.

C. Horizontally rotating table—The work table can swing around on the column to accommodate different shaped work. Or to completely clear the table to mount extremely tall work on the base plate (Fig. 10-28).

D. Quill depth control (also called a quill stop)—Provision is made to limit and control the hole depth. Most quill stops (Fig. 10-29) are graduated with a vertical rule and division lines around the circumference of the stop nut. One full turn often equals a common increment such as a $\frac{1}{16}$-in. change in quill depth, or a 2-mm change, for example. Take the time to investigate what those graduations equal on your lab's press as that knowledge can save a lot of guesswork when drilling to an exact depth.

E. Some standard presses feature a tilting table that tilts clockwise and counterclockwise when viewed from the operator side.

Figure 10-29 *The quill stop will allow the spindle down another 20 mm. One full turn of the stop equals 1.0 mm.*

Power Feed Drill Press

Evolving up in size and strength, this heavier machine features a power feed and a larger quill to accommodate bigger drills (Fig. 10-30). These presses are less common in schools, but are well suited to industry, where they are put to work drilling holes from 12 mm up to 30 mm or greater ($\frac{1}{2}$ in. up).

Due to the extra force they must withstand, many of these presses do not use belt drives. Instead a series of gears is used to shift speeds. For that reason, you might also hear them called "gear head presses."

In addition to the features of the standard press, these presses feature:

A. Varying power feed rate selections—The quill can be pushed down by hand, as with the two preceding presses, or the operator can pull the hand lever sideways to engage the power feed. This feature not only prevents operator fatigue, but it also imparts consistency to drilling operations. It's challenging to consistently keep the right down feed by hand pressure.

B. Micrometer quill stop—Halts both hand and power down feeding.

C. Coolant systems and reservoir—An option, but very common on these presses.

D. Larger spindle, usually #3MT.

E. These machines have a reversing spindle to allow power tapping. Some feature automatic stops to reverse the motor at a preset depth.

F. Gear head to create slower speeds but greater rotational torque (force) for tapping and reaming.

The power press is designed for drilling large lots of parts. Typical of all manually operated equipment, gear head presses are being displaced by CNC mills and lathes. That's also true of the next press. However, they are strong, accurate machines and will be in shops for years to come. They still do the job when the CNC machines are loaded with work!

Figure 10-30 *A power feed* drill press *or sometimes called a* gear head drill press.

Radial Arm Drill Press (Radial Drills)

This machine is generally the most powerful of the three (but there are crossover exceptions). It is usually larger than the previous two. In industry, there are **radial arm drill presses** large enough to drive an automobile on the work table. There are also *radials* with two complete drilling heads at each end of the work table to enable reaching extremely large work.

Radial Arm Presses Take the Head to the Work

The presses discussed thus far require the operator to push or tap the work into vertical alignment with the drill, then clamp it down. However, as work gets very big or bulky, that's difficult or even impossible to do. Using the radial arm capability on these large presses, after swinging the arm out of the way, the operator can move the work directly

Radial Arm Drill Press

Column
Sliding drill head
Elevating rail
Revolving arm
Coolant reservoir
Tilting table

Figure 10-31 *A radial arm press can position over the work table or swing to allow the crane over the table.*

over the table with an overhead crane, to set it down. The crane is removed and the work clamped firmly in place on the table (Fig. 10-31). The operator then positions the arm and drill head back over the work, by moving in or out along the rail and in a circular arc about the column. Once the drill is in position over the work, an electric lock (or a manual one on smaller radials) is actuated.

In addition to the features of the power press, radial drill presses (Fig. 10-32) have these features.

A. The entire radial head moves up and down on the column, by hand crank or by power.

Major Functions

Quill hand feed (rotate)
power feed (pull out)
Radius lock
Radius in/out hand wheel
Elevate rail
Lock/unlock button column rotation
Spindle clutch forward/reverse

Figure 10-32 *The main parts and features of a radial arm drill press.*

The table does not move up or down. Some tables can tilt, others cannot.

B. The motor reverses by clutch action.

C. A true clutch is included that allows gently starting the spindle rotation and reversing by slipping the clutch. This is a great feature for power tapping and other operations requiring operator control such as spotfacing (coming up).

D. Extra large work table is included with tee-slot-type hold down capability.

Drill Press Safety Always

Because drill presses appear to be simple, and harmless, many machinists both new and experienced, let their attitude slip when running them. We instructors see the following three accidents occur far too often—don't let them happen to you!

Unclamped Parts Spinning Out of Control

It's tempting to put a workpiece on the table and drill it without clamping, or even dumber, to try to hold it by hand! Don't do it! The problem is that as the drill breaks through the far side of the unclamped work, it grabs the workpiece, pulling it up the drill bit much like a screw in wood, then it spins and goes out of control on the bit. Depending on the bit's RPM, the wildly rotating part can hit your hand several times before you can react. But the accident isn't over yet!

As the unbalanced object spins out of control on the bit, the bit often breaks, then the part flies out of the drill press. It's a "fire-drill" easily avoided by taking a few moments to secure the work in a vise or to the table with clamps before drilling.

If due to incorrect clamping, the work comes loose during drilling, your actions are to immediately turn off the press and step back. There are no heros in machine shops, only smart people. Don't try to grab the wildly spinning work with your hand! See Fig. 10-33.

Pay close attention to instruction on how to clamp work the right way to completely avoid this accident!

Figure 10-33 *Warning! This accident happens when work is not clamped!*

Hair or Clothing Caught in the Spindle

The spindle and bit can grab and instantly wrap up and pull any loose item that gets too close. This happens for three reasons. One is that the friction of drilling causes a static electric buildup in the tool, similar to a comb held to your arm after combing your hair. Items like hair or clothing are attracted to the tool. Once contact is made, they wrap up lightening fast!

The second reason is that there are air currents created by the spinning tool. Like a fan, air is slung out and away. If it spins out, more air must rush in elsewhere to replace it, bringing along the hair or cloth.

This accident commonly occurs when a machinist with long hair leans over to see how the drill is doing.

KEYPOINT
Wear a hat or hair band and appropriate clothing for the shop environment. No long sleeves or jewelry while operating any machine, but especially a drill press!

Chips Flying Out (and Rarely a Broken Tool)

The third way to get caught by the drill press is the long stringy chips they produce

Figure 10-34 *Unbroken chips can be dangerous.*

(Fig. 10-34). If not broken by operator action, they fling out in an expanding radius as they grow in length. Protect yourself from them by wearing safety glasses, no long or loose clothing, no fabric gloves that will be caught by the chips, and clear away anything on the table that might get caught up by them. If, as in the upcoming Shop Talk, long chips become a real problem, use an in and out motion on the feed lever, called *chip breaking* or **peck drilling.**

Drilling to a Controlled Depth

Clearing and Breaking Chips During Drilling This is a good exercise to practice for getting to know drill operations and limits before writing programs. As drills progress deeper into the hole, at first, small pecks safely break the chip, but at a given depth (learned from experience), you must begin chip-clearing pecks. Get the feel for the need to peck drill and practice it on the manual drill press. The only way to go much deeper than six to eight times the drill's diameter using a standard twist drill is to perform peck drilling.

KEYPOINT

As a rough guide, switch to peck drilling when hole depth is five times hole diameter.

Small pecks or even no pecks at all, work well for hole depths up to around five times

the drill's diameter, depending on alloy and coolant supply. A $\frac{1}{8}$-in. drill, for example, will not need to clear chips until drilling down to around $\frac{5}{8}$ in. or so, depending on the kind of metal being drilled. But drilling deeper requires switching to big pecks. If you don't, the drill flutes will jam up with chips and will not be able to deliver them out of the hole, called *clogging* or *loading.* If not stopped, the loaded drill will quickly lead to the bit breaking or burning up due to friction between the loaded bit and the hole. The jammed chips can also roughen the finished hole.

Coolants help lubricate the chips and prevent clogging. But as the depth progresses, chips become packed more tightly in the drill's flutes and prevent the coolant from getting down to the cut area. Again, to avoid burning the bit, big pecks momentarily allow coolant to flow down to the contact area to cool the drill and prepare for the next peck.

Drill Press Setups

Now let's look at four common methods of holding the work in a drill press: vises, clamps, plates, and fixtures. Each has it's own purpose. All are employed on automated equipment except drill press vises (see the Trade Tips).

Drill Press Vises

These specialized vises are made differently from standard milling vises in two ways (Fig. 10-35). The most obvious feature is the fast open and close feature for quick size adjustment and turnaround time (changing from one part to another).

The second feature that sets drill vises apart is the shelf in the jaws that enables suspending and holding work up off the floor of the vise to drill entirely through the work without the drill touching the vise bed.

KEYPOINT

Clamp the Vise to the Drill Press!
Once the work is secured in the vise, you aren't done with the setup. Next, clamp the vise to the table in one of the ways described next.

Typical drill press vise

- Pivot joint
- Half nut quick release
- Lift for quick movement
- Shelf holds work up off vise floor

Figure 10-35 *A drill press vise has two unique features.*

Trade Tip

Drill Press Vises Cannot Hold Work for Milling!
Never use drill press vises for operations that produce a side thrust such as milling. They are designed for vertical pressure only and will be dangerously underrated for the duty. Learn the difference between drill vises and milling vises and use them correctly.

Clamping Parts

There are a variety of clamping tools for drill press work. Some are used on other machines and a few are made expressly for drill press work since they would not withstand forces other than vertical drilling.

Using Table Clamps The most common clamp used throughout the shop is a simple bar with a slot through which a bolt is placed (Fig. 10-36). It is supported at the back with a heel block. There are a variety of modifications made to these clamps, which you'll see when studying milling. There is a right and wrong way to use table clamps.

Table clamps

- Protective shims
- Sturdy washer
- Heel block

Figure 10-36 *Two kinds of table clamps.*

Figure 10-37 *What is wrong with both clamps?*

KEYPOINT

To concentrate the pressure on the work, not the heel block, place the bolt close to the work, not to the heel block. Also, set the heel block such that the clamp is level or tilted slightly down toward the work, which prevents pushing the work out from it.

Test Your Powers of Observation

In Fig. 10-37, there are four things wrong with the clamp on the right. What are they? But there is also something wrong with the clamp setup on the left, too. What is it? Answers are found just before the unit review.

Other table clamps are used exclusively on drill presses as well. A typical quick clamp example is shown in Fig. 10-38. Common

Figure 10-38 *A quick-action table clamp for a drill press.*

Drill bit aligned to table clearance hole

Clamp not shown for clarity

Elevating parallels

Figure 10-39 *Two ways to protect the drill press table.*

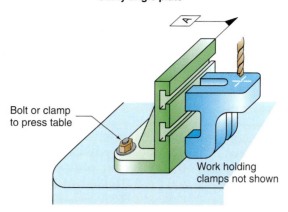

Utility angle plate

Bolt or clamp to press table

Work holding clamps not shown

Figure 10-40 *Drilling parallel to a datum or surface.*

"C" clamps are also used on drill presses at times.

Prevent Drilling the Vise or Table!

When locating and clamping work directly on the drill table, or any machine, be certain that the setup provides cutting tool limits that will not allow it to contact the table surface. There are two solutions on drill presses.

◆ *Use the clearance hole* (Fig. 10-39). Place the workpiece with the intended hole location directly over the clearance hole in the machine table. Before clamping, first test the location with the drill bit to the clearance hole. Rotate the table if needed, such that the drill passes through the hole.

◆ *Elevate the work* up off the table by clamping over parallel bars (Fig. 10-39) a block of metal or even a sheet of plywood to provide drill clearance to the table. In this case, use the quill stop on the press to prevent the drill from reaching the table as it breaks through the work.

Work Holding Plates Create Vertical Datums

When holes or other cuts must be parallel to a datum, or when work is difficult to clamp in a vise or directly to the drill press table (Fig. 10-40), one solution is the shop grade, right-angle plate. Recall that these utility plates are different from the layout version that are not to be used during machining. They are heavy-duty, work-holding plates with holes for bolts or tee-slots for clamping work. The

clamps we've discussed could be added to Fig. 10-40 to firmly hold the part on the plate.

Drill Jigs

In manufacturing, the word **jig** means a holding tool that secures parts that are otherwise difficult to hold for machining, welding, or assembly. Jigs ensure accurate location of the workpiece on the machine and accuracy of the operations to be performed. They also make it faster to exchange one part for the next during production. In Fig. 10-41, the top of the jig slips off two pins, to exchange parts. This jig locates the face of the part on datum A, similarly to the angle plate setup of Fig. 10-40. By using the left-right position of the part in the jig, datum B is obtained by slipping the central slot in the part over a locator tab in the jig. When in position, a clamp is screwed against the workpiece. Once in position, the hardened slip bushing ensures that the drill will not wander.

Drill and ream jig

Turn lock groove

Reamer diameter busing

Hardened interchangeable slip bushing

Datum B locator

Guide pins for removable top plate

Guide bar keeps jig from spinning

Part clamp

Figure 10-41 *A drill jig holds the work and guides the drill.*

Drill jigs take on two forms:

◆ *Custom made* tools are specifically built for a single part or perhaps to hold a family of very similar parts. Figure 10-41 is a custom-made jig consisting of four steel plates fastened together.

◆ *Universal kits* enable the machinist to make up a jig using components. These are sometimes called box jigs, as they start with several sizes of square boxes. A fourth plate on the near side of Fig. 10-41 would simulate a box jig. To the basis box, work support points are added to create datums, locators, and clamping screws. Bars with bushing holes are bolted to the box as required, and finally, legs are put on where the jig must be held above the table to allow drilling through the work.

Interchangeable Guide Bushings Where drilling and reaming must be at a precise location, but the work will be done on a manually operated press, drill jig guide bushings are used in Fig. 10-42—called **slip bushings.** First the drill size bushing is inserted in the fixture with a one-quarter twist that

locks it in position. Then the hole is drilled. Next, the drill bushing is exchanged for one that fits the reamer. Each bushing guides each tool to an exact location.

Observation Answers

The right-hand clamp is tilted down, it will push work out or slip off. There is no washer, the nut is touching on corner only. There is no shim, so it will mar the workpiece. Finally, the bolt is too far from the workpiece.

The left-hand clamp has no protective shim, it might mar the workpiece!

Turn to
lock bushing

Figure 10-42 *After drilling, the removable guide bushing will be rotated to unlock if from the jig, then exchanged for the reamer bushing.*

UNIT 10-2 REVIEW

Review the Key Terms

Bushing (slip bushing)
A hard steel guide to ensure drill and reamer location.

Fixture (jig)
A work holding devise that accurately secures odd shapes and shortens part turnaround time.

Peck drilling
The in-out-in action used to break long chips into short segments and to clear chips during deep drilling.

Quill
A cylinder that contains the spindle bearings and spindle. The quill slides in and out to create the Z axis of most machines.

Radial (arm) drill press
A drill press for large work wherein the head is moved over the intended hole using the radial movement of an arm.

Sensitive drill press
A lightweight, high-speed drill press for small drills.

Replay the Key Points

◆ Adjust RPM on variable speed transmissions only when the spindle is turning.

◆ When setting up any machine, be certain the cutting tool will not contact the machine table.

◆ Drill presses can be dangerous even though they look simple and safe.

Respond

1. What is the normal upper limit of diameter of straight-shank drills? For the standard drill press? Metric limit?

2. Which drill presses are equipped with a reversing motor to accommodate power tapping?

Figure 10-43 *Identify this industrial drill press.*

3. What is different about a drill press vise compared to a standard vise?

4. True or false? For safety, you must either clamp your part to the table or angle plate or put it in a vise. The vise must also be clamped down. If this statement is false, what will make it true?

5. Name the type of drill press shown in Fig. 10-43.

6. What four methods exist for holding work on a press?

7. Which method in Question 6 runs the risk of drilling the table?

8. How do you prevent the problem of Question 7?

9. Name a method of preventing long stringy chips on a drill press.

10. Describe the four methods of work holding for a drill press.

UNIT 10-3

Speeds and Feeds for Drilling

Introduction: Setting the correct RPM and feed rate is necessary for good finish, tool life, and size accuracy on all machines. All the methods we'll discuss determine the best *initial* values. After testing them on the machine, they are almost always adjusted up or down to suit the particular situation. For example, does the setup tend to chatter, how is the particular tool performing, is coolant available, and so on? The skill of fine-tuning a setup improves with experience. It's nearly intuitive at times. It's also critical to recog-

nize safety limits for both speed and feed rates, and not to exceed them.

When practicing at the manual drill press observe closely the result of changing RPM (cutting speed) and feed rates, down pressure. That vital knowledge will help ensure success in your CNC programs!

KEY TERMS

Empirical data
Data collected by experimentation rather than calculation. All speed and feed charts are empirical.

Feed rate
Cutter advancement through the workpiece, expressed in inches of decimal parts of a millimeter, per minute.

Surface speed
The speed in feet per minute or meters per minute, of the drill rim.

Drill RPM

As discussed previously, there is an optimum surface speed when drilling a given material and that translates into the RPM of the various tools in your setup.

For example: from Appendix III in this book, a surface speed of 100 feet per minute (F/M) for drilling steel is recommended and 175 F/M when drilling brass. Or 30 meters per minute. *But that is not the RPM! It is the rim speed of the object that is spinning.*

KEYPOINT

A. There is a "best" **surface speed** for any material being drilled.
B. Different diameter drills must be set to turn at different revolutions per minute—such that their rim is at the correct surface speed.
C. Smaller drills require *faster* RPMs, while larger drills must go *slower* to achieve the same surface speed at their rim: big drill = small RPM and little drill = big RPM.

Five Methods for Accessing RPM

There are five ways to determine the right RPM for a drill or any other spinning tool.

◆ *RPM chart* based on material being machined and tool diameter.

◆ *Formula* using surface speed and tool diameter as the variables.

Estimating RPM We'll study RPM charts and formulae here in Section 10, but while using them, you'll also be developing your built-in sense of the right speed. Even though "knowing when it's right or wrong" is not exact, it is very much a part of the tradesman's competencies. The journeyman is able to look at a spinning tool and instantly know whether it's close to right or dangerously wrong. Typical of most life lessons, this knowledge will probably be gained more from your mistakes than from your successes. But it's good to develop it before attempting to write programs. The CNC machines are faster, stronger, and less forgiving than the manual ones.

◆ *Insight* based on experience.
◆ *Dedicated speed and feed, slide-rule calculator.*
◆ *CAM*, built-in RPM calculation within the software.

The dedicated S&F calculator will be learned in Section 12. CAM calculated RPM, the last method, is incorporated into programming softwares both off-line or at the machine control. You'll have a chance to use this ability in Section 25. If given the tool diameter and material hardness, modern softwares such as Mastercam®, contain the formula we're about to explore. They can, if commanded, automatically compute tool RPM.

Drill Speed Chart

This is the simple way to obtain tool RPM. You'll find them on the wall in most shops, on plastic pocket charts, metal plates on the machine itself, and in reference books. The chart is entered with two cross-referenced, variables: *drill diameter* and *type of material*.

KEYPOINT

Drill charts assume the tool is made from HSS (typical of most shop quality drills). But, if using a carbide-tipped drill, the RPM can be increased by a factor of three or more.

There's not much training required to learn to use an RPM chart. The fastest way is to try a few practice problems.

Check Your Understanding 10-3A

Use Appendix III, the drill speed chart. Answers are found at the end of Section 10.

1. How fast should a $\frac{3}{8}$-in. drill be turning when drilling mild steel?
2. What is the correct RPM for a $\frac{1}{4}$-in.-diameter drill in aluminum?
3. What is the RPM for a 1.0-in. drill in brass?

Critical Thinking

4. RPM changes as drill diameter changes. Look at Appendix III, under aluminum and compare RPM for a $\frac{1}{8}$-in. drill to a $\frac{3}{16}$ inch bit. (diameter difference = $\frac{1}{16}$ inch). Now compare the RPM for a $1\frac{1}{8}$-in. to $1\frac{3}{16}$-in.-diameter bit (also $\frac{1}{16}$ inch difference). Explain why the RPM difference is so great between the two smaller drills but the change is very little between the two larger drills.
5. Looking at the chart, what can you say about the machinability of carbon steel, compared to annealed stainless?
6. Not covered in the reading: What should you logically do if the RPM you find isn't available on the machine you are operating? (This happens on manual machinery such as step pulley drill presses where there are only eight speeds available but never on CNC where spindles are computer controlled.)

Calculating RPM

Drill charts are specific, don't cover diameters that lie between two common drills, and they cover only a range of sizes, so any drill larger or smaller than the chart must be calculated. On the other hand, formula results cover any diameter, large or small. Also when using the short formula coming up, the calculations can often be performed in your head, thus there is no need for a chart. Formulas use two variables:

◆ Diameter of the drill (*DIA*)
◆ Recommended surface speed (for the metal being drilled) (*SS*)

Another advantage of a formula is that it can be programmed into your calculator with stops for each variable (surface speed and tool diameter). Since the calculator is always by your side, it's a ready method, but you must remember recommended surface speeds, or have an SS chart handy or store them in memory.

Surface Speed To use either formula, long or short, one must start with a recommended surface speed chart. Appendix IV provides the student biased numbers. It also features rounded numbers that are easily recalled. This should aid in developing the journeyman's feel for RPM.

RPM Formulae—Long and Short Version The short is derived from the long. Both produce acceptable results with only 5 percent difference in RPM. That's well within acceptability, since initial RPM is a ballpark (best estimate) process anyway.

When working in the inch system, surface speed is entered in feet per minute while diameters are given in inches, which requires the factor of 12 to equalize units. This formula is derived from the circumference of the tool rotating in one minute:

> **Long RPM Formula (Diameters in decimal inches, PI = 3.14159)**
>
> $$\text{RPM} = \frac{\text{Surface speed} \times 12}{\pi\ (\text{PI}) \times \text{Diameter}}$$

Example calculation: A $\frac{3}{8}$-in. drill cutting mild steel. Turn to Appendix IV to obtain the surface speed.

$$\text{RPM} = \frac{100\ \text{F/M} \times 12}{3.14159 \times 0.375\ \text{DIA}}$$
$$= 1018.6\ \text{RPM}$$

Now compare this result to the Appendix Drill RPM Chart at 1066, just slightly faster. The long formula is the version you would program into your calculator and the one found in CAM softwares.

The short version shown next, is useful because it can often be performed in your head. It's derived by shortening PI to 3.0 not 3.14159. That reduces the formula to

> **Short Formula (Diameters in decimal inches)**
>
> $$\text{RPM} = \frac{\text{Surface speed} \times 4}{\text{Diameter of rotating object}}$$

Now, resolve for the $\frac{3}{8}$-in. drill example using this formula. Try it in your head first.

It's 400 divided by a number slightly less than 0.4—the result will be close to, but slightly larger than 1,000.

$$\text{RPM} = \frac{100 \times 4}{0.375}$$
$$= 1066\ \text{RPM}$$

Metric Short Formula If the spinning tool diameter (or lathe workpiece) is dimensioned in millimeters and the feed rate in meters per minute, substitute 300 for the 4 in the preceding formula.

> **Metric Short Formula (Diameters in mm and SS in meters/minute)**
>
> $$\text{RPM} = \frac{\text{Surface speed} \times 300}{\text{Diameter of rotating object in mm}}$$

Reamer RPM

In theory, the reamer RPM should be the same as an equal-sized drill. Since both tools are HSS and the diameter difference is very little, the same surface speeds and resulting RPM should be nearly the same. However, in real practice this is usually not so.

Reaming brings the additional challenge of chatter due to the thin chip they produce and the reamer's multiple flutes. To stop reamer chatter, we usually slow them to less than half the drilling RPM; sometimes even lower. The goal is to find the nearest RPM to the correct speed, that does not chatter. If the reamer does vibrate, the hole will be oversize and of poor finish.

Drill Feed Rates

Drill infeed, the **feed rate** at which the drill advances into the work, is expressed in *thousandths of an inch per revolution* or in *decimal parts of a millimeter per revolution*. When pulling the feed lever by hand, infeed on a standard manual press is a matter of feel, sound, and observing the chip—all quickly learned with practice. Feeds on power presses and on CNC machines must be selected before drilling.

> **KEYPOINT**
>
> Drill feeds range from a light 0.0005 in. (0.1 mm) per revolution to a heavy 0.040 in. (1.0 mm) per revolution, or even higher on large radial arm machines.

Selecting the feed rate is a three-way balance between tool life, cycle time, and machine finish. Here's a list of the factors to consider:

1. **Depth of Hole—Ability to Deliver Chips Out of Hole**
   As the hole gets deeper, the removal of chips becomes a problem. Often, you must slow the feed or begin peck drilling at a given depth.

2. Material Type

Some materials produce chips that tend to feed up and out of the drill hole easily: cast iron, free-cutting steel, hard aluminum, and some brasses are good drilling materials. On the other hand, softer aluminum and the tougher stainless steels are difficult to drill because the chips tend to build up in the flutes of the drill.

3. Coolant Availability

Flood coolant (a stream pumped out by the machine) not only reduces friction during the rake phase, but it also lubricates the chip's passing up and out of the flutes. If a flood stream isn't available, typical of most small drill presses, then a cup of cutting oil brushed on the tool is an adequate second choice, or a spray bottle with synthetic coolant can be used.

4. Size of Drill

As shown in the sample chart coming up, smaller drills require lighter feeds to avoid breakage.

5. Strength of Setup

6. Expected Finish and Accuracy of Hole

Pushing harder causes rougher, less accurate holes.

Drill Feed Rate Chart

Drill Size	Feed Rate (Range) in IPR
$\frac{1}{16}$ to $\frac{3}{16}$	0.001 to 0.004
$\frac{3}{16}$ to $\frac{3}{8}$	0.002 to 0.006
$\frac{3}{8}$ to $\frac{1}{2}$	0.003 to 0.008
$\frac{1}{2}$ to $\frac{3}{4}$	0.004 to 0.010
$\frac{3}{4}$ to $1\frac{1}{2}$	0.005 to 0.015

Student note: Remember, these are moderate rates.

UNIT 10-3 REVIEW

Review the Key Terms

Empirical data

Data collected by experimentation rather than calculation. All speed and feed charts are empirical.

Feed rate

Cutter advancement through the workpiece, expressed in inches of decimal parts of a millimeter, per minute.

Surface speed

The speed in feet per minute or meters per minute of the drill rim.

Replay the Key Points

- There is a "best" surface speed for any material being drilled.
- Different diameter drills must be set to turn at different revolutions per minute—such that the rim is at the correct surface speed.
- Smaller drills require *faster* RPM's while larger drills must go *slower* to achieve the same surface speed at their rim.
- Drill RPM and feed rate charts assume an HSS tool. If the tool is carbide, rates may be increased.
- Reamer RPM is often adjusted to half the drill RPM to stop chatter.

Respond

Calculating Speeds and Feeds

Unless directed otherwise, find surface speeds in Appendix IV.

1. Use the long formula to solve correct RPM for a 2.0-inch diameter drill in brass.
2. Use the short version (no calculator): a $\frac{1}{2}$-in. drill in annealed stainless steel.

Critical Thinking Problems

3. A line in a program includes the code *S400*, meaning the programmed RPM is to be 400 (spindle at 400). It's written to drill $\frac{1}{2}$-in. hard stainless, using a $\frac{7}{8}$-in. drill. Is it correctly written? *Assume a cutting speed of 70 FPM for this material.* Use the short formula to analyze this in your head before calculating it.
4. Use both formulae to compute RPM. Then compare percentage difference between the faster and slower result: Letter "A" drill in aluminum?
5. Calculate the RPM for a 12-mm drill cutting steel at 30 m per minute surface speed.

Extended Reference

6. Using the short formula, and *Machinery's Handbook* or CD to find recommended surface speed, compute the RPM for a Letter J Drill, cutting red brass. (*Hints:* The main ingredient in brass is? (Unit 5-2). You need speed and feed tables for drilling in order to access the correct surface speed.)

7. Using *Machinery's Handbook* data and the long formula, compute the correct RPM for a 2.0-in. drill cutting AISI alloy steel, Type 4140 (a chrome-molly steel discussed in Section 15). This material is in it's softest annealed state.

8. Still drilling the material in Question 7, you must now switch to a 36-mm diameter drill. It should be turning at _____ RPM. Which formula should be used?

9. A 0.625-in. access hole must be drilled and reamed in the side of a cast iron transmission case. On a separate piece of paper, fill in the setup facts listed next. Use only data found in this book and the short RPM formula.

   Reamer drill diameter

   Reamer drill RPM

   Drill feed rate range

   Reamer RPM

10. Using *Machinery's Handbook* and the long formula determine speed and feed for a 1.0-in. drill, cutting 1010 alloy, plain carbon steel in its annealed (softest) state.

    Recommended surface speed

    Drill RPM

    Drill feed rate range

Figure 10-44 *A CNC machining center can position, center drill, drill, and ream to tolerances below one thousandth of an inch.*

KEY TERMS

Center-finder
An adjustable pointer used to pin point, the exact center of a rotating spindle.

Core drill
A multiflute cross between a drill and reamer, that must follow a pilot hole. Used mostly to drill out rough cores in castings.

Counterbore
A round opening surrounding a hole, usually to receive a bolt head. See also spotfacing.

Countersink
A cone-shaped opening at the entrance of a hole, or the tool that produces this shape.

Engagement (thread)
The amount of contact between a bolt and nut, compared to the theoretical 100 percent; 75 percent engagement is common.

Gun drills
Specialized tools for drilling extra deep holes.

Spade drill
A flat cutting tool with two cutting edges, used on more rigid machines than drill presses.

Spotface
A smooth surface around a hole, created for a washer or nut.

Step drills
Drill press tools that cut multiple features and diameters at one time. May create angular transitions between diameters.

Wiggler
A set of utility holding devises that can be used as an indicator holder, an edge-finder, and a center-finder.

UNIT 10-4

Drilling Per Print and Secondary Operations

Introduction: Now we investigate producing the hole in the right position on the work, within the size tolerance (Fig. 10-44). We'll explore a couple of standard drill press methods in Unit 4, where we move or tap the part into alignment with the spindle. One produces a repeatable location accuracy of 0.030 in. and the other around 0.010 in. But that's not good enough for many jobs.

But once the tool is positioned correctly over the part, there still remains the challenge of getting the drill to bite in and drill at that location. Even when the part is perfectly aligned and clamped down, drills tend to wander if the upcoming steps are skipped.

Two Drill Press Location Methods

The first method is fast but not very accurate. The second improves accuracy but takes more time to perform. The best accuracy for either method starts with good, sharp layout lines with high contrast between the line and dyed metal. Since you'll be using your eyes to determine alignment in both cases, good lighting is also very important.

Layout/Punch Method— Repeatability Approximately 030 in.

Step 1   Layout the Hole Locations

Always reference part datums as a basis for the layout lines (if they are shown on the drawing) (Fig. 10-45).

Step 2   Pilot Punch (Prick Punch)

Use a very light tap of the pilot punch while holding it perpendicular to the work surface (Fig. 10-46). After checking the accuracy of the mark, a second slightly harder tap is then performed.

Step 3   Center Punch

Follow the pilot punch with a broader deeper center punch mark (Fig. 10-47). Again, be certain the punch is held vertically. Remember, don't do any of this hammering at the layout table!

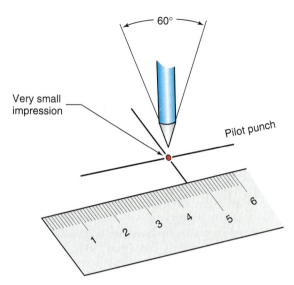

Figure 10-46 *A pilot (prick) punch creates a small, 60° indentation at the layout location.*

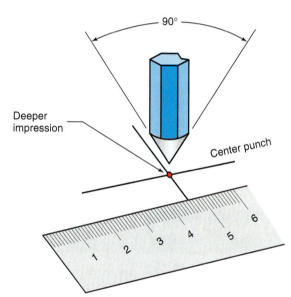

Figure 10-47 *Following the pilot, the 90° center punch drives the mark deeper and wider.*

Step 4   Visually Locate and Spot Drill to Test Location

Before pilot drilling, test the work location to be certain the drill is lined up over the layout lines. With the work lightly clamped in place, a center drill is mounted in the drill chuck and spun slowly by hand as the part is tapped into alignment with a plastic hammer. Tighten the clamps a bit more when the work appears directly beneath the spindle.

Then spot the location. Turn on the spindle and very lightly touch the work surface with the center drill's pilot. This is not a drilling operation—do not remove metal yet. Stop

Figure 10-45 *The layout must be clear and accurate.*

and read the shiny round spot created. If it's right on target, a bright ring will be created around the center punch mark. Now tighten the work clamps one last time and center drill to the depth shown in Fig. 10-49.

The objective of a spot drilling is to remove no metal just layout dye—at the exact place where the center will start when fed down.

Center drills are best for this operation as they are short, rigid tools. They feature a precision ground body and a very small pilot on their tip. Center drills were originally designed for lathe work, which we'll explore later, but they work best for spot drilling the crossed layout lines in Fig. 10-48. Center drills are supplied in four sizes #1 smallest to #4 largest. On the drill press we generally use a #2 unless the pilot is bigger than the intended drill diameter.

Trade Tip

A center drill's RPM should be computed based on the tiny pilot. As such, it will require a high RPM to ensure good location accuracy. However, the correct RPM will often exceed the drill press's range of speeds. When it does, select the highest possible RPM and push down very gently on the quill feed lever. Center drills are harder than drill bits and the tiny pilot breaks easily.

Figure 10-48 *To spot the location, touch the work surface with the spinning center drill, then read the marks.*

Correct center drill depth not to the full diameter

Figure 10-49 *A center drill does not drill below this distance.*

Step 5   Pilot Drill to Correct Depth

Change to the pilot drill then follow the center drill dimple with the pilot. Drill deeply enough (half way through the work or so) such that the next drill will tend to follow it (Fig. 10-49).

Step 6   Drill to Size

Check the sharpness and geometry of the drill before chucking it. Set the right RPM on the press then test the location by gently touching the hole rim with the spinning drill bit.

Trade Tip

Pulling a Mislocated Drill Back on Location A started drill that has wandered off location can be pulled back toward the correct location with this trade trick. But it only works, if you detect the problem before drilling downward to the drill's full diameter (Fig. 10-50).

With a hammer or chisel create deep marks in the angular cone surface that is the started hole. The marks are made on the side to which you wish the drill to return. They catch on the cutting edge and draw the drill toward themselves. They can bring the drill location back on target.

Punch marks draw the start back toward correct location as it drills deeper

Correct location

Started cone shows the drill is wandering off location

Figure 10-50 *It's possible to pull a mislocated drill back on location, if the start isn't yet at full diameter.*

Drill Press Method 2— Layout/Center-Finder Pickup—Repeatability Approximately 0.010 in. or Better

This method uses an alignment tool, the **center-finder,** to refine location (Fig. 10-51). A center-finder is a two-piece pointer whose tip can be gently tapped or nudged onto a wobble-free, perfect center on its rotating base. Theoretically, a precision ground, solid metal pointer could be used, but in practice, every chuck and spindle have some wobble. The center-finder eliminates any hole misalignment from spindle or chuck wobble.

Some steps and photos are omitted because they duplicate steps in method 1.

> To use the center-finder correctly, no punch marks are made in the layout.

Step 1   Layout

Step 2   Rough Position the Work
Add the holding method tighten lightly only.

Step 3   Center-Find Location With the center-finder held lightly in the drill chuck, set the drill press at a moderate RPM from 400 to 900 maximum. If the RPM is set too fast, the center-finder can be damaged by centrifugal force. The adjustable point is held on the body by a spring. It can be stretched beyond usefulness if the point is flung out from excessive RPM.

Center the Point Lightly nudge the wobbling tip until it settles on the exact axis of the spinning spindle. Use a pencil, rule, or other small but hard instrument for this purpose.

Step 4   Align the Work Under the Rotating Center-Finder With the spindle rotating, bring the quill down until the point is approximately $\frac{1}{16}$ in. from the layout surface. With light taps to the workpiece with a dead blow mallet or plastic hammer, move the part until the point is right over the intersection of both layout lines (Fig. 10-52). Be sure to position your eye in direct line with the X direction line then the Y, to avoid parallax error.

> The center-finder never touches the work. *It does not leave a mark!*

Step 5   Clamp the Work in Position
Use the center-finder again to see if the clamping has disturbed location.

Step 6   Spot Drill to Test Work Alignment

Figure 10-51 *A center-finder is used to improve spindle location over layout lines.*

Adjustable edge-finder used on milling machines

Body for holding in a chuck

Adjustable center-finder point

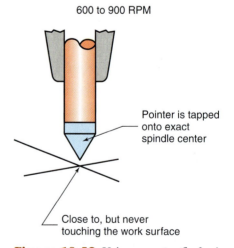

600 to 900 RPM

Pointer is tapped onto exact spindle center

Close to, but never touching the work surface

Figure 10-52 *Using a center-finder is easy.*

Drilling and Taping Threaded Holes on the Drill Press

Selecting the Correct Tap Drill

Recall from Section 5, that there is a designated drill for a specific tap. If not predrilled to the correct size, the tap will break if the hole is too small while a hole too large will produce weak or nonexistent threads. When you read a print callout such as:

Drill and Tap $\frac{1}{2}$-13 UNC

or

$\frac{1}{2}$-20 UNF

or

M8-1.25 ISO

it's specifying not only a certain tap, but also a specific drill size. That size selection is found on the drill chart or in standard machinist's references.

Using a Tap Drill Chart

Turn to Appendix I, the drill chart. What is the correct size drill for a $\frac{1}{2}$-13 UNC tap?

Answer: a $\frac{27}{64}$-in. drill

Now look for the tap drill for a $\frac{1}{2}$-20 UNF tap.

Answer: $\frac{29}{64}$ in. (0.4531 in.)

KEYPOINT

Even two different pitches of the same size thread require different tap drill diameters. Every thread designation has it's own tap drill size.

Why the Specific Tap Drill Size?

You may see a note on some tap drill charts that the recommended drill produces a 75 percent **engagement** thread. That's the standard fit. Here's what that means. The drill size is calculated such that a bolt threaded into the correctly drilled and tapped hole will touch on 75 percent of their thread faces; 75 percent of what?

ShopTalk

Wrong Tap Drill A common error made by trainees is to choose the drill size equal to the thread size. For example, to drill a $\frac{1}{2}$-in.-diameter hole for a $\frac{1}{2}$-13 tap. If you do this, the tap drops through because there is no metal left for the threads.

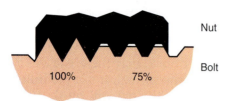

Figure 10-55 *The correctly chosen tap drill produces internal threads with a 75 percent engagement.*

A theoretical thread, with pointed triangle tops and zero mechanical clearance between the bolt and nut, would produce a 100 percent engagement. But, as we've discussed before, the thread form triangle is truncated, clipped at the tiny points. Using the recommended chart diameter forms the truncations on the inner diameter of the thread, as shown in Fig. 10-55.

Tapping Methods

There are three common methods used on drill presses for tapping threads.

Hand turning   Power turning
Tapping heads

Hand-Turning the Tap Using the Drill Press and a Tap Guide

This method is used for single or a limited number of tapped holes (Fig. 10-56). To begin, a pointed center is mounted in the chuck. It will be used as a guide to keep the tap perpendicular to the work. The spindle motor is not used in this method. With one hand on the quill lever to maintain light down pressure against the tap, turn the tap handle with your free hand. Be sure to use a tapping compound or cutting oil.

Turning the Tap by Power

This operation is accomplished on a gear head or radial arm drill press, with a reversing

Tap guide

Hand-tapping wrench

Pilot hole in tap

Figure 10-56 *Hand-turning the tap in the drill press is useful for one or two parts.*

Figure 10-57 *Tricky to do, this skilled machinist has the tap mounted in a chuck, within a reversing gear head press.*

spindle and lower speed ranges (Fig. 10-57). Standard drill presses cannot be used this way.

Safety Precautions Here the press screws the tap in and out while you keep light pressure on the feed lever. When power tapping on a drill press, be sure you do not accidentally engage the power feed when the tap is in the hole. The power feed will advance the quill down at a different rate than the tap must lead into the work. Pressure will instantly build up and the tap will break or the part must move to accommodate the different advancement rates.

Another factor to control is the fact that the spindle of most gear drive presses will coast some distance with the power turned off. When nearing the bottom of the threads, be sure to allow for the extra distance the tap will advance while the spindle decelerates and reverses. Power tapping requires practice and a demonstration by your instructor! The chances of breaking the tap are near 100 percent, at least for trainees! This method would be chosen to quickly tap one or two holes, but would never be used for long production of many tapped holes if a tapping head was available.

Tapping Head Attachments

Tapping heads are good solutions for driving taps on any drill press with or without reversing spindle motors (Fig. 10-58). When many parts must be tapped, this attachment

Figure 10-58 *A tapping head adds several advantages when used in a standard or gear head drill press.*

◆ Adjustable torque friction that allows the tap to slip rotationally in the holder if it jams during the cut. (Not found on all tapping heads.)

Counterboring and Spotfacing Operations

Two specialty operations are performed on the drill presses using cutting tools designed for this purpose (Fig. 10-59). These operations follow a drilled hole and produce a flat, smooth, circular surface around the hole. They are different terms for the same operation. The depth of the cut is their only difference. A **counterbore** is a deep cut, while a **spotface** is a shallow cut. The cutting tools are correctly called either a counterbore or spotface.

In Fig. 10-60, the computer symbol for counterboring/spotfacing is a wide "U" and the depth is the down arrow symbol. The computer symbol is shown in the upper callout, while the older written version of the callout is below.

Spotfacing a shallow smoothed circle for a washer or bolt head is usually required around a bolt hole in a rough casting. Spotfacing depth is commonly called out as in the illustration as *minimum clean up* or sometimes abbreviated to C/U.

Counterboring is machined to a controlled depth. A typical use of a counterbored hole is to receive the head of a bolt below the surface of the part. When counterboring, you must set a quill stop to achieve depth accuracy.

There are three versions of counterbores. Some are single-piece cutters with a solid pilot, but the more popular version features interchangeable pilots (Fig. 10-61). By exchanging the pilot, these counterboring tools can follow a range of hole sizes. The third tool that counterbores is a custom-made step drill (coming up next). They are mainly used

prevents tap breakage and operator fatigue. It turns any drill press into an efficient production tapping machine.

A tapping head adds four or five features to a standard drill press:

◆ Geared down RPM—safer for tapping.

◆ Quick reversing mechanism—As the tap reaches its lower stroke limit, the operator holds back on the quill lever arm and the head instantly shifts to reverse without reversing the motor on the press.

◆ Efficient collet type holder for taps (not a chuck)—Often the collet is rubber isolated such that it absorbs much of the shock that occurs during tapping, to avoid breakage while rotating and keeping the tap centered in the spindle.

◆ A "floating" tap holder mechanism—A major advantage of tapping heads, this feature permits the tap to move up and down in the spindle but not to move sideways. This results in some forgiveness to over-run of the spindle motor and operator lack of skill.

Counterbore with inter-changing pilot
(also called a spotfacing tool)

Multi-fluted core drill

Double ended centerdrill
(also combination drill and countersink)

Step drills-many shapes

Countersinks

Back spotfacing tool
for machining inside walls

Turn lock mandrel

Figure 10-59 *Secondary drill press tools you'll need to recognize, some used for CNC operations.*

⌴ 1.00 ⌴ 0.750 ↧
⌴ 1.00 ⌴ min C/U

Spotface 1.0 dia
minimum clean up

Counterbore
1.0 dia × 0.75 deep

Figure 10-60 *Compare the CAD symbol (top) for spotface/counterbore to the previous word call-out below.*

to drill and counterbore standard bolt and head sizes all in one operation.

All three feature a central pilot that ensures that the cutter stays centered in the hole. The pilot also ensures that the zero degree lead on the cutting edges doesn't wander or grab as the cutting edge touches the work surface.

Multiflute counterbore

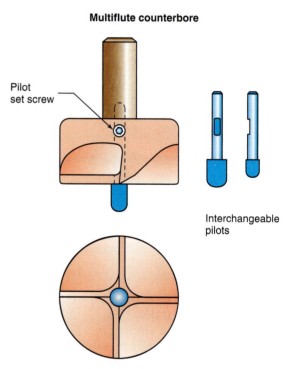

Pilot
set screw

Interchangeable
pilots

Figure 10-61 *Counterbores are also used for spotfacing.*

Countersinking

This is a drill press operation that produces a cone-shaped opening around a drilled hole, for several reasons:

◆ To remove sharp edges and burrs (recall the term: *breaking the edge*)

Figure 10-62 *Countersinks: closed face (left), open faced, and micrometer stop (right).*

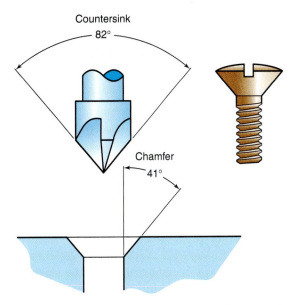

Figure 10-63 *An 82° countersink produces a 41° chamfer.*

- ◆ To provide a starting slope for tapping threads
- ◆ To finish the raw edge of a hole for functional requirements or for appearance.
- ◆ To provide a conical nest for a flat head screw (as shown shortly in Fig. 10-63)

Countersinking is similar to spotfacing. The tool must be turned at a lower rate than normal cutting speed (50 percent or less) to avoid chatter. The closed-face countersink (Fig. 10-62) is the least prone to chatter and tends to produce the better finish, but since it has only one cutting edge, it doesn't last as long as the others before regrinding is necessary.

Eliminating the Confusion— Countersinking versus Chamfering

You might see either term used on a print (Fig. 10-63). While they are related, they are different terms. A chamfer is a small angular cut at the intersection of two surfaces and it can be around a hole entrance.

KEYPOINT

A chamfer is specified as the angle of the chamfer as compared to a vertical axis line. A countersink is called out as to the included angle of the cone.

The two common included angles for countersinks are 82° and 90°, although others are used in industry. Flathead and oval-head screws (Fig. 10-63) are made with 82° shoulders.

KEYPOINT

An 82° countersink produces a 41° chamfer around a hole.

Micrometer Stop Countersinks

When countersink depth must be controlled within a tighter tolerance than is possible using the quill stop, on a drill press, we use the micrometer stop countersink (Fig. 10-64). It allows the countersink to extend down through the contact skirt to a preset depth. This tool is extra useful where material thickness varies or

Figure 10-64 *A micrometer stop countersink accurately controls depth.*

is rough, yet the countersink depth must be held to a close tolerance. Countersinks can also be used away from the drill press, in hand drill applications.

Specialty Setups

You'll soon come to appreciate that making many setups is similar to constructing a Lego® block masterpiece. You must work with basic components to assemble something custom fit to the part shape and the task at hand. I believe setups are the fun stuff of machining. Here are a few suggestions that work on a drill press.

Drilling Though Round Work

The challenge is to get the drill to start on the center of the round surface. First, a pair of layout lines is needed. One from the end of the object, 0.750 in. in the example (Fig. 10-65).

Next, a centerline is laid out with the work clamped in the vee block but with it lying on its side. Touch the upper surface of the round work with the scribe. Now, with the work clamped in the vee block, turn it upright and set it in the drill press vise. The marked intersection will be vertical and ready for center-finder location.

Note that a second vee block is supporting the work on both sides of the hole. The pair of vee blocks can be held in a drill press vise, Fig. 10-66 or clamped directly to the table.

Figure 10-66 *Supporting the work in two vee blocks for drilling.*

Drilling Holes at an Angle to a Work Surface or Datum

Here are two ways to drill at an angle. The simple method is to rotate the drill press table as shown in Fig. 10-67, if the drill press is so equipped. If not, the work is tilted and held at the correct angle, in a tilting vise.

The challenge in either setup is to start a drill location accurately on a tilted surface (Fig. 10-68). The bit tends to walk down hill before it bites in. To prevent this, start with a number 1 center drill set at a high RPM. Lightly touch the surface with the drill, then back it away immediately. Repeat this action with a series of light pecks, each just a few thousandths deeper than the last. Once the

Radius distance down from top

0.750

Figure 10-65 *Laying out a centerline starts the process.*

Tilting table   Tilting vise

Figure 10-67 *Two angle drilling setups.*

Figure 10-68 *Starting a drill on location on an angular surface, without wandering.*

Extremely light pecks very high RPM

entire pilot of the drill has entered the work, you can increase the force.

If the surface is slanted beyond 15° relative to the drill's axis, nothing can prevent wandering, especially if the work material is hard. The illustrated surface is at 12°, and you can see the problem—without a pilot hole, the drill is going to slip to the right. Depending on the kind of material, hard or soft, even that much slant might not be possible to drill accurately on a press unless a drill guide bushing can be set up to prevent wandering.

In these cases, we move the job to a vertical milling machine with a more rigid spindle. There, a circular landing spot can be created that's perpendicular to the intended hole's axis—called prespotting the hole. The cutter is a rigid, flat-bottom, two-flute tool called an end mill. It's similar to a drill with 0° point angle (Fig. 10-69). We'll study these cutters in detail in Section 12.

Specialty Drills

There are four types of drills beyond the common twist drill that you might encounter in your first year of employment but not likely in tech school. The spade drill, the gun drill, the step or pilot drill, and the core drill. Here is a brief description of each.

Spade drills are used to drill large holes above 2 in. in diameter or for odd sizes. A spade drill can remove large quantities of metal at a high rate. It has a removable cutting edge that can be carbide or HSS. These drills are not common on standard drill presses due to need for rigidity, but they are used in production CNC turning, on larger radial arm presses, and occasionally on manual lathes. Spade drills can drill deeper than twist drills because of their chip removing ability.

Very precise changes in hole diameter are possible by custom grinding the inserted cutter (Fig. 10-70).

Gun drills are specialized tools and processes designed specifically for drilling extra deep holes. When drilling deeply, standard twist drills tend to wander into a curved axis—and they eventually break in the hole. Also, standard twist drills are poor at removing chips at any great depth because of the increasing friction between chip and helical flute.

A gun drill corrects these problems by using a three-point suspension head that supports itself as it cuts Fig. 10-71. A gun drill has a single carbide cutting edge and two support pads spaced at 120° from the cutter. They also feature a straight flute that better sends chips out using pressurized cutting oil or fluid injected down to the tip, to force the chips

Prespot level for pilot hole

Two flute end mill

Tilt work

Figure 10-69 *Sometimes it's necessary to create pre-spots to allow the drill to bite on location.*

Spade drill

Chip relief

Two cutting edges

Interchangeble carbide or high-speed cutting tips

Figure 10-70 *An insert cutter, spade drill.*

Gun drill

Figure 10-71 *A gun drill can cut extremely deep holes.*

Typical step drill shapes

Figure 10-74 *Various step drills.*

out. The fluid is pumped through a gallery (oil hole) that runs through the tool's body. It is injected right to the cut.

Gun drilling can be accomplished on regular lathes using special oil injection pumps and attachments. However, if a lot of gun drilling is needed in the shop it is usually assigned to specialized machines having the ability to pump the high-volume cutting fluid and retrieve it into the sump again (Fig. 10-72).

Core drills are a combination of a drill and a reamer (Fig. 10-73). The main purpose is to machine holes in castings that already have a rough cast hole—called a *cored hole*. A core drill features a very strong web and has several flutes (three, four, six, or more for larger tools). It will not start a hole without a core or a pilot hole in the metal. These drills tend to produce a fine finish, often nearly as good as a reamer.

Step drills are designed to accomplish two tasks at one time (Fig. 10-74). They can drill and chamfer, drill and counterbore, or any combination of pilot, drill, counterbore, and chamfer, depending on their ground shape.

Figure 10-72 *A gun drilling machine in industry.*

Multi-Flute Core Drill

Figure 10-73 *A core drill is designed to follow rough holes in castings. It resembles a reamer but with deeper flutes.*

UNIT 10-4 REVIEW

Review the Key Terms

Center-finder
An adjustable pointer used to pin point, the exact center of a rotating spindle.

Core drill
A multiflute cross between a drill and reamer, that must follow a pilot hole. Used mostly to drill out rough cores in castings.

Counterbore
A round opening surrounding a hole, usually to receive a bolt head. See also spotfacing.

Countersink
A cone-shaped opening at the entrance of a hole, or the tool that produces this shape.

Engagement (thread)
The amount of contact between a bolt and nut, compared to the theoretical 100 percent; 75 percent engagement is common.

Gun drills
Specialized tools for drilling extra deep holes.

Spade drill
A flat cutting tool with two cutting edges, used on more rigid machines than drill presses.

Spotface
A smooth surface around a hole, created for a washer or nut.

Step drills

Drill press tools that cut multiple features and diameters at one time. May create angular transitions between diameters.

Wiggler

A set of utility holding devises that can be used as an indicator holder, an edge-finder, and a center-finder.

Replay the Key Points

◆ The objective of spot drilling is to remove no metal just layout dye—at the exact place where the center will start when fed down.

◆ No center punch marks are made when picking up hole locations from layout using a center-finder. The center-finder never touches nor marks the workpiece surface.

◆ Two different pitches of the same size thread require different tap drill diameters. Every thread designation has its own tap drill size.

◆ A chamfer is specified as the angle of the chamfer as compared to a vertical axis line.

◆ A countersink is called out as to the included angle of the cone.

◆ An 82° countersink produces a 41° chamfer around a hole.

Respond 10-4

1. Using Appendix II, what is the correct tap drill for a $\frac{5}{16}$-24 UNF thread.

2. True or false? Two common methods of locating a part exactly on the axis of a drill press spindle are the center-finder (or wiggler) and the layout punch method. If it is false, what will make it true?

3. Name a third, more accurate method of locating a drill on a standard press.

4. Draw a sketch or describe the method used to "pull" a drill back toward the right location if found that it has wandered off position. At what point in the drilling can this be accomplished?

5. True or false? To mount a Morse taper-shank reamer in a drill spindle requires a shell adapter. If it is false, what will make it true.

6. Name two methods of tapping by power in a drill press capable of reverse rotation.

7. Draw a sketch or describe the colinear method of aligning a drill bit to a layout line. What tool and processes would you need?

Figure 10-75 *Identify the three drill press tools from left to right.*

8. Name the operations, in the sequence needed to *ream a hole* in undrilled material (the punch mark method).

9. True or false? Machining a chamfer around a hole is called flatspacing. If it is false, what will make it true?

10. What is the difference between a *spotface* and a *counterbore*?

11. Identify two drilling methods for cutting extra deep holes.

12. Of the above two methods, which requires a special machine or attachment and why?

13. Identify the tools shown from left to right in Fig. 10-75.

UNIT 10-5

Sharpening Drill Bits

Introduction: While there are easy-to-use manual sharpening jigs and automatic machines available to sharpen drills, (Fig. 10-76), we will concentrate on hand-sharpening drill bits. This is an ability no machinist should be without. You'll use it on the job and in the home workshop, too. More than useful, hand grinding a drill is one of the fastest and least inexpensive ways to increase tool geometry knowledge.

But, there's an even greater reason for Unit 10-5. The techniques and hand coordination skills required to grind a drill bit will apply to touching up or hand-grinding many other

Figure 10-76 *A drill sharpening machine in industry.*

tools in the shop. We try to avoid hand-ground tools in production machining, but occasionally, they become the only short-term solution to keep the machine moving. Handmade tools are more common for the tool and die industry and especially, the mold maker. Even today in an era of CNC tool-grinding machines, every so often, the only way to produce some wild-shaped tool is to do it by good old hand skills.

Before sharpening a bit, it's important to recognize if it's time to grind it or not. Grinding reduces its shop life span. While nonproductive down time can be created by the machinist stopping too soon to sharpen or replace a dull tool, waiting too long, pushing the tool beyond normal wear, usually results in disaster both for the workpiece and the cutting tool!

KEY TERMS

Dead center (drill)
The noncutting portion of a drill point—caused by the central web.

Overtempering
Reducing the hardness of any HSS tool bit by heating it during grinding.

Scatter shield (upper grinding wheel guard)
The metal grinding wheel guard above the grinding wheel, designed to contain a broken wheel if it explodes.

Recognizing a Dull Bit

There are two ways to detect a dull bit.

Inspection before Mounting

Check the cutting edge especially at the outer corners and along the margins near the cutting edge. Look for chips in the cutting edge and rounded corners at the margins. But not all dull drills are so obvious. So, for the normally worn bit, hold it up to catch the light. The cutting edge will appear as a bright line. Compare a sharp and a dull drill—you'll see the difference (Fig. 10-77).

Observing Signs while Drilling

There are five signals that indicate the bit is failing.

◆ Excessive noise—squeal, vibration, and chatter.
◆ Chips jamming in the flutes. This can also be caused by deep drilling, lack of coolant, and/or wrong speeds or feeds.
◆ Yellow, brown, or blue chips emerging from the hole when drilling steel. Discolored chips can also be caused by excessive speed or feed rates but whatever the cause, it will soon lead to a dull drill.
◆ The drilled hole measures too large.
◆ Excessive heat—steam if using coolant or smoke if using cutting oil.

Not recognizing any of these symptoms can result in a ruined tip or total destruction of the bit (Fig. 10-77).

Damage

Figure 10-77 *This drill cannot be easily reground. Notice the margin damage.*

Regrinding the Bit

There are two features to be controlled while sharpening a normally worn standard twist drill:

1. **Point Angle (lead angle)**
   Both cutting edges are of equal length and angle.
2. **Clearance Angle**
   Six to ten degrees.

Holding the Bit for Sharpening

As shown in Fig. 10-78, hold the bit with the cutting edge *horizontal* (parallel to the floor) with the cutting edge facing up.

KEYPOINT

The cutting edge will remain horizontal the entire time it's in contact with the grinding wheel.

Your right hand supports the front of the bit and your left is on the shank. Note, hand positions can be reversed by preference.

Trade Tip

Hand Position Practice Here's a safe "dry run" (CNC term meaning to try out the operation or program without cutting metal) to develop the hand position and movements needed to sharpen a drill. Use a large bit around a half inch or bigger that is already correctly ground.

Place the cutting edge at position 2, at the top of the grinding stroke, against the wheel as shown in Fig. 10-79—*Safety note!!! The grinding wheel is not spinning.* Now, while pressing the bit lightly against the wheel, lower the bit tip to position 1, as in Fig. 10-80, and in so doing carry the wheel down with the movement. As you do so, keep the cutting edge horizontal—do not twist it! Now, move the bit (and wheel) back to position 2. Repeat this motion, always keeping the cutting edge horizontal.

Safety Note—the eye guard is not illustrated in these figures for clarity, only; be sure to use this important guard in actual grinding practice. Repeating this up and down motion is the movement needed to grind the bit.

Why is the horizontal factor for the cutting edge, so important? Turn to Fig. 10-86 later in the unit to see why.

Figure 10-78 *Hold the cutting edge level to the floor during the entire grind motion. Guard removed for clarity.*

Safety Check List

☐ You have on safety glasses.

☐ Make sure the tool post is adjusted close to the wheel ($\frac{1}{16}$-in. to $\frac{1}{32}$-in. space)

☐ Both the eye guard and **scatter shield** (upper guard) are in place.

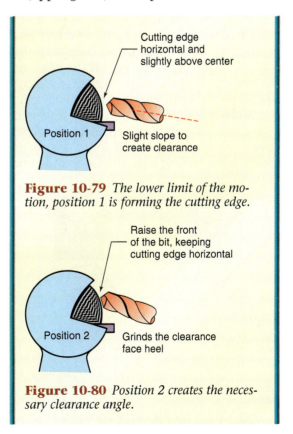

Figure 10-79 *The lower limit of the motion, position 1 is forming the cutting edge.*

Figure 10-80 *Position 2 creates the necessary clearance angle.*

☐ Make a visual inspection of the wheel for cracks or broken out chunks. Also inspect to see that the grinding wheel surface is straight and not clogged by soft metal.

☐ Make sure the water pot is ready—heat will be generated.

Now with the grinder running place the bit up at position 2 close to but not touching the wheel. Move it forward and with a light touch until very light sparks show up. Touching on the heel of the bit at position 2 ensures that you won't accidentally damage the cutting edge.

Then immediately lower the bit from position 2 down to position 1, keeping the cutting edge horizontal, then back up to position 2 again. Repeat the action several times, then stop and do the same on the other cutting edge. Do not force it, let it just touch the wheel at first to get the feel of the action.

KEYPOINT

The goal is to grind equal length cutting edges and equal angles relative to the drill's axis.

Control by Reading the Sparks

The spark pattern can be used as a visual control of the cutting edge angles you are producing (Fig. 10-81). Sparks will be equally distributed across the entire edge if the angle on the bit is being reproduced. To change the cutting angle, concentrate the sparks on one edge or the other until the surface is fully reshaped.

To control the clearance angle on the down stroke, you must stop at position 1. Do not go lower or the cutting edge will be rounded off

Figure 10-82 *Read the sparks to create a perfect cutting edge!*

Correct clearance   Rounded off cutting edge

Incorrectly rotated below position 1

Figure 10-83 *Lowering below position 1 creates a drill that will not cut!*

and the bit will not cut. The spark pattern can be used to gage when the drill is lowered to position 1 (Fig. 10-82). Pivoting the bit below position 1, rounds the cutting edge clearance and the bit will not cut (a common error made by beginners). See also Fig. 10-83.

At the instant the cutting edge is at position 1, sparks will peel off the upper side of the bit across the rake surface as a chip would. Before reaching position 1, the sparks will go past the cutting edge—straight down.

KEYPOINT

When the sparks peel off the top rake surface, you are at position 1—do not go any lower.

Trade Tip

A Visual Template of 59° If you have difficulty obtaining the 59° cutting angle, here is a hint that can help. Using a protractor, mark a line at 59° on the tool rest of the grinder. Sight down this guideline as you hold the bit. The correct angle will soon become natural to reproduce (Fig. 10-84).

Equal spark pattern   Changing the angle
indicates equal pressure

Figure 10-81 *Read the spark pattern to control the cutting angle.*

Figure 10-84 *A 59° guideline helps maintain the right angle.*

59°

Guide line drawn on tool rest

On center
unequal angles

Off center
equal angles

Figure 10-85 *Two typical drill point problems.*

Checking and Correcting the Shape

During the grind, use a drill point gage (Fig. 10-85) to check the cutting angle and centering of the point.

Wrong Angle

Two problems can arise. Different angles on each cutting edge causes early failure due to only one edge contacting the work and thus doing all the cutting. Equal angles but off center causes an enlarged hole (Fig. 10-85). The longer edge forms the radius of the hole.

Checking the Dead Center As discussed previously, the web of the twist drill creates an inefficient chisel that only scrapes metal away. If you have held the drill correctly during grinding, with the cutting edge horizontal, the web will be at it's shortest length. The angle will be approximately 115° relative to the cutting edge (Fig. 10-86). However, if you have rotated or tilted the cutting edge, the **dead center** appears as the bit on the right. The bit on the right can have its center thinned but the better fix is to regrind it.

Long Dead Center on Short Drills You may produce a dead center that looks like (Fig. 10-87). The angle is right relative to the

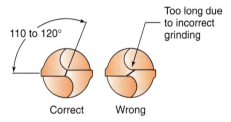

110 to 120°

Too long due to incorrect grinding

Correct

Wrong

Figure 10-86 *The dead center looks like this when the drill isn't held horizontal while grinding.*

Trade Tip

The enlarged hole made by an off-center point is sometimes deliberately used by machinists to produce a slightly larger hole than the drill. This "trade trick" will work only when there is no pilot hole. Without a pilot hole, the off-center drill pivots around the dead center, which produces a hole not quite equal to twice the longest cutting edge.

Keep the Bit Cool Keep dipping the bit in the water pot while grinding. Don't wait for it to get too hot to hold. Never let it discolor at all. If you do heat it up, the hardness can be reduced and will become dull more quickly. This is called **overtempering** the metal.

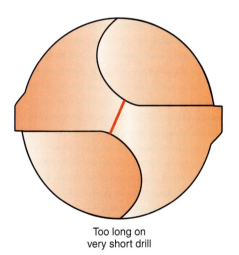

Too long on
very short drill

Figure 10-87 *A long dead center appears on a drill that's shortened from many resharpenings.*

cutting edges, but the length is too long. You did nothing wrong. For added strength, the twist drill's web is made progressively thicker closer to the shank. On bits that have been reground many times and have lost over half their total length, the dead center becomes extra wide. You must thin this dead center for all drilling circumstances (Fig. 10-87).

UNIT 10-5 REVIEW

Review the Key Terms

Dead center (drill)
The noncutting portion of a drill point—caused by the central web.

Overtempering
Reducing the hardness of any HSS tool bit by heating it during grinding.

Scatter shield (upper grinding wheel guard)
The metal grinding wheel guard above the grinding wheel, designed to contain a broken wheel if it explodes.

Replay the Key Points

◆ During drill bit sharpening, the cutting edge will remain horizontal the entire time it's in contact with the grinding wheel.

◆ The sharpening goal is to grind equal length cutting edges and equal angles relative to the drill's axis.

- When grinding a drill bit, when the sparks peel off the top rake surface, you are at position 1—do not go any lower.

Respond

1. Explain why you must hold the cutting edge of a drill bit horizontal during grinding.
2. When sharpening a drill bit, which face or surface is contacting the grinding wheel?
3. To produce a standard drill point, at what angle are you grinding the cutting edge, relative to the drill's axis?
4. How do you know when you are reproducing the angle on the bit or changing it during grinding?
5. On the down stroke toward position 1, how do you know to stop lowering the bit?

Unit 10-1

We learned that the drill and reamer combination are the best method to produce a hole that's round, smooth, and of consistent accuracy. We learned that there are two ways a drill or reamer is held in a drill press, either straight or taper shank, and we saw that there are several modifications that can be made to the drill point shape to achieve a special goal.

Unit 10-2

Standard drill presses divide out by their sizes and drilling heads. Starting with the light-duty, sensitive drill press used for extremely small holes, we looked at presses in advancing horsepower and size all the way up to the giant radial arm press big enough to drive a car onto the table. Perhaps the greatest demarcation line between presses is their ability to rotate their spindle forward and backward. Without that ability, a tapping head is required. We also looked at the ways to safely hold work on the press.

Unit 10-3

There are several ways a machinist might determine drill and reamer RPM: charts, calculating formally or in your head, from built-in speed calculators in CAM software, and from good old intuition based on experience. That last category will come in handy not only when setting up manual machines but when trying out a brand new program. A good operator must be able to see the tool's spin and know instantly whether or not action is required.

Unit 10-4

By now, if you've had any lab experience on the machines, you've seen that setups are a lot like working with Lego® blocks. Every new part shape and cut sequence brings some new puzzle to solve as to how to hold the object safely and reliably, using a few components such as vises, clamps, angle plates, parallel blocks, and such, we only touched on a few of the myriad possibilities. Along with planning the cut sequence, great setups are the most complex and challenging part of our trade!

Unit 10-5

Grinding a drill bit that looks right and works well takes practice. Since you undoubtedly have a machine or jig that also does the job, was it worth the effort to learn to grind them by hand? Yes. Even in the age of computer machining, there are many situations where a fully competent journeyman must be able to modify or quickly sharpen a tool by hand. Not just drills, we sometimes find ourselves whipping up a special lathe or mill cutter too.

Questions and Problems

Test your total recall about drilling. The answers follow the review.

1. True or false? The best material for an industrial twist drill is high-carbon steel. If it is false, what will make it true? (Obj. 10-1)

2. You see the following callout on a technical print:

   Drill and Tap $\frac{5}{16}$-18 UNC

   What drill bit must you select for this hole? (Obj. 10-1)

3. Now you have checked out the supposedly correct drill bit for tap drilling the $\frac{5}{16}$-18 UNC thread of Question 2, name the five remaining factors you should examine on the bit before mounting it in a press. (Obj. 10-1)

4. A Morse taper shank has a taper ratio of ____ inch per foot. (Obj. 10-1)

5. Explain why machine shops use tapered-shank drills for holes above ____ in., even though these tools are more costly? (Obj. 10-1)

6. Identify the drill bit features A through G in Fig. 10-88. (Objs. 10-1 and 10-5; and from previous discovery in Section 5)

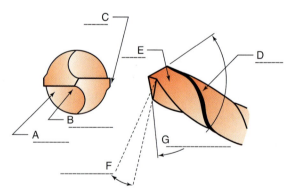

Figure 10-88 *Identify the main features of a standard drill bit.*

Figure 10-89 *Identify the drill press vise by its features.*

7. In Fig. 10-89, the same part is setup to drill in two different vises.
   A. Which vise is designed to be a drill press vise in Fig. 10-89?
   B. What features set drill press vises apart from the other type.
   C. For what safety reason, is it important to know the difference in the two vises shown?
   D. What's wrong with the setup, using the upper vise? (Obj. 10-2)

8. What two variables are used to enter a drill speed chart? (Obj. 10-3)

9. Use the short formula and the recommended cutting speeds here in the Appendix IV, to calculate the correct RPM for a 0.4375 in. ($\frac{7}{16}$ in.) drill bit cutting cast iron. (Obj. 10-3)

10. Use the long formula to solve Problem 9. Will the answer be higher or lower by how much? Use Appendix IV to access surface speeds. (Obj. 10-3)

11. Name the two different methods drilling a hole to a drawing location on a standard drill press. List each method's expected repeatability. Describe each briefly. (Obj. 10-4)

12. A center-finder can be used in two ways to pick up a location: As an accurate spindle *pointer* or as an *optical colinear* locator. Which method is the more accurate? What is the repeatability of each? (Obj. 10-4)

13. There are two methods of *power tapping internal threads* some drill presses. But, only one on a press that has no spindle reverse.
    A. Identify the tapping method that drives the tap in then out on presses without reverse?
    B. Which presses have reversing spindles? (Obj. 10-4)

14. True or false? A counterboring tool requires a pilot to keep it from chattering and to keep it concentric to the hole. A spotfacing tool needs no pilot since it cuts less deeply than the counterbore.
    If this statement is false, what will make it true? (Obj. 10-4)

15. A standard flat head screw has an included angle of ____ degrees. What countersink must you select to produce a cone shaped location for that screw head
    (A) 41°, (B) 82°, or (C) 59°? (Obj. 10-1)

16. When drilling soft brass or sheet metal, how can you modify the bit for safety? (Obj. 10-5)

17. After regrinding a standard twist drill, the dead center is inefficient.
    A. What modification can be made to the drills point geometry to improve its cutting action?
    B. What two improvements does this modification add to the use of the drill?
    C. Under what three conditions would this modification not be needed? (Obj. 10-5)

18. In Fig. 10-90, something is wrong with one or more of the bits. Name the problem and tell what the result will be if it is used to drill. Describe what needs to be done to correct each bit.

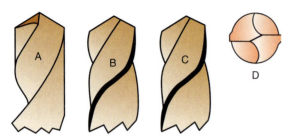

Figure 10-90 *Identify the problem with each drill, and the result.*

19. A new drill bit is checked out from the crib and it has a golden-colored plating on the cutting portion but not the shank. What might that added feature be for? See Fig. 10-91. The solution was not covered in the reading but you have the facts to solve it.

Figure 10-91 *Why would this coating be on the drill bit?*

CNC Questions

20. On CNC machines that perform drilling operations there are two kinds of cycles that aid in drilling deep holes. Name and describe them.

21. Solving this may require discussion with fellow students—but it is an important difference in CNC machines and manuals for tapping operations. Why might a CNC machine not need a tapping head to cut internal threads using power feed?

Section 10 Answers

ANSWERS 10-1A

1. Size #13 at 0.185 in. and size #12 at 0.189 in.
2. The sizes get smaller as the numbers get bigger. This leaves possibilities open-ended as the need arises for smaller and smaller size drills. It is easy to add a size 81, for example.
3. A size #80 at 0.0135 in. (Note that most standard number drill sets stop at #60.)
4. #9 smaller size; #10 first larger size

5. A "Z" size drill. It is 0.413 in. in diameter. That's the last letter size drill.
6. It is true. That's the way both the number and letter sets are organized on the chart.
7. A letter A drill is 0.234 in.—just under $\frac{1}{4}$ in. at 0.250.
8. Letter size drills are larger.

ANSWERS 10-1B

1. $0.1875 = \frac{11}{64}$-in.-diameter reamer drill
   0.250 $\frac{15}{64}$
   0.3125 $\frac{9}{32}$
   0.375 $\frac{11}{32}$
   0.4375 $\frac{13}{32}$
   0.500 $\frac{15}{32}$
2. 11.0 or 11.5 mm
3. Any above $\frac{1}{2}$ in., 0.5625 and 0.625 in.

4. 118°
5. 135° (A flatter point geometry)
6. It creates a stronger edge with more bulk.
7. The drill will not start easily, it requires a pilot hole.
8. Leave it as is, harder to feed through but will drill; drill a pilot hole; modify the dead center (thinning).

ANSWERS 10-2

1. $\frac{1}{2}$ in. or 12 mm in both cases
2. Power presses and radial arm presses
3. A drill vise features quick open and close plus a shelf in the jaws to support work up off the floor of the vise.
4. True
5. It is either a radial drill press or radial arm press.
6. Vise, angle plate, fixture, direct clamp
7. The direct clamp

8. Clamp over the drill clearance hole or use parallel bars to elevate the work up off the table.
9. Peck drilling is an in and out interrupted down feed.
10. Vise, a standard drill press vise that holds off the vise floor; direct table clamp, either over the clearance hole in the table or elevated up on parallels; angle plate, designed for machining duty; fixture, a special tool made for the purpose.

ANSWERS 10-3A

1. 1,066 RPM
2. 4,000 RPM
3. 700 RPM
4. The difference in the two small drill's diameters is great when considering percentage of diameter. A $\frac{3}{16}$-in. drill is 150 percent bigger than

a $\frac{1}{8}$-in. drill. Therefore its RPM will be a lot slower. A $1\frac{3}{16}$-in. drill is only 5 percent bigger than a $1\frac{1}{8}$-in. drill. Not much difference in RPM is required. This is an important concept to help develop a sense of what is the correct RPM for a given tool.

5. Carbon steel and soft stainless machine about the same as far as speed is concerned. At 90 F/M. See Appendix IV for surface speeds.

ANSWERS 10-3B

1. 334 rpm

$$\frac{175 \times 12}{3.1416 \times 2.0}$$

2. 720 RPM

3. "Probably." It's spinning faster than the formula result at approximately 320 RPM but it is within reason. The setup has most likely been sped up after the initial calculation. Flood coolant might have been added.

4. A. Short formula

$$\frac{250 \times 4}{0.234} = 4,273 \text{ RPM}$$

  B. Long

$$\frac{250 \times 12}{\pi \times 0.234} = 4,081$$

$$4,273/4,081 = 1.047\% \qquad (5\% \text{ faster})$$

5. 750 RPM

6. All brass alloys are copper based. The correct surface speed is *140 F/M*. See Table 22A under "Recommended Feeds and Speeds for Drilling Copper Alloys" in *Machinery's Handbook*.

$$\frac{140 \times 4}{0.277} = 2,022 \text{ RPM}$$

7. The fastest cutting speed listed is *90* F/M—that would be for the most machinable (softest) material (Table 17 in *Machinery's Handbook*—drilling speed and feeds).

$$\frac{90 \times 12}{\pi \times 2.0} = 172 \text{ RPM}$$

8. Any one of the formulae can be used as long as the units are correct—all require conversion.

6. The safe approach is to choose the next lowest RPM that is available on the machine then test the results. If all goes well, it might be possible to shift up to the next higher RPM.

Machinery's Handbook contains a good conversion chart.

  A. *Convert 36 mm to inches* then use the 90 F/M surface speed in the short formula:

$$\frac{90 \times 4}{1.41732 \text{ in.}} = 254 \text{ RPM}$$

  B. *Convert 36 mm to inches* then use the long formula:

$$\frac{90 \times 12}{\pi \times 1.41735 \text{ in.}} = 243 \text{ RPM}$$

  C. *Convert 90 F/M to meters per minute* and use the metric formula

$$\frac{27.432 \times 300}{36 \text{ mm}} = 229 \text{ RPM}$$

KEYPOINT

The metric short formula yields speeds about 10 percent lower than the long formula.

9. Reamer drill diameter   0.5938 ($\frac{19}{32}$ in. for $\frac{1}{32}$-in. excess)

Reamer drill RPM   674
Drill feed rate range   0.004 to 0.010 in.
Reamer RPM   337   (half drill RPM)

10. Recommended surface speed   100 F/M
Drill RPM   382
Drill feed rate range   0.011 to 0.021 IPM

1. Letter I at 0.272 diameter.
2. It is true.
3. Drill jig
4. Punch marks on the face of the started cone on the side toward which the drill needs to be moved. This is possible only before the drill has drilled to full diameter.
5. False. It might require a taper sleeve adapter depending on the spindle and tool shank size.
6. Power turn in and out; tapping head
7. The tool would be a center-finder. The sketch would show the layout line being compared to its own reflection. When these two lines coincide, the part is aligned.
8. Layout lines, pilot punch, center punch, center drill, drill, and ream. Note more than one drill might be used.

9. False. Machining a chamfer around a hole is *countersinking*.
10. They are both enlarged, flat-bottom holes concentric to a drilled hole. A spotface is shallow, while a counterbore is cut to a specific depth.
11. Gun drilling and spade drilling
12. Gun drilling due to the lubrication that is forced through the drill shank to blow the chips out of the hole.
13. A. A closed-face countersink
    B. A counter boring tool (spotfacer solid body type)
    C. A keyless chuck

1. To produce the shortest possible dead center
2. The clearance face
3. 59°
4. Equal distribution of sparks or concentrated on one side

5. The sparks peel off onto the rake face. Before they stay on the grinding wheel and go straight down.

Answers to Section Review Questions

1. It is false. High-carbon steel is the budget metal of home-workshop drills but not industrial, quality high-speed steel.
2. You would check out a letter "Q" drill which has a 0.257-in. diameter.
3. A. Measure the drill on the margins. It may not be the correct size!
   B. What shank type is the drill?
   C. Is it sharp on the cutting edge and margins?
   D. Are the cutting edges the correct length and angle?
      Check this with your drill gage.
   E. Is the dead center in need of thinning?
4. This ratio is $\frac{5}{8}$ in. per foot. Here in drilling, this is not a critical number to remember but it will become more important when study lathes and turning tapers. Get a jump up, learn it now!

5. Due to rotational forces, chucks can't squeeze the drills tightly enough to prevent slipping. Taper-shank drives do not slip and the taper ensures an accurate—on center—mount every time. (Unless a careless person has mounted it without cleaning both mating surfaces, which can damage the shank.)
6. The parts of a drill bit (Fig. 10-92).
7. A. The drill press vise is on the lower right of Fig. 10-89.
   B. It features a quick release/advance and shelves to hold work above the floor of the vise.
   C. Because the drill press vise is not made to take any force other than downward pressure, it should never be taken to other machines.
   D. The drill cut into the bed of the vise

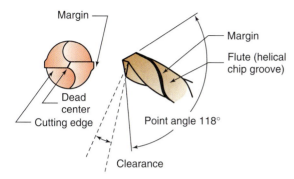

Figure 10-92 *The features of a drill bit.*

8. Drill size and material type* to be drilled.
9. $\frac{7}{16}$-in. drill in cast iron = 914 RPM
10. 5 percent slower at 873 RPM
11. Both methods include moving the part into alignment with the spindle then clamping it in place. *Method 1*—layout—punch pickup—0.030-in. accuracy. Using a punch mark on the layout line intersection, to create a dimple or dent where the work is to be drilled. *Method 2*—layout—center finder pickup—0.010-in. accuracy. Locating an adjustable pointer over the intersected layout lines. (No punch marks!)

Critical Thinking

18. See Fig. 10-93.
    A. Rounded clearance from dripping below position 1 when grinding. Would rub and not cut at all.
    B. Unequal angles. The cutting edge on the right would do all the cutting and a slightly oversize hole would result if no pilot hole was used. This bit would become dull too soon.
    C. Dead center off center (angles OK). This bit would drill an oversize hole if no pilot hole was used.
    D. Was held wrong for sharpening—dead center too long. Would drill but it would require much pressure to penetrate plus too much frictional heat.
19. It is a hard coating that reduces friction on the rake surface during the rake phase. That coating is called titanium-nitride abbreviated to

12. The colinear method is accurate to 0.005 in. with practice. (Lining up the layout line with its reflection.)
13. A tapping head attachment reverses rotation automatically. The radial arm and gear head (power feed) press normally feature reversing spindles.
14. A spotfacer and a counterbore are the same tool and operation. Both operations require a pilot to keep the cut concentric (centered to) the drilled hole and to also stop chattering as the tool starts the cut.
15. B. 82° countersink
16. Grind small negative rake surfaces on the drill bit.
17. A. Thin the web. (Grind two tiny cutting edges out of the dead center.)
    B. It reduces down feed cutting pressure and frictional heat.
    C. It is not needed if a pilot drill is used previously, since the dead center has no metal to cut, if the press has power feed and is able to force the bit through, or if the material can be drilled with hand pressure such as aluminum or brass.

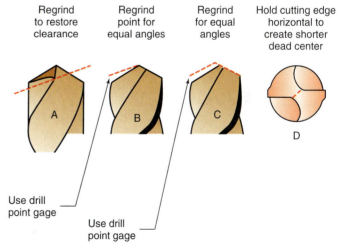

Figure 10-93 *The repair needed for each drill.*

*Recommended surface speed might be substituted as one variable instead of material type on some charts.

TiN. It is found on many HSS tools to increase their performance and tool life.

20. Small pecks or chip breaking cycles drill down a given distance then halt forward progress backup a few thousandths then continue feeding. Big pecks or chip clearing cycles drill down a given distance then withdraw all the way out of the hole. They then reposition back to the original depth and continue drilling.

21. This was a deep technical question—don't be dismayed if you didn't solve it. Here's the answer, which touches on the highest level of CNC program and operation skills.

   The modern CNC machine can power feed the tap at the exact rate that the thread leads in or out. In other words, as the tap advances its pitch distance, with each revolution into or out of the work, the machine can push or pull the spindle at the same rate and thus not break the tap. CNC controls that perform tapping, sense their exact RPM and can coordinate their feed axis advance rate at the exact ratio with the RPM. That's the theory! It works most of the time!

   But now—here's the rest of the truth. Depending on the situation, there are variables that sometimes require a tapping head even on the best CNC controls with absolute feedback (sense of) of both RPM and feed rates. Why? Two reasons: Cutting edge efficiency on the tap, a minor problem, and spindle acceleration curves, a major problem that affects other cutting operations, too.

   Taps don't always start cutting the instant they touch the work, sometimes there's a slight lag between touching and cutting. The spindle must push harder to get the tap to bite in and start cutting. That cutting lag increases as the tap dulls. Also, it changes with differing alloy hardness. These things cannot be sensed by the standard CNC control. Also, even though they start and stop very efficiently, CNC spindle controls must always deal with the deceleration and acceleration curves of the spindle and axis drive motors and the total mass in motion. That mass changes, depending on the drag on the tap and the total weight being started and stopped in the setup.

For example, compare tapping the end of a long steel bar, with a two inch tap on a CNC lathe—the tap is big and drags greatly in the cut, plus the weight of the chuck and workpiece moving on the lathe add up to a big acceleration challenge. Compare that to a half-inch tap cutting aluminum on a mill, which has nearly no acceleration problems other than the machine spindle drive itself— very different acceleration characteristics. Tracking and coordinating these kinds of

motion problems requires a supercomputer in terms of calculation speed. Sometimes the machine looses track of the exact difference in tap lead and axis advancement. Conclusion: When writing programs that tap, check with experienced programmers concerning when to use a tapping head and when it's not needed.

CNC tap holders   When tapping on a CNC machine, the lag difference between tap and axis lead is never great, even for difficult acceleration situations. So, CNC tapping heads are very different from standard drill press versions. They provide a bit of in-out slippage of the tap relative to the axis advancement but do not include the reversing and rotational slip functions of the standard drill press tapping head. These specialized tool holders are called *tap holders* rather than tapping heads.

Section 11

Turning Operations

Introduction

Some time ago, manually operated lathes were considered the heart of the shop and the primary machine for two reasons:

Products of the time were mostly round; for example, axles, gears, and shafts.

In theory, lathes were at the base of the machine pyramid. Theoretically one of those ancient lathes could make another working lathe.

But that's changed where you'll be working. Two factors put turning operations behind milling. First, turning spins the workpiece, not always a good idea compared to milling, where it's only the cutter spins. Also today's CNC mill can create round surfaces of bearing quality in roundness and smoothness. But the CNC mill can also machine a variety of shapes not possible on a standard CNC lathe. Still, there are times when turning is the efficient (or only) way to get the job done. Knowing when it's right or not can make a big difference in quality, safety, and profit.

There's another reason to learn turning: machine evolution isn't over! Today's computing power has given us fifth-generation machines that multi-task. The hard line between lathe and mill is fading. Today, we find CNC lathes with milling heads and feed axes. A second universal lathe employs bolt-on cutting tools attached to the lathe tool turret. Hydraulically powered, they rotate milling cutters, drills, and other spinning cutters, and move them to cut the work held in the lathe chuck—called live tooling lathes. On the other side of the fuzzy line, new multi-task mills equipped with powerful auxiliary rotary axes can spin the work when it's the better way to cut the feature.

These multiplex machines open up new operations for job planning. Operations can now be written that don't simply mill or turn; they combine them. Rotating the work as a lathe would, while contacting it with a spinning mill cutter, leads to new ways to shape metal unseen previously—called *generating* the shape.

Here's an example in Fig. 11-1. A hex head is machined onto a spinning round shaft. Using a cutter with the exact tooth spacing needed, the CNC control ensures that the cutter's RPM is

Coordinated RPM
mill-turning a hex head

Figure 11-1 *On the frontier, simultaneously turning and milling.*

coordinated to the work speed such that the teeth pass in a straight line creating a six-sided object. In just seconds the hexagon is machined. As these machines become accepted in industry, and grow even more capable, new ideas will arise and you'll be there to invent them!

But before you can write or manage programs that use these fantastic possibilities, you've got to clearly understand the basics of turning. Section 11 is about spinning the work when it's the right process, and how to do it efficiently and safely.

Objectives

UNIT 11-1

Basic Lathe Operations

Introduction: Unit 11-1 introduces the basic ways lathes shape metal. It defines the name of the operation, the nature of it, a typical cutting tool required, and in some cases, a bit of basic information about the setup required. The purpose is familiarization before lab training; Unit 1 introduces what a lathe can do. In the review problems, you'll be given a part print and asked to identify the various lathe operations needed to complete it, and in what logical sequence they might be machined.

Manual Verses Programmed Lathes This unit assumes it's likely you'll be practicing operations and setups on a manually operated machine before trying them on a CNC. But there are some differences in how these two kinds of lathes perform them. When it's important to know the difference, it will be pointed out. These operations are not everything that can be accomplished on a lathe. But by mastering them, you become ready for your industrial experience.

KEYPOINT

Keep in mind as you scan, the goal is an understanding of what can be done, not how to do it! That comes later.

Accuracy Comparisons Some operations are more accurate than others when performed on a manually operated lathe. On CNC lathes, the accuracy and repeatability is far better than that described here and it's more uniform throughout. The numbers are my relative estimates, provided for comparison of the various operations at the beginner level—they are not *a standard*. The first time, you might find them challenging but a good goal. In most cases, a skilled machinist would be able to produce results finer than those described here. The age and condition of the lathe on which you will be practicing are also factors.

KEY TERMS

Boring
A single-point tool operation (boring bar) used to machine a cylindrical hole, but can be used for other internal features.

Chamfer
An angular surface usually specified in degrees with respect to the work centerline, on a corner of the workpiece.

Drive dog
The clamp-on accessory used to transfer spindle rotation to a workpiece.

Facing
A lathe operation that cuts the end of the work to be flat and perpendicular to its axis.

Included angle
Specifying a conical surface from one side to the other—twice the half angle. Sometimes called the whole angle in lathe work.

Knurling
A rough pattern deliberately produced for grip or to increase size.

Parting
A relatively narrow groove that severs the work from the stock remaining. Accomplished with a special tool designed for this purpose.

Taper
A slightly changing diameter specified in change per foot or degrees. Might also be specified in half angle degrees or whole angle degrees.

Operation 1— Straight Turning

During straight turning, the diameter of the work is reduced to a precise diameter and probably also to a length dimension, creating a cylinder. It's accomplished by positioning the cutting tool inward along the X axis (Fig. 11-2) to a given diameter, then moving the tool along the Z axis (left-right motion as you face the lathe) to cut the metal away.

Straight Turning

Feed direction
Z axis

Figure 11-2 *Straight turning machines a cylinder.*

During a turning operation, the work rotates at the calculated RPM, while the cutting tool is forced laterally at a specific *inches per revolution* feed rate past the work (for example, 0.006 IPR). On both manual and CNC lathes, the *Z* axis moves to your left/right. The *X* axis moves inward and out. On a manual lathe, the *X* and *Z* tool motion is provided either by a hand crank or by a power feed. Exact position is achieved by hand dial with micrometer graduations or by digital readouts.

Figure 11-4 *Some lathes feature a micrometer stop to assist Z axis depth stop control.*

Lathe feeds are normally specified in inches per revolution (for example, 0.006 IPR). Lathe axes are *Z*, left/right, and *X*, in/out, on the diameter of the work (Fig. 11-3). The *Z* axis of a lathe (or any machine) is parallel to the axis of the main spindle.

Turning Accuracy

On most manual lathes, there are differences in the repeatability between the two axes. The *X* axis is cranked into position with an accurate micrometer dial graduated in thousandths of an inch. Expect a repeating accuracy of around 0.001 to 0.0015 in., depending on the lathe's quality and condition. That means a repeatability on the work diameter at 0.002 to 0.003 in. plus or minus.

Because the *Z* axis is not moved by a precision lead screw as the *X* is, but by a simple spur gear pulling the tool on a rack gear, the *Z* axis repeatability is not as accurate unless the lathe is equipped with electronic digital positioning. Using the *Z* axis hand dial, position repeatability varies from 0.005 in. on the

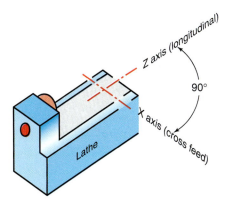

Figure 11-3 *The X and Z axes on a lathe.*

better models up to around 0.030 in. on everyday lathes.

Z Axis Stops

Lacking the digital positioner, there are two ways to improve tool *Z* position and part length results for the *Z* axis. A precision micrometer stop attachment (Fig. 11-4) is bolted to the lathe bed. Its probe stops *Z* axis movement and can be adjusted in 0.001- or 0.0005-in. increments, depending on its construction. Alternately a dial indicator mounted on a magnetic base can also be used to control fine movements of the *Z* axis. Your instructor will show you how.

On manually operated lathes, there are differences in the mechanical method by which each axis is located, creating different repeatabilities. This problem doesn't exist on CNC lathes.

Diameter to Length Ratio Determines Setup

During straight turning, work holding is an important issue for both safety and accuracy. We'll devote Unit 11-4 to the subject. For now, the way the work is held for straight turning will vary, depending on the length of the raw stock compared to its diameter. Shorter parts are held on the chuck end only (Fig. 11-5).

Anything over an approximate five-to-one ratio (length to diameter) usually requires extra support

ShopTalk

When discussing manual lathe motion, machinists might refer to *Z* axis motion as longitudinal (or long axis), while *X* is often commonly called cross feed.

ShopTalk

How Fine Can You Go? In contrast to manual lathes, all CNC lathes feature a resolution well below 0.001 in. (usually 0.0001 to 0.0002 in.), with repeatabilities near 0.0005 in. or better. A few superlathes have resolutions of 0.00005 in.—they work at 50 millionths in. resolution! No kidding.

**Work support determined by
length to diameter ratio**

Up to 5/1 — Chuck support only

Up to 10/1 — Add tailstock

Add center supports

Above 10/1

Travel and steady rests—
any combination

Figure 11-5 *The amount of support needed depends on work length.*

at the outboard end. Parts beyond 10 to 12 compared to 1, will also require support at both ends and in the middle too. The end support shown in Fig. 11-5 is called a tailstock. It ensures that the medium-length shaft doesn't bow away from the tool, bend, or vibrate out of control. Heavier chip removal rates determine which holding chuck will be used and how much material must be reserved for gripping the work. This *grip* issue is a huge part of lathe planning.

KEYPOINT

Length to diameter ratios beyond 5 to 1 require tailstock support. Ratios beyond 10 to 1 require midpart support. These are rules of thumb, there are many variables. Early setups of long stock must be checked by a journeyman or instructor before taking a cut with them.

Notice that in Fig. 11-6, a turning tool features the three basic tool angles: lead (cutting angle), rake, and clearance plus a nose radius to improve finish.

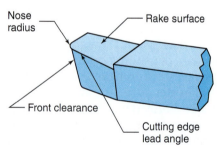

Nose radius

Rake surface

Front clearance

Cutting edge lead angle

Figure 11-6 *A tuning tool features the four universal geometry features.*

Operation 2—Facing

Here the tool cuts across the end (face) of a part, moving on the *X* axis. The objective is to make it flat, perpendicular to the *Z* axis and to a given length. Or, as shown in Fig. 11-7, to machine a step in the work to a given dimension from a starting reference—usually, the outer face of the work. During **facing,** the *X* axis moves, while the *Z* axis is locked in position. On manual lathes, the *X* axis can be hand cranked inward (or outward) toward (or away from) the center of the work. It can also be power fed.

On manual lathes, the repeatability of facing work to a specified dimension can be from 0.002 to 0.030 in., depending on how the lathe is equipped for *Z* axis positioning of the tool. The dial indicator attachment previously shown can be used, however, a digital readout positioner is the best solution, but these options are not always found on manual lathes.

Facing to a Step Dimension

In Fig. 11-8, we need to cut a face step 0.125 in. from the work face. The first step is to coordinate the cutting tool to the position readout

Figure 11-7 *A facing cut is taken by moving the X axis.*

Z = 0.0000

Set to zero on readout, dial, or indicator

Tool touching work face

Z = −0.1250

Facing the step to the dimension

Figure 11-8 *Prepping for a face cut, the machinist touches the work then sets the axis to zero.*

(digital or indicator). Either the tool is just touched to the outer end of the work, or a light face cut is taken. Then the position is set to read zero. The hand dial numbers would be set to read zero as well, as a double check of the tool's position.

Second, the tool is moved to the first cut position at Z-0.080 in., for example. This cut, called a roughing pass, takes the majority of the metal to be removed. Next a semifinish pass at Z-0.100 is taken. This cut might also be called a *spring pass* because it neutralizes tool or work deflection from the rough pass. Finally, the tool is positioned at Z-0.125 and the finish pass is taken.

Digital Positioning

An attachment added to the lathe, positioners report tool location relative to any given zero point preset by the operator, usually to a 0.0005-in. resolution (or finer).

KEYPOINT

Learning to use digital tool positioners and tool to work coordination, are excellent preparation for CNC training.

RPM—The Need for Speed

One big challenge to facing operations is that the effective rim diameter is constantly changing as the tool moves in or out along the X axis. Thus, a change in RPM is needed to maintain consistent cutting speed. Some manual lathes feature a variable speed drive, their RPM can be increased or decreased during the cut. However, most feature gear driven spindles to accommodate heavy work. They must be shifted to a single RPM and they cannot be shifted on-the-fly (while the spindle is turning). So the cutting speed for facing is chosen as a compromise. It will either be too slow or too fast during some part of the face cut.

Trade Tip

CNC lathe controls feature a constant surface speed command that, when invoked, continuously changes RPM as the facing cut progresses.

Operation 3—Drilling and Reaming

On manual lathes, drilling operations are accomplished using the tailstock with its sliding quill. It's always hand fed by crank. The tailstock holds cutting tools on the center of the work and drives them forward by a sliding quill. The quill features a Morse taper hole (the same as on drill presses). Differing from a drill press, the cutting tools do not rotate, it's the work that spins here (Fig. 11-9). Drills and reamers are mounted either in chucks or directly into the Morse taper.

Diameter Accuracy

There are four factors affecting diameter accuracy when drilling on any lathe. Tool sharpness and size are critical. The other factors are the rate at which the drill is fed inward and heat/coolant issues. Diameter repeatabilities of around 0.005 in. for drilling and 0.0005 in. for reaming are typical. Coolants are a must in deep holes, but they tend to spray off the rim of the chuck. As shown in Fig. 11-10, when drilling parts in a chuck, coolant must be

Figure 11-9 *The lathe tailstock holds and drives the drill tools.*

Figure 11-10 *Without a chuck guard the coolant would be flung out everywhere!*

contained in a chuck guard not only to keep coolant from flying out but to safely contain chips too.

Hole Depth Accuracy

Hole depth is an accuracy challenge, it will repeat from 0.005 to 0.030 in. (0.1 to 0.5 mm), depending on tailstock dial graduations. Better lathes feature graduations as fine as 0.005 in. (Fig. 11-9).

KEYPOINT

Always Check Tailstock Centering Before Drilling!
For taper turning operations (coming up later) the tailstock can be offset sideways—off the horizontal center of the spindle. However, to not break drills and reamers and to create accurate hole sizes, the tailstock must be set back on center for drilling.

Starting the Hole on Center

Even though the work is spinning about the spindle center and the tailstock is centered, a drill can and does wander at initial contact. To prevent this, a face cut should be taken to clean up the work flat and perpendicular, as long as there is material to allow it. Then a small pilot or a center drill should be used to spot the exact center, the same as for drill press location.

Safety Note—Chuck RPM Limits

When the workpiece is held in a standard lathe chuck, sometimes RPM cannot be shifted to the correct speed for a small drill, even though the lathe may be able to achieve it. Most shops limit RPM for chuck work. In my shop, it's 1,000 RPM and that assumes the work is well balanced! Due to centrifugal forces and vibration, cast-iron chucks can explode at their rim! Poorly held parts can and do fly out of the lathe! The worst accident I have seen in my entire career was caused by a large part flying out of a lathe. It not only hurt the machinist but went on to tear up a big part of the shop before coming to a stop many yards away!

KEYPOINT

Find out about RPM policy. Though I'm repeating myself, always have your setups checked!

There are chucks made for both manual and CNC lathes that safely withstand greater RPM limits, as we'll see examples in Unit 11-4. However, on a CNC lathe operating under constant surface speed mode, a dangerous condition can develop when a facing cut is going toward the center of the work, or when using a very small drill. The control will attempt to speed up the spindle beyond safe limits. To counter that potential, a red-line code word can be invoked that limits RPM to a specific maximum.

Operation 5—Turning Angles and Tapered Surfaces

Conical work (changing diameter as a cone, not a cylinder) may be called tapered, chamfered, or when the conical surface connects two other surfaces, might be called a *transition* (Fig. 11-11).

When turning tapers, there are two related characteristics to verify within tolerance:

The *diameter* of the surface at the large and small ends

The *angle* specified in degrees or sometimes in taper amount per foot

Angular Feature Expressions

To cut tapers one must understand two terms found on engineering drawings:

Chamfers (steep transitions) are specified in degrees relative to the work centerline, for example, 45° chamfer or a 45° transition.

Shallow **tapers** are usually specified with diameter change per length or in degrees, for example, Morse tapers are $\frac{5}{8}$ in. per foot of shaft length, or it can be expressed as an angle at 1.49° taper.

Be Certain It's Either the Included or Half Angle When reading a taper callout on a print, determine which angle is being shown, the half or full angle (Fig. 11-12).

Figure 11-11 *This taper shank adapter exhibits four angular surfaces.*

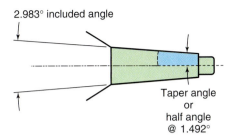
2.983° included angle

Taper angle
or
half angle
@ 1.492°

Figure 11-12 *Tapers are noted in either the* half *or* included *(whole) angle.*

KEYPOINT

The *half angle* (sometimes called the taper angle), is measured from one side of the work with reference to the centerline. In contrast, the **included angle** compares one side of the part to the other side—the overall taper angle of the cone.

Four Angular Setup Methods

On CNC lathes there is no difference in the way any taper cut is set up and made, since the computer can move the X axis relative to the Z axis at any specified feed ratio. However, on manual lathes, there are four different setups, depending on how steep and long the taper is to be.

Method 1—Swiveling the Compound Axis Manual lathes feature an auxiliary axis that can be rotated to any desired angle with respect to the X-Z framework. Hand-cranked only, the compound provides a simple way of producing a taper of moderate accuracy. Note the perspective for these drawings is from above, looking down on the lathe (Fig. 11-13).

Tool compound axis rotated
(viewed from overhead)

Compound set to
the half angle

Hand
crank

Cutting
tool

Figure 11-13 *The* tool compound rest *allows machining steep angles by rotating it to the correct half angle.*

Must use complimentary or supplementary angles When swiveling the compound a common mistake is made by beginners. Most compound bases are graduated in 90° quadrants with a few supplied in 180° segments. When setting the angle, be on the watch for predictable complimentary and supplementary errors.

KEYPOINT

The compound axis on most manual lathes is at 0° when it is parallel to the X axis (Fig. 11-14). However, engineering drawings usually specify the part angle off the centerline Z axis.

On most lathes, you will be setting the compound to the complimentary angle of that to be machined. However, on a rare few, 0° on the compound occurs with it parallel to the Z axis, and you may be setting the compound to the supplementary angle. Confused, see the Trade Tip.

Using the compound to cut angular surfaces, can be as accurate as $\frac{1}{2}$ degree and 0.002 in. on the diameter. But, since both the X axis micrometer dial and compound dial are involved, repeatability must be assumed not better than 0.005 in.

To produce the half angle

Set at the complimentary angle

Zero parallel to X axis

Figure 11-14 *Avoid complimentary angle problems associated with the compound axis.*

Method 2—Forming the Angle This short, steep-tapered surface is machined with a tool ground for the purpose. We'll see other versions of the form tool coming up (Fig. 11-15). Forming a cone or radius is accurate and fast in production situations. Form tools are even used for tricky CNC lathe work, where the control could duplicate the shape but the form tool does a better job.

Accuracy For both diameter accuracy and repeatability, this method is the best of the four, as long as the form tool is set to the right angle. The tool is cranked inward along the X axis to the desired size and therefore repeatability is as good as the lathe's X axis.

Method 3—The Taper Attachment Once it's set up, turning with a taper attachment is similar to straight turning. The lathe is engaged to move along the Z axis in the usual way, by power feed or hand-cranking. While it moves laterally, the angled taper attachment moves the tool in or out along the X axis at the ratio set on the attachment (Fig. 11-16).

Accuracy The taper attachment can be adjusted for any taper angle between 0° and around 25°. The angular resolution is around $\frac{1}{2}$ degree using the attachment's graduations alone, but can be improved greatly with sine

Forming the Angle

Form ground bit

Figure 11-15 *Steep angles can be formed. The tool bit is ground and setup to the shape of the angle required.*

X axis and cutting tool follow the taper attachment guide ways

Figure 11-16 *The tool is moved in two axes at the same time, producing the taper. Not all lathes have a taper attachment.*

bars, dial indicators, and vernier protractor tricks you'll learn later.

Although diameter is determined by X axis position, repeatability isn't as good as the lathe's X repeatability because the taper attachment includes a mechanical clearance to allow it to be operated freely. Expect around 0.005-in. repeatability, but that can vary depending on the quality and maintenance of the attachment. The taper angle will repeat perfectly once the attachment is adjusted to the correct angle and locked in position.

KEYPOINT

Remember, this is a visual task inventory, it's unimportant that you fully understand the attachment right now, just that it is one of the options for producing a taper on a long workpiece.

Method 4—Offsetting the Tailstock This method makes very shallow taper angles. You would choose this method because of the ease with which the part can be removed and quickly replaced in the setup with no disruption to accuracy or because the work is very long. Notice there is no chuck, but rather the part is held between pointed centers that we'll discuss more in Unit 11-3 (Fig. 11-17).

Drive plate | Drive dog attached to the workpiece

Offset to spindle axis

Figure 11-17 *Looking down, the tailstock is moved off center to make a shallow taper.*

In this method, these centers are not parallel to the Z axis by *half* the taper amount.

Using the offset method of producing a taper, the tool action is the same as for straight turning except now the work is *not parallel* to the Z axis of the lathe. The work is revolved by a clamp-on driver called a **drive dog** because it has a tail. It's tail is inserted into a slot in the drive plate.

Operation 6—Boring

Boring is an accurate method of machining an inside diameter or another internal feature on the work. Bored holes are usually straight (cylindrical), but they can also be machined with a taper using the compound or the taper attachment (Fig. 11-18). This lathe operation requires a cutting tool called a boring bar.

Reasons to Bore a Hole

The advantages of boring over reaming are twofold. One is that any size can be produced with a repeatability of 0.0005 in., assuming the lathe's X axis is accurate. Second, a boring tool will machine to the exact centered location without wandering as a drill/reamer combination might.

Two or three passes (light sequential cuts) of a boring bar will correct a misaligned drilled hole. During boring, the tool is either hand or power fed, usually inward along the

Boring an Inside Diameter

Boring bar

Figure 11-18 *This lathe is set up to bore a hole with a boring bar.*

Z axis. Reversing the Z feed, outward boring is also possible and less risky as the cutter is powering away from the bottom of the hole.

There are three challenges to boring operations:

1. **Boring Bar Deflection**
   This causes inaccurate sizes and chatter due to the long, thin boring bar.

2. **Ejecting the Chips Out of the Hole**
   Coolants help, but chips tend to wind around the boring bar and jam at the bottom of the hole. Operators must always watch and listen for chip build up when boring. The jammed chips rub and ruin the newly machined surface and/or break the cutter.

3. **Blind Operation**
   The operator cannot see the cut as it proceeds. The depth of the bore must be controlled without being able to see the tool.

In addition to producing internal cylinders and cones, other operations are performed with a form-shaped boring tool. Features such as corner fillets, round or square internal grooves and threads are made with a formed boring bar (Fig. 11-19).

Operation 7—Forming

Forming is used more on manual than CNC lathes to create special shapes that are impossible to produce using tool axis motion. The tool bit is ground to reproduce the desired shape (Fig. 11-20).

We seldom use a form-ground tool on a CNC lathe for two reasons. First, it is a custom ground tool, which we try to avoid. Second, the controller can produce most shapes by driving standard tools in the shaped tool path.

Internal operations using a boring bar

Internal groove

Internal threads

Figure 11-19 *Internal grooves and threads are produced with a boring bar but using a shaped bit.*

Forming a Complex Shape

Form ground bit

Figure 11-20 *A forming tool making a complex shaped cut.*

KEYPOINT

When forced against the rotating workpiece, the form tool produces a reverse image of itself. Cutting fluid and slow RPM are vital for success in forming, and due to the large tool contact area, chatter is always a potential problem. Accuracy is difficult to determine for forming due to the complex shapes, but within an overall 0.0015 in. might be expected.

Operation 8—Threading

Cutting threads can be accomplished in three ways on the manual lathe: taps mounted in the tailstock, dies in regular handles, and using a forming tool to cut the helical groove between threads, called *single-point threading*. Single-point threading (*single pointing* in shop lingo) is a vital lesson in using the manual lathe and, in your first year machining course, is likely to be the most complex operation to

setup and to perform. It's time consuming and prone to errors due to the many details and actions the machinist must handle.

Still, it's a vital competency to have because any custom or standard outside or inside thread can be produced using single-point methods. There are times when single pointing the thread is the right choice; perhaps one thread is needed when a die or tap isn't available, or the CNC lathe is doing other things. This is the full journeyman's fallback method!

The setup uses a cutting tool ground to the correct thread form with large clearance and rake angles. Since the spindle will be rotating slowly, these extra angles help to improve the finish (Fig. 11-21). As with all threading operations, cutting fluid is an absolute must. As we've just seen, single-point threading can also be performed on internal threads using a boring bar and a form-ground cutting tool.

Single-point threading can repeat at around 0.001 in. if the lathe is accurate and well maintained. But, since it requires the use of two hand dials, the X and the compound axes, care must be taken to control axis backlash. We've got a bang-up trade trick coming up in Unit 11-7 to make this aspect a snap.

Tapping and Die Threading

These are two less complex options for producing threads at the lathe.

Taps in the Tailstock To produce internal threads, regular taps can be mounted in a drill chuck in the tailstock. The spindle is placed in

External threads

Thread relief

Single-point threading tool

Internal threads

Thread relief to stop threading

Chamfer for better thread start

Figure 11-21 *A threading tool is a type of form tool.*

neutral then rotated by hand to cut threads where tap breakage might be an issue. For sturdy taps of $\frac{1}{2}$ in. or greater, the spindle can be placed at a low RPM then started forward under power. When using power, it's wise to slip the clutch and never remove your hand from the start lever. If the lathe can be reversed, the tap can be removed under power too. If not, the spindle is placed in neutral and rotated backward by hand to reverse the tap. Alternatively it can be unchucked then unscrewed with a regular tap handle.

KEYPOINT

The tailstock hand wheel is not used for this kind of tapping on the manual lathe. Advancing the tap into the hole (and out) is done by hand pressure on the entire tailstock. Left unlocked on the ways, it's slid forward by hand pressure to engage the tap with the work (Fig 11-22).

As the tap nears full depth, the spindle is stopped and then reversed. At the same time, the machinist pulls the tailstock outward. It's a coordinated action requiring a demonstration before you try it yourself.

Handheld Dies (and Taps) It is possible to thread with a handheld tap or die, mounted in a handle, pushed against the work as the chuck is rotated slowly by hand—*not by power!* If it is power driven, there's a chance to pinch hands between the handle and lathe bed or saddle. Also, it's prone to alignment error of the cutting tool to the work axis. However, if your instructor approves of this method, ask for a demonstration. It can be used to quickly make utility threads, although they are not customer quality—we call this a "quick and dirty" method. See the Trade Tip.

Production Taps and Dies In contrast, there are times when large batches of parts are required with high-quality threads and the CNC machine isn't available. Then, on any lathe, manual, automatic screw cutting, or even the CNC, we turn to the collapsing tap or automatic die head. These are designed to cut the threads in one pass, on the inward stroke. Then using a triggering action, withdraw their cutting edges so they can be pulled back quickly rather than unscrewing them from the thread, which tends to dull the cutting edges in some materials. They are more

commonly used on automatic lathes made expressly for threading, but if mounted in the tool post or tool turret and centered to the spindle, both vertically and in the X axis, *they can cut threads faster than any other method* with repeatabilities in the subthousandth range.

Another advantage to these kinds of threading tools is that they use inserted HSS or carbide cutters that are easily exchanged to a different thread. The die inserts can be removed from the head and quickly resharpened. This threading setup would be chosen when hundreds of parts are to be threaded and tolerances are tight. Dies and taps of this kind pay for themselves in large batch production (Fig. 11-23).

Figure 11-23 *Trip dies and taps speed up production threading but would not be common in school shops.*

Operation 9—Parting (Cutting Off)

The **parting** operation is primarily used to cut the finished product away from the raw material remaining in the chuck, or occasionally to produce an extra deep groove in the cylindrical side of the work. Parting tools are special, thin blades that produce a narrow, deep groove (Fig. 11-24, top view).

The parting tool is either hand or power fed in along the X axis. Parting tools are supplied in either high-speed tool steel or as brazed carbide tips and carbide insert types.

Accurate work length control is the objective. Repeatabilities vary, depending on Z axis positioning of the tool from 0.030 in., using hand wheel graduations to 0.002 or 0.003 in. if digital readouts or long range DTIs are used. Besides positioning the cutting tool then locking it in place, another challenge to length control is the perfect alignment of the tool to the X axis, and its tendency to wander to one side as it dulls. Coolants, well-calculated RPMs, and moderate feed rates are essential in a parting setup.

KEYPOINT

Using the chuck face or the lathe's cross slide as a reference, the parting tool must be perfectly parallel to the X axis to maintain accurate cutoff lengths. Keep parting tool overhang to minimum!

Parting Operation Challenges

In addition to part length, there are several factors making this a tough operation to perform consistently and safely. Chip removal, chatter, and tool digging and/or breakage are a concern.

Part Length To achieve the required length, the tool's left edge is touched to the work end (dashed lines in Fig. 11-24), then after withdrawing it from the workpiece, the Z axis is moved the distance required (tool width plus part dimension). Most parting tools are made to an exact width to help with the math. But be aware that as the HSS version is reground, the effective width may change.

Chips Wider Than the Groove Due to the flowing deformation of the metal, the chips grow slightly wider than the groove made by the tool—they rub in the groove and often coil up into a tight mass. Coolants are a must both to alleviate heat and to lubricate the chip so it can be ejected from the groove. (See the Trade Tip, Fig. 11-25.) Ideally, coolant should be induced from above and below the cut to get it to where it's needed. However, it's uncommon for manual lathes to feature dual spigots, so it must be flooded generously into the groove. A chuck guard is then necessary.

Chatter and Tool Shattering It's likely you'll be using HSS parting tools initially when learning parting. While the carbide insert or carbide-brazed tip parting tool outperforms the HSS by a large factor, the tougher HSS tool forgives most of the digging and chattering problems associated with parting. Most—but not all!

KEYPOINT

Caution!
Parting tools do shatter when the cut is deep and the tool is extended far from the holder. They dig in, pull down, then snap into dangerous projectile shards! Be sure to set the tool with just enough overhang to do the job and use conservative feeds.

Constant Surface Speed Needed Similar to facing, the need for increasing RPM becomes critical as the tool approaches center. Due to the tight deformation problems of the chip piling up within the groove, surface speed is a must-do challenge. This requires a variable drive spindle or CNC control in constant surface speed mode to do it right. Sometimes, as the tool creates an ever smaller diameter, the workpiece will break prematurely, leaving an unwanted protrusion as shown shortly in Fig. 11-26. However, RPMs closer to the right speed without exceeding the chuck limit, will help reduce this problem.

Parting the finished
workpiece off the bar

Figure 11-24 *During parting, chip removal is a challenge.*

Chip Folding Tool To help the wide chip made by parting come out of the groove, it's possible to grind a small vee on the top rake surface of the tool. The chip flows over this surface creating a disturbance line in the chip that tends to fold it, thus it does not hang in the groove so tightly. Many commercial carbide parting tools have this feature built into their top surface geometry. The groove is put in using the corner of a well-dressed grinding wheel—have your instructor demonstrate this modification (Fig. 11-25).

Grinding a Slight Lead Angle on the Parting Tool
If many parts are to be cut off a bar, then a small, unwanted point can be left behind on each finished part after the parts disconnect (Fig. 11-26). It must then be removed later, adding non-value-adding time to the product. To solve this problem, grind a face lead angle around 10° maximum on the parting tool, with the more pointed corner toward the work (Fig. 11-26). The pointed tool tends to create a weak spot in the bump that allows it to break away or leaves it on the raw stock not the work. The cutting tool then proceeds forward to cut the bump off the raw stock.

Chip folding vee groove

Figure 11-25 *A top groove in the parting tool helps to crease and fold the chip.*

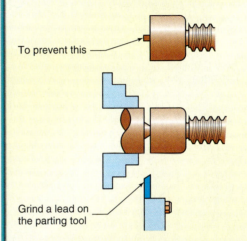

To prevent this

Grind a lead on the parting tool

Figure 11-26 *Modifying the lead, the raw stock retains the small protrusion rather than the workpiece.*

Dropping the Part Refer back in Fig. 11-24. There is a potential problem—can you see it? As the final cut is made, the part is going to fall down. But the support center is still in the right end. This can lead to real pinching problems for the loose part and chuck, unless the lathe is stopped instantly. There are a couple solutions but . . .

KEYPOINT

Safety
None of the solutions involve reaching in with your hand to catch the part as it is finally severed from the stock! (The key words here are *severed* and *hand!*)

Stop and Withdraw the Center Slightly. Do this before recommencing to fully disconnect the object. Then when it does disconnect, let the part fall into a catch basket or fall to the chip bin if damage to the part isn't a critical issue.

If the Part Is Nearly Through with the Support Center in Place. *Stop the spindle* then *hand saw* the object off the stock or bend it off. This is the safer solution especially for beginners.

Operation 10—Grooving and Operation 11—Face Grooving

There are two types of grooves made on the lathe. The first is similar to a parting tool cut, a groove into the outer round surface of a part, while the second is on the face of the work. Tools used for outside diameter grooving are similar in tip shape but shorter than parting tools. Face-grooving tools must be slightly modified with extra side clearance to fit within the outside curvature of the groove the tool makes (Figs. 11-27 and 11-28).

Expect a repeatability of 0.0005 in. for the width of the formed groove, which is often a critical feature. Diameter repeatability depends on the positioning method of the lathe and is from 0.002 to 0.0005 in.

Side groove Face groove

Trepanned face

Figure 11-27 *Face grooves require a modified tool bit shown in Fig. 11-28.*

Face Grooving Trepanning Tool

Extra side clearance required

Figure 11-28 *A face grooving tool requires more side clearance on one side to allow for the shape of the groove produced.*

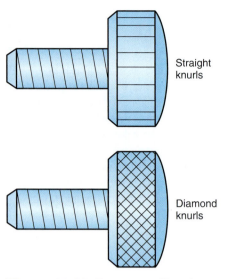

Straight knurls

Diamond knurls

Figure 11-29 *Two types of knurls on a shaft, straight and diamond.*

You have seen this diamond or straight line pattern on handles of tools or on thumb screws (Fig. 11-29). Grip and appearance are the objectives for knurling, so size is not the usual objective. Expect size repeatabilities of no better than 0.010 in.

Knurling is called out for two purposes. First to improve hand grip or bonding qualities. Second, to make a shaft bigger. Because the displaced metal is shoved up, knurling actually increases the effective size of a shaft. Suppose a worn shaft no longer holds a press-fitted bearing in place. It can be knurled then turned back to the correct diameter to hold the bearing again. It won't be as good as the original since there is far less metal to form the surface, but it does work! (See Fig. 11-30.)

A face-grooving tool can also be used to remove material from the face of a part to a flat-rimmed dish or pan. This is often called *trepanning*, and is probably a reference to even older wood-turning lathes.

Operation 12— Knurling

Knurling (rhymes with *sterling* silver) produces a deliberate rough pattern on a workpiece surface. Knurling does not remove metal but rather displaces it. With the work rotating slowly, a knurling tool creates small, sharp dents in the work. The displaced metal is deformed (shoved) up next to the dent, a little bit at a time, into the rotating forming dies.

Figure 11-30 *Knurling requires a rotary tool comprised of two dies that displace but do not remove metal.*

Operations 13 and 14—Hand Filing and Hand Sanding

We try to avoid handwork on machines as much as possible—because we're machinists. Still, there comes a time when just a bit can make the difference in

Minute size reductions. Filing or sanding can take smaller bites than a cutting tool, for example, sanding to remove a tenth to exactly bring a diameter into tolerance.

Surface finish and appearance. Sanding can create finishes not possible with cutting tools. It is useful for final details such as burr removal with a file, corner radii, or chamfers.

When filing and sanding, machinists achieve close to 0.0002-in. accuracy (repeatability is not appropriate here). But we always seek other solutions, since hand operations are not only slow and on the low-tech side, they also bring your hand and arms dangerously close to the chuck.

Safety Precaution

Because you will be placing your hand close to a rotating spindle, there are specific safety precautions for both filing and sanding. Figures 11-31 and 11-32 show the right way to

Figure 11-32 *When sanding work, always hold the media as shown. Why? Answer found in the Unit 11-1 quiz.*

file and sand. Hand and arm positioning are critical as is your braced stance in front of the lathe! As a critical thinking exercise, you'll be asked to reason why and how in the review. Study the illustration closely.

Figure 11-31 *This machinist is using the safe, skilled hand position for filing, along with a chuck guard and no loose sleeves or jewelry.*

UNIT 11-1 REVIEW

Review the Key Terms

Boring
A single-point tool operation (boring bar) used to machine a cylindrical hole, but can be used for other internal features.

Chamfer
An angular surface usually specified in degrees with respect to the work centerline, on a corner of the workpiece.

Drive dog
The clamp-on accessory used to transfer spindle rotation to a workpiece.

Facing
A lathe operation that cuts the end of the work to be flat and perpendicular to its axis.

Included angle
Specifying a conical surface from one side to the other—twice the half angle. Sometimes called the whole angle in lathe work.

Knurling

A rough pattern deliberately produced for grip or to increase size.

Parting

A relatively narrow groove that severs the work from the stock remaining. Accomplished with a special tool designed for this purpose.

Taper

A slightly changing diameter specified in change per foot or degrees. Might also be specified in half angle degrees or whole angle degrees.

Replay the Key Points

◆ Lathe feeds are normally specified in inches per revolution, for example, 0.006 IPR.

◆ Lathe axes are Z, left/right, and X, in/out, on the diameter of the work.

◆ The Z axis of a lathe (or any machine) is parallel to the axis of the main spindle.

◆ Because there are so many variables as to how to hold work, it's critical that your setups be checked by your instructor before starting them.

◆ For accurate drilling operations, and to not break drills and reamers, the tailstock must be set on center.

◆ Know shop RPM policy and always have setups checked!

◆ The *half angle* (sometimes called the taper angle) refers one side to the work compared to the centerline, while the *included angle* refers to side to side—the overall taper of the object.

◆ The compound axis is at 0° when it is parallel to the X axis on most machines.

Respond

Do you recognize the basic lathe tasks? Complete the following questions to find out.

1. On a piece of paper, name the operations needed to turn this axle from raw stock (Fig. 11-33).

2. Grooving the side and face of a part can be accomplished using a parting tool. Is this statement true or false? If it is false, what will make it true?

3. Without referring back to the text, list the basic operations performed on a manual lathe. Describe each briefly.

4. Identify the operations required to make 25 of the fixture locator pin in Fig. 11-34. They will be machined from a 2.25-in.-diameter bar of aluminum. In this single-view drawing, the circles with the cross lines indicate diameters.

Figure 11-33 *Name the lathe operations required to turn this axle.*

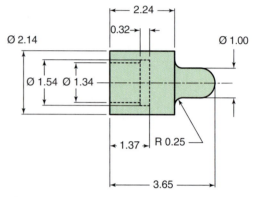

Figure 11-34 *Identify the operations needed to make this pin. The bar is long and many will be made.*

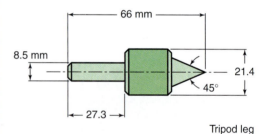

Tripod leg

Figure 11-35 *Name the operations needed to machine this tripod tip.*

5. Name the operations required to turn the camera tripod tip shown in Fig. 11-35.

6. List the potential safety hazards presented in turning on a lathe not found on other machines.

Critical Thinking for Safety

7. Return to Fig. 11-31 and list the reasons for the hand and body positioning. Note, safety is a matter of common sense in most cases. This was not covered in the reading.

8. Look again at Fig. 11-32 and identify the reason for two hands on the sandpaper instead of pinching it into a loop with only one hand.

How the Lathe Works

Introduction: Now we assemble the major components of a screw-cutting lathe, one part at a time, and observe how they operate. You might also hear this machine called an engine lathe because they were one of the earliest of machines to be powered with their own drive motors rather than by an overhead drive shaft powering the entire shop. We'll start with the most critical part of the machine, the bed ways and frame. Then we'll build it, one piece at a time.

KEY TERMS

Apron
The front, vertical part of the carriage where most of the operator controls are located.

Bed ways
The precision spine of the lathe on which tool motion slides.

Carriage
The assembled apron and saddle make up the carriage.

Chasing dial
The indicator of when to engage the half nut lever to ensure that each new pass will fall in the same thread groove as the last.

Dovetail
The guides upon which the cross slide and compound slide.

Gibs
Small wedge-shaped shims that are used to compensate for wear and minimize unwanted motion in tool slide.

Half nut (lever)
The devise that closes the clam shell half nut upon the lead screw to start pulling the carriage for threading.

Saddle (lathe)
The major component that rides upon the bed ways.

Tailstock
The far end support and drilling accessory.

Tool compound (lathe)
The pivoting axis of the lathe used to machine angular surfaces.

Building a Lathe One Component at a Time

Bed Ways and Frame

The **bed ways,** often shortened to *ways*, are the precision heart of the lathe. They are usually made of fine-grain cast iron or of alloy steel. Their surface is hardened and ground flat and parallel. The ways are attached to the frame, which must also be flat and parallel to the shop floor. Frames vary in their construction from single castings to welded components and assembled combinations of both. On some lesser quality lathes, the frame and the ways are combined into a single casting, but that makes replacement of damaged ways impossible.

All sliding-axis components are supported by the ways and frame. The tailstock clamps to them too. If a turned shaft is to be perfectly cylindrical, the lathe must have ways in top condition.

Longest Supported Bar The length of the ways determines the longest bar that can be supported for machining (Fig. 11-36). It's the longest bar that can be held between the two centers one in the tailstock and one in the driving headstock.

Professional Care There are four vital items of care that must be given to bed ways.

Keep the ways clean and oiled. After use, wipe the chips, coolant, and dirt off the ways with a shop cloth, then rewipe on a thin coat of machine lubricating oil. This oil is important because it provides initial lubrication for the next movement of the lathe after cleaning. The lathe will

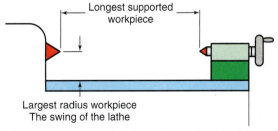

Lathe Size

Longest supported workpiece

Largest radius workpiece
The swing of the lathe

Figure 11-36 *Lathe size is determined by longest supported part and largest radius to be rotated without hitting the bed ways.*

then lubricate itself but not immediately. The oil also prevents rust during idle periods. A reminder from Section 5, while compressed air is used in many production shops to clean chips off the lathe, it is unwise to use air to clean the ways, use a brush or a rag. Chips can be forced into the dirt seals and wipers of the **saddle** (the device that slides on the ways). These seals are not designed nor are they able to hold out compressed air.

Never lay machine accessories, work material, or hand tools on the ways. The lathe bed seems just right for temporary storage but it's a poor practice.

When adding accessories to the lathe, protect them from any falling object. In keeping this policy, you will find a special wood block, called a cradle, made for protecting the ways when changing chucks (Fig. 11-37).

Always check and use the lubricating system on your lathe. Lathes either feature an automatic system or a manual oiling pump(s). Determine which and make sure the reservoir is above the add mark with the right type of oil for sliding friction ways (Fig. 11-38). Then pump the manual plunger system on a schedule. Not sure how long between pumps? Other than a little extra oil in the sump, you can't do it too often. Pumping way lube two or three times per shift is usually considered OK. Read the operator's manual or placard on the machine or ask someone.

Figure 11-38 *If the lathe does not have automatic slide lubricators, then it will feature a manual pump like this.*

Figure 11-37 *Do it right—use a wood cradle when changing chucks.*

The Headstock

Adding on to the lathe frame, the next major component, the headstock contains the necessary gearing, shifters, and spindle bearings (Fig. 11-39). The spindle bearings support and rotate the spindle, chuck and work.

The headstock case is bored to support the spindle in a set of highly engineered bearings. They must allow smooth rotation of often heavy work and the chuck yet prevent any other movement, side or end. They must also absorb the shock and force of machining. This demands exceptionally strong bearings.

Headstock Lubrication Is Critical To keep these special bearings alive and well, there will be a headstock lubricating procedure for your lathe separate from the ways (Fig. 11-40). Most likely, an oil level sight glass

The Headstock and Gear Case

Figure 11-39 *The headstock supports the work, drives the spindle, and contains the shifters for various RPMs.*

Figure 11-40 *Opening the outer cover, user-serviceable components are seen, but many more are beyond this within the inner headstock case.*

or dipstick will be present. There are several types of spindle oil, so be certain to read the placard or operator's manual for that machine. Keep the oil above the add mark but do not overfill, as that can sometimes blow out seals.

The Spindle and Spindle Taper

The external part of the spindle is the *spindle taper*, also called the *spindle nose*. It features a

There are two lubrication systems on lathes: one for the ways using a viscous (thick) way lube and one for the spindle and gearbox—called spindle oil. Do not mix them!

taper on which precision chucks or other work holding attachments such as face plates are firmly mounted. The nose will feature a set of locks to pull and lock the chuck or other holding devise on the taper during turning (Fig. 11-41).

KEYPOINT

It's beyond critical that you never attach any spindle accessory (chucks, face plates, and adapters) without first cleaning both spindle taper and the mating surface carefully (Fig. 11-42). Dented spindles destroy the accuracy of a lathe! Putting them together dirty is a sure way to ruin the lathe and to become unemployed.

Bar Feeding Through the Spindle The spindle continues from the operator side, through the headstock bearings, to the far

Figure 11-41 *The spindle nose taper and chuck locks work together to securely hold chucks and other accessories on the lathe.*

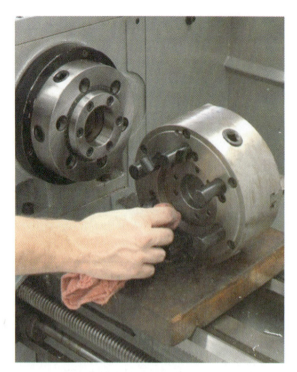

Figure 11-42 Critical! *Always clean the tapers on the spindle nose and mating surface on the chuck before mounting and tightening.*

side of the headstock. It's a hollow tube so that long bars can slide completely through. The diameter of that tube is another way we rate the size of lathe—the largest bar that can be slid through the spindle. Lathe spindle holes commonly range from 1.0 in. up to around 3.0 in., in training facilities and much larger in industry.

Making a part from a long length of metal is called a *bar operation*. A single part is machined then cut off the bar. Then with the lathe stopped, the bar is unchucked, slid out, rechucked, and the next part is made. When a bar is within the spindle, several precautions must be taken.

Safety Note about Long Bars Long bars running entirely through the headstock can present a serious safety problem. When holding and machining these bars be absolutely sure that the far end (out of the headstock into the shop) is guarded and supported. If this step is not performed:

The Immediate Danger! The spinning bar could hurt other machinists as they walk by your lathe. Place a bar support and some form of barrier around the headstock or better yet, select a bar length that does not protrude at all! Fig. 11-43.

Danger!

Figure 11-43 Danger! *A long, unsupported bar is a death-causing accident ready to happen! See Fig. 11-44.*

The Whip Far more dangerous, when the lathe is started rotating with a thinner unsupported bar such as in Fig. 11-44 protruding from the headstock, it will first vibrate then wobble, then quickly bend to a 90° angle, which violently whips the bent rod in a very large circle.

Safety Key Point The force of this out-of-balance condition can actually move lathes right off their foundation. The danger to yourself and others is tremendous, from the bar tip swinging in a large circle with its tip moving at close to the speed of sound!

Danger!

Figure 11-44 *This is a real example. The instructor started the spindle and here's what happened!*

Third feature for lathe size

Spindle hole-largest bar through

Figure 11-45 *Size is also determined by the biggest diameter bar that can be passed through the spindle hole.*

Lathe Swing—Largest Radius Part The third lathe size rating is the *swing*. The swing of a lathe is the largest part radius that can be rotated over the bed ways. It's the distance from center of the headstock spindle to the ways. Common swing radii are from 8 to 22 in. in schools, while swing capacities of several feet can be found in industry (Fig. 11-45).

Lathe size is determined by swing radius, longest supported bar, and largest diameter bar through the spindle.

Lathe Carriage

This lathe component is made up of two parts joined together, the horizontal saddle and the vertical apron (Fig. 11-46). The entire mechanism adds the left/right Z axis motion. The **carriage** slides upon the bed ways to provide the Z axis of the lathe. It is moved either by hand or by power for straight turning. It also supports the X axis, then the compound and ultimately the cutting tool. When

The Carriage Includes Saddle and Apron

X motion

Saddle

X axis dovetail way

Z axis motion

Z axis hand wheel

Feed and thread clutch levers

Apron

Figure 11-46 *The saddle attached to the apron forms the carriage to provide Z axis tool movement.*

Lathe Controls on the Saddle

Thread engage lever (half nut lever)

Feed engage levers

Z axis handwheel

Axis select feed X or Z

Spindle start

Foot brake (optional)

Figure 11-47 *The apron contains most of the operator controls used during turning.*

Z axis motion must be stopped, the saddle features a positive lock.

The Apron

The vertical, front of the carriage is called the **apron.** The controls used during turning to start and stop the action are located here (Fig. 11-47). We'll look at them individually in a moment.

Cross Slide

Now assemble the cross slide over the carriage dovetail to create the X axis, perpendicular to the Z axis (Fig. 11-48). It can be moved by hand crank or by power feed for facing or parting and features a micrometer dial for accurate positioning. The X position can be locked on most lathes, to prevent tool slippage during turning, for example.

Diameter Graduations The lathe's X axis is graduated for diameter values. For example, after taking a cut, then dialing the X axis forward by another 0.010 in. and taking a second cut, the measured part will be 0.010 in.

Cross slide

Dovetail ways

X axis micrometer dial

Figure 11-48 *The cross slide moves the tool in and out along the X axis. It is held to the saddle by a dovetail way.*

ShopTalk

Radius Lathes Rarely, one might encounter a very old, indirect reading lathe that moves the tool in or out at the dialed amount. That bit of questionable engineering changes work diameter at twice the dialed amount! These confusing machines were known as *radius calibrated* and they've mostly been scrapped or converted to diameter calibrated dials. I've only run one in my entire career and it was a real nightmare to use!

Direct reading or diameter calibrated *X* axis

Tool is advanced 0.005 in.

0.010 in. smaller diameter

Figure 11-49 *If the cross slide and tool are advanced 0.005 in., the diameter becomes 0.010 in. smaller. The amount dialed equals the amount of diameter* change.

Figure 11-50 *Adjustable gibs minimize unwanted movement of the cross slide and the tool compound too.*

smaller in diameter. Similarly, CNC lathes are usually programmed with *X* coordinate values representing diameters.

This useful feature is known as a diameter graduate lathe or sometimes as a direct reading lathe. So, when dialing the *X* axis in to a smaller diameter, by 0.010 in., for example, what is the actual tool movement? The tool goes forward half the amount (0.005 in. in this example), which reduces the radius 0.005 in. But that translates to a 0.010-in. diameter change. See Fig. 11-49.

Motion and Positioning Adjustments and Backlash

Lathes are made such that both backlash and unwanted axis movement can be adjusted to a minimum. What is "backlash"? See the Trade Tip coming up.

Gibs for Slide Adjustment

To minimize unwanted side or up and down movement of the *X* axis cross slide, which would destroy accuracy and the finish on the workpiece, small adjustable wedges are used in the **dovetail** part of the slide (Fig. 11-50). These wedges are called **gibs.** The gib screw requires periodic adjustment to compensate for wear. You'll see gibs on many other machines that feature sliding actions.

With your safety glasses on and with permission, take a moment to view the gibs on both the cross slide and the upper compound axis of a lathe in your lab.

Trade Tip

Understanding Backlash! Also called "mechanical play" or mechanical clearance, machinists must understand and control backlash. It affects the accuracy of all machines in the shop where a tool, machine table, or slide is positioned or moved during machining. Uncontrolled, backlash not only spoils accuracy, it can cause an accident!

For example, backlash is present in the cross slide of a lathe because it's moved by feed screw, turned by the *X* axis handwheel. Turning the handwheel moves the axis drive nut (and cutting tool) fastened to the cross slide.

The axis is either pushed or pulled, depending on the direction the hand wheel is rotated.

There must be a necessary, small space between the nut and screw or the system will bind. That minimum required clearance is responsible for the backlash. Over time, the nut wears and the backlash increases. With regular lubrication, machinists minimize this wear. On most lathes it can be adjusted to a minimum. But not all backlash can be eliminated with this nut and screw system. Later in Section 17, we'll see another kind of axis positioning and drive called a ball screw, used on CNC machines. They eliminate backlash problems but they are too expensive for manual machines. So, how does backlash affect your actions when running the manual lathe or mill? Answer this (Fig. 11-51).

The answer is that for a short time, the tool does not move even though the reversed crank and micrometer dial are indicating that it should be. That brief stall in movement, when the crank direction is reversed, is the backlash. It also means that for a short time the axis has no positive control should the cutting action grab the tool—it can jump forward or back! That uncontrolled jump can lead to a broken cutter or worse.

It is caused by the clearance between screw and nut—reversing. This might be as little as 0.002 in. or as great as 0.030 in. or more, depending on wear and maintenance. A minimum amount of backlash is necessary, but is problematic on machines using standard nut/screw systems.

Compensating When referring to a micrometer dial position on a machine, you must always know in which direction the handle is turning with respect to the graduations and backlash. Your instructor or a journey-level machinist will demonstrate how to deal with backlash on your lab's machines. Remember—you must know how to compensate for backlash in all setups where a tool is positioned. Digital positioners eliminate mispositioning the cutting tool, but they do not overcome potential slippage when the tool is pulled across the backlash gap.

What happens when rotation is reversed?

Figure 11-51 *After cranking inward, what happens to tool movement when it's reversed (cranked outward)?*

Antibacklash split nuts

Adjust apart then lock to the cross slide

Figure 11-52 *Adjusting split nuts apart, maintains backlash at a minimum.*

Split Nuts Adjust for Backlash

There are many ways to maintain a minimum amount of backlash in an axis screw. The split nut is the more common method used on engine lathe cross slides (Fig. 11-52). By adjusting the gap between the two nut halves, a minimum amount of play is maintained, minimizing backlash in both directions. We'll explore antibacklash devises in more detail later as controlling it is part of a machinist's bag of tricks.

Compound Axis

The compound axis adds versatility to the lathe. Mounted on top of the cross slide, it has no power feed. It can be rotated to any angle with respect to the *X-Z* axes then locked in position. The **compound** is used mostly for turning angular surfaces. The compound also features a slide lock for the times when movement is not needed, and they have a micrometer dial on the hand crank. The compound slide is fitted to its base using dovetails and gibs to control unwanted side movement (Fig. 11-53).

Tool Compound Axis

Tee slot mount

Micrometer dial

Rotation lock bolts

Angle graduations

Figure 11-53 *The tool compound is rotated by loosening the nuts.*

The compound is subject to backlash error in the positioning screw too. If diameter accuracy is important, then lock the compound axis at the supported position shown in Fig. 11-54, not hanging out over its dovetail.

Trade Tips

Lathe Compounds *Keep it supported for general work.* In Fig. 11-54, you see a common error. It is possible to crank the compound slide out quite far from it's supporting dovetail. Occasionally, extending it this far is necessary for special or delicate work. However, avoid this position for everyday work. The slide is too weak when extended this far! It will deflect down, breaking tools, ruining work, and occasionally breaking off the compound itself!

KEYPOINT

When not using the compound axis, always keep it drawn in to the minimum extension and locked as shown in Fig. 11-54.

Overextended weak point

Correct support

Figure 11-54 Caution! *Overextending the compound can break it at the tee slot—the weakest point.*

Angle graduations on the compound. Most lathes (but not all) mark the angular graduations such that the compound is at *zero degrees* when it's parallel with the *X* axis. However, when the compound is swung to machine a steep angle, recall that the drawing usually specifies the needed angle with respect to the *Z* axis centerline (Fig. 11-55).

KEYPOINT

Set and lock the compound to the complimentary angle.

Figure 11-55 *To machine the 20° half angle (reference to the Z axis), the compound is rotated to 70° from the X axis.*

Lubricating the Compound

Some lathes include the compound in the total saddle lubrication system, however, on others, the compound is lubricated using a hand oil can. Lube it with way oil, on the schedule your instructor specifies or once per shift. With this addition to our lathe, the axis and tool positioning system is complete.

The Gear Box

The area directly under the headstock is the gear case and the electric motor. The features you need to understand for now are the "quick change" shifting gears (Fig. 11-56). These gears provide control over four major setup selections:

Feed rate amount in inches per revolution

The feed rate direction—*X*, left/right, or *Z*, in/out

The RPM

Selection of feeding action or threading

Trade Tip

Lathe Gear Changing Shift like a "pro." Never attempt to shift these gears while the lathe is running. A single ruined cluster gear (the main selection gear) can cost as much as a month's wages or more for a machinist!

Second, when changing gears, you may need to rotate the spindle by hand just a bit to "drift" the gears into mesh. Some lathes (other machines too) feature a jog button for this purpose. When you tap it briefly, the spindle motor jumps forward to gently bump gears into mesh.

Feed & lead selection

RPM selection

Figure 11-56 *The quick-change gear case contains most of the setup shifters and gear selections.*

The lathe will feature a built-in chart (Fig. 11-57) on the headstock to show which gear positions result in which feed rate or thread pitch.

KEYPOINT

When you jog the spindle motor, it will coast and stop. When it's nearly stopped, ease the gears into mesh.

Tailstock

The **tailstock** is an integral part of the lathe although it's a bolt-on component. It's not an accessory.

Three Tailstock Movements In addition to the quill in/out function (the drilling axis) the tailstock features two more movements. Both are for setup position only, and must be locked into place during turning.

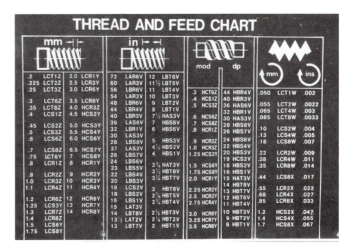

Figure 11-57 *Most screw cutting lathes feature a feed rate and thread lead gear selection chart similar to this.*

Z position The tailstock slides on the bed ways. This is used only for rough positioning. It is then locked for drilling or for supporting work. When sliding the tailstock, always wipe chips and dirt away from the ways to prevent scoring the bed ways.

Quill feed Note in Fig. 11-58 that the quill is graduated for drill depth control. However, these graduations are ruler accurate only. The hand wheel also has graduations of varying resolutions. A 0.005-in. resolution is common on better machines.

Lateral offset Mentioned previously, the tailstock quill frame can be moved sideways on its base such that it is no longer in line with the headstock, for tapering operations. It is shown set to the back in Fig. 11-58.

Tailstock Functions

Figure 11-58 *The tailstock has three possible movements: the setup positioning parallel to the Z axis, the lateral positioning off the centerline for taper operations, and the quill slide positioning parallel to the Z axis for drilling.*

Section 11 Turning Operations ♦ **401**

Tailstock Off Center If you move the tailstock off center for taper turning, then return it to alignment before leaving the lathe. (Or leave a note.) Should the next operator not detect the misaligned condition, the surprise taper produced might result in a scrap part. When finishing a job, do not leave the machine with the tailstock off center unless you leave a note that it is off (Fig. 11-59). When setting up the lathe, always check its centering.

In addition to the tailstock position scale, a quick eyeball check is made by sliding it all the way forward to compare it's center point to one placed in the headstock.

An indicator is required for perfect alignment (Fig. 11-60). Precisely centering the tailstock requires the use of a dial indicator. The indicator holder is mounted on or in the chuck and spun around the tailstock quill.

Alignment scale may be a vernier

Figure 11-59 *Recentering the tailstock, the offset graduations below the crank are also a good rough guide.*

Recentering the Tailstock

Magnetic holder or clamp on

Spindle gears in neutral

Adjust to zero front to back

Figure 11-60 *The best way to align the tailstock to the exact spindle center.*

Work Support Accessories

Extra long work cannot be safely supported by the chuck and tailstock alone. To prevent the central portion from springing away from the cutter, or from riding up on it, one or more center supports are needed. There are two types and they can be used individually or together.

Follow Rests

Because it's bolted to the cross slide or carriage, the follow rest (also sometimes called a travel rest), moves along with the tool in the *Z* axis direction. The follow rest is a two-point support opposing the work's tendency to spring up and away from the cutter. The follow rest can be bolted ahead of the cut to ride on the uncut metal or behind the cut to ride on the cut just made. Figure 11-61 shows the rest from the tailstock side.

Better follow rests feature roller bearings, but they also can use consumable, brass rub pads that contact the work. If using the brass pad version, a heavy pressure grease lubricant is used to aid in reducing friction and to prevent smearing of the brass onto the work. Watch out for heat and too much pressure against the work with these type supports (Fig. 11-62).

Placing the Follow Rest Ahead of the Cut If you wish the cut to be concentric to an existing diameter, then the follow rest must be set to contact the work ahead of the cutter, on the original surface. In order to do that, the original diameter contacting the support must be smooth and round. Then,

The Follow Rest

Adjustable brass pads or rollers

Cutting tool

Cross slide top

Figure 11-61 *The travel rest is a two point support that moves with the saddle.*

Figure 11-62 *On smooth stock with no runout, the travel rest leads the cut to machine the work concentric to the original surface.*

due to a changing diameter, the supports must be readjusted after each pass of the tool.

Placing the Follow Rest Behind the Cutter This rest position is used when the original material is rough not round, or deliberately not on center, as in turning a cam roller. For this setup, it's necessary to create a smooth place to contact the rest pads to the work surface. A short cut is taken when the cutter is near the tailstock or chuck. Then the lathe is stopped and the supports are brought into contact. Here too, the rest must be reset after each pass (Fig. 11-63).

Figure 11-63 *To support and cut long stock with runout, such as this hexagonal bar, the travel rest is set up to follow the cut.*

Steady Rest

Figure 11-64 *A steady rest clamps to the bed ways and does not move during turning operations.*

KEYPOINT

A follow rest can be placed ahead of the cutter or behind, depending on the work condition and effect desired. Two follow rests can be used, one on either side of the cutter.

Steady Rests

These three-point supports clamp to any position along the bed ways. They do not move with the tool (Fig. 11-64). One (or more) are used to support long work between cuts.

To use a steady rest, the work must be round and smooth. Where the raw stock isn't round or smooth, its necessary to prepare it first by chucking the shaft near the center portion (with half its length remaining inside the spindle tube), then with a tailstock supporting the outboard end, cut a small section to any diameter that cleans up. It's then pulled out and the rest supports are brought into light contact. Do not overpressurize them, they should be brought just in contact then stop.

Controls Runout of the Work

As long as the work is round and smooth, runout of the workpiece before contacting the rollers is not an issue. As the rollers or pads are brought together, they will remove the wobble. (See the Trade Tip.)

KEYPOINT

Sometimes the only way to get the job done is to setup both steady and follow rests working together for very long or slender parts. Use caution as either work support can force the work off center if not set up correctly.

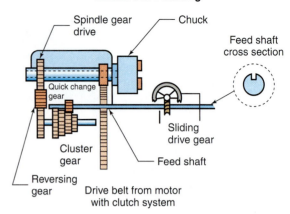

Manual Drive Gearing

- Spindle gear drive
- Chuck
- Feed shaft cross section
- Quick change gear
- Cluster gear
- Reversing gear
- Sliding drive gear
- Feed shaft
- Drive belt from motor with clutch system

Figure 11-65 *A simple sketch of the feed system on a lathe.*

The feed shaft rotates at the rate selected with the quick-change gears, at a specific ratio with respect to the spindle. It passes through the apron, where a takeoff gear slides upon it and is driven through a key in the shaft key way—but this happens inside the apron out of sight (Fig. 11-65).

Feed Direction

Manual lathes feature a selector that determines the direction of feed. The X axis can move either toward center or away, and the Z toward the headstock or away. The direction selector is usually found on the gear box near the feed shaft output, but on some lathes it might also be located on the apron.

Axis Drive Systems

The final lathe component we'll investigate are the power feed and threading systems. Due to a difference in backlash control and precision, there are two different axis motion systems on a lathe: one for moving the cutting tool in the X and Z axes for general machining and one for driving the threading tool along the Z axis only.

Feed System

For general machining, the spindle is connected through gears, to a shaft that runs just below the ways—called the feed shaft. Take a moment to go to the lab and see it along the front frame extending out to the tailstock end. It's a round shaft with a square groove in it called a key way. In mechanical devises, this kind of shaft is sometimes called a *jackshaft*.

Differences in Lathes

These are a couple of details to discover when operating a lathe for the first time:

1. **Single or Double Engagement Levers** On most larger lathes, there will be two feed engagement levers, one to start feeding in the X axis and one for the Z axis. However, on smaller compact models, there may be only one feed engagement lever. If so, you must then select whether the feed will start the X or Z using a selector lever.

2. On-the-Fly Reverse or Not

On some, the feed direction can be reversed without stopping the spindle, on others you must stop the lathe. Know the difference or grind gears on the lathe that must be stopped, or even break them!

Lead Engagement System for Threading

Due to cutter's travel ratio with respect to spindle turns, a very different tool movement system is necessary for threading. To make precision threads the cutter is pulled along the carriage using a screw action rather than gearing. This system would wear out quickly if used for daily machining, therefore it's reserved for threading. It's a long screw, called the *lead screw* running next to the feed shaft, seen in Fig. 11-66. You might identify it on a lathe in your lab, now.

When setting up for threading, the *feed direction must be set to neutral and the lead screw must be engaged.*

Once the lathe's lead screw is selected for threading, a new engagement lever is used to start the tool moving, the **half nut** *engage lever* (Fig. 11-66), located on the right side of the lathe apron. On most but not all lathes,

Thread-chasing Dial Mechanism

Figure 11-67 *The chasing dial indicates when it's time to start threading.*

the half nut lever is locked out until the feed is set to neutral. The lead screw system imparts tool motion in the *Z* axis direction only.

Chasing Dial

The **chasing dial** tells the machinist when it's time to start the threading tool moving. During threading, several passes of the cutting tool are usually necessary. The chasing dial is a gear-driven indicator, as shown in Fig. 11-67. It indicates when the cutting tool will start in the same thread groove as the previous pass. The chasing dial is located near the half nut lever. We'll investigate its use in Unit 11-7.

Figure 11-66 *For threading only, we use the* lead screw, half nut engagement lever, *and* chasing dial.

UNIT 11-2 REVIEW

Review the Key Terms

Apron
The front, vertical part of the carriage where most of the operator controls are located.

Bed ways
The precision spine of the lathe on which tool motion slides.

Carriage
The assembled apron and saddle make up the carriage.

Chasing dial
The indicator of when to engage the half nut lever to ensure that each new pass will fall in the same thread groove as the last.

Dovetail
The guides upon which the cross slide and compound slide.

Gibs
Small wedge shaped shims that are used to compensate for wear and minimize unwanted motion in tool slide.

Half nut (lever)
The devise that closes the clam shell half nut upon the lead screw to start pulling the carriage for threading.

Saddle (lathe)
The major component that rides upon the bed ways.

Tailstock
The far end support and drilling accessory.

Tool compound (lathe)
The pivoting axis of the lathe used to machine angular surfaces.

Replay the Key Points

◆ For power drive of an axis, you must select either feeds or threads—two different systems.

◆ Direction of feeding or threading is selected on the headstock gearbox. This causes either a right-hand spiral or left-handed threads. Or when feeding in/out or right/left selection.

◆ When threading, a separate drive shaft and engagement lever is used. The engagement lever is not a clutch. Called the half nut lever, it is either engaged or not, but it will not slip as a clutch would.

◆ To engage the half nut lever at the right time, the lathe features a chasing dial that indicates when to drop the lever in gear.

Respond

1. On a sheet of paper, identify the indicated items in Fig. 11-68. Then with permission from your instructor, with the correct safety eyewear, identify the same items on a lathe in your lab. Quiz a fellow student as to each item's purpose. Note, do not interfere with machinists operating the machine.

2. With a partner, identify and describe the function of these parts on a manually operated lathe.
   Bed ways
   Saddle—tool carriage
   Apron
   Tool compound
   Quick change gearing
   Z axis hand wheel
   X motion
   Tailstock lock
   Tailstock quill
   Axis direction selector
   Dovetail and gibs (cross slide or compound)
   Chasing dial for threading

3. The compound is an adjustable axis that has a micrometer dial. Is this statement true or false? If it is false, what will make it true?

4. What danger is there in extending the tool compound out to the end of the stroke when machining? When is overextension of the compound necessary and justified?

5. Of the two different work rests available, which has three point suspension?

6. Describe a "follow rest" and it's use in machining. Be complete.

7. Name the three possible movements of a tailstock and a brief description of each.

8. What is the name for the coasting process that helps to engage gears while shifting a lathe?

9. Name the moving objects used during turning operations, both rotating and sliding, that could be a safety danger. Things that could catch, cut, or pinch the operator. (Not control levers, hand wheels, or dials.)

10. Head stock spindles are hollow on lathes to save weight and reduce the horsepower required to drive them. Is this statement true or false? If it is false, what will make it true?

11. Name two reasons for wiping oil on the bed ways after cleaning them.

12. Match the axis with the sliding devise
   A. X axis Compound
   B. Z axis Saddle
                     Cross slide

Figure 11-68 *Identify the following lathe parts.*

Figure 11-69 *Identify this lathe operation used on both manual and CNC machines.*

13. Screw cutting lathes feature two complete systems for moving the cutting tool. These systems are the _____ and the _____ systems.

14. The splitting dial is used to ensure the tool runs down the same thread groove each time. Is this statement true or false? If it is false, what will make it true?

15. Gibs are small wedges used to minimize unwanted movement in the saddle, cross slide, and compound slides of a lathe. Is this statement true or false? If it is false, what will make it true?

16. The axis direction selector is found on the lathe _____.

17. What is the name of the operation shown in Fig. 11-69.

UNIT 11-3

Work-Holding Methods

Introduction: There are six different methods of holding material for turning. The choice depends on several factors. The first is, which of them do you have? Your objective at this stage is to know the difference in their abilities. Which is safer for the job and holds best?

Work holding is a big issue with turning. Keep in mind that the work-holding devise you select will keep the spinning mass under control. If the job is for a CNC lathe, your choice is even more critical as the speeds and forces multiply. Here in Unit 11-3, we explore the options for work holding on manual lathes. Most, but not all, apply to CNC as well.

KEY TERMS

Cam locks
The most common method of holding chucks on a lathe, they use rotating cams in the lathe spindle that grip pins protruding from the chuck.

Collets
Sometimes called spring collets due the way they are made, these precision hardened and ground griping tools hold the work by being pulled into a tapered hole.

Dead centers
A support center that has no bearings for rotation. They rotate with the work when used in the headstock but rub when used in the tailstock.

Fixtures
Custom-made tools for holding special shaped work not possible to reliably hold by any other method. Sometimes called jigs.

Grip
The comparative amount of bite a holding method is able to impart to the work.

Independent chucks
Chucks that feature jaws able to move individually.

Live centers
Support centers used only in the tailstock, they have bearings to eliminate friction.

Scroll plate
The helical face thread that moves all three jaws simultaneously in a three-jaw chuck.

Soft jaws
A custom-machined three-jaw insert used where concentricity or part shape is an issue.

Decision Factors

In planning your work holding setup, there are five aspects to consider:

1. **Work/Operation Requirements**
   To choose wisely, above all else, you must know the operations that are to be performed. That includes how much raw stock must be removed and how? The remaining choice factors depend on your understanding of what you will be doing in this setup. Find the facts on the drawing and work order, and by comparing the raw stock to be turned to the finished shape and size. Determine the nature of the work. Here are a couple of examples where the choice might be different. (A) It's a rough HRS bar that will be completely machined away or (B) it's a brass forging and must not be marred by the holding method since some of the original

surface remains on the finished part. Stock has or doesn't have excess for gripping inside the chuck. Is it a long bar job—through the headstock? What are the job tolerances?

2. **Grip**

How much holding pressure is needed and possible without marring or crushing the work due to how much material will be machined away or how strong or delicate is the object to be held?

3. **Turnaround Time**

Is this a one-off part or are you setting up to make several parts? How quickly should the setup be rechucked? This becomes a greater issue with CNC lathe work.

4. **Runout (discussed next)**

This is an accuracy issue. What are the runout tolerances for the job? What is the typical surface wobble method?

5. **Special Characteristics**

Each method offers special advantages or a definite type of work for which it is best suited.

We'll look at each of these categories as we examine the various holding methods.

Runout

From Section 4, runout is the geometric characteristic of surface wobble on rotating objects. It is always present in lathe work but unwanted. Each work holding method exhibits various amounts of runout. Your objective is to match the holding method to the tolerance.

Measuring Runout

In Fig. 11-70, an indicator is touched to the work surface while the object is slowly rotated 360 degrees (usually by hand—placing the spindle gears in neutral). The indicator is

Testing Runout

Figure 11-70 *The total movement detected is the runout—a geometric term for surface "wobble" on a rotating object.*

adjusted to read zero at the lowest or highest point in the rotation. Next, rotate the part and find the greatest difference in values (recall that to be FIM or TIR). Both ends of a shaft should be tested for runout, and possibly the middle also, if the shaft is long. It is possible for one end to show very little TIR while the other end might show a large runout.

KEYPOINT

Workpiece runout can be caused by irregularities of the work surface, but the point is, it can also be introduced into the setup by the holding method itself. It's usually a combination of both.

With some turning jobs, you need not give much consideration to initial runout; for example, when holding a rough metal bar to turn its entire length to finished diameters. Then when complete, it is parted off the remaining bar gripped in the chuck. Any wobble in the initial work surface is removed with the machining—the accuracy of the chuck isn't an issue.

However, now suppose you wish to machine a new feature on a previously machined axle. Functionally, the new diameter must be concentric to the original shaft's centerline and surface. Here runout control of the holding method is of the highest concern.

The Work-Holding Methods

Here are the possibilities.

Three-Jaw Universal Chucks

These chucks are simple to use and have an excellent grip on work where heavy cuts are to be made. The main advantage to three-jaw chucks is that the jaws close simultaneously, similar to a drill chuck. As such, three-jaw chucks feature moderately fast turnaround times.

As shown in Fig. 11-71, turning the chuck wrench gear rotates the **scroll plate** inside the chuck body. The scroll plate is a flat disk with a helical groove. Moving jaws simultaneously gives rise to the term *universal jaws*, as opposed to chucks whose jaws open or

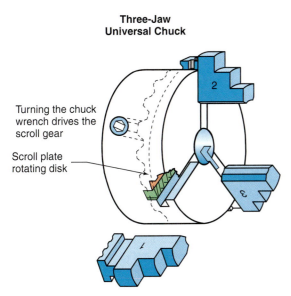

Three-Jaw Universal Chuck

Turning the chuck wrench drives the scroll gear

Scroll plate rotating disk

Figure 11-71 *A universal* three-jaw chuck.

Outside Grip

Inside Grip

Figure 11-72 *Three-jaw chucks feature* inside *and* outside *grip jaws.*

close independently. In Fig. 11-71, one jaw has been removed to show the scroll plate beneath.

To replace the removed jaw in Fig. 11-71, all three jaws must be removed completely, then inserted onto the scroll, in their numbered sequence. Otherwise the chuck will not close on center. Have your instructor demonstrate this task.

Reversed Jaws for Extra Large Work

Figure 11-73 *Here the chuck jaws have been reversed to nest grip a large diameter object.*

The jaws can grip on the outside or the inside of objects (Fig. 11-72). However, when objects become relatively large, the jaws of many chucks (both three- and four-jaw chucks coming up next) can be reversed such that they nest-grip, as shown in Fig. 11-73.

Reversing Chuck Jaws

There are two ways chucks might be manufactured for reversing jaws:

Two-piece jaws (more expensive but more useful)

Total reversing jaws (remove from chuck carrier slot, turn around, and reinsert)

Two-Piece Jaws To reverse these industrial quality jaws, two bolts are removed (Fig. 11-74). The jaw is tapped lightly with a soft hammer to remove it from its carrier, which remains in the chuck. Next, both the jaw and carrier are cleaned of dirt and

Two-piece Reversing Jaws

Soft jaw blank

Hardened steel jaws

Keyed jaw carrier

Figure 11-74 *Two-piece jaws are reversed by removing two bolts.*

chips. Then the jaw is rebolted in the reverse position.

Custom soft aluminum jaws are an option that bolts in place of the hardened steel jaw. We will look at the uses of soft jaws later.

Solid Jaw—Total Reversal Here the jaws cannot be taken apart from the carrier. To reverse these jaws, back them all the way out of the chuck. Reverse then reinsert while turning the chuck wrench.

Trade Tip

Jaw Reversing Sequence After cleaning, make sure to replace the jaws in the correct order in the chuck. Notice that both the jaws and the slots in the chuck are numbered 1, 2, and 3. Insert and recatch jaw 1, on the scroll plate in slot 1. Then rotate the chuck and start 2 in 2 then 3 in 3 last.

Why Choose a Three-Jaw Chuck?

A high-quality, new three-jaw chuck might produce as little as 0.001 to 0.003 in. runout when gripping a perfectly round object. However, as they are used for heavy machining, runout degrades due to overtightening, crashes, and normal wear of the scroll plate and spindle mount taper.

Runout   Not the best choice to control runout. Expect from 0.005 to 0.010 in., depending on the history of the chuck. Of all five methods, three-jaw chucks are the least capable of producing zero runout.

Grip   Excellent on solid round work. They grip a large range of part sizes both outside and inside. Three-jaw chucks with steel jaws can mar the work.

Turnaround   Medium fast.

Special Characteristics   Besides its universal nature, the three-jaw's main advantage is the ability to create custom-shaped soft jaws to fit odd-shaped work not easily held in other chucks.

Trade Tip

Flying Wrench Rule   When tightening or removing work from a chuck, *never leave the wrench in the chuck!!*

On the very day I wrote this unit, a student in the next lab to mine, forgot that the wrench was in his chuck when he turned on the lathe. It hurled past him and three other terrified students, on its way to a plate glass window into the hall, where a group of amazed high school visitors stood watching. The student machinist was speechless for several hours. A bunch of visitors decided right then they would not enter the trade! Fortunately there were no injuries—this time.

To avoid this problem *never* leave a chuck wrench in the chuck for any reason. **DO NOT** *take your hand off the wrench when it's is in the chuck. Remove that wrench—always!!!* (See Fig. 11-75.)

Figure 11-75 *Leave it, and you'll launch it!*

Four-Jaw Chucks

Four-jaw chucks feature individually operated jaws. Each jaw is cranked in or out independent of the other three. Thus their second name, **independent** jaw **chucks.** As shown in Fig. 11-76, each jaw need not remain concentric to the chuck, therefore odd-shaped work can be gripped. To reverse these jaws, each is backed out, then reversed and reinserted, one at a time in any order since each is independent of the others.

Independent Four-jaw Chucks

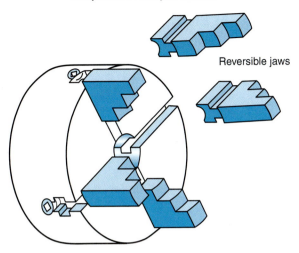

Reversible jaws

Figure 11-76 *Each jaw is moved independently thus 100 percent of all runout may be adjusted out of the setup.*

Why Choose a Four-Jaw Chuck?

Runout   There is no inherent runout in the chuck—it depends on how skillfully the machinist tightens them. Using an indicator, the machinist can adjust a perfectly round object such that all the runout is gone—perfectly centered. This is called "indicating the work" or "truing the work."

KEYPOINT

Because of its adjustable nature, the four-jaw chuck is the only method that can eliminate 100 percent of the runout when gripping a perfectly round object.

Grip   A very strong grip can be had on round and square objects.

Turnaround   Although there are a couple of trade tips that help speed up changing from one part to another, the four-jaw is one of the slowest to turnaround because each jaw must be tightened, usually while watching an indicator to center the work. Since they are such a slow part exchange method, they're avoided in production CNC setups if possible.

Special Characteristics   Four-jaw chucks are used where runout is of the highest concern for the job. They are also good at holding square and odd-shaped work, something a three-jaw chuck cannot do.

Trade Tip

Indicating Work in a Four-Jaw Chuck   Here are several trade secrets that save time and frustration when indicating a job in a four-jaw. The indication procedure has three phases: (1) rough centering, (2) indicator centering, and (3) tightening. By following this sequence the entire procedure will take only minutes.

Rough Quick Centering   Use the concentric lines scribed on the chuck face to coordinate centering the work. Using this trade trick, the object is placed close to center initially. Crank an edge of each jaw to the same line (Fig. 11-77), a little bigger than the workpiece. Then turn the wrench for each one-half turn inward, for each jaw. Then one-quarter turn, then another quarter until the jaws grip the work loosely. The jaws all come closed approximately on center. This should yield concentricity within about 0.050 in. or so, which is usually good enough to begin using the dial indicator, next.

Indicator Centering   With the spindle in neutral, place the indicator probe on the workpiece and rotate the object until the lowest point is under the probe. (The lowest point is farthest away from the indicator.) Zero the dial, then very slightly loosen the jaw closest to the low point. Rotate the object 180° and tighten the opposite jaw an equal amount. Keep doing this until the runout is close to being within tolerance for the setup. But you are not done—at this point the grip is not tight enough for machining.

Tightening the Chuck on Center   At a point learned from experience, you will switch from loosening the low jaw to increasing the grip. Only tighten the high side (Fig. 11-78). This adds holding pressure while bringing the object right on center.

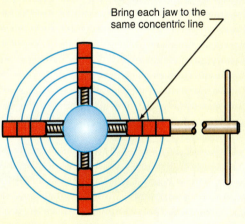

Bring each jaw to the same concentric line

Figure 11-77 *Trade tip—Use the centering guide lines on the chuck face.*

Figure 11-78 *Centering the workpiece occurs in stages:*
1. *Rough-in (eyeball to lines on chuck)*
2. *Tightening & loosening opposite jaw*
3. *Final tightening of all jaws to gain grip.*

Make a Final Check   Be sure you tighten all four jaws. Remember, two opposing jaws may be perfectly centered but not fully tightened. Test all jaws. The right amount of jaw pressure is a complex issue learned from experience. Factors are

1. How much metal removal is planned?
2. How solid is the work being chucked? Could it be crushed as with a thin-walled tube?
3. Are you chucking on finished surfaces or on excess material? Tight jaws will cause marks in the metal.
4. How long is the work and is the end supported at the tailstock? Do you expect vibration? This can loosen a chuck.

Adding Leverage Precaution   Lever tightening by placing an extension on the chuck wrench is not a recommended practice. However, in high-removal situations where the work absolutely must not slip in the chuck, we all do it sometimes. This is a risky procedure as

Centering Work That Isn't Round

Figure 11-79 *Centering an object that is not round is a matter of averaging opposing dial readings.*

it can mar the work and distort the chuck itself, and even bend the wrench. Your instructor or supervisor may hold a different opinion on using a chuck wrench extension bar. Ask before using one.

Indicating Work That Is Not Round   Occasionally you'll need to center work that is not round (Fig. 11-79). After roughly centering the work, set the indicator to zero then rotate the spindle 360° and observe. If the work is not round, it will show multiple high and low points. Compare this to the runout of a round object, which would produce a single high and low point. During indication of this kind of object, balance the highs and the lows on the average.

Collet Chucks

These precision chucks grip by pulling the tapered collet into a tapered hole, which squeezes its inside diameter closed on the work. To allow it to close, then spring open again the collet is split along its axis in three or four locations around the rim, thus their other name, *spring collets* (Fig. 11-80). **Collets** are a precision devise made of hardened and ground tool steel.

Why Choose a Collet for Your Setup?

Collets are the best gripper of small-diameter rods, where light cuts are expected. On manual lathes, they are a poor choice if heavy turning is to occur—they are best suited to light machining situations. However, they are the most common gripping method on CNC lathes where hydraulic or pneumatic force is used to pull them closed, thus they grip better than the manually closed version.

Grip   Medium to light **grip** on manual lathes depending on the pulling method for closing the collet. However, on CNC

Spring Collets

Figure 11-80 *A collet grips the work when it's sprung closed by being pulled into a precision tapered hole.*

lathes, they grip about as good as any method we'll discuss.

Runout Quality collets in good condition feature the least built-in runout of all the methods presented. They are guaranteed to have less than 0.001 in. of runout when new, with some higher quality versions offering as little as maximum 0.0003-in. factory runout.

Turnaround Very fast. In fact, on most production CNC lathes, the spindle needn't be stopped to open the collet chuck if the lathe is equipped with a bar feeder. While still rotating, the collet opens enough that the bar can slide forward a preset distance, then the collet closes again. But even with manual collets, they are the fastest turnaround holding method.

Special Characteristics Collets are designed to hold small-diameter work. They are also chosen where runout is an issue. They can be purchased in regular shapes such as squares and hexagons or as blanks, whereby the shop machines the shape needed. Each collet has a small range of sizes that may gripped without distortion to the collet or marring the work. The usual range is $\frac{1}{32}$ in. or 1 mm for each collet. This means that the shop must have 24 collets to grip a total range of sizes from $\frac{1}{8}$ in. diameter of up to 1.0 in.

Hex Collet

Figure 11-81 *Collets are supplied in regular shapes, as shown, and in blank form ready for custom machining.*

Setting Up Collets There are two kinds of collet chuck accessories for the manual lathe: front closing and rear closing (Fig. 11-82). The front-closing version is a self-contained bolt-on accessory, while the rear-closing collet requires extra components mounted on the lathe to pull the collet closed. Front-closing sets are more expensive, but they are faster to use since one needn't walk around the lathe to operate them, so they are common in production work. There are two kinds of front-closing collets—spring and rubber isolated.

KEYPOINT

Rubber collets are different from spring collets. Their jaws float in flexible rubber such that each collet has a larger grip range than spring collets and can accommodate small part irregularities better (Fig. 11-82).

KEYPOINT

The grip surface is round at the specified size and distorted slightly when squeezed smaller. Never squeeze the collet beyond its intended range—the fingers of the collet break when this is done! The closer the part size is to the collet, the better the grip will be. It's best to use rubber collets when part size is more than 0.015 in. from collet size.

Lathe Collets

A rubber-flex collet chuck has a wider range than other collets

Soft and Custom Shapes Collets that haven't been hardened yet can be purchased, then custom machined to a special work shape. Hardened and ground collets can also be purchased with several different grip shapes other than round (Fig. 11-81); for example, to make bolts from preshaped hex bar stock. After turning and threading, the bolt would be parted off and another length slid out to begin another part.

A draw-in collet mounted in a lathe spindle — Handwheel — Drawbar — Spindle-nose cap — Collect — Taper sleeve

A spring collet chuck

Figure 11-82 *Front-closing rubber collets grip a wider range per collet and cannot be broken so easily.*

The rear-closing types are more common in schools as they seem to be a standard accessory for manual lathes. The setup requires four components (Fig. 11-82):

1. **Spindle Collet Adapter**

   A solid metal sleeve that's lightly pressed into the spindle similar to a drill sleeve. The adapter converts the spindle taper to the correct internal angle matching the collet.

2. **Draw Bar**

   A hollow round shaft that fits through the spindle. It pulls the collet into the taper adapter. The collet screws into the headstock end of the draw bar.

3. **The Collet**

Note that there are 12 different standardized collet shapes. Several might be used in a single machine shop. Be sure the collet chosen matches the spindle adapter.

4. **Collet Closer**

   This will be either a lever attached to the back of the headstock or a hand wheel devise. Both pull the draw bar into the headstock, which in turn pulls the collet into a tapered hole. They then remain in the pull position until they are reversed to release the work.

Threaded and Expanding Mandrels

Sometimes the work cannot be gripped in a chuck or collet. But, it if has a hole drilled through it, it can be held on one of two kinds of *mandrels*. A mandrel is a round bar that's inserted into a hole in the workpiece. There are times when they are the only solution. Here's an example—suppose you must reduce the outside diameter of 20, $\frac{1}{2}$-in.-diameter washers. How would you hold them in any of the methods above? The answer: you can't.

Threaded Mandrels The solution in this case is the threaded mandrel (Fig. 11-83). It's not unlike a $\frac{1}{2}$-in.-precision shoulder bolt. The washers are placed on the mandrel, then the nut is tightened upon the stack. The grip force is the friction between washers, so the nut must be tightened very well.

These kinds of mandrels are often machined on the spot, as needed. A bar of steel

Threaded Mandrel

Parts to be turned — Center support

Figure 11-83 *Threaded mandrels are used to hold small parts with holes drilled through.*

is gripped in a three-jaw chuck, then the portion that's needed to fit the washers is turned down, leaving a shoulder, as shown, to push the washers against. The threads are machined, then the mandrel is ready to go and it's perfectly on center since it was machined on the spot.

Tailstock Support When making or using a mandrel to machine work, the 5-to-1 ratio of diameter to length dictates whether to use a tailstock support. However, it's not a bad idea to use a live center in the nut end, unless it interferes with very small work.

Trade Tip

When one small item requires mandrel turning, an everyday bolt can be used, if concentricity of the hole to the finished turning isn't a big issue. If its head is gripped firmly in the three-jaw chuck or a collet, then nuts placed on both sides of the object will allow light turning. For support and safety, you may need to center drill the bolt then insert a live center.

Why Choose a Threaded Mandrel?

Choose one of these when you have small work with a central hole, when no other method will do the job.

Grip Restricted to light duty, small work

Runout Depends on situation. Can be near zero if the mandrel is a precision shaft or when it's machined on the spot. Overall runout is a function of the mandrel's condition, how it's held in the lathe, and how well it was made.

Turnaround Time Medium slow but considering it's probably the only way to hold the work, this isn't often an issue.

Special Characteristics Suitable for work that can be held with side pressure from the nut, as in the stack of washers. Could be used for a single washer in the example using two nuts.

Precision Expanding Mandrels

These are ground and hardened precision holding devises. They work somewhat like an internal collet. The two-piece mandrel is inserted in the hole in the work, then the tapered shaft is driven into the expanding sleeve with a soft-face hammer until it fits tightly within. The work-holding force is friction against the hole's wall.

Once it is tightened inside the work, the mandrel and work are usually placed between centers on the lathe. But the mandrel shank can also be held in a chuck or collet when centers are needed.

KEYPOINT

The 20-washer job would be impossible with this mandrel due to variation in their hole sizes. Only the smallest inside diameter (ID) washers would be held. The rest could spin freely.

Machining a bushing (Fig. 11-84), is a good example of an expanding mandrel job. Suppose it's outside diameter (OD) must be reduced by 0.030 inch and that new OD must be concentric to the ID within 0.001 in. The expanding mandrel is the right choice.

Grip Restricted to extra light duty, small work.

Precision Two-piece Expanding Mandrel

Single part to be turned

Expansion slots

Tapered section

Drive dog flat

Centerdrill hole

Tapered hole

Figure 11-84 *Expanding mandrels work like internal collets.*

Runout 0.0005 plus or minus depending on condition and method it's held in the lathe.

Turnaround Time Very slow since it must be removed from the lathe to load and unload. But considering it's the only way to hold the work, speedy turnaround isn't often an issue here either.

Special Characteristics Suitable for work with a precision hole and where concentricity is a prime requirement. Will fit a modest range of hole sizes. Must have several sizes to cover holes from $\frac{1}{8}$ in. up to approximately $1\frac{1}{2}$ in.

Lathe Fixtures

In production, we might need to hold and turn castings, welded parts, and forgings that are an unusual shape—often an unbalanced object. Figure 11-85 is the perfect example.

Lathe **fixtures** are custom made expressly for a single part or family of similar parts. They feature a set of clamping devises to ensure safe and accurate part holding and some means of nesting (locating) the part accurately with respect to the center axis of the lathe. Fixtures are used in industry when part shapes are impossible or unsafe to hold by any other means.

Fixtures Are Potentially Dangerous

Two challenges are created: out-of-balance rotation and swiftly passing corners, clamps, and edges. Those spinning protrusions are

Figure 11-85 *A custom fixture is being used to hold this odd shaped part in a production lathe.*

Face plate or CNC mill,
which is the safer setup?

Large bored hole ——

Figure 11-86 *Boring the holes could be accomplished in a lathe fixture but it's safer on a CNC mill.*

— Tee-nut slots

— Direct bolt
better solution

Screw-in
heel block

Figure 11-87 *The transmission bolted to a face plate.*

scary to watch go whizzing by! Often the fixture features counterweights and guards, but they don't cancel all the dynamic problems or danger. Compounding the balance problems, as the object is machined, its center of mass changes. You will require further training in using a fixture and will probably not use one in school (Fig. 11-86).

Fixture Grip   Very good as long as the machinist has paid attention as he clamps! Correctly made to withstand the forces, fixtures can be used where heavy machining is to occur.

Runout   Since a large, odd-shaped object is mounted in a fixture, runout is difficult to quantify. It will vary with each fixture.

Turnaround   Again, since every fixture is different, some are quick to use, while most are very slow for part removal and replacement. All locating surfaces must be cleaned, then checked to be certain the part is nested on them correctly and that all clamps are tight.

Face Plates

Face plates are a kind of universal fixture. They are large flat disks, mounted in place of the chuck, upon which we clamp and bolt the workpiece. They feature slots or tee-slots for bolts and clamps, in a radial pattern (Fig. 11-87). Work is clamped directly to the plate using high-strength setup bolts, nuts, and clamps similar to the drill press or milling machine—with one critical exception.

KEYPOINT

When using clamps to hold work to a face plate, use only screw-in heel blocks, never use loose heels. Even well clamped down, during the heat, impact, and force of machining plus centrifugal action, separate heel blocks can be flung out thus letting go of the work. (See Section 12-4 for clamp training.)

Using the transmission case example, bolts could be put through the mounting ears if their holes are already drilled. That way, the part is secured well to the plate. If not, safe planning might include drilling them first. However, as shown, clamps are a necessary evil at times with face plate setups.

Last Resort Choice Face plates are only chosen to hold objects that are impossible to chuck otherwise and cannot be machined elsewhere. Even more than fixtures, which usually feature work nests, pins, and custom-made clamps to retain parts, face plate setups are dangerous due to clamping and holding

Trade Tip

Safety If machining isn't required to the center of a job, then a solid wood block can be forced against work on a face plate using a live center in the tailstock. This adds an extra protection from work coming loose.

Wood block support for setups too. The wood block support can be used when centering work on the face plate or in the four-jaw chuck—before the clamps or jaws are tightened. The forced block holds the work in place but allows light tapping to move the workpiece closer to center.

issues. I do not recommend that they be attempted until you have much experience with turning. *Always have someone double check your face plate setup!* See the Trade Tip.

Why Choose a Face Plate Setup?

Grip   Medium well if all clamps are well tightened. Face plates can be used for heavy machining by an experienced machinist but that's not a recommended practice.

Runout   Runout is a function of the care with which the work has been indicated or located on the plate. There are two runout issues with faceplates. *Axial runout* concerns having the work accurately centered around the spindle axis. Accurately centering the work is usually accomplished by roughly clamping it to the faceplate, then tapping it toward center while testing the object with a DTI. For *radial runout* the face of the plate itself must not wobble such that one of the work datums will be perpendicular to the lathe's axis. Sometimes, with use, the machinist may need to take a light face cut off the face plate surface to return radial runout to zero. Never cut the face plate without asking!

Turnaround   As slow as it gets. Don't even think of hurrying a face plate setup!

Special Features   Used to hold odd shaped work but must have one flat surface to place against the face plate.

Holding Work Between Precision Centers

This final holding method isn't common in industry for lathe work but it is used on precision cylindrical grinding machines. Even though we try to avoid this old method of work holding, there are times when the work has center support holes already drilled and it's the right choice to get the workpiece quickly aligned to the lathe. In Fig. 11-88, the work is drilled on both ends, with center holes. The holes are made with a center drill, which is the purpose for which center drills were originally created.

The center drilling can be accomplished on the lathe with the work held in a three-jaw chuck or collet or it can be done on the drill press with the work held in a vee block.

Next, remove the chuck and place one pointed center in the spindle and one in the tailstock (Fig. 11-88). Four accessories are added to the setup to positively center and rotate the part;

A *lathe dog* is then bolted directly on the work. The "tail of the dog" is then placed in one of the slots in the drive plate.

A *drive plate* is mounted on the spindle nose taper. The drive plate is a small face plate with one or two slots that receive the drive dog tail. (A face plate can also be used to drive the dog's tail but they are too big for the purpose.)

A **dead** (solid metal) 60° **center** is placed in the headstock end. This center rotates

Turning Between Centers

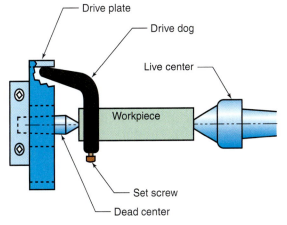

Figure 11-88 *Four components for holding work between centers. Note, center drill holes must be present.*

Lathe Centers

A 60° lathe center

A half-center

Roller and cage assembly

Rotating cover

Spring ring

Spindle

Needle roller bearing

Screwed retainer

"O" ring

End cap

Locking screw

Body

Removable point

Figure 11-89 *Two types of lathe centers* dead *or solid and the popular* live center.

along with the work (Fig. 11-89, left side).

A **live center** (roller bearing) is placed in the tailstock (Fig. 11-89). The live center point rotates with the work while its shank remains stationary in the tailstock.

Grip Shown in Fig. 11-88, the work is rotated by a clamp on the drive dog. It transfers spindle torque to the workpiece via a set screw. That's not a very efficient grip. Between center work is reserved for light cuts only.

Runout Very low—less than 0.001 in. as long as the drilled center holes are round and the centers are in good shape.

Turnaround Once the initial setup is made, this is a very fast way to replace parts in the setup. Working between centers is especially useful when the part must be removed and replaced back on the exact center again.

Special Characteristics Used where a student part might not be finished during your training shift, so it can be removed and replaced exactly back on center next time. Turning between centers is a good way to bring old parts that were originally turned between centers, back onto the lathe axis. Also used where other means of holding the work are not possible.

About Centers When the turning tool must come close to the work's center as in facing or threading a small object, the relieved center, as shown in Fig. 11-89 is used; but only at these special times. Due to the smaller support surface, it is the wrong choice for general turning. Also, since it has no bearings, it requires constant lubrication and attention as the work heats from machining.

The center drill, sometimes called a combined countersink and drill, was originally invented to prepare work for this holding method. To use it correctly, drill in to the work about $\frac{3}{4}$ of the distance up the cone portion (Fig. 11-90). This leaves a cone-shaped receptacle for the lathe center.

While live centers in the tailstock are far more useful, sometimes a dead center is used when runout is a prime concern or because a live center is unavailable. Live centers are moving objects themselves and can add small amounts of runout.

When setting up a dead center in the tailstock, friction becomes a problem between it and the work. Grease must be used to minimize heat and wear. The pilot part of the center drill not only relieves the hole for the center point, but it also forms a reservoir where special high-pressure grease is placed before putting the part into the setup.

Correct Center Drill Depth

2/3 drill cone

Do not drill beyond the full diameter

Forms the center point relief and a reservoir for pressure lube

Figure 11-90 *Use about three fourths of the cone portion but do not drill beyond the cone.*

Mounting Chucks and Other Accessories on the Lathe

Chucks that screw onto the lathe spindle have been phased out of industry due to their inability to turn the spindle backward without the dangerous risk of unscrewing from the lathe. They will not be considered here, but you might encounter them in home or school shops—if you do, get extra instruction!

Cam Lock Chucks

The **cam lock** *spindle* is safe, strong, and universally accepted for mounting chucks and face plates on a lathe, both manual and CNC. The lock system uses hardened steel pins extending from the back of the chuck or face plate (Fig. 11-91) that are firmly gripped by rotating cams in the spindle nose of the lathe (Fig. 11-92). Correctly locked, the chuck holds equally well for forward and reverse spindle rotation (Fig. 11-93).

Disassembly of Cam Locks Follow this procedure to remove a cam lock chuck or face plate. Caution! Before proceeding be sure the machine is locked out—the main power is off, not just turned off.

1. Place a wood cradle under the chuck for protection of your hand should it slip, and the chuck.

2. If an overhead shop lifting system is available, screw an eyebolt into the chuck—ask for instruction on this step. If overhead lifting isn't available, ask a fellow machinist to stand by to assist. As the chuck is removed, do not extend your back out over the lathe bed—do not get into the wrong lifting position!

Figure 11-92 Repeated Warning! *Clean cam holes, pins, and tapers before assembly, because one tiny chip left behind will dent them due to the great pressure between them!*

3. Using a special square or hexagonal wrench, turn all locks counterclockwise to the unlock position, as shown in Fig. 11-93.

Trade Tip

They will pop up just a bit at this position.

4. Now for safety, turn one cam back toward the hold position. But leave it loose. This restrains the chuck from falling free as you pop it off the spindle taper.

5. To break the holding force of the taper, using a soft brass hammer, strike the chuck at the rim with a moderate blow. The chuck will pop off but not fall away due to the single cam that is holding it.

6. With the chuck supported from above or with two machinists holding it, one in front and one behind the lathe, unlock the final cam, lift the chuck away and store it in its rack or shelf.

Figure 11-91 *Cam pins firmly mount chucks and face plates, allowing safe forward or reverse spindle rotation.*

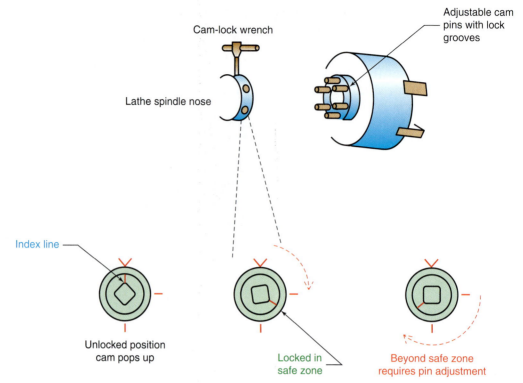

Figure 11-93 Trade Tip—*The index line shows the lock status.*

KEYPOINT

During this entire operation, keep your hands out from under the heavy chuck.

Trade Tip

On the final tighten, read the lock index lines per Fig. 11-93. With normal use or especially after a serious crash, the cam pins can become out of adjustment for a safe lockup. If the index lies within the safe zone it's OK to proceed. If, with the cam locked, any lie beyond (shown on the right), then remove the chuck and adjust the cam pins as needed by removing the set screw, then turning them in one turn. Note that the groove will always be toward the inner circle.

Assembly of a Cam Lock Chuck (Pay Close Attention to Steps 2, 5, 6, and 7)
Again, be certain the machine is locked off or the main power is turned off.

1. Place a wood cradle under the spindle.
2. Clean all mating surfaces on the cams, chuck, and spindle first with a shop towel then with a final wipe of your clean hand.
3. Turn all cams to the unlock position.
4. Lift the chuck into place using a lifting device or a helper as a second choice.
5. Lightly turn each cam until it just starts to snug up. Do this in an across and back pattern to equalize the initial seating of the chuck; softly and not all on one side at first.
6. Retighten in this same pattern. Always turn the cam wrench in a clockwise direction as viewed from above. This can be

repeated twice more with a bit more force each time. Apply 40 or 50 pounds of force on the wrench for the final tighten.

7. Do a final test "around the circle" this time to ensure all cams are tight.
8. Remove the lifting aids, wrench, and the wood cradle.

Reading the Cam Lock Indicators
When tightening a cam lock device, there are indicator lines on the lathe spindle nose that show the lock position. They surround the

hole where the cam lock wrench is inserted in the lathe (Fig. 11-93). There are three indicator lines:

The cam position index line

The safe lock line is on the rim

The unlock position indicator—often a vee shape to distinguish it from the lock position

KEYPOINT

If the lock tightens beyond the safe line, the chuck pins are incorrectly adjusted. Do not use a chuck if the cam lines show as on the right in Fig. 11-93. Ask your instructor for help in resetting the pins.

Soft Jaw Work Holding

There are two different versions of soft jaws for two different purposes:

1. Protectors on regular jaws to prevent marring finished work
2. Specially machined aluminum jaws to hold difficult shapes and/or to better align the work on center

Protectors

Usually cut from sheet aluminum, but any soft metal or even plastic would work, these little sheet metal pads are held with your hand or with tape until the chuck pressure keeps them in place. Since the chuck pressure distorts them, they are consumed in the setup. We usually shear several dozen at one time.

Machined Soft Jaws

Used often in real manufacturing but seldom in school, custom-machined **soft jaws** are chosen when runout is a prime issue or when the work can't be gripped by any other means (Fig. 11-95). For example, this clevis (a round end on a rod) must be held to cut the threads. Soft jaws could be used. They are often created for a special job on a CNC lathe.

Soft jaws begin as blank aluminum, or occasionally mild steel for extra-demanding production of many parts. The backside is premachined such that it will key and bolt in place of the regular hardened jaws. Once mounted on the chuck, they are custom machined to fit a given workpiece. Soft jaws are used on three-jaw chucks. Bored in place, on

Trade Tip

Equal Thickness When using protectors be certain they are all three of equal thickness when used with universal three-jaw chucks. If not, the work will be held off center! If using them with a four-jaw chuck, thickness isn't as critical as you will be centering the work with a DTI anyway (Fig. 11-94).

Figure 11-94 *Soft sheet metal protects a finished shaft in a chuck.*

the chuck, they bring the work perfectly onto center every time.

Exact Centering on a Three-Jaw Chuck
Cutting soft jaws correctly, on the same three-jaw chuck in which they will be used to hold the workpiece, ensures they then close on the work very close to center each and every time. But only as long as they aren't removed from the chuck and the chuck is not

Figure 11-95 *This clevis would be difficult to hold without the custom soft jaws.*

removed from the lathe. Using them without disturbance, runout of the work can be as repeatable as 0.0005 in. or better.

Soft jaws are often removed from the chuck to be used later, to hold the same job or similarly shaped objects. But, once they are removed then remounted at a later time, the centering must not be trusted—perhaps degrading to an estimated 0.005 in. plus or minus. That's due to refastening pressure and general alignment of the jaws to the chuck. So, when reusing them, a light skim cut of the grip surfaces restores their near-perfect centering capability.

Once made, soft jaws can become a shop setup tooling. They are stamped with the part number and stored for later use. When this occurs, if the jaws are made to fit a three-jaw chuck, the jaw number must be stamped on each (jaw 1, 2, or 3). This helps to eliminate degrading runout when remounting them. And the chuck ID number helps too, if many chucks are used in the shop.

Ensuring Chuck Grip Bias

To ensure perfect centering with each opening and closing of the custom-machined jaws, it's necessary that the chuck be forced tight against some object when the custom shaping of the jaws is being done. This brings the chuck jaws to the right pressure and alignment bias with each closing. If the chuck isn't stabilized by closing force, the machining will not be correct and it will be outright dangerous, as the jaws will move as the cutter hits them.

If the work is going to be gripped from the outside, then the chuck must be closed downward toward center, as shown in Fig. 11-96. A hex nut or a round plug can do the job. See the Trade Tip.

Countersunk bolts and thick side of jaw to be machined

Adjustable spider

Figure 11-96 *A spider helps pressurize the jaws correctly during their custom machining.*

Trade Tip

Making Your Own Spider A handy self-made tool for boring soft jaws with inward bias is the adjustable three-point spreader called a spider. Drill and tap three holes in a large hex nut. Then add the right length bolt to fill the space. Sometimes, when the entire jaw must be machined, the spacer can be placed against the jaw carriers farther into the chuck.

When the soft jaws will be used to grip the work from its inside—with an outward chucking force—a solid metal ring or other retaining devise must be placed around the outer surface of the jaws to stabilize and bias the grip during custom machining. Check with your instructor of foreman as to what is available. There are a few trade tricks on how to do this, but I'll leave them to your individual shop policy.

UNIT 11-3 REVIEW

Review the Key Terms

Cam locks
The most common method of holding chucks on a lathe, they use rotating cams in the lathe spindle that grip pins protruding from the chuck.

Collets
Sometimes called spring collets due the way they are made, these precision hardened and ground griping tools hold the work by being pulled into a tapered hole.

Dead centers
A support center that has no bearings for rotation. They rotate with the work when used in the headstock but rub when used in the tailstock.

Fixtures
Custom-made tools for holding special shaped work not possible to reliably hold by any other method. Sometimes called jigs.

Grip
The comparative amount of bite a holding method is able to impart to the work.

Independent chucks
Chucks that feature jaws able to move individually.

Live centers
Support centers used only in the tailstock, they have bearings to eliminate friction.

Scroll plate

The helical face thread that moves all three jaws simultaneously in a three-jaw chuck.

Soft jaws

A custom-machined three-jaw insert used where concentricity or part shape is an issue.

Replay the Key Points

◆ Workpiece runout can be introduced to the setup by the holding method itself.

◆ Solid reversing jaws can be reinserted in the wrong order, resulting in chuck closure at some point other than center.

◆ A collet grip is round at the specified size and distorted slightly when squeezed smaller. Never squeeze the collet closed beyond its range.

◆ In using cam lock chuck mounts, do not use a chuck if the index line goes beyond the safe lock position.

◆ Each and every time, clean the spindle taper and chuck before mating them. This is one of the most irritating brain lapses that beginners can make. *Just one trapped chip ruins expensive equipment!*

Respond

Choose the right chuck and support accessories for the following jobs: Note, there is often more than one correct answer. Note that, although parts may be similar in shape, other conditions such as batch size, raw material, and tolerances dictate choices. Check each answer as you go. Compare yours to those of a fellow student.

1. Ten, expensive A arm, titanium steering castings, designed for an undersea rover must be turned round on the pointed end. They resemble a bent, 12-in.-tall letter A.

2. Fifteen, 2.5-in.-diameter shafts must be turned to 0.750-in. diameter then threaded with 1.0 of threads. They are precut to 8.250 in. long from a CRS bar—no finishing of the 2.5-in. diameter will be required. Runout requirement is 0.005 in. between the shaft and new threads. See Fig. 11-97.

3. Five hundred, finished 15-mm-diameter twist drills must be turned down to 0.350 in. on the soft part of the shank (to make turned-down shank drills), with a runout requirement of 0.001 in. to the original diameter. They are 100 mm long and require about 12 mm length of turned-down surface.

4. A photocopy machine needs its main roller just cleaned up to remove scratches, then repolished to the

Figure 11-97 *Which holding method would be best for this CRS cut bar job?*

original 8-μI finish. (*Hint:* Runout is tightly toleranced.) See Fig. 11-98.

5. You are assigned to turn 300 aluminum window screen holder pins per work order and print. (*Hint:* They must be turned very fast for good cutting action.) See Fig. 11-99.

Now, answer the following questions from what you know about chucks.

6. True or false? Due to manufacturing tolerances, industrial-quality four-jaw chucks will have a certified maximum runout when new, however, with use and abuse, runout can degrade due to stretching of the scroll plate. If it is false, what will make it true?

7. What are the special characteristics of a three-jaw chuck?

8. What are the advantages of a collet chuck? What are the disadvantages?

9. What type of work is best suited to three-jaw chucks?

Figure 11-98 *Turn one paper roll for minimum cleanup of scratches only, then repolish to a fine finish.*

Make from aluminum

Figure 11-99 *Turn the 19-mm aluminum rod to 8.0 mm, chamfer, then part off.*

10. Name two reasons for choosing a between-centers setup.

11. Why would you use a face plate setup?

12. True or false? The three-jaw chuck is the most common chuck because it grips well and is quick to use, but it produces a moderate runout. If it is false, what will make it true?

UNIT 11-4

Turning Tool Basics

Introduction: Another important aspect of setting up a lathe is choosing the cutting tool and the tool holder. Often, depending on the tools on hand, several might work OK but only one or two might be best for the task.

KEY TERMS

Cemented carbide
Tungsten carbide grains cemented together with a cobalt binder and a term sometimes used to describe a carbide tip is brazed on a steel lathe shank.

Chip breaker—chip dam
A groove in a lathe cutting tool (breaker) or metal protrusion (dam) used to snap off the advancing chip into safe "C" shapes.

Cobalt
A tough metallic element added to HSS steel bits for extra toughness or used as the "glue" in a sintered tungsten carbide.

Indexable insert
An insert carbide (or other cutting material) tip that features more than one cutting edge per insert. When the insert is indexed to a new position, it doesn't loose the tool's position on the cutter holder.

Inscribed circle (abbreviated IC)
The size of a carbide insert, the largest circle that can be drawn on the insert. The tool is working efficiently when it is cutting a depth of $\frac{2}{3}$ the IC.

Side rake (also compound rake)
A lathe tool featuring a rake angle that has a bias to cut left or right.

Sintering
Powered metal is pressed into a mold then fired to form a solid composite with characteristics different from any cast metal.

Choosing and Setting Up the Right Tool for the Job

The factors to consider when choosing the right cutting tool are

Tool Shape
Will it be able to cut the surfaces and shapes required?
  Often more than one cutting tool will be needed to complete the part shape. If so, then the kind of tool post (cutting tool holder) becomes an issue too.
Positive or negative rake
  Depends on the alloy to be cut and the amount of metal to be removed.
Nose radius and lead angle
Right- or left-hand bias
  Turning, threading, and facing tools are usually ground or selected such that they cut in one direction.

Tool Construction
Standard Technology:
  Solid HSS
  Carbide insert tooling
High-tech cutters for high production or for cutting extremely difficult alloys
  Polycrystalline cutting edges
    Artificial and natural gem stones embedded into the tool's cutting edge
  Ceramic or ceramic/metallic composite cutting tools
    Aiming at harder cutters, some are made of nonmetal materials. While they cut very well, they are also brittle.
Brazed-tip carbide (mostly as form tools)
  A steel shank with a shaped piece of tungsten-carbide brazed on to form the cutting edge, they are being displaced due to carbide insert technology and CNC abilities to drive standard tools to create shapes rather than form them.

While all types of cutting tools can be found in industry, the standard technology options are the more practical choices for training. We'll concentrate our explorations there.

Tool Shapes

Other than the drilling, reaming, and tapping from the tailstock, the basic operations can be accomplished with the following five utility shapes. There are many variations within each type, including HSS and carbide insert tooth varieties. We'll look at both, but for now let's discuss their shape.

Turning Tools

Although there are neutral bias tools, they usually do not produce good finishes. They are used for utility setups where several different kinds of cuts must be made by the same tool or where tool changing is slow or difficult. But for best results, we generally choose a tool with a cut bias. Depending on how the lead and rake angles are ground, turning tools are biased to feed to the operator's left or they can face inward toward the work center, or to the right and face outward on the work. There are a nearly infinite number of combinations of rake, lead, nose radius, and tool composition. In fact, like two artists choosing their favorite brushes, often one will work best for one machinist, while a different combination is chosen and used with good results by another (Fig. 11-100).

Parting and Grooving Tools

Similar in shape and function, the difference is the depth to which these cutting tools must drive into the work (Fig. 11-101).

Figure 11-101 *A parting tool and a grooving tool are very similar.*

Trade Tip

Since parting requires a long thin tool that must overhang the tool support, HSS is the best choice for learning. It can flex without breaking. But in industry, the insert carbide parting tool can cut faster and last longer.

Form Tools

Form tools (Fig. 11-102) are custom made mostly from solid HSS, or for longer tool life, with steel shanks and brazed-on carbide tips. Today, however, CNC lathes are reducing the

Forming Tool

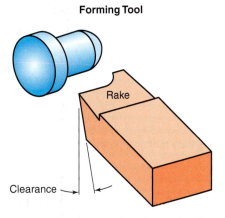

Figure 11-102 *A forming tool will make an outside corner radius.*

need for custom-shaped form tools. Programming allows complex movements of off-the-shelf, standard tools to reproduce custom shapes. Additionally, using standard cutting tools eliminates the challenges of slow cutting, chatter, and tearing of the surface associated with forming.

Threading Tools

Threading tools (Fig. 11-103) are purchased in preformed versions or as insert-type carbide tools. But it's likely that your first threading lesson will include grinding your own using a blank HSS bit. It must be ground with the

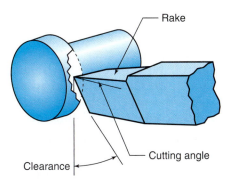

Figure 11-100 *A turning tool is used for other kinds of operations too.*

Threading Tool

Figure 11-103 *A threading tool is a type of form tool.*

included angle of the thread (60° for Unified Threads) and small flat section on the tip, to duplicate the thread form.

Boring Bars

Boring bars are found in machine shops in three varieties and a great many sizes. Many are used on the milling machines as well as the lathes, both manual and CNC. Many of the concepts we'll discuss here also apply to boring on the mill, where it's the tool that spins, not the work.

Preshaped Solid Boring Bars One popular version for small holes is the "Swiss" type (Fig. 11-104) that's mounted in a tool holder block for lathe work. Although relatively expensive to purchase, they produce excellent results when boring diameters below 1.0 in.

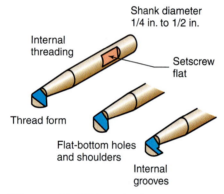

Solid Boring Bars for Small Holes

Shank diameter 1/4 in. to 1/2 in.

Internal threading

Setscrew flat

Thread form

Flat-bottom holes and shoulders

Internal grooves

Figure 11-104 *Boring bars for small diameter holes are usually the preshaped "Swiss" type.*

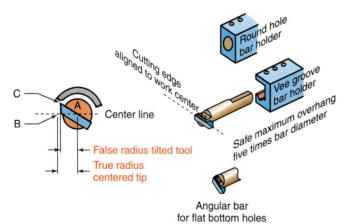

Universal Boring Bars

Round hole bar holder

Cutting edge aligned to work center

Vee groove bar holder

C

A

B

Center line

Safe maximum overhang five times bar diameter

False radius tilted tool

True radius centered tip

Angular bar for flat bottom holes

Figure 11-105 *Shop-made boring bars are used on manual lathes.*

Universal Boring Bars These shop-made bars are common for tooling and production work. Using set screws to hold the cutter in a vee groove, they accept standard HSS tips, ground to any shape by the user (Fig. 11-105). Or a carbide cutter can be used as well. Machining the bar tip at an angle can be useful for reaching the hole bottom before the bar face rubs the workpiece.

KEYPOINT

To ensure correct cutting action, when making your own boring bar be certain that the upper edge of the tool bit slot is near the center of the bar's diameter.

Setting Up a Boring Bar

Four factors must be considered when setting a boring bar.

Factor 1—Clearances There are four clearances to be checked for a successful boring bar setup:

1. **Bit Length**
   First, as shown in Fig. 11-105, the replacement bit might be too long. It might rub the backside of the hole on the first cut. The bit cannot protrude out the back of the bar beyond the radius of the drilled hole.

2. **Chip Clearance Between Bar and Hole**
   While the goal is to use the largest bar diameter that will fit into the hole, it cannot be so big that chips cannot eject. If they bunch up (a common problem during boring), then coolants cannot reach the cut. Heat builds rapidly with tool failure to follow.

3. **Cutting Edge Clearance**
   Next, as shown in Fig. 11-106, more cutting tool clearance than normal must be ground such that the cutter's lower edge does not rub on the hole's curvature.

4. **Tool Holder Clearance**
   As Fig. 11-107 shows, after setting the depth control, be sure the tool holder does not touch the face of the work when the cut is at full depth. That space is necessary to allow coolant in and chips out.

Factor 2—Depth Control and Overhang When choosing a boring bar, remember the guideline of five-to-one overhang ratio, length to diameter (Fig. 11-106). Long, small-diameter boring bars tend to dig and chatter. Many boring bar holders allow the bar to slide in and out—use that function to set a minimum overhang.

Depth control Since boring is a blind operation, you cannot see when to stop cutting inward. You must then build in some form of positive stop to ensure depth control of the bored hole (Fig. 11-107). If the lathe isn't equipped with a digital positioner, here is a way to set it up. With the bit's cutting edge touching the work face, set a zero against a gage block stack equal to the bored hole's depth. This can be on an indicator, as shown, or use a micrometer stop if the lathe has one. On many lathes, the microstop features a trip that halts *Z* axis movement as it contacts the stop.

A set of gage blocks can be used to coordinate set the depth (Fig. 11-107). After zeroing the dial, remove the stack and the saddle will come to the preset zero at the correct depth. Note that the full depth clearance must be checked too, as shown.

Figure 11-106 *Factors for a correct boring bar setup.*

Factor 3—Vertical Axis Alignment This tool setup parameter is important to ensure size control when boring. The cutting edge must be on the vertical center of the workpiece. Centering of the tool can be accomplished in one of three ways. The best solution is using the quick change vertical adjustment screw. The next choice uses shims under the tool in the holder, as with an open-sided tool post. The least desirable, is by rotating the bar in the round hole mount because that changes the effective rake angle as well.

Study the end view inset in Fig. 11-105 to see why. Distance A to B is the true radius

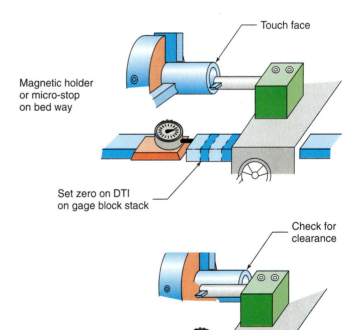

Figure 11-107 *Setting depth control and checking tool holder clearance.*

distance from work center to the inside of the hole, parallel to the X axis of a correctly centered tool. If the bit is set correctly, cutting edge movement will be along that horizontal line—so when making a diameter change on the X dial, the same amount of diameter change occurs in the hole.

Distance A to C represents the effective, smaller radius of the incorrectly tilted tool. A given change in the X axis would produce a false result. The farther the cutting edge is misplaced from the horizontal X axis, the greater will be this shrinking effect. Remember this concept. It's important when aligning boring bars on milling machines too. The cutting edge must be at the work centerline to achieve diameter accuracy and correct form tool shape as well.

Figure 11-108 *Carbide insert tooth boring bars with triangular inserts such as this are common in industry.*

KEYPOINT

To setup a boring bar the cutting edge must be at the work centerline to achieve diameter accuracy and form tool shape. Use the largest bar that will fit into the drilled hole yet allow coolant in and chips out. Be certain the setup has adequate clearance in four ways: cutter in hole, total length, bar, and tool holder.

Carbide Insert Boring Bars

The third kind of boring bar is mostly used for larger holes and heavier cuts. However, they are also supplied for holes as small as $\frac{3}{8}$ in. (Fig. 11-108). We'll study the insert geometry and makeup in Unit 11-6. For now, their main advantage is the quick changeover to new sharp inserts without disturbing the setup position of the bar itself. Every cutter we'll discuss here can be purchased as a carbide insert version.

Direction of Tool Cut

The next setup choice is the *hand* of the tool (the direction it's designed to cut—to the left or to the right). Many turning, facing, and threading tools are given a compound rake angle—with a bias toward cutting on one edge (Fig. 11-109). This

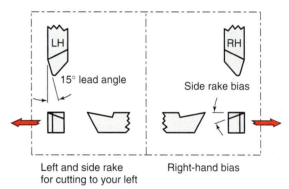

Left and Right Hand Cutting Tools

Figure 11-109 *Lead and rake angles combine to create cutting tool hand bias.*

Figure 11-110 *With the shield open, note the tool turret located behind the chuck on this CNC lathe.*

tool geometry modification is called **side rake** or sometimes it's referred to as **compound rake** (see Shop Talk). The lead angle is also ground onto the tool to cut toward either the left or the right.

Neutral Rake

Lathe tools can be ground or selected with no side rake bias such that they cut in both directions, but they usually don't perform as well as those with a bias. In addition to a potentially better finish, chip ejection becomes better with a rake bias.

KEYPOINT

Tool hand bias is composed of lead and rake angle combinations. Positive bias is a must on threading tools ground for the manual lathe, because they must cut at an extraslow speed. The positive rake helps the slow moving chip deform.

Cutting Tool Material

The third choice is the material from which the cutter is made. While there are several high-tech choices such as embedded diamond crystal tips or ceramic-metal composites, their cost is justified only in special application production. For now, we'll limit our investigation to the standard choices found in most tech schools and the average shop: HSS and carbide cutting tools.

High-Speed Steel Tools

For this lesson, the HSS tools also include two slightly harder but similar metals called cast alloy and **cobalt** HSS tools. As a beginner, the difference isn't significant. They do perform slightly better than plain HSS when cutting harder work but their cost puts them beyond normal school budgets.

The similarities between cast alloy, cobalt, and HSS tools are significant—all of these tools are meant to be ground in your shop, using a standard grinding wheel. They are usually supplied blank or partially preformed, with the machinist grinding the final shape and regrinding them when dull or when a different cutter geometry is required.

Why Choose an HSS Tool? They are the lowest initial cost cutting tool. While they dull quicker in service, compared to carbide, HSS tools offer an extremely long overall

Figure 11-111 *A blank HSS bit, a ground bit, and a new parting tool.*

shop life because they can be reground many times, to any practical shape. The most important feature is that they are the toughest cutting tool. Toughness does not mean a cutting tool that can machine the most metal. It does mean it can withstand vibration, chatter, physical and thermal shocks (hit and miss coolants), and flexing without breaking (Fig. 11-111).

KEYPOINT

Choose HSS tools over carbide when an *interrupted cut* must be taken. These hammering situations (a noncontinuous surface where the cutter must enter and come out of the cut with each revolution of the workpiece) break harder tools.

On the down side, HSS tools withstand the least heat from machining, therefore the cutting speed is the lowest of the two everyday choices. The cobalt and cast alloy tools perform a bit better at higher temperatures compared to plain HSS in this respect—but not by greater than 15 percent.

Carbide Insert Tooling

This is the tool of choice for industrial production. These tools are a combination of a precision, hardened alloy steel shank with a replaceable carbide tip—called an *insert*. Most inserts feature more than one cutting edge (Fig. 11-112). When the tip dulls it can be rotated or reversed to another edge. Inserts that have this feature are called **indexable inserts.**

Carbide—It's Not a Metal as You Know It!

When we surround the heavy metal tungsten with carbon, then sustain it at high temperature without oxygen, it forms tungsten-carbide (commonly called carbide). The first firing creates a material that cannot then be formed into cutting tools in the standard ways because it melts at around 6,000°F (approximately three times the temperature of steel).

So, to make carbide cutting tools, the coarse carbide crystals are broken down into a uniform, fine powder using ball mills to crush them. It's then mixed well with a second metal powder made of cobalt. Next the mixture is pressed into a mold. The soft inserts have no strength until they are placed in a high-temperature furnace. During the elevated firing, the tough cobalt melts and glues together the hard carbide grains. The process called **sintering,** forms a composite that is both hard and tough—more or less like concrete where the rocks are held together with cement (Fig. 11-114 on next page). In fact, another name for carbide is **cemented carbide.**

Hardness/Toughness Ratios
By changing the ratio of carbide to cobalt in the insert, the hardness verses the toughness is altered. Higher cobalt content yields a tougher carbide tool but one that is less hard. More carbide relative to cobalt, creates a harder tool but it's also more brittle and prone to chipping.

Insert tooling is more costly for initial purchase compared to HSS, since the tool shanks must be purchased as well as a supply of inserts (Fig. 11-113). But, for the following reasons, they are very economical over a long period of use.

Why Choose Insert Tooling?
Insert tools have several big advantages over HSS.

Quick Replacement The inserts feature guaranteed repeatability, usually to a tolerance of less than 0.001 in. The dulled cutting edge can be quickly rotated to a new edge or the entire insert can be replaced with very little size change in the setup. That means a minimum of down (nonproductive) time when the insert is indexed or replaced.

High Production Carbide's hardness allows cutting speeds at least three times faster than standard HSS and much higher in some metals.

Multiple Choices Another big advantage is that many different combinations of hardness/toughness ratios, harder coated inserts, rake angles, and corner radii are available in the same shape insert. For example, if the insert is chipping before it becomes dull, then try a tougher insert with slightly less hardness. That change takes less than a minute to do. Or suppose the tool is chattering. Then change to a smaller nose radius insert (this will change work size). These changes are quick since they require changing only the insert, not the entire cutting tool.

Tool Geometry Control

Insert tooling adds flexibility to your setup. Inserts are supplied in a wide variety of sizes and shapes. Within a single insert tool holder, you can quickly change the nose radius, chip breaker, hardness, and even the rake angle in some cases by simply replacing

Figure 11-112 *A variety of insert tools used for turning.*

Common Insert Shapes

Diamond Square

Chip-breaking groove

Triangle Round

Figure 11-113 *Indexable inserts are found in many shapes and sizes. These are the four most common.*

the insert in the tool shank. Today, nearly all lathe tooling needs can be met with quick change insert tools with the single exception of custom-shaped forming bits.

Understanding Carbide Insert Tooling Numbering

When selecting an insert tool for your setup you must know industry standard numbering. The system includes data fields that identify the size, hardness, and shape of the insert. A similar system exists for selecting the tool holder.

Figure 11-114 *Carbide is a powdered metallurgical product with fine grains of hard carbide cemented together by other tougher metals.*

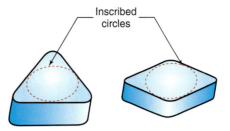

Inscribed circles

Figure 11-115 *The inscribed circle describes insert size.*

When making a manual lathe setup, you have some latitude as to the cutting tool size and shape chosen, but when making a CNC lathe setup, most program setup sheets call out exactly which tool holder and insert are to be used. They will probably be called out using the following number system or they will be in picture form—often it's both. At that time, you'll need access to this information from supplier catalogs, wall charts, or reference books such as *Machinery's Handbook*© or the *Machinist's Ready Reference*©.

Insert Size This decision is about how big the insert shape should be. It's based on the tool post capacity, the availability of sizes in the shop, and the shape and size of the work. The amount of metal to be removed is a factor too. Bigger tooling is better for strength, but remember that the larger tool is more costly. Make sure your setup will fully utilize it.

KEYPOINT

Choosing a large insert for small cut depths is not economical. But choosing a small insert that is overused is not smart either.

The Inscribed Circle For comparative reference, the largest circle that can be drawn on the inset face, is used for the size called the **inscribed circle** and abbreviated **IC**. Ideally, choose an insert such that it will be cutting at about $\frac{2}{3}$ of it's full IC cutting depth for roughing cuts (Fig. 11-115).

Insert Shape Cutting tool shape is a more complex issue than size. Keep in mind a CNC document will probably call out exactly what insert shape must be used because there are issues you'll learn in programming as to how the cutter must approach and cut the material.

KEYPOINT

For CNC work, setting up the wrong shape (not what the program thinks it will be) can make scrap parts and even cause a crash!

The insert shape and the tool holder shape are linked. In the following three charts, note how the holder is chosen. It's not critical to understand it all at this time. You should be aware of the system and where to find it for future reference (see Figs. 11-116 to 11-118).

Similar to the holder, the insert itself has a numbering system. Figure 11-119 is an example of a standard number chart.

Trade Tip

Chip Breaking Breaking the pesky long chips into harmless "C" shapes is a competency that improves your setup and work day as well. Insert tools usually feature a **chip breaker** groove or a protruding bump that bends the chip violently enough as it flows past that it breaks into small, safe, easy-to-handle packets. However, when you grind your own bit it is possible to add a chip breaker groove.

In place of grinding a breaker into the tool, machinists often set up a small piece of metal known as a **chip dam** or chip trap that also bends the chip so tightly that it breaks off (Fig. 11-120 on page 436). It is simply clamped in with the tool, then adjusted to work best yet release the chip after breaking it.

Toolholder Identification System

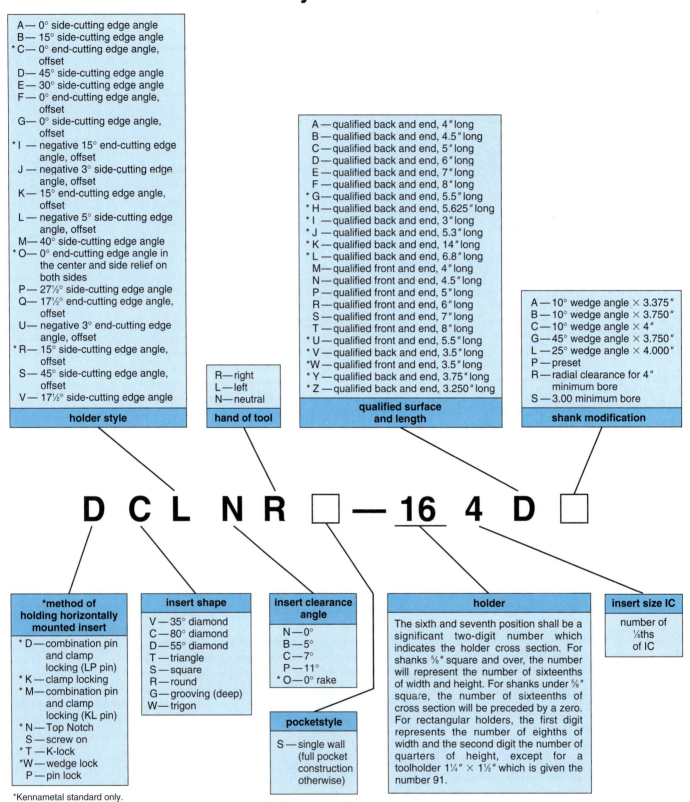

Figure 11-116 *Standard lathe tool selection identification chart.*

Metric Toolholder Identification System

holder style

A—straight shank with 0° side-cutting edge angle
B—straight shank with 15° side-cutting edge angle
C—straight or offset shank with 0° end-cutting edge angle
D—straight shank with 45° side-cutting edge angle
E—straight shank with 30° side-cutting edge angle
F—offset shank with 0° end-cutting edge angle
G—offset shank with 0° side-cutting edge angle
J—offset shank with negative 3° side-cutting edge angle
K—offset shank with 15° end-cutting edge angle
L—offset shank with negative 5° side-cutting edge angle
M—straight shank with 40° side-cutting edge angle
N—straight shank with 27° side-cutting edge angle
O—offset shank with 30° negative end-cutting edge angle
R—offset shank with 15° side-cutting edge angle
S—offset shank with 45° side-cutting edge angle

rake

N—negative
P—positive

hand of tool

R—right
L—left

holder length

H—100mm
J —110mm
K—125mm
L —140mm
M—150mm
N—160mm
P—170mm
Q—180mm
R—200mm
S—250mm

M C L N R □ - 25 25 M 12 Q

method of holding insert

(horizontally mounted)
K—clamp locking
M—with locking pin
(insert with hole in center) and clamp
N—Top Notch clamp
S—screw down
W—wedge locking

insert shape

C—30° diamond
D—55° diamond
K—55° parallelogram (Top Notch)
R—round
S—square
T—triangle
V—35° diamond
W—trigon

holder height

A significant two-digit number which indicates the holder height in millimeters.

holder width

A significant two-digit number which indicates the holder width in millimeters.

qualified

pocket style

S—single wall, (full pocket construction otherwise)

insert cutting edge length in millimeters		¼" 6,35	⅜" 9,32	½" 12,70	⅝" 15,88	¾" 19,05	1" 25,40
IC							
triangle		11	16	22	27	33	44
square round			09	12	15	19	25
55° diamond				15	19		
80° diamond				12		19	25
35° diamond			16				

D —insert IC
L—length on insert cutting edge

Figure 11-117 *Standard lathe tool selection identification chart.*

Turning Insert Identification System

symbol	insert shape	shape	nose angle (degree)
S	square	square	90
T	triangular	triangular	60
C			80
D			55
E		rhombic (diamond)	75
F			50
M			86
V			35
W		hexagonal (trigon)	80
H		hexagonal	120
O		octagonal	135
P		pentagonal	108
L		rectangular	90
A			85
B		parallelogram-shaped	82
K			55
R		round	—

1 shape

tolerances

IC: theoretical diameter of the insert inscribed circle
T : thickness
B : see figure below

tolerance class	tolerance on "IC"		tolerance on "B"		tolerance on "T"	
	in	mm	in	mm	in	mm
C	± .0010	± 0,025	± .0005	± 0,013	± .001	± 0,025
H	± .0005	± 0,013	± .0005	± 0,013	± .001	± 0,025
E	± .0010	± 0,025	± .0010	± 0,025	± .001	± 0,025
G	± .0010	± 0,025	± .0010	± 0,025	± .005	± 0,13
M	see tables at right		see tables at right		±.005	± 0,13
U	see tables at right		see tables at right		±.005	± 0,13

3 tolerance

example

	position 1	position 2	position 3	position 4
inch	C	N	M	G
(metric)	C	N	M	G
position	1	2	3	4

N—0°
A—3°
B—5°
C—7°
P—11°
D—15°
E—20°
F—25°
G—30°

relief angle 2

4 insert type

symbol	hole	shape of hole	chipbreaker	shape of insert's section	ordinary system	IC less than ¼"
N	without		without		N	
R			single sided		R	E
F			double sided		F	
A		cylindrical hole	without		A	
M			single sided		M	
G			double sided		G	
W		partly cylindrical hole, 40–60° countersink	without		A	
T			single sided		M	
Q	with	partly cylindrical hole, 40–60° double countersink	without		A	D
U			double sided		G	
B		partly cylindrical hole, 70–90° countersink	without		A	
H			single sided		M	
C		partly cylindrical hole, 70–90° double countersink	without		A	
J			double sided		G	
X		special			X	X

*inch system only.

5 size

inch	IC		metric cutting edge length						
	in	mm	C	D	R	S	T	V	W
1.2 (5)	5/32	3,97	—	04	03	03	06	—	—
1.5 (8)	3/16	4,76	04	05	04	04	08	08	—
1.8 (7)	7/32	5,56	05	06	05	05	09	09	03
—	0.236	6,00	—	—	06	—	—	—	—
2	¼	6,35	06	07	06	06	11	11	04
2.5	5/16	7,94	08	09	07	06	13	13	05
—	0.315	8,00	—	—	08	—	—	—	—
3	3/8	9,52	09	11	09	09	16	16	06
—	0.394	10,00	—	—	10	—	—	—	—
3.5	7/16	11,11	11	13	11	11	19	19	07
—	0.472	12,00	—	—	12	—	—	—	—
4	½	12,70	12	15	12	12	22	22	08
4.5	9/16	14,29	14	17	14	14	24	24	09
5	5/8	15,88	16	19	15	15	27	27	10
—	0.630	16,00	—	—	16	—	—	—	—
5.5	11/16	17,46	17	21	17	17	30	30	11
6	¾	19,05	19	23	19	19	33	33	13
—	0.787	20,00	—	—	20	—	—	—	—
7	7/8	22,22	22	27	22	22	38	38	15
—	0.984	25,00	—	—	25	—	—	—	—
8	1	25,40	25	31	25	25	44	44	17
10	1 ¼	31,75	32	38	31	31	54	54	21
—	1.260	32,00	—	—	32	—	—	—	—

NOTE: inch sizes in parenthesis for "alternate sizes" D or E (under ¼ inch IC).

Figure 11-118 *Standard numbering system for carbide inserts.*

IC		± tolerance on "IC"							
		class "M" tolerance						class "U" tolerance	
		shapes S, T, C, R & W		shape "D"		shape "V"		shapes "S, T & C"	
inch	metric	inch	mm	inch	mm	inch	mm	inch	mm
5/32	3,97			—	—	—	—	—	—
3/16	4,76			—	—	—	—		
7/32	5,56	.002	0,05	.002	0,05	.002	0,05	.003	0,06
1/4	6,35								
5/16	7,94								
3/8	9,52								
7/16	11,11	.003	0,06	.003	0,06	.003	0,06	.005	0,13
1/2	12,70								
9/16	14,29								
5/8	15,88	.004	0,10	.004	0,10	.004	0,10	.007	0,18
11/16	17,46								
3/4	19,05								
7/8	22,22	.005	0,13	—	—	—	—	.010	0,25
1	25,40			—	—	—	—		
1 1/4	22,22	.006	0,15	—	—	—	—		

IC		± tolerance on "B"							
		class "M" tolerance						class "U" tolerance	
		shapes S, T, C, R & W		shape "D"		shape "V"		shapes "S, T & C"	
inch	metric	inch	mm	inch	mm	inch	mm	inch	mm
5/32	3,97			—	—	—	—	—	—
3/16	4,76			—	—	—	—		
7/32	5,56	.003	0,06	.004	0,11	—	—	.005	0,13
1/4	6,35					—	—		
5/16	7,94					—	—		
3/8	9,52					.007	0,18		
7/16	11,11	.005	0,13	.006	0,15	—	—	—	—
1/2	12,70					.010	0,25	.008	0,20
9/16	14,29					—	—	—	—
5/8	15,88	.006	0,15	.007	0,18	—	—	.011	0,27
11/16	17,46					—	—		
3/4	19,05					—	—		
7/8	22,22					—	—		
1	25,40	.007	0,18					.015	0,38
1 1/4	22,22	.008	0,20	—	—	—	—		

Figure 11-119 *Standard numbering systems for carbide inserts. (Continued)*

— 4 3 2 □ □ □

12 04 08 □ □ □

⑤ ⑥ ⑦ ⑧ ⑨ ⑩

⑥ thickness

thickness		symbol	
in	mm	inch	metric
1/32	0,79	0.5 (1)	—
1/16	1,59	1 (2)	01
5/64	1,98	1.2	T1
3/32	2,38	1.5 (3)	02
1/8	3,18	2	03
5/32	3,97	2.5	T3
3/16	4,76	3	04
7/32	5,56	3.5	05
1/4	6,35	4	06
5/16	7,94	5	07
3/2	9,52	6	09
7/13	11,11	7	11
1/2	12,70	8	12

NOTE: inch sizes in parenthesis for "alternate sizes" D or E (under 1/4 inch IC).

hand of insert (optional) ⑧

R L

⑦ corner radius

corner radius		symbol	
in	mm	inch	metric
.004	0,1	0	01
.008	0,2	0.5	02
1/64	0,4	1	04
1/32	0,3	2	08
3/64	1,2	3	12
1/16	1,3	4	16
5/64	2,0	5	20
3/32	2,1	6	24
7/64	2,3	7	28
1/8	3,2	8	32
round insert (inch)		—	00
round insert (metric)		—	M0

⑨ & ⑩
cutting edge condition or chip control features (optional)

T — negative land

K — light feed chip control, double sided Kenloc insert

M — heavy feed chip control, deep floor Kenloc

N — narrow land Kentrol insert with chip control on one side

W — heavy-duty chip control, wide land Kenloc insert one side

J — polished to 4-microinch AA (rake face only)

UF— ultra-fine finishing

See Technical Section for additional conditions and chip control features.

Do it Yourself Chip Control

Chip breaker groove

Bolt-on steel chip dam

Figure 11-120 *A chip breaker or chip dam, aids in creating safer, easier to handle "C" chips.*

Rocker Tool Post

Tool post

Tool holder

Rocker shim

Rocker spacer

Tee-nut washer

Compound

Figure 11-121 *Six rocker post components: Tee-nut washer, tool post, rocker disk spacer, rocker, clamp bolt, and tool holder.*

Figure 11-122 *Left-, neutral-, and right-hand rocker post tool holders.*

Manual Lathe Tool Posts

A tool post holds the cutting tool on the lathe compound, there are four types. Each offers some advantage.

Industrial Varieties

These include

    Solid tool post
    Quick change
    Indexing turret
    Rocker tool post

Not often used in industry, the rocker tool posts are the most universal but the least strong.

Tool Post Strength

With each tool post, I'll provide a relative depth of cut and feed rate. The descriptions are meant to provide comparisons between the various kinds, not to set a standard. Each tool post is different in terms of sturdiness, and that's a prime factor in choosing the right one for your setup. If the work requires light cuts, then all of the following can be chosen. But as the cut forces increase, some posts should not be used.

Rocker Tool Posts and Tool Holders

This is the weakest of the four presented but is also the lowest cost and the most universal. The rocking action allows easy vertical and right and left positioning. (Vertical positioning is discussed after tool posts.) Rocker posts hold standard HSS cutting tools. The cutting tool is clamped in a tool shank by tightening a square headed set screw (Fig. 11-121). Then the pair are clamped into the post. Due to their tendency to move and slip, rocking posts are not often used in industry especially where heavy cuts are made, but when the standard tools don't do the job, they can be useful in some situations.

In order to use a rocker tool post, the shop must have a selection of the tool holders (Fig. 11-122) made for the post. These holders have a "hand" similar to tool bits. This allows the tool more "reach" in tight machining situations and provides better cutting angles in some setups. They are either right, left, or neutral (pointing straight ahead). They are not suitable for jobs requiring several tools to complete the shape.

Why Use a Rocker Tool Post? The main reason is that nothing else is available. However, rocker tool posts are able to rotate through 360 degrees, and the tool can be pivoted up or down on the rocker. This makes them simple to setup and very universal. Carefully used, a rocker tool post can meet most of your training needs. The down side is that heavy cuts or a dull tool often cause them to slip. Also, they are slow for changing to a new cutting tool.

Solid Tool Posts

Also called open-sided tool posts, they are the opposite of rockers: heavy-duty brutes. Solid tool posts can hold solid HSS bits, or the tool holders for rockers, or carbide insert tools, as shown in Fig. 11-123, but only as long as the tool's cutting edge can be shimmed to the vertical center of the work. They are used for roughing work where extra heavy metal removal rates are needed. They are a good tool post to use during knurling due to the force on the tool.

But because the tool shank is held down with set screws, they are not meant for changing tools. Set these tool posts up when you expect to use a single tool throughout the turning of the product.

While they are easily rotated to point the tool in any direction, vertical adjustment is

Open-sided or Solid Tool Post

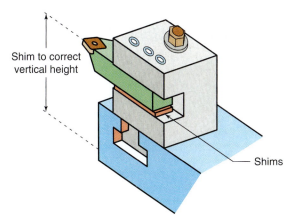

Shim to correct vertical height

Shims

Figure 11-123 *A solid tool post (open sided) is the strongest and simplest of the four types presented.*

possible only by adding shims. Shimming means placing thin metal sheets under the tool to raise it to the spindle center. As always, be mindful of the tool's overhang from the post—keep it to the minimum.

Quick Change Tool Posts

Also called cam lock tool posts, their purpose is to allow several different tools to be set up then quickly exchanged—one at a time. Each tool is mounted on it's own tool block. When the tool block with it's cutting tool is removed then returned to the post again, the cutting tool tip returns to its original position relative to the X and Z axes, within a repeatability of perhaps 0.0005 in. or less. But that's only if the tool post and holders are undamaged and cleaned.

The quick change tool post features a dovetail method of removing and replacing the tool blocks in the same position every time. The vertical height of the tool is set with a locator pin and adjusting screw. They can hold either HSS cutting tools or carbide insert types. Note in Fig. 11-124 that a drill or reamer can be held in the quick change tool post rather than the tailstock, but you are then faced with positioning it on the exact X axis center of the work.

Why Choose a Quick Change Tool Post? Set up a quick change tool post where several cutting tools are required to finish parts. They are especially useful when running batches of parts with multiple operations. After initial setup of the cutting tool in its own tool block, exchange time tool-to-tool, is a matter of just seconds.

ShopTalk

Standard Load Tools In both CNC and manual lathe work, we often leave the most common turning tools mounted in the turret or in a set of tool blocks all the time. Then the required custom tools are added in other blocks (or turret stations) for a particular setup.

A Quick-change Tool-Holder
and various types
of cutting tools

Figure 11-124 *Quick change tool posts with pre-set blocks reduce tool changing time plus they repeat position accurately when replaced.*

Quick change tool post sets (Fig. 11-125) are supplied in different sizes to accommodate various lathe swings.

Indexing Tool Turret

This accessory is another way of setting up multiple cutting tools, but here, they stay on

Figure 11-125 *A quick change tool post in action.*

the tool post. A screw, cam, or lock pin action allows the operator to unlock the turret, rotate it to several discreet positions relative to the work, (usually 8 or 16 positions), then securely lock it in position.

Why Choose a Turret Tool Post? These holders are the fastest for production work if the job doesn't require more than four cutting tools. When indexed to the next tool, they repeat position within 0.0005 in. or better.

Although the turret will hold the tool shank for which they were designed, close to vertical center, most will require shimming. Thus they are slow to set up initially. It's also necessary to ensure that the other cutting tools not being used, do not interfere with the work. For example, in Fig. 11-126, the carbide parting tool (upper right) might touch the tailstock center. If it does, then it must be moved to another position or the turning tool on the left could be slid farther out. Most CNC turning center (production

Indexing Tool Turret

Figure 11-126 *A square turret offers multiple positioning for several permanently mounted tools.*

lathes) present the same challenge, except their tool turrets hold many more tools.

Tool interference is one of the many items that must be cautiously and slowly checked when the program is tried for the first time.

Why Choose a Turret Tool Holder? This tool post is designed for runs of many parts where two to four cutting tools are needed to turn a part. The turret post is the fastest for changing tools during turning operations and the second most solid holding method behind solid tool posts.

Cuts of around 0.090-in. depth of cut and 0.025 IPM are possible.

They can hold carbide insert and HSS cutting tools. Often, a standard load of tools is set up and left in the holder all the time. The entire tool load can be left in the turret and then can be removed from the lathe. The entire turret is quickly bolted to the lathe for a superfast setup. For example, a right-hand turning tool, a left-hand tool, a parting tool, and a chamfer forming tool could be left on the turret, then custom tools could be added as needed. Turrets are costly but rugged with a long life in the shop.

Indexing Quick Change Tool Posts

Combining the best features of both production tool posts, the quick change indexing tool post shown in Fig. 11-127 offers removable tool blocks, but it can hold four blocks on

Figure 11-127 *An indexing quick change turret features 15° discreet positioning and quick change blocks.*

the turret at any time—one per side. Then the turret can be indexed to 24 positions—every 15°. This adds great flexibility to the setup.

Positioning Turning Tools on Vertical Center

When mounting a tool in any tool post, it's important to set the cutting edge at or near the vertical center of the work. As shown in Fig. 11-128, it will cut OK if positioned slightly below, but the low cutting tool can dangerously encourage thin work to bend and then climb up over the bit. Also, low tools tend to distort form shapes.

KEYPOINT

If the tool is very far above center, the clearance face will rub and the cutting edge will not contact the work at all (see the Trade Tip on page 440).

Tool Post Application		
Type	**Relative Removal**	**Best Suited for**
Rocker tool post	0.045 in. at 0.010 to 0.015 IPR	Light work
Open sided	0.100 in. at 0.025 to 0.035	Roughing and simple work
Quick change	0.085 in. at 0.020 to 0.025	Complex batch requiring multitools
Indexing turret	0.090 in. at 0.025 to 0.030	Rough/finish, medium number of tools
Indexing quick change	0.090 in. at 0.025 to 0.030	Very complex setups for batches

Tool Vertical Height

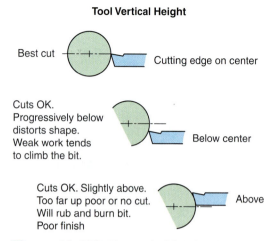

Best cut — Cutting edge on center

Cuts OK.
Progressively below distorts shape.
Weak work tends to climb the bit.
— Below center

Cuts OK. Slightly above.
Too far up poor or no cut.
Will rub and burn bit.
Poor finish
— Above

Figure 11-128 *The vertical height of the cutting edge must be set.*

Trade Tip

There is disagreement among masters with regard to tool centering. Some recommend that a lathe cutting tool should be set just a bit above center because the cutting forces will pull it downward during heavy cuts, thus it ends up on center. I do not agree, as it is a guessing game as to how far above to place it. I place it on center.

Three Methods of Tool Centering

Here are the ways we ensure the tool's cutting edge is on center of the work.

1. **Dedicated Height Gage (Best Method)**
   The surest method is a simple shop-made, bent rod pointer. The rod is welded or screwed into a heavy steel base. The rod end is sharpened as a divider point, with a flat surface (Fig. 11-129). Set upon the

Dedicated Tool Height Gage

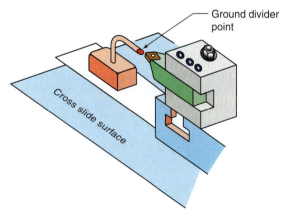

Ground divider point

Cross slide surface

Figure 11-129 *A shop-made height gage points out the spindle centerline.*

Testing Tool Height with the Tailstock Center

Figure 11-130 *The cutting tool can be compared to the height of the tailstock/headstock center point.*

top surface of the cross slide or bed ways (whichever is more convenient on that lathe), it makes tool alignment a snap.

2. **Tailstock Center**
   Sliding the tool and tailstock (or headstock) center together can also point out vertical centering (Fig. 11-130).

Trade Tip

A surface gage used for layout work or a small vernier height gage, set to the correct vertical distance, can also be used to align cutting tools on the lathe.

3. **Captured Flat Object (Quick and Dirty Method)**
   This method is a Trade Trick recommended only if you take it easy. Here, the cutting tool is cranked forward against the work surface to trap and hold any flat object (a wood popsicle stick or if you are extra careful, your pocket ruler) *lightly* against any centered round surface, the workpiece, the chuck side surface, or tailstock quill. Never use enough pressure to mar objects! The slant of the trapped object indicates vertical position (Fig. 11-131). Not all instructors recommend this method, especially when trapping a pocket rule, for it is inexact, potentially damaging to the ruler and the cutting tool's sharp cutting edge, plus it can chip brittle carbide tools if too much pressure is brought to bear. I do use it, as it's a quick way to check rough tool height.

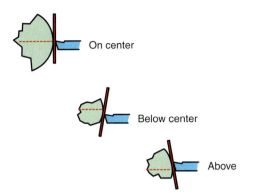

Capturing a Straight Edge

On center

Below center

Above

Figure 11-131 *Capturing a flat object between the bit and round workpiece roughly shows tool height. Caution, can chip carbide bits.*

UNIT 11-4 REVIEW

Review the Key Terms

Cemented carbide
Tungsten carbide grains cemented together with a cobalt binder and a term sometimes used to describe a carbide tip is brazed on a steel lathe shank.

Chip breaker—chip dam
A groove in a lathe cutting tool (breaker) or metal protrusion (dam) used to snap off the advancing chip into safe "C" shapes.

Cobalt
A tough metallic element added to HSS steel bits for extra toughness or used as the "glue" in a sintered tungsten carbide.

Indexable insert
An insert carbide (or other cutting material) tip that features more than one cutting edge per insert. When the insert is indexed to a new position, it doesn't loose the tool's position on the cutter holder.

Inscribed circle (abbreviated IC)
The size of a carbide insert, the largest circle that can be drawn on the insert. The tool is working efficiently when it is cutting a depth of $\frac{2}{3}$ the IC.

Side rake (also compound rake)
A lathe tool featuring a rake angle that has a bias to cut left or right.

Sintering
Powered metal is pressed into a mold then fired to form a solid composite with characteristics different from any cast metal.

Replay the Key Points

◆ Choose HSS tools over carbide when an *interrupted cut* must be taken that breaks

harder tools (a noncontinuous surface where the cutter must enter and come out of the cut with each revolution). They also work better where flex or chatter breaks harder cutting tools.

◆ Choosing a large insert for small cut depths is not economical. But choosing a small insert that is overused is also not economical either.

◆ For safety, limit cuts to around 0.050- to 0.075-in. depth of cut and 0.008 to 0.012 IPM when using a rocker tool post—or to the limits set by your instructor.

◆ Never try to exchange a quick change tool while the spindle is turning or rotate tool turrets. *Stop the lathe to change tools!*

◆ Always have someone check your setup as you learn.

Respond

The CNC setup sheet notes that Tool Number 5 must be a KTMNR-164 tool holder. The following five questions refer to that holder. For outside research, you might go to *Machinery's Handbook* rather than the charts here.

1. What is the size of the insert?
2. What is the "hand" of this holder?
3. What size is the shank?
4. What is the shape of the insert used on this tool?
5. This is a compound question. What is the recommended maximum working depth of the insert for this tool?
6. A cemented carbide tool is chosen when carbide is a must but some custom shape is needed that is not available in insert tooling. Is this statement true or false? If it is false, what will make it true?
7. A rocker tool post is chosen because of its ability to swivel and pivot in any direction and take deep or heavy cuts. Is this statement true or false? If it is false, what will make it true.
8. Identify the tool post in Fig. 11-132.

Figure 11-132 *Identify this production tool post.*

9. Name the three tool posts most often used on standard lathes in industry (this excludes rocker tool posts) and give a brief description of each.

10. You have an order of 250 spring pins. They require a large amount of straight turning, facing and, a chamfer on the corner. Which tool post would be best for this production run?

11. Modifying the job in Question 10, you now need to knurl the work, make a deep groove on one end and also perform an internal bore to a depth of 4 in. as well as the other operations required. Which tool post?

12. On the very first cut for roughing away the excess metal on the pins in Questions 10 and 11, the HSS tool gets very hot, the finish is very poor. What is wrong and what can you do?

Note that there are many factors here. Take your time in thinking this out as to the possibilities.

UNIT 11-5

Lathe Safety

Introduction: Coming up in Units 11-6 and 11-7, we'll look at setups. From the start, it's beyond critical that you plan and execute every setup with safety as the prime directive. Here, in Unit 11-5, we'll consider some specific safety tips that pertain directly to lathe operation.

KEYPOINT

Unit 11-5 is short but the whole thing, 100 percent, is *key points*.

This unit's solutions are by no means comprehensive, they are the common problems we instructors observe beginners experiencing. Each setup and operation presents some challenge that must be recognized and solved by good common sense. There are two overshadowing traits you must develop:

1. Safety first and always—it's easy to get too busy to remember safety! (For me too, for all of us!) It's common for beginners to start out with an extrasafe approach but, as they get comfortable with the equipment, to let that attitude down.

2. If it looks dangerous, it most likely is. Discipline yourself to take the time to do it right.

From all the instructors who helped plan this book, here is our condensed best advice on how to stay in full control of the potential danger of turning.

Best Practice Safety on Lathes

Know Your Equipment

Accidentally engaging the wrong axis or moving the tool the wrong direction is arguably the most common source of ruined student work and accidents. Here are two different aspects to equipment skills to develop.

1. Review the various dials and levers before early machining exercises, always ask yourself "what will happen when I pull this lever or turn this dial?" Think out your actions before performing them.

2. Have an emergency crash plan ready and practice it! It's common for a beginner or even an experienced machinist, to do the wrong thing during a developing accident. When something goes wrong, a part slips in a chuck or a tool breaks, for example, you must have quick reflexes to stop the machine. That takes practice. With a delayed or, even worse, a wrong action, accidents are amplified beyond their initial danger *by the operator!* (see Fig. 11-133).

Practice your actions. Stop the lathe as quickly as you can! Many lathes feature a combination spindle stop and brake lever. In addition to that lever, the brake will be duplicated on the floor with a bar that is foot operated. If your lathe has one, train your foot to use it.

Instantly Recognize Three Classes of Danger

Class 1 You hear or see a sign that something might not be right. Stop, investigate, and fix it, before it does degrade. Examples include changing sounds, degrading finishes, smoke, steam, or blue chips when they weren't previously seen or heard. In this case, you might let the cut finish before stopping.

Class 2 No doubt about it, you hear or see something coming loose or

Hand
spindle
lever

Foot
brake

Figure 11-133 *Never start the lathe without a review of how to stop it quickly and safely—the spindle lever or the foot brake.*

breaking. The danger is real and you must act quickly to prevent it from degrading into a class 3 crash. Stop now, investigate, and fix the problem. This one probably warrants a new setup.

Class 3   This is the catastrophe. It's a fast developing all-out crash—there is nothing to do but get away from the machine as fast as you can. Once it's under control and everybody is safe, confer with your instructor or supervisor to be certain this never happens again.

The point is, there's no time to think out your actions when a class 3 accident occurs. Like a Kung-Fu master, be mentally prepared to act. This goes hand-in-hand with your crash plan. Prepare now, for the eventuality when there is nothing to do but duck and run! We'll review the three classes again in CNC machining, for there the crashes are more potent!

KEYPOINT

Build in Safety with Every Setup

Setups that are not well planned are a common cause of accidents and ruined work. Choose work holding and cutter tooling that's rigid with the absolute minimum overhang. Test cutting tools for chuck interference by rotating the spindle in neutral before engaging power turning.

In Fig. 11-134 how many things can you find that are wrong with setup B compared to setup A? Not only is setup B's tool and holder set too far out from the turret, but the turret itself is mounted out beyond the lip of the compound. Adding to the overhang, the bit is out too far from the holder. This is a weak, long lever arm just made for trouble.

What's wrong with the setup shown in Fig. 11-135?

Setup B

Setup A

Figure 11-134 *The tool on the right extends too far. It will flex down and vibrate, perhaps digging into the work.*

Figure 11-135 *What is wrong with this setup?*

Figure 11-136 *Now it's safe to turn.*

Again, if you have even the slightest doubt that the setup is solid enough, stop and rethink the rigidity (ability to resist flexing and bending). Don't give it a try to see if it "might work" (Fig. 11-136).

Make Sure Chuck Guards Are in Place

Don't forget to add the chuck guard to your setup. The chuck guard does three things (Fig. 11-137).

Figure 11-137 *This lathe has a chuck guard keeping the chips and coolant from flying out.*

1. It keeps coolants inside the lathe rather than letting them fling off the chuck into the shop.
2. It prevents chips from flying out to hit you or others nearby.
3. The guard helps contain the workpiece should it accidentally be flung out.

Guards are even more critical in CNC turning, where the forces and speeds are accelerated. Most CNC turning centers feature a full containment safety shield that surrounds the entire machining area. If that guard isn't in place, the lathe will not start except by a major adjustment to the control's software (called a parameter change).

Control Chips

Long stringy chips are a multiple problem. They cut and grab and, they also catch and wrap around chucks with an amazing speed causing a tangled mess, called a *rat's nest*. (Fig. 11-138). There are two steps you can take to prevent chip problems.

Break Chips into "C" Shapes Do this by using better feed rates and tool geometry and by adding a chip breaker on top of the tool (Fig. 11-139). Ask your instructor to show you how to break your chip if it starts to dangerously curl up and out of the setup. If it's a roughing cut, you can briefly stop the feed movement then start it again. Don't stop it long enough to let the tool dwell and rub, but the quick stall causes a weak place in the chip that breaks it. There are also times when without disengaging the feed, the handwheel can be moved forcefully forward,

Figure 11-138 Danger! *This setup is producing a sharp, hot, and very strong wire!*

Figure 11-139 *Broken "C" shape chips indicate correct tool geometry, good cutting speed, and feed rate.*

called *crowding the wheel,* then when it is let go, that also causes a weak place in the chip.

Keep Long Chips Cleared Away Sometimes it isn't possible to break chips, especially on light cuts. When this happens use the chip hook shown in Fig. 11-140 to keep stringy chips pulled away from the chuck and the work. Also keep piles of chips swept away from the machine. They can cause feet to slip on floors, which then leads to further injury if you fall toward the moving lathe.

Calculate Speeds and Start Moderately, Then Work Up to Maximum Rates

As a beginner, do not guess at the correct speeds.

Figure 11-140 *Use a chip hook to safely pull long chips out and away from the lathe.*

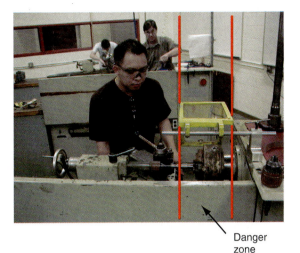

Danger zone

Figure 11-141 *Stand to one side of the danger zone, away from the rim of the chuck and workpiece where the chips could fly out of the setup.*

Don't Stand in the Danger Zone When Operating the Machine

Remember, when filing use your left hand on the file handle and during all turning operations, stand to the right of the area where chips will be ejected (Fig. 11-141).

Mind That Chuck Wrench!

Remember, the forgotten chuck wrench left in the chuck is transformed into a missile when the spindle is turned on. Never take your hand off the wrench when it's in the chuck.

Dress for the Job

Also a reminder, no jewelry or loose clothing. Control your hair with a hat. Wear eye protection even if the machine isn't moving and protect your hearing with plugs or muffs if noise is uncomfortably high.

UNIT 11-5 REVIEW

Replay the Key Points

- Know your equipment.
- Consider your actions before trying them.
- Have an emergency crash plan ready and practice it!
- Build in safety with every setup.
- Instantly recognize three classes of danger.
- Make sure chuck guards are in place.
- Build in chip control. Use chip breakers. Keep dangerous stingy chips cleared away.
- Always turn off and lock-out the power to the lathe or open main breakers for chuck or other accessory changes, such that the

---

lathe cannot be accidentally started. This is super critical for CNC setup work!

- ◆ Calculate speeds and feeds—start out moderately.
- ◆ Don't stand in the danger zone when operating the machine.
- ◆ Remember that chuck wrench!
- ◆ Dress for the job.

Respond

1. Name the four ways a pro controls chips on a lathe.
2. What operator action can you take when a class 1 event occurs?
3. What is the main characteristic of a class 3 crash?
4. What's a good rule to prevent flying chuck wrenches?

Critical Thinking

5. What would you say is the most common cause of accidents on a lathe?

UNIT 11-6

Lathe Setups that Work Right— Troubleshooting

Introduction: After a brief discussion about calculating RPMs for lathe work, we'll look at setups and trouble shooting. Then, after completing a demonstration of the lathe from your instructor and any planned checkout activities, you'll be ready to setup and operate a lathe.

KEY TERMS

Coordinating the axis
Setting the micrometer dial or digital readout to represent the true diameter of the work.

Cp (capability of the process)
The normal amount of variation in any process.

SPC (statistical process control)
A method of detecting normal and trend variation in any process.

Calculating Lathe RPM

There is an additional consideration for correct RPM on lathes beyond good cutting action: *maximum safe limits!* Since many jobs are held in chucks, often the correct cutting speed for the work is too fast for the chuck rim. Chucks,

face plates, and other setups can fly apart due to centrifugal forces. Recall that general limits for chucks are usually around 1,000 RPM or less on manual lathes. However, please check with your instructor for your shop limit.

KEYPOINT

Your shop will have its upper limit RPM for certain holding methods.

Computing RPMs for turning is similar to the process for drills. To compute the RPM for a lathe setup,

1. Find the suggested surface speed based on the tool alloy and metal to be cut. To find a recommended surface speed, use the student chart in Appendix IV, a commercial Recommended Surface Speed chart, or *Machinery's Handbook.*
2. Now calculate the RPM such that the work outer rim will run at the recommended surface speed if it is safe. Using the short formula:

$$RPM = \frac{4 \times Surface\ speed}{Diameter\ (of\ rotating\ object)}$$

For example, a 3.0-in.-diameter mild steel part is turned using a HSS turning tool. In the Appendix IV chart, we find the recommended surface speed to be 100 F/M.

In the formula

$$\frac{4 \times 100}{3} = 133\ RPM$$

Calculating the RPM is critical to achieve good finishes, accuracy, and safety. Remember, the smaller object must rotate faster to achieve a given surface speed.

Trade Tip

Constantly Changing Diameter Work Discussed in Unit 11-1, a problem arises in RPM selection when facing or parting because the tool moves inward toward center (or out) and thus has a changing rim diameter. Good cutting action would require a constantly changing RPM. The spindle should go faster as it advances inward.

There are three possible solutions:

1. (Commonly Used) Use a compromise RPM—Select an intermediate RPM that is initially a bit high but as the tool approaches center will be too slow.

Planning Your Setup

Even a simple job such as the one in this exercise has many steps. Take a few minutes and give it a try, see how you would plan and execute this job. This exercise would be more beneficial if worked with a partner of equal skill. Not all of the steps have been covered in the reading but you have the knowledge to work them out. You will need to extend and read the Trade Tips along the way too. We'll be following this W/O.

> **Work Order** U11-6-142
>
> Make Qty 15—Steel Shear Pins, Per Dwg 11-142 (Fig. 11-142)
>
> Rough Stock: Cold Rolled Steel 1.375 in. Diameter
>
> Stock has been presawn into 9-in. lengths with excess for grip. The ends are rough. Machine this order using HSS tools only

Instructions

Use a sheet of paper to plan every step in the setup and in the machining operations. The goals are safety, quality, and efficiency (in that order). Compare your finished setup and operation guide to the one provided. In making your setup choices, be aware of the print tolerances for various features of the part.

Here is a check list of the things to be considered:

☐ Holding method
☐ Cutting tools
☐ Tool post type
☐ Work support
☐ Operation sequences

General notes
Make from 1 3/8 in. CRS
Gereral tolerance +/−0.3 in.

Shear pins, steel
P/N 11-139

Figure 11-142 *Plan and produce 15 of these shafts.*

My Suggested Setup

This is one solution. There are other combinations that would produce good results.

1. **Holding Method—Three-Jaw Chuck**
   There's lots of metal to remove so runout is not important but good grip is. In making 15 parts, a reasonable turnaround time is needed. The parts have excess length to hold inside the chuck.

2. **Cutting Tools**
   Right-hand turning tool—cuts toward the chuck.
   Facing tool (left hand turning), cuts inward toward center.
   Parting tool
   Chamfer form tool to remove sharp edges per drawing
   Center drill in chuck in tailstock

3. **Tool Post**
   Multitool setup. Either a cam lock or a turret is right for this job.

4. **Work Support**
   Tailstock, live center. You will have about 7 in. of material protruding from the chuck. Thus it's beyond the 5-to-1 ratio for diameter to length of unsupported objects.

Suggested Operation Sequences

Step 1 End Prep

The part protrudes out of the chuck a short distance—1 in. or less. At 500 RPM, face the end—cleanup a minimum amount for a true face using a right-hand (RH) facing tool. At 1,000 RPM, center drill per Fig. 11-143 (or your shop's upper chuck RPM limit).

Figure 11-143 *Ready for the center drilling of the bar.*

Grouping Operations It is sometimes more efficient to perform the same operation to all 15 parts before moving on to the next step (Fig. 11-144). In this case, all could be chucked then faced on one end and center drilled. Another example would be in the actual turning of the diameter. Each part might be rough turned to within 0.030 in. of finished size then set aside to cool while rough turning the remaining parts.

Fifteen parts faced and center drilled

Figure 11-144 *To achieve efficiency, all 15 parts could be faced and center drilled without changing the setup.*

First cut measures 1.234 in.

Set/Lock dial at 0.2340 in.

Coordinating the micrometer dial

Figure 11-145 *Coordinating the micrometer dial to represent the turned results is a time saver.*

example, after making the test cut, the shaft measures 1.2340 in. Without disturbing the axis position, set the position 1.2340. Coordinating the axis accounts for backlash and tool spring.

Step 2   Rechuck for Turning

There are 7 in. protruding from the three-jaw chuck. Setup tailstock live center support.

Step 3   Reshift the Lathe for Turning

Set RPM to 400. (Turning steel with a HSS tool.) Set feed rate 0.015 in./rev.—a roughing rate for the OD (outside diameter).

Step 4   Turning

The first action will be to roughly coordinate the micrometer dial and/or digital position readout to represent the diameter the tool produces.

Just touch the tool bit to the OD of the work then set the dial or digital readout to 1.3750 (0.375 on micrometer dials), then lightly lock the dial. Next take a short Z axis test cut of about $\frac{1}{4}$ in. long and from 0.030 in. to 0.050 in. deep (the rough pass amount).

Stop the lathe and measure the result with a micrometer. Now refine the X axis position on the readout or dial to represent the amount remaining on the shaft until final size. This is an important step often overlooked by beginners. It saves time in getting a final result as quickly as possible. This act is known as **coordinating the axis** (Fig. 11-145). For

Achieving Finished Size While coordinating, the X axis dial is important for any single part, it's usefulness increases for production when multiple parts are to be produced. However, to use this Trade Tip, the setup requires a cam lock or square turret tool post such that the turning tool can be put back in place in the same position each time it is used.

When machining the second part, if the axis is coordinated, the dial should read "Zero" when you reach the finished size. But be aware, other factors can come into play; heat and spring in both the tool and shaft can cause the first setting to be close but not exact. The coordination can change.

Differences in the depth of cut taken will affect the size reproduced too. Taking more material tends to spring the tool away from the work. Continue to monitor size on succeeding parts.

After a few parts, the setup settles down. That is, the variation stabilizes. As a skilled machinist, you will learn to minimize the time taken for this to occur. Once the variables are accounted for, the micrometer dial and digital readout will remain accurate for a long period of time until the cutting tool begins to dull.

Rough and Finish Tools   To stabilize the setup even further many machinists would make this setup using two cutting tools, one roughing and one for finishing cuts. Most of the hard work is done by the tool that does not take the last pass. This is also a CNC practice.

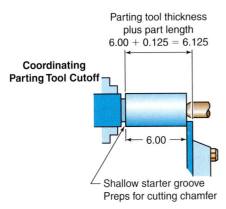

Parting tool thickness
plus part length
6.00 + 0.125 = 6.125

Coordinating Parting Tool Cutoff

6.00

Shallow starter groove
Preps for cutting chamfer

Figure 11-146 *Touch the parting tool to the right end of the work, then move it to the left a distance equal to the part length plus the tool thickness. Make a groove for the chamfer tool.*

Step 5   Chamfering

To chamfer the part on the end in the chuck, first make a groove with the parting tool. To obtain the right parting tool Z axis position, use the digital readout or a micrometer stop attachment if the lathe is so equipped. First, touch the parting tool to the right edge then move it to the left a distance equal to the blade thickness plus the part length (Fig. 11-146). This can be accomplished with the Z axis hand wheel if it is graduated finely enough.

Trade Tip

Most parting tools are manufactured to an exact thickness of 0.100, 0.125, or 0.1875 in.

Step 6   Parting

After chamfering, remount the parting tool and position it at the groove. Lock the saddle and shift the X axis to feed at 0.005 per revolution—inward. Use coolant for parting to lubricate the chip. This is so even with metals that require no coolant. Lubrication helps the chip come out of the deep groove.

Be ready! The instant the part is severed from the parent metal still in the chuck you might need to stop the lathe. Have your foot or hand on the stop lever!

Operator Control

Problem Solving When Things Don't Go Per Plan!

As a skilled machine operator, you need the ability to reason out how to improve a setup or fix a problem. On the previous job the 1.00-in. diameter's tolerance was plus/minus 0.005 in. If, after running 10 parts, you find the diameter varies by, say, plus or minus 0.001 in., then your setup is in control and producing consistent results. In the process, you have a normal variation of 0.002 in., spread evenly around an average size, say, 1.002 in. Parts range from 1.001 to 1.003 in.

The challenge now is to get the process variation centered around the target size, in this case, 1.000 in. To do so requires a final adjustment of the micrometer dial on the machine. Then monitor successive parts to see that the sizes remain centered around the target value.

You'll experience a natural variation from heat, mechanical play in the machine, your own actions in controlling the backlash, and other factors. These factors add up to an over-all limiting repeatability of the setup called the **capability of the process** abbreviated to **Cp.**

Your responsibility as a journey machinist is to recognize how much variation is built in to the setup and machine and how much is the result of factors that require control. If, for example, you observe a trend of parts growing bigger in diameter by 0.0002 in. with each new part, what would you suspect is the cause?

Tool wear would be the number one suspect, but could there be other factors? How about the tool slipping in the holder? The micrometer dial could be slipping too. So, your job is similar to that of a detective—to reason out the clues and decide the most likely cause.

Troubleshooting Steps

Using the clues, decide what the possible causes might be.

Test each one in the setup to eliminate each.

Find solid solutions that won't allow the problem to arise again.

Keep in mind, sometimes the problem can be caused by more than one factor.

Statistical Process Control

Beyond safety, the machinist's main duty is to find and control variation, to make parts as close to nominal as possible—that's the foundation of quality work. To do so, you must understand the nature of variation in the setup and cutting process. It divides into three types: common cause (normal), trend, and special cause. Common causes are the amount created by the equipment and process that cannot be eliminated altogether. Special variation is that created by one time only events that might lead to a scrap part, but there is little that can be done to the process to avoid it. For example, every day at 12:45 P.M. a train goes by the factory, causing the machine to vibrate. One needs to detect the cause and avoid it, but you would not change the setup. That leaves trends as the real enemy to quality—something is changing or has changed away from normal variation.

A very good tool for finding and tracking all three kinds of variation is called **statistical process control. SPC** is a math/graphic tool that works especially well on production runs in machining, but it can be applied to any process that has a quality outcome, machining or otherwise. We'll investigate it in detail in Section 27.

SPC does two things: it determines what is a normal amount of variation in the process, then it graphically tracks, in real time, variation in the work to show if there are trends developing away from normal. It can be done on a PC at the machine, or many modern CNC controls have SPC built into their software. SPC doesn't explain why the variation is occurring, that's up to the skills of the machinist. SPC does assist the machinist with an early warning of developing trends. Then, the skills take over in finding the root cause(s) and determining the permanent fixes. The upcoming problems require that kind of detective work.

UNIT 11-6 REVIEW

Review the Key Terms

Coordinating the axis
Setting the micrometer dial or digital readout, to represent the true diameter of the work.

Cp (capability of the process)
The normal amount of variation in any process.

SPC (statistical process control)
A method of detecting normal and trend variation in any process.

Replay the Key Points

◆ Your shop will have their own upper limits on certain holding methods—RPM.

◆ Operations are frequently grouped into batches. The decision to run parts that way depends on turnaround time for the holding method.

◆ Coordinating the micrometer dial saves lots of extra measuring and testing.

◆ When a part is severed from the raw stock, do not try to catch it with your hands!

Respond

Taper in a turned shaft is frequently a problem. Here are some critical thinking challenges. The detective work and fixes are typical of the daily challenges we face. Anybody can run a machine when everything is going OK. It's the skilled master that can keep it making good parts when things go awry!

1. *Taper in First Part After First Roughing Pass.* Turning a length of $\frac{3}{4}$-in.-diameter CRS stock down to 0.625 in., after the first rough cut you've found a problem. Using a micrometer on both ends and the middle, plus testing it with a straight edge, you find a straight taper (not barrel shaped) with the tailstock end being the largest size. The difference end to end is 0.018 in. (Fig. 11-147).

Figure 11-147 *This part was produced on the first rough cut. What could the problem be?*

A. List what the five possible causes might be.

B. Describe how you might test to eliminate each possibility.

C. Assuming the cause is as described in the answers, find the specific solution. For a quick fix method, see the Trade Tip in the answers.

2. (Same job as problem 1) *After Five Parts with No Taper, the Sixth Shows Tapering.* Similar to the last problem, the right end is bigger than the left. List the possible causes.

3. *Poor Surface Finish on Rough and Semifinish Passes.* Solve this problem before taking the final pass.

UNIT 11-7

Single-Point Threading

Introduction: Continuing setups, we look at single-point thread cutting on a manual lathe. In my opinion, single-point threading is the most complex activity in first semester machining in terms of both setup complexities and operator actions.

You will not be able to read this material then go to the lathe and make single-point thread. Scan it now, for a basic understanding. Then, attend the demonstration by your instructor. Finally, take the book with you to use as a reference when setting up and machining the first thread. Pay special attention to the Trade Tips for threading, they work!

KEY TERMS

Center gage
The small arrow-shaped tool used as a setup template and thread checking guide.

Chasing dial
The indicator of when to engage the half nut lever to make each pass in the same thread groove as the last.

Half nut
The split nut engagement devise for pulling the carriage along at the right pitch ratio to the spindle. The half nut lever engages the threading action.

Plunge threading
Single-point threading where the compound is not rotated. The thread is cut on both edges of the bit.

Relief (thread relief)
A groove formed in the shaft in which the thread bit stops cutting.

Shoulder thread
A thread that allows the object to be screwed in up to the very end of the thread.

The Setup for Threading

For this exercise, we'll produce a single thread on this part (Fig. 11-148).

Task 1   Holding Method

Runout of the thread to the existing diameter is tight (0.003 in.) so either hold the finished shaft in a four-jaw chuck, a collet, or between centers, or machine the thread diameter from a larger shaft. A three-jaw is probably not accurate enough. Always consider a tailstock support for threading.

Task 2   Part Prep

First, be certain the shaft is at the major diameter for the thread. Next, machine a lead-in chamfer and a **thread relief** (if the design will permit it). Your actions will be different as the tool reaches the end of the thread, depending on the way it terminates (Fig. 11-149).

The thread on the top of Fig. 11-149 is called a **shoulder thread,** meaning that in use it can be screwed completely to the flat end—the shoulder. Doing so requires a thread relief. The threading action is simplified with a relief because it provides a place to stop the tool's Z axis travel, without ruining the thread. I strongly recommend that your first few threads be relief types if you have the choice.

However, it's sometimes necessary to thread without a relief, as shown on the bottom object. The unrelieved, *pull out* thread requires a coordinated movement of both hands! As the

Material: mild steel CRS

Figure 11-148 *You are required to set up and thread one part.*

Figure 11-149 *Your actions change depending on how the threads terminate.*

Checking Bit Shape with a Center Gage

Off center

Slanted shank

Figure 11-150 *Grind the bit to 60°. The relationship of the point can vary to the shank as shown.*

bit reaches a mark on the part or a position on a dial indicator, you must quickly crank the *X* axis back so the tool clears the thread while simultaneously lifting the half nut lever to stop the carriage. That's a white knuckle time!

If you miss this simultaneous move, the thread is ruined. The pull out thread then has a short section where incomplete, tapered threads are produced as you crank the tool back, while threading to a relief allows simply stopping the carriage when the bit reaches the relief. Then crank the tool back.

KEYPOINT

If given the choice, always thread to a relief.

Task 3   Grinding the Thread Bit

For this exercise we will use the Unified Thread Form that you already know and assume you aren't using preshaped threading tools (even though they save a lot of time). The thread form features a 60° included angle template and a small flat at the tip. You will need to custom grind the bit to produce the UNC 1-8 thread form.

First, form grind the bit to a 60° included angle using the center gage.

Trade Tip

When grinding your own bit, the point need not be in any specific relationship to the shank and is often slanted to one side to allow threading to a shoulder. In other words, it isn't required that the point is on the center of the bit nor that it points straight off the bit (Fig. 11-150).

The top rake is ground as a compounded angle—both to the side and back to improve the cutting action of one edge only. The tip flat is a different width for each thread pitch. Larger threads mean a larger flat. It will be one-eighth of the thread pitch distance (thread-to-thread distance).

For our example, Fig. 11-151, 8 threads per inch (TPI), the pitch distance would be 0.125 in. Dividing that by 8 yields the flat width of 0.0156 in.

1-8 UNC thread

Pitch distance = 0.125 in. (thread-to-thread)

Flat width = 0.125/8 = 0.0156 or rounded to 0.016 in. flat

This is so for ISO metric 60° thread forms too. So a metric thread example would be M8-1.25 (Pitch distance = 1.25 mm already supplied)

$$\frac{1.25}{8} = 0.16 \text{ mm tip flat}$$

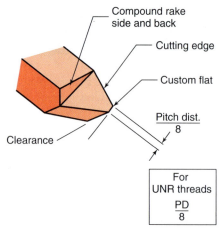

Compound rake side and back

Cutting edge

Custom flat

Clearance

$\dfrac{\text{Pitch dist.}}{8}$

For UNR threads $\dfrac{PD}{8}$

Figure 11-151 *Notice the combined side and back rake makes this a right-hand bit—cutting toward the chuck using the left edge only.*

Use a Center Gage to Check the 60° Angle Center gages are small, flat, arrow-shaped templates that have three purposes while threading. (Watch for two more uses coming up.) Shown in Fig. 11-151, it is being used as a template for checking and grinding the included angle on the bit.

Trade Tip

Hold the center gage template on the tool bit up, such that a light source is behind the pair. Any shape irregularities will show very well.

Task 4   Setting the Compound

Single-point threading requires several passes of increasing depth, to complete the total thread depth. To do so, we swivel the compound to an angle slightly less than half the included angle of the thread. For a Unified Thread of 60°, turn the compound from 29° to 29.5°—in the counterclockwise direction (Fig. 11-152).

Cutting edge

29.5°

Figure 11-152 *For a UNC thread, the compound is set at 29.5 degrees, counterclockwise.*

Trade Tip

Why 29 to 29.5 Degrees? Because threading on a manual lathe requires several simultaneous, quick operator actions, the spindle is slowed far below efficient cutting speeds, otherwise a human couldn't keep up with the hand-eye motions. But that means the cutting speed is far below good deformation. To aid in machine finishing at these low speeds, the tool must be as efficient as it can be. Therefore, a large positive, compounded rake to the side and back is used.

Then to take advantage of that tool geometry, only one edge of the bit cuts best. The compound is turned such that each new pass advances the cutting edge into the work. In other words, the good edge does the cutting. If not, both edges would cut the thread resulting in lower quality finish. (This is called **plunge threading**.)

Threading progresses by cranking the compound forward—not the cross slide—for a reason. By moving inward along the compound axis, the bit is cutting with the left-hand edge only—the edge that is ground to cut correctly. But why not 30° instead of 29.5°?

Using a 29.5° angle rather than exactly half the thread included angle ensures that the back side of the thread is smoothed after each new pass. A very small amount of metal is removed by the back side of the bit to ensure that small steps aren't produced.

However, when threading on the CNC lathe where the controller starts and stops tool motion, the cutting speeds are restored to full efficiency. Then the single-edge cutting action described in Figs. 11-152 and 11-153 is not required. In CNC threading, there are advantages and disadvantages to both single-edge cutting and plunge threading.

30 degree thread angle

Successive passes—transition smoothed

Cutting edge

Figure 11-153 *Each pass is taken by the best edge of the tool. The back edge smoothes the transitions of each pass.*

Task 5   Positioning the Bit

Then after aligning the compound, there are two items to check when mounting the tool in the tool post:

Vertical center height—as usual

Bit angle relative to the axis of the work—checked using the thread center gage

Tool Alignment with a Center Gage

Align gap

Hold gage against tool

Aligned to work axis

Figure 11-154 *Holding the center gage firmly on the bit, move them both to the workpiece and compare.*

Coordinating Micrometer Dials

Tool touching work

Set both to zero

Figure 11-155 *After removing backlash, lightly crank the bit to the work then set both micrometer dials to zero.*

Vertical Height Centering First, check vertical centering of the bit. For threading, this task is more important than just the cutting clearance/action. If the threading bit was positioned low, it would cut OK, but the 60° shape would be distorted.

Setting the Cutting Angle Using the center gage a second time to align the tool to the *Z* axis of the lathe, there is a right and wrong way of doing this task. For best results, *hold the gage firmly against the bit,* then move the *X* axis inward close to the work. Now, compare the gap between the gage edge and the work or chuck side (Fig. 11-154). Using the gage this way, the alignment is easy to see. Do not hold the gage against the work then put the tool in the template groove. That action makes it difficult to detect any misalignment, which would only show between the gage vee and the bit.

Task 6 Coordinate the Micrometer Dials—*X* axis and Compound

With the bit correctly positioned in the lathe and the compound swiveled and locked down, dial it forward to remove backlash in both the *X* and the compound axes. Check to see that the compound is not overextended at that position, then continue the *X* forward and touch it lightly to the workpiece. With the tool just touching the workpiece, zero both micrometer dials, compound and *X* (Fig. 11-155).

Note, it is common to use the dials as you have now coordinated them to make the thread, however, there is a much more skilled version of coordination coming up in a great Trade Tip, don't miss it.

Task 7 Gear Selections for Threading

Next, set the gearing on the lathe. There are five items to check off:

Select thread drive Shift such that the lead screw is engaged not the feed system.

Right-hand spiral The tool will be moving toward the chuck when the half nut lever is engaged.

Thread pitch levers (Two or three shift levers) Set for a pitch of 8 for this example (maybe four shifters for dual metric/imperial lathes). Refer to the chart on the lathe front.

RPM Set for a low RPM (300 plus or minus). Although too low for efficient cutting, it is not possible to react fast enough at higher spindle speeds during single-point threading. Threading is difficult to control if the RPM is very high.

Engage the **chasing dial** This is done by loosening a screw and pivoting it into engagement with the lead screw.

Making the Thread— Your Actions

OK—the lathe is ready to go. There are three actions you must take to be in control of threading. While I'll give a brief explanation, you'll need a demonstration before the solo:

Marking and checking the first pass
Completing the finished thread passes
Measuring the thread

First Pass—Positioning the Tool and Axes

Since you'll be working with the compound axis, which has no digital readout, even if the *X* does, use digital *X* as a backup reading only. Concentrate on the micrometer dial readings for threading.

With the *X* drawn away from the work 0.050 to 0.150 in., the saddle is moved to the right, off the work. Then the *X* axis is re-cranked back to the original zero position.

Next, the compound axis is cranked forward beyond zero, just enough to cut a light test cut on the work. From 0.003 to 0.010 in. are reasonable amounts. It is a good idea to snug the axis lock on the compound when making a threading pass. Not locked, just snug.

KEYPOINT

On the first pass, you aren't cutting the thread, but rather, making a scratch that can be tested to see if the gear selections yield the pitch distance required.

After the first pass test the thread for correct pitch. Use one of the methods shown in Fig. 11-156 or a caliper set to the correct pitch distance. This will indicate that the gearing is correct for the intended thread. Note that center gages have two or four scales, one of which will usually be divisible by the thread pitch. In this example, 8 threads per inch will divide into 24 every three divisions but the 32-division scale could be used too; 8 is prime to either scale.

Half Nut Lever Engagement Before cutting that first pass, I strongly recommend practicing the following actions, with the tool drawn away from the work, a few times to get the feel of single point threading.

Your left hand's first action is on the *Z* axis, saddle hand wheel rim, lightly. This is to hold

back on the wheel and keep the backlash to the right against the lead screw. This is the way to get good repeatability of tool position on the lead screw. Once the thread is cutting, you'll be moving that hand position.

Your right hand is on the **half nut** lever. When the chasing dial gets close to lining up, drop the lever down lightly before the index lines completely coincide. This will ensure that you do not miss engagement, which must be made smartly on time. Doing so, you will feel the half nut touch, pause, then drop down into engagement. (See the chasing dial rules—next.)

KEYPOINT

When learning to thread (Fig. 11-157), it's wise to keep your right hand on the half nut lever throughout the pass! Then, if a problem occurs, you don't need to fumble for the correct lever.

Left hand second action is as the thread begins to cut, transfer your left hand to the *X* axis dial ready to crank it back.

KEYPOINT

Remember to stop the *Z* axis motion, pull up on the half nut lever when the bit reaches the relief. Then crank back on the *X* axis.

Figure 11-157 *Left view: Threading begins: left hand on the Z hand wheel, lightly holding back, right on the half nut lever. Right view: With the half nut engaged, after the saddle moves, switch left hand to the X cross feed dial.*

Two Methods to Check Screw Pitch Distance

Figure 11-156 *The witness marks are compared to the standard lines on the center gage or to a screw pitch gage.*

Caution: When pulling up on the half nut lever before the thread relief, with the tool within a cut thread, you will produce an unwanted groove around the shaft, ruining the thread. So follow the next procedure.

Crash Procedure To stop threading somewhere in the middle in an emergency, first, using your left hand, crank away on the *X* axis hand wheel, then quickly pull up on the half nut lever with your right hand. This disengages the action without making any unwanted groove in the threads.

Threading Passes

To position for a new pass, with the *X* axis still cranked back, return the *Z* axis to the start position, to the right end. Then dial the *X* back to the zero then the compound in a reasonable amount of around 0.020 to 0.050 in. depending on the size of the thread, ready for the next pass. Watch the chasing dial for engagement timing and keep repeating these actions. For the best cutting action, paint cutting oil or even better yet, thread oil on with a brush. Flood coolant isn't recommended as it obscures your vision of the cutting tool, which must be seen to stop exactly at the relief.

When to Engage the Chasing Dial

The half nut inside the lathe saddle is a clam shell that closes on the lead screw when the lever is engaged. With some thread pitches, it can be engaged in the wrong position, which cuts the wrong groove and ruins the thread. A standard lathe features an eight-position chasing dial to prevent this from happening (Fig. 11-158).

Its purpose is to show the operator when the half nut lever can be engaged such that the tool will fall into the same thread groove with each pass of the tool. The rules for its use are based on the number of threads per inch you wish to produce. The outer index line will coincide with one of the eight lines on the dial to indicate correct timing to start the thread. Look back at Fig. 11-67 for an example of the chasing dial.

Which line or lines you can use is based on pitch. If the thread has an

Even number threads—for example, 16, 18, or 20 TPI

Drop the lever on any of the eight lines.

Odd number threads—for example, 13 or 27

Rules for engaging the split-nut for thread cutting

Threads per inch to be cut	When to engage split nut	Reading on dial
Even number of threads	Engage at any graduation on the dial 1 1 1/2 2 2 1/2 3 3 1/2 4 4 1/2	
Odd number of threads	Engage at any main division 1 2 3 4	
Fractional number of threads	1/2 threads, e.g., 11 1/2 engage at every other main division 1 & 3, or 2 & 4 other fractional threads engage at same division every time	

Figure 11-158 *The eight possible engagement positions for a chasing dial.*

Drop the lever on the numbered lines only.

Fractional threads—for example, $27\frac{1}{2}$ TPI

Drop the lever on any two opposing numbered lines—for example, 1 and 3, or 2 and 4 (see the Trade Tip.)

Trade Tip

When you must drop the half nut lever on selected lines, it's a good idea to use a highlighter to clearly mark the target line(s). Or, if there's any doubt, just use the same line every time, that guarantees correct alignment. (I use line 1 when doing this.)

Metric Threads

When cutting metric threads on older lathes (pre mid-1980s), it's necessary to open the gear case and add or change drive gearing between the spindle and the lead screw. However, many modern manual lathes are made to easily cut both imperial and metric threads by use of an extra quick change gear selection on the headstock. That complicates the chasing dial and rules. To solve the chasing dial issue, some feature two interchangeable chasing dials, one for imperial pitches and one for metric. However, other dual-system lathes are equipped with chasing dials with more than eight positions. For these, find and follow the manufacturer's instructions or use my Trade Tip and drop the lever on the same dial position for every pass.

Internal Threading Option 1

Cutting motion inward

Thread relief

Crank compound outward each pass

29.5°

Crank *X* inward to retract bar Outward to zero for each pass

Figure 11-160 *Boring an internal thread option 1—cutting inward.*

Trade Tip

Coordinate the Thread Depth on the *X* Axis— Save Calculating and Extra Measuring This tip simplifies cutting the thread, to the correct depth. The classic method of dialing forward along the compound axis until the full depth is reached, means moving the tool into the work along the long (calculated) side of a triangle, to create the full depth—or alternatively cutting and measuring until the depth is reached. But, adding these two simple steps after coordinating the *X* and compound dials to zero, makes thread depth a snap!

Step 1 From the coordination position (bit tip touching—both dials at zero) *back away on the compound axis about three times the thread depth* (Fig. 11-159).

Step 2 Now, with the bit off the work, *crank the X axis forward a distance equal to the thread depth.*

That distance is easy to find in *Machinery's Handbook* or to compute, depending on the series and root shape. For example, the depth of a $\frac{1}{2}$-13 UN thread = 0.54127 × Pitch distance = 0.0416 in.

Keep in mind that on diameter-calibrated lathes, that means you will actually dial in twice the thread depth to move the bit forward the real depth. For this example, that means cranking 0.0832 in. The tool will physically advance the required 0.0416 in. *Now reset X to zero.*

With this coordination correctly performed, you will still move the *X* in and out to its zero position for each pass, and will still be infeeding along the compound axis. But you will stop when both dials reach zero. Presto, the thread will be at the right depth!

Thread Depth Trade Tip

Dials set at zero tool touching as usual

Step 1 back away using compound

Step 2 dial *X* axis forward double thread depth

More than double thread depth

Reset *X* to zero

Figure 11-159 *Use this trade tip to simplify threading.*

Setting Up to Cut Internal Threads

Cutting an internal thread is a bit more complex compared to the external setup (Fig. 11-160). You cannot see the tool as it approaches the end of the cut. That means a DTI or other means of showing when the half nut lever must be disengaged, at the exact *Z* axis position each time. Digital readouts can be used but require you to look away from the action. There are two choices:

Option 1 Progressing Inward To cut an internal thread with the cutting tool going to your left, the compound is rotated 29.5° in a clockwise direction (opposite normal rotation). Operator actions are similar to cutting external threads with two exceptions. After the thread pass is completed, you must crank the *X* axis inward to clear the tool of the work for the next pass— not out! (See the thread stop Trade Tip.) Second, to take successive passes, the compound axis is cranked outward, not in, until the full depth is achieved.

Option 2 Threading Outward Here are two versions of a safer setup whereby each pass is started with the boring bar all the way in, at the bottom of the far end of the hole, then the cut feeds outward—to your right, away from the bottom of the hole (Fig. 11-161). This creates less operator stress since the half nut lever is disengaged after the bit is safely out of the hole, where you can see it. Both setups are based on simulating unscrewing the cutter while it cuts a right-hand thread therefore the lathe must be able to run in reverse.

Internal Threading—Two Options

Cutting motion outward →
Reverse spindle rotation both setups

Bit upside down

Cutting edge upward

Crank compound out each pass

Crank *X* inward to retract bar
Outward to zero for each pass

29.5°

Crank compound in each pass

29.5°

Crank *X* toward you to retract
Away to zero for each pass

Figure 11-161 *Safer setups—boring outward: version 1, bit upside down, in front, and version 2, bit up, on far side of hole.*

Bit Upside Down

In version one, the bit is turned over in the boring bar. The cutting tool is positioned on the near side of the hole.

Bit on Far Side Right Side Up

Here the bit is upright on the far side of the hole. In some cases this means that the machinist can see it at full depth, when it's parked in the relief, an added advantage.

KEYPOINT

For either setup, a thread relief is necessary. If a relief isn't possible, then neither method can be used since there is no place to park the cutting tool until the thread pass begins.

Cam Lock Chucks Only for Reverse In order to achieve a right-hand helix, the lathe must rotate backward. That means a screw-on chuck cannot be safely used.

Trade Tip

Thread Stops Many industrial lathes feature a special *X* axis setup provision that sets hard limits on the amount and direction the *X* axis handwheel can be cranked. They are very useful when making both internal and external threads. If the lathe has a thread stop, look for a small thumb wheel on the perimeter of the *X* axis collar, behind the crank. Ask your instructor to explain how they are used.

KEYPOINT

Thread stops correctly setup, prevent accidentally cranking the *X* axis the wrong direction at the wrong time!

UNIT 11-7 REVIEW

Review the Key Terms

Center gage
The small arrow-shaped tool used as a setup template and thread checking guide.

Chasing dial
The indicator of when to engage the half nut lever to make each pass in the same thread groove as the last.

Half nut
The split nut engagement devise for pulling the carriage along at the right pitch ratio to the spindle. The half nut lever engages the threading action.

Plunge threading
Single-point threading where the compound is not rotated. The thread is cut on both edges of the bit.

Relief (thread relief)
A groove formed in the shaft in which the thread bit stops cutting.

Shoulder thread
A thread that allows the object to be screwed in up to the very end of the thread.

Replay the Key Points

◆ Given the choice, always thread to a relief.

◆ On the first threading pass, you aren't cutting the thread, but rather making a scratch that can be tested to see that all the gear selections yield the pitch distance required.

◆ To cut an internal thread the compound is rotated 29.5° in a clock-wise direction to advance the bit's cutting edge into the work

◆ Chasing dial rules:
Even number threads—on any of the eight lines
Odd number threads—on the numbered lines only
Fractional threads—on any two opposing numbered lines

Respond 11-7

Threading is complex to explain on paper but it will become clear with hands on practice. Let's see what retention occurred here. On a sheet of paper, list the steps to setting up a thread and the actions you must take to make one. Questions A through D are critical thinking and there are no answers provided. Double check yours against the reading or better yet, talk them through with a partner.

A. The Setup

   1)   2)   3)   4)   5)

B. Operating the Lathe

   1)   2)   3)   4)   5)

C. Hand Positions

D. After positioning the bit for a pass what are the starting actions:

A} Watch the _____ _____

B} Push [ ]Up or [ ]Down on the _____ _____

C}

The stopping actions

At the relief?

A)   B)

Before the relief? (Usually an emergency stop)

A)   B)

Positioning the tool for a new pass:

Answer the following about threading. Answers found at the end of the section.

1. To what angle must the compound be turned for a Unified Thread?

2. Explain why you swivel the compound in Question 1.

3. What is the width of a thread bit, nose flat for 16 threads per inch?

4. Engaging the lead screw is one of the five setup steps for threading. Is this statement true or false? If it is false, correct it to be true.

5. How would you make a left-hand thread? This is not covered in the reading—think it out.

6. To make 20 threads per inch, where can you engage the chasing dial? Where for 19 TPI? For $27\frac{1}{2}$ TPI?

7. What is the depth of thread for a UNF $\frac{1}{2}$-20 screw?

the thread until it reaches the full pitch depth. But since the deepest feature, the bottom crest of the triangle called the **root diameter** or **minor diameter,** includes a clearance amount to allow the screw to assemble with a mating nut, it isn't the key feature that must be measured. Plus it's hard to reach the root with a measuring tool.

Measure the Pitch Diameter The function being controlled by the dimensions and tolerances is the fit between the faces of the screw and nut. To do so, we need to measure somewhere within the thread form, not at the root or top crest of the thread. The crest (or major diameter) is the surface of the original shaft that hasn't been machined away during threading. So, we measure at a theoretical point (Fig. 11-162), at the center of the thread faces, called the **pitch diameter.**

The pitch diameter is actually the center of engagement between the mated screw and nut. Therefore it isn't exactly at the center of either component. It occurs at the center of their interface. However, for convenience, it can be envisioned as the physical center of any given screw thread since it's within a few tenths of a thousandth of the middle for screw sizes smaller than 1 in.

KEYPOINT

The pitch diameter is a theoretical measure of the diameter of the overall thread.

Unit 11-8 introduces several options for pitch diameter measurement that might be found in a tech school. There are more sophisticated methods to be learned on the job. Methods beyond those presented here are industry specific, depending on the use to which the products will be put. They range from optical projection comparitors through electronic probing equipment and jigs.

UNIT 11-8

Measuring Threads

Introduction: When machining a screw thread, we make several, light cuts called **passes,** until the thread is at its full depth in the shaft. Each pass reduces the diameter of

Figure 11-162 The pitch diameter *compared to the other thread dimensions.*

Completely measuring threads can be a complex issue depending on the use to which they will be put and their tolerances. Not only is pitch diameter to be checked, but the thread form must be verified and the lead measured over a great distance too. Here, we'll limit our investigation to ways of measuring the pitch diameter. These tests represent the larger part of thread measuring until the threads become very closely controlled. Check with your instructor to learn which methods are used in your lab.

KEY TERMS

Passes (thread)
Each cut to deepen the thread toward the pitch diameter.

Pitch diameter
The theoretical center of engagement between internal and external mated threads.

Pitch gage (snap gage)
A horseshoe-shaped gage with two pairs of pitch rollers. If the first pair slips over but not the second, the thread should assemble.

Root diameter (minor diameter)
The smaller diameter on the thread form—internal or external threads.

Thread ring gage
A precision nut that can be set to test the pitch diameter of threads.

Measuring Pitch Diameter

As we've learned previously, dimensions can be measured or they can be functionally gauged. Of these methods, two are measurements and two gages of fit.

Gage 1   Testing Diameter Against a Nut Standard

The bottom-line, functional test finds if a nut assembles with the screw. If it goes on without dragging or excessive wobble, it's Ok for some utility applications. This is the least technical method. But when a thread must be quickly produced for tooling or other noncustomer purposes, it works! Both this test and the next verify thread lead to some extent but only over a short distance—the width of the nut.

Gage 2   Adjustable Ring Nuts

A more technical version of the nut gage, is the *adjustable* **thread ring gage.** Simple to use, sometimes one gage nut is set up at the smallest size allowable and a second is set at the largest size to become a go, no-go pair.

Measurement 1   Measuring with a Thread Micrometer

This is a special micrometer featuring pointed tips that fit into the Unified Form thread grooves (or other thread form if the mic has interchangeable tips). To use this method, refer to the data chart in the mic's kit or in standard reference books such as *Machinery's Handbook*. It will list the size that the mic will read when the thread is at the correct pitch diameter.

This is a fairly accurate method of determining pitch diameter, but doesn't actually verify thread form or lead. A thread can pass this test yet not fit to a mating part! However, if the thread form is under control and the lathe produces accurate leads (the usual situation), then thread mics are a good option.

Thread mics are more universal than test nuts, since they can measure any thread. There are reference tales to determine the pitch diameter for a given standard thread and a formula to calculate the pitch diameter for special threads not found in the chart.

Measurement 2   Thread Wires

This is an accurate, universal method, but slow. Three wires are held in place as per Fig. 11-163. A standard micrometer then measures over the wire tops. A data chart (or formula found next) is needed to determine the exact measured distance over the wires when the pitch diameter is correct.

While they are supplied in handy sets (Fig. 11-164), thread wires can be any set of three round pins of the same diameter. They are selected such that the wire diameter contacts the thread form near but not perfectly at the pitch diameter for that thread's pitch. All that's required is that they protrude slightly above the major diameter to contact the micrometer face.

Trade Tip

When measuring nonstandard threads or you do not have a three wire set as shown in Fig. 11-164, here's three useful formulae from reference books:

To find *M*, the measurement over the wires at the correct pitch diameter, within a standard 60° thread form:

Major Diameter Known (from *Machinists Ready Reference©*)

$$M = D + (3 \times W) - \frac{1.515}{N}$$

Pitch Diameter Known (from *Machinery's Handbook©*)

$$M = E - 0.86603P + 3W$$

Best Wire Diameter The closest, standard wire size that contacts thread nearest the pitch diameter in 60° threads.

$$W = 0.57735 \times PD$$

where

M = Measurement over wires in thousandths of an inch

D = Thread major diameter (from a reference book)

W = Wire diameter

N = Number of threads per inch

P = Pitch diameter of thread

PD = Pitch distance = 1/TPI

These formulae do not compensate for the compounding of the measurement caused by the thread's lead angle. They are close enough for class 3 threads in everyday manufacturing. Closer formulae that do compensate plus a more complete explanation are found in ASME B1.2-1992. Read further in *Machinery's Handbook©*.

Figure 11-164 *A set of wires—three of each diameter. Note the clay holding the wires until they are trapped by the micrometer.*

Gage Test 3 Using a Pitch Gage Standard

The **pitch gage** (Fig. 11-165) is an adjustable go, no-go set of thread rollers. If the first pair can fit over the thread but not the second pair, the thread is within the calibrated tolerance. These gages offer one advantage over the previous methods, they functionally simulate mating with several threads, thus check a small part of the lead as well as the pitch diameter.

This gage is accurate and quick to use thus it is used most often in production machining. It's sometimes called a **snap gage** due to the way it drops on the test piece, it requires calibration against precision threaded standards.

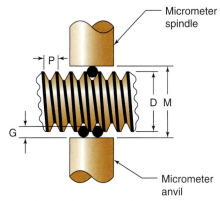

Three wires arranged correctly for measuring 60° threads

The screw thread micrometer measures the pitch diameter of a thread

Figure 11-163 *Two common methods of measuring threads.*

Figure 11-165 *A pitch gage slipping over the first rollers, but not the second, shows the thread is within tolerance.*

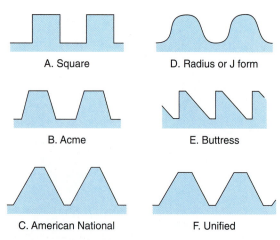

A. Square   D. Radius or J form

B. Acme   E. Buttress

C. American National   F. Unified

Figure 11-166 *Other thread forms you might be called on to machine.*

Other Thread Forms

We will study technical screw threads as a subject in Section 14, but for now, you should recognize the other thread forms used in mechanical devises (Fig. 11-166). They are cut in similar fashion as the Unified form and measured similarly, with the exception of the square form thread, which must always be functionally gauged.

UNIT 11-8 REVIEW

Review the Key Terms

Passes (thread)
Each cut to deepen the thread toward the pitch diameter.

Pitch diameter
The theoretical center of engagement between internal and external mated threads.

Pitch gage (snap gage)
A horseshoe-shaped gage with two pairs of pitch rollers. If the first pair slips over but not the second, the thread should assemble.

Root diameter (minor diameter)
The smaller diameter on the thread form—internal or external threads.

Thread ring gage
A precision nut that can be set to test the pitch diameter of threads.

Replay the Key Points

- The pitch diameter is a theoretical measure of the diameter of the overall thread.
- The pitch diameter can be measured or gauged.

◆ For measurements over given wire sizes, look in *Machinery's Handbook.*

Respond

1. Which methods of measuring pitch diameter use a micrometer?

2. What is the deciding factor for choosing the size of the three wires used to measure pitch diameter?

3. Describe the pitch diameter for a thread.

4. True or false? Pitch diameter can be verified by functional gage and measuring. If it's false, what will make it true?

Critical Thinking

5. Of the four methods we've just discussed, which verify thread form? If they can't, then how might you think thread form can be evaluated?

6. From *Machinery's Handbook,* what wire size is best for testing American Standard Unified threads (60°) with 16 threads per inch? (*Hint:* The best wire diameter contacts the thread at the pitch diameter but there is a range of wire sizes that will do the job.)

7. Measure an American Standard $\frac{1}{2}''$-13 UNC screw thread. From the *Handbook,* what formula would you use to determine the pitch diameter if you have the measurement M over three wires of known size W?

8. Can wires of 0.050-in. diameter be used to measure a American Standard Unified thread form of 13 TPI?

9. Using *Machinery's Handbook,* you have three wires of 0.050-in. diameter, to measure the pitch diameter of a $\frac{1}{2}$-13 UNC thread. What should the measurement M be, if the thread pitch diameter is correct?

10. For the thread of Question 9 the measurement M (over 0.050-in. wires) is found to be 0.5235 in. diameter. How much more must be machined away to make it per specs?

Section 11 Review

Unit 11-1

Before learning how to perform the basic operations, we needed to define the target—what a lathe can do. Keep in mind, all of Section 11 is offered not only to learn how to recognize and safely perform operations on the manual lathe but to get a feeling for them such that you can write reasonable program statements that do them right on the CNC machine as well.

Unit 11-2

To run and maintain a lathe correctly, it's critical to understand the way they are put together. While there are differences in the way they are made, they deliver the same range of functions. For example, some budget lathes put the lead and feed functions together into a single shaft with a key way and lead threads both. Other lathes feature two axis clutches, one for *X* and one for *Z*, while less costly lathes feature a single clutch with a selector for which axis is to be actuated on the apron. If you understand the range of functions a lathe must perform, then it's easy to adapt to the specific machine.

Unit 11-3

Do you clearly understand the differences in the six ways to hold work and how to choose the right one? If not, review Unit 11-3 again. In my career, I have seen two spectacular crashes involving misheld work on the lathe. In one, the operator was injured seriously. In the other, we were pretty white knuckled as the part went spinning throughout the shop causing all sorts of damage to things but not people—lucky that time! If you don't recall this unit clearly, then go back and do it again. Education by disaster is unacceptable and inevitable if you don't understand work holding.

Unit 11-4

Like a lot of subjects in this book, we've delved into the subject to the point that you have a working knowledge. You now know about basic cutting tools to set up and perform the 15 operations. However, we've only seen the tip of the iceberg. Cutting tools are evolving as I write this book. Stay informed through supplier catalogs and seminars, by going to tool shows, and by subscribing to trade publications such as the *American Machinist*©, the *Machine Tool Digest*©, *CNC Machining*©, or several other fine magazines. Ask your instructor for a recommended publication for your area.

Unit 11-5

I hope I've made my case for extra caution when setting up and running a lathe. They are powerful and dangerous. You must stand in front of them while they spin the work. Chips can fly out like bullets, as they are flung off the spindle. Uncontrolled, they can be as sharp as a razor and strong too. Then there are the heated chips and work, and finally the misheld part flying out. All can be controlled with professional caution and skills.

Unit 11-6

Although CNC lathes have displaced manual machines on the front lines, to be a true journey-level machinist working in a tooling or job shop, you must be able to deliver the quick part made the good old standby way. Manual machines will never be eliminated from the factory or especially from tool and die, repair, or other small shops.

Unit 11-7

Almost as difficult as rubbing your stomach with your right hand, while patting your head with your left, the hand-eye coordination for threading takes practice! It is a good idea to watch a demonstration then practice the actions with the tool pulled back a safe distance. This too is a skill a competent journeyman must understand even though we make most modern threads using programs or automated equipment, not hand action.

Unit 11-8

The real trick here is the same as for all measuring tasks—determine the tolerance then find the tool that delivers the repeatability and resolution needed. Then practice.

Questions and Problems

1. Of the basic metal cutting operations on a lathe, which are performed from the tailstock? (Objs. 11-1 and 11-2)

2. Facing on a manually operated lathe presents a problem that is easily overcome on a CNC lathe. What is it? (Obj. 11-1)

Complete Part Setup Problems

3. In Fig. 11-167, name the operations in sequence, required to complete the spur gar shaft before the gear teeth are cut on it. It comes to your lathe as a 2.500-in.-diameter bar of CRS—cut to 8.0-in. lengths. (Obj. 11-1)

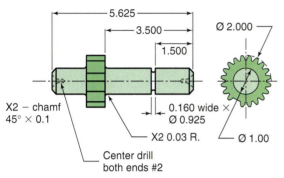

Figure 11-167 *Write a plan to machine this spur gear blank before the teeth are milled on it.*

4. For Fig. 11-167, what holding method would be correct? Explain? (Objs. 11-3 and 11-5)

5. Would a tailstock support be *required* when turning (Fig. 11-167)? (Objs. 11-2 and 11-6)

6. To complete the gear shaft on the lathe, which tool post would be ideal? Explain. (Obj. 11-3)

7. To complete Fig. 11-167, which cutting tools should ideally be carbide insert? Which would probably be HSS? (Obj. 11-4)

8. To machine Fig. 11-167, calculate the correct RPM to machine the 2.0-in.-diameter gear blank and the 1.00 shaft assuming carbide cutting tools. Use Appendix IV, the SS chart, and the short formula. (Obj. 11-6)

9. The RPM calculations were based on finished diameters. Would they be correct for the roughing passes? You have two options. Explain. (Obj. 11-6)

Critical Thinking Problems

10. The pilot on the center drill is 0.125 in. diameter. What is the correct RPM for center drilling? (Objs. 11-5 and 11-6)

11. If there were 20 parts to be made per Fig. 11-167, is there a more economical way to machine them on the manual lathe? Explain. (Objs. 11-2 and 11-3)

12. Assuming you adopt the method suggested for making 20 parts in Question 11, is there a special safety consideration? There are two options. Explain. (Objs. 11-3 and 11-5)

13. True or false? The center gage is used to center threads on the uncut shaft. If it's false, what will make it true? (Obj. 11-7)

14. True or false? When facing work it's a good idea to lock the *X* axis to avoid cutting tool slippage. If it's false, what will make it true? (Obj. 11-1)

15. Explain why a face-grooving (trepanning) cutting tool must have a slightly different geometry than a tool meant to groove the side of the workpiece. (Obj. 11-4)

16. Of the four thread-measuring tools and methods, which is the most accurate at measuring the thread pitch diameter and lead for several threads? (Obj. 11-8)

17. What is the pitch diameter for a $\frac{7}{16}$"-14 UNC screw thread? (Obj. 11-8)

18. What is the recommended wire diameter and measurement over the wires to check the $\frac{7}{16}$"-14 UNC thread of Question 17? (Obj. 11-8)

CNC Questions

19. Why must you set the RPM so far below the normal recommended surface speed when threading on the manual lathe yet they can be cut at the correct spindle RPM on CNC lathes? (Objs. 11-2 and 11-7)

20. Beyond the actual computer control of the various functions, what is the basic mechanical difference between a CNC lathe and a manually operated machine? (Obj. 11-2)

Section 11 Answers

1. This axle required:

   A. Straight turning
   B. Taper (by tailstock offset or taper attachment since the length was long)
   C. Drilling (center drill in this case)
   D. Face cut
   E. Chamfer (form tool or turn compound)
   F. Threading
   G. Grooving in two places

2. No it can't. False. A parting tool does not have the side clearance required for trepanning.

3. Here are the 15 first year skills you need to be able to setup and perform on a lathe: *straight turning* (short, medium, and long shafts), reducing the diameter of a shaft parallel to its axis; *facing*, Machining the end surface of a part perpendicular to it's axis; *drilling/reaming* (no explanation necessary); *cutting angles*, short angular surfaces such as a chamfer or transition; *tapers*, precision angular surfaces such as a drill shank taper; *boring*, enlarging a hole parallel to the work axis, can be tapered, and is accomplished with a boring bar; *Forming*, cutting with a shaped bit; *threading—single point*, cutting threads with a tool ground with the thread groove shape; *parting*, Making a groove deep enough to cut the work away from raw stock; *grooving*, similar to parting using a grooving tool; *face grooving*, similar to a parting operation but on the face of the work; *knurling*, forcing a pattern into the metal for non-slip bonding and size expansion; *filing and sanding*.

4. Note, the round nose pin would probably be machined with one lathe tool. The face, turn, and fillet radius could be made with a turning tool featuring a nose radius of 0.25 in., which would form the fillet as it turns and faces (Fig. 11-168).

5. See Fig. 11-169. Note, depending on how this job is planned, there might also be a parting operation if it's made from bar stock.

6. The problems are twofold. The workpiece is only poorly held, and can either fly out or bend and whip around. The second is that the operator must stand very close to the danger zone. Long stringy chips are a problem associated with lathes. They tend to grab and cut operators who are not paying attention.

7. Hands placed such that the file end never points at your body should it catch on the work. Body is upright, not leaning over the lathe. Note: Never use a file without a handle. The pointed tang (the portion inside the handle) can become a dangerous spear.

8. Here's the result of holding sandpaper in one hand. The pinched paper can wrap around the shaft, catch your thumb, and pull it into the paper and shaft. You will get a nasty sanded thumb or a dislocation (Fig. 11-170).

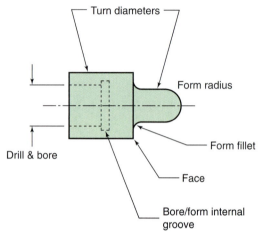

Figure 11-168 *The operations for Problem 4.*

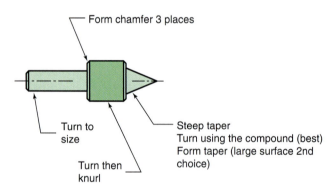

Figure 11-169 *The operations for Problem 5.*

Figure 11-170 *Warning: holding sandpaper or cloth in one hand can catch and wrap your thumb into the fold!*

ANSWERS 11-2

1. Lathe components are identified in Fig. 11-171.
2. *Bed ways,* Support entire carriage; *saddle, Z* axis movement over ways; *apron,* carriage front with operator controls; *compound,* swivelling axis for tool mounting; *quick gears,* feed/thread selection; *Z handwheel,* left/right movement by hand; *X motion,* in/out movements?—cross slide; *tailstock lock,* stop tailstock quill movement; *tailstock quill,* main spindle for in/out motion; *axis dir. select,* in or out or left or right movement; *dovetail and gibs,* guide and adjust out axis clearance; *chasing dial,* indicates when to engage half nut lever.

Figure 11-171 *The major components of the lathe.*

3. True

4. It may snap off due to the weak spot created by the tee slot groove for mounting tool holders. Extending is justified only when the operation is delicate or special.

5. A steady rest has three support points for the work.

6. A two-point support for work that moves along with the tool.

7. A tailstock can move the quill in and out for drilling, slide the entire accessory forward and back for setups, and move sideways to machine tapers.

8. Moving the spindle either by hand or bumping it with power is called drifting or jogging.

9. The following objects rotate: spindle, workpiece with chips, feed rod, lead screw. The following slide and could be a pinch point: X and Z axes under power. Others such as the quill are unlikely to be dangerous because they are slow and hand operated.

10. False. Spindles are hollow to allow long bars through.

11. Initial lubrication upon first move after cleaning: prevent rust

12. X axis = cross slide; Z axis = saddle (The compound is an auxiliary axis that can be turned to any angle.)

13. Feed; lead

14. False; the chasing dial

15. False. Gibs are used on the cross slide and compound only.

16. Quick change gear box (most common) (There are a few lathes that have this function on the apron.)

17. Parting off or shortened to parting

ANSWERS 11-3

1. This job justifies the cost of a special lathe fixture.

2. This could be a typical three-jaw job if a good chuck is used—to meet the 0.005-in. runout requirement. The job requires completing multiple parts requiring a relatively large amount of metal removed with a runout tolerance of a normal three-jaw chuck. The three-jaw has quick turnaround and a good grip. The end will require support, as the object is 8 in. long—too long for unsupported rough machining. If the chuck can't meet the runout requirement, a four-jaw will be used. Special note, be cautious of marring the original material surface when chucking.

3. This job would be easily accomplished in a 15-mm collet. The metal removal is small and the drill is actually a cylinder if held over a large enough area. Note, excess pressure could mar the drill margins. A collet would have a quick turnaround time for this quantity of parts. Alternate ideas: This job could also be accomplished in a soft jaw setup on a three-jaw chuck. Or, a standard three-jaw chuck would hold the drills on the fully round part near the shank, however, the runout requirement might be risky.

4. The center holes are important for this job. This roller could be turned in one of three setups:

   A. A collet and tailstock support if the existing center hole is OK and the collet is in top condition to meet the runout requirement.

   B. A four-jaw chuck and tailstock support—indicate the datum surfaces to remove all runout. Be careful to protect existing surfaces.

   C. Between centers, if the original center holes are in good shape. Be cautious of marking the end where the dog is clamped on.

   All three methods could meet the exacting runout tolerance. The job requires very little metal removal so grip isn't an issue. Ask your instructor how this job might be accomplished on the equipment you have.

5. This is primarily collet chuck job with no end support needed as the 19-mm part protrudes less than 25 mm so it is safe without support. A three-jaw chuck might work but the small-diameter aluminum would require very high RPMs and would be beyond safe limits for a large chuck. Use caution when machining the 60° angle on the end of the thin pin—it could bend. It may be necessary to put it on first.

6. False. Four-jaw chucks have only as much runout as the machinist permits. They can be adjusted for 100 percent accuracy. This statement is true for three-jaw chucks. Four-jaw chucks need no scroll plate because their jaws move independently.

7. Three-jaw chucks have good grip for heavy work and quick turnaround time, and can hold both inside and outside diameters.

8. Advantages of a collet: very quick turnaround, excellent control of runout, best holder for small round work—sometimes small rods cannot be held by any other means; *disadvantages:* weak grip—light, precision cuts only; the shop must own many sizes, which becomes expensive; can be damaged by overtightening or chips not cleaned away during assembly.

9. A three-jaw chuck is best for work where runout is not a major concern, although good three-jaws can have as little as 0.001 in. runout when in perfect condition.

10. Where a long taper is to be produced by moving the tailstock center off center. Or where several machinists must share a single lathe— the correctly prepared part may be removed and reinserted with perfect realignment.

11. Face plates are used only for part shapes that cannot be gripped in any other method. Fixtures are better than face plates because they are custom made to fit the part. Fixtures have better grip—face plates are universal versions.

12. True

ANSWERS 11-4

1. $\frac{1}{2}$ in. inscribed circle (four times $\frac{1}{8}$ in.)
2. Right hand
3. 1-in. shank size (16 times $\frac{1}{16}$ in.)
4. It is a triangle
5. About $\frac{2}{3}$ of the inscribed circle, which is $\frac{1}{2}$ in. $\frac{2}{3} \times \frac{1}{2}$ in. $= \frac{1}{3}$ in. or 0.333 in.
6. True
7. The statement is false. A rocker post is not recommended for heavy cuts due to its tendency to move during heavy metal removal.
8. A quick change or also called a cam lock.
9. *Solid tool or open-sided tool posts,* a solid block with a side opening used for setups where heavy metal is removed and/or a single tool can complete the entire setup; *quick change or cam lock tool post,* a set of tools, a dovetail post with a cam lock devise to hold special blocks in place on the post, chosen where several tools are needed in a setup; *turret tools,* also used where several tools are needed, the tools are permanently mounted on the turret. Care must be taken in mounting tools such that those not in use do not interfere into the work or chuck. More solid than cam locks but slightly limited to the variety of tools that may be used.
10. A turret is first choice and a cam lock second. The turret would be fast to index new tools and solid enough for the heavy metal removal required of this job. The three tools are simple and would not interfere in the turret.

11. This job would require a quick, cam lock tool post because some tools are long and there are too many for the turret. The feed rate and removal rate would need to be slowed down a bit and several smaller cuts taken because this tool post isn't quite as strong as the turret. Note that for efficiency, you might chuck the parts twice. Once for a roughing cut using a solid tool post where most of the excess metal is removed leaving just enough for the second setup using the cam lock post.

12. Rubbing or wrong tool geometry are the most likely since it occurred so soon in the operation. Your next choices would be that the surface speed was too fast and the tool burned up quickly or that the tool was not sharp to begin with. Check tool height using the rule capture method. Next inspect the cutting edge for rounding and regrind if needed. Recalculate the RPM and use coolant. You should also change to carbide tooling for the roughing cycle as it will be more efficient, but this then requires a new RPM.

1. Chip breakers to break chips into "C" shapes; chip dams to break chips into "C" shapes; chuck guards; use a chip hook to clear stringy chips away.
2. Stop and investigate—fix the problem if there is one.
3. Run!
4. Never take your hand off it. That way you won't leave it in the chuck.
5. There are two possible answers: Incorrect operator actions due to unfamiliarity or poor setups.

1. *Taper.* Five possibilities to check are

   The tapering attachment is engaged from a previous job.

   The part is not parallel to the axis of the lathe. That means the tailstock is off center.

   Using the second clue, the work is bigger on the right end, it could be flexing away from the tool. That could only occur if the tailstock has slipped and isn't truly supporting the work.

   It could be that the tool wears or slips in its holder/tool post as it is cutting.

   Eliminate tool wear with a check of the tool's cutting edge and accidental taper attachment engagement by checking the engagement nut to see if it's tight (engaged) or loose. Check all tool holding nuts and setscrews to ensure it isn't slipping.

   That leaves work flex and centered alignment as potential causes—both tailstock problems. The quick test is a hand spin of the live center. If it spins freely, it needs to be replaced in the hole and locked in position. If it is correctly touching, then the problem has to be tailstock misalignment. The taper must be caused by an incorrectly adjusted center. You have found it is tailstock misalignment.

2. *The Sixth Part Shows Taper.* The setup is OK, tailstock on center, taper attachment isn't engaged, and so on. This is a trend. However, it has to be something wrong at the tailstock. There are two possibilities: It has loosened on the previous part. But here's a real possibility: poorly chucked raw stock could also be the

Trade Tip

Realign it quickly using this process (Fig. 11-172).

1. Measure both ends of the part.
2. Calculate the difference in sizes.
3. Put a dial indicator on the tailstock near the work.
4. Adjust the center such that the work moves half the taper amount in the correct direction (toward or away from the operator).

Removing the Taper

Probe at part end

Adjust half taper amount toward the front

Figure 11-172 *Realigning the off-center tailstock.*

problem—it might not have been tightened fully in the chuck and it has been pushed back into the chuck by the cutting tool's pressure, leaving the tailstock support not touching. This is the most likely problem since tool wear would produce a cylindrical part (not tapered) but of the wrong diameter.

3. *Rough Finish.* This could be caused by six factors: incorrect tool shape (nose radius, clearance, and lead angles), incorrect vertical

centering of the tool, vibrations in the work/ tool, lack of coolant, and finally, the RPM or feed are wrong. Poor finish is often a combination of these factors. After a quick inspection of the tool for loose mounting, signs of wear, correct shape (left or right hand, shear, and rake angles) and correct vertical centering, if no problems are found, change the RPM to solve this problem. If the finish does not improve, change the tool shape to a smaller nose radius to correct chatter or a larger nose radius if no chatter is present.

ANSWERS 11-7

1. 29.5°
2. To use the efficient cutting edge of the bit to remove metal—yet still take a very small amount off the back of the thread to smooth out the transition between passes.
3. Tool nose flat formula: Pitch distance/8; pitch distance is 0.0625; 0.0625/8 = 0.0078 in. rounded to 0.008 in. flat.
4. True
5. Shift the direction lever such that the lead screw moves the tool to the right when the half nut is engaged.
6. To make 20 threads per inch, you can engage the chasing dial on any line graduation. For 19 TPI, engage at the numbered lines only. At line 1 would be best.
7. $E = M + 0.086603 \times P - 3 \times$ Wire dia.

ANSWERS 11-8

1. Thread mics and/or three wires
2. They contact the thread close to the pitch diameter and protrude above the major diameter of the thread form.
3. The center diameter of the engagement between a mated screw and nut.
4. True
5. Even the gages show only that the thread will engage with a counterpart. They do not show that the thread form is the right shape. The optical comparitor can project the magnified thread form onto a screen, to be measured for angle and crest/root width.

6. Contact at pitch diameter = 0.0361 in. (0.0562 in. Max to 0.0350 in. Min).
7. Pitch diameter = major dia.(nominal) + (0.086603 × Pitch distance) + (3 × Wire Dia.). From *Machinery's Handbook, Formulas for Checking Pitch Diameters of Screw Threads*, where *M* is the measurement over the three wires and *W* is the wire diameter are known.
8. Yes, 0.050-in. wires are acceptable.
9. From the table, *Dimension Over Wires of Given Diameter for Checking Screw Threads*, W = 0.050 in. M = 0.5334.
10. None—it's undersize now. This thread is scrap!

Answers to Section Review Questions

1. Drilling and reaming.
2. The need for constantly changing RPM to maintain constant surface speed.
3. Part protruding a small distance from chuck (assumes spindle hole is bigger than 2.5 in.). *Opp 5—Face one end*, holding in three-jaw chuck; *Opp 10—Center Drill #2* (one end only), Pull stock out with 2 in. in chuck for grip-tailstock support; *Opp 15—*rough and finish turn 2.000; *Opp 20—*rough turn 1.000 diameter on both sides of gear—leave excess 1.060 dia and 0.060 on 0.75 dim; *Opp 25—*face both sides of gear to 0.750 width at 3.500 in. dimension, requires two different facing tools to reach both sides of gear; *Opp 30—*finish turn 1.000 diameter—note full radius on cutting tool; *Opp 35—*cut 0.16-in. groove at 1.5 in. dimension × 0.925 dia.; *Opp 40—*chamfer right end 0.1 × 45°; *Opp 40—*part from stock at 5.625 in. dimension.

Note I would not part the finished object completely off the stock. I would stop with just a bit still connected, then remove the part from the chuck, reverse it and using collets, hold it by the 1.000 diameter and cut away the last part that was originally in the raw stock within the chuck. That way the work doesn't fall, only the excess. I would then chamfer and center drill the other end since it may be needed to cut the gear teeth.

4. Three-jaw chuck. There is lots of excess on the finished diameter so runout is not an issue. The cuts will be relatively heavy so good grip is an issue. The bar has length excess so gripping in a chuck is the way to go.

5. Yes support is needed. Note that the length-to-diameter ratio of the raw stock isn't the factor to be considered, it's the diameter that comes out beyond 5-to-1 length-to-diameter ratio.

6. The quick change since several tools are required. A square tool post would be awkward due to the need for two facing tools.

7. Carbide insert tools for turning and facing; HSS for center drilling; grooving and parting probably HSS but carbide would be OK too, if in the inventory.

8. CRS is mild steel to be cut at 300 F/M SS; 2.0 in. dia. at 600 RPM; 1.0 in. dia. at 1,200 RPM.

9. No. The rough diameters are larger by a half inch for cutting the 2.0 diameter and a full inch for the 1.0 finished shaft—now turned to 2.00 diameter. *Best solution:* Shift to an intermediate RPM. *Possible solution:* Test cut to see if the tool will withstand faster speeds. Remember, the speed charts in Appendix IV are student numbers (low for safety).

10. Maximum RPM for chuck *1,000 RPM* or shop set limit! Calculated RPM for 0.125 in.-diameter HSS tool at 100 F/M = 2,400 RPM.

11. Yes. Use a long bar through the spindle and machine each part by pulling it out enough for

one part then part it off. This saves the chucking amount on each cut bar.

12. The initial bar will be approximately 10 ft long for the 20 parts. It will protrude out the back side of the headstock. *Solution 1:* OK. Use a long bar support and protect others from walking into it. *Solution 2:* Better. Use two bars around 5 ft long. They don't protrude from the headstock.

13. False. The center gage helps determine the shape of the threading bit and to align it to the shaft. To align its centerline parallel to the lathe's X axis.

14. False. The Z axis is locked. It's the X that must feed in or out to face.

15. The face-grooving tool must have more side clearance on its outer edge. In truth, it needs little or no clearance on the inner side because the cut slopes away from the cutting edge. However, if it does have side clearance on both sides, it can be used as a regular grooving tool.

16. The thread pitch gage

17. 0.3911-in. pitch diameter for $\frac{7}{16}$-14 thread (from "Coarse Thread Series UNC" in *Machinery's Handbook*).

18. 0.050 wires at 0.4793 in. over the wires (from "Dimension Over Wires of Given Diameter" *Machinery's Handbook*).

19. The manual lathe requires the operator to drop the half nut at the exact point on the chasing dial, with each thread pass, therefore RPM must be low to do it right. The CNC lathe starts the Z axis in exact synchronization with the thread groove, thus it can turn at the most efficient SS.

20. The CNC lathe has no gearing between the spindle motor and the X and Z axes. They are driven independently by the controller.

Section 12

Mills and Milling Operations

Introduction

At the time this text was compiled, according to the National Machine Tool Builders Association, nearly 80 percent of all new machining equipment sold in North America was machines that mill. They shape metal by rotating a cutter and moving it along an axis through the workpiece. Clearly, milling is a central issue in machining.

Although their prime duty is to create flat and square surfaces, mills can perform a wide range of tasks. Skilled machinists can cut angles, form radii, bore holes, and using attachments, make circles or parts of circles and helices (spiral cuts—plural of helix). Using special cutters, mills can cut gear teeth and make dovetails and tee slots. The most amazing thing is that all of these shapes can be made on manual machines, without CNC control!

Now, add the computer to drive the feed axes, and the possibilities are almost limitless! Using CAD/CAM-generated programs, there's practically no 3-D shape that cannot be created. Today, if it can be drawn onscreen it probably can be milled. Here are the units of learning in Section 12 to gain this valuable milling background.

Objectives

What Does a Mill Do?

Introduction: Unit 12-1 provides a background in the operations expected of a beginning machinist. It provides the ability to look at a part print or a finished item and then identify the basic operations required to make it. We will also discuss a few basics about the way the setup is made and which cutter might do the job. We'll explore setups and cutters in greater detail in Units 12-3, 12-5, and 12-6.

Like all others in this book, Section 12 assumes you'll be practicing these operations on manual mills then transferring the experience to CNC machines. When differences are significant, they will be pointed out. Most of the 12 operations are common on programmed machines as well.

KEY TERMS

Arbor
A precision shaft used to support wheel type cutters.

Ball nose end mill
A cutter featuring a full radius at the tip.

Boring
Machining precision holes by rotating a single point tool in a circle.

Bull nose end mill
A cutter featuring radiused corners and a central flat face.

Climb milling
Peripheral cutting that amplifies the cutter feed action—the work feed agrees with the cutter rim and if not controlled, can lead to serious accidents.

Conventional milling
Peripheral cutting where the cutter opposes the work feed direction.

Depth of cut (DOC)
The distance from the work surface to the cut surface.

Fillet
An inside corner radius between two surfaces.

Indexed (spaced) work
Machining multiple features in a circular array; for example, a bolt circle.

Keyway (key)
A square slot or groove in an object in which a key is inserted to prevent rotational slippage on the shaft.

Peripheral/profile cutting
Machining around the outside edge or surface of the work.

Plunge
Drilling into the work with an end mill that is ground for the purpose.

Pocket milling
Machining down inside a part using both face and peripheral cuts.

Radial engagement (RE)
The amount of cutter diameter contacting the work compared to its full diameter; 66 percent RE is ideal.

Ramp
A multiaxis movement of the cutter into the work.

Straddle (gang) milling
Using two or more wheel cutters on a single arbor.

Milling Operations

The 12 common operations divide into four types;

Facing—cutting one flat surface using the end of the cutter teeth

Profiling—cutting the outer rim of the work, using the side of the cutter

Drilling

Boring—creating round internal surfaces

Operation 1—Face Milling

Arguably the most basic operation of all is a face cut. The cutter is lowered below the surface of the work to the **depth of cut (DOC),** then set to engage the work at a given portion of its diameter, called **radial engagement (RE)** (Fig. 12-3). The RE is expressed as a percentage of the cutter's diameter that is in contact with the work. The upper illustration in Fig. 12-3 has a 2 to 3 radial engagement (66 percent). The cut width is about two thirds of the total it could cut.

There are several reasons to setup a cut with less than 100 percent RE. We'll reason out

Face Milling a Flat Surface

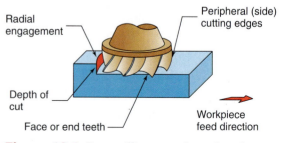

Radial engagement

Peripheral (side) cutting edges

Depth of cut

Face or end teeth

Workpiece feed direction

Figure 12-1 *Face milling uses the end teeth to create a flat surface.*

3 = 66% RE
2

Better tool life
but why?

100% RE

Not recommended,
however, will work

Figure 12-3 *Radial engagement is an important part of the face mill setup.*

Figure 12-2 *This high-tech, 1½-in.-diameter carbide insert cutter blasts through this steel block.*

why, as we go. However, there are times when we must use 100 percent RE because of the lack of a bigger cutter or from other machine limitations, but the better solution is to set up a bigger diameter cutter when a large area is to be faced, but why? Read on.

Step Cuts

It's common to make both face and peripheral cuts simultaneously as in Fig. 12-2, where a high-volume cut is progressing. Note the thick blue chips the cutter is taking from the steel workpiece using a 40 percent radial engagement but a very large depth of cut.

KEYPOINT

While it's possible to use a 100 percent radial engagement for a face cut (Fig. 12-3), for tool life and finish, during face cutting it's desirable to keep RE below 66 percent.

Critical Thinking

Study Fig. 12-3 carefully, and see if you can solve the puzzle of why there is the difference in tool life and finish for the 66 percent RE compared to the 100 percent RE. The answer won't be fully discovered until the end of Unit 12-5.

Removal Rate—In Cubic Inches Per Minute

The combination of depth of cut, radial engagement, and feed rate create the overall removal rate—stated in cubic inches (or cubic centimeters) of metal removed per minute.

The removal rate translates directly to horsepower required to make a cut. We'll delve into calculating these things in Units 12-4 and 12-5.

KEYPOINT

Mill feed rates are specified in inches of travel per minute (*IPM*) or millimeters per minute. That's different than drill presses and lathes, where it's in inches per revolution.

Three Cutters That Face

Face milling can be accomplished using three different type cutting tools. Learn their names and general characteristics.

Face Milling Cutters These cutters are made for the purpose. Supplied in either HSS (Fig. 12-1) or carbide (Figs. 12-2 and 12-3), face mills remove the most metal per minute of the three types. The carbide insert cutters shown in Fig. 12-4 are by far the most common

Figure 12-4 *A variety of insert tooth face mills.*

in industry due to their high removal rates, quick replenishment of cutting edges, and easy control of cutter geometry.

Face mills produce a medium fine finish around 64 microinches (μIn) or better under ideal conditions. While face mills are costly compared to the next two cutter types (a 10-in.-diameter, 12-tooth insert cutter can cost two weeks wages for a journey-level machinist), remember they are the fastest metal removing tool of the three. Considering productivity, they often pay for themselves in the first month of use!

End Milling Cutters (Endmills) These cutters are also used to face cut because they have cutting teeth both on their end and sides. Also, when their face teeth are ground to do so, end mills can **plunge** straight down into the work, similar to drilling. Face mills cannot plunge, therefore they must be set at the DOC when they are not in contact with the workpiece.

KEYPOINT

Not all end mills can plunge—they must have cutting ability to their face center. Only the plunging end mill can begin an inward hole or shaped pocket in a workpiece surface (Fig. 12-5).

End mills with two or four teeth are common. Three-tooth cutters are also used in some free-cutting metals such as aluminum. Limited numbers are made of cobalt HSS, which is a bit harder and tougher than standard HSS, but cobalt cutters are more costly. The extra cost

for cobalt HSS is not often justified, except in cases where carbide can't do the job because it's cutting edge chips due to chatter but plain HSS dulls too quickly due to work hardness.

End mills are also supplied in solid carbide, which is more costly than HSS but cuts three times faster. They are also made as carbide insert tooth versions. The choice of which end mill to use is a basic skill that you'll explore in the shop—the right choice can make all the difference in accuracy, tool life, and production!

End mills produce a medium fine finish and when facing, remove metal at a relatively slow rate, compared to face mill cutters. Using lower feed rates, end mills can produce a surface finish of around 64 μIn.

KEYPOINT

To cut softer metals, choose the least flutes, harder metals warrant more teeth.

Fly Cutters are a single tooth, facing cutter, used more in tooling work than production due to very slow cutting (Fig. 12-6).

End Mills

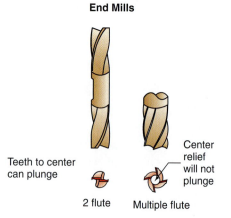

Teeth to center can plunge

Center relief will not plunge

2 flute   Multiple flute

Figure 12-5 *End mills feature both face and side teeth. Some are double-ended for cost savings; one body with two cutters.*

Fly Cutter

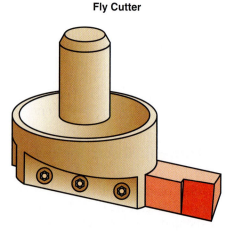

Figure 12-6 *A fly cutter creates fine finishes.*

Figure 12-7 *A fly cutter makes light face cuts only, but creates a very flat surface with good finish.*

Figure 12-8 *This block is being profile milled using the peripheral (side) teeth of the cutter.*

They are correctly setup where a very large surface must be exceptionally flat or smooth and have an excellent finish of 32 μIn or better. The inserted HSS or carbide bit in a fly cutter is ground with a shape similar to a lathe bit.

While fly cutters are inexpensive and can be shop-made, they are very slow metal removers with some serious safety considerations. Tool chatter and centrifugal force at the end of the extended fly bar are challenges for fly cutting (Fig. 12-7).

KEYPOINT

Caution!
The out-of-balance nature of fly cutting makes maximum RPM, feed rate, and DOC all critical issues. A suggested limit might be 0.015-in. DOC, a 3 in. per minute feed rate, and an RPM under 700 for a 5-in.-diameter cut. *Check with your instructor when setting up a fly cutter—some have different limits.*

Operation 2—Peripheral Milling

Peripheral cutting (also called **profiling**) means cutting around the outside or inside surface of a part using the side teeth of the cutter. The most common cutter for profiling is the end mill (Fig. 12-8). The top on this block is being profile milled by a two-flute end mill.

Climb verses Conventional Milling a Critical Issue!

Profiling has it's own set of safety rules and challenges, which we will study in Unit 12-4, but the most important is knowing the differences in the forces, depending on the feed direction. When cutting with side teeth, the cutter may be fed forward along the work as a wheel driven upon a road—called **climb milling.** During climb milling, the cutter rotation in the workpiece tries to pull the milling machine feed axis and work through the cut. Or you can feed the cutter the opposite direction along the perimeter, with the teeth opposing the feed axis direction, called **conventional milling** (Fig. 12-9). It's critical for safety, tool life, and finish to clearly understand the difference.

Conventional

Climb milling

Figure 12-9 *It's important to know the difference between climb and conventional milling.*

Critical Thinking

Here are four questions to answer. Between climb and conventional cutting,

Which creates the best finish?

Why?

Which requires extra strong setups and backlash control?

Which provides the best tool life and why?

The truth is related to radial engagement. These questions will be answered soon, but reason them out now.

KEYPOINT

There is a big difference in the forces created when climb milling compared to conventional milling. They make a difference in outcomes and especially in the setup.

Operations 3 and 4—Step and Pocket Milling

Often a combination of face and peripheral milling is required in one cut. A *pocket,* shown in Fig. 12-10, is a depression or cavity milled down into the metal. In CNC programming, a second pocket milled within the floor of another is said to be *nested.* These operations are called steps, or pockets.

Plunging to Cut Depth

To start the pocket operation, a hole must be opened to the rough pocket depth—a third operation similar to drilling—called *plunging* when it's done with an end mill. While it is possible to use a twist drill to start the pocket, it's faster to plunge the end mill to depth, then mill sideways to make the pocket. To do so, examine the end mill. It must have the

Figure 12-10 *A cutaway view of two pockets with the second nested within the first.*

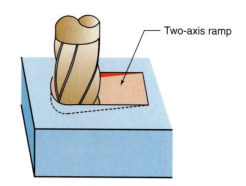

Figure 12-11 *Ramping down to depth allows chips to escape and coolant to enter far better than plunging.*

teeth ground with one or more cutting edges completely to center—similar to a drill, but with a zero degree point angle (Fig. 12-11).

Three- and four-flute cutters can plunge if correctly ground with one or more teeth that cross the face center. Most two-flute cutters can plunge, but not all! Some feature hollow faces at their center, without cutting edges.

KEYPOINT

End mills must be ground with one or more teeth that cross the center face, such that they remove all metal during a plunge. If they are not, the unremoved center material will stop the plunge and probably burn or break the cutter.

Ramping to Cut Depth

See Fig. 12-11. Feeding the end mill sideways while plunging down is known as a ramp plunge or shortened to just **ramp.** This operation is not often used on manual mills but it's common on CNC because it makes a slanted groove that releases chips better than a drilled hole, plus it allows coolant to reach the cut better. Programmers almost always choose ramping over plunging to cutter depth, while manual mill operators usually use simple plunging. Spiraling the ramp down, in a helix rather than in a straight line, offers a couple of advantages. First, it keeps the plunge area smaller, and it can terminate tangent to the wall of the pocket, thus causing fewer tool marks in the finish. That move is called a *spiral ramp.* Here, the cutter moves in a circular path while it plunges downward.

Layout line above
the vise a small distance

Figure 12-12 *Tilting the part using layout is a simple method with repeatability around 1°.*

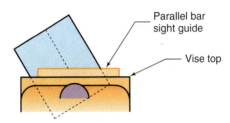

Parallel bar
sight guide

Vise top

Figure 12-13 *Improving alignment using a parallel bar.*

this setup at around one to one and one-half degree—good only for semiprecision work. Using a sine bar setup improves accuracy to within a few seconds variation.

Trade Tip

Pocket Secrets **Pocket milling** tosses a couple of serious challenges your way, both caused by backlash in the feed axes (mechanical clearance between drive nut and screw). As the machinist makes the final cuts around the inside of the pocket, each axis must be reversed while the cutting tool is in contact with the work. It can be pulled into the work surface by the cutting action when the clearance between the axis screw and nut is slack. It also leaves an unwanted indent in the surface often called an undercut. This problem exists on both manual and CNC milling machines, but its far worse on manual ones due to the type of axis drive screws they have. CNC machines have nearly no backlash compared to manual machines.

Changing directions in the corners, the cutter tends to dig in (pull beyond the intended path) into the corners as the backlash is taken up on the axis. The second challenge exists only on manual mills lacking digital positioners, because the micrometer dials are valid only when cranking one direction. Yet you need to crank to locations on both sides of the pocket. If the machine has digital readouts, this problem is solved for cutter position, but not from the corner digging. Without digital position, use layout and measuring tools to track progress.

To avoid the backlash corner digging, first rough within 0.015 to 0.030 in. of final size. Then take two light finish cuts around all surfaces. Second, always use climb milling as it tends to push the cutter away from the work rather than dig in. Third, on manual mills, always tighten the lock on the unused axes (*X* or *Y* and the *Z* quill). Finally, use digital readouts over mechanical dials whenever possible because they are not subject to backlash errors.

Operation 5—Angle Cutting

On manual mills, angular cuts can be set up several different ways, by either tilting the cutter or tilting and rotating the part. Angles can also be cut using form ground cutters. Today the most common method of making an angular surface, is to feed two axes simultaneously through CNC control.

Figures 12-12 through 12-17 show five manual mill setups that could be used to machine the angle on a drill gage blank. Each offers different mixes of accuracy and required setup time.

1. Tilting the Part

Using a layout line and the parallel bar (Fig. 12-13). Using the Trade Tip, repeat

Trade Tip

Improving Layout Alignment (Fig. 12-14) To improve accuracy when tilting the work in a vise, lay a ground parallel on the top of the vise jaw, then level the layout to it.

Tilting the Part Using a Sine Bar

59°

Vise floor

Figure 12-14 *Tilting the part using a sine bar with repeatability around 0° 0' 15."*

Figure 12-15 *Tilting either the A or B axis (shown) is a simple way to cut an angle.*

2. Tilting the Head

The spindle head of many smaller vertical milling machines feature one or two axes that can be set at an angle with respect to their table (Fig. 12-15). On their flange rim (Fig. 12-16) they feature angular graduations with a resolution of either $\frac{1}{2}$ or 1 degree.

Critical Thinking Question In Fig. 12-16, the setup is made to mill the 59° angle on the drill gage. So, what angle do you set on the sine bar? (The answer is found just before the unit review.)

3. Rotating the Part

There are a few ways angular work can be setup parallel to the mill table, then rotated either to the 59° angle or its compliment, depending on the setup accessory: a rotating vise, a protractor, or slot block parallels (Section 8) and a sine bar. To finish the setup in Fig. 12-17, the work would be clamped to the table with a protective space between it and the table

Tilting mill head
Set to correct angle
Refine rotation until zero at both positions
Mill table

Figure 12-16 *The tilt angle can be refined beyond the machine's scale, using a sine bar and DTI.*

Corner radius left after cut
31°
Mill table (top view)
Slot block parallels

Figure 12-17 *Rotating the work using slot blocks and a sine bar.*

top. However, for the drill gage example, milling the angle with the side of a cutter presents a problem—the sharp corner will not be milled correctly due to the cutter diameter (Fig. 12-17).

KEYPOINT

Laying the work on the mill table and rotating it relative to an axis works particularly well for large parts too big to tilt. When doing so, never clamp the work directly on the table, always provide a space between the two with shims.

Operation 6—Forming Cuts

Radius Cutting

On manual mills, inside or outside radius cutting (rounded corners) is accomplished using form-ground cutters. Here the pocket shown previously now has received several radii (plural of radius). Inside corner radii are also called **fillets** on drawings (Fig. 12-18).

The teeth of radius cutters have either a convex curve or a concave curve ground onto their corner (Fig. 12-19). As with lathe tools, creating a shape in the work using a shaped

External radius (corner)
Internal radius (fillet)

Figure 12-18 *A radius corner has been added to the pocket and the outside corner. The concave radius is often called a fillet.*

Fillet Radius

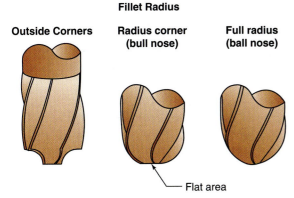

| Outside Corners | Radius corner (bull nose) | Full radius (ball nose) |

Flat area

Figure 12-19 *Three form ground radius cutters: an outside corner radius, a bull nose, and a ball nose end mill.*

cutter is known as *forming*. The two types of end mills ground to form concave corner radii are the **ball nose** (full end radius) and **bull nose end mill** (a small flat on the end).

Custom Form Shapes

When the shape is unusual, the simplest form cutter to make in the shop is the fly cutter. Its HSS bit can be custom ground to nearly any shape without special equipment. In Fig. 12-20, an angle and corner radii are put on this object. Figure 12-21 shows two types of purchased form cutters for special shapes, the tee slot, and the dovetail.

CNC Tool Grinding

When cutter shape exceeds normal cutter grind methods, they can be form ground on CNC tool grinding machines. Today, practically any form tooth shape can be reproduced on CNC programmed tool grinding equipment, shown in Section 13. However,

Figure 12-20 *The simplest shop-made form cutter is ground into the bit for a fly cutter.*

Form Cutters

| "T" slot | Single angle |

Figure 12-21 *Many different angles and several other shapes can be purchased preground onto end mills.*

CNC mills have replaced much of the need for complex-shaped cutters.

Operation 7— Drilling/Reaming

We perform hole drilling on the milling machine when the hole's location tolerance is tighter than drill press accuracy, at around 0.015 in. or less. Using micrometer accuracy, mills can locate their tables from within 0.001 to 0.003 in. in the *X-Y* plane, and with digital positioners do even better. The bits and reamers are held in a drill chuck or a taper adapter, but they can also be held in collets, which tend to hold the drill on the spindle axis with less runout.

Drill tools can be hand or power fed into the work. While mill spindles are far more rigid than drill press, the drills can still wander off location as they start, so a pilot drill (or center drill) is required to aid location accuracy. **Boring** after drilling is the best location solution, which we'll explore next.

Hole Accuracy—Location Repeatability

Manual mill location repeatabilities of 0.001 to 0.003 in. are typical using *X-Y* micrometer dials and finer using electronic digital positing, but only when the hole is bored on location.

> **KEYPOINT**
>
> To achieve both location and diameter accuracy, sometimes it's best to center drill on location, drill the rough hole, bore it within 0.015 in. of finished size, then finish the size with a reamer.

Diameter Repeatability

Diameter accuracy of drilled and reamed holes on milling machines repeats no different than on a drill press. You can expect around 0.003

in. for drilling alone and 0.0003 in. for reamers. However, on mills we can add an adjustable boring head to machine any diameter with 0.0002 in. repeatability, if a quality head is used along with good speeds and feed, and sharp boring tools.

Coolants, Accuracy, and Tool Life The constant challenge for tool life and the accuracy associated with sharp cutting tools is the delivery of coolants down to the cut area. When hole depths exceed three to five times the drill diameter, the chips being wedged up and out of the helical flutes (or plunging end mill) stop the coolant as it tries to penetrate down into the hole. There are three solutions.

Peck Drilling As discussed previously, there are two types. *Chip breaking* is used for medium deep holes, whose depth is up to eight times diameter, whereby the infeed is halted and the drill pulled back from around 0.010 to 0.050 in. then sent back to drill. Beyond the eight-to-one ratio, a second peck is called *chip clearing* because the cutter is pulled all the way out of the hole then sent back down to drill again. That allows most of the chips to get out and a burst of coolant to flow down to the bottom of the hole.

High-Pressure External Coolant Here a focused stream of coolant is sprayed down the tool flutes and can penetrate to depths of five or six times drill diameter. But it too becomes blocked by the outgoing chips. The high-pressure coolant system is an add-on feature for all manual machines and most CNC machines, but it's a good solution for many machining problems including medium-deep hole drilling.

High-Pressure Coolant Through the Tool The winner for chip clearing, accurate holes, and tool life features an injection hole through the tool center (Fig. 12-22). High-pressure coolant is forced down to the action area. Special sealed machine collars surround the tool shank and holder and force the coolant down to the point of application. Not only does this cool the cut, thus extending tool life while also aiding diameter accuracy, it also pressure forces the chips up and out of the hole, thus eliminating the need for pecking. Gun drills using this center coolant concept can drill to a nearly unlimited depth. This is an add-on accessory and requires special tools that feature the coolant galleries through which the stream is pumped (Fig. 12-22).

Drilling on Ram and Turret Mills

It's common to drill on the smaller universal mills, called ram and turret machines, often called Bridgeport-type mills (see the Shop Talk on next page). Typical reasons might be close location and/or diameter tolerances

Figure 12-22 *The coolant delivery holes shown in the blown up end view of this carbide tipped deep hole drill.*

within 0.003 in. or less, or when no reamer is available for a special diameter so we must bore the hole. Creating accurate angular holes is also a typical ram and turret operation because the milling head can be tilted in two directions relative to the table.

Ram and turret mills are unquestionably the most universal manually operated machines in any shop. In schools and job and tooling shops they are the standard for small milling machines (Fig. 12-23). When we use their sturdy sliding quill and quill feed system, which is similar to a power feed drill press, they improve and simplify drilling operations compared to drill presses and other kinds of mills not well suited to drilling.

Ram and turret mills are very flexible in terms of setup configurations and the range of the operations they can perform. The reason for their formal name, *ram and turret,* can be seen in Fig. 12-23, where atop the solid turret column, sits a sliding ram that can also pivot around the turret. At the outer end of the ram is the machine head, which pivots on two axes.

Many other power accessories fit on the rear end of the ram, which can be unlocked and pivoted 180° to bring it to the front of the machine. These accessories can perform medium-duty milling, drilling, boring, and all the other cuts shown in this unit. Using

attachments, they can cut slots with reciprocating heads (ones that stroke in and out) and if we add grinding heads, they can grind precision surfaces. Many other operations are also possible.

KEYPOINT

Due to their flexibility and popularity in training labs, we'll be using Bridgeport mills in our examples throughout the manual milling section.

Operation 8—Boring

Precision holes are bored on mills using a rotating, adjustable cutter, called a boring head. This specialized tool allows outward micrometer positioning of a single-point cutter (Fig. 12-24). A rough-out hole is first drilled, or milled, then finish passes are taken by dialing the boring tool outward to a bigger radius, then machining a pass through the hole.

Using the single-point tool and several passes to eliminate out-of-roundness and mispositioning of the rough hole, the boring process tends to machine the hole exactly on location, very round, and straight through the work, with no wandering from location. Boring creates a round, smooth hole of 32 μIn or better with size (Fig. 12-25).

Figure 12-23 *The ram and turret mill is popular for offload support work, tooling, and training.*

Micrometer Adjustable Boring Head

Micrometer dial

Slide lock screw

Boring bar

Figure 12-24 *A typical adjustable boring head using the Swiss type boring bar.*

Typical Operations using a Boring Head

Boring

Facing

Chamfering

Turning

Figure 12-25 *Several operations are possible with a boring head and the right cutting tool!*

Operation 9—Arbor Cutter Work

Sometimes part features such as slots or tabs are cut using *arbor cutters*, sometimes called *wheel cutters* in shop lingo. They might also be called *plain mill cutters* (Figs. 12-26 and 12-27). These cutters are made to cut on their rim teeth, as shown in Fig. 12-26. Like a wheel on an axle, they are mounted on a precision shaft called an **arbor** that is held in the mill's spindle taper. The arbor features a square **keyway** along its entire length. The cutter has a similar square groove so that when a square steel key is inserted it positively rotates the cutter without slipping. Precision-ground spacers are used to position the cutter in place along the shaft. These cutters are either HSS or carbide insert tooth (Fig. 12-28).

Two Kinds of Arbors—Stub (Short) or End Supported (Long)

Depending on the setup, a short or long arbor might be called for. When selecting an arbor for a setup, always keep the minimum overhang

Arbor-Mounted Tools

Stub arbor

Precision spacers

Drive keyway

Figure 12-26 *Wheel-shaped cutting tools are mounted on arbors.*

Slitting Saw Cutter

Figure 12-27 *A thin wheel cutter is called a slitting saw.*

Figure 12-28 *The arbor, HSS wheel cutter, and overarm support are shown in this setup. Carbide insert cutters (inset) might also be used on the arbor.*

principle in mind. There are two types of arbors from which you can select.

Stub arbors (Fig. 12-26) are used where *reach* isn't great. Reach is the distance from the machine spindle out to the cutter. This is also called extension or overhang (Fig. 12-27). Because it is supported only on the machine end, the stub arbor has a limited reach before the arbor flexes, the cutter chatters, and the arbor is at risk of deflecting and bending. Stub arbors are set up on both vertical spindle mills and on horizontal spindle mills.

End-support arbors are used when the required reach exceeds the stub capability, or when the need arises for multiple cutters on a single arbor (Fig. 12-29). Long arbor support

Gang or Straddle Milling

Figure 12-29 *Long arbors feature pilots that are supported at the far end.*

Typical Indexed Features

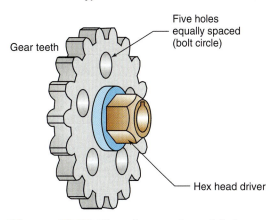

Figure 12-30 *Three features that might be indexed on a manual milling machine.*

is not possible on vertical mills, however, many horizontal machines are designed for this purpose. On the horizontal mill equipped for arbor work, long arbors are solidly supported at the outer end as well as the near end. These machines feature an *over-arm* sliding ram designed for this purpose. An arbor support attachment bolts to the ram, to support the pilot on the arbor (Fig. 12-29).

Straddle or Gang Milling

Cutting on two or more sides of a feature is accomplished using two (or more) wheel cutters with precision spacing between. This production setup is known as **straddle milling,** as shown in Fig. 12-29, where the cutters finish the width of the tab. When several cutters are used to finish several features at one time, it's called a **gang milling.** Both the depth of cut and side spacing is ultrarepeatable, because they are rigidly held on the arbor with no variation other than that due to cutter wear. Due to the heavy cutter load and the high cost of a gang of cutters plus the fact that gang milling is most useful where large batches of parts are milled, you probably won't see this done in school. Another fact is that with CNC

Indexing and CNC While most instructors believe that you definitely need indexing skills to be considered a complete journey machinist, and to be able to make your own tooling, the need for manual indexing is diminishing greatly due to the abilities of CAD/CAM programming and controlled rotary axes on CNC machines. Today's shops cannot compete with CNC methods in production work using manual indexing methods. Today these skills are best used in tooling and one-off type work.

capabilities in most production shops, gang milling has diminished greatly.

Operation 10— Indexed and Spaced Work

Indexing or sometimes called **spacing** is an intermediate skill you will use to produce work where duplicate features are machined in several locations around a circular pattern. Examples might include gear teeth, holes drilled in a circle (called a bolt circle), or machining the flat sides on a hex head bolt, as shown in Fig. 12-30.

Using an accessory called an *indexing chuck* or *indexing head* (coming up), the work is rotated then held in the new cutting position for each feature. After drilling the first hole in the example drawing, the work would be rotated $\frac{1}{5}$ of a full revolution, then the next hole would be drilled. Then it's rotated and drilled again. The indexing device is also useful for layout and drill press work (Fig. 12-31).

Operation 11—Rotary Table Work

When precision arcs (curves less than a full circle) must be machined on a manual mill, another accessory called a *rotary table* must be used (Fig. 12-32). The part drawing depicts a stamping die to be machined that will cut the bill of a baseball hat

Making the Hat Bill Punch Although the baseball hat die could be produced on a rotary table by a skilled toolmaker, today it would more than likely be made using a CNC mill. To make it manually, the blank would need to be positioned with the exact center of each curve directly under the spindle, then one axis would be moved until the correct cut radius is reached. All of which requires lots of math and time and includes a large risk for error!

Then after a time-consuming setup, each of the four curves would be machined by hand cranking. I estimate the process would take around 3 hours (for setups, math calculations, and cutting the die). But, using computer methods it would take minutes to write the program and to machine the first part. But that's a single part comparison. Making more than one part would widen the time gap. Therefore rotary accessories are fading from the production scene, but are still used in tooling work. Even as a tool and die maker, I'd still choose a CNC mill to machine the ball cap die, if the CNC mill was available.

Making the Hat Bill Punch (*continued*) If the drawing is a computer file as is this one, then the programmer would use CAM (computer-assisted machining) software to import it, take away unwanted layers of dimensions and notes, then generate the program based on the part geometry. But even if the drawing didn't exist, the programmer could create the part shape using a draw utility such as found in Mastercam (Section 25) in less than 2 minutes—no math, no errors! Once the tool path was created then evaluated onscreen (another few minutes), it would be downloaded to the CNC controller. We're into the process about 15 minutes thus far. The setup would take maybe 20 minutes and the actual machining maybe twice that. You can do the comparison. But there's also the advantage of CNC accuracy.

This kind of single-part production called "one-off" was once considered impractical using program methods but not today. Using the graphic programming methods you'll learn in Section 25, the time savings would be close to 300 percent over manual methods (my estimation) for just one part. But that's not all. CNC doesn't just turn hours into minutes and reduce human errors; it makes it possible to machine shapes difficult or impossible to make on the rotary table or by any other manual part movement during machining! Once the file is proven, hundreds of hat dies could be machined today, and anytime in the future when the program is called back to use again!

Figure 12-31 *A dividing head (shown) or indexing chuck (simpler version) rotates the work an exact amount then locks in place for machining.*

Rotary Milling Table

- Degree graduations
- Rotating table
- Vernier scale
- Clamp lever
- Hand wheel

Figure 12-32 *A rotary table is a circular feed axis for milling or for indexing.*

Hat Bill Die

R 11.77
R 5.00
6.16
15.58

Figure 12-33 *This hat die could be made using a rotary table on a manual mill.*

from flat material. The die is to be machined from flat steel. This is a typical shape that might be made using a rotary table (Fig. 12-33).

The rotary table can be used both as an indexer (moving parts from position to position) and to move the part in an arc during machining. Used in that capacity, it becomes a feed axis that rotates by hand cranking. The machinist can work from layout lines on the part or by setting datum stop points on the graduated table. The precision rotary table may be moved as fine as 0.2° using a vernier scale adjacent to the handle.

Sine Bar Answer The complimentary angle at 31°. Return to Fig. 12-14, the same logic also applies there.

UNIT 12-1 REVIEW

Review the Key Terms

Arbor
A precision shaft used to support wheel type cutters.

Ball nose end mill
A cutter featuring a full radius at the tip.

Boring
Machining precision holes by rotating a single point tool in a circle.

Bull nose end mill
A cutter featuring radiused corners and a central flat face.

Climb milling
Peripheral cutting that amplifies the cutter feed action—the work feed agrees with the cutter rim and if not controlled, can lead to serious accidents.

Conventional milling
Peripheral cutting where the cutter opposes the work feed direction.

Depth of cut (DOC)
The distance from the work surface to the cut surface.

Fillet
An inside corner radius between two surfaces.

Indexed (spaced) work
Machining multiple features in a circular array; for example, a bolt circle.

Keyway (key)
A square slot or groove in an object in which a key is inserted to prevent rotational slippage on the shaft.

Peripheral/profile cutting
Machining around the outside edge or surface of the work.

Plunge
Drilling into the work with an end mill that is ground for the purpose.

Pocket milling
Machining down inside a part using both face and peripheral cuts.

Radial engagement (RE)
The amount of cutter diameter contacting the work compared to its full diameter; 66 percent RE is ideal.

Ramp
A multiaxis movement of the cutter, into the work.

Straddle (gang) milling
Using two or more wheel cutters on a single arbor.

Replay the Key Points

◆ Mill feed rates are expressed in inches of travel per minute (IPM) (or millimeters/minute).

◆ Not all end mills can plunge.

◆ To cut softer metals, choose the least flutes, for harder metals choose more teeth.

◆ Safety—There is a big difference in the forces of climb milling and conventional milling—it makes a difference in the setup.

◆ Safety—The out-of-balance nature of fly cutting makes maximum RPM, feed rate, and DOC all critical issues. Check with your instructor for shop limits.

◆ When placing work on the mill table, never clamp it directly on the table, always provide a space between the two with shims.

◆ To achieve location and diameter accuracy, it's best to center drill on location, drill the rough hole, bore it within 0.015 in. of finished size, then finish the size with a reamer.

Respond

1. Of those listed here, identify the operations that can be accomplished with an end mill.

   Facing

   Peripheral cutting

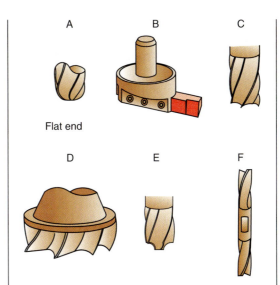

A   B   C

Flat end

D   E   F

Figure 12-34 *Identify the cutters.*

Step milling

Pocket milling

Cutting angles

Forming—inside and outside radius

Drilling, reaming

Boring

Straddle and gang cutting

2. True or false? When drilling down to form the start of a pocket, a two-flute end mill must be modified such that it will nest. If this statement is false, what makes it true?

3. Identify the cutters shown in Fig. 12-34. For what operation would each be used?

4. Name the three types of facing tools for milling machines. Please note, there are more cutter types but these are the basic three experienced in your first year.

5. Milling with two cutters mounted on an arbor that are spaced an exact distance apart, is called _____ milling.

6. Why might you move a hole-drilling operation from a drill press job to a vertical mill?

7. Explain and compare a ramp to a plunge cut.

8. Objective: modify this brake disk in Fig. 12-35 by drilling cooling holes. Name the operation that will be used. What is the name often given to a pattern of holes such as shown in Fig. 12-35?

24 holes × 12 mm dia
equally spaced

R 180 mm

Figure 12-35 *What operation is needed to drill these cooling holes in this disk brake? What is the name of the pattern produced?*

Critical Thinking

9. When setting up to drill holes on a milling machine, the operator must first locate the spindle directly over a datum then move it to each hole location. There is a vital step missing in this statement—what is it?

Figure 12-36 *What is the name of the operation being performed and the advantages over drilling and reaming?*

Name the operations

C

B — R 0.25

A

D

Figure 12-37 *Name the operations and setup needed to make this object.*

10. Describe the two advantages of using the operation shown in Fig. 12-36 over drilling and reaming a hole.

11. Using the operations we've discussed, identify the four used to make the part sketched in Fig. 12-37 and the possible setup/cutter combination to make it. There may be more than one way to accomplish the feature.

UNIT 12-2

How Does a Mill Work?

Introduction: Similar to what we did with lathes, here we'll assemble then examine a ram and turret mill, one feature at a time. But first, we'll take a quick look at the difference between horizontal and vertical milling machines.

KEY TERMS

Knee
The second largest component mounted vertically on the column. The knee provides the Z axis motion on a vertical mill.

Linear axis
Straight-line motion. May be used as a motion for machining or limited to positioning during setups only.

Orthogonal axis set
A 90° set of linear axes found on all CNC machines, which may have two or three primary linear axes. Their orientation to the world varies, but they always remain the same to each other.

Rotary axis
A motion or position axis that moves in an arc about a pivotal axis (partial or full circle).

Saddle
The mill component that mounts on the knee providing the Y axis motion on vertical milling machines. The lathe component that fits over the ways.

Work envelope
The space where machining can occur bounded by axis limits.

Comparing Modern Vertical and Horizontal Spindle Mills

Vertical milling machines, which include the ram types, are very popular in industry and schools due to the ease of setup and use. Our

Figure 12-38 *Vertical and horizontal milling machines are similar.*

Figure 12-40 *Placing the work in a vertical mill work envelope.*

early investigations will focus on them. However, horizontal machines have their purpose especially in heavy industry. You should know when a job is better suited to a horizontal than to a vertical spindle mill. Let's briefly examine the differences.

Both machines shown in Fig. 12-38 have horizontal work tables, carried on a knee that runs vertically, up and down on a dovetail column. That forms one axis of movement for both—but it's the *Z* axis for the vertical and it's the *Y* axis for the horizontal—we'll see why when we examine axis conventions.

A second version of the horizontal mill features a vertical surface for mounting the work. The vertical table, horizontal mill is generally the larger of the two and is usually found only in shops that machine massive work (Fig. 12-39). But both horizontal machines are better suited to heavy or extra large work compared to the vertical. We discover why next.

Horizontal Mill Advantages over a Vertical Mill

1. Part Loading
Large parts can be suspended from above with a crane, then easily loaded directly down onto either type of horizontal table. No special crane rigging is required. In comparison, it's difficult to load a vertical

mill with large parts, since its milling extends out over the table (Fig. 12-40).

2. Chip Clearing
When large volumes of chips are made on the vertical table horizontal mill, they simply fall away along with coolants. They flush easily out from deep pockets with coolant injection.

3. Larger Parts
Because there is no overhead frame, very large parts can be loaded onto either version of a horizontal mill, but the vertical table version can securely hold literally tons of work within it's work envelope.

Work Envelope

The **work envelope** is a term used to describe the boundaries of space in which machining can occur (on any machine—not just mills). The envelope is bounded by the extent of spindle and axis travel. On mills, the work envelope becomes a cubic space bounded by the *X*, *Y*, and *Z* axis limits (Figs. 12-40 and 12-41).

One of the greatest distinctions between vertical and horizontal mills is that on the vertical mill, the workpiece must be able to fit within the work envelope—the *Z* axis limits (Fig. 12-42). In other words, inside the

Figure 12-39 *This is the larger version of the horizontal mill.*

Figure 12-41 *Placing the work in a horizontal mill's work envelope.*

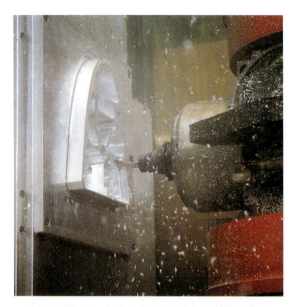

Figure 12-42 *Chips fly in every direction as this horizontal-spindle hypermill rips through a large aluminum billet. Note the vertical table.*

Figure 12-43 *A horizontal spindle machining center with horizontal table.*

machine. However, on horizontal machines, the work sits on or clamps to the table, but the work size is not limited to the work envelope (Fig. 12-43).

Vertical Milling Machines

That said, in most of today's shops the vertical milling machine is the more popular equipment (Fig. 12-44). It is easy and quick to setup, fast to turnaround to another setup, and very versatile for work up to a maximum of 2 or 3 ft long. From here on we'll be discussing a vertical mill.

Column and Base

The spine of the mill is the solid metal column and base that compares to the bed ways

Figure 12-44 *The MVP in most machine shops— the vertical machining center, milling machine.*

Figure 12-45 *The base and column of a typical vertical mill.*

on a lathe. It features a vertical, precision dovetail that guides and supports the knee— the up and down sliding axis that supports the table. The dovetail sliding upon the column becomes the Z axis (Fig. 12-45).

Lubrication of the dovetail ways is essential for long life and precision. Your machine may have automatic or manual lubrication— always know which. The lube reservoir and pump are probably mounted beside the base; use them on a daily schedule if your machine is so equipped. Take a minute to look at the mill in your lab to determine which lubrication system is provided.

The Knee

The next most massive part of the axis system is the **knee** (Fig. 12-46). The knee mounts on the column dovetail using gibs similar to lathe axes. The knee on most mills can be hand cranked up and may feature a power feed also. The Z axis will feature a micrometer dial and an axis lock. The first of three axes is now complete; two remain to be added.

Knee and *Y* Axis Dovetail

Vertical (*Z*) handwheel

Y axis dovetail

Knee

Vertical elevating screw

Figure 12-46 *Fitting over the column, the knee slides up and down on dovetail ways. It's moved by hand and/or power feed. The knee forms the Z axis and supports the Y axis.*

Table and Saddle

The mill now receives the *Y* axis **saddle** that moves in and out with respect to the operator. It slides upon a dovetail machined to the knee's top surface. Then upon the saddle, the table dovetail forms the right/left *X* axis (Figs. 12-47 and 12-48). Each of these axes has an

X Axis Saddle Added Over *Y* Axis Dovet

X axis dovetail

Saddle

Y axis micrometer dial

Figure 12-47 *The mill has now received the saddle and the X axis.*

The *X* Axis Table Completes the Set

X axis handwheels and micrometer dials

X axis lock

Y axis lock

Y axis handwheel

Figure 12-48 *The table slides on the X axis dovetail to complete the axis set for a vertical milling machine.*

adjustable micrometer dial and axis lock. Some feature power feed, while all feature hand cranks. Each also has adjustable gibs to minimize unwanted axis motion. Each axis has a given amount of backlash in the positioning system due to mechanical clearance between its drive screw and axis nut. Backlash must be accounted for by the machinist, as we've discussed previously.

Industry Standards for Axis Notation

Axis movement is either straight line—called **linear**—or in a circle—called a **rotary** motion. Some rotary axes can move during machining, while others are adjustable for setups only and are then locked in position during the cutting. These are called positioning axes. This first lesson will be followed in further sections where axis identification becomes a hypercritical issue when writing programs or setting up the machine.

When two or three linear axes are at 90° to each other, such as in the milling machines in Fig. 12-49, they are called an **orthogonal axis set** (the same root word as for orthographic projection). This 90° relationship of the axes is constant from machine to machine, worldwide. Depending on the machine type, the set might be in a different position relative to the floor—for example, the *Z* axis between horizontal and vertical mills—however, their orientation to each other within the set remains a constant.

In Fig. 12-49, compare the vertical mill's axis set to the set for a horizontal mill. The horizontal mill's set has been rotated such that *Z* axis is now parallel to the floor.

KEYPOINT

Note that according to convention, the *Z* axis will always be the linear axis parallel to the main spindle centerline.

Standard Axis Nomenclature

Y   *Z*

X   *X*

Z   *Y*

Horizontal milling machine

Figure 12-49 *Compare the orthogonal set for a horizontal milling machine to a vertical.*

Rotary Axis Identification A, B, and C

A relationship exists for rotary motion letters too. If the center of the rotary motion revolves around a line parallel to the X axis, then it is an A axis, motion around Y becomes B motion, and that around Z is then designated C.

KEYPOINT

Rotation around a line parallel to X is the A, axis, around Y is B, and around Z becomes C. In rotary motion, A rotates around X, B rotates around Y, and C rotates around Z.

You will soon see that the ram and turret mill features six auxiliary movements. Two rotary motions of the head that are for setup position only. Next, the quill is power or hand fed during machining. The ram rotates about the turret, and it slides in and out for setup positioning. So, other than the quill feed, all are limited to setup only—they cannot be used as machining motions.

KEYPOINT

On CNC mills, the A and B axes are common fourth and fifth machine rotary feed motions.

TRY IT

KEYPOINT

Although learning the axis nomenclature doesn't make a whole lot of difference for operating manual machines—right/left, in/out, and up/down would be OK—this is an essential preparation for working with CNC machines.

A. From what we learned in Section 11, is this statement true for lathes? *The Z axis is parallel to the main spindle axis.* Draw a sketch to assist—sketch both the X and Z axes.

B. On an R&T (ram and turret) type mill (Fig. 12-50), the two rotary positioning movements of the spindle head can be called A and B according to the industry standard notation. In looking from the side of the mill table, what rotary axis positioning would this be, A or B?

Identify This Rotary Axis

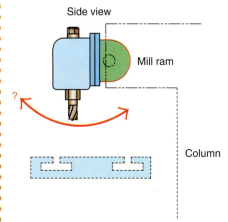

Figure 12-50 *Name this setup rotary axis.*

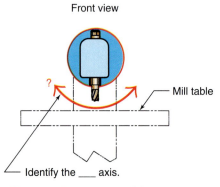

Figure 12-51 *Name this rotary setup axis when the mill is viewed from the side.*

C. Then viewed from the front of the vertical mill, according to the industry standard, what rotary axis is shown in Fig. 12-51?

D. One variety of *horizontal* milling machine, known as a universal mill, features a horizontal rotating work table that is positioned then locked. Looking back at the horizontal mill axis set, what rotary axis would this be?

ANSWER

A. Yes. The Z axis is the lathe saddle, which moves parallel to the spindle axis. That shouldn't be a surprise, these are world standards!

B. See Fig. 12-52.

C. The rotary axis is an A axis because it rotates around a line parallel to the X axis (Fig. 12-53).

D. The rotary table on horizontal mills would be a B axis because it rotates around a line parallel to the Y axis in the orthogonal set.

This rotary axis is an "A" because it revolves around a line parallel to the X axis.

Figure 12-52 *Name this setup axis viewed from the front.*

The "B" axis rotates around Y.

Figure 12-53 *The B rotates around the Y.*

Machine Table and Feed System

If the mill is equipped with a power feed, then the power feed gear case is found on one end of the table, as seen in Fig. 12-54. Feed selections for the X axis will range from below 1 in. per minute to 20 in. or more per minute (entry level rates). Note, only a few small mills, such as the ram and turret, do not feature a power feed on all axes.

Feeds range from below 1 in. to around 20 in. per minute on small mills.

The mill table forms the X-Y plane and features *tee slots* used for bolting work or accessories (Fig. 12-55). The upper groove in the slots can be used to align work in several ways. Using the slot correctly can save a lot of extra time and add rigidity and precision to your setups.

X axis drive

Figure 12-54 *The X axis feed gear case can be seen on this CNC Easytrack mill.*

Precision slot

Slot block parallel lightly press fit in the mill table

Tee nut

Figure 12-55 *The tee slots are a precision part of the machine.*

Tee slots are a *precision part of the table.*

Pro Table Care

In Fig. 12-56, something is very wrong! With all your setups, the table must be protected from dents, scratches, and especially from machine cuts! That includes the sides of the tee slots. Do not hammer on the table or store material or

Figure 12-57 *Improving the setup in Fig. 12-56, parallel bars are under the work to keep the cutter away from the table.*

tools on it. Any nicks must be lightly honed away using a fine grain flat stone.

X-Y-Z Digital Position Readouts (DROs)

A superhandy attachment is often added to milling machines to increase the accuracy and to help deal with backlash. The electronic display shows the position of the table with reference to the operator-chosen datum zero (Fig. 12-58). This is similar to adjusting the axis dials to zero at a datum.

This ability to zero the axis position at any point within the work envelope is called *full floating zero.* You'll find the very same concept in CNC work. Modern digital positioners also convert between imperial inch and metric units, and most can store and recall significant points then count down the distance to go to that point, thus making it easy to return to a location when machining several parts in a batch.

Using electronic location rather than micrometer dials, the axis positions are displayed independent of mechanical backlash. So, without the stall of a mechanical micrometer dial, drilling becomes more accurate. However, backlash still remains a challenge to be dealt with when moving the axis during machining due to the looseness of machine motions.

Figure 12-58 *A digital readout displays the X-Y-Z position with reference to a movable zero point chosen to represent the datum basis of the workpiece.*

Figure 12-59 *Comparing the milling head on the standard vertical machine in the foreground to the ram and turret in the back, you can see it's made for heavier work but it's less universal.*

Figure 12-60 *The head and ram are extended to reach a part further out on the table.*

The Machining Head

The machining head features the quill, quill feed rates, automatic and manual quill depth control, RPM adjustments, and feed levers and stops. Before looking at these features, let's examine the way in which the head is mounted on vertical machines.

Standard Vertical Mill Head

On a heavy-duty vertical mill, the head is an integral part of the column (Fig. 12-59). It is positioned out from the column above the table. The work envelope is then limited to the area over which the head can be positioned.

Universal Ram and Turret Head

On these machines, the head is mounted on the ram. Following the minimum overhang principle, the machining head and ram must be withdrawn into its slide to the minimum that reaches the work. In Fig. 12-60, the ram has been extended to reach over a large part, but it may need to be withdrawn again for your setup. (Also see Fig. 12-61.)

Figure 12-61 *To machine this bicycle assembly jig we had to rotate the turret and extend the ram, extending the work envelope.*

KEYPOINT

Always draw the ram extension back to the column for normal work and lock it securely in position.

Features Found on the Machining Head

The R&T spindle head has the following features:

◆ RPM selection (step pulley or variable speed transmission)

◆ Spindle start switch

◆ Spindle brake and lock to stop rotation

◆ Quill feed by hand similar to a drill press, by power feed by engaging a clutch, and by manual powering of the power feed system. By shifting the motor drive gear into neutral, then engaging the feed clutch, one can turn a hand wheel to power the quill feed system. This mechanical feeding action provides a very controlled method of drilling down.

◆ A and B axis tilt (position only not for feeding)

◆ Quill lock and quill stops for production drilling and boring. These allow stopping automatic feed and provide a solid stop for quill movement.

◆ Quill feed rate selection in inches per revolution. Ram and turret mills commonly offer selection of 0.006 IPR, 0.003 IPR, or 0.0015 IPR feed rates, where IPR is inches per minute, as in drilling rates.

◆ Cutter holding draw bar

◆ Ram slide and turret rotation—lock bolts and crank for the slide.

You will need a demonstration and must learn the appropriate response for each feature from your instructor before setting up and operating the machine. Go to the lab if possible, and using Fig. 12-62, identify these functions on your machines.

Figure 12-62 *Learn these features and functions on the ram and turret mill.*

Spindle speed control

RPM range
hi – low

Quill feed
engage

Hand quill feed

Micrometer
quill stop

Quill lock

Figure 12-62 *(Continued)*

- Feeds on an R&T mill range from below 1 in. to around 20 in. per minute.
- On a milling machine, tee slots are a *precision part of the table.*
- Always draw the ram extension back to the column for minimum overhang setups then lock it securely.
- *Caution! Never machine close to the table with an end mill, face cutter, or drill bit.*

Respond

Note that no answers are provided for these questions.

1. With a partner, identify the following milling machine components on a vertical mill in your shop. If you have a horizontal mill, find as many as you can on it also. Remember to put on safety glasses and avoid interfering with others working in the shop.

   - ☐ *X* axis
   - ☐ Spindle start
   - ☐ Feed rate selector
   - ☐ RPM selection (May be a set of levers, a dial, or a step pulley belt)
   - ☐ *Y* axis motion lock
   - ☐ Tee slots
   - ☐ *Z* axis and *W* axis if it exists
   - ☐ Feed start lever
   - ☐ Lubrication system (May be automatic, in which case there will be nothing visible except an oil reservoir and an oil level indicator)
   - ☐ Spindle brake (May be a function of the start clutch—when it is pulled down, it engages the brake. It may be a separate lever as on a Bridgeport mill. The separate brake should never be actuated while the spindle is turned on except in extreme emergencies.)
   - ☐ Quill lever (Present on smaller ram and turret vertical mills only)
   - ☐ *A* axis pivoting head—lock bolts and crank point, if so provided
   - ☐ Ram slide and lock bolts
   - ☐ Boring feed rate selector (Found on the head of Bridgeport machines and many others)
   - ☐ Cutter draw bar
   - ☐ *B* axis lock bolts

2. From the shop manual or from measuring, what are the limits of the work envelope of a typical machine in your shop? With the ram and turret centered?

Introduction: In this unit, we look at mounting cutters on the machine and selecting the correct RPM. Both safety and quality depend on making the right choice.

KEY TERMS

Collet—DIN standard
A small flat-nosed spring collet used in end mill holders for manual and CNC machines.

Collet—R-8
A popular shape mill spindle shape designated as type 7 by the ASTME.

Drawbar
A long bolt that pulls the cutter into the mill spindle. There are two types: direct pull and tension nut pull.

ISO/ANSI mill tapers
A world standard mill spindle taper featuring $3\frac{1}{2}$ in. per foot. Formerly called the "American Standard Mill Taper." Common sizes are #30, #40, and #50.

Shell end mill
A hollow cutter that mounts on a pilot arbor.

Mill Setups that Work Right

Setting up Cutters

There are five major methods of mounting and holding a cutter in a spindle. Each has a particular purpose and advantage. All five feature a ratio of taper per length. One major difference between them is how the cutting tool is held in the taper.

Milling Machine Tapers

Machine tapers divide into shallow and steep versions. Shallow tapers such as the Morse tapers (studied in Section 11) are used for drill work. Once driven into their tapered receiving hole, they tend to stay in place, thus their name—self-holding. But they stay in place under loads from the end.

KEYPOINT

Self-holding tapers are not good at holding side loads, which is exactly what milling forces create. They will not stay in place. Therefore, Morse tapered tools are used only for drill work when used on mills, never for milling!

Drawn Tapers for Milling

To resist milling vibrations and side-load forces as well as end forces, we employ two of the five possible standard machine tapers defined by the American Society of Toolmaking Engineers. *Both are not self-holding.* Both must be pulled into their receptacle with a draw bolt or draw bar, that remains a permanent part of the setup.

Two kinds of milling machine tapers are commonly used on mills, the R-8 taper, used on smaller machines and the Standard Mill Taper used on heavier machines. You may hear veteran machinists refer to them as *American Standard* mill tapers.

1. **The ISO/ANSI Standard Milling Machine Taper**

   As shown in Fig. 12-63, most larger milling machines feature a spindle hole with a taper ratio of $3\frac{1}{2}$ in. per foot. Similar to the various Morse taper sizes, different sections of that $3\frac{1}{2}$ in. per foot cone are used for varying machine sizes. They run from #30 (small end around $\frac{5}{8}$ in.), #40 (small end 1 in.), to the #50 (small end around $1\frac{1}{2}$ in.). Figure 12-63 is proportionately drawn for three sizes.

2. **Torque Transfer**

   Spindle torque is considerable for mill cutters compared to drills. It is also an intermittent load because the cutter teeth enter and leave the cut with each revolution, causing a hammering action that would stall the rotation if they are not positively driven. For milling, spindle torque is transferred to the cutter through drive keys that protrude from the milling machine spindle flange (Fig. 12-64) and mate with slots on either side of the taper flange.

Figure 12-63 *These three ISO/ANSI standard milling machine tapers all have the same taper, $3\frac{1}{2}$ in. per foot, but they are different portions of that cone.*

Figure 12-64 *The drive keys show on the flange of this milling machine spindle.*

While a few cutters are a solid unit (the taper is part of the cutter) as the upper end mill in Fig. 12-63, it's far more likely you'll find four other types in use on both the manual and CNC mills (Fig. 12-65). They are the collet end mill holder (Fig. 12-66), the plain end mill holder, the pilot arbor (Fig. 12-67), and the stub arbor (Fig. 12-68). The pilot arbor, also called a C style arbor, can be loaded with hollow center cutters called **shell end mills** and with shell face mills.

Figure 12-65 *A variety of different size end mill cutter holders with standard milling machine tapers.*

Drive keyway

Spacers

Quick Change Adapters

Rotating lock collar

Figure 12-68 *Stub arbors are used to hold slitting saws, wheel cutters. A quick change holder is useful on CNC machines that have no automatic tool changer.*

Typical R-8 Mill Tools

Drill chuck

Spring collet

Pilot arbors

Drive keyway

Stub arbors

Solid end mill holders

Boring head

Figure 12-69 *An array of R-8 cutter and tool options.*

R-8 Max Capacity = $\frac{3}{4}$ in.

The largest standard size R-8 spring collet is 0.750-in. (19-mm) diameter. However slightly larger collets can be special ordered.

KEYPOINT

- All collet type end mill holders transfer torque to the cutter by squeeze action alone, they are not meant for heavy machining. When loads increase, change to solid end mill holders.
- All mill tools use some form of precision taper. Always clean both components, the taper and the receiving hole, before assembly. The pressure created by a single trapped chip can permanently dent both the taper, and worse, the milling machine spindle.
- *Caution!* Larger diameter **R-8 collets** break very easily when sprung even slightly beyond their range capacity!

Trade Tip

Snap It in! Never put a collet into a **DIN** or Erickson type end mill holder (Fig. 12-66) without first popping the collet into the extraction snap ring in the closing nut. If you fail to take this step, the collet will not be removable after tightening the collet on the cutter.

KEYPOINT

If this step is forgotten, the only way to get the collet out of the hole usually requires destroying the expensive collet!

Collet End Mill Holders

Extraction snap ring groove
See the Trade Tip

Closing nut

Snap ring

DIN or Erickson spring collet

Figure 12-66 *A collet end mill holder is useful on manual or on CNC machines.*

Solid End Mill Holders

Pilot

Cutter drive keys

Pilot cutter arbor

Figure 12-67 *Pilot arbors mount hollow center end mills called* shell end mills *and also face mills.*

R-8 Machine Taper

Most small mills feature a spindle nose taper designated style 7 by the ASTME, but it's commonly called an R-8 taper (Fig. 12-69). There are a wide array of tool holders and accessories featuring R-8 shanks (Fig. 12-69).

Drawbars

Drawbars (also called draw-in bolts, or draw-bolts) are true to their name, they pull and keep cutter holders in the mill spindle taper during machining. On manual machines and a few smaller CNC machines, you will encounter two slightly different types—recognizing each and using them correctly is a safety issue.

For safety you must understand the difference between direct pull and tension nut drawbars.

1. **Direct Pull**

   A direct pull drawbar is truly an extra long, high-strength bolt. Using your hand and then a wrench, it is screwed in until the right pressure is developed to pull and hold the cutter in the machine spindle (Fig. 12-70). About 40 pounds of pressure on the wrench is a good guideline, but have your instructor demonstrate the amount of wrench torque needed.

2. **Tension Nut**

   The difference for this drawbar is that the final cinch is accomplished by the nut not the drawbar. The nut pulls the tool into the spindle by shortening the drawbar. The nut type is more heavy duty and is found on larger machines. See the Trade Tip.

Drawbar Engagement

While a rule of thumb is to screw the drawbar into the tool holder, a minimum distance equal to the thread's diameter (for example, a $\frac{1}{2}$-in.-diameter thread screws in $\frac{1}{2}$ in. deep), deeper is better to maintain a constant pull without stretching the drawbar threads. But to avoid an accident, be sure you do not bottom out the tension nut type. See the Trade Tip.

Trade Tip

Tightening Cutters Right Two drawbar details, if missed, can cause serious accidents.

◆ First, if the nut type is mistaken for the direct pull type, and it is simply screwed in until it bottoms out,

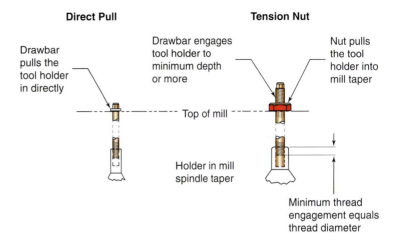

Two Types of Drawbars

Figure 12-70 *Cutters fly out violently when these two drawbars are not recognized. The nut type is used more often on larger mills.*

using it as if it were the pull type, then there is no tension on the cutter. The nut type must first be screwed into the holder a minimum distance equal to its diameter but 10 turns or more is usual. Do not bottom it out—use the nut to pull.

◆ The second accident I've seen more than once is caused by not aligning the torque drive keys on the cutter taper with the flange slots on the mill spindle. From the top of the mill, the unseen cutter feels as though it is tightening since the flange is bottomed against the key tops, as would be the case in Fig. 12-71, if it were tightened in that position. However, with the first touch of the cutter to the work, the cutter spins, loses its hold, and a serious disaster ensues. Always check that the keys are in the drive slots!

◆ *When Tightening Set Screws, Do the "JOGGLE"* Sometimes, when tightening the set screws in solid-end mill holders, the flat spots on the cutter aren't perpendicular to the screw. Tightened just once, they seem OK, but they loosen instantly when the cutter

Figure 12-71 *Caution, the keys can miss the slots!*

hits metal. To avoid this potential accident, do the joggle. First tighten one screw, then release $\frac{1}{8}$ turn or less. Now move the end mill in and out and rotate it around as you slowly retighten the screw. You can feel the cutter fall into alignment. But do it again. The second time, just loosen the set screw until the cutter can joggle a tiny bit. Then on retightening, it should be perfectly aligned to the set screw. Not clear? Ask for a demonstration of this simple but effective trick.

Drills and Reamers

Drill tools can be held in mills using three different methods.

1. Straight shank tool directly in collets.
2. In chucks drawn into the mill. The chuck may have an R-8 taper or it may be a straight shank itself and held in a collet.
3. Morse taper adapter—drawn into the mill spindle. This sleeve converts the mill spindle to accept standard drill tapers (Fig. 12-72).

Caution!
Drills can vibrate or even pull out of Morse adapters during heavy use or during chatter. Unless there is no other option, they are not recommended for drills or reamers in mills.

Selecting RPM

Depending on the alloy being machined, and how hard it is, and the operation being per-

R-8 taper adapter

Morse taper

Figure 12-72 *Morse taper tools must be used for drilling and reaming tasks only, never for milling which creates side thrust and will dislodge the Morse taper!*

formed and the kind of cutter that will do the job, there are a range of mill cutter RPMs that produce a good finish, efficient metal removal, and tool life. But as we've discussed before, there are usually trade-offs between these three goals. How wide that RPM range is depends on the alloy and how difficult it is to machine. In general, start out with conservative speeds for roughing cuts, then increase RPM for finish work. There are three convenient ways to learn RPM selection.

Chart RPM

There are tables in all machinist's reference books that provide mill RPM selections. They are useful but, as with any chart, have range and discretion limits.

Calculated RPM

The short and long formulae we learned in Sections 10 and 11 in the drilling and turning setups also apply to milling cutter speeds. The diameter variable is always the spinning object, so for milling it's the cutter. Let's review.

The long formula is

$$\frac{\textbf{General}}{\textbf{RPM}} = \frac{\textbf{Surface speed} \times \textbf{12}}{\textbf{3.1416} \times \textbf{Diameter of cutter}}$$

The variables are cutter alloy, cutter diameter, and work material.

The short formula is

$$\textbf{RPM} = \frac{\textbf{Surface speed} \times \textbf{4}}{\textbf{Diameter of cutter}}$$

Remember, using our short formula, the RPM will be within 5 percent of that calculated from the general formula. And similar to drill and lathe work, the recommended surface speed is found in a chart in Appendix IV, in reference handbooks, and in cutter manufacturer supplied charts. The chart is entered with two arguments: the cutter type (carbide, HSS, ceramic, or composite) and the work material.

Example

This setup requires a $\frac{1}{2}$-in. HSS end mill. You will be cutting aluminum. Start by finding the surface speed in Appendix IV.

$$SS = 250 \text{ feet per minute}$$
(HSS cutter in aluminum)

$$RPM = \frac{250 \times 4}{0.5}$$

$$RPM = 2,000$$

TRY IT

Find the following RPMs.

A. For a carbide cutter of 6-in. diameter is cutting mild steel, find surface speed and RPM.

B. When machining brass with a 2-in.-diameter HSS face cutter, find SS and RPM.

C. When a pocket must be milled into a carbon-steel forging using a $\frac{3}{4}$-in. HSS end mill, find SS and RPM.

ANSWERS

Keep in mind there are many other factors that affect your final RPM selection. For example, what speeds are available on the machine? Is coolant available and how rigid is the setup? Factors such as these will suggest either more or less RPMs to the experienced machinist.

A. SS = 300, RPM = 200.

B. SS = 175, RPM = 350.

C. SS = 90, RPM = 480.

Slide Rule Calculations

Perhaps the most useful tool for RPM selection is the dedicated slide rule calculator. These are provided by cutter manufacturers. Slide rules make the long formula a snap to solve by printing the arguments and results on a sliding index card. Take a look at examples later in Fig. 12-97 in Unit 12-5.

Carbide Insert Mill Cutters

Similar to the case for lathe insert cutters, the ASTME/ISO have designated milling machine cutters standards to define cutter shape, size, and hardness of the insert tooth.

It's not necessary to learn all the facts now, however, at this time, please turn to the *Machinery's Handbook* section on indexable inserts and scan the information and illustrations. Also look over the industrial chart shown in Fig. 12-73. A similar reference will be used where you work, if the shop uses progressive tool technology. We'll return to it for a few insert selections in the unit questions and in Section 12 Review.

Insert Carbide Tooling Advantages and Disadvantages

While cutters similar to that shown in Fig. 12-74 are popular in industry for both manual and CNC work, they aren't ideal in every situation. Knowing when carbide tools are right for the task and when they are not makes a big difference in successful setups. Let's compare them.

Carbide Insert Cutter Advantages

1. **Faster Metal Removal**
   Carbide insert cutters are usually three times faster in most steel-milling situations compared to HSS. They may increase speeds many times that in milling other metals.

2. **Faster Turnaround**
   Dull or broken inserts are easy to exchange and due to the cutter construction, the new insert nests back within 0.001 in. or better. Thus all machining will resume at nearly the same size.

3. **Quick Change Geometry**
   By exchanging inserts, the machinist can change hardness, nose radius, and rake angle in just moments (Fig. 12-75).

4. **Long Tool Life**
   Correctly applied, carbide tooling extends tool life and reduces down time many times over HSS.

5. **Can Eliminate the Need for Coolants**
   Some materials cut better if no coolant is used when using the right cutter.

Disadvantages

1. **High Initial Cost**
   Insert tooling costs many times that of HSS cutting methods. However, they become quite economical when put into long-term industrial machining.

2. **Vulnerable to Breakage**
   Used incorrectly, carbide tooling lacks the forgiveness of HSS and any excessive shock loading will chip and break the inserts. Often when an interrupted, hammering action breaks carbide cutters, HSS can withstand the punishment.

3. **Less Sharp**
   Because carbide is so brittle, it must have a tiny edge radius to avoid chipping. In comparison, HSS cutters can be honed to a very fine edge when extremely sharp cutters are required. Carbide cutters then, cause more force and heat compared to HSS.

For all these differences, most training facilities teach students to use HSS cutters first.

Figure 12-73 *Industry standard carbide insert identification. Review this information and know where to find data of this kind when the job requires it.*

Figure 12-74 *An advanced design, Chip-splitter™ carbide insert, mill cutter, ripping through steel at high speed.*

Figure 12-75 *When exchanging inserts, it's critical to clean out the tiniest chips and dirt before installing the new tooth.*

UNIT 12-3 REVIEW

Review the Key Terms

Collet—DIN standard
A small flat-nosed spring collet used in end mill holders for manual and CNC machines.

Collet—R-8
A popular shape mill spindle shape designated as type 7 by the ASTME.

Drawbar
A long bolt that pulls the cutter into the mill spindle. There are two types, direct pull and tension nut pull.

ISO/ANSI mill tapers
A world standard mill spindle taper featuring $3\frac{1}{2}$ in. per foot. Formerly called the "American Standard Mill Taper." Common sizes are #30, #40, and #50.

Shell end mill
A hollow cutter that mounts on a pilot arbor.

Replay the Key Points

◆ *Be sure to snap the end mill collet into the snap ring. If this step is forgotten, the only way to get the collet out of the hole usually requires destroying the expensive collet!*

◆ All collet type end mill holders transfer torque to the cutter by squeeze action alone, they are not meant for heavy machining. When loads increase, change to solid end mill holders.

◆ All mill tools use some form of precision taper. Always clean both components before assembly. The pressure created by a single trapped chip can permanently dent both the taper and worse, the milling machine spindle.

◆ *Caution!* Larger diameter R-8 collets break very easily when sprung even slightly beyond their range capacity!

◆ There are two kinds of drawbars—be sure you know the difference.

◆ From Section 11, when exchanging inserts, be sure to clean the insert seat.

Respond

1. From *Machinery's Handbook,* what does a number (single or double digit) in the fifth position in a carbide insert call out, for example, TNMG682 (assume a nonmetric insert)?

To get a feel for carbide selection standards, but using *Machinery's Handbook,* describe these inserts.

2. TNMG-432

3. DPMP-631

4. The top of a cast-iron transmission case must be machined flat using a 4-in.-carbide face mill. What are SS and RPM?

5. A peripheral cut must be made to an aluminum bar using a 0.25-in.-diameter end mill. Identify SS and RPM.

6. Cut a slot in a stainless steel forging using an expensive solid carbide 1.0-in. end mill. What are the SS and RPM?

7. You have seen three different methods of selecting RPM, what are they?

UNIT 12-4

Avoid These Errors—Great Setups and Safety

Introduction: Mills are generally safe machines, but they do have potential dangers, including fast rotating cutters, flying chips, and incorrectly held work that can shift in the setup, resulting in broken cutters and part ejection. Unit 12-4 lists the common errors that new machinists commit on milling machines, but more, it presents ways to avoid these mistakes. Similar to the "hit list" for lathe training, these potential accidents are gleaned from other's experiences (mine included). Each accident described here can be avoided.

The top item in terms of frequency is presented first. This list isn't a summary of accidents but rather a list of the *preventative skills* you'll need plus some Trade Tips to get it right! In fact, you might envision this entire unit to be a Trade Tip.

Unit 12-4 is not general, it does not include dangers such as cut or pinched fingers, eye injuries from not wearing glasses, and so on. It highlights specific mill problems, but most importantly, it *offers solutions*.

KEY TERMS

Backlash compensation (compensator)
A one-way brake that is set to hold back the X axis of some commercial duty mills. Compensators help avoid problems from aggressive feeding during climb milling.

Deflection (cutter)
Bend in an end mill (or any log tool) due to a low rigidity ratio and machining forces.

Packing block
The portion of the clamp setup that supports the back of the clamp.

Rigidity
The relative ability to resist movement—can apply to the work, setup, and the cutter.

Solid jaw (vise)
The datum surface used for nesting work surfaces.

Tangential cut
A movement of the face cut to one side such that the cutting force is lengthened and shifted.

Common Problems and Their Prevention

Work Not Held Properly

During machining the part shifts.

Results

Cutter digs in and the part is ruined

Cutter digs in and breaks into flying fragments

The part is flung from the setup

The Solutions

Use table clamps and vises the right way—they serve as a lever.

Use clamp leverage to your advantage—see the Trade Tip.

Use vises the right way—see the Trade Tip.

Trade Tip

Using Table Clamps There are many versions and combinations of table clamps and the support **packing block** at the back. All must follow a few simple physical rules to hold right. Use your built-in sense of mechanical devices and compare the two clamps in Fig. 12-76, to determine what is wrong with the one on the right.
Clamp Answers (Fig. 12-77).

1. *The clamp on the right has the bolt too far from the work*—the larger portion of the holding force is concentrated on the packing block rather than the work where it is needed.
2. *The clamp on the right is tilted back, thus all its force is concentrated on the corner of the work.* This not only damages the work, it tends to tilt the workpiece up off the table. Plus it will push itself off the work when the cutting begins—its greatest danger. The one on the left is tilting slightly down to avoid this problem even when it is tightened hard enough to bow the clamp down, it's level.
3. *The left clamp is setup with a thin, soft metal shim to protect the workpiece. The clamp on the right has none.*
4. *The differing packing blocks are equally OK*—the special step block and clamp on the left or the flat top block and plain clamp work fine for this application. You may need to shim the flat block to get the right height.

What's wrong with this picture?

Figure 12-76 *Compare these clamps. Can you spot four errors?*

Missing washer unstable platform for the nut

Missing soft shim will mar the work

Packing too low

Concentrated force lifts part and pushes clamp off the workpiece

Tee nut bolt too far from the work

Figure 12-77 *Did you find all these problems?*

Solid jaw

Mill vise datum surface

Figure 12-79 *The solid jaw is the* datum surface. *Align it to the machine axes and always place important edges of the work against the solid jaw.*

Trade Tip

Correctly Using Mill Vises

Rule 1—Make sure the work is down low, close to the vise with plenty of grip.

Rule 2—Make sure it is true to the solid (nonsliding) vise jaw. An uneven part that is not parallel will not hold well because all the force will be concentrated in one location.

Critical Question What's the best fix for the uneven part in Fig. 12-78? (The answer immediately precedes the unit Review.)

Using the Solid Vise Jaw The jaw that does not move on the mill vise is the gage surface. It forms an important datum for milling. *Use it correctly by placing any work requiring alignment to the vise, against the* **solid jaw** *always* (Fig. 12-79).

Dangerous pressure concentration

Equalized pressure

NEVER!

Better but not a perfect solution

Why?

Figure 12-78 *Sometimes a round pin can help to equalize work pressure against the solid jaw—but not this time. It might mar the work but there's another reason. Why?*

Trade Tip

Using a Mill Vise Whenever parts are loaded into a vise, it's imperative to test how well they are seated down against the vise floor or supporting parallels. If, after closing the vise, the parallel bars supporting them can be moved, the part has lifted up due to vise closure. This is a predictable error that often causes undersized parts. One way to detect this problem is to place a piece of paper under the parts, on both sides of the vise. Then, if they are seated correctly, the paper will not move when tugged after the vise is tightened (Fig. 12-80).

But if the test shows the parts have lifted up when the vise was closed, here's what to do. Release then snug the vise pressure about two-thirds of final pressure. Now tap the part down with a deadblow (sand-filled) plastic hammer (Fig. 12-81). Now close it and give the part one last firm hit with the deadblow.

If the part won't seat, then check it's shape. A part whose sides aren't square to its base can't be seated

Figure 12-80 *Two paper feelers show whether the workpiece has lifted off the vise floor when tightened.*

Figure 12-81 *A deadblow hammer tends to cancel the rebound that happens when driving parts with a brass face hammer.*

down to full floor contact. In this case, either accept side contact parallel to the floor but not fully seated, or do not use a vise, use clamps and parallels to hold the work.

Vise Quality Another pro tip for vise work is that mill vises aren't all alike. Some work just fine for utility work holding, but they aren't precise enough for the job in Fig. 12-80. Their jaws aren't hardened and ground perpendicular to the vise's axis, their construction is loose, and they tend to lift parts up. Other vises are precision ground machine accessories (Fig. 12-82) featuring well-guided moving jaws that do not tip when they are closed. Due to the way they are engineered, the mill vise of Fig. 12-82 places $\frac{1}{2}$ pound down pressure on the work for every pound of closure pressure. Their moving jaw guides are adjustable to compensate for wear. Learn to recognize which vise is which in your shop, then use the better vise on the precision jobs.

Soft Jaws When work shape becomes impossible to hold in a flat jaw vise, we exchange the hardened jaws for soft jaws that can be custom machined to the right shape. Soft jaws also aid in holding batches of parts in just the right location. More coming up in Unit 12-5.

Figure 12-82 *The AngLock precision vise is built to 0.001-in. tolerance for size and squareness. When it's closed, it puts $\frac{1}{2}$ lb down pressure on the workpiece, for each pound of closing pressure. It can deliver thousands of pounds closing pressure.*

Poor Cutter Choice

Too long, too weak, or poorly held work. (Fly cutters can be a problem in this area.)

Results

The cutter spins in the collet, ruining the collet and work piece

The cutter chatters and breaks with flying fragments

The Solutions

Always use the shortest, most sturdy cutter to do the job without **deflection** (bending) (Figs. 12-83, 12-84, and 12-85). Long cutters tend to chatter and break.

Always use the joggle method of tightening set screws.

Double-check the draw bar during setups and often during operation.

Question: What is wrong with the setup of Fig. 12-83?

Wrong Spindle Speeds

(Too high or too low)

Results

Too high—heat, chatter, cutter dulling, potential work hardening

Too low—poor finish, cutter digging, and breakage

The Solution

Always compute the spindle speed and use coolant whenever practical.

Figure 12-83 *Explain what must be done to this setup to make it safe.*

Deflection
due to extension and diameter

Extension

1/4"   3/4"   1/4"   3/8"

Twice the extension   Half the extension
only 1/3 the depth   3 times the depth
of cut   of cut

Diameter

1/4"   1/2"   1"

Rigidity of 1   Rigidity of 8   Rigidity of 64

Figure 12-84 *The rigidity of the cutter increases exponentially as cutter diameter increases.*

Vertical edge
no deflection

Long
thin
cutter

Cutter deflection
causes tapered
side

Figure 12-85 *Poorly chosen, the cutter will deflect (bend) and can even break!*

Dull Cutter

Either the student begins cutting with a dull cutter or it dulls during machining.

Results

High heat causes excessively hot chips, which burn the operator.

High heat burns up cutter or work hardens the material (steel, bronze, and other metals prone to hardening)

Cutter loads up on chips and breaks—common in aluminum (Fig. 12-86)

The Solutions

Examine the cutter before mounting it in the machine

Use coolant

Figure 12-86 Stop and solve it before proceeding! *Your cutter is* loaded, *probably from dullness or inadequate coolant, or possibly due to excessive spindle speeds.*

Use correct speeds and feeds

Look and listen for signs of progressive dulling.

Incorrect Operator Action

Initiating the wrong movement at the wrong time

Results

The machine moves the wrong direction and the cutter contacts either the part, vise, machine table, or fixture in the wrong location. Ruined part, fixture, and cutter. Possible broken and flying cutter.

One common action is to start the cutter rotating backward—no kidding it happens regularly!

The Solutions

Know and review the function of each machine control. Think out actions. *Know and practice how to stop all actions in an emergency.*

Unrealistically High Feed Rates

Results

Rough finish on parts

Broken cutters or possibly shifting parts in setup

The Solution

Pay close attention to feed rate training coming up. Start low and increment up to a maximized rate.

Climb Milling Aggressive Feed

Without proper part and machine axis control

Results

Broken cutter

Part begins to self-feed and may be ripped out of setup

The Solution

Climb milling is a better way to peripheral cut but you must know and control the danger. It tends to pull the part along which can get out of control. (More training coming up)

On manual mills, lock the machine axes not being used for feeding movements and set a drag on the axis being used. Note, using axis locks to hold back aggressive feeds causes wear to mill dovetails and to the lock pads, but sometimes, it's the only solution to use climb milling safely. Larger milling machines feature **backlash compensaters.** They are one way brake designed to retard aggressive feed without excessive wear to machine parts.

Leaving a Wrench on the Drawbar Nut While Starting the Spindle

Results

The wrench hits the machine frame, flies off into the shop.

The Solutions

Never take your hand off the wrench when it's on the machine.

Always check before turning on the machine that the wrench has been cleared from the drawbar (Fig. 12-87).

Figure 12-87 *Dents in the motor housing show how dangerous leaving the wrench on the draw bar can be!*

Short abrupt "hammering" action

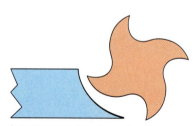

Longer more shearing action

Figure 12-88 *Which cut will vibrate less?*

KEYPOINT

Both cutters in Fig. 12-88 are at 50 percent radial engagement, but there is a big difference in the cutting action.

The short hammering cut can be improved to a longer action known as a **tangential cut** (cutting on the tangent not directly across). Much of the cutting force is transferred to the longer axis of the work and these tend to be far less vibration.

Trade Tip

Work Entry Angle and Forces Milling is an intermittent action. Each cutter tooth enters and exits the work, thus causing a hammering action on both cutter and work. However, in Fig. 12-88 we see good and bad ways to make a cut. The lower cutter has simply been shifted off to one side of the part, to cause a longer shear action of the chip.

Control the chip ejection path

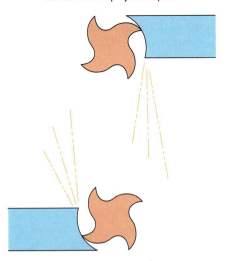

Figure 12-89 *Top view. The direction of cut can control the direction of the chip ejection.*

KEYPOINT

Controlling Chips

Chip ejection can be aimed. Figure 12-89 shows two cuts of equal efficiency, but one sends the chips out at the operator and one sends them harmlessly into the machine frame. A bonus is that the better setup forces the work toward the solid jaw of the vise—a good practice!

KEYPOINT

Safe, Strong Setups

A setup is a physical demonstration of mechanical concepts—physics in action. Use your judgment, if it looks weak or dangerous, it probably is. Be skeptical of your setups and never machine without evaluating for **rigidity** and tightness. Always look for better, more rigid solutions.

Answer to Critical Question For the uneven part shown in Fig. 12-78, machine the block parallel, if possible, then put it back in the vise. If it can't be machined parallel, then hold it in some other way, for example, with down pressure from table clamps.

UNIT 12-4 REVIEW

Review the Key Terms

Backlash compensation (compensator)
A one-way brake brought against the X axis of some commercial duty mills. Compensators help avoid problems from aggressive feeding during climb milling.

Deflection (cutter)
Bend in an end mill (or any log tool) due to a low rigidity ratio and machining forces.

Packing block
The portion of the clamp setup that supports the back of the clamp.

Rigidity
The relative ability to resist movement—can apply to the work, setup, and the cutter.

Solid jaw (vise)
The datum surface used for nesting work surfaces.

Tangential cut
A movement of the face cut to one side such that the cutting force is lengthened and shifted.

Replay the Key Points

- Setups are complex demonstrations of physical science. Be on the lookout for ways to make them more rigid and safe.
- Cutter actions can be rearranged to improve setups.
- There is a right and a wrong way to use machine vises.

Respond

1. On a sheet of paper, sketch a safe table clamp. There will be seven components to each clamp setup.
2. Why are there parallel bars under the work in Fig. 12-90? What is missing from the setup?

Critical Thinking

3. To machine this pocket and drill the hole through the part, the plan calls for holding it in a vise (Fig. 12-91). Sketch your ideas for holding it against the solid jaw. Ask others in your training group how they would do it. This is a typical setup problem, although it is not directly covered in the reading.
4. What is good and bad about a *tangential* cut?
5. After cutting an aluminum bar with an end mill, for a short while, the cutter

Mill table

Parallel bars

Figure 12-90 *Why the parallel under the part? What's missing here?*

Machining this pocket
drilling this hole through

Protrudes farthest
rounded end

Datum surface

Figure 12-91 *Sketch or describe your solution to holding this part correctly in the vise. Invent the solution from simple shop components.*

snaps off! What could have been the problem(s)? What is(are) the solution(s)? Identify at least five.

6. A. What is dangerous about *climb* milling?
   B. What must you do to control the machine during climb milling?

UNIT 12-5

Pre-CNC Mill Setups

Introduction: This unit contains tasks that are critical to operating both CNC and manual milling machines. They involve computing feed rates and aligning work and work-holding vises and other accessories on the mill table.

KEY TERMS

Coordinating axes
Positioning the spindle or cutter in an exact relationship with the work or holding accessories, then setting axis registers (dials or digital readouts) to zero. Creates the RZ or PRZ.

Coaxial indicator
A vertical orientation DTI for finding the axis of a hole by rotating its probe while the dial face remains stationery.

Edge-finder
An alignment tool that wobbles until it comes into tangent contact with a surface, at which time the spindle's centerline is exactly 0.100 in. in from the edge.

Feed per tooth FPT (chip load)
The recommended amount of advancement one tooth can withstand per revolution.

Jig bore indicator
A vertical orientation indicator whose dial faces up when the probe is on an edge or within a hole.

Program reference zero (PRZ)
A starting point reference for a program that must be coordinated on the workpiece or tooling where the axis registers are set to read zero.

Reference zero (RZ)
A starting point reference on a workpiece where the dials and axis registers are set to read zero. All subsequent cuts will be made with dimensional reference to the RZ.

Wiggler
A universal edge and center finding tool set that also includes an articulated indicator holder.

Ten-Point Checklist for Mill Setups

The following roadmap can be used to make certain you've not missed anything in your setups. Some points will be familiar, while some are new. It might be a good idea to copy then check the items for your first few setups. There are a few more tasks needed for setting up a CNC that we'll add later.

- ☐ **1.** Mount the cutter
- ☐ **2.** Calculate and select the RPM
- ☐ **3.** Mount and align the work-holding accessory
- ☐ **4.** Bolt the work to the table using clamps
- ☐ **5.** Calculate and select the feed rate
- ☐ **6.** Check all safety guards
- ☐ **7.** Lock all unused axes including the quill lock/backlash compensator
- ☐ **8.** True the Universal Head (If equipped)
- ☐ **9.** Align the spindle with the part datums
- ☐ **10.** Test the coolant before starting the cut

Using Climb and Conventional Milling

We need to look more closely now at the chip being made to understand the differences in the two cutting methods (Fig. 12-92). The first fact we uncover is that *conventional* cutting isn't the convention in the modern world of CNC. There, the machine axis drives have nearly zero backlash and the tiny bit they do have is controlled automatically. CNC machines don't aggressively self-feed out of control, due to the climb cut's force multiplication, as happens on hand-cranked manual machines. But the part and the cutter are still at risk from these forces on any milling machine.

When hand cranking a climb cut using a manual mill, the crank handle will move nearly on its own! But that can (and sometimes does) get out of control. If allowed to self-feed, the axis movement can accelerate until either the cutter breaks or the part is pulled out of the setup. Also, if the part isn't held strongly enough, it can begin to move into the cutter, which also spells disaster.

Compare the Action

Take a moment to compare Fig. 12-93, which shows climb cutting, to Fig. 12-94, where a conventional cut is being made. Note that the climb cut starts with a thick chip as each tooth enters the work. That's good, it makes a nonrubbing entry into the work. Then as the tooth exits the metal, the chip becomes progressively thinner. That's good too. The machined surface is formed by a final, sheared slice, from inside the material toward the outside. Both work finish and tool life are excellent when compared to the way a conventional cut progresses.

Notice first off in Fig. 12-94 that during the conventional cut, the cutting edge doesn't start into the metal until some distance past the theoretical entry point. For a short

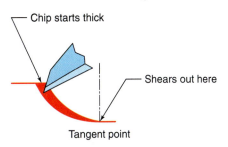

Figure 12-92 *The forces of climb verses conventional milling.*

distance, as the machine feeds forward, the tooth builds up pressure until the pressure is great enough for the cutting edge to dig in to start cutting. Yet, that rubbed zone is what

Figure 12-93 *A climb cut produces a better finish and longer tool life.*

Figure 12-94 *The conventional cut starts by rubbing until the tooth builds pressure to dig in, producing poor finish and dulling the cutting edge.*

will become the finish left behind! The tooth edge also suffers as it rubs on the material, especially if the material is hard, such as carbon steel or bronze. Using conventional cutting, surface work hardening will occur faster in a material prone to it, and the dulling/work hardening accelerates as the edge begins to fail.

KEYPOINT

During a peripheral conventional cut, the edge rubs for a short distance before it builds enough pressure to dig in.

Trade Tip

Exceptions—When to Use Conventional Cutting On manual or CNC mills, it is acceptable to cut back and forth using first conventional then climb milling when reducing a surface with peripheral cutting, especially on softer materials such as aluminum. Just keep in mind the different actions produced—snug the brake for the climb pass then release for the conventional cut. However, to produce the best finish, choose climb milling for the final pass.

When to Use Conventional Cutting There are two times when conventional cutting is called for.

1. *Hard Crust* Cast iron often forms a hard thin outer surface, but it has a soft interior (Fig. 12-95). Here, by correctly choosing conventional cutting, the tooth starts under the crust until it breaks upward, pushing the crust away. In this case, climb milling would tend to dull the cutter more quickly because each tooth would start it's cut in the hard crust.
2. *When the Setup or Part are too Thin or Weak* Sometimes, the aggressive action of climb milling will flex, bend, or lift a part in a setup. In this case, try conventional cutting to stabilize the work and the setup.

Crust

Cutter lifts up under hard crust

Figure 12-95 *Conventional cutting is correct for work with a hard crust.*

Computing Mill Feed Rates

Recall that mill feeds are expressed in inches per minute IPM (or meters per minute). Reasonable student rates vary from under 1 in. per minute to around 20 in. per minute depending on the work material, setup rigidity, and the machine. Another big factor is your skill level; stay on the conservative side while learning. Industry rates are in the hundreds of feet per minute and are accelerating as machines and controls evolve. Feed rates of 100 IPM or greater are everyday stuff with CNC rates at 300 to 700 IPM and faster where the computer can control the action and still stay on course! (See Fig. 12-96.) There is a specific study of maximized machining, called high-speed machining, that we'll look at in Section 26.

To Find a Feed Rate in IPM

The process of finding a feed rate is a three-step multiplication.

Determine FPT

From a chart, find chip load for one tooth making one revolution

It's the recommended bite for one tooth— **feed per tooth (FPT)**

FTP is also called the **chip load.**

Figure 12-96 *This horizontal machining center is removing aluminum at 750 IPM.*

FPT is found on the chart by crossing two arguments: the cutter material (HSS or carbide) and the work material on the chart.

From experimenting, researchers have found minimum and maximum amounts of forward feed, for one tooth, making one revolution. The range provides good finish and tool life. It's then compiled into a recommendation chart.

FPT × Number of teeth = Feed per revolution

Multiply FPT times the number of teeth in the cutter

Determines feed advancement for one single revolution of the cutter = FPR

FPR × RPM = Feed Rate

RPM has been precalculated

Quick Review—Calculating RPM

After finding the recommended surface speed from a chart, it is combined in either the long or short formula with the cutter diameter. The SS is determined with two arguments: work hardness and cutter hardness (carbide or HSS).

Example Problem We'll be using *Machinery's Handbook* or CD for this exercise, however, commercial slide rules designed just for milling are available and quick to use. Please turn to the section on "Speeds and Feeds—Feed Rates for Mills."

Calculate the feed rate for milling cast steel, using a $\frac{3}{4}$-in.-diameter, four-tooth, HSS end mill cutting to a depth of 0.250 in. with the RPM at 480

Find the chart "Recommended FEED in inches PER TOOTH for HSS Milling Cutters."

Question? What is the total range of tooth feeds for cutting cast steel?

The range is 0.001 in. to 0.012 in. depending on cutter type, depth of cut, and hardness of work material. We'll assume the steel is the least hard (top row). The rate is?

Answer 0.003 in. per tooth if at 0.250 DOC.

Next multiply times 4 teeth, which equals 0.012 in. per revolution.

Now, multiply times the RPM at 480.

480 × 0.012 = 5.76 in. per minute feed rate

Feed rate = FPT × number teeth × RPM.

TRY IT

From *Machinery's Handbook*,

A. Find the feed rate for milling plain carbon steel (lowest hardness) using a $\frac{1}{2}$-in. two-flute HSS end mill cutter at 0.050 in. depth of cut. Calculated RPM = 800.

B. Cutting gray cast iron (hardest this time), with an eight-tooth HSS face mill. RPM = 200. This time give me the range of possible feeds.

ANSWERS

A. Feed per tooth = 0.002 in.; 0.004 in. per revolution; *3.2 in. per minute.*

B. Feed per tooth = 0.002 to 0.008 in. per tooth (range of possibilities); 0.016 to 0.064 per revolution. From *3.2* IPM to finish cut up to *12.8 in.* per minute for roughing.

Trade Tip

Speeds and Feeds With experience, you'll develop a "feel" for the right rate. Experienced machinists often select initial feed rates from their head, but not always! There are times when a special material or unusual cutter is used; it's then that we all fall back on calculating.

The two pocket slide rule calculators shown in Fig. 12-97 are a convenient way to quickly find speeds and feeds. They are supplied by cutter manufacturers, usually free of cost. Check with your local tooling company or go online and request one. You'll find they follow the same formulae and input arguments we've already discussed for both RPM and feed rates.

Figure 12-97 *Speed and feed calculators should be a part of your tool box.*

Setting Up Milling Machine Vises

Now we turn our attention to the nuts and bolts aspect of the setup, literally!

Clean It First

When mounting vises, always wipe the bottom with a clean lint-free rag, then give it a final wipe with your clean hand. Always clean the mill table with equal care. This palm trick will detect any chips or nicks protruding. The final hand swipe is a Trade Tip.

Aligning

If the vise has slot alignment keys, do not set it directly on the mill table then slide it across to location. Set the vise on a piece of plywood or other protective material then slide it into position—tilt the vise up to remove the wood, then lower the keys into the slots, making sure no chips or dirt has fallen on the table while you were doing so.

Indicating Vises

There are two cases when alignment keys are not used and we turn to a dial test indicator to setup the vise (Fig. 12-98).

The vise has no keys—yet the setup requires perfect alignment to a machine axis (common).

The setup requires the vise at some angle to an axis.

Naturally, there are a few trade tricks!

KEYPOINT

Always indicate the *solid jaw* of the vise, never the moving jaw. The solid jaw will become the reliable datum for your part. A ground parallel held in the vise averages out jaw damage or irregularities.

Set Up a Pivot Bolt

To begin, lightly tighten one tee nut bolt, then snug the second bolt just beyond finger tight. That simplifies moving it as the vise

Seeking zero reading both ends

Tap vise base with soft hammer

Machine axis

Pivot bolt

Figure 12-98 *The indicator is against the nonmoving (solid) jaw.*

Collet Spindle clamp Magnetic base to mill frame

Quill housing mount

Figure 12-99 *Five ways to hold your indicator on the mill.*

pivots on the tighter bolt during final alignment. Now, with an indicator mounted on the mill frame or spindle, place the probe against the pivot bolt end, zero it, then move the mill axis back and forth. While it moves, tap the vise side with a soft hammer, until a zero reading is obtained on both ends of the vise jaw (Fig. 12-98).

Mount the indicator on the mill, in one of five possible ways (Fig. 12-99).

KEYPOINT

Avoid flinging your indicator across the shop, shift the spindle to neutral for the first three indicators shown in Fig. 12-99.

Trade Tip

Positioning Mill Vises and Aligning Work Here are a few handy tips that simplify the task of positioning a vise in mill setup!

Read the Trend! Instead of cranking back and forth, end-to-end to indicate the vise, try this: with practice you can whip a vise into alignment in seconds! Start the power feed at a moderate rate around 2 IPM. As it moves away from the pivot bolt, watch the dial movement to determine which way the vise is misaligned. Start lightly tapping the vise as the indicator travels across the jaw. The indicator's needle movement will slow then stop as the vise approaches alignment. Using this technique, one can often complete an alignment within 1 minute!

Balance It. Move It. Given a choice, never mount a vise at the extreme end of the mill table (manual or

CNC). This creates an unbalanced load on the table and excessive wear on the gibs. However, it is good practice to move different setups to various locations somewhat near table center (Fig. 12-100), especially on CNC machines where moves repeat many times in production runs. That avoids long-term concentrated wear to the drive system.

Angle It. Here are three methods of positioning a vise at an angle to the *X* or *Y* axis. Each is progressively more accurate but requires a bit more time to perform.

1. *Rotating angular vises* with graduated bases are useful when angular tolerances are no tighter than 1° accuracy (Fig. 12-101).
2. *Indicating a protractor* is the most convenient method. Lightly clamp the base of the protractor in the vise, then indicate the blade. Expect 10 minutes or better accuracy, which is 12 times as accurate as method 1. Don't forget to clamp the second ear down to the table as shown in Fig. 12-102.
3. *Indicating a sine bar*, is the most precise method available (Fig. 12-103). Accuracy will be within 0.0005°. A major improvement!

Quick Alignment Here's another tip that will save you much time, use the table slots and slot blocks to align a workpiece datum to the *X* axis (Fig. 12-104).

Figure 12-100 *If possible, mount the vise near the table center to balance wear on dovetails and gibs.*

Figure 12-101 *Some vises feature a swivel base. The better vises feature angular graduations as on this example.*

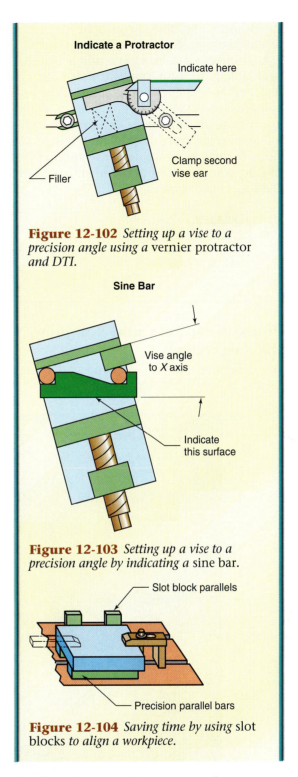

Figure 12-102 *Setting up a vise to a precision angle using a* vernier protractor *and DTI.*

Figure 12-103 *Setting up a vise to a precision angle by indicating a* sine bar.

Figure 12-104 *Saving time by using* slot blocks *to align a workpiece.*

Indicating a Universal Head Machine

When drilling or boring vertical holes on mills, the head must be perpendicular to the table. If not, the hole produced will not be vertical. On rigid vertical mills, the head cannot be out of alignment. However, when using a ram and turret mill, or one with an *A-B* head auxiliary, the *A* and *B* axes must be aligned for precise drilling, boring, and many other operations.

Aligning the *A-B* axes perpendicular to the mill table is accomplished with an indicator fastened to the spindle. The indicator and spindle are rotated by hand, in a circle, against the machine table or a ground parallel to average irregularities and bridge tee slots. You will require a demonstration of this task to completely understand it. There are the three major phases.

Phase 1—Testing and Roughing the Alignment

Graduations First Look at the graduations on the head (Fig. 12-105). Both the *A* and *B* axes have them. Any gross misalignment will be apparent.

Using a Square and Alignment Notches in the Head Now, using the small notches machined into the machining head, refine the alignment. As shown in Fig. 12-106, these notches are parallel to the quill and can be used against a master square sitting on the mill table.

Trade Tip

Paper Feelers to Eliminate Guessing To get a good "feel" at both contact points between square and pad, hold a piece of paper at each contact point and tug on each. Any difference in pressure will be easily felt. If this is done right, the head error can be almost removed without using the indicator.

B axis crank
B axis lock bolts (4 bolts)
B axis angular index line
A axis crank
A axis lock bolts (3 bolts)
A axis index line

Figure 12-105 *The A and B axis angular graduations are a rough indicator of alignment.*

A Axis   B Axis

Figure 12-106 *Equalizing the drag on paper feelers against alignment notches in the head roughly aligns the A/B axes.*

Phase 2—A Axis Alignment (Front to Back)

Leaving the lock bolts finger tight, use the front to back swing of the dial test indicator sweeping the table to determine perpendicularity (Fig. 12-107). Your objective is to adjust the head until the indicator reads zero-zero, front and back. Once the *A* axis alignment is accomplished, lock the bolts a little at a time then retest to see if locking disturbed the position.

There is a complexity in aligning the *A* axis. Note in Fig. 12-108, that the indicator pivots at the end of a long arm. That means that, as the head's tilt is adjusted, the sweep of the indicator moves in an arc at the end of the arm. Both the front and the back of the sweep move in the same direction. That fact makes a difference in how you adjust the head toward alignment. Here's an example to clear this up. In Fig. 12-109, the head is not perpendicular to the table. At position 1 the indicator shows a difference of 0.020 in. farther away from the table, in the front. From that, you can deduce that it's tilted outward.

So, with the indicator in the front, we crank the axis down 0.020 on the dial. But after doing so (Fig. 12-110), we still find a 0.009 front to back difference. The adjustment didn't

Aligning the "*A*" Axis

90.00°

Clamp on holder spindle in neutral

Swing indicator front to back

Parallel bar bridges tee slots

Figure 12-107 *Correct the A axis error first.*

bring the head into alignment due to the indicator being at the end of the long arm. Both the front and the back swung down closer to the table with the first adjustment.

Figure 12-108 *The indicator is at the end of a long arm.*

Position 1
Misalignment Detected

Indicator
sweep plane

0.020" less DTI
interference in front

Figure 12-109 *Position 1—0.020-in. error read on indicator.*

Position 2
After Correcting 0.020"

0.009" remaining

Figure 12-110 *Position 2—0.009-in. error remaining.*

KEYPOINT

The correction turns out to be a ratio of *amount taken* compared to *amount achieved* on the indicator dial.

The ratio is a function of the long indicator arm from the head's pivot point to the table contact of the probe. That distance changes from setup to setup, depending on indicator holder and quill position.

The ratio, in this example, is a little over half—in cranking out 0.020 in. on the indicator; 0.011 in. of error was removed. Discovering that with the first adjustment, reposition in the front and dial down say 0.017 in. to take out the 0.009-in. needed correction. The ratio changes as the probe approaches parallel, so, a final adjustment is probably going to be required, but it will be a small correction. Have your instructor demonstrate and explain this concept. It's easy to understand when you are doing it, but difficult to put into words.

Trade Tip

Use a Ground Parallel to Smooth Out Table Irregularities When performing the DTI alignment of the universal head, place a large ground parallel on the table to account for the tee slots and any dents that might disturb the probe as it swings. A fine stone run across the table will remove protrusions but the DTI needle can still trip in dents. A parallel on the table solves that.

Phase 3—*B* Axis Alignment

The *B* axis is easier to align than the *A* axis (Fig. 12-111). This also is accomplished by sweeping the indicator in a circle on the

A axis

B axis

Figure 12-111 *Rough align the A axis first, then the B axis. Then align them a second time and check the alignment after tightening lock bolts.*

Figure 12-112 *Use a sine bar and DTI to set precise angles.*

table while the right-hand reading is compared to that on the left. The head is adjusted until these readings are equal, zero-zero. But this time there's no ratio challenge. All one must do is determine the right-left difference then adjust away half the amount needed. One side goes up while the other comes down.

Sine Bar for Angles For setting up precision angles of the *A* or *B* axes, after roughly adjusting the head using the graduations on the axis, use a sine bar and sweep it for zero, zero on a DTI (Fig. 12-112).

Datum Locating a Milling Machine

For many manual mill and all CNC setups, it becomes necessary to locate the spindle over a unique place on the part:

◆ An intersection of two edges
◆ Centered over a hole
◆ On the intersection of layout line's—center-finders are covered in Section 10

With the spindle located at that point, the position registers (micrometer dial or readouts) will be set to zero. This is a prime CNC setup skill.

KEYPOINT

That *X-Y* datum reference point is called the **program reference zero (PRZ)** in CNC programs and setups. On manual machines we call it the datum point, or the **reference zero (RZ).** It's symbolized by a bull's-eye target.

There are three tools used to position the mill's spindle at a desired RZ.

Tool	Repeatability
Edge-finder	0.003 to 0.001 (Using Light Flash Trade Tip coming up)
Center-finder	0.005 to 0.003 (Using Co-Linear Trade Tip in Section 10 Unit 4)
Dial test indicator	0.0005 or better

Edge Finding

For this job (Fig. 12-113), we will be drilling five holes in a part that is already shaped. The hole location tolerance is plus/minus 0.010 in. so an **edge-finder** is appropriate. The RZ is at the intersection of datums A and B, the upper left corner. Once we position the mill spindle over that corner, we'll zero both micrometer dials (or *X* and *Y* readouts) then crank out, in turn, to each of the five locations and drill the holes.

KEYPOINT

In Fig. 12-114, the RZ is at the intersection of two primary datums. The vise has been indicated true to the mill's *X* axis. The part is loaded into the vise with datum feature A against the solid vise jaw.

In Fig. 12-115, note the target symbol in the corner of the part. That's the symbol for RZ or PRZ on a setup document. I'll use it often in

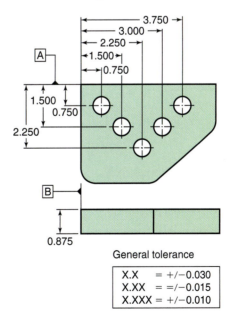

General tolerance

X.X	= +/−0.030
X.XX	= =/−0.015
X.XXX	= +/−0.010

Figure 12-113 *We need to drill this part to a 0.010-in. location tolerance.*

Figure 12-114 *The intersection of datums A and B will become the RZ.*

Figure 12-115 *The symbol used to denote RZ or PRZ in setup documents.*

Combination Center and Edge-Finder

Pointer for finding lines and points

Tension spring

0.625 dia body
(0.500-in. single end)

0.200-in.-diameter
edge finder probe

Figure 12-116 *The edge finder helps place the spindle axis over a part edge within 0.003-in. repeatability.*

later parts of this book as do programmers and others in manufacturing.

Using the Edge-Finder This handy little tool makes edge datum location a snap (Fig. 12-116). It's a priority 1 for your tool box purchase plan. We'll be using the cylindrical probe. In theory, any round object of known diameter could be used for edge finding, except that there's always some runout in the way it's held in the spindle. But, due to the probe's ability to float on the end of the body, it

centers on the spindle axis independent of spindle runout.

Mounted in a drill chuck or collet, it's spun at around 200 to 300 RPM but no more. The spring can be stretched if the RPM is set too high. The objective is to start it with a deliberate amount of runout, then, watching the wobble, approach the edge with the probe touching the work, until the runout disappears.

The objective is to stop moving the spinning probe (0.200 in. diameter) toward the edge at the instant that's it's perfectly tangent. At that location, the spindle axis is 0.100 in. from the edge (half the probe diameter). All that remains to do is set the position register at plus or minus 0.100, depending on which side of the edge the probe is relative to the work. In Fig. 12-117, it's to the left—in the minus direction compared to the edge.

KEYPOINT

The radius of the probe is 0.100 in. By moving the adjusted axis to zero in Fig. 12-117, the spindle would be directly over the X axis edge and the backlash would be accounted for. It's now ready to position over the upcoming hole locations.

Setting the dials or registers to represent the spindle position relative to a RZ or PRZ is called **coordinating the axes.** After coordinating the X register, do the same to the Y register to complete the RZ.

Right and Wrong Edge-Finder Use A common error made by machinists performing edge finding is to watch for the probe to

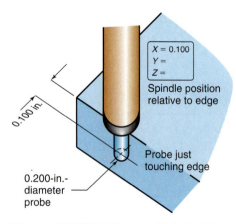

Figure 12-117 *The spindle is 0.100 in. to the left of the work edge as the probe just touches.*

snap sideways then assume that location is the perfect tangent location—it is not. In order for the probe to snap, friction must be present and that only happens once the edge-finder is in forceful contact with the edge—too far past perfect contact! That probe snap location will reproduce reliably at the same location every time, but it's beyond perfect alignment by around 0.001 to 0.002 in. It can be used for rough edge finding but it doesn't produce the best results. See the Trade Tip.

Using a DTI to Coordinate an Edge This is a vital skill for coordinating reference points when tolerances are finer than edge finding can deliver—below 0.001 in. The objective is to place the DTI on the edge, then hand rotate it in an arc (with the spindle in neutral). When the numbers seen on the dial are zero-zero on both sides of the edge, the spindle is located. You'll need to hold a ground parallel on one side for this test, as shown in Fig. 12-118.

Edge Finding with a DTI

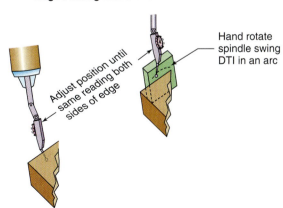

Figure 12-118 *Locating over an edge using a DTI.*

How Does It Work? With the indicator set just off the spindle axis, swing the spindle and indicator in an arc as you crank the axis toward the edge. As it touches the edge, the indicator readings will increase until a highest interference point (the *null* point) is reached then drop off again.

Return to the null position, set the indicator to zero. Now pull it up to clear the edge, turn it 180°, and move the probe to the other side of the edge. Read the difference in this null value compared to the zero. Determine by the readings (plus or minus from the zero), which way the spindle is from the edge, then move $\frac{1}{2}$ the difference toward alignment. Test again as it usually requires a second and even a third fine adjustment.

The Wiggler Set As shown in Fig. 12-119, the **wiggler** and its accessories are used in a number of tasks for alignment and indicator holding. You must have either an edge-finder or a wiggler in your tool box. When picking up lines, wigglers aren't as accurate as center-finders because they do not reflect the line's image. The pointer is used for eyeball location.

Squaring a Blank Part Using the Datum Process

This is an absolute bottom-line skill for milling and very often the start of a job plan. We'll square up the drill gage blank to the print dimensions. The four steps are shown next, but not everything is explained. Questions will be asked at the end based on what

Figure 12-119 *The wiggler set is used similarly to an edge-finder to place a spindle over an RZ.*

you see in the drawings. The objective is to mill a 3.000- × 5.000-in. rectangle with 90° corners.

We'll start with 0.100-in. machining excess on all four edges. The setup includes a vise true to the *X* axis of the machine with a $\frac{1}{2}$-in. end mill in a collet. *Hint:* Watch Datum edge A as it progresses through the operations.

KEYPOINT

The only way this process will produce accurate results is to clean chips and remove burrs between each operation!

Step 1 Create the Datum Feature Edge

Based on the priorities on the print, start with the most reliable, longest, or most important edge of the part. In this case, the long edge is also Datum A on the print (Fig. 12-120), but that may not always be the case. It might be necessary to create an intermediate surface on the way toward making datum feature A.

Step 2 Machine the Opposite Edge Parallel and to Size

Reversing the part in the vise and testing to see that datum feature A is fully seated on the vise floor, cut the opposite edge parallel to datum A. Take a small test cut and measure the result. Coordinate the *Z* axis position (readout/micrometer dial) to your measurement, then take roughing passes, check it again, then make the final pass according to the coordinated reading.

The first operation makes Datum feature A

Figure 12-120 *Machining Datum A—Minimum clean up to produce a clean surface free of saw marks.*

KEYPOINT

Do not skip this micrometer dial coordination, it assists and speeds precision. Note that just a bit of the part was left protruding from the vise, to facilitate measuring it (Fig. 12-121).

Second Cut Parallel to A

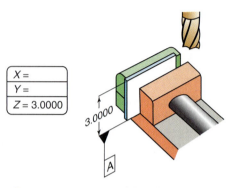

X =
Y =
Z = 3.0000

3.0000

A

Figure 12-121 *Machine the opposite edge to size. Measure the result on the part protruding from the vise.*

After the coordination it is done, as long as the quill remains locked in place, subsequent face, step, and pocket cuts with that end mill will produce the same *Z* axis depth relative to the vise floor datum.

Step 3 Machine One Corner Square

Lay the part down in the vise on parallels. Machine one edge for clean up using the *Y* axis this time. *Notice that datum edge A is against the solid jaw of the vise* (Fig. 12-122).

Step 4 Machine Final Edge to Size

Slide the part such that the other end protrudes out of the vise but keep datum edge A against the vise jaw. Now machine the final

Third cut 90 degrees to A clean up

A

Two parallel bars support the work

Figure 12-122 *Note that the original datum A edge is against the solid vise jaw.*

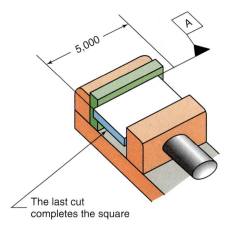

The last cut completes the square

Figure 12-123 *Machining the final end to size, square to Datum A.*

Jig Bore Indicator

Figure 12-124 *Locating over a hole using a jig bore (vertical) DTI.*

end to size (Fig. 12-123). There are two other possible ways to make the final edge. It could be accomplished with the part standing up and datum A on the vise floor, as in step 2. Or the part could be flipped over keeping datum edge A against the vise and cutting the part exactly as in step 3, with the cutter on the same side of the vise.

Indicating a Hole for Datum Point Location

Sometimes the RZ is a hole's axis rather than intersecting edges. Often a CNC mill fixture will have a precision bored hole that represents the setup program reference zero for the fixture. Both the X and Y axes must be coordinated over this hole. Use an indicator in the mill spindle for this setup skill.

Vertical Indicators

To locate over a hole, use an indicator, as shown in Fig. 12-124. Although an every day DTI can be used for this task, there are two indicators made for the task of overhead location: a **jig bore indicator** has the dial facing up when the probe is vertical. They are also called *vertical indicators*. They speed up this operation by presenting their dial faces in such a way that they can be easily read without parallax.

Jig Bore Indicators With the spindle in neutral, hand rotate the indicator while adjusting spindle position until a zero-zero reading occurs in the X and Y axis direction. A jig bore (or jib borer) indicator would be a priority 2 for your tool box. You can get along without them, still they are so very handy to have when setting up over holes.

Coaxial Indicators A **coaxial indicator** is also designed for this task. Its dial also faces up, but this time the machine revolves the probe under power at around 100 RPM or so, but the dial remains stationery by holding back on a handle.

First the X spindle position is dialed in by holding the handle parallel to X. In that orientation, the swings of the needle represent X misalignment. Adjust the spindle position relative to the hole while watching the swings diminish until no movement of the dial hand is detected—it's on center for X. Then the indicator handle is rotated parallel with Y and the task repeated. This takes just seconds. These indicators are complex, precise mechanisms, and as such they are relatively expensive. They are normally shop owned.

Specialty Vise Setups

Vise work can be very creative. The fun part of the challenge is to find and set up the right combination of holding tools to keep a complex part stable, repeatable, safe, and in the right datum relationship. Here's a few tips as to how to proceed—but there are literally an infinite number of ideas remaining to be discovered.

Vise Soft Jaws Often on CNC mills and occasionally on manual mills, it becomes

Figure 12-125 *The hard jaws are removed.*

Figure 12-128 *One example of the infinite possibilities.*

Figure 12-126 *Two examples of custom shaped vise jaws.*

necessary to mill special shapes into sacrificial aluminum (or steel) jaws. The blanks are mounted in the vise, then cut to fit one or more odd-shaped parts (Figs. 12-125 and 12-126). Note that in Fig. 12-126, the jaws on this fixture vise can be rotated to accommodate the round part.

Vise End Stops Sometimes it's necessary to load a series of parts into the vise, with each stopping in the same datum relationship. Datum A is against the solid jaw, while datum B touches the stop. If the vise isn't equipped with these kinds of stops (Fig. 12-127), then a stop can be bolted directly to the table.

Compound Rotate and Tilt Vises The possibilities are limitless. Figure 12-128 shows just one of many. I consider these kinds of challenges to be less like work and more like fun.

Cutter Question Answered—Why Longer Tool Life with 66 Percent Radial Engagement?

The answer divides into two parts: Chip heat and cutting edge entry angle. To solve it, I need to ask another question: Study the two magnified views of tooth entry, and decide which is the better way to setup a face mill cut, positive or negative entry (Fig. 12-129). *Hint:* remember chip formation has two phases.

Figure 12-127 *This vise is setup with swing-away end stops.*

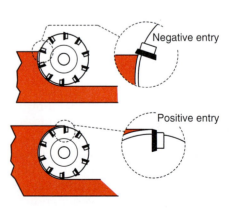

Figure 12-129 *Compare entry angles of the teeth.*

Next clue: the positive entry cutter starts the cut first, then, as the cutter feeds forward, the chip thickens and the rake begins. The negative entry cutter starts the rake before the cut.

Contrary to initial logic, negative tooth entry is better. Why? First off, it would seem that the face contacting the work before the cutting edge wouldn't cut, that this would break the tooth, especially if it's brittle carbide. But, in truth, there's little more pressure on the rake startup than for any rake at any time in the cut (just a bit more until the plastic flow begins, but that's almost instantaneous). In fact, the recommended feed per tooth force is chosen such that the rake won't break the tooth. On the other hand, the positive entry cut contacts the delicate sharp edge first, which tends to chip the edge, and it wears until a sufficient rake builds up to move the boundary zone up the bit face. Heat builds up quickly, but the chip cannot conduct it away as it's ultrathin at the start of the cut and at the end.

Strange as it may seem at first, cutting tools last longer if a negative entry is used.

UNIT 12-5 REVIEW

Review the Key Terms

Coaxial indicator
A vertical orientation DTI used expressly for finding the axis of a hole. Its probe rotates while the dial face remains stationery thus reporting wobble in the axis in question X or Y.

Coordinating axes
Positioning the spindle or cutter in an exact relationship with the work or holding accessories, then setting axis registers (dials or digital readouts) to zero. Creates the RZ or PRZ.

Edge-finder
An alignment tool that wobbles until it comes into tangent contact with a surface, at which time the spindle's centerline is exactly 0.100 in. in from the edge.

Feed per tooth FPT (chip load)
The recommended amount of advancement one tooth can withstand per revolution.

Jig bore indicator
A vertical orientation indicator whose dial faces up when the probe is on an edge or within a hole.

Program reference zero (PRZ)
A starting point reference for a program that must be coordinated on the workpiece or tooling where the axis registers are set to read zero.

Reference zero (RZ)
A starting point reference on a workpiece where the dials and axis registers are set to read zero. All subsequent cuts will be made with dimensional reference to the RZ.

Wiggler
A universal edge and center finding tool set that also includes an articulated indicator holder.

Replay the Key Points

- Climb milling is a better process than conventional cutting, producing a finer finish and longer tool life, but it also produces a multiplication of forces that must be anticipated.
- During peripheral cutting, with a conventional cut, the edge rubs for a short distance before it builds enough pressure to dig in.
- Always indicate the *solid jaw* of the vise, never the moving jaw.
- Avoid flinging your indicator across the shop, shift the spindle to neutral!
- Nearly all parts are loaded into a mill vise with datum feature A against the solid vise jaw.

Respond
Answers follow—check your understanding of milling

1. In the square up exercise in this unit, what two types of cutting were used?
2. When the peripheral cuts were being taken on the part ends (Fig. 12-122), which axis lock must be left snug during a climb cut but able to move?
3. Describe two advantages to climb milling compared to conventional. Explain.
4. True or false? Coordinating the milling machine (or lathe) requires positioning the axes to RZ then testing the distance between work and cutter. If this statement is false, what makes it true?
5. Compute the feed rate for the following facts:

    Cutting aluminum with a two-flute, 0.750-in. diameter HSS endmill

    Recommended FPT = 0.007 in.

    Recommended SS = 400 F/M

Section 12 Review

Unit 12-1

The focus of this unit was to first identify and then to learn the nature of the basic milling operations. These are the critical skills you'll probably be practicing at the manually operated milling machines. By doing so, you'll be safer and more efficient at learning on the CNC mills. But there is also a good chance you'll need to run manual mills during your first lab assignment or your first employment. That's because another theme in Unit 12-1 was that milling rules the shop in the modern world.

Unit 12-2

We learned how the orthogonal axis system fits together and how these axes relate to all other machines—worldwide. I promise, we'll be discussing axes a lot before reaching the end of this book. We also explored the flexibility of work envelopes and that the ram and turret (Bridgeport) mills have an amazing range of setup configurations and capabilities that make them unique to the shop. That, even though most manually operated mills have been relegated to supportive and tooling roles, they will never be shoved to the scrap heap by advancing technology.

Unit 12-3

Setting up a mill means making lots of choices. One of the most basic is that of which cutter to use and how to mount it in the milling machine. The right cutter can make all the difference in quality, pro-ductivity, and tool life. Getting the RPM close to the optimum is another aspect of a good setup but the point was made that there is a range of speeds that will do the job.

Unit 12-4

Why not learn from other's mistakes? It saves a lot of embarrassment. This unit presented a compilation of the most frequent problems we instructors see every semester. Raising your level of conscious about them wasn't our main objective. Our goal was for you to know the solutions. It should start you on the way to being able to walk past a setup and know whether or not it's rock solid. No exaggeration, I can turn my back on the shop and know when things are going right or wrong on most of the machines, just by sound! I suspect most instructors would tell you the same.

Unit 12-5

This is one of my favorite subjects to teach. Every new part brings a set of puzzles to solve. How to hold the work? Which cuts to make first? What's the most efficient and safest way to plan the job? What tooling would be the best? Each unique part might be planned out and setup in at least 50 different ways—each offering some advantage and disadvantage. After pondering dozens more of the skills, the committee and I felt these were the essentials. They are the ones you need for the manual machines but transfer onward to the CNC.

Questions and Problems

1. There are two types of plunging. Name and describe them. Which is the better entry process? (Obj. 12-1)

2. The table tee slots are used for two purposes—identify both and list the items that are used in the slots. (Objs. 12-2 and 12-5)

3. What vertical and horizontal milling machine component is held on the column dovetail? (Obj. 12-2)

4. The feed rate for mills is expressed in inches per minute. Is this statement true or false? If it is false, what will make it true? (Objs. 12-2 and 12-5)

5. Tools, collets, and milling machine arbors are held in the spindle with what two devises working together? (Obj. 12-3)

6. Although there are others used in industry, name the two common milling machine tapers shown in Fig. 12-130. (Obj. 12-3)

7. Describe four necessary table clamping practices. Be complete. (Obj. 12-4)

8. What is a normal feed rate range for a student machinist? (Obj. 12-5)

9. When a pocket is milled within the bottom of another pocket, it is called a normal pocket. Is this statement true or false? If it is false, what will make it true? (Obj. 12-1)

10. Briefly describe the advantages of *climb milling*. (Obj. 12-5)

11. What is dangerous about climb milling? (Obj. 12-5)

12. On a sheet of paper, draw a sketch of conventional and climb milling. Be sure to include the nature of the chip formation. (Obj. 12-5)

13. When is conventional milling acceptable and when is it necessary? (Obj. 12-5)

14. Compute the speed and feed rates for the following job.
    Use *Machinery's Handbook* for the feed rate and Appendix IV for the RPM
    Cutting cast aluminum—as cast
    Using a $\frac{3}{4}$-in.-diameter end mill with four teeth
    Depth of cut is 0.250

15. Identify the tools and holders used on milling machines, as shown in Fig. 12-131. (Objs. 12-1 and 12-3)

A   B
C   D

Figure 12-131 *Identify these tools and holders.*

Figure 12-130 *Identify these two mill spindle tapers.*

16. See Fig. 12-132. You need to machine (8), $\frac{1}{4}$-in.-thick aluminum spacers to 4.000 in., removing

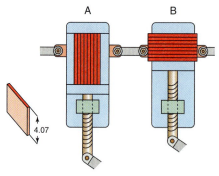

Figure 12-132 *Which vise load is smart? Which is not?*

0.070 in. on one side only. You've discovered that they will fit in the mill vise either way (A) or (B). One way is safe the other dangerous. Explain.

17. Describe how you could use an end mill to locate a spindle over the edge of an object. In doing so, what would be the major challenge to accuracy? How does an edge-finder overcome this problem?

18. How could you go about aligning a spindle over a hole that was not round? There are two ways. Describe each and the geometric result of each. This is a tough question.

CNC Questions

19. In Fig. 12-133, identify the *X*, *Y*, and *Z* axes.

20. How is it possible for a CNC mill to make a spiral ramp? Describe the axis moves.

Figure 12-133 *Identify the three linear axes on this CNC vertical mill.*

Section 12 Answers

ANSWERS 12-1

1. Facing, peripheral cutting, pocket milling, step milling, and cutting angles (setup dependent or a form ground cutter); radius cutting uses a modified end mill.
2. False. Cutters must be relieved to plunge, not nest.
3. A. Bull nose end mill (radiused corner end mill)—Form inside corner fillet radii while cutting steps and pockets.
   B. Fly cutter—Large flat surfaces, light machining. Can be used to form cut with shaped bit.
   C. Four-flute end mill—face, profile, step but usually not plunge unless specifically ground to do so.
   D. Face mill—Cut large flat areas; heavy machining.
   E. Corner radius end mill—form radii on outside corners.
   F. Two-flute end mill—profile, plunge, and face.
4. Face mill, end mill, and fly cutter.
5. Straddle milling (or gang milling).
6. To gain better hole location accuracy.
7. Plunging is straight, single-axis drilling using an end mill. Ramping is a multiaxis move that creates a tapered entry slot that allows chips to escape and coolants to enter.
8. Indexing or spacing (either term)—the pattern is often called a bolt circle.
9. The datum position had to be set at zero on the micrometer dials and/or digital positioner.
10. The boring bar machines the hole to the exact location of the spindle. Boring can create a cylinder of any size; drilling and reaming cannot.
11. A. Angle—with a face mill or end mill—tilt the part or tilt the cutter.
    B. Pocket with $\frac{1}{2}$-in.-diameter end mill—plunge to depth.
    C. Outside corner radius (not a fillet)—using form-ground end mill or fly cutter.
    D. Slot—bull nose end mill to cut near side and square corner end mill on the other side or a wheel cutter with radius ground onto left side and no radius on right.

ANSWERS 12-3

1. It denotes the size of the inscribed circle, in $\frac{1}{8}$'s of an inch.
2. TNMG-432: T means triangular; N means neutral clearance 0°, insert tips to create clearance, makes it negative rake but both sides are usable; M means moderate repeatabilty; G means a straight hole with double-sided cutting edges; 4 indicates $\frac{4}{8} = \frac{1}{2}$-in. inscribed circle; 3 means $\frac{3}{16}$ in. thick; 2 means $\frac{1}{32}$ in. corner radius.
3. DPMH-631: D means 55° diamond; P means 11° relief; M indicates 0.002 to 0.004 in. tolerance on repeatability; H means countersunk hole, chip groove one side only; 6 means $\frac{3}{4}$ in. IC ($\frac{6}{8}$); 3 means $\frac{3}{16}$-in. thick; 1 indicates $\frac{1}{64}$ in. corner radius.
4. SS = 300, RPM = 300
5. SS = 250, RPM = 4,000 (not possible on many mills).
6. SS = 200, RPM = 800
7. Transmission selection, step pulley, and variable speed drive.

ANSWERS 12-4

1. Bolt closer to the work than to the packing—puts pressure on work not packing; slight tilt to clamp; packing slightly higher than the work; shim under clamp to protect work; tee nut and a solid washer.
2. To avoid machining the machine table; a washer and a protective shim are missing.
3. Place *parallels* under both edges, and use a solid block, often called a *filler block,* between the

Filler block

Close vise against filler

Matched parallel bars

Figure 12-134 *By placing parallel bars under the object to clear the drill bit, then a filler block to push it against the solid vise jaw, this oddly shaped part can be machined.*

two protrusions (Fig. 12-134). This distributes the force without damaging the work.

ANSWERS 12-5

1. *Face cutting,* for datum edge A and the opposite edge, *peripheral cutting,* for both ends, probably climbing but conventional would work.
2. Snug the *Y* axis if the cut is climbing to stop self-feeding (*X* and *Z* and quill are fully locked).
3. Climb milling produces a better finish due to the cutting edge shearing out the thin chip as it exits the metal. It gives longer tool life due to not rubbing the cutting edge.

Answers to Section Review Questions

1. *Straight plunging,* drilling straight down; *ramp plunging,* drilling down while moving sideways, creates a slot into which coolant enters and chips come out.
2. To hold back on tee nuts for clamping and for accessory and work alignment using slot blocks and vise keys
3. Knee
4. True
5. A draw bar pulling a taper into the spindle hole
6. An R-8 on the left and an ISO standard (American Standard) milling machine on the right
7. A. The bolt should be as close as possible to the object to be clamped, not to the packing block.
   B. Place a soft shim under the clamp nose to protect the work.

4. *Good,* it lengthens the chip thus reducing the hammering; *bad,* it transfers some of the machining force against the solid jaw to a sideways action rather than against the solid jaw of the vise.
5. *Loaded cutter,* use coolant and/or slow the RPM; *dull cutter,* obvious; *excessive feed rate,* obvious; *extra long cutter,* obvious; *part moved in setup,* check and recheck; *climb milling without a brake or other axes weren't locked,* see Unit 12-5 for the solution.
6. Climb milling makes a better finish as you will soon see, but it will also aggressively pull the part along, which can get out of control. Lock all unused axes and set a drag or backlash compensator on the moving axis.

4. False. We move the cutter or spindle to a known position relative to the work (the intersection of two edges or the axis of a hole) then set that position to be zero.
5. 2,133 RPM (short formula) $0.007 \times 2 \times 2,133 = 29.86$ IPM.

C. Never have the tail of the clamp below the front—this places all the clamp pressure on the edge of the work causing damage, tending to tilt the part and making it unstable.
D. Always use a strong washer between nut and clamp.
8. From below 1.0 IPM up to around 20 IPM (while learning).
9. False—a nested pocket.
10. Climb milling produces a better finish because of the way the tooth shears the chip off the metal—inside to out. Climb milling also produces longer tool life.
11. Climb milling can self-feed due to the agreement of the cutting force and the feed direction.
12. Compare your sketch to Fig. 12-92.

13. Conventional cutting is acceptable when reducing soft metal such as aluminum. It's necessary when the work has a hard crust and soft interior or when the setup is not rigid enough to withstand climb milling forces.

14. Using the short RPM formula and a chip load of 0.004 in. per tooth.

$$RPM = \frac{250 \times 4}{0.75} = 1,333$$

Feed rate $0.004 \times 4 \times 1,333 = 21.3$ IPM (a moderately fast feed).

15. A. R-8 taper with micrometer boring head
B. Standard mill taper stub arbor with wheel cutter
C. Quick change tool holder, chuck, and drill bit
D. Collet end mill holder with standard taper

16. Vise B is safe, each part is being squeezed with an equal amount of pressure. Vise A will put holding pressure only on the longest one or possibly two parts. The rest might not be held at all unless they are all perfectly the same width—an impossibility!

17. By just bringing the end mill to the edge until it is tangent. That position is the radius distance off the edge. Note, this is known as touching off in CNC work alignment. The problem is runout in the holder and end mill. The edge-finder works independent of the tool's runout.

18. There are two. The easy method is to put a test pin in the hole that just slips in thus accounting for hole irregularities. Then indicate the pin not the hole. This would place your spindle over the largest cylindrical space available.

The other method is to indicate the hole by balancing out irregularities similar to indicating nonround object in a four-jaw chuck on the lathe. This would put your spindle over the average volume of space, which could very well be a different location than the pin (Fig. 12-135). In Fig. 12-134, the hole has a flat spot on the right and a bulge on the left. You can see that balancing out readings on the DTI would position the spindle to the left (red centerline) of the location in which the largest pin would reside (blue centerline).

19. X to the right and left; Y in and out; Z up and down.

20. The computer moves the X and Y simultaneously in a circle while moving the Z straight down. We'll learn that a simultaneous X-Y circle is called *circular interpolation*.

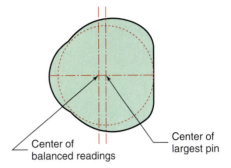

Center of balanced readings

Center of largest pin

Figure 12-135 *The different location methods yield different results geometrically.*

Section 13

Precision Grinding Operations and Machines

Introduction

Figure 13-1 *Mirror finishes require vibration-free spindles, the right grinding wheel (Unit 13-1), balanced and dressed right (13-3), and good setup parameters (13-4).*

Figure 13-2 *This spray of sparks is actually microchips elevated above 2,500°F and reacting with room oxygen.*

Precision grinding is an operation that usually follows turning, drilling, and milling. It improves surface finish, and accuracy, plus it shapes metal that's hardened beyond the capability of standard cutters. Grinding produces finishes as smooth as 16 μIn with repeatable tolerances near 0.0001 in. (0.003 mm). However, given experience and quality machinery, far finer results are possible (Fig. 13-1).

Removing metal with a grinding wheel seems different from cutting it away; it's even called an abrasive process. The Latin root, *abrade* means to scratch or wear away, not cut. The leftover *slurry* produced by grinding doesn't look anything like chips. It's a dust, or mud if coolant is used.

But examined closely, the waste particles reveal microscopic chips along with a small amount of broken abrasive grains and bonding agent torn from the wheel. Before those grains broke away, they were actually tiny cutters, each with its sharp edge, making one chip per revolution as it whizzed by the work surface. This means that the physics of chip making applies here too. For example, each grain is a negative rake cutter producing the usual external and internal friction and heat. But in this case, the heat is so great and the chips so well supplied with room oxygen as they're flung out of the grinding wheel, that metals that oxidize ignite, resulting in sparks (Fig. 13-2).

The work heats too. The solution is familiar: add coolant and choose the right grinding wheel for the task. Stopping to resharpen the wheel before it's too dull is another heat control. So, in grinding work, coolants perform the same antifriction and heat conduction services they do elsewhere in machining. But with grinding, they also lubricate the chip and fling it out of the wheel.

Different from drilling, turning, and milling, grinding isn't performed on a single machine. Every machining process has its specialized machine that grinds. For example, cylindrical grinding machines take on some of the lathe's capabilities, producing internal and external round surfaces, tapers, and contoured shapes. But the manual

Figure 13-3 *This third-generation cylindrical grinder is a true high-speed machine. Using advanced abrasives and CNC control, it can rough turn, face, and make finished contours as well as redress it's own wheel, all at removal rates unheard of in the past.*

versions can't cut threads, so the thread grinding machine assumes that task. On the other hand, CNC cylindrical grinders can cut threads. Jig grinding machines parallel the vertical mill or the jig borer's duties, precisely locating and finishing nearly perfect holes. They use small-diameter grinding wheels mounted on steel mandrels. Due to their small diameters, those wheels are driven by air turbines screaming up at an amazing 80,000 to 150,000 RPM and higher!

In the past, grinders lagged behind in computerization, but today every grinding machine is supplied in a programmable form. Tool and cutter grinders and gear grinders are leading the race, but high-speed CNC cylindrical machines are now in service that rough cut in steps just like a lathe, dress their wheel, then make final cuts to shoulders, tapers, and contours. Their manual counterparts require time-consuming setups when going from one operation to another, but this is not so on the computerized versions (Fig. 13-3).

We're here not to learn how to run those machines but rather to understand how a grinding wheel removes metal, then to practice solid setups on a surface grinder.

Objectives

UNIT 13-1

Selecting the Right Grinding Wheel for the Job

Introduction: Just like changing cutting tools to achieve a different result, different grinding wheels return varying rates of removal, finishes, and tool life. Beyond the simple bench grinder wheels we've discussed previously, there are a lot of variations in the abrasives, the way they are held together, and how the abrasive grains and bonding agents are structured in the wheel.

Adding CNC Control Grinding wheel selection is an example of how much CNC has changed the way we plan and set up jobs. When working on manually operated equipment, wheel dressing robs production time to stop and redress the wheel; but we must do it for accuracy and efficiency. This is especially so when form grinding, where the wheel is shaped as a counterpart of the desired contour

Figure 13-5 *Manually dressed form grinding wheels must be the perfect counterpart to the desired shape.*

(Figs. 13-4 and 13-5). But depending on the process chosen, the wheel dress can take nearly as long as the grinding itself!

So, for efficiency on the manually operated grinder, the wheel selected should hold its shape for the longest period of time. But in choosing a hard wheel for shape holding, we trade away the ability to remove larger amounts of metal—we'll see why in just a bit. That means the work has to be milled very close to the finished shape before grinding, due to the low removal rate of the finishing wheel, usually within 0.003 to 0.005 in. of the desired size. So, pregrind machining is a costly, time-consuming operation itself.

But CNC equipment dresses itself automatically and quickly. The dress time factor becomes far less critical (Fig. 13-6). A fast-cutting, softer wheel will work more efficiently without creating heat, so the extra cost of very close tolerance machining can be dropped, and more excess can be left on the work to grind away. And, even though the wheel will need dressing more often, it's not a problem. It can be scheduled into the program as often as needed.

2-D Compensation for Wheel Shape But beyond faster cycle times, there are three more advantages gained with CNC control. First and foremost, we no longer need to dress the wheel into the exact shape. Using the CNC surface grinder shown in Fig. 13-7, the grinding wheel corners need to be radiused only with a known arc (Fig. 13-8). Then by keeping that radius tangent to the

Figure 13-4 *The wheel guard has been removed to show a typical form grind operation using a shaped wheel.*

work surface, a complex shape can be ground. Also, that nonspecialized wheel can be used in many other ways—as is. In contrast, manual grinder shaped wheels must be completely reshaped to grind another job or they must be put on a shelf to await a similar part!

3-D Compensation The final ability brought by CPU driven machines is to produce three-dimensional ground shapes. By controlling all three axes simultaneously, contours like that shown in Fig. 13-9 are possible. That's something we could not do previously.

Extreme Safety Planning—No Kidding! The theme throughout grinding must be based on the orbital speeds of the grinding wheels. They range from a bottom line of 4,000 feet per minute up to 12,000 feet per minute or more! That's bullet speed! So, while these speeds are right for the way a grinding wheel cuts metal, they also dictate the speed of an accident! Need proof? Go look at the left scatter shield on your lab's surface grinder—undoubtedly my case will be proven by chipped paint, craters, and dents!

Common grinding accidents occur more often due to poor setups that do not anticipate the dynamic forces and heat of grinding. Fewer are caused by incorrect operator actions at the wrong time.

Review We're going to learn a ton of new facts about abrasives and grinding wheels, but before we do, do you recall these facts?

◆ For normal day-to-day grinding, identify the two common types of abrasives.

◆ What are the two methods of bonding them together into grinding wheels?

Answer *Aluminum oxide* is the common, economical abrasive for everyday duties, and *silicone carbide* is used for harder work like grinding carbide cutting tools. You may have also experienced man-made or natural diamond abrasives to grind extra hard materials or for finishing work finer than 32 μIn.

The bonds are *vitrified,* clay fired to the hardness of ceramic, and *resinoid,* the low-heat process of gluing grains together with flexible plastic compounds and sometimes reinforcing them with fibers.

Figure 13-6 *A CNC grinder dresses its own wheel faster and more accurately.*

Figure 13-7 *A CNC grinder can perform many tasks more efficiently and economically due to wheel selection options.*

KEY TERMS

Alumna
The semi-refined version of bauxite, aluminum ore, from which aluminum oxide abrasive is

CNC-Dressed Wheels

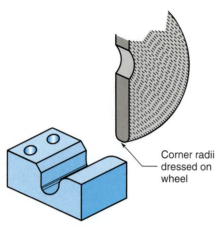

Corner radii dressed on wheel

Figure 13-8 *CNC grinders contour the work by keeping the known wheel radii tangent to the desired work shape.*

Three-dimensional contours possible only with CNC

Figure 13-9 *By controlling all three axes simultaneously, CNC grinders can create 3-D shapes.*

Similar to cutting tool geometry and selection, abrasives and their application are a set of very technical facts that are evolving quickly. Machinists don't always agree on what works best. Creativity comes into play in this arena.

For example, Appendix V in this book offers a handy wheel recommendation chart for different applications, supplied by the Norton Company (a leader in abrasives and grinding wheel manufacturing). Wheel applications can also be found in *Machinery's Handbook*©. Not surprisingly, you'll find differences between these two definitive sources. Today's journeyman stays informed about evolving grinding products and applications through manufacturer's catalogs, websites, trade shows, magazines, and published engineering data distributed by sales people.

made or further refining produces the metal aluminum.

Bond (wheel)
The material that glues the abrasive grains together.

Boron nitride (cubic boron nitride)
A man-made abrasive next to diamond in hardness. It's crystal structure looks cube shaped. It tends to stay sharp and to fracture well.

Friability
The ability to shear away small flakes to self-refresh an abrasive's sharp edge before it shears out of the bond.

Glazing
A grinding wheel condition where the grains have not self-dressed, yet they have no sharp edges left. The wheel must be dressed.

Grade (grinding wheel)
A relative index of the hardness of the bond not the abrasive—the hardness of the wheel.

Grit size
The number of screen squares within a square inch. The larger the number is, the finer the grain size.

Post (grinding wheel)
The column of bond material linking individual grains.

Structure (grinding wheel)
An index of the relative amount of porosity in a grinding wheel relative to grit and bond. Structure runs from 1 = dense to 16 = very open.

Self-dress (self-refresh)
New sharp edges are created by either the dull grain fracturing from pressure or the bond letting go.

Truing (grinding wheel)
Dressing the wheel to reshape it and to remove runout.

Five Factors for Grinding Wheel Selection

Thus far, we've learned that wheels are changed after prolonged use and dressing that makes their diameter too small for efficient rim speeds or when they are possibly fractured from abuse.

KEYPOINT

Due to the cost of maintaining an inventory of different grinding wheels, premounted and balanced on quick change hubs, a school budget may not allow many wheel change options, but in industry it's expected that the machinist include correct wheel selection during the setup of most precision grinders.

For example, one wheel is best for roughing, while another lasts longer for finish work, or one type is good for hard steel but another is best for bronze or aluminum. One removes metal quickly, but a second creates a 16 μIn finish. Sound familiar? It's the trade-off game based on skillful use of knowledge and experience.

To assist in identifying the wheel's makeup, we use the industry standard ANSI code to tell us several vital parameters about it. It's found on the paper gasket glued to the wheel, or may be stenciled directly on the

Figure 13-10 *An identification number similar to this will be stenciled on the wheel or printed on its paper gasket.*

wheel. There will probably be six or seven columns, separated by dashes or spaces (Fig. 13-10). However, the first and last columns are manufacturer/product specific, so we'll investigate the five universal data in the order in which they are found. These are your control factors for all grinding wheels, no matter where they are used, and there's a lot of technology behind them.

Abrasive type   Grit size   Bond hardness structure   Bond type

KEYPOINT

The culmination of this unit is a cause-and-effect chart. It's there to begin grinding wheel selection skills.

Decision Factors for Wheel Choice

Choosing the right grinding wheel for the job will be based on

1. *Amount to remove*—Is it a lot (over 0.015 in.) or very little metal?
2. *Metal and hardness*
3. *Required finish*
4. *Batch size*
5. *Cycle times and tool life*

For each of the five wheel characteristics, I'll lay out a simplified baseline for correct application. See the Shop Talk.

On the practical side, unless your shop specializes in technical grinding, where wheel choices can be limitless, most shops don't stock a huge inventory of different kinds of grinding wheels. Your options will normally be two or three kinds of abrasives at the most,

in several grits and structures. However, when those universal wheels don't deliver, then the following information must be clearly understood.

Abrasive Types—Column 1 (not considering the prefix column) (Fig. 13-11)

There are four basic kinds of abrasives in use today. The first three are produced by chemical reactions between two elements when elevated to high temperatures in an electric furnace:

1. **Aluminum oxide (A)**
2. **Silicon carbide, (C)**   (Also called carborundum, an early brand name.)
3. **Boron nitride (B) or cubic boron nitride (CBN)**   It is called cubic due to its box-shaped crystal structure.

and

4. **Diamond (D)**   Both natural and the man-made (synthetic) abrasives are used.

Differences in hardness and crystal structure (which contributes to its ability to fracture away in small flakes—a desirable quality), overall grain strength, and resistance to edge dulling make each abrasive type best suited for a particular application.

Man-Made Diamonds Are a Machinist's Best Friend
They are created by super high compression of carbonaceous graphite in a high-heat environment. The man-made version is five times stronger than the naturally occurring form due to its uniform microstructure.

KEYPOINT

Similar to cutting tools differences, the main abrasive property is hardness but with extra abrasive hardness, purchase price also generally goes up. Machinists soon learn when grinding difficult materials that the least expensive may or may not be the economical choice for the job.

Super Abrasives

Most advanced abrasives designed for specialized tasks are subcategories of these four. But there are also combinations of two or more along with ceramic/metallic compounds. For example, if we add an oxide of the metal zirconium to the aluminum oxide grain mix, the wheel gains efficiency in cut speed and grain durability. The ratio of zirconium oxide to aluminum oxide determines specific applications, but in general all combinations are faster cutting than plain aluminum oxide

Standard Marking System Chart

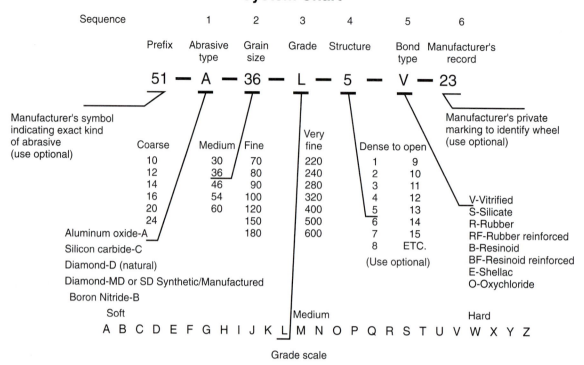

Sequence		1	2	3	4	5	6
	Prefix	Abrasive type	Grain size	Grade	Structure	Bond type	Manufacturer's record
	51 —	A —	36 —	L —	5 —	V —	23

Manufacturer's symbol indicating exact kind of abrasive (use optional)

Manufacturer's private marking to identify wheel (use optional)

Coarse	Medium	Fine	Very fine	Dense to open	
10	30	70	220	1	9
12	36	80	240	2	10
14	46	90	280	3	11
16	54	100	320	4	12
20	60	120	400	5	13
24		150	500	6	14
		180	600	7	15
				8	ETC.
				(Use optional)	

Aluminum oxide-A
Silicon carbide-C
Diamond-D (natural)
Diamond-MD or SD Synthetic/Manufactured
Boron Nitride-B

V-Vitrified
S-Silicate
R-Rubber
RF-Rubber reinforced
B-Resinoid
BF-Resinoid reinforced
E-Shellac
O-Oxychloride

Soft   Medium   Hard
A B C D E F G H I J K L M N O P Q R S T U V W X Y Z

Grade scale

Figure 13-11 *The ANSI numbering system for grinding wheel selection features seven columns of information.*

abrasives. These abrasives are beyond the scope of this unit. Learning to apply the other four will provide a good working knowledge.

Abrasive Friability

One valuable abrasive property is it's ability to stay sharp and to refresh it's own sharpness between dressings. An abrasive that has good friability allows it to shear away tiny flakes, many times before the whole grain breaks completely out of the wheel. For example, the more modern versions of boron nitride produce flakes measured in millionths of an inch! This is known as the **self-dress** or **self-refresh** friability of the grain (Fig. 13-12). The wheel outlasts other abrasives several times over. But its life also depends on its hardness and how long each sharp edge lasts. We'll start with the softest abrasive and work toward diamond.

Aluminum Oxide

The workhorse of abrasive grains, aluminum oxide is made by combining **alumna** (semirefined aluminum ore) with oxygen at very

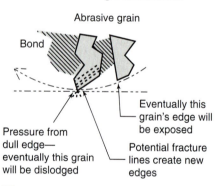

Self-Refreshing Characteristic

Abrasive grain

Bond

Eventually this grain's edge will be exposed

Pressure from dull edge— eventually this grain will be dislodged

Potential fracture lines create new edges

Figure 13-12 *As one grain dulls it either fractures to create a new edge or the bond shears to expose another grain behind.*

high temperatures. It's the universal choice, providing good economical results over the widest range of work materials. Aluminum oxide stays sharp when grinding high-tensile alloy and tool steels. The chances are most of the grinding wheels in your lab are aluminum oxide.

Purity and Hardness There is a range of hardness **grades** for aluminum oxide, depending on the purity of the process by which it's made. Purity relates to hardness. As

Figure 13-13 *A variety of abrasive wheels used in manufacturing. Not all are for precision grinding.*

the grain's hardness goes up, its edge lasts longer but overall toughness goes down.

Dark gray 94 percent pure This is the tougher version. But its sharp edge doesn't last as long, especially when grinding hard metals and cast iron. While it's the abrasive of choice for pedestal grinder wheels due to it's toughness, this abrasive is not used on precision grinders. It is shown in the center of Fig. 13-13 (yellow label).

Correct Application
- Good results when grinding plain and medium hard alloy steels.
- Not recommended for cast iron because aided by the wheel's impurities, the aluminum bonds to the iron and the wheel dulls quickly.
- Can grind some nonferrous metals—not the best choice. However, when doing so, wax or oil helps chips release from the wheel when it has cleared the cut.

Light gray 94 to 97 percent purity See the top yellow and orange label in Fig. 13-13. This is the harder version but is more costly. This popular abrasive represents an estimated 50 percent of all wheels on surface grinders and cylindrical grinding machines.

Correct Application
- Good for soft through hard alloy steels up to fully hardened tool steel.
- Acceptable for most other metals—will grind cast iron but not the best choice.

White (or stained) 97 percent See Fig. 13-13 (upper right corner) and 13-1,13-2, and 13-4. The purest form of aluminum oxide is bright white, but it's often stained a bright color to distinguish it from lower cost grinding wheels.

Correct Application
- Good results on extra hard tool and alloy steels.
- Can grind softer grades of carbide, but not the best choice.
- Best for general tool and cuter grinding.

Silicone Carbide

This material is made by combining common silica sand with coke (refined carbonaceous coal) in furnaces purged of oxygen. The result is an abrasive that's about 33 to 50 percent harder than aluminum oxide. Silicone carbide's hardness combined with a tough but friable crystal structure means that it creates sharper cutting edges than aluminum oxide, thus it cuts faster. But the crystals tend to shear in bigger flakes, thus grains shear completely from the wheel sooner.

Observed closely, grains of silicone carbide appear as a deep-colored background reflecting rainbow colors. There are varying hues of this abrasive running from black through green—each with slightly different properties. You've seen silicone carbide on high-quality wet or dry sandpaper and cloth—the deep black abrasive paper with a sheen.

Correct Application
Black and Gray SC (Fig. 13-13—blue/white label)
- Excellent for cast iron.
- Good over a wide range of steels, but the expense isn't usually justified over aluminum oxide.
- Good for grinding aluminum, copper brass, and bronzes.
- Acceptable results grinding some stainless steels and titanium.
- Acceptable results on tungsten carbide tools.

Green SC (Not shown)
- Preferred for grinding tungsten carbide cutting tools.
  See the Trade Tip.

KEYPOINT

Green wheels are usually reserved for *nothing but carbide tools*. They are costly by comparison to aluminum oxide bench wheels, and due to the abrasive and binder's properties, they wear away too fast for everyday use.

Boron Nitride

Boron, a mineral somewhat like silicone, is combined with nitrogen to form an abrasive that's twice the hardness of aluminum oxide and the closest man-made abrasive to diamond hardness. The process by which it's created is costly, however, boron nitride's edge-holding ability cuts chips extremely well compared to the two previous choices. Since it's best used for extra tough grinding jobs, the cost is often balanced against efficiency.

KEYPOINT

While it's a superior abrasive, in general, due to cost, we do not use boron nitride unless aluminum oxide or silicone carbide cannot do the job. However, its extreme hardness, fast yet cooler cutting action, along with less dressing of the wheel sometimes combine to make it the best choice on a tough job.

Correct Application

◆ Best for form grinding of extremely hard metals.
◆ Excellent on very hard metals and other materials such as glass.
◆ Performs well on carbide tools but expensive if SC can do the job.
◆ Excellent on high nickel/cobalt superalloys.
◆ Excellent for long abrasive life and high-demand applications such as CNC high-speed grinding.

Figure 13-14 *A diamond wheel is ideal for finishing these slots.*

Diamond Abrasive (D) Natural (MD) Manufactured (SD) Synthetic (Fig. 13-14)

Now were talking expensive! However, there are some tasks that only diamond can do. Its greatest advantage beyond hardness is that diamond cutting edges tend to remain sharp longer than all other abrasives, thus it cuts with a great deal less heat. That property alone makes it the right choice for fine tool and cutter grinding where we cannot risk overheating thin cutting edges.

Correct Application

◆ Excellent, fine finish on carbide tools.
◆ The only choice for grinding exceptionally hard materials such as ceramic or glass.
◆ Excellent for hardest grades of tungsten steel tools.
◆ Best where heat sensitive work may harden—diamond cuts with less friction and heat.
◆ Excellent for fine finishing.

Grit (Grain) Size—Column 2

Grit size is a simple issue but an important one for success. In general, the finer the abrasive, the finer the finish it will produce since it makes smaller cuts in the workpiece. But with small abrasive size, removal speed goes down and heat becomes a greater issue as the tiny cutting edges become dull. But there's a

Relative Grain Sizes

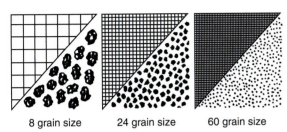

8 grain size   24 grain size   60 grain size

Figure 13-15 *Grain size is based on the number of squares per inch, through which a single grain can pass.*

Wheel Grades

Weak "posts"   Medium-strength "posts"   Strong "posts"

Figure 13-16 *Wheel grade (hardness) is directly related to the wheel bond.*

unique solution to these challenges found in the way the grains are bonded together—called the wheel **structure.** We'll look at that after bonds.

For now, let's get an initial feel for grit sizes. You've already seen grits specified as coarse (18 to 24 grit), medium (26–46), and fine (46–80), when selecting pedestal grinding wheels. That means that after crushing the large abrasive crystals down to small grains, they are sifted through a grid with the specified *number of spaces per square inch*—also called *screens* (Fig. 13-15).

> **KEYPOINT**
>
> Note in the application chart, that the three basic grinding machines are listed in an order. Each uses a slightly finer grain for it's general duties. (There are some exceptions to this guideline.)

Grade—The Hardness of the Bond—Column 3

In selecting the right wheel for the job, it's important to consider wheel hardness expressed as a letter grade, with "*A*" being the softest grade. Wheel hardness is a combination of the ratio of **bond** relative to abrasive grains and the air spaces between them and the hardness of bonding agent itself. The bonding material holding the grains in the wheel forms links between grains called **posts** (Fig. 13-16).

Varying the amount and hardness of the posts causes grinding wheels to perform differently, even though the abrasive grain hardness is kept a constant. More bonding material tends to hold the grains longer and stronger—a harder wheel. Sometimes a harder wheel is just right, when form grinding, for example, but it's not always a good thing. When grinding harder work materials, for example, grain edges dull more quickly. In these situations, it's best if the grains release sooner. If not, a condition known as **glazing** can occur.

A glazed wheel is one that has lost its cutting ability. It has been used long beyond the point where it should have been dressed. When this is done, the heat and friction create a glass-like surface that rubs and does not abrade away material. A glazed wheel will reflect light and is easily recognized. If not dressed to clear the problem, it burns the material rather than cutting it. Wheels glaze quickly if they are poorly chosen for grinding hard alloys.

> **KEYPOINT**
>
> Wheel glazing can be blamed on (1) a wheel grade that's too hard, (2) an abrasive that's too soft, (3) using it beyond the point where it should have been dressed (most likely), and (4) choosing grinder feed rates too low to cause the self-dress (not a common reason).

Self-Dress Property

Working together, the bond hardness (grade) and the abrasive's ability to fracture continuously refresh the wheel's ability to cut cleanly, between dressings. But only up to a point where it must be redressed to restore wheel surface runout and shape due to random release of grains. The time between dressing the wheel is a function of several factors:

◆ Work material type and hardness

◆ Amount of material to be removed

◆ Type of grinding operation

◆ Coolant application

◆ Wheel grade

◆ Abrasive type

Knowing when to stop grinding and dress the wheel is an essential skill learned from experience.

Grinding wheel self-refreshing action is both good and bad. While it renews the cutting action for a limited time, it also means the wheel is eroding away. So, how much self-dress is right for the job? How hard should the wheel bond be?

KEYPOINT

When grinding a hard metal, *use a* soft wheel, *and when grinding a* soft metal, *use a* harder wheel.

In general, softer bonds are usually needed for harder material to cause more self-dressing, but abrasive hardness also affects this choice. Softer materials grind more economically with a harder bonded wheel.

Grades From A (soft) to Z

Bond hardness varies from very soft "A" through very hard "Z." Within a span of perhaps two or three grades, wheels deliver nearly the same result. For example, a wheel of F bond hardness will perform similar to an H bond.

Correct Application—Bond Hardness

Bond Grade	Application
A through F	Carbide and extremely hard tool steel, glass
G through P	Large contact area processes Rapid material removal General hardened alloy steels (85 percent of all jobs)
Q through S	Nonferrous materials— CI with crust removed
	Small narrow contact areas where wheel is fragile
	High-force-demand applications
T through Z	Nonferrous, very soft materials, extreme RPM

Harder bond wheels are also chosen for final grinding operations, where very little metal is to be removed or for form grinding where the wheel is dressed into a special contour shape. Here, it is desirable that the wheel hold that shape longer so the self-dress is less important—assuming, of course, that it's a manually dressed machine.

KEYPOINT

Wheel hardness is a combination of bond material and amount of bond within the ratio—it has nothing to do with abrasive hardness.

Wheel Structure—Column 4

Open space between grains also has a lot to do with wheel performance. A wheel is truly composed of abrasive grains, bond material, and *empty space*. Figure 13-16 also shows the effect of wheel structure. Space is necessary in every wheel for three reasons:

1. **A Temporary Place for the Chip to Form**
   When the grinding wheel is in contact with the work, the forming chip must have someplace to stay until the wheel clears the work, to eject it.

2. **Create Coolant Induction Channels**
   Properly directed at the wheel/work interface, coolant is injected into the air space just before it contacts the work. The coolant is trapped in the air space as the wheel rotates onto the work. Thus, it is carried into the action area right where the chip is forming. Then on leaving the cut, centrifugal force ejects the coolant and chip.

3. Cools the Wheel

During noncontact time, while the wheel rolls over for another cut, air swirls in the space to cool individual grains. This is an especially important property when coolant is not possible and room air carried in and out of these spaces becomes the wheel's coolant. So, a dry setup might warrant a more open structure wheel compared to one where coolant will be injected.

Structures from 1, Dense, though 16, Open (more space in the ratio)

Within a range of perhaps two or three units, wheels perform similarly.

Correct Applications of Structure

1 through 5	Regrinding cutters, form grinding, removing less than 0.010 in. overall material from the workpiece
5 through 10	General grinding operations, removing from 0.010 to 0.020 in. overall
10 through 15	Rough grinding, or where heat is a problem, removing above 0.020 in.

Porous Wheels

The overall ratio of spaces compared to solids (grains and bond) can be created by larger individual air space holes or by multiple smaller holes between the grains. If a wheel is deliberately created with large, singular pores, its designation will end in a letter P. That's a wheel expressly made for high-volume, rough grinding.

A-48-G-12-V-P

Bond Types—Column 5

There are three basic bonds used in conventional precision grinding wheels, with variations within each:

Vitrified   Resinoid   Metallic

We've previously discussed the first two, so that leaves the metallic nickel-copper-based alloys used to embed grains of diamond, CBN, or other expensive crystal grains onto tool steel disks, as shown on the diamond wheel in Fig. 13-14.

They are precisely assembled such that approximately $\frac{1}{3}$ of the grain's height is above the bonding agent and all are equally spaced, creating a highly aggressive and cool cutting wheel (Fig. 13-17). These wheels are made by electrodeposition (depositing metal with electricity) of the bond metal around the grains.

A Metallic Bonded Face Grinding Wheel

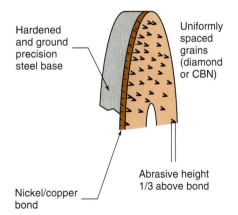

Figure 13-17 *A metallic bonded, diamond, or CBN face wheel cuts very quickly and creates less heat than other types.*

Correct Application of Bonds

Vitrified

- Used where coolants or oils are used—glass-like material is unaffected by water, oil, and most other chemicals.
- For general purpose grinding where exceptional mechanical shock is not going to occur.
- Can withstand exceptional heat.
- Used where rim speeds do not exceed 6,500 F/M.

Resinoid

- Used where high-speed CNC grinding demands rim speeds above 12,000 F/M.*
- Used where mechanical shock may be a factor.*
- Not recommended for use around certain solvents and acids.

Metallic Bonds

- For extreme long life where dressing is impractical.
- To hold expensive grains firmly for a long time.

- For very high speed grinding.
- For certain electrical and chemical applications where the wheel must conduct electricity.

*Requires fiber reinforcement for this duty.

Control Factors

The following is a compilation of basic problems associated with grinding, the target goal for improving them, and the action needed to do so. But as usual, something gained often means something lost—most adjustments other than adding coolant are trades in time or cost or some other downside element lost. For example, changing from aluminum oxide to silicone carbide then to boron nitride results in improved production but costs rise.

Memorizing these facts is not the objective. Understanding why each works is your goal. Note that when a sharper abrasive is indicated, it means a harder abrasive that holds its edge better.

Symptom	Target	Action	Trade-Off
Slow cutting	Increase production	Coarser grain	Rougher finish
		Harder abrasive	Wheel cost
		Softer bond	Increases wheel wear
		More open structure	Increases wheel wear
Dressing too often	Increase production	Harder bond	Glazing/loading
	Less down time	Harder abrasive	Expensive wheel
		Increase cutting	More force against work
		Pressure-feed	Self-dresses better but requires strong setup
		Add coolant	None
Poor finish	Improve quality	Finer grain	Heat buildup
		Dress more often	Production time
		Harder bond	Possible glazing
		Rebalance wheel	Down time
		Slow cutting rate	Extra time
		Sharper abrasive	Cost
Parts overheating	Protect workpiece	Add coolant	None
		Dress more often	Wheel wear
		More open structure	Coarser finish
		Coarser grain	Coarser finish
		Sharper abrasive	Cost
		Focus coolant	Overspray
Wheel glazing	Increase dress life	Coarser grain	Coarser finish
		More open structure	Faster wheel wear
		Sharper abrasive	Cost
		Add coolant	None
		Dress more often	Wheel wear and time

(Continued)

Early wheel breakdown	Increase dress life	Harder grade	Potential glazing
		Sharper abrasive	Cost
Wheel loading	Increase dress life and production	Less feed pressure	Longer cycle time
		Denser wheel	Glazing/heat
		Inject coolant	No downside
		More open structure	Fast wheel wear
		Dress more often	Wheel life shorter
		Spray mist if no coolant	Breathing protection
		Softer bond	Breaks down faster
		Sharper abrasive	Generally costlier
		Take lighter cuts	Slower production

UNIT 13-1 REVIEW

Review the Key Terms

Alumna
The semirefined version of bauxite, aluminum ore, from which aluminum oxide abrasive is made or further refining produces the metal aluminum.

Bond (wheel)
The material that glues the abrasive grains together.

Boron nitride (cubic boron nitride)
A man-made abrasive next to diamond in hardness. It's crystal structure looks cube shaped. It tends to stay sharp and to fracture well.

Friability
The ability to shear away small flakes to self-refresh an abrasive's sharp edge before it shears out of the bond.

Glazing
A grinding wheel condition where the grains have not self-dressed, yet they have no sharp edges left. The wheel must be dressed.

Grade (grinding wheel)
A relative index of the hardness of the bond not the abrasive—the hardness of the wheel.

Grit size
The number of screen squares within a square inch. The larger the number is, the finer the grain size.

Post (grinding wheel)
The column of bond material linking individual grains.

Structure (grinding wheel)
An index of the relative amount of porosity in a grinding wheel relative to grit and bond. Structure runs from 1 = dense to 16 = very open.

Self-dress (self-refresh)
New sharp edges are created by either the dull grain fracturing from pressure or the bond letting go.

Truing (grinding wheel)
Dressing the wheel to reshape it and to remove runout.

Replay the Key Points

◆ Predictable accidents occur due to incorrect setups that do not anticipate the dynamic forces of grinding. Some happen from incorrect operator actions.

◆ Grinding wheels are exchanged to suit the job.

◆ Green, silicone carbide wheels are usually reserved for nothing but carbide tools.

◆ In general, due to cost, we use the softest abrasive that will do the job.

◆ Hard metal = soft wheel and soft metal = hard wheel.

◆ Wheel hardness is a combination of bond material and amount of bond within the ratio—it has nothing to do with abrasive hardness.

◆ A wheel with more space is needed for heavy removal grinding while less space is required for finish work.

Respond

1. Interpret this grinding wheel.

   A-46-H-7-V

   Describe what each data field means and what would be a typical application for this wheel. You may want to refer to *Machinery's Handbook* for this question.

2. List three reasons a grinding wheel might not cut the work material and a possible solution to each.

ShopTalk

Nuts! When a grinding wheel is produced, the powdered clay bond and abrasive grains are pressed into a mold then removed. At that stage, it's easily broken with a light tap. That unfired wheel is furnace heated to vitrify (melt the bond) into a porcelain-like substance. On cooling, a precision lead or plastic hub is pressed in the center hole. Finally, after a factory dress, the wheel is carefully packaged and shipped to the customer. But, the question is: given that the raw material was smashed into a mold under great pressure, how does the wheel end up with air spaces?

Interestingly, the original mixture starts out with three ingredients: abrasive, bond, and an organic filler. Among other things the filler can be made of specially ground and sized walnut shells! When the wheel is fired to create the bond, the filler turns to vapor and a tiny bit of ash; it disappears leaving spaces in the wheel.

3. In general terms what might this grinding wheel be best suited to do?

C-80-R-3-V

4. True or false? It is more economical to use the softest grade of aluminum oxide wheel that will do the job because as the hardness of the grains increases, so does the cost. If this statement is false, what will make it true?

5. On a piece of paper, fill in your selections for a roughing wheel cutting mild steel. You needn't be exact—provide a range for the characteristic needed.

UNIT 13-2

Reciprocal Surface Grinding

Introduction: In a lab or on the job, your first experience with precision grinding will probably be a small surface grinder. They create precise flat surfaces with finishes smoother than 16 μIn, finishing parallel bars, for example. They're an indispensable machine for precision tooling work and many other tasks supporting production. With just a bit of setup and operation training, very rewarding results can be yours with these little gems.

Safety is a big item even on these little machines. Though they are simple to understand and operate, there are several specific precautions because the work is often held by a magnet. While cylindrical or tool and cutter grinders may or may not be included in

your first-year experience, it's certain that surface grinders and magnetic chucks will be. For that reason, they are the central focus of Units 13-2 and 13-3.

KEY TERMS

Electromagnetic chuck
Grip can be switched on or off and residual magnetism can be reduced to zero.

Hydrostatic bearing
A predictably thick oil film separates sliding components.

Permanent magnetic chuck
Sets of magnets either oppose or agree with each other inside the chuck. A lever rotates one set relative to the second set, which either cancels the pull or causes it.

Pick feed (grinder down feed)
An adjustable ratchet feed that cranks the Y axis down at the end of each X stroke of a surface or cylindrical grinder.

Reciprocating
A back and forth action.

Reciprocal Grinders

There are two different types of surface grinding processes used in industry. The first, **reciprocating** (back and forth) uses the rim of the wheel and moves the work in a straight line under it. By far, these are the more common grinders due to their simple construction and they are the ones we'll concentrate on.

Rotary surface grinders, the other type, use a spinning chuck whose face is horizontal to hold and move the work past the grinding wheel. The grinding wheel cuts with its face rather than its rim—similar to a face mill. In general, these high removal machines are a specialty process not found in every shop and rarely found in schools—we'll cover them in Unit 13-5.

Manual or Tool Room Surface Grinders

See Figs. 13-18 and 13-19. These machines grind by moving the worktable back and forth with hand motion only. They also feature manually actuated motions of the wheel head Z axis and Y axis saddle, within a work envelope of approximately 18 to 24 in. long by 8 in. wide by 10 to 18 in. tall. As seen in the lower right corner of Fig. 13-18, most small grinders feature a fine movement for

1. Column
2. Saddle
3. Table driving handwheel—X axis
4. Saddle driving handwheel—Z axis
5. Electric cabinet
6. Wheel guard
7. Grinding wheel
8. Table
9. Micro up/down feed device
10. Wheel head driving handwheel—Y axis
11. Base

Figure 13-18 *A small surface grinder has three movements* X, Y, *and* Z.

Figure 13-19 *The X axis rests and rolls on non-lubricated, precision-matched sets of ball bearings.*

Figure 13-21 *Always wear eye protection and a dust mask when dressing the wheel or running a grinder without coolant.*

the *Y* axis—to enable removing a very small amount of material (#10 in the figure).

To impart the nearly friction free action, the *X* axis glides on roller bearings. Differing from milling machines, the table is not held down with a dovetail. It sits atop the frame, suspended by antifriction rollers. The rollers are held together in sets called carriers. Note, the rollers are sealed from dirt and slurry, but the protective covers are removed in the drawing to illustrate concept only.

As such, these machines are designed for light loads only. The table is usually not attached to the ways; it can be lifted straight up for cleaning and maintenance. The table *X* axis is moved left and right by a gear on a rack, using hand pressure only (Fig. 13-20).

Two more hand-operated, dovetail slide motions move the wheel head up and down, the *Y* axis that controls work thickness, and the *Z* axis saddle in/out relative to the operator, stepping the grinding wheel sideways across the work.

KEYPOINT

Breathing Protection When Grinding
The air around most grinding machines can be polluted with fine metallic and wheel dust particles and fine mist coolant flung from the grinding wheel (Fig. 13-21). This is especially true when dressing the wheel.

Manual Surface Grinder Functions

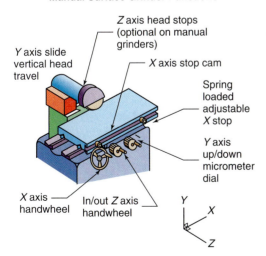

Figure 13-20 *The surface grinder functions and setup options.*

Trade Tip

Learning to Use a Hand-Operated Machine Operating one of the small machines is like patting your head while rubbing your stomach with the other hand. The coordinated movements require practice to get them smooth and consistent. One must move the table back and forth with the left hand, while moving the *Z* axis in/out, at the end of each *X* stroke with the right. Then as the wheel comes to half its width off the work, the *Y* axis must be cranked down for another pass across the work.

Here's how. First, using your left hand, move the *X* axis back and forth against the spring-loaded stops. With no other movements, just repeat that motion for a while. Next, your right hand cranks the *Z* axis in and out, moving the grinding wheel sideways half its width each time, at the end of each *X* stroke.

Moving the wheel only when it is off the part at the end produces straight grinding marks. Once this two-axis motion is mastered, using your right hand, crank the *Y* axis down no more than 0.001 in. only at the inner or outer edge of the work. Do not crank down at the same edge each time. Alternating the down feed equalizes the wear pattern on the wheel and grinds the work flatter.

Safety Note The predictable operator caused accident results from confusing the *Z* hand crank with the *Y* axis "down." Both are usually actuated with your right hand. The accident occurs by forgetting which hand wheel you are grasping! This can cause a big "down" when "across" was the intention. The unhappy result is a burned or undersized workpiece, or a stalled wheel against the work, possibly breaking the wheel or gouging and dislodging the work.

When first learning, try placing a piece of tape or tying a short strip of rag to the *Y* axis hand wheel. It's there to remind you that this is the down handle, just until you get comfortable with this three-movement operation. In Unit 13-3, we'll learn how to set up these machines.

Hydraulic Production Grinders

The big brother of the reciprocal surface grinders, these machines have work envelopes of from around 24 in. in *X* to over 20 ft or even more. For example, in Fig. 13-22, the table length is 6 m (nearly 20 ft).

Beyond its hydraulic actuated *X* axis, its *Y* and *Z* axes are also power fed. Most power-fed grinder functions can be operated automatically or by shifting gears, by hand too. The *X* axis on very large grinders does not have a hand motion as it's too big to move without hydraulic force. Using adjustable stops for all three axes, these machines often feature totally automatic or CNC operation.

Since removal rate and heat are constant challenges, these grinders always feature powerful coolant systems! Particulate masks are even more important here as well as very good safety glasses.

Figure 13-22 *Larger surface grinders such as this and the one shown in Fig. 13-23 feature power-driven X, Y, and Z axes.*

Safety Note Awesome accidents on grinders happen faster than you can react. They are not a first-timer's experience. In my shop, students learn to operate a the small grinder first.

Due to their heavy-duty nature, the table no longer rolls but rather they are pushed along with a hydraulic cylinder. The table slides on a **hydrostatic bearing** (Unit 1-3).

Warming Up a Hydraulic Machine While any machine surface that slides usually requires a lube film between the moving and stationary components, true hydrostatic bearings force a predictable thickness of oil between the two. When they are warmed up to full pressure, no metal-to-metal contact occurs, and the components are separated by an exact gap filled with oil. The Trade Tip applies to manual, large automatics, and all CNC machines.

Trade Tips

Exercising Most Machines Similar to a long-distance runner or weight lifter, it's important to "stretch out and warm up" hydraulic and CNC machines before heavy demands are placed on drives and bearings. After a long period of down time, this is an operator's start-up responsibility for accuracy and machine life. Here's a three-step procedure for a surface grinder but other machines are similar.

After checking and servicing the hydraulic oil level if needed, first start the pump (without moving the table) and let the machine idle for 2 or 3 minutes while the pressure builds up and the table lifts up on its oil film cushion.

Pressure Build—Exercise Warm Up—Full Stroke Then lightly "exercise" the slides by hand crank or power feed. This entails short stroking the *X* axis several times—around 4 to 6 in. Finally, to distribute the oil evenly and to bleed any trapped air in the ram cylinder from a long period on nonuse, run the table to its full left and right limits. *Caution!* Your foreman or instructor will need to show you how steps 2 and 3 are accomplished on the machine in your lab, especially the third step, where you will be taking the machine onto its end stops.

To exercise CNC equipment, I recommend a warm up program that loops through several long and short axis moves. It must take into account any setup on the machine so as not to hit the grinding wheel or cutter against work or tooling. Load and run it while you open your toolbox, get the apron on, and read the instructions for the day.

Setting Up a Reciprocal Surface Grinder

Setting up a reciprocal surface grinder entails four tasks:

1. *Setting the stroke and feed limits*—X length, Z width, and stepover amounts; Y down-feed if the machine has this feature.
2. *Setting safe work holding.*
3. *Coolant supply.*
4. *Mounting and dressing the grinding wheel* (Unit 13-3 coming). Is the wheel right for the job and dressed true and sharp?

Correct Stroke Length and Width

Setting up a surface grinder requires three or four features depending on type of grinder:

1. **X Stroke Limits Right/Left**
   Set beyond the workpiece surface by at least $\frac{1}{2}$ in.
2. **Z Axis Limits—In/Out Full Travel of the Wheel**
   Set to go beyond the edge by half the wheels width.
3. **Z Axis Stepover Amount Per X Stroke**
   Distance the wheel is moved sideways for each new X stroke.
   Set to step in/out half the wheel's width with each stroke.
4. **Amount of Y Axis Down-Feed— Situational**
   From 0.0001 to 0.0005 in. depending on work, wheel grit, and structure and strength of the setup. These are conservative amounts for learning. Manual machines do not feature autofeed.

Surface grinder setup stroke and feed limits are set depending on the type of grinder—manual, automatic, or CNC. The setup rectangle (X length and Z width) must be bigger than the workpiece surface by a given amount. Figure 13-24 depicts a typical setup—a rectangle of stroke length and width.

All surface grinders feature adjustable stops for the X stroke length, and most also have Z axis width stop cams. On manually operated surface grinders, the Y axis down feed is then a hand function. The down feed on automatics is a ratchet that engages a tooth gear on the Y axis hand wheel, called a **pick feed**. It can be disengaged or swung onto the Y

Figure 13-23 *A typical size surface grinder.*

handwheel when automatic feed is needed. The pick is adjusted to pull a given number of 0.0001- or 0.0002-in. clicks down at the end of each stroke. On CNC machines, all these functions are menu driven on the controller (Fig. 13-23).

Guidelines for Setup and Operation

Setting X and Z cam stops entails moving the grinding wheel beyond the work in each axis (Fig. 13-24). It is there, when the wheel is not in contact with the work surface, that lateral and down movements occur. The right/left X stroke should clear the work by at least a half inch minimum. See Fig. 13-25.

Surface Grind Setup Rectangle

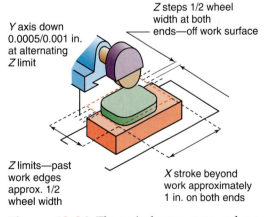

Y axis down 0.0005/0.001 in. at alternating Z limit

Z steps 1/2 wheel width at both ends—off work surface

Z limits—past work edges approx. 1/2 wheel width

X stroke beyond work approximately 1 in. on both ends

Figure 13-24 *The typical setup stops and motions for any surface grinder, manual, automatic, or CNC.*

Figure 13-25 *The same limit rectangle and steps become program entries (called parameters), on this CNC surface grinder control.*

KEYPOINT

The extra *X* stroke length beyond contact allows good coolant flushing, but it also improves professional looking results. It is the zone where the wheel is moved sideways in *Z* and the down feed occurs. See the Trade Tip.

Z Axis Steps and Limits

The *Z* axis has two functions: sideways movement for each stroke and full limit at each edge of the work.

KEYPOINT

Z limits are set such that the wheel travels half its width with each stroke and off the work edge at each side.

Trade Tip

A Bigger *X* Stroke Is Safer Other than cycle time, there's no problem in setting the *X* length much larger than the object. In fact, it's safer for your first few setups. However, in production the goal is to set the *X* limits such that the wheel clears just enough to avoid stepover (*Z* axis) tracks in the finish produced.

Coolants for Grinding Operations

Two Ways: Low-Pressure Flood and High-Pressure Injection

At the extreme rim speeds required for grinding, friction and heat become huge is-

sues. On surface grinders equipped with coolant systems, to cool the work, in general, a low-pressure flood is used. On many surface grinders, the wheel guard contains the flood system or it can be a nozzle. Either way, it's a general stream of coolant that protects the work from overheating. While some of that flood also becomes trapped in the air spaces in the wheel, so that it reaches the action area where the chip is being deformed, but it's not enough nor is it consistent enough.

Coolant Injection

To reach the point where the chip is being abraded from the work, a focused stream is directed at the contact point where the wheel begins its pass across the work. That traps coolant in the wheel's air spaces. Trapped coolant performs two services:

1. **Lubes Deformation**
   It reliably lubricates chip formation as each tiny negative rake cutter (abrasive grain) deforms the chip.
2. **Chip Ejection**
   After the chip has been formed, the trapped coolant keeps the hot chip from bonding to the abrasive grains. Also as it sprays out of the wheel on leaving the cut, its mass helps to fling out the chip.

What to Do When the Machine Has No Coolant

Most small tool room and manual grinders do not have flood coolant. If so, you have two options:

Mist Coolant A coolant spray accessory works well. For best results, it should be aimed at the wheel contact area. The spray that doesn't become trapped within the wheel will help to cool the work, but the important aspect is to get coolant to the chip formation zone. Also, if a mist isn't available, then plain air can be directed at the wheel and can work to keep it cool, but it will have no effect on chip deformation.

Create a Cooler Setup or Operation There are several things one can do when no coolant is available:

1. **Choose a Cooler Grinding Wheel**
   One with a more open structure, or larger grain size, or a harder abrasive that tends to hold its edge longer.

2. Grind Less Material/Less Removal per Minute

This is the last option. Let the material cool naturally by conduction to room air.

A few materials can be successfully ground with no coolant (cast iron, for example) but if available, coolants always improve grinding in four ways:

1. Reduce Heat Production

They lower friction and there is thus less deformation heat.

2. Remove Unwanted Heat from the Work and Holding Accessories

On precision grinders, this is a critical issue since tolerances are small. Thermal expansion of the part or work holding devise can be a real accuracy spoiler.

3. Extend Wheel Life by Cooling the Abrasive Grains

Because the mineral grains are exceptionally hard, coolants extend their lives by stabilizing unwanted temperature fracturing from hot/cold/hot cycles. But that's only when an adequate supply of coolant is injected into the interface between wheel and work directly at the contact point (Fig. 13-26). Flooding the general area helps cool the work, but it can cause hit-and-miss cooling of the abrasive grains (Fig. 13-26). The best coolant injection on any grinder setup occurs by forcing a stream into the wheel as it contacts the work.

4. Chip Ejection

Coolants trapped in the air space with the chip help to fling them out of the wheel during the split second when the wheel isn't in contact with the work. If not, the chip remains embedded to cause wheel loading (Fig. 13-27). So, as the wheel clears the work, due to centrifugal force, the trapped coolant is slung out, carrying

Two Effective Coolant Deliveries

Figure 13-26 *Flooded coolant reduces general work temperature, but it lubricates the cutting action only on a hit-or-miss basis. A better source is the nozzle on the right, focused on the wheel work interface.*

Correctly Injected Coolant

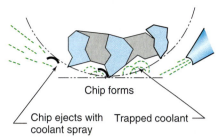

Figure 13-27 *Correctly injected coolant is entrapped in the wheel pores until it is flung out, helping to clear the chip from the wheel.*

the small chip with it. To cause this effect, coolants must be injected forcefully at the point of wheel/work contact.

KEYPOINT

Correctly used, grinding coolant creates a fine spray out the far side of the wheel but that indicates the right use of the fluid (Fig. 13-28). This is true for cylindrical grinders as well.

Coolant-to-Water Ratios for Grinding

Most shops use the same coolant syrup as that which is used throughout the shop. But for grinding, it's mixed in a thinner ratio of water to syrup, at or near 50 to 1. Read the directions for the product. Shops that do lots of grinding or where special alloys, such as titanium or magnesium, are ground use specialized coolants designed for abrasive work only. See the Trade Tip.

Trade Tip

Mixing Coolant Always read mixing instructions for coolants carefully. They aren't all the same. Follow specific recommendations for the particular application. Ratios at or above 50 to 1 are common for grind work. Beyond diminishing performance, the upper limit to the mix is not to exceed the coolant's ability to protect the machinery from overnight rust.

Also, many manufacturers recommend *adding water to the concentrated syrup while stirring constantly*, not syrup to the water. Why? It has to do with the way the molecules bond. A more complete mix is obtained in this way.

Figure 13-28 *Coolant is correctly injected into the wheel-work interface on this automotive cam grinder.*

Six Work-Holding Methods

There are six options to hold the workpiece for surface grinding. We'll do an overview of the two most common first, then consider the rest in the next unit.

Precision Grinding Vises

Work can be held in a small vise, much the same as on a milling machine. But for grinding, there is a more precise version reserved for surface grinding and tool room operations only (Fig. 13-29; shown in use in Fig. 13-4).

Figure 13-29 *A typical precision grinder vise.*

The vise skills you already have also apply in grinding.

◆ Always use the solid jaw to nest datum surfaces.
◆ Remember for every load of the vise, there's sure to be slurry dust left on the precision surfaces. Wipe them before loading the vise.
◆ Use a reboundless hammer to lightly tap the work until firmly seated.

KEYPOINT

For any setup, in a vise or work held with some other method, the wheel's action forces the work to the operator's left (see Fig. 13-30). Be certain no person stands to the left of the grinder. This will become more important with more complex setups and for larger reciprocal surface grinders.

Magnetic Chucks

Work is often held on surface grinding machines with a magnetic chuck, which might be an **electromagnetic chuck** (Fig. 13-31), where the grip can be switched on or off. But more likely on a small machine, a **permanent**

No – Danger   Yes – Safe

Figure 13-30 *By incorrectly moving to the left of the wheel (first photo), the operator has put himself in danger while the machinist in the second photo is in the safe zone.*

Figure 13-31 *A typical electromagnetic chuck.*

magnetic chuck is used (Fig. 13-32). No electricity is involved (Fig. 13-33).

Similar to a magnetic-base indicator holder, they are activated by a mechanical lever that either opposes or aligns magnets within their interior. When they oppose, the force is canceled, but when they agree, the attractive force can be great—great enough to pinch finger tips!

Ensuring a Safe Setup

Always check that the magnetic chuck is turned off before setting any object on it.

When setting work upon a magnetic chuck, never place your fingers under the object! While it should be turned off, if the chuck is accidentally left on, when the work gets close to the chuck, the downward force can be great enough to smash and cut finger tips! See Fig. 13-32.

After the magnet is on, test it by grasping the work then push and pull it to be certain there can be no movement sideways or any tendency to rock or tip. There is so much to learn about these devises for both safety and successful setups that Unit 13-4 is dedicated to setting them up. Have the foreman or instructor check your first few setups.

Figure 13-32 *Never have your fingers under an object as you put it upon a magnetic chuck. If the chuck is accidentally left on, the downward force can smash and cut them!*

Figure 13-33 *A permanent magnetic chuck.*

UNIT 13-2 REVIEW

Review the Key Terms

Electromagnetic chuck
Grip can be switched on or off and residual magnetism can be reduced to zero.

Hydrostatic bearing
A predictably thick oil film separates sliding components.

Permanent magnetic chuck
Sets of magnets either oppose or agree with each other inside the chuck. A lever rotates one set relative to the second set, which either cancels the pull or causes it.

Pick feed (grinder down feed)
An adjustable ratchet feed that cranks the Y axis down at the end of each X stroke of a surface or cylindrical grinder.

Reciprocating
A back and forth action.

Replay the Key Points

◆ *Breathing protection!* Beyond flying work and wheel fragments, grinders present a breathing danger from dust and flying coolants—wear a mask or filter.

◆ Coolants help to fling the tiny chips out of the wheel during the split second when the wheel isn't in contact with the work.

◆ Correctly used, grinding coolant is injected into the wheel-work interface. This creates a spray out the other side of the wheel, but indicates the right use of the fluid.

◆ For any setup, in a vise or held by some other method, the wheel's force against the work is to the left. This will become more important with more complex setups and when using larger reciprocal surface grinders.

◆ *Never put your fingers under the object!* Accidentally left on, when the work gets close to the magnetic chuck, the downward force can be great enough to smash finger tips.

The activities for Unit 13-2 are combined with those for Unit 13-3.

UNIT 13-3

Setups That Work Right

Introduction: Safety and solid setups blend together, especially when using magnetic chucks and high-speed grinding wheels. I cannot stress this enough that using grinders requires individual instruction and guidance.

Using safety guards and eye protection, even if you are not actually removing metal is a good start—why? *Wheels can and do explode even when they are revolving without touching the work.*

Magnetic chucks add a new twist to setup safety. While magnets produce precise results, their relative grip compared to the positive gripping chucks, clamps, and vises of milling machines, make them a subject for special focus. When magnetic chucks are incorrectly applied, parts tip, slip, and sometimes fly out of the machine. That almost always leads to a broken grinding wheel—and flying objects!

Strap clamp (magnetic chuck)
Used when stop blocks aren't secure enough or they are too thick to clear the grinding wheel. This U-shaped clamp holds with set screws against the chuck side.

Truing a grinding wheel
Dressing a grinding wheel to make it perfectly cylindrical and concentric to the grinder spindle.

Mounting, Dressing, and Balancing a Grinding Wheel

The first setup step is to mount the wheel on the grinder. Here's the step-by-step procedure.

Removal of the Old Wheel

Use a wood block or other protective material on the chuck to protect both the outgoing wheel and the chuck/table. Loosen the center spindle nut $\frac{1}{4}$ turn. Tap the spindle hub with a plastic hammer to break the self-holding taper, then remove the wheel and hub. (If the grinder features a simple **flange** and nut wheel mount similar to a bench grinder, then the tapping isn't necessary.) Caution—grinders often use a left-hand thread for flange nuts!

After removing the outgoing grinding wheel, never roll it or lay it where other items might be stored upon it.

KEYPOINT

If you find it's a damaged wheel, show it to your instructor or foreman then take it directly to the garbage can—don't take chances that it will accidentally be reused at a later time. Break it up completely!

Check the Identification, RPM, and Size

Be certain the replacement wheel is the right makeup, diameter, and shape using the standard number selection for the machine and the job. Look for the maximum RPM label and check that against the machine on which it is to be mounted.

Figure 13-34 *Can you see the crack in this wheel? Always inspect wheels for visible signs of damage and check that it has paper washers in good condition or that new ones are ready to use.*

Task 1   The Right Wheel and Condition

Visually inspect the wheel for cracks and chips and that it has two new paper gasket washers.

Paper washers perform two vital services for grinding wheels (Fig. 13-34).

1. **Equalize Flange Clamping Pressure**
   Without the washers, the clamping force is concentrated on individual, high grains, promoting wheel breakage if the flange is clamped too tightly. Remove a used washer to see the imprint of the grains on the paper. The soft paper spreads the load over a safer area of the wheel.

2. **Absorb Shock**
   They prevent wheel breakage by acting as a damper when a wheel is accidentally overloaded or some other accident occurs such as work slipping or tipping.

Never attempt to use a chipped or damaged wheel. While it's possible to trim dress a wheel beyond the damaged portion, that's to be done by supervision, you are not authorized to do so at the beginner level. Just don't

Wheel supported on a finger

Figure 13-35 *A ring test helps find hidden cracks.*

use a wheel that looks anything other than perfect. But, as Fig. 13-35 shows, you can't always see damage so don't skip task 2.

Task 2   Ring Test the Wheel

A broken vitrified wheel sounds much the same as a cracked coffee cup, it will thud rather than make a clear ringing sound. To **ring test** the wheel, suspend it with any convenient hard object in the mounting hole (or your finger will work), then tap it in four places around the wheel with a plastic screwdriver handle or some other hard nonmetal object. Rotate it about a fourth of a turn and do it a second time (Fig. 13-36).

But you can't always hear cracks that don't leave a long fracture in the wheel, so, after mounting the wheel perform the acceleration test (coming up).

Task 3   Mount the Wheel On the Grinder

There are several details to see to when mounting a precision grinding wheel. Grinding wheels mount on machines in one of two

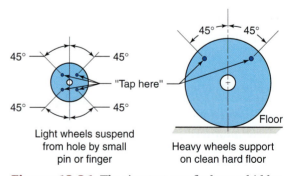

Light wheels suspend from hole by small pin or finger

Heavy wheels support on clean hard floor

Figure 13-36 *The ring test can find most hidden cracks if they are extensive.*

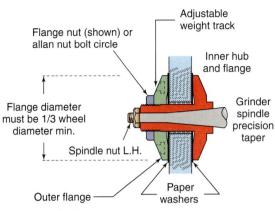

Precision Quick Change Grinder Hub

Flange nut (shown) or allan nut bolt circle

Adjustable weight track

Inner hub and flange

Flange diameter must be 1/3 wheel diameter min.

Grinder spindle precision taper

Spindle nut L.H.

Outer flange

Paper washers

Figure 13-37 *A cross section of a typical precision grinding wheel flange and quick change taper mount.*

different ways: directly or on quick change hubs.

Wheel Flange Mount Styles—Direct or Quick Change The direct mount grinder spindle is similar to a bench grinder's. The center hole in the wheel fits the machine's spindle shaft (Fig. 13-37). The wheel is held in place by a pair of steel flange washers and a flange nut (usually left-handed) (Fig. 13-38) that screws onto the grinder shaft. This system is used on some tool and cutter grinders and small surface

Figure 13-38 *The guard has been removed to show this typical small grinder wheel flange.*

grinders. However, simple, direct mounted wheels have two problems. First, the machine's spindle shaft is subject to damage when an accident happens that is severe enough to break the wheel. Second, these mounts do not include wheel-balancing provision.

The quick change hub is more expensive, but it is the better solution for production wheel mounting where setup turnaround time matters (Fig. 13-37). Grinding wheels mounted on a hub can be balanced using sliding weights.

Shops usually own many of these flange sets so they can keep an inventory of often used wheels mounted, balanced, and true—ready for quick turnaround. Hubs of this type also protect the machine's spindle from damage during an accident.

KEYPOINT

Grinding wheel balance is a critical issue not only to achieve good finish and accuracy, it's also necessary for long life of the grinder's spindle bearings. We'll see how in just a bit.

Task 4 Torque the Wheel Flange Nut

When mounting the wheel on the **arbor,** tighten the nut or the circle of Allen bolts to the manufacturer's specifications—if available. Usually, someone will need to show you how much is acceptable. Too much is as dangerous as not enough! Larger grinding wheels require more pressure on the outer flange, but any wheel can be cracked from overtightening.

Torque Patterns If the outer flange is tightened with a series of bolts in a circle as in Fig. 13-39, then use a cross-over torque pattern and procedure. Tighten slowly, a bit at a time in the order shown. Never tighten on one side first. Distribute the force equally.

Task 5 Primary Wheel Balance

Wheels placed on arbors require a balancing operation in one step (or a finer two-step process) for two reasons:

1. **Wheel Density Differences**
   Quality wheels are uniform in their density but they are not perfect.

Typical Torque Pattern Sequence

Flange tighten

Spindle mount nut

Figure 13-39 *To avoid concentrating force in any one spot, bolt circle wheel flanges require tightening a bit at a time following a pattern, until they reach the specified torque.*

2. Mounting Clearance

There is a necessary clearance between the grinding wheel hub and the spindle arbor, which will affect balance.

Static and Dynamic Balancing There are two ways a wheel may be balanced. You've seen both processes in your local tire dealer's workroom.

Trade Tip

Be on the watch for a stenciled arrow on some industrial grinding wheels. It will say "Up" or "Down" (Fig. 13-40). Mounted this way, the wheel should come closer to initial balance. That was the way the wheel was mounted, balanced, and trued at the factory.

UP

Figure 13-40 *Mount the wheel according to the arrow if one is stenciled on the wheel or washer.*

Figure 13-41 *A static balance fixture consists of two pairs of knife-edge, hardened steel wheels with very free running bearings.*

Static balancing is the simpler method, but it's not the best. It's a gravity test made by supporting the wheel and mounting the flange on a nearly friction free set of rollers (Fig. 13-41). The heavy side of the wheel will tend to roll down. Weights are added and shifted until the wheel has no net tendency to roll. This is the method you will most likely experience in school. But static balancing doesn't eliminate all the problems of balance when it's spun up to cutting speed.

Dynamic balancing rotates the wheel at its normal operating speed then detects through sensors where and how the weights must be placed. To see why, consider a bicycle pedal. It should balance out statically, as long as both pedals weigh the same. Taking the chain off the sprocket to allow free rotation, it shouldn't have a heavy side. But now suppose it's spun at 1,000 RPM. You can see that, although statically balanced, the weight isn't distributed about the axis of the pedal. It will have a tendency to wobble, but the bearings prevent it, and that translates to vibration. Similarly, to balance a grinding wheel, weights must be placed on both flanges of the wheel, inner and outer, to cancel dynamic balance problems.

Dynamic balancing is the preferred industrial process for close tolerance grinding (Fig. 13-42). It's also necessary to preserve machine bearings and deliver the very finest finishes. Newer CNC machines include dynamic

Figure 13-42 *A dynamic balance sensing unit is attached to the wheel while it's spinning. A computer determines where and how big the weights must be.*

balancing as part of their internal routines. We'll see this concept again when we get to high-speed machining (Section 27) where extreme RPM cutters in their holders must also be dynamically balanced.

Task 6   Mount and Dress/True the Wheel

Dressing the wheel is a resharpening of their edge by deliberately fracturing them and/or exposing the new ones. **Truing the wheel** is making its rim concentric to the machine's axis and making its grind surface perfectly cylindrical. They are done at the same time.

Grinding wheels must be dressed when they are loaded (see Fig. 5-49 in Section 5) and when they are **glazed.** Glazing occurs when grinding hard work for a long time beyond where the grains retain any sharp edge. Heat causes the wheel bond to smear with the resulting surface something like smooth glass. Normal wear has made the wheel face out of true.

Precision Wheel Dressing Tools Assuming we are dressing an aluminum oxide, or silicone carbide wheel, for precision surface grinder and cylindrical work we almost always use diamonds (Fig. 13-43). A single diamond crystal mounted in copper/nickel is called a **nib.** It's not a gem-quality diamond, but nevertheless, it is expensive. Alternately, less expensive diamond chips can be used when they are embedded in soft

Figure 13-43 *A diamond nib in a surface grind block.*

materials such as copper or hard resins (Fig. 13-44).

As seen in Fig. 13-43, either diamond dresser can be mounted in a block that adheres to the magnetic chuck for surface grind dressing. It is placed in the end of a round rod for dressing cylindrical wheels. If the mounting block is a solid cube, the nib can be mounted on its side when it's necessary to dress the side rather than the rim of a grinding wheel.

**Embedded Diamond Chip
Wheel Dresser**

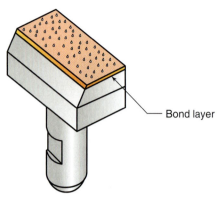

Bond layer

Figure 13-44 *A diamond chip dresser works nearly as well (for most dressing situations), but costs less.*

15 to 20 degrees

Steel dresser block

Magnetic chuck

Figure 13-45 *The dresser is set 15° beyond the wheel's center—can you see why?*

Trade Tip

Dressing Here are a few tips on the right way to dress a wheel.

The diamond is set at about 15° to 20° beyond the wheel's center, as shown in Fig. 13-45.

Make sure the X axis is locked or stopped from moving.

Using the Y axis, the wheel is dropped until it just touches the diamond held on the magnetic chuck. Then crank it down about 0.005 in. more for the first pass.

Using either the Z axis hand wheel or power feed, it is drawn across the diamond, moderately slowly at about 1.5 to 2.0 in. per minute.

To know that the wheel is true, look for consistent color across the face and listen for intermittent sounds turning to a solid sound. Bringing it to full true may take two or more passes of around 0.003 to 0.010 in.

Do not take a deep pass, for the possibility of grinding away the nib mounting material exists. The diamond is never found once it's ejected! (I've never found it! Good luck.)

The final pass is taken slightly slower at around 1.0 in. per minute. There are two mistakes made here, both of which only dull the new grains. Going too slow such that the new sharp edge hits the diamond a second time on the next revolution or taking an extra pass back across without lowering the wheel again "just to be sure."

The final tip is to rotate the nib in its holder from time to time. That keeps it a uniformly pointed cone and prolongs its life in the mount by wearing it away on all sides.

KEYPOINT

When dressing a wheel take one final pass slightly slower than when truing, but not too slow. Be sure the X axis can't move.

Side-Wheel Dressing Grinding a workpiece surface using the side grinding wheel (not the rim) isn't a beginner operation. But to finish steps in the work and for other situations that do arise, it's sometimes required. We often side-wheel when doing tool and cutter grinding. But there we use a wheel made for the task. On a surface grinder, where the wheel's rim is intended to do the job, heat is the greatest problem. When we contact the work along an arc, too much wheel surface is in contact with the work at any time. When side-wheel work is necessary, use Fig. 13-46 to guide the dress setup and seek help for the first time.

Side Wheel Dressing

Wheel guard

Cubical dress block

Figure 13-46 *Side wheel dressing requires extra caution.*

The Fine Balancing Act When using static balancing, to achieve the best possible finish and a long wheel life, after the first balance and initial truing, remove the wheel and hub from the grinder, then balance it a second time. You will find that the truing operation has slightly changed the initial settings in many cases.

Task 7 Warm Up the Wheel

With all guards in place, start the grinding wheel in accelerating stages. Turn on the spindle for a count of two then switch it off before it gets to full speed. Then turn it back on for a count of four and off again. Finally let it go to full speed but stand out of the path. Let it spin freely for a minute or two as a test. A wheel with hidden damage will normally blow during that time!

This acceleration test is a good practice after a shift change or when the machine has not been used for some time. On machines that use coolant, the fluid will collect at the bottom of the wheel in the air spaces as it slows to a stop and this slow start method will sling out the unbalancing liquid.

Task 8 Always Inspect a Mounted Wheel

If the wheel is already on the machine, rotate it slowly and inspect it for cracks and chips before you use it. The ring test is not valid here because of the paper washer and flange pressure. Report any damage immediately—do not try to dress it away.

Safely Using a Magnetic Chuck

The magnetic chuck can be used in countless ways. We'll look at a few basic methods, but you'll soon find other creative ways to combine lessons here, with parallels, vee-blocks, and other items used to set up mills and other machines. It can be used to even grind nonmagnetic material such as aluminum or brass, which won't adhere to the magnet. To do so, the work is held by trapping it in place, using **stop block** packing, to completely contain the work (Fig. 13-47). Shown shortly in Fig. 13-50.

Figure 13-47 *Anticipate and prevent movement based on the forces of grinding.*

The bottom line for success is to analyze the forces and the work's potential for unwanted movement. Then set up stops, clamps, and other holding tools to stop it dead! *If the setup looks even slightly weak, it can't be used!*

Critical Safety

Potential movement due to the grinding wheel's force is to slide to the left and/or to tip over to the left (Fig. 13-47).

Critical Question Which movement is more dangerous, sliding or tipping, and why?

Answer Both are totally unacceptable, but the tipping is worse because it pivots the block on the leading edge, which jams its back edge into the wheel. That always leads to a severely damaged wheel or a total explosion. Tipping also ruins the workpiece and gouges the chuck too!

Now, examine Fig. 13-48, where both blocks on the chuck are steel. *Which is dangerous?*

That one was easy, the tall block will tend to tip over more readily. But here's a key point: *even the short one isn't well setup.* It too will tend to tip or slide to the left. One key is to appraise the amount of hold area relative to the grinding forces and height. If you can tilt the work with your hand, it's definitely not safe!

But How About This Pair? Each is the same steel block. In Fig. 13-49, one is a better setup—why?

While both blocks will hold to the magnet with nearly the same attractive force, block A has its long axis braced against tipping. Because its longer face is held at right angles to the grinder force, it presents less resistance to tipping. But the important point is that both blocks could possibly tip. Neither is ready to grind! They must be trapped with stop blocks.

Which is dangerous?

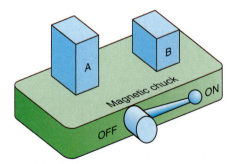

Figure 13-48 *Which steel block is dangerous in this setup?*

Which is dangerous?

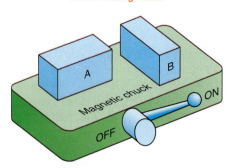

Figure 13-49 *Placement can be right or wrong too!*

Stop Blocks

When setups include tall work such as in Fig. 13-48 or 13-50, tipping problems can sometimes be solved in a couple of ways. Stop blocks of steel are the more immediate solution. One can use parallel bars and burr-free steel bars of any size and shape as stops, as long as they are shorter than the workpiece. They can't be rough or they might damage the chuck (Fig. 13-50). The larger surface area they have in contact with the magnet, the better they will hold.

Trade Tip

After setting the magnet to *on,* use a reboundless hammer to tap stop blocks firmly against the workpiece. If there's any doubt put paper feelers between the blocks and the work. By tugging on it after setting the blocks, the paper will show if they are firmly set. Then just leave the paper in the setup after the blocks are tapped up tight. If the paper is well trapped between work and blocking, the coolant will not affect the paper fibers enough to change its thickness.

Using a firm push with your hand, always test the work to simulate the grinding action before starting the machine. Test to see that the work cannot move.

Stop Blocks Prevent Movement

Figure 13-50 *Correctly placed steel stop blocks prevent unwanted movement.*

**Fully Contained Magnetic
or Nonmagnetic Work**

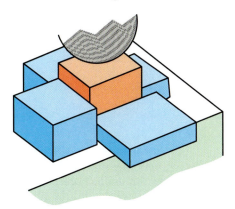

Figure 13-51 *Four stop blocks fully contain the work—it cannot move laterally and the wheel above stops it from coming up.*

Full Containment—Float Grinding

In Fig. 13-51, the work is fully stopped on the chuck, on all four sides. If, using the hammer and paper trade tricks, the packing is tight against the work, it cannot move in any horizontal direction. This setup will allow grinding of nonmagnetic work. Think about it—once the wheel is directly over the work, it's totally trapped. It floats in the setup but cannot move—thus, the name **float grinding.** However, heat expansion then contraction can loosen the packing. Grind slowly, use coolants if possible, take small down feed passes and check it often to be certain it remains tight.

Note in setups similar to that shown in Fig. 13-51, that the front block must be nearly as tall as the work to prevent tipping—no less than 75 percent of the work's height. The back block need not be as tall.

Precision Angle Plates

Utility plates can be magnetically chucked then used to hold work using clamps or bolts. This is the best solution for work that is too tall or otherwise unsuitable for direct chucking (Fig. 13-52). But even though they adhere to the magnet very well, the angle plates can be tipped. Blocking them in place is necessary too.

Critical Question

Can you see another way to rearrange the setup of Fig. 13-52 to ensure squareness in two axes?

Using Angle Plates and Clamps

Figure 13-52 *A combination of angle plates, clamps, and stop blocks is set up here.*

Trade Tip

When using C clamps on grinders, a bit of tape to hold the tee handle from bouncing into the wheel is a very good idea!

Answer

Move the rear (to the right) angle plate to the work's side so that it's trapped in a pair of angle plates that are 90° to each other. Then move the rear stop block forward (to the left) against the work, to trap it. Still use a C clamp or other holding method to pull the work to angle plates.

Magnetic Parallels

Odd-shaped work can be held using magnetic parallel blocks that transfer the force up to a part requiring some space from the table. We discussed them back in Section 8 but here's where they truly come to the forefront of setups. *Magnetic parallels* are made of alternating pieces of steel and any nonmagnetic material—brass, aluminum, or even plastic. This

Trade Tip

Safety Note The cores of magnetic parallels and other magnetic transfer tools must be placed *parallel* to the lines of force from the magnet below them or they will cancel most of the holding force! Try an experiment. Note the magnetic cores on a chuck in your shop. They run front to back in the Z direction. Now place a magnetic parallel one way with a steel object on top of it, then turn on the magnet. Try to move the object. Now, turn the magnet off, rotate the setup 90°, and try again. That's the proof you need.

Magnetic Parallels

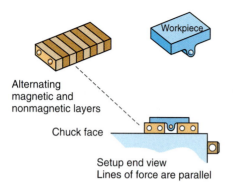

Figure 13-53 *Magnetic parallels transfer the chuck's force although they diminish it to some degree.*

Back Guide

Figure 13-55 *Most chucks are fitted with back guides to align work, the vise, or other accessories to the X axis, and to block work against the back of the chuck.*

creates a series of spaced magnetic cores that pass the force of the chuck through to the object being held. But be aware, they weaken it to some degree.

In Fig. 13-53, a steel hinge base needs grinding on the bottom. It is easily held using two matched magnetic parallels to accommodate the hump. Even though this looks to be a pretty secure setup, I strongly recommend adding additional packing around the work to be certain it cannot move.

Chuck Strap Clamps

Sometimes blocks alone are not secure enough or they aren't thin enough to clear the grinding wheel. Under these circumstances, a *chuck strap clamp* accessory can be used as a front and possibly as a backstop too. They are a simple U-shaped steel bar with set screws in the front (Fig. 13-54). **Strap clamps** work best when the magnet is turned on to keep the central thin portion from bowing up due to the set screw force, but they can be used without the magnet.

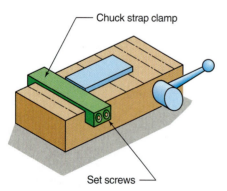

Figure 13-54 *A chuck strap clamp can take the place of thin blocking.*

KEYPOINT

Strap clamps are mainly used as very low front blocks.

Chuck Back Guides

To aid in alignment of work to the axis of the machine, when grinding slots of form grinding, for example, most chucks feature a back guide. Its height is adjustable within a small range (Fig. 13-55).

KEYPOINT

Don't assume the chuck and back guide are aligned to the grinder's X axis. They only bolt in place and can be far from parallel. Accidents can also misalign them. Mount a DTI on the wheel guard and run along the length of back guide to test its alignment.

Angle Grinding Sine Plates

Expensive and precise, these holding tools are used to setup and grind angular surfaces (Fig. 13-56).

Maintaining Magnetic Holding Tools

At long intervals, the shop supervisor will choose to regrind the top of the chuck itself or other magnetic holding tools. Although regrinding is not an everyday maintenance operation, the chuck is made with enough

Magnetic Sine Plate

Figure 13-56 *Magnetic sine plates tilt the work to a precise angle and also hold it for grinding.*

Figure 13-57 *This chuck control includes automatic demagnetization. It also includes cycle buttons and a rotary switch to diminish the chuck's holding force—a useful feature when plates are warped.*

top metal to regrind it many times over a long period of years. With professional use, the chuck should not need grinding more than twice a year maximum, and then only 0.002 to 0.005 in. will be removed.

Between grinding, small nicks and dents can be flat filed and honed away. Only after many of these imperfections have occurred, is chuck grinding justified.

Never leave coolant sitting on the chuck at the end of a shift and be certain the magnet is turned off. If the electrical types are left on, they will heat up, and the permanent types will weaken over a long period of being left on.

Demagnetizing Work

Any work material that is drawn to a magnet (iron, steel, some stainless steels, some nickel alloys, and a very unusual alloy of aluminum-nickel and cobalt called Alnico), if left on the magnet for a period of time, will themselves become magnetized. Usually, that residual magnetism must be canceled to get the object off the chuck. And it must be canceled to send it to the next work station. It's difficult to measure magnetized metals in some cases.

Demagnetizing on Electromagnetic Chucks

To cancel residual magnetism, electro chucks often feature some method of cycling the force down and thus automatically canceling the magnetism. Switching these magnets to the off position begins the cycle, which lasts about 1 minute. Using the chuck itself, the demagnetizer control creates alternating fields of slowly diminishing force, that drain the work metal of it's residual magnetism, **demagnetizing** it. An indicator will tell you it's OK to remove the work (Fig. 13-57).

Demagnetizing on Manual Chucks

Manual chucks require a two-step process. First, switch the magnet off and move the work to a position that is 90° to its original if you can. Quickly throw the chuck lever two or three times—on then off. This usually cancels most but not all of the magnetism. If all residual magnetism is to be removed, the work is then placed on or in an electric demagnetizer accessory.

Trade Tip

If your shop doesn't have a demagnetizer to use with the permanent magnetic chucks, take your work to the electro chuck!

Grinding a Plate Flat and Parallel

The most basic operation on a surface grinder is to reduce the height of an object to a specific size. Other than setting the trip stops and motions, the actual grind procedure requires no discussion beyond your instructor's demonstration of the machine's features.

Figure 13-58 *Step 1—Lay the work with the dome up, then shim it. Turn on the magnet and grind about 75 percent of the surface flat.*

Figure 13-59 *Release, demagnetize, reverse, and shim the work on the opposite side. Now grind it for 75 percent clean up.*

Thin and Warped Plates

But when the given plate is warped or flexible, it cannot simply be chucked down and ground. As it is released, it will spring back from flat on the chuck to its original pregrind shape. Here's how to solve that problem. The secret lies in a trade tip called "shimming and flipping."

The real tip is in this first step (Fig. 13-58). Lay the plate with the curvature upward and shim the middle such that the magnet does not pull out the warp when it is turned on—this is a critical step. With no shim, the plate would spring down under magnetic pull, then spring back after it is released.

With the shim(s) in place, turn on the magnet and grind the plate for a minimum cleanup of about 75 percent of the surface. Next, release it, demagnetize it, flip it over, and shim the other side at the outer edges (Fig. 13-59).

Again the shims prevent magnetic change from its natural position. Then grind about 75 percent cleanup on this side. Release, flip, and repeat steps 1 and 2 a second time and presto, the plate will be flat and parallel. Then it can be held directly on the magnetic chuck with no warp for final size grinding.

Operating a Surface Grinder

Each surface grinder is different. This instruction is general and will require specific information and a demonstration of the grinder in your lab. Here's a set of guidelines for success.

Always Clean All Surfaces

Loose grit is the enemy of precision! Always wipe the chuck surface with a rubber squeegee then a clean rag, then give it a final test using *your hand* (Fig. 13-60). Your hand is the best grit detector! One single piece of grit left on the table between the work and chuck will be 5 to 10 times the expected repeatability of the operation!

Depth Feed—How Much?

A common error made by beginners is to attempt to remove too much metal per stroke. Each setup, work, wheel, and machine combination will be different. Similar to milling and turning, there is no hard and fast rule, but here are some guidelines to remember for setup and operation of the small surface grinder:

Cutting	Down-feed for pass across the work	Suggestions
Steel mild	0.0005 to 0.003	Use coolant.
Steel hard	0.0005 to 0.001	Use coolant and dress often.
Cast iron	0.002 to 0.006	Coolant not required—But it helps keep dust under control. Cast iron dust will contaminate coolants.
Aluminum	0.0005 to 0.005	Wheel is open structure and coarse. Be on the watch for loading and heat build up, especially when no coolant is available.
Nonferrous metals	0.0005 to 0.003	
Stainless	0.0005 to 0.001	Use coolants.

Figure 13-60 *After cleaning with a squeegee, then a rag, then do a final wipe with your clean hand. It's the only way to guarantee that no grit is left behind.*

Removing Work from the Chuck

Removing work from a magnetic chuck can be challenging because of both the air seal created against the chuck and the residual magnetism.

Safety Always

Never attempt to take any work off the chuck before the wheel completely stops turning.

Save That Chuck!

To avoid gouges in the precision chuck surface, do not slide the work if you can tilt then lift it. A paper gasket placed under the work at setup time will prevent scratches when trying to remove work. The secret here is an old telephone book: its thin paper is perfectly parallel and will stay so even when the coolant soaks it.

UNIT 13-3 REVIEW

Review the Key Terms

Arbor
The precision hub, shaft, and flange mount for a grinding wheel.

Demagnetizing
Removing residual magnetism from magnetic workpieces exposed to the chuck.

Dressing
Reexposing sharp grains to unclog, sharpen, and true a grinding wheel.

Dynamic balancing
Rotating grinding wheels and cutters at working servos while removing all out-of-balance conditions.

Flange (grinding wheel)
The wide sides of the wheel mount designed to spread the load over a large area. Must be $\frac{1}{3}$ the wheel diameter or greater.

Float grinding
Holding and grinding nonmagnetic material using blocking and a magnetic chuck.

Glazed grinding wheel
An overused wheel that's lost its cutting edges. Heat causes the bond to smear and forms a nearly glass-like surface that will not cut.

Nib
A mounted diamond on stick—used for dressing a grinding wheel.

Ring test
Tapping the unmounted grinding wheel in several spots while listening for a clear bell sound. A broken wheel produces a thud.

Static balancing
A gravity test of grinding wheels to shift weights to neutralize roll down.

Stop blocking
Extra blocks that adhere to the magnetic chuck used to guarantee the work will not move.

Strap clamp (magnetic chuck)
Used when stop blocks aren't secure enough or they are too thick to clear the grinding wheel. This U-shaped clamp holds with set screws against the chuck side.

Truing a grinding wheel
Dressing a grinding wheel to make it perfectly cylindrical and concentric to the grinder spindle.

Replay the Key Points

◆ Grinding wheel balance is a critical issue not only to achieve good finish and accuracy, it's also necessary for long life of the grinder's spindle bearings.

◆ When dressing, take one final pass slightly slower than when truing, but not too slow. But do not go back over the wheel to be sure. Be sure the X axis can't move.

◆ When mounting a grinding wheel
  ◇ Ring and visual test the wheel.
  ◇ Double-check wheel size and RPM.
  ◇ Always use paper washers.
  ◇ Torque the nut(s) per specs or instructor demonstration.
  ◇ Balance the wheel, dress it, then rebalance it.
  ◇ Warm up the wheel with a short run before grinding.
  ◇ Always visually inspect a mounted wheel before using it.

◆ Potential movement due to the grinding wheel's force is to slide to the left and/or to tip over to the left. Tipping is more dangerous.

◆ Appraise the amount of hold area relative to the grinding forces and height. If you can tilt the work with your hand, it's definitely not safe!

Respond

1. List the seven steps to mounting a new wheel on a grinder.
2. How is a grinding wheel balanced?
3. Which of the two equal sized, steel blocks shown in Fig. 13-61 is safest for grinding the top? Describe how to improve the setup of both.

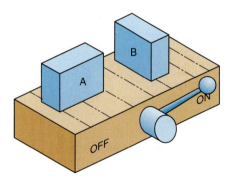

Figure 13-61 *Which is more safe and why? How can this setup be safer?*

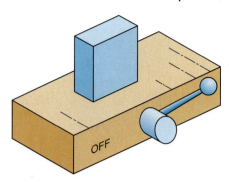

Figure 13-62 *What is needed to make this setup safe?*

4. On a sheet of paper, sketch the correct blocking for the cast iron object in Fig. 13-62 to be ground safely. Or describe it in words.
5. Would blocking and grinding aluminum on the surface grinder be correctly called float, flat, or finish grinding?
6. Name at least five other options beyond blocking that can be used to secure a steel workpiece on a surface grinder.

Critical Thinking Question

7. When dressing the grinding wheel, it is important to place the diamond at least 15° to the left of the wheel (beyond center) (see Fig. 13-45). We didn't discuss this, however, you should be able to answer why.
8. When must a demagnetizer be used?
9. Name two times when a demagnetizer would not be needed.
10. For what three reasons is dressing with a diamond performed?

UNIT 13-4

Other Grinder Operations—Visual Inventory

Introduction: Our objective here is to provide a background idea about other types of grinders found in industry that may or may not be in a technical school curriculum. They will most likely be introduced in an industrial apprenticeship or on-the-job training. When any new grinder is to be learned, the skills gained in wheel selection and metal removal by surface grinding will apply.

Unit 13-4 is offered so you can describe and recognize these technical processes. As such, it does not contain safety or skill training. We'll just take a brief look at the machine and the kind of work it can do.

Manually Operated Grinders Live On

Although each machine described is found in both manual and CNC versions, the manuals are still doing their job in many shops. These have been slow to be replaced by CNC versions because, in the past, grinding machines were envisioned to be single-operation equipment—the part is put in and a single surface is ground. For a long time, job planning was written with that image in mind. Each machine does its particular specialty very well and won't be pushed out of the shop for some time to come. But it will happen someday.

That one-at-a-time image is changing for grinders in general due to CNC control. It has already changed for tool and cutter and gear grind machines. In Fig. 13-63, a CNC tool grinder is finishing a worm gear on its OD.

CNC grinders can perform multiple operations, dress their own grinding wheel, change grinding wheels, dress the new wheel into a contour if needed, and make complex shapes not possible by manual grinding (Fig. 13-64). We even see grinding heads attached to CNC

Figure 13-64 *Looking closely, the Aluminum Oxide wheel shows when the safety door is opened on the CNC cylindrical grinder. Otherwise, it might be taken for a turning center.*

machining and turning centers and tool grinders as part of the machining center's capability. Here are a few selected industrial grinders.

KEY TERMS

Blanchard grinder
The common name for a rotary table, vertical-axis-wheel-head, surface grinder featuring segmented wheels.

Centerless grinder
A cylindrical grinder that does not hold the work, but rather contains it in a three-point suspension between a grinding wheel, support blade, and rubber regulating wheel.

Cylindrical grinders
One of two kinds of grinders that create round objects: the center and the centerless types. Only the centerless type can grind the entire length of a shaft. Only the center type can grind a taper.

Front feeding
Putting cylindrical work through a centerless grinder by tilting the regulating wheel. This grinds the work along its full length.

Regulating wheel (centerless grinder)
The wheel that rotates and controls the workpiece in a centerless grinder.

Through feed
Feeding work from front to back on a centerless grinder.

Figure 13-63 *A CNC tool and cutter grinder finishes the OD of this worm gear.*

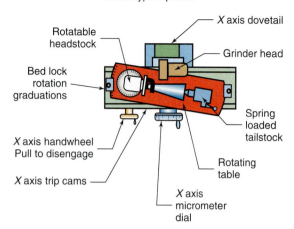

Cylindrical Grinder
Center type-top view

Rotatable headstock

X axis dovetail

Grinder head

Bed lock rotation graduations

Spring loaded tailstock

X axis handwheel Pull to disengage

Rotating table

X axis trip cams

X axis micrometer dial

Figure 13-65 *Overhead view of basic manual cylindrical grinder.*

Grinding Tapers on Cylindrical Grinders

Pivoting headstock

Pivoting grind wheel

Figure 13-66 *An overhead sketch of two different kinds of taper setups. Both functions can be used together to achieve special operations.*

Figure 13-67 *This cylindrical grinding wheel has been angled to widen its face and to reach the work shoulder (top view).*

Cylindrical Grinders

There are two general types of machines for grinding precision round objects: **the center type** and **centerless grinders.** Each machine grinds round objects but in a radically different way.

Center Type

These grinders are similar to a lathe. They hold and rotate the work by the same means as a lathe: collets, chucks, and between centers.

Basic Construction Much of your lathe training will be invaluable here especially for work holding. But there is a difference between the grinder frame and the lathe.

Pivoting Bed When adjusting to cut a straight cylinder shaft on a lathe, the tailstock is moved laterally until the shaft runs parallel to the lathe's Z axis. The lathe frame remains stationary. Not so on a cylindrical grinder. The entire bed pivots at its center, which includes the headstock and the tailstock. That flat bed supporting the headstock and tailstock is clamped at each end to a saddle that sits atop bed ways. The bed and ways reciprocate in the Z (right-left) axis under hand cranking or power feed (Fig. 13-65).

In the overhead view, the bed is set to grind a long taper with the work held between centers. To adjust away taper, the locks at each end are released and the entire bed with head and tailstock are rotated relative to the saddle.

Other Setup Options—Grinding Steep Tapers The cylindrical grinder can cut short tapers and grind shoulders in two different ways (Fig. 13-66).

Pivot the headstock or the grinding wheel The headstock can be moved on the bed ways— forward and back similar to a tailstock. And it can be turned and locked relative to the bed, as shown on the left of the drawing, or the grinding wheel can be pivoted, as shown on the right. In both setups, the bed X axis is locked and the work is held and rotated in a front closing collet. Figure 13-67 shows a typical use of these pivoting functions. Here the grinding wheel has been pivoted, then dressed parallel to the machine's axis again. This allows grinding up to a sharp corner shoulder without touching the face of the work—only the diameter.

Universal internal attachment Some cylindrical grinders feature a second grinding head

Figure 13-68 *The internal head is set up on this cylindrical grinder.*

Figure 13-70 *Center type cylindrical grinders can hold and grind objects that aren't round.*

that allows internal grinding operations (Fig. 13-68). Here, the work is held in a chuck or on a face plate. The external wheel and guard has been removed. The internal head is mounted and driven by a belt from the spindle motor. A mounted stone is being spun in the internal head. These stones rotate at high rates—from 10,000 to 30,000 RPM on cylindrical grinding machines. But, on dedicated internal grinding machines, air turbines drive them to 150,000 RPM or higher (Fig. 13-69).

Inside-Diameter-Mounted Grinding Wheels

Figure 13-69 *High-RPM grinding wheels for internal diameters are mounted on mandrels.*

Why Choose a Center Type? Cylindrical grinders can hold repeatable tolerances of 0.0003 to 0.0001 in. with finishes easily within 16 μIn or finer. The advantage of these machines over the centerless type is that they can grind work that is not initially round and can round areas on work of any size or shape that can be swung within the work envelope: for example, a bearing on a camshaft, crankshaft or a shoulder diameter on a hex bolt (Fig. 13-70).

Centerless Cylindrical Grinders

These unique cylindrical machines can grind a round shaft along its full length—something a center type cannot do. Work is not chucked. It floats and is trapped in a three-point tunnel

between the steel support blade, the rubber regulating wheel, and the grinding wheel (Fig. 13-71). The centerless grinder's main purpose is to reduce diameters. They do not finish tapers or shoulders. A centerless is best suited to pins and shafts, but they grind them accurately and very fast. Repeatable tolerances of 0.0001 to 0.0002 in. with finishes if 16 μIn are common (Fig. 13-72).

KEYPOINT

Due to the unsupported work and the nature of inserting it into the tunnel, it is unlikely that you'll encounter a centerless grinder as a beginning machinist. They require experienced operators.

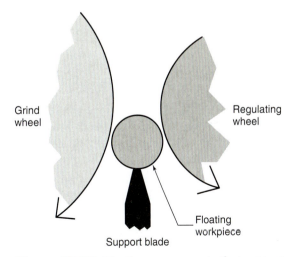

Figure 13-71 *The three components that support the work on a centerless grinder.*

Figure 13-73 *A simplified sketch of through feeding on a centerless grinder.*

A **through feed** setup is made where a large number of parts are to be ground to a specific diameter, without taper, along the entire length. Figure 13-73 is a simplified diagram of a through feed setup. This is a very fast method. The work can be loaded, one after the next, as fast as the operator can insert them! They fall out to a catch pan.

Front Feed (Insert and Remove) In **front feeding,** the regulating wheel is parallel to the grind wheel, and the work is inserted in the opened tunnel, bigger than finished size. A combination stop gage and part pusher rod is at the far end of the tunnel to ensure the work goes in to a preset depth and that it stays inserted at that depth (Fig. 13-74).

Then the grinding wheel is dialed toward the regulator until it reaches final diameter. This action is automatic or hand cranked. Upon

Figure 13-74 *A simplified sketch of front feeding on a centerless grinder.*

Figure 13-72 *Both CNC and manually operated centerless grinders are found in industry.*

To use a centerless process, the work has two requirements:

1. It must be round and smooth initially.
2. Its diameter must fit into the three-point work tunnel.

Through Feed or Front Feed There are two ways the centerless grinder is set up to grind round work.

Through feed for large batches By slightly tilting the rubber **regulating wheel** axis relative to the grind wheel, work can be inserted in the front to eject from the back into a part-catching tray. The grinding wheel is set at the correct diameter. It is not fed inward so a very limited amount of material can be removed in this setup (0.003 to 0.005 in. maximum).

completion, the grinding wheel is pulled back to open the tunnel. The part eject lever is pulled, pushing the work out of the tunnel, toward the operator where it's removed. This setup allows grinding to a step, as shown, which cannot be done with through feed.

KEYPOINT

Centerless grinders do not hold the work, but rather, they trap it in a three-way suspension between support blade, grinding wheel, and regulating wheel.

Tool and Cutter Grinders

End mills, reamers, lathe tools, face mills, and all other cutting tools are sharpened on a universal tool and cutter grinder. These machines are exceptionally flexible in their setup configurations. The manual machines feature a fully tilting and rotating head plus a horizontally rotating table. It's likely that your school will have one of these machines.

On the manual versions, there are literally dozens of accessories available to hold, rotate, and grind tools. Additionally, many kinds of grinding wheel shapes are available (Fig. 13-77) as well as different kinds of straight line and radius dressing tools. Learning to operate a tool and cutter grinder is a trade in itself.

Internal Cylindrical Grinders

Three varieties of machines finish the inside diameter of work. First are **jig grinders** (Fig. 13-78) which locate and grind holes to exact size and position. Next **internal diameter (ID) grinding machines** (Fig. 13-79) which are similar to cylindrical grinders with internal heads, rotate the work while stroking the mandrel and stone in and out of the hole. They can grind large objects. The third is an internal attachment for cylindrical grinders that we've already seen.

KEYPOINT

Mandrel Wheels Must Achieve Extreme RPMs
In order for these relatively small diameter grinding wheels to achieve the 4,000 minimum F/M surface speed required for efficient grinding, their RPM must be from 10,000 to 150,000 RPM, depending on the diameter of the wheel. Be extra careful and double-check RPM setting on machines that feature multiple setting.

Figure 13-75 *Sharpening the spiral peripheral teeth on an end mill requires screwing the cutter forward by resting it on a thin support finger, and pushing it past the grinding wheel.*

Figure 13-76 *Sharpening the end teeth on an end mill.*

Jig Grinding Machines

The objective of a jig grinding machine is final machining of holes to the best possible cylindricity, finish, size, and position tolerances. They are considered extra precision equipment. Operation of this machine is often considered a specialty within the machinist's role.

Jig grinders feature an ultraprecise positioning table to locate holes within 0.0001 in. repeatability. The grind head functions something like an adjustable boring head, in that the radius of swing of the mandrel grinding wheel can be dialed out to any radius, in 0.0001-in. increments. Constructed somewhat like a vertical mill, jig grinders feature a vertical spindle usually powered by an air turbine (Fig. 13-78).

Grinding Wheels for Tool & Cutter Grinding

Type 1 Straight

Type 6 Straight cup

Type 11 Flaring cup

Type 12 Dish

Type 13 Saucer

Figure 13-77 *Several different grinding wheel shapes are used on tool and cutter grind machines.*

Figure 13-78 *A jig grinding machine features a very accurate X-Y locating table and the ability to dial in hold diameters within 0.0002 to 0.0001 in. accuracy.*

ID Grinding Machines

Inside diameter grinders are much like an OD machine in purpose but not in construction (Fig. 13-79). Having short ways and extra heavy headstocks, they are made to hold and rotate large and heavy work while finishing a single hole. Their purpose is extra precise control of the diameter and its taper.

The Rotary Surface Grinding Processes

These machines are high-volume metal removers compared to reciprocal machines. The large grinding wheel rotates in a horizontal plane (vertical axis). Different from regular grinding wheels, they are composed of replaceable segments bolted to a base plate. This type of grinding wheel is called a composite. It's similar to a large, insert tooth, face mill cutter. The advantage of segments is that one is easy to handle at a time and one or more damaged segment can be changed at any time.

The work is magnetically held on a rotating, horizontal table. These two

Figure 13-79 *An internal diameter (ID) grinder resembles a cylindrical grinder but it's designed to rotate larger work and the smaller grinding wheels must rotate at much faster RPM.*

disks, the chuck and wheel, are rotated off center to each other while the distance between them is closed to control work thickness. They both spin the same direction (counterclockwise), which means at their contact point

Rotary Head/Rotary Table Grinders
(Blanchard grinders)

- Bolt-on wheel segments
- Containment safety shield
- *Z* axis micrometer dial
- *X* axis dovetail
- Work envelope
- Horizontal rotary magnetic chuck

Figure 13-80 *A Blanchard grinder rotates the work on a horizontal table while grinding it with an overhead vertical axis wheel.*

the wheel action opposes the chuck and work (Fig. 13-80).

It doesn't take too long to see that only an expert ought to run one of these machines. The forces are tremendous.

UNIT 13-4 REVIEW

Review the Key Terms

Blanchard grinder
The common name for a rotary table, vertical axis wheel head, surface grinder featuring segmented wheels.

Centerless grinder
A cylindrical grinder that does not hold the work but rather contains it in a three-point suspension between a grinding wheel, support blade, and rubber regulating wheel.

Cylindrical grinders
One of two kinds of grinders that create round objects: the center and the centerless types. Only the centerless can grind the entire length of a shaft. Only the center type can grind a taper.

Front feeding
Putting cylindrical work through a centerless grinder by tilting the regulating wheel. This grinds the work along its full length.

Regulating wheel (centerless grinder)
The wheel that rotates and controls the workpiece in a centerless grinder.

Through feed
Feeding work from front to back on a centerless grinder.

Replay the Key Points

- Cylindrical grinders can hold repeatable tolerances of 0.001 in. with finishes easily within 16 μIn or finer. The advantage of these machines over centerless is

that they can grind work that is not round initially.

- Centerless grinders do not hold the work but rather they trap it in a three-way suspension between support blade, grinding wheel, and regulating wheel.
- Be extra careful and double-check RPM setting on machines that feature multiple setting.
- Mandrel-mounted wheels need to run at very high RPMs to achieve correct cutting speeds.

Respond

1. Name three machines that grind the inside hole diameters.
2. Beyond the fact that it cuts with a grinding wheel, how is a standard cylindrical grinder different from a lathe?
3. True or false? Only the jig grinding machine can finish several holes on one part. If it's false, what will make it true?
4. Which machine uses segmented grinding stones? What is its shop lingo name?

Critical Thinking Question

5. Approximately how fast (RPM) must the rim of a $\frac{1}{2}$-in.-diameter, mandrel-mounted grinding wheel go to equal the rim of a 14-in.-diameter cylindrical grinder, doing 1,714 RPM? What RIM speed is that? (*Hint:* The short RPM formula helps solves this.)

Figure 13-81 *A Blanchard grinder can remove metal efficiently using a high powered, horizontal grinding wheel (yellow with stone segments below) while holding close tolerances.*

ShopTalk

Full Circle Roles (continued) them out to be sharpened in contract tool grind shops. Today, only the largest shops maintain their own grind facility.

But hold on—there's sure to be more change! CNC tool grinding is returning to the machinist in the small shop, in two different ways. First, CNC tool and cutter grinders are lightning quick to setup due to universal tooling and their stored cutter routines. But even though they take just minutes to finish the job, the operator would still need to go to the grinder to sharpen tools while their machine ran unattended. So, the obvious next step is multitask CNC milling and turning machines able to grind their own tools using accessories such as shown in Fig. 13-77! Machinists of the future will once again be their own tool grinders!

ShopTalk

By Any Name Like the Bridgeport mill, these machines are often referred to as **Blanchard grinders** due to a popular manufacturer (Fig. 13-81). These brutes can often remove more metal per minute than a milling machine! That's not an exaggeration, using a 48-in. chuck Blanchard, I have regularly made fairly large "C" shaped, blue chips when grinding steel! Rotary machines are especially good for premachining where one or two sides of a group of parts must be made flat or to size. The accuracy is within 0.001 in. repeatability or better.

Unit 13-1

Like so many other skills we've discussed, we concentrated on the basics of wheel selection to begin your education. I'm sure you realize there will be dozens of new abrasives types and wheel combinations found out in industry—it's going to be far more complex. But the skills you've already gained on lathes and mills, combined with surface grinding experience, will be enough preparation to transfer to many different kinds of grinders. The only missing piece lies ahead in the CNC sections and lab practice.

Unit 13-2

Upon completing Unit 13-2 you should have gained a fairly good idea of how a reciprocating surface grinder is constructed and how it works. You are ready to see the demonstration and try a small job on the tool room version. You are ready to proceed to Unit 13-3, to make setups that overcome these forces and safely grind most materials.

Unit 13-3

An understanding of magnetic chucks along with grinding wheel skills of Unit 13-1, is the heart of Section 13. You now know how to analyze the forces and to anticipate how to put together setups that guarantee safe grinding.

Unit 13-4

There are more grinding machines than the ones we've seen in Unit 13-4. Other than thread and gear grinders, which require a great deal of training, they all follow the same principles we've learned: choosing the right grinding wheel, anticipating the heat and forces, and making setups that overcome them will be fairly simple. Using the surface grinder experience, you will have the background to transfer knowledge to other machines in shop.

Questions and Problems

1. Interpret the grinding wheel shown in Fig. 13-82. How will it be best used? (Obj. 13-1)

2. What variable does the bond hardness control? Explain the grade selection when grinding a fully hard tungsten steel tool bit (HSS). (Obj. 13-1)

A - 68 - F - 4 - V
RPM 1,650 Max

Figure 13-82 *In general, how might this grinding wheel be used? For what duty and material?*

3. True or false? In general, coarser grit wheels with less air space are needed to perform heavy-duty removal grinding—rough grinding. If it is false, what will make it true? (Obj. 13-1)

4. True or false? In general, coarser wheels with less air space are needed to grind soft metals such as aluminum. If it is false, what will make it true? (Obj. 13-1)

5. What duties does coolant perform for grinding? Identify the extra duty beyond those it performed for chip cutting. (Obj. 13-2)

6. What are you seeking during a ring test? (Obj. 13-2)

7. What does a *paper washer* do? Name two things. (Obj. 13-2)

8. True or false? A mounted diamond used for wheel dressing is called a *nib*. If it is false, what will make it true? (Obj. 13-2)

Critical Thinking Question

9. While we didn't cover it in the reading, why is a diamond nib set 15° beyond wheel center when dressing a wheel? (See Fig. 13-45.) Why not place the diamond on the other side of the wheel? (Obj. 13-2)

10. Using Fig. 13-83 as a guide, describe or sketch the chuck accessory that would prevent slipping of this thin steel object. (Obj. 13-3)

11. Describe and explain the final pass when dressing a grinding wheel. (Obj. 13-2)

Set screws

Figure 13-83 *The setup.*

12. On a sheet of paper, sketch or describe in words, the stages of grinding a warped plate— to become flat and parallel. (Assuming it is thin enough that the magnetic chuck will pull it down flat to the chuck.) (Obj. 13-3)

13. Describe briefly or sketch the difference in the two kinds of cylindrical grinders. (Obj. 13-4)

14. On a sheet of paper, sketch the work tunnel in a centerless grinder. Include the regulating wheel, the grinding wheel, and the support blade. Indicate the direction each wheel rotates. (Obj. 13-4)

15. Explain or sketch the difference between lathe frames and manual cylindrical grinders. (Obj. 13-4)

Critical Thinking Problems

16. What's wrong with this surface grind setup shown in Fig. 13-84?

17. This aluminum workpiece shown in Fig. 13-85 is not ready to grind. Why? Describe or sketch the improvements needed in the setup.

18. You are surface grinding a particularly hard tool steel die component with a white aluminum oxide grinding wheel. After a short

Figure 13-85 *What can be done to improve the grind setup of this aluminum workpiece?*

Stop blocks

Steel workpiece

Figure 13-84 *How can this setup be improved?*

time, the wheel begins leaving a dark band of burned metal as it strokes across the work. List as many reasons as you can think of that might have caused the problem.

19. You've dressed the wheel from Question 18, made sure the coolant is directed right, and then began grinding again. It starts grinding well, but the burn marks soon return. In sequence, what might you do to troubleshoot this setup?

CNC Problem

20. Identify the advantages that CNC surface grinders have over manual ones.

ANSWERS 13-1

1. A = aluminum oxide abrasive; 46 = medium grit at 46 screen; H = medium hard bond; 7 = average air spaces and bond posts; V = vitrified. See Fig. 13-86. Probably used for general purpose grinding of alloy steels of moderate hardness. For semi- and finish grinding.

2. A wheel won't cut if

   The abrasive is not hard enough—obvious fix.

Figure 13-86 *Problem 1.*

The wheel is glazed—dress the wheel and choose a softer grade.

The wheel is loaded—dress the wheel and use coolant, a more open structure and coarser abrasive grains.

3. This wheel is designed for very little metal removal probably on hard steel requiring a very fine finish. It features silicone-carbide abrasive, a clue that it's for harder steel or cast iron. It also has fine grains (80), a high density (3), and a relatively hard bond (R). It would be best used for finish work or manually dressed forming.

4. It is true, but only to a point. The harder wheel holds its shape longer and may turn out to be the better choice on a case-by-case basis.

5. Aluminum oxide 20 to 30; grit, F to J; 5 to 8, vitrified.

ANSWERS 13-3

1. Here are the seven steps to remember:
   Ring and visual test the wheel.
   Double-check wheel size and RPM.
   Always use paper washers.
   Torque the nut.
   Balance the wheel, dress it, then rebalance it.
   Warm up the wheel.
   Always inspect a mounted wheel.

2. By mounting the wheel on its arbor then placing it in a balance fixture and moving small weights spaced around the mounting flange.

3. Workpiece A is safer but still could use stop blocking due to its height.

4. This is a very tall workpiece. A pair of angle plates or very large stop blocks with a C-clamp holding them together is strongly advised.

5. Aluminum is nonmagnetic—float grinding.

6. Work can be held in a vise, or on an angle plate; work can be held on magnetic parallels; work can be made more secure using the back guide; blocking can be improved with a chuck strap clamp; angular work can be held with magnetic sine plates.

7. Because if something slips, it's sent out of the wheel without gouging or digging. If it's placed on the right, it will dig in if it moves and damage both the wheel and nib.

8. When the material is magnetic and retains some residual magnetism from its contact with the magnetic chuck.

9. When the material is nonmagnetic, such as aluminum; when the chuck itself features a demagnetizing cycle.

10. Dressing reexposes sharp edges and removes loading and glazing; dressing trues the wheel and creates a new flat surface.

ANSWERS 13-4

1. ID grinders, jig grinders, and cylindrical grinders with ID attachments.
2. The entire bed pivots on the saddle. The headstock can be rotated and moved forward and back. The grinding head can also be rotated.
3. True. All the other machines studied have no way of positioning the work or grinding head to some other location. (A CNC cutter grinder might be able to grind several holes using a mandrel-mounted wheel, but it's not the intended application.)
4. The rotary table, horizontal head surface grinder, often called the "Blanchard."
5. They are doing 6,000 F/M rim speed. To achieve this speed, the $\frac{1}{2}$-in.-diameter stone must turn 48,000 RPM!

Answers to Section Review Questions

1. A fairly fine, soft, dense wheel, best suited for finish work on hard steel.
2. Bond hardness controls self-dress (along with grain friability)
   A soft bond would be chosen from around D through G to improve self-dress.
3. True for the grit but false for the structure More space is needed to hold the chips.
4. Same as answer 3.
5. Reduce friction, conduct heat away from work and wheel, help fling the chip out of the wheel when it's no longer in contact with the work.
6. A clear bell tone indicating the wheel has no invisible cracks, which produces a thud sound.
7. It averages out grain differences between the flanges to avoid pressure concentrations and it absorbs shocks to the wheel.
8. True
9. If the block slips or the table moves, the nib is harmlessly pushed out of the wheel. Otherwise it would dig in and break the wheel.
10. A chuck strap clamp. A thin bar clamped to the left of the work to stop it from shifting on the chuck. It's used when stop block parallels aren't strong enough or they are too tall and interfere with the grinding wheel.
11. One moderately slow pass that exposes new grains but does not contact them a second time. *Not* a second pass "to be sure."
12. *Step 1—warm up,* shim center and grind 75 percent of the surface flat. Step 2—flip over, shim ends and edges and grind 100 percent flat. Step 3—flip over and shim center if needed (may be able to skip this step). Step 4—flip over, and grind to size if plate lies flat without magnet on.
13. Center grinders hold work similar to lathes— between centers, chucks, and collets. Centerless grinders do not hold the work but rather trap it in a three-component tunnel.
14. See Fig. 13-87.
15. The entire bed of a cylindrical grinder (headstock and tailstock) pivots relative to its X axis ways.

Critical Thinking

16. There is nothing wrong with placing a high stop block on the right—it can be left as is. However, a high block on the left is the missing element to keep the work from tipping to the left. Adding side blocking would also be a good idea.

Regulating wheel

Floating workpiece

Grind wheel   Support blade

Figure 13-87 *Problem 14.*

High block in front

Figure 13-88 *Problem 17.*

17. First the block needs to be revolved 180° so the front can be blocked high (Fig. 13-88). Blocking needs to be placed on all four edges. Alternately, the back gage could be raised for the back stop, as shown. As an added measure of safety, a C-clamp could be added across the end stop blocks as long as it doesn't interfere with the surface grinder. You probably ought to turn on the magnet too!

18. A. It was used beyond the time when it should have been dressed. It's probably glazed now.
  B. The down feed was cranked down too much. Too big of a bite.

C. Coolant wasn't forced into the work/wheel interface and the work began heating until it reached the oxidization temp.

19. *If it fails a second time, it must be the wrong grinding wheel.*
  A. *Change wheel grade/structure*—softer more open structure.
  B. *Change abrasive*—a coarser grit wheel, but that sacrifices finish.
  Try a silicone carbide wheel then boron nitride if that doesn't work.

20. Automatic dressing saves time and wheel life since grinding wheels don't need to be shaped to the exact counterpart of the work; 3-D contours are possible; faster cutting wheels can be used since dress time is a lesser consideration.

Section 14

Technical Screw Threads

Introduction

In the mechanical world, threads perform a wide range of duties beyond fastening. They forcefully drive aircraft control surfaces or quickly move and accurately position CNC axes. Threads also join pipes that carry high-pressure fluids or hold scaffolding together—each a similar but different duty. Threads are used to lift objects as heavy as your car or to secure the lid on a mayonnaise jar. Each function has its unique thread, different from the standard versions we've thus far studied. Knowing how to machine them is one indicator of journey-level skills, because they require an understanding of their functional differences and the skillful use of data sources.

After a discussion on the technology of specialty threads, we'll practice using information found in *Machinery's Handbook*©. You'll need a copy available either on CD or hardbound to complete problems in this section. When scanning the *Handbook*, you'll quickly see that we're only touching the highlights. Any edition or version of a machinist's reference should yield the same results, but the needed information can be found in a slightly different place than described in the questions here.

Machining technical threads also requires in-depth knowledge of how to set up the machinery. When prepping manually operated lathes for threading, you'll need to find the specs to grind or select the tool bit's shape and size, to set the lathes lead selectors, and to make other setup changes. When programming CNC lathes to cut any thread, we use prewritten routines called *canned cycles* or *fixed cycles*, a generic template capable of cutting any thread, except it's missing the numbers (parameters). It's a matter of filling in blanks or leaving them off when they don't apply (see Fig. 14-1). The point is that operating a manual or CNC lathe, you'll need to understand thread formulas.

Today, few shops could be competitive making batches of threaded parts by any method other than an automated or CNC lathe. Still, when the job is *hot* and your shop needs one or two specialty threads made right away, it's the top machinist who is capable of whipping them out on the old reliable manual lathe. Here's the background you need to be that machinist, making specialty threads on any lathe, CNC or not.

Objectives

UNIT 14-1

Technical Thread Variations

Introduction: Threads perform three distinctly different task families (Figs. 14-1 and 14-2). Some designs require a thread to perform a combination of duties. Based on the target function, there are subtle differences. Those small details make a difference in the machining and measuring of a specific thread. Before we can discuss the hard data, let's look briefly at the ways threads are put to work. Take a moment to read the terms.

Figure 14-1 *A CNC thread routine makes thread cutting a snap.*

Three Task Families

Figure 14-2 *Screw threads have three different functions that sometimes overlap.*

Acme form
A thread built on a 29° triangle.

Axial force (screw thread)
The lateral force of a screw upon a nut, parallel to the screw's axis.

Ball-screw (recirculating)
Split nuts with a strong preload forcing them apart on the screw. Round balls fill the hollow space between the nuts and screw. The ball-screw has zero backlash and very little friction.

Buttress form
A thread designed for thrust in one direction only.

Galling (rhymes with falling)
Deformation of metal under heavy loads. Causes smearing and lockup of threaded fasteners.

Helix angle (threads)
The angle created in a thread, whereby the triangle has its opposite side equal to the thread's lead and it's adjacent the circumference of the thread.

Lead (pronounced "leed," long E)
The distance a thread translates in one full revolution.

Multiple start
Threads featuring more than one helical groove.

Radial force (screw threads)
The outward push of the screw against the nut, tending to keep the nut locked in position on the screw when tightened.

Square form
Designed for multiple translations—not commonly used today.

Stripping
Shearing away of the threads from excess force.

Stud bolts
Bolts with no head—threads on both ends.

Thrust
The aspect of threads that is the *push* along the axis of the thread.

Translation
The function of a thread that moves an object.

The Three Task Families

Task 1 Fasten and Assemble

These are the most common uses of threaded objects. A common bolt fastening a component on your car's engine is a one-time-only assembly. It may be removed later, but it's designed to be permanent. On the other hand, bolts that locate and fasten an inspection cover on a hydraulic fluid reservoir on a high-pressure pump must be screwed down tight to prevent leaks. But they are going to be removed and reinserted many times throughout their service lives. They will be made with more precise threads and be of higher carbon steel—perhaps hardened to an SAE Grade 8 (Unit 14-3).

Still others are fasteners on which lives depend, so they must not vibrate loose and/or must be made of extra strong metal. For example, the bolts and nuts holding wheels on trains and airplanes. Their *root* (bottom of the thread shape) will be rounded to distribute stresses better. They too would be made from hardened steel but the Grade 8 would be too brittle—perhaps a slightly tougher but a little softer bolt, a Grade 7, would do the job better? Using the wrong bolt in this application could lead to disaster.

Task 2   Impart Motion and/or Thrust (called *Translation*)

The feed screws that move the milling machine table in *X*, *Y*, or *Z* are bidirectional **translation** screws (**lead** screws). They are designed to move their axis hundreds or even thousands of times per day, for years. They are machined with one of four very different kinds of threads, depending on how much force is imparted, called the **thrust** *load*, and also on how fast the translation must occur. Another great divisor between translation screws is the amount of acceptable backlash (Unit 14-1).

As you've discovered in the lab, manual mills and lathes have a given amount of lead screw backlash—reversing the axis crank causes a stall in motion until the clearance between the nut and screw reverses. However, that kind of action is unacceptable in precision CNC machinery. Requiring zero backlash, CNC axis drives make use of a highly engineered thread called a ball-screw. Then there are thread and nut combinations that fall between the simple lead screw and the ball-screw, with just a bit of backlash, called a preload antibacklash screw (Unit 14-1).

On the less accurate side, the screw in a bench vise or on a screw jack for lifting an automobile are both translators, but the force is in one direction only and backlash is of no consequence. But between them there are sub-differences. A screw jack must translate its load farther with each turn of the crank than a bench vise, otherwise it would take too long to lift the car. So, we modify the thread for each function. The vise must also remain tightly closed during vibration and cutting forces so it fastens temporarily as well as moves the vise

jaw. Those two duties require two different versions of a form called an **Acme** (Unit 14-2).

Task 3   Form a Seal and Join

Threads joining pipes where leakage is not permitted may carry low-pressure water, gasoline, or high-pressure gasses. Each requires a similar but slightly different pipe thread.

Some pipe threads are tapered so they seal when tightened. As the tapered surfaces tightly contact each other, we call that bringing the thread *home*. The everyday pipe thread is less precise. It needs a sealant compound between the two joining pieces to make a tight seal when it's home. A second version is more precise in its form, because it must seal without compounds, to avoid chemical reactions with the material carried within the pipe—called a dry seal pipe thread.

KEYPOINT

Threaded mechanical devices *fasten and assemble, translate and thrust,* and *seal and join.* Some threads are called upon to perform a combination of these duties. Each specialized duty has a specific modification for the task.

Thread Forms for Various Functions

Unified Threads

Before discussing other thread forms, we need to take a closer look at some of the terms and formulae that define the Unified standard thread. Take a moment to study Fig. 14-3. Throughout Unit 14-1, we'll be comparing other threads to the standard 60° triangular form to highlight functional differences.

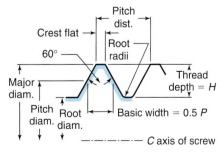

Unified Thread Terms

Figure 14-3 *Basic screw thread terms found in ANSI/ASME B1.7M (revised).*

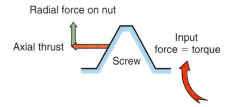

Figure 14-4 *When the face of the screw pushes on the nut, two forces are created: the axial force is parallel to the screw's axis, and the radial force is outward against the nut.*

Radial Verses Axial Thrust

There are two kinds of forces acting on a thread, one along the threaded shaft's long axis—**axial thrust** (force). The other is outward, known as radial thrust. Much of the difference between threads lies in how much radial thrust is needed to maintain fastness.

> **KEYPOINT**
>
> Increased radial thrust means the thread will remain fastened better, it won't loosen or vibrate loose. But radial thrust robs axial thrust from the total input torque put upon the thread.

The next three forms are designed to push laterally. In other words, compared to a 60° triangular thread, they convert more of the rotary torque put upon the screw into axial thrust. Radial thrust is the characteristic of a screw thread to translate some of the torque put upon it, into force outward against the nut (Fig. 14-4). For a fastener, radial thrust is a good thing. When a standard nut is tightened on a bolt, it locks on the thread due to the wedging effect of **radial thrust** (force), such that it resists loosening under vibration, changing loads and temperatures. On the other hand, radial force becomes lost power and undue friction in a translating screw.

> **KEYPOINT**
>
> In all screw (and gear) systems, the input force never perfectly equals the output forces due to friction loss. But the objective is always to use as much of the force in a useful direction, as possible, known as effective force.

Acme Threads

Evolving from the 60° thread, the Acme thread (Fig. 14-5) is also triangular and

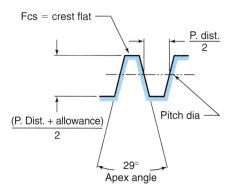

Figure 14-5 *Acme screw threads impart bidirectional thrust and moderately accurate translations (ASME/ANSI B1.5 (revised)).*

truncated. But its triangle isn't isosceles, it's equilateral with an apex angle of 29°. In Fig. 14-6, a standard 60° thread is compared to an Acme. When we observe their shapes, it's obvious the Acme is better suited to axial thrust.

Acme threads are designed to push harder—their faces are more vertical with larger contact areas, thus they tend to wear less and convert more torque to axial thrust. They last longer when constantly translating, as a milling machine's lead screw would.

> **KEYPOINT**
>
> Functionally, Acme threads resist stripping and wear better during constant use. They are specified for devices that must repeat translations and thrusts—in both directions.

But, due to steeper faces, Acme threads do not stay fastened as well as 60° threads due to reduced radial force. The 14.5° faces (29°/2) on Acme threads would make poor fasteners. They loosen due to temperature changes and vibration.

There are several subforms and classes of Acme thread, depending on the specific application. Figure 14-6 depicts two: the standard and the stub (truncated farther down the

Comparing Forms

Figure 14-6 *A comparison of Unified and Acme forms reveals their functional differences.*

triangle, making shorter less deep threads). The choice depends on the size of the shaft upon which they will be cut. To avoid stripping, the stub thread must deliver a slightly diminished trust force compared to the taller version. But the stub doesn't weaken the shaft on which it's machined nearly as much. It creates a more compact assembly.

For the same reasons as an Acme, based on function, most other thread forms can also be specified as stub threads.

Buttress Threads

These threads perform similarly to Acme threads, delivering repeated translation thrusts, but here, in one direction only. They feature an even sturdier form and vertical face compared to both Acme and Unified forms (Fig. 14-7). It's obvious which face is the thrust direction. They efficiently translate torque into nearly 100 percent axial thrust (less friction loss).

The screw for a drill press vise shown in Fig. 14-8 or a screw type car jack might be made with a buttress thread. Since there is a momentary thrust in the opposite direction to open the vise, the Acme thread might also be specified, especially for higher demand vises such as in milling. The car jack's thrust will always be up and it's a perfect application of the **buttress form.**

Buttress Threads

Type A   Type B   Type C (radius root)

Figure 14-7 *Buttress threads are designed to thrust axially—in one direction.*

Figure 14-8 *A drill press vise screw might be made with a buttress thread.*

Buttress threads are designed to translate and thrust in one direction.

Keeping to our theme of subtle differences, there are three subcategories of buttress thread forms. Type A, straight vertical faces, and Type B, 5° slope to the load face, causing a bit of radial thrust and a radius root for extreme duty applications machined upon hardened screw shafts.

Square Threads

These are the strongest threads for multiple translation and thrust loads in both directions. But, totally lacking radial thrust, they work loose the easiest of all the thread forms so they're never used where permanency is of concern. **Square form** threads offer the advantage of large surface contact area between nut and screw, thus they last a long time during repeated use (Fig. 14-9).

Not Used in Engineered Devices

Even though they offer translating advantages (strong, long lasting, and bidirectional), square threads are the least popular for precision applications due to the difficulty in measuring their pitch diameter. Their straight sides make it impossible to gauge pitch diameter (PD) other than measuring the *root diameter* then calculating PD based on it. Also, due to their broad, flat *crests* (the top of the thread form), they are more difficult to produce with a tap than all other thread forms. There is no practical measuring device (other than optical comparitors, which require removing the work from the lathe or by functionally screwing them into a comparison nut), that tells when they are machined to the right diameter.

Square Threads

Figure 14-9 *Square threads are the best at imparting thrust but it's difficult to measure their effective diameter.*

The bolt is tested against an acceptable nut. Conversely, while machining a nut, an acceptable screw is the functional test. If the gauge assembles, not tight or loose, the threads are complete. Due to this lack of size control, square threads are usually found on crude devices such as the clamps in Fig. 14-9.

Pipe Threads

Pipe threads fall into two major types: mechanical fastening and sealing types. Each of these types is then subdivided into two categories that could truly be called loose and precision fitting. Radial thrust becomes important again, since these threads must not only stay permanently assembled but must also seal. So most pipe threads feature tapered internal and external threads—they tighten as they are screwed in (Figs. 14-10 and 14-11). Therefore to gain radial thrust, pipe threads use the familiar 60° triangle on both straight or tapered pipe threads.

Nominal Differences for Pipe Threads

Pipe threads share one unique difference from all other threads. Their measured physical size is much larger in diameter than their

Figure 14-10 *Because pipe threads are designated by the ID of a standard wall pipe, they are physically much bigger than their nominal size.*

Figure 14-11 *Tapered pipe thread terms.*

nominal size. (Recall *nominal* means the target or design size.) In the past, when pipe pressures were low due to the technologies of the day, pipe threads were based on the inside diameter of the pipe, not the major diameter of the thread. So, a $\frac{1}{2}$-in. pipe thread was close to $\frac{7}{8}$-in. in diameter because the hole in the pipe was $\frac{7}{8}$ in., the determining factor (Fig. 14-10).

That relationship remains true for standard wall thickness, wrought steel pipes today (the metal and plastic pipes you would find at the hardware store). However, for heavy or extra heavy wall pipes, the difference between inside diameter and thread size is even greater.

> **KEYPOINT**
> Pipe sizes are based on capacity not physical diameter. Pipe diameters are bigger than the nominal size.

The National Pipe Thread Designations

In North America, pipe threads fall within the American Society of Mechanical Engineers (ASME) (formerly ANSI) standards B1.20.1-revised. Similar to the unified system, their designations look similar to other threads. Here are the three most common, but as seen in *Machinery's Handbook*, there are a bunch of subcategories:

NPT (National Pipe Tapered)

Example: $\frac{1}{8}$-27 NPT. This is a tapered thread commonly used for joining water pipes in your home, carrying low pressure fluids. The seal between the male (outside thread) and female thread is accomplished when the conical thread comes home with a sealant compound between them. Taps and dies are available to cut these threads (Fig. 14-12).

> **KEYPOINT**
> Pipe thread pitches are different from common fasteners such that no other thread might accidentally be assembled with them.

NPTF (National Pipe Dry-Seal)

Example: $\frac{3}{4}$-14 NTPF. A tapered pipe thread that seals without compound or sealing material. This is the more precise thread used for

Figure 14-12 *A set of tapered pipe taps and dies.*

gasses and high pressures. The F is used to designate fit-up dry.

NPSM (National Pipe Straight Mechanical)

Example: $1\frac{1}{4}$-$11\frac{1}{2}$ NPSM. Designed for general assembly where no internal pressure is required. Notice that the pitch is eleven and *one-half* threads per inch to prevent assembly to other threads, which could lead to disasters, as in scaffolding struts and bars, for example.

Critical Application Threads and Fasteners

When the thread must be rigidly controlled such that it will not fail, we must turn to standardized quality. These parts are made according to very rigid rules and specifications *not found in Machinery's Handbook,* but in government and professional association documents such as the ASME-ANSI or military specifications (Mil-Spec).

These documents are published at several levels, national government (Mil-Spec) or International Standards Organization (ISO), professional association standards (Society of Automotive Engineers, American Society of Mechanical Engineers), and individual companies publish their own specifications.

Where a national or international standard is in existence, lower authority specs such as company standards may expand on but not override or change them. They can supplement the specification to

ShopTalk

The Bad and Good News About Tapered Threads!

Without a die, outside tapered threads are not commonly made on the manual lathe. However, they can be machined using a tapering attachment—we'll see how in Unit 14-4. Since that setup is very time-consuming compared to making them on a special pipe thread lathe using dies, tapered threads are not considered a normal manual lathe task. The dedicated thread lathe is a portable machine that can be moved to a job site and quickly produce straight or tapered outside diameter pipe threads. When they are occasionally requested in the machine shop, they are usually produced with tapered dies and taps, if available (Fig. 14-13).

Figure 14-13 *A portable pipe threading lathe is used at construction sites.*

Critical Threads

1/2–13.UNCJ

1.70
Complete
threads

2.00

XX ▷ XXXX XXXXX

YY ▷ YYY – YYYYY

ZZ ▷ ZZ.ZZZZ

11 ▷ Per mil-spec S-8879

Figure 14-14 *When threads must meet standard specifications, the control document will be noted on the drawing. You must meet the specs if the thread is to be legal.*

tailor it to the individual industry, but that expansion can never dilute a higher authority.

These closely controlled specifications are not normally called out on the drawing—but they are noted usually with flags that tell the reader what specification to use (Fig. 14-14), flag 11 in the drawing. You'll need further training in this aspect of machining. It's usually covered in a complete or advanced print reading course at a specific company.

KEYPOINT

If a thread spec is shown on the drawing but the information is not on the drawing (the usual condition) then you must obtain the data and guarantee that your work follows it exactly. To be legal and certified, the parts must meet those specs.

Controlled Radius Root Threads

These are a good example of a standardized thread form for a critical application. Sometimes the straight-sided 60° triangular form must be modified due to the hard intersection of the thread face and root diameter. The sharp corners create points where forces can concentrate and start a failure in the bolt. The bolt's strength is improved by distributing the forces over a rounded corner, by adding closely controlled root radii (Fig. 14-14). Radius roots can be called out on any thread form, but when they are critical, a specification will be noted.

Where that radius must be controlled for exactness, the thread callout on the drawing will include the letter "J." For example, $\frac{1}{3}$-13 UNCJ. You must be able to determine the exact radius on bolt or nut. There are questions coming up

Figure 14-15 *A Helicoil® thread insert creates strong, stainless steel threads (or other alloys) within a softer metal workpiece.*

in this section's response problems so you can practice this kind of data access.

Threaded Inserts

Standard tapped internal threads work fine in softer metals such as magnesium or aluminum, for many applications. But when great demand is put upon bolts screwed into these softer materials or where they must be assembled repeatedly without failing, standard tapped threads can strip or deform.

So, when the assembly must not fail, we create a stronger, harder internal thread than the parent material by inserting 60° cross section, stainless steel or other hard alloy wires wound into a helical coil (Fig. 14-15). The insert wire is threaded into specially tapped holes to create a standard threaded hole that will accept a standard bolt.

KEYPOINT

Threaded inserts are used in soft and hard metals and even in plastics, to create stronger and more reliable threads.

Installing Threaded Inserts

To install a threaded insert, first a specific diameter hole is drilled. Then a special tap is driven into the material, creating the thread that accepts the insert. It has the right pitch distance but a larger diameter than the standard size thread.

Figure 14-16. Then to insert the coil, an installation tool is placed over a driving tang that protrudes from the coil's outer end, to drive it into the tapped hole. Then once installed, the tang is removed from the coil by striking it with a rod.

Advantages After insertion, the coil expands outward with a spring-like action, to

Figure 14-16 *Installation is a four-step process: drill, thread the hole with special tap, then screw-in the insert, last break off the drive tab.*

Grade 1	Grade 5	Grade 7	Grade 8
Med-low carbon steel Everyday use No heat treat	Medium carbon steel heat treated	Medium carbon alloy steel heat treat med hard	Medium carbon alloy heat treat 15% harder

Figure 14-17 *SAE bolt grade markings indicate the material from which it's made, the heat treatment (if used), and relative hardness of a bolt.*

anchor it permanently in place. The key advantage to threaded inserts is that the coil distributes the assembly load over the full length of the thread, thus providing greater overall strength. They not only resist stripping, they wear better for repeated use. Threaded inserts are common in aircraft, automotive, and many other manufacturing industries. The most common manufactured variety is Heli-Coil®.

Fastener Grades

Steel bolts are classified by their strength grade. Hardness determines strength. Bolt grades span from 1 (softest) to 8 (hardest) (Fig. 14-17).

But as the bolt's tensile strength (rated strength during stretching) increases, the ability to flex and bend without breaking goes down. The hardest bolt isn't always the best choice. Here's a baseline for application.

ShopTalk

Home and Automotive Applications of Threaded Inserts Accidentally stripped spark plug holes in your lawn mower or car engine can be re-tapped with specially made tools found at automotive supply stores. They come in kits with the right insert to fit the original spark plug. When tapping the hole, be extra careful to not let any of the chips fall into the cylinder, which would destroy the engine. The best solution is to remove the cylinder head. With small engines, you can sometimes plug the hole with a rag and then turn the entire engine upside down during tapping, such that chips tend to fall out—downward. Careful removal of the plug afterward helps prevent unwanted chips from entering the engine.

Correct Application

Grade 1 = Everyday applications—the bolts and nuts found at the hardware store. Assembling a backyard barbecue or child's wagon.

Grade 5 = Medium strength—fastening noncritical but demanding items such as a universal joint or a wheel on your car. Often found in automotive stores.

Grade 7 = More demanding applications— a cylinder head on your car or a jet engine part. Might be at the hardware store but usually bought through companies specializing in fasteners. May need to be certified to meet specs.

Grade 8 = Highest demand applications— milling fixtures, dies, and a part on a CNC lathe. Usually purchased through certified suppliers if they will be assembling critical items where safety is concerned.

KEYPOINT

Bolt grade is determined by code marks on the bolt head. Grade is normally specified in the drawing notes—not in the thread callout.

Thread Variations

Now let's examine two unusual "twists" that can be added to any of the thread forms we've studied: left-handed and multiple-start threads.

Left-Handed Threads

These threads tighten in the opposite direction to those we've thus far studied. They have a functional purpose: sometimes the right hand helix spirals the wrong direction. They might loosen under load and vibration.

For example, the nut holding the grinding wheel on the left side of a pedestal grinder would loosen from the grinding forces if it were a right-hand thread but if we make it a left-handed spindle and nut, it remains tight. Studs holding the wheels on some cars are

Left-hand
thread helix

Tighten

Figure 14-18 *Left-hand threads tighten in the reverse direction.*

left-handed on the left wheels. You can see the LH stamped in the end of such **stud bolts** (bolts without a head—threaded on both ends).

A third example is a turn-buckle joiner for a cable (Fig. 14-18). Its purpose is to tighten the joint between two cables or rods, when the center buckle is rotated. As it is turned, the right-hand thread pulls from one end while the left-hand thread also pulls inward on the opposite end (Fig. 14-19).

Any thread can be specified as left-hand spiral. For example,

$$\tfrac{3}{8}\text{-24 UNF LH}$$

or

$$\text{M12-1 LH}$$

Left-hand taps and dies are available but it's not common to find them in most shops unless they make many parts requiring left-hand threads. In Unit 14-3, we will explore two manual lathe methods of making these threads. Naturally, left-hand threads are a snap to produce on CNC lathes, but the trade tips we'll be learning to produce them on manual lathes will be useful.

Multiple-Start Threads

These threads are something of a mystery to beginning machinists. They feature more than one helical groove on the screw and

Right hand

Left-hand thread

Figure 14-19 *A turn-buckle is an example of left-hand thread use.*

on the mating nut. At first, they might be difficult to envision—more than one thread groove on the same shaft? Their function is to increase the rate of translation per revolution.

KEYPOINT

Multiple-start threads move the nut farther in one revolution compared to single-start threads.

The term **multiple start** means there are two or more individual thread grooves in the object. To explore them further, we need to define the term **lead** (long "E") and compare it to pitch distance.

Lead is the distance translated by the nut in one revolution.

Pitch Distance is the repeating distance between threads.

KEYPOINT

Lead and pitch distance are the same on single-start threads but different on multiple starts.

A double-start thread translates twice the pitch distance in one revolution compared to a single-start thread. Look at Fig. 14-20 to see a double-start thread not yet complete. As yet, it has only the first thread groove completed. Notice the blank space between grooves, ready for the second start. Then in Fig. 14-21, it's finished with both grooves. Placing an imaginary

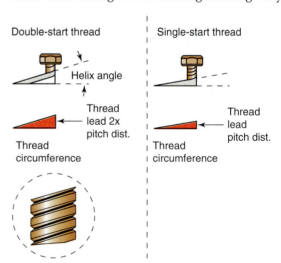

Figure 14-20 *A double-start thread has two separate helical grooves. Only one groove has been machined in the magnified view.*

Both Starts Complete

Start #1

Start #2
180° opposite

Figure 14-21 *Both thread helices are now machined on this double-start thread.*

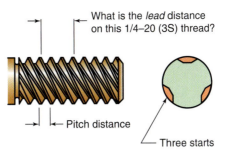

What is the *lead* distance on this 1/4–20 (3S) thread?

Pitch distance

Three starts

Figure 14-22 *Calculate the lead.*

pencil in a groove, then tracing it one full turn of the helix, it would lead twice as far as would a single-start thread of the same thread pitch—it has a steeper (larger) **helix angle.**

The relative stripping strength of the double-start thread remains about the same as a single-start thread of the same pitch; the finished thread has the same number of threads per inch as a single start and it has the same amount of thread surface carrying the load. The difference is the steeper angle of the individual threads.

Examples of multiple-start threads can be found all around you. For examples, the screw-on top of a glass mayonnaise jar or on a quality writing pen cap or your gas cap on your car. With one quarter turn of your wrist, it's assembled or removed. They are either three- or four-start threads.

Calculating Lead The lead of any thread is equal to the pitch distance multiplied by the number of starts.

Lead = Pitch distance × Number of starts.

For example, what is the lead for this thread: $\frac{1}{2}$-20 UNF 2S (Double Start)?

Find the pitch distance.   1.000/20 = 0.050 in.

Multiply PD × 2   0.100 in.

In Fig. 14-22, it follows that, if the same thread were a triple start, what lead would it have? Three times the pitch distance at 0.150 in.

Trading Mechanical Advantage Using multistart threads in a device, creates a trade-

off of mechanical force for distance traveled, compared to single-start thread. With the inclined plane (wedge effect) becoming a steeper helix angle, the amount of mechanical force created per unit of torque applied is directly reduced as the lead increases. In other words, it takes a harder twist on the wrench to achieve the same tightening force as the multiple starts go up.

Backlash Control Threads

In some mechanical devices, backlash must be controlled to a minimum or eliminated completely, based on function. There are three common kinds of screw-nut systems that control backlash: split nuts, loaded split nuts, and ball-screw split nuts. Each is a design evolution toward less backlash.

An axis drive on everyday machinery can be as simple as an Acme screw and a suitably well-fit bronze nut. That system must have some mechanical clearance, even when new, otherwise it will bind. The best fit screw and nut will wear over time, even when well lubricated. Backlash can be a problem in these devices and it gets worse with time. There is nothing to do but replace the nut when backlash exceeds practical limits. The nut is generally made of a softer material (brass or bronze) to deliberately wear out before the more costly lead screw. But there are better systems.

Split Nuts—Adjustable

Most industrial-quality machines are fitted with bronze split nuts—two nuts running on the same screw shaft—with a small space between them. We saw an example in Section 11 on the *X* axis, cross feed on most lathes. Backlash is adjusted to a minimum by moving one nut relative to the other—then locking its

Adjustable Backlash Lead Screw

Fixed nut   Adjustable nut
(parallel to
axis only)

Hardend
steel
Acme lead
screw

Figure 14-23 *A simplified adjustable split nut backlash screw and nut.*

position on the pair's carrier. Adjusted correctly, it can have very little backlash, but zero plus it will still wear. If adjusted to zero, the system would have too much friction and binding would be a problem (Fig. 14-23).

Manual mills feature similar devices on their *X*, *Y*, and *Z* axes. These systems require periodic adjustment to compensate for wear.

Preload Split Nuts— Self-Adjusting

The next higher backlash design also uses the split nut concept, but now the two nuts have a strong spring (or hydraulic piston) placed between them. It allows the adjustable nut to move laterally (within a very small distance), relative to the other. The spring (red) pushes the sliding nut (orange) away from the fixed nut. In this system, pressure is exerted against both sides of the thread face, creating zero backlash for thrust loads up to given amount (Fig. 14-24).

They continuously compensate for wear in the screw, even when the wear isn't uniform. The problem is that they often wear more in their center of travel than at the two ends. The spring load nut system can compensate over

Preload Antibacklash Lead Screw

Fixed nut

Controlled slide
(Parallel to axis only)

Hardened
steel
Acme lead
screw

Preload force

Figure 14-24 *A simplified preloaded split nut system uses spring or hydraulic pressure to force nuts apart.*

these changing wear patterns for backlash, because it constantly adjusts. But accuracy is lost.

The rigid split nut system cannot accommodate uneven wear. Adjusted correctly for the central worn section, it will tighten and possibly jam on the ends of the axis screw.

But the preloaded split nut has two limitations. First, the spring action accelerates wear on the nut. The nuts must be replaced on a schedule based on use. The other is that the spring can be overcome by extreme machining loads, making the axis jump or bounce; not a good thing during a cut. As such, spring-load axis screws are limited to small, light-duty CNC machines only. One solution to the wear problem is to use Delron® or Teflon® nuts or some other high-density plastic that slides extremely well on the metal screw. But they too cannot withstand heavy machining loads.

To increase the amount of machining load withstood, designers can add hydraulic pressure between the two nuts. However, the nuts are then subject to even more rapid wear. The final solution for the big gun, industrial-strength machines is the ball-screw.

Ball-Screws—Continuous Feed

While other axis drive devices have been tried for CNC equipment, we keep coming back to **ball-screws** as the translation system of choice. Again, a preload is used between two split nuts, but this time the pressure is so strong that no machining force can overcome the parting force on the nuts. Under that much side load, the friction would be so great between the nut and screw that the system would bind or the nuts would wear out almost immediately. The friction solution is to forsake the Acme thread and machine and precisely grind thread grooves into half-round spaces on both sides—nut and screw (Fig. 14-25). This creates a round space between the two components.

Then precision balls are introduced between the nut and screw. They roll and take the thrust load while reducing friction close

Ball-screw and Nut

Figure 14-25 *A ball-screw and single nut.*

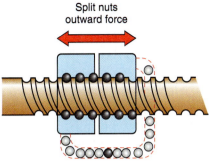

Recirculating Ball-screw

Split nuts
outward force

Recirculation Channel

Figure 14-26 *A ball-screw is the most popular CNC axis drive translating screw.*

to zero. But, in designing a ball-screw, one more problem had to be overcome—if the axis were to move farther than the width of the nuts, the balls would roll out of the nut. So, a recirculation channel surrounds the nut pair (Fig. 14-26). It can be as simple as a tube leading from the opening on one end of one nut to the opening on the far end of the other nut. As the screw translates over a great distance, the balls roll out into the channel to be pushed into the far end, thus continuous recirculation of the balls.

Single and Multiple Start Ball-Screws

To take advantage of the fast central processors in newer CNC controls, ball-screws feature multiple-start threads. Older CNC machines and other devices such as truck steering systems, use single-start ball threads.

UNIT 14-1 REVIEW

Review the Key Terms

Acme form
A thread built on a 29° triangle.

Axial force (screw thread)
The lateral force of a screw upon a nut, parallel to the screw's axis.

Ball-screw (recirculating)
Split nuts with a strong preload forcing them apart on the screw. Round balls fill the hollow space between the nuts and screw. The ball-screw has zero backlash and very little friction.

Buttress form
A thread designed for thrust in one direction only.

Galling (rhymes with falling)
Deformation of metal under heavy loads. Causes smearing and lockup of threaded fasteners.

Helix angle (threads)
The angle created in a thread, whereby the triangle has its opposite side equal to the thread's lead and it's adjacent the circumference of the thread.

Lead (pronounced "leed," long E)
The distance a thread translates in one full revolution.

Multiple start
Threads featuring more than one helical groove.

Radial force (screw threads)
The outward push of the screw against the nut, tending to keep the nut locked in position on the screw when tightened.

Square form
Designed for multiple translations—not commonly used today.

Stripping
Shearing away of the threads from excess force.

Stud bolts
Bolts with no head—threads on both ends.

Thrust
The aspect of threads that is the *push* along the axis of the thread.

Translation
The function of a thread that moves an object.

Replay the Key Points

- Lead of a thread is equal to the pitch distance times the number of starts. These two numbers are the same on single-start threads.

- Pipe threads are specified with pitches different from common fastener threads such that no other nonpipe thread might accidentally be assembled with them.

- The Acme thread form is a truncated 29° triangle offering more face area compared to 60° threads.

- Buttress threads are specified when translation or thrust in one direction is the function of the device.

Respond

1. True or false? A *buttress* thread is designed for thrust in one direction only but has less contact surface between the nut and bolt, compared to a standard 60° triangular thread. If it is false, what will make it true?

2. Identify the thread form shown in Fig. 14-27.

Figure 14-27 *Identify this thread form.*

3. In Fig. 14-28, the screw has 8 TPI. What is the lead?

4. Write the correct drawing callout for the pipe threads in Fig. 14-29. Requires looking in the *Handbook*. What thread form is used on this thread?

5. From *Machinery's Handbook*, what is the correct *pitch* for this tapered pipe thread?

$$2\tfrac{1}{2}\text{-}\underline{\quad}\ \text{NPT}$$

6. How would the callout change for the pipe thread of Problem 5 if it were to seal without a sealant compound?

7. How big is the inside diameter of the pipe shown in Fig. 14-29 in Problem 4?

8. Describe the three major functions of screw threads.

9. True or false? The difference in helix angles between a single- and double-start thread of the same diameter and pitch, is that the double-start thread is greater (larger). See Fig. 14-20. If it is false, what will make it true?

10. A. Name two reasons a designer might specify a left-handed thread.
    B. Identify one functional reason to use multiple-lead threads over single-lead threads.

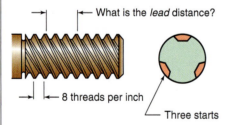

What is the *lead* distance?

8 threads per inch

Three starts

Figure 14-28 *Calculate the lead.*

1.050 dia

Figure 14-29 *Write the thread callout for this pipe thread (standard wall pipe).*

Getting and Using Thread Data from References

Introduction: You will need to make calculations when setting up to machine and measure a specialized thread. They will be based on reference books, specification charts, and company data. Here we'll concentrate on *Machinery's Handbook©* but many other sources should yield the same results. Keep in mind that this unit intends to provide a working knowledge of the kinds of information needed and how to find it when you need it—do not try to memorize the data.

There is little instruction needed, only practice is required. Next are a few examples with answers found at the end of this section. Check them as you go.

UNIT 14-2 REVIEW

Respond

Solve the following puzzles which typify a custom thread setup in industry.

Problems 1 and 2 refer to this general purpose thread:

ACME 1.0-8

1. What will the *smallest* possible *major diameter* be for the bolt (external thread)? This is the lower size limit of the shaft before threading. (*Hint:* Look under screw threads—Acme, in the index or search words if using the CD version.)

2. You are to set up and bore an internal Acme thread to the specification shown here. How wide should the crest flat be when this thread is to its finished size?

3. You must spotface a minimum cleanup flat area on a drill jig to accept a $\tfrac{1}{2}$-in. standard wing nut. What diameter counterbore should you obtain? See the hint in the answers if you get stuck.

4. You have been assigned to turn a type B buttress thread. To do so, you will need to grind the included angle of the bit to _____ degrees?

5. You are to rethread a spark plug and need information about the various dimensions. You suspect it is a metric thread. Where will you look?

6. There are actually two common types of American Standard Acme threads, what are they?

7. *Finalize Your Scan.* From scanning the *index*, list the various major topics in which you could find information about threads. See if you can find more than I did in 3 minutes or less.

8. Turn to two subjects from your list or mine in which you have no information at all, such as rolling threads or Aero threads. Read just enough of the technical data there to satisfy your curiosity then stop.

There are no new terms nor key points to review in Unit 14-2. Move on to Unit 14-3.

Figure 14-30 *Die heads (trip dies) produce quality, external threads by cutting forward in one pass, then snap open for rapid travel off the part.*

UNIT 14-3

Manual Lathe Setups for Technical Threads

Introduction: This unit discusses how to set up a manual lathe to produce various technical threads. However, much will also apply to CNC lathe setups. Before proceeding, there is a threading option we've not yet discussed, the **tripping die** or collapsing tap head. Their main purpose is long-run production of high-quality threads. Mounted on the lathe tailstock or tool turret, they cut threads quickly and accurately with very repeatabilities within 0.001 inch.

As shown in Fig. 14-30, they are supplied with interchangeable jaws that can be configured to cut nearly every possible thread variation and form we have discussed. While being fed inward, cutting the thread, they lock at thread size, until touching a trip lever that opens the die or collapses the tap. Then they can be withdrawn at high speeds rather than unscrewing them.

They make high-quality threads very fast. Once the die or tap is set up to the correct thread diameter and lathe at the best RPM, the setup is supplied with plenty of flood coolant or thread cutting oil (always essential when cutting threads) and these production dies reproduce high quality time after time.

KEY TERMS

Inverted tool
A left-hand thread setup whereby the lathe cutter is turned upside down to allow cutting toward the chuck with reversed spindle rotation.

Reverse travel
A left-hand thread setup whereby the cutter travels away from the chuck during a thread pass.

Tripping die
A production thread cutting tool that cuts on the inward stroke but collapses to allow rapid removal on completing the thread.

Left-Hand Thread Setups

There are two options for modifying the normal, single-point threading setup (Section 11):

Inverted tool   Reverse travel

Inverted Tool Method for Left-Hand Threads

Here the thread cutting tool is mounted upside down in the tool holder (Fig. 14-31). The spindle rotates clockwise when viewed from the tailstock—backwards from normal turning so the cutting action is up rather than down (see the Trade Tip).

In this setup, during the cut, the tool travels to the operator's left, in the normal direction, toward the chuck. This means that the threading tool can be safely and conveniently started off the workpiece as with normal thread turning, thus providing a bit of time and space to engage the half-nut clutch. Also,

Left-Hand Threads—Inverted Tool

Reverse spindle rotation

Left-hand helix

Tool travel

Solid tool post

Compound swung CCW Half angle 29.5°

Figure 14-31 *The tool is mounted upside down to machine these left-hand threads.*

at the end of the thread pass, it can be withdrawn to create a pull-out thread (incomplete threads at far end). So this method can be used when no thread relief is permitted in the work.

KEYPOINT

The inverted tool method provides the escape distance needed in case you miss the half-nut lever engagement mark! It must be used when no thread relief is allowed on the workpiece.

Although the setup is a bit more complex, it does make a good part. The next setup, reverse travel, offers no room for engagement error. Using it and missing the half-hut mark means a ruined part!

Trade Tip

Safety Considerations *Cam Locks Only* Before running a lathe backward (clockwise), be certain the chuck is mounted on the spindle using modern cam locks, not a screw-on spindle nose. On a rare few older lathes, threaded chucks can dangerously unscrew due to the opposite forces of reverse turning. Most using threaded spindles will not rotate backward but—.

Solid Tool Posts An open-sided or solid tool post must be used because it's able to withstand the upward force. The post chosen must allow the cutting edge to reach the work's vertical centerline. Quick change tool posts are not allowed, as the action tends to take the tool block up off the post.

Left-Hand Threading Tool The pre-shaped threading tool with the correct thread form is ground with a left-hand side rake if that tool was positioned upright (cutting to your right normally).

The Tool Compound Angle Similar to cutting right-hand threads, to ensure that the cut occurs on the correct cutting edge of the tool bit, the compound is swung to slightly less than half the included angle of the thread form. In this method, it is swung counterclockwise, into the same position as if the threads were right-hand helix. This method works for internal threads too, but the compound direction is reversed.

Reverse Tool Travel Method for Left-Hand Threads

This method is a bit simpler to set up since the cutting tool is mounted in the standard position, with a left-hand rake—to cut to your right. The tool starts at the headstock end of the work. For this method, shift the tool travel direction such that the lead screw sends the tool away from the chuck when the half-nut is engaged, with the spindle turning forward, in the normal rotation.

Thread Relief Required If no thread relief is possible on the part, then the reverse tool travel method cannot be used. The tool must start from a static parking spot while you wait for the chasing dial to come to lock on position. To do so, a relief groove must exist in which to park the tool (Fig. 14-32).

Trade Tip

Forsake the Chasing Dial Rules! In both methods of machining a left-handed thread, the chasing dial will be rotating backward to the direction you have been accustomed to seeing. Although the rules are still valid for left-hand threads, for a beginner, it would be a good idea to simplify engagement by dropping the lever on the same mark every time. That takes away one small variable in an otherwise oddball setup. I mark the line I intend to use on the dial with a red marker just to be sure.

Multiple-Start Thread Setups

Here's how to set up and machine multiple-start threads on a manual lathe.

The trick in this setup is to build in a way to accurately index (rotate the part a fraction of

Left-Hand Threads—Reverse Travel

Normal spindle rotation

Thread relief

Tool travel

Solid tool post

Compound
swung clockwise
Half angle 29.5°

Figure 14-32 *Left-hand threads can be produced by reversing the tool travel as the spindle rotates normally.*

a turn) forward after machining each complete start, ready to machine another start. If it's a double-start thread, the part must be indexed half way around after completing the first thread. It follows that it must be indexed one third of the way for each groove of a three-start thread, and so on.

The setup is made two different ways.

Method 1   Holding Between Centers

If the number of starts are two or three, a drive plate can be used to create the correct spacing between thread starts. Placing the drive dog tail in one slot, one complete thread is made. Then without disturbing the drive dog on the work, the tailstock is withdrawn, the dog is pulled from the slot, and the part is rotated to the next slot in the plate representing the multiple. Most drive plates feature three slots equally spaced with a fourth opposite one of them, to create two-start threads. For greater multiples, use a face plate with six equally spaced drive slots.

In the example in Fig. 14-33, a three-start thread is being produced. Common multistart threads are two, three, and four.

Multiple-Start Threads

Drive or
face plate

Complete
first thread

Place tail
here for
second start

Figure 14-33 *This setup will produce a triple-start thread. The tail of the dog is placed in each slot then a single thread is produced with the correct lead.*

KEYPOINT

Multithreads

A. For each revolution of the spindle during threading, the tool must move laterally, a distance equal to the lead not the pitch distance. *Set the quick change gears to the lead not the pitch.*

B. The bit must be ground with more side clearance on the cutting side than for normal threads due to the increased helix angle of the thread. The bit must be extra relieved on the cutting edge face (similar to a face grooving tool—more side clearance on one side than the other).

C. After a thread is begun, never change gears for RPM or set the spindle in neutral. If you do so, the coordination between spindle and tool will be lost and the threads will be damaged.

Method 2   Indexing the Whole Chuck

Equally spaced cam locks on the back of a lathe chuck can also be used to index work for multiple threads. This little tip works especially well for work that is difficult to hold between lathe centers. Leaving the work tightened in the chuck, unlock the cams the remove it from the lathe spindle nose. Index it the amount needed and remount the chuck. Before setting this up, count the number of cam locks to be certain your number of starts is prime to the spindle's number (usually six cams).

Other than the indexing operation, every other aspect of producing the multiple thread remains the same including measuring the thread depth. A reminder, using the special X axis thread depth coordination makes the depth control easy too (recall the Trade Tip in Unit 11-6 on coordinating the dials).

Producing Tapered Threads

From time to time, machinists must produce a custom-tapered thread. It can be cut on a manual lathe featuring a taper attachment or using the tailstock set off center.

Taper Attachment Method

This setup is chosen for work suitable for holding in a chuck, or not easily adapted to tailstock taper turning (Fig. 14-34). As explained in Unit 14-1, you must begin the setup by calculating either the taper angle or taper per inch to set the angular difference

Shallow Tapered Threads

Taper attachment

Short straight distance—no taper occurs

Figure 14-34 *A tapered thread can be made with a taper attachment or by setting the tailstock off center. If using the taper attachment, allow for the flat spot.*

between the bed ways and the attachment. Most attachments are graduated in both units. When using angles keep in mind that you must compute the half-angle, one side compared to the work's axis, not the full included angle.

Trade Tip

A taper attachment has considerable mechanical clearance. Therefore, when the tool starts to move in the Z direction, the X action stalls for a while until the mechanical clearances are taken up and the tool begins to move in or out. That means that there is a short distance where *no taper* will be produced—the tool moves in a straight line. To solve this, make sure that the tool is engaged far enough off the work to accommodate the straight portion before it cuts metal.

Once the tapered portion of the shaft is machined and verified correct, producing the thread is no different from straight threads. The compound is turned to slightly less than half the thread's included angle and the micrometer dials are adjusted to zero. If a thread stop is available on the lathe, to help limit in and out cranks on the X axis, it is a good idea to use it. The thread stop prevents starting the tool at the wrong position during successive cuts.

Offset Tailstock

Work held between centers with the tailstock offset to half the taper angle, can be used for tapered threads. There is no mechanical play for the tool—the threads are easily produced without concern for a straight section.

Trade Tip

If using an offset center tailstock method, when two parts of slightly different length are threaded, the taper angle will be slightly different. Draw a pencil sketch if this is not immediately obvious to you.

UNIT 14-3 REVIEW

Review the Key Terms

Inverted tool
A left-hand thread setup whereby the lathe cutter is turned upside down to allow cutting toward the chuck with reversed spindle rotation.

Reverse travel
A left-hand thread setup whereby the cutter travels away from the chuck during a thread pass.

Tripping die
A production thread cutting tool that cuts on the inward stroke but collapses to allow rapid removal on completing the thread.

Replay the Key Points

◆ For each revolution of the spindle, during multistart threading, the tool must move laterally, a distance equal to the lead not the pitch distance.

◆ When cutting multiple-start threads, the bit must be ground with more side clearance on the cutting side than for normal threads.

◆ There are two methods of single-point machining a left-hand thread: Toward the chuck with the tool upside down and the spindle running backward or with the tool upright and spindle running forward but moving from the chuck to the far end of the work.

Respond

1. When setting up a lathe to cut a multiple-start thread would you shift the gearing to the pitch or lead of the thread? Why?

Critical Thinking

2. A tapered thread with an included angle of 34° must be machined. How will you make the setup?

3. What is the lead of this thread? $\frac{3}{4}$-12-*Triple Start*

4. Can an object with a 12.5° included angle tapered thread be made on a manual lathe equipped with a tapering attachment?

5. If Question 4 is possible, to what angle must you set the taper attachment?

Section 14 Review

Unit 14-1

After looking through *Machinery's Handbook,* I'm sure you realize that we've only touched the high spots on technical thread variations. In my apprenticeship, we studied them for an entire semester! The rest of your education will happen on the job when the nature of your shop's business determines which specialty threads you'll need to understand and make. For example, an aerospace machinist in California will be making radically different threads compared to a tool and die maker in Detroit.

Unit 14-2

Although working with charts and large published specification books, and all the numbers they contain may seem boring, clearly management and engineers must have that deep understanding of their usage. Digging into them deeply is an investment in your own future!

Unit 14-3

Few modern machinists would recommend producing large batches of technical threads using a manual lathe. That would be a business plan sure to go quickly out of business. However, in a job shop environment or in a tooling shop supporting CNC production, your skills in this area will be essential. It's very rewarding to know you are the only person around who can make the difficult threads on the manual lathe!

Questions and Problems

1. Other than fastening, name the two other functions of threaded devices. Describe each. (Obj. 14-1)

2. Which thread form is built on a 29° triangle? (Obj. 14-1)

3. Which thread form features a radius root? (Obj. 14-1)

4. For what engineering reason was the radius root thread invented? (Obj. 14-1)

5. For what functional reason do we sometimes machine left-hand helix threads? (Obj. 14-1)

6. Briefly describe four ways a left-hand thread can be made. (Obj. 14-3)

7. A steel hex head bolt has three lines on its head, equally spaced. What class thread is it? (Obj. 14-2)

8. For what function is the buttress thread designed? (Obj. 14-1)

9. What advantage does an Acme thread offer over a 60° thread form? (Obj. 14-1)

10. Identify two possible ways to index a three-start thread on a manual lathe. (Obj. 14-3)

11. When single-point turning a tapered thread, there are two possible setups. Identify them. (Obj. 14-3)

Problems

12. The pitch distance of a given thread is 0.07692 in. but when a nut is turned upon it, it advances a little over 0.300 in. Explain. (Obj. 14-2)

13. What is the number of threads per inch pitch for the screw in Problem 12? (Obj. 14-2)

14. Explain what is different about the following thread compared to a standard unified form thread: $\frac{7}{16}$"-18 UNCJ. (Obj. 14-2)

Critical Thinking Questions

The two bolts shown in Fig. 14-35 both have a problem in their manufacture.

On Bolt A, the pitch diameter is smaller than it should be by 0.010 in. However, the major diameter is right. In other words, the round rod from which the bolt was made, is the right diameter but the threads themselves are too small.

On Bolt B, the major diameter is undersize by 0.010 in. but the pitch diameter of the thread checks out to be correct.

15. When each of these bolts are assembled with a correctly machined nut, how will they fit?

16. How will each bolt perform when used? What are its weaknesses?

17. To setup and machine a $\frac{1}{4}$"-20 UNC thread, you must know the pitch diameter. From *Machinery's Handbook,* identify the minimum and maximum PD value for a Class 2 screw thread.

18. A $\frac{7}{16}$-18 UNC screw is measured and all dimensions are found to be within tolerance including the pitch diameter. Then the outside diameter of the shaft and the thread crests are damaged. To repair it, the diameter of the entire shaft length is turned down by 0.015 in. Then a chasing die is run back over the threads (Section 5). How much did that rework change the pitch diameter?

Figure 14-35 *How will a correctly machined nut fit on each problem bolt?*

CNC Questions

19. Briefly describe the most popular CNC axis feed screw and nut. Why is it used?

20. Name three or more reasons to choose CNC lathes to produce specialty threads whenever possible.

21. What is the one thread a CNC lathe might make that a manual lathe cannot?

Section 14 Answers

ANSWERS 14-1

1. It is false—the buttress thread features more contact area than a 60° thread.
2. It is a *left-hand,* Acme thread. (Not important if you missed the hand helix.)
3. 0.375 or $\frac{3}{8}$ in. Calculated: Pitch distance is $\frac{1}{8}$ in. = 0.125 × 3. The lead is three times the pitch.
4. From "Basic Dimension—American Standard Tapered Pipe Threads" $\frac{3}{4}$-14 NTP (National Standard Tapered Pipe Thread). Uses a 60° triangle form.
5. 8 threads per inch $2\frac{1}{2}$-8 NPT

6. $2\frac{1}{2}$-8 NPT*F* (fit up dry)
7. $\frac{3}{4}$ in. (assuming it is a standard wall pipe)
8. Assembly (single or multiple), translation (moving objects), sealing (liquids and gasses).
9. It is true—steeper helix angles translate farther.
10. A. The left-handed thread is chosen when function dictates that the right-handed thread won't perform correctly. Could be for safety or to change the thrust to direction turned.
    B. The multiple thread is chosen to increase lead.

ANSWERS 14-2

1. 0.995 in.
   1.00 in. − 0.05 × Pitch distance = 0.125 × 0.05 = 0.006
   (but never more than 0.005 in. less than nominal)
2. (*Hint:* Found in Basic Formulas for General Purpose Acme Threads.)
   Answer = 0.0463 in. basic truncation.
   Formula: F_{CN} = 0.3707 × Pitch distance
3. (*Hint:* Look under heading of *nuts* in the index or subject search.) The counterbore must be $\frac{3}{4}$ in. minimum diameter. The *D* factor on the chart.
4. The bit will be ground to 50°.
   (*Hint:* Screw Threads—Buttress Form.)
5. Found in Screw Thread Systems Spark Plugs (both imperial and metric).
6. Acme and Stub Acme

7. Here are fifteen obvious places
   Aero Threads
   ANSI Standards (The American National Standards Institute)
   Automatic Screw Machines
   Bolts and Nuts—Another major source of information
   British Standards
   Cutting Threads
   Dies Threading
   Gages for Measuring Threads
   Machine Screws
   Nuts and Bolts (bolts and nuts)
   Pipe
   Screw Thread Systems—The major source
   Taps
   Thread Cutting (cutting threads)
   Thread Rolling

ANSWERS 14-3

1. To the lead because that's the distance the tool bit must move laterally in one revolution.
2. On the CNC lathe, neither the taper attachment nor the tailstock offset methods will produce such a steep taper.

3. $\frac{1.0}{12}$ in. × 3 = 0.25 in. lead
4. Yes
5. $\frac{12.5}{2}$ = 6.25

Answers to Section Review Questions

1. Translate—to move a device; thrust, force a device with axial thrust; Seal, keep fluids within a pipe; join, connect pipes.
2. Acme threads
3. It's used most commonly on the standard 60° triangular thread but it could be put on any thread.
4. To prevent concentrated forces at the base of the thread at sharp corner intersections.
5. The right-hand thread doesn't function correctly in the application.
6. Left-hand taps and dies; CNC lathes; reverse travel on manual lathe; inverted tool, reverse spindle on manual lathe.
7. It's probably a class 2 since most bolts are, but the three lines don't tell you that. They do tell you it's an SAE, Grade 5 bolt.
8. To translate more force in one direction than the other.
9. Reduce wear from friction by increasing the load-carrying surface.
10. Using a drive/face plate with three slots or the cams on a six-cam-lock chuck.
11. Set over the tailstock half the taper distance; set the taper attachment to the correct half angle or taper per foot.
12. 0.3077/0.07692 = Four starts
13. 1/0.0769 = 13 TPI
14. To reduce breaking of the threads or shaft at the sharp corner intersection of the thread face and root.
15. How will each fit?
    Bolt A will fit loosely—the clearance between the undersize bolt and correctly machine nut are too great.
    Bolt B will fit correctly with no excess looseness.

Figure 14-36 *Review Question 15.*

16. What are their weaknesses?
    Bolt A will tend to strip more easily as it has a weaker thread triangle.
    Bolt B will not tend to strip very much more than a full diameter thread. Supportive fact: Recall that the normal truncation to cause 75 percent engagement of threads only reduces the stripping resistance of a normal thread, by 4 percent. Bolt B is truncated a bit more.
17. 0.2164 Max to 0.2108 Min from, "Unified Series Screw Threads"
18. No change—turning the outside diameter down affects only the thread crest but not the thread.
19. The ball-screw is a pair of split nuts with a great preload between them. The round space between nut and screw is filled with recirculating balls to eliminate backlash and reduce friction.
20. By changing program lines and the cutting tool, the control can make a multiple-start, left-hand, or tapered threads. A CNC lathe can make variable pitch threads. RPM may be set at or near the correct cutting speed. Thread finish, tool life, and accuracy improve. Time is reduced greatly.
21. Variable pitch threads are only possible on CNC lathes.

Section 15

Metallurgy for Machinists—Heat Treating and Measuring Hardness

Introduction

This section rounds out your background in metals. Its primary goal is to understand why heat treatment of metal is often necessary in a job plan, and to learn the basics of do-it-yourself heat treating of steel and how to test hardness in metals.

There are five reasons that the various implements in Fig. 15-1 might have their physical conditions changed before, during, or after machining, forming, and welding. To

Make it harder for better wear or strength characteristics.

Make it softer to increase machinability or to improve long-term durability.

Relieve stresses created during its original formation at the steel mill, or to relieve stress created by machining, welding, and other processes.

Change chemical properties, which then permits further heat treatment.

Stabilize the metal from future changes in shape.

Figure 15-1 *Various items are heat treated to various states of hardness and tempered durability.*

Level of Accuracy

Similar to machining, heat treating can be performed to close tolerance using exacting equipment and standardized processes. But it can also be done to less demanding requirements. When traceability, product quality, and function are critical, processes are performed by specialists (Fig. 15-2). But utility heat treatment of shop tooling is another matter. In many shops, there's a need to harden or soften jig and fixture components and other items such as steel vise jaws. These are things we make in-house to keep the work moving along, but they won't be sold to customers. Using a small electric furnace or a hand-held torch, the tradesperson is able to perform the basics.

Learning to perform minor heat treatment is a major goal in Section 15. But you'll also gain an appreciation of the skills of specialists and the work they perform. They must understand the use of powerful high-heat gas and electric furnaces and other thermal equipment. Once those energy-hungry beasts are up to temperature, they stay energized! So, unless a shop has a constant need for heat treatment, it's not practical to maintain a facility. Today, most shops send their certified work to an outside contractor that serves an industrial area (Fig. 15-3).

Measuring hardness is a basic requirement for machinists. You should be able to take your work to the hardness tester to determine if it's right according to the job plan, or to test mystery metal to determine correct cutting speed.

Objectives

Figure 15-2 *In modern manufacturing, most shops contract their certified heat treating to outside specialty shops that have the high-energy specialty equipment and skilled heat treat people.*

Figure 15-3 *Machinists with the ability to do their own heat treatment of shop tooling, short run production, or repair work deepen their career potential.*

UNIT 15-1

Steel and Other Alloys

Introduction: As we learned in Section 5, steels are supplied in large families known as the alloy steels, tool steels, and stainless steels. These groups could also be called consumer production steel, technical application steels, and noncorrosive steels.

There are dozens of families and subtypes within each. Adding to the list, many steel manufacturers offer proprietary (their own patented formula) alloys. So, similar to other specifications we've studied, memorizing data is not the objective. Knowing where to find the information and how to use it is.

Some references create a fourth classification for steels, *plain carbon steel* or sometimes they are called the *carbon steels*. These are the simple formula steels used for car bodies, lawn mower handles, sheet roofing, and many other applications. Depending on their carbon content, those above a given percent (Unit 15-2) can be heat treated to improve physical hardness. These are steels with nearly no extra metals or minerals added to create superior characteristics other than hardness. Carbon steels don't bend, flex, or take changing loads as well as those steels alloyed with further ingredients. In this text, we will consider the plain steels to be the base material among the alloy steels.

When performing heat treatment, it's critical to identify the exact alloy. This unit is an introduction to the way steels (and other metal alloys) are specifically designated. Based on a given steel's formula, heat treating temperatures, times, and cooling media vary.

KEYPOINT

Heat treat temperatures and cooling methods vary depending on the alloy's major ingredient and carbon content.

KEY TERMS

Alloy
A specific formula for a metal.

Alloy steel (SAE steel)
A group of steels tailored to production part making.

Corrosion resistant steels (CRES)
Stainless steel used in medical, food, and other applications where rust is a problem.

Quench
Rapidly cooling a metal from high to lower temperature to create hardness.

Tool steels
A family of custom-designed steels used for extreme duty.

The Numbering System

In the past, steels (and most other commercial metals) were identified with one of several numbering systems, depending on the branch of manufacturing in which it was being used. Each professional organization or association tailored a system to be useful to them.

AISI	American Iron and Steel Institute (used most for machining)
SAE	Society of Automotive Engineers
ASTM	American Society for Testing Materials
ASME	American Society of Mechanical Engineers
ANSI	American National Standards Institute
AWS	American Welding Society

Plus there are other military and government organizations that classified metals.

Combining systems, the ASME created the Unified Numbering System *UNS*, that we'll look at in a moment.

TRY IT

Please turn to *Machinery's Handbook,* under *AISI, SAE and UNS Numbers for Plain Carbon, Alloy and Tool Steels.* Note the similarity in these systems. Note too that the UNS system adds a letter to help identify alloy steel's group, similar to tool steels. Keep a marker in that section—problems coming up next.

Steel Classifications—Main Modifying Ingredient First

Within the AISI numbering system, the major family is identified by the first digit or first two digits of the series. They denote the second most common ingredient beyond plain carbon steel. That provides a general idea of how the **alloy** will perform and how it should be heat treated.

For example, all steels having the heavy metal molybdenum as a secondary ingredient, begin with a 4. 4340 steel then is a particular kind of alloy steel found in mountain bicycle frames, for example. 4130 is also a molybdenum steel, but with slightly different properties. A bit less costly, 4130 is used for high-demand automobile parts such as axles.

To start your vocabulary, here are four useful groups to learn:

Number	Type
1xxx	Carbon steel (including mild steels)
2xxx	Nickel steels
3xxx	Nickel chromium steels (stainless)
4xxx	Molybdenum steels

TRY IT

To get a feel for the system and where this information is found, answer the following questions from the *Handbook*. Quickly scanning the section named *STANDARD STEELS,* you will see that there are three kinds of information about the various alloys:

Numbered alloys with ingredients
Word descriptions of group characteristic
Recommended applications

The following three questions will take you to those three areas of the handbook.

A. You are given a work order to make snap rings (a circular retaining pin). They must be made of the correct steel. Which steel alloy would be recommended?

B. As you will see in Unit 15-2, carbon is the major factor for selecting heat treating temperatures. What is the amount of carbon in the specific alloy of Question 1?

C. The print calls for using a stainless steel of the 30316 alloy. Can it be hardened by heat treatment? If you can't find this information see the hint in the answers.

ANSWERS

A. 1060 steel
B. 0.55 to 0.65 percent carbon. (*Hint:* Look under **steels—corrosion resistant** abbreviated to **CRES.**)
C. No it cannot be made harder. This group of stainless steels are very resistant to corrosion but they cannot be hardened by heat treatment to change their characteristics. They can be stress relieved, however.

Tool Steel Identification

The second group of steels are custom designed for extra demanding applications. In general, they are more expensive than **alloy steels,** but there are exceptions. Due to cost and their extreme properties, they aren't often specified in consumer products but are used more for tooling applications. The tool steels are generally higher in carbon content, alloying minerals and metals. They are custom tailored for uses such as dies, punches, wood lathe tools, drills, and end mills. Common tools such as hammers, files, and concrete chisels are not usually made from these tool steels, but rather from the AISI alloy steels.

Tool steels are identified by a letter followed by a number. The letter not only identifies the alloy family, but it also identifies the general procedure for heat treatment. When steel is cooled rapidly, from a high temperature to a much lower level, it becomes hardened. That's called **quenching** the steel. So, the W class of tool steels are quenched in water. But water would be too severe and would shatter some steels so a few are correctly quenched in oil (the O group), which removes heat a little slower than water. Still others are supersensitive to the quench rate and they harden by exposure to moving room air, the A tool steels.

For now, be aware of the following types with a more complete chart in the handbook:

Designation	Application	Quench Media
A	Die components for medium duty	Air (very slow quench)
D	Heavy die components	Oil/salt (medium quench)
M	Metal cutting tools (HSS)	Melted salts (slow quench)
M,T	Tungsten metal cutting tools	Melted salts
W	Pins, bushings, and shafts for medium duty	Water
O	Pins, bushings, and shafts requiring more toughness	Oil

We will look at tool steels again after some discussion of heat treating processes coming up next.

A. Each type requires a given kind of quench and each alloy within the type requires different temperatures before and after the quench.

B. Alloy steels for general production applications are designated with a series of four, five, or six numbers. The first one or two digits identify the group's major added ingredient(s) and its specific formula. Further digits identify specific alloys.

C. Tool steels for heavy-duty applications are identified by a letter followed by a number. It tells not only the type, but also something about the needed heat treatment.

UNIT 15-1 REVIEW

Review the Key Terms

Alloy
A specific formula for a metal.

Alloy steel (SAE steel)
A group of steels tailored to production part making.

Corrosion resistant steels (CRES)
Stainless steel used in medical, food, and other applications where rust is a problem.

Quench
Rapidly cooling a metal from high to lower temperature to create hardness.

Tool steels
A family of custom-designed steels used for extreme duty.

Replay the Key Points

◆ Due to popular use in the machining field, in this book we'll use the AISI steel classification system.

◆ Each alloy requires different heat treat temperatures and procedures.

Respond

Questions are deferred until Unit 2.

UNIT 15-2

Heat Treating Steel

Introduction: Heat treating steel requires changing workpiece temperatures of the material, in three stages, to permanently alter its hardness. Some transitions are slow and some are very fast. Some require high temperatures—500° to 2,700°F—while some are relatively low—from 275° to 500°F. Performing any of these operations requires the handling of superhot metal! Safety must be considered! There's a very well-defined set of heat treat protective gear we'll cover in Unit 15-2.

Safety
Regular shop safety attire is not adequate to perform heat treating of metals. Your instructor will show you exactly what equipment is required in your lab.

Here in Unit 15-2, we'll learn to perform three in-shop processes on heat-treatable steel—quench hardening, thermal tempering of hardened steel, and thermal annealing. The objective is to change its hardness and physical condition.

KEY TERMS

Annealed
The softest state of steel—100 percent of the hardness has been drawn out.

Annealing
Drawing 100 percent of the hardness from a steel, taking it back to its softest state.

Austinite
A fine-grained solution of carbon and iron that is not stable. It will revert back to rough layers of soft metal if left to cool slowly.

Carbon stripping
Surface disturbance of steel whereby carbon molecules have migrated to the surface to form a black crust. Caused by overheating too long.

Critical temperature
The point at which carbon begins to dissolve into the iron crystals and austinite forms in steel. See transformation temperature.

Drawing (heat treat term)
To reduce hardness down from a full hard, brittle condition by heating to a temperature below the critical point.

Martensite
A desirable, permanent crystal structure created by freezing austinite structure through quenching.

Normalized (steel)
Steel that has been heated just below the critical point and cooled slowly to leave it stress free but to retain a bit of its physical hardness as well. A subset of annealing.

Thermal shock
The fast reduction of heat through quenching. Each alloy requires a given level of thermal shock to lock the crystal structure into full hardness.

Transformation temperature
See critical temperature.

Three States of Steel

A little bit of carbon makes a big difference in steel! To be heat treatable (the ability to be hardened), steel must have a minimum of $\frac{3}{10}$ of 1 percent (0.3 percent) up to a maximum of around 3 percent. Some exotic steels may contain 4 percent. But in general, carbon above the 3 percent, has little or no effect on hardness.

Mild (low carbon) steel below the 0.3 percent threshold cannot be hardened without first changing its carbon content—which we'll look at in Unit 15-3, case hardening. Heat treatable steel can be placed in three physical states.

1. **Annealed**
   The softest, most machinable state—the easiest to machine.

2. **Full hard**
   The hardest condition of steels capable of being hardened. Most steels are too brittle to use at this point. And often too hard to machine in any way except grinding or electrical discharge erosion (Section 26).

3. **Tempered**
   A lower, useful hardness state. The condition the product will be put in for its life as a useful object.

The Three-Step Hardening Process

To perform your own hardening of steel, you must take the metal through the three steps. Here's a quick run-through. We'll discuss each in more detail right after.

1. **High Heat**
   Heating thoroughly to a specific prequench temperature of 1,300° to 1,900°F, depending on the alloy (700 to 1,000°C). Temperature varies by alloy. It's called the **critical temperature** or also the **transformation temperature** (Fig. 15-4).

2. **Quench**
   A rapid cooling to create desired full hardness. Plunging into water, saltwater brine, oil, air, or melted salts. Once it's cooled, the metal is too hard for practical use. It tends to crack if used at that hardness. It can even crack during the quench, or soon after, due to internal stresses created during the quench (Fig. 15-5).

3. **Temper**
   Reheat and hold at a controlled lower temperature to reduce the brittleness of full hard metal hardness to a usable condition

High Heat-Prequench

1,300° to 1,900°

Furnace controller

Figure 15-4 *Heat to the recommended conversion temperature.*

Tempering

Holding parts at specific temperature for 30 minutes or more.

Lower Temperature 375° to 1,400° F.

Figure 15-6 *After quenching, the parts must be returned to the furnace as soon as possible, at a preset, specific lower temperature.*

Quench

Plunge below surface, swish until cool.

Water or oil

Figure 15-5 Quench *to freeze the structure creating the desired full hardness.*

(Fig. 15-6). Nothing is then done to the steel other than letting it cool slowly in room air or even more slowly by wrapping it in a insulating blanket for some tool steels.

Step by Step

Here are the expanded facts you'll need to harden steel.

Step 1 Elevating to the Hardening Temperature Elevate the metal, slightly above (50° to 100°F) the *transformation* state. At that point, the steel's crystals change from coarse layers separated by nonuniform layers of carbonaceous metal, into fine crystals with the carbon in uniform solution. This state is called the **austinite** *structure and the steel is called austinitic.*

When it is held at this elevated temperature until the metal is uniformly hot (called soaking), the internal structure forms a desirable, fine-grained, well-distributed condition. The change in structure isn't instant, it takes time. So, the steel must soak at that elevated temperature for while. A rough guideline is about 1 hour for every inch of thickness.

Then, if quenched while converted to this state, the crystal structure locks into the hardest condition it is capable of reaching.

KEYPOINT

Slightly exceeding the transformation temperature and holding it at that temperature to soak, creates austenitic structure: fine, well-mixed grains. But that state is not stable, it reverts back to a coarse separated structure when left to cool slowly.

Austinitic steel is a good condition, but it's not stable. If left to cool slowly, the austenite grains separate back into layered materials that are not hard. But if frozen quickly enough with the right quench media, the grains remain fine and become locked into full hardness. Then in this state, the steel is known as **martensite**—the objective of the process.

Here are a few general conversion temperatures to memorize, to impress your friends! Note the inverse relationship of carbon to temperature.

Carbon Percentage	Transformation Temperature
0.65 to 0.80	1,450° to 1,550°F
0.80 to 0.95	1,410° to 1,460°
0.95 to 1.10	1,390° to 1,430°
1.10 and above	1,380° to 1,420°

Step 2   Quenching To create the necessary locking action, different alloys require varying amounts of **thermal shock,** the rate at which heat is conducted away from the hot metal. Too little shock and the steel doesn't reach full hardness. But the hardness that does occur is nonuniform and nearly impossible to predict. It's softer than full hard and has soft and hard spots within—a very undesirable condition. It will tend to break along demarcation lines between the areas. Thin sections harden more. The only sure way to achieve complete, uniform hardness is to hold the metal at the prescribed high heat then quench it at the prescribed rate (Fig. 15-7).

Various quench media conduct away heat at various rates. For example, cold water absorbs heat far faster than hot water. Between the two quenches in terms of shock rate, we use brine, water mixed with salt. All the water quenches are too fast for many high-carbon steels, so heat treat oil is commonly used in the small shop or even exposure to room air for air hardening steels.

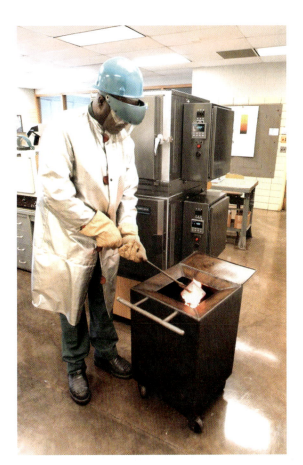

Figure 15-7 *To avoid warping or stressing the work, plunge it completely into the liquid then stir it to quench all sides evenly.*

too much shock. Charts are supplied by the manufacturer of the particular steel. Many steel companies supply a complete reference book for their line of steels.

KEYPOINT

Different alloys require different rates of quench—not too fast or too slow.

To bring steel to the full hard state, you must know?

What is the right transformation temperature for this alloy?

How long must it remain at the transformation state?

What is the correct quench media?

To quench right is to immerse the steel in the quench media then move it about until it stops boiling the liquid. Do not remove it from the quench just to see if it's cool—hard and soft spots can result.

Too much shock! Quenching too fast can be harmful to metal. Some alloys shatter if given

Trade Tip

Quenching *The right way to plunge hot parts* (Fig. 15-8). Plunge the hot workpiece into the quench media with it's broadest surfaces vertical, which then surrounds the workpiece with cold liquid as close to simultaneously as possible, called a vertical plunge. Then keep it well below the surface and move it constantly in a figure eight pattern to flush the boiling gas and expose all sides of the object to fresh liquid. Also keep moving it up and down in the tank. Do not just drop it to the tank bottom, as the side that's down will not be exposed to fluid.

It's important that no area of the workpiece cools at a different rate due to the boiling action of the media around the hot part or in the way it was plunged into the quench. Not plunging right and swishing results in warpage, cracks, soft spots, and unnecessary internal stress.

Figure 15-8 *The right and wrong way to quench a part.*

Combo-Quenching Sometimes parts with thin or complex sections will crack or warp excessively when quenched in water even though it's the right media and the plunge is performed right. If so, try this: quench them for a few seconds in the oil, then transfer them quickly into the water while still hot. If you do so, the sharpest edge of the shock will be taken away. *Caution!* When moving the part from the oil to the water, oil on the surface will vaporize and may flare up. Wear a complete set of heat treat protective clothing, use long tongs to hold the work, and work only in a fire-preventive ventilated heat treat facility. Do not try a double quench without supervision.

Step 3 Tempering the Quenched Steel

OK, the steel is at the full hard state, but that's too hard to be useful. Tempering the steel down to a useful hardness where it won't crack under load involves heating it to an exact temperature well below transformation. It's done by placing the work in the furnace (or the torch flame), then holding it there until it is uniformly well heated throughout, usually one-half hour or more per inch, in an electric furnace, depending on the total volume of metal, the desired temperature, and the alloy.

Tempering is also called **drawing** because it draws out the brittleness.

KEYPOINT

The higher the draw heat, the lower the resultant hardness will be. Higher heat = lower hardness. Draw temperatures range from around 400°F to over 1,200°F. To avoid the formation of stress cracks, full hard steel should be tempered immediately after quenching.

Here too, it's critical that all sections of the work achieve uniform temperature. After the work is heated and soaks, it is usually placed on a wire rack so that air can circulate around all sides, then allowed to cool slowly. Everyone understands to not touch work with their hands when it's in that location. Some sensitive steels might be placed in an insulating blanket or steel box to safely cool them.

KEYPOINT

After the draw is finished, there is little or no difference in hardness results in the way the work is cooled. The metal needs to slowly cool to room temperature without a steep transition.

Draw Temperatures and Hardness

In Unit 15-5 we will learn how to measure hardness and what the resulting numbers mean. For now, I need a scale to illustrate the inverse relationship of draw temperature to hardness. One of the more popular systems for comparative hardness is the Rockwell index. When a Rockwell test is made on a typical hard, tool steel block for example, it will rate as high as 63 to 65 at full hardness. At its annealed (softest) state, it will be near 20. For reference, mild steel, which cannot be hardened, would measure very close to 0 (zero). Draw charts (Fig. 15-9) or draw temperature tables like the one shown next, provide half the information needed for doing your own heat treat.

Figure 15-9 *A typical hardness curve as draw temperature is raised.*

Typical Draw Temperatures and the Application

Draw Temperature (degrees Fahrenheit)	Hardness Rockwell Index	Typical Application
375 to 400	58 to 62 (highest useful without brittleness)	End mills
450	56 to 58	Slitting saw
500	54 to 56	Punches
500	52 to 54	Bending dies
600	48 to 52	Drill bushing
1,300 (1 hour—slow cool)	Normalized (discussed next)	
1,400 Slow cool in furnace	Annealed	

Annealing, Stress Relieving, and Normalizing Steel

All three of these heat treat processes are subsets of drawing. They are all accomplished by putting the metal into a very high draw temperature. Each process softens the metal but each has a different objective.

Annealing Takes it to the Softest State

Annealing is a 100 percent draw back to the steel's softest state. It's done to make the metal machinable or to remove work hardening spots. To anneal steel, the work is put in the furnace set at *just above transformation temperature,* then left to cool very slow (usually by turning off the furnace without opening it). But it can also be wrapped in an insulating blanket, buried in dry sand, or put in a closed container.

Stress Relieving Internal stress can be induced into metal in several ways: by machining, welding, through bending and forming, and even at the rolling mills where it is originally made. To relieve that stress the metal is heated close to but *just below the transformation point.* It's then held there for 1 or 2 hours, depending on thickness, then it is cooled on an extra slow schedule of furnace temperature reductions.

Normalizing Also a stress relieving process, this time the top temperature is *farther below the transformation point prescribed for each alloy*—slightly less than the stress-relieving temperature. During normalizing, the metal must be cooled on a schedule. Normalizing might be seen as very long, very severe draw. It relieves stress but doesn't quite take all the hardness from the object, leaving it nearly

stress free but slightly hard. The difference between normalizing and stress relieving is that normalizing can be performed only on heat treated metal whereby plain steel unable to be hardened can only be stress relieved to become **normalized steel.**

KEYPOINT

All three processes require a slow, controlled cool down.

A. Annealing occurs above transformation temperature.
B. Stress relieving occurs just below transformation temperature.
C. Normalizing occurs at a prescribed temperature farther below transformation.

Trade Tip

High Heat *Caution!* This unit is not about correctly using an oxygen/acetylene torch. You must be trained to use it correctly. Ask for a demonstration and safe operation test from your instructor.

Uniform Heating While the best results are accomplished in a controlled furnace, you might be called on to perform a heat treatment using a torch. If so, it's important to heat the object uniformly. Concentrate the flame on the thicker parts while "playing the flame" on other parts (Fig. 15-10). Apportion the heating (where you aim the flame), based on volume in the work. This helps create an even heat throughout the part.

Carbon Stripping At the high temperatures required for heat treating, carbon in the steel can move due to the high

Playing the Heat for Uniformity

Thin cross section

Thick

More time here

Figure 15-10 *When using a torch, be sure to play the heat by concentrating on thick sections.*

molecular kinetic energy. Near the transformation temperature, carbon can migrate to the surface of the workpiece—called **carbon stripping,** a highly undesirable condition (Fig. 15-11). It's seen as dark spots on the white hot steel surface, deliberately overheated. After cooling, the stripped carbon lies in flakes on the surface. Beneath those flakes, the surface is rough. The destroyed soft shell can be as deep as 0.060 in. Without reintroducing more carbon, it cannot be hardened due to its lack of carbon.

Carbon-Rich Flame To help prevent stripping, a torch flame can be adjusted richer (higher) in carbon than a normal welding flame. To do so, reduce the oxygen to acetylene ratio in the mixture (Fig. 15-12). The rich flame appears less focused, more orange, without the blue cone of the neutral flame on the bottom. It isn't hot enough for other duties, but during heat treat work it will strip less carbon due to the higher levels surrounding the hot steel.

Time and Temperature Also, to control stripping, do not overheat the work as was done intentionally in (Fig. 15-11). Bring it uniformly up to the prescribed high temperature (dull red through glowing orange, depending on alloy) then quench it immediately. If you hold it at or above the critical point for extra time, carbon will continue to migrate to the surface.

Figure 15-12 *The rich flame above isn't as efficient, but it tends to not strip carbon during heat treat operations.*

KEYPOINT

Stripping Is a Function of Temperature and Time
To avoid stripping, do not overheat and do not hold parts at high temperature any longer than required for uniform heating.

When You Must Use a Torch—Utility Temperature Controls

Commercial furnaces can be set to any temperature between 300°F and approximately 2,600°F. They hold temperatures within 2 percent of the target and graph and record temperature profiles in order to create a certificate of the process (Fig. 15-13). For a general shop heat treat, the electric furnace controls we use range from 5, 7, or 10 percent, depending on the quality of the control.

But, when controlled furnaces aren't available, or the work is too big to fit in the chamber, the next best temperature control is the *pyrometer* (Fig. 15-14). A pyrometer is a high-heat measuring devise. In industry, there are two types: those that must touch the part with a probe, similar to a meat thermometer, and those that measure optically, and thus do not need to touch the hot part. They are used mostly where the heat is too great to come near the parts.

Carbon flakes   Lost carbon

Figure 15-11 *Dark carbon flakes can be seen on this overheated steel as it passes the critical temperature. The result shows more clearly after it cools—free carbon stripped from the steel.*

Figure 15-13 *A commercial furnace can control the heat within 2% accuracy.*

Figure 15-14 *A pyrometer can tell the temperature of hot metals.*

Pyrometers are common in heat treat shops, but it's unlikely you'll have one in a small shop. When both furnace and pyrometer aren't available, we fall back to a set of trade tips to control temperatures. They aren't accurate enough to perform a critical heat treat but they work well for utility purposes.

High Heat Control Using Color Brightness—Steel Glow

As steel molecules (and those of many other metals) heat, they begin to radiate energy as visible light, called iridescence. Beginning with a dull red, the metal glows brighter as the energy builds toward a higher temperature. With practice, that color brightness can be used as a rough gauge of temperature (Fig. 15-15). This sequence in the figure was shot as the part heated.

1,800°+F

1,650°F

1,450°F

1,300°F

1,100°F

Figure 15-15 *Learn to identify temperature by glow.*

To Gage Steel's Temperature	Approximate Temperature
Dull red	1,100°F
Bright red (called *cherry red*)	1,300°F
Orange	1,450°F
Bright orange	1,650°F
White orange (*white hot*)	1,800°F

Note that carbon steel melts at 2,000 to 2,200°F.

KEYPOINT

I estimate temperature discrimination by glow to be no better than ±200°F, and only after some guided experience.

The only way to get a grip on this skill is to work with a journey-level craftperson and see a demonstration.

Low-Temperature Controls for Drawing

There are two ways to know how hot metal is becoming before it glows; the melt gauge crayon and surface oxide colors.

1. Melting Gages

These handy little pocket crayons are specially formulated to melt at different temperatures (Fig. 15-16). They can be purchased in 50° increments from 250°F up to 1,200°F. When the draw temperature is be-

low 700°F, several lines are drawn on the cold metal, at various places so different cross-section masses can be tracked as they heat. For temperatures above 700°F, have three crayons bracketing the desired temperature ready. Swipe each on the hot surface. The middle temperature should melt but not the higher temp.

2. Avoid Overheating

Not exceeding the draw temperature, especially on thin sections, is the main challenge. Once it's melted, the crayon indicates no further information.

Oxide Color for Various Temperatures (Carbon Steel)	
Color	**Temperature**
Pale yellow	430°F
Straw yellow	450/460°F
Dark yellow	480/490°F
Brown-yellow	500°F
Brown-purple	520°F
Light purple	530°F
Purple	540°F
Dark purple	550°F
Blue	560°F
Dark blue	570°F
Light blue	600°F and above

Figure 15-16 *A Tempilstick® melting crayon gage can be used to indicate draw temperature from 250° up to 1,200°F.*

Trade Tip

Surface Oxide Color and Temperature Here's another tip that helps determine draw temperature when no other low-temperature gauge is available. Clean, shiny steel will react with room oxygen as it is heated. The oxides change color as the temperature increases (Fig. 15-17). The shiny steel in the figure was heated on one end. Starting from room temperature, it shows a light straw yellow at around 400°F, then slowly progress up through a series of color changes, to a light bluish color before it reached the dull red glowing point. The problem here is that the surface temperature doesn't always tell the core temperature. Also, different alloys vary slightly in their color spectrum so this hint, although useful, must also be

Figure 15-17 *Various oxide colors can be used to indicate temperatures of steel.*

regarded as a rough estimate. Still, oxide color can be helpful when no other temperature control is available. Clean the metal surface to bright steel, where you need to determine temperature. It's smart to error on the cool side of the temperature. If the finished product is tested and found to still be too hard, there is no serious harm in retempering it to a slightly higher temperature, causing a lower hardness.

However, heating it too hot the first time is a one way trip! There's no way that the overtempered part can't be increased in hardness a bit! It must be completely hardened then redrawn back.

Safety Equipment Requirements

Heat treatment of metals requires the right protective equipment and clothing (Fig. 15-18).

1. **Extra Eye Protection**
   Safety glasses plus a full-face shield made for high temperatures. It may also be partially mirrored to reflect heat.

2. **Fire Extinguisher**
   Immediately available.

3. **Protective High-Temperature Coat**
   Made for heat treating and with long sleeves.

4. **Gauntlet Gloves**
   Long to overlap coat sleeves, insulated for high temperatures. *Caution!* You will not handle hot parts with these gloves. *Hot parts are always handled with long gripping tongs.* Always have two pairs of *tongs* available.

5. **Steel Shoe Protectors**

6. **Other Precautions**
   Be certain there is someone within voice range when heat treating. Memorize where the nearest *eye bath* is located.

Hot Parts Become Slippery!
Highly heated parts above the transformation temperature become slippery on their surface. Handle them with extra care.

Hard hat with face shield and rear neck protection

Safety glasses under face shield

Alumanized heat reflective coat or leather coat/apron

Gauntlet gloves

Long tongs

Leather high top boots with steel top shields

Figure 15-18 *This craftperson is correctly dressed for safe heat treat work.*

ShopTalk

Liquid Salt Molten salts are used in several heat treat applications, at both low and high temperatures. They are efficient media for both heating and quenching. Thus far, we've discussed liquefied salt as a quenching media, but it can also be used for high temperatures (Fig. 15-19). But a common question arises—is it the same salt I use on my food?

The answer is no. The definition of a salt is a neutralized combination of an acid and an alkali base. Before melting, the granulated salt does resemble table salt. To identify different kinds of heat treat salts, note that they are sometimes stained various colors. They are definitely not the familiar spice on your potato chips!

Additionally, it's wise to have a backup person, correctly dressed for heat treat, standing by during transferring of hot parts. A second pair of tongs can often save the day.

Quenching Media for Heat Treating

In professional shops many different liquid materials are used to quench work. But in the tooling and small shop, there are three:

Plain cold water (can be heated to prevent overshocking).

Brine, water with salt added to slow its conductivity.

Heat treating oil, a slower, more gentle media—less thermal shock.

Heat treating oil is specifically formulated to not ignite unless the entire vat is heated beyond its flash point. But it will flare when the hot part is accidentally removed before fully cooling.

Melted Salt, Air, and Compound Quenching

To further minimize shock, heat treat shops use liquid salt quenches. Molten quench salt becomes a liquid at around 350°F. Maintained in a electrically heated tank, this quench is very gentle in terms of thermal shock.

Figure 15-19 *A high-heat, liquid salt bath at 1,300°F.*

KEYPOINT

Safety Using Quench Oil

Flame Control Once a hot part is plunged into the heat treat oil, never remove it until the boiling action ceases unless performing a double quench. Otherwise, if the hot oily part contacts room oxygen, it will flare up. If you accidentally do remove it, plunge it back, and the flame will extinguish. Never heat treat with any oil other than one specified for quenching! Regular lube oil cannot be used.

Do Not Overheat the Oil Vat Do not use a small volume of quench oil to heat treat many parts. If the vat is small, with each quench, it will become hotter, which reduces its shock value progressively, and it increases the danger of flare up. If using a small furnace as in Fig. 15-3, this isn't usually a problem as not enough parts can be loaded and heated to overheat a normal sized oil vat (4 or 5 gallons).

Air Hardening Tool Steel— Air Quenching

In terms of gentleness, moving air provides enough heat removal for one family of high nickel tool steels known as the air hardening tool steels (Fig. 15-20). Designated by the letter

Figure 15-20 *Compare microphotographs of 4140 steel in its annealed state, coarse and layered (upper photo) to the same alloy at 54 Rockwell C scale (R_C 54), where the structure is fine and well distributed.*

A, examples commonly used are A-2 and A-6 die steels. Once heated to their transformation temperature, which is generally higher than that of other tool steels, they are removed from the furnace and set on racks that allow air movement to surround the hot parts. This provides enough freeze rate to lock the austinitic crystal structure to martensite.

UNIT 15-2 REVIEW

Review the Key Terms

Annealed
The softest state of steel—100 percent of the hardness has been drawn out.

Annealing
Drawing 100 percent of the hardness from a steel, taking it back to its softest state.

Austinite
A fine-grained solution of carbon and iron that is not stable. It will revert back to rough layers of soft metal if left to cool slowly.

Carbon stripping
Surface disturbance of steel whereby carbon molecules have migrated to the surface to form a black crust. Caused by overheating too long.

Critical temperature
The point at which carbon begins to dissolve into the iron crystals and austinite forms in steel. See transformation temperature.

Drawing (heat treat term)
To reduce hardness down from a full hard, brittle condition by heating to a temperature below the critical point.

Martensite
A desirable, permanent crystal structure created by freezing austinte structure through quenching.

Normalized (steel)
Steel that has been heated just below the critical point and cooled slowly to leave it stress free but to retain a bit of its physical hardness as well. A subset of annealing.

Thermal shock
The fast reduction of heat through quenching. Each alloy requires a given level of thermal shock to lock the crystal structure into full hardness.

Transformation temperature
See critical temperature.

Replay the Key Points
- Heat treatable steel runs from 0.3 percent up to around 3 percent maximum carbon.

- Slightly exceeding the transformation temperature creates an austenitic structure: fine, well-mixed grains. Quenching it freezes the grains into desirable martensite.
- To harden a particular alloy you must know the specific hardening temperature for this alloy and the correct quench media.
- The higher the draw heat, the lower the resultant hardness will become. It is critical that all sections of the work achieve a uniform temperature.
- Normalizing, stress relieving, and annealing are similar processes, requiring elevating the work at or near the critical temperature then cooling slowly.

Respond
From the reading and from *Machinery's Handbook*.

1. List the steps required to bring a steel to full hardness.
2. List the various quench media used to heat treat steel.
   A. for the in-shop heat treating
   B. for professional heat treating
3. An optical pyrometer is one method of monitoring the temperature of steel for high heat. Is this statement true or false? If it is false, what will make it true? Name two more methods.
4. In order for steel to be heat treated, it must have a carbon content above 3 percent. Is this statement true or false? If it is false, what will make it true?
5. Define annealed steel.
6. To temper steel what must be done to it?
7. Why is steel not used in full hard state?
8. Name the three missing reasons to heat treat metal: Harden it, soften it, _____, _____, and_____.
9. Describe the transformation state for steel.
10. After tempering a steel part at 500°F, it is tested and found to be too hard for the application. What must be done?
11. Starting with annealed steel, what must be done to temper it to a prescribed hardness?
12. List the four temperature monitoring methods for *tempering,* in order of their exactness. Which are in-shop methods?
13. As the tempering temperature is elevated, the temper becomes harder in most tool steel. Is this statement true or false? If it is false, what will make it true?

Case Hardening Steel

Introduction: Mild steel can be modified with extra carbon, such that its surface can be made nearly as hard as tool steel. This process creates a shell of high carbon steel (called the *case*) from one to two millimeters deep. After adding the carbon case, the steel's surface can then be hardened in the normal fashion.

Like other forms of heat treating, there are methods for exact case depth control and shop utility processes. In this unit we'll discuss other methods but concentrate how you can do it in the shop.

Steels Are Case Hardened for Two Reasons

To harden mild steel This is usually done for the economy of not using expensive alloy steel for a product yet having a surface hardness able to withstand abrasion and wear.

To change the properties of alloy steels Others need to be very hard on the outside and tough on the inside. For example, an automobile axle, made of heat treated and tempered alloy steel can withstand flexing and heavy loads but it must also be very hard at the bearing surfaces (Fig. 15-21). In other words, the steel has a double requirement for surface hardness and interior toughness. This is possible using the case process over a medium high carbon alloy steel.

Figure 15-21 *The axles on this muscle car must be heat treated for toughness to withstand 800 horsepower without snapping, yet they must also be hard on their outer shell for bearing purposes. (See Fig. 15-22.)*

KEY TERMS

Case hardening
Several processes that modify the surface of steel by adding extra carbon then hardening that shell.

Carburizing
Adding carbon to the outer surface of steel. Accomplished with liquids, solids, and gasses.

Nitriding
A gas hardening process that adds nitrogen rather than carbon to a thin shell around specially formulated steels.

The Case-Hardening Process

If we were to cut open the axle of Fig. 15-21, it would appear as the drawing of Fig. 15-22. The outer surface has been modified with extra carbon then quenched to lock the fine grains. The case hardened surface creates a set of characteristics for the axle's function. The core must be flexible and not prone to stress cracking, but its outer surface must be hard to resist wear. This is possible using case or surface hardening.

The case zone extends down from 0.015 to 0.200 in., depending on the process used (shop or commercial). A small boundary zone occurs where carbon content falls off rapidly, leading to the core of the original unmodified structure. That core can be alloy steel that is also heat treated but to a much lower hardness. This process can also be used to modify low-carbon steel incapable of heat treatment otherwise.

Before hardening, the carbon content increases by elevating the steel's temperature

Figure 15-22 *A simulated microphotograph of a cross section of an axle shows fine, high-carbon crystals on the surface, then a gradual increase toward coarse structure below the case.*

from 1,500° to 1,650°F. That's at or above what would be the transformation point assuming the steel had enough carbon to convert. At that point, due to thermal energy, carbon molecules begin to migrate outward—we see it as the carbon crust that forms on the outside of the work (Fig. 15-10). But once the carbon moves outward, it can also migrate inward if there is a carbon-rich material on the outside of the metal.

Once the steel is heated slightly above critical temperature, it is then immersed in a carbon-rich material—solid, gaseous, or liquid. See the Shop Talk.

Shop Case Hardening

For everyday shop tooling work, we use the pack-hardening process by surrounding heated metal in powdered carbon made just for utility **case hardening.** It produces a shell depth of, from 0.015 in. to 0.045 in. maximum, depending on the number of times the steel is heated and coated with fresh compound. Compared to commercial processes, carbon cannot penetrate very far. But, other than shallow depths, it offers several advantages:

It can be performed with torch or small furnace.

It is a relatively simple, quick way to harden a steel item.

Specific areas can be spot hardened while others are left soft.

The compound is inexpensive and if used conservatively, will harden many parts.

Heating then packing the object in a case-hardening compound produces from 0.015 in. (one hot immersion then quench) up to around 0.045 in. case depth (immersion, clean, reheat, and reimmerse repeatedly up to three or four times, then quench). This more thorough process also produces about a 20 percent harder surface.

Three-Step Process to Raise Carbon Content

Adding carbon to the steel's surface is not complex. It involves heating the metal then exposing it to the compound.

Step 1   Heat to 1,500°–1,650°F

Figure 15-23 *Case hardening compound is rich in carbon. It's made to transfer carbon but to not flare up.*

Step 2   Surround with Carbon Compound (Fig. 15-24)

Step 3   Absorption Time— Sustaining at Temperature
This last step has two forms, depending on whether a furnace or torch is used.

Torch process If the shop has no furnace or if a shallow case of less than 0.5 mm is required, simply burying the heated work in the carbon compound then leaving it to cool naturally will produce results. After cleaning the spent compound from

Figure 15-24 *After removal from the furnace nearby, using long-handled tongs, the heated part is quickly buried in case-hardening compound.*

ShopTalk

In the past, metal workers used charcoal, powdered coal, and even charred wood chips to induce carbon into hot metal. But you can see the problem with those materials—fire. The work had to be buried far enough within it that room oxygen didn't make it to the hot interface. Today, we use specially formulated prepared compounds (Fig. 15-23) that look like coarse, black sand. They are stabilized carbon and will not ignite or give off dangerous vapors.

the work, this process can be repeated to add extra depth but never to the depth of the next method.

Controlled furnace A deeper case is gained by burying the heated workpiece in a steel tray filled with compound, back in the electric furnace and holding it at the absorption temperature for 1 hour our more.

Step 4 Quench Quenching of the case should be performed soon after the case has been created.

If using a torch, the metal is removed from the compound then cooled enough to handle. Then the crust of spent compound is removed. The metal is reheated to 1,500° to 1,650°F and plunged in cold water.

If using the furnace absorption method, it's possible to move the hot metal directly from the furnace to the quench tank. The problem is the crust of spent compound will contaminate the water. Unless you plan to change the water, it's better to remove the work from the furnace, cool it, clean it, then reheat it and quench it. But to avoid surface cracks, this should be done soon after the work is cased.

The Professional Surface-Hardening Methods

While these are not machinist tasks, you should know a bit about them.

Gas Carburizing

Gas **carburizing** is a highly controlled case-hardening process where a deep uniform layer is required, up to 3 mm. This process does not distort the surface of the work as much as pack hardening (Fig. 15-25). After placing the parts in a sealed furnace, the furnace interior is purged of oxygen by displacing it with nitrogen. Then the furnace and parts are heated to the critical temperature. At that point preheated natural gas is flooded into the furnace to displace the nitrogen. The carbon-rich gas is kept at a positive pressure for a length of time, depending on desired case depth. The gas contributes the carbon but does not burn since there's no oxygen.

Figure 15-25 *An electric/gas* carburizing furnace *must be operated by a trained expert only.*

The depth of the case is controlled by pressure and exposure time. The case can reach a maximum of around 3 mm deep.

Liquid Salt Case Hardening

Using salt compounds heated electrically to a liquid state, similar to salt quenches, but to the critical temperature required to move carbon, an enriching environment can be created by adding carbon compounds to the liquid. This process brings several advantages for case hardening over other methods.

Faster Cycle Time—The heated liquid transfers heat and carbon to the work fastest of all the methods.

Less Distortion

Selective Hardening—see the copper Trade Tip.

Trade Tip

Using Copper Plating as a Carbon Barrier
Copper will not melt at the temperatures required for heat treating steel. Copper plating on steel will stop carbon movement in or out of the surface and is used to protect the surface during regular heat treating of high-carbon steel. It is also used to selectively mask off areas of the workpiece that are not to be cased.

This tip can be used to protect precision parts from carbon stripping during normal heat treating. For example, a complex-shaped, finished tool steel die can be preplated with copper, then heat treated. Once the copper is stripped back off the work after heat treat, the finish is near the original! The copper formed a barrier to carbon stripping.

Copper plating is also used during liquid carburizing where selected areas of the work are required to remain unmodified—no carbon. Only the raw surfaces without the copper shield will absorb carbon from the salts; similar to stenciling, the copper shields selected areas of the work.

Nitride Gas Hardening

This method is unique in that it modifies the steel in a different way. Steels formulated for this process will produce a shallow hardness of a few thousandths when their structure is modified by nitrogen not carbon. The process is similar to gas carburizing except the heated gas is ammonia. Ammonia gas requires extra precautions, therefore this process is performed in specialty heat treat shops only.

Nitriding is relatively quick to perform compared to the other surface-hardening processes and produces a very hard but thin case on the work.

UNIT 15-3 REVIEW

Review the Key Terms

Case hardening
Several processes that modify the surface of steel by adding extra carbon then hardening that shell.

Carburizing
Adding carbon to the outer surface of steel. Accomplished with liquids, solids, and gasses.

Nitriding
A gas hardening process that adds nitrogen rather than carbon to a thin shell around specially formulated steels.

Replay the Key Points

◆ Steel is given a case to modify and harden the surface of mild steel and to change the properties of alloy steels.

Respond

1. Using the solid pack method, the hardness of the case is dependent upon the length of *time in the furnace,* while the metal in question is exposed to the carbon compound. Is this statement true or false? If it is false, what will make it true?
2. List the steps required to create a case hardened steel workpiece.
3. Using the solid pack method, when would you not replace the packed work in the furnace?
4. Name and briefly describe the professional case hardening processes.
5. Describe why we can case harden steel.

UNIT 15-4

Aluminum Alloys and Heat-Treated Conditions

Introduction: Due to its low weight-to-strength ratio, and its availability in our earth's crust, aluminum is a very common metal in modern manufacturing. It too is heat treated to improve physical characteristics, but the way heating and quenching are used are exactly opposite that for steel (see the Section Review).

Some aluminum alloys can be hardened for strength and durability. They can also be softened to increase malleability, to easily form or stretch it without cracking. Like steel, aluminum can also hold internal stresses caused by casting, mill rolling, machining, welding, and other processing. Aluminum can be relieved using thermal heat treatment, but it is accomplished with a very different set of processes compared to steel.

No aluminum alloy will accidentally work harden, nor can it be deliberately hardened by heat treatment, beyond the machinability range. Thus annealing isn't used to bring it back to machinability. But it is annealed to better bend, stretch, and form it.

Unit 15-4 provides a background about the possible heat treat conditions of aluminum. When the engineering drawing specifies a given heat-treated aluminum, the material must meet that callout. But it's not a how-to-do-it unit. Aluminum heat treating is a specialty usually not performed by machinists.

KEYPOINT

Aluminum specifications usually include not only the alloy identification, but the heat treated condition too.

Because the heat treat process has many variations, a complete designation will include not only the alloy, but also its exact heat treat condition and often even the way in which it was put in that condition.

KEY TERMS

Condition
The heat treat hardness level, and process for aluminum.

Wrought
Any process that forms (but does not cut) a metal when it is not in the liquid state.

Figure 15-26 *The alloy, hardness, and method of tempering can be identified on this aluminum bar.*

Aluminum Alloys

Similar to steels, the alloys of aluminum are specified by number (Fig. 15-26). The first digit identifies the major alloying ingredient. The remaining numbers identify specific ingredients in the subgroup. For example, the alloy 5052 is modified with the metal magnesium.

Aluminum is supplied in cast alloys and mill rolled bars and sheets. All alloys that are cold or hot worked after they are formed at the smelter are called **wrought,** referring to the working of the metal after it is solidified. Here are the major groupings:

Type	Major Added Ingredient (Metal or Mineral)
1000	Iron and silicon
2000	Copper
3000	Manganese
4000	High silicon
5000	Magnesium
6000	Magnesium/silicon
7000	Zinc/magnesium/silicon/copper

Heat Treat Conditions

The temper (also called **condition**) is indicated by a letter and a number, or often a series of numbers that follow the alloy number—separated by a dash. For example, 5051-T6 is hardened to condition T using process 6 (found in the *Handbook*).

This is a magnesium aluminum and it has been put in the tempered condition T6. Each letter denotes a certain physical state. These you should remember.

Letter	Condition
F	As fabricated including work and weld stresses (not often used in this condition)
O	Annealed—softest state usually created for upcoming forming operations
H	Hardened and will be a specific process; for example; H1 = strain hardened and annealed
W	An unstable soft condition whereby the aluminum will be formed then hardened
T	Heat treated to a useful hardness by a specified process. This is the common category for finished products. Numbers following indicate how hard, then which process was used. For example, 5051-T65 is hardened to level 6 using process 5.

UNIT 15-4 REVIEW

Review the Key Terms

Condition
The heat treat hardness level, and process for aluminum.

Wrought
Any process that forms (but does not cut) a metal when it is not in the liquid state.

Replay the Key Points

◆ Aluminum specifications include the alloy, heat treat condition, and the process by which it arrived at the condition.

◆ When a specific aluminum is called out on the drawing, all criteria must be met, including the process by which it was placed in a given condition, if it is to be a legal part.

◆ Aluminum heat treating requires hot and cold baths, thus the specialty equipment isn't normally used as an in-shop process.

Respond

Using *Machinery's Handbook* as a reference:

1. Describe an aluminum alloy denoted to be 7075-T6-51.
2. List the various temper conditions of aluminum.
3. Which conditions would be soft and stable?
4. How much copper (abbreviated CU) and silicon are in a typical 6061 aluminum alloy?

UNIT 15-5

Measuring Metal Hardness

Introduction: Beyond verifying our in-shop heat treatment, testing hardness is sometimes necessary for production work as well. Even though it's bad planning, occasionally a job arrives at our machine with an unknown alloy or maybe its composition is known but the hardness isn't. It is possible to use a file to roughly test the machinability of that metal, but the best way to select cutter types, speeds, and feeds is a true hardness measurement.

KEY TERMS

Brinell
Testing hardness by reading the diameter of a ball penetrator mark.

Elasticity
The property used by the Shore system to test hardness.

Rockwell
Testing hardness by reading a penetrator depth.

Scleroscope
The devise that performs a Shore test, which measures the rebound of a diamond-tipped hammer off the test piece.

Shore
Testing hardness by measuring the bounce of a diamond-tipped hammer.

Properties Used to Test Hardness

As metal becomes harder, or softer, its physical properties change. We will study these properties further in Unit 15-6 but for now, we are interested in two. The three tests we'll look at use these two properties.

1. **Malleability**
   This is metal's ability to be deformed without tearing or cracking. From our study of chip making, you'll recall that harder metal resists deformation more so gaging this property can provide a relative hardness value.

2. **Elasticity**
   Elasticity is the ability to stretch without breaking. It can be measured by bouncing an object of known hardness off its surface. The harder the work is, the higher the test object will bounce.

The Rockwell Hardness Test

The **Rockwell** is a widely accepted method for both soft and hard metals. This system gauges malleability by measuring the depth a pointed probe of known shape and size will penetrate into the material given an exact amount of force upon it. Due to Rockwell's range it is the most popular test in tooling and small production shops and training labs.

Rockwell Numbers

There are several different scales within the Rockwell system, which we'll look up in the *Handbook* later. For this discussion we'll use the Rockwell C scale, correctly used on hardened steel. The C scale can be said to start at 0 (annealed steel) and runs up to 68, harder than a HSS tool bit, near that of a carbide tool. It is symbolized by a larger R with the scale subscript.

$$R_C$$

KEYPOINT

Rockwell C numbers rise as the hardness rises. Rockwell tests cover a wide range of material hardness.

The Two-Step Rockwell Test

To perform a Rockwell test, a part with a clean smooth surface is placed in what could be described as a very strong, large micrometer frame or precision arbor press (Fig. 15-27).

Step 1 Calibrate Load The test object is set upon the lower anvil such that it's stable and won't move when pressed down from above. Next, a cone-shaped diamond penetrator is brought into contact then driven

Figure 15-27 *A Rockwell tester has a strong frame with a dial to indicate diamond penetration depth into the workpiece: deeper penetration indicates softer metal.*

into the metal to a predetermined pressure of 10 kilograms (22 pounds). That causes the conical point to sink into the metal from 0.003 to 0.006 in. This is the initial calibration load. At that time, a large dial indicator is rotated to read zero (Fig. 15-28).

Step 2 Test Load Then with the calibration pressure upon the penetrator and indicator set to zero, a second, additional 10 kg test load is added. As the diamond sinks farther, its added depth is translated to the dial, but in an inverse relationship. The deeper the diamond penetrates, the softer the metal tests, therefore the lower number that must appear on the dial face. Inversely, when the point can't go very deep, the metal is hard and registers higher on the dial face (Fig. 15-29).

KEYPOINT

Step 1 Touch surface, initial load, set dial to zero. *Step 2* Add second load—measure penetration depth.

Calibration Load

Figure 15-28 *The initial load is placed on the penetrator then the dial set to zero.*

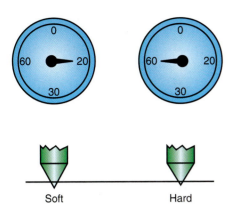

Soft   Hard

Figure 15-29 *Then by hand or by automatic action, the test load is added to the penetrator. Shallow penetration shows hard metal. Increasing depth shows softer metal.*

The Brinell Hardness Test

A second popular hardness test, the **Brinell** test, is very similar to the Rockwell system in that a *penetrator* is forced into the sample, however, here the measured gauge is the *diameter* of the dent made by penetration of a hard steel ball of known size, into the workpiece surface. Hardened tool steel balls are used for testing softer material, while a carbide penetrator ball is used to test harder metals (Fig. 15-30).

   Due to the upper hardness, limiting factor of Brinell balls, this test is correctly used as a test of soft to medium hard metals—it is used more for customer products than for hard tooling.

KEYPOINT

Brinell is correctly used for relatively-soft to medium-hard, metals. As such, most production drawings normally specify hardness ranges in Brinell.

Brinell Scale Numbers

The scale runs from 160 for annealed steel up to approximately 700 for very hard steel. They offer three advantages over Rockwell:

1. Brinell hardness numbers relate to the metal's tensile strength. (Which we'll investigate in Unit 15-6.) Briefly, tensile strength determines how much force is needed to permanently deform one square inch of the metal, by pulling it. Tensile strength is expressed in pounds per square inch, required to deform one square inch of metal by pulling.

Brinell Penetrator Ball

Impression width

Figure 15-30 *The Brinell hardness test measures the diameter of the round impression made by a hard steel ball. Brinell discriminates well in softer metals but isn't applicable for very hard metal.*

KEYPOINT

Brinell readings are approximately twice the tensile strength in 1,000 pound units. For example, a metal reading of 260 Brinell, means that a given square inch would require 130,000 lb pull to deform it permanently.

2. Brinell numbers are greatly expanded compared to Rockwell numbers, allowing more discrimination between softer metals of only slightly different hardness. This can be useful where a small difference in hardness can make a big difference in performance of the material.

3. The Brinell system has a single scale covering its entire range of hardness, while the Rockwell system must be switched between ranges as softness range varies.

Trade Tip

Test It, But Don't Scrap It! The Rockwell and Brinell systems are a destructive test. They create a permanent mark in the metal. It is a common error for entry-level people to make the test in the wrong location on the work. Tests should be performed only on nonfunctional surfaces or in the exact location noted by a flag on the drawing.

The Shore Hardness Test

The **Shore** test is very different from the previous two; it gauges elasticity not malleability. If a precision test probe is dropped from a known distance, upon the work surface, its bounce height can be measured. The harder the metal, the higher the probe will bounce. For this test, a precisely rounded, diamond-tipped hammer of exact weight is dropped 255 mm to the work's surface. It bounces back up a scale that retains the highest reading.

The Shore test has two advantages over Rockwell and Brinell:

No permanent deep marks; this test is nondestructive.

The **scleroscope,** the instrument that makes the test, is small and portable (Fig. 15-31). This test can be taken to the work, while the work must come to the test for both Brinell and Rockwell systems.

The Shore test has two requirements.

1. **Horizontal, Flat Test Surface**
The work to be tested must have a flat surface big enough to support the scope and it must be able to be positioned horizontally such that the scope tower is oriented vertical and perpendicular to the work.

2. **Solid Sample**
The work must be solid enough to cause a true reading of the bounce. Any superficial case hardening, work movement, or flexible surfaces will falsify the reading, as will round or unusual shapes which absorb some of the bounce.

Hardness and Your Machinist Work

Just today a student brought a piece of stainless steel to the lab from which he intended to make a model jet engine rotor. He bought it at a local scrap dealer and didn't know the

Figure 15-31 *Shore Scleroscope, a portable hardness tester that measures rebound height of a diamond tipped, drop hammer.*

alloy or hardness. He asked, "What cutting speed do I use?" A hardness test provided the setup information.

Recall that in finding the recommended cutting speed for any setup two arguments were crossed:

Tool bit hardness   Work hardness

Below is a baseline chart of Rockwell C and Brinell hardness samples. When in doubt, always look it up.

Critical Question

The student's stainless did adhere to a magnet and tested R_C, 40. We then assumed it was in the 400 series (martensitic stainless steels), probably 416 as there was a lot of ma-

R_C/Brinell	Typical Product	Machinability	Speeds in ft/min HSS	Carbide
20/226	Soft carbon steel	Good	100	210
30/286	Drill shank	Fair	80	180
40/371	Axe head	Poor	50	120
45/421	Chisel, wood	End of HSS range	X	80[#]
50/475	Chisel, metal	End of carbide range	X	40
60/625*	HSS end mill	Grind only	X	X

Cutting Speeds Based on Work Hardness

[#]Considered the end of practical machinability—tool life and work hardening become major issues below this line. Extreme machining conditions indicate grinding.
*Average number—600 is beyond Standard Ball Brinell Test.

rine ship building in the region. Since he was in the Engineering I Lab, he had HSS cutters only. Based on the chart just given, what cutting speed did we setup?

Answer

50 feet per minute using coolant, HSS tools, and watching the cutting action like a hawk. The hardness of the metal was close to the end of the HSS machinability range. One incident of dull tool, and work hardening would occur, and the part would be scrapped or need to be annealed. He knew his tools would dull far faster than normal and must not be allowed to go beyond the first stage of breakdown. Gaining this kind of awareness is a major objective for the following problems.

UNIT 15-5 REVIEW

Review the Key Terms

Brinell
Testing hardness by reading the diameter of a ball penetrator mark.

Elasticity
The property used by the Shore system to test hardness.

Rockwell
Testing hardness by reading a penetrator depth.

Scleroscope
The devise that performs a Shore test, which measures the rebound of a diamond-tipped hammer off the test piece.

Shore
Testing hardness by measuring the bounce of a diamond-tipped hammer.

Replay the Key Points

◆ Rockwell C numbers rise as the hardness rises. Rockwell tests cover a wide range of material hardness.

◆ The Rockwell test: step 1—touch surface, initial load, set dial to zero; step 2—add second load, measure penetration depth.

◆ Brinell hardness is correctly used for relatively soft to medium hard, metals. As such, production drawings normally specify hardness ranges in Brinell.

◆ Test it, but don't scrap it!

Respond

1. What two properties are used for metal testing?

2. List the three common tests for metal hardness and describe their method.

3. Of the three hardness tests, which do not mar the part?
   ☐ Rockwell
   ☐ Brinell
   ☐ Shore

4. Briefly summarize the cutting speed table with respect to work hardness. From *Machinery's Handbook,* hardness testing sections.

5. A workpiece tests to be 42 on the Rockwell C scale. What would it's Brinell hardness be?

6. Using the workpiece of Problem 5, what tensile strength would the workpiece exhibit?

7. From the *Handbook* we find that there are several Rockwell scales. Identify the correct Rockwell scale for:
   A. Extremely soft metals
   B. Very hard metals
   C. Medium hard metals
   D. Case-hardened work

Trade Tip

Note, each of these scales uses a different amount of test pressure and might require a penetrator change.

8. A Brinell test is performed and read to be 344, what is the equivalent Shore reading?

9. A milling machine is already set to cut a pocket using a high speed end mill. The work material tests at 328 Brinell. The RPM for the $\frac{3}{4}$ in. cutter is 400. Is this correct? If not what is the correct speed? Use the Cutting Speed Based on Work Hardness chart in this unit then convert to RPM.

UNIT 15-6

Physical Properties of Metals

Introduction: In addition to cutting or abrading metal from a workpiece, most metals can be formed by bending, hammering, pushing, rolling, pulling, casting of liquid metal, and extruding of semiliquid metal (squirting softened metal through a die).

There are subset methods within each category. Before any process is chosen to machine or form a metal, the planner must know the way it "behaves" while being worked. The master property that predicts its cooperation or refusal to be shaped, is *malleability,* its ability to be permanently moved from one shape to another without breaking. But malleability is only one of several characteristics a given metal will exhibit. You should know something about each.

There are three groupings of properties attributed to metals: physical, chemical, and electrical. Each has significance to manufacturing. Chemical and electrical properties are important for design work, however, in Unit 15-6 we'll limit exploration to physical characteristics since they affect machine setups.

In choosing a metal for a design, engineers balance factors of hardness, wear, and corrosion resistance, and toughness against cost, machinability, availability, and **weldability** (can it be welded). All tied to physical characteristics.

KEY TERMS

Ductility
The measure of the ability to be drawn out to thinner cross section, in one axis, without breaking.

Elastic limit (elasticity)
The measure of the ability to be stressed by bending, pulling, twisting etc. without permanent deformation.

Eutectic
An alloy that melts at a lower temperature than any of it's constituent ingredients.

Malleability
The measure of the ability to be deformed in multiple directions without breaking.

Newton
The SI standard unit of force = gravitational pull of 1 kg.

Tensile strength (ultimate strength)
The final strength of a metal in lb/in.^{2} or Newtons/cm^2. Metal fractures at this point.

Weldability
Some metals can be welded and some cannot. Some can be welded but only using special materials and processes.

Yield stength
The point at which the elastic limit is exceeded.

Metal Characteristics

Malleability

A metal is malleable when it can be deformed by pushing, rolling, or hammering without cracking, separating, or breaking. A metal that exhibits good **malleability** can be deformed in all directions. Gold is so supermalleable that it can be rolled or hammered out to sections of an amazing 0.00001 in.! (a hundred-thousandth of an inch!). Aluminum is close to gold in this trait. We use the foil rolled from aluminum everyday in our homes.

In other words, harder metals of the same alloy cannot be deformed as much without cracking. This ratio is the major contributing factor to the machinability of a metal. A less malleable metal resists deformation more, and thus creates more heat and requires lower cutting speeds.

However, metals that are very malleable are sometimes also difficult to machine. For example, lead is downright difficult to machine to a dimension because it's so soft that it "squirms" while being cut. It won't stand up against the cutter force. It is similar to cutting a sponge with a spinning cutter. Softer alloys and heat treat states of aluminum also tend to be "gummy." Instead of forming clean chips that eject from drills and end mills, they smear across the tooth face and clog flutes, stopping further cutting action. We've discussed this gummy property previously.

KEYPOINT

Malleability is the ability to be deformed in *all* directions and it is inversely related to hardness.

Ductility is a subset property of malleability. A metal that is ductile can be drawn out by pulling or rolling—but in a *single direction.* Soft copper is a ductile metal. It is easily drawn into wires and tubes.

KEYPOINT

Ductility is the property that allows metal to be deformed in one direction only. It is a subset of malleability.

Amazing Alloys Combining metals yields some bizarre results. For example, an **eutectic** alloy is one that oddly melts at a much lower temperature than any of the ingredients from which it was made. Solder is a eutectic. Made of lead melting at 621°F and tin melting at 449°F, solder melts at the far lower temperature of 350°F! Combining different amounts of each metal affects the melting temperature and that effect can be graphed. The eutectic melting point is at the bottom of a funnel-shaped line, the lowest melting point combination is called the eutectic point.

Another and even stranger fact arises when adding the metal bismuth to the solder combination. Bismuth is a pinkish metal melting at 520°F, which further lowers the eutectic point. The new alloy not only melts far below the temperature of boiling water, a truly amazing 149°F for the eutectic, but in a particular ratio of bismuth to the lead and tin, it exhibits the totally unique property of remaining a constant volume from liquid to solid. No other alloy does this. This characteristic makes it useful in complex, technical machining.

Commercially named Cerro-True, it is used when delicate materials must be supported to be machined without bending, flexing, or vibrating. Liquified, this amazing metal is cast around and inside delicate workpieces, to support them during machining. Due to the zero shrink factor, the liquid metal remains supportive of the workpiece after it cools to a solid. Both work and supporting Cerro-True are machined together. After the cutting is complete, the chips and work are both put in hot water where the eutectic melts away and can be drawn off the bottom of the tank to be reused!

A second custom-designed version of this metal is called Cerro-Bend. Using slightly different proportions, it expands just slightly when it solidifies from the liquid state! This property is used to fill then support metal tubing to prevent wrinkling and collapse when bending them around relatively small radii. After bending, the removal of the solder-like metal is easy with the hot water tank.

Greater than Their Parts! Another startling fact: Many alloys achieve greater properties than any of their individual ingredients. An exceptional example lies in the metal beryllium-copper, commonly used for electrical conducting springs. While it is 97 percent soft copper, with only 2 percent beryllium and 1 percent cobalt, it is possible to heat treat the alloy to a R_C 50, hard enough to machine soft steel! No kidding.

Tensile, Yield, and Elastic Strength

These three properties are measures of the ability to withstand stress (push, pull, bend, and so on) without damage. Each is directly proportionate to the heat treat condition of the alloy.

The Three Stages of Measurable Strength

To understand the different stages, lets perform a test on 1 in.^{2} or 1 cm^2 of a given metal. This is a **tensile strength** test—meaning we'll put tension on the metal by pulling it apart. We'll increase tension progressively until it breaks and three stages of strength are observed (Fig. 15-32).

Stage 1 Elastic Strength Until a given pressure is reached nothing permanent happens to the test piece. It will stretch a tiny amount, but will return to its original length when the force is removed. The force remained within the **elastic limit** of the test metal. A safe design ensures that the alloy chosen will perform well within a safe margin below its elastic limit.

Stage 2 Yield Strength Then at a greater force, which is different for each alloy and heat treat condition, it would start to stretch permanently. It would not return to its original shape if the strain was ended. It has exceeded its elastic limit. That permanent deformation point is noted in pounds per square unit, as its **yield strength.** The yield point must be known by the tool maker if

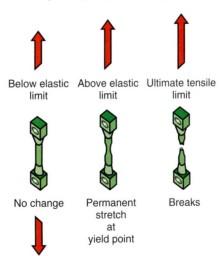

Testing Strength by Increasing Tension

Below elastic limit   Above elastic limit   Ultimate tensile limit

No change   Permanent stretch at yield point   Breaks

Figure 15-32 *A tensile strength test finds three stages of stressed metal: below, then above the elastic limit, then at the ultimate limit, where it breaks.*

Force Units Tensile, yield, and elastic strengths are expressed in one of two possible units. Imperial: pounds per square inch or SI: newtons per square centimeter.

A newton is the international basic unit for force. To reproduce a newton, you would measure the amount of gravitational pull on 1 kg of mass. In other words, 1 newton of force is how much pull a 1-kg item exerts toward the earth's center.

Remember: A newton is a standard unit of force equal to just over two pounds. It is the force of gravity on 1 kg of mass.

building a die that forms by bending a given metal. It indicates in pounds per square inch how much force will be required to cause a permanent formation of the metal.

Stage 3 Tensile Strength Then as the pull continues to increase, the resistance to stretching begins to decrease on an accelerating curve. It starts to pull apart rapidly then snaps. It reached its *ultimate strength* expressed as its **tensile strength** in pounds per square inch or **newtons** per square centimeter (see the Shop Talk).

It should be noted that some alloys have little or no ability to stretch thus they go from the elastic limit directly to the breaking point. Many cast alloys exhibit such a trait.

UNIT 15-6 REVIEW

Review the Key Terms

Ductility
The measure of the ability to be drawn out to thinner cross section, in one axis, without breaking.

Elastic limit (elasticity)
The measure of the ability to be stressed by bending, pulling, twisting etc. without permanent deformation.

Eutectic
An alloy that melts at a lower temperature than any of its constituent ingredients.

Malleability
The measure of the ability to be deformed in multiple directions without breaking.

Newton
The SI standard unit of force = gravitational pull of 1 kg.

Tensile strength (ultimate strength)
The final strength of a metal in lb/in.^{2} or Newtons/cm^2. Metal fractures at this point.

Weldability

Some metals can be welded and some cannot. Some can be welded but only using special materials and processes.

Yield stength

The point at which the elastic limit is exceeded.

Replay the Key Points

◆ Malleability is the master physical property. It is the ability to be deformed in all directions. It is inversely related to hardness.

◆ Ductility is the property that allows metal to be deformed in one direction only.

◆ There are three stages within tensile strength test: elastic limit, yield point, and ultimate strength.

Respond

1. Using ten words or less, define: malleability, ductility, and elasticity.

2. A newton is a unit of mass that describes the tensile strength of metal. Is this statement true or false? If it is false, what will make it true?

3. Which will be the greater values for every alloy, the tensile or yield strength?

4. How are stress forces expressed for various alloys?

5. You need to make a shaft for a food processing machine and think that a stainless steel would be the best material for the application. It must withstand at least 40,000 pounds per in.^{2} tension with a 5 percent safety margin. Would 30302 stainless steel be acceptable?

6. Many car enthusiasts brag about their "Mag" (magnesium wheels). But in actual fact, most of those wheels are made from aluminum! Is aluminum's tensile strength stronger or weaker than cast magnesium?

7. Of the two alloy steels 4130 and 4340, which has the stronger yield strength?

Section 15 Review

Unit 15-1

Alloy and tool steels and the many varieties of other metals we machine, are only the tip of the subject. Every time I read about metals, I find surprises. In doing the research to write this section, I learned perhaps 10 new things about some of the oddball metals and their properties.

For enrichment only, see where you might obtain lab samples of some of the metals and alloys you might be called upon to machine, yet they are not common. Compare their weight to steel. Find out how rare their ingredients are in the earth's crust. Are they an alloy or a pure element? Research what it takes to extract them from the earth. Some metals are actually abundant but so tightly bonded to other elements that they are too expensive to use. Where are they found and what nations control them? This is a big issue, as many are needed here in North America to keep manufacturing going, yet they are mined thousands of miles away where the supply is controlled by other governments. Here are a few suggestions for your list:

Magnesium	Beryllium-copper
Pure beryllium (lighter than aluminum and more costly than gold)	
Titanium	Tin
Tantalum	Molybdenum
Hastalloy®	Cobalt
Nickel	Pure tungsten
Chromium	Manganese
Inconnell	

Unit 15-2

Similar to the special ability to cut technical threads using a manual lathe, the ability to do your own heat treatment of tool steel is a skill that separates the machine operator from the journeyman machinist. Today, it isn't required of the CNC machinist. However, if your career plans include progressing to tool making, heat treating then becomes an essential skill. A thorough knowledge is also needed to become a job planner, engineer/designer, or programmer.

Unit 15-3

All the other heat treat processes change the physical properties of metal. Case hardening changes steel's chemical composition by adding more carbon to the alloy. However, using the shop process of packing the carbon compound around hot steel. The carbon can migrate only so far between the hot austinetic crystals even if the process is repeated two or more times. After quenching, it forms a shallow shell that's 0.045 inch thick at the best. To *push* carbon into the steel further, requires technical processes performed by a specialist.

The area where this process most effects the machinist is that of planning to grind a job after it has been case hardened using the packing process. The challenge is to plan rough machine cuts that leave just the right amount of excess to be finished after case hardening. If the case is not deep, (0.030 in. average), it can easily be ground completely away!

Unit 15-4

Heat treating aluminum is a very different process from heat treating steel, and a good example of the strangeness of metallurgy. It's exactly backward in every respect when compared to steel. When aluminum is heated then quenched, usually in very cold water, it enters its softest state, not the hardest. But if left alone, over time it will return to a harder state by a process called aging. But using a lower heat, similar to tempering steel, the process can be speeded up. So here too it's opposite, heating it after quenching speeds up and increases aluminum's hardening.

Although it's the third most abundant element in the earth's crust, behind oxygen and silicone, aluminum is a difficult metal to extract because it forms tight bonds to other minerals and metals. Found in many compounds in the ocean and on most continents, there is only one form, bauxite, that can be economically refined. But even then the process requires huge amounts of electrical energy.

Perhaps one of its more interesting properties is that in it's pure state it forms an oxide that's nearly impervious to further reaction or corrosion. Pure aluminum literally creates its own anticorrosion coating. But aluminum without alloys is far too malleable to use in products. So, to take advantage of the oxide property, aluminum manufacturers coat strong alloys with a thin layer of pure aluminum, called *cladding*. It then oxidizes and stops further reaction to outside elements that would corrode the alloy without protection.

Unit 15-5

Unlike heat treating, measuring hardness and understanding Rockwell and Brinell index numbers aren't added skills for the modern machinist. They are an integral part of the job. There are other systems and methods, but the two, Rockwell and Brinell, are by far, the most popular ways we test and rate hardness. And they are the way we communicate how fast a metal is to be machined. This is important information for high production today, but as computers and the machines of the next generation speed up, using hardness information will become ever more critical.

Unit 15-6

It's useful to understand the various physical properties of metals when machining, heat treating, and measuring hardness. This becomes especially true when we maximize production and cutter life during CNC machining.

However, if entering the tool and die field, an in-depth understanding of those properties is critical. The tools must be based on calculating forces and they are expensive. For example a three station, progressive die (one that shapes and cuts metal in stages) can easily cost $15,000 or far more. The die maker must calculate the exact shear property of the metal being formed in order for it to work right. If not the punch doesn't fit the die opening (both too much clearance or too little won't work) and it will fail catastrophically or constantly need resharpening.

Questions and Problems

1. What facts are found in the AISI numbering system to help indicate how a given alloy must be heat treated? In other words, what makes the difference in critical temperatures and quenches? (Objs. 15-1 and 15-2)

2. Identify and describe the three general groupings in which we divide steels. Some references include a fourth group, what is it? (Obj. 15-1)

3. In terms of carbon, what is the dividing line between mild and heat treatable steel? (Obj. 15-1)

4. What AISI family includes the mild steels? (Obj. 15-1)

5. Using *Machinery's Handbook,* what is the critical temperature required for heat treating for AISI 1019 steel? (Objs. 15-1 and 15-2)

6. What is the critical temperature for AISI 1095 steel? (Objs. 15-1 and 15-2)

7. Identify and describe the three stages of heat treating steel. What must be done to put the steel in each state? (Obj. 15-2)

8. When the steel is heated above the critical temperature, then quenched, what occurs? (Obj. 15-2)

9. Following quenching, why must we then temper the hard steel? (Obj. 15-2)

Critical Thinking Questions

10. Why don't we save time by heating steel to a temperature just below the critical point, then quench it? Wouldn't it save the tempering step? (Obj. 15-2)

11. What is another name for tempering steel? (Obj. 15-2)

12. What are the correct quenching media for these tool steels? A-2, O-1, and W-2 (Obj. 15-2)

13. What is the objective of heating steel above the transformation temperature (critical temperature) then quenching it? (Obj. 15-2)

14. Identify two reasons we might case harden a steel part. (Obj. 15-3)

15. Describe the case hardening process you can perform in your shop. (Obj. 15-3)

Reference Book Problems

16. Completely describe the information in this aluminum specification: 7075-T651. (Obj. 15-4)

17. What is the tensile strength of the 7075-T651? What is the expected Brinell hardness index for this aluminum in this temper condition? (Objs. 15-4 and 15-6)

18. Name the two most common hardness tests in North America and breifly describe each. (Obj. 15-5)

19. List and describe at least five physical properties of metal. (Obj. 15-6)

20. True or false? Other than when performing CNC, HSM (high-speed machining), we normally do not look at the heat treat condition or alloy of the aluminum. If the statement is false, what will make it true?

Section 15 Answers

ANSWERS 15-2

1. A. Heat to *prescribed high temperature*.
   B. Quench in *prescribed* media.
2. A. Water, air, heat treat oil.
   B. Water, air, heat treat oil, molten salt.
3. It is true. A controlled furnace is the better way. Color of the glow is the least useful.
4. It is false, the steel must have carbon between 0.3 to 3 percent.
5. Steel at its softest state
6. Tempering is a reduction in hardness down to a controlled, useful level. It is heated to a specific temperature below the critical temperature then left to cool.
7. Full hard steel is brittle.
8. To change the chemical makeup—add carbon; to remove residual stress created by machining, heat treating, forming, or welding; to stabilize it from long-term changes.

9. Short acceptable answer: The temperature at which the steel's structure becomes fine and will be hard if frozen at that state. The longer answer: The temperature at which the steel transforms into austenite, a fine even-grained condition that is not stable. If frozen by quenching, the steel becomes hard martensite.
10. It must be retempered at a *higher* temperature than 400°F to further reduce the hardness.
11. High heat, quench, then temper
12. Controlled furnace, pyrometer, temp stick, surface oxides. Lacking a controlled furnace, temp sticks and oxides are used.
13. False—just the opposite relationship—As the temperature rises, the hardness decreases.

ANSWERS 15-3

1. It is false. The metal is not hard after packing, it is ready to be hardened. However, the depth of modified steel is dependent on time at temperature.
2. Heat to high temp; surround with carbon compound (solid form); hold at temperature for a given time; clean and reheat to the transformation temperature; quench; temper if required.

3. Only when the case is to be very shallow or when there is no controlled furnace available.
4. Gas carburizing; nitriding; liquid salt; exposing heated metal to a carbon- or nitrogen-rich atmosphere to change the thin shell of metal.
5. At the transformation (critical) temperature, carbon molecules can migrate in or out of the steel. More go in than come out if there is a carbon-rich material surrounding the steel.

ANSWERS 15-4

1. 7075 is a high strength alloy of aluminum with a high ratio of added ingredients. The T6 denotes it has been solution heat treated, the 5 in the last field shows it has been stress relieved, and the 1 in the 51 shows it was relieved by stretching. (This is a common condition for aircraft application.)

2. F, O, H, W, T
3. O condition
4. 0.25 percent copper, 0.60 silicone (found under wrought aluminum compositions).

ANSWERS 15-5

1. *Ductility*—the ability to deform; *elasticity*—bending without deforming and bouncing objects.
2. A. Brinell—Testing hardness by reading the diameter an impression made by a ball penetrator.
   B. Rockwell—Testing hardness by reading a penetrator depth.
   C. Shore—Testing hardness by measuring the bounce of a diamond-tipped hammer.
3. The Shore test although a very shallow mark can be produced, it is considered nondestructive.
4. Harder metals require slower cutting speeds with a Rockwell C of 45 being the upper limit for practical machining.

5. 390
6. Approximately 180 KSI (180,000 lb/in.^{2})
7. A. E scale
   B. A scale
   C. B scale
   D. D scale
   The C scale accounts for most of the usage of Rockwell. Remember: "C" for deep, hard testing.
8. Shore 51
9. It is correct. A Brinell of 328 is about half way between 286 and 371. Extrapolating a recommended surface speed of 75 ft/min.

$$\text{Using the short formula RPM} = \frac{\text{SS} \times 4}{0.75}$$

$$\text{RPM} = 400$$

ANSWERS 15-6

1. Malleability, the ability to be deformed in any direction without breaking; ductility, the ability to be deformed in one direction without breaking; elasticity, the ability to be stressed without permanent deformation.
2. False. A newton is a unit of *force* not mass. It is the force gravity exerts on a kilogram of mass.
3. The tensile is the ultimate strength of the metal and will always be the higher value.

Answers to Section Review Questions

1. The main alloying ingredient and the amount of carbon in the alloy.
2. Alloy steels, used for consumer products; tool steels, used for high-demand applications, in general, more expensive; stainless steels, used where corrosion is a problem. Some references include plain carbon steels as a group.
3. Around 0.3 percent (3/10 of 1 percent)
4. The group that begins with 1xxx
5. Carbon content = 0.15 to 0.20 percent. Sorry, this was a trick question! 1019 steel is mild steel, with carbon below 0.3 percent. It cannot be hardened.
6. Carbon content 1095 = 0.90 to 1.03 percent. The range then falls between 1,390 to 1,430°F. The average temperature for conversion would be 1410°F.

4. In units of force per square area. Pounds per square inch or newtons per square centimeter.
5. Yes, the tensile strength of 30302 is 85K. From *Strength of Materials—Strength of Iron and Steel.*
6. Yes and no, depending on the heat treat condition of the aluminum; Aluminum 19K lb/in.^{2} minimum to 48K max; magnesium 22 to 40; heat-treated aluminum might be a stronger choice and it's far less expensive too.
7. 4340

7. *Annealed,* softest state—heat it to the critical temperature and cool slowly; *full hard,* hardest condition throughout, heat to the critical temperature then quench; *tempered,* reduced hardness to a useful condition based on application, heat full hard steel to a lower temperature for a specific time.
8. The austinitic crystals freeze into a fine-grained structure with evenly distributed carbon content. It becomes full hard.
9. Because the steel will be too brittle in most cases, for use. It can even crack due to internal stress built up by quenching.
10. Because the hardness will not be complete, it will be nonuniform and is unpredictable. The steel will be harder in thin sections and subject to failure.
11. Drawing

12. A-2 = air quench; O-1 = oil quench (heat treat oil); W-2 = water quench.
13. Thermal shock to freeze the austinite structure
14. A. To create a hardenable shell around the outside of mild steel
    B. To create a hardenable shell around alloy steel that must be tough on the inside and hard on the outside
15. The steel is heated to 1,300 to 1,600°F; remove from furnace, surround with Casenite compound.
    A. Let cool for a very shallow case.
    B. Return to furnace to reheat along with Casenite for deeper case. Reheat then quench to finish the hardness. Temper (may not be required)
16. Alloy 7075 = modified with zinc, silicone, and magnesium; T6 = solution heat treated and artificially aged; 51 = stress relieved by stretching. Note answers 16 and 17 are from *Machinery's Handbook* under Non-ferrous alloys.
17. 83,000 lb/in.^{2} (83 ksi); Bh = 150
18. Rockwell, a probe is driven in and the depth of penetration is measured. For harder materials, the probe is a pointed diamond cone. For softer, it is a hard steel ball. Rockwell is useful through soft to very hard metals.
    Brinell, a hardened carbide ball is driven in and the width of penetration is measured. Brinell is used mostly on moderately hard, production materials.
19. *Malleability:* the ability to be deformed (stretched, compressed, pushed, and so on) in every direction, without breaking or cracking; *ductility:* a subset of malleability, the ability to be deformed in one direction; *elasticity:* the ability to withstand bending, stretching and compressing with no permanent change to shape; *tensile strength* (ultimate strength): The point at which an alloy will stop stretching and break; *yield strength:* the point (in PSI) at which the alloy's elastic limit is exceeded; *weldability:* some alloys can be welded, some can't. Some only using special processes and materials.
20. It is true. Except that when machining the newer, extremely high silicone aluminum alloys, we must choose diamond-coated carbide tools that can withstand the abrasion, and they are most efficient if the exact RPM is selected. This was not covered in the reading.

Section 16

Job Planning

Introduction

Even making a single workpiece with the simplest of shapes requires careful choices before putting it into work. More to the point, not planning the simplest or any job, easy or complex, is a surefire guarantee it will be difficult to complete, of poor quality, unprofitable, or even dangerous!

One of the highest skills within our trade, the machinist who excels in planning has great opportunity for advancement. But it's a hard won competency that's different from previous subjects we've discussed, for four reasons.

First, planning solutions are often matrix-like in that any single job could take many different routes through the shop on its way to completion. Some sequences are faster, while others lead to more repeatable accuracy, some toward cost or time savings, and on it goes. Every job presents a puzzle with a nearly infinite set of solutions, each with trade-off advantages and disadvantages. There is no single or perfect answer.

Second, planning often requires inventive thinking beyond conventional solutions. Some new designs create a new challenge: how to hold an oddly shaped object as it nears completion, or how much material is just enough for roughing but not too much for economical manufacturing.

Third, plans are never quite finished. Even the best change after the work is in progress. As safer ways are discovered or as new ideas occur we edit them as part of normal operations to upgrade capabilities. In industry, I've seen the same job go through the shop dozens of times, and before each run, improvements are edited in based on the last time we ran it.

Fourth, writing plans requires a full understanding of all the machine types and the operations they can perform. If you've followed the book outline, you are at that point now. You now see the bigger picture—that the shop functions as a unit.

Here in Section 16, after a few hints for success, we serve up six part-making puzzles for which our solutions (reviewed by several instructors) are offered. Four jobs have not been planned yet, and two are already done but they might need improvement. You'll need to compare your solution to ours, discuss it with others, then decide for yourself if your plans are the better solution?

Three Kinds of Planned Success

In the shop, there are three different kinds of planning. They work together to keep everything running smoothly:

- Job flow
- Shop flow
- Operation sequence

Job Flow Decision #1 This is the route any single part or batch takes through the shop. For example, do we surface grind the work before or after heat treat? Does the job go to the mills first, then the lathe? These plans must occur first.

Shop Flow This is the day-to-day operation of the shop. It schedules and moves jobs through the shop simultaneously. Unplanned, they arrive at critical workstations (example: CNC mills or surface grinders) at the same time. When they clash, one must be given priority while the others are stopped, re-routed, or sent to another machine shop. Keeping the total workload moving efficiently might also be called *shop load* or *shop floor planning*. As a management task, shop flow is beyond the objectives of Section 16 but it's a big part of operations.

Shop flow is vital. Most large companies employ specialists in the field. They maintain job boards and wall charts where they move tags to track it all. But even the experts often turn to management software when visual aids can't solve the matrix. PC-based management predicts bottleneck points, and schedules and adjusts sequences to avoid them.

Shop flow is a perfect description with a river of jobs flowing through. New ones enter upstream daily. Each has its own priority and potential to affect those already in work. Often, a job's priority will change halfway through the loop. Suddenly it goes from normal to *hot:* the customer needs it now! That changes everything in its routing. It causes turbulence, which then disturbs other lower priority work. You can see that only a computer with specific programming can reschedule the flow when things don't go as planned—which is almost always!

Absences and Breakdowns The two worst dams in the flow occur when the machinist doesn't make it to work that day or a machine breaks down, especially if that station is one of the bottlenecks! If available, another machinist can be substituted or the job could be moved to "bump" another job. But then, a whole series of disturbances ripples out.

Operation Sequences The third kind of planning is the machinist's problem, which is logically sequencing operations. They're step-by-step instructions, aimed at achieving the drawing dimensions and specs.

One of the highest skills within our trade, the machinist that excels in planning has great opportunity for advancement. Here's what we'll learn in Section 16 to improve that ability.

Objectives

Planning Jobs Right

Introduction: Learning by doing—it's the only way to learn to plan. With some guidance, in Unit 16-1 you are to plan out four jobs, each more complex. Then on compiling your best ideas, compare yours to mine. That's the key here, trying then comparing.

KEY TERMS

Grip excess
Excess used to hold a bar in a lathe chuck beyond that which is needed for the part.

Temporary datum
Planning a cut sequence that creates a reference to be cut away when no longer needed.

The Prime Questions

A solid plan must answer six or seven questions, depending on whether or not it's to be machined by CNC methods. All must be answered at the same time or nearly so. They are intertwined, each affecting the others. Use the following as a checklist for writing your plan.

☐ *Which machine types first?*
The answer to this will depend on the number and complexity of parts in the batch and the kinds of machines available for this job. Job priority affects this decision too, since a hot job can bump others from certain machines.

☐ *How is the work to be held and located during machining?*
Often overlooked by beginners, this must be solved right after you decide which machine gets the job first. An incorrect choice can lead to needless variation being introduced into the job.

☐ *Does the job need work-holding and reference excess material?*
Well-planned excess material to hold onto or to create temporary reference surfaces can save money and time.

☐ *Do the first cuts lead to reliable reference?*
The first vital cuts must create references that will be used for moving the parts between setups.

☐ *Is the overall sequence efficient?*
Can several operations be accomplished in one setup?

☐ *Are specialized holding or cutting tools justified?*
This decision is based on two factors: batch size and the probability of completing the job using only standard holding and cutting tools. For example, are soft jaws for the vise or chuck justified? This decision is often based on part quantities—more parts in the batch justifies specialized tooling.

☐ *Reference point (CNC jobs only).*
Since our plans will be limited to manually operated machines, we won't be using this factor here in Section 16. But it must be in the list of questions when planning a job that's to be machined from a program. There must be zero point chosen, a place where all coordinates are set to $X = 0.0$, $Y = 0.0$, and $Z = 0.0$.

That reference point applies to the program, to the holding tooling, and to the machine setup. Everybody in the loop must know where it is located on the part or on the holding method. Choosing that point wisely is a prime decision that must be made at the time when the other questions are decided.

The upcoming challenges focus attention on these issues. Here's few tips about how to answer them.

Excess Material

Planned for Holding and Intermediate Operations

Although excess must be made into chips, which consumes money and time, sometimes by planning the right amount in the right place, the added material costs are outweighed by improved safety and quality, or the excess provides a far more efficient way to hold and cut the part. Thus far in this text, excess has been presented as a minimum allowance to completely machine a surface. But there are two more aspects.

KEYPOINT

There are times when it is very cost effective to plan excess metal beyond a clean-up cut minimum. It's used for *work holding of difficult objects* and *a planned temporary datum to start the job right.*

Work Holding Excess

Here's an example of work-holding material for a lathe job (called **grip excess** *material* for lathe work).

Figure 16-1 *There are several different ways to plan operation sequences for this Plum Bob.*

You are to turn five (5), brass surveyor's plumb bobs (swinging pointers to indicate a vertical line) (Fig. 16-1). They're made from $1\frac{1}{2}$-in.-diameter brass bar stock, a relatively expensive material. This stock doesn't fit through the lathe's spindle hole, so bar feeding and parting off isn't an option. That leaves two others:

1. **Sawing Blanks to Length** (Fig. 16-2) Turn half the shape, reversing to finish the job.
2. **Sawing with Grip Material** (Fig. 16-3) Holding an extra length, machine the entire shape in one setup.

Both plans would require the bar to be cut into lengths before machining. But the grip excess (plan 2) would include 2 in. extra material per part length. Let's compare.

Second Operation

▶ Two separate operations
▶ Added cost of special tooling
▶ Potential concentricity problems

After first operation

Figure 16-2 *Option 1—The second of two setups.*

1. Cut to 3.100 in. long. We'll first turn the point and knurled section, leaving a small shoulder between the bar stock and 1.437 in. diameter. This could be done in a standard three-jaw chuck. Then reverse it by pushing the shoulder against soft jaws (Fig. 16-2). The soft jaws would help ensure concentricity of the second cuts and to not mar the finished knurls. With the temporary shoulder nested against the jaws, it's possible to finish the shape.

2. On the other hand, how about cutting the bar into 5-in.-long pieces? (See Fig. 16-3.) That leaves a minimum of $1\frac{1}{2}$ in. stock to grip in the chuck so the entire shape can be turned then parted off at the final operation. No special tooling is required and the whole profile can be turned in one setup.

Critical Question Is plan 2 more economical? Probably, even though a length of brass is lost for each part. It's going to go faster and be safer, and the shape is going to be more concentric end to end. But the extra brass stock is costly too. The final answer would be a time study verses current material cost. (See the Shop Talk.)

ShopTalk

In plan 2, a small but significant fact to be considered when batches become large, is that the excess will be parted off as a solid chunk. A solid piece of metal commands a better price when it's recycled and it will offset some of the added cost of excess.

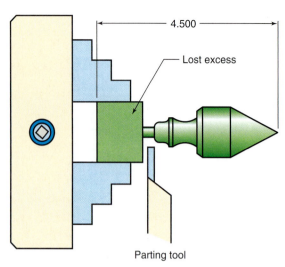

4.500

Lost excess

Parting tool

Figure 16-3 *Option 2—Cut all features in one setup, then part off the finished work.*

KEYPOINT

In this example, plan 2 is likely to be more economical, but there are times when plans similar to plan 1 might work better.

First Cuts Based on Design Datum Basis

Related to excess material is the decision of how to create a reliable reference datum(s) at the start of the job. Unless the parts can be finished in one setup, as plan 2 was, the parts will need to be moved from one setup to the next, or to another machine. The shoulder in plan 1 that allows us to reverse the parts and be able to locate them in the soft jaws with reference to the pointed end, was a **temporary** reference **datum.**

Temporary Reference

So the third reason for planned excess is to create a temporary basis for the next cuts. That extra step is used when one can't immediately cut the actual datum to begin the job. It's machined early in the plan, then sequential setups or cuts are referenced to it until an actual high-priority datum can be cut. The temporary surface then simulates a high-priority datum until one can be created. Then, after shaping the part, where it's no longer needed, the temporary surface is removed. Of course, the objective is always to minimize that planned amount.

KEYPOINT

Good plans begin with vital, first cuts that create sequential control throughout the remaining operations. That first set of operations might be as simple as squaring two sides of a block or a lathe face cut, but it is the starting point reference from which all sequential operations originate.

Recall squaring a part on the milling machine (Section 12). After establishing a reliable datum feature, all sequential cuts were made by placing that surface against the solid jaw of the vise.

Holding Tooling and Method

The next prime question is how will the job be held on that machine to make the first cuts? This must be answered before writing cut sequences or a program. The obvious reason is safety and efficiency. But the subtle reason is to control variation. Does your choice of holding method eliminate or cause variation in the job? Rather than discuss it here, one problem was expressly created to challenge your thinking in this aspect of planning.

KEYPOINT

Always analyze the holding method and ask whether it cause needless variation.

Note too that each problem asks you to machine different numbers of parts. The quantity in the batch changes some of the thinking, especially on holding tooling. Larger part runs may justify special tooling (soft chuck or vise jaws or fixtures), where small batches are more economically solved using standard vises and chucks, even though they take longer to use.

Problem 1   Cast Iron Angle Plates

OK—it's your turn. Plan and write a set of operations to machine these four precision angle plates including semi-finish milling and final grinding on all surfaces marked with a 64-μIn callout (Fig. 16-4).

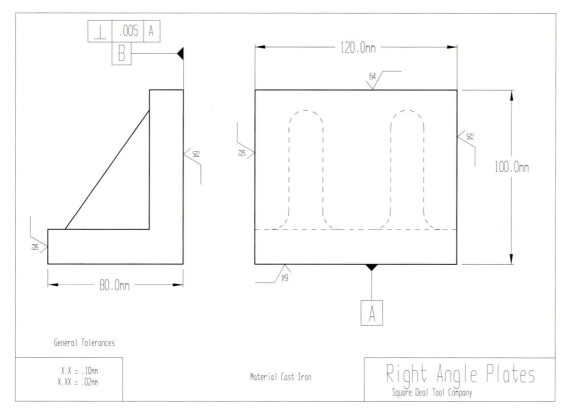

Figure 16-4 *Planning problem 1—Right angle plates.*

Instructions

◆ Machine only features that have surface finish dimensions.

◆ 6 mm excess has been cast on every face of the casting to be machined.

◆ Leave 0.4 mm grind excess after milling on every face.

Using the model here, write a complete machining plan on a sheet of paper. Add more sequence numbers if required. Then compare your final version to mine.

Job Planning Sheet

Job Name: Right Angle Plate   **Number:** 4 Machined in matched set
Material: Cast Iron
Special Tooling or Holding:

Sequence	Operation	Comments
005		
010		
015		
020		
025		
030		
035		
040		
045		
050		

Figure 16-5 *Planning 2—Valve dome.*

Problem 2 Valve Domes

These valve domes (Fig. 16-5) are to be machined from precision brass forgings. Each has 0.10 in. excess on all dimensions to be machined. Include operations to make all holes and slots. On completion, turn to the answers and compare your plan to mine.

Instructions

◆ Write your plan to produce (Qty. 150).
◆ Plan this job without the benefit of CNC equipment.
◆ How would you change the plan if CNC mills and lathes were available?

Problem 3 Tap Wrench Nuts

You are to machine five of these tap wrench nuts from a 1.00-in.-diameter 4130 steel bar preheat treated to Rockwell-C 28 (Fig. 16-6). Study the print and write your plan on a sheet of paper.

Instructions

◆ Make (Qty. 5) five of these nuts.

Problem 4 Locator Bolts

These special fixture bolts (Fig. 16-7) might need more than one setup to complete, or do they? Note they're made from mild steel but no raw material size or shape has been specified.

Instructions

◆ Make 25 in this batch.
◆ *Choose the raw material* shape and size.
◆ You may need to look up standard steel shapes and sizes in *Machinery's Handbook*.
◆ Problem 4 is a time-verses-quality challenge.

Figure 16-6 *Planning problem 3—Tap wrench nut.*

Figure 16-7 Planning problem 4—Locator bolt.

Troubleshooting

The most skilled planners cannot foresee every trap that arises as the machining commences. Plans are improved regularly as part of normal operations. However, as your instructor in print, I'm compelled to add a bit of advise: as a new person, exercise restraint in criticizing a plan or instructions. If safety and quality are at risk, then go ahead, but do it with finesse.

Seeing problems with work orders and improving them is a skill needed for competency in both manual and CNC machining. Problems that cause part errors and safety hazards must be eliminated when they are found. Here's an example (Fig. 16-8).

KEYPOINT

In every shop, there will be an official revision procedure that keeps everyone that it affects informed of the changes. Never write on or alter a work order/plan without permission.

Problem 5   Lock Keys

After reviewing the plan offered, edit it or reroute it altogether to solve sequencing problems if they exist. On completion, compare your improvements to mine. (*Hint:* There are at least four big errors in this plan.)

Figure 16-8 Planning problem 5—Lock keys.

Job Name: Lock Keys
Number of Parts: 15
Material: 1.0 by $\frac{3}{8}$ in. CRS Bar 45 in. long
Special Tooling or Holding: None

Sequence	Operation	Comments
005	Saw to lengths 3.0	0.125 in. excess each end
010	Layout Angle	
015	Rough Saw Angle	Leave 0.030 Machining Excess
020	Mill 18.0 Degree Angle	Tilt Mill Head
025	Mill 2.75 × 0.875 rectangle	
030	Case Harden R_c 48 min	
035	Drill $\frac{1}{4}$-in. hole and Chamfer 2 sides	

Problem 6   Double-Arm Bracket

You are to make a batch of double-arm brackets (Fig. 16-9) from a solid block of aluminum. After reading the original plan, edit it with improvements or scrap it altogether and start over, if you feel justification.

Now write your own plan for the double-arm bracket. It need not take the path I've chosen here.

Figure 16-9 *Planning problem 6—Double-arm bracket.*

Job Planning Sheet

Job Name: Bracket Double Arm
Material: Aluminum: 6061-T651 Aluminum
3.0-in.-Diameter Bar
Sawed to 15-in. lengths
Qty 17
Tooling or Holding Soft Jaws Lathe/Mill Vise

Sequence	Operation	Comments
005	Turn 2.750 Dia (2 bars)	4 Jaw Chuck & Tailstock Center 14 in. long turned portion. Excess for grip & facing parting
010	Face and Part off 1.5 lg.	Soft three jaw chuck—tailstock support Creates Datum Feature A Excess for milling
015	Mill 2.20 × 0.50 Deep	Soft vise jaws on (Fig. 16-10) 2.750 dia—Datum B
020	Reverse Part Mill 1.3 × 0.25 R	Standard Vise on 2.20 Dim 1.300 Dim Across Ears
025	Mill 0.31 Ears	
030	Drill 0.375 through	Vise with end stop
035	Layout Ear Profile	
040	Rough Saw	Leave 0.060 Excess
045	Mill Angle Sides	
050	Mill 1.25 Dim	
055	Remove all burrs and sharp edges	

Operation 015

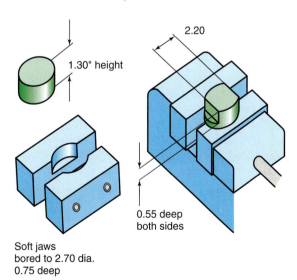

1.30" height

2.20

0.55 deep
both sides

Soft jaws
bored to 2.70 dia.
0.75 deep

Figure 16-10 *Operation 015 for the double-arm bracket.*

Section 16   Our Answers

Problem 1   Right-Angle Plates

Justification

I first machined a temporary surface (E), which will lead to a reliable milling of face A, then B. To mill E, each plate was first nested on uncut face A, against the mill table (Fig. 16-11). Then an 80 percent clean up cut was made creating a temporary datum that represented the "average Surface A." Surface E will be milled to the pregrind 100.8 mm later, after datum A is complete.

Why the intermediate step? The important issue was to machine face A completely to about 5.6 mm depth, leaving 0.4 mm for grind.

To machine face A parallel to its rough cast average surface might be difficult to do as a first cut because it needed to be level to the cutter's plane. However, creating the temp surface makes it easy.

To understand the reason for this extra step, imagine there was very little original excess on all sides of the raw casting instead of the generous 6 mm. The challenge would be positioning unmachined surface A as horizontal as possible, accounting for all casting irregularities, so it would clean up completely and still leave grind material. With so little original excess, the intermediate step would be absolutely necessary to orient the average surface A when it was cut (Fig. 16-12).

Following the Datum Milling Process

After surface A was machined, the remaining operations are simplified. The next cut would be surface B machined relative to surface A, nested on a vise or mill table. Then with the critical 90° surfaces under control, both the 80.0-mm dimensions could be ma-

Figure 16-11 *Machining reference surface E on the angle plates.*

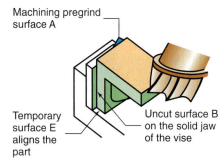

Figure 16-12 *Machining datum feature A relative to surface E.*

chined to a pregrind size of 80.8 mm (0.4-mm grind stock each side), and the 120-mm dimension would be milled to 120.8 mm.

The grind operations would follow a similar sequence, except no temporary datum is needed. The job could start by grinding B relative to A since they now are reliable. Then A would be finished at 90° to B. Then the 80- and 120-mm dimensions would be completed.

005	Cut temp. surface E	Part nested on raw feature A Cut E 80% cleanup
010	Nesting on E—Cut Face A	
015	Mill face B	Nest against face A
020	Mill 100.8 mm	Nest on A—finishes Surface E 0.8 Grind excess (0.4 each surface)

025	Mill 80.8 mm	Nest on B
030	Mill 120.8 mm	Nest on A—Index to B
035	Grind surface B	Nest on A
040	Grind A	
045	Grind 100 mm	
050	Grind 80 mm	
055	Grind 120 mm	
060	Break all sharp edges—remove burrs	

Problem 2   Valve Domes

Operations 005 Justification

My manual lathe/mill plan would make two sets of soft aluminum jaws for a three-jaw chuck. Then for the first operation 005, chuck the part inside raw datum bore B. The dome would be outward (Fig. 16-13). The part would nest against raw surface A, parallel to the chuck face as shown.

Machining 0.1 on each surface (flange face and outside diameter), an intermediate datum frame is created, composed of the new 2.50 diameter and the faced flange. Different from the casting problem 1, the flange surface and diameter can be machined to final size because forgings are more accurate in their raw state while castings often have irregularities. I would machine all 100 parts in this setup first.

These intermediate surfaces were to be machined anyway but they now represent the average of raw datum features A and B. They will then be used to create A and B.

Operation 010

Now reverse the part to machine datums A and B. Reversing the dome and holding on the newly created intermediate surfaces allows machining of the two critical datums in perfect perpendicularity to each other. Bored aluminum soft jaws would be a good idea to guarantee concentricity and to prevent marring the outside of the domes (Fig. 16-14). At that time, the 0.375 hole would be drilled undersize, bored for concentricity, then reamed to ensure the position tolerance to B and size.

Operation 005

Soft jaws—3 jaw chuck
2.00 in. dia.

A

B

Turn
2.50 in.

Face to 0.37 in.
0.25 in. radius

Figure 16-13 *Operation 005—Valve domes.*

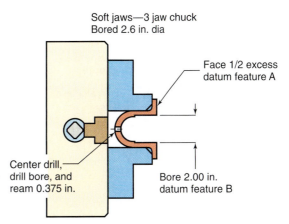

Operation 010

Soft jaws—3 jaw chuck
Bored 2.6 in. dia

Face 1/2 excess
datum feature A

Center drill,
drill bore, and
ream 0.375 in.

Bore 2.00 in.
datum feature B

Figure 16-14 *Operation 010—Valve dome plan.*

Machined in this sequence, each part would repeat exactly and the relationship between datums A and B would be guaranteed from machining them in the same setup. For this large batch, special tooling was justified—lathe soft jaws, two times.

Critical Question After establishing the datum features, there are two possibilities for proceeding with holes and the slot. One is more accurate, while one is faster. Carefully analyze both versions of my plan, each has a subtle potential problem.

More Accurate Plan—Two Setups

The slot centerline has a basic angularity relationship of 30° to datum B with a fairly tight 0.003-in. tolerance. In this version of the plan, we drill the holes first with the dome side up, positioning the part on a datum B locator (1.9995 in.), face A down. The part is now located on the actual datums (Fig. 16-15).

Placing the part with datum bore B over a 1.9995 locator pin, and against a side pin to ensure the holes are correctly drilled in the flange enlargements, the holes could be jig or CNC drilled with reference to datum B.

Now reversing the part, face A up, holding on the original 2.50 diameter and locating over a pin in one drilled hole, mill the slot. This might require vise soft jaws or a flat plate cut out to receive the dome.

Faster Solution—One Step The slot and holes could be machined in a single setup (Fig. 16-16). By locating on the original 2.500 diameter and flange face, and against a pin on the side of a flange, both

Drill and/or Mill Slot

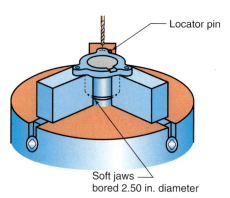

Figure 16-16 *Drilling and milling the slot using the lathe chuck with soft jaws, to hold the work.*

the holes and slot could then be machined with the dome down.

Troubleshooting

Both plans to drill the holes and mill the slot have their own potential errors. Can you find it? Can you correct it? Make your investigation and choices now. Then read the following.

Two-Step Plan—Potential Error In the two-step plan, the 2.00-in. bore, datum B, is located over a precision pin with a MMC clearance of 0.0005 in. in the tightest-fit case, to drill the holes. Datum B, the bored hole has a diameter tolerance of plus 0.002 in. The difference in the bore B size and the locator will allow movement in any direction, which will be passed on to the hole locations—degrading their location accuracy. How can that problem be solved and keep this plan?

A solution might be instead of a 2.00 locator pin, inside gripping soft jaws to hold the part perfectly concentric to datum B while drilling the holes.

Simple Drill Jig

Figure 16-15 *Drilling the valve domes with a shop made jig.*

One-Step Plan—Possible Error Holding on the 2.500-in. surface is not the datum, remember, *it represents it.* If there is any axial or perpendicularity error in these two intermediate surfaces, relative to A and B, that error will then be passed on to the holes and slot. The probability of this occurring is low, however.

However, if this problem occurs, there is no solution. If the error is great enough to affect the hole and slot position, this plan must be scrapped! On the other hand, if the error between the temp surfaces and datums A and B is zero or very small, this plan is faster by far, plus it ensures the relationship between the holes and slot since they are made in a single setup. This solution should be considered first since both features are machined in a single setup.

CNC Solution

If a CNC milling machine were available, this job would have progressed differently. The first setup would use the external soft jaws in a vise or mill indexing chuck to hold the work with the dome inward. Then Operations 005 through 025 would machine datums A and B all holes and the slot. Cutting all critical features within a single setup would guarantee the relationship between features. A strong advantage of CNC mills is that they can generate the round datum A plus mill or drill all other features too. After the first setup, the part could be off loaded to a lathe to finish the subordinate, flange, and 2.500 diameters relative to the datums.

Problem 3 Tap Wrench Nuts

Justification

This plan is a single-setup completion of all dimensions and features without rechucking the parts. It ensures concentricity plus a quick turn-around time.

All features will be machined in one chucking, except the 0.030-in. chamfer on the 0.50-in. hole. To solve that, the last step is one of those creative ideas—see the Trade Tip following the plan.

Job Planning Sheet

Name: Tap Wrench Nut
Number of Parts: Qty = 5
Material: 1.0 dia 4130 Bar R_c 28
Tooling or Holding

Sequence	Operation	Comments
005	Clean Up Face	Three Jaw—1.5 in. Protrudes from chuck
010	C'Drill & Drill 0.375	
015	Turn 0.857 diameter	Carbide TNR-16
020	Knurl	Med. Diamond
025	Turn Outside Taper	Rotate Compound 15°
030	Bore 0.573 Diameter	Minor Thread Dia.
035	Bore Internal Taper	Same Taper Angle 15°
040	Form Chamfers ID, OD	0.65 × 45°
045	Bore Thread Relief	Internal groove tool
050	Thread $\frac{5}{8}$-14	Internal Thread Tool
055	Part Off To Length	1.75 long
060	Form 0.03 Chamfer	Hold on threaded mandrel

Trade Tip

A Threaded Mandrel See Fig. 16-17. Holding the finished nuts for the internal 0.03-in. chamfer was the only issue with my plan. A collet can grip over knurls but there is a potential for damage to the high points formed by the tips of the diamonds. However, if a right-hand-threaded object such as this is screwed onto a threaded mandrel, the cutting action will tend to tighten it. If this mandrel is made with a shoulder such that the work cannot overtighten, then the chamfer can be formed with little danger of marring the work. Be sure to screw the workpiece on tightly before machining so that it doesn't ram tighten! Gripping it with sandpaper and pliers will unscrew it.

But how about a creative setup that finishes all details before parting off? Figure 16-18 shows how.

Threaded Mandrel

030 × 45° chamfer

Stop shoulder

Figure 16-17 *Using a threaded mandrel to machine the final chamfer.*

Boring the Last Chamfer

1.00 in. dia barstock

1.75 part length

Figure 16-18 *Another option, boring the chamfer before parting the nut off the bar might save making another setup.*

Problem 4   Locator Bolts

Justification

To commence my plan, I chose 1.0-in. *square* CRS bar cut into two lengths, long enough to bar feed turn five parts, one at a time. I left excess for parting and for holding the final, fifth part too. Each bar was five parts long, so I could lay each on the small surface grinder to grind the 0.875 dimension.

Read this plan and see if you can improve on it—there is a better solution that you might have already found.

To start the job, *I finished the square datum surfaces while the two bars were long.* Using the datum mill and grind process, I completed the 0.875-in. dimensions square. I then took them to the lathe for turning while being held in custom made soft jaws or in a square collet. It is possible to make a square hole on the center of the chuck jaws. A two-jaw universal chuck with custom soft jaws could also be used.

A four jaw chuck would work, but there would be a lot of indicating to get the square on the spindle center each time. Plus the surfaces of the datum would need to be protected from chuck marks. Since each part had to be turned on both ends, this extra tooling would be justified for even 10 parts.

Did your plan turn the parts with the tapped hole facing out (Fig. 16-19)? If it did, then you could have made all operations on the part in one setup, which was faster than mine. I turned the part around for each end, which makes length of features more difficult to control. Using a one-step plan would be acceptable for CNC turning as well.

7/8" precision square collet

Utility relief (not on print)

0.65 dia. × 60° C-sink

3.00 part off length

Figure 16-19 *Using a square collet to establish datum Axis A-B.*

Job Planning Sheet

Job Name: Locator Bolt
Number of Parts: 10
Material: 1.0 Square CRS—2 Lengths, 19.0 in. Long
Tooling or Holding 0.875 square collet or 0.875 soft jaws square $\frac{5}{8}$-18 threading tool

Sequence	Operation	Comments
005	Datum Mill 0.9 × 0.9	0.025 Excess for grind
010	Surface Grind 0.875 Square	
015	Turn 0.625 Dia	Lathe with 0.875 collet
020	Form thread relief	
025	Thread $\frac{5}{8}$-18	
030	Part Off	
035	Reverse in Soft Jaws	Use material stop in spindle hole
040	C'Drill—Drill $\frac{5}{16}$	
045	Countersink	
050	Tap in Lathe	Caution—do not hand hold tap handle
055	Break Sharp Edges	

Job Planning Sheet 16-5 (My Improved Plan)

Job Name: Lock Keys
Number of Parts: 15
Material: 1.0 by $\frac{3}{8}$ in. CRS Bar 48 in. long
Special Tooling or Holding None

Sequence	Operation	Comments
005	Cut stock into three equal 16-in. lengths	
010	Stack mill stock to 0.875 tall	One setup
015	Surface grind stock to 0.312 thick	One setup
020	Saw to lengths 2.85	0.05 excess each end
025	Mill 2.75 × 0.875	
030	Drill 0.25 Chamfer both sides	
035	Mill 18.0 degree angle	Tilt mill head
040	Case harden R_c 48 min.	
045	Break sharp edges	

Problem 5   Lock Key

Here are the errors. Did you find all five?

1. **Stock Issue Problem**

   The 45 in. wasn't enough total metal—there was no kerf allowance built in for the saw cuts!

2. **Dimension Missed**

   There was no provision for the 0.312 dimension—the keys were left 0.375 thick.

3. **Sequencing Problems**

   Milling the angle first was a big mistake. The rectangle should have been finished first. By milling

0.875 and grinding 0.312 dims, in bulk stock many part loads were saved. By doing so, the angle could have been cut without a layout operation.

4. **Machining After Heat Treat Problem**
Drilling after case hardening is difficult or impossible. Drill while in the rectangular shape—easy to hold and locate.

5. **Missed Operation**
There was no part finish operation.

PART 3

Introduction to Computer Numerical Control Machining

To be a master, yesterday's machinist needed many skills and traits. But even though the knowledge bank was large and took years to gain, once journey level was reached, it remained relatively stable. Sure, processes and technologies did change over time, but the rate of change was nothing compared to how change progresses today! Not even close!

For a career to flourish today, machinists must add two more skills to their list. We'll start with *adaptability*. It's more than a willingness to change, it's about looking forward to evolution, seeking it and embracing it. Adapting will surely be your greatest challenge and your biggest career asset.

In that early machinist's world about three quarters of the price of an average manufactured item was in the labor cost to make it, but not today. Due to the efficiency of computers and software, and to fierce international competition, the wage portion is being driven down and down. Today, any business wishing to have a future, foresees and adapts to technologies that deliver faster, better, and less costly products. That goes for the machinist too.

The workplace is evolving too. Previously, programmed machines were used for long production runs. Their great cost, slow setup, and programming time justified purchasing them only when the shop expected to make hundreds or thousands of similar parts. But today, these machines are so easy to get going that batches of 5 or 10 parts are normal and economical.

Progressive shops today consider one part to be a batch! Instead of completing a large group of parts before moving the job to another station, these shops move single parts or small batches through in a continuous flow, from machine to machine, and from station to station. It's a flash quick turnaround environment.

In some of these shops, management is changing the way they see their floor plan too. We no longer see a rigid, unchanging group of bolted down machines (called *monoliths*). Instead machines are mounted on thick steel pallets that can be skidded to a new location by forklift.

After conferring with planners and lead machinists, on the weekend, they reconfigure the machines to match the most efficient part flow for the upcoming workload (Fig. P3-1).

Today's toolmakers are using CNC to quickly and accurately make one-of-a-kind molds, dies, and other implements. Not only are they using computers to do their work more quickly, they can create unique items that were impossible to machine previously. Today I can't imagine anyone doing competitive tool and die work without a CNC mill and a CNC EDM machine.

The point is that in this sea of change the attitude that floats cannot be a monolith. As I write this, the tasks required to process a normal job from the engineer's drawing to final inspection are those of the

Figure P3-1 *These machines are not monoliths.*

- Designer/engineer
- Programmer
- Planner
- Setup/lead
- Tool maker
- Machine operator
- Quality assurance (previously called the inspector), possibly using programmed coordinate machines

But in the next generation, these tasks will merge, evolve, or even disappear! For example, that last job title, inspector, has already been transformed and it will continue to evolve. Many shops are seeing that the responsibility for part quality must lie within the machinist's role. In today's teamwork environment, the inspection department became quality assurance, and the "inspector" now functions more as an auditor, investing more time toward improving and monitoring the quality system and less inspecting the machinist's work. Everybody works together to ensure the product is right. Today some of the inspection equipment itself is automated and programmable and the machinists as well as inspectors have access to it. Tomorrow, this equipment will surely be even more capable (Section 26).

So, as technologies advance, the wage portion of a product's cost will continue to drop. You're here to gain the knowledge base from which adaptation to that challenge will be a snap rather than a struggle.

Better communication skills are the second new survival trait. Yesterday's craftsman had a print, a machine, and perhaps a planning sheet. For the most part, they made and executed most of the decisions as to how and when the part would be machined. But today, the planner, programmer, tool maker, lead machinist, and operator are all involved in the final outcome. In the new shop, they need to share program data, design planning, and setup information.

Without communication much of the advantage of automation can be lost and the process can even become dangerous! Today conservative feed rates are 300 in. per minute with mill spindles spinning at 8,000 RPM—double or triple those figures for high-speed machining! Many accidents can be traced back to misunderstood directions or missed setup statements. We'll weave the ability to read and write concise instructions into many of the sections and units.

So, welcome to the world of programmed machine tools. It's an adaptable, communicative place. Today is an excellent time to be an entry-level craftsperson.

Section 17

Coordinates, Axes, and Motion

Introduction

The bulk of a program is comprised of point coordinates that refer distances to axes. Along with statements that determine how the machine is to use them, those X, Y, and Z coordinates are used either for tool motion or positioning tools on the work, or for reference. The axes they refer to are the same ones we've already learned on mills and lathes. But here in Section 17, we'll formalize them.

These four units are bedrock knowledge for everything in CNC and most of the remaining textbook.

Objectives

ShopTalk

EIA-RS267-B Axial motion and position are standardized by the Electronic Industries Association (EIA) here in North America in their Recommended Standard EIA267-B. There is also a parallel standard through the ISO (International Standards Organization). These standards actually include 14 defined axes for motion and position, however, using the 9 described here, covers all normal CNC equipment.

World Axis Standards

Introduction: There are nine standard axes universally used in CNC machining. Three are the familiar *primary linear* (straight line) movements X, Y, and Z. Three *primary rotary axes* (A, B, and C) are used to identify arc or circular movements such as a programmable turntable, lathe spindle, or an articulating, wrist action milling head (rotary motion but not in a full circle).

Finally, we have three secondary, straight-line, motions called the *auxiliary linear* axes (U, V, and W). They are added to the system for multiaxis production, sometimes called *multitasking*.

KEY TERMS

Articulate
A wrist-type action that moves in an arc but not a full circle. Mill heads articulate.

Discreet planes
One of three unique planes defined by the axes that lie on it: X-Y, Y-Z, or X-Z.

Five-axis mill
A vertical or horizontal machine with a head that articulates in the A and B axes.

Orthogonal axis set
Three axes lying at mutual 90° angles.

Right-hand rule
Used to determine the letter identification of axes within an orthogonal set.

Rule of thumb
Used to determine the sign value of rotary axes.

World orientation
The relationship of a machine's axis set with respect to the floor and operator.

The Primary Linear Axes X, Y, and Z

The basic axes used to define a three-space (three-dimensional) envelope lie at 90 degrees to each other, and as such are called an **orthogonal axis set.** Using the same root word as orthographic projection, the set is comprised of axis lines at mutual 90° angles,

Figure 17-1 *The three primary linear axes of* X, Y, *and* Z.

intersecting at a common reference point (Fig. 17-1).

Three Primary Planes

Combining any two primary axis lines defines a flat plane. There are three planes: X-Y, X-Z, and Y-Z (Fig. 17-2). For example, when viewing a part placed on a vertical milling machine, the table represents the X-Y plane, while a lathe object is viewed in the X-Z plane—usually from overhead.

When the machine control is capable of cutting curves in more than one of these three **discreet** *(unique)* **planes,** the programmer must add a code word to define in which plane the motion is to occur. We'll learn the codes later.

Three Discreet Planes

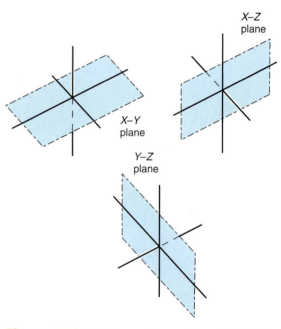

Figure 17-2 *The three primary planes* X-Y, X-Z, *and* Y-Z.

Axis ID on a CNC Machine

When facing a new machine for the first time, the **world orientation** of its axis set (relationship to the floor and to the operator) can often be identified this way, in this order (Fig. 17-3).

- Z The axis parallel to and opposing the main spindle
- X Usually the longest axis, *usually* parallel to the floor
- Y The axis perpendicular to both X and Z

These are conventional guides, not a standard. For any given CNC machine, the set needn't be in any given relationship to the world. While the axis set remains orthogonal and the axes are in the same position to each other, the axis set can be rotated to any world position.

Try it yourself. Use the right-hand rule for a mill in your shop. Depending on the world perspective of the axis set, you may find your hand in any position, but it will be found to fit the rule. There are almost no violations to

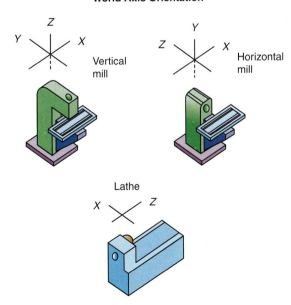

World Axis Orientation

Figure 17-3 *The primary axes as they apply to three familiar machines.*

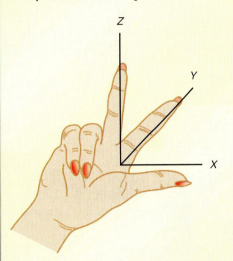
the orthogonal set, but it can be found lying in some odd positions relative to the world, especially on advanced CNC equipment and robots.

The most common example of skewed world orientation, shown in Figs. 17-5 and 17-6, is a *slant bed lathe,* where the X axis has been tilted relative to the floor. This modification makes chip and coolant ejection more efficient and improves operator access for setting up tools.

CNC Slant Bed Lathe
(viewed from tailstock)

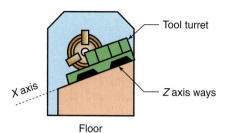

Figure 17-5 *A slant bed lathe features a tilted X axis relative to the floor, to improve chip ejection and operator access to tools.*

Figure 17-6 *The X axis of slant bed lathe is tilted relative to the floor to improve coolant and chip ejection and to aid operator access during setup.*

The Primary Rotary Axes *A, B,* and *C*

Some CNC machines feature programmable axes that rotate or **articulate.** According to the EIA267-B standard, there are three primary rotary axes:

A, B, and *C*

Each is identified by the central primary axis around which it pivots. The

A axis rotates around a line parallel to *X*
B axis rotates around a line parallel to *Y*
C axis rotates around a line parallel to *Z*

The primary rotary axes are identified by their central axis.

Question

To identify a rotary axis, first find its central axis. For example, a lathe equipped with a positioning spindle that can stop at any given rotation angle, facilitates drilling holes spaced around the perimeter of a turned part (Fig. 17-7). What is this axis called?

Answer

The indexing chuck turns about an axis parallel to the lathe's *Z* axis, therefore it's a *C* rotary axis. If it can stop only at a given location, not feed through an arc during machining, it is a positioning axis.

That same *C* axis could also be capable of rotating at programmed, slow feed rates, which would then enable milling operations as well, assuming a milling head was added. It is then known as a *feed axis.* To facilitate the milling operation, a powered tooling head is added to the tool turret (Fig. 17-8).

Mill Rotary Axes

When angled or warped surfaces are required, but they can't be cut with standard milling equipment, we turn to rotary axis machines. There are two ways a CNC mill might employ a programmable rotary axis: by rotating the part or by rotating the cutter head, or a combination of the two.

Articulating Mill Heads These machines are equipped with a spindle head that can rotate in one or two planes during a cut (*A* or

Figure 17-7 *A lathe spindle capable of positioning to an exact location, would be designated to be a "C" axis according to the EIA standards.*

Figure 17-10 *This high-speed turbine is being machined by an* A/B *axis tilting mill head.*

Figure 17-8 *Multitasking lathes feature powered mill cutters in the tool turret, often called live tooling.*

$A + B$). The articulation is similar to the two Bridgeport type mill head rotations, but here it's empowered to feed in an arc (Fig. 17-9).

Five-Axis Cutter Heads Machines with this capability are known as **five-axis mills** with axes X, Y, and Z and A and B (Fig. 17-10).

Plus or Minus Rotary Motion— Rule of Thumb

To define which direction a rotary motion is to occur, clockwise or counterclockwise, we use a plus or minus sign on the coordinate. Try it yourself. Use the rule of thumb to solve the following questions about cutting a warped surface.

TRY IT
Warped Part Critical Questions: The part shown in Fig. 17-11, must be profile milled, by climb cutting (right to left on the page). To make this warped surface, an A/B articulating head must follow the changing warp as the mill moves on the X and Y axes. Use the rule of thumb (Fig. 17-12) to determine which direction the head must rotate, in the plus or minus A and B directions.

A. Before cutting, is the A axis tilted in the plus or minus direction from vertical? (This is shown on the right of Fig. 17-11.)

B. Which way does A move during the cut?

C. The B axis starts at a zero degree tilt, does it end in a plus or minus tilt at the end of the cut? (See the left side of the figure.) Answers found on page 677.

A-B Articulating Mill Head

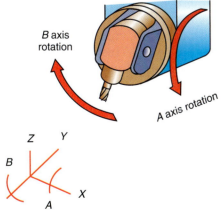

Figure 17-9 *A five-axis mill features two articulating motions—A and B.*

Milling a Curved Warped Surface

Figure 17-11 *To climb mill this part requires both* A *and* B *axis articulation. What direction are they (plus or minus)? Use the rule of thumb to solve it.*

Figure 17-12 *The* rule of thumb *helps to identify rotary axis sign value, direction sense.*

Rule of Thumb—Rotary Axis Sign Value

To identify whether the rotary axis direction is positive or negative (clockwise or counter-clockwise), use the rule of thumb (Fig. 17-12).

First, identify the positive direction for the central axis around which the rotation occurs (+X, +Y, or +Z). Then, pointing the thumb of your right hand along that positive direction, your fingers will curl in the positive rotary axis direction. Negative rotary motion would be against your fingers.

Part Rotating and Indexing

The second way a machining center or CNC mill might achieve rotary axis cutting and indexing is by moving the part rather than the cutter head. This can be accomplished on a programmable *A* axis accessory (more common), as shown on the right of Fig. 17-13, or a built-in axis in the middle of the table. There are also several other versions of programmable part index and rotate accessories.

Part Indexing A very efficient way to index parts is a *four-sided, tombstone table*

CNC Rotary Axes

X axis

Figure 17-13 *Programmable rotary axes can be add-on accessories or built-in parts of the machine.*

Figure 17-14 *Production tombstone fixtures rotate the work to create safe, efficient access to more than one side of the workpiece.*

(Fig. 17-14). Here parts are loaded on four sides of a sturdy vertical, work-holding tower. Each side can be indexed and locked toward the spindle. For example, the part receives all face milling cuts at one station then the tombstone indexes and peripheral, and side cuts are made on the second face. Drilling and boring is done on the third. Three sides of the part can be machined in one setup. Another advantage of this kind of holding and indexing tooling is that the parts can be exchanged safely, on the fourth side, opposite the spindle, while the machining is in progress. The third advantage is that whole towers can be preset with the correct holding fixtures such that when *turning the machine around* (setting it up for a new job), it takes just minutes to remove one tombstone and replace it with a second, completely ready to go with parts preloaded! This is a good example of ways to keep the chips flying for profit! We'll see more in Section 18.

KEYPOINT

For program purposes, part rotation accessories and some index accessories must be assigned an axis letter identification. When this is done, the identification follows the rule of thumb, based on the right-hand rule.

The Secondary Linear Axes U, V, and W

The lathe drilling attachment in Fig. 17-7 illustrates another axis set. CNC machines occasionally receive secondary, straight-line axes to add auxiliary tool slides or boring quills and other machining functions to their capability.

Secondary Linear Rule

To identify the secondary linear axes, determine the primary linear parallels (X, Y, or Z). If the secondary axis is parallel to

X, it is the U axis

Y, it is the V axis

Z, it is the W axis

For example, the small vertical milling machine shown in Fig. 17-15 has its programmable quill designated as the Z axis, therefore the knee, which also moves vertically, becomes the "W" axis according to EIA267-B.

- **TRY IT**
- **Warped Axis Answers**
- A. The program starts with a $+15°$ (positive) A axis tilt.
- B. Then the head starts pivoting in the negative A direction.
- C. The B axis ends with a negative tilt.

- Did you miss this one? Remember, the thumb points along the positive central axis—opposite each X and Y part edge.

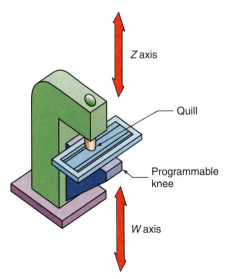

Auxiliary Linear

Z axis

Quill

Programmable knee

W axis

Figure 17-15 *The programmable milling machine knee would be called a* W *axis since it moves parallel to the Z axis.*

UNIT 17-1 REVIEW

Review the Key Terms

Articulate
A wrist-type action that moves in an arc but not a full circle. Mill heads articulate.

Discreet planes
One of three unique planes defined by the axes that lie on it: X-Y, Y-Z, or X-Z.

Five-axis mill
A vertical or horizontal machine with a head that articulates in the A and B axes.

Orthogonal axis set
Three axes lying at mutual 90° angles.

Right-hand rule
Used to determine the letter identification of axes within an orthogonal set.

Rule of thumb
Used to determine the sign value of rotary axes.

World orientation
The relationship of a machine's axis set with respect to the floor and operator.

Replay the Key Points

- To change from one discreet plane to another, a code word must be included in the program.
- The rotary axes are identified by their central axis.

Respond

1. On a sheet of paper, using the right-hand rule, complete the orthogonal set for the sketch in Fig. 17-16.

2. Which primary linear axis is parallel to the main spindle axis? Which axis is central to a B axis move?

3. Identify the rotary axes on this five-axis mill in Fig. 17-17 and the direction they are moving ($+$ or $-$).

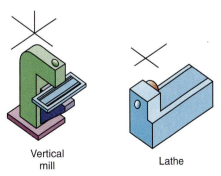

Vertical mill

Lathe

Figure 17-16 *Identify the axis set using the right-hand rule.*

Figure 17-17 *Identify the two rotary motions on this five-axis, articulating mill head.*

4. A multitasking CNC lathe capable of cutting both sides of the work features two programmable tool turrets. Each can move individually in two directions. Identify the missing axes on this machine and their sign value (+ or −) (Fig. 17-18).

5. The primary axes correspond to each other. Fill in this chart.

Primary linear		Y
Secondary linear	U	
Primary rotary		C

Twin Turret Lathe

Figure 17-18 *Identify the primary and two auxiliary axes of this lathe.*

Figure 17-19 *Is this a positive or a negative A and B axis motion?*

6. This five-axis mill in Fig. 17-19 is rotating its cutter head in the direction indicated. Is this a positive or negative A or B axis movement according to the rule of thumb?

7. Complete the orthogonal set for the CNC horizontal mill in Fig. 17-20.

Name These Axes

Figure 17-20 *Complete the axis set for this horizontal mill.*

UNIT 17-2

Coordinate Systems and Points

Introduction: The Spanish map maker Cartes (Car-tez) used a grid of lines crossing at 90° to identify any point on his geographical maps. Four centuries later we're still using them on street maps. Way beyond his intention, mathematicians and machinists adapted them to programming and called them *Cartesian coordinates* to help us make CNC machining possible. Two Cartesian coordinates (X and Y) identify points on two-dimensional planes. By adding a third (X, Y, and Z) we can identify any point in three dimensions—called the *three space*. They are also called rectangular coordinates.

The point identified might be used for reference as with the center of a circle. It might also be used as a location where a hole is to be drilled or to define one data point on a surface. Or a coordinate might be the target for a motion statement such that the tool drives to it at a feed rate.

Cartesian coordinates don't serve all the point identifications required. Sometimes information comes to us on an engineering drawing that's not in X, Y, and Z. Like a rocket leaving the earth, we sometimes think of point identification in terms of how far it has traveled from the origin (radius = R) and at what angle (A). These are known as **polar coordinates.** For example, a bolt circle is much easier to define using polar coordinates rather than X by Y locations.

KEYPOINT

In CAD drawing and CNC programming by hand or by CAM software, we mostly use rectangular coordinates, but every so often it's easier to switch to polar coordinates. They can save a lot of math when drawing dimensions are provided in terms of radius and angle rather than rectangular distances from datums.

Unit 17-2 is about three kinds of coordinates used in CNC. It's also about the way we assign value to them to determine direction.

KEY TERMS

Absolute coordinates
Values based on the origin of the axis set—the PRZ.

Cartesian coordinates
Rectangular coordinate set that refers points to reference axes.

Coordinating (CNC axes)
Physically moving the cutter to a known position relative to the part geometry then setting axis registers to represent that position.

Default (CNC condition)
An expected or preset value or mode.

Entity
A straight line or curved arc with a start and end point coordinate.

Full floating reference
The ability to place the PRZ anywhere within the work envelope (and sometimes beyond in special situations).

Homing (the CNC machine)
Driving the machine to a fixed position on the machine, machine home.

Incremental value (relative value)
Coordinates based on the previous entry. Jump-to-jump values. Sometimes called relative coordinates because each entry is based on the last.

Machine coordinates
A set of coordinates that always refer to M/H not to the PRZ.

Machine home (M/H)
The never changing, utility position used for safe parking, accuracy refreshment, and setup purposes.

Null entry
A command, mode, or coordinate that has been set previously and can be optionally omitted from a program statement.

Polar coordinates
Point identification using radial and angular displacement from the origin.

Program reference zero (PRZ)
Master origin for program and part geometry. Chosen by the programmer, PRZ is coordinated by the machinist during setup. It can be placed anywhere within the envelope on modern machines.

Quadrant
One of four possible 90° segments lying on a flat plane, created by the intersection of the two axes that define the plane. Absolute coordinate values depend on the quadrant in which a point lies.

Three space
A three-dimensional work envelope defined by X, Y, and Z coordinates or by spherical polar coordinates.

Incremental Dimensioning

0.75 0.75 0.75 0.75

B ▸ 0.75
      1.50
      2.25
      3.00

A

Absolute Dimensioning

Figure 17-21 *Examples of absolute and incremental (also called relative) dimensioning.*

Absolute and Incremental Value Coordinates

There are two different ways to use both Cartesian and polar coordinates. How we use them depends on whether the coordinate refers to the origin (more common) or takes its reference from the previous point called incremental coordinates. Figure 17-21 shows examples of two different kinds of dimensioning. Note that in the real world these would not be mixed, as in the example drawing, because the tolerances would conflict.

KEYPOINT

Normally, the designer would use one type or the other—absolute or incremental dimensioning.

Definitions

Absolute Value Coordinates

An **absolute coordinate** is one where each entry represents the point's distance from the master origin: $X = 0$, $Y = 0$, and $Z = 0$. We use absolute point identification more often in CNC work because prints usually refer features to origins on the print (Datums A and B, for example). They are more popular because absolute values have

become conventional for CAM-generated programs.

Incremental Value Coordinates

These refer to a coordinate set where each entry represents the identified point's distance from the previous point. These points can be envisioned as jumps from the present location to the next. Using incremental values often saves math and time during setups and editing of programs.

Absolute Value Coordinates

Now let's develop a working ability using coordinates. We'll start with the most commonly used: the absolute value rectangular.

Referenced to the Origin— Program Reference Zero

All absolute coordinates refer to a singular starting place—the origin. In CNC work it's called the **program reference zero (PRZ)**, *the origin of the grid.* The PRZ is the master reference point on which the program and the setup are based. Selecting the PRZ location relative to the part's geometric features is among the first critical decisions for planning.

KEYPOINT

The PRZ must be based on geometric priorities whether or not the print is GDT.

Setting the PRZ on the Machine

When the machine is setup, the PRZ must be located at the same position on the physical part as it was selected for the program. This is a critical setup task on both mills and lathes. We'll use a mill example.

The action is similar to setting something up (Fig. 17-22) to be drilled on a manual milling machine. First, the vise is indicated true to the machine's axis and bolted in place. It does not matter where the vise is located on the table because the PRZ can be set at any position within the work envelope on modern equipment. This is known as **full floating reference.**

When the machinist sets the PRZ relative to the part, vise, chuck, or fixture, the action is called **coordinating** the machine. The spindle is located over one corner of the work in this example. The next action is to set the digital readout and/or micrometer dials to

Co-ordinated PRZ

Absolute position registers
set to zero

Pos Abs	X 00.000
	Y 00.000
	Z 00.000

Tool touching top at
datum corner of work

Figure 17-22 *Coordinating the PRZ to the physical part requires setting the axis registers to zero when the tool is at the PRZ position.*

read zero at that position on the manual mill. On the CNC, it's the same, the control registers are set to X0.0 and Y0.0. Then, after coordinating the *X* and *Y* axes, the cutter would be touched to the top of the work and its register set to Z = 0.0000, assuming that's where the *Z* axis PRZ has been chosen for the program (a common practice).

KEYPOINT

The tool just touching the work surface is the simplest way to illustrate the *Z* zero position. It works fine for programs that use just one cutting tool. Later we'll learn that *Z* zero in a setup with several cutting tools may be at machine home or some other height off the part to allow for different length tools.

With the PRZ coordinated, all absolute moves of the machine's axes would then refer to that point. On a mill part, the PRZ is commonly placed on a corner of the part as in Fig. 17-22, while on a lathe part the PRZ is often placed at the outer tip and centerline of the work.

The PRZ can sometimes be placed outside the work envelope, under special circumstances. First the spindle or tool is positioned on the part. But this time the axes aren't set to zero, rather their position relative to PRZ is written into the machine registers. Not all controllers will permit this action. However, it is possible since the machine refers coordinates to it, but needn't actually move the spindle or cutting tool to it.

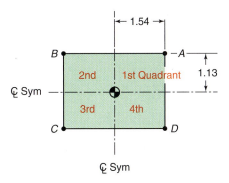

Figure 17-23 *Absolute values in the four quadrants.*

By engineering convention, the PRZ is shown by the bulls' eye circle shown in Fig. 17-23.

Absolute Value in the Four Quadrants

Depending on which side of the PRZ a coordinate lies, its value will be either positive or negative. Using the *X-Y* plane for the example, the intersection of the two axes creates four possible zones called **quadrants.** They are numbered from the upper right in a counterclockwise direction (Fig. 17-23). Each ordinate (*X* or *Y*) combines to make a coordinate and each has a positive or negative value.

ShopTalk

Depending on local custom, you might also hear the PRZ called *program home, program datum point, program zero,* or *X-Y-Z zero.* Whatever you call it, remember that it's the point from which all absolute coordinates originate or to which they refer.

KEYPOINT

Slightly different from math statements, for CNC coordinates, we place the minus sign after the letter ordinate to indicate negative value. We do not use a plus sign. No sign indicates positive value.

X− is negative   *X* is positive

Each *X* and *Y* value is made unique by the quadrant in which it is located. If, as shown in Fig. 17-23, point *B* is within the second quadrant, then its coordinates are

$$B = X-1.5400, Y1.1300$$

In Fig. 17-23, what are the absolute coordinates for points *C* and *D*?

Answer

$$C = X-1.5400, Y-1.1300$$
$$D = X1.5400, Y-1.1300$$

PRZ symbol

Pos Abs
X 02.250
Y 01.000
Z-01.000

Figure 17-24 X *and* Y *are positive absolute value, but* Z *is negative.*

Lathe Quadrants

When programming work for the lathe, the workpiece is viewed in the *X-Z* plane—from overhead. The PRZ is usually placed at the outer end centerline (but not always). If it is located in this easily coordinated place, the entire path to be machined lies within the second quadrant, making all *Z* values negative but all *X* values positive, as seen in Fig. 17-25.

In Fig. 17-25, the coordinates for point *A* are

$$A = X1.0000, Z0.0000$$

Figure 17-25 *The quadrants for* X *and* Z *axis values on lathe work.*

What are the coordinates for points *B*, *C*, and *D*?

Answer

$$B = X1.000, Z-2.7500$$
$$C = X1.8750, Z-2.7500$$
$$D = X1.8750, Z-4.0000$$

Diameter Programming

A second lathe convention is illustrated in Fig. 17-25. In nearly all lathe programs, the *X* axis coordinates are based on diameter.

On some controls and CAM systems, it is possible to program in radius values, but only after specifying radius-valued *X* coordinates; *X* = diameter is a **default** condition for most turning centers. (Default here means the fallback value or condition.)

Points for Geometry and/or Reference

Coordinate points fall into two categories by the way they will be used in the program. Some points can do double duty. In Fig. 17-22, the PRZ was also on the part as a geometry point to be machined.

Geometry Points

Geometry points are those you intend to use for position or motion—a location to be machined. They are the end of a cut or a center location for a hole or an intermediate point on a surface (Fig. 17-26).

Figure 17-26 *Geometry points occur at significant locations.*

Figure 17-27 *Points can be coaxial (stacked on top of each other along a primary axis).*

Panel instrument	Suitable for the use of local references

Figure 17-28 *The math required to write coordinates for this instrument panel's features could be simplified by using a temporary local reference for each cutout pattern.*

Identifying Geometry Points

A geometry point occurs at the junction between any two **entities** (Fig. 17-27). An entity is a familiar term for those with CAD training. They are individual straight lines or arcs. Each has a start and end point coordinate. An arc must also have a center point reference point. Where any two entities join, become tangent, intersect, or cross, there is also a unique geometry point common to both.

Reference Points

In addition to the PRZ, there are two other kinds of reference points used in CNC. One not normally used for program execution is **machine home (M/H).** M/H is not a floating point, it lies at an exact fixed (repeatable) location within the machine's envelope, usually with all axes fully withdrawn away from the chuck or table. Driving the machine to M/H is called **homing.**

It's a utility location used mostly for setup functions and for accuracy coordination in older machines lacking the ability to retain their position when they are turned off. We'll set M/H aside until we look at operator responsibilities in Sections 19 and 20.

Local Reference Points When writing a program it's possible to switch from the PRZ to a temporary *local reference zero point (LRZ)* that lies at a known distance from the PRZ. Once the features are machined that relate to it, the LRZ is canceled to again refer to the PRZ. Local references are used for a couple reasons.

1. One reason might be that a feature, or group of features, refers to a reference other than the PRZ (Fig. 17-28). Many calculations would be required to write absolute value coordinates for each cutout referring back to the PRZ for each coordinate. But setting a local reference point, at the center, then referring to it for that group saves math and programming time when writing the program without CAM.

2. An LRZ can also be used to repeat the entire program or a part of a program in another location within the work envelope. The repeat could be on one part as the instrument cutouts, or we could be machining two of the same parts in two vises. The second part would use a local reference, at a given distance from the first (Fig. 17-29).

3. We also use an LRZ on very large parts such as wing spars, for example, where the master PRZ might be hundreds of feet away. A temporary point can be established that's far more convenient for setup work. The bed of the gantry mill in Fig. 17-30 is 200' long.

Trade Tip

In CNC work there's almost always more than one way to do a job using the power of the codes.

Subroutines that contain all the commands for one instrument cutout can be used to simplify the instrument panel program. Such would be written using incremental coordinates (coming up next). Then we need only move to the center location to begin the subroutine. We'll learn about subroutines in logic commands in Section 23.

Another method of writing the two vise setup of Fig. 17-29, is called a *fixture offset.* Most controls can remember several PRZ points within the work envelope. Given the right code, the registers can switch from PRZ1 to PRZ2 and so on. Most controls store from 6 to 10 PRZ locations with reference to the machine home. The codes commonly used are G53, G54, G55, G56, and so on, depending on the manufacturer. Once the new fixture offset code is read, the machine begins acting on the reference location stored on the fixture offset page of the control. Many shops load G53 with all zeros so it's an easy way to return to (or refer to) machine home. G53 then cancels any fixture offset. So, for the two vise routine, the logic would be G54 = vise 1, G55 = vise 2. We'll learn more about fixture offsets in Section 22.

LRZ a known distance from PRZ

PRZ

Figure 17-29 *An LRZ might be used to run the same program in two vises.*

Figure 17-30 *Machine home and PRZ could be 20 feet apart (left) or 200 feet on this giant, 3 spindle, 5 axis, gantry mill.*

Incremental Coordinates

Incremental **Cartesian coordinates** are a useful, second method of identifying a point. They might be called *relative* coordinates because each movement of the tool is related to the last spindle or tool position. The coordinate is the distance from the present position to the next. **Incremental values** are sometimes chosen for several reasons. For example, when

The *drawing* or a portion of it, is incrementally dimensioned and the program will be written without CAM software. In that case, using absolute values will require lots of math. On the other hand, the incremental values are all there—no calculations to make.

Manual *keyboard entries* were used for example. A brief tooling program written at the control to shape soft jaws, or to move an indicator probe an exact amount using keyed in entries, would both be simplified using incremental coordinate entries.

Incremental solutions are needed. When a solution to a programming calculation yields an incremental value. In other words, solving for the next position or local reference, creates the incremental distance. To compute the absolute distance, further math would be necessary.

Incremental values are required. A few specific commands must be in incremental values. We'll see curve commands where the center of the circle must be identified with incremental coordinates.

Extra moves are to be edited in. A written program could require editing. It's often easier to slip in incremental moves to fix something in a program otherwise written in absolute values.

Incremental Motion Values

The axis value, plus or minus, describes the direction in which the tool movement is to be made from the present position to the next. The center of the illustration in Fig. 17-31 is not necessarily the PRZ, but should be envisioned as the present location of the tool. For example, at this position, the coordinates $X10.0$ $Y1.0$ take the tool to position 1. Both are positive movements. However, an $X-10.0$ $Y-1.0$ would take it the opposite direction.

• TRY IT
Critical Question *Starting at point 1, determine incremental coordinates to move from point 1 to point 2, then onward to points 3, 4, and 5 in sequence. Note the 1-in. scale on the drawing.*

ANSWER

To Point	Incremental	Coordinates
2	$X-1.600$	$Y1.000$
3	$X-0.600$	$Y-1.400$
4	$X1.200$	$Y-0.600$
5	$X1.400$	$Y0.000$

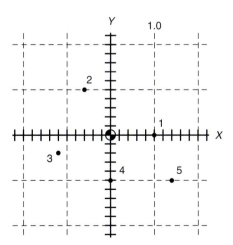

Figure 17-31 *Identify these points incrementally, in sequence.*

KEYPOINT

Note that the point 5 coordinate X1.400, Y0.000 can be shortened to X1.400 with no Y entry since there is no movement of the Y axis from point 4 to point 5. That's called a *null entry.* Null coordinates can be left off or left in.

Trade Tip

A good way to envision incremental coordinates is to think of the present position as a mini-PRZ, "just for now." That means that the coordinate values (plus or minus) of the next point follow the quadrant principle, but using the present position as the reference not the absolute origin.

Metric Coordinates

Metric value coordinates are as simple to enter as those in inches. But here too, the right code must be provided to alert the control whether the units being entered are metric or imperial. Although it would be poor practice, both units can be mixed in the program. That would only happen if a wacky drawing mixes them!

As you know, a millimeter is 25.4 times smaller than an inch. So, coordinates are usually carried only to three decimal places; for example, $X3.750$.

KEYPOINT

There is a code for metric-valued coordinates and one to tell the control the program is in imperial (inch) units. It can, for example, be G20 = Imperial and G21 = Metric, or G70 = Imperial or G71 = Metric.

G21 X2.000
This line sets metric values and the entry is 2 mm.

G20 X2.0000
This line tells the control to return to imperial values and that the entry is 2 in.

Modern controls allow complete flexibility of units and values. Most can accept either absolute or incremental and metric or imperial values with ease, as long as they are accompanied by the right code. Again, while it's a bad idea, the units and value (absolute = G90 or incremental = G91) can be changed within the program or even within a single command with no loss of accuracy. Behind the readout video screen, all modern controls keep track of tool position absolutely, using metric values, disregarding the program values and what you see on the screen. Changing units and values only changes the way you see the data.

Which Coordinates Are Correct?

This isn't an issue if you are programming with CAM software. The software will convert dimensions on the CAD drawing into equivalent distances in whatever units and values you have selected for the output program (almost always absolute values). However, when programming by keyed data entries, as at a PC or on the controller keypad, the smart programmer selects the coordinate identification that requires the least math—thus the least chance for error. Math is sometimes unavoidable in programming, but when a large number of the calculations are popping up based on converting point locations into absolute or incremental values, or to Cartesian from polar (to be studied next) or visa versa, then change units and values to stop making conversion calculations.

KEYPOINT

One of the three types of coordinates will be found to identify most of the points on the drawing: incremental Cartesian, absolute Cartesian, or polar.

Trade Tip

Incremental coordinates can be useful when several parts of nearly the same size or shape are programmed (Fig. 17-32). Using one base program, the modification can be inserted where needed. For example, the three handles could be made from a single program with two additional *X* moves added to adjust the part length to machine the dash 1, 2, or 3 version. This wouldn't work using absolute coordinates because the next sequence would go right back to the original shape. For example, by adding a couple of incremental moves of 0.400 in., the dash 1 is changed to the dash 2 part.

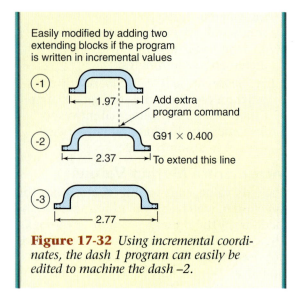

Figure 17-32 *Using incremental coordinates, the dash 1 program can easily be edited to machine the dash –2.*

Coordinate Policy

For uniformity, many shops have a specific policy about coordinate selection that restricts values to a single type.

Conventions in Program Commands

Eliminating Nulls (Repeated Absolute Coordinates)

While it's correct to represent each geometry or reference point in a program with a complete *X*, *Z*, or *X*, *Y*, and *Z* coordinates, not every coordinate must contain all entries to be complete data.

Absolute Null Rule

If a given absolute coordinate value has been entered in a previous command statement, then that axis ordinate entry might be skipped. Called **null entries,** the duplicated portion can be left off for convenience and to create shorter programs.

For example, compare the two sets of coordinates shown next, where the mill cutter starts 1.0 above PRZ (*X*0.0, *Y*0.0, *Z*1.0). It then positions down, in absolute value coordinates, to touch the work. It then positions laterally in *X* to point *A*, then changes *Y* and *Z* positions to point *B*. Relate the full coordinates to the abbreviated set that follows where the nulls are omitted. The machine actions remain the same.

Complete Set

From start to PRZ	*X*0.0 *Y*0.0 *Z*0.0
To point *A*	*X*2.5 *Y*0.0 *Z*0.0
To point *B*	*X*2.5 *Y*−2.0 *Z*−1.0

Nulls Removed

$Z0.0$
$X2.5$
$Y-2.0\ Z-1.0$

Null Elimination

If the position register (screen readout) for a particular axis motion is not going to change from one entry to the next, then that portion of the upcoming coordinate (X, Y, or Z) can be omitted. This is so for both absolute and incremental coordinates.

Trade Tip

To Null or Not to Null When writing the program without CAM assistance, leaving off nulls works especially well for incremental programming. Since each new command statement is the amount and direction of the next move, removing nulls shortens long programs such that they create less data sent to permanent storage. This becomes important for complex 3-D shapes. It's a matter of choice in some cases.

But there are times when nulls are better left in. As a beginner, there is a chance your work will require some debugging. Nulls that are left in help with this task. There are other times when you will choose to leave them in the program as well. For example, when programming curves (Section 22), nulls occur in the references and geometry points, yet they are best left in for good record keeping of your calculations.

The following procedures and conventions apply to using both absolute and incremental coordinates. The CNC controls of today are flexible in their data input requirements both for entry order and for decimal placement, but that hasn't always been so. Also there are a few commands where the order of entry is rigidly defined, no matter the generation of control.

Entry Order

Modern controls can be programmed in any X-Y-Z order with acceptable results. The coordinate, $X1.0$, $Y2.0$, $Z3.0$ could be entered $Y2.0$, $X1.0$, $Z3.0$, or ordered $Z3.0$, $X1.0$, $Y2.0$ too. Most controls do not require data order as long as it is complete within one command. However, ordering it alphabetically is the norm.

Leading and Trailing Zeros

On modern controls, decimal number entries need only have their decimal in the right place, and the significant digits correctly arranged. For example, $X.37$ and $X0.37000$ are both acceptable. However, there are three reasons that program entries should be ordered conventionally. The program is often easier to edit. Compare these

Unstructured
$X1.37\ Y1$
$X.256$
$X.037\ Y1.125\ Z.5$
$Z-0.2500$
$X.037\ Z-.05$

Structured
$X1.3700$ $Y1.0000$
$X0.2560$
$X.03750$ $Y1.1250$ $Z0.5000$
                                   $Z-0.2500$
$X0.037$ $Z-0.0500$

Both sets of coordinates convey the same data. However, in the second set, the columns align, and they extend to the accuracy of the machine. All the zeros aren't necessary on but they set a uniform, easy-to-scan format. Ordering the program is a matter of choice and shop policy.

KEYPOINT

When you are assigned to a new control find out what conventions are required for data entry. *Older controls often require a fixed format* with rigid ordering.

When Whole Numbers Are Required

Some entries must be entered as whole numbers without a decimal. For example,

Line numbers never have a decimal point
    N001 cannot be N001.0

Tool numbers
    T10 cannot be T10.0
    But they might be T0010 if the control holds many tool definitions.

Number of passes to complete a shape
    P12 cannot be P12.0

These situations will be learned on an individual control basis. Read the control manual.

UNIT 17-2 REVIEW

Review the Key Terms

Absolute coordinates
Values based on the origin of the axis set—the PRZ.

Cartesian coordinates
Rectangular coordinate set that refers points to reference axes.

Coordinating (CNC axes)
Physically moving the cutter to a known position relative to the part geometry then setting axis registers to represent that position.

Default (CNC condition)
An expected or preset value or mode.

Entity
A straight line or curved arc with a start and end point coordinate.

Full floating reference
The ability to place the PRZ anywhere within the work envelope (and sometimes beyond in special situations).

Homing (the CNC machine)
Driving the machine to a fixed position on the machine, machine home.

Incremental value (relative value)
Coordinates based on the previous entry. Jump to jump values. Sometimes called relative coordinates because each entry is based on the last.

Machine coordinates
A set of coordinates that always refer to M/H not to the PRZ.

Machine home (M/H)
The never changing, utility position used for safe parking, accuracy refreshment, and setup purposes.

Null entry
A command, mode, or coordinate that has been set previously and can be optionally omitted from a program statement.

Polar coordinates
Point identification using radial and angular displacement from the origin.

Program reference zero (PRZ)
Master origin for program and part geometry. Chosen by the programmer, PRZ is coordinated by the machinist during setup. It can be placed anywhere within the envelope on modern machines.

Quadrant
One of four possible 90° segments lying on a flat plane, created by the intersection of the two axes that define the plane. Absolute coordinate values depend on the quadrant in which a point lies.

Three space
A three-dimensional work envelope defined by X, Y, and Z coordinates or by spherical polar coordinates.

Replay the Key Points

◆ Absolute coordinates always refer to their X, Y, and Z distance from the PRZ.

◆ Incremental coordinates identify the distance and direction to the *next* point using the present point as its reference.

◆ Coordinate points are used either for geometry or reference, and one point can be used in both capacities at times.

◆ The PRZ is chosen from its datum basis on the print.

◆ The PRZ also exists on the physical envelope of the machine and setting it is a setup task.

◆ If the machine reads $X0.0$, $Y0.0$, and $Z0.0$, and the cutter is positioned at PRZ, it is coordinated.

◆ Each primary plane has four quadrants, which determine absolute coordinate sign values.

◆ Null values may or may not be omitted by the programmer's choice.

◆ Modern CNC equipment offers certain entry flexibilitys which can be used on a choice basis.

◆ There is no difference in accuracy using either absolute or incremental or metric or imperial values as the control tracks progress internally, and always absolutely.

◆ We can switch freely between absolute and incremental as long as the right command code is given to the control so it knows which is being used.

◆ We often use incremental coordinates for editing where we must insert an extra move or to write quick setup or tooling programs at the control.

Respond

Resolve any misunderstandings or get assistance with answers that are incorrect or unclear. Do not go beyond this unit without 100 percent comprehension.

Absolute Coordinates

1. Coordinates that refer to the PRZ for their values are known as absolute rectangular coordinates. Is this statement true or false? If it is false, what will make it true?

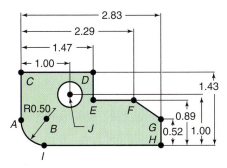

Figure 17-33 *Mill questions and problems.*

Point	Absolute coordinates	
A	X2.00	Y0.83
B	X0.83	Y2.00
C	X0.83	Y2.00
D	X-2.00	Y0.83
E	X-2.00	Y-0.83
F	X-0.83	Y-2.00
G	X0.83	Y-2.00
H	X2.00	Y-0.83

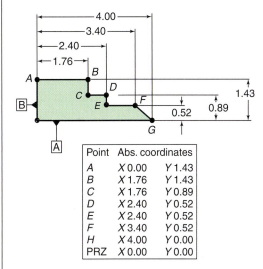

Figure 17-35 *Use graph paper to create a program grid like this.*

2. Name in their order of importance, the two guideline rules for selecting the PRZ.

3. Draw a sketch of the PRZ symbol on a sheet of paper. Where should it be chosen for mill job shown in Fig. 17-33?

4. Sketch of the correct location for the PRZ on the lathe job in Fig. 17-34. Student note: Often only half of the profile is shown on lathe work.

5. Write a set of *X-Y absolute value* coordinates for the indicated points in Problem 3 (Fig. 17-33). Record them in the sequence shown, then check your answers against those provided. Note, *Z* values are not needed for this exercise, but would be necessary in a program.

6. Write a set of absolute value coordinates for the lathe job of Question 4 (Fig. 17-34). Note, typically, half the part is shown for symmetrical lathe work. The *Z* axis is horizontal on the page and the *X* is vertical.

7. On graph paper, draw a graph similar to that shown in Fig. 17-35, then plot these points and label them in order. Then connect the dots to form a familiar shape. All lines will be straight. Some may be diagonal on the page.

Critical Thinking

8. This problem (Fig. 17-36), illustrates a major difference between absolute and incremental coordinates. We will rerun

Point	Abs. coordinates	
A	X 0.00	Y 1.43
B	X 1.76	Y 1.43
C	X 1.76	Y 0.89
D	X 2.40	Y 0.52
E	X 2.40	Y 0.52
F	X 3.40	Y 0.52
H	X 4.00	Y 0.00
PRZ	X 0.00	Y 0.00

Figure 17-36 *Plot these points to see what the error does to the shape. Save your drawing for later analysis.*

it a second time in the incremental problem set. The set of *X-Y*, absolute mill coordinates in Fig. 17-36 have a deliberate error that *does not produce the shape shown*. Using engineer's graph paper (0.1-in. or 0.2-in. grid), plot and connect them in sequence using a contrasting color, to see what happens. Save your drawing.

KEYPOINT

To enrich this exercise, plot the points on a CNC simulator software or using the geometry creation part of a CAD/CAM system.

9. Find the null values in the coordinate set of Problem 5. Copy the page then use a highlighter or circle those that could be dropped from the program.

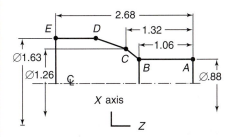

Figure 17-34 *Lathe questions and problems.*

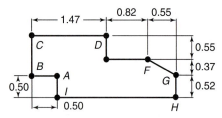

Figure 17-37 *Write a set of incremental coordinates for this part.*

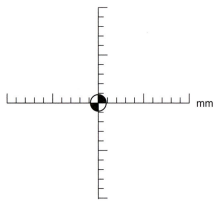

Figure 17-38 *Your plot grid needn't be any particular scale or actual size. Choose a scale that fills the page with the part dimensions.*

Problems in Incremental Coordinates

10. Write a set of incremental coordinates for the shape shown in Fig. 17-37. Note that the drawing scale needn't be actual size. Assume the tool is positioned at the PRZ (lower left corner) and proceed to point *A*, then on around the part clockwise. Close the shape from point *I* to point *A*. No *Z* axis movements are needed for this exercise.

11. Plot the following incremental coordinates with a pencil, in sequence. Choose a scale on graph similar to that shown in Fig. 17-38. The program begins with the tool (pencil in this case) 5 mm above the paper at the PRZ. It touches to make a line when it's at $Z-5.0$ mm. Nulls have been removed.

Sequence Number	Coordinates
N005	$Z-5.0$ (touches the paper to make a line)
N010	Y10.0
N015	$X-5.0$ Y10.0
N020	Z5.0

N025	X10.0
N030	$Z-5.0$
N035	$X-5.0$ $Y-10.0$ (makes diagonal line)
N040	Z5.0
N045	X20.0 Y10.0 (diagonal)
N050	$Z-5.0$
N055	$X-10.0$
N060	$Y-10.0$
N065	X2.5
N070	$X-2.5$ (retraces the previous line)
N075	$Y-10.0$
N080	X10.0
N085	Z5.0
N090	X10.0
N095	$Z-5.0$
N100	X10.0
N105	Y10.0
N110	$X-10.0$
N115	Y10.0
N120	X10.0
N125	END

Trade Tip

In the coordinate set of Problem 11, sequence numbers have been added. They are an organizational tool you should include in most programs. However, in many cases, they are optional. Numbering them five apart, leaves space for future editing.

Sequence numbers, are sometimes required when certain programming logic tools called loops, subroutines, and fixed cycles are used. These statements jump or bridge the program forward or backward out of the normal inline sequence. This is called *branching logic*, which we'll study later. In these instances, sequence numbers become a necessary address for the jump and/or for the return back to the normal program sequence on completion of the branched commands.

Point	Incr. coordinates
A	Y1.44
B	X1.76
C	Y-0.55
D	X0.64 Y-0.37
E	Y-0.37
F	X1.00
G	Y0.60 Y-0.52
PRZ	X-4.0

(Sum 0.52 + 0.37 + 0.55)

Figure 17-39 *Once again, the program errors and does not produce the shape shown. Analyze the difference in absolute and incremental errors.*

Critical Thinking

12. After verifying your understanding of Problem 11, return and modify only the first five lines by inserting the null coordinates that have been omitted. Note, this is a critical thinking point to teach yourself. Read the answers carefully.

13. The set of *X-Y, incremental* coordinates in Fig. 17-39 have an error that does not produce the shape shown—once again. Start the PRZ. Plot and connect them in sequence using a contrasting color, to do the same problem in absolute value. What happens to the shape? Solving this puzzle will point out a big difference between the two kinds of coordinates when an error is made.

UNIT 17-3

CNC Machine Motions

Introduction: Whether setting up, operating, or programming a CNC machine, the various kinds of movements they are capable of making must be understood. We will look at CNC machines from the simplest to the more complex. Understanding CNC motions and the part shapes they produce will improve your understanding of writing and editing commands and performing setups.

For safety, it's important to know how your CNC machine is going to respond to various types of motion commands. There can be a difference, depending on the age and processor speed. Older controls with slower computer power move differently in some cases, when driven by up-to-date high-speed computers. Where it will matter, these differences will be pointed out. Then the savvy operator must determine which version of the movement their control will produce.

KEY TERMS

Circular interpolation
Machining an arc in two or more axes, at the specified feed rate.

Default (value or condition)
A fall-back number to which the machine will refer when entries exceed axis or controller limits.

Linear interpolation
Machining in one or more axes in a straight line at the specified feed rate.

Rapid travel (rapid)
The fastest axial speed that the machine can produce. Used as a positioning move for efficiency.

Retract height (R)
The safe position to which a tool is positioned toward at rapid, or backed away from the work before rapid positioning.

2-D control
A control capable of machining a circle using two axes at one time only.

$2\frac{1}{2}$-D control
A control capable of circular interpolation in any two axes while performing linear interpolation in the third axis.

3-D control
Performing circular interpolation in three axes at a specified feed rate. Not limited to three planes for arcs.

Axis Moves

There are four ways tool motion occurs for lathes or mills.

Rapid Travel

Rapid travel, usually shortened to simply *rapid,* is the fastest speed the machine can produce. Rapid is an efficiency move to reposition for another cut or to change tools or parts. Common speeds range from a slow 100 in. per minute (IPM) on training machines to a blazing 1,000 IPM on industrial equipment, or more. Rapid is a one-speed entry—no velocity is specified. If it's too fast for safety, the machine operator can slow it down.

KEYPOINT

Rapid travel can be manually adjusted below full speed using the *rapid override* function (Fig. 17-40). Rapid cannot be adjusted above the 100 percent level (in comparison, feed rate moves can be adjusted above 100 percent).

The operator chooses to override a rapid move during a new program tryout to provide time between moves for safe tracking of the cutter. Some controls use keypad override rather than a physical dial as shown in Fig. 17-40.

Rapid Height or Distance Off In planning a program, one must determine the closest rapid approach movement toward the work. The tool must be slowed to a feed rate at that distance from the work material. The safe distance must take into account clamps, vise, chuck jaws, and other holding tooling. As a beginner programmer, stop cutting tool rapid travel when it's 12 mm or $\frac{1}{2}$ in. from the work surface. This is large for industry but safe. It can be modified later when you are more confident in your programs.

That safe travel height above the work/tooling is called the **retract height** on milling machines (abbreviated to **R**) where the tool must pull back several times, then move over the work surface as in drilling a series of holes for example. On lathes it might also be called the *departure point* in certain commands where the tool must retract back to a safe starting place before taking another pass. Retract moves away from the work are usually made at rapid travel.

Linear verses Nonlinear Rapid The path a CNC machine takes while performing rapid travel falls into two versions depending on the speed of the microprocessor in the controller (Fig. 17-41). It's important to know which path your machine follows. Older machines take a dog-leg path. If the odd path goes unaccounted for in a program, the cutter can run into the part or the holding tools.

KEYPOINT

All newer machines with 16-bit or higher microprocessors follow the true linear motion.

In the illustration, two controls reposition the tool from *A* to *B* at rapid speed. Note the difference between the slower CPU dash line and the newer solid line. The slower CPU cannot output motor control data fast enough to portion the two drive motors to move in a straight line from *A* to *B*.

When given a rapid command, the older control rotates both axis drive motors at their

Will now travel at one-half rapid speed

Rapid % override

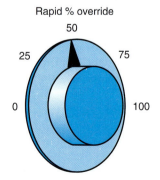

Figure 17-40 *A typical override control can reduce rapid travel below the machine's maximum velocity.*

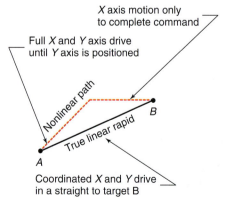

Figure 17-41 *Older controls do not follow a true straight line path from point to point, while in rapid travel mode. It's critical to know what path your control will take!*

full speed until one of the axes reaches its destination, producing a 45° angle for a distance. Then the other axis continues on at full speed to its destination.

The faster CPU control portions out the motor control speeds such that both X and Y arrive at the destination at the exact same time—producing a straight line from A to B. That true linear movement requires more computer speed.

Linear Interpolation—Straight-Line Motion at Feed Rate

Linear motion occurs at the specified feed rate in the program. Like twiddling the dials on an Etch-A-Sketch® or both X and Y handles at differing rates on a mill, you could almost produce the diagonal line shown in Fig. 17-42. The result would not be very smooth or accurate. Additionally, achieving the desired feed rate would be an impossible challenge.

As in Fig. 17-42, when the CNC machine is commanded to move in **linear interpolation,** one, two, or three axes must move at coordinated speeds to arrive at the destination, in a straight line, at the tool feed rate specified in the program.

From the drawing you can see that when more than one axis motor is driving the tool, each must be turning at some subrate below that which is specified. Their combined effect produces tool movement at the specified rate. This is known as *interpolation* (interpolation means to find a value between two others). For example, to linear interpolate a straight line at 20° going from A to B, the Y axis must rotate at 36.3 percent of the X axis speed (the tangent ratio of the line's slope).

If the programmed feed rate is 400 mm per minute (mm/M), for example, the X axis must drive at 375.87 mm/M while the Y axis turns at 136.81 mm/M. The CPU controls each axis drive to achieve the result.

Linear Interpolation

Figure 17-42 *The combined action of differential* X *and* Y *feed rates creates the linear path at the programmed feed rate.*

Suppose the absolute coordinates of B are $X2.5000$, $Y.8750$. Then the following command line would move from A to B.

G21 G01 F400. $X2.5000$ $Y0.8750$

G21 = metric values

G01 = linear interpolation

F400. = feed rate = 400 mm/M

Again, don't get too hung up on memorizing codes as yet. We'll concentrate on them later.

Feed Rate Override Similar to rapid movement, feed rate movements can be overridden by the operator to fine-tune the cutting action. But there's a difference between rapid and feed override control. The usual adjustment range for feed rate is from 0 percent up to 150 percent of the programmed rate. Unless the programmed rate is near or at the machine's fastest rate already, the feed can be increased above the programmed rate.

KEYPOINT

On most CNC controls, feed rate can be increased above the program value to a maximum increase, usually 150 percent of the programmed rate.

Circular Interpolation

Circular interpolation is the next higher CNC motion. It produces a full circle or a part of one (arc), at the programmed feed rate. Circular motion adds an additional challenge for the computer beyond linear interpolation. When producing a true circle, the control must continuously change the ratios between the drive motors for each increment of arc.

Each axis is interpolating below the target feed rate, but their combined action creates the correct tool velocity. Keep in mind that in addition to the axis motor control calculations, the CPU is also comparing resultant feedback to keep everything on track (Fig. 17-43).

Default Rates During circular interpolation, due to the high number of calculations per second, some older (slower CPUs) must limit feed rates to a maximum **default** *rate.* These controls cannot produce circles at the same speed as straight lines. For example, an older control can perform linear interpolation at 200 IPM but must slow to 140 IPM for arcs. Even though the program is at a higher

Circular Interpolation

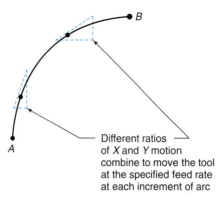

Figure 17-43 *By constantly varying the X and Y rates, a true circle can be produced at the programmed feed rate.*

Two-D controls
Primary planes only

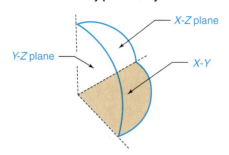

Figure 17-44 *A 2-D control can produce a circle in one primary plane at a time only.*

rate, the control will fall back or default during the circular movement. Newer 32-bit (or higher) CPUs have no such limits until extreme feed rates are encountered.

The term default will be used elsewhere in CNC. It means a fall-back or preset bias built into the control. A default also occurs when a choice must be made by the control when other information is not available.

Axis Combinations for Milling Machines

CNC lathes can perform the tool motions of linear and circular interpolation as already described in the primary plane of X-Z only. This discussion is about mills only. Three-axis milling machines are further categorized by their ability to perform various circular moves:

Two-dimensional motion

Three-dimensional motion

Two-and-one-half-dimensional motion

The difference is significant.

Two-Dimensional Motion

The lower level **2-D control** is restricted to moving two axes while producing a circle. That means arcs in the primary planes of X-Y, X-Z, or Y-Z only—one plane at a time (Fig. 17-44).

The program must include the command G17 to indicate circular interpolation in the X-Y plane, G18 to produce a circle in the X-Z plane, and G19 for Y-Z circular interpolation.

Three-Dimensional Motion

A 3-D mill requires a fast processor or very slow feed rates. It must interpolate and drive three axes simultaneously to produce arcs. Perfect

A Three-Dimensional Control

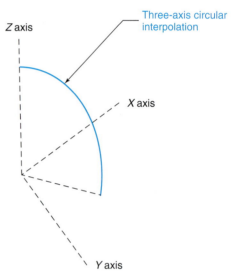

Figure 17-45 *A 3-D control can produce an arc using three axes simultaneously. That means arcs in any plane can be machined.*

circular tool motion involving X, Y, and Z axes at the programmed feed rate is a great challenge for the processor. **3-D controls** are not common in industry (Fig. 17-45).

Two-and-One-Half-Dimensional Controls

Most controls today fall between the preceding two, producing circular motion in one of the primary planes X-Y, X-Z, or Y-Z, while moving the third axis in a straight line at the same time. That motion, called a $2\frac{1}{2}$-D move (two axis circular while one moves linear), produces a spiral similar to a thread—always parallel to a primary axis (Fig. 17-46).

Machines with $2\frac{1}{2}$**-D control** are able to machine the spiral ramp pass we've discussed previously in milling. Today, in many cases we do not need CNC controls that feature

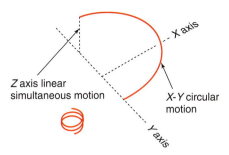

A 2.5-Dimensional Control

Z axis linear simultaneous motion

X axis

X-Y circular motion

Y axis

Figure 17-46 *Most CNC controls can produce $2\frac{1}{2}$-D motion.*

Figure 17-47 *Most CNC controls are capable of $2\frac{1}{2}$-D motion.*

high-level interpolations as long as a capable CAM system backs it up for program commands (Fig. 17-47). The CAM system can approximate complex surfaces using either 2- or $2\frac{1}{2}$-D movements.

Trade Tip

Conic Interpolations Not all regular curves are parts of circles. The regular curves shown in Fig. 17-48 are called the conic sections. When a cone is cut at angles other than 90° across its axis, first an elliptic cross section is produced, then tilting the slice further, a parabola and then a hyperbola are produced. Each is a curve with a slightly different characteristic. Some controls can interpolate these curves using the same 2-D and 3-D rules outlined. You'll learn of them on the job when they apply to a special shape. Machining conic curves is beyond the scope of Unit 17, but you should be aware of their existence.

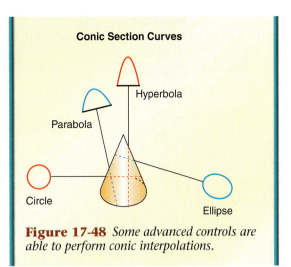

Conic Section Curves

Hyperbola

Parabola

Circle

Ellipse

Figure 17-48 *Some advanced controls are able to perform conic interpolations.*

CAM-Generated Shapes

Because most machine controls are limited to linear and circular motions, CAM software cannot simply generate an irregular surface. Instead, the CAM software breaks the surface into tiny arcs or straight lines, then links them together to approximate the contour.

The length of these approximation lines is defined by a set tolerance away from the perfect shape, and how steep the curvature is. Because the CAM-generated, three-dimensional shape is a series of tiny lines and arcs, the program's data becomes long, especially when surfaces are complex. Advanced software such as Mastercam, our example CAM program, can reprocess the program once it has been generated to create more efficient arcs that fit the shape better, but there is still a lot of data compared to a true 3-D process.

However long the data might be, the CAM process for complex shapes is the method of choice today. The only problem it brings is data storage, which is easily managed. Down to the microscopic level the surface, the difference between true 3-D generated shapes and CAM-approximated ones is indistinguishable. Additionally, using CAM, the CNC machine is far less costly and complex.

UNIT 17-3 REVIEW

Review the Key Terms

Circular interpolation
Machining an arc in two or more axes, at the specified feed rate.

Default (value or condition)
A fall-back number to which the machine will refer when entries exceed axis or controller limits.

Linear interpolation

Machining in one or more axes in a straight line at the specified feed rate.

Rapid travel (rapid)

The fastest axial speed that the machine can produce. Used as a positioning move for efficiency.

Retract height (R)

The safe position to which a tool is positioned toward at rapid, or backed away from the work before rapid positioning.

2-D control

A control capable of machining a circle using two axes at one time only.

$2\frac{1}{2}$-D control

A control capable of circular interpolation in any two axes while performing linear interpolation in the third axis.

3-D control

Performing circular interpolation in three axes at a specified feed rate. Not limited to three planes for arcs.

Replay the Key Points

◆ Rapid travel can be manually adjusted below full speed using the *rapid override* function that allows reducing velocity only.

◆ Rapid cannot be overridden above the machines 100 percent level.

◆ Older slower CPUs may produce a dog-leg kind of rapid travel, which must be accounted for in programs.

◆ Feed rate can be increased above the program rate up to the maximum rate of the machine, usually to 150 percent of programmed rate.

Respond

Critical Thinking

1. When might a shop need a higher level control than 2-D for a lathe?

2. During a circular interpolation, the feed rate is specified with the code word G21 F600, meaning the programmer has set the metric values (G21) and the feed rate at 600 mm per minute. However during actual operation the machine is observed to actually cut at 400 mm/M under that command. Offer two possible explanations.

3. A hemisphere must be machined on the top of an odd-shaped casting. The shop has a CNC lathe and 2-D CNC mill. Describe ways in which it might be machined.

4. The move illustrated in Fig. 17-49 is called a *helical ramp plunge or a spiral*

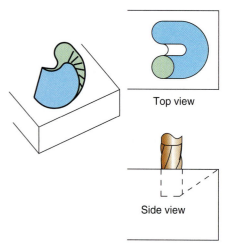

A Spiral Ramp Plunge

Top view

Side view

Figure 17-49 *What kind of CNC control can make a spiral ramp, 2-D, $2\frac{1}{2}$-D, or 3-D?*

ramp. It is used in CNC milling to plunge down into the work to start milling a pocket. What type of control is needed to perform such a move? Referring to your manual milling machine skills, why would the illustrated move be superior to straight plunging and to a linear down ramp? This is a critical thinking question not directly covered in the reading.

5. Name a CNC mill type and action that can produce a smooth version of the hemisphere of Question 3.

6. Describe the actions of the three levels of mill controls: 2-D, $2\frac{1}{2}$-D, and 3-D.

7. Two control limitations described in this unit were default feed rates and types of circular interpolations. To what are both due?

8. What type machine movement performs the ramp down into a pocket shown in Fig. 17-50?

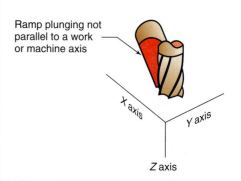

Ramp plunging not parallel to a work or machine axis

X axis

Y axis

Z axis

Figure 17-50 *What control level can produce this ramp plunge?*

UNIT 17-4

Polar Coordinates

Introduction: In addition to the *X*, *Y*, and *Z* rectangular coordinates, a unique point can also be identified using its radial and angular position relative to an origin or a local reference. The two-dimensional polar universe is a series of concentric circles with one radial line extending out, used for the angular reference. Distance out from the origin determines the radius ordinate, while angular displacement from the **polar reference line (PRL)** completes the pair of coordinates. Polar coordinates are expressed as R, A (radius, angle) (Fig. 17-51).

KEY TERMS

Absolute polar (coordinates)
Point identification using radius and angle ordinates, based on the PRZ.

Incremental polar (coordinates)
Point identification coordinates based on another polar point, not the PRZ.

Polar reference line (PRL)
A line radiating out of the PRZ that indicates zero degrees.

Spherical coordinates
A three-space coordinate using radius, angle, and angle.

Absolute and Incremental Polar Values

Absolute Polar Coordinates

If the central reference is the PRZ point for the part geometry and the angular reference line is horizontal (zero degrees), then the coordinate is an **absolute** value **polar** coordi-

Incremental coordinates

Figure 17-52 *Some drawing dimensions are polar rather than rectangular.*

nate. For example, in Fig. 17-52, point *A* is absolutely identified in the flat plane as:

R7.16 A52.00

That coordinate identifies point *A* as unique. Many (but not all) CNC controls can act on it as reference or geometry, as easily as a Cartesian value.

Incremental Polar Coordinates

Sometimes the point cannot be absolutely identified from the PRZ but it can be usefully located using another known point as a reference. If the radial and angular dimension is local, then an **incremental polar** coordinate can do the job. In Fig. 17-52, point *B* is identified using the center of the radius as its origin and the radial line to point *A* as it's angular reference.

R2.38 A−142.00

Positive and Negative Angles In Fig. 17-52, note that the 142° angle is expressed as a negative number. That's polar coordinate convention.

Figure 17-51 *Point A can be defined using its radial and angular displacement from the PRZ and PRL.*

KEYPOINT

Following the Rule of Thumb
If the angular displacement is clockwise from the PRL or from a local reference, then it is a negative angle. If it is counterclockwise it is a positive angle. Identifying point *A* using point *B* as an incremental reference, the coordinate would be

R2.38 A142.00

This is true for both absolute and incremental angles (Fig. 17-53).

Polar
Angular sign value

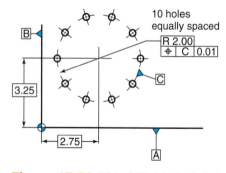

Figure 17-53 *Polar motion clockwise is negative, counterclockwise is positive.*

Similar to incremental coordinates, Cartesian coordinates, an incremental polar point is identified from the present location and a local center reference, used just for now. Most CNC controllers feature absolute polar coordinates. Only a few make use of incremental polar ones but where they do, nearly all trigonometry is eliminated when writing programs from the engineer's drawing via keyboard entries.

If the drawing has the part geometry defined, then most significant points are dimensioned with some form of rectangular or polar reference. By switching between coordinate types and values, all of the calculations required to convert can be eliminated. All remaining calculations are needed only to define significant points not dimensioned on the drawing.

To use incremental polar coordinates, the control must feature a convenient way of fixing local reference points for the center of the local system (Fig. 17-54). It will be in effect as

Figure 17-54 *This drilled bolt circle is dimensioned using polar coordinates.*

long as the coordinates refer to that point only. For example, fixing the center of the pattern as the local reference:

Hole 1 = Radius 2.0 Angle 0.00
2 = R2.00 A36.0
3 = R2.00 A72.0

Trade Tip

Polar *Read the Control Manual—Avoid Calculations and Potential Errors* Do yourself a big favor, read the control manual to see how polar coordinates are used on a particular machine. Nearly all modern CNC controls make use of absolute polar coordinates, but to varying degrees. Not all will feature the ability to use incremental polar values or easily place a local reference as described above.

However, on those that do, program math is reduced greatly if you've caught on to this trick of the trade. Having the choice of either absolute or incremental polar and Cartesian coordinates to identify any single point on the part geometry, most can be programmed directly from the drawing with no calculating (Fig. 17-55).

Figure 17-55 *Most scientific calculators include a polar-to-rectangular converter.*

Then, Read Your Calculator Manual When working on a control that doesn't feature polar coordinates, here is a second trick to help speed up polar-to-rectangular (P-R) conversions (Fig. 17-55). Your standard scientific calculator probably features a conversion function between polar and rectangular coordinates: P-R and the reverse, R-P. Learn the polar-to-rectangular conversion function. This simplifies conversion trigonometry!

Spherical Coordinates—Three-D Polar

Points can be identified in three-space using a second type of radius-angle-angle grid called

spherical polar **coordinates.** They are not used on CNC controls and beyond the scope of this unit. When three-dimensional problems arise, programmers fall back to X, Y, and Z Cartesian coordinates or to CAM systems to make the calculations for them.

However, looking into computer-integrated manufacturing, there are certain industrial robots that move their bodies around a central pivot and feature two angular axes for the arm. Therefore, the terminal gripper moves about a hemispherical envelope. These spherical types are occasionally programmed with a language similar to a CNC machine, using radius, angle, and angle values to describe unique positions of the arm.

UNIT 17-4 REVIEW

Review the Key Terms

Absolute polar (coordinates)
Point identification using radius and angle ordinates, based on the PRZ.

Incremental polar (coordinates)
Point identification coordinates based on another polar point, not the PRZ.

Polar reference line (PRL)
A line radiating out of the PRZ that indicates zero degrees.

Spherical coordinates
A three-space coordinate using radius, angle, and angle.

Replay the Key Points

◆ If angular displacement is clockwise from a local line, or from the PRL, it is a negative angle.

Respond

1. Complete the polar coordinate chart for the 10-drilled-hole, bolt circle problem shown in Fig. 17-55. Retain null values in your coordinates.

2. Rewrite the coordinate set for the 10 holes of Figure 17-55, but using the negative polar circle this time. All CNC controls that use polar coordinates would understand either value, positive or negative.

Figure 17-56 *Define point 3 using absolute polar coordinates.*

Critical Thinking

3. Point 3 must be identified for a program (Fig. 17-56). Select the correct coordinate values. Explain your choice.

4. State factors for coordinate selection in a program in their order of importance to the programmer.

5. Define a polar coordinate in your own ten words or less.

Critical Thinking

6. Review Problem 1, consider how much trigonometry would be required if the control was limited to X-Y Cartesian values. (No CAM available.) If you have a R-P conversion on your calculator, try it now on Problem 1. Convert each polar position to rectangular.

Trade Tip

Bolt Circle Routines Many controls have a third solution to finding rectangular coordinates for Problem 1, it's called a bolt circle routine (canned cycle). To use it, after entering the code word that starts the routine, you supply the parameters for drilling at the given radius from the local reference center. Additionally, the routine needs angular distance from PRL to the first hole and the number of holes in the circle—presto, the control computes the coordinates for each hole. The final facts concern how the holes will be drilled, in big or small pecks, total hole depth, and how far to retract the drill for rapid travel between holes.

Section 17 Review

Unit 17-1

The next level of machine evolution is just getting underway. Whole new species are on the market that have blurred the distinction between cutting and grinding, turning and milling, they are complete, one-stop machining centers. That trend will continue since it reduces the cost of manufacturing. That means that being able to identify machine axes on new machinery will become more complex and necessary.

Those multitasking machines will feature many slides and rotary motions. All must be assigned axis names and directions. It's predictable that we might exceed the nine EIA standards we've studied.

Unit 17-2

I don't ever foresee a time when the ability to write keyboard-entered tooling programs or to edit existing CAM-generated programs won't be a job requirement for a lead machinist. To do so, a good working ability with both absolute and incremental Cartesian and polar coordinates and command codes will remain essential job skills. However, mainstream programming is under evolutionary change.

While complex drawings that already exist as wire frames or flat 2-D patterns will never be converted, today most new CAD drawings are provided as solid objects with surfaces. Given a solid model, CAM programming becomes more automatic. With a defined 3-D target and the alloy and size of the raw billet known, the software selects most of the operation sequences, number of cuts, speeds and feeds, and the right tool to use.

Since PC-based CNC controls are capable of holding and applying CAM programming software, many jobs with solid model drawings can go directly from design to CNC machine, where the controller writes the program! We're talking about job title evolution by reducing man-hours.

Unit 17-3

The ability to perform in-depth data manipulations at the machine is one of the major factors separating operators from machinists. The machinist can control and improve the process, the operator controls the running of the machine.

CAM-generated programs don't use high-level machine motions nor do they use logic loops and subroutines. The programs they generate are comprised of very close approximations of surfaces using arcs and straight lines. Therefore an understanding of the ways a machine is able to interpolate is needed for the same reasons as already outlined, to improve and try out programs and to write mini-tooling-routines for your machine.

Unit 17-4

I admit that knowing and using polar coordinates for CNC machine work might be closer to a trade tip than a fundamental subject. Not many machinists use them on a daily basis and they aren't standardized within industry as are Cartesian coordinates. Some controls don't accept radius and angle entries, and between those that do there are wide differences as to what level they use them.

All controls that do accept polar coordinates use absolute values based on the PRZ. However, fewer use incremental polar because they require fixing an LRZ. That may or may not be a function of the control. However, if your CNC control does accept R/A entries, a working understanding of them will simplify many of your daily tasks. Read your operator's manual to find out.

KEYPOINT

But, while they are optional in CNC work, when drawing part geometry in CAD software, polar coordinates are an indispensable tool.

Questions and Problems

1. We've studied nine axes that define motion and position in CNC work. Describe them. (Obj. 17-1)

2. True or false? When an axis moves with a wrist action, not in a complete circle it is said to be interpolating. If it is false, what makes it true? (Objs. 17-1 and 7-2)

Critical Thinking

Problems 3 through 6 refer to Figure 17-57, and the following planning.

Job Planning Part Number PLC 17-1

005   Cut aluminum blanks 5 × 2.75 × 1.25 (thickness excess for holding in vise)

010   Profile mill all details × 0.900 total depth (0.875 + 0.025 extra)

015   Reverse parts and fly cut the thickness excess to 0.875 in. dimension.

The final operation will be to turn the parts over and hold in soft jaws and cut away the excess. This allows the parts to be 100 percent profile milled at one time (Fig. 17-58).

3. In Fig. 17-57, note that part PLC 17-A is not dimensioned geometrically. Where should the PRZ be placed on the drawing to write a profile milling program? Explain. (Obj. 17-2)

4. Figure 17-59 depicts four possible positions we could use to setup a vise to hold the part for milling of the PLC 17-A. In which quadrant does each setup lie? (Objs. 17-1 and 17-2)

5. How many circular interpolations will be required to mill the profile of the PLC 17-A? (Obj. 17-3)

6. Of the four possible setup orientations of the PLC 17-1 (Fig. 17-59), which could be used to machine the part, A, B, C, or D? Which would

Figure 17-58 *Here's how we'll hold and profile mill the part.*

Possible Quadrant Orientations

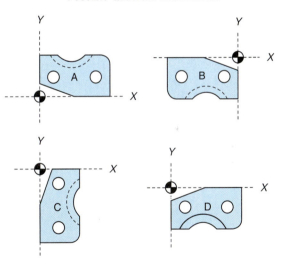

Figure 17-59 *Setup options in various quadrants.*

be the most efficient A, B, C, or D? Explain. (Obj. 17-2)

7. With the part in position D (Fig. 17-59), what are the absolute coordinates for the center of the two 0.812-in. holes? (Obj. 17-2)

8. Using the drawing orientation (Fig. 17-57), what are the incremental coordinates of point K (arc center) using point D as the present reference location? (Obj. 17-2)

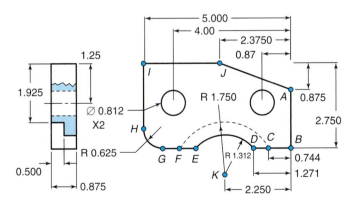

Tol. general +/−

X.XX = 0.30	
X.XXX = 0.005	PLC 17-A

Figure 17-57 *Problems 3–10.*

9. What are the absolute coordinates of point *K* with the part in the *D* orientation? In Fig. 17-59 (Obj. 17-2)

10. With the part in the drawing orientation, write the absolute coordinates of the geometry. Fig. 17-57 points in the order given. Leave null entries off. You are not writing curve commands just listing the coordinates. (Obj. 17-2)

11. To perform a spiral ramp plunge, what level machine motion must the control be capable of performing? (Obj. 17-3)

12. How do CAM programs produce three-dimensional shapes? (Obj. 17-3)

13. True or false? An incremental polar coordinate uses the PRZ as its center reference and a local line as its angular reference. If this statement is false, what will make it true? (Obj. 17-4)

14. In Fig. 17-60, where should the PRZ be placed to turn this pin chuck collet? In what orientation should it be machined? Explain. (Obj. 17-2)

15. Are there any circular interpolations required for the pin chuck? Why? (Fig. 17-60) (Obj. 17-3)

Pin Chuck Collet

Figure 17-60 *Turning problem.*

16. Since the 50° dimension is angular, would polar coordinates be used to define the tapered surface? (Fig. 17-60.) (Obj.17-4)

17. Assuming the part is 180° rotated,* what are the geometry point, absolute coordinates, for turning this part? Start from PRZ then turn toward the chuck. You are not writing a program, just listing the significant geometry points required to turn the OD. (*Hint:* See Problem 18, then check the answers.) (Objs. 17-1 and 17-2)

*With the taper to the right

Critical Thinking

18. The second geometry point for Fig. 17-60 was the diameter at *X*0.6840, *Z*0.0000. It's not on the drawing. How was it calculated?

19. Figure 17-61 depicts a part requiring two wrench flats milled on the end of a round inspection cap. Would the coordinates be more efficient in Cartesian or polar? Explain. (Objs. 17-2 and 17-4)

20. What are the absolute polar coordinates of points *A*, *B*, *C*, and *D*, at the ends of each wrench flat? (Fig. 17-61.) (Obj. 17-4)

Figure 17-61 *Inspection cap.*

ANSWERS 17-1

1. See Fig. 17-62.
2. The *Z* axis; the *Y* axis
3. The *A* axis is moving in the plus direction; the *B* axis is moving in the plus direction.
4. The secondary axes would be a *U* and *W*. Note, for programming consistency, the *W* axis is negative toward the work. As is the *X* axis.

5. *X   Y   Z*
   U   V   W
   A   B   C
6. It is a negative *A* motion.
7. See Fig. 17-63.

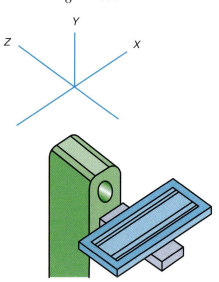

Figure 17-63 *Answer to Problem 7.*

Vertical mill   Lathe

Figure 17-62 *Answer to Problem 1.*

ANSWERS 17-2

1. True
2. By the *datum* basis always—actually the *only* rule. Place it in the first quadrant if possible—a minor consideration
3. See Fig. 17-64.
4. See Fig. 17-65.
5. See Fig. 17-64.
6. See Fig. 17-65.
7. Not a closed octagon. See Fig. 17-66.

Trade Tip

Did you close the octagon shape for Problem 7? If so, you assumed where the program was going. Be cautious of these assumptions during program tryout. Don't assume the cutter will do what you want it to do. Read and analyze the data.

Point	Abs.	Coordinates
A	*X* 0.00	*Y* 0.50
B	*X* 0.50	*Y* 0.50
C	*X* 0.00	*Y* 1.43
D	*X* 1.47	*Y* 1.43
E	*X* 1.47	*Y* 0.89
F	*X* 2.29	*Y* 0.89
G	*X* 2.83	*Y* 0.52
H	*X* 2.83	*Y* 0.00
I	*X* 0.50	*Y* 0.00
J	*X* 1.00	*Y* 1.00

Figure 17-64 *Answer to Problems 3 and 5.*

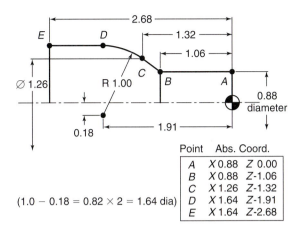

Figure 17-65 *Answer to Problems 4 and 6.*

$(1.0 - 0.18 = 0.82 \times 2 = 1.64 \text{ dia})$

Point	Abs. Coord.
A	X 0.88 Z 0.00
B	X 0.88 Z-1.06
C	X 1.26 Z-1.32
D	X 1.64 Z-1.91
E	X 1.64 Z-2.68

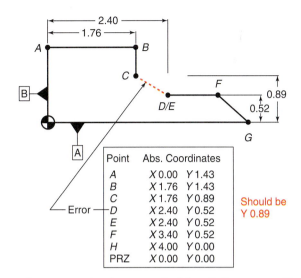

Point	Abs. Coordinates
A	X 0.00 Y 1.43
B	X 1.76 Y 1.43
C	X 1.76 Y 0.89
D	X 2.40 Y 0.52
E	X 2.40 Y 0.52
F	X 3.40 Y 0.52
H	X 4.00 Y 0.00
PRZ	X 0.00 Y 0.00

Error — D   Should be Y 0.89

Figure 17-67 *Answer to Problem 8. Error caused by wrong absolute coordinate.*

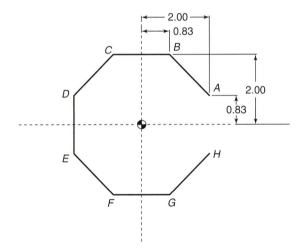

Figure 17-66 *Answer to Problem 7.*

Point	Abs. Coordinates	
A	X 0.00	Y 0.50
B	X 0.50	Y 0.50
C	X 0.00	Y 1.43
D	X 1.47	Y 1.43
E	X 1.47	Y 0.89
F	X 2.29	Y 0.89
G	X 2.83	Y 0.52
H	X 2.83	Y 0.00
I	X 5.00	Y 0.00
J	X 1.00	Y 1.00

Figure 17-68 *Answer to Problem 9. Drop these null entries.*

8. The error causes an angular line from point *C* to *D*. Since point *D* is incorrectly impressed over point *E*, line *D-E* is missing. However, all else is correctly shaped beyond that error (Fig. 17-67).

9. The null values that can be dropped are as shown in Fig. 17-68.

10. Here are the incremental coordinates:

N005	X.50 Y0.50	(Move to point *A*)
N010	X−0.50 Y0.0	(Y0.0 is an incremental null value—no *Y* motion)
N015	Y0.94	(All nulls removed from here down)
N020	X1.47	(Point *D*)

N025	Y−0.55	
N030	X0.82	
N035	X0.55 Y−0.37	
N040	Y−0.52	(Point *H*)
N045	X−2.34	
N050	Y0.50	(Closes shape to point *A*)

11. They produced a block style "YES" 50 mm long and 10 mm tall.

12. Here are the null values for the program:

N005	X0.00 Y0.00 Z−5.0
N010	X0.00 Y1.00 Z0.00
N015	X−0.50 Y1.00 Z0.00
N020	X0.00 Y0.00 Z.50
N025	X1.00 Y0.00 Z0.00

Explanation: In each case, the null coordinate had to be a "zero" for any other value would have caused movement. A null value would maintain the present location of the tool in that axis. Another way to envision this is that there was *no movement expected* for each null—a zero.

13. The angular line was made again between *C* and *D*. But *D* now becomes the relative origin for the next point *E*, which is now shifted 0.37 lower on the page, as are all other entries after the error. In other words, even though the entries after the error were the right amount, they inherit the positional shift too. While both parts would have been scrap, the error becomes far worse in incremental-valued coordinates where the cutter might have hit the vise or holding fixture (Fig. 17-69).

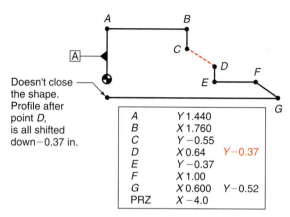

Doesn't close the shape. Profile after point *D*, is all shifted down −0.37 in.

A	Y 1.440	
B	X 1.760	
C	Y −0.55	
D	X 0.64	Y −0.37
E	Y −0.37	
F	X 1.00	
G	X 0.600	Y −0.52
PRZ	X −4.0	

Figure 17-69 *Answer to Problem 13. Error's effect using incremental coordinates.*

ANSWERS 17-3

1. Additional linear slides added as milling options cause the need for possible higher levels of tool movement.
2. Either the *feed rate override* is being used at about 66 percent or the control has a circular feed rate *default limit* of 400 mm/M.
3. Using a face plate or fixture, the CNC lathe could accomplish this with ease simply performing circular interpolation while rotating the part (Fig. 17-70).

There are two mill options:

Machine many circles using a *ball nose* end mill (Fig. 17-71). Produces a rough surface finish depending on how many circles are programmed. A slow method, but sometimes a necessary one.

Add an additional rotary axis table and spin the part while performing slow circular interpolation. This then is the same action as the lathe but the work rotation speed is safer for large or odd shaped objects.

4. It is a $2\frac{1}{2}$-axis move. Making a circular approach to the work surface while plunging down in

Turning the Hemisphere Using
***X-Z* Circular Interpolation**

Figure 17-70 *Turning the dome shape.*

Ball Nose Approximation

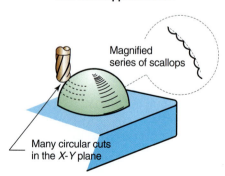

Magnified series of scallops

Many circular cuts in the *X-Y* plane

Figure 17-71 *Milling the dome shape using 2-D circular interpolation.*

the Z axis. A 3-D control can perform this move as well. It is better because

A slot is made for chip and coolant transfer. The cutter comes into contact with the finished surface in a tangential path producing a smooth transition from ramp to intended surface with less tool marks at the point of contact.

5. A *five-axis* CNC mill could produce the hemisphere by continuously keeping the cutter tangent to the surface at all times (Fig. 17-72).

6. 2-D—circular interpolation in a primary plane only; $2\frac{1}{2}$-D—circular in a primary plane with the third axis performing linear interpolation; 3-D—circular interpolation using all three axes.

7. The speed with which the microprocessor can output control data. Older controls were too slow for full-speed motion during these moves.

8. Since it's only a three-axis linear move, it's possible on all three types of mill controls!

Five-Axis Mill Solution

— Continuously keeping the cutter tangent to the sphere

Figure 17-72 *Milling the dome shape using X-Y circular interpolation while keeping the cutter tangent to the surface.*

ANSWERS 17-4

1. Positive polar values (note, the R2.00 has been omitted for point 5 on). On some controls the repeating radius can be dropped while on others it will need to be included when working with polar.

4 R2.00 A108.0	8 A252.0
5 A144.0	9 A288.0
6 A180.0	10 A324.0
7 A216.0	

2.

1 R2.00 A0.0	6 A-180.0
2 A-324.0	7 A-144.0
3 A-288.0	8 A-108.0
4 A-252.0	9 A-72.0
5 A-216.0	10 A-36.0

Answers to Section Review Questions

1. Primary Linear X, Y, and Z (straight line)
Primary Rotary A, B, and C (circular or arc motion)
Secondary Linear U, V, and W (linear parallel to X, Y, and Z)

2. False—it is articulating.

3. Intersection of upper and right edges—all dimensions originate there.

4. A = 1st quadrant
B = 3rd quadrant

3. There were two options:
Absolute Cartesian   X32.5   Y27.50
Absolute polar   R42.573   A49.764
The rectangular dimensions are single decimal place numbers, while the radius and angle appear to be calculated from them. Cartesian coordinates might be a slightly better choice due to rounding errors. Either would do the job.

4. *First*, company policy (not stated in the reading but a reality in industrial programming); *second*, control capability; *third*, least math conversion if you know the rules of the control.

5. Position coordinate identification in a plane using radius and angle.

C = 4th quadrant
D = 4th quadrant

5. Five circular interpolations. According to planning *all* details must be profile milled. The 0.75-in. radius and the 2.500- and 2.675-in. arcs must be circular interpolated. But the 1.312-in.-diameter holes must also be milled, not drilled.

6. They all could be used since the PRZ is in the right position relative to the work. However, position D allows the 2.675-in. counterbored radius to be machined while the other details are cut. Setups A, B, and C could be used, but the counterbore would need to be machined when the part is flipped over to remove the excess.

7. X 0.8700   Y−1.2500
 X 4.0000   Y−1.2500

8. X−0.979   Y−0.925

9. X2.250   Y−3.675

10.

Point A	X0.0000	Y−0.8750
B		Y−2.750
C	X−0.744	
D	X−1.271	
E	X−3.229	
F	X−3.756	
G	X−4.375	
H	X−5.000	Y−2.125
I		Y0.000 (zero is not a null)
J	X−2.375	

11. Two-and-one-half dimensional ($2\frac{1}{2}$-D) circular in two axes and linear in the third.

12. By approximating the surface with short straight and curved lines (best fit within a tolerance).

13. It can be true if the tool is located at PRZ! However, normally an incremental polar coordinate uses a local center reference and the present location point, as it's angular reference.

14. At the centerline and at the tip of the 50° face (Fig. 17-73). The part should be turned around 180° from the drawing orientation, with the

Pin Chuck Solutions

Figure 17-73 *Pin collet chuck solutions.*

50° point facing the tailstock. In that position, all details can be machined and the PRZ is easily verified at the outer tip of the material.

15. No, it would have been formed due to its small size.

16. No, the radius distance between points would need to be calculated—that's extra work.

17. Coordinates are in *X* and *Z*. Remember, all *X* coordinates are diameter values for most lathe programs (Fig. 17-73).

X0.0000	Z0.0000	(PRZ at centerline)
X.6840		
X1.5000	Z−0.8750	(Top of taper)
	Z−2.1250	(Corner of relief—note, you would turn to the 2.125 shoulder then return to form the relief)
X1.7500		

18. See Fig. 17-73.

19. Polar (assuming control accepts them) since both radius and angle are known. To use rectangular, the *Y* dimension would need to be computed, not a big deal, but why bother when it's right on the drawing in polar?

20.

R1.125	A 28.5	
	A151.5	(180−28.5)
	A208.5	(180+28.5)
	A331.5	(360−28.5)

Section 18

CNC Systems

Introduction

Section 18 provides the background concerning the machines and the data they use necessary to manage CNC equipment. We'll look at three aspects of CNC systems:

Axis Drive Mechanics   Building on Section 17's explanation of the kinds of axis moves a CNC machine can make, here we'll investigate the mechanics of how they make them. We'll learn how they can make a near perfect surface when cutting into and out of changing material loads. Even with varying excess from one part to the next, every part comes off the machine a near perfect duplicate of the last!

Machine Variations and Evolution   While today's machines are far superior to anything in the past, it's a sure bet they aren't the ultimate. So, our goal is an understanding of the drive behind machine evolution such that the future is easier to see. Reading the time line carefully, you might foresee a big change just around the corner!

Programs   We'll also examine program creation and data management. Section 18 isn't about writing programs, it's about where they come from and how they are managed.

Objectives

A CNC Axis Drive

Introduction: The biggest differences between a manual machine and a modern industrial CNC is the way it moves its axes. A CNC axis shares a lot of similarities with using your arm to move and place an object. Your brain sends signals though connective nerves to the synapses that trigger muscles. It also receives reports back that compare the command to the result—called a kinetic sense. Without looking, you always know where your arm is located and how fast it is moving. Then, if the item doesn't budge, you can add more power to get it going or at some point decide it's beyond your ability. If the object moves easily, you can back off on the force.

KEY TERMS

Absolute positioned system
A CNC drive system that retains its grid position even when turned off, then on. These machines feature linear feedback devices.

Ball-screw
A backlash compensating, linear drive system in which a split nut is forced laterally both directions against a screw. Balls roll through the circular channel between nut and screw.

Closed loop
A circuit that includes monitors to send a progress signal back to the CPU.

CPU—central processing unit
The component that decodes the program, sends the drive commands to the servo motors, and monitors and adjusts progress.

Feedback
The signal returning from the drive system to the CPU.

Hard limits
Physical switches that verify location of an axis position.

Homing the machine
Sending all axes to limit switches in order to achieve a zero base for the encoders. Generally required only on older machines.

Initialization
Turning on the control, depending on sophistication, routines must be followed to prepare for machining. Newer machines load software and ready themselves; older units need more preparation.

MCU
The master control unit see CPU.

Machine home
A never-changing position where the machine must be driven to refresh its feedback signal at initial start-up.

Open loop
A budget drive system that does not feature any provision for feedback signal processing.

Servo error
An unwanted condition in a closed-loop system where the axis progress exceeds the tolerance from the expected result.

Servo motor
A highly controllable motor with predictable power, speed, and acceleration curves based on input energy.

Soft limits
Optical sensors that verify proximity to hard limits switches.

Stepper motor
A drive motor that moves a given part of a rotation, given one pulse of energy.

The Five Components

Using five electrical/mechanical components, a CNC drive performs the same tasks as using muscles to move and position your arm. The controller can keep a cutter hugging the true programmed shape from within 0.002 or 0.003 in. for heavy cuts down to an amazing 0.0002 in. on light loads, depending on machine construction and the nature of the shape and volume of metal being removed.

The control constantly compares physical tool position to the expected position. If the cutter's progress doesn't fall within a preset lag tolerance, the controller has one or two options, depending on basic architecture within its operating system.

More commonly, it can wait a preset time, until the physical cutter location catches up to the expected position before issuing another axis drive command. Second, on more powerful machines, the control can increase the axis drive force when the cutter enters heavier cuts and starts to lag behind, and it can back off on the axis drive as the cutter breaks into lighter work sections. The system performing these marvels is a **closed-loop,** *linear, servo axis drive.*

That mouthful is shortened to a CNC axis drive! It's responsible for the cutting motions

Figure 18-1 *A closed-loop CNC axis drive system.*

and for the positioning accuracy. Like your muscle movements, five components are required (Fig. 18-1):

CNC Drive	Muscle Motion
Controller unit	Brain—processing in/outgoing data
Axis drive relay card	Synapses—triggering muscles
Translation axis screw	Tendons pulling
Controllable drive motor	The muscle
Feedback device	Sensory nerves reporting progress

In order to achieve the maximum results and understand limitations, the action of each component should be generally understood by the machinist, which is our goal for this unit.

Three Kinds of Repeatability

When buying a CNC machine, one of the first facts to inquire about would be its axis repeatability. We begin with resolution (see Fig. 18-2).

Machine Resolution This is the smallest movement the axis drive is able to make, almost always, 0.0001 in. or 0.02 mm. It's either a mechanical limit, based on the finest movement the axis drive motors are able to make, or it can also be a sensor limit of how fine the machine can read movement and send a signal back to the controller—one or the other, usually the mechanical limit.

Axis Repeatability Next up we discuss the capability of the axis drive to move the spindle or tool away from a specific location then

Figure 18-2 *Super precision CNC machines such as this Super-Mini Mill, have resolutions below 0.0001 in.*

drive it back to the same place within a close tolerance. The approach test must be made from every part of the work envelope and at varying speeds. But, the return test is never under a cutting force load. CNC axis repeatabilities are usually around 0.0002 in., again with exceptions. But repeatability doesn't tell the whole story. By itself, repeatability can be misleading.

Shape Repeatability—The Final Factor
Machinists sometimes think of axis repeatability as the tolerance limit for holding to a curved path, however, that isn't always so, especially as removal rates increase. Many factors combine to create the shape repeatability. Depending on the overall strength of the machine's axis drives and the nature of the machining, a few machines are able to reproduce shapes within the specified repeatability while most tolerate more deviation.

There are extreme accuracy machines designed for specialty work that feature 0.00005-in. resolution (fifty millionths inch). These machines can reproduce parts within 0.0001 in. each and every time.

Physical Servo Errors

Similar to discovering your arm hasn't the strength to push the work along, when a program demands more than the machine or the cutter can deliver, or when a cutter dulls or loads up, the machine can become out of sync. When its physical path exceeds a preset deviation limit, it enters a condition known as **servo error.** On screen and possibly with flashing lights, the control will signal the operator and, at the same time, halt all progress when it cannot bring the axis back into coordination within a given time lag.

Trade Tip

As the operator, you have three main weapons against servo error. First, use the feed and speed override controls to fine-tune cutting forces (editing in permanent changes to the program if needed). To do so machinists must understand the forces and physics of chip making and the limits of their machine. Closely monitoring cutter sharpness and coolant delivery can also help prevent a lot of servo stalls.

There are also programming methods that can prevent servo and shape errors. They are the same skills learned on manual machines, like taking a light final pass with a full-sharp cutter reserved for finish work. Your manual machine experience really counts here!

Mechanical Axis Translation

The heart of the precision drive is mechanical. We've already seen it in Section 14, the *backlash eliminating, recirculating ball-screw, and preloaded split nut.* Its purpose is to move, position, and reverse axis motion with no backlash. For a reminder of how they look see Fig. 18-3.

Controller

The CNC brain is sometimes called the *master control unit* (**MCU**), or the **central processing unit** (**CPU**). This component has four or five functions, depending on whether or not it is connected to an outside computer. For our study of axis systems, the controller's CPU converts stored program commands or

Recirculating Ball-screw

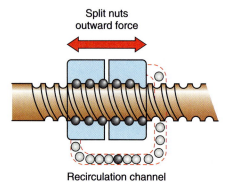

Figure 18-3 *Ball-screws are the mechanical heart of the drive system.*

keyboard entries by the operator into drive commands, then sends them to the axis motor relays.

KEYPOINT

Axis drive commands do not go from the CPU to the drive motors, they go to intermediate relays.

The CPU issues a low-energy command (usually 12 V DC) to relay devices that work as electricity amplifiers and switches. At the same time, it is also processing incoming feedback from the axis drive sensors to determine what action is required to maintain the programmed feed rate.

Axis Drive Relays

An individual CNC controller can be used to drive a variety of different CNC machines. To do so, the controller doesn't directly command the motors to move. It commands an axis drive relay that's custom tailored to the specific drive motor. Drive relays make it possible for a single control model to move any kind of motor, big or little, servo or stepper (studied next). One control is able to operate a small offload mill or it can be wired into a giant horizontal mill with over 100 hp per axis, because of the axis relays.

Wired between the controller and the motor, the intermediate relay card receives the low-energy control command, then amplifies it and configures it to the motor's needs. At the relay, the control command triggers high-energy electrical current configured into a direct current stream or as pulses, depending on the kind of motor being driven.

Drive Motors—Servo or Stepper

Different from manually operated machines, that often use one motor to drive several functions through gears, belts and clutches, CNC drive motors are highly controllable and predictable. Given an exact amount of input energy, they reliably move the same way each time. They accelerate and decelerate on an exact time/load curve.

> CNC machines use one drive motor per axis and each motor is very different from those on manuals.

There are two kinds of popular drive motors, **servo motors** and **stepper motors.** (A few machines such as precision grinders, use hydraulic motors powered by pumps but they are far less common.)

Direct Current Servo Motors—More Power for Industrial Machines Electric servo motors are stronger but also more costly. They produce accurate RPM based on continuous input energy. As the input energy increases, servo motors spin faster or drive harder when pushing axis drives.

KEYPOINT
> The DC servo motor is by far the most common.

Steppers—Moving in Small Increments
Steppers are the second most common drive muscle. They are generally used in smaller machines. Steppers do just that, they move in tiny jumps rather than spinning as servos do. Grasping the shaft of an unmounted stepper and spinning it, you'd feel it click or pulse as the motor rotates through small magnetic arcs.

Using the CPU's clock function to coordinate timed pulses to the drive relay, after reading the programmed feed rate, the controller determines how fast the pulses must be sent. The faster they are issued, the faster

the motor rotates (up to its maximum speed).

Each pulse from the axis relay, rotates the motor an exact arc. For example, common steppers rotate through one full circle with 15, 30, or 45 pulses per revolution, depending on their construction. So the motor moves 24°, 12°, or 8° per pulse. That, in turn, translates to an exact rotation of the ball-screw.

Since the movement of a stepper is predictable, light-duty machines can be constructed with no feedback system, called **open-loop** *circuitry.* Open-loop machines are budget equipment only. With no feedback, they do not sense or control variables, nor do they repeat shapes very well. In fact, an open-loop axis can be stopped completely, yet the control will continue blindly issuing program commands, with no movement resulting. With no kinetic sense, the control never knows it's out of sync!

KEYPOINT
> Open-loop machines are best used for detail and offload work or for woodwork, but they aren't practical for heavy machining.

The Feedback Device

The fifth component is the watchdog in the system. Monitoring the axis progress and location, the **feedback** device is responsible

ShopTalk

Precision ball-screws are so critical to the manufacture of CNC machines that a certain foreign machine building company wished to make high-quality machines here in the United States. Before they even considered building a factory, they sent a study team to determine if there were more than one ball-screw manufacturer that offered the accuracy required. Not until they knew that single fact would they commit to thinking about the investment (Fig. 18-4). Go to the Internet and write www.wedin.com to see some interesting ball-screw manufacturing articles.

Figure 18-4 *This giant ball-screw is one of the largest ever made in the United States.*

for detecting and reporting position in real time back to the processor. There are two general families:

1. **Direct Linear Feedback**
   Sometimes called **absolute positioning,** these systems make use of various kinds of calibrated distance sensors. Although more sophisticated, the scales are similar to digital positioning systems on manual machinery. The controller tracks axis position using signals from the scale. Direct feedback provides the advantage of acquiring position instantly during initial powering up. Machines equipped with direct feedback reacquire their axis and tool positions when the control is first turned on.

2. **Indirect Feedback**
   These devices monitor the amount of drive screw rotation as shown in Fig. 18-5, called *encoders (counters).* About the only advantage of these encoding devices is their low cost compared to absolute positioning systems. The biggest disadvantage of encoders is that they can become lost either from depowering or from heavy machining loads. Since the CPU must count pulses to track position, it can lose the count when the control is switched off or when the cutter encounters

sudden heavy cuts. The system that's built in this way, must be homed by parking it on relay switches **(hard limits)** or optical sensors **(soft limits)** to regain a sense of where it is. Machines of this type are set up such that they will not operate until they are homed after being powered up in the morning, so they have a zero base from which to begin positioning.

Initializing (Turning on a CNC Drive)

When **homing the machine,** or **initialization** (powering up the system) of encoding machines, each axis must be driven up to a reliable start point, usually against axis limit switches. That position is known as the **machine home** or *axis home.* Once parked at that never-changing hard-wired zero base location, usually at the far end of each axis, the controller then has a datum origin for its axis count.

Trade Tip

On older encoding drives, it's a good idea to insert a home command into the program, after a long or difficult roughing cut, before taking the finish passes. To simplify the action, homing can usually be performed with a program command, rather than manually by the operator. Then with the axis coordination refreshed, the final pass can be taken with improved accuracy. It's often convenient during homing to program a tool change to a finish cutter since the tool change position is often at or near the home position on both mills and lathes. Presto—the machine has a sharp cutter and a refreshed axis position.

There are other kinds of linear and rotary feedback drive mechanisms, but scales and encoders (Fig. 18-5) represent most of the possibilities for common equipment. Still more advanced systems use reflected laser light as their axis sensing mechanism.

KEYPOINT

No matter what kind of device, feedback's purpose remains the same: sense the exact position of the machine in X, Y, Z, A, B, and C axes then feed that information back to the CPU.

Simplified Encoder—Indirect Feedback

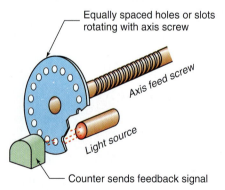

Equally spaced holes or slots rotating with axis screw

Axis feed screw

Light source

Counter sends feedback signal

Figure 18-5 *An indirect encoder counts pulses as the axis screw rotates.*

UNIT 18-1 REVIEW

Review the Key Terms

Absolute positioned system
A CNC drive system that retains its grid position even when turned off, then on. These machines feature linear feedback devices.

Ball-screw
A backlash compensating, linear drive system in which a split nut is forced laterally both directions against a screw. Balls roll through the circular channel between nut and screw.

Closed loop
A circuit that includes monitors to send a progress signal back to the CPU.

CPU—central processing unit
The component that decodes the program, sends the drive commands to the servo motors, and monitors and adjusts progress.

Feedback
The signal returning from the drive system to the CPU.

Hard limits
Physical switches that verify location of an axis position.

Homing the machine
Sending all axes to limit switches in order to achieve a zero base for the encoders. Generally required only on older machines.

Initialization
Turning on the control, depending on sophistication, routines must be followed to prepare for machining. Newer machines load software and ready themselves, older units need more preparation.

MCU
The master control unit see CPU.

Machine home
A never changing position where the machine must be driven to refresh its feedback signal at initial start-up.

Open loop
A budget drive system that does not feature any provision for feedback signal processing.

Servo error
An unwanted condition in a closed-loop system where the axis progress exceeds the tolerance from the expected result.

Servo motor
A highly controllable motor with predictable power, speed, and acceleration curves based on input energy.

Soft limits
Optical sensors that verify proximity to hard limits switches.

Stepper motor
A drive motor that moves a given part of a rotation, given one pulse of energy.

Replay the Key Points

- There are five components in a modern CNC axis drive system: controller, axis relay, servo motor, ball-screw, and feedback device.

- Servo errors usually happen when cutting forces suddenly build beyond the axis drive's ability to push; for example, when an end mill loads up and stalls.

- Open-loop machines are best used for detail and offload machining or for woodwork, but never for industrial machining.

- The direct current servo motor is by far the most common.

- The feedback device senses the exact position of the machine in X, Y, Z, A, B, and C axes then sends that information back to the CPU.

Respond

1. On a sheet of paper, roughly sketch the five components of a CNC drive system—identify each component by name. Check your sketch against Fig. 18-1.

2. Describe in less than 10 words:
   - A. Direct feedback
   - B. An open-loop drive (critical thinking question)
   - C. A servo motor
   - D. An axis drive relay

3. Describe in less than 20 words, how a ball-screw aids CNC accuracy.

Critical Thinking

4. When during a cut, an axis drive is suddenly loaded, the MCU senses the lagging progress and adds extra energy to the drive to compensate. Is this statement true or false? If it is false, what will make it true?

5. Some budget controls lack _____, therefore commands are not verified by the CPU.

UNIT 18-2

Industrial Machining and Turning Centers

Introduction: In Unit 18-2 we'll take a look at the kinds of machines you might find on the job. We'll look at the features and accessories of modern machinery, many of which may not be present among tech school lab equipment due to their expense and because they consume far too much material per hour to fit into a training budget!

KEY TERMS

Bar cutter/loader
A turning center accessory that saws bar-stock into blanks ready to load then turn.

Bar feeders
A turning center add-on that contains a long spinning bar and advances it when a new length is required.

Bar loaders
A turning center accessory that places new bars of turning stock into the bar feeder or bar cutter.

Chip-to-chip (CTC) cycle time
The time between cutting with one tool then a second tool.

Downtime
Nonproductive time when no work is being done.

Live tooling
A lathe tool that can spin, making milling and drilling operations possible on turning center jobs.

Pallet changer
A machining center with twin workstations—each able to rotate into the work envelope while the other is outside being loaded with new parts to be machined.

Part-to-part cycle time
The time period between finished parts in a multipart run.

Process control robot (PCR)
An in-process inspection robot that can measure size during a program. It can also feedback data to the control to adjust for variation.

Qualified tooling
Cutters inserts and holders that can be replaced with very little variation from one to the next.

Random tool storage
The ability to store an outgoing cutting tool into any tool drum position, with its tool number and new position being tracked by the control. This improves CTC time.

Tombstone
A vertical square work-holding fixture often used on pallate changing machining centers. The tombstone allows access to three sides of the workpiece in a single setup.

The Evolutionary Time Line

Evolution toward the perfect machine is driven by four factors:

New products	Not possible previously
Error elimination	Remove human intervention variability thus reduce scrap
Increased production	Reduce labor cost and lead time
Higher quality	More reliable reproduction

To clearly see into the future, we take a brief look backward. While it's enriching to know our trade's family tree, this information has a greater purpose: to deepen understanding of the reasons behind machine evolution.

The Big Clues

The first stride up from manual machines began because there "had to be better way"! But at a certain point, the drive changed from doing what couldn't be done to doing it better and ever more quickly. Also notice in the time line that soon after the machines were born a big branch of the changing technology had to do with programming systems—where and how they were written. Note too that the person, place, and method of writing a program has changed several times in the time line, and will predictably change again in the future! The next phase might lay the job back in your hands, skipping the programmer. Here's an overview of unfolding dates.

1946	**Aerospace spin-offs following World War II require complex shapes.** To machine helicopter rotor blades, John Parsons' Corporation experiments with tables of X-Y-Z movements called out to one human operator at each axis crank of a manual mill. You can guess how that went?
1949	**Parsons set up a demonstration of his early solution for the U.S. Airforce.** It showed it was possible to automatically generate complex shapes using his X-Y-Z data as a program that drove the machine axes.
1952	**First true programmed machine.** Joint effort put forth by the Airforce, Parsons, and the MIT Servo-Mechanisms Lab yielded the first true programmed

machine tool. It worked! Driving a Cincinnati Hydrotel®, the electrical relay cabinets were larger than the machine tool itself (Fig. 18-6), but it worked!

1955 **Tape drive machinery is born.** The Airforce awards a $35 million contract to produce 100 numerical control (NC) machines for aerospace manufacturing. Machinery is reserved for military purposes. Note, there's no "C," just NC.

KEYPOINT

These machines have no memory, they follow numeric programs on punched tape or cards, a lot like a music box.

1960 **Private industry gets NC machinery.** Machine tool manufacturers begin supplying simple two axis tape drive drills, lathes, and milling machines, to non-military industry. Machines begin the ascent toward the "best machine". Programming remains a matter of calculating and entering commands by hand.

KEYPOINT

At this stage, time required to write the program is usually many times longer than the time needed to run the program! Math challenges and entry errors are an ever-present challenge.

Standardized code word vocabulary for programming machine tools becomes necessary. It's organized by the Electronic Industries Association (EIA) RS-267-B code vocabulary is born. Programs are written by highly skilled specialists with an in-depth math ability.

1963 **Word routine, automatic programming language (APT) is born.** Using subroutines in a language called APT, programming can be done by writing word statements such as *"goto PT 1 [point 1]"* or *"goaround R2.5 [2.5 radius]"* then feeding that into large mainframe computers to calculate the coordinates.

Figure 18-6 *The first programmed machine tool had larger electrical cabinets than the machine!*

Owning a mainframe is beyond most budgets, other than the largest corporations! So, if using APT, we used teletype machines (tape-reading typewriters that can also punch new tapes) to send raw APT programs to the mainframe located several states away, which then returns the finished program commands! *There has to be a better way to program machines!*

1970 **CNC is born.** Machines with RAM (random access memory) become commercially available. Programs can be edited and written right at the CNC machine but that means it isn't running while it's being programmed.

1975 **Desktop PC computers become affordable and programming becomes a simpler, more affordable issue.** Programs can be written for many different machines, using a single PC workstation. Desktop APT becomes a reality and programming gets a big efficiency boost. CAD drawings begin.
CAM begins. Experiments begin by creating software that solves some of the programming math for *X-Y-Z* coordinates. As the math solution software becomes evermore capable, it becomes obvious that the software might be able to write programs directly from the part CAD image template.

However, due to computer software expense, CAD drawings are rare. Programs are still written by specialists, but far less math is required.

KEYPOINT

Note that at this point, evolution is on three ladders, each affecting the other: CAD/CAM software systems, PC chips, and the physical machines, each on their own path. Each driving the others and improving the process.

1980	**Full-functioning graphic programming.** **Affordable CAD desktop software.** **Chip Technology increases calculation speed.** Multitasking CNC machines. **Graphic programming systems become common on machine controllers.** Some programming is performed on the shop floor on the control, by the machinist. Onboard graphic programming systems add great expense to CNC controllers. Most new machines have their own CAM system. "There has to be a better way."

KEYPOINT

Now, the physical machines begin to change rapidly due to the CPU's ability to multitask.

1990	**PC-based CAM systems move most programming away from operators to the office environment.** **Solid CAD drawings become more common (Unit 18-3).**
1995	**CAM systems operate on solids.** Wireframe drawings (Unit 18-3) are still used but solid drawings begin to make gains in industry.
2000	**Automated CAM software takes solid drawings into programs automatically.** **PC-based controls and lightning-fast programming based on solid models make shop floor programming by the machinist a possibility once again.**

What's Driving It Now?

From the mid-1980s, machine evolution has been mainly driven by the need to improve efficiency (make more parts per hour). To compute manufacturing costs, machine time is divided into two broad categories: *uptime* (making chips), also called *WIP* (work in progress), and **downtime** (not making chips). Due to their great cost and potential for profit, CNC machines must deliver uptime production in the area of 90 to 95 percent in order to pay for themselves, plus wages, overhead, and profit. Uptime is challenged by five kinds of nonproductive downtime that drive machine evolution and the shop team to solve.

1. **Part-to-Part (PTP) Turnaround (Cycle Time)**

   This is the time required to remove the finished part then reload a second. This includes cleaning the setup of dirt and chips and carefully reinserting and clamping or chucking the next part. Well-thought-out operator actions are a key element in this battle, but the machines themselves have been changed to minimize PTP time too, as we shall see. CNC evolution has created a whole group of part loading options: bar loaders and feeders for lathes, pallet-changing systems for mills, and robot part loading arms for both kinds of machines.

2. **Ongoing Maintenance Downtime**

   This involves replacing dull cutters, maintaining lube and coolant levels, and other tasks that do not make parts but must occur, even though they slow production. If safe, many of these tasks are best dealt with while the machine is operating, but not all. Some can be performed safely only with the machine stopped. There are some computer solutions to this downtime, too. When machines have the tool capacity, two tools with the same shape and size can be put into the tool storage drum. Then after using the first for a given number or cycles, the machine automatically switches to the second sharp tool. At that time, the first can be removed from the setup to be sharpened or replaced. This is known as *scheduled tooling*.

3. **New Setup Turnaround**

   These are tasks required to get a new setup up and running. Writing and inserting a new program; unbolting the old, then aligning and bolting the new work-holding tools; and inserting new cutters in the turret and trying out the program on a first part run are just three of the many machine nonproductive turnaround times.

4. **Chip-to-Chip (CTC) Tool Change Cycle**

This is the period of time required to change cutting tools during a program. It includes stopping the spindle, rapidly traveling the machine to a tool change position, exchanging tools, starting the spindle again, and repositioning it back at the work with chips being removed once again.

5. **Inefficient Programs**

This category is complex. It includes unneeded traveling of the cutter without touching metal, or changing tools or moving parts between setups too many times. Later we'll examine some ways to find these time bandits, and how to get them locked up through smart editing of programs.

Today's machine evolution is driven by the need to maximize WIP.

So, now we investigate the turning and machining center functions and accessories that trim away nonproductive time.

Vertical Machining Centers

Unarguably the backbone machine in industry today, the vertical machining center (Fig. 18-7) is capable of not only changing its tools and parts, the word "center" indicates that multioperational work can be finished on them. They can thread internally and externally, bore, drill ream, and mill round objects that would have been machined on a lathe in the past.

Machining centers are able to machine three-dimensional shapes when they are programmed with CAD. If we add an A-B head and/or a rotary table, vertical CNC mills can machine nearly any shape that can be drawn on a computer screen.

Tech Specs for Vertical Mills— Machining Centers

Here are the capabilities these machines bring to the workplace.

Tool Changing Most machining centers can store, sort and automatically change, a minimum of 16 or 18 cutting tools, up to approximately 36 within standard equipment. Some feature optional tool carriage extensions or conveyor systems that hold an unlimited number of cutting tools.

Chip-to-Chip Time, CTC Cycle Time, or Sometimes Called Tool-to-Tool (TCT) The speed with which these machines can withdraw from cutting metal to change then return to fully making chips with a second cutting tool, is called the chip-to-chip tool change time. Cycles of from 2.5 to 10 seconds, are common. The cutting tool flange is gripped by a robotic arm shown in Fig. 18-8. This efficient tool arm is known as an end-for-end changer. It has the incoming tool in one gripper, on one end of the arm, and has gripped and pulled the outgoing tool from the spindle at the other end.

Figure 18-7 *The backbone of today's CNC shop, a vertical spindle machining center (VMC).*

Figure 18-8 *An end-for-end robot tool change arm pulls the outgoing tool and has the incoming on the opposite end.*

Figure 18-9 *Tool information stored on one page of the controller tool size, tool offsets, and current location.*

CNC Tool Management Most modern CNC tool management systems perform **random tool storage** (older CNC machines do not). Random storage means that any tool may be restored at the closest storage drum hole, or tool storage conveyor position at any time during the program run (Fig. 18-9). At the first

Figure 18-10 *A typical tool storage drum on a machining center stores up to 36 tools (varies with machine).*

insertion, the operator informs the control that a given tool number is in a particular drum position (Fig. 18-10). Then the control keeps an accounting of where each tool is located in the drum after each tool change. The point is, the drum needn't rotate to put the tool away in the same hole each time. After use, it places the tool in the first available open hole.

KEYPOINT

In setting up autoload cutting tools, in a machining center, there are two numbers that must be coordinated; The *tool number* assigned by the programmer (found in the setup document) and the *tool position number on the machine's tool drum*, chosen by the operator at setup.

Once this tool load coordination is performed, the computer tracks and moves the tools for efficiency.

KEYPOINT

Older CNC machines with tool changers may require the tool number to match the tool storage drum position, called an assigned location. On these machine we try to put tool #1 in hole #1 if it's possible.

Trade Tip

Drum Rotation as Soon as Possible Smart programmers write a command to rotate the tool drum or conveyor to the next tool to be used before it's needed. For example, while cutting with tool 3, the drum is commanded to rotate to tool 4. Then, when a tool change command is issued, the tool is in the change position ready and waiting to improve chip-to-chip time.

CNC Cutter Holders If the CNC mill has no automatic tool changer, then standard mill tapers and flanges can be used to hold cutters in the spindle. However, to facilitate autotool changing, cutters must be mounted in holders with three modifications compared to manual mill tool holders (Fig. 18-11).

In many cases, tool holders are interchangeable between machines. But there are some differences between the following tool holder features. Although undesirable, some shops must use a variety of tool holders based on

Three Differences in CNC Mill Tool Holders

Pull stud

Robot arm gripper groove

Deep/shallow drive slots or other code mark recognizable by machine

Figure 18-11 *CNC mill cutter holders have three modifications from manual mill holders.*

There are variations of each of these three modifications. Be certain the tool holder you intend to use fits the specific CNC mill tool changer.

the tool changer arms on various machines. In these situations, it's dangerous to mix holders between machines.

1. Spindle Orientation Coding

Spindle orientation has two purposes: to ensure that wide tools don't interfere with each other in the storage drum and to stop the spindle in a given position during machining. Spindle codes can be a dot or barcode on the flange, seen by the robot. It is often a difference in the drive slots and lugs, one being deep and the other shallow (Fig. 18-11). Stopping the cutter at a specific rotational orientation during a program can be useful. For example, withdrawing a boring bar from a finished hole, can scratch the surface if it's pulled straight up. However, if the programmer commands the machine to orient the cutting tool tip with respect to the hole's axis, he can add a tiny *X* or *Y* axis retract away from the wall, then pull the boring bar up with no mark.

2. Automatic Draw Studs

The CNC tool change cycle includes an automatic tightening into the taper. The CNC holder is fitted with a draw stud as shown in Fig. 18-11. The end lobe of the stud is gripped and pulled to hold the taper in place. Many modern CNC machines include an interlock sensor that will not allow the spindle to start unless the tool is correctly nested in the spindle.

3. Robot Grip Slots

The flange rim is modified with a groove machined to nest the gripper claw of the robot arm. There are a couple of different patterns for this and not all tool changers require this modification.

Work Pallet Changing

To reduce part-to-part cycle time, machining centers are fitted with twin rotating work surfaces on which *pallets* are located. A pallet is similar to an interchangeable mill table or **tombstone.** This feature may not be present on school machines as it is an expensive option. However, in production machining, a **pallet changer** greatly reduces part-to-part cycles, plus improving operator safety. The ingoing part is preloaded and ready to rotate in when the current part being machined is done. Pallets are usually exchanged in 15 seconds, chip-to-chip time (Fig. 18-12).

Preset Pallet Fixtures (Tombstones)

Additionally, to reduce setup time, CNC shops might own several pallets that can be readied before they are needed. The new setup can be assembled on a pallet off the machine then quickly exchanged when the new program is loaded.

High-Speed Accurate Spindles

Most small to medium CNC machining centers feature spindles capable of higher RPMs than standard milling machines. From 10,000 to 15,000 RPM is common but high-speed machining (HSM) spindles achieve revs at

Figure 18-12 *A pallet-changing mill speeds part-to-part cycle time and improves safety.*

20K or faster. These spindles accommodate the high RPM demands of small cutters and/or high-volume metal removal of materials such as aluminum.

Temperature-Controlled Spindle Oil

To maintain stable accuracy and long life lubrication, many machining centers feature temperature stabilized bearing oil. As the work forces heat the spindle oil, it is circulated through a cooler/filter before reentering the bearings always at the same thickness (viscosity). This not only stabilizes accuracy

Figure 18-13 *Spindle oil is cleaned, cooled, and recirculated by this stabilizing unit.*

Figure 18-14 *Twin coolant streams flood this end mill cut.*

but it imparts long life to exceptionally high-quality bearings (Fig. 18-13).

Coolant Systems

Chip ejection and cutter lubrication become even more important at the high removal and spindle rates used in CNC milling. To meet this challenge, machining centers include high-volume coolant systems. Some feature more than one programmable outlet such that coolant can be directed from two or more sides of a cut. Coolants are often "blasted" more than flooded over the cut area (Fig. 18-14).

Programmable Coolant

Some industrial CNC machines feature coolant nozzles that can swivel in two axes. Using a joy stick or two axis knobs, the operator aims them during the first run of the program. The coolant positions are attached to the tool number memory, then recalled during subsequent runs of the program.

Mist Coolants

A second type of coolant is the air-driven mist. Combining both air and coolant streams, mist coolant not only blows chips away but it often reaches deep into pockets and bores where flood coolant is driven back by flying chips (Fig. 18-15).

At the correct time the programmer might call on mist or flood coolant. But more likely, these decisions will be the machinist's.

KEYPOINT

If program editing is allowed by management, coolant improvements can be edited into the program at tryout time.

Figure 18-15 *Mist coolant is often more efficient than flood.*

Figure 18-16 *High-pressure coolant can be forced into deep pockets and holes or through the tool. Photo Courtesy ChipBLASTER, Inc.*

High-Pressure Coolants

Usually an added accessory, many machining tasks can be improved if the coolant can be injected forcefully at or into the cut. Figure 18-16 shows a high-pressure coolant system that injects a strong enough stream to remove wire burrs from aluminum!

Thorough the Tool Coolants

Technology borrowed from the gun drilling department, high-pressure coolant can be injected through special cutters featuring central galleries to channel the coolant right to the cut area. They help force chips up out of deep holes and also get to the action where they are needed (Fig. 18-17).

Automatic Safety Features

Autodoors

A common feature often found on CNC turning centers as well as machining centers is the automatic closing safety door (Fig. 18-18), many with sensors to detect and protect anything in their path. These doors are more prevalent on pallet-changing machines since they require larger front space to pass the rotating pallets, but they are also found on standard vertical machining centers and turning centers.

KEYPOINT

Containment screens and doors found on machining and turning centers not only stop flying chips, they also retain expensive coolants in the machine.

Remote Monitoring

Using replaceable video cameras rather than human eyes, the operator can display the

Figure 18-17 *Using special cutters HP coolants can be injected through the tool, to the point of greatest need. Photo Courtesy ChipBLASTER, Inc.*

machining action on his control screen (Fig. 18-19) as opposed to putting his face in harm's way. Here we see a split screen with program and distance to go screens plus the actual machining. This is useful where safety guards, flying chips, and coolants obscure visibility from the operator side of the machine.

Automated Chip Conveyors

High chip volume, coolant containment, and fast-moving machinery make it difficult to remove chips by hand. Automated conveyors are added to machining and turning centers to serve the purpose. A chip conveyor not

Figure 18-18 *Automatic Safety doors are found on many newer* lathes *and* mills.

Figure 18-20 *For high-volume machining, chips are conveyed away by automatic systems.*

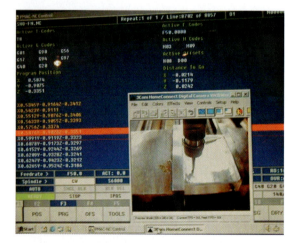

Figure 18-19 *The operator has chosen to split the screen with program data and video monitoring of the machining.*

only aids ongoing maintenance, but improves operator safety by making it unnecessary to reach into the machine to clean away chips. As shown in Fig. 18-20, the chips are sent to a cart that can be taken to recycle bins when it's convenient. Larger systems send their chips directly to dumpsters, to then be handled by forklift trucks as part of the factory's recycling program.

Gated Control for Peripheral Devices

The next two advancements aren't machine accessories, but rather, they are programmed machines themselves. They work in conjunction with one or more CNC machines. They load or inspect parts at appropriate times, using their own internal controllers and programs. When it's time for their assistance, the CNC control halts progress then signals them to proceed—it opens the gate for the go-ahead. Completing their task, they signal the CNC control, which then closes the gate switch to be certain of coordinated activity. The CNC control is the master control for gating.

Automated Part Loading (Robotic) Where work is heavy or the turnaround demand is constant and/or fast, part-loading robots can improve cycle times and reduce operator strain and boredom and improve safety. When placed in a cell of two or more machines as shown in Fig. 18-21, they can keep groups of machines working without human intervention.

Automated Part Inspection When in-process part quality is critical and inspection must be performed at special times within the PTP cycle, the **process control robot (PCR)** also greatly enhances PTP cycles and quality (Fig. 18-22). Results from the PCR are fed back to the control to enable automatic adjustments to cutter offsets to maintain tolerances. Cutter offsets are next and are discussed in detail in Section 22.

Figure 18-21 *This robot efficiently and safely moves heavy engine parts between CNC machines.*

Figure 18-22 *This production control robot, (PCR) automatically inspects critical operations during the cycle.*

Automated Cutter Offsets

This popular accessory has become a near-standard on both machining and turning centers. When they are first loaded into the machine, the exact cutter diameter and length must be entered into the control's tool memory.

While a program is written with a certain cutter length and diameter in mind, the actual tool needn't be an exact match, just close to the one for which the program was written. If the capable control is informed of the real cutter size and shape, it can compensate for the difference between programmed and actual size.

These differences are called tool offsets and loading them into the control is an operator responsibility. Using offsets saves time and money since any good, resharpened cutter can be used. The offsets must either be determined by hand measurement and then the data entered by hand, or they can be automatically determined by probes or optical sensors.

Automated Tool Grind Systems

Using either laser vision inspection and/or scheduled tool exchange, CNC controls are capable of checking tool sharpness, grinding tools, then putting the tools back into the tool drum. The laser can determine how much grinding will remove the dull edge and

it can then communicate that to the tool storage page to change the offsets for that tool. With just a bit of preplanning and set up, the machine needn't stop production! To do so, it's sometimes necessary to keep two tools in the drum depending on the cycle time.

Operator Warning Lights— Yellow Announcement and Red Warning

To aid operator monitoring, many CNC controls feature bright flashing lights placed high for visibility. If enabled, the yellow signals the need for operator intervention such as for a clamp change or the end of the program. This feature allows operators to manage more than one machine within a cell. The red light indicates something is seriously wrong such as low coolant, a program fault that cannot be processed, or a servo error.

Operator beacons (Fig. 18-23) are especially useful when one operator is monitoring more than one machine in a cell of several.

Tech-Specs for CNC Turning Centers

Since lathes had already evolved into very sophisticated automatic versions of turret lathes, multiple-spindle lathes, and screw machines, long before programmed machine tools came on the scene, the CNC version represents a lesser degree of evolution than

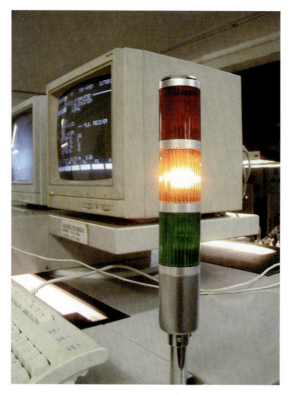

Figure 18-23 *Operator beacons warn and inform of major events.*

Figure 18-24 *Tool offsets can be determined by laser vision or touch probes.*

milling machines, up from their manually operated ancestors. Still, they are changing rapidly today. Many of the cycle time and safety improvements discussed for machining centers are standard or available on turning centers as well:

Chip removal conveyors

High-volume coolant systems with low-level warnings

Automatic safety door closure with interlock switches

Video monitoring systems on split screen controllers

Automatic cutter offsets where by touching the cutting tool to a sensing probe (Fig. 18-24) or passing it through a laser vision system, the control stores the size and position relative to the setup

Individual part-loading robots

Automatic Bar Feeding

For machining many parts from a single long bar of material, the **bar feeder** pays for itself, often within a month or two. As shown in Fig. 18-25, the bar feeder supports and safely contains a rapidly spinning bar, up to 20 ft long.

Upon parting off of the finished work into the part catcher (another automated acces-

sory), the bar feeder pushes the bar forward an exact amount to make another part.

Bar feed is accomplished in two different ways. On better feeders, the bar is located forward an exact distance by the feeder advancement. On less expensive versions, the bar feeds forward until touching a programmed stop mounted in the tool turret.

Nonstop Feed

If collets are used to hold the work, using both types of bar feeders, the spindle need not be stopped to advance the work unless it protrudes very far from the spindle where part whip would be a problem.

Through the program, without stopping the spindle, the collets open and the feeder advances the material. This ability allows long unattended operation, freeing the machinist to perform other duties such as writing future programs or burring parts.

Changing Acceleration Loads

Bar feeding is by far the more common way to automate long runs but it has one problem: if the job requires RPM changes, the entire bar

Figure 18-25 *A bar feeder (left side of machine) safely contains and advances the long bar.*

Figure 18-26 *A bar cutter loader takes the strain off the spindle motor by placing short cut parts in the chuck.*

must be accelerated and stopped. The answer is the bar cutter changer (Fig. 18-26).

Sometimes when the work is closely toleranced, the differences in spindle acceleration of a long bar that become continuously shorter with each part can make a difference in final shape, finish, and size. Speeding up and stopping long bars also loads the spindle motor. To help with that, some bar feeders offer part rotation drives to assist the spindle but even though the bar is rotating in an oil bath, all vibrations caused by the long bars in the feeder cannot be damped. The bar cutter is the answer.

Automatic bar cutter/part loaders The **bar cutter/loader** is an automatic accessory that is a saw and loading robot combination. The saw cuts the work into slices, then automatically loads them into the stopped spindle. The controller opens the chuck or collet, then sends a gate signal to the loader to pull the present part and load a new blank.

Automatic bar loaders If the production run requires many lengths of bar to be consumed, in both a bar feeder and a bar cutter, an automated **bar loader** is the answer (Fig. 18-27). It holds and inserts one bar at time into the bar feed system. Using this feature, it can be literal shifts before the machine requires human attention.

Multiple-tool turrets Most new CNC turning centers feature multiple-station tool turrets (one or two opposing turrets). Many different cutting tools can be mounted on the turret as well as individual coolant nozzles (Fig. 18-28).

Figure 18-27 *An automated part loading arm.*

Figure 18-28 *This standard tool turret can accept from 10 to 18 tools depending on size.*

Figure 18-29 *A programmable tailstock (left end) is needed when work is long.*

Programmable tailstock Many but not all turning centers use a programmable center support. The addition of a tailstock depends on the kinds of products being made (Fig. 18-29). If they are long, an autotailstock is a good addition.

Live milling and drilling tooling When specially equipped, some CNC lathes can spin mill cutters, drills, and reamers from the tool turret. This is accomplished using hydraulic powered motors. Figure 18-30 shows a lathe drilling a hole using **live tooling.**

Some newer lathes feature a third axis of motion (*Y*) for the turret. That allows milling of shapes off center to the workpiece. These

Figure 18-31 *This CNC turning center features a Y axis to allow curved mill cuts.*

machines can complete an amazing array of part shapes (Fig. 18-31).

Hydraulic Chucking

CNC turning centers include automatic collets and chucks that speed up PTP turn around time. Hydraulic powered chucking ensures fast, consistent grip from one part to the next when compared to manually tightened chucks. The amount of grip pressure is adjustable.

Trade Tip

The close-open-close chuck joggle trade tip learned in manual lathe training will be useful here too. When just placing the work in the three-jaw chuck and closing it, the work can dangerously be pinched off center between only two jaws. To avoid this problem, after gripping the work, the chuck is opened a bit then reclosed while jiggling the part. It will usually find the true center.

Safety Foot Pedal Switches For safety and convenience, hydraulic chucks are usually activated by a foot pedal, so the operator can use both hands to support the work. Using a combination toe lift plate, then a down-activated switch farther back beyond that, the chuck cannot be accidentally opened at the wrong time (Fig. 18-32). Safety features are also built into the program and control to ensure that the chuck cannot open during a cut.

Figure 18-30 *Live tooling accessories allow milling type cuts using lathes.*

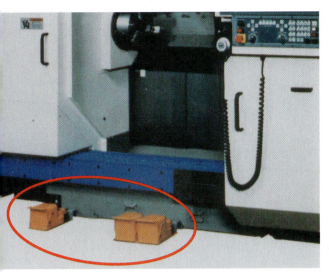

Figure 18-32 *Safety, chuck foot switches are interlocked through control gates and double trip foot action, to prevent accidental opening at the wrong time.*

Qualified Tooling

Toward the goal of more WIP time, the CNC lathe takes quick change tooling to new levels of efficiency and accuracy with **qualified tooling.** Improvements in machines and cutter tooling combine for faster cycle and setup times.

The turret is made such that sturdy tool holders can be quickly mounted and any replacement holder within the same system will position in exactly the same position relative to the work and the turret. Additionally, the tools themselves feature close tolerance, replaceable inserts. These features mean quick exchange of tool holders and cutting tip inserts with little disruption to production. The alignment pins of this turret are shown Fig. 18-33.

Tool Interference and Turret Capacity

Since lathes must hold in the turret all the tools required to finish a job, two issues arise that effect both program planning and program tryout. First, placing long tools such as drills and reamers next to each other (for example, T001 = Drill and T002 = Reamer) can create a situation where during drilling, the reamer also touches the workpiece, near its rim if the work's diameter is large. It's the machinist's job to test for these situations during program tryout. Be certain the turret withdraws far enough to rotate safely to the next position, then after indexing to the next tool, slow the rapid approach back to the

Figure 18-33 *Qualifying pins on the turret ensure cutter tooling can be replaced with minimum variation.*

workpiece, to be certain that adjacent tools do not touch the work.

The second issue arises when the program plan requires more cutting tools than can be loaded into the turret. Many cutting tools can be loaded into large turning centers (16 to 24, depending on turret size and the size of the cutting tools) since CNCs feature big turrets. However, on smaller turning centers, where capacity is limited, programmers can turn to multipurpose tools, such as shown in Fig. 18-34. They can face, turn, and

Figure 18-34 *Multipurpose cutting tools can accomplish several operations without indexing the turret. They reduce tool inventory and cycle times.*

part off, as shown, in making this control rod ball end. Multipurpose tools also reduce turret rotation and total cycle time, but they aren't as strong for heavy cuts compared to standard tooling. The decision to use multi-purpose tools is the program planners, and must be built into the program and tool numbers must be assigned.

UNIT 18-2 REVIEW

Review the Key Terms

Bar cutter/loader
A turning center accessory that saws bar-stock into blanks ready to load then turn.

Bar feeders
A turning center add-on that contains a long spinning bar and advances it when a new length is required.

Bar loaders
A turning center accessory that places new bars of turning stock into the bar feeder or bar cutter.

Chip-to-chip (CTC) cycle time
The time between cutting with one tool then a second tool.

Downtime
Nonproductive time when no work is being done.

Live tooling
A lathe tool that can spin, making milling and drilling operations possible on turning center jobs.

Pallet changer
A machining center with twin worksta-tions—each able to rotate into the work envelope while the other is outside being loaded with new parts to be machined.

Part-to-part cycle time
The time period between finished parts in a multipart run.

Process control robot (PCR)
An in-process inspection robot that can measure size during a program. It can also feedback data to the control to adjust for variation.

Qualified tooling
Cutters inserts and holders that can be re-placed with very little variation from one to the next.

Random tool storage
The ability to store an outgoing cutting tool into any tool drum position with its tool number and new position being tracked by the control. This improves CTC time.

Tombstone
A vertical square work-holding fixture of-ten used on palate changing machining centers. The tombstone allows access to three sides of the workpiece in a single setup.

Replay the Key Points

- *NC evolution* began because skilled ma-chinists using manual machinery, could no longer deliver advancing aerospace designs.
- *CNC machine evolution* toward faster more accurate production continues today.
- *CAM software evolution* now carries the torch of programs that machine creative new shapes impossible to machine oth-erwise.
- When loading tools into a machining center, there are two numbers that must be coordinated: *tool number* in the program and *tool position number* on the machine.
- Accidents equate to downtime as well as hurt workers and disabled machinery. CNC machines feature several improve-ments to avoid them.

Respond

1. When was the first true programmed machine tool invented? And who were the three partners in the consortium?
2. List as many improvements as you recall, without looking back, that evolved milling machines into CNC machining centers.

   Identify those that require computer direction.
3. Name three major differences between a CNC turning center and a manual lathe.
4. On a turning center, name the accessory that contains long bars and pushes them forward when another length of stock is required.
5. What was the original, main reason that programmed machine tools were invented? What drives machine evolution today?

Critical Thinking

6. Extending the time line forward, what advancements in CNC turning and machining centers might we see in the future for mills and lathes?

UNIT 18-3

Program Creation and Data Management

Introduction: This unit provides an overview of the way in which program data is created and managed. It will help to clear up some of the data management terms specific to CNC work. It's not about writing programs.

KEY TERMS

Bulk load
A version of DNC whereby large packages of program code are sent to the controller, then when these are nearly consumed, another package is sent down.

DNC (direct numerical control)
The host computer assumes the role of RAM sending program data to the control as it is needed. *Used mostly where CAM programs exceed the CNC control's RAM.*

Download
Sending a program from an offline PC to a CNC control.

Drip feed
A version of DNC whereby small packages of data are sent from PC to controller in just enough data for the present command and the next.

Hand load
Bringing a completed program to a CNC control via diskette.

Imported image
Bringing a part geometry file from a CAD system into a CAM system. Or bringing any computer file from one software to another.

Part geometry
The image of the part to be machined. Can be drawn online or imported from a CAD system.

Postprocessing
Converting a generic tool path program into a machine-readable version.

Tool path
A generic program used by the CAM computer on which a postprocessed machine-readable program is constructed.

Program Forms and Origins

Programs originate in one of three possible ways.

Figure 18-35 *Graphic interface (GI) programming at the CNC control allows the operator to create a program while the machine is cutting another part.*

1. **Handwritten Codes at a Keypad**
   When no CAM or graphic interface is available. Handy for setup, warm-up, and tooling routines.

2. **CAM Software—Graphic Driven**
   The program is created using CAD file.

3. **Graphic Interface Onboard the Machine Controller**
   The part shape is drawn on the control unit.

Graphic interface (GI) programming (Fig. 18-35) puts the responsibility for program creation in the hands of the machinist. It is usually performed on a control capable of multitasking so that the new program can be created while another is running. Otherwise the machine must sit idle while it is programmed during setup. The use of GI programming depends on management philosophy.

EIA/ISO Codes

There are two forms in which program data might be compiled to suit a specific control. By far the more common is the EIA/ISO code words we've discussed previously. A complete vocabulary is found in Appendix VI. There are exceptions.

ShopTalk

Not Perfect Standardization
While EIA language is a standard, there are differences in code application from control to control. From the time line, we saw that codes were invented for the NC era, before CNC. Typical of a technology explosion, the inventors didn't see how many words would eventually be required, so they assigned 100 general machining words (G0 through G99) and 100 miscellaneous utilities such as coolant on/off, or tool change words (M0 through M99).

But, of those original 100 words in each category, 50 were left unassigned as to their specific meaning. Those were left to the control manufacturers for custom functions. Today, however, using CNC, the control can read an unlimited number of code words. Additionally, there are a few slight differences in EIA words from ISO words.

(continued)

Conversational Language A few controls use conversational language specific to the unit. Conversational languages are effective and easy to learn but they are dedicated to one specific controller family. Once the language is mastered, the concepts transfer to other conversational controls but not the words. As such, EIA/ISO, *G code* language is the more common program method. (See the Shop Talk.)

KEYPOINT

> Within the first 100 G and M words, many will be specific to the machine. Words 100 and above are also machine specific.

Why Learn Codes at All?

You might ask what purpose learning the codes serves since most programs are created by CAM. The answer is that they are necessary for three critical reasons.

1. *To have the ability to read and control a program* generated by CAM. For example, if a tool breaks during a long program, on a machine that does not have mid-cycle halt capability, you must stop and replace the tool. Then you need to intersect the program at the same point or go back to the very beginning and run it from line 1. With the ability to read the CAM program, you can control the restart mid-program.

2. *The ability to write mini-programs* for tooling and setup functions or when a CAM system is not available.

3. *The ability to edit* graphically generated programs for maximum efficiency or to correct errors.

Three Types of Part Geometry

CAM and GI programming systems create the code (or conversational) words based on an accurate drawing of the intended job, called the **part geometry.** The CAM software will follow the drawing lines or surfaces to create the program codes.

Part geometry falls into three categories, depending on whether it is going to the lathe or mill and how complex the milled shape might be.

Flat Pattern Line Drawings

For simple parts, a flat pattern works fine for both lathe and mill work. Similar to a pencil drawing, this geometry is a profile, two-dimensional drawing composed of straight lines and arcs. Flat patterns are the only kind of drawing needed for *X-Z* lathe work (Fig. 18-36). Both the full and half part profiles are commonly used. Since lathe parts exhibit *Z* axis symmetry, all the dimensions needed are represented on one side of the centerline, and the half profile drawing is commonly used.

To employ a flat pattern for an *X-Y-Z* mill program, at the time of program creation the user must assign *Z* axis depth to various cuts (Fig. 18-37). In the two-view orthographic, the top view would be used to produce the program but each element would need to be assigned a *Z* axis depth as the elements are chained together. If that process became long, the best geometry would then be wireframe.

Typical Lathe Flat Pattern Drawings

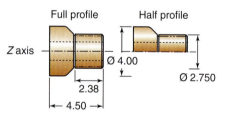

Figure 18-36 *Lathe part geometry can be two-dimensional.*

Figure 18-37 *The top view, if drawn correctly (per Fig. 18-41 coming up), can be used as a flat pattern geometry.*

Wireframe Geometry

Figure 18-38 *A wireframe consists of lines and arcs in X, Y, and Z dimensions.*

Wireframe Geometry

Used for mill work, a wireframe is still a set of lines, but they have depth, as shown in the perspective view of Fig. 18-38. If the geometry is in wireframe form, then the CAM system can assign *Z* axis cut depths automatically. Wireframes also help to show part features not easily visualized with a flat pattern drawing. But wireframes do not indicate surfaces, just lines. The user must still tell the CAM system to which side of the line the cutter must go. When shapes are complex, that becomes difficult. Then the next higher geometry in terms of usefulness for CAM programming is the solid model.

Solid Model

A solid model looks as though the part was made from a piece of clay. It's composed of computer-generated surfaces rather than lines. To create a solid part geometry, one begins with a large rectangle (or other billet shape) then unwanted pieces are removed just as though it was machined from raw stock. Shaped pieces can also be added to the geometry as it takes shape (Figs. 18-39 and 18-40). Solid model part geometry offers several advantages:

- Easy pictorial visualization of all part features. Note that features are not easy to see in the wireframe form.
- Near automatic programming capability on complex shapes with compound curves (software dependant).
- Complete control of every side of the object.

Program Creation

Importing Geometry

Once the geometry is drawn in the CAD, it's ready to **import** (bring the image into the CAM software as a readable file). It may need

Figure 18-39 *A solid model starts as a machined product would, with a billet of raw material.*

Figure 18-40 *Because it represents the finished product, the solid model can be viewed from any side.*

to have dimensions and other details that clutter the drawing removed. If the drawing was correctly drawn for CNC, those details will be on layers that can be removed or made transparent (masked) while the program is being created.

Next, the user begins the process of turning geometry into program codes. This procedure can be grouped into two general types.

KEYPOINT

We'll devote much training to actually writing CNC programs later, both by hand and by CAM. This material is an overview of how programs come into existence.

Section 18 CNC Systems ◆ **735**

Flat Pattern or Wireframe Element Selection

Starting with a flat pattern or wireframe of the geometry, individual elements are pointed out by the user using the mouse, in the sequence and direction they are to be machined. The programming software does not know they define a surface, it only sees the elements as lines and arcs, so the user must indicate to which side of the line the cutter must go.

Then, either before or after element sequencing, the user assigns machining data such as cutter size and shape and number of roughing and finishing passes to the sequence. Then with the elements linked together in a sequence, the software then uses the machining data to create a complete **tool path.** Or sometimes called an *element chain.*

Differences in Software

Adding Machining Data Before or After Path Selection Depending on the software, machining data such as cutter size and number of roughing passes are either added before the path is selected or some require the data to be entered after the chain is complete.

Pre- or Postcustomizing into Codes for a Given Control All programs must be custom translated into a specific machine's conventions. Control-to-control, there are differences in format as well as usage of the unassigned code words. The CAM software has a library of these control specifics. The various templates for each control are called *posts* (**postprocessing** *templates*) or sometimes *filters.*

When a CAM system has the ability to create the custom program for a given control, it is said to *support* it (write specific codes for a specific control).

Some CAM software requires that the specific template be selected before element chaining, while others postprocess the tool path for the machine after the path is selected.

Geometry Associative Programs When the design changes, so must the program. Most modern CAM software is able to automatically update the program if any element in the geometry is changed. Some less-sophisticated software requires the user to go back and rechain and process the program if any part of the geometry was changed.

Figure 18-41 *For programming purposes, CAD drawings must be accurately sized and have each intersection or tangent point trimmed perfectly.*

Geometry Drawing Accuracy To be of use for a CNC program, the person doing the drawing must ensure absolute accuracy of elements (Fig. 18-41). Lines must be drawn in the exact length desired on the work and line intersections must be trimmed exactly. Solid models can't simply look like the part, they must be an exact size and shape replica of the real part. If these rules aren't followed, the CAD drawing might look right to the eye, but it will not generate the right part in CNC. That accurate picture can be the original CAD drawing or it can be developed within the CAM system.

Sending Programs to the Controller

Data Transfer

Once a machine-readable program is ready, it is sent to the controller. This act is known as **downloading** (Fig. 18-42).

Active or Permanent Memory

The CNC controller stores programs in one of two locations: active memory ready to machine or permanent storage for later use. Permanent storage is where the downloaded program goes initially. The total capacity of a control to hold programs in both locations is an issue.

Downloading CNC Data

Figure 18-42 *Downloads send entire programs to the controller's permanent memory.*

Direct Numerical Control

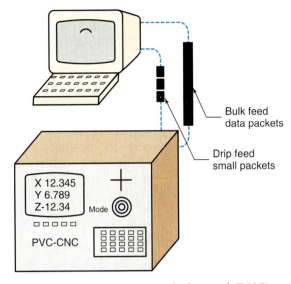

Figure 18-43 *Direct numerical control (DNC) sends packets to the control's active RAM just in time for execution.*

Transfer Types—Hand, Drip, and Bulk

The program can be transferred to the CNC memory in one of three ways:

1. **Standard Disk—Hand Loading**

   The program is copied to a standard computer diskette and hand carried to the controller. This is sometimes called the *sneaker net (using your shoes to transfer the data, or a foot-shake routine).*

2. **Direct Download**

   Through a communication wire, the PC host sends the entire program to the controller. The word "download" infers that the program is coming from a higher or master source. *Uploading* occurs in the reverse direction, program data is sent to the host or PC from the control.

3. **Direct Numerical Control (DNC)**

   The third download method is used when program data exceeds the memory capacity of the control. CAM-generated programs use many small lines to approximate a surface, thus programs for complex shapes become exceptionally large. The solution is to send it in packets—just in time to be used. There are two ways in which this direct numerical transfer can occur (Fig. 18-43).

   Drip Feed Transfer. The data is sent in small units just in time for the control to translate the codes and make the machining movements.

   Bulk Load Transfer. The data is sent in large packages that nearly fill the controller's RAM. As the data is consumed, the host computer sends the next package of data at the need time.

> **KEYPOINT**
> A download occurs when a CNC program is transferred from a master computer to a dedicated CNC controller.

> **KEYPOINT**
> DNC is used mostly for the times when the CNC control's RAM is inadequate to hold an entire CAM-generated program. Working together, the host supplies the data, and the controller continues to drive the machine.

UNIT 18-3 REVIEW

Review the Key Terms

Bulk load

A version of DNC whereby large packages of program code are sent to the controller, then when these are nearly consumed, another package is sent down.

DNC (direct numerical control)

The host computer assumes the role of RAM sending program data to the control as it is needed. *Used mostly where CAM programs exceed the CNC control's RAM.*

Download

Sending a program from an offline PC to a CNC control.

Drip feed

A version of DNC whereby small packages of data are sent from PC to controller in just enough data for the present command and the next.

Hand load

Bringing a completed program to a CNC control via diskette.

Imported image

Bringing a part geometry file from a CAD system into a CAM system. Or bringing any computer file from one software to another.

Part geometry

The image of the part to be machined. Can be drawn online or imported from a CAD system.

Postprocessing

Converting a generic tool path program into a machine-readable version.

Tool path

A generic program used by the CAM computer on which a postprocessed machine-readable program is constructed.

Replay the Key Points

◆ Within the first 100 G and M words, many will be specific to the machine. Words 100 and above are also machine specific.

◆ For a CAD drawing to be of use to the CAM system, it must be drawn to perfect accuracy and must be trimmed with perfect intersections between elements.

◆ A download occurs when a CNC program is transferred from a master computer to a dedicated CNC controller.

◆ A DNC occurs when a host computer feeds a program in small or big packets, to the control to use as needed.

Respond

1. Test your data vocabulary use. How many of these CNC data terms can you define in your own words?

Bulk load	Download	DNC (direct
Drip feed	Hand load	numerical
Offline	Tool path	control)
Taxonomy		Imported image
		Postprocessing

2. In your own words, describe the two ways CAM software creates a program using part geometry.

3. Define solid model part geometry.

4. Extending the terms you have thus far learned, can you describe a CNC *upload?*

5. Given the time line, who might be responsible for programs in the future?

Section 18 Review

Unit 18-1

We keep coming back to ball-screws. Other drive mechanisms have been tried for CNC axis work. The linear drive is one experiment that held some hope. A linear is somewhat like a stepper motor unwrapped into a straight line. Given a pulse of energy, the sliding component moves forward one magnetic section along the stationary driver. They are fast and accurate. However, the problem is there is no mechanical advantage as with screw drives. In order to develop the amount of thrust needed, the linear magnets must be huge and consume too much energy!

Unit 18-2

In addition to cutting metal, as we do on CNC lathes and mills, we'll soon see machines grinding in conjunction with cutting metal. The technical challenges to integrate them into classic machinery aren't great.

Many other operations are CNC driven and can be added to most machines, too. For example, laser cutting, waterjet cutting, electrical discharge eroding and deposition of metal using focused lasers and powered metal (Section 28). These and several other metal cutting and shaping processes are also CNC driven. Look for them to be added to the complete machining center sometime soon.

Unit 18-3

While the software war rages on as to the leading CAD and CAM system, we the machinists can only benefit. As software becomes more automatic in its ability to master programming, the responsibility may be returned to the machinist.

Questions and Problems

1. Which component in a complete CNC axis drive sends low-energy commands to make the servo motors move? (Obj. 18-1)

2. True or false? The axis encoder makes it possible for a CNC control to drive several different kinds of axis drive motors. If this statement is false, what makes it true? (Obj. 18-1)

3. What kind of axis drive motor is the most popular for heavy-duty CNC machines? (Obj. 18-1)

4. What condition can occur within a closed-loop feedback system when the program demand exceeds the ability to thrust the axis? With what result? (Obj. 18-1)

5. What factors and controls does the operator have to prevent servo errors? (Obj. 18-1)

Critical Thinking Questions

6. Why can't a stepper motor machine with no feedback go into servo error? (Obj. 18-1)

7. True or false? CNC machines lacking encoders must be homed when initialized. If this statement is false, what makes it true? (Obj. 18-1)

8. True or false? CNC machines with direct reading scales need not be homed when first powered up. If this statement is true—then why? (Obj. 18-1)

9. For what reason do manufacturers use stepper motors in CNC axis drives? (Obj. 18-1)

10. Why are CNC machines evolving today? (Obj. 18-2)

11. Why were NC machines invented originally? (Obj. 18-2)

12. Name the three separate branches of evolution that intertwine to improve CNC production. (Obj. 18-2)

13. Identify the feature that speeds up automatic tool changing by not requiring a given tool to be placed in a specific drum location. (Obj. 18-2)

14. When setting up a CNC mill, what two tool numbers must the control know initially? (Obj. 18-2)

15. Identify the kinds of part geometry, and describe each. (Obj. 18-3)

16. In order for a CAD drawing to be of use in CNC work what must be done? (Obj. 18-3)

17. Describe DNC data transfer. Why is it necessary? (Obj. 18-3)

18. On a sheet of paper, sketch the five CNC components in a closed-loop axis drive. Briefly describe each. (Obj. 18-1)

19. Which component in Question 18 is responsible for reporting position back to the control? (Obj. 18-1)

20. What is one main reason for DNC capability? (Obj. 18-3)

Section 18 Answers

ANSWERS 18-1

1. Per Fig. 18-1, your sketch has control unit, relay, servo motor, ball-screw, and feedback device (indirect or direct).
2. A. Direct feedback—Position signals provided by scales similar to manual digital positioners.
   B. Open-loop drive—Blind command, no feedback from the axis motion.
   C. A servo motor—Highly controllable motors producing consistent rotation and acceleration.
   D. Drive relay—A translation device that amplifies the control signal and configures the energy sent to the motor.
3. Ball-screws minimize backlash when screws are reversed by pushing both directions on the rolling balls in the split nut.
4. This statement can be true for some CNC machines that feature powerful axis drives. More commonly, CNC controls wait until the lag catches up to the command.
5. They lacked feedback.

ANSWERS 18-2

1. 1952; MIT Servo Lab, Parsons' Corporation, and the U.S. Airforce
2. Underlined answers require computer direction.
   <u>Automatic tool changing with random storage</u>
   High-volume coolant systems
   High-speed spindles
   <u>Automated safety equipment</u>
   Chip removal systems
   Warnings for low vital supplies (oil/air/coolant)
   <u>Part-loading options</u>
      <u>Pallets</u>
      <u>Tombstones</u>
      <u>Robotic part loading</u>
   <u>Automatic cutter offsets</u>
   <u>Programmable coolants</u>
   <u>Automatic tool grinding</u>
   <u>Scheduled tooling</u>
   Remote monitoring of machining
3. Any combination of the following: hydraulic chucking, automated safety doors, high-volume coolant systems, automated part catching—parted off items, programmable tailstocks, qualified tooling, remote monitoring of progress, bar feeders and loaders coordinated to program, automatic tool offsets
4. A bar feed(er)
5. Original: There had to be a better way. Making parts too difficult for manual methods. Today: More efficient part and machine turnaround cycles.
6. Machines will continue to become more efficient—they will be able to accomplish more operations in one setup, with less minutes in the cost of manufacturing.

1. See the Key Terms at the beginning of Unit 18-3.
2. *Element selection:* starting with a line drawing of the geometry, elements are selected and machining data provided; *surface selection:* starting with a solid model of the part geometry, the sequence of surfaces are indicated. The CAM system determines cut sequences automatically given desired conditions.

Answers to Section Review Questions

1. The controller (MCU)
2. False. The axis drive relay does this task.
3. Direct current servo motors
4. Servo error. The control cannot bring the real position of the cutter to the commanded position within a time limit. The machine stops and alarms the operator.
5. Sharp cutters, good coolant delivery, monitoring, and overriding cutting speeds and feeds
6. Open-loop circuitry has no kinetic sense; it never knows when it's out of sync.
7. False. Machines *with* encoders must be homed when initialized.
8. True. They wake up knowing where they are located.
9. Budget price
10. More efficiency—more WIP
11. To make products that skilled machinists using manual machines could not produce.
12. The machines, PCs, and CAD/CAM software
13. Random tool storage

3. Solid model part geometry—computer-generated define all sides of the object.
4. Uploading is the sending of data from a CNC control to the host CAM computer.
5. To eliminate man-hour costs, the engineer's solid model drawing can be sent to the machine with full multitasking CAM software. The machinist can guide the software to write the program.

14. Tool number, from the program, and tool location in the drum
15. Flat pattern, lines and arc in a single plane; wireframe, lines and arcs in X, Y, and Z; solid model, continuous surfaces
16. Features must be exact size and have intersections trimmed perfectly.
17. To send program data to the CNC control as it is needed. Because the program is larger than the control's RAM.
18. *Controller:* sends out low-energy axis drive signals; *axis relay:* translates control signal to motor commands; *drive motor:* drives spindle or axis ball-screws; DC servo, powerful continuous energy; stepper, less powerful pulses cause finite arc of rotation; *axis drive ball-screw; feedback device:* sends position signal back to control, direct reading, scales, indirect, encoders
19. Feedback device encoder or scale
20. Because a CAM-generated program is too big to fit the control's memory.

Section 19

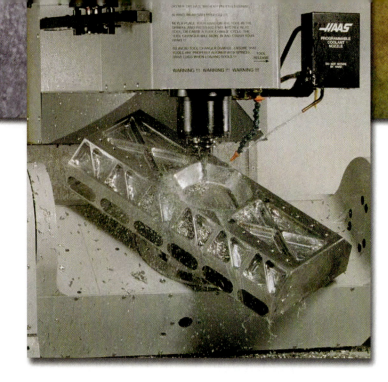

CNC Controls

Introduction

Before studying this section, you might elect to move to Section 20, then return afterward. The choice depends on the course outline and your career path.

Employed CNC Operator Sequence

Often in industry, one's career begins as a CNC operator, performing guided production work. Then gaining some experience you'll be instructed in setups, adjustments, and edits—increasing responsibility.

If you are now employed in a progression something like that, then the best choice might be to remain here in Section 19, followed by Section 20, then onward in the sequence presented.

Training Program Sequence

In most tech schools, with no pressure to produce parts, the learning sequence is reversed. We start with entry-level programming and simulation machining. These early, safe experiences increase the effectivity and safety of our classrooms. Later, you'll be guided through the first download and try out one of your programs. If that fits your needs, go to Section 20 and come back here afterward.

Objectives

Starting a CNC Machine

Introduction: Depending on the sophistication of the control and the machine's complexity, cold starting it may be simple. You'll need to check vital supplies such as air pressure, lubes, and coolants, then be sure safety guards and emergency buttons are in the operate position. Last, with the main breaker switch on, touch a green button. The rest of the procedure is automatic.

These higher-level industrial machines wake up knowing their axis locations, which tool number is in the active position, and where the others are in the tool changer or turret. They load their operating system software automatically. In just a few moments, green lights and a video message indicate the machine is ready to exercise its axes and spindle bearings before it's put to work.

On the other hand, lower-level machines will require *initialization* (a formal start-up procedure). It may be complex, requiring several steps, in a defined order, or it might be semiautomatic. Accuracy and safety will depend on performing all steps correctly. Many controls that require a start-up procedure won't allow full use or program execution until the initialization is done right.

KEYPOINT

Although older CNC machines are less sophisticated in their operating systems, they are accurate, efficient metal removers. They are well distributed in the real world and the chances are good that one might be your first machine!

KEY TERMS

E-stop button
The big red button that depowers all machine functions. May require several actions to restart after touching this button.

Initialize (initialization)
A process through which some CNC machines are brought up to full operational readiness. May be semiautomatic or may need intervention.

Lockout key switch
A multilevel switch that with the key removed, allows or disallows certain control functions and authorities.

Midcycle stop (midauto halt)
A way to quickly stop a program run without loosing physical position or program data.

Overt (action)
A deliberate and often difficult move to ensure it isn't done by accident.

RIS (reduced instruction set) controller
Reduced instruction set controls using icons and double function buttons to create a super-compact control pad.

Soft limits
A switch that halts or slows rapid axis motion, before reaching hard limit switches, to prevent accidental overtravel of axes into full mechanical limits.

Which CNC Control?

When an author writes a general CNC text such as this, there's a conflict of which control panel to use for examples. Should it be the best/latest control? (Nearly impossible to choose, and it changes quickly.) The most widely distributed in industry might be the better choice. (That's easy, it's a Fanuc® control, but there are many models and levels plus not every school has them.) Finally, the most affordable CNC trainer could be the best example, since it's likely to be in many tech schools. You see the problem? To solve it, we invented our own generic controller (Fig. 19-1); introducing the plain vanilla control (PVC) unit!

The controls found on the PVC are universal. They may have a different name or be actuated differently on a given unit, but they'll be there in some form because they're basic to setup and operation of the machine. Throughout the CNC chapters of this book, we'll refer to this base control.

Kinds of Control Units

CNC control panels vary widely. Here are the three most common examples.

Standard View Screen, Dials, Keys, and Switches Figure 19-2 shows a group of typical CNC controllers. They feature a vid-screen, control dials, switches, buttons, and indicator lights to show conditions and modes. Representing the largest share of control types, these units are made specifically for industrial machines.

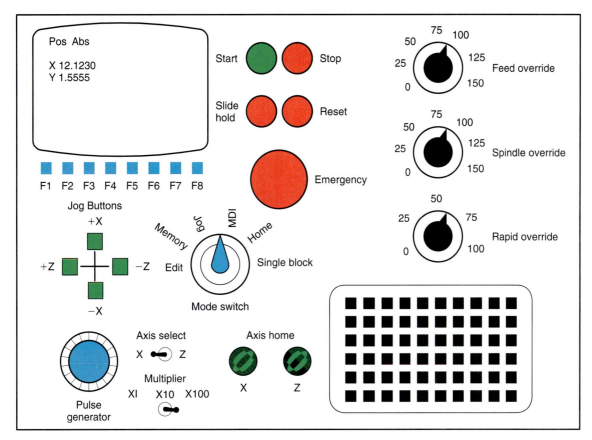

Figure 19-1 *A generic CNC control panel.*

Figure 19-2a, b, c *Bridgeport, Fanuc, and Hass control panels.*

Figure 19-3 *A RIS (reduced instruction set) control performs all operation functions but not program editing.*

RIS Controls

Figure 19-3 shows a graphic **RIS** (*reduced instruction set*) **controller** touch pad, sometimes called an operator pendant. While it takes a bit of training to know what all the symbols mean, once learned, these highly compressed control panels work with simple beauty. The pad can be taken in hand to any viewpoint around the machine. Some are even wireless!

RIS units are dedicated to setup and operation modes only. To keep the panel palm sized, they have no keypad. Editing and program writing tasks are performed on an interfaced PC located nearby. Some RIS panels feature a very limited key set similar to a scientific calculator.

KEYPOINT

On many industrial controls, a handy portable pendant with just the manual setup functions is an accessory added to the main control unit. This allows easy viewing of dial test indicators and other alignment items not easily seen from outside the work envelope. These pendants are not RIS controls, they cannot manipulate setups or access data. Pendants are for setup motions only (Figs. 19-4 and 19-5).

Operator pendant

Figure 19-4 *An operator pendant is often added to high end industrial controls.*

Figure 19-5 *A pendant provides manual axis movement from any view perspective.*

PC-Based Controls

Finally (Fig. 19-6), there is a controller based solely on a standard PC. Following the axis drive discussion in Section 18, you understand that any fast processor and hard drive could store and read a program then send a signal to the axis drive relays. Many machine manufactures offer PC-based systems as an option.

Familiar Operating Systems Using everyday operating systems such as Windows® or Unix®, PC-based CNC controls offer

Huge hard drive capacities to hold long programs.

Standard hard drive and computer components that bring interchangeability and easy repair when necessary.

CAD/CAM that can be performed at the CNC machine if the software is loaded.

Part inventory using bar code scans.

General shop communications such as e-mail, work order, or engineering drawing downloads.

The ability to perform statistical process control charting and all sorts of other tasks right at the CNC machine.

Most PC-based controls include an operator pendant to assist with setup and program-run functions.

Lockout Switches

To protect machines, setups, programs, and people, many CNC controls feature a locking switch that's similar to setting a user access limit on a PC. With the switch set and the key removed, various levels of responsibility are allowed while others cannot be accessed. Figure 19-7 shows such a **lockout key switch.**

1. **Operate** (lowest responsibility level)
   Allows halt, start, stop, and speed/feed functions and *disables* edit and set up information changes.

2. **Setup** (medium level)
   Allows operate functions plus process changes such as tool diameter, length, or PRZ location and *disables* program edit and load capability.

3. **Edit** (full control responsibility level)
   Allows access to all functions, including changing setup information, writing new programs, and changing existing programs in storage.

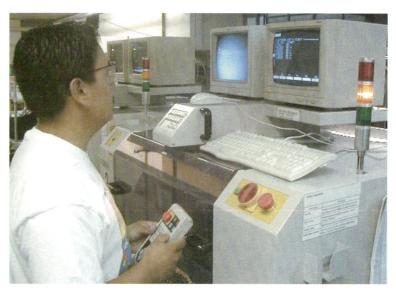

Figure 19-6 *A CNC machine with PC control.*

Figure 19-7 *The key-operated protect switch on a control allows various levels of operator responsibility.*

Starting a CNC Control and Machine

Prechecks

Many machines will only come up partially then halt if the correct actions or necessary supplies are low or missing. There are three or four "under the hood " checks to be performed before powering up any machine.

1. **Air Pressure**
   Many but not all machines require clean compressed air. If the machine does, it must be supplied at the required pressure, 50 PSI or higher. It's found

ShopTalk

At the full-edit level, given the entry commands to open the door, the control's operating system can be edited or erased! Here you might change items such as the safety door interlock setting such that the machine will operate with the guard open, or move the **soft limits** to a slightly different location, or change how the tool changer sorts tools or rotates. *Never attempt these things without guidance and permission since operator safety can be compromised or a total control wipeout is possible!*

on machines that blast tool holders to clear chips during tool changing or to power lubrication systems or to activate clamps and other auxiliary functions.

2. **Lube Oil**

This is required by *all* machines, but it is not always an autosensed function. Remember, there will be more than one product: spindle oil and way lube. In general, newer machines will not operate if the lube reservoirs are low. Many retrofit (manual machines made into CNC) or older model CNCs might run without lube oil, damaging the machine.

3. **Hydraulic Oil**

Not all machines require this, but usually lathes with hydrochucking do. Check the oil level before initialization, and after the machine is up and running on a schedule, *check* that the correct pressure is set for the chuck. There will be an adjustable dial on the back or side of the turning center with a gage. Check the setup document or ask the lead machinist for the correct pressure for the specific job. Too much chuck pressure mars or even crushes work and can bend tooling; too little holds the work weakly and dangerously.

4. **Low Coolant**

Nearly all machines require coolant, but it may or may not cause an alarm if it's low. It can be a real disaster if coolant runs out during a heavy cut. Many controls sense when coolant is at the add mark, but will not stop the machine until a current cycle completes.

The Major Start-Up Controls

Next to bring the control up to a run condition, check the emergency-stop buttons and put them in the run position (Fig. 19-8).

Emergency-Stop Buttons

Rotating collar   Rotating button

Figure 19-8 *The largest red button is for final shut down and emergencies.*

Initialization Checks

Read the specifics from the operator's manual before you **initialize** the machine. You might need to

Index tool turrets or tool changer drums to position 1.

Close chucks and safety doors.

Start hydraulics at a specific time—before starting controls.

Home the axis drives to reset the position registers.

Coordinate the cutting tool inventory.

Plunger Safety Switches—The Big Red Button

As seen in Fig. 19-8, these buttons are universally used as an emergency safety stop, **E-stop.** When depressed during normal power-down or in a real emergency, the button stays locked off until the base collar or the button itself is hand rotated to unlock it. This **overt** (deliberate) **action** creates a reminder of clearing the problem, if one existed, before running the machine. Depressing *Ctrl-Alt-Del* on your PC is an example of an overt action.

KEYPOINT

> Depending on axis feedback and control sophistication, depressing the E-stop during a program run can have an effect on the parts, setup, and program operation. Other than normal power-down, it is only used when, in your best judgment, an emergency stop is the only right action to take.

Never use the E-stop button midcycle unless all other halt actions would be too slow to be safe. In addition to stopping all axis moves, this button stops the spindle, coolant, and any other auxiliary functions being used. On older CNC machines with encoders, the PRZ location is lost because this action removes power to the feedback system, resulting in a lost register count. Later we'll learn several halt and **midcycle stop** techniques, one of which will be a better choice for situations that could become emergencies but aren't yet.

KEYPOINT

> E-stop is for all-out emergencies. It may disable all safety brakes, allowing the spindle to coast to a stop.

Multiple E-Stop Locations

For quick access on larger machines, there might be more than one E-stop button. Machining centers might feature one on the front panel and one at the tool turret. Lathe operators might find one at the bar feeder and one by the coolant/chip system as well as the one on the main control panel. An E-stop is often part of the pendant too. If any single switch is depressed, the machine won't start.

KEYPOINT

Safety

As an operator, it's a good idea to review the location of all E-stops occasionally so you can react quickly in an emergency.

If, during machine use, the E-Stop button is hit, you must reinitialize the machine, repower the axis drives, then manually jog (see the key terms in Unit 19-2) the cutter away from the work or danger area, then start the program over.

Trade Tip

Here's Your Wakeup Call! In extra precision situations, you will find that machined sizes sometimes change as the machine warms up, especially after being down for an extended time. Metal expansion within its frame and ways and oil viscosity can be such a critical issue that some shops prefer to leave the machine on standby to idle overnight rather than turn them off.

When starting a cold machine, it's a good idea to exercise the axes and spindle, especially after long down periods. This entails moving the machine to the soft limits for each axis—in both directions, turning the spindle on and off several times and letting it idle for 3 minutes or so at a moderate RPM. That action ensures all moving or sliding and rotating surfaces receive an equal coat of lubricant and the bearings are warmed up slightly.

While the warm-up can be performed manually, try this hint to save some time. Write one or more custom "wake-up programs" that automatically put the machine through the muscle flexing. Different routines can account for setup interference (one that uses no Z axis move, for example, and one that does). Then, the control can put itself through the warm-up while you read the tie-in book, put on your apron, and get tools out. Save these miniprograms in memory and call them up when needed. In Section 23, we'll learn to use a program logic technique called a loop to speed up these programs. Don't miss it!

UNIT 19-1 REVIEW

Review the Key Terms

E-stop button
The big red button that depowers all machine functions. May require several actions to restart after touching this button.

Initialize (initialization)
A process through which some CNC machines are brought up to full operational readiness. May be semiautomatic or may need intervention.

Lockout key switch
A multilevel switch that with the key removed, allows or disallows certain control functions and authorities.

Midcycle stop (midauto halt)
A way to quickly stop a program run without loosing physical position or program data.

Overt (action)
A deliberate and often difficult move to ensure it isn't done by accident.

RIS (reduced instruction set) controller
Reduced instruction set controls using icons and double function buttons to create a supercompact control pad.

Soft limits
A switch that halts or slows rapid axis motion, before reaching hard limit switches, to prevent accidental overtravel of axes into full mechanical limits.

Replay the Key Points

◆ Know and review often where the emergency-stop buttons are located.

◆ Depressing the E-stop button during a production run nearly always has a marked effect on setup and program operation.

◆ To start certain machines index tool turrets to position 1, close chucks and safety doors, start hydraulics at a specific time—before starting controls, home the axis drives to reset the position registers, and coordinate the cutting tool inventory.

Respond

1. Describe why some CNC machines must be homed before they can be used to cut material.

2. _____ a machine may be fully automatic, requiring a single button touch, or it may require several steps in a given order.

3. Why does the E-stop button remain depressed when it is touched?

4. You need to change a tool diameter in a vertical mill setup. You *cannot* do that from which modes?

5. How is an operator pendant different from an RIS controller?

UNIT 19-2

Controls for Setup and Operation

Introduction: OK, the machine is up and ready. Unit 19-2 explains the five basic controls that all operators must know to load a program, set up the machine, and get the whole thing going. Again, they may be slightly different from machine to machine, but each must be present in some form.

For example, on a given control, the three individual override controls (spindle RPM, feed, and rapid travel) discussed here are found on a single dial with the operator selecting which action is to be controlled.

KEY TERMS

Distance-to-go screen
A video display of the amount of travel yet to occur to complete the current command.

Hard keys
Dedicated entry keys on the control panel.

Hard limit
The emergency end of axis travel.

Jog
A manual movement of a CNC axis, instituted by the operator.

Manual pulse generator (MPG)
A rotating dial that has many finite positions (clicks) representing one jump of the axis selected.

Multiplier Controls (rate control)
A three- or four-position switch that sets the amount of jog feed per jump or pulse.

Override
A dial control to fine-tune feed rate, rapid travel, and spindle PRM.

Slide hold
The little red button that halts all axis motion with no consequence to other functions or the program.

Soft keys
Control entry keys positioned just below the video screen that assume different functions depending on the mode and page selected.

Soft limits
The switch that stops the normal range of axis travel. There is no consequence for touching this switch other than axis stoppage.

Mode Selection

Operation modes are used to simplify and make tasks more efficient and safe by putting the functions needed at the forefront. Each mode allows some functions while banning others. Usually, this prime control is a rotary switch as in Fig. 19-9, but it can also be a screen selection. Selecting the mode of operation (what tasks are to be done) is usually the first action to take.

Depending on the design of the control these universal tasks may be arranged slightly different than the illustration.

Possible Modes

Home Home provides access to the automatic homing cycle or to manual performance. Home may or may not be present on newer machines with constant positioning capability.

Jog A **jog** allows the operator access to manual motion functions discussed next. For safety, this mode disallows automatic program functions as the operator is usually observing indicators, tapping vises, and generally within the work envelope.

CNC This is the program run selection. May also be called *auto*. When executing a full program run, the mode will be set here.

Single Step (Single Block) This operates one command line at a time then waits for another touch of the (green) cycle start button.

Mode Switch

Figure 19-9 *A rotary mode selection switch is common on a CNC control.*

A single step is used mostly during program tryout (Section 20).

Manual Data Input (MDI) MDI allows one or more command lines to be written at the keypad. They can then be executed, if so desired, using the cycle start button.

Edit This mode enables full program memory and storage access except to the program currently in the active buffer (the one being used at the time).

Memory This mode is for file management, calling a program from permanent storage, or performing a download or upload from a host PC. Memory is also used by some controls when writing a program from the keypad.

Hard Keys Versus Soft Keys

There are two types of entry keys on a control panel (Fig. 19-10).

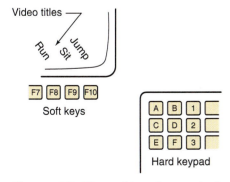

Figure 19-10 *Hard keys have a single or possibly a dual purpose, soft keys have multiple purposes based on the mode and video screen selected.*

Hard Keys These are normal keys associated with computers. **Hard keys** may be plastic (as on your PC), or they may be membrane keys (as on some cell phones) under a sealed layer to protect them from coolants and chips. Each hard key has a dedicated primary purpose and possibly a selectable second or even third function similar to those on calculators.

Soft Keys Depending on mode, the selection of **soft keys** expands the possible uses of function keys at the top of the keypad. They work like the F-1 through F-9 or F-12 keys on your PC; by adding video titles above them as a mode or screen page is selected, they assume new roles.

Manual Movement of CNC Axes

Homing Axes

During initialization, older machines or open-loop systems of any age, require a homing cycle to zero base their axis system. However, newer CNC machines generally do not. When initialized, these machines come online automatically recognizing their axis positions and tool locations. However, there are still various reasons to home almost all CNC machines, for setup and tooling purposes. Here's how: After selecting the home mode, the operator drives all axes to their home position, either manually using controls, or with an automatic cycle, but only after the cutter is jogged clear of setup obstacles. Then the axis is slowly driven upon the hard limits and the position registers are set to zero. Doing so requires an understanding of limit switches.

Limit Switches

The work envelope of a CNC machine is protected and contained by limit switches that stop the axis drives before hitting the physical end of travel. Damage would occur if an axis were driven onto its full mechanical limit—the end of the ball-screw. There are two kinds of limit switches you must understand.

Soft Limits (Normal Operation End)

Figure 19-11 depicts a simplified version of a **soft limit.** It's an optical cam device that signals the control to slow down then stop. On most CNC machines, the soft limit switch

Soft Limits

Figure 19-11 *A simplified soft limit stops axis movement with no consequence to the program or setup.*

can be contacted at full rapid travel speed with no consequence to program or position due to its deceleration function. The axis will simply stop traveling fast when the soft limit is reached.

KEYPOINT

Once an axis is on the soft limit, a command in the reverse direction requires no special prefix or action. Soft limits are the boundaries of the normal work envelope seen in Fig. 19-11.

Trade Tip

Parking on the Soft Limits A programming trick is often used when parking a machine for a part change. The programmer or operator commands the *Z* axis to rapid away 100 in. An easy figure to write, but most machines haven't that much *Z* axis envelope. The objective is to simply drive the machine upon the soft limits, where it stops.

When machining extremely heavy loads, on older indirect feedback machines, it may be necessary to refresh the coordination of the axis registers due to mechanical slippage and mathematical rounding in the control during rough cuts. After roughing is done, add a homing command into the program, then do the finish cuts with refreshed axis coordination.

KEYPOINT

Caution!
Do not try this without determining if the machine features dual soft/hard switches. Hitting hard limits at rapid speed is a crash!

Hard Limits (Crash Stops and/or Axis Refresh)

A **hard limit** is a physical contact switch beyond the soft limits, but before the full mechanical limit of the axis travel. This switch is the absolute end of powered axis travel. The hard limit is a never-changing position used to zero axis registers. Contacting the hard limit during feed or rapid movement stops axis motion and puts the machine in an alarm condition.

KEYPOINT

Hitting a hard limit during program operation is considered a minor crash.

Backing Off Hard Limits Once on the hard limits, the operator must make some overt action such as touching the jog and slow buttons at the same time, or the reset and jog together, to allow the machine to back off the limit toward the open work envelope again. The action will be different depending on the control, but it ensures that the axis is moved slowly and correctly.

Manual Axis Setup Movements

There are four ways for moving machine axes for setup work, for homing them, and for driving a cutter (with no program) through material to machine a fixture or chuck jaws. One or two methods must be part of every control.

KEYPOINT

Warning
The force behind the axis is great and a very clear understanding of the following controls is necessary before attempting to use them.

1. **Continuous Jog**
   This is a steady axis movement as long as the jog button is depressed (Fig. 19-12). The rapid feed override will select the speed at which the axis moves (discussed later).

2. **Incremental Jog**
   The machine will jump a given amount with a single push of the button at the speed selected on the override. A selector switch, called a **multiplier** (Fig. 19-13) is used to select the distance covered in one

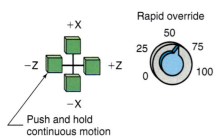

Jog Buttons

+X

Rapid override
50
25   75
0   100

−Z   +Z

−X

Push and hold
continuous motion

Figure 19-12 Continuous jog *moves the axis selected while the button is depressed. The rapid override is used to determine the speed at which the axis moves.*

Select axis

X1 X10 X100   +   −   X Y Z

Multiplier   Manual pulse generator

Figure 19-13 *The* resolution multiplier *applies to either* incremental jog *or to* handwheel (MPG) *movements. For either movement, the axis and multiple of the resolution must be selected.*

jump. For example, if the machines resolution is 0.0001 in., then selecting

Multiples of the resolution
X1 = 0.0001 for each push
X10 = 0.001
X100 = 0.01
X1,000 = 0.1

Incremental jog is used during setups to move the machine an exact distance in multiples of the machine's resolution. If the control has this feature, the same buttons are often used for both continuous and incremental jog. Not all CNC controls feature incremental jog.

Manual Pulse Generators

A common method of moving an axis, this control might also be called a "handwheel." As the relatively large dial is rotated by hand, small clicks can be felt. Each represents one jump of the selected axis. The **manual pulse generator (MPG)** is a handy combination of incremental and/or continuous motions. If the dial is rotated continuously, a fast motion is obtained, but one jump is also possible by turning it one click. The jump amount is determined from the multiplier switch and the speed by the override dial.

In the illustration, the *X* axis will move 100 times the resolution (normally 0.0001 in.) per pulse = 0.010 in. Rotating the dial counterclockwise produces positive *X* motion. On machining centers, to aid in setups, the MPG is often placed on the pendant to be taken into and around the machine to improve viewing angles.

Trade Tip

MDI Axis Movement (manual data input) Here's a handy fourth method of moving one or more CNC axes a given exact distance. It's used by hand writing one-time-only, miniprograms at the controller. Full edit capability must be unlocked in order to use it!

After selecting the correct mode, the operator keys the needed commands. Then with a push of the start button, the machine will obey that single command block, once only.

All jog actions described thus far cause single axis, straight-line moves, while this method expands the possibilities. Using an MDI statement, any motion within the capability of the control is possible: multiaxis linear and circular.

Machine Halt Controls

Making the Right Halt Choice

When an emergency happens, the need for prepracticed action is critical. Many crashes get worse due to inaction or worse, the wrong action. Practice is the real trade tip. Make sure to know your stop and halt options. It's important to be prepared to make a quick decision between the big and little red button. Remember, the big button ends the program and possibly disturbs the setup PRZ. However, never hesitate to hit it if safety is an issue. Reviewing the possibilities for stopping and halting, from time to time, is a good idea.

In Section 20, we'll investigate your operator actions when stopping the machine in

KEYPOINT

There is a difference in stopping and halting. Halting implies that the action is temporary and normal: to measure a key dimension, to move a clamp, or to add coolant or brush away clogged chips. After halting, machining will resume. Halting is an operator responsibility and the emergency-stop button shown in Fig. 19-8 is *not used in this capacity.*

Start

Slide
hold

Emergency
stop

Figure 19-14 *The three main operator control buttons are color coded: start (green), slide hold (red), and emergency stop (red).*

situations ranging from in-process adjustments to an all out emergency. Here in Section 19, we'll discuss various controls used to halt the machine during a program.

Here are the various methods of halting program actions.

1. **Depress *Slide Hold***

   This control may also be called *feed hold*. It is present on most CNC controls (Fig. 19-14). The red slide hold button is smaller than the E-stop button. It immediately halts all axis motion with no other consequence to the program. Spindle and coolant and any other secondary functions such as clamps, lights, and so on, continue as programmed or set from MDI. With a second touch of this button or possibly the cycle start button (depending on control), the program resumes normal motion.

Trade Tip

Halting While there are no data, setup, and/or program consequences to halting, there may be physical machining effects. With the spindle continuing to spin, the cutting tool might be dwelling on the work. This can lead to work hardening, dull cutters, or unwanted marks in the material. Most controls allow a spindle stop during a slide hold.

2. **Turn the *Feed Rate Override* to *Zero***

   Here the halt is made by reducing feed during a cut to nothing. Because the spindle or milling cutter will continue to rotate with the cutter possibly in contact with the work, this method is not recommended. But small adjustments such as aiming the coolant can be quickly made in this way. See rapid override for another method of stopping the machine for small adjustments.

3. **Switch the *Mode Control* to *Single Block* or *Cycle Stop***

   The machine will finish the current command then halt. The cycle start must be pressed to resume activity. Leaving the control mode in single block, it is possible to allow one command at a time using the start button, until the cutter is clear of the work then fix the problem.

Trade Tip

Program Interrupt Many controls feature the useful ability to halt machining then take further axis actions such as moving a dull cutter off the work or stopping the spindle then replacing it. Using program interrupt allows all jog and manual motion actions plus other functions normally disallowed during slide hold.

Then, with the problem resolved, with a touch of the green button, the control will reset all machining and logic conditions within the program, then slowly move the cutter back to the exact position where the interrupt took place. It will then continue normally.

Video Screen Selections

Depending on mode choices and tasks the video screen can be set to display a wide range of information. Most allow operator selection of split screens with more than one selection (shown in Fig. 18-19). Here are the four basic screens often selected for setup and operation.

Tool Path Graphics Screens

This selection (Fig. 19-15) can be used in three very different ways and is often selected for one half of a split screen:

1. **Program Monitoring**

   This process tracks the position of a cutting tool using a picture of the progress. When coolant, chips, and safety shields obscure the machining, this is one way of tracking progress.

2. **Program Dry Run**

   The first test of a program can be made on the video screen with all axes locked out—using the slide hold or axis lock switches. The graphics will provide a rough picture of the program path with no material being machined.

3. **Graphic Programming**

   An option on more advanced controls, the program is developed by drawing

Figure 19-15 *When you cannot see the cutting tool, monitor progress with screen graphics.*

elements on the screen similar to CAD drafting. This option will be learned in advanced training.

Coordinate Position Screens

There are three kinds of position screens.

KEYPOINT

Verify a clear understanding of the following position screens. This knowledge is absolutely necessary for efficient, safe setup and operation.

1. **Machine Coordinates Screen (Mach Coords)**

   This screen shows the position of the spindle or tool with respect to machine home independent of any program reference position that's been set. Machine coordinates are mostly used for setup functions, locating the PRZ, and aligning holding tools. The mach coords screen is available even on machines that do not require a homing operation. Machine coordinates are used to check major items such as clearance to tail stocks on lathes or approaching full travel range of an axis. It is unlikely that you will use machine coordinates to monitor a program in progress because the distances refer to the machine position rather than the tool to work position.

2. **Work Coordinate Position Screen (Work Coords)**

   This is the screen most likely to be displayed during a program run. The current position of the tool is shown relative to the selected PRZ. This, along with the **distance-to-go** (DTG) screens, are the most common screens.

3. **Distance-to-Go Screen**

   The amount of travel left to complete the current command line. This screen is extra useful when trying out a new program. Used in conjunction with the single block mode and feed and rapid override controls, a block of commands can be read and started, then halted at a questionable position. Then with the remaining distance to go displayed, the operator can physically measure the gap between the tool and the chuck, for example.

Split and Combined Screens

Various combinations of these screens may be selected by the operator. For example, during a proven program run, operators will often display machine and DTG and work coordinates along with graphic progress.

For example, a program is being tried out for the first time. On the left of the screen, the tool has been commanded forward 2.000 in. in the Z axis, toward the chuck. To check the command for safety, the operator halts the tool motion and selects distance to go on the right side of the screen.

Halting the axis and spindle motion, the screen shows 1.575 in. remains to complete the block. Then using a ruler, the machinist measures the gap between tool and chuck to determine that more than 1.575 in. exists before restarting the spindle and axis drives, allowing completion of the command.

Override Controls

There are three functions on a CNC control panel that allow fine adjustment to spindle RPM and Axis Movement Rates. We've already discussed the rapid **override** which adjusts the jog rates and also the rapid travel rate during a program.

Notice that the rapid override can be set all the way to *zero* (a trade trick method of stopping a CNC machine when it encounters the next rapid movement command in the program). Turning the override to zero, the machine will complete the current feed rate commands until it encounters a rapid travel command, at which point it stops moving

KEYPOINT

Caution
In Fig. 19-16, stop the spindle before attempting to measure physical distance to go.

Split Control Screen

Command Line	Dist to Go
	X 00.000
G01 Z-2.0000	Z-1.5750

Linear movement of *Z*-2.00 in.

Distance to complete the current command

Actual tool position

Does more distance than 1.575 in. exist?

Figure 19-16 *This CNC lathe has been commanded to move 2.0 in. minus along the Z axis—left screen split. The right screen shows 1.575 in. remains to complete the command.*

Figure 19-17 *The rapid travel override control is used to slow programmed rapid moves and manual job commands.*

Figure 19-18 *The spindle speed override can reduce RPM from a minimum of 5 percent to 10 percent of the programmed rate or increase it to 150 percent.*

KEYPOINT

Rapid Override
While the fastest possible rapid moves are desirable to trim cycle times, the rapid speed is slowed during a first tryout of a program.

until the override is turned up again (Fig. 19-17).

Spindle Speed Override

This control works during manual use of the spindle and during programs (Fig. 19-18). Notice that it not only decreases RPM below the programmed rate but it can increase it from 150% to 200% above that which has been entered.

Feed Rate Override

This dial works similar to the spindle RPM control. It can adjust down to *zero* and up to *150 percent* of that which has been programmed (unless the machine is at its limit feed rate already).

UNIT 19-2 REVIEW

Review the Key Terms

Distance-to-go screen
A video display of the amount of travel yet to occur to complete the current command.

Hard keys
Dedicated entry keys on the control panel.

Hard limit
The emergency end of axis travel.

Jog
A manual movement of a CNC axis, instituted by the operator.

Manual pulse generator (MPG)
A rotating dial that has many finite positions (clicks) representing one jump of the axis selected.

Multiplier controls (rate control)
A three- or four-position switch that sets the amount of jog feed per jump or pulse.

Override
A dial control to fine-tune feed rate, rapid travel, and spindle PRM.

Slide hold
The little red button that halts all axis motion with no consequence to other functions or the program.

Soft keys
Control entry keys positioned just below the video screen that assume different functions depending on the mode and page selected.

Soft limits
The switch that stops the normal range of axis travel. There is no consequence for touching this switch other than axis stoppage.

Replay the Key Points

◆ On some controls, edit mode is used when creating a new program from the control keypad. On others, program entry is done in MDI mode.

◆ Once an axis is on the soft limit, a command in the reverse direction requires no special prefix or action.

◆ Soft limits are the boundaries of the normal work envelope seen in Fig. 19-11.

◆ *Caution!* Do not try moving an axis to the limits without determining if the machine features both soft and hard limits. Hitting hard limits without the deceleration of soft limits, at rapid speed is a crash!

◆ The force behind the axis drive is great and a very clear understanding of the controls is necessary before attempting to use them.

◆ There is a difference in stopping and halting. Halting implies that the action is temporary and normal. For example, measuring a key dimension, moving a clamp, or adding coolant.

◆ Knowing the difference between the three coordinate sets in a CNC control—machine, work, and distance to go—is absolutely vital to setup and operation.

◆ Always stop the spindle before attempting to measure physical distance to go.

◆ While the fastest possible rapid moves are desirable to clip cycle times, the rapid speed is slowed during a first tryout of a program.

Respond

1. Describe the difference in the big and little red buttons?

2. What does an MPG handwheel do?

Critical Thinking

3. Other than running a stored program, what is the only axis movement method that can make curves or move two or more axes simultaneously?

4. Identify the master control that allows certain functions while banning others.

5. Of the three override controls, which cannot go beyond 100 percent? Why?

6. Why do some machines not require an initialization?

7. A program is being tried out for the first time on a vertical machining center. During a rapid move, the slide hold is touched, and the axes are read on the DTG screen $X1.5065$ $Y0.0055$ $Z1.0030$. With the spindle stopped, the cutter is found to be 1.030 in. above the work surface. Is it safe to allow this command to continue?

8. What does the *mach coords* screen tell the machinist?

9. Your machine refuses to start up. There is no action at all when touching the start button. What might be the problem?

10. Manually jogging an axis can be in three forms. Name them.

Section 19 Review

Unit 19-1

While we've tried to describe the various controls in a general way, every machine is different. Even two from the same manufacturer might have different Rev levels in their operating systems. As unlikely as that might seem, I've been in just such a situation. Moving from one machine to the next, I had to remember which of the two I was operating. One required a separate hydraulic start-up that had to happen after initializing the control. One started its hydraulics automatically. The first had no low-lube warning, while the second did. There were other differences too, even though they were identical machines outwardly.

Unit 19-2

Problem 15 coming up will ask you to go to the shop and identify the basic controls on a unit in your lab or shop. This is more than an academic exercise. You'll be doing the same quick orientation on new equipment throughout your career.

While these universal controls are sure to become even more compact than the RIS unit shown previously, they have carefully chosen functions that must be present on any machine in the foreseeable future. Learn to find them first, they are the controls you'll need to be safe and efficient as soon as possible.

The best policy is to first read the start-up and control sections of the operator's manual. Then ask your lead or another machinist who is familiar with the machine, if you don't know for sure.

Questions and Problems

1. On a sheet of paper, identify the CNC controls A through H in Fig. 19-19 and describe the function of each. (Obj. 19-2)

2. Describe hard limits and soft limits in 10 words or less. (Obj. 19-1)

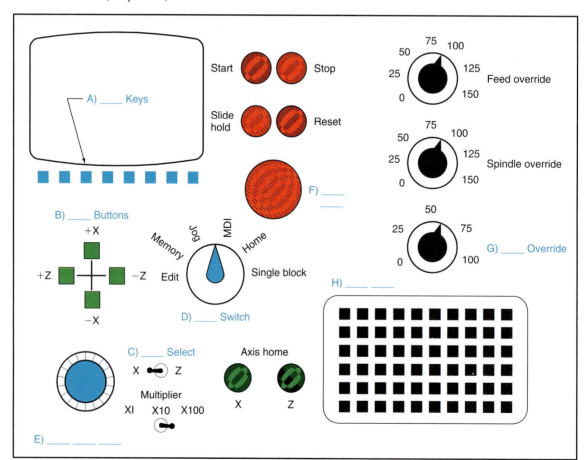

Figure 19-19 *Identify the lettered controls and features.*

Critical Thinking

3. True or false? A CNC milling machine must be homed in order to set the video *X*, *Y*, *Z* work coordinate position to zero. If it is false, what will make it true? (Objs. 19-1 and 19-2)

4. Before starting a CNC machine name four items that should be checked. (Obj. 19-1)

5. To perform a program *edit,* name two tasks that must be accomplished. (Obj. 19-2)

6. Without looking back, name the other mode selections beyond edit. (Obj. 19-2)

7. Name five possible methods of moving a CNC machine other than using a stored program. (Obj. 19-2)

8. What button stops all axis motion with *no consequence* to the program? (However, there may be a consequence to the work.) (Obj. 19-2)

9. What would happen during a cut in progress, if you push or turn
   A. rapid override to zero?
   B. feed rate override to zero?
   C. spindle speed to 125 percent?
   D. emergency stop?
   E. single-step mode? (Obj. 19-2)

10. What do hard limits prevent? (Obj. 19-1)

11. True or false? In using the MPG on a machine with a resolution of 0.0001 in., you wish to increase the rate from 0.001 to 0.010 in. per click. You would select the *X*10 multiplier from the mode switch. If it is false, what will make it true? (Obj. 19-2)

12. What CNC functions may be overridden by the operator? (Obj. 19-2)

13. Identify the three CNC controls in Fig. 19-20 that work together and explain their purpose. (Obj. 19-2)

14. Not directly covered in the reading: You are to try out a new program. From what you have learned, name the panel controls and video screens you will use. (Obj. 19-2)

15. Wearing correct eye protection, observe one or more CNC machines in your lab and list the controls and functions you recognize as compared to the generic control. Note, there may be different names and other control not discussed here. (Obj. 19-2)

Figure 19-20 *Identify the name and purpose of these three CNC controls that work together.*

ANSWERS 19-1

1. Older machines lacking absolute positioning or those without a feedback system must achieve a zero balance for axis drives and tool turrets before they can start operating.
2. Initializing (starting the machine)
3. To force the operator to clear the condition for which it was originally hit, if it was an emergency

4. The *operate* level will not allow setup changes. CNC or autocycle—but slow all axes.
5. The pendant is for manual movement of the machine during setups—the RIS is a full CNC controller but very compact.

ANSWERS 19-2

1. The big red button is an emergency stop. It permanently depowers the machine. The little red is for slide hold (feed halt). It is for a temporary, normal halt before resuming machining.
2. It is a method of manually moving an axis by turning the handle.
3. An MDI command line written at the keyboard
4. The *mode* selector
5. The rapid control, because that's all the machine can do
6. They have full-time position awareness (absolute positioning), therefore they wake up knowing where their axes and tools are located.
7. No. The cutter won't hit the work surface (this time) since it will stop rapid travel with 0.027 in. to go before hitting the work. However, that's too close for a safe program. Some editing might be in order.
8. Axis/spindle distance from machine home
9. E-stop still down (most likely); main breaker off
10. Continuous, incremental (jumps), and handwheel (manual pulse generation—MPG)

Answers to Section Review Questions

1. A. Soft keys—variable function entry keys
   B. Jog buttons—manual continuous or incremental motion of an axis
   C. Axis select—set the axis to be jogged
   D. Mode switch—select the operating activity
   E. Manual pulse generator (MPG)—hand dial the selected axis at the rate selected on the multiplier switch
   F. Emergency stop
   G. Rapid override—lower the rapid travel rate from 100 to 0 percent
   H. Hard keys—dedicated entry keys

2. *Hard limits*—the full end of axis travel; *soft limits*—the end of the normal work envelope
3. This statement is *false* for two reasons! Some machines don't require homing since they always know their machine coordinate position. Also, homing has nothing to do with setting work coordinates to zero. They are coordinated to the current PRZ within the envelope.

4. Previous to starting check: air, coolant, oils, and emergency-stop switches that must be released.

5. Select edit from the mode switch. Load the program to be edited into the edit buffer.

6. Various control modes: home, jog, automatic/program/CNC, single step, manual data input, edit, memory

7. From the MPG, continuous jog, incremental jog, continuous feed, from an MDI entry

8. Slide hold or feed rate override

9. A. Rapid override to zero: The machine would continue machining until a rapid command was encountered in the program then it would stop.

   B. Feed rate override to zero: The machine would stop all axis motion. However, the cutter would remain spinning in contact with the work.

   C. Spindle speed to 125 percent: The spindle would turn 125 percent of its programmed rate.

   D. Emergency stop: All activity would halt and the program would be reset. The machine would be turned off.

   E. Single step: The control would complete the command line it was working on then wait for another start command from control panel.

10. Hard limits prevent the machine from driving into the full mechanical limit of travel on a linear axis.

11. It is false for two reasons: The multiplier is not on the mode switch and the multiplication would be 100 not 10.

12. Most CNC controls allow operator override of spindle RPM, feed rate, and rapid travel rate.

13. The manual pulse generator (MPG) moves the selected axis by small increments or feeds it continuously be turning the dial. The axis select. The multiplier selects the rate of multiples each pulse will jump.

14. First a *dry run* using the graphic screen with the axes locked or on slide hold. Then with a cutter in the machine, select the distance to go screen to indicate how far each axis will travel to complete the current command; the single block mode selection; the start button to start each new block; and the feed and rapid override switches to slowly approach the work and setup.

Section 20

Operating a CNC Machine

Introduction

Taking command of a production machine is a big responsibility. It means producing quality work at a rate that creates profit. But there's more, you must keep yourself safe, keep records of production, and maintain the machine and work area. This section and the next, provide the information and skills needed. But they are also designed to teach more than an operator needs to run production, they are aimed at the career-minded trainee.

Typically in industry, new people are given the job of putting parts into a setup already proven by a lead person. You'll touch the green button to wait until the program ends to replace the workpiece with another. That's known as the *button-pusher* phase of your career. While it may not be the greatest challenge, don't underestimate the potential for learning, there's plenty of valuable knowledge and experience to be gained.

No matter where your career might be aimed in manufacturing or engineering, the time spent operating a CNC machine is worthwhile. Programmers, toolmakers, setup people, and supervisors have a more complete background and can do their jobs more effectively if they know how to get the machines up and running. The following units do both: they teach you how to run the machine right and how to deepen your knowledge for career advancement.

Objectives

CNC Planning

Introduction: In a small shop or tech lab situation, you'll plan and execute your own program. That means you'll know all the facts about the setup and the tool path when trying it for the first time. From the beginning, you'll know what cutting tools will be used and what their tool numbers are, where the PRZ is located, and the sequence of events. Before writing the program, you've planned it all. The point is that there are no facts to find out.

But in industry, before the job reaches the shop floor, several people work on it. The planner, the programmer, and setup and tooling people all make choices that affect the job; all that communication must be clearly understood or disaster is sure to follow.

In most shops, the setup and operation facts are recorded in a document similar to a work order. In some shops, the CNC document can be part of the work order but it's usually separate because it changes far more often than a normal work order.

KEY TERMS

Compensate
To adjust a program to accommodate a given cutter radius.

Compensated program
A program prepared to look for offsets.

Cove
An internal detail that sets the upper limit for cutter radius.

Cutter centerline (cutter path)
A less common compensated program, where a given cutter radius has been built into the path but can be compensated with offsets. See minus comp.

Flank interference
A compensation challenge on lathes whereby the control must keep the sides of the tool from touching the workpiece.

Generating
A curved cutter path that creates a part radius larger than the cutter radius. The motion preferred to forming.

Look ahead
The controller's ability to see upcoming part geometry shape interference problems with cove entry by the cutter.

Minus compensation
A negative offset number applied to a cutter centerline program to bring the cutter closer to the part geometry.

Offset
A variable number entered into controller tool memory by the operator. The program refers to the offset by the tool number or from a code word in the program.

Optional stop (opt stop)
Given the right halt code in the program, the operator can choose to halt at certain places or to switch them off and run through the halt.

Part path program
A program based on positive compensation offsets away from the shape of the part. The more common program type.

Radius offset
The tangent distance away from the part geometry for lathe or mill cutters.

Remarks
Notes embedded within the program.

Tool bias/approach vector
Defines the lathe tool's orientation in the setup thus the way it must move toward the workpiece and away during compensation.

Tool orientation
The direction the lathe tool is pointing.

The Vital Information You Must Have

A CNC document conveys two kinds of information:

Machine setup facts	Program run instructions

Those critical instructions are found in two possible places:

Hard copy document

Embedded remarks within the program

Differences from a Standard Work Order

First, you'll notice that the step-by-step operation sequences are gone. Those facts exist within the structure of the program. In their place, you'll find details about the overall sequence in which sets of cutter operations are to occur and there will be a specific cutter that will carry them out. For example,

T01 = 6″ carbide insert tooth face mill,

Opps #N015 through N240,

Face cuts to top surface.

In addition to the usual work order information found in the heading of the document—

print, part and revision numbers, material type and heat-treat condition, grain and type of material (castings, forgings, bars, etc.), and part counts—the CNC document lists

- *Program identification code with revision level* (see Trade Tip)
- *Program location*—central computer, disk, tape, and so on
- *Specific machine for which it is processed*
- *Special jigs/fixtures* with ID numbers for this job
- *Raw material size, alloy, and heat-treat condition*
- *Standard tooling required* along with their tool number in the program
- *Custom tooling required* and their tool number
- *The PRZ location in the setup*
- *Specific location and kind of chucking/holding/clamping tooling*—This is a critical factor because the program expects them at a given place and height above the work.
- *Event sequence*
- *Any special instructions particular to the job* For example,
- *Clamp changes* at midprogram halts
- *Warnings* about a large piece of excess material that will fly off the work at a given point in the program
- *The part must be rotated* or *repositioned* in the chuck at a certain halt point
- A key dimension that must be measured a given halt point

Operation Remarks within the Programmed

Most dedicated controls and all PC-based controllers allow **remarks** to be embedded in the program where they can be emphasized in bold or flashing characters (if commanded to do so) on screen, at the right time. In addition to making them appear at the best time to be noticed, embedded remarks

Trade Tip

Program Revisions Editing and tracking program updates is a huge issue in CNC work. We often improve a program several times when it's run for the first few times. There will be a very well-defined policy about who edits them and when they are changed, and especially how the data is managed. Small edits such as feed rate changes might be allowed from the machinist or the lead person. But those changes must be communicated with the programming department and the revision level for the program must be updated if they are incorporated. *Large changes are usually done only by the programmer, and they must be formally documented.* Due to potential data mix-up, some companies do not allow machinists to edit.

When a program is edited, the revision level must be included in the program header and also within the setup document. The revision level of a program can be tracked by date and time or by a Rev number/letter as on drawings, or it can be a serial number.

can easily be updated by the shop as better ideas are found. (Check policy to edit them.)

Each control has its own way of setting these messages aside without attempting to read them as program commands, here are the two most common methods:

1. **REM:** *This is a special note* ___ ___ **END REM:**
   The control displays but ignores characters between REM and END REM. The control recognizes the REM as a bracket.

2. **(***This is a special note*** ___ ___)**
   The control displays but ignores all characters between the parenthesis.

```
-----
:%
(PROGRAM 256-56b - HYDRAULIC FITTING - PROGRAM 2)
(REV JAN 12 - 2006)
(PRZ @ LLC, TOP OF WORK)
G80 G90 G40 G20
S2000 F20.0
N005   T01 M6 (.5 EM - 2 FLUTE)
N010   M3 M7
N015   G0  X1.00  Y2.500  (TO TOP OF RADII)
N020   Z.25
N025   G01  Z - .250
```

Figure 20-1 *CNC instructions can be embedded and emphasized in the program.*

The remarks are put in the program as warnings or flags of special events or they might describe better or safer ways to run the program (Fig. 20-1). For a simple job, a complete CNC document may be nothing more than a few lines of remarks in the program, perhaps accompanied by a sketch similar to Fig. 20-2. But the documentation can also be several pages of critical information. Now, let's examine each category.

The PRZ Location

At the top of the list, in pictures and/or in words, the PRZ location will be noted. Also, the method of locating multiple parts in a batch, such that each nests in the setup with its datum basis and the PRZ in a constant repeating position.

Hill Machining - CNC Documentation			
Dwg # *106-54-309*	Rev *A*	Prog # */106-54*	Prog Rev *22*
Part Count *50*	Mat *7775* *Alu*	Programmer *Fitz*	Date Orig *2-2004*
Holding Tools/Fixtures		Notes	
T01 *.5 ⌀ EM-2FL*	T02 *5/16" Drill*	T03 *3/8" Reamer*	T04
T05	T06	T07	T08
T09	T10	T11	T12
T13	T14	T15	T16
Special Instructions *Parts are pre-cut with datum edges finished Program profiles contour*			

PRZ Setup Sketch

Figure 20-2 *Typical simple CNC setup and run document.*

The specific setup that creates the PRZ may be part of the instruction, or it may be left up to the machinist. Figure 20-2 shows two pins on one side of the part, and a third on the other along with a clamp location. If you miss those details, one of two problems might happen:

1. **Caution!**
   An accident can occur as the cutter hits the wrong location on the part or even tries to cut the holding tooling. Because this is an obvious danger, it is an uncommon mistake.

2. **Introduced Variation**
   This is a far more subtle and common problem for beginners whereby needless tolerance is consumed. To understand this potential, answer the critical questions.

Introduced Variation Questions

The job is to CNC drill, three holes in a batch of parts as shown in Fig. 20-3. On the print, note that the three holes are to be located relative to two datums and their intersection should be set up as the PRZ location for this job. For this exercise, the work has been premilled square and the radius is already machined on the left corner.

The 15.0-mm and 20.25-mm positions are basic to the hole location. The overall length of 75.0 mm has a loose tolerance of 1 mm.

The setup in Fig. 20-3 will accomplish the goal. The solid vise jaw provides datum A, while the end stop establishes datum B. However, by incorrectly setting up a material stop on the wrong end of the work, as shown in Fig. 20-4, the operator has misplaced the B locator relative to the PRZ. The datum A locator is correct but datum B is now on the wrong end.

The Correct Setup

Figure 20-3 *The part datum locators are placed right. Every part will reproduce the PRZ position.*

The Wrong Setup

Incorrect datum B stop location

A

Figure 20-4 *What variation does this cause?*

Question What are the consequences of this error? No accident will occur because the part is nearly symmetrical. But in this setup, by how much could the 15-mm end space be missed with respect to datum B?

Answer The end spacing to datum B can be shifted by as much as the overall length tolerance for the bar, of plus or minus 1.0 mm. That's greater than the hole end spacing tolerance. The result is that the hole pattern will shift left or right with the overall length variation of the parts.

KEYPOINT

With the Datum B locator incorrectly placed, the holes will now be referenced from the right end, not the critical left, datum feature B.

But that kind of error is obvious, this next one is not. In an attempt to continue using the wrong setup, the length of all the blank parts was premachined to the 75.0-mm length within an overly tight 0.01-mm tolerance before the CNC drilling operation.

Question The raw parts are now very close to 75 mm. So now, can't that right end locator setup be used? Think this through and you'll have gained an important CNC concept.

Answer No matter how close the raw parts were machined to the 75-mm dimension, length variation still exists. Each part loaded starts with introduced left-end variation whereby, if the PRZ was correctly located on the left side, the only variation for the hole positions would be due to the drilling alone.

In trying to make the wrongly planned setup work, the cost per part was needlessly driven up by extra operations while the quality was

reduced! Some parts will be potential scrap due to the combined effect of drilled and induced end variation. Normally, these PRZ issues will have been resolved by the programmer and toolmaker before an operator receives a job.

KEYPOINT

PRZ location and holding method are the most basic facts.

Types and Sizes of Cutter Tooling

In most cases using a modern CNC control, the cutter tooling outlined in the CNC document is a guideline for size and shape. The actual tool selected can vary within small limits. After some shop talk about the savings gained from this ability, we'll explore the operator's role in cutter selection based on the CNC document.

Cutter Selection Based on Program Type

Compensated Programs When a program is written, the programmer has a certain cutter in mind for turning or milling. But in most cases, that which is set up in the machine need not be the exact diameter, radius, or length originally chosen.

To be able to use this adaptive ability, the program must be correctly prepared as a *tool* **compensated program.** For now we're concerned with reading the CNC document and selecting cutters that will work right from those available. In Section 21 we'll look at finding those compensating factors and entering them in the control. We'll look at mill examples, then lathes.

Two Kinds of Tool Compensation When a program is written it is in one of two possible types: cutter centerline or part path.

KEYPOINT

It's essential that you know which program path is used, since the cutter value you must enter into tool memory will be quite different depending on the type.

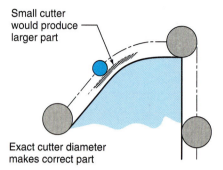

Cutter Centerline—No Compensation

Small cutter would produce larger part

Exact cutter diameter makes correct part

Figure 20-5 *A noncompensated cutter centerline program will produce the correct part only if the exact cutter size is used.*

Program Type I Cutter Centerline—Built-In Compensation

A specific cutter size is built into the program either by the CAM system or by computing it then handwriting the program codes. The **cutter centerline** program generates the dotted line shown in Fig. 20-5. This path is calculated at the time it is written.

KEYPOINT

A cutter centerline program moves the tool on a path that is the tool radius away from the part geometry.

Compensated or Uncompensated If the cutter centerline program is written with compensation commands, any cutter can be used that's close to the size for which it was written (as long as it can fit into all coves in the part shape).

If the cutter path isn't set up for compensation, then it will produce the right size and shape only if the exact cutter for which it was written is used. Uncompensated, if any other cutter radius is used, the resulting part will be either larger or smaller than expected—the blue cutter is undersized and makes a part too large, by the radius difference in the two cutters.

Further, using the correct cutter, as it wears, the uncompensated program subsequently makes part surfaces that become progressively larger due to deflection. There are two fixes: change to a new, full-sized cutter or edit the program to include compensation.

ShopTalk

In shop lingo, we often shorten the term "cutter radius compensation" to "cutter comp" or just "comp."

Computing the Tool Path Centerline

Figure 20-6 *Point A on the part geometry becomes point A' on the cutter centerline. It is shifted a small amount both in X and Y.*

Writing Centerline Programs—Point Shifts Writing a manually compensated program such as in Fig. 20-5 is a good exercise for beginners. To do so, each coordinate point on the part geometry must be shifted out to the center of the cutter's radius (Fig. 20-6). To write the cutter path using a 1.00-in. end mill, point A, on the part geometry, had to be shifted up to point A' (A prime) to compensate for the 0.50 radius distance off the part geometry.

KEYPOINT

All program compensations, whether mill or lathe, are based on keeping the cutter radius tangent to the part geometry.

Minus Compensations to Centerline Programs The centerline program is adjusted by entering the difference in the intended cutter radius for which it was written, and the actual radius of the real cutter put in the machine. As such, it's sometimes called a **minus compensation** or *minus comp program* since the real cutter is almost always smaller than the full-diameter cutter chosen by the programmer.

If the exact cutter size for which the program was written is used, there is no adjustment needed. But for example, in Fig. 20-7, if a 0.725-in.-diameter end mill is used for a program written for a cutter of 0.750-in. diameter, the compensation, based on the radius, is entered in the control as a *minus* 0.0125 in. (the required radial change to come closer to the geometry line by 0.0125 in.).

Question For a compensated cutter centerline program, what adjustment must be entered in the control for a 1.060-in.-diameter

Cutter Centerline
(computer compensated)

Offset amount
0.0125 in. minus

Compensated
for 0.725-in.-
diameter cutter

Centerline
0.750-in.-
diameter
cutter

Figure 20-7 *A 0.75-in.-diameter cutter centerline program using a 0.725-in.-diameter cutter is compensated 0.0125 in. toward the part geometry.*

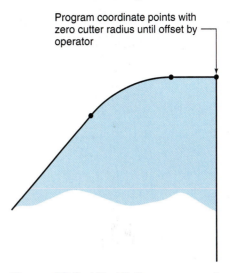

Part Path Program

Program coordinate points with
zero cutter radius until offset by
operator

Figure 20-8 *A Part Path program creates a cutter path of zero offset—zero diameter until told differently with a compensating tool radius offset.*

cutter, if the program was originally written for a 1.00-in.-diameter cutter.

Answer The actual cutter is bigger than the basic cutter—not the usual condition. So, it must be adjusted by *plus 0.030 in.* based on the radius. The new cutter must back away to allow for the larger diameter.

> **KEYPOINT**
>
> An uncompensated cutter centerline program can be edited to create a compensated version.

Program Type II Part Path (Part Geometry Coordinates)

This second kind of program is by far the most common industrial version—the program path is the part shape. As shown in Fig. 20-8, the part coordinates are programmed directly from the print without any cutter radius shift off the shape.

Positive Compensation To use the part path program, the operator enters the actual cutter radius (or sometimes the diameter depending on control type) into the control tool memory—as a positive compensation to the program line.

> **KEYPOINT**
>
> Using programs based on part path, the control creates the cutter path by moving away from the geometry a distance equal to the cutter **radius offset** it finds in the control's tool memory. The machinist enters that offset.

In effect, compensated centerline and the part path programs are the same except a part path program is written for a zero radius cutter. It is subsequently compensated to some other real radius when used on the shop floor.

> **KEYPOINT**
>
> A **part path program** is composed of coordinate points shown on the print. While it can be a different size, there is a cutter size listed in the CNC document. It is the target choice to produce best results.

Part Geometry Programs Are Common

If the program is not a part path program, that fact should be noted in the CNC document. Otherwise, assume a positive offset away from the part geometry.

> **KEYPOINT**
>
> To set up a CNC machine, you must know what kind of program is used—cutter centerline or part path.

ShopTalk

Offsets **Offset** is a CNC term used for the variable that adjusts a program for real conditions impossible for the programmer to know. Offsets are used to account for variance in end mill diameters and lengths, and for lathe tool nose radius and the tools shape.

Offsets are entered in the control memory by the machinist. Offsets are not in the program, they exist in the control's tool memory. The correctly prepared program calls up these saved variables to compensate.

The offset being discussed is for tool radius compensation. You'll see others as well—for example, tool length offsets (Fig. 20-9). For now, remember, an offset is a controller-stored variable that adjusts a program for a specific condition. Using offsets adds great flexibility to CNC work.

Figure 20-9 *Tool length and radius offsets are written into their tool number.*

For safety reasons, most shops use one or the other, most likely a part path program since they feature less math. Shops normally do not mix the two kinds of programs, however, some shops accept jobs that come with the program.

Lower Limits on Cutter Variation

1. Minimum Lathe

Tool nose radius is a matter of *good cut qualities* up to and including zero radius. Here is another area where manual lathe practice is valuable. By now you should have experimented with varying nose radii to get different finish and antichatter results.

2. Minimum Mill Cutter, Based on Cutter Strength

The lower limit for end mill diameter is strength. The cutter must hold up to the demands but otherwise, a smaller cutter will always produce the correct shape if it is correctly compensated and long enough.

Upper Limits for Cutter Radius

To understand the upper limits of cutter radius, we need to define two new terms.

Generating a Curve Versus Forming
When the cutter moves in an arc, producing a curved surface, it is called **generating** the shape. Compare that to forming in Fig. 20-10, where the cutter simply leaves behind its own shape.

Coves Borrowed from carpentry, the term **cove** means any internal detail that must be produced by profile milling or turning. A slot is a good example of the internal shape, as

Forming   Generating

Figure 20-10 Generating *versus* forming.

In most cases, generating is the superior machining process in terms of finish and avoidance of undercut corners, but requires a cutter with a radius smaller than the part geometry.

seen in Fig. 20-11. The cutter chosen must be small enough to enter and exit the cove without touching the opposite side.

Cutter Diameter/Radius Rule (Mill/Lathe)

When selecting a cutter size or shape, its radius cannot be larger than the smallest inside (concave) radius, or cove, on the part geometry. Of the three cutters shown in Fig. 20-12, only two will be able to machine the correct geometry. One will form, while one will generate the shape. The third will not work! But the result (scrap or uncut part) will depend on the capability of the control to see the problem.

Controls with Various Levels of "Look Ahead"

Various controllers react differently when we try to use a cutter radius too large for the cove. For example, using the 1.000-in.-diameter cutter in Fig. 20-12.

Using a very low-level control, when the 1.0-in. cutter follows the slanting line down, it will overrun the upcoming vertical surface, and undercut it. A cutter that is too large to fit the cove will make an incorrect part.

The cutter must either form or generate these radii

Figure 20-11 *An internal cove is a set of details requiring the cutter to enter and exit without touching opposite walls.*

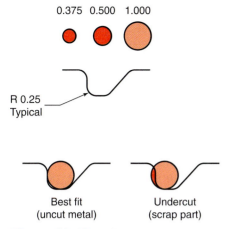

R 0.25
Typical

Best fit
(uncut metal)

Undercut
(scrap part)

Figure 20-12 *Only one cutter can generate the corner radius, one will form it. The third will either cut a best fit leaving uncut metal, or the control will alarm and halt with a (control-dependent) geometry error.*

Most CNC controls today can see that upcoming interference event by examining the part geometry and the cutter radius. Depending on the level of software in your control it will

Error out—some will cause an error code and stop machining.

Be best fit—some will take the larger cutter as far into the cove as possible then out, without undercutting the opposite side of the cove. The part won't be the right shape but it won't be scrap either.

Be scrap—lower level controls won't see the problem at all. They would produce scrap a scrap part if cutter C was used in Fig. 20-12.

Trade Tip

Avoid Forming if Possible When selecting cutters larger than the programmed size, it is poor practice to change up to a size that's equal to a cove radius on the part. If the cove radius is the same as that of the cutter, then the generated curve exists but it has a zero radius. The cutter forms the cove radius. The control will be able to deal with the change. It will produce a tool path curve with "zero" radius, as shown in Fig. 20-13, but forming tends to pull the cutter into corners leaving undercuts and to also produce tool chatter.

Generating cutter   Full form cutter

Generating arc   Zero radius arc

Figure 20-13 *The compensation software calculates a curved path for the cut on the right, but the path has zero radius.*

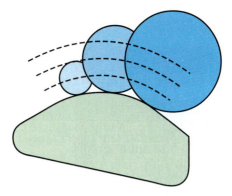

Figure 20-14 *For outside shapes with no coves, any cutter diameter is able to produce good results if the program is compensated.*

Let's sum this up:

◆ The lower limit is cutter weakness for mills, and finish produced for lathes.

◆ If there are no internal coves within the part shape, any cutter larger or smaller will work (Fig. 20-14) as long as the program is cutter compensated.

◆ The best choice is close to the original size for which the program was written.

◆ If the program is not written for compensation, the cutter must be the exact diameter for which it was written, but that program can be edited for comp.

Lathe Cutter Selection

When selecting lathe tools from the CNC document, then loading tool information into the control, one additional issue that arises is tool **flank interference.**

Tool Flank Interference Compensated cutter paths for lathe work must keep the cutter tangent to the part geometry as on mill compensation. But lathe controllers must also consider the sides of the cutting tool. The control must know which direction the tool is pointing and how wide the tool is in order to create a useful tool path following a part geometry.

In Fig. 20-15, the 30° diamond-shaped cutter is set up in the straight position. It could

Possible heel drag interference

Figure 20-15 *Lathe tool shank size, orientation, and cutter shape must be known to the control to avoid approach interference problems in the compensated path.*

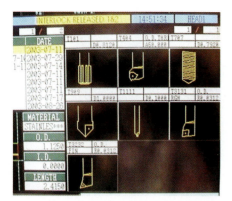

Figure 20-16 *A standard tool library selection.*

make the thread relief, while the 45° insert in the 45° holder would drag on the heel as it cuts the cove slope. CNC terms used for this lathe **tool orientation** are **tool bias** or **approach vector.** There are two ways in which this tool bias information can be entered in the control's tool memory:

Tool Library Many modern CNC lathe controls contain a standard library of the common tools. During the setup, when a tool number is loaded into the tool memory, it is attached to a screen-selected image that represents the correct tool shape and the direction in which it's pointing in the setup. For example, after selecting turning tools from a menu, the image shown in Fig. 20-16 might appear. The machinist selects the tool shape and orientation that suits the setup. Then the shank size and nose radius offsets will be attached to that selection. Most tool libraries of this type, are customizable. New tools are supplied on disk by manufacturers or they can be custom drawn using control definitions.

Tool Vector Selections Without the tool library function, controls attach an approach vector to a tool selection. The vector numbers are determined from a chart, such as shown in Fig. 20-17. For example, after selecting *grooving tools,* an entry from one to eight will be requested. The vector number represents the direction in which the grooving tool is oriented. During tool path compensation, it tells the controller how to approach and pull away from part surfaces.

Tool Vector Numbers

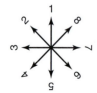

Figure 20-17 *This control features eight possible* approach vector *numbers.*

It's your job to avoid these problems during tool selection for the setup and tryout. Call on your manual lathe tool selection experience to avoid this problem. Vector tool bias selections do not guarantee interference problems are solved. Try out new programs carefully.

Phantom Tool

A puzzling problem occurs when using nose-radius-compensated programs for turning but the wrong radius is stored in tool memory. Similar to a mill program with the wrong cutter diameter in memory, a distorted part will be produced. But with lathes an odd twist occurs. Even with the wrong radius entered, or none at all, as in the example (Fig. 20-18), straight diameters and face distances are machined right, yet curved, and angular surfaces are machined wrong!

Think this through—it's a bit of a brain teaser. Lathe compensation programs follow the same logic as a mill, in that the control keeps a round cutter tangent to the programmed shape (Fig. 20-18). If no nose radius is entered, the control assumes a zero radius, thus keeps a pointed tool tangent to the geometry. That works fine if the tool really is pointed.

In this example, the control has a zero nose radius in memory. However, the radiused tool has been coordinated for both *X* diameter and *Z* face distance from PRZ. Its radiused nose is

Part Distortion from Wrong Tool Nose Radius

Figure 20-18 *Without the nose radius offset in the control, a curved surface is altered.*

touching the work at tangent points *X* and *Z*. As long as the tool cuts only straight diameters and perpendicular faces, it will make correct dimensions. It will form the tool nose radius at internal corner intersections.

However, when cutting curved surfaces, as shown on the right in the figure, the control perceives that it has a pointed tool, known as the phantom tool point, thus it pulls back to keep the phantom point on the geometry. That makes a distorted shape. This occurs only when the operator forgets to associate a nose radius with a cutter memory. Given the wrong nose radius or no entered radius, distortion occurs on angular surfaces as well.

KEYPOINT

When mounting a lathe tool, both the tool radius and the tool shank shape must be entered into the control's tool offset memory.

Understanding Cutting Tool Numbers

Tool numbers in the CNC document must be correctly identified in the control (Fig. 20-19). If the machine is a lathe, the task is simple, tool 01 goes in turret position number 1. Tool 02 goes in position 2 and so on. Some mills require the same tool load as lathes; T01 goes in storage drum hole 1, T02 in hole 2, and so on. These are generally older mills lacking random tool storage ability. More modern CNC mills allow the storing of any tool in any drum hole, called random tool storage. But due to random tool storage, the tool-loading task is bit more complex. For example, the mill operator is storing T01 in open drum position 08 in Fig. 20-20. His next action will be to immediately go to the control and enter tool 1 at position 8 on the tool tracking page.

Tool numbers are arranged in sets of two-digit data fields, in one, two, or three columns. Various controls use different tool number sets. Some take up the cutter offset information when reading the tool number, while others require a separate code to get and use offsets.

Separate Code Example
T01 H43 D1

Use tool #1 and apply length offset (H43) = to the number stored in D1 (diameter 1)

Figure 20-20 *After storing random tools, write the tool number (T01) into the turret position (hole 8) on the tool storage page.*

Figure 20-19 *On CNC lathes, tool position on the turret is also the tool number in the program.*

Single Column

T01

Tool number (Dropping leading zero, can also be T1)

Tool Radius Attachment— Two-Column Format

T0101

Tool number/Offset number

Using this format, radius offset is the second column. In the example, radius offset value 1 is applied to tool 1. For consistency, offset 1 is usually attached to tool 1, but there are times when a second offset might be placed on tool 1 by calling it T0102.

For example, a roughing program uses T0102, which has a deliberate false offset bigger than the tool radius, which then roughs out the part, leaving a bit of finish metal. Then the offset is exchanged for the true compensation and the program is rerun with T0101 to finish the shape—the same cutter but different offsets.

Tool Radius and Length Numbers—Column 3

On some controls, the tool number includes a third data filed to include tertiary offsets. The most common is the length offset used on machining centers:

$$\textbf{T010203} = \text{Tool number/Radius offset/Length offset}$$

Understanding Mill Tool Length Differences

Tool length offsets make both the programmer's and the operator's jobs much easier. We'll study how to determine and enter them in Machine Setups in Section 21. For now, you need to know that they apply to real tools of varying lengths.

KEYPOINT

During mill setup, the operator must determine and enter the tool length offset differences as determined by gage measuring the tools.

When the program is being written, the programmer assumes that all tools in the setup are the same length (it doesn't matter how long, just that all are the same). He does not need to adjust Z axis moves, to account for differing tool lengths.

During setup or tool replacement, the operator determines length differences in the real tools by touching a reference surface with each, then reading the Z position on the screen. Next, the difference in each is stored as the tool's Z length offset (Fig. 20-21).

So, using the example of Fig. 20-21, under tool number T03, length offset is a plus 2.450 in. for the end mill compared to tool 1, the drill. Any time, under length offset 03, the program calls the end mill to a Z axis position, it combines the axis move with the offset distance. The Z axis must move an additional 2.450 in. to reach Z zero using the end mill, when compared to the drill.

Tool Length Differences

Figure 20-21 *If the drill is set to touch the work surface at Z = 0.00, then the end mill requires a length offset of plus 2.450 in. in order to also go to Z = 0.00.*

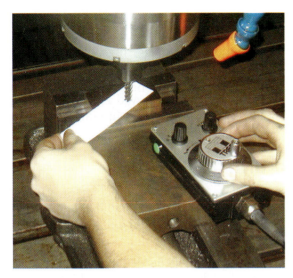

Figure 20-22 *Touching the bit to the work surface is a common way to coordinate Z axis reference.*

Figure 20-22 shows an operator using a feeler gage to test the drill bit. There are several versions of this task that will be covered later in Section 21.

Presetting Tool Lengths Many high-production shops set tool lengths such that every tool number 1, the drill in Fig. 20-21, is the same length before it is given to the mill operator. That way, a dull tool can be quickly replaced with no length testing by the operator. Another solution is the automatic probe shown in Figs. 20-23 and 20-24.

While there are manual methods of determining length offsets, both at the machine and at the preset bench, as seen in Fig. 20-22, the automatic probe has some advantages in

Figure 20-23 *The Z axis position of cutting tools can be coordinated using an auto probe that communicates contact to the control.*

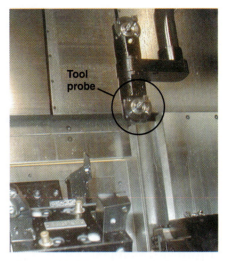

Figure 20-24 *This CNC lathe features an automatic probe to determine tool position and offset.*

that it senses then automatically loads the offset for the operator.

KEYPOINT

Using one of several methods, before a multiple-tool program can be run on a milling machine, length offsets must be determined and stored for each cutting tool.

Cut Sequence and Special Instructions

Sequence of Operations When trying out a new program the operator absolutely must know the following facts from the CNC document.

Which cutter will be used in which sequence.

What machining operations occur in what order.

Any special intervention required (list coming up).

Sequence information may be listed in the document or embedded in the program as comments, often both. For example,

T02 [Tool 2—$\frac{3}{4}$ in. EM will rough three pockets and datum C]

T03 [Tool 3—$\frac{1}{4}$ in. drill–deep drill watch coolant stream]

If we know this information, adjustments such as feed and rapid travel override can be used as they are needed.

Special Instructions Events that would otherwise be a "big surprise" to the operator should be made clear in the documentation beforehand. Here's a list of the unusual events that can happen.

Lathe surprise events

1. Cutoff Operations

The part or excess material becomes unattached and may become a potential projectile. Instructions for avoidance might be included.

2. Extreme Machining

The coolant stream is critical such as heavy roughing cycles or deep drilling.

3. Tooling Challenges

Experience shows that the cutter might fail at a particular point in the program thus extra attention must be given to be able to halt operation instantly.

4. Part in Chuck Changes

5. Optional Stops

These are critical places where the programmer has provided a halt if it is needed. For example, every 10 parts must be measured on a critical dimension. Between halt times, a toggle switch, such as the one shown in Fig. 20-25, is set to ignore the **optional stops** or **opt stops.** There may be more than one opt stop switch such that various intervention places can be turned on or off, at the operator's discretion.

Typical mill surprise events

1. An Exceptionally Deep Operation

Coolant and close monitoring are required here.

2. Vibration and Chatter May Occur

The operator may need to override speed and feed controls as the machine cuts through a critical area.

Figure 20-25 *Optional stop switches recognize or ignore halt points in the program by the operator's choice.*

3. Midprogram Clamp Changes or Additions

Sometimes as machining progresses, the part becomes progressively weaker and requires extra support. Or to machine an area that was shadowed by a clamp, at a given point it must be shifted.

UNIT 20-1 REVIEW

Review the Key Terms

Compensate
To adjust a program to accommodate a given cutter radius.

Compensated program
A program prepared to look for offsets.

Cove
An internal detail that sets the upper limit for cutter radius.

Cutter centerline (cutter path)
A less common compensated program, where a given cutter radius has been built into the path but can be compensated with offsets. See minus comp.

Flank interference
A compensation challenge on lathes whereby the control must keep the sides of the tool from touching the workpiece.

Generating
A curved path that creates a part radius larger than the cutter radius. The motion preferred to forming.

Look ahead
The controller's ability to see upcoming part geometry shape interference problems with cove entry by the cutter.

Minus compensation
A negative offset number applied to a cutter centerline program to bring the cutter closer to the part geometry.

Offset
A variable number entered into controller tool memory by the operator. The program refers to the offset by the tool number or from a code word in the program.

Optional stop (opt stop)
Given the right halt code in the program, the operator can choose to halt at certain places or to switch them off and run through the halt.

Part path program
A program based on positive compensation offsets away from the shape of the part. The more common program type.

Radius offset
The tangent distance away from the part geometry for lathe or mill cutters.

Remark
Notes embedded within the program.

Tool bias/approach vector
Both terms for the lathe tool's orientation in the setup.

Tool orientation
The direction the lathe tool is pointing.

Replay the Key Points

◆ The CNC documentation foresees potential problems, eliminates guesswork, and gets the job going right with the least amount of experimentation.

◆ By matching the program's header Rev level to the CNC documentation, data crossing can be avoided. There will also be a policy about dumping old programs from permanent memory.

◆ When a datum locator is not placed correctly, needless variation will be introduced into a production run.

◆ PRZ location and holding method are among the most basic decisions.

◆ Program compensations (mill or lathe) are based on keeping the cutter radius tangent to the part geometry.

◆ An uncompensated cutter centerline program can be edited to create a compensated version.

◆ When using programs based on part path, the control creates the cutter path by moving away from the geometry a distance equal to the cutter radius offset it finds in the control's tool memory. The machinist enters that offset.

◆ A part path program is composed of coordinate points shown on the print.

◆ You must know what kind of program is used—cutter centerline or part path.

◆ In most cases, generating an arc is the superior process compared to forming it, in terms of finish and avoiding undercut corners, but generation requires a cutter with a radius smaller than the part geometry.

◆ Vector tool bias selections do not guarantee that interference problems are solved. Use caution when trying new programs.

◆ When mounting a lathe tool, tool radius and the tool flank shape and size must be entered into the control's tool offset memory.

◆ During mill setup, the operator must enter the Z axis tool length offset differences as determined by gaging and comparing each tool against a reference.

◆ Using one of several methods, before a multiple tool program can be run on a milling machine, length offsets must be determined and stored for each cutting tool.

Respond

1. List the three kinds of compensated programs and give a brief description of each. Which is the most common and why?

2. As parts are run on a lathe, their outside diameter becomes progressively bigger and their length increases. What can the operator do?

3. Using Fig. 20-26, to profile mill this part; select the best mill cutter from these; 0.25, 0.375, 0.500, 0.625, 0.750, and 1.00 in. diameters.

Critical Thinking

4. You are to set up to CNC mill the internal pocket and two drilled and tapped holes (Fig. 20-27). Sketch the setup tooling if the part has been premilled to a finished square and the 0.750-in. slot has also been machined into the block.

5. Don't look back. List as many important items as you can, that might be listed in a CNC document. The book identifies 14.

Figure 20-26 *Select the best cutter from the list of Problem 3.*

Figure 20-27 *Describe the setup holding and alignment tooling for this part per Problem 4.*

UNIT 20-2

Controller Alarms

Introduction: Many potential problems are detected by an alert operator. But others are announced by the control itself, in codes or words, called an alarm condition. When the control finds a problem, it takes certain actions. It may keep on cutting, but flash a screen error message or a yellow warning light, or both. It also might just stop. Something must be corrected. Depending on the control type, the error message can be easy to read or sometimes it's cryptic and requires a bit of detective work to solve.

These messages are never welcome, because they not only announce something requiring operator (or programmer) intervention but they often halt the machine until solved. We'll cover a few of the more common ones. On the job, you'll need an operator's manual that lists them all.

KEY TERMS

Branching errors
The programs logical sequence is in error. It wanders off and can't return to the original address from which it branched.

Look ahead
The controls ability to analyze and compensate program commands ahead of the active line. This ability can cause some frustration when looking for the cause of an alarm.

Servo errors
The machine has become uncoordinated between physical position and recorded position.

Syntax error
Any entry that doesn't appear in the right order or form for the control to understand.

Alarm Classes

A few controls provide alarms in readable words. But many flash alpha-numeric error codes (Fig. 20-28). The operator is then left with the task of decoding what went wrong. Given a little experience, it's usually easy to find and fix, but sometimes an alarm can be a real headache to sort out.

Take a moment to scan the following example alarms to get a feel for the various ways a control can stall. The main objective is to understand the kinds of alarms within a

Figure 20-28 *Control alarms can be simple or a real challenge to solve, but they are never welcome!*

KEYPOINT

Alarms fall into general groups. Recognizing what the group indicates helps speed the solution.

group, thus you'll know where to start looking for the problem, inside the program data, in the machine setup, or in the machine itself.

Physical and Hardware Errors

These errors can occur when running any program new or old. Here the problem lies outside the program!

◆ Machine supplies low
   Air pressure, hydraulic, coolant, and lube oil
   Each may have its own code or all will be grouped into a single alarm condition. For example, alarm 01 = low fluids.

◆ Hard limit overtravel

◆ Violating a safety switch
   Door open, chuck switch, bar feeder, part loader, and so on

◆ **Servo errors**
   Due to exceptional action, extra heavy machining, or a crash, the machine has lost its position count on the feedback system.

Syntax and Program Errors

These problems lie within the program—but only within new programs. They are fixed by solving math or logical order errors, then editing them out. See the Trade Tip on look-ahead controls.

◆ **Syntax errors**—switching a letter "o" for a 0 (number zero), incorrect wording of statements, incorrect ordering of a rigid set of variables

- Missing data—no feed rate or RPM (very common for beginners)
- Math errors—curve or other computations do not make sense
- Extra data—two, mutually exclusive, actions are requested in the same command. For example, both rapid travel and feed rate commands are given in the same line. The machine might alarm or it might accept only the last command it reads on that line.
- Limited power—two switching commands have been given in the same line but, due to exceptional power drain, only one is allowed on any given line. For example, turning on the coolant and work light and spray mist all at one time might be disallowed on older controls.
- Radical commands—the programmer has asked for a rate beyond the ability of the machine. For example, 200,000 RPM when 2,000 was the intent.

- Branching logic—a loop, bridge, or subroutine is used (program tricks to save extra writing) but they aren't addressed such that after completion, the control can return to the main program and keep going. Control becomes confused by such a **branching error.**

Operator/Controller Errors

Review of previous actions or entries.

- Compensation impossible—also common for beginners. Here, a cutter has been loaded such that the control cannot use it to compensate the part geometry. See Fig. 20-29.
- Erroneous offsets—impossibly big or small offsets, usually a decimal point error.
- Machine not initialized correctly—the homing sequence that has not been performed.

Trade Tip

Look Ahead Controls On modern controls with **look ahead** capability, one cannot assume the problem is on the line where the control has stalled. Depending on the level of intelligence, most controls look ahead from a minimum of three command lines up to the entire program, as they attempt to perform tool compensation. They read and compensate several lines ahead of the active command line. If not, tool interference such as that shown in Fig. 20-29 can occur.

The control is machining line *A-B*, but it has foreseen a problem coming up as it rounds the corner, it will hit point *E*, the cutter won't fit into the cove. It has *detected the problem five command lines ahead in the program*. It errors and halts. In this example, after finding out where the problem might be, the fix is to install a smaller diameter tool.

Interface area several blocks ahead

Figure 20-29 *Looking ahead, the control may see the interference problem depending on how many blocks it analyzes ahead of the current command from A to B.*

UNIT 20-2 REVIEW

Review the Key Terms

Branching errors
The programs logical sequence is in error. It wanders off and can't return to the original address from which it branched.

Look ahead
The controls ability to analyze and compensate program commands ahead of the active line. This ability can cause some frustration when looking for the cause of an alarm.

Servo errors
The machine has become uncoordinated between physical position and recorded position.

Syntax error
Any entry that doesn't appear in the right order or form for the control to understand.

Replay the Key Points

- Alarms fall into groups. Recognizing them helps find and fix the problem.
- Due to look ahead capability, the problem may be several lines ahead of the line on which the control has stalled.

- An alarm may be in flashing words that are easily understood, or it may be a code number that must be decoded from a list in the operator's manual.

Respond
The activity for alarm codes is to take a moment to scan the manual of the machine you are about to operate. Or your instructor may have a handout prepared on the same subject. Notice the types of errors and the way they are communicated on your specific control. Be aware that the explanations are often translated from the language of another country and as such can be difficult to understand.

UNIT 20-3

Operator Actions (During Production Runs)

Introduction: Everything is set up and tested. Production is ready to go. This brief unit describes the activities that a pro-manager performs to be in full control. It's been put into a unit because alertness is such a vital subject for safety.

It's easy to sit back and do little or nothing during a run when everything *seems* to be going right. But it's just then, when you are in "potato mode," that something goes wrong suddenly. The solution is to be ready, but I admit, sometimes that's not easy to do when the run is long and has worked perfectly for some time.

Practicing Emergency Plans
By now you have seen or heard an accident on a manual machine. And you also know that crashes occur at fantastic speeds in CNC, especially on turning machines.

KEYPOINT

The best operator action during emergencies always results from preplanning and practice. Review and practice emergency moves at the beginning of your work day.

When starting a machine, do a quick mental review of the following three-level occurrence list. The real skill objective is to analyze quickly at which level the problem is and take the right action—at the right time. Some problems can wait until a natural division in cutting action while others must be halted ASAP, right now! The challenge is, there isn't much time to sort it out sometimes. Another challenge is that different machines feature varying kinds of halt controls.

Notice that as the emergency level increases, that the action options decrease, yet they must be performed evermore quickly!

Level 1—Something Isn't Right
The Event As yet this is a nonemergency that might get worse. Examples include a sound that hasn't been present or a finish that seems to be deteriorating.

The Action Plan Memorize the methods that halt the machining at a convenient time, with no consequences. Some of these actions are not immediate, the machine will continue to run for a while, then halt at a natural time when the cutter is not touching the workpiece (see Section 19).

- **Switch mode to single block—**
  Will stop upon completing current command.
- **Turn rapid override to zero now—**
  Will stop as the next rapid travel code is read.
- **Turn rapid override to zero after a rapid move commences—**
  This halts the action with the cutter not touching the work.
- **Touch halt/slide hold at a tool change—**
  Will stop with tool off the work.
- **Midcycle halt—**
  Will stop right now, but then the cuter is dwelling upon the work. The spindle must be stopped.

Level 2—A Problem Is Obvious and Will Degrade—Immediate Action
The Event A low-level emergency such as chip buildup exists. It must be solved efficiently before something much worse occurs.

The Action Plan Memorize the controls available to stop the machining quickly.

- Slide hold
- Feed rate override to zero
- Midcycle halt

Level 3—An All Out Crash—Fully Dangerous to Machine and Operator

The Event Parts of the setup and cutter shards are flying out, axes are diving into and stalling on machine tables and chucks, it's the big one! Stop everything right now!

The Action Plan Above all else, be certain you are out of danger! Only if it's safe to do so, hit the emergency stop with lightning quick reflexes.

> **KEYPOINT**
>
> **Urgent!**
> Level 3 situations occur too fast to think about your actions. You must have a preplanned action and have practiced it enough that it's instantaneous. Know the location of all E-stops. If the machine has more than one E-stop button, practice turning toward it, and away from the danger.

Problems and Emergencies Caused by Operators

Here are the four main actions of operators that cause problems. They are organized in their rough order of occurrence (author's opinion) within a long production run.

1. Not loading the part right.
   Misaligned, loose clamping pressure or part upside down or backward.
   Parts shift, breaking cutter. Or parts fly from setup.
2. Not removing chips well enough during cut.
   Keeps coolant away from the cutter—loading and breakage.
   Work hardening and poor finish.
   Causes chip loading in mill cutters.
   Can catch and destroy work finish on lathes.
3. Not keeping coolant streams aimed at critical cuts (mills only).
4. Using cutters too long.
   Avoidance of changing dull cutters too long.
   Results in vibration and poor finish with a possibly oversize part.

Boredom—The Enemy of Efficiency (and Promotion) If you allow it, long or repetitive CNC work can lull you into a less alert state. Here are a few suggestions as to how to avoid doing so:

Monitor the progress.

Improve part turnaround time (information next).

Remove burrs from finished parts.

Measure a critical dimension.

Record measured results in computer database.

Stack it in the outgoing pallet or tub.

Pick up next part from the "in pile" ahead of time to use it.

Blow off new part.

File and remove burrs on new part.

Place prepped part ready for next load.

Turnaround Time Until given the responsibility of editing the program, you can't do much about inefficient movements in the program. But you can analyze your own actions. Many new operators seem to be working furiously for a short time then doing nearly nothing for the remaining cycle time. Look for tasks that can be accomplished while the machine is running.

It may seem trivial at first, but 10 seconds cut off the cycle time can make a big difference in delivery schedules and job costing. Remember, one of the main advantages of CNC is the rapid cycle times.

> **KEYPOINT**
>
> Your job is to refine cycle times without compromising quality or safety. Don't rush, work efficiently.

Debugging Was Once Real Finding and correcting errors is called debugging the program. We assume the problems are called bugs—right? History proves differently.

Instead of silicone chips and transistors, the first monster computers controlled electronic pulses using vacuum tubes roughly three times the size of your thumb. The control mechanism within each was a pair of flat metal plates. Heating one changed the conductivity of the other but the heat needed also caused one of the plates to glow with a small orange light.

Literally hundreds of these vacuum tubes were required to operate the computer but their collective, happy glow was just too much for congenial insects! Every so often, technicians had to open them and using a brush and vacuum truly debug the system of suicidal guests.

UNIT 20-3 REVIEW

Replay the Key Points

◆ Know the location of all E-stops. If the machine has more than one E-stop button, practice turning toward it, away from the danger.

- The best operator action during these times always results from preplanning and practice. Practice every so often.
- Your job is to refine cycle times without compromising safety. Don't rush, work efficiently.

Respond

1. Describe the nature of a level 1 problem and give an example using a mill cutter.
2. Identify several ways to stop machining to investigate a level 1 situation.
3. Describe a level 2 situation and give an example that might require action.
4. What must be done in a level 2 situation?
5. Describe an example of a level 3 situation and what you must do.

Figure 20-30 *Ears are the best method for detecting an early problem. Personal music is a bad idea.*

UNIT 20-4

Monitoring and Adjusting

Introduction: This unit explains several ways to stay in control for safety and efficiency. After exploring machine monitoring, we'll look at compensating and changing tools.

KEY TERMS

Control chart
A statistical method of detecting trends requiring offset adjustments.

Critical dimension (key dimension)
One or more dimensions chosen by function, that represent all dimensions made by a single tool.

Position page (screen)
The control monitoring page that displays axis positions and distance to go information.

Monitoring Production

Absolutely the number one method of knowing how the cut is progressing is the sound it makes. That's why personal ear-sets are almost always banned in machine shops, while a few permit radios on benches if not turned up too loud (Fig. 20-30).

Next, we keep an eye on cutter progress, the finish produced, chip ejection, and coolants. But due to safety shields and to flying coolants and chips, visual monitoring isn't always possible, so we turn to technical solutions.

Alternate Monitoring Methods

There are three options of tracking what's going on inside your machine when you can't see the action directly:

Camera   Controller position page
The graphic monitoring page

Video Camera An add-on accessory, not all machines feature video cameras, however, they work well where machining is obscured behind screens covered with coolant and chips.

Controller Position Page This is arguably the choice most often selected during normal program monitoring and unarguably the best screen for program tryout. The **position page** can be arranged on most controls to show one or all three displays:

1. The present coordinate position and kind of motion being made
2. The destination coordinates
3. The distance to go (DTG) to complete current command

Controller Graphics Page Modern CNC controls offer varying levels of graphic sophistication. The latest show color differentiated cutters, raw work, and finished cuts in animated real time. Older graphic screens depict the tool path similar to wireframe geometry.

Monitoring by Measuring

An essential way to monitor production is to measure certain carefully selected **critical dimensions** called **key dimensions** and

A Control Chart

Figure 20-31 *Control charts track progress and highlight trends that need operator intervention.*

surface finish. They indicate the condition of the cutting tools. If this information is fed into a computer program capable of creating a chart called a **control chart,** as shown in Fig. 20-31, the operator can get a good picture of tool wear and other factors such as heat buildup in the setup. For more information on this ability, read Section 27: Statistical Process Control, SPC.

From the example chart, it's easy to see that if the operator takes no action, soon the parts will go beyond tolerance. Using this kind of data, an operator can also predict how many parts might be made before it's time to change cutting tools.

Finding Key Dimensions Often in CNC work, when a single tool machines several details on one workpiece, the operator need not measure every dimension it produces. For example in Fig. 20-32, a representative diameter and length can be chosen such that if they remain in control, all others normally follow if machined by the same tool.

If the object has varying tolerances based on function, then the key dimensions should be chosen from the tightest.

Single Tool Machines—All Details

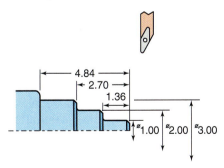

Figure 20-32 *Key dimensions represent many features as shown here, or can be the most critical features of the part.*

Sometimes the engineering drawing will note which dimensions are to be used for this tracking purpose and there may be a symbol for which dimensions are key to the function.

Maintaining Size and Shape

Offsets As shown in the control chart (Fig. 20-32), if a trend shows in key dimensions produced by a single cutting tool, the likely cause is tool wear. You'll have two options.

◆ Adjust the tool offset (plus or minus depending on type of program)

To fix a trend like Fig. 20-31 we normally change the offset. That causes more tool pressure against the workpiece, so as a fix it has limits beyond which the trend will degrade even faster. By changing the offset you are pushing the cutting tool closer to the work to return the dimension back to the target. So, there's more friction and heat size control, and tool failure begins to accelerate. Experience and SPC will tell you how many adjustments can be entered before changing tools (usually no more than two). Still, it is a standard practice.

◆ Exchange the tool for a sharp cutter

Trade Tip

Settling Down—Initial Tool Break-In Veteran machinists know there is an initial period when cutting tools wear away the fresh new edge then stop deteriorating for a long period (Fig. 20-33). This is especially true for new carbide tools. When the change stops, the operation is said to "settle down." Then at a given time, the changing dimension will begin again as the tool dulls to a given point. The real trade trick is the ability to predict how many cycles that tool will last before unwanted events occur.

Using commercial inserts, an edge prep hone operation has already been performed during manufacture. But on shop ground carbide tools, you can stabilize the break-in period by skillful use of a diamond hone. Remove from 0.003 to 0.010 in. from the cutting edge. As strange as that seems, this dulling improves tool stability because, using correct feeds, the tool's cutting edge is being depressurized by the rake face lifting the chip away from the parent metal. The extremely sharp edge tends to chip, while the slightly rounded edge stands up well to the action.

Initial break-in

From 0.003 to 0.010 radius
cutting edge-carbide tools

Figure 20-33 *Initial size changes as carbide cutting tools break-in. Then it stabilizes for an extended period as wear slows.*

Solving Dimensional Error When monitoring progress in an established run that was producing good parts, we often discover dimensions that begin to wander away from nominal. While changing cutter offsets will usually bring the dimension back to acceptable, you may be chasing a symptom of a problem, not attacking the cause. Before you can solve the root problem, you need to sort out what caused the dimension change. Here are seven general causes. There are others based more on specific situations.

◆ Dull tool
◆ Tool deflection—weak bending tool
◆ Part or setup flexing
◆ Mislocated PRZ
◆ Misaligned holding tooling or misaligned in the holding tooling
◆ Incorrect information in tool memory (wrong radius, offset, or length)
◆ Incorrect program

The final analysis might show a combination of causes. The upcoming activities present an opportunity to test your detective work in this area. Use this list to solve situations where parts are not coming out right.

Calculating Offset Change Amounts

Before adjusting tool nose radius, the following must be clearly understood.

For Lathes

1. **Reducing Nose Radius Values**
   Using smaller tool nose radius adjustments moves the cutter closer to the part geometry to remove more material on the workpiece. It produces smaller outside diameters and bigger bore diameters.

2. **Increasing Nose Radius Values**
   Adjusting nose radius values upward moves the cutter away from the workpiece to create larger OD dimensions and a smaller ID.

But, you can only uses nose radius offsets to solve dimensional errors, when *all diameters* produced by the same cutting tool are universally in need of adjustment.

KEYPOINT

Adjusting the nose radius cannot correct Z axis distance problems (critical thinking problem coming up).

But when you do change the nose radius, it's similar to changing the *X* axis micrometer dial on a manual lathe—the amount of change required is determined by dividing in half, the difference in the nominal dimension and the measured result.

For example, by entering a 0.003-in. smaller nose radius, the tool moves in 0.003 in. closer to the work but produces a 0.006-in. reduction in diameter.

Nose radius change value =

$$\frac{\textbf{Actual dimension} - \textbf{Target dimension}}{2}$$

Positive or Negative Nose Radius Change The formula just given produces a positive value when the work is too big compared to the nominal target—therefore the stored nose radius value must be reduced by the result to bring the tool closer to the work. If the formula produces a negative value, the actual size is smaller than the target and the tool nose radius value must then be made larger in the control by the calculated amount.

Mill Offset Change—Critical Question For profile milling cutters, it takes a bit of detective work to think out whether the correction needed is positive (away from the geometry line) or negative (toward the geometry

Calculate the Offset Value

Oversize 0.017"   0.017" oversize

Over by 0.017"

Figure 20-34 *Calculate the offset.*

line). Figure 20-34 shows three critical features, on three different parts, a tab, a slot, and a step. Each feature measures to be 0.017 in. too large. What radius offset correction is needed for each? Compare your answer to the response found at the start of the Respond for this unit.

Note that while the change in offset is truly a radius change, many controls work from mill cutter diameter. In such a control, the offset change must be doubled to equal the diametric change the new offset cutter diameter entered directly.

UNIT 20-4 REVIEW

Review the Key Terms

Control chart
A statistical method of detecting trends requiring offset adjustments.

Critical dimension (key dimension)
One or more dimensions chosen by function, that represent all dimensions made by a single tool.

Position page (screen)
The control monitoring page that displays axis positions and distance to go information.

Replay the Key Points

◆ Personal music machines are a bad idea when running a CNC machine.

◆ Offset changes must be calculated by measuring the key dimension and comparing them to the nominal target.

Critical Question Response

The tab requires a negative correction of 0.0085 in.—toward the line—half the amount measured oversize. The slot requires 0.0085 in. away—positive offset. *No offset can correct the problem on the third part.* Think this through, the problem can only be a mistake in the program. Changing the

offset would move the step right or left depending on whether it's a positive or negative correction, but it would not change the wall to wall dimension.

Respond

1. You are operating the CNC turning center on a first part run. After measuring four dimensions, you see a need for offsetting two tools. In Fig. 20-35, the actual measured results are shown in brackets. T01 cuts the 1.00- and 2.000-in. diameters and faces between them but does not face the end of the work. T02 cuts the larger face out to the dimensional limits of the stock size. Using Fig. 20-36 as a guide, enter your calculated offsets, a simulated tool memory page. Notice in Fig. 20-36 that since this is a first part with fresh tools, there are no initial offsets—only the nose radius for compensation.

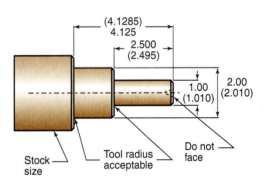

Figure 20-35 *Problem 1.*

Lathe Tool Memory

Tool	Name	Tool radius	Radius offset
T01	Turn–1	0.06	
T02	Turn–1	0.06	
T03	Part		
T04			
T05			
T06			
T07			
T08			

Note, the–1 behind the turning tool name indicates that the tools are both type 1 straight turning tools with a 15 lead. This ID number comes from the tool library

Figure 20-36 *Typical tool memory page.*

Critical Thinking

2. In monitoring a long run on the CNC vertical mill, you find the condition for a nominal 0.750-in. slot width as shown in Fig. 20-37. It is being profile milled with a 0.5000-in. end mill, T0101. This is a mature setup and has already been offset, as shown in Fig. 20-38. Calculate and rewrite the radius offset required to

Figure 20-37 *Problem 2.*

Mill Tool Memory Page 1

Tool	Name	Tool Diam.	Drum Location
T01	EM	0.500	13
T02			
T03			
T04			

Note, this number changes as the control tracks random tool storage management

Mill Tool Memory Page 2

Offset	Amount	Offset	Amount	Offset	Amount
01	−0.007	09		18	
02		10		19	
03		11		20	
04		12		21	

Figure 20-38 *Note the initial 0.007-in. offset on this tool memory page.*

bring the dimension to nominal again. Explain your calculated answer.

3. Explain why one tool might use two different offsets in a single program, for example, T0101 and T0102.

4. What does the DTG screen tell the machinist?

5. How is the DTG screen used for program tryout?

Section 20 Review

Unit 20-1

I've worked in a shop that produces airplane wing spars. The blank aluminum is several inches thick and 150 ft long. Each billet is worth over $5,000 before the job begins! It weighs over 2 tons when it's laid down on the big gantry mill bed. At first, due to the great weight, it needs no clamps to keep in place for the first few program runs. The raw material goes onto the machine by overhead crane with special lifting devices, but it comes off by three people picking it up at the ends and in the middle! The finished part weighs less than 200 pounds. To get to that stage, it took several programs and setups. The point is that work like that takes good documentation and communication. One error caused by misunderstanding a direction can cost thousands of dollars!

Unit 20-2

Troubleshooting a problem is simplified if you become familiar with the classes of alarms. Then step back and look at the facts. Was the program CAM generated or by command entry at a keyboard. Is the part geometry complex or simple? Then apply PEP logic (predictable error point). For example, if the program was written by hand key entry, always suspect a letter "o" was accidentally substituted for the number 0. If the geometry is complex and the program is CAM generated, suspect cutter shape interference. However, if it's handwritten, suspect a calculation mistake and so on. Before jumping right into scanning the program lines to find the problem, try a little logic on what kind of errors a particular program might have.

Unit 20-3

If you remember only one safety tip from Section 20, don't forget to practice what to do in level 2 and 3 events. Many times, I've seen operators turn a level 2 into a level 3 by their incorrect actions or by paralysis (no action). It's not enough to know the dials and knobs on the control. When a problem develops suddenly, you must move almost automatically. To be ready, one should practice the moves at the start of a shift or during a long job. Review how you'll halt and how you'll stop the machine.

Unit 20-4

The chances are that as an entry-level machinist you won't be changing offsets for a while. However, when the lead machinist does so, ask questions and ask if you can do it while he watches.

Questions and Problems

1. Safety question: Why is it so vital that the PRZ location be clearly documented by the programmer and understood by the machinist? (Obj. 20-1)

2. Why are program revision levels important to verify using the setup document and work order? Where is the program revision level found, in the document or in the program header? (Obj. 20-1)

3. What three ways are program Rev levels documented? (Obj. 20-1)

4. Identify the two kinds of information a CNC document conveys. That information can be found in two different places—where? (Obj. 20-1)

5. Describe a tool-compensated program. (Objs. 20-1 and 20-4)

6. Name the two kinds of programs that can be compensated. (Objs. 20-1 and 20-4)

7. Describe the nature of a level 2 incident and actions you must take. (Obj. 20-3)

8. What operator actions are required for a level 2 incident on a CNC machine? (Obj. 20-3)

9. True or false? The operator actions for a level 2 incident can also be used for a level 3, except they must be taken immediately rather than waiting until a natural place to halt. If it's false, what makes it true? (Obj. 20-3)

Critical Thinking Problems

Use this list of possible factors that might be at fault to solve the following problems in a CNC part run. (Obj. 20-4)

Clue List (partial—there may be other causes, too)

- Dull tool
- Tool deflection
- Part or setup flexing
- Mislocated PRZ
- Misaligned holding tooling or misaligned in the holding tooling
- Incorrect information in tool memory (wrong radius, offset, or length)
- Incorrect program

10. Using a compensated, part path program and a 60° diamond insert cutting tool with a 0.060-in. nose radius, shank setup at vector position 4 as shown in Fig. 20-39. After coordinating the setup and turning one part, the following dimensions are measured. (Objs. 20-3 and 20-4)

Print	Actual
1.5000	1.4995
2.7500	2.7505
0.500 R	0.565
45° × 0.125 chamf.	45° @ 0.190 each leg
2.3125	2.3123

Is there a problem? If so what might it be?

11. Using Fig. 20-39, the part is machined and the 1.5000-in. dimension is measured at 1.507 and the 2.7500 one comes out as 2.7568 in. The radius, chamfer, and the 2.3125-in. dimensions are all measured to be right on target. What correction(s) are called for? (Objs. 20-3 and 20-4)

12. In Fig. 20-39, the first run of the program shows the following turned dimensions. What's the problem and how can it be corrected? The tolerance for diameters and length is ±0.003 in. (Objs. 20-3 and 20-4)

Print Size	Actual
1.5000	1.5003
2.7500	2.7503
0.500 R	0.501
45° × 0.125 Chamf.	45° @ .123 each leg
2.3125	2.2135

Figure 20-39 *Problems 10 through 12.*

13. Per Fig. 20-40, the part shows the drawing measurements on the 1.500-in. straight diameter after making 74 parts. List the possible problems in their order of likelihood.

(*Hints*: Testing one part in every five, 40 parts were made found to be OK. Then the setup was assumed to be settled so no further measurement was done until the 75th part. All other measurements are found to be OK on that part.) (Obj. 20-4)

Figure 20-40 *What is the most likely cause?*

14. See Fig. 20-41, Horseshoe Fitting 204-56B. A setup is made according to the document sketch (Fig. 20-42) of the fitting blank.

Instructions

It has been precut 3.625 × 1.25 × 5.25 (excess for machining right side)

The program is a compensated part path
Using T01

    End mill the 0.875 deep pocket including the 0.50 radius

    Profile mill end to 5.125 dimension

T02

    Drill the hole.

Setup Sketch 204-56B (hyd. fitting)

T01 = 0.875-in., two-flute end mill
T02 = 1.000-in.-dia drill

Use parallel bars under part for drill clearance

Figure 20-42 *CNC setup sketch part number 204-56B.*

Figure 20-41 *Problems 14 through 17.*

On the first part, the condition shown in Fig. 20-43 is found to be true. What is needed to bring the four key dimensions exactly on target? (Obj. 20-4)

15. Using the tools as outlined in the setup document (Fig. 20-42) will the program for the 204-56B Hydraulic Fitting form or generate the 0.50 radius at the left side of the pocket? What radius will the cutter path follow? (Obj. 20-1)

16. In machining the 204-56B, you have set up a 1.50-in.-diameter end mill in place of the 0.875 and reentered the cutter diameter offset for it. What will be the problem if any? (Objs. 20-3 and 20-1)

Key Dimensions on First Part
204-56B hyd. fitting

3.25

0.7513

2.8735

5.010

0.7513

Figure 20-43 *Is there more than one reason behind these measurement errors?*

17. On the program tryout for the 204-56B fitting, the condition shown in Fig. 20-44 is discovered. What is the problem and fix? (Obj. 20-1)

18. Monday morning, you touch the green button to start your machine. What's wrong in each situation? (Obj. 20-2)

A. It will not start at all.

B. It starts to a certain point, the red light flashes, and further initialization is impossible.

C. It comes fully up and is a machine that does not require initialization. But now it will not move in jog or automatic modes.

19. Why would the control alarm out (stop machining) as you touch the cycle start button to make the 20th part after things have been going well? (Obj. 20-2)

20. What is a tool path *look ahead?* Describe how it might complicate finding an alarm condition. (Obj. 20-2)

Key Dimensions on First Part
204-56B hyd. fitting

5.126

3.2525

0.751

0.753

0.748

5.128

Figure 20-44 *Why might this part be tapered on the edge?*

Section 20 Answers

ANSWERS 20-1

1. *Manually compensated,* generates a cutter centerline for a specific cutter radius; *cutter centerline,* generates a cutter centerline with a target size cutter radius but can be adjusted plus or minus depending on real cutter used; *part path—part geometry,* generates a cutter centerline of zero radius offset until a positive value offset is entered in the control's tool memory. The most common program type due to entering the print information in the program.

2. After checking for a loose tool, enter a negative adjustment to the radius offset to account for tool wear. Or change to a sharp tool and re-enter the radius offset.

3. The 0.375-in.-diameter cutter produces a *0.1875-in. radius,* which is the largest size below forming the 0.25-in.-internal radii inside the slot.

4. The real problem is this was poorly planned! It should have been sequenced by holding the blank, then cutting the Datum C slot along with the details referenced to it, *at the same time.* However, to locate the center of the Datum C slot, on these parts, I put in a Datum C rotating diamond pin (see Fig. 20-45). When it's turned, it centers the slot over the pin's center. I then clamped the work to the Datum A locator block to do the machining.

Figure 20-45 *One possible solution—but the best would have been a better job plan sequence!*

5. Here are fifteen items of interest to the operator, found in the CNC document: print, part, and revision numbers; material type and heat-treat condition; grain and material form (castings, forgings, bars, etc.); part counts; the program identification; program location (central computer, disk, tape, and so on); the specific machine for which it is postprocessed; jig and fixture ID numbers for this job; any custom tooling required; the setup location of the PRZ; the tool numbers with types and sizes; the kind and location of all clamping/chucking or holding tooling; the sequence of events; any specific instructions particular to the job; hints and tips learned from previous runs.

ANSWERS 20-3

1. A change that may or may not be turning into a problem but warrants investigation. A cutter may be getting dull.

2. Let the machine complete the present cut then stop during a rapid travel move or at a tool change.

3. A problem that if not corrected will become more serious. The cutter is loading up with chips.

4. Halt the machining now but not with the emergency-stop button.

5. An all out emergency stop is a must. Protect yourself first, and stop the machine if that action is safe. Touch the big red button as you turn away from the machine.

1. Tool T01 required a minus offset of 0.005 in. Both diameters were 0.010 in. over, while the length was 0.005 in. short. By moving 0.005 in. closer to the geometry, this offset will solve both errors. T02 required a plus offset of 0.0035 in. to decrease the length problem.

Tool #	Nose radius offset
T01 0.06	−0.005
T02 0.06	0.0035

2. The cutter is making the slot 0.011 in. too big. The cutter was set *too close* to the geometry line. The current offset of −0.007 in. was too much. It needs to come *away* from the geometry line by 0.0055 in. to correct the 0.011-in. error. (Calculate −0.007 + 0.0055 = 0.0015 new offset.)

Tool #	Tool Dia.	Number	Offset
T0101	0.500	01	0.0015

3. One for roughing then one for a finish pass?
4. The distance to go to complete the present command?
5. Stop and verify that there's room to complete the distance to go before letting the machine complete the move.

Answers to Section Review Questions

1. Because all coordinated machine moves refer to the PRZ. To misplace it in the setup would mean a certain crash where cutting tools contact the stock or holding tools in the wrong location.
2. To avoid making the wrong part by running the wrong version of the program even the though the part numbers match. The program revision level can only be found in the program header.
3. By Revision Letter, by Serial Number or by Date
4. Setup facts; program run facts. Documentation is found in hard copy and embedded as remarks in the program itself (programmer's option but operator's might edit them in).
5. One that can accommodate varying tool sizes and shapes using offsets
6. Cutter centerline (cutter path) and part path (part geometry); length offset; diameter offset if required
7. A clear but moderate emergency that will get worse if action isn't taken. Touch the slide hold, or the midcycle interrupt, or turn feed or rapid overrides to zero, or turn the mode to single block.
8. Some form of halt—turn rapid travel to zero, select single block, or even slide hold.
9. False. A level 3 means hit the emergency stop right now.
10. The X and Z axis dimensions are in close control, only angular and curved features. Most likely, the control has the wrong nose radii stored for compensation. Since angular and curved surfaces are about 0.060 large, the operator probably forgot to enter a nose radius. If that's not the problem, then the program is written with the wrong dimensions for the 0.50 radius and 45° chamfer.
11. In this case, it's highly likely that it's a setup coordination problem. The X axis position of the turning tool, needs to be shifted inward 0.0035 in. (half the diameter change needed).
12. A small nose offset change (−0.0001 in.) might be used to take away the 0.0003-in. oversize diameters, however, since it's the first part, and only three-tenths over doesn't violate the tolerance, I wouldn't deal with it just yet. However, the length error is clearly a program problem. The 2.3125 dimension has been transposed to 2.2135 in. and must be edited. No setup change or radius offset can correct this.
13. Something is degrading. The tapered part could only be from
    Part deflection due to dull cutter.
        Change insert tips. Add insert change more often to operator's routine.
        Note, adding tailstock support will help reduce the taper as cutters become dull. It will extend the length of time between tool replacement but it will require a center drill cycle and slow the cycle time.

14. There are two items wrong here. *Program wrong:* Since the 0.750 wall on the left is only off a bit, yet the 5.125 dimension is very wrong, the problem must be the program data. *Needs cutter diameter offset adjustment:* The offset must be reduced by 0.0013 in. The three dimensions 0.75, 0.75, and 2.875 all show the cutter is 0.0013 in. too far away from the part geometry.

15. It will generate it with a cutter path radius of 0.0625 in. 0.500 part radius − 0.4375 cutter radius = 0.0625 in cutter path radius.

16. A control alarm for geometry interference. The larger cutter cannot generate or form the 0.50 concave radius, inside the pocket.

17. The part is rotated counterclockwise. There are two possible causes: vise is not indicated true to the X axis or the part was loaded on a chip or burr, causing it to be rotated.

18. A. Main breaker or depressed emergency-stop button.

B. Low supplies of air or oil or it could be on a limit switch.

C. Feed and rapid override dials were turned to zero when it was last turned off.

19. On the 19th part cycle, some vital supply became low. The machine completed the cycle but won't let you make the next part without replenishing it.

20. The control analyzes the tool path several blocks ahead of the current command to see if there are problems in the command logic, command syntax, or cutter interference. The control will halt on a command line that's several blocks ahead.

Section 21

Program Planning

Introduction

For me, writing and executing a successful CNC program is the ultimate computer game! I can still remember the first one I wrote. We instructors understand that students are more than anxious to do the same—create their own, then test it on metal. But experience proves over and again that while it's fun to do, starting there is a giant mistake!

Unguided, beginners often compile the program then try to solve the details of carrying it out afterward. Here in Section 21, we learn that a modest investment in planning before writing returns big dividends in productivity, quality, and safety. Along with organizational skills, Section 21 is also about communicating instructions to others. To do both, you will call heavily upon your manual machine experience. There are just two units required to plan out CNC work.

Objectives

The Final Challenge

In the review we'll tackle a couple jobs, planning them from start to finish. The last problem has two answers: one provided by myself, then a second by the students and instructor of the Machining Technology class at Bates Technical College in Tacoma, Washington. When I wrote my job plan I had no idea if our plans would agree or diverge, but I expected they would diverge. Both plans will make good parts. The point is, although a geometric science, this is very much a creative endeavor.

UNIT 21-1

Selecting the Origin, Quadrant, and Axes

Introduction: Selecting the axis orientation, the PRZ location, the holding methods, and the cut sequences are inseparable decisions. Each choice affects the other and they must all be made at the same time. Will we look at each as individual units only for book organization. The order in which these choices are made is so intertwined that there is no clear order in which to make them.

In some simple cases, the choice of holding tools can be made later at setup time (standard chucks, vises, and clamps). But many times in real CNC work, they must be preselected and sketched. Why? To avoid situations where your program accidentally machines through vise jaws when bar clamps would have been the right choice, or to avoid throwing parts from the wrong kind of lathe chuck. Clearly, everything must be well thought out before writing the program and those choices must be documented.

KEY TERMS

Coordinate shift
A programming technique used on both mills and lathes to temporarily shift the PRZ for safety or convenience.

Tooling reference—set point
An alternate method of establishing PRZ using a block attached to the fixture.

Touch method (touch off)
A physical method of setting a Z axis PRZ at the machine by touching the cutter lightly to the work surface.

Work stop/spindle stop
An adjustable stop to locate raw material in the same place with each part.

PRZ Selection

The choice of PRZ is usually a relatively easy decision even though it's critical.

KEYPOINT

The PRZ must be based on functional priorities and geometric datums if they are within the design. If you are not using a geometric design, then the most reliable and easily located surfaces should be used.

In some cases, PRZ selection is based on temporary surfaces aimed at machining the higher priority functions and features as soon as possible. Some of the PRZ choices involve work and/or axis orientation. We'll start with lathe work, the simpler issue, since it's a two-axis machine, then look at mills. Here are two guidelines for PRZ selection for either.

Choosing the PRZ Location— Lathe or Mill

Primarily, PRZ must reflect the datum/ dimensioning basis of the design.

Second, choose some physical location that's easy to locate during the setup—a large surface, a well-machined hole, etc.

PRZ Location for Turning Work

X Axis on Centerline To control diameters, the program reference zero always lies on the work center for the X axis. Diameters are always referenced from the lathe's centerline for turning work.

Z Axis PRZ While the Z axis location is often at the far right tip, as shown in Fig. 21-1, away from the headstock, occasionally, function

Typical PRZ

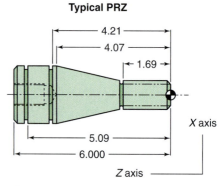

Figure 21-1 *A typical PRZ location for turning work.*

A different function places the PRZ elsewhere

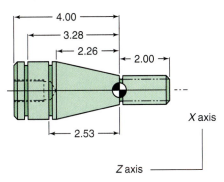

Figure 21-2 *Function occasionally places the PRZ in some location other than the work tip.*

will dictate *Z* location other than the tip, as shown in Fig. 21-2, where it's obvious where the PRZ should be. In this case, it's easy for the operator to locate it—2.00 in. inward from the outer tip, plus some excess material for machining, say 2.050 in. (0.050 in. extra for a face cut).

KEYPOINT

For Fig. 21-2, the shifted PRZ is a critical fact (the PRZ is not at the "usual" location) and must be absolutely clear in the CNC document.

In the case shown in Fig. 21-2, during setup the machinist will touch the facing tool to the right, outer tip of the work then set the controller position registers at *Z*2.050—a plus *Z* distance from PRZ. That means the *Z*-PRZ is 2.050 in. farther into the part. If that setup detail is missed, the program run would be a total disaster.

Work Reversal

Often, work must be machined on both ends, as shown in Fig. 21-3. Bar feeding can only make the external details and the internal features on one end. For example, it's going to require a second setup to machine the threads. There's two ways to solve this.

1. Machined Temporary Datum
   Machine features in from one end, to a selected *Z* axis position. Then remove, reverse chuck the part nested against your temporary reference, then complete the final features.

Requires machining on both ends

Figure 21-3 *This job requires machining on both ends.*

Figure 21-4 *A spindle work stop provides a Z axis locator.*

2. Spindle Work Stop
   If the work is being machined from cut stock with no chucking excess, one solution to controlling a reverse *Z* axis position is to use a **spindle stop** or **work stop,** as shown in Fig. 21-4. After machining the features onto every part in the batch, each is reversed and rechucked. The adjustable stop sets the *Z* position of the work based on the far end.

   When we use a work stop, the PRZ remains the same physical location on the work, but it's now inside the machine spindle. The machinist must set the *Z* position registers to represent a PRZ that the tool cannot reach!

KEYPOINT

Once again this must be crystal clear in the document. A sketch is usually the best solution.

Trade Tip

Chuck Jaw Faces A simple work stop can be the face of the chuck jaws in many cases. Often when machining custom jaws, a step can be faced into the jaws to function as a work stop.

To coordinate such a setup, with the PRZ inside the lathe headstock, the operator writes where the Z-PRZ is located relative to the cutting tool, even though the tool cannot travel there.

Caution! It's common programming practice to rapid travel the tool back to Z0.0 or close to it, when the next roughing or finish pass is to be taken. Forgetting that the PRZ is inside the headstock would result in smashing the tool into the chuck—a disaster! There is a convenient planning fix for these situations using a programming tool called a **coordinate shift.** *(See Trade Tip next.)*

Trade Tip

Shifting Coordinates There are times when the physical location for the PRZ on the work puts it in an unreachable location to coordinate it. It falls inside a chuck or beyond the work envelope on large parts, for example. In these cases, we can use a coordinate shift command. This program command sets up a temporary reference location, based on the master PRZ location. For example, in Fig. 21-5, positioning the tool 6.5 in. from the master PRZ, the programmer has moved the tool 0.5 in. away from the far tip of the finished work length—6.50 in. from the master work stop. All programmed Z axis moves are still relative to the original PRZ so no variation will be introduced. They now include the 6.5 in. The programmer can write the program using print dimensions then shift the reference point out to a location that's useful and safe.

From that command onward, all positions relate to the temporary reference (TPRZ), yet that reference fully

Shifting the coordinates

Figure 21-5 *Shifting coordinates can be a time saver.*

represents the original, 6.0 in. farther into the spindle. This saves tons of extra programming math and potential accidents. You will see many possible applications of this time-saving tool in both lathe and mill work.

Mill Program—Reference Zeros

With mills, we add one more guideline for PRZ selection, beyond the datum basis and easy position to setup on the part:

> Try to use the first quadrant if possible. This is a secondary issue of rotating the work such that it lies in the first X-Y quadrant with the best axis advantage (Fig. 21-6).

Top Surface—Lower-Left Corner The most common selection for mill PRZ is the top surface, lower-left corner shown in Fig. 21-6. We try to position the work such that the longer side is parallel to the X axis but that's only for convenience. Located that way, all absolute value coordinates on the geometry will be positive (it's not a big deal but it eliminates minus signs).

But that's not the only choice, as shown in Fig. 21-7. In each orientation, the datum basis of the design is preserved on the work. The only difference is the axis orientation and positive or negative values resulting when absolute coordinates are used.

Rotating the work to some other quadrant is often chosen to accommodate holding tools or work envelope space available, as shown in Fig. 21-8, where the vise has been turned sideways such that it will completely fit within the chip guard with the door closed. The long axis of the work is now parallel to the Y axis on the vertical mill.

Typical PRZ location
top surface lower-left corner

Figure 21-6 *Top left corner is a common PRZ location.*

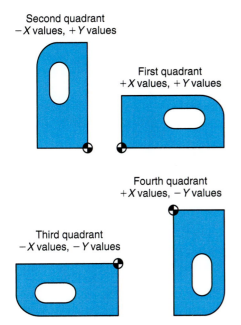

Second quadrant
−X values, + Y values

First quadrant
+X values, + Y values

Fourth quadrant
+X values, − Y values

Third quadrant
−X values, − Y values

Figure 21-7 *The same workpiece with the same PRZ, but located in each quadrant.*

Typical setup with work in the fourth quadrant

PRZ

Table tee slots

Y axis

X axis

Vise handle clears door

Safety door closed—top view

Figure 21-8 *This part orientation was chosen to fully contain the vise within the safety shield. This places the work in the fourth quadrant.*

KEYPOINT

Axis orientation and PRZ are preprogram decisions. Once the program is written, the setup must be made as shown in the document.

Question In Fig. 21-8, in which quadrant is the work placed with the vise rotated as shown? (See the caption for the answer.)

Touch Method for Setting Tool Reference Selecting the top surface of the work for the Z axis reference is a common method. It simplifies the task of coordinating the Z axis position of tools and for determining length offsets. It's done by just touching the top of the work with the tool tip—lightly with no hard contact. Better yet, the Z reference can be set by staying a given distance off the work then setting the Z axis position register to represent that height off.

This is known as the **touch-off method.** It's used on machines without tool probes or camera vision. We also touch the tool to the work on lathe setups. Touching a cutting tool to the work by jogging is not without risk. To do so, you must drive the tool toward the work using machine controls that have no sensitivity. Servo drives produce hundreds of pounds pressure without any sense of over-contact, which mars the work and breaks the tool! Don't miss the Trade Tip!

Trade Tip

Feelers for Tool Touching Here are two methods that improve touching off.

Use a paper feeler (Fig. 21-9). Slipping it back and forth as the nonrotating cutter is slowly jogged toward the work, contact will be obvious when the paper drags. The thickness of the paper is then accounted for in setting the tool position, usually 0.003 in. must be added to the Z axis register position.

An even better contact detector is a precision setup feeler (Fig. 21-10). They are usually made of a soft material such as plastic or aluminum to protect the work and cutter. Make your own to some easy math size such as 10 mm or 1.00 in.

Figure 21-9 *A paper feeler is a good practice when touching tools to work for Z axis touch methods.*

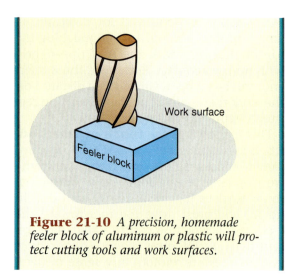

Figure 21-10 *A precision, homemade feeler block of aluminum or plastic will protect cutting tools and work surfaces.*

Tooling References—PRZ not on the Workpiece

Fixtures are custom-made holding devices that

- Hold odd shapes that cannot be held safely or reliably using standard methods.
- Provide quick part-to-part turnaround time.
- Reduce variation in production due to reliable loading and holding.

For now, we're concerned with a special type of reference provided by the fixture. Figure 21-11 is an example of an odd part shape that requires a fixture. In these cases, the raw stock is usually a casting, forging, or weldment, with no previous machining. In this

Figure 21-11 *The design has a datum basis, but until the rough casting is machined, there is no reliable place to indicate a PRZ. This part is ideally held in a mill fixture.*

Figure 21-12 *A mill fixture solves both holding and reference point location. Note the set point on the fixture is relative to the PRZ—it is not the PRZ.*

starting condition, none of the surfaces provide a reliable location on which the physical PRZ can be located.

In the example (Fig. 21-12), to cut the bottom pads to create datum A, and bore datum B, this part must be held in a fixture. Then the PRZ reference is planned and located on the fixture. These planned references are called **set points** or *fixture reference points*.

KEYPOINT

The PRZ is not the set point. The PRZ is a known distance from the set point at datums A and B. In Fig. 21-12, when the spindle is directly over the set point hole, it is then at Y2.000 relative to the PRZ.

Using a DTI to position the spindle over the hole in the set block, the operator then sets the axis registers to represent the position of the set point relative to the part. This information must be clearly sketched in the document.

The set point is usually a precision hole bored into a tooling block bolted to the fixture. The hole's axis is the location for the X-Y set point, and top of the block is used for the Z reference.

Trade Tip

Soft Jaw Vises and Chucks Custom-machined, soft jaws are handy shop-made fixtures. For Fig. 21-11, can you see a way to hold the part in custom-machined soft jaws? They could be machined to custom fit the work. If the part shape is simple, the tooling program is written by the operator or toolmaker. But there's another trade tip coming up in Unit 21-2. A part program can be edited and used to cut the soft jaws by switching cutter compensation to the other side of the geometry line. One program can be modified to make both the workpiece and a fixture shape that holds the part!

UNIT 21-1 REVIEW

Review the Key Terms

Coordinate shift
A programming technique used on both mills and lathes to temporarily shift the PRZ for safety or convenience.

Tooling reference—set point
An alternate method of establishing PRZ using a block attached to the fixture.

Touch method (touch off)
A physical method of setting a Z axis PRZ at the machine by touching the cutter lightly to the work surface.

Work stop/spindle stop
An adjustable stop to locate raw material in the same place with each part.

Replay the Key Points

- When choosing the PRZ location
  - Primarily, it must reflect the datum/dimensioning basis of the design.
  - If possible it should be some physical location that's easy to reproduce in the setup—a large surface, reliable machined hole, and the like. But sometimes, that's not possible, so tooling reference becomes necessary.
  - When a PRZ is in a location other than what might be expected, it must be made crystal clear in the setup document.
  - Keep in mind that axis orientation and PRZ are preprogram decisions. Once the program is written, the part must be placed as shown in the document.
  - With a set point, the PRZ is normally still on the part, not on the fixture. The set point represents a known distance from the PRZ.

Respond

1. What is the main guideline for choosing a PRZ location?
2. What is the secondary guideline for choosing a PRZ location?
3. Why would you choose to plan a set point for establishing PRZ?

Critical Thinking

4. What could happen if the PRZ was not coordinated correctly on the machine relative to its program position? (Two potential problems.)
5. Explain why, if possible the PRZ is chosen such that the entire part to be milled lies within the first quadrant.

UNIT 21-2

Selecting the Holding Methods and Cut Sequences

Introduction: Unit 21-2 builds upon setup skills gained in lab assignments using manual machines. But with CNC work, the method chosen must be rock solid to withstand the thrust and heat.

Single-point threading is a perfect example. On manual lathes, threading occurs at an artificially low rate to accommodate the physical drop of the half-nut lever. That results in light cuts because the speed is so low.

But on CNC lathes, the computer starts and stops the cutting tool in perfect coordination to the spindle. The CNC spindle can rotate near normal cutting speeds, therefore the cutting tool moves at amazing rates! The chuck grip and work rigidity must be able to withstand that force.

KEY TERMS

Sacrifice plate
A metal plate bolted above mill tables to protect them and provide a utility surface for drilling and tapping threaded holes for clamps and grooves for parallels and pins.

Subplate
Same as sacrifice plates.

Machining Center Holding Tooling

We start by investigating the differences in CNC holding on mills. We'll limit our investigation to vertical CNC mills, where there are three common holding selections for your plan:

- Precision vise or vises—with hard or soft jaws
- Direct bolt down—over a sacrifice plate or set up on parallels
- Fixture—shop or professionally made

Precision Vises

We need say nothing about hard jaw vises at this point in your training. However, with CNC work, soft jaw vises are used more often because we need to hold odd-shaped work and also provide a locator for a PRZ.

The vise jaws are custom-machined to fit the part using the CNC control. Often, the finished soft jaw set (Fig. 21-13) is given a tooling number and saved for future applications of the program. This fact must then be added to the CNC document under comments—that tooling already exists and need not be made each time the program is run.

Figure 21-13 *A shop-made soft jaw vise is custom-machined to fit an odd-shaped part, difficult to hold by any other means.*

Trade Tip

Cutting Both Sides of a Line Cutting a set of soft jaws on either the lathe or the mill is simplified if you

know this trick of the trade. By modifying the radius compensation commands in the program written for the workpiece, the cleaver operator can create a reverse image program. Here's how.

In machining the box cutter cams shown at the top of Fig. 21-14, for the second set of operations the need arises to hold their perimeters to machine the pocket and drill the hole in the bottom. To cut the vise jaws, the outer profile program can be modified by switching the cutter to the opposite side of the geometry line (shown at the bottom of the illustration).

The cutter is traveling along the left side of the geometry to cut the workpiece. But it's switched, by editing, to the right side to machine the reverse image vise jaw. Using program data already in existence saves time!

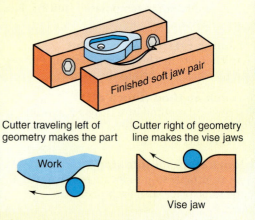

Cutter traveling left of geometry makes the part

Cutter right of geometry line makes the vise jaws

Work

Vise jaw

Figure 21-14 *Using first left-hand, then right-hand compensation, both the work and the vise jaws can be machined using the same program.*

KEYPOINT

Caution!
This is an advanced editing challenge that requires more training to perform. There are several modifications to be made to the program to eliminate unwanted tool motions, tool changes, and so on.

Direct Bolt Down—Sacrifice Plate

Bolting directly to the machine table is less common in CNC work mostly because it's slow to replace production parts but to also keep fast moving cutters away from the machine table. We use bolt-down holding when the work won't fit into a vise or to gain a bit

Figure 21-15 *This* sacrifice plate *on the machining center protects the table and also adds useful setup features.*

more *Z* axis work envelope for an extra tall workpiece.

For protection during bolt-down holding and for other setup utility reasons, a protective flat plate called a **sacrifice plate** or sometimes a **subplate** is usually bolted to the CNC mill table. These thick, aluminum plates (shown in Fig. 21-15) serve two purposes in CNC work, they

◆ Protect the machine table.
◆ Provide a soft utility surface for creating setups.

High-density, reinforced plastic is occasionally used as a sacrifice plate, too.

These plates often feature premachined, precision slots to fit vises or slot blocks, saving alignment time. Precision round dowel pin holes are sometimes bored to align work directly. They also feature a matrix of tapped holes, usually $\frac{1}{2}$-13 UNC the common setup stud size. They are spaced regularly in *X* and *Y* directions.

Custom Machining the Subplate A quick fixture can be made on top of the plate, as shown in Fig. 21-16.

> **KEYPOINT**
> Not all shops allow machining the subplate to create a special setup. Ask first!

While they are considered consumable, these utility plates are costly to replace. Make every effort to use features already machined on them. As cuts are made to the plate, it becomes an unstable platform, so there comes a time to resurface them with a fly cutter or replace them when they become too thin.

A quick fixture

Figure 21-16 *A typical quick, shop-made fixture features a slot fitting a pair of slot block parallels, a bored end alignment hole, and an array of tapped clamp holes.*

Figure 21-17 *A heavy-duty shop-made fixture in use on this twin spindle Cincinnati Vertical CNC machining center.*

Machine Fixtures

Discussed previously, fixtures are a commonly used work-holding tool in industry (Figs. 21-12 and 21-17). These custom-made holding tools range from inexpensive shop-made to the designed, professionally made fixtures. Costing from a few hundred to thousands of dollars, fixtures are a good solution for some CNC plans, but not for others.

> **KEYPOINT**
> When using a designed fixture in your job plan, be sure there is no other less costly way to hold the work—and deliver the needed repeatability. Custom fixture cost is usually not justified unless they are to be used more than once or there is absolutely no other way to hold the work safely or efficiently.

ShopTalk

Creative Solutions Observe solutions in your shop. Collect creative ideas and think up others. Mandrels, rotary tables, and rotary chucks are all used—use the building blocks gained in manual machining. I've even been forced to use double back tape to hold a very thin workpiece on a mill table!

Vacuum Chucks One useful holding tool for mill work is the vacuum chuck, as seen in Fig. 21-18. The mill sacrifice plate just mentioned can often be used to make a vacuum chuck by plugging the bolt holes then machining the sealing details into the plate. A vacuum pump made for the purpose is used to hold the work tightly to the seal O-rings cut into the plate. For CNC router woodwork, vacuum holding is the universal choice (Fig. 21-19).

Vacuum chuck

Machined O-ring groove seals vacuum between part and plate

Vacuum inlet pipe thread hole

Figure 21-18 *A vacuum chuck is made by machining an O-ring groove in a plate with a vacuum channel to distribute the holding force. A vacuum pump supplies the energy to hold the work to the plate.*

Figure 21-19 *This giant gantry mill is producing three wing spars at one time, all held down with vacuum chucks. Notice the large chip volume.*

Turning Center Work-Holding Tools

Similar to manual lathe work, there are the familiar selections of four-jaw chucks, mandrels, and face plates, but they are normally used for one-off work or when a three-jaw or collet chuck cannot be used. The three-jaw and collet chucks are used for production. With both, there are differences from those on manual lathes. In all cases, long work is supported by a manually set tailstock center, or the faster programmable tailstock.

◆ Common
  Three-jaw chucks—hydraulic with hard or soft jaws
  Collets—hydraulic
◆ Less useful
  Face plates/turning fixtures—bolted to face plates or chucked up. With these any out-of-balance condition is very dangerous thus the speed must be kept low.

Four-jaw chucks can be mounted on CNC machines equipped with cam lock spindles. They are only used in very short run situations or for single parts.
  Mandrels

Hydraulic Three-Jaw Chucks

There are five differences in three-jaw chucks made for CNC turning centers, compared to those on engine lathes. First, the chuck is made for extreme duty including much higher RPM. Second, the scroll plate is replaced by a hydraulic pull yoke, inside the spindle housing. CNC chucks aren't tightened with a wrench, because doing so would be too slow and the grip would vary depending on the operator. We've previously looked at the safety foot switch that actuates the chuck—the third difference. It's interlocked such that once the chuck tightens and the program commences, the chuck cannot be accidentally opened.

KEYPOINT

Reminder
The force developed by a hydraulic chuck can crush hollow parts and must be adjusted up or down by the operator. Experience will show what pressure is enough to hold, but not destroy or mar, the work.

The fourth difference is the jaw travel range from open to closed. Limited by the hydraulic action on CNC chucks, the open-close range of about 5 mm is adjusted to fit the work diameter. That's accomplished by bolting the jaws in place along a serrated carrier (Fig. 21-20). Commonly, each serration represents 1 mm of radial change for the jaw.

Figure 21-20 *The hydraulically powered chuck's grip radius is adjusted through precise serrations on the back of the jaw.*

Position the jaws such that they grip the work at about half their travel range to allow for raw part variation.

Consider Soft Jaws
For many reasons, they are a creative solution for CNC work holding.

Closing Sense The fifth CNC chuck difference is that on most industrial turning centers a closing sense for inside or outside gripping action must be preselected by the operator. Set for outside grip, they close inward and interlock upon reaching the full pressure set by the operator. Only then, with the green grip indicator on, will the spindle start (Fig. 21-21).

Safety Note! To set up a three-jaw hydraulic, the operator must place all jaws at equal radii such that they will close on the work at about *half their full travel*. If not, when there is an unusually small diameter part within a batch, it might not be held as the carriage runs out of travel.

Hard or Soft Jaws Soft aluminum jaws are very common, but we also use steel jaws for CNC production work. Correctly bored, these jaws can serve the function of a work stop while gripping the odd-shaped part. Another advantage arises when tight concentricity is an issue. By boring the jaws in the setup in which they will be used, the work centers on the lathe's axis with nearly zero runout.

When custom jaws are unbolted then reused at a later time, they are normally lightly rebored, due to the impossibility of refastening them in the exact location of the original bore.

Hydraulic Collets In our manual machining discussions, we considered collets a medium-grip work holder. While CNC collets don't grip as securely as a chuck, they hold strongly enough due to hydraulic closing, to do moderately heavy roughing of work up to around 1.5 in. in diameter.

CNC spring collets are heavier than their manual counterparts and stronger, but they can be broken. Be certain the bar stock to be gripped is within the range of the collet. Work that is too small will bend or break the collet leaves when flexed beyond their working range. Be certain to adjust the hydraulic pressure based on the size and strength of the work. Like chucks, collets can crush tubing. CNC collets are also supplied (or custom machined) to fit odd-shaped work.

Most lathes require the grip switch set to outside when using collets.

Turning Fixtures Due to safety issues and expense, lathe fixtures are considered a last-resort decision. You would choose a fixture when the work could not be held any other way and the job justifies the expense.

Due to centrifugal forces combined with cutting forces, CNC lathe fixtures must include bullet-proof clamping and extra-good balancing. Fixture making and use is beyond the scope of this unit. However, their care is something we must discuss.

Fixture Care Toolmakers work for days on fixtures—they are precise and costly. The tolerances used are below one-third that of the critical dimensions of the intended work. They *are expensive and must be treated with the greatest respect.* Here are the seven commandments of fixtures:

1. Never modify them without permission. If modified to any great degree, a note in the CNC document is necessary.
2. Keep all clamps, pins, and other attachments with the fixture when you are through with it.

Figure 21-21 *A grip sense switch tells the machine which orientation is the safety close position. If no closed feedback exists, the machine will alarm out and not start.*

3. Oil all steel components before long-term storage.

4. Always clean fixtures before mounting on the CNC machine and after use.

5. When loading parts, do not use hammers, pry bars, and so on. Fixtures are tough but can be damaged!

6. *Know the fixture before using it!* If the part can be inserted wrong, determine which way is right. Also study the clamping. It might be possible to distort the work with excess clamping force.

7. Read and check all instructions provided and be certain the part and fixture numbers match the work order!

How to Get Started Writing Your Plan

To begin review the engineering drawing for these five major factors:

1. **Critical Features and Dimensions**
   Look at tolerances and datums.

2. **Scan the Overall Work Shape and Special Features**
   Do these details create special cuts or holding considerations?
   They may require special cutters or holding tools.

3. **Know the Material Type, the Raw Billet Cut Size, and Consider Excess**
   Remember excess material may save money in the long run.

4. **Number of Parts**
   Doing this may justify certain decisions about tooling.
   Do future runs of the same part justify better tooling costs?

5. **Raw Material and Finished Part Value**
   May justify cutting and holding tooling decisions.

As you solve the problems coming up, consider these five factors. It's rare that a perfect answer shows itself. It's entirely possible that your plan will have improvements over mine. With each, there will be a blank CNC document to copy then fill in.

Don't Dismiss Manual Machining

Due to book organization, we've presented machining as though it was either manual or CNC, while in industry, machining is simply machining. In other words, a job might efficiently start on the manual machines, squaring up or rough turning, for example. Then it might be moved on to the CNC area for the main machining, then back for offload operations. For example, the bar stock of the first problem has been premachined by facing one end plus center drilling on both ends. Those operations could be done on a manual lathe, if the CNCs were jammed with work.

If they can deliver the work at the required rate, using manual machines often improves the price of the final product. The manual machines are far less costly to own and maintain. Use them to allow expensive CNC equipment to do the work, only they can do.

UNIT 21-2 REVIEW

Review the Key Terms

Sacrifice plate
A metal plate put on mill tables to protect them and provide a utility surface for clamps and grooves for parallels and pins.

Subplate
Same as sacrifice plates.

Review the Key Points
- Not all shops allow machining the subplate to create a special setup.

- Use a designed fixture in your job plan only if you are positive there is not a less costly way to hold the work.
- The force developed by hydraulic chucks can crush hollow parts and must be adjusted up or down by the operator.
- Always consider soft jaws in CNC work.
- Depending on the need be sure to select the interlock inside or outside grip.

Respond

1. When would the machinist choose a fixture for a CNC plan?
2. Based on Unit 21-1, what program reference is most likely when using fixtures?
3. What is different about a three-jaw chuck on CNC lathe compared to a manual chuck? (There are two facts.)
4. True or false? When setting up a CNC three-jaw chuck, one must set the jaws such that they close at the middle of their grip range.
5. In addition to the midtravel range, what other consideration is needed for setting CNC three-jaw chucks?

Section 21 Review

Unit 21-1

On the Space Shuttle there are liquid oxygen booster turbines that cost in excess of $15,000 each and you could hold them in your hands! The program that made them had to be right! But no matter whether it's a high-value part like that or an assignment for training, all programs need to run right. Yet for any given part geometry there are probably over a hundred different plans that all lead to a finished part. How do you know yours is among the best? There are three gages in this order:

Operator and machine safety

Part quality—variation is minimized

Cost of manufacturing

　　Over-all cycle time

　　Value- and nonvalue-adding time (no chips) in the cycle

How much handling—more than one setup?

Does it cause offload processing and burring?

Cutter tool life and cutters costs

Excess raw material consumed

Unit 21-2

Real experience is the only teacher of this skill. Observing others will help but doing a plan that matters is the real test. To get it going, after a few questions, there are two challenge problems in the review, one for turning and one for milling. Neither is easy. Go at them as though the job was real! In the workplace, I always consult others on a difficult plan. Do the same here—bounce your ideas off another student. Because the CNCs can do so many more operations in one setup, compared to manual machining, every job plan for a new shaped part is radically different. The possibilities are nearly unlimited.

Questions and Problems

1. Before planning out a job there are five factors that should be considered. Name them. (Obj. 21-2)

2. When would a coordinate shift be used? Why? (Obj. 21-1)

Critical Thinking

3. A milled part is very large. Its end must protrude out beyond the area where the spindle can be brought over the physical PRZ can be coordinated? Would a coordinate shift or a fixture reference solve the problem? (Obj. 21-1)

4. True or false? In setting up a part you discover the vise protrudes out beyond the safety door. It won't close. But you can turn the vise 90° (with the part) and it closes. You can edit a coordinate shift into the program now to make everything work. If the statement is false, what makes it true? (Obj. 21-1)

5. Name at least three important details that are different about setting up a CNC hydraulic three-jaw chuck compared to a standard three jaw? (Obj. 21-2)

6. Name an improvement to aid in the touch-off method of coordinating cutting tools. (Obj. 21-1)

7. Is it desirable to place milled parts within the first quadrant? Why? What is the advantage? (Obj. 21-1)

8. What are the major factors in PRZ selection before writing a program? (Obj. 21-1)

CNC Planning Problems

9. **Turned Stub Arbor** (Objs. 21-1 and 21-2)

    Per the work order instructions on the next page and the part drawing, Fig. 21-22, plan the holding methods, cut sequences, PRZ location and special instructions such as rechucking or part reversal.

Be sure to sketch any details you feel the operator will need. A blank CNC document is included on page 814 to be copied then used for this and the next assignment. There is also a blank master in the Instructor's Guide for this book.

Figure 21-22 *Problem 9.*

Tolerances	Notes			
X.XX = +/−0.010 X.XXX = +/−0.0005 X.XXXX = + 0.0000 − 0.0005	All dia not specified to be concentric to Ⓐ 0.010 45° × 0.125 chamf × 4 Break all sharp edges	**CNC planning**	**Stub arbor**	FA-2122-1

Drawing labels and dimensions:
- 5/8-18 UNC
- Detail X3
- 1.75
- 2.5 drill depth
- American std #40
- Ⓐ
- 1.75
- 0.50
- Stock dia.
- 4.95
- 4.20
- ◎ ⌀0.001 A
- 1.00
- 0.7500
- 1.00
- 0.5000
- ◎ ⌀0.001 A
- 5/8-18 UNC
- 0.125 × 0.062R
- 3.75
- 2.50
- 3.45
- Rev A Add section view
- Rev B Updated concentricity

Work Order—Stub Arbor
Make Qty-50 Part Number FA-2122-1 Rev B Sht. 1 of 1
4340 Bar 2.50 dia stock—Cut to length plus 3.0" chucking excess
Prefaced one end and center drilled—both ends
Stress Relieved and Heat-treated to R$_c$-24/48
Lot 1 of 50 lots*

*More parts to be made over many months.

Step One—Organizing the Major Factors

From the print, make notes about the five job characteristics:

☐ Critical tolerances and functions
☐ Overall shape and condition
☐ Material available
☐ Number of parts
☐ Value and complexity

Step Two

Now, select the type of turning tools needed for the job and assign them tool numbers on the document (Fig. 21-23).

Step Three

Now, with a pretty good feel for the part and program, plan out the PRZ, holding tooling, and cut sequences on the document. If, in your opinion, another operator might not understand printed word instructions, draw sketches. Assume this may be a long-term instruction used repeatedly by others in the shop.

McGraw Machining-CNC Documentation

Print Number FA-2221	Part Number - 1	Rev Level B	Sheet 1 of 1
Date	Part Count 50 of 500	Material 4340 HT	

Holding Tools/Fixtures		Name-Stub Arbor	
	Cutter Tooling		
T01	T02		
T03	T04		
T05	T06		
T07	T08		

Event Sequence and Special Instructions

PRZ Setup Sketch

Figure 21-23 *Photocopy and use this blank for your planning sheet.*

10. Valve Body (Objs. 21-1 and 21-2)

This is a product typical of small parts encountered in high-tech, electromechanical devices. Use Fig. 21-24 and the work order.

Hint: The end mill that will fit between the pockets, walls, and the $\frac{3}{4}$-in. post is 0.3125 ($\frac{5}{16}$). See the Shop Talk next for some vital background information. Note that the cutter must also have a 0.125-in. corner radius to form the internal radii between walls and pocket bottom.

Divergent Answers

To illustrate the different ways products can be made, I will offer my solution and for comparison, so will another instructor (Mr. Bob Storrar). Neither is superior. Compare yours to ours. Have your instructor or another journey-level master review your solution.

Work Order—Valve Body

Make Qty—20 Part Number FA-2124-2   Rev NEW
Sht 1 of 1

Aluminum Blocks—Grain direction not critical

Saw and premachine flat and square to 3.0 × 3.0 × 2.0

A one time only order—not to be repeated

Figure 21-24 *Problem 10.*

Section 21 Answers

ANSWERS 21-1

1. Datum basis on drawing or based on main functions of object
2. Reliable easily reproduced location on object
3. There is no way to locate it on the part, which is too rough or odd shaped. The reference is placed on the fixture, which may be the PRZ but more likely represents a given distance from it.

ANSWERS 21-2

1. When there is no other practical way to hold the work
2. A set point location relative to the PRZ on the odd-shaped work
3. A. CNC chucks have a limited range and must be adjusted to fit the work.
   B. They are *usually* hydraulic.

Answers to Review Questions

1. Critical features and dimensions with tight tolerances

   Overall work shape and special features

      Holding considerations
      Special cutters or holding tools

   Material type, the raw billet cut size, and consider excess

      Excess material may save money in the long run.

   Number of parts

      How many will be run this time?
      Do future runs of the same part justify better tooling costs?

   Raw material and finished part value

      May justify cutting and holding tooling decisions

2. When the physical location is beyond reach to coordinate (inside a chuck or outside the work envelope). The coordinate shift allows the program to refer to the original location using

Answers to Planning Problems

9. **Stub Arbors**

   This solution is a good example of the need to make special instructions clear to the operator. Read this closely, as there is a potential problem in loading the material.

4. Poor quality or scrap parts; a crash where the tool hits objects that weren't planned to be cut.
5. It's for convenience—by placing the part within the first quadrant, all *X-Y* coordinate values are positive.

4. False. The jaws are set such that they grip the work in the middle of their travel range to allow for raw stock variation.
5. The jaw radius is set by moving them in or out by 1-mm increments. Be sure they are all three at the same radius!

print dimensions even though it's beyond reach.
3. Either would work. However, the fixture would need to be built.
4. Very false! Rotating the part 90° would change the work axes relative to the program axes.
5. A. Pressure must be regulated
   B. Jaws must be positioned
   C. The interlock sense switch must be set for internal or external.
   D. Foot switch with interlock—no chuck wrench
6. Plastic or aluminum feelers or paper
7. No, it's only a convenience in that the part coordinates are always positive in the first quadrant.
8. *Primary*—must reflect the datum/dimensioning basis of the design; *secondary*—a reliable critical feature that's easy to reproduce in the setup—large surface, machined hole, and so on.

The 0.75- and 0.50-in. diameters and their concentricity to datum A, the taper, are the critical features for this program. The best guarantee of concentricity is to machine these details at one time. This requires one program and one

Figure 21-25 *Space grooved out to start right-hand turning tool.*

Turret-mounted work stop

After turret positions, pull work firmly up to stop

Figure 21-26 *The turret-mounted work stop in place to locate the material.*

chucking with no special tooling required. By facing the back to length, the chamfer and the tapped hole can be efficiently accomplished with an offload to a manual lathe by chucking the work on the 1.00-in.-diameter shoulder.

Tool List

T01 = Left-hand (LH) turning tool—roughing
T02 = LH turning tool—finish
T03 = Right-hand (RH) turning—roughing
T04 = RH turning tool—finish
T05 = 0.125 wide—radius nose grooving tool
T06 = Threading Tool—18 TPI
T07 = Grooving/turning tool (To open up space for RH turning tool for taper Fig. 21-25)
T08 = Material stop—in turret

PRZ

Big end of taper—4.95 from work stop—right end. Use work stop on turret as first program sequence to establish part position.

Holding Method

Three-jaw—no soft jaws needed, all diameters turned together. Use tailstock support.

Event Sequencing

REM: Grip work with stock farther within chuck than final position.
Cycle Start
  T08—Index work stop (Fig. 21-26)
  Advance turret to PRZ
  Program halt (yellow light)
REM: Open chuck
REM: Pull stock to right to touch material stop—establishes PRZ
  Close chuck
  Close safety shield

Cycle Start
  Withdraw work stop
  Tailstock forward
  T01—Rough out thread and shaft diameters—0.030 excess
  T07—Groove and side cut relief to start taper. (Note, if a side cutting groove tool is unavailable, make multiple grooves side by side. Cut beyond left end of part for tool shank clearance and part off on manual lathes.)
T03—Rough pilot and taper—0.030 finish excess
T04—Finish taper and pilot—do not chamfer
T02—Finish shaft diameters and chamfers
T05—Form thread relief
T06—Thread $\frac{5}{8}$-18 UNC external thread

Critical Question? There is a real potential for a crash with this sequence. It exists in the first few events establishing the position of the material in the chuck. What is the problem and what is the fix? Reread the operation sequence to determine an improvement. Answer following directly but think it out first.

Critical Answer

The Problem—The material could be misloaded.

If the operator didn't place the material far enough to the left initially, the material stop could hit the protruding bar as it advanced into position.

The Fix—Rewrite the load sequence:
Before starting the cycle, place the material in the chuck but untightened.
  Touch Cycle Start

Indexes material stop in turret then moves forward

This action pushes the loose material into chuck

Halt

Now push material against material stop.

10. Valve Body

(Fitzpatrick—Solution, compare to Bates Solution next)

Scanning the drawing, the critical functions are

The relationship of the central hole and the small hole in the flange, to Datums B and C

The angular surface to Datums B and C

The thickness

All these can be machined at one time by holding on the material below the flange (see Fig. 21-27). I choose to make the part in two setups. First with the island side up locating on prefinished edges. This setup will produce datum features A, B, and C on the work.

Program 1—Island Side Up
Tool List

T01 = 4-in. face mill—cuts datum feature A
T02 = $\frac{5}{16}$-in. end mill with 0.125 corner radius
T03 = Center drill
T04 = $\frac{11}{32}$-in. drill (ream size, for central hole)
T05 = 0.25 drill (directly drilling 0.015 position tolerance)
T06 = 0.375 reamer (finish central hole)
T07 = 0.25 Dia countersink—to break edges 0.010

PRZ

The intersection of Datums B and C at the top surface of the work. Note top surface will be cut for a minimum clean up in this program, which establishes datum feature A. This setup places the work in the third quadrant.

Holding Method

Standard, hard jaw vise. Datum B parallel to X axis. Locator on right edge.

Event Sequence

Coolant on

T01 Face upper surface half of excess to 1.875 dim
T03 Center drill for both hole locations
T04 Drill large hole through entire block—peck drill
T05 Drill small hole
T06 Ream 0.375 REM: make sure coolant is on reamer
T02 Rough and finish pocket and outer perimeter 0.05 in. below flange
T07 Chamfer 0.010 in., all upper surfaces

Trade Tip

Chamfering Edges In the preceding sequence, a chamfering tool finishes all machined edges with a small chamfer. While that adds a few seconds to the program run time, it saves hours of offload or bench/hand finishing of the sharp edges. Additionally, this creates a uniform edge. Chamfer depth and width are controlled using tool radius and length compensation.

Program 10—Island Down Phase
Tool List

T01 = 4.00-dia. face mill—finish top
T09 = 0.50-dia. end mill—roughing pocket
T10 = 0.375-dia. end mill—finishing pocket
T06 = 0.25-dia. chamfering tool

PRZ

Datum intersection B and C with datum A—down.

PRZ for Z axis is 1.875 above vise floor—requires sketch for operator.

Work is in fourth quadrant in this position

Program 1
Top face, profile, holes, and pocket

Material stop

Edge against vise solid jaw

Cut profile below half depth but above vise surface

Figure 21-27 *Machining the island side of the work first, locating material against Datums B and C.*

Figure 21-28 *The second program is held in a soft jaw vise with the square portion up.*

Program 2

Custom fit
soft jaw

Workpiece
contour down

Bates College progression setup

Fixture offsets

G54
G55
G56

Dovetail vise

Figure 21-29 *Problem 10, solution 2. This setup sketch shows three stations in a progression.*

Holding Method
Vise with moving jaw cut to accept rounded corner and flange (Fig. 21-28).

Event Sequence
T01 Finish top surface to 1.875 in. from datum A

T09 Rough pocket—rough and finish perimeter

T10 Finish pocket

T06 Finish edges and large chamfer of 0.375-in. hole

10. Solution 2 (Bates College Solution)
New terms used
Dovetail fixture (Fig. 21-30)
A work-holding method whereby a dovetail is precut upon the raw stock, then held in dovetail jaws.
Fixture offsets
Stored reference points that, when called in the program, fix the PRZ with reference to machine home.

As predicted, the instructor and students of Bates College devised a very different plan from mine. They chose to move parts through a three-vise progressive setup, whereas I used batch production.

Fixture Offsets To easily make this setup, the three vises are indicated parallel to the X axis, then bolted to the mill table in any convenient position (Fig. 21-29). Each vise's PRZ position is found using an indicator. It is then entered as a fixture offset under G54, G55, and G56, in the controller's offset page. The program refers to these reference points when machining the part at that station. Using this method, the vises can be bolted and indicated true to the X axis, anywhere on the machine—then when the program calls each using G54, for example, it works from that vise's reference until moving to vise 2 using G55 then vise 3 at G56.

Before reading the comments, carefully compare the two plans, to see the differences. Discuss this with your class and decide which you would use and why? Doing so will provide a lot of insight into job planning.

Dovetail fixture

End stop
flush with fixture

Can also be vise soft jaws

Figure 21-30 *Problem 10, solution 2. A dovetail fixture can be very useful where details must be machined on the entire profile.*

Problem 10—Job Plan
Two Vises Stations 1 & 3
Dovetail Fixture Station 2—sketch Figs. 21-29 and 21-30

Material:

3-in. square bar stock

prep> 2.22 long (1.875 for stock + 0.25 for dovetail + 0.1 kerf)

Cutters

Dovetail cutter	T01
3-in. face mill	T02
$\frac{1}{2}$-in. 2-FL Em 2.5-in long	T03
$\frac{3}{8}$ Bullnose EM	T04
Center drill #2	T05
$\frac{1}{4}$ drill	T06
0.375 drill/reamer	T07
$\frac{1}{4}$-in. chamfer	T08

Event Sequence

Dovetail	Station 1 G54
Face	Station 2 then St. 3
Profile + ear top	Station 2 Ear Bottom St. 3
Center drill/drill	Station 2
Ream 0.375	Station 2
Pocket	Station 2 then St. 3
Chamfer	Station 2 then St. 3

Three lessons gained in their solution:

Progressions versus Batches   They used a three-station progression that produced one good part per cycle of the program. After each program cycle, the parts are move forward one station. Figure 21-29 is a typical sketch of such a setup. Whereas my plan would complete all the parts in the batch, on one side, then make a second setup in soft jaws to finish the batch. (Mine could easily be converted into a two-station progression.)

A progression is better suited to onward production since the first part off the machine can go directly to the next workstation thus smoothing out job flow. However, when something goes wrong in a multistation progression, every station down the line from the hiccup must slow or stop unless a few buffer parts are suspended between stations to smooth out unexpected events. Progressions are an efficient shop structure but require more job flow planning. They cannot be easily mixed with batch production. It's usually one or the other, and both have advantages.

Critical questions—dovetail grip excess   Stations 1 and 3 are standard vises while the center is a *dovetail fixture* (Fig. 21-30). Dovetailing is a useful version of grip excess that uses specially shaped vise jaws or a dedicated dovetail fixture as shown. At station 1 a dovetail is made on the part. (Note that takes non-value-adding time to the part cost—is it worth it?) That allows a complete profile cut to be made at station 2. The Bates plan required an extra 0.25-in. raw material to create the dovetails. Again, that's usually an economical decision. Is it in this case?

Fixture offsets   The third lesson is their use of fixture offsets *G54, G55, and G56.* That allows them to mount and indicate each vise (station 1 and stations 2 and 3) where they are practical on the machine. They don't need to be lined up as in the drawing! After bolting them down true to the *X* axis, the operator then entered their machine coordinate position for each. That circumvents the nearly impossible task of placing three stations in an exact preplanned relationship to each other.

Each plan has its advantages and disadvantages but both would lead to quality products.

One final comment, their plan makes a complete profile cut around the perimeter at station 2, other than the shadow under by the small ear. That portion is completed when the dovetail is cut away at station 3. In contrast, my plan necessitates a profile cut on both sides—joining half way down the outside wall. Mine would load faster per cycle, since there are only two stations, but mine presents a chance of a mismatch between the two profiles. In this case, that's no big deal, but it could be on other parts and might require extra deburring to smooth the line where the two profiles join.

Section 22

Level-One Programming

Introduction

In this book, we'll approach programming in three stages, writing codes by hand first, then working up to CAM programming in Section 25. Here in Section 22 we study arranging code words into complete statements that prepare for or cause a machine action, called commands or command blocks. Often these sentences are shop-shortened to blocks. Then we will put the blocks together to write two kinds of programs:

Manually Compensated Creating a cutter centerline tool path without assistance from CAM software or graphic interface controls.

Tool Radius Compensated Writing a part path program with commands that tell the control to compensate for a given cutter size and shape.

To complete the shape assignments, you'll learn to write arc commands.

After Section 22, you should be able to write programs for most of your basic school assignments with the exception of branching logic statements and special routines (Section 23—second-level programming) and complex surfaces only possible with CAM assistance (Section 25).

Why Manually Compile Program Codes at All?

Today in industry nearly all production programs are generated using CAM software. It's a matter of efficient use of your time and staying up with the competition. CAM programming brings even more to the shop than turning hours of calculating and code writing into minutes, it also programs shapes not possible any other way.

So the question is valid. If it's time consuming and math intense, why not skip writing codes manually and go directly to CAM? The answer is fourfold.

The first factor is that to be in complete control of a machine tool, one must be able to read the program! Being able to read code means you'll know what the machine is about to do before it does it! It's also necessary to read code to solve what's wrong when an error message appears onscreen. Writing codes is the fastest way to gain these skills.

Second, the CAM program (or any program) is rarely a perfect work. There are places where it can be edited for safety or efficiency. With code and program structure background, someone else needn't be called in to make changes. You will have the skill to do them.

Third, during a machine setup the need often arises for a quick program to make soft jaws, a fixture, or some other provision for holding the work. These tasks you can do at your keypad if you can write code.

The last reason is that code writing is a backup skill. When a programming system isn't available or the PC goes down, the competent journeyman can still get the machine up and running.

Here are the units of learning that provide the fundamentals of *BAM* (brain-assisted machining).

Objectives

Code Words and Program Conventions

Introduction: To begin, we'll take a look at a code word vocabulary. It's not a full set, but these words form a good beginning. Then we'll investigate data conventions: how to make entries the control can accept. We'll then learn how to join the code words together into complete command blocks.

The main objective is not to memorize codes, although doing so will accelerate mastery. Here, we'll be focusing on the possibilities of the words. If you know the function exists, you can look it up when you need it.

Upon completion of Section 22, you'll be able to write and read most basic programs for lathes and mills, with the exceptions outlined earlier.

KEY TERMS

Alphanumeric code
A combination of a prefix letter with trailing numbers.

Code groups
Similar words that are mutually exclusive within any single command line.

Command (command line)
A complete statement causing a machine action.

Command word (word)
A part of an instruction that causes action or prepares for action or some utility function. The basic building blocks of programs.

Conversational
A CNC language based on spoken words.

Default
A fall-back or a preset condition built into the control.

Machine assembly language
The underlying, operating commands that your program causes. This language works behind and is caused by CNC programming.

Modular command
A word that stays on until canceled or superseded by any other word from its group.

One-shot codes
Words that effect the line in which they are written but have no further effect.

Prepatory codes (G codes)
Codes that cause or prepare for a machine action.

RS274-D (EIA Recommended Standard 274-D)
The standardization of code words for CNC programming.

Programming Languages

Three Ways to Compile Programs

When the control reads your program commands, transparently, without your knowing of it, it translates each word into a more basic set called **machine assembly language** (MAL). MAL works at the on/off level to process and send signals to actuate drives and start or stop various machine functions. Although MAL is the universal language working within all machines, writing programs in it would be very slow and clumsy for day-to-day use.

To speed and simplify the task, friendly languages have been invented. Similar to computer programming languages, each word in the vocabulary is actually a routine in itself. The MAL understands and breaks each code word down into machine-readable signals.

Today there are three different methods of compiling a machine-readable program.

1. **Graphic Interface**

    This method generates program commands from screen pictures. Mentioned previously, many control manufacturers offer this feature. While graphic programming is easy to learn, in fact it's designed with ease in mind, each system is designed for a particular control. Transfer of knowledge is simple but there's no universality between systems. Therefore we'll not study it here. Today, most graphic controls also convert their programs into EIA codes to facilitate editing and data exchange to other machine systems.

2. **Conversational**

    These controls are based on the spoken language of the country in which it is being used. If it's English, for example, then command words might be *go-to, go-around,* and *spindle on and arc.* Like graphic programming, conversational input is also easy to learn. But it too is controller dependant. What you know on one control will transfer to another manufacturer's product, but not verbatim. Even in the same language, they don't speak the same "dialect"! It is less common for a conversational control

Figure 22-1 *A typical conversational control is programmed using machinist language.*

to convert programs to EIA codes since editing is easy in the original form (Fig. 22-1).

3. EIA Codes (ISO)

This is the closest we come to a universal language set. The basic vocabulary is standardized by the Electronics Industries Association, EIA—Recommended Standard RS274-D, here in North America. Worldwide it's overseen by the International Standards Organization (ISO). ISO and EIA codes are virtually the same at the beginner level, and there are only a few differences at any level.

While it's close to universal, you will find code differences from control to control. Being over 30 years old, the code language was originally written for lower-level machine tools than those of today. But the originators foresaw the need for future expansion. They interspersed defined words with blank words that had no meaning, to be defined by the manufacturer of the control. Thus, the language had room for growth.

Today, every manufacturer uses the originally recommended core of 50 motion statements and 50 miscellaneous statements of a possible 100 within each set (one set for turning and one for milling). But, we've outgrown the total 200-word vocabulary some time ago.

Industry then created new words, starting with the 50 unassigned words within the basic set. Then as computers grew in power and speed, many more beyond the original 100 level were added.

Many of the first 50 new words have become universally accepted for CNC work simply because they became popular on Fanuc® CNC controls, a world leader in control distribution. These words could also be argued to be core words.

So the core remains a near-constant throughout the industry, with a few differences that are easily understood. The best news is that core commands comprise nearly all programs except the most complex of shapes and program structures.

Input Conventions

The flexibility of today's microprocessors and text editors makes it easy to write programs. Many rigid rules and conventions of the past have been relaxed.

For example, while it's good policy to put complete numbers into the program along with correctly placed decimal points, you needn't worry about leading and trailing zeros on entries:

10. or 10.0 are both acceptable input as much as 010.0000 (past practice).

Without the capabilities of today's CPUs, previously you would have needed to write:

0010060 *[tab]* 005000 *[tab]* 0200090
[end of block]

While today, you can write:
X1.006 Y5.0 Z20.009
Or you could also write:
X1.0060 Y5.0000 Z20.0090

If all necessary code and math components are present and the decimal point is in the correct place in each, the statement will be understood by the control. That said, flexible input and edit conventions may or may not be present on the machine you run. It depends on your machine's age and level of technology. Find out by reading the program manual.

Program Organization

Sometimes it's useful to arrange data in a form that makes reading it easier. Scan the two example side-by-side programs that convey the same data. Don't be concerned with what these programs do, assume they both do the same thing. Look at the data organization below.

Notice in the outline form, that all motions using T01 are set up so it's easy to tell how many axis moves and codes pertain to that tool. Also the tool is noted in brackets. It's easy to see how many lines T01 covers: from N020 down to N055 where the program changes to T02. Also notice that the same applies to each feed or rapid travel mode. N030 is a rapid move and so too is N035. Then on Line N040, the G01 mode, covers lines 40, 45, and 50. On line 45, it's easy to expect no Y tool motion.

Line by Line				
N005	G80 G40			
N010	G94 F20. G70			
N015	G17 G90 S1000			
N020	T0101 M6			
N025	M3			
N030	G0 X.5 Y0.0 Z.5			
N035	X.1 Y−0.1 Z0.0			
N040	G1 X0.0 Y−0.5 Z−0.5			
N045	X3.0 Z−1.5			
N050	Z−3.0			
N055	G0 X100. Y10. Z100			
N060	T0202 M6			

Outline				
N005	G80 G40			
N010	G94 F20. G70			
N015	G17 G90 S1000			
N020	T0101 M6	(0.5 End Mill)		
N025	M3			
N030	G0	X.5	Y0.0	Z.5
N035		X.1	Y−0.1	Z0.0
N040	G1	X0.0	Y−0.5	Z−0.5
N045		X3.0		Z−1.5
N050				Z−3.0
N055	G0	X100.	Y10.	Z100.
N060	T0202 M6	(#3 C'Drill)		

When given an outlined program, some controls will rearrange the data into line-by-line form as the program is downloaded, depending on the sophistication of the control's text editor. Others accept the data as-is and leave it in outline form.

CAM software generates line-by-line programs. However, tool and operation notes can be inserted and they can be arranged into outline form by editing.

Positive and Negative CNC Values

As a reminder, in CNC programs, a positive value is indicated by the number without a + sign. $X10.0$ is positive. However, the minus sign must be shown if the value is negative ($X-10.0$).

The value sign occurs differently than it would in standard math notation. Place the minus sign between the axis letter and the significant numbers. That order first indicates the axis or other code to which the negative value applies then the numbers follow.

Code Words

A code word is a single statement that

Sets a mode.

Causes an action.

Prepares for an action (sometimes called **prepatory codes** or words).

Starts or stops a function (coolant or spindle, for example).

Complete Commands

Program words are arranged into sets of **command lines** (shortened to **commands**). Or sometimes they're also called *command blocks*. Whatever you call them, a command is a complete instruction containing words and/or coordinates for one action. If the action includes axis movement, it's sometimes called a *motion statement*. The command might be as short as G00 $X1.0$, which has one word and one coordinate, causing a rapid positioning to $X1.0$. Or the command may contain several words, numbers, and coordinates longer than one entire screen line.

Controller Difference— End of Block

When the command line is complete, most controls accept the ⏎ (Enter on a standard keyboard) to signify the end of that command. For example,

G01 $X1.5000$ $Y-0.375$

After the 0.375 entry, I touched the same enter key, and the text editor moved down to the next line. The CPU inserted a transparent end-of-line code that signifies a finished command that the MAL understands. However, some controls require an *end-of-block* symbol entered at the end of the command. For example,

G01 X1.5000 Y−0.375**:**

Baseline Vocabulary

Code Word Prefixes

We've already seen a few words in previous chapters comprised of a prefix letter followed by numbers (called **alphanumeric code**). Your objective for now is to become familiar with six common word prefix letters plus a set of selected words. You'll find the entire vocabulary in the appendix of this book.

Question While scanning the code words, notice that they are in groupings represented by horizontal lines. See if you can figure out why.

G Codes These are the preparatory and the motion words. Since they start with G, they are also called *general* codes. G words with axis coordinates form the heart of the program.

Group 1

G00 Move at rapid travel—one or more axes
G01 Linear movement at feed rate—one or more axes
G02 Circular motion—clockwise at feed rate 2 axis limit
G03 Circular motion—C-clockwise at feed rate 2 axis
G04 Dwell—A stall for a specified time

Group 2

G17 Machining in the *X-Y* plane—mills only
G18 Machining in the *X-Z* plane—mills only
G19 Machining in the *Y-Z* plane—mills

Group 3

G70 Inch input (ISO = G20 and sometimes in the United States)
G71 Metric input (ISO = G21)

Group 4

G40* Cancels tool radius compensation
G41 Compensate cutter radius to left of geometry line
G42 Compensate cutter radius to right of geometry line

Group 5

G80* Cancels previous routines
See "Modular and One Shot Words"
G81 Drilling cycle
G82 Drill cycle with pecks
G83 Drill and bore cycle

Group 6

G90 Absolute value coordinate input
G91 Incremental value coordinate input

Group 7

G94 Feed at inches/millimeters per minute—usually used for mills but lathes understand G94
G95 Feed at inches/millimeters per rev (lathes and drilling)

*On some controls M02 on M30 also cancel compensation and routines.

Group 1

M00 Program halt—utility stop for operator actions
M01 Optional stop—operator selected or ignored
M02 Program end—reset to start
Can also use M30—stop and rewind tape—from days when we used punched tape as data storage media

Group 2

M03 Spindle start forward (lathe or mill)
M04 Spindle start reverse
M05 Spindle stop
M13 Spindle forward + coolant on
M14 Spindle reverse + coolant on

Group 3

M06 Tool change

Group 4

M07 Coolant on—spigot B (often assigned to mist coolant)
M08 Coolant on—spigot A
M09 Coolant off

M Code Words Miscellaneous functions or actions are the second most common words used in the program. They cause utility functions such as tool change, spindle on or off, and coolant.

Control Differences On older controls, to prevent electrical current surge and CPU data overloading, only one M code is permitted per command line. For example, entering M3 and M8 on the same command line would cause a program syntax alarm. Today's controls have no problem with multiple M functions. But that's not the reason for the groupings—keep guessing.

Notice that nearly all of this short list, other than M14, M15 and M30, can be abbreviated to a letter and a single number.

T Code Words—Tool Numbers and Time T words have two applications. By far, the more common is to denote tools as previously discussed. For example, T0101 M6, rotates the mill storage drum to tool 1, then changes to that tool and takes up offset 1 from tool memory. T words are also used for time functions in dwells.

Control Differences When Using T for Time Most controls accept G4 T10.5 to indicate a dwell of 10.5 seconds. On them, when the T follows the G4 (or some other command requiring a time span), the control accepts the entry within the same command line, to indicate time. However, others require some notation to indicate the T word is time not a tool: M4/T10.5 (for example, a slash is added to denote time).

F Word—Feed Rates

G94 G70 F20.0   Specifies feed rate motion at 20 IPM.

G95 G71 F.5   Specifies feed rate at 0.5 mm per revolution.

Controller Differences for Metric Feeds Note, most controls accept metric feeds in parts of millimeters, which would change the last example to F0005. Read your control manual for the specific form.

S Words—Speeds Causes the spindle (lathe or mill) to rotate at the specified XXXX RPM. For example, S2000 M03 starts spindle at 2,000 RPM forward rotation.

N Words—Line Number Addresses or Number Entry N code words are used in two ways:

N001 = Line Number Addresses Line numbers are used for program organization. In most cases, they can be omitted, although most programmers use them.

N20 = Numeric Value or Repeat in a Routine
For example, in a milling pocket routine, the cutter is to cut down N20 times by dividing the total depth into twenty small plunges downward.

Why Codes are Grouped?

Mutually Exclusive Words—Code Groups It's common and often necessary, to have several code words **groups** on the same command line. Each is needed in combination to complete the command. However, some G and M words are incompatible with each other and cannot be within the same command. They are *mutually exclusive,* meaning each overrides, conflicts with, or cancels the others from the same group just seen.

For example, you cannot ask a controller to rapid travel and feed an axis, all in the same command line or command it to turn the coolant on and off simultaneously!

Controller Differences Safety Warning!

Accidentally writing two exclusive G or M codes in the same command line will cause one of two possible events depending on the control:

Not dangerous—The machine will hang up on a syntax error code for you to sort out.

Dangerous!—A few controls will ignore all but the last code of that group, on that line. That could be a disaster if it wasn't the intended action!

Modular and One-Shot Code Words

Called **one-shot codes**—some codes affect only the line on which they appear. For example,

N005 M01

N010 G01 *Y*1.0000

At line N005, an optional stop has no bearing on line 10 following. These are called a *one-shot* or *one-time-only* words. We'll see others when we scan the EIA set in the appendix.

Modular codes or **modular commands** work as follows. Upon reading some command words, the control retains their setting. They stay in effect (turned on), until they are canceled by another word in a future command. The cancel word can be

A program end

Another word from the same group

A specific cancel word—The master cancel words are G80 (end all canned cycles) and G40 (end tool radius compensation), and some use G49 to cancel tool length compensation.

These *modal* words are similar to a two-position, toggle switch: if it's turned on, it stays on until turned off (canceled). Thus their

other common name: *toggled* commands. For example:

N005 G00   *X*20.0000

N010   *X*30.0000

N015   *X*45.0000

After line 5, the rapid travel is still in effect. All three moves occur at rapid, but

N005 G00 *X*20.0000

N010 G01 *X*30.0000

N015   *X*45.0000

would be quite different. Here, the G01 on line 10 overwrites the rapid mode of line 5. So, what action will occur on line 15? The machine will feed in a straight line to *X*45.0. The G01 is the mode.

Default Conditions

The word **default** has two similar but slightly different meanings in CNC work:

1. **Best-Guess Value**

   This is a fallback value that the control chooses when incomplete or impossible information is supplied. For example, single-point threading on a turning center. Suppose a previous RPM was set at 3,000. The next command line initiates threading of a large shaft with a fast lead. The machine cannot possibly coordinate the starting and stopping of the tool at the high RPM so it will fall back, or "default" to a lower RPM at which it can control the threading action.

2. **Preset Value**

   When first powering-up some machines, certain conditional values or modes will be assumed. They are the "default values" for that control; a bias to one condition or another, built in. For example, a given control always starts in absolute value mode until commanded differently. Each control exhibiting default values has it's own list.

Most newer controls have no default values. They start up in the modes and values set in the previous program. Therefore, make it your policy to set all expected modal conditions such as feed type, metric/inch values, and absolute/incremental coordinates at the start of all programs no matter whether the control has defaults or not. We'll learn to do this in Unit 22-2.

Outside Research

This activity will expand your list but again, it's for exposure to the possibilities. Ask your instructor for the handout listing codes of the machines in your lab, if available. If not, ask for permission to use the programming manual.

First, compare the codes in your lab to the short list in this unit. If you find differences, note them and mark them on the real vocabulary you will use in your lab. That's what any machinist needs to do when moving to a different machine, be certain of the vocabulary of noncore words.

Now, turn to Appendix VI to see the full EIA list of code words. Please notice that columns three and four contain check marks as to whether the code is "Function Retained"—modular, or "Affects One Line Only"—one shot. Compare that list to the codes used in your lab. They will be used for future reference as you move between controllers.

UNIT 22-1 REVIEW

Review the Key Terms

Alphanumeric code
A combination of a prefix letter with trailing numbers.

Code groups
Similar words that are mutually exclusive within any single command line.

Command (command line)
A complete statement causing a machine action.

Command word (word)
A part of an instruction that causes action or prepares for action or some utility function. The basic building blocks of programs.

Conversational
A CNC language based on spoken words.

Default
A fall-back or a preset condition built into the control.

Machine assembly language
The underlying, operating commands that your program causes. This language works behind and is caused by CNC programming.

Modular command
A word that stays on until canceled or superseded by any other word from its group.

One-shot codes
Words that effect the line in which they are written but have no further effect.

Prepatory codes (G codes)
Codes that cause or prepare for a machine action.

RS274-D (EIA Recommended Standard 274-D)
The standardization of code words for CNC programming.

Respond

1. List the five code word, letter prefixes and describe their usage.
2. What actions will occur on these three lines?
   N025 G01 G94 F19. *X*30.0000
   N030   *Y*1.0000
   N035 G00   *X*0.0 *Y*0.0
3. What will occur given this command?
   N001 G0 *X*1.000 G01 *Y*15.0
4. Working with a partner machinist, without looking back, list as many G code words as you can remember.
5. Now list the M words you remember.
6. Why are CNC codes arranged into groups?
7. True or false? In the example lines of Problem 2, line N030 is not a command, it is a word only. If it's false, what will make it true?

UNIT 22-2

Outer Program Structure

Introduction: This unit shows how programs start and end. There are a few commands needed to ensure that unwanted actions do not occur. In industry, there will most likely be a specific template for start and end program structure that will be pasted into every program. When CAM programming, you can compile the a program header and finisher, then save them to be pasted in at the right place.

Special Communication and File Symbols Before any safety codes are written, a few controls may require a special symbol set as the first two lines or a program start command word. Where program data start symbols are needed, they alone must proceed any remarks or code words. For example on a Fanuc® control:

%
O2345

For a Fanuc® control the % (program start) then letter O followed by a program number (O2345). To determine if your control requires a file symbol, read the program manual or call up and scan any program stored in memory.

M99 (program start code word for another control in my lab)

N005

Air cut
(Shop lingo) An inefficient time when the cutter is moving at feed rate but not making chips.

Departure point
The safe place from which a program starts and then parks the tools at completion. Also the safe place from which certain cycles begin and end.

Retract position
The safe distance off the work beyond which no rapid moves toward the work should occur.

Start-up commands (safety commands)
The first few command words that set expected modes and cancel unwanted actions or values.

Program Start Words

There are two categories of **start-up command** information that must be in the lines after the program start and file name:

Cancel words   Mode-setting words

As stated before, not all programs need all these preconditions, but there is no harm in including them unless they are not wanted. For example, I usually start my programs with the G70 code to signify it's in inch values or G71 if it's metric. Normally, the control would not be set to metric mode—except that one time when I assume the control mode was inches—but the last was metric. Big surprise, my incorrectly headed program tried to start in millimeter mode! So you see, there are exceptional "trap" situations that can be disastrous without the right start-up codes.

Here is another example. Suppose a clogged cutter has caused the operator to halt mid-program to clean the chips out. Then he chooses to restart the program from the beginning. If, somewhere within the original run through the program, a mode had been set just before the halt, the control will restart in it, unless cancel words begin the program.

KEYPOINT

A complete program start, including safety cancel lines and mode settings, helps avoid exceptional traps.

Setting and Canceling Modes

Required modes can actually be written into the program anywhere, as long as they are entered before or on the first line in which they are needed. For example,

Sample 1
N010 G94 G70 F20.
N015 G1 *X*2.00

But

Sample 2
N015 G94 G70 F20. G1 *X*2.0

will work equally well. The feed method G94, G70 inch values and feed rate were both available at the time they were needed, in each sample. Sample 1 had all modes fixed on line 10, before the linear interpolation. Sample 2 would work equally well, however, it's crowded with data.

Suggested Start Codes and Modes

Here is a list of those modes and values that should be set at the start of your program:

General Modes

G94 or G95	Feed type (per minute or per revolution)
G70 or G71	Inch/metric values (may be G20 or G21)
G90 or G91	Absolute/incremental values (usually absolute to start)
G17, G18, or G19	Mills only—working plane *X-Y* or *X-Z* or *Y-Z*

On both lathe and mill start-up commands, most programmers spread out the commands over two or three lines and they might also write a bracketed note to explain.

Specific Machining Data

FXXX	Initial feed rate
SXXX	Spindle RPM

The feed rate and spindle speeds must be entered before any machining can occur—but they need not be in the start section of the program. Again, there would be no harm in placing them there.

Canceling Modes for Safety

To ensure that a program starts cleanly without trying to execute some unwanted mode from a previous command, we put cancel codes in the start-up. There are two for now.

Others will be needed as your programs become more complex.

G80   Cancels all previously used canned cycles (Section 23)

G40   Cancels any previously used radius compensation

Here's a suggested generic startup template.

N005 G80 G40   (Cancel modes)

N010 G70/71 G94 (Initial feed)
  (or 95) FXXX

N015 SXXX G90   (Coord system
  (or 91)   abs./incr.)

Then, if it's a mill program

N020 G17/18/19   (Milling plane)

Secondary Start-Up Commands

Now that the modes are right, there is another start-up consideration—closing the safety guards and starting spindles and coolants and the initial motions. If your machine features automatic enclosures, there will be an M code word that closes the door. It should be the very next command after the mode setting lines.

Next are the physical tool motions. Every program has a staging area called the **departure point.** It is the place where your program parks the machine to end the previous cycle and where it also begins the next cycle. On a mill, it is probably the tool change position. On a lathe it is a full *X* and *Z* retract which is also machine home.

Yesterday in our lab, a student learned the hard way about setting a homing command as part of her program start-up. She had a bolt that held her work to the subplate. Her mill program start departed from PRZ with the cutter 1 in. above *Z* zero. But she had jogged the cutter away from PRZ to tighten the bolt. When she started the program, the cutter rapid traveled right through the bolt on its way to the first cut! If there was a homing command in the start-up, that wouldn't have happened.

Safe Rapid Travel Toward or Above the Work—The Retract Position

The next item in the sequence would be to change to the first tool (if the machine has a tool changer), start the spindle and coolant then rapid travel toward the work. The rapid travel destination is called *retract height* or *retract distance* (Fig. 22-2).

This **retract** or **R position** is the safe coordinate distance of the cutting tool from the work beyond which it should not rapid travel. It's also the safe distance a tool would retract away to move to a new location on the part—thus the name. At the R distance, you will switch to feed rate to complete tool contact with the work.

A drill cycle on a mill is a good example. One of the data fields in the G81 drill routine is the R coordinate. It is the jump height of the drill between holes. The safe distance above the work for a rapid travel move to a new location. The letter R is used in many cycles to indicate this safe distance from the work (Fig. 22-2).

Departure point
fully retracted *X, Y,* and *Z*

First rapid move

R distance = retract height
The safety limit of rapid travel

0.5 in. or 12 mm for
student programs

Figure 22-2 *A program typically starts from a departure point then moves at rapid travel speed to the safe retract height (or retract distance on lathes).*

Safe R Distance Verses Program Efficiency

For student programs and even your first few professional programs as well, I strongly recommend R heights of 0.5 in. or 12 mm (or more, as your shop policy sets). While that's safe, it requires relatively long **air cuts** with the tool, making its final approach at feed rate, without making chips. Later, these distances can be shortened for efficiency if shop policy allows it.

Lathe tool changing falls into this safety concept as well. Professionals retract the turret just enough to safely rotate it without interference from any tool in the turret.

KEYPOINT

Depending on which direction the turret indexes, the safe retract distance to change tools may need to be far enough from the chuck to clear every tool in the machine. (See the Trade Tip.)

But close tool changing is not a good idea for beginners. For student lathe programs, pull the turret as far back as it will go, usually to the home position, the farthest possible distance from the work, then change tools.

KEYPOINT

For beginners, it's a very good idea to run the turret all the way back in Z and out in X (home position) before changing to another tool.

With experience both lathe and mill retracts and lathe tool change positions can be shortened for efficiency.

Program End Commands

Ending the program is a simpler issue, but it must still have a few vital features to be safe and efficient. Modes that are in effect should canceled, *even though they will be canceled by the start section* when the program is run again.

This is especially true for tool compensation programs. As the tool completes the final cut, it's pulled off a safe distance from the work, then rapid traveled back to the departure point at which time the compensation is ended, when the tool is safely away from the work surface. That way, when it moves from the tool radius distance away from the geometry to center over the geometry line, it won't accidentally touch the work or chuck or vise. Tool compensation is canceled with the G40 code. We'll explore this more thoroughly coming up.

As the tool is pulled off the work, stop the coolant M9. The M30 or M2 command will also turn it off, but it's poor practice to do so at the very end. That generally slings coolant everywhere! Not only does it cause a slippery mess, the cost of coolant lost to the shop floor robs profit as well. This problem occurs mostly on smaller mills without full enclosure. Turn it off at the end of the final feed movement. Another convenient program end is to put a blank tool in the spindle or rotate the lathe turret to a blank position (see the Trade Tip).

Safe to home

Must retract to R
before homing

Figure 22-3 *Rapid moves must always be made to and from a safe location.*

Safe Rapid Travel Away from the Work

A potential accident can occur when pulling the cutting tool away from the work for a tool change or to park the machine at program end. The total path must always be considered.

In Fig. 22-3, a drill has completed drilling to depth. To simply home the machine from the bottom of the hole would be a double crash. It would start moving in *X*, *Y*, and *Z* axes at rapid travel. It must be pulled straight out of the hole then sent to home. But when homing it right from the R position, it could also hit a vise jaw or clamp if not retracted far enough above the work first. The drill must be retracted straight up in the *Z* axis, above the vise, then sent to home.

Always pull directly off the work, at feed rate, to a safe distance, then analyze the path back toward the departure point before homing the cutting tool!

UNIT 22-2 REVIEW

Review the Key Terms

Air cut
(Shop lingo) An inefficient time when the cutter is moving at feed rate but not making chips.

Departure point
The safe place from which a program starts and then parks the tools at completion. Also the safe place from which certain cycles begin and end.

Retract position
The safe distance off the work beyond which no rapid moves toward the work should occur.

Start-up commands (safety commands)
The first few command words that set expected modes and cancel unwanted actions or values.

Replay the Key Points

♦ A complete program start, including safety cancel lines and fixing needed modes, helps guarantee success.

♦ For beginners, it's strongly suggested that all axis moves toward the work stop rapid traveling at least 0.5 in. from the work.

♦ For beginners, it's strongly suggested you withdraw all tools to their farthest (safest) position before changing parts or changing tools.

♦ Always pull directly off the work, at feed rate, to a safe distance, then analyze the path toward the departure point before homing the cutting tool!

Respond

1. A program begins with:
   %
   O5432
   What does it mean?

2. True or false? Modal commands are those that cancel or override all previous modes set in the program. If it's false, what makes it true?

3. Why do we add the word *G80* to the beginning of a program?

4. Explain the two elements of good program start.

5. Identify at least two modes that should be set at the start of a program.

6. Why must a tool retract path always be analyzed before sending the tool back to the departure point?

UNIT 22-3

Manually Compensated, Linear Cutter Paths

Introduction: OK, it's just about time to give program writing a go. At the end, I'll offer a solution program in the answers. Check it against yours. If your lab has CNC simulation software this is the time to learn to use it. If so, write your program in the text editor then define the raw material size, cutting tools, and the PRZ location on the material. Then machine the screen-blank using your program.

Not only is it rewarding to see the codes turned into cyber-parts, it's the fastest way to test your program. At this stage of learning simulators are by far the best learning tool. If you have no simulators, you can also test a program by backplotting it in the lab's CAM system or by using controller graphics but with the machine safely locked up in dry run mode. (You'll need some guidance for that test.)

After a discussion of how to calculate the tool path, we'll make a CNC mill part, start to finish. You will write the plan, then the codes. I'll provide a few math hints for coordinates. Notice that there are no arcs. Curve commands are coming up in Unit 22-4. Note, you will need trigonometry and the Pythagorean theorem to complete some of these activities.

Calculating the Tool Path

Here we will take a look at *X-Y* shifts for a profile mill, tool path but they could also be *X-Z* shifts for turning a workpiece using a radius nose cutting tool.

Sketching the Tool Path

You are going to calculate coordinates then write a tool path program for the latch lock (Fig. 22-4). To do so, this program will climb mill from point *A* to Point *I*, using a 0.500-in.-diameter cutter (0.25 distance off the geometry).

Draw a Sketch

The first step is to draw the tool path at the needed tool radius offset distance. The centerline is blue in Fig. 22-5. We don't do this

Sample tool path sketch

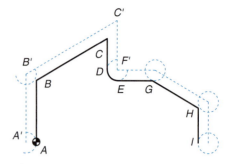

Figure 22-5 *Sketching the tool path.*

on a day-to-day basis, but it helps at the start. Use a contrasting color to draw a dotted line to represent the cutter centerline. A circle template often assists with this process.

KEYPOINT

Keep in mind that while the circles are easy to envision as end mills profile milling a part, they could also be the nose radius on a lathe tool. The concept is the same—a radius tangent to a geometry line.

Four Kinds of Compensation Shift

Now, label each significant point on the part geometry then look for its counterpart on the tool path. Points lie at each end of lines, at line intersections, and were we using arcs, at their tangent points and centers.

KEYPOINT

The coordinate that will be entered in the program is not on the geometry. It is the cutter center, on the cutter path.

In this book, I'll refer to the shifted coordinate point on the cutter path, as a prime of the point on the part geometry. For example, point *A* on the part geometry becomes point *A'* on the cutter path. There are four distinct situations illustrated by the latch lock and Fig. 22-5:

◆ *Straight Shift Point A'* Because the cutter is tangent to a vertical line, Point A shifts to the left by a minus 0.25 cutter radius to become point A' in the tool path.

◆ *Compound Shift Point B'* Point B shifts in both *X* and *Y* to become B'. This requires

Latch lock (0.25 thick MS)

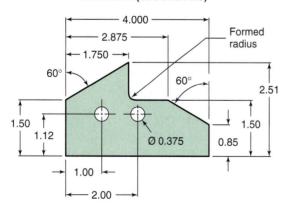

Figure 22-4 *The latch lock to be programmed.*

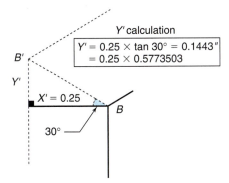

$$Y' \text{ calculation}$$
$$Y' = 0.25 \times \tan 30° = 0.1443''$$
$$= 0.25 \times 0.5773503$$

Figure 22-6 *Solving for the X' and Y' shifts. Each triangle is different based on the part geometry.*

a triangle solution. There are others on the part geometry.

- *Tangent Point F'* Point *F* is a special situation. Points *D* and *E* on the geometry disappear because this particular tool path will form the 0.25 part radius. The cutter will cut in a straight line, to point *F'* then turn 90° to cut to point *G'*. If the cutter was smaller in diameter, we would need to write a curve command based on points *D'* and *E'*.
- *Centers Don't Shift* Center references, such as drilling holes, are not shifted.

The Shift Triangles

Now sketch the shift triangles where they are needed, as shown in Fig. 22-6. Then solve for the *X'* and *Y'* distances. It will be useful to blow them up to some greater size. Continue the color scheme from the parent sketch.

Trade Tip

Finding and Solving the Coordinate Shift Triangles

1. Label the cutter radius first. It will be the distance the centerline lies from the part line.
2. Label the 90° angle.
3. Using basic geometry, look for lengths or angles that are known. In the example, the 60° angle on the print.

   Or, if you do not have the math skills to solve these problems, try drawing the geometry and tool path accurately in CAD then analyze the elements. This is how the CAM system derives point coordinates.

When these coordinate shift calculations are complete, it's time to write the program codes.

UNIT 22-3 REVIEW

Respond

Program Problem—Latch Lock

Calculate and write a cutter centerline program to mill Fig. 22-4, the steel latch lock. Use the following program sheet as a model. If you get stuck on any math, you will find hints in the answers.

Job Instructions

1. Work in absolute values.
2. Calculate spindle speeds with 100 F/M surface speed. Conservatively feed the cutter at 10 in. per minute for milling and 0.003 in. per revolution for drilling.
3. Upon completion your program may not be in the exact order of the answer, but they are OK as long as all data is correct and all present. The simulator will show if the program works or not.
4. Use a 0.50-in. diameter, four flute cutter and a 0.375-in. drill. For this first program, center drilling is not necessary.
5. Cut the profile using climb milling—just one finish pass around. (Although a rough and finish pass would produce more consistent results if this were a real part.)
6. The lower edge has been premachined. Place your clamps along that edge and start machining along the left-hand edge which has 0.030-in. machining excess.
7. Use absolute value coordinates.
8. If your machine uses codes that differ from the book list, use them.
9. Work on a program sheet similar to the following.
10. The end mill and drill may cut into the sacrifice plate.

Program Sheet

N005	N100	N195
N010	N105	N200
N015	N110	N205
N020	N115	N210
N025	N120	N215
N030	N125	N220
N035	N130	N225
N040	N135	N230

N045	N140	N235
N050	N145	N240
N055	N150	N245
N060	N155	N250
N065	N160	N255
N070	N165	N260
N075	N170	N265
N080	N175	N270
N085	N180	N275
N090	N185	N280
N095	N190	N285

UNIT 22-4

Writing Arc Commands

Introduction: Programming arcs is easy with a bit of training. There are actually three methods, two of which we'll consider here.

The Radius Method

The Center Method (*IJK* Method)

The Polar Arc Method (Not studied)

An arc is a polar entity, thus writing arc commands is simple enough that it requires little instruction. However, not all controls feature this ability. Unique to this method, the entries include the degrees in the arc.

Depending on how the dimensions are provided, using one of the three methods eliminates nearly all math required to program arcs.

KEY TERMS

Center point arc method (*IJK*)
The arc command that requires curvature code, end point coordinates in *X, Y,* or *Z* and center point coordinates in *I, J,* or *K*.

Curvature codes
The code word that causes CW or CCW arc motion.

Polar arc method
The arc command that requires curvature direction, radius of arc, and number of degrees in the arc.

Radius arc method
An arc command that requires curvature code, end point, and radius entries.

The Five Arc Facts

To draw any defined arc on paper using a compass, there are five parameters required. They are the same facts the controller will need (Fig. 22-7).

Arc radius	**Start point**	**End point**
Center point	**Direction clockwise or**	
	counterclockwise	

Default Facts

All five facts are not needed in each command. Given any three, then two will be automatically known. For example, if the start point and the center points are known, then the radius is the distance between them—it is automatically available.

Or, given the arc's start and end points, the center point will lie on a line perpendicular to a line connecting them (Fig. 22-8). The control can derive the center point in this case if it's given the radius and direction, clockwise (CW) or counterclockwise (CCW).

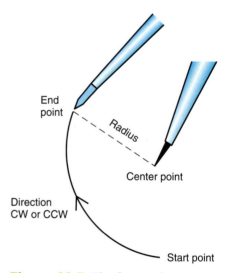

Figure 22-7 *The five arc facts.*

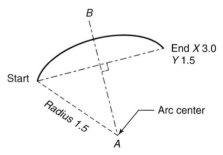

Figure 22-8 *Given the start and end points, the control can derive that the center point lies along line A-B.*

The difference in the two arc commands, center identification or radius method, is which facts must be entered and which are determined by default.

Current Tool Position— Always the Start Position

In all arc methods (polar included) one fact is always automatically known, the start point. So, with the start point known, only two more facts must be provided in either command.

The Radius Method for Arc Commands

The **radius arc method** is the simpler of the two nonpolar commands, as long as the arc's radius is known. Factors in the command are

Curvature Code, End Point Coordinates, Radius

G2 X3.0 Y1.5 R1.5

This example moves the tool clockwise (G2) from its present position to X3.0 Y1.5 with a radius of 1.5 in.

There are two direction codes:

G02 = Clockwise circular interpolation
G03 = Counterclockwise interpolation

The end point coordinates may be absolute or incremental per the information available (Fig. 22-9).

Arcs Larger Than 180°

There are actually two possible arcs that could connect the start to the end point, given the factors outlined. One is less than or equal to 180° and the other is greater than

Two radius method commands

G2 X 0.5 Y 0.94 R 0.5
Using
absolute values

G3 X 0.38 Y 0.17 R 0.5
Using
incremental values

Figure 22-9 *Two example radius method arc commands.*

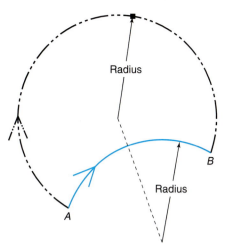

Figure 22-10 *Two possible arcs given start, curvature, end, and radius factors.*

180°. In Fig. 22-10 both arcs move clockwise from *A* to *B* and both have the same radius. The solid blue arc is the default entity, less than 180°. To program the arc greater than 180° using the radius method, the control must be given an additional symbol.

All controls (of which I'm aware) default to the lesser arc unless a special symbol is added to the command, to indicate that the arc is the larger segment.

Controller Differences— Greater Arcs

The symbol that signifies the arc larger than 180° varies from control to control. Here are two examples:

Default arc :   G02 *X*14.0 *Y*12.0 R16.0

Greater arc :   G02 *X*14.0 *Y*12.0 R−16.0

The minus sign placed on the radius, indicates that the arc is greater.

/G2 *X*14 *Y*12 R16

The slash indicates the larger arc on our generic control. Read your manual.

The Math

When the arc is 90° on square part edges, the radius command is ultrasimple; just write the **curvature code,** end coordinates, and radius. However, when the arc is not a simple corner, there are some computations to be made. Each geometry presents its own set of conditions.

Make a sketch then fill in what you know. Use the Pythagorean theorem and trigonometry to solve for the unknown coordinates. The upcoming activity will give you some practice at calculating and writing arcs using both methods.

The Center Point (*IJK*) Method for Arc Commands

For the **center point arc method (*IJK*)**, this command is the one CAM systems use because it can denote an arc of any number of degrees (zero up to a full circle), with no confusion between lesser and greater arcs. Factors in the command are

Curvature Code, End Coordinates, Center Coordinates
G02 X.75 Y1.75 I.75 J0.0

The example moves the tool from the present position to the *X-Y* position using a center point that's 0.75 in. from the start in the *X* direction and right on the *Y* line (*J* = 0.0).

For this command method, the end point coordinates may be absolute or incremental depending on information at hand, however, the *center point coordinates are normally incremental* using the start point as their reference. Read that last statement again—center point identification is incremental and identified with alternate axis letters.

Why Use *I, J,* and *K*?

When working in an *X-Y* plane on a mill, for example, there are two sets of coordinates within the arc command, the *X* and *Y* end point coordinates and the *I* and *J* center coordinates (Figs. 22-11 and 22-12).

G02 X0.75 Y1.75 I0.75 J0.0

Figure 22-11 *An arc command using the center point method.*

G2 X0.5 Y0.94 I0.5 J0.0

Absolute ID of point *B* — Incremental ID center from point *A*

Figure 22-12 *A second example.*

If a lathe program is being written, then the center is identified with *I* and *K*.

Controller Differences—Other Uses of *I, J,* and *K* Depending on the control type, other uses are made of the *I, J,* and *K* prefix in commands other than for arcs. Again, they are used as alternate *X, Y,* and *Z* coordinates, usually as incremental values. For example on a given lathe control, *I* and *K* are used as incremental moves of the *X* or *Z* axis without changing to incremental mode.

Fanuc Lathe Controls and Incremental Coordinates When you wish to use incremental coordinate values on many Fanuc lathe controls, there is a special convention, G91 is not required nor used to set the incremental mode. These controls default to the absolute mode and have no incremental mode. When you wish to use an incremental coordinate, the *X* and *Z* are replaced with a *U* and *W*. So, on most other controls G91 *X*10.0 would be an incremental jump of 10 units in the plus *X* direction. However on Fanuc, *U*10.0 would cause that action.

UNIT 22-4 REVIEW

Review the Key Terms

Center point arc method (IJK)
The arc command that requires curvature code, end point coordinates in *X, Y,* or *Z* and center point coordinates in I, J or K.

Curvature codes
The code word that causes CW or CCW arc motion.

Polar arc method
The arc command that requires curvature direction, radius of arc and number of degrees in the arc.

Radius arc method
An arc command that requires curvature code, end point, and radius entries.

Replay the Key Points

◆ In all arc methods (polar included), one fact is always automatically known, the start point.

◆ Between the *IJK* and radius methods, the radius is the simpler to calculate and enter as long as the given data supports it.

◆ There are two arc direction codes G02 = CW and G03 = CCW.

◆ Arcs less than 180° are the default value. Arcs greater require some symbol.

◆ In CNC work *I, J,* and *K* are used as incremental, alternate forms of *X, Y,* and *Z.*

Respond

Calculate and write a command for the given arcs. For the first two problems, assume *the control is in incremental mode.*

1. See Fig. 22-13.
   A. Write the default arc command using the radius method.
   B. Now write the same arc command using the center method.

2. Using the information in Fig. 22-14, write three different arc commands: the default command from *A* to *B,* the arc greater than 180 degrees from *A* to *B*

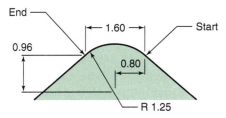

Figure 22-13 *Problem 1—Write a radius method command then a center method command.*

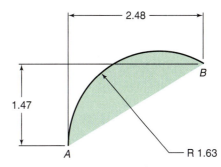

Figure 22-14 *Problem 2—Write the radius method command.*

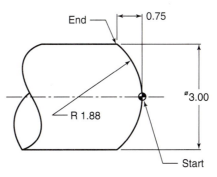

Figure 22-15 *Problem 3—Write the radius method command.*

(use the minus radius symbol for greater arc), and the default arc from *B* to *A.*

3. See Fig. 22-15. Write an *IJK* arc command for this turning job, using absolute value coordinates. Then write the radius arc command.

4. Now write a metric program to turn (in one pass) then part off, the aluminum bullet nose pin in Fig. 22-16. This is a bar feed job held in collets. Do not turn the stock diameter. Use a 2-mm, nose radius, lathe tool and write your program in absolute values. You may use either

Figure 22-16 *Write a manually compensated program to turn this bullet nose pin using a 2-mm nose radius tool.*

RPM See Fig. 22-16. Since the tool will start in the center then work outward along the profile, the spindle should start at a very high RPM, then continuously slow as the effective diameter at the tool gets smaller. There is a way to add constant surface speed (CSS) to your program to initiate this continuously changing RPM in such cases. We will study this ability in Section 23, but there is no harm in reading ahead in the programming manual in your lab and using the correct code to implement CSS.

arc command method—both will be shown in the answers along with some math hints if you get stuck. Do not write roughing passes—this is an exercise to compile codes and arc commands only. Use 1,000 RPM and feed at 0.1 mm per revolution.

UNIT 22-5

Writing Compensated Programs

Introduction: Now we write part path programs with commands that cause the control to calculate the tool path. That means no more X'-Y' shift calculations to make! The programmed coordinates are part definitions taken from the drawing. They may still need to be calculated, but they are easier to determine. From the previous activities, you can see, this skill is a real time saver.

Also, a compensated program allows some latitude as to which cutter will be used. With all factors correctly written into the program and the offset stored in the control, the cutter's rim for peripheral milling or nose radius for turning, will remain tangent to the work along the entire part geometry. In other words, the cutter's outer radius will follow the part path. The cutting tool can be exchanged for another with a different radius and still produce an accurate part as long as the new tool radius has been correctly entered in the control.

KEY TERMS

Dummy move
A line of any convenient length, that transitions to the ramp.

Lock on
The cutter has arrived at tangency to the line and will remain so until told differently.

Off ramp
The short line used to smoothly transition off the work and to sometimes create a phantom buffer before turning off compensation.

On ramp
A short added line used to lock on to the part geometry before chips are made.

Preparing a Program for Compensation

Three items are required in the program to initiate compensation.

1. **Which Side of the Geometry Line—G41 Left or G42 Right?**
   The compensated program defines to which side of the upcoming geometry line is the controller to move the cutter. That happens on the ramp.

2. **The On Ramp**
   Similar to an automobile starting onto the highway, compensated programs require a distance to get started—sometimes called the **lock on** *distance*. This ramp is an added geometry line or series of lines that begin away from the workpiece but lead onto it in a logical place. The cutter is commanded to a departure point at the beginning of the ramp to get a compensation sense. Then as it follows the ramp, the cutter becomes tangent to the line before cutting metal.

KEYPOINT

The cutter becomes tangent to the geometry before touching the metal to be cut.

3. **How Far Right or Left—Radius Offset**
   The control will move the cutter left or right of the first ramp line, by the distance stored in the tool radius offset memory. The G41 or G42 is entered in the program, on or just before the first ramp line where tangency is to begin.

The complete command to compensate includes G41 or G42 plus an entry to tell which offset to use

G41 D01 (cutter left, use offset stored in D1)

G42 T0101 (cutter right, use tool 01 with offset 01)

Cutter Right or Left

The tangency designation is as though you were behind the cutter looking in the direction of the cut (Fig. 22-17). If the metal is on your right (to the right of the cutter) it is cutter left and vice versa.

In the illustration, the mill cutter is climb cutting a profile—cutter left and conventional cutting when on the right. Adding either word to the program causes the cutter to move to a tangent relationship with the part geometry *on the next axis move after the code, in the current program plane.*

Figure 22-17 *Compensating left and right.*

The command varies slightly with different controls. Here are two examples in a CNC mill program.

G17 G41 D1

X-Y plane, cutter left using stored tool offset 1

T0101 G17 G42

Tool 1 offset 1, *X-Y* plane, cutter right

Current Program Plane on Mills

The G17 code tells the control to compensate in the *X-Y* plane. Under the G17, *Z* axis motion has no effect on compensation on modern controls.

Controller Differences There are a few older CNC controls that will unlock from compensation, if given a *Z* axis move.

The G17 should already be present in the program start lines on mills. No plane code is needed for lathe programs, since they operate only in the *X-Z* plane.

Then, once the cutter radius is correctly locked on to the geometry (tangent and following connected lines), it will remain so along the entire part path until the G40 command is given to end compensation or the program is ended.

Designing Ramps

The on-ramp can be a single line that leads onto the geometry, or it may need to be multi-element (Fig. 22-18). On the left side of the

Figure 22-18 *The ramp line provides a direction sense for the control to lock on to the shape. There's a slight chance it might nick the corner.*

drawing, the cutter is first positioned at point *A*. It is uncompensated, centered over the line. Then the comp codes are read and the cutter commanded to point *B*. As it moves, the cutter becomes tangent to the line at point *C*. Following that angular path, there's a chance the cut path will touch the corner of the part.

KEYPOINT

It actually moves to point *C*—gradually (typical of most modern controls).

So, one improvement is to preposition the cutter back a bit and change the approach angle as shown on the right side of the drawing. Another improvement involves two or even three connected lines, of any convenient length (Fig. 22-19). The multielement ramp joins with the geometry better, to begin the cut. They provide a smooth transition into the cut to avoid cutter marks and nicks in surfaces and corners. The cutter is fully

Two-part improved ramp

Figure 22-19 *The double ramp line reduces chances of nicking the part.*

tangent to the line before it touches metal, at point *C*.

Program Example Codes

Here's a typical set of commands to write a tool compensated program:

```
(Demo Mill Comp Start)
N005 G80 G40
N010 T0101 M6                (Control has radius offset 01)
N015 S1000  G94 G90 G17      (The program is in the X-Y plane)
N020 M3 G0 X−0.5 Y−0.5 Z.25  (Rapid positioning—point A)
N025 G41                     (Cutter left on next X, Y, or X-Y move)
N030 G1 Z−0.25               (Not X-Y plane—no comp yet)
N035        X0.0 Y0.0        (Planar move, becomes tangent at ramp end)
N035    Y1.0                 (Now ready to follow remaining geometry commands)
```

Ending Compensation

When the cutter has completed the shape, it's very important that the path moves off the work smoothly, again with little or no abrupt transition. The off ramp is easily created by adding an extension line to the geometry. It usually continues in the same direction as the cut as long as chucks and clamps are not in the way.

KEYPOINT

Only after the cutter is well clear of the work, turn off compensation with the G40 code. *Warning:* Only under very special conditions do we ever turn off compensation while the cutter is in contact with the work! Normally the tool is withdrawn safely off the surface.

Controller Differences—Safety Notes on Compensation As CNC processors became faster and competition forced features on manufactures, many of the bugs in compensation have been fixed. However, in industry or tech school, there's every chance you'll be running two different machines that behave very differently during the lock on, while staying locked on, and during the unlock phase of compensation.

Lock on Modern controls commanded to become tangent, will gradually move away at an angle to the line until they are tangent at the far end as shown in Figs. 22-18 and 22-19. But on a few older controls, they move all the compensation distance before starting along the ramp—at point *A*. Given the command to move from point *A* to point *B*, they move directly sideways, 90° to the upcoming line,

Trade Tip

Comp Tricks Here are a couple tips that might be useful and a warning.

Ramp Lines Can Be Ultrashort There are times when the cutter can't be positioned very far from the surface where the cut will begin—a vise jaw, tight bore, or opposite wall of a narrow pocket, for example. There are two solutions: on a mill, begin the *X-Y* ramp above the surface in the *Z* axis, then after tangency to the ramp is established, drop the cutter down to a short line leading onto the surface to be cut. Modern controls will remain locked on to compensation during a motion not in the present plane. So in G17, a *Z* motion has no effect on the comp. *Warning*—a few older controls unlock during that *Z* motion. Read the program manual.

A second solution drops the cutter to the *Z* level surface at the start of the ramp line but just a short distance off the geometry. Then a very short first ramp line is programmed; for example, 0.001 in. long. The control follows it while becoming tangent. This is known as a **dummy move**. What actually occurs is that the cutter moves forward 0.001 in. but also sideways by the radius amount. It's still following line *A-B* 0.001 in. long, but going to *C*. With the cutter now locked on the dummy, it connects to a second line that leads onto the part geometry. Voilá, no nicks.

Rapid Lock Most modern controls are capable of locking onto the ramp line while in rapid travel! It depends on CPU speed. To save time, you can initiate compensation while moving from the program departure point toward the work. Be careful to make the rapid motion continue onto the work surface in the same general direction, otherwise the direction sense might be reversed and the cutter end up on the wrong side of the line (Fig. 22-20).

Warning! Do Not Reverse the Cutter Direction While Locked on If a program is written to move from point *A* to point *B*, while in G41, for example, then commanded back to point *A*, a big problem occurs. Think this through. On the return move, the cutter must jump back across the line to stay on the left as shown in Fig. 22-20. This jump usually results in a crash. The control does not "see" a metal part on one or the other side of the line, it simply sees a line connecting points *A* to *B* then a new line from *B* to *A* and it will go to the left side of each!

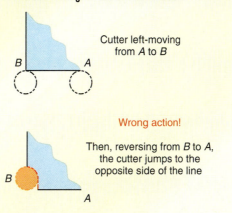

Figure 22-20 *Wrong action! Reversing cutter direction while locked on results in a jump across the line.*

then start along the ramp. Read the manual or ask before starting a real compensation program! In general, that same control is incapable of initiating comp during a rapid travel move.

Third axis moves mills only As mentioned already, some lower-level controls will unlock even though not commanded to do so, if an axis move outside the program plane is given. That bug made it complex to drop the cutter down for sequential passes when roughing a pocket, for example. That was fixed very early on and few controls today have this handicap.

Unlocking Seeing the G40 cancel code, most new controls gradually move away from the off-ramp line on the next physical line command within the current plane. But an older control will unlock the same as it locked on—90° directly sideways, then follow the line!

Buffer off ramps Some controls require a compound **off ramp** that has at least three line elements beyond the actual part geometry. Since the control is looking ahead at the program line, that control must not see the G40 command within three off-ramp moves ahead, while still making real moves. If the three-part, off ramp is not present to act as a buffer, the control will stop compensating while still on the part geometry—a disaster in many cases. These extra buffer elements can be dummy length.

UNIT 22-5 REVIEW

Review the Key Terms

Dummy move
A line of any convenient length, that transitions to the ramp.

Lock on
The cutter has arrived at tangency to the line and will remain so until told differently.

Off ramp
The short line used to smoothly transition off the work and to sometimes create a phantom buffer before turning off compensation.

On ramp
A short added line used to lock on to the part geometry before chips are made.

Replay the Key Points

- During compensation, the distance the cutter stays right or left is determined by the radius offset stored in the tool memory. The command to compensate is in the program, but the distance is not.

- After the compensation code, the cutter becomes tangent at the far end of the next line (for most modern controls).

- Do not reverse cut direction while locked on to geometry—the cutter will jump across the line.

◆ Turn off compensation with the G40 code *only* after the cutter is well clear of the work. Never turn off compensation while the cutter is in contact with the work!

Respond

Note: We'll write compensated programs in the review problems. Answer these questions to determine if you understand the concept.

1. What advantages are there to using cutter compensation in a program?

2. True or false? To compensate to the left side of the geometry line one must enter M41 on or before the next move within the current plane. If it's false, what makes it true?

Critical Thinking Problem

(Not covered in this reading but mentioned previously.)

3. I have a previously written *cutter center* line program that's long. I don't want to start over to write a part path program. It was specifically written for a 0.750-in.-diameter end mill, exactly. But I have only a 0.735-in.-diameter cutter.

   A. What happens if I use that cutter?

   B. Can I edit the program to include compensation for the different cutter?

4. Describe at least two things the on ramp does for compensation.

5. Why must you clear the cutter well off the part geometry before entering G40?

Unit 22-1

How many code words do you now know? Day by day, you'll add more to your vocabulary. But occasionally we all need to look one up in the index for a given control!

Unit 22-2

Programs aren't unlike a painting. Every artist has his or her own techniques that work for them, as does the programmer. That said, don't take too much time trying to make the data look like a masterpiece of organization. Most CAM systems generate a line-by-line arrangement so it's the standard for data.

Unit 22-3

One item probably overstressed in Unit 22-3, is to know the control's way of doing things when initiating compensation. In general, most of the odd differences are gone in today's controllers. They all operate pretty much the same, locking on and off gradually, at rapid travel if needed and staying locked until told differently.

Modern operating systems can be upgraded easily via comm wire or by disk. So as improvements are written they can easily be added to the control's bag of tricks. But the older CNC control had its operating system hardwired on ROM chips. Upgrading was possible but expensive and not without bugs. But, unlike the constant obsolescence of desktop PCs, those expensive machines aren't replaced when new versions are introduced and they aren't often upgraded either. Old CNC machines are very common in industry and in schools. They make excellent parts as long as the machinist and programmer understand their foibles.

Unit 22-4

Unless one calculates and writes a few arc command lines, they remain a mystery, especially the *IJK* version. You should be able to do it, but today most of us solve our computations through the CAM system. CAD drawings can also be analyzed for distances. That's how I checked the problems here—I drew an accurate version of the problem then analyze the database for distances and coordinates. Today, however, arc coordinate computation is a backup skill I wouldn't want to be without.

Unit 22-5

Program verification is a big issue in CNC work. You'll be comparing yours to mine in the review coming next. But in the real world, that's not the way it's done. It's just not possible to read all the lines of even a modestly long program to check it. Nobody can do it unless it's short and simple. So that means verification by another means. We have three options.

In a CAM system like Mastercam® in Section 25, one of the utilities is called *Tool Path Verification*. It's a playback of the cutter path. But watching it can determine gross errors on your part, for example, if the cutter was sent in the wrong direction or it went through a clamp on a rapid move. Still you'll use it often, it helps to see the results of your selections. Higher up the test scale, programmers use surface verification programs (Fig. 22-21) that create complete cyber parts along with tool marks and errors if they exist. Note the yellow rapid travel mistake in the program. This software compares the CAD drawing surface to the projected machined results. Go to the Internet and search for CNC Tool Path Verification Software or enter *MetaCut Utilities* in a search.

Test Parts the Final Proof Finally, if the part is very expensive, we machine material of lesser value then measure it. This can be a special reusable wax or foam made for CNC program tryout. It cuts easily yet holds dimensions. Wood can also be used but it's not recommended, as the fine dust tends to clog machines not sealed against it.

Figure 22-21 *The tool path and surface are analyzed by the software and an error is found—see the yellow rapid travel flag where the cutter ruined the work.*

Questions and Problems

1. Identify the six common code word prefixes. (Obj. 22-1)

2. If you saw an entry of T10.5, would you assume it to be a tool or time word and why? (Obj. 22-1)

3. True or false? CNC codes are grouped together for easy reference. If this statement is false, what makes it true? (Obj. 22-1)

4. True or false? An entry of *G71 X-042.32* would be rejected by most modern CNC controls due to the leading zero before the 42 in. and the trailing zero beyond the 0.32. (Obj. 22-1)

5. After the file number and communication symbol (if needed) your program should have two kinds of safety start data—what should they do and why? (Obj. 22-2)

6. When pulling a drill back to a safe distance beyond the work, to what level is it withdrawn? (Obj. 22-2)

7. Write a cutter centerline coordinate for point *A′*, use Fig. 22-22, assume 0.125 radius on the tool nose. Use absolute values. *X* axis values are diameters on the drawing. See the blow-up sketch. (Obj. 22-3)

8. See Fig. 22-22. Complete the centerline, cutter path coordinates (not the program) from PRZ to point *E′* using absolute values and *X* diameters. Do not write the curve command (see Problems 9 and 10). (Obj. 22-3)

9. See Fig. 22-22. Write the curve command from point *C′* to point *D′* using the radius method and absolute values. (Obj. 22-4)

10. See Fig. 22-22. Write the curve command using the center point method (*IK*). (Obj. 22-4)

Problems 11 through 16

Latch lock—cutter compensation program

See Fig. 22-23. Write a complete tool radius compensated program to profile machine the latch lock. (Obj. 22-5)

◆ Planning:
  ◆ Climb mill from the left corner of the 60° angle (point *A*) to the far end of the 50° face. One pass only.
  ◆ Be sure to include on and off ramps for compensation. The on ramp is a double line.
  ◆ The stock is $\frac{3}{16}$-in.-thick steel plate. It has been sheared with 0.030 in. excess on the 2.51-in. dimension. It is 4.00 in. wide.
  ◆ The stock will be clamped over the lower edge on a sacrifice plate that will be slightly cut by the end mill.
  ◆ Use absolute values and the radius method command for the 0.25-in. inside corner radius (Obj. 22-5)

Problems 17–20

Write a tool radius compensated program for the familiar bullet nose pin (Fig. 22-24). (Obj. 22-5)

Figure 22-23 *Review Problems 11 through 16.*

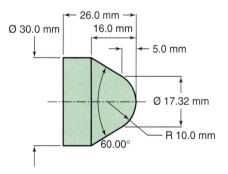

Figure 22-24 *Review Problems 17 through 20.*

Figure 22-22 *Review Problems 7 through 10.*

ANSWERS 22-1

1. G—prepares for or causes an action; M—miscellaneous actions; T—tool codes; S—spindle RPM; F—feed rate; N—line address
2. A linear movement to position X30.0 at a feed rate of 19 in. per minute, then to Y1.00 at feed rate, then to X0.0 Y0.0 at rapid rate.
3. An error. Either the program will stop the machine or it will obey only the G01.
4. Check your answer against the reading.
5. Check your answer against the reading.
6. To list those that cannot appear on the same command line.
7. The statement is false. It is a complete command line because the rest of the command is modal and preexists in N025.

ANSWERS 22-2

1. These are the program start and the program file number
2. False. The modal command cancels or changes only others from its group!
3. To cancel previously used modal cycles (drilling or threading, for example)
4. A safety cancel line ends surprise modes; expected modes are set.
5. *Coordinate values* (absolute or incremental); *feed rate type* (IPR or IPM); *unit values* (inch or metric)
6. Unanalyzed, the rapid travel move could hit work or setup. Pull directly away from the work at feed rate, then rapid toward home when you are sure the tool is clear of all objects.

ANSWERS 22-3

Note, short form codes are used when possible. Some null coordinates have been omitted. The work is clamped down to a sacrifice plate for milling and drilling. PRZ is lower left corner, top surface.

```
(Latch Lock PRZ LLC-Top—cutter CL)
N005 G80 G40
N010 G94 G70 F10.
N015 G90 S800 G17
N020 T01 M6 (0.5 EM NO OFFSET)

N025 M3
N030 G0 X−0.25
N035   Y−0.25 (BEYOND EDGE)
N040   Z0.50 (R HEIGHT)

N045 M8
N050 G1 Z−0.275   (BELOW BOTM)
N055        Y1.6443
N060   X2.0    Y2.943

N065        Y1.750
N070   X2.942
N075   X4.250 Y0.9943
N080        Y−0.25

N085 G0 Z100. M9 (TO LIMITS)
N090   X100. Y100.    M5
N095   T0202 (0.375 DRILL)
```

```
N100 G0 X1.0 Y1.12
N105   Z.5
N110 M3
N115 M8

N120 G95 F0.006 (DRILL FEED)
N125 G1 Z−0.4
N130 G0 Z0.5
N135   X2.0

N140 G1 Z−0.4
N145 G0 Z100. M9 (ON LIMITS)
N150   X100. Y100.
N155 M2 (END)
```

ANSWERS 22-4

1. G3 $X−1.16$ $Y0.0$ R1.25 (radius method > 180)
   G3 $X−1.16$ $Y0.0$ $I−0.80$ $J−0.96$ (center method)
   Note, the $Y0.0$ could have been dropped as null entries as there is no incremental difference in start and end point with regard to Y value. However, it is recommended that you leave nulls in arc commands for checking the math.
2. G2 $X2.48$ $Y1.47$ R1.63
   G2 $X2.48$ $Y1.47$ R−1.63
   G2 $X−2.48$ $Y−1.47$ R1.63
3. G3 $X3.0$ $Z−0.75$ $I−1.88$ $K0.0$
   Remember, lathe X axis coordinates are normally given in diameter values.
   G3 $X3.0$ $Z−0.75$ R1.88
4. See Fig. 22-25.
   (Bullet Nose Pin)
   N005 G80 G40
   N010 G95 G71 F.1

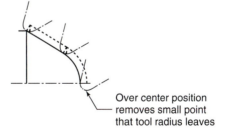

Over center position removes small point that tool radius leaves

Shift triangle

2.0 mm nose rad   30.00°

X' radius shift
cos 30 × 2 = 1.732 mm
Doubled = 3.464 diameter shift

Z' shift
sin 30 × 2 = 1 mm

Figure 22-25 *Solving the shift triangle.*

N015 S1000
N020 T0101 (LH turning tool 2 mm NR)
N025 M3
N030 G0 X0.0 Z10.0 M8
N035 G1   Z0.0 X−2.0 (See answer sketch for explanation of X pos.)
N040   X0.0
N040 G3   X20.784 Z−4.0 R10.0
N040 G3   X20.784 Z−4.0 I0.0 K−10.0
     (*IJK* alternate—note center is 2 mm from present tool position)

N045 G1 X33.464 Z−15.0
N045 G0 X100. Z100. (Tool change position)
N050 T0202 (parting tool)
N055 G0 35.0 Z−29.0 (allows for 3-mm parting tool thickness)
N060 G1 X−0.1 M8
N065 G0 X100. Z100. M9
N070 M2

ANSWERS 22-5

1. Program coordinates are part geometry, thus saving lots of math. Cutters radius can vary yet still produce the right dimensions.
2. False. It's G41 not M41.
3. A. The part will be too large by 0.0075 in. on every side.
   B. Yes, the radius offset becomes a negative number—0.0075 in. closer to the geometry.

Answers to Section Review Questions

1. G, M, T, F, S, N
2. A time code since tool are whole numbers.
3. False. They are grouped together to show mutually exclusive codes.
4. False for three reasons: It's a metric value not inches (G71). Also, the number is a complete number with the decimal in the right place. The zeros, leading or trailing, have no bearing on their acceptance.

4. It provides a smooth transition onto the part geometry; it provides a line on which to get the sense of direction. Which then shows the control how far to go left or right of the line.
5. To avoid marring the work when the cutter unlocks from the geometry line, G40 cancels compensation.

5. Cancel unwanted modes to avoid traps. Examples include G40 and G80. Set needed modes and conditions to begin, for example, G90, G70, and G95.
6. The retract height.
7. Point *A′* lies at X2.7693 Z0.1250. Remember the X values are diameters. To calculate point *A′* find new diameter using (2.625 + 2 × [Tan 30° × 0.125])

8. Point *A'* X2.7693 Z0.1250
   Point *B'* X4.2462 Z−1.1540
   Point *C'* X4.2462 Z−3.6250
   Point *D'* X4.7462 Z−3.8750
   Point *E'* X5.7500 Z−3.8750
9. G2 *X4.7462 Z−3.8750 R0.125*
10 G2 *X4.7462 Z−3.8750 I0.125 K0.000*

Problems 11 through 16
(Radius comp—latch lock program)

(PRZ lower left corner)
N005 G80 G40 G70
N010 G94 G17 G90 F10.
N015 S550
N020 T0101 M6   (0.75* Dia end
                                  mill).

N025 G0   *X*−1.0 *Y*1.0 M3
N030   *Z*.5 M8
N035 G41
N040 G1   *Z*−0.200 (0.025 in. Into
                                       sacrifice plate)
N045   *X*−0.500 *Y*1.25   (Ramp—locks
                                    on)
N050   *X*0.000 *Y*1.500
N055   *X*1.75 *Y*2.51
N060   *Y*2.125
N065 G3 *X*2.375 *Y*1.500 R.625 (.625 corner
                                     radius)

N070   G1   *X*2.875
N075   *X*4.00   *Y*.375   (end of 45°
                                        geometry)
N080   *X*4.500   *Y*0.0   (off ramp)
N085   G0   *Z*100.   (*Z* soft limits)
N090   G40 *X*100. *Y*100.   (Park on soft
                                        limits)

N095 M2

Problems 17–19
(Bullet nose pin)
(Metric values)
N005 G80 G40 G71
      G95 F.1 S1000
      T0101
      M3 G42 (Comp on)
N010 G0 *X*0.0 *Z*5. (Rapid ramp on)
N010 M8
N015 *Z*0.0
N020 G3 *X*17.32 *Z*−5.0 R10.0
N025 G1 *X*30.0 *Z*−26.0
N030 *X*40.0 (Off ramp)
N035 G98 (Home turret**)
N040 G40
N045 M2

*Your cutter could be larger or smaller, in which case the RPM will be different.

**This is sometimes used for homing codes but it's not universal.

Section 23

Level-Two Programming

Introduction

Next we examine the computer logic programming that is used extensively in handwritten programs and to a lesser degree within CAM-generated programs. Using these techniques one can simplify programming math and reduce total data. But more, it's just plain fun to do these things!

We'll start with fixed cycles. Used for milling and turning, all CNC controls contain a few within their capability. Many feature a library of *canned cycles* (everything is contained within the command including the formulae that make the calculations to carry out the routine).

Custom Cycles

Some advanced controls allow custom-written (or purchased) specialized cycles. They support the manufacturing of a particular feature or unique product. For example, a shop makes custom chain sprockets. While each sprocket is different, they vary in just three characteristics. So a tailored cycle that profile mills the chain teeth is a real time saver.

After entering the code to begin sprocket milling, let's say it's G151, the cycle would require a pitch parameter for the sprocket P (which sets the diameter), and W the width of the chain (depth to plunge the end mill), and the N parameter to determine the number of teeth, plus the X-Y center location.

G151 *P*.5 *W*.1875 *N*23 *X*1.500 *Y*0.000

Presto! Supplied with just those variables, using the formula within the cycle, the control calculates the size and shape of each tooth at the correct angle with respect to the axis of the sprocket, then makes the correct axis moves. So, with no other programming, the G151 roughs then finish mills the sprocket. The magic happens transparently within the controller after it reads the G151.

In comparison, the same program generated by CAM software, would be hundreds or even thousands of command lines long. The CAM software does not generate one continuous line, it breaks the graphic image of every tooth profile into tiny segments. It then calculates the best fit arc or straight line for each segment. The control then writes axis commands for each. Down to the microlevel, the profile is identical to the canned cycle shape—it's just the data to make it, that's different. We won't investigate writing these custom routines (sometimes called *parametric cycles*), they are beyond the beginner's level. Here we will explore the standard cycles all machinists should understand.

Then after canned cycles, we'll look at some higher programming functions:

◆ Branching logic with loops and subroutines

◆ Mirror imaging

◆ Scaling

◆ Constant surface speed

Most are program tools that CAM software doesn't use, but that the hand programmer can employ to great advantage. Here are the units that come close to an art form and fun.

Objectives

Fixed Cycles

Introduction: A fixed (canned) cycle begins with a code such as G81, which starts the drilling cycle. The code is followed by several variable entries called *parameters*. By entering the parameter once, the *modal cycle* retains them to avoid repetition. Recall that a modal command means it's a mode to be switched on by the G81, and it stays on until canceled by the G80—cancel cycle code—or by another word from the same group such as G83, the peck drill cycle. M2 (end of program), also cancels cycles on most controls.

KEY TERMS

Bolt circle
A pattern of holes or other features, distributed about a central datum.

Canned cycle
A second common name for a fixed cycle.

Departure point
The initial position from which a given routine is started. Often designated with the letter I in programming manuals.

Parameter
A blank requiring an entry to customize a fixed cycle.

Shape address numbers (shape definition)
The line numbers often defined by P and Q parameters in canned cycles.

Tiled routines
Overlapping and connecting mill routines to machine away a given area of complex shape.

How Fixed Cycles Save Time

Follow this example G81 for drilling a series of holes on a vertical mill:

N019 T01 H1 G43 M6
   (G43 = tool length offset,
   H1 = apply length offset 1)
N020 G81 X1.00 Y2.00 Z−0.5
   R.375

Same Language, Different Dialects! Keep in mind as we look at various cycles for mill and lathe, and at their parameter designators, that these are evolved words that came into existence after the original assigned core words. While the cycles operate similarly from control to control, the code and the parameter letters may be different than those used here. For example, a unit in our lab uses *D* for the hole depth parameter, not *Z* as discussed. The parameters and codes sited herein are those I'm familiar with; it's impossible to know them all!

Command line 20 tells the control to go to the X-Y location and drill a hole to a depth of $Z−0.5$ then retract at rapid travel to 0.375 above the surface of the work. In this two-line program, I used G43, a common code to initiate tool length offsets. We'll look at how length offsets are setup and used in Section 24.

The G81 code is modal. It's still in effect unless canceled.

Critical Question?

So on the next line, if I enter:

N021 X1.75 Y2.5

What will happen? Don't read the answer, think it through.

Answer

The control will move in rapid travel to the new location and repeat the modal cycle because it's still in effect. Now, what happens on line 22?

N022 X3.00 R.75

It goes to the next location moving the X axis only (Y was a null entry) to drill the hole to $Z−0.5$ but this time retracts to a different R height. You can see, a modal **canned cycle** can be useful. All one needs to do is enter the parameters that change, hole-to-hole.

Fixed cycles are used in two different but closely related ways.

1. **Abbreviate Repeating, Sequential Commands**
   The drill routine above is a good example. To program 50 holes without a cycle would require writing all the X, Y, and Z axis movements 50 times!

2. **Simplify Otherwise Complex Operations**
   Some operations would be hard or nearly impossible to write with standard commands. The threading cycle on the lathe is a perfect example (coming up). In some cases, there are no standard commands that produce a perfect thread. But by changing **parameters** within the various threading cycles, a variety of threads can be easily programmed including, internal,

left-handed, and tapered versions—as easy as entering a code word then filling in five parameters!

Lathe Cycles

Threading Cycles

To understand CNC threading we must first compare manual to CNC lathes. There's a major mechanical difference. On a manual lathe, when the half-nut lever is engaged, a geared connection between spindle rotation and tool lead is locked in. But on a CNC lathe, there is no mechanical connection. The spindle has its drive motor and the tool turret has one drive motor for X and one for the Z axis.

The computer determines the moment to start the tool moving and at what axis speed it must move to produce the thread lead. It does it perfectly, hitting the mark every time.

KEYPOINT

But that creates four critical checks for CNC operators and programmers using a threading cycle.

Any Lead Is Possible

With computer control, any thread lead can be produced: metric or imperial, left- or right-hand, fractional, tapered, and even nonconstant lead threads.

Acceleration Distance Required

Without gearing, the tool cannot instantly start moving in the correct relationship to the spindle. The Z axis must accelerate over a short distance before it's up to the correct lateral speed. Depending on the power of the drive system, some acceleration distance must be built in by starting a short air cut—not contacting the workpiece. Otherwise imperfect threads will be made during the acceleration. Read the manual for that machine to see how much acceleration zone might be required.

On a few older machines with less responsive drive systems (Fig. 23-1), there is also a deceleration distance at the end of the thread as well but it is always shorter than the acceleration zone.

Single-pass threading cycle

Figure 23-1 *The* thread cycle *requires a departure point outside both the X and Z positions relative to the finished thread.*

Lack of Operator Control

During threading, the demand on the control to keep everything in coordination, is intense. In order to govern the RPM-to-lead ratio, once the thread cycle begins, the control defaults to a specific RPM and also locks out these operator functions during threading:

Single-step mode   Spindle speed override
Feed override   Rapid override

KEYPOINT

Because the control must maintain rigid control of the tool to spindle, once the thread cycle begins there is no altering speeds, stopping, or even slowing the lathe except with the emergency stop—it goes on autopilot!

That makes for some nervous first part try-outs when a thread cycle is in the program!

The Departure Point

The prepositioning of the tool before the control reads the thread cycle command is critical. The tool must be parked outside the diameter (or inside for internal threads) and beyond the start length of the thread in Z because that's the position to which the tool will rapid return to begin successive passes. The departure point, sometimes labeled "I" for initial position, creates the retract position in the cycle. This prepositioning is also important for several other lathe cycles coming up.

But it must not be too far away either. It's that position that determines where the first pass begins. Positioning the tool too far away in the X diameter can create several air passes the full length of the thread before metal is cut. To see this effect, experiment with changing departure points then viewing the results on a screen graphics dry run or on a PC simulator or in a CAM toolpath backplot.

There are four or five parameters for the thread cycle depending on which cycle is chosen—single-pass or multipass threading. For example, a $\frac{3}{4}$-10 UNC thread:

G33   X.75 Z−1.5 H.866 A60.0 P.1

X = Thread diameter 0.75

Z = Final length 1.5 from PRZ

H = Thread depth (single depth—might be a "D" parameter on your lathe)

A = Thread form angle—60 (This parameter determines infeed angle similar to rotating the compound on a standard lathe.)

P = Pitch distance accurate to five decimal places. (Your lathe might accept number of threads per inch for this parameter.)

Ta = Parameter varies but it equals taper amount (Used with G33 and G34 nonconstant thread leads. If this parameter is left off, the thread will be straight.)

The Multipass Threading Sequence (Fig. 23-2)

1. The tool will start from the initial departure point. That's the preposition of the tool from the previous command.

2. It will rapid to the first X diameter for the first pass. Note that you do not make this calculation, the control will do it.

3. Then cut to full Z length.

4. Withdraw in X axis rapid, to the initial X diameter, then rapid to the Z departure point.

5. The control then repeats all these steps, in coordination such that the tool cuts the same groove each time, progressively deeper until a complete thread is made.

Material Roughing and Finishing Cycles on a Lathe

These are exceptionally useful programming tools for removing a large volume of metal. Without them, many roughing passes would need to be written. These three cycles (Fig. 23-3) are commonly numbered as

G70 Finish

G71 Z direction roughing (turning)

G72 X direction roughing (facing)

Combined, they first rough out in efficient straight-line passes, leaving a stepped rough

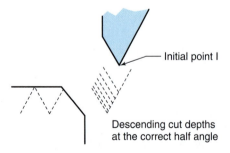

Multipass threading

Initial point I

Descending cut depths at the correct half angle

Figure 23-2 *The sequence of passes.*

G71-roughing   G72-roughing

G70-contour finishing

Figure 23-3 *The roughing and finishing cycles efficiently remove volumes of metal.*

surface. The G70 cycle then finishes the shape in a smooth contour cut, based on your parameters. Often it's more efficient to rough cut using a straight line and then switch to contouring to remove the final material.

These cycles work similar to tool compensation in that they follow the part geometry. They will work in conjunction with (or without) the G42 or G41 tool compensation commands.

When one of the roughing cycles is read by the control, using the "D" depth of cut parameter, the control divides the removal zone into straight line passes that stop short of the finished surface. That shell of excess left for the finish pass is determined by the I and K parameters (Fig. 23-4). I equals distance away from the final shape, along the X axis while K equals distance off in Z.

These cycles will make their feed passes either inward or out, depending on where the programmer places the tool before evoking them. In other words, if the tool must progress from its departure point, toward the chuck as the cycle begins, the control interprets that to mean the passes are inward. All feed passes will be inward and all rapid returns will be outward back to the departure point.

Shape Definition Addresses

To use these three roughing/finishing cycles, the parameters include a program line number on which the cycle is to start working and a line number on which it is to stop. These are called the **shape definition addresses.** The program entries between the start address P and end Q, are the part geometry program lines that the cycle will follow.

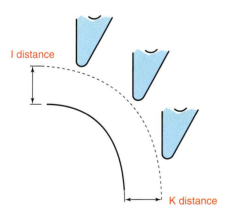

Figure 23-4 *The "I" and "K" parameters determine how much excess is left in the X and Z directions after roughing.*

KEYPOINT

The finished shape of the object lies between these two line numbers in the form of regular program commands and coordinates.

For example,

N0015 G71 *P0020 Q0035 I.5 K.3 D2*

This X axis, roughing command will start cutting from the departure point written on the previous line, close to (but leave excess) over the contour between line number 0020 through N0035. It will divide the total removal into X axis passes of 2 mm (*D2*) depth and leave 0.5 mm excess away from the finished shape in the X direction and 0.3 in the Z.

Controller Difference

Note, your control may require a slightly different form for the "D" parameter as in *D2000* equals two millimeters depth.

After the roughing cycle is finished, the control moves to the next line where it reads:

N0016 *G70 P020 Q035*

Then on line 16, the control reads the finish cycle command, it will contour cut the shape between lines N0020 and N0035.

KEYPOINT

A. These roughing and finish cycles are the first program commands in this book, where line numbers are essential to implement the command. The cycle must include parameters for the line number to start and to end the contour.

B. Roughing and finish cycles are non-modal. Once the control branches back from the shape end address, it continues on into the program with the cycle no longer in effect. (See the next paragraph for branch definition.)

Controller Differences—Order of Entry
On most CNC controls, the shape definition (line numbers 20 through 35 in the example) must be placed beyond the command line in which the cycle is implemented, not before. The search for definition then occurs forward into the program structure—that's called a *branch*. However, a few controls will also branch backward to find the needed line number addresses, that's often called a *loop*.

As usual, read the programming manual to find out; however, if the program is structured with nothing but forward branches, it will work in all cases.

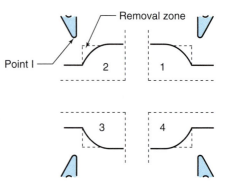

Figure 23-5 *Four cases for roughing cycles are determined by the departure point I.*

Trade Tip

Using Roughing and Finishing Cycles *Writing the Definition*   It's difficult to know the exact *P* and *Q* numbers until the entire program is written. To solve that small dilemma, write the program with the rouging and finishing cycle commands in their correct location within the program, but leave the "*P*" and "*Q*" parameters as XXX and ZZZZ (or blank). Then when the program is finished and you know exactly which line numbers contain the final shape definition, go back and fill in the shape addresses.

Two Cutting Tools   For better surface finish and longer tool life, rough with one cutter under G71 or G72, then change tools to a finish cutter before commencing the G70 cycle. Using one finish pass, sometimes the square corners of the roughing cuts leave disturbances in the final shape as the cutter hits large and small cut depths. To solve it, take two passes with the finish cutter. On the semifinish pass, set the radius offset such that the cutter backs away and leaves 0.5 mm or 0.015 in. for the finish pass. Return and cut again with the correct offset but now there are no bumps in the excess.

Modal Feed Rates

In reading the program manual, you will see that a provision for a feed rate is made within most canned cycles. If an *F* parameter is given within the cycle command, that rate is modal along with the cycle. It is in effect as long as the code is in effect but as soon as the G80 cancel cycle command is read or another cycle implemented, the original feed rate, before the mode was commanded, is reinstated.

The Departure Determines Quadrant Sense

For many cycles the departure point is a critical consideration in your planning. It must be slightly outside the contour in both *X* and *Z* directions to provide a quadrant sense, as shown in Fig. 23-5. There are four cases, with each determined by the relationship of the initial departure *point I* with respect to the work. Each quadrant sense then causes the positive direction for *I* and *K* excess as well. Quadrants three and four would be applied for internal roughing with a boring bar.

Trade Tip

Combining Cycles   With both lathe and mill roughing cycles, it is often necessary to use more than one cycle to completely rough the work. (Using them in overlapping blocks to machine the entire shape). This program trick is known as tiling, rhymes with smiling (laying routines next to each other as you would floor tiles), so that we have **tiled routines.**

KEYPOINT

If a G70 command follows either the G71 or G72, the control will first rough, then contour finish, the work with no extra programming other than the shape definition and line numbers.

Mill Cycles

Similar to lathe finishing cycles, these are time savers. However, there are two notes to consider.

Simple Shapes Normally

A few high-end controls do feature complex shape routines but they are not universal control to control. We can look at only three canned cycles that are close to common leaving full investigation for outside research.

No Island Geometry within Pockets

Due to the complex shapes encountered in milling, they are usually provided as rectangle and round pocket clearing routines on all controls but not normally will they finish a contoured geometry with islands nested within. However, as the Trade Tip describes, both rectangular and circular clearing cycles can be combined to machine a given area.

Listed next for three fixed cycles are the common parameters. Their function will be the same on all controls. For example, each must be provided a total depth, number of passes, and overall shape size but the letter designator might be different in your lab.

Rectangular and Square Pockets (Fig. 23-6)

X and *Y* coordinates for one corner of the pocket

L and *W* parameters for length and width

Plus or minus values to designate direction from the reference corner

D parameter for the total depth

H parameter for the distance to the top of the pocket from PRZ

N parameter for the number of vertical rough passes

P or *R* parameter for the number of lateral stepovers

Note, this routine mills away all the material from the top to the total depth of a rectangle in both vertical and horizontal steps. It can be applied to the outside of an object as well as a pocket.

Round Pocket Routines

X and *Y* coordinates of the center

D parameter for total depth

H parameter for the height of pocket top from PRZ

R parameter for the pocket radius

P or *R* for the number of radial movements inward or outward

N parameter for the number of vertical passes

Rectangular pocket cycles

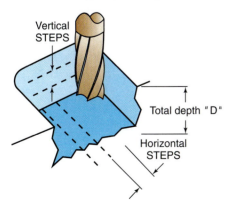

Figure 23-6 *A rectangular pocket routine.*

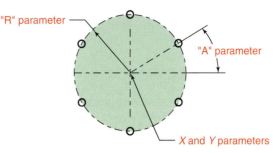

Six-hole bolt circle
"N" parameter = 6 (holes)

"R" parameter

"A" parameter

X and *Y* parameters

Figure 23-7 *A bolt circle fixed cycle.*

Note, these round area routines are often capable of both clockwise or counterclockwise circular passes and may start in the center and work out or on the rim and work inward.

Bolt Circle Routines (Fig. 23-7)

A third common mill cycle provides a quick way to calculate and drill holes in a pattern distributed about a datum, called a **bolt circle.** While there is no standardization, this routine is fairly common on most controls.

The **departure point** is usually the center of the pattern. It is either set as a local reference or there is a *X-Y* parameter for the pattern center within the routine.

G88 *N*12 *A*12.5 *D*−0.5 *P*.125 *R*.5 *H*0.0

N = Number of holes in the pattern

A = Absolute angle from the horizontal reference line to the first hole. This establishes the pattern relationship to the *X* and *Y* axes.

D = Total hole depth

P = Parameter for pecking action

R = Parameter for the retract height between holes

H = Parameter for the top of the hole from PRZ

UNIT 23-1 REVIEW

Review the Key Terms

Bolt circle
A pattern of holes or other features, distributed about a central datum.

Canned cycle
A second common name for a fixed cycle.

Departure point
The initial position from which a given routine is started. Often designated with the letter I in programming manuals.

Parameter
A blank requiring an entry to customize a fixed cycle.

Shape address numbers (shape definition)
The line numbers often defined by P and Q parameters in canned cycles.

Tiled routines
Overlapping and connecting mill routines to machine away a given area of complex shape.

Replay the Key Points

◆ Many cycles are modal. They remain in effect unless canceled (G80) or a program end (M2) is read.

◆ Once the thread cycle begins there is no altering speeds, stopping, or even slowing the lathe except with the emergency stop!

◆ Roughing and finishing cycles require line number address.

Respond In Your Lab

1. Why is the departure point and/or initial point critical for many cycles? Give three reasons.

2. Are canned cycles appropriate for complex shapes on milled parts?

3. Ask for the handout describing the lathe and/or mill fixed cycles in your training lab. If this is unavailable, ask to read the programming manual for the control. The object here is to know the possibilities. Look too at the parameters and compare them to those given here.

4. Write a program that uses fixed cycles to machine an object. If you have graphic simulators, try the program out. A program using cycles and tool compensation is fairly high level, and it is anticipated that there will be bugs to work out.

5. When using a thread cycle on the lathe, why must we park the tool slightly back in the Z axis before the thread starts?

UNIT 23-2

Branching Logic–Loops and Subroutines

Introduction: In Unit 23-1 when we used the P and Q addresses in both the rough and finish routines, the control had to leave the command line, where the roughing cycle command is read, and move forward in the program to the line number address designated by the P parameter. Then after completing the shape definition string, it returned to the branch origination. It would then continue onward in the normal program.

This nonlinear program flow can be shown in a graphic form called a **logic diagram.** If you have any experience in computer programming, many of the logic manipulations described will be familiar.

We will investigate two useful techniques, but there are others beyond the scope of this unit. Within those advanced techniques left, for outside research are the ability to

Measure the work using interfaced equipment and automatically adjust offsets up to a limit.

Sense the cut pressure and change to a new sharp cutter when the pressure exceeds a preprogrammed limit. This ability is called *adaptive control.*

Count and record the number of parts produced per hour and per day and the total run count, then stop at the number required—this is common on bar feed turning centers.

Communicate with part loading robots and outside control databases.

KEY TERMS

Branch
A departure from a linear format whereby the program flow jumps to a designated address to create a loop or to find and operate a subroutine then return to the origin of the callout.

Logic diagram
A graphic tool used to visualize the syntax of branching logic.

Nesting
The use of a loop or subroutine within a loop or subroutine.

Smart loop
A programming tool that turns the program flow back upon itself a given number of times.

Subroutines

Miniprograms Called within Programs

The purpose of a subroutine (often called subs) is to avoid repetitious programming or to put aside groups of commands to be used again at a later time. Clever programmers sometimes save a number of subs then paste them together to create programs quickly. For example, an instrument shop makes custom panels that hold various instruments, customized to each request. For each instrument one could save the panel cutout shape as a subroutine. Then, in the program locate the mill spindle over the needed hole and call the sub.

A subroutine is nothing more than another program. It has a start, a body, and an end. But different from the programs we've been studying, it is stored in memory then called up and used by an active, master program. The actual commands within the subroutine are those we have studied including fixed cycles and tool compensation. Nearly any program command can be used within a subroutine.

Controller Differences

Logic programming was introduced long after codes were invented, thus there's a great deal of variation in the codes and parameters used to start and end subroutines. Some controls call a subroutine from the main program using a G code, while others use an M98 code and still others use the logic statement *Gosub*.

To simplify explanations, the words GOSUB *NNNN*, SUB *NNNN*, and ENDSUB are used to represent the **branch** commands along with their line number addresses (*NNNN*). These are words borrowed from conversational controls (also from computer programming languages), however they covey the meaning easily.

Outside Research—In Your Lab

To use subroutines, you will need your control's equivalent words that

1. **Call the Subroutine from the Main Program (GOSUB *NNNN*)**
Causes the program to branch out to the stored file name given in the command.

2. **File Name (SUB *NNNN*)**
The control searches memory for this identification then upon finding it, operates it as a program.

3. **End Subroutine (ENDSUB)**
The last line of the subroutine. This causes the control to end the subroutine and branch back to the next command line in the master program, *after* the call subroutine command. Several controls use M99 for this command.

These can be shown in logic diagrams (Fig. 23-8).

A single subroutine logic diagram

Figure 23-8 *A simple logic diagram of a subroutine.*

Solve this logic challenge

Figure 23-9 *This instrument plate has four repeating patterns.*

Critical Thinking Challenge

A subroutine could be used to program the drilling of the sixteen, 0.250-in.-diameter holes in Fig. 23-9. Look at the detail view and see if you can think of how this would be accomplished. Diagram or write the program structure. Then read the program provided just before the Unit 23-2 Review, to see one possible solution. The real trade secret lies in the subroutine.

KEYPOINT

Caution!
When swapping modes (abs/Incr) as in this example, be certain to reinstate the original mode (G90) before leaving the subroutine. Otherwise the next move after the last subroutine has run might be a big surprise!

Nested Subroutines

It is possible to have a subroutine call out a second subroutine. That second branch is then said to be **nested** within the first. For example, the circular cutout in Fig. 23-9, could be a second subroutine, called out in the drill pattern routine. The second subroutine is also stored in memory and called when needed.

Some controls limit nesting of subroutines (and loops—discussed next) to three levels deep, a subroutine within a subroutine within a subroutine. Other more capable controls feature a higher nest limit, all the way up to no limits. However, nesting much beyond three deep, usually ends in a tangled mess, especially when something goes wrong or editing is required!

Loops

A logic loop is a useful technique that also makes efficient use of repeating data, but this time it's located within the main program. There are two kinds of loops, the *counted* or **smart loop** and the *unending, dumb loop*. The counted loop is of prime interest as a programming tool, while the dumb loop has limited use.

Smart Loops

A smart loop includes the branch statement and a count function. It branches back to repeat a selected group of command lines within the program, a given number of times, then completing the loop, moves on to the next command line after the loop start.

There are actually two different structures to these loops, as shown in a logic diagram (Fig. 23-10), dependent on the control type, the branching statement is placed at the

front of the loop or at the end. This is a controller difference in words only.

The overall effect is the same, the program branches backward to repeat a group of commands a given number of times (Fig. 23-10). Both forms repeat the indicated commands six times, until the count is satisfied then move forward onto line N50. Notice the difference in repeat requests based on the loop structure. The second structure needed five more repeats to complete six times through the loop, because it has already completed the commands one time as the flow reaches the branch back.

KEYPOINT

The number of repeats is an issue—depending on where your control picks up the number of times to repeat a loop. If on your control, the loop repeat statement occurs at the *end of the loop* then the *requested* repeats will be one less than the number desired $N = R - 1$ Why? Because one has already been made. But, if the repeat statement occurs at the *loop start*, then $N = R$.

Dumb Loops

A dumb loop is a logic branch that does not count but turns back on itself forever if not stopped by the operator (Fig. 23-11). The only way to stop a dumb loop is to include a program halt or optional stop within the command lines or by an operator intervention such as feed hold, cycle stop, or emergency stop. An uncounted loop is actually a syntax error. It never reaches the end of the program. Dumb loops have only two applications.

Two logic loop structures

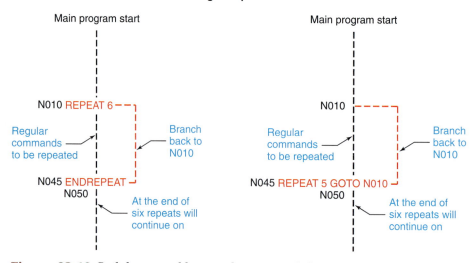

Figure 23-10 *Both loops would repeat the commands between line numbers 10 through 45, six times, then skip forward in the program.*

An unending dumb logic loop

Main program start

N010

N045 GOTO N010

M2

Figure 23-11 *A dumb loop diagram.*

Warm Up Routine

They can be a stored program that can be quickly called up to "exercise" all axes of a cold machine before it is put into production.

(Exercise Loop—Departure from Machine Home)

N005 G0 X50. (run to limit switches or end of travel)

N010 Y50.

N015 Z−50.

N020 X0.0 (or a machine home statement)

N025 Y0.0

N030 Z0.0

N035 GOTO N005

N040 M2

Never Ending Program

Dumb loops can be a program that runs continuously, whereby the operator places the part in the machine then depresses "cycle start," which overcomes the halt command on line N025.

(Simple Drill and Repeat Loop)

N005 M3 S2000 G95

N010 G0 Z.5

N015 G1 Z−0.5 F0.003

N020 G0 Z100. (limit home)

N025 M0   (program halt)

N030 GOTO N005

N035 M2

Neither dumb loop ever reaches the M2 command, so it's actually not needed to run it. However, most controls will make a syntax check as the program is loaded into memory. Without the correct program end code, the control will error out.

Answer to Critical Thinking Challenge

To simplify, most of the usual start codes and modes have been eliminated, study the logic and subroutine only.

The trick was to position the tool at the center of each pattern using absolute coordinates, then switch to incremental values within the sub. In effect, that makes the center position a mini-PRZ or local reference for the first move, within the subroutine.

Instrument Plate Main Program (From Fig. 23-9)	Instrument Plate Subroutine 01
T01 M6 ($\frac{1}{4}$ drill)	SUB 01
G90 M3 S3000	G91 (incremental mode)
G0 X2.0 Y2.0 (center of 1st group)	G81 X.5 Y.5 R.25 D−0.5
GOSUB 01	X−1.0
G0 X4.0	Y−1.0
GOSUB 01	X1.0
G0 Y4.0	G80 G90 (absolute mode again)
GOSUB 01	ENDSUB
G0 X2.0	
GOSUB 01	
G0 Z100 X100 Y100 (park at limits)	
M2	

UNIT 23-2 REVIEW

Review the Key Terms

Branch
A departure from a linear format whereby the program flow jumps to a designated address to create a loop or to find and operate a subroutine then return to the origin of the callout.

Logic diagram
A graphic tool used to visualize the syntax of branching logic.

Nesting
The use of a loop or subroutine within a loop or subroutine.

Smart loop
A programming tool that turns the program flow back upon itself a given number of times.

Replay the Key Points

◆ When changing modes within nested routines or loops, be certain to reinstate the original mode before leaving the routine otherwise the next move afterward might be a big surprise!

◆ For loops, the $n$umber of $r$epeats will be either $N = R$ or $N = R - 1$ depending on where the repeat statement is located in the loop.

Respond

1. Can you invent a way to drill the grid plate shown in Fig. 23-12 using a loop? Try it yourself but use the commands of your equipment this time, if available. If not, use the words shown in the logic diagrams. There are several solutions to this problem including subroutines but for now, use only one loop.

2. Now, can you see a way to improve your program using a nested loop?

Grid plate

Figure 23-12 *Program this* grid plate *using* loops.

Special CNC Programming Tools

Introduction: This unit lays out a collection of tools of varying degrees of usefulness. The goal here is to introduce the concepts. Each will require a bit of reading within the programming manual to implement. They are all highly specialized computer functions, therefore none are standard code words.

Unquestionably, the most useful is the constant surface speed (CSS) command which allows the CNC lathe control to maintain a selected cutting speed on operations where the cut diameter is constantly changing, as when facing the work.

KEY TERMS

Mirroring
A program code or draw utility whereby geometry is reversed.

Obverse (shape)
The counterpart shape required to mold a correct feature.

Scaling
A program factor (or draw utility) whereby dimensions are expanded or contracted by a percentage.

Shrink factor
The amount of percentage change of part geometry.

Turning Center—Constant Surface Speed

This command is easy to write and demonstrates a powerful difference in CNC and manual machining where, due to changing radius, the selected RPM is always a compromise when face cutting or parting are done.

Before instituting CSS, the tool is first moved to an initial point. That X diameter will then serve as a model for the surface speed. The RPM is set at that point and the CSS code is given:

G0 *X*2.0 *Z*.5   (Rapid to Initial Point)

G96 S2000   (G96, is a common code used for CSS, but the code on your control might vary)

After the model surface speed is set with the G96, any change in X axis diameter will cause a corresponding increase or decrease in the RPM according to the familiar RPM formula.

The RPM Limits

You can see that, using CSS, as the lathe cutting tool faces inward, as it approaches the work center, the calculated RPM will approach infinity—super high! For example,

X Axis Diameter	RPM
2.00	2,000 (the model)
1.00	4,000
0.50	8,000
0.25	16,000
0.0125	32,000 (No way!)

The G50 RPM Limit Code

A common code used to stop RPM at a safe ceiling is G50 (on many controls but not a core assigned word, so it may vary—see the Trade Tip).

    N015   G0 X2.0 Z.5
    N020   G96 S2000
    N025   **G50 S4500**

The RPM in the preceding constant surface speed program will not exceed 4,500.

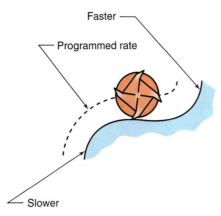

Nonconstant surface speed

Faster —

— Programmed rate

— Slower

Figure 23-13 *As the end mill cuts around curves, its feed rate varies at the rim.*

Under normal milling conditions this velocity differential has little effect, and we need not do anything about it in the program. However, when milling metals with a narrow machinability window such as hardened stainless or phosphorous bronze, for example, or when performing high-speed machining, which pushes limits and creates a very narrow window for surface speed, then CSS can be useful. An uncommon code, the word and application varies from control to control.

Milling Cutter—Constant Feed Rate

CSS is very common in lathe programs but it also exists for profile milling. As an end mill cuts around an object (Fig. 23-13), the feed rate at the contact rim varies slightly depending on whether it is producing an outside or inside curve, as shown in Fig. 23-13. That's because the rate is predicated on the cutter centerline, the dotted line.

Axis Scaling

This programming tool is used to expand or shrink machined results of a program. It can be applied to all axes or to a single axis, depending on the desired effect. It's used where a complex geometry and drawing would make it difficult to calculate the expansion or contraction of the workpiece, so the print dimensions are used for program coordinates—then the shape is scaled up or down using the power of the computer. If you are familiar with CAD drawing, then you've probably already used **scaling.**

One good example is that of milling a precision, plastic injection mold cavity. On these molds, the plastic is forced at elevated temperature to flow in and conform to the mold. But as it cools, it also shrinks. The mold must be slightly larger than the finished part by a predictable ratio.

That **shrink factor** is calculable and simple to edit into the program. After writing the finished part geometry, a positive factor is inserted at the start. After the code word (we'll use G100 on our generic control), a percentage of increase is included with each axis:

$$G100 \ X1.05 \ Y1.05$$

That would cause all X and Y axis movements beyond that command line, to be 5 percent larger. In this example, the Z axis commands would be unaffected.

Another use of scaling is found in high-volume (big chip loads, not HSM) machining where deformation heat expands work. Then as it cools, the parts are found to be undersize. Using part scaling, a slightly expanded part can be made that accounts for thermal expansion, such that the cool part is the right size. If the problem occurs on one axis of the work because it is long, then the scaling can be applied to only that axis. The ratio can be adjusted by the operator as conditions change. For example a new tool causes 1.5 percent expansion but as it dulls 2 then 3 percent, at which point the cutter must be changed to a sharp tool again and the ratio is returned to normal in the program.

CAD/CAM Scaling and Mirroring

The scaling and mirroring commands (next) are used when editing a previously written program or handwriting a new program. However, CAM and CAD software also have these capabilities. One can easily expand or contract the part geometry by a given ratio before creating a program. It can also be turned into a mirror image just as easily.

The effect of mirroring axes

Mirrored X axis makes opposite part

As programmed (or drawn) no mirroring

Mirrored X and Y same part rotated

Mirrored Y axis makes opposite part

Figure 23-14 *Results of mirroring.*

Axis Mirroring

Similar to scaling, **mirroring** modifies all axis movements beyond the command, until it's canceled by a flip/flop of the same command—one time turns on mirroring, a second time turns it off.*

*On the controls of which I'm aware, there is no cancel code for mirroring or for scaling, but there could easily be one—read the manual! To my knowledge, they are turned off by reversing the effect of the command, for example, entering G110 X0.0 scaling of the X axis is ended by returning to a scale factor of zero.

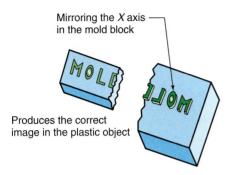

Mirroring the X axis in the mold block

Produces the correct image in the plastic object

Figure 23-15 *Mirroring a letter mold.*

Given a mirror axis command, the axis drives the opposite direction to the programmed command. One or more axes may be mirrored depending on the affect desired. Mirroring is another unassigned code, we'll make it G110 on our generic control.

A good example of mirroring is that of machining a mold or die. In this case, it's making plastic letters (Fig. 23-15). The mold must be the mirror image of the finished part. Often, it's far easier to program the letters or other features from the print, in their normal form, and then edit and insert an axis mirror command before the geometry begins.

$$G110\ X$$

Presto, that's all the command required to reverse the X axis but leave the Y and Z alone.

KEYPOINT

Mirroring only *X* is the needed command to perform the letter reversal of Fig. 23-14, called an **obverse** mold. Mirroring *X* and *Y* axes wouldn't produce a mirror image part, it would change its location only.

Double Axis Mirroring

Be aware that mirroring both axes does not produce the obverse effect. Shown in Fig. 23-15, mirroring both the X and Y axes on a mill program will only serve to shift the part location.

UNIT 23-3 REVIEW

Review the Key Terms

Mirroring
A program code or draw utility whereby geometry is reversed.

Obverse (shape)
The counterpart shape required to mold a correct feature.

Scaling
A program factor (or draw utility) whereby dimensions are expanded or contracted by a percentage.

Shrink factor
The amount of percentage change of part geometry.

Replay the Key Points

◆ The objective was to learn the possibilities of these commands, not to learn how to use them. Each requires more study in the individual programming manual.

◆ Mirror only one axis to obtain an obverse geometry.

Respond

1. Describe a *subroutine* compared to a *loop*.
2. True or false? Constant surface speed is a lathe programming tool used to maintain a model cutting speed as diameters vary. If it is false, what will make it true?
3. When G96 is used in a turning program, what safety factor should be included?
4. Describe canned cycles. How are they used? What is another name for them? What are the blank variables called within?
5. Compare a smart loop to a dumb.
6. Describe the difference in G71 and G72 roughing cycles for the lathe.
7. In the following command lines, what purpose does the G80 code perform?

N075	G81	X1.0 Y1.0 D−0.5 R0.2
N080		X1.5
N085	G80	
N090	G0	X0.0

8. Describe axis mirroring and axis scaling.

Outside Research

From a handout or from the programming manual,

9. How many threading cycles does your lathe feature?
10. Determine the recommended acceleration distance for a thread cycle on the lathe in your lab.
11. What codes, if your control features them, are needed to cause RPM limits, axis scaling, and axis mirroring?
12. Identify the area-pocket roughing fixed cycles found in your milling machine programming manual.
13. Does the CNC milling machine in your lab feature constant feed rate?

Section 23 Review

Units 23-1, 23-2, and 23-3

All of Section 23 was introductory, to whet your appetite for the higher functions when the opportunity arises. In every unit of this section, we only tipped the iceberg as far as the special things that this generation of CNC controls can do, especially the PC-based versions! But what about tomorrow? The real frontiers of this special kind of programming lie within algebra.

At some point in your career, if you wish to use the power, it would be a good idea to learn the logic statements and program flow of MS Visual Basic® (VB). VB is more than a language to write games and screen images. Learning the way data can be manipulated and branched is also the fundamentals of the CNC subjects we didn't touch on.

Using smart programming, CNC controls can do algebraic manipulations within programs. For example, consider the custom chain sprocket routine discussed in the introduction. It was easy to describe how it might work. But actually writing the formula that defines the tooth shapes when each complex curve lies at a different angle to the central axis of the blank requires a great deal of math and logic.

Without these skills, you could solve the problem by writing a single tooth profile and use a rotary C axis to index the blank to mill each tooth. But writing the routine to mill every tooth without revolving the part, would be a far greater accomplishment. Then any CNC mill could make the sprockets.

Questions and Problems

Problems 1 through 6 refer to the following command:

$$N0195 \ G71 \ P0155 \ Q0185 \ I0.03 \ K0.03 \ D0.05$$

1. Describe the machine type and the action. (Obj. 23-1)

2. How much material will be withheld for a finish pass? (Obj. 23-1)

3. How much material will be taken with each advancement of the X axis? (Obj. 23-1)

4. Write a finishing command to follow the example command. (Obj. 23-1)

Critical Thinking Questions

5. What does this command tell you about the logic of the control on which it is to operate? (Obj. 23-1)

6. How could you write a set of program commands to semirough the shape leaving 0.015-in excess, then return a finish the shape? (Obj. 23-1)

7. When performing a threading routine on the lathe, which operator control functions are disabled. Explain? (Obj. 23-1)

8. Write a threading routine (use the code your lathe uses or G33) for a 12-mm, 1.75-pitch, 24-mm-long standard 60° thread form. (Obj. 23-1)

9. True or false? A loop is actually a syntax error because the control never gets to the end of the program. If it is false, what makes it true? (Obj. 23-2)

For problems 10 through 12, the following looped program is written to drill the holes for Fig. 23-16. Assume a single tool program ($\frac{1}{4}$-in. drill) with Z0.0 at the work surface. The tool is parked above PRZ 5.0 in. The material is 0.375-in. thick and the work is clamped over a 0.50-in. thick sacrifice plate that will be drilled.

Figure 23-16 *Review Problem 10.*

10. Does the loop drill the five holes? (Obj. 23-2)
11. Is there any danger with the program? (Obj. 23-2)
12. If the program needs editing, repair it to run correctly. (Obj. 23-2)

    (Final review loop program—Problems 10 through 12)

    | N005 | G80 G40 | |
    | N010 | S2000 G94 | |
    | N015 | T01 M6 | ($\frac{1}{4}$″ drill—no offset) |
    | N020 | M3 | |
    | N025 | G0 X1.0 Y1.0 | |
    | N030 | Z.25 M8 | (rapid to retract height) |
    | N030 | G91 G1 | |
    | N035 | Z−0.700 | (drill into sac. plate—0.1) |
    | N040 | Z0.700 | |
    | N045 | REPEAT 005 GOTO N030 | |
    | N050 | G0 Z5.00 | (park spindle) |
    | N055 | X0.0 Y0.0 | |
    | N060 | M2 | |

13. Define a subroutine. How would it be used? (Obj. 23-2)
14. What is different about a subroutine compared to a master program? (Obj. 23-2)
15. Define a nested loop subroutine. How deep can one normally nest branches? (Obj. 23-2)

Critical Thinking—Not Covered in the Reading

16. Can mirroring be used to machine soft jaws to fit a contoured part? In other words, can you edit mirroring into an existing program to machine a counterpart of the object to be held, into the soft jaws? (Obj. 23-2)
17. Describe the purpose of G96 for turning. What precaution must be used? (Obj. 23-3)
18. True or false? An axis-scaling code would not be found in a program created by CAM. If this statement is false, what makes it true? (Obj. 23-3)
19. Describe when and why constant surface speed might be used for milling profile work. (Obj. 23-3)

Critical Thinking Problem

20.
A. In the following lathe commands, what Surface Speed (not RPM) has been set as a model?
B. At what diameter will the RPM stop accelerating?
   Hint: The solutions require manipulating the long RPM formula:

$$RPM = \frac{SS \times 12}{Circumference}$$

The needed formulae can be found in the answers.

N005	G0 X2.500	(Rapid tool to model diameter)
N010	S1000 M3	
N015	G96	
N020	G50 S3000	
N025	G1 X0.0	

Section 23 Answers

ANSWERS 23-1

1. To avoid excessive air cuts before the real cut begins; to determine the quadrant in which the cycle will occur; to provide a safe retract location; to provide an acceleration distance for threading.

2. No, CAM is a better solution.
3. Answer found in your lab research
4. Answer in lab
5. To avoid imperfect threads

ANSWERS 23-2

1. One solution is a loop that moves the tool 0.50 in. to the right then drills a hole, repeated nine more times. That loop is used three times after positioning the tool at the start of each horizontal row.

   N005   T01 M6— ($\frac{1}{8}$ Drill—starting at PRZ)
   N010   G0 $X0.0$ $Y.5$ $Z.25$ (Departure point for first row)
   N015   G91 M3 S4000   (Incremental mode)
   N020   G81 $X.5$ $D-0.25$ $R1.0$ (drills first hole)
   N025   **REPEAT 5**
      GOTO N020   (Drill next five holes)
   N030   G90 G80   (Restores original modes)
   N035   G0 $X0.0$ $Y1.0$ $Z.25$ (Departure for second row)

   N040   G91 G81 $X.5$ $D-0.25$ $R1.0$
   N045   **REPEAT 5**
      GOTO N040
   N050   G80 G90
   N055   G0 $X0.0$ $Y1.5$ $Z.25$
   N060   G91 G81 $X.5$ $D-0.25$ $Z0.25$
   N070   **REPEAT 5**
      GOTO 060
   N075   G90 G80
   N080   G0 $Z100$
   N085   M2

2. The three horizontal rows can be placed in a nested loop that moves the tool up 0.5 in. in Y and back 5.0 in. to the left in X, each time, before calling the nested drill loop.

ANSWERS 23-3

1. A subroutine is a custom written, miniprogram called up from storage by a main program. A loop exists within the main program and is repeated.

2. The statement is true.
3. An RPM limiter as the tool approaches zero diameter
4. Canned cycles or fixed cycles are used to shorten programming of repetitive, complex, or otherwise impossible shapes. They are groups of commands that require only a code and completed parameters to make up an entire packet of programming.
5. A smart loop repeats selected commands to satisfy a count then moves beyond the loop farther into the program. A dumb loop repeats forever.

6. G71 divides the removal zone into straight turning passes in the Z axis direction while G72 divides the zone into X axis facing cuts.
7. When using a G81 drill cycle, it is a good idea to enter a G80 command at the end of all drilled holes because it is a modal cycle. If not canceled, the next X or Y coordinate will cause the machine to position then drill, when positioning was the sole intent!
8. Mirroring is the reversal of one or more axes to make opposite hand parts or to shift position within the work envelope. Scaling is expanding or shrinking one or more axes to change the size of the finished product.

Answers to Section Review Questions

1. It's a lathe routine to rough using Z axis straight turning passes.
2. The cutter will stop 0.30 in. from the final shape.
3. 0.05 in. per pass = 0.100 in. smaller diameter
4. G71 $P0155$ $Q0185$
5. It will branch backward (many won't). It is at line N0195 but the shape definition addresses start at N0155.
6. By using two finish routines and tool radius compensation. Change the radius offset to a larger value than that which is actually on the cutter for the semifinish pass.
7. All operator functions for speed, rapid and feed plus single step mode. It's because the control on the lathe must maintain rigid control of the Spindle rotation to the cutting tool position.
8. G33 $X12.2$ $Z-24.0$ $H1.515$ $A60.0$ $P1.75$ (The routine includes clearance to avoid tool drag upon rapid return; $H = 0.866 \times$ pitch distance.)
9. It's False. A dumb (uncounted) loop is a syntax error.
10–12. There are three problems with the program.
    A. It drills only the first hole, then it will go up and down at that location. It has no stepover move!
    B. The program doesn't cancel the G91 incremental command after the loop. If we run it a second time, it will begin in incremental mode. Note that a G90 was added twice, once after the loop then in the start lines.
    C. It needed to be repeated *four* more times as there are five holes total. Or alternately the loop repeat could have been moved to the front of the loop and kept at five repeats.
13. A program called from memory, from a master program. Subroutines shorten programs by eliminating repetition.
14. Nothing in the actual commands but they must be identified with a file number that will be called out in the master program, and they must have an end-subroutine word at their end.
15. Nesting is writing a branching logic statement within another branching statement. Generally three levels deep or more.
16. No. It produces opposite hand parts, but not the nest that holds the part. Switching tool compensation codes G41/G42 is the trick used to make soft jaws.

17. By setting a model for surface speed, any change in X diameter causes a proportionate change in RPM. For safe RPMs it should be limited when facing at or close to the center (G50).
    (ANS. Problem 10–12 Five Hole Program—Rev 1)
    N005 G80 G40 G90
    N010 S2000 G94
    N015 T01 M6   ($\frac{1}{4}$-in. drill)
    N020 M3
    N025 G0 $X1.0$ $Y1.0$
    N030 Z.25 M8   (rapid to R)
    N030 G91 G1
    N035 $Z-0.700$
    N040 Z0.700
    N041 $X0.625$ (Stepover)
    N045 REPEAT *004* GOTO N030
    N050 G90 G0 Z5.0.   (Abs. val. + Park)
    N055 $X0.0$ $Y0.0$
    N060 M2
18. It's probably true. If the object needed scaling, up or down, when using CAM, most likely the geometry would be resized before the program was written. However, there's a slight possibility the scaling code was edited into the CAM program based on the 100 percent size. In other words, the CAM program is 100 percent but it was scaled up or down later.
19. When the window for surface speed is narrow, to improve tool life and avoid work hardening, while machining difficult material or during maximized machining such as HSM.
20. N005 G0 $X2.500$ (Rapid tool to model diameter)
    N010 S1000 M3
    N015 G96
    N020 G50 S3000   (reaches 3K @ 0.83 Dia.)
    N025 G1 $X0.0$

    A. Formula: $SS = \dfrac{RPM \times (\pi \times Dia)}{12}$

    $SS = \dfrac{1{,}000 \times (7.854)}{12}$

    $SS = 654.5$ @ Model X Diameter

    B. Formula: $Dia = \dfrac{SS \times 12}{\pi \times RPM}$

    $Dia = 0.833$ in.

Section 24

Setting Up a CNC Machine

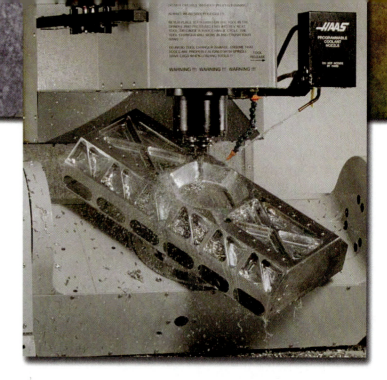

Introduction

In Section 24 we assume the planning and program are finished, but untested as yet. So it's time to focus on getting that first CNC job up and running. Following the plan, you'll mount work-holding devises, set up and coordinate the PRZ, add cutting tools, then load data into the control. OK—now it's time for that white-knuckles, first part run!

How white the knuckles are depends on a complete understanding of the goals of Unit 24-3, where we'll discuss 12 different ways to ensure success—9 of which occur before touching the cutter to the metal. How many program examinations are used to prove it works, varies by situation. But clearly, recognizing the tricky part runs then taking the best proof and tryout action is a worthwhile skill to have!

Section 24 is about managing a CNC machine. While much of it is deep, study it to the max! In my opinion, knowledge of setting up a CNC machine is the center of this book.

Machine and Shop Differences

While the setup tasks of Section 24 have a common objective, the way each shop and CNC control performs them varies. Your instructor will have a particular way of setting up a PRZ, while your shop foreman may set a policy that's completely different. (Note the discussion of master tool verses master position methods for setting PRZ.) In some cases, newer controls from the same manufacturer will do things differently than the older version. For each task, we'll discuss what the planning committee and I believe is the range of common procedures. The key point is that we'll discuss the range of possibilities, but no shop mixes them, it would be too confusing. They'll have one procedure.

Lead/Setup Machinists

To be considered a journey-level machinist, you must be capable of all the tasks in Section 24. However, for efficiency, many shops employ lead machinists to perform them. They are capable of setting up, debugging, and running every machine in the shop. While they occasionally fill in when the workload is exceptional or a key machinist is missing, generally they perform no production work. They circulate from machine to machine, setting up, helping, and troubleshooting problems.

After making a first part that passes inspection, they might remain until critical dimensions stabilize and normal variation can be determined. Shop leads also anticipate cutting tool replacements before they are needed. CNC lead people also work with toolmakers to fabricate jigs and fixtures or, in some shops, they make the tools themselves. Lead work is a challenging, well-paid career and a step toward management or owning your own shop! The following goals teach how to get your machine going and perhaps get a career up and running right as well!

Objectives

Aligning and Coordinating One Cutting Tool in Machining Centers

Introduction: First we'll make the work-holding setup, then we'll synchronize its physical location within the work envelope to the axis registers in the control. We'll start with the simplest case, a single cutting tool, first in a CNC mill then in a lathe. Then in Unit 24-2, we'll progress to setting multiple tools, where a lot of the variation between procedures arises.

KEYPOINT

> In Unit 24-1, after making the physical setup, we'll locate a single cutting tool at the PRZ location, then set the axis registers to read X0, Y0, and Z0 if it is a mill program. That's called the master tool method for setting PRZ. There is another way described in Unit 24-2, called master position.

This section assumes previous experience with dial test indicators and edge- and center-finders to align tooling to machine axes. If you don't have that experience, return to Sections 7 and 12 for a review. Previous practice with digital axis positioners on either lathes or mills will also be useful.

KEY TERMS

For brevity I often state that a concept applies to a "mill," which means a vertical machining center probably equipped with automatic tool changing. A lathe means a turning center with the cutting tools mounted in an indexing tool turret.

Axis preset command
A code that writes machine coordinates into work coordinate registers based on the present tool position.

Calc (position)
Modern controller function whereby axis values can be written into work coordinate registers or offset memory.

Coordination (coordinating)
Physically aligning holding and cutting tools then writing axis registers to represent that position as the PRZ.

Fixture offset
A method of setting PRZ. A code word (usually G54 through G59) that refers axis position relative to PRZ, using

stored data in memory. Fixture offsets can be changed by editing memory.

Master tool method
A setup method whereby one cutting tool is used as a model for all other tools in the setup. All other cutting tools are given offsets so that they perform as though they were the length of the master.

Touch-off method
Bringing a cutting tool to the work until it touches a feeler of known thickness.

Machining Center Vise Alignment

Controller Screens Needed

To perform CNC setups, you'll need to clearly understand the difference in and how to access and write data into the following control screens (sometimes called pages) on your control. On most modern controls, two or more of these screens may be combined by choice within split-screen displays. Here's a brief summary of what each represents and how it's used for setup and operation:

Machine Coordinates
  Mills
  The absolute *X-Y-Z spindle position* with reference to machine home
  Lathes
  The *X-Z tool turret position* relative to M/H
    Used for
    Alignment and setup reference locations and holding tools, to machine tooling such as vise and chuck jaws and as a utility reference during fixture offset setups
    Not Used for
    Machine coordinates are absolute and fixed, they cannot be changed, nor will they be used during program operation. They do not directly show a relationship with the program PRZ.

Work Coordinates
  Both Lathes and Mills The *position of the cutting tool with reference to the current PRZ* that has been established during the setup.
    Used for
    Basic data for *monitoring program tryouts and during normal program runs*.
    Not Used for
    Work coordinates are not used until the setup is synchronized to the physical location to the PRZ. They are useless until that final step since they represent a previous setup no longer on the machine. A

main objective for CNC setup is to coordinate the *X* and *Y* (and *Z*) axes to represent the actual tool position in these coordinates. You will be changing this screen to represent the setup.

Work coordinates aren't complete until the setup is finalized.

Distance to Go (DTG)

These are countdown incremental values that tell the operator how far the cutter will move in each axis, *to complete the current command.* They are used when clearance is in question during cuts, either during program or MDI commands. Often we halt the motion with override or slide hold functions, measure how far the cutter is from danger, then compare that to the DTG values.

Used for

They are used extensively *during program tryout;* for example, a lathe tool starts to rapid toward the chuck, to be certain it won't hit the chuck, touch slide hold, stop the spindle rotation then measure the distance the tool can travel. Then we compare that to the DTG screen.

Not Used for

Machine or PRZ reference—they report how far the cutter is going to move to finish the current axis command. *DTG values cannot be changed* by hand data entry.

Tool Memory
Used for

Discussed previously, this screen compiles the tool number, its tool changer location along with its compensating offsets for diameter/radius and length. Lathe tool shape or approach vector are also stored here.

Fixture Offset—Mills

Fixture offsets *locate the PRZ relative to machine home* (usually on mills but lathes can use them too). They are used to locate one single PRZ or multiple PRZs. Common codes for fixture offsets are G54 and G55, up to G59. G53 is commonly reserved to identify machine home in the program, but it can be used as a fixture offset as well.

Used for

In industry, several vises and fixtures can be prelocated to reduce setup time. As they are mounted, keys or pins fix them in an exact location on the sacrifice plate. In one quick step, it's aligned and at a precise known machine reference location within the work envelope. Then saved fixture offsets can be implemented to establish the PRZ. On most controls, a minimum of seven locations can be set. With little chance for error other than chips accidentally left behind, the holding tooling and *X-Y* reference for the setup is ready to go.

A quick method to return reference to machine home, and to move the tool to M/H, is to write G53 X0 Y0 Z0, where the G53 offset has been preloaded with X0, Y0, and Z0—it has been reserved for machine home.

Floating Vise Setup

One of the more basic setup tasks is to locate a single vise anywhere on the CNC mill table. With some practice, you should be able to complete this step in less than a half hour. Here's how.

Work Orientation First

If you didn't do the planning, start by determining the how the raw stock must lie relative to the machine then find the holding tool(s) to make it so. In most cases, the vise, clamps, or fixture can be placed near the center of the table, but not always. Long or odd-shaped parts sometimes need the setup made elsewhere.

Next clean everything; the vise and sacrifice plate or mill table must be free of chips.

X-Y Alignment Step by Step

☐ **Load the Indicator Holder— Manually or MDI**—Many shops keep an indicator holder as a permanent part of the tool

Figure 24-1 *Many CNC mills maintain a permanent indicator holder as Tool Number 16.*

drum load. For example, in our shop it's always T16 = indicator (Fig. 24-1).

Preloaded Holder If so, switch the control to MDI mode (manual data input—may also be edit mode on some controls) then enter the tool number and the M6 code. Then switch to single block and touch >cycle start<. Normally, the indicator is not kept on the holder, so once the changer arm has safely retracted, mount it in the holder.

No Holder Loaded If there is no holder in the drum, switch the control to manual mode and follow the manual tool change sequence for that control. Change to a blank tool position, then use the draw bar to pull in an indicator holder, or a suitable tool holder upon which a universal indicator holder can be clamped, or a drill chuck with an edge-finder.

☐ **Shift the Spindle to Neutral**
☐ **Select the Manual Axis Movement Method**
  Pulse generator—hand wheel or jog buttons

☐ **Alignment Indication Procedure—** At this point, the procedure is the same as on manual machines. Use either the end-to-end or trend methods of aligning the datum surface (Section 12) to the machine axis. Remember, after the final bolt tightening recheck the alignment.

Trade Tip

Reminder For better alignment of vises, a ground parallel held in the jaws averages imperfections and provides a smooth datum surface upon which to run the indicator probe.

X-Y PRZ Controller Coordination

The vise is bolted in place and its axis alignment is rechecked. The next task is to synchronize the *X-Y* PRZ in the axis registers, called **coordinating** the setup. We'll look at two alignments—PRZ over a setup hole and PRZ at an intersection of two surfaces. The *Z* axis position can be coordinated only after cutting tools are in the machine (Unit 24-2).

Parking the Spindle Over a Hole Either the part has a precise hole that's at a known distance from the PRZ or it might be the PRZ, or the setup includes a setpoint hole bored in a block that's at a known relationship with the work PRZ.

Step by Step (Fig. 24-2)
☐ **Shift Spindle to Neutral to Rotate the Indicator by Hand**
☐ **Rough Alignment—**With the mounted indicator probe just above the hole, switch to manual movement, then jog until the probe swings visually concentric around the hole's rim, through 360°. Try to get it within 0.015 in. or so the indicator will be able to read the full amount when it's put in the hole.
☐ **Hand Tilt the Probe Safely Beyond the Hole's Edge then Drop the *Z* Axis**
☐ **Final Alignment** Hand rotate the spindle while jogging until TIR = 0.0.

Y axis

X = 00.0000
Y = 00.0000
Z = (Not Yet)

Work coordinates

X axis

Figure 24-2 *Aligning the indicator over a setpoint hole.*

Note, after alignment, it's a good idea to drop the probe deeper into the hole then test the *X-Y* alignment a second time, just in case there are roundness or angularity issues. The second test will detect them. If the two tests don't agree, then try placing a precision pin in the hole or alternately try averaging the two locations.

KEYPOINT

The spindle is now at the exact *X-Y* PRZ location or it might be at a known setpoint position relative to the PRZ, according to the planning. *But the axis registers do not represent that position yet.*

Loading Registers (Fig. 24-2)

☐ **Switch to Work Coordinates Screen**—There are three ways in which a number can be entered into an axis register in the control.

1. **Axis Zero**

   The simplest is to use a reset button that zeros the registers. However, on some controls that also resets other data. You may not want to loose it all. Some controls don't have a reset function, but where they do, it's useful at times if one knows how it effects all other data.

2. **Manual Input**

   The universal method is to place the cursor over the column and write the digits, one at a time, from the keypad.

3. **Calculate and Code Functions**

   Beyond the MDI input, there are differences in axis loading protocols from control to control. Here are a couple of representative examples:

Preset Code Word Not all controls feature this code word, especially those with fixture offsets. It is a word that sets the axis registers to a given value. All the data is in the command, none is stored in memory as with fixture offsets.

If your control has a preset code word, it can be a useful way to set a PRZ. It can also be used in a program, but exercise caution in using it, as traps arise when one switches the PRZ coordination. Sometimes the PRZ can be permanently lost unless its machine home coordinates are written down.

KEYPOINT

An **axis preset command** writes whatever value it contains into the axis registers at the time it reads it in the program.

In order to use this feature, the tool/spindle must be at a known position relative to the PRZ. If by accident it's not at that expected position, then the PRZ is lost until reestablished.

Suppose our generic control uses G60 as its axis preset code, and we're parked at a setpoint 2 in. positive, from the PRZ along the *X* axis. With the work coordinates page displayed, turn to MDI mode and key in

G60 *X*2.00 *Y*0.0

>cycle start< in single block mode.
The registers will now read *X*02.0000 *Y*00.0000.

(Calc Function) Calculating Position Most modern controls provide a way to load register information directly into registers or to use math in data entries. Go to the work coordinates page and touch the "CALC" soft key below the screen.

If the present location is the PRZ, then enter the calc function and then touch it a second time indicating that all the numbers should be zero. The register then reads *X*0.0000, *Y*0.0000 (note, there may be control differences in how this function works on your machine).

Trade Tip

Many controls allow the use of math functions when entering data in the calc mode. For example, suppose an edge finder is just tangent to a fixture surface that's a plus 2.625 in. from *X* axis PRZ. Then enter 2.625 − 0.1 (the minus 0.1 accounts for the radius of the probe). The register will then read 2.525.

Parking Over a Corner Intersection X-Y PRZ Datum Point

This alignment procedure is a bit more complex since you'll be aligning to two edges, one at a time. Using an indicator, one can position the spindle at an intersection point between two edges, near or at the repeatability of the indicator.

One datum edge for mill work is usually the solid vise jaw or a parallel held against that jaw, to become datum A. The second is probably the raw stock held against a material stop to establish Datum B. It does not matter which is coordinated first. In Fig. 24-3, we see such a setup, requiring X-Y datum alignments.

Step by Step

☐ **Rough Align**—Begin with an edge-finder or center-finder. Jog both the spindle into eyeball location over the X-Y point. At that time, it can be useful to zero out the work coordinates and use them for final alignment—however, machine coordinates can also be used until the alignment is finalized.

☐ **Final Indicator Alignment, One Axis**—We'll start by placing the spindle directly over the X axis line (vise jaw), which *zeros the Y axis.*

Aligning the spindle over the edge parallel to the *X* axis, establishes the *Y* axis zero.

Figure 24-3 *This setup requires physically parking the spindle over the intersection of edges A and B to establish PRZ in the X-Y plane.*

Locating over an edge

Reads zero in both positions

Ground parallel forms reverse counterpart

Figure 24-4 *Establishing the* Y *axis zero. The indicator reads* zero *on both sides of the X axis, edge A. To do so requires a ground parallel held against the edge as shown.*

☐ **Indicator Null Over Vise Jaw—Parallel to X Axis**—The objective is to jog until the same indicator reading (usually zero) shows on both sides of the line, as shown in Fig. 24-4. It's done just like on the manual machines, jogging the axis in small amounts until it's centered over the edge.

☐ **Work Coordinates Page, Set Y Axis Register to Zero**—Disregard Y axis position for now. An axis reset could be used here.

☐ **Null over the Y Axis Edge—Set X Zero**—The operation is repeated against the surface parallel to the Y axis, thus establishing the X axis zero. PRZ in the X-Y plane is then complete.

To zero the second axis, do not use the reset function as that will erase the previous axis coordination.

Trade Tip

There are three hints that simplify edge alignment:

1. *Rough Alignment* Using a precision ground pointer or better yet a center-finder, one can bring the position quite close to alignment (Fig. 24-5). Sometimes, if there's enough rough stock on the work, this may be accurate enough to position the spindle. If not, this operation saves time by prealigning for the more accurate methods.

Rough alignment

Center-finder

Figure 24-5 *A center-finder (or even a precision pointer) saves time.*

2. *Datum Hole Accessory* This little homemade gem can save time during precision datum alignment. It features a hole precisely bored directly in line with a ground test edge. When clamped against the edge in question, indication is simplified by spinning the probe in the hole. You can make this tool as a shop project! (See Fig. 24-6.)

3. *Edge-Finders* More accurate than center-finding but less than the indicator methods, edge-finders are often employed to align axis registers in CNC setup work. In one step per edge, the test cylinder is brought to the exact tangent relationship with the edge in question. At that position, the spindle is 0.100 in. if the probe is 0.200 in. diameter (most common). The axis register is then set to represent that coordinated position at plus or minus 0.1000 in., or the calc function can be used to account for the probe radius.

Make this handy indicator target

Bored hole exactly centered over edge

Figure 24-6 *An indicator target speeds the use of DTI over an edge.*

Using Fixture Offsets Instead of Setting Work Coordinates

The preceding procedure involved a single vise with a single *X-Y* PRZ coordination, which required changing *work coordinate* entries to represent that PRZ datum point. But there is another way the setup is synchronized. By loading the *machine coordinates* of the PRZ into a given fixture offset memory.

KEYPOINT

To use fixture offsets for PRZ synchronization, the program must contain a fixture offset command (G54 through G59). Then you must ensure that the correct offset memory is loaded with the machine coordinates of that offset—you will determine those values during setup.

To use fixture offsets, we begin the same way, by locating the spindle at the PRZ. But here we read the *machine coordinates* of that location. The machine reference values are then written into fixture offset memory under G54 through G59 (most controls). That can be done manually or some controls will load them automatically using the calculation (calc) function.

Then when the control reads a fixture offset command in the program, all subsequent axis moves are shown in work coordinates relative to that location as PRZ. From the moment the offset command is read, work coordinates will be changed relative to that new fixture offset PRZ.

KEYPOINT

From the moment a fixture offset command is read in the program until it is canceled, work coordinates will represent the new PRZ.

Fixture offsets are easier to use compared to the older preset axis codes, because the PRZ location data is stored in control memory not in the program lines. They can be easily edited. To shift a preset location, one must edit the program itself. Also, since presets assume the axis is at the intended location, it's easy to *get lost* if the machine isn't actually at that position when the code is read.

Fixture Offset Uses These programming tools can be used in three different ways:

To setup a single PRZ

To setup and use more than one machining station in a single program—multiple vises, for example

To setup and save several different PRZs for different programs

Offsets for a Single PRZ In some shops, to maintain consistency, fixture offsets are used to set up every PRZ. Every program header starts with the fixture offset code, where the coordinates of the reference point are stored.

Control and Policy Differences—Canceling Fixture Offsets Most newer controls feature offsets starting at G53. So G53 *X*0.0 *Y*0.0 cancels any offset.

On the other hand, some controls have a specific fixture offset cancel code. G28 is one popular command but there are others. Where the fixture cancel command exists, G53 can then be used as a fixture offset. Be sure to learn how your particular control cancels a fixture offset.

Edge Intersection Step by Step

☐ **DTI Position at the PRZ location—** Using the methods already outlined, position at PRZ #1.

☐ **Go to *Machine Coordinates* Page—** Record the absolute position of the PRZ with respect to M/H. Then switch to the fixture page and enter the location under the particular fixture memory line (G54 through G59).

Auto-Loading Most CNC controls feature autoentry of machine positions into fixture offset memory. Once the spindle is parked at the PRZ position, a button touch will load that position into the highlighted fixture offset line. If so use it to avoid transposing digits from hand entry. Those controls also provide calc entry when the setpoint location of the PRZ is different than the spindle's position.

TRY IT

Got It? Test Your Understanding: In Fig. 24-7, vise 1 PRZ is stored under G54 and vise 2 under G55. The dimensions are the machine coordinates discovered with an indicator during setup. They are relative to M/H. Each was entered into fixture offsets. Next the program calls

N025 G54, G0 *X*0.0 *Y*0.0
(no *Z* axis movement)

Figure 24-7 *These machine coordinates are stored under G54 and G55.*

A. Where is the cutter going to go?
B. After the move, what will the work coordinates then read?
C. What will the machine coordinates read?

OK, that was easy but now, command line 30 is read;

N030 G55

D. What then will the *work coordinate* registers read?

ANSWERS

A. Above the PRZ for vise 1.
B. Work coordinates: *X*0.0 *Y*0.0.
C. Machine coordinates: *X*13.3767 *Y*−06.5566
D. Work coords: *X*−5.3240 *Y*5.9567
   The cutter won't move, only change references. It's the distance difference in the two vises, *X-Y* location.

![KEYPOINT]

Fixture offsets are memory stored locations relative to machine home. They are called by the program, but the numbers are not in the program. Fixture offsets are modal until canceled usually with a G53 or G28.

Z Axis Coordination

OK, the *X-Y* PRZ is finished, now the single cutting tool is loaded into the spindle and it's time to coordinate the *Z* axis zero. Continuing with our simple case, it's an end mill. The plan states it's at *Z*0.0 when it is just touching the workpiece surface. Here too, there are several variations so it's critical to understand the range of possibilities.

Caution!
Not setting Z height correctly is a common cause for tryout crashes!

Step by Step

☐ **Load Cutting Tool**—Either draw the cutter and holder into the spindle manually or if it's in the tool drum, switch to MDI then enter the tool number and M6 code, then execute the tool change in a single block.

☐ **Change to Manual—Jog Mode (without autoprobe)**—Jog the Z axis toward the work until the end mill's tip is at an exact relationship with the planned PRZ location—the top surface of the work, in this example. Again, read the setup planning—the Z datum point could be elsewhere.

Approaching the surface slowly, the objective is to hand sense contact, known as the **touch-off method.** Touching a cutter to work material isn't the best policy, the tool is manually rotated while being jogged toward contact with the work. Just until it drags a bit!

KEYPOINT

Do not rotate the cutter backwards to avoid cutting edge damage.

Gaging Z Height An improvement over touching the work surface with the cutter, is the gage block or paper feeler method previously discussed (Fig. 24-8). Either object will be of known thickness and prevents the tool from touching the work directly. It also acts as an improved sensitivity gage.

☐ **Set Z Axis Register in Work Coordinates**—Upon physically positioning the Z axis, the control is then set to the correct positive value relative to the tool position above the work or to Z = 0.0000 if it's touching.

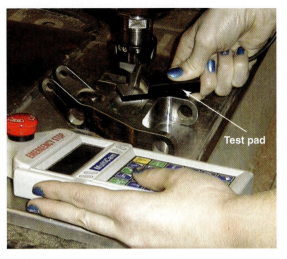

Test pad

Figure 24-8 *Using a tool-setting probe to establish Z axis position.*

Probes and Machine Vision

The safest, most reliable and fastest method of setting Z axis coordination is an autoprobe (Fig. 24-9) or a tool vision system. Easy to use, the tool probe has a known thickness saved within the control. Once the tool probe cycle is commenced, it slowly drives the tool toward the probe lying upon the reference surface. Then at the exact location that it touches, it halts motion and sends a signal to the control to set Z zero.

Controller Differences for Tool Sensing and Setting Probe sensors range from small electronic pads, hand located between the datum surface and the tool, up to sophisticated laser vision cameras on swing-away arms. After the probe senses the tool's length, loading the axis registers varies too. At contact, the position might be hand loaded using the probe's known thickness. But more likely on these sophisticated controls, the Z position

Figure 24-9 *This probe is sensing cutter position.*

will be automatically loaded into work coordinates. If there ever was a time to read the manual, this is it!

Coordinating PRZ on a Turning Center

Here too, we start with a single turning tool, then build upon that with offsets and hints in Unit 24-2. Similar to the mill, the cutting tool must be positioned in a known relationship to the intended PRZ in both *X* and *Z* axes. Then one axis at a time, the PRZ is loaded into the work coordinate registers.

Coordinating the *X* Axis

Normally, it matters not which axis is coordinated first. We'll start with the *X*.

Differing from the mill procedure, it is not possible to accurately position a turning tool exactly at *X* zero—the work centerline. The procedure is to position it at a known diameter then load that into the *X* axis position. This is especially true when using a tool with a nose radius, since it's difficult to detect when the tip is exactly at center of the work. Assume the work-holding device with a length of material is already in place.

X Diameter Coordination

Touching a Known Diameter Using the touch-off method, the cutting tool is slowly jogged into contact with any utility, round stock of known size, held in collets or a chuck (Fig. 24-10). To facilitate a feel sense, the spindle is hand rotated in neutral and a paper feeler is used. (See the Trade Tip.)

Better Accuracy—Machining a Test Diameter Here we use manual movement of the cutting tool to machine a short test cylinder. The setup bar or workpiece is machined to any cleanup diameter.

That measured diameter becomes the *X* position of that turning tool.

Figure 24-10 *Coordinating the* X *axis involves making or touching off a known diameter.*

Creating this temporary diameter with the turning tool eliminates potential runout and feel inaccuracies of touching the workpiece, and it reduces fine-tuning of the setup due to work and tool deflection experienced during actual machining. But it doesn't eliminate them altogether, as the manual cutting won't perfectly match program movements and metal removal—you may need to refine the *X* PRZ during the first few parts to be made with that cutting tool.

Step by Step

Because metal will be cut in this procedure, there is danger of a crash. *Be certain you understand the control and have instructor assistance.*

- ☐ **Select Manual Mode**
- ☐ **Index the Tool Turret—Mount Cutting Tool**—Be certain that the turret is fully withdrawn from the chuck to ensure tool clearance as it rotates.
- ☐ **Select RPM—Moderate**
- ☐ **Start the Spindle**—Check the workpiece first to be certain it is tightly held.
- ☐ **Select Jog or MPG mode**

- ☐ **Move the Cutting Tool $\frac{1}{2}$ in. from the Work**
- ☐ **Slow the Axis Motion**
- ☐ **Move X Axis to Make an Estimated 0.015-in.-Deep Test Cut**
- ☐ **Using Moderate Jog Rate or Continuous Feed, Machine the Test Diameter**—Double-check which axis is selected when you begin the machining pass. You will be moving the Z axis perhaps 0.25 in. or so into the workpiece to create a test diameter.
- ☐ **Withdraw the Cutting Tool in the Z Axis Only—Stop the Spindle**

Warning!
Do not alter the X position when withdrawing. Then before reaching in to measure the test diameter, lock out accidental axis movements.

- ☐ **Measure the Test Diameter**
- ☐ **Load the Test Diameter into X Work Coordinates**

Z Axis Coordination

Setting the Z axis PRZ value is a one step touch-off procedure performed on the work material not on a test bar. Or for improved accuracy, a test cut can also be made.

For production work, where batches of parts are to be made, a material stop or bar feed step-out length must first be set up to establish the work's Z axis location, before coordinating the PRZ.

Step by Step
- ☐ **Chuck the Work Material**—Place it in the position in which is to be machined.
- ☐ **Jog the Turning tool into End Contact**—(Or using MDI/JOG, machine the end for a minimum cleanup.)
- ☐ **Withdraw in the X Axis Only, Then Stop**—The tool is withdrawn in X to preserve the tool's Z0.0 position.
- ☐ **Load Z Axis Registers**

Probes and Tool Vision Systems

Similar to the mills, the most accurate and fastest method of coordinating a lathe setup is to autoprobe the cutting tool's position relative to the machine coordinates (Figs. 24-11 and 24-12).

Work Coords
X 01.2500
Z 00.0000

Lightly touch workpiece

Figure 24-11 *Coordinating the Z axis.*

Figure 24-12 X *axis measurement or tool coordination is easy using a probe.*

Master Tool Method

Now begins the great argument among CNC machinists—follow it closely. This discussion is critical for a deep understanding of CNC setup procedures. It's at the root of variation from shop to shop. So far, examples of setting up a single cutting tool at the physical PRZ location (or at a known distance from it) then writing that location into work coordinates, is known as the **master tool method.** It's simple and easy to explain for a first experience—move the tool to PRZ then set the registers to represent PRZ. But a master tool brings complexities to making setups with several cutting tools.

Length Offsets Compare to Master

Using a master tool, when additional cutting tools are used, they undoubtedly have different lengths compared to the master. Those differences must be accounted for as tool length offsets in tool memory. It works fine as long as the master tool remains in the setup. Each tool's offset makes it perform as though it had the same length as the master. Any replacement tool must be offset against the master so they all touch the workpiece when each is commanded to go to Z zero.

KEYPOINT

Using the master tool method, each cutting tool is given a length-compensating value, but the original master's value is zero until it must be replaced. All other tools are length-compensated relative to the master.

Calling out tool 2, you would add on a command to use length offset 2, which either holds back or adds Z movement to compensate for the difference in it and the original master.

Example Let's say that offsets are commanded with a G43 on our mill using H lines in tool memory. Suppose tool 2 is 1.5375 in. longer than the master. Its H value would be −1.5375 in offset memory, not in the program. (It's longer so it must be held back by the difference in it and the master.)

T02 M6
G43 H2   (use length offset $H2 = -1.5375$)
G0 X0.0 Y0.0 Z0.0

Holding back by the value found in H2, it then moves to Z PRZ exactly as would the master. At the moment the offset command is read both work coordinates and DTG screens represent true tool to work distance.

Replacing the Master

The big problem arises when the master becomes dull or broken. Its replacement must be exactly the same length, or values must be changed in the control or the following happens.

Critical Question In a mill setup where the master cutting tool (an endmill) touches the work top surface at Z0.0, after setting the PRZ, you then exchange a dull master with a *longer replacement!* Then without changing values in the control, command:

T01 M6
G0 X0.0 Y0.0 Z0.0

What will happen?

Answer The longer tool will try to rapid travel below the work surface, at the PRZ location, a serious crash!

To solve this problem, there are four options if using master tool:

Option 1   Offset the New Master Here the programmer put a length-compensating command in the program for all tools including the master originally, or it can be edited in. But until the master is replaced, it's compensating value remains zero.

Then at replacement, the machinist uses the touch method to find the difference in the first and second masters, then stores the new compensation in tool length memory. Using this solution, the new master tool is compensated to perform like the original master.

The problem now is that all other tools have no physical master for comparison and replacing them becomes a problem. It can be done, but it's a complex process. To do so, one must first take up the Z axis offset for the master so that as the machine is moved, it represents the work coordinates of the new master tool. Next, touch off the replacement tool and read its work coordinate position relative to the new master. That becomes its offset (but with sign reversed). For example, you have a new master in the machine and its offset has been determined and written under H1 in tool memory. You are now replacing Tool #2.

MDI enter G43 H1 T02 M6 (loads tool 2 in the spindle but takes up offset for the master). Now manually touch tool 2 to the work surface or Z axis reference. Read its Z position in *work coordinates*—that's its difference from the new master. Reverse the sign (plus/minus) and put that value into tool 2 length offset. At this point, T02 G43 H2 will then be synchronized to the new master.

Option 2   Reset All Tools The next solution is to resetup the new master's PRZ. Touching off the new master, a new Z PRZ is written into the control. Then, all other tools in the setup must be offset relative to it again. It's starting the setup over again with a new master tool—very time consuming unless there are only two or three tools! This option is not recommended, but it does work.

Option 3   Preset Duplicate Tools Some shops use fixtures to set the cutting tool's *gage distance*—the distance it protrudes from the mill spindle taper. This requires special cutter holders that allow adjustment of tools with respect to their length. Using preset tools, since each replacement is the same length, no length offset change is needed for master or other tools.

Using a preset replacement, the new tool can even be loaded into the storage drum before it's needed, then switched into the production based on a replacement schedule. Or at a predictable time, the switch can be triggered by either part count (common) or due to increasing tool pressure beyond a set limit (uncommon). With both options, there is zero down time. Presetting tool length is the best solution for accuracy, efficiency, and operator simplicity. But it requires a big investment in presetting equipment and tool holders. Note that presetting is also used with master position methods.

Option 4 Dummy Master Solution A final solution for tool replacement under the master tool method is to plan and load a dummy tool in the drum or turret. It's touched to the workpiece, then the Z PRZ is set to it. All other tools are offset to it. Not used as a cutter, it never wears out or looses its master status. Usually when a dummy master is used, its tool number is one of the higher numbers in the machine so that real cutting tools occupy the lower numbers, T18 or T20, for example. No reference is made to the dummy in the program—it is not called up in the program at all.

OK, the single tool setup is ready to go on both machines, lathe and mill. Review carefully, since we'll be making more complex setups in Unit 24-2.

UNIT 24-1 REVIEW

Review the Key Terms

Axis preset command
A code that writes new numbers into work coordinate registers based on the present tool position.

Calc (position)
Controller function whereby a tool's position, other than at the reference zero, is loaded into work coordinate registers or offset memory.

Coordination (coordinating)
Physically aligning cutting tools to the work then writing axis registers to represent that position.

Fixture offsets
A code word that refers to stored data in memory, to set the PRZ relative to machine home. Can be changed by editing memory.

Master tool method
A setup method whereby one cutting tool is used as a model for all other tools in the setup. It has no offsets initially. All other tools in the setup are given offsets so that they too touch zero.

Touch-off method
Bringing a cutting tool to the work until it touches a feeler of known thickness.

Replay the Key Points

- Work coordinates aren't complete until the setup is finalized (coordinated).
- When setting the PRZ, the spindle or cutting tool must be at the exact X-Y or X-Z PRZ location (or it might be at a known setpoint position relative to it per the planning).
- The axis preset command writes whatever values it contains into the axis registers.
- When aligning over the intersection of two edges in mill setup, zeroing over the edge parallel to the X axis, establishes the Y axis zero.
- Fixture offsets are memory-stored *machine coordinate positions* of individual PRZ locations—G54 through G59; called by the program, but the numbers are not in the program; modal until canceled usually with a G53.
- Setting a primary cutting tool's tip at the PRZ, is known as the master tool method of setting Z axis coordination.

Respond

1. Identify and briefly describe the controller screens needed for setup work.
2. True or false? When aligning over a setpoint hole, you cannot use work coordinates since they do not represent the PRZ as yet.
3. If Question 2 is true, then what position screen can you use?
4. List the main steps to coordinating a mill setup, in their order of occurrence. Use one cutting tool using the master tool method.
5. List the main steps to coordinate a lathe setup using the master tool method and one cutting tool.
6. What do machine coordinates tell the setup person?
7. In the planning you are told that a mill job has a fixture setpoint that is $X-3.00$ from PRZ and $Y-2.00$ in. What is the setup procedure after bolting and aligning the fixture to the mill table?
8. True or false? The touch off method is one way to coordinate X and Z tool

position relative to the PRZ, for turning setups. If false, what makes it true?

9. Beyond eliminating feel inaccuracies, what advantage does turning a test cylinder bring to the touch-off method?

Critical Thinking Question

10. Using the master tool method of setting length offsets, suppose the master tool T01 breaks and is replaced by a new end mill. What can the operator do to get the mill setup running again? (not a program change)

   Facts: The replacement end mill is 0.0377 in. longer.

   The program uses three cutting tools: T02, an end mill, and T03 a drill.

   The master tool had no length offset at the time it broke.

In Your Lab—Problems 11 through 20—CNC Setup Problem—The Mystery Block Challenge

Here's a test to perform on your CNC mill that verifies understanding. The objective is to align and coordinate a ground parallel on a CNC mill. But there's a special twist to the game.

Instructions

After coordination is complete, using only the indicator probe and axis position registers, determine the length of the block to within a half-thousandth! Your instructor may have a set of blocks prepared for this activity. However, any parallel of a size not known to you, will be OK (Fig. 24-13).

KEYPOINT

Caution!

If this is your first manual movement of this machine, be sure to reduce all axis override functions to 25% or less and lock the spindle in neutral for this activity.

Step by Step

After lightly clamping one end, align the parallel block using a test indicator. Aim for an end-to-end accuracy of 0.0005 or less.

Use an edge finder first, then DTI, to establish the PRZ following the previous edge intersection procedure.

Load the PRZ position into the work coordinates.

Mystery block challenge

Figure 24-13 *Align and coordinate the block using DTI. Then determine its length using only the indicator and the work coordinates.*

Critical Thinking Now, see if you can think through how to use the axis position in work coords and the DTI, to measure the length of the block.

Upon completion, check your result with a micrometer or against the instructor's answer sheet. You should be within 0.0005 inch or closer.

UNIT 24-2

Multiple Cutting Tools— Determining and Setting Offsets

Introduction: Now we investigate setting up several cutting tools in a tool-changing machining center and in a turning center with an indexing turret. We're going to determine and enter two kinds of offsets:

   Tool radius for keeping the cutting tool tangent to the geometry—assuming the program is compensated.

   Tool length differences—for both turning tools and milling

KEY TERMS

Mach coords (abbreviation)
Shop lingo for machine coordinates with reference to machine home.

Material shift offset
Allows a small shift in the PRZ to accommodate abnormal workpieces.

Work coords (abbreviation)
Shop lingo for work coordinates.

Determining Tool Offsets

Tool Radius

Tool radius offset is a simple issue for both lathe and mill. All that's required is to know the kind of tool path (cutter centerline or part path) then enter the right offset in tool memory.

Radius Values Entered in Memory

For mill tool path compensation, tool radius is the standard protocol over tool diameter. The reason is that changing radius distance away or toward the geometry simplifies the effect. The change in radius is the amount the cutter moves closer or further away from the geometry—therefore, it equals the difference in size change. However, I have encountered one mill control that used tool diameter then divided that value in half for compensation. Lathe tool nose offsets are always expressed in radius values.

Why Change Radius Comp for Profile Milling or Turning?

Radius offsets might be adjusted by the operator, to compensate for:

1. **Tool Wear**
   Up to a limit where a new tool must be mounted.
2. **Machine Warm Up**
   Heat and bearing characteristics can effect size.
3. **Tool Replacement from Failure**
   While qualified insert tools index within a guaranteed tolerance from one insert to the next, they all have some variation. Sometimes it's a good idea to back the tool path out just a bit (0.0003 in. or so) when putting a new insert in the holder.
4. **Change a Dimension**
   Remember this will change all dimensions made by that cutter.
5. **Adjust for Deflection**
   Or for other factors such as vibration.

Which Kind of Path?

Even with just one or two cutting tools in the setup, it would be unusual for an industrial program to be written without tool compensation. However (Comp) is even more certain if multiple cutting tools are used. But even though the chances are near 100 percent that it is compensated, it's a good idea to verify it. Look for a check block on the plan.

[ ] *Tool Comp*

Or Look for a Note in the Planning

If the path type isn't indicated in those ways, then scan the program for G41/42 and comp codes in the first few lines after the first tool change.

Part or Cutter Path Program?

Next, verify whether the program is a *cutter centerline,* version, in which case the cutter radius offset will be the small difference between the recommended cutter envisioned by the programmer and the actual cutter put in the machine.

[ ] *Cutter Centerline Path*   [ ] *Part path*

For example, the planning states T01 = $\frac{1}{2}$″ End Mill and the program is written for cutter centerline, as checked above. You determine that the actual cutter is 0.4950 in. diameter. The radius offset entered will be a −0.0025, shifting the slightly smaller cutter's path closer to the part geometry. The offset value is the difference in tool radii.

Trade Tip

In using Mastercam, the program supplied with this text, you will be given a choice [ ] *Comp in Computer* or [ ] *Comp in Program* when writing a program.

Choosing "comp in computer" generates a cutter centerline program based on a cutter radius or diameter selected by the programmer. This path requires small negative offsets for undersize cutters.

Choosing "comp in program" generates a part path program requiring large positive offsets equal to the real tool's radius.

But suppose the program is a part path:

[ ] *Cutter Centerline Path*   [ ] *Part Path*

Then the tool radius offset value will be the whole cutter radius that's required to keep it tangent to the geometry. Using the same

To avoid confusion, machine shops normally do not mix program types regarding tool compensation. All programs will be either part path or tool centerline—*but that's when they write their own programs.* However, if it's a shop that contracts outside work, then many jobs are accompanied with a drawing and a program! In these shops, be on the watch for tool path differences.

A problem occurs in using the master tool method for length offsets. If due to wear or damage, the master tool must be replaced. The standard for offsets is lost unless it's replaced with an exact duplicate (often the case using preset tooling). However, without preset tools, using a new master of a different length can be used, if it's given its own offset, to make it perform like the old master. But everything gets complex from there on.

So, there's a simple solution to work around the problem. A dummy master is added to the tool load (Fig. 24-15). The dummy master is never used to cut metal or called out in the program, thus it is never absolutely necessary to replace it. In fact, the dummy master needn't be a real cutting tool. It can be an aluminum or plastic cylinder that's mounted as a cutter, then used for touching off more safely since it won't mar the work surface.

0.495-in. end mill, the offset will be plus 0.2475 in.

Tool Length Compensation—Mills

Using three cutting tools in a milling machine, we'll compare the two methods of setting the length offsets: *master tool* then *master position.* Each protocol has it's own particular advantage and disadvantage. Most shops use master position.

> **KEYPOINT**
>
> The choice of length offset method is a setup decision only—not a planning item. The difference lies in how you set the values in tool memory.

Possible Algebraic Controller Difference

We assume in these examples that the control combines the commanded axis move with the length offset by the rules of algebra (algebraically). That is, it adds plus value offsets and subtracts minus values from the command distance in the program. That's the common usage, but it's possible that a given control may combine the offset opposite from this. Read the programming manual to be certain or ask someone.

Mill Tool Length Offsets— Master Tool

In Fig. 24-14, the master just touches the work surface at $Z0.0$ (PRZ zero). Then, T02 and T03 length offsets represent the physical length difference in them and the master. Compared to T01, T02 is a shorter tool and requires an additional 1.376 in. to make a move to Z zero, a positive offset.

> T02
> G43 *H2* G0 Z0.0   (with 01.3760 stored in length offset *H2*)

The offset allows the program to call each tool to $Z0.0$ (or any Z

Figure 24-14 *Under the master tool method, every other tool receives an offset equal to the difference in length except the master tool.*

Figure 24-15 *One solution is to use a dummy master that never gets broken or worn.*

position relative to the work disregarding length differences between tools).

Finding and Storing Length Offsets
Under the master tool method, first the master is set to read $Z0.0$ when it's touching the reference surface.

> **KEYPOINT**
>
> **Caution!**
> Before setting up the master tool's Z zero position, go to tool memory and be sure it has no length offset *HXX* = 0.0000 (zero), for the initial master! *XX* = the master tool number. Also be certain no other offsets are in effect in the control.

Then each successive tool is brought to the touch point (or shim or probe) and its Z axis value read in *work coordinates.*

Critical Question If after setting the $Z0.0$ for T01, with no length offset in effect, then T02 is loaded into the spindle and touched to the work surface in Fig. 24-14, what Z axis position will be displayed on the *work coordinate screen?*

Answer $Z-01.3760$—The shorter tool had to move 1.376 beyond Z zero to touch the work.

A tool shorter than the master, such as T02, will need to go farther down to touch—beyond the Z zero point. A longer tool such as T03, will be $Z0.9200$ above Z zero in work coordinates when it touches, thus $H3$ will be a negative 00.9200 offset. Any Z axis move of T03 must hold back 0.92 in. to synchronize it to the master.

TRY IT

Got the Idea? Here are two critical questions to see (Fig. 24-14).

A. With T01 just touching the reference surface and the Z axis register zeroed, with no offset in effect, the work coordinates read $Z00.000$. Then the following command line is read.

G43 H2 (No tool change or axis motion—just the correct offset for T02, is read and used.)

What will the work coordinates read? Plus or minus 1.376 in.?

B. OK—Tool 2 is loaded into the spindle then brought down to just touch the work surface. No change in offsets $H2$ is still in effect. What do the work coordinates then display?

ANSWERS

A. $Z01.3760$. Why? Because the Z axis' quill position will be 1.376 above where T02 would reach zero.

B. $Z00.000$. The tool is touching and its length offset is in effect, so it's now at Z zero.

Controller Differences— Example Commands

In addition to the H word, some controls use the tool ID line to denote the length offset. The following command lines would compensate tool 2 to perform as though it had the length of tool 1—the master:

T0202 M6 (Load tool 02 using offset line 02 from tool memory)

G43 G0 Z.5 (Now under length comp 02, rapid this tool to a position 0.5 above the Z axis PRZ. For Fig. 24-14, this tool will move 1.376 in. more than it would without compensation.)

Master Position Method for Length Offsets

Here's the method that's more common in industry for multiple tools. In this procedure, all tools are assigned an offset from the beginning, *there is no master*. Offset values are each tool's machine coordinate distance from machine home to the datum reference (Fig. 24-16).

Using the same tools from the previous example, note that the offset values still represent the physical length differences but this time they are large numbers.

Using the MP setup method, when any tool is commanded to a Z position, the control must always *add the offset* to the movement. An uncompensated Z axis move would

Figure 24-16 *Compare the master position method to the master tool. Every tool receives a positive offset from machine home to the datum surface.*

Figure 24-17 *Imagine a very long, dummy tool—all other tools would need a positive offset to equal its length.*

actually be no movement at all. No motion would occur until the length offset is added on. Using this protocol, length offsets in tool memory would now look like this:

Offset Number	Length Offset	Radius Offset	Drum Posit
01	20.000		
02	21.376		
03	19.080		

The length differences between the three tools are still there but they represent their distances down from machine home (master position) to the datum surface. The offset then is the positive correction placed upon the real tools such that each will drop to the work surface ($Z = 0.0$) by their offset distance when commanded G43 *HXX* G0 *Z*0.0.

Another way to envision master position is to compare it to a master tool setup. In a master position setup, envision that there is an imaginary dummy tool but it's so long that it just fits between the work surface zero with the spindle retracted to machine home. That imaginary tool is at *Z* zero when the spindle is all the way retracted at machine home. Because the real tools are all much shorter, they all require positive length offsets. So in reality, that very long, master is actually replaced by a master position (Fig. 24-17).

Confused? Don't worry. Machine experience will help clear it up. And remember, no shop mixes the two methods—and they usually choose master position since it simplifies replacing cutting tools.

Lathe Multiple Tool Setups

Like mills, lathe offsets and PRZ setup can be made using master tool or master position. Differing from mills where only length varies, between different lathe cutting tools there are both *X* and the *Z* axis differences.

Both *X* and *Z* Differences

If the turret is rotated with no axis motion, the turning tool, the facing tool, and the drill all index into a different position relative to the workpiece (Fig. 24-18). Those *X* and *Z* differences must be stored as tool offsets—either as master tool or master position numbers.

X and Z tool differences

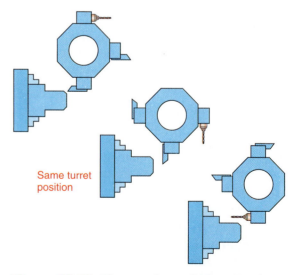

Figure 24-18 *Three cutting tools show X axis and Z location differences when the turret is indexed but not moved relative to the spindle.*

And again this is a setup decision only—either method will work.

X Axis Offsets and PRZ

We'll start with the X axis since it's a fairly simple issue for most setups with multiple tools. When cutting tools such as drills, reamers, and taps are used, X axis PRZ and offsets are normally based on them. They are tools that must be positioned on the spindle centerline to cut correctly—called centering tools.

X Axis Constant—Turret Position on Center

As shown in Fig. 24-19, centering tools are mounted in bored mounting blocks. Each block places the tool the same X distance from the turret center. Therefore, the X axis machine coordinate position of the turret, when it centers these tools, becomes a constant for X setup.

How that crucial machine coordinate position is used depends on whether it's a master tool or master position method, regarding X axis offsets.

Master Tool Normally all center cutting tools have no X axis offset. They are *all* the master for X position. Replacing a centering tool has no effect on its X offset since the replacement goes right back on center.

If there are no center cutting tools, you have two options: use a dummy centering tool or select some other master tool.

Master Position Every cutting tool is offset from M/H to the X PRZ. All center cutting tools receive the same X axis offset—to position it from machine home, to X center. It will be the sum of the turret and tool distances shown in Fig. 24-19.

Keep in mind for CNC lathe work, X axis numbers are diameters, even machine home. Think of machine home as the largest diameter the turret can reach.

Finding the Centered Position If you do not know the X-centered position, it can be found by placing any cylindrical object in a center cutting tool block (a 1-in. test pin in the drawing), then using the touch method, bring its rim into tangency with a turned test cylinder of known diameter (Fig. 24-20).

KEYPOINT

At that tangency location, the turret is at the sum of the two *diameters* away from the center of the spindle, or as shown (R1 + R2) * 2, double the sum of the two radii.

Critical Question Suppose in Fig. 24-20 that the *machine coordinate* is −18.303 in. with the test pin touching the test cylinder. The test pin is 1.0-in. diameter and the turned cylinder is 3.500 in. diameter. What is the X center constant on that CNC lathe?

Answer It will be 18.303 + 1.000 + 3.500 = 22.803 in. The sign of the 18.303 was reversed to make it a positive distance.

That machine coordinate of the centered turret position never changes. It's usually hard wired into the machine home routine.

X axis centered constant distance

Figure 24-19 *The X constant is the sum of two diameters.*

Finding the X constant for centering tools

Figure 24-20 *The constant X position is the sum of three numbers.*

That *X*-centered machine coordinate will be recorded in one of several places:

In the operator's manual

In the tie-in book

Z Axis Offsets

Compensation for tool differences with respect to the Z axis PRZ is very similar to mill setup and depends on whether it's a master tool or master position method.

Z Comp Master Tool

For a master tool setup, after setting the *Z* axis zero in work coordinates each subsequent cutting tool is brought up until touching the *Z* reference (usually the work face) then the **work coordinate (work coords) position** is noted. That's the *Z* axis difference. The sign value is reversed to become the saved offset value for that tool. (assumes algebraic combining). The offsets will be either positive or negative depending on whether the tool in question is longer or shorter compared to the master.

Z Comp Master Position

To offset tools in the *Z* direction, each is touched to the reference surface, and its **machine coordinate (mach coords) position** relative to machine home, is noted and saved as its offset—always positive values.

Material Shift Offsets

Most modern CNC lathes (and mills too) offer an offset screen that allows the operator to shift the PRZ a small amount but not loose the original PRZ location to provide a **material shift offset.** This handy operator control is designed to deal with variations in raw stock, castings, and forgings, or when previous machining leaves the work close to not having enough excess on one side, to cleanup, but there is plenty on the other side. Sometimes the stock is slightly different from the norm, a little longer or shorter or maybe warped for example. The part can still be made but only by shifting the PRZ just a bit

to equalize excess or keep the machining within the odd shaped part. It's probably not going to be a permanent shift, but applied to a limited number of parts, then canceled.

UNIT 24-2 REVIEW

Review the Key Terms

Mach coords (abbreviation)
Shop lingo for machine coordinates with reference to machine home.

Material shift offset
Allows a small shift in the PRZ to accommodate abnormal workpieces.

Work coords (abbreviation)
Shop lingo for work coordinates.

Replay the Key Points

◆ For milling

◆ *Master tool* Using the MT protocol, the length offset value of a given tool is found on the *work coordinate* screen, as the physical difference in the master at *Z* zero and the current tool touching the work surface (or feeler or probe).

◆ *Master position* The length offset under master position, for each tool is the *machine coordinate Z* position when the tool touches the datum surface—but with the sign reversed to positive. Remember that the offset is determined from the machine coordinate screen.

◆ For turning

◆ *Master tool* Normally all center cutting tools have no *X* axis offset. They are *all* the master for *X* position. Breaking one has no effect on *X* offsets since the replacement goes right back on center.

◆ *Master position* Every center cutting tool receives the same *X* axis offset—to position it from machine home to *X* center.

Respond

The following critical thinking problems involve setup of PRZ and offsets.

1. A rectangular workpiece that is faced on its top and a pocket milled into its top surface is measured. The pocket is the right depth but the part is too tall by 0.005 in. In this program, a single end mill faces the part and cuts the pocket. The program is written with tool length compensation. Is the problem with the program or the setup? How can the problem be corrected? (There are two methods.)

2. See Fig. 24-21. After machining and measuring the profile of this object, the

Cutter centerline
tool path

1.625

2.750

Figure 24-21 *Compute the offset change needed.*

machinist finds the two dimensions to be 2.7536 and 1.6286 in. The program is a compensated, cutter path that was written for a 0.500-in.-diameter cutter. What offset correction is needed?

A. What will the new offset be for the cutter path program?

B. What would the offset need to be if this was a part path program and this part was produced with a 0.500-in.-diameter cutter offset?

3. Is this statement true or false? In a setup as in Problem 1, one end mill with the program written with tool compensation, the control does not require a length offset. If it is false, what will make it true?

4. Describe the coordination of the PRZ on a vertical mill CNC machine, when the alignment is at the intersection of two datum edges on the work or tooling? Assume the rough position has already been performed.

5. When performing the PRZ coordination of Problem 4, the controller will be on the tool memory page. Is this statement true or false? If it is false, what will make it true?

6. After turning, an object shows two diameters to be oversize by 0.05 mm while two are exactly right. All *Z* axis dimensions are too short by 0.1 mm. What must be done to correct this single tool setup and part run? The program is *not written* with *tool comp offsets.*

7. Describe the master tool method of setting up a lathe or mill.

8. Describe the master position method of setting up a lathe or mill.

9. How are material shift offsets used?

10. Which tool compensation method is the less complex when replacing work or broken tools? Explain.

UNIT 24-3

Proving a Program—Tryout

Introduction: OK, the setup is synchronized to the workpiece. Cutting and holding tools are in place and ready to go. The objective is to get the program proven and working as soon as possible, but safely. There are 12 testing methods presented here in a hierarchy, each exposing a bit more about the program. We begin the process tests that present zero risk then work toward those that move machine axes without cutting material so there's a bit of risk. Finally, the ultimate tests are those that make real parts. There are three different ways to cut metal, with lessening degrees of test control. Which of these 12 methods are selected, depends on

The number and cost of the parts to be run. When making many parts each with a relatively low value, the first needn't be perfect. Called a *tryout part,* it can be measured and corrections made to produce exact dimensions on the next part. However, when making only one or two parts from expensive forgings, for example, more testing is required before cutting metal.

The history of the program (and programmer). Has it been run before? If so, you might skip most steps and do a single step run until the cutter seems to be going OK, then turn to full autorun and let it go but still keep a close eye on the progress.

The nature of the object. A complex part will predictability need more analysis.

Your experience level. First time? Use every trick available!

KEY TERMS

Associativity
Linked elements in a program. Each is associated with the others. Highlighting or changing one changes all linked windows.

Axis disable
A tryout method whereby the operator can lock an axis but the control displays as though it was moving.

Dry run
Any one of several program testing methods whereby the machine does not make chips but goes through all or part of a full autocycle.

Optional stop (opt stop)
A controller function whereby certain codes may be recognized or ignored within the program.

Single-step tryout
Evaluating a program by fully running it one code at a time. The last test before a full autorun.

Tryout material
Soft or inexpensive material used to machine a shape without the danger of wasting real material or breaking cutters.

Program Tests

It would be unusual that all these options would ever be used in any program test. However, each is a tool that should be understood, so that the right ones can be chosen for the situation.

Safest Testing

Test 1   Reading Programs This test is best used when trying to solve an error in the program or a small shape glitch. It's not an effective method for testing an entire program. Trying to decipher the data has little value especially when a typical CAM-generated program is many lines long, or the shape is complex. Program scanning is mentioned here for the rare situations when there is no graphic capability for evaluation.

Tests 2, 3, and 4   Graphic Evaluations
One or more of these tests are a must-do in modern machining. They are found in the shop in three varieties:

Test 2   CAM tool path evaluation
Test 3   Specific tool path utility software
Test 4   CNC control graphics

CAM tool path evaluation Also called a *program plot or backplot,* if the program is readable by the CAM system (nearly all are), then it will have the ability to replay the tool path generated (Fig. 24-22). Most CAM software offers the ability to display the work material and the cutting tools too and to view the results progressively, from varying angles to see cut depths and other programmed movements not visible from a simple top view. Figures 24-22 and 24-23 show two views of the same program. One is a wireframe of the tool path with blue line showing G01 moves, while G00 are yellow.

Figure 24-22 *A sample CAM backplot shows a wireframe tool path in blue and rapid moves in yellow.*

Figure 24-23 *Testing further, the view has been zoomed in where a problem is suspected, with the cutter displayed.*

Depending on user selections the backplot can display the cutting tool (Fig. 24-23), and the work object and view it from many different perspectives during and after the screen-machining is complete. Where there is a need to zoom in close, details of the machining can be displayed.

But a graphic backplot cannot tell the user if dimensions are correct or if the program has danger points where the cutter might hit bolts, vise jaws, or other objects in the setup unless they are included in the material designation. Testing for these things is yet to come.

● **TRY IT**
The best way to learn this test option is to do one. We'll be using Mastercam but if your class has different software it will work similarly. If you haven't done so, please install your Machining Technology CD or go to a class PC with the program.*
The following is the simple way to do a backplot, but on completion experiment with different views and display options.

*Can be either a fully implemented program or the demo version, as on your CD.

CAM Backplot Step By Step
Start MasterCam Mill
[ ] **Files,**
  Get,
    2-D or 3-D
      Click any file—The part geometry will appear on screen
[ ] **Main Menu**
[ ] **NC Utils** (Utility tools for programming and testing)
  Backplot,
  Run
      Blue lines are cutter paths, yellow are rapid moves.
  or Step
      Click to read each command block.

Try different options Now change the various backplot options to see different view perspectives, and different cutting tool displays.

In Fig. 24-23, I made four changes. First, using G-View-Dynamic (Get View at the bottom of the control window), the geometry was mouse rotated to show the corner of the part better, then zoomed in to a specific feature. Next, the tool display was set to shaded. Then I single-stepped through until the cutting tool reached the windowed area in question.

> Change tool display
>
> Single-step to see one block at a time
>
> Change perspectives and zoom in

Test 5 Toolpath Utility Software The ultimate graphic test, these software programs are created especially for testing and editing CNC programs. In most cases, surface and tool path software is a separate program from the CAM, but working in conjunction with it. Our example, Metacut Utilities® is such a program.

We've chosen Metacut to demonstrate tool path and part evaluation as part of the program tryout process (see the Shop Talk). A student demo copy is found at www.Metacut.com. Go there to download a free trial copy of their software.

In Fig. 24-24, using Metacut software, I've split the screen into four quadrants to show some of the ways it can help find and fix program bugs and prevent crashes.

> No modern CNC shop should be without some form of tool path and surface analysis system. It can save time and improve both safety and quality and avoid potential scrap work.

Similar to CAM, Metacut can show a wire frame of the cutter path or a shaded solid model of the product (Fig. 24-24), from any view perspective. Zooming in to see small details both during the cyber cut, or after, makes it possible to analyze small details such as mismatches and uncut transitions. It even displays tool marks. It can detect events that are crashes, for example, when in rapid travel mode, the cutter touches the surface. In the "Quick Tour" demo, you'll see this flagged in yellow. It can display the cutting tool as the cut progresses and will differentiate rapid moves from feed rate cuts.

Total Screen Associativity and Surface Analysis

For evaluating, debugging, and editing programs, a most useful feature is forward/

Figure 24-24 *Using tool path utility software, many details of the program can be revealed.*

reverse **associativity.** This means that all windows of the program are linked. Cutting tools, program code lines, and cut elements in each window can be highlighted to show their part in the program.

For example, as you click on a line of code, the cut element that it makes is highlighted, plus the cutting tool that made it will be highlighted too. Or if you click on a cut area in the graphic window, or on a single cut line, then the line(s) of code that made it will be highlighted as will the tool. Finally, if a tool is highlighted, then the playback shows those cuts made by it along with the command codes. Highlighting one, in any window, instantly connects it to the other two windows. That makes program investigation and editing a snap, especially when it's many lines long as are most CAM programs.

• TRY IT
Open the Metacut program. The demo opening screen will ask if you would like to take the tour—do it! If your lab has Metacut installed, then go to HELP—TOUR.

As you will quickly see, this is a powerful CNC proving and improving tool. After the tour is over, try opening a file saved in the software, then try the various manipulations yourself. Use the on-screen aids and the excellent, printable tutor found under the HELP menu. Later, in Section 25, we'll be putting a program you write in Mastercam into Metacut to evaluate final results.

It's not uncommon for a CAM-generated program to be hundreds or even thousands of code lines long for complex shapes. When something is wrong on such a program, it's nearly impossible to find the exact line on which the problem occurred unless using specific software designed for the task.

Test 6 CNC Controller Tool Path Graphics Once the program has been downloaded into the control, it's a good idea to give the program a dry run test with all axes locked out. This test can determine if the program works in the particular control and that the tool path is that which you expected—not a wrong program.

Machine Motion Tests But that's the end of graphic methods, it's time to use the CNC machine to further test the program. There are four levels to which one might go.

A Dry Run

A **dry run** is an actual run of the program with one or more machine functions disabled—no chips will be made. There are several different ways to perform a dry run.

Test 7 Axis Disable If the control offers an **axis disable,** you can first lock up all axes, then run through the program in dry run status. This test shows that tool changes are OK and that the program syntax is acceptable. If the control has screen graphics it will also show the tool path and the real times required for each cut (Fig. 24-25). During the run, the work coordinates will change as if the machine was moving and all modes and active codes will also show onscreen. It's a rough outer test but it's quick and has zero risk.

Next, using a mill for the example, unlock X and Y axes, but withhold the Z up at machine home (or withhold X or Z if it is a lathe). Now the X-Y spindle motions can be analyzed for danger points where a rapid move might hit a clamp or the vise. You might then move the Z axis a little farther toward the work, then lock it again and dry run it again.

Another addition to the dry run is to allow tool changing during the run, but only after a lockout test. Here, the tools index into position and out at the correct time in the program but the axis slides remain locked. This feature may not be available on all machines.

Test 8 Cutting Tool Withheld Using this method, all the axes and tool changer are allowed to move, thus all motion can be

Figure 24-25 *Many CNC controls can display a tool path analysis from many different angles.*

observed for danger points. A disabled tool can be setup in three different ways:

1. Remove the cutting tool from the machine.

2. Use offsets to force the cutting tool back a great distance. Don't forget how far back so it can be reversed for the real program run.

3. Better suited to multitool setups, is to shift the PRZ in the *Z* axis, up away from the work. Again don't forget the amount of shift so it can be returned afterward.

In each of the cutting tool withhold tests, there is some risk as axes will be moving.

Test 9 Missing Material This is another form of dry run, but it has more risk since the cutting tools will be coming to their intended height with respect to the vise, chuck, or clamps even though the work material is removed for now.

Test 10 Tryout Material Here everything is enabled, cutting tools, tool changing, and all axis motion. However, the part is not the real material. Instead a **tryout material** that is less expensive and machines more easily is used. This test creates the first dimensional examination of the product, and is usually performed because the work material is expensive or in scarce supply; for example, if the job requires machining three expensive titanium blocks and no more can be had! Here the first part must be right as well as the following two! A test material run always follows a series of dry runs.

Test material can be special wax designed just for a CNC tryout. It too isn't cheap but the chips can be remelted and cast back into a block! See the Trade Tip.

Special, high-density foam that is also made for the purpose can be used, for tryout parts (Fig. 24-26). Sometimes when nothing else is available, we use wood or even aluminum. Foam and wood, however, can be dust problems.

Figure 24-26 *Tryout material is a safe way to obtain real part dimensions.*

Trade Tip

Reusing Wax Before a wax tryout, first clean the machine's chip bin so as to not mix metal chips with wax. But 100 percent cleanup isn't necessary if you follow the upcoming hint. After we make the cuts, the chips are picked up with a small battery powered, handheld vacuum that we reserve just for reclaiming wax. Then melt the reclaimed wax in a pan over a hot plate. The liquid is then poured through a tea strainer to remove stray metal chips, when it's recast in the mold. After several recycles the wax's properties begin to degrade, so a special conditioner is added to the liquid before recasting. It restores dimensional stability.

Dry Run

Test 11 Oversize Radius Offsets This test only works if the program is tool radius compensated. Here, radius offsets are faked such that the cutter backs away from the work by a large amount, thus little or even no contact is made on profile cuts.

Be sure to record the offset changes in a multitool setup to guarantee they are written back correctly when the test is complete.

Using radius offsets, one can machine the work into an oversize version then measure it to help ascertain that the true offsets will produce a good part! Be aware, the next part after the offsets are restored, may not be of correct size due to tool deflection differences in the light test cut and the full removal cut.

Test 12 Single Step OK—it's show time! Here we make chips but not at the rate they would be made under autocycle. This test is universally used on all CNC controls with which I am familiar—old or new.

For a **single-step tryout,** first select the single-step mode with the setup complete, program loaded and work material ready. It's a complete setup and part run with only axis motion slowed. Then, with rapid travel turned down to 25 percent or lower, and the feed rate backed down too, touch the >Cycle Start< button, the control will complete the given command block then await another push. The problem is the spindle will dwell while the control awaits a second touch of the start button. Dulling the cutter against the work is a potential problem here.

The real trick to using this test is to analyze the program block under the cursor. Then watch the position registers and the distance to go screen. *Caution! Look and think before touching the start button.* Ask yourself "Is the next move reasonable?" Don't assume anything—you are especially vulnerable to this mistake if you wrote the program yourself.

Using DTG During a single-step test, when it's questionable that a cut or rapid move is safe, allow most of the commanded to occur then turn the override function to zero (rapid or feed depending on the command), stopping the axis move midway through the command. Next, stop the spindle and use a rule or some other means of measuring how far the tool can go safely. Then compare that to the "Distance to Go" screen on the control. *Caution!* Do not put your hand near a spinning cutter or lathe chuck to make this test! Turn off the spindle.

In Run

Test 13 Optional Stops To use this test, the codes to cause an optional stop must be built or edited into the program. Optional stops halt the program at specific points when the operator selects them, such as key measurements that might be performed on every twentieth part for example. But between, no

Figure 24-27 *Optional stop 1 has been enabled by the machinist while the others remain disabled.*

halting is desired, therefore the switch is left in the open position and the **optstop** code is ignored (Fig. 24-27).

This test is used for ongoing part evaluation but it can also be used for a first part run if the stop points are well thought out.

Many CNC controls indicate an optional stop by a yellow flashing light.

Notice in Fig. 24-27 that there are multiple switches. Various levels of intervention may be created. One optional stop for a problematic dimension or supercritical finish requiring a lot of operator attention then a second for a larger range SPC sampling, for example.

Comparing the Methods

The objective is to choose the best combination from each category to match the job. Here's a quick overview of the possibilities:

Prerun Tests
  Program reading—for spot checks of error points
  Graphic evaluation—a must-do if available
  CAM PC graphic evaluation—if available (best choice)
Dry Runs
  An axis disable
  Tool or material withheld
  Oversize offsets—no cutter contact
Test Runs
  Tryout material

Fake offsets—reducing until final size achieved

Single-step testing

Halt Run

Optional stop—One or more custom intervention points. *Must be edited or written into the program.*

Single-step cycle Switching to this mode the control will complete the block it is currently reading then wait.

Feed/speed override—Correctly used only during a non-contact movement of tool with work. Any other time, this is a low level emergency stop procedure.

UNIT 24-4 REVIEW

Review the Key Terms

Associativity
Linked elements in a program. Each is associated with the others. Highlighting or changing one changes, all linked windows.

Axis disable
A tryout method whereby the operator can lock an axis but the control displays as though it was moving.

Dry run
Any one of several program testing methods whereby the machine does not make chips but goes through all or part of a full autocycle.

Optional stop (opt stop)
A controller function whereby certain codes may be recognized or ignored within the program.

Single-step tryout
Evaluating a program by fully running it one code at a time. The last test before a full autorun.

Tryout material
Soft or inexpensive material used to machine a shape without the danger of wasting real material or breaking cutters.

Replay the Key Points

- Tryout material can be used to make a test part that can be measured.

- Shifting the PRZ away from the material only works if the setup is a master tool PRZ.

- Using oversize radius offsets, one can machine the work into a blown up version of the part then measure it to help ascertain that the true offsets produce a good part!

- To best use a single-step test, analyze position and distance to go. Ask yourself "Is the next move reasonable?"
- Optional stops are used for ongoing part evaluation but it can also be used for a first part run if the stop points are well thought out.

Respond

1. Which program tests have no risk?
2. Of the graphic program tests with no risk, which is the best for program testing? Why?
3. What are the main issues a backplot can reveal?
4. What cannot be determined by a backplot?
5. How would you go about moving all three axes of a vertical mill, yet not make chips?
6. In order to change the PRZ in a multitool mill setup, to try out a program, what kind of setup must be made?

Critical Thinking

7. Your assignment is to CNC mill an expensive stainless steel forging (Qty 1) into a complex, close tolerance shape. Which program tests would you perform?
8. True or false? All graphic tests of programs have zero risk.
9. Describe a single-step test through a program. At what stage of testing would you perform it?
10. True or false? Optional stops are more useful for part checking during the run, than for program testing.

Special Research

If you haven't already done so, it's time to install the Machining Technology disk and run the tutorial for the Metacut Utilities. Note the following:

11. All aspects of the screen image are associated.
12. Changing any line of program code changes the part shape when the screen is refreshed.
13. You may mouse select any element or code line and they are strung together.
14. As the program is run, each element and its direction is indicated by a chain snake.

Section 24 Review

Unit 24-1

When making a setup on any machine, it's critical to consider the forces that will be brought to bear on the work. By now, you've experienced these forces in your lab and perhaps even seen an accident occur. (I hope not but . . .!)

That's not news and neither is the fact that when it's to be a CNC job, the safety issue is multiplied. All work must be held in the very best grip possible. Be certain that studs, bolts, tee-nuts, and clamps are in good condition and that they are grade five—made for machine work. All bolts must be torqued down but not overstressed. All chucks must be tested and retested for grip. Leave nothing to chance before the program begins.

The worst industrial accident I've seen occurred when a 40-pound, aluminum wheel casting came out of a CNC lathe chuck. With a bit of investigation we believe the tool hit a sand pocket in the casting then broke. It dug in and the part was ripped out doing near 400 RPM. It not only broke through the safety screen, it knocked the operator's tool box off the work bench then went careening through the shop. From that awful experience, the program was reworked to include tailstock support during roughing cuts!

Unit 24-2

The chances are, you'll work in more than one shop in your career. When moving from one to the next, you'll experience differences in the way they do things with regards to setups, offsets, and PRZ setting. For example, in teaching in an overseas shop, the standard policy was that PRZs on lathes were set up using the same command codes we would fixture offsets on a mill. Their reason was they felt it was consistent with the way the mills were set up.

In particular, you'll find people who insist that either master tool or master position is the *only way* to setup a PRZ. It depends on the way they were taught originally. The savvy tradesman learns to adapt through understanding the possibilities. You'll also find combinations of the above three methods: master tool, master position, and fixture offsets. If you understand the difference in each, then adapting will be easy. That was the theme of Units 24-1 and 24-2.

Unit 24-3

There are dozens of combinations of the 12 possible testing methods. I wouldn't begin a program run without at least two tests no matter how proven it was in the past. Beyond training situations, you'll always be expected to get the machine up and running as soon as possible. "Time truly is money." However, the time cost of an accident is far greater than the testing beforehand. Program testing, especially at the beginner level, is a time investment that pays dividends.

Overall CNC Review

We're at the end of the CNC-specific sections. Next we take a look at CAM programming, then at other advanced technologies. As such, the following problems and questions are a final review of your understanding of the whole subject, Sections 17 through 24: foundation knowledge, monitoring production, editing, operating CNC machines, programming, and machine setup and testing. Each question is keyed to the section from which it was taken. Restudy might be indicated if you miss the target on two or more from any given section.

1. Based on Fig. 24-28, identify the five axes of a CNC vertical mill and the sign value for each (plus or minus). (17)

2. If the spindle of a CNC turning center is programmable, that is it can make controlled angular movements as would an indexing head, it is identified as the ＿＿ axis. (17)

3. Is this statement true or false? According to the *rule of thumb*, by pointing the thumb of your right hand, along the positive axis of rotation, your fingers will curl in the positive direction of rotary motion. If it is false, what will make it true? (17)

4. In handwriting the program coordinates for Fig. 24-29, what would be an easy method of avoiding calculations for points *E, F,* and *G*? (23)

5. In programming Fig. 24-29, the correct answer was to switch to incremental coordinates at

Figure 24-29 *Review Problem 4. How can these points be identified?*

point *D*. What EIA code made that change? (22)

6. The modal code required in Question 5 for incremental values will remain in effect until _____. (22)

Figure 24-28 *Identify these CNC axes.*

Critical Thinking

7. In handwriting a program for Fig. 24-30, the bolt circle holes must be drilled. Name at least three possible methods of writing this portion of the program other than graphic/CAM. What is the fastest of all three? (17, 23)

8. Is this statement true or false? Cartesian coordinates that refer to the PRZ for their values are known as Absolute Cartesian coordinates. If it is false, what will make it true? (17)

9. Name in their order of importance, the two guideline rules for selecting the PRZ. (17)

10. Where should the PRZ be chosen for the mill job shown in Fig. 24-31? (17)

11. A hemisphere must be machined on the top of an odd shaped casting. The shop has a CNC lathe and 2-D CNC mill; describe at least two simple ways in which it might be machined. (17)

12. Identify and describe three electrical CNC drive components. (18)

13. What is the function of a ball-screw in a CNC machine? (18)

Figure 24-30 *Describe several ways to write the program code without CAM software.*

Figure 24-31 *Problem 10. Where should the PRZ be located?*

14. Is this statement true or false? DNC can be accomplished in bulk or drip feed modes from the CNC control to the host computer. If it is false, what will make it true? (18)

15. Some CNC machines require a *home cycle* when initially turned on, while others do not—why? (19)

16. There are four methods of *halting* a CNC autocycle during a program with no consequence to the run other than the tool might be idling upon the work. With each method the autocycle may be restarted with a simple action. Name them and the restart action required. (19, 24)

17. During a setup, you wish to position the tool turret of a turning center at an exact position within the work envelope with respect to machine home zero. Name three operator methods of accomplishing this move. (19)

18. Name and describe three position screens commonly found on a CNC control. (19)

19. To correctly use the MPG, one must place the control in _____ mode and also use two other operator controls, name and describe them. (19)

20. Label the CNC controls A through H on the PVC controller shown in Fig. 24-32 and describe the function of each. (19)

21. What would happen during an auto cycle if you push or turn (19)
   A. Rapid override to *zero?*
   B. Feed rate override to zero?
   C. Spindle speed to 125 percent?
   D. Emergency stop?
   E. Single step?

22. List as many categories of vital information found in a CNC setup document as you can recall. These are vital to both operator and setup person. (19)

23. List two kinds of compensated profile mill programs and give a brief description of each. Which is the most common and why? (20)

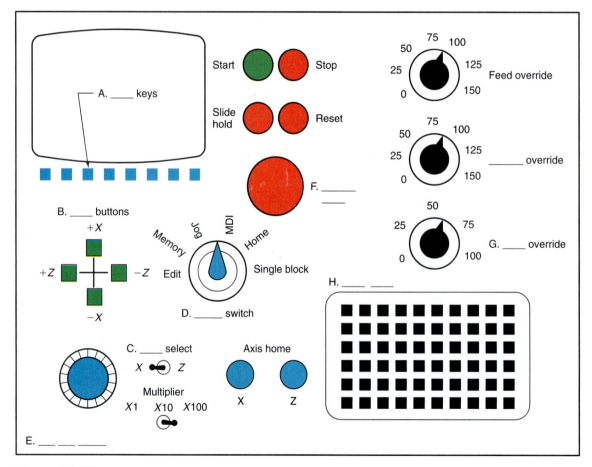

Figure 24-32

24. During an autocycle, several errors can cause alarms. List and describe them. (20)

25. Safety question—*vital*. Name and describe the three levels of operator intervention for problems. Under each level, name one or two methods of stopping the run to make the fix. (20)

26. Name at least three or four methods of monitoring a production CNC machine run. (20)

Critical Thinking

27. You have produced the part shown in Fig. 24-33 on a CNC machining center with T0101 = 0.500-in. diameter, two flute end mill for the pocket, and T0202 = 0.375-in. drill Actual measurements after machining are shown in red (brackets). What must be adjusted to bring the part into target nominal, PRZ or offsets? (Note no tolerance is given for this problem.) (20)

28. There are many factors to consider when planning out an efficient, safe CNC program. Beyond machine type, name the three most critical. (21)

29. Define the following planning and setup terms: (21)
coordinate shift
tooling reference—set point
tooling points
touch method

30. Is this statement true or false? Soft jaws are a problem for machinists. Made of the wrong material, both vises and chuck jaws deteriorate from use. If it is false, what will make it true? (21)

31. Name the three input methods of generating a CNC program. Which is more common? (22)

Figure 24-33 *Problem 27. What can be done to correct these dimensions?*

32. Fill in and describe as many of the following codes as you can remember. Note, the object was never to memorize codes but to learn the possibilities. See how many you do remember. (22)
G00
G01
G02
G03
G04
G17 (mill code)
G18
G19
G70
G71 (might be G21)
G40
G41
G42
G80
G90
G91
G94
G95 (mostly on lathes and drilling)
M00
M01
M02 (can also use M30)
M03
M04
M05
M06 (mill only)
M07
M08
M09

33. Why are G and M codes grouped into blocks? (22)

34. What actions will occur on these three command lines? (22)
N025   G01 *X*30.0000 G94 F19. G90 G70
N030   *Y*1.0000
N035   G00 *X*0.0 *Y*0.0

35. What will occur given this command? (22)
N001 G0 *X*1.000 G01 *Y*15.0

36. You are in G17 on a vertical CNC mill. Write a part path, *IJK* method (center ID) command block for the arc of Fig. 24-34. Then write the same arc command using the radius method. (22)

37. To initiate cutter radius compensation, what considerations must be (22)

   A. in the program planning?

   B. in the program?

   C. in the controller?

38. How are *fixed cycles* used? What is another shop term for these programming tools? (23)

39. Safety question. What is the singular danger of using the threading cycle on a CNC lathe/turning center as compared to all other fixed cycles? (23)

40. Is this statement true or false? A bolt circle is a threaded fastener that is bent into a round shape. If it is false, what will make it true? (23)

41. With respect to fixed cycles, define the following CNC terms: departure point, parameter, and tiled routines. (23)

42. Define the following setup terms: coordinating, PRZ—datum point, and master tool (24)

43. A machinist wishes to evaluate a CNC program *graphically* but the CNC control has no graphic capability. Name at least two of three possible solutions. (*Hint:* Think globally.) (24)

44. What is the best test before loading the program into the control and performing a dry run? (24)

45. How many program tests evaluate part dimensions? Describe them? (24)

46. What are the factors that determine how much program testing is advised for a given job? (24)

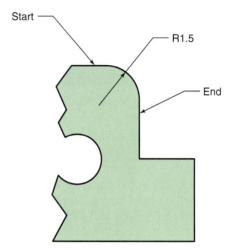

Write this arc command

Start

R1.5

End

Figure 24-34 *Problem 36. Write these arc commands.*

47. You wish to machine a cone shaped hole in an aluminum block. It has been predrilled to 1.00 in. diameter (Fig. 24-35). List the various ways it can it be machined, both Manual and CNC. Which would be the most accurate with the best machine finish?

48. A program has machined castings correctly for several hours and made 25 good parts. The 26th part is removed from the fixture and it shows that one side hasn't cleaned up all the way. What could be the problem?

49. You are attempting to do a dry run, graphic evaluation of a handwritten program for a lathe control. It halts on the highlighted line (*), with error code #34B. Looking that up in the operator's manual you find it means a syntax error has occurred. What's wrong? How do you fix it? See Fig. 24-36.

```
:O12345
(Final CNC Quiz—Turned Pin)
(2.000 CRS Stock—Cut to length—PRZ outer
Tip—Per Fig. 24-36)
N005   G80  G70  G90  G40
N010   G0 X100.   Z100.
(Home turret to soft limits)
N015   T0101
(0.03NR, RH turning—offsets 01)
N020   S650           G95   F0.005
N025   G0    X2.100  Z0.250   M03      (Rapid to departure)
N030   G41   X2.00                     (First ramp)
N035   G1    X1.750  Z0.050            (Ramp on)*
N035                 Z−3.000           (First pass)
N040   G0    X1.800  Z0.250            (Rapid back for second pass)
N045   G1    X1.375  Y−3.000           (Second pass—finish 1.375 diameter)
N050         X1.875                    (Face shoulder)
N055   G3    X2.000  Z−3.0625  R.0625  (Corner radius)
N060   G1    X2.125                     (Pull off contour)
N065   G0    X100. Z100.0
N070   G40
N075   M2
```

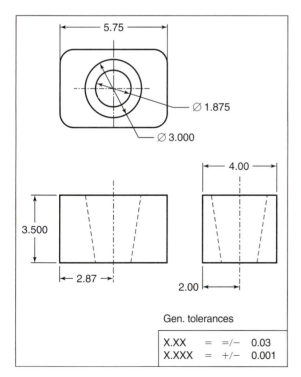

Figure 24-35 *How many ways are there to cut this tapered hole?*

50. You now have the syntax problem repaired in the program. It dry runs OK but the graphics show an odd shape at line N045—not per drawing. What's the problem? How do you fix it?

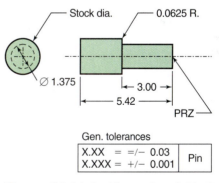

Figure 24-36 *Problems 49 and 50. Compare the program to the drawing and edit out errors if they are found.*

ANSWERS 24-1

1. *Machine coordinates* absolute position relative to M/H; *work coordinates* represent the coordinated tool position; tool memory contains the tool number along with its compensating offsets; the fixture offset page locates the PRZ relative to machine home.

2. False. Since you are only using the screen to follow relative DTI movement, work coordinates can be used with equal ease to machine coordinates.

3. See answer 2.

4. A. Clean, bolt, and align work-holding devise.
   B. Coordinate *X-Y* PRZ (order does not matter).
   C. Load cutting tool.
   D. Coordinate *Z* PRZ.

5. A. Chuck test material or workpiece.
   B. Mount cutting tool.
   C. Start spindle and machine test cylinder.
   D. Measure and write into *X* axis register.
   E. Touch off *Z* axis PRZ and coordinate in axis registers.

6. The absolute position relative to machine home.

7. Indicate over the hole then write $X-03.0000$ $Y-02.0000$ into the work coordinate registers (or calculate that difference into the machine location in fixture memory).

8. It's true.

Mystery Block Challenge

You should have been able to use a test indicator zeroed over an edge, to detect the size of the block within 0.0003 in. of its micrometer size. The *trade*

9. By actually turning a test piece, tool deflection and runout issues are reduced, which aids in overall accuracy of the setup. Less adjustment is needed after making the first part.

10. Here are your options—none involve changing the program other than being certain that all three tools have length compensation commands.
    A. Be certain the new tool's length is 0.0377 longer when it is mounted in the spindle, then set a -0.0377 offset in its length offset value. It then performs as though it was the old T01. The other two tools need no change using this solution.
    Reestablish the new tool's *Z* zero, by touching off the work, then:
    B. Add 0.0377 in. more to the other tool's length comp offsets. They then move farther down to pace the longer master. This can be done MDI or by using the calc function.
    C. Or, retouch the other tools then set their offsets to compare with the master tool. This solution is essentially starting over.

From Problem 10, you can see that with any further tool replacement beyond this point, and subsequent offsets become a real headache using the master tool method! In Unit 24-2 we'll learn a much better method for multitools.

tip is to first null the indicator over an edge then not disturb the probe. When the probe reaches zero null, the spindle is directly over the edge.

ANSWERS 24-2

1. Because *all Z* axis features are too tall, the problem lies with the setup, not the program. Change the tool length offset to a greater, negative number. Change the *Z* axis PRZ deeper (farther down) by 0.005 in.

2. The object is uniformly oversize by 0.0036 in. on both dimensions. The amount per side to be reduced = 0.0018 in.
   A. Cutter path offset will be -0.0018 in., moving the cutter closer to the work.

B. Part path offset will be 0.2482 radius, or 0.4964 if the entry is a diameter.

3. The statement is true, although an offset may be used to adjust cut depth easily. This could also be accomplished by shifting the PRZ down or using material offsets.

4. Making an datum edge-edge intersection PRZ alignment: Center the spindle over the edge parallel to *X*. Set the *Y* axis at zero position in the control. Center the indicator over *Y* edge. Set the *X* axis at zero in the control.

5. It is false. The controller will probably be on a position page showing the work coordinates but up to the time where work coords are changed it could be on mach coords too.

6. There are two different problems. Since the *X* diameters are nonuniform in their errors and there is only the single turning tool, the problem is either the program (to be edited) or the cutting action/sequence of the tool. It can not be fixed with offsets or PRZ adjustments.

The *Z* axis problem is corrected by shifting the PRZ back (away from the chuck) by 0.1 mm.

7. Your answer should make the following points:
   A. A single tool is chosen as the model. All other tools are offset to perform as that tool.
   B. The master is brought into a known position relative to the intended PRZ, physical position on the work or fixturing.
   C. That position is then loaded into the work coordinate page of the control as *X*0.0 *Y*0.0 *Z*0.0 with no length offset.

8. All tools receive a positive value offset. It is the large distance of the individual tool's machine coordinate position from M/H to the datum surface.

9. To accommodate odd-shaped work where the part can be made but the overall shape must be shifted within the workpiece.

10. The master position method assumes all tools have an offset and therefore makes it very easy to change any tool without math challenges.

ANSWERS 24-3

1. All graphic tests; all axes disabled—no movement.
2. A toolpath utility program because it is made for the purpose of testing.
3. Gross shape errors and dangerous rapid travel moves.
4. Dimensions and interference with tooling on the machine.
5. A. Remove the cutter.
   B. Remove the work.
   C. Move the PRZ.
   D. Fake the offsets (radius and length).
6. Master tool
7. There is no perfect answer. However, I'd use the backplot function as the program was

being developed. Then a toolpath utility test viewing from several angles. Finally, I'd do a couple of dry runs. Next, I'd choose to make a test material part. Finally, I would not do a single-step run, as I only get one chance to make the part.

8. It's pretty much true unless the graphic test accompanies a dry run on a machine.
9. A single-step test involves reading the codes on screen before allowing them, and analyzing each using position registers and distance to go.
10. Mostly true, however, they can be inserted into critical points to check measurements during a first part run.

Answers to Section Review Questions

1. Five axes are shown in Fig. 24-37.
2. Since the spindle rotates around the *Z* axis it is the *C* axis.

3. It is true.
4. Switch to incremental coordinate values when identifying point *E* from point *D*.

Figure 24-37 *The five axes with one negative sign.*

5. It is G91.
6. The modal code (G91) can be canceled by any other code from its group—G90, in this case. Note, many modal codes will be canceled by the M02 (M30) command but on modern controls, these two codes stay in effect after the program has ended.
7. This bolt circle can be easily programmed using:
   A. *Fastest*—A bolt circle, canned cycle which usually includes drilling routines. If not, by using the G83 drill cycle within the bolt circle routine.
   B. Temporarily shifting the coordinate set to center upon the pattern center. Calculate the positions relative to the center of the pattern. Then use a drilling fixed cycle.
   C. A local reference zero. Calculate positions. Use a drill cycle (if the control has such a provision) and a drill fixed cycle.
   D. A fourth potential solution lies within polar coordinates, if you use them. Radii and angles are the basis of polar values. However, they have not been studied in this book and the full answer would be lengthy.
8. It is true.

9. *Primary:* The datum basis of the print; *Secondary:* Within the first quadrant (if possible—mills only).
10. Your symbol should have been placed at the intersection of the right and lower edges based on the dimensions. However, programming the part in this orientation would produce all negative X axis coordinates, since the part lies in the second quadrant. You might consider rotating the part before programming it.
11. A. The easy method is on the CNC lathe if the object can be held and turned safely.
    B. Using a ball nose end mill and making horizontal circles around the dome.
12. The three electrical components in a CNC drive are
    A. Servo motor
    B. Feedback (positioning) device
    C. Central processing unit
13. To drive and position a linear axis while eliminating backlash by preloading the drive axis in both directions.
14. It is false. Reword: From the host computer to the CNC control. (Moving data from a CNC control to a PC is an upload not a direct numerical control.)
15. Feedback counters must be reset to an absolute zero with respect to the machine work envelope. They do not retain absolute position when the machine is off while other systems do retain absolute machine position when powered down.
16. *Override to zero*—return to its original value; *single cycle*—Switch mode to auto and touch cycle start; *slide hold*—touch cycle start; *optional stop*—touch cycle start
17. Manual pulse generator/position screen (hand dials); jog buttons/position screen; MDI—writing a small command to go to the given position
18. Work coordinates; position with respect to PRZ; machine coordinates; position with respect to machine home; distance to go; amount of axis travel left to complete current command.
19. First, the control must be in *jog mode; axis select* switch activates given axis; *multiplier* sets how much movement per pulse, 1 or 10 or 100 times the machine resolution.
20. A. Soft keys—variable function entry keys
    B. Jog buttons—manual continuous or incremental motion of an axis

C. Axis select—set the axis to be jogged
D. Mode switch—select the operating activity
E. Manual pulse generator (MPG)—hand dial the selected axis at the rate selected on the multiplier switch.
F. Emergency stop
G. Rapid override—lower the rapid travel rate from 100 to zero percent
H. Hard keys—dedicated entry keys

21. What would happen if you push or turn:
    A. The machine would continue machining until a rapid command was encountered in the program then it would stop.
    B. The machine would stop all axis motion. However, the cutter would remain spinning in contact with the work.
    C. The spindle would turn 125 percent of its programmed rate.
    D. All activity would halt and the program would be reset. The machine would be turned off.
    E. The control would complete the command line it was working on then wait for another *start* command from control panel.

22. Here are 10 major categories but there may be more on your list:

    The program identification and location—central computer, disk, tape, and so on
    The specific machine for which it is postprocessed
    Jig and fixture ID numbers for this job
    Any custom tooling required
    The setup location of the PRZ
    The types and sizes of all cutter tooling and their assigned tool numbers within the program
    The kind and location of all clamping/chucking or holding
    The sequence of events
    Specific instructions particular to the job—for example, clamp changes at midprogram halts; a warning and instruction about a large piece of excess material that will fly off the work at a given point in the program; the part must be rotated or repositioned in the chuck at a certain point. Is the program compensated or not?

23. *Manually compensated cutter centerline*—a program written for an exact cutter size; *part path compensated*—a program written to follow the part geometry with any acceptable cutter within given limits. By far the most common program.

    Note that there is a third program type generated by CAM software, the compensated, cutter centerline program. This program is written for a specific tool radius but is also capable of compensation. It can be adjusted toward or away from the part geometry by adding plus or minus offsets.

24. Depending on the controller, this is a very limited list. Most controls produce many subsets of these and other error types. If your answer approximates the following, give yourself credit for your answer.

    Physical and hardware errors—Operator action is required to correct some condition. Examples include axis limits have been hit, low oil or air, safety switches, misfeeds, and chuck closures.
    Servo errors—Exceptional action, extra heavy machining or a crash, the machine has lost its position count on the feedback system.
    Syntax and program errors—math or syntax problem fixed by program editing, for example, switching a letter "o" for a zero; incorrect wording of statements; incorrect ordering of a rigid set of variables; missing data, or radical commands, logic problems with branching programs.
    Operator/controller errors—previous actions or entries, offsets cause impossible tool path; machine not initialized correctly.

25. Again, this is a minimum response. There are many ways to stop and correct problems based on operator judgement by situation. Do you understand the general concept of the three ascending levels of operator intervention and one or two methods of halting the run?

    Level 1—Something isn't right. May be dangerous. Noise or finish deterioration. A *halt* run action is called for within a given time. Switch to single-cycle mode and wait for end of current block. Turn the rapid override to zero and wait for next rapid move after completing current block.
    Level 2—A problem is obvious. It is dangerous or serious, but a crash is not occurring yet but eminent if action is not taken quickly. May require a cycle end—full

stop or may be fixed with a less critical *halt*. But it must be fixed quickly as it will deteriorate.

Level 3—An all out crash. A surprise event with dangers to you, the machine, and the work. Fast response includes emergency stop while leaving the danger area. Practice this move and memorize alternate E-stop buttons away from the danger zone.

26. Monitoring a CNC production run can be accomplished by

Listening to the sound of the machining—primary method.

Watching the cutting action and finishes produced as well as the chips flow and characteristics.

Measuring completed parts.

Watching graphic and position pages on the control. Not discussed in the training but a valid symptom, watching the horsepower meters on the control. If they trend upward without changing feeds or speeds, the tool is dulling.

27. T0101　Reduce radius offset by 0.003 in. to cause the cutter to go toward the line thus making a larger pocket. Oversize by 0.006/2 = 0.003 rad reduction. Add 0.0075 in. to length offset value (to machine deeper in the *Z* axis).

　　 T0202　New drill bit—wrong size. Decrease length offset value by 0.030 in. (to drill shallower).

Note, shifting the PRZ is not a potential fix here as one tool is too deep, while one is too shallow.

28. Three major areas are select the PRZ and axis orientation of the raw stock; select the holding methods; plan the cut sequence which includes the kinds of cutting tools.

29. *Coordinate shift* A programming technique used on both mills and lathes to temporarily shift the PRZ for safety or convenience.

Tooling reference-set point An alternate method of establishing PRZ using a block attached to the fixture.

Tooling points A support pad that is part of a mill fixture to hold the work.

Touch method A physical method of setting a *Z* axis PRZ at the machine.

30. It is false. Soft jaws are usually made of aluminum or soft steel, such that they may be machined into a custom holding shape by the CNC machine itself.

31. *Conversational languages* Based on the spoken language of the country (English in the USA)

Graphic input Based on accurately drawn pictures of the work to be performed. This method is also used on most CAM systems.

EIA-ISO codes Based on alphanumeric codes assigned to various programming functions. The most common system, most graphic systems are able to translate their programs into a code set.

32. G00　Move at rapid travel—one or more axes

　　 G01　Linear movement at feed rate—one or more axes

　　 G02　Circular motion—clockwise at feed rate two axis limit

　　 G03　Circular motion—counterclockwise at feed rate two axis

　　 G04　Dwell—a stall for a specified time

　　 G17　Machining in the *X-Y* plane—mills only

　　 G18　Machining in the *X-Z* plane—mills

　　 G19　Machining in the *Y-Z* plane—mills

　　 G70　Inch input (not universal, sometimes G20)

　　 G71　Metric input (not universal, sometimes G21)

　　 G40　Safety—cancels previous tool radius compensation

　　 G41　Compensate cutter radius to left of geometry line

　　 G42　Compensate cutter radius to right

G80   Safety—cancels any previous special routines

G90   Absolute value coordinate input

G91   Incremental value coordinate input

G94   Feed at inches/millimeters per minute (usually mills)

G95   Feed at inches/millimeters per rev (lathe and drill)

M00   Program halt—utility stop for operator actions

M01   Optional stop—operator selected or ignored

M02   Program end—reset to start (can also use M30)

M03   Spindle start—forward (lathe or mill)

M04   Spindle start—reverse

M05   Spindle stop

M06   Tool change (mill only—in conjunction with tool code)

M07   Coolant on—spigot A

M08   Coolant on—spigot B (often assigned to mist coolant)

M09   Coolant off

33. The groupings divide mutually exclusive codes. Only one from any group may be commanded on any given line.

34. N025 G01 X30.0000 G94 G90 G70 F19.
   The X axis will feed in a straight line to position X30.0 in. from PRZ, using 19 in. per minute feed rate.
   N030 Y1.0000
   The Y will now position to Y1.00 in.
   N035 G00 X0.0 Y0.0
   The X and Y will now rapid back to the PRZ.

35. This line is wrong. It may cause an error code (most common) or the machine may ignore the rapid command and execute only the G01 linear interpolation. The problem is that two codes from the same group have been commanded on the same line.

   N001 G0 X1.000 G01 Y15.0

36. G91 G02 X1.5 Y−1.5 R1.5
   G91 G02 X1.5 Y−1.5 I0.0 J-1.5

37. *The planning* must include a ramp on/off move and physical space to do so.
   The program must include the ramp moves the comp code G41 or G42 and the offset call out with the tool.
   The controller must be loaded with the correct tool radius (and tool vector/shape if it is a lathe program).

38. Sometimes called "canned cycles" these are program time savers that require the completion of data fields variables for commonly repeated events such as peck drilling or roughing out work.

39. Because the lathe must control the thread lead to an exact degree, the speed/feed ratio can not be varied or interrupted. Therefore both *override and single-step functions are disabled* during this cycle only.

40. It is false. A bolt circle is a series of evenly spaced holes in a circular pattern, that may be a fixed canned cycle on your control.

41. *Departure point*—The initial position from which a given routine is started. Often designated with the letter I in programming manuals. *Parameter*—A blank requiring an entry to customize a fixed cycle. *Tiled routines*—Overlapping and connecting mill routines to machine away a given area of complex shape.

42. *Coordinating*—The setup task of physically positioning the tool relative to the work PRZ then loading the axial values into the controller. When the coordinated tool is commanded to 0,0,0 it is physically at the PRZ.
   Datum point—The physical location on the work or fixture that is either the PRZ or a known distance from the PRZ.
   Master tool—A setup policy whereby one tool in the setup is the template for PRZ. It will be at PRZ when commanded X0 Y0 Z0 on mills for example. But to position all other tools to this position, they must be given an axis offset.

43. A CAM backplot tool path evaluation (most immediate); a tool path utility software (best solution); repost (process) the file to a different control with graphic capability (last resort)
44. Graphic tool path utility software.
45. Machining test material—soft wax, foam, or sometimes wood/aluminum. Using oversize offsets which makes a large version of the part. A single-step run makes the part but the cutter dwells on the work surface.
46. Number, availability, and value of parts to be run; history of program and programmer; your experience level; complexity of work to be done/part shape
47. Boring bar on lathe
   Good finish and accuracy
   Note—this could be CNC or on manual lathe with taper attachment.

End mill on five axis—CNC mill
   Moderate finish due to long extension— may chatter.
CAM-generated ball nose set of circular cuts approximate cone.
   This would leave cutter mark, waves. The more cuts taken the finer the waves. Very time consuming but might work if it's the only machine available.
48. Strongest possibility is that it's a bad casting.
   Check all other dimensions—if OK, then it is the casting.
   Second possibility is that you loaded the casting into the fixture incorrectly or a chip got between it and the locators.
   Check to see if machined feature dimensions have shifted relative to the uncut surfaces of casting, if yes, then it was your loading.

The fixture has slipped due to vibration. Check fixture alignment and retighten.

Tool breakage might be the problem since the problem occurred suddenly.

This should have been obvious—wear isn't the problem as that would have shown slowly over several parts.

49. Hint—before reading on, remember the control is in compensation and looking ahead into the program lines.

Answer—The problem was an error on Line N045

Change N045 G1 X1.375 *Y*−3.000
        N045 G1 X1.375 Z−3.000

No *Y* axis on lathe

50. The movement made a tapered object. Add line N046

N045 G1 X1.3750 (To finished diameter)
N046   Z−3.000 (Make final cut)

PART 4

Advanced and Advancing Technology

Fast technical evolution is the very reason for this book. But it's also your greatest challenge as a craftsperson of today and tomorrow. Some of the subjects in Part IV are new, while some have been around for some time. For example, cutter geometry and composition or statistical process control are not new, but today due to multiplying technologies (one affecting the other) they are undergoing more change than at any time in the past. Still others are just showing up in the shop, while this book is being written.

Part 4 has two very different goals. The first is to introduce

Part 4 is more about adapting to change than it is about the technologies themselves, and this is our second and more critical goal. By graphing the trends of the recent past, one can easily see that your career span will encompass far more change than that which has already happened. You will need to be a very different machinist than those of us who compiled and wrote this book. You will need to sort through changes to find the good ones and reject the whizbang that will drop away. You'll need to look forward and see the changes coming and to be ready—To do otherwise will be to choose to be a dinosaur.

Computer-Assisted Machining (CAM) Programming

Introduction

Today most CNC programs are created with CAM software, except for tooling, or graphic programs compiled at the control. Compared to handwritten codes, CAM brings seven advantages to manufacturing:

Universal Programming System Using PC workstations, any CNC controller can be programmed. Most existing programs can be quickly reprocessed by CAM to be read by any other CNC machine. Data is managed better since programs are saved in a central location, using familiar file storage media. This frees CNC controller memory to serve machining rather than program storage.

Speed Programs are written in a fraction of the time. Adding efficiency, newer softwares feature *associativity*. Changing one or more elements on the drawing template can automatically update the program. Design revisions and other production changes require a fraction of the time they once did.

Writes Shapes Impossible Otherwise CAM can create supercomplex shapes that cannot be defined using standard commands. We would use it for this reason alone.

Effortlessly Solves Math, Eliminating Potential Errors When CAD lines and arcs exist on-screen, then their properties and common points are known to the computer. For example, in Fig. 25-1 by drawing the circle in its exact *X-Y* location with the correct radius, then trimming lines *A-B* and *D-C* to it, their intersection points are known to the computer. While every machinist must know how to compute those trim points using trigonometry and geometry, it takes time.

Toolpath Verification Tool The ability to *backplot* the tool path during its creation and during final evaluation is a valuable safety feature. It shows that the cut is going to go where expected—graphically, without risk!

Existing Templates The geometry template often exists as a CAD drawing—a real time saver for complex parts! That drawing can be sent from the customer to the contracting shop anywhere around the globe. The finished program and backploted rendered image can be sent back the same day for approval.

Office Environment Away from the noise, and interruptions of the shop, programs are easy to create. Also, learning CAM software is simplified since they are driven by familiar operating systems such as Windows® or Unix®.

Step 1
The circle is drawn, then lines *A-B* and *C-D*

Step 2
The arc is trimmed to lines *A-B* and *C-D*

Figure 25-1 *Solving math using CAD.*

The Four CAM Tasks

There are four distinct phases to create a program using CAM. We'll practice all four in Section 28.

1 Geometry Acquisition

A graphic template must be acquired first. If it already exists, it can be *imported* (brought in from outside the CAM software). Or the geometry can be drawn using the utilities within the CAM software. Finally, it could be drawn in a CAD program then imported to the CAM.

2 Tool Path and Machining Parameters

Based on the geometry, drawing elements or surfaces are selected in the order in which they are to be machined. They are then given cut parameters. For example, where to stop rapid travel toward the work, depth of cut, and how many roughing/finishing passes are required. That creates a tool path but it's not yet a program.

3 Tool Path, Program, and Shape Evaluation

While it's not required, graphically observing the resultant tool path for errors is a good step to take, especially when the workpiece is expensive or you have only one piece of material or when safety is in question.

4 Program Management

Converting the toolpath into a specific program for a specific CNC control, called *postprocessing* in our example software Mastercam®. At this stage, operator comments and safety and/or efficiency moves are sometimes edited into the program. It is also then saved where it can be downloaded for use by the machine. Creating shop floor management documents such as setup sketches and tool lists is also part of program management.

Which Software? Why?

For a usable practicum, our committee chose Mastercam® based on a survey of schools. Although a great product, this is not an endorsement of one software. There are many fine systems available. Here in Section 25, we're discussing the four basic tasks, not how to use Mastercam, therefore, many of its features will be left for a formal course or for self-exploration.

Objectives

CAM Task 1–The Part Geometry Model

Introduction: The process starts with a template, a drawing of the part to be machined. It can be drawn using Autocad®, Cadkey®, or CATIA®, for example, or any CAD program, then the template is transferred to the CAM. Or it can be drawn within the CAM software using its draw utility. The drawing cannot be created using everyday draw programs found on home computers because many do not generate vector entities. Vector drawings are comprised of unique points, lines, arcs, surfaces, and solids that can be manipulated and used for CNC work. We'll discuss this more coming up.

If the CAD graphic file is prepped into a form readable by the CAM software, then transferred to it, it's said to be *exported*. When the foreign drawing is brought into a CAM program from an outside source, it's being *imported*. In Unit 25-1, we'll practice creating our own CAD drawing. Then we'll import and modify Fig. 25-2 within the CAM software. If you've had CAD or designer experience, many of these tasks will be familiar.

The upcoming exercises may be easy or may require a bit of experimentation if it's your first time. At times, you'll need to use critical thinking to solve the problem before moving onward. That's what we all must do when learning unfamiliar software. Using help screens to assist in solving problems like those here in Section 25, it will begin to make sense. It's also good learning to refrain from asking questions until you are sure you are really stuck. Solving your own dilemmas will enhance your learning. To speed things along in the problems, step-by-step instruction is provided and where it is, I'll provide a graphic of what yours should resemble. However, sometimes to enhance learning, you'll be cut loose to figure out how to achieve a goal on your own, using information you've just learned.

The Tasks to be Performed In Unit 25-1, based on Fig. 25-2, we'll be getting a top view ready to profile mill, cut a pocket, and drill a 0.375-in.-diameter hole.

Figure 25-2 *Master drawing for Section 25.*

KEY TERMS

Boss
A thickened feature on a machined or cast part, usually around a hole.

Branch points (Mastercam)
Points where more than two elements share a common X-Y-Z coordinate. Indicated by a red arrow in Mastercam.

Entities
Points, lines, arcs, and surfaces in a CAD drawing.

Graphic format
The file form in which drawing graphics are saved.

Layers (CAD/CAM drawing)
Placing drawing geometry on one level with details, notes, and dimension overlaid on one or more other clear levels.

Masking
Temporarily making invisible, details and entities on a given level or layer.

Supported (file format)
A CAD geometry file format that the CAM system can import directly.

Transforming (Xform Mastercam Menu)
A main menu pick that begins changing entities by moving, scaling, rotating, mirroring.

Translating (entities and/or geometry)
Moving selected entities from point to point.

Trimming
Making two or more elements have shared endpoints in three-space.

Vector (entities CAD/CAM)
CAD drawing geometry lines and arcs with a start, end, and direction, necessary for elements to be manipulated.

CNC-Ready Geometry

In modern manufacturing, the CAD drawing should be created with CNC in mind. That is, all entities (arcs, lines, and surfaces) are drawn accurately and trimmed perfectly. But the designer often can't envision the machining sequence, or which details will be cut first and which will be set aside for later; for example, adding temporary tabs, or cutting intermediate datum surfaces (see Fig. 25-7 later in the section), or the methods by which the work will be held and machined. Those are details the planner adds and the programmer executes. Therefore, even the very best CAD drawings often require work before they are CAM program-ready. Here's a few things that must be done:

1. **Masking/Erasing**
   Notes, dimensions, and non-geometry clutter must be removed. In Fig. 25-3, I've masked layer 2 to show only the object lines found on layer 1. Before writing the profile mill program, we'll erase the front and end views too, since they are also on layer 1, they cannot be masked. (Information coming up.)

2. **Establish Part Orientation and PRZ—Critical Step**
   The $X0$, $Y0$, $Z0$ point on the CAD drawing isn't necessarily where the PRZ should be for the particular CNC program. During practice sessions, we'll rotate the geometry to orient the X axis per our job plan then **translate** (move) it such that the PRZ is in the correct location on the part according to the plan.

3. **Temporary Utility Lines/Surfaces**
   These lines have several purposes:

   Tie bars to kill vibration or to hold large chunks of metal safely as they are nearly cut away from the stock.

   Temporary datums per your job planning.

   Roughing surfaces—for example, leaving excess for heat treatment then a final grind later.

   Added ramp lines when a very specific lock on/off path is required for safe tool compensation.

 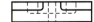

Figure 25-3 *Layer 2 is masked to show only essential geometry.*

Containment lines where the cutter path must be directed to another surface or where a pocket milling routine would be the most efficient way to reduce the part but the geometry isn't actually a closed pocket. In that case, an added fence can be drawn to close the shape.

Masking Details on Layers

Extra details on a CAD drawing can either be erased or they can be **masked.** Masking makes unwanted lines, dimensions, and details invisible temporarily. To use this convenience, the original drawing must have been created with different kinds of information on separate **layers** (*also called levels in some software*). Layers are as though the drawing lines were on one sheet of clear plastic, while the dimensions and notes are on a second overlay (many layers can be used). In Fig. 25-2, I drew the part geometry on layer 1 and the section view, dimensions and notes on layer 2. In Fig. 25-3, I temporarily masked layer 2 to show only the orthographic projection.

Erasure Selection Methods

If the drawing isn't layered, then unwanted details must be erased one by one, by the programmer. Although it takes longer than masking, erasing doesn't take long once the various ways the toolbar utilities and menu selections are memorized, along with the ways to grab and select entities to be erased. The easiest is the window selection. Click on one corner then drag the window to the opposite corner. Only entities and notes fully inside the window will be blown away. Windows can also be selected to pick differently. Others wipe out items completely outside the window, or partially inside.

They can be created using multiline shapes to surround otherwise difficult to pick entities while avoiding others. Mastercam makes it easy because elements that have been selected to go change color to flag the upcoming wipeout, but even so mistakes happen! In that case, look for the *unerase* button on the tool bar that brings back the last step, and the undelete selection within the Delete menu that brings back every element one at a time! That one is very useful for the really big mistakes—it all comes back!

Exact Geometry

For any CNC template, whether for lathe or mill, lines and arcs must be drawn to the exact length then **trimmed** perfectly. Trimming means creating two elements with shared endpoints in three-space X, Y, and Z.

For example, when drawing Fig. 25-1, I could have sketched the circle's center approximately where it belonged rather than using its exact X-Y coordinates. Using the right diameter and approximate location, it would have *looked right* but been slightly inaccurate, not making the exact part.

Additionally, the horizontal line C-D was trimmed exactly to the circle, at the microscopic level, no gaps or overruns. See Fig. 25-4 where two examples of bad trims would lead to problems when creating the tool path.

In both cases, the errors in Fig. 25-4 might be less than 0.001 in., not enough to make much difference in the part, yet they may not

Bad intersection trims

Overrun —⟋   ⟍— Gap

Figure 25-4 *Two examples of sloppy drawing technique.*

work as a CAM model; the line elements are discontinuous. Most software has a selectable tolerance for bridging discontinuous geometry, but avoiding it is by far the best policy.

It's frustrating to get part way through a toolpath routine only to discover that elements aren't right and the gap is too big to jump. Progress stops to find and fix the problem. It's far worse to measure a machined product to find it to be dimensionally wrong due to the original CAD drawing size.

KEYPOINT

It's usually no more time consuming to draw part geometry accurately than to sketch lines and arcs approximately.

Three-Dimensional Milling Models

When working with *X-Y wireframe geometry* (connected lines and arcs resembling wires that define part edges and surfaces), adding three-dimensional *Z* depth may or may not be a good idea. We'll see in the exercises that it isn't always required.

If the *Z* depth is included in the geometry model, then cut depth will be automatic (Fig. 25-5). However, a flat *X-Y* top profile template can be useful at times, with no depth at all. Using a flat pattern for profile and pocket cuts, we set the *Z* cut depth when defining the

parameters for that segment of the toolpath. This is a nonissue for lathe work were a flat *X-Z* template is all that's required unless working on a multiplexed lathe with added axes.

KEYPOINT

When writing the CAM milling program, *Z* depth can be acquired from a three-dimensional geometry for automatic cut depth, or depth can be written during the parameter assignments (examples coming up).

Shown in Fig. 25-5, the 3-D wireframe drawing has depth. It was created by translating top view lines down to the required cut depths. During Mastercam programming, when an element is selected to machine, the cutter will automatically go to that depth or rough down to it in steps.

Trade Tip

Translating Depth into the Drawing To give a drawing *Z* depth, draw details in the *X-Y* plane then transform by translating them to their correct *Z* depth. By doing so, the various line elements will be at the intended *Z* cut depth.

Program Reference Zero Translation

Another task is to ensure that the drawing *X-Y-Z* origin is at the intended PRZ location per the plan. It isn't always where you might expect it to be. For example, in Fig. 25-6, the CAD drafter chose to originate the drawing at the center of the hole, a natural place to start this shape. But the hole doesn't exist in the raw stock. A better location for the first program would be the corner, to contact the vise

Creating the PRZ origin by translating

X0, Y0 Natural start point for CAD drawing

PRZ translated to work corner for CAM

Figure 25-6 *The drawing is translated to a new reference point.*

Figure 25-5 *A wireframe geometry of the part shows Z depth as well as X-Y flat elements.*

jaw and a stop. Here's a typical menu tree to make it happen:

Transform *(Xform Mastercam menu)*

Translate   *A specific point on the geometry will be selected as a control point to be moved to a new X-Y-Z location, taking all the selected entities with it. This is known as point-to-point translation.*

Window or Individual Select the elements to be translated

Done

Move *(or copy if that's what is required)*

Control Point Method/Mouse Pick

This sets the point to be moved to the origin.

Origin

Sends picked elements to X0, Y0, Z0.

Trade Tip

Cleaning Up Before Exporting While masking is easy in Mastercam and other software, it's usually a good idea to load the copied program file while still in CAD, then erase unneeded layers and details *before* exporting the file to the CAM system. This avoids transferring excess data that may or may not be understood in the CAM system. A clear example will be seen when you bring over the master CAD drawing in the student files into the CAM software—a few details won't make the transfer intact. Details like notes and dimensioning aren't needed for programming and are usually best left behind.

Modifying Geometry

There are lots of reasons that we change the original drawing. Figure 25-7 shows an example. A tie bar of stabilizing metal has been

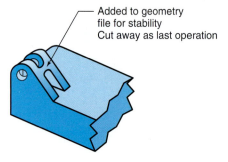

Added to geometry file for stability
Cut away as last operation

Figure 25-7 *A typical add-on to stabilize the part temporarily.*

---

created and left untouched during profile machining and drilling. It will control flexing and vibration. It then will be machined away as an offload operation.

Understanding CAM Graphics

To move files between software programs, it's necessary to understand a little about how they are formatted within each. Wouldn't it be great if there was only a single standard for graphics files? But, of course, each CAD/CAM software has its own internal way of defining drawing vector entities to save them as a file. File format differences are true of bitmap graphics as well. The arrangement and way graphic data is saved is called the **graphic format.** The good news is that there are industrywide standards such as IGES (International Graphic Exchange System) or the newer ACIS format for 3-D and solid geometry. Just like on starships, these universal translators help aliens talk to each other!

KEYPOINT

When importing or exporting a geometry file between software programs, you must determine if the file is in a format that can be directly transferred (**supported**) by the CAM or must be translated (converted into a readable format) before export to it.

Supported Formats = Direct Transfer The desirable condition whereby the CAD file is defined and saved in a form that can be understood by the CAM directly. For example, under the Autodesk® selection in Mastercam, you'll find .DXF (Drawing Exchange Format) or .DWG. Both are Autocad® protocols and they are supported by Mastercam and all other software. This is true for Cadkey's CDL® format as well.

Unsupported Formats = Intermediate Translation While still in the CAD system, the file is translated into a universal form such as the early universal format, IGES. Once translated, the CAM system can then import it and then retranslate it into its native format.

Figure 25-8 shows the possible formats that Mastercam can import and use.

DWG   Autocad—native drawing file

DXF   Autocad—drawing exchange format

Figure 25-8 *Most CAM systems can import various file formats.*

CDL   Cadkey drafting software format
IGES   International Graphics Exchange System format
ASCII American Standard Code for Information Interchange
ACIS   Solid model format found in most 3-D systems, based on a computer chip and using the ACIS kernel

Standard Exchange Formats

The world standard is the International Graphic Exchange System, IGES, format. IGES is best for line and arc, wire frame, and simple surface geometry, but the newer ACIS works better for surfaces of complex design created in solid modeling software that use the ACIS kernel chip.

Trade Tip

For Complex Geometries Even though a CAD file might be in a supported format, sometimes there are advantages in translating it into a universal form before exporting it to the CAM. This step might prevent programming glitches and complexities. For example, specific font letters to be engraved where the letter subroutines do not come across correctly unless universally translated.

Vector Geometry

CAD/CAM softwares deal with screen graphics differently than everyday computer images. To be usable, all CAD line and arc entities must be unique, that is, they have a start and end location in *X*, *Y*, and *Z* and a center point, if it's an arc. These are known as **vectors** (a start point, direction, and an end point). Defined in that way, they can be moved, trimmed, and manipulated because they are defined with math formulae.

By contrast, everyday screen images such as clip art or photos are composed of a series of colored dots called bitmaps or sometimes a raster screen. In a bitmap, there are no defined entities, just a bunch of dots. A raster is a scanning line, similar to radar. Every time it whizzes past, it paints the dots one at a time. The dots can be changed one by one, but there is nothing that links them together other than common color or general area on the screen. OK, let's make this really confusing, to paint a vector entity on a computer screen still requires the monitor to raster.

KEYPOINT

CAD/CAM geometry consists of vector entities (mathematically defined lines and surfaces that can be manipulated).

Drawing Geometry

Ok, you made it through the reading, now the software! We'll be drawing two kinds of **entities,** straight lines and circular arcs in Unit 25-2, called *primitive elements*. However, as we'll see in Unit 25-3, entities can also be nonregular curves and complex surfaces.

Defining Entities

While you'll be working within Mastercam, the way entities are created, trimmed, and manipulated is not different from any CAD/CAM software. The essential skill here is learning how to define unique points in three-space. There's a lot of technique and tricks involved and they aren't learned from reading!

KEYPOINT

A vector entity is defined by defining unique points.

Ten Possible Ways to Define an Entity Point

A new entity can be created that starts and ends at

Coordinate position (*X*, *Y*, and *Z* coordinates relative to the origin)

End point—of another existing entity

Middle point—along another entity

Relative point—*X*, *Y*, and *Z* distance from any other known point

Polar position—angular and radial distance from any known point

Parallel distance—on a line a given distance from another entity

Arc center—center of existing arc

Tangent or Perpendicular—to another entity

Point—selects a unique point previously drawn

Mouse sketch approximate—no finite location until after it's placed on the drawing. At placement, the software will show that position in *X*, *Y*, and *Z*.

Note that sketching is used to create points somewhere near the location, using a cursor select. Points, arcs, and lines defined in this way are usually drawn as temporary trim lines or ramps for tool compensation for example.

Learning the 10-point definitions is **the** skill for computer drawing. You'll only get good at it by practicing. Experiment with different ways to define entities in the exercises coming up.

Critical Question

Selecting which definition to use for an entity point depends on whether or not other entities exist and their relationship to the one to be drawn. For example, given the following information (Fig. 25-9), but starting with a blank screen (no lines drawn yet), how many ways can you draw the blue line *AC*? Line *AC* lies in the *X-Y* plane for this example—no *Z* depth. In other words, in how many ways can you use the 10 definitions to draw the line exactly as it's defined?

OK, now expand the possibilities. What if only line *A-B* was already drawn on the screen? What if both lines *A-B* and *B-C* were already drawn?

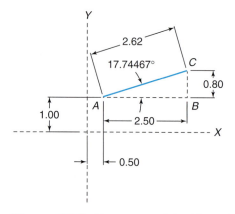

Figure 25-9 *How many of the 10 methods can define Line A-C with no other elements drawn on the screen? With only line A-B? With lines A-B and B-C?*

TRY IT

Using the 10-point definitions, it's time to draw your own geometry.

OBJECTIVES

To draw a flat *X-Y* template of Fig. 25-2, top view only. It will then be used to create a milling program in Unit 25-2. We'll do it in three steps.

Step 1   Draw Flat Geometry—X-Y top view only.

*Step 2   Translate Elements to Z Depth—*Makes pocket and perimeter machining depths automatic.

*Step 3   Rotate and Translate Entire Geometry—*Make datum A parallel to the *X* axis and place the PRZ at the intersection of datums *A* and *B*.

INSTRUCTIONS

Follow the step-by-step instructions until told to proceed without them. Begin by drawing two temporary centerlines—crossing at the drawing origin. The 0.375-in. hole will be at the *X-Y* origin for this exercise.

Draw the left half of the geometry from the vertical centerline (*Y* axis). We'll mirror copy it at the end to learn a valuable time saver.

Each bullet begins a set of menu picks to follow. Boldface type are menu selections or required data. Italicized text are helpful comments.

Plan enough time to complete the assignment as the file cannot be saved within the student program. There are file copies of each major step saved on your

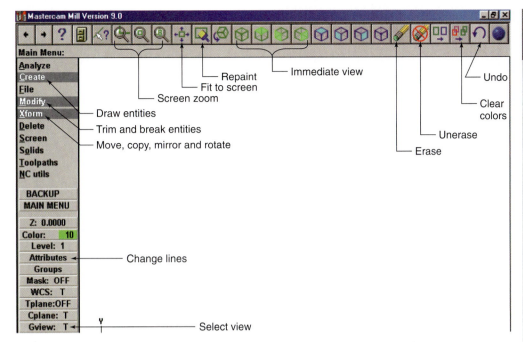

Figure 25-10 *The menu selections you'll be using to draw the part template.*

Machining and CNC Technology student disk. Use them when time runs out in one session or as a check comparison to your geometry.

KEYPOINT

There are many creative ways to draw this geometry. As you progress, envision and practice the possibilities. Experiment!

In Fig. 25-10, you'll be using the following highlighted menu selections to draw, trim, and transform and modify entities to create Fig. 25-11. Note and use the selected icons on the tool bar shown in Fig. 25-10. They are useful for setting perspectives and for other immediate functions such as zoom and undo.

ShopTalk

Mastery Software Some CAM systems, but not all, are capable of analyzing tool paths and the objects they produce, then maximizing and filtering the program data for efficiency. Called mastery software, these systems can implement cycles, subprograms for repeating features, and other geometry efficiency improvements such as examining the finished tool path then replacing tiny straight lines with arcs to better approximate surfaces and reduce data volume. This not only creates smaller programs of greater significance, it improves shape and finish, removes mismatched surfaces, and can reduce machine cycle time. Our example software, Mastercam, can perform these improvements and compressions.

Figure 25-11 *Step 1—draw the flat geometry.*

Step 1 Drawing Part Geometry

Start Mastercam Mill *(Using Version 9 but any ver. will be similar)*

> **Gview: T** *Get View = Top (lower left menu). That's the X-Y plane on the screen. The program should default here but check to be sure. You can change view perspective at any time.*
>
> **Cplane: T** *Construction plane = Top—should also default to top view but always check. These two items should be checked when starting a top view.*

Attributes *(change line style/width to draw the centerlines)*

> **Centerline**
> > **Thinnest Line** *(choose the thinnest line option)*
> > > **OK**

Create
> **Line**
> > **Vertical**
> > > **Origin** *Starts Line at X0, Y0*
> > > > **Sketch** *Click anywhere above Y0. Click again to accept at cursor location—line turns color (probably green).*

You have now drawn a vertical centerline starting at the origin. It does not matter how tall.

Backup
  Horizontal
    Origin
      Sketch *Click to left of CL any length.*
      Click again to accept.
      You have now drawn two center-
      lines that intersect at the origin.
Main Menu *(prep to draw circle)*

Trade Tip

F9 At any time, to display the axes and origin, touch F9. Axes will appear as very light lines along with a distance scale for the current zoom level. Touching F9 again turns the axis display off.

Attributes *prep to draw object lines—#2 is wider, pick solid lines*
  Solid Line/Width #2
Main Menu

Create
  Arc
    Cent Pt + Dia.
      Defines arc using center point and a diameter
      .375 *enter diameter at bottom of screen*
        Origin *to place center at X0, Y0*

Trade Tip

Mastercam will remain at that menu level and continue to draw parallel lines without going back, just click the centerline again to draw the next vertical line.

Main Menu—*go back to main by selecting Main Menu or by touching esc.*

Create
  Line
    Parallel
      Side + Distance
        Mouse Click Vertical Centerline
        Line flashes to show pick
          Click Direction
          Choose left side for now
            1.35/2*
            Half 1.35 **Boss** *Width*
            <ENTER>

Click Vertical Centerline/Click Left Direction
  2.35/2 *Half pocket width*

Click Last Line Drawn—Click Left Direction
  .75 *Incremental distance from last line*

Go to top tool bar and perform a screen fit.
Or you can touch Alt-F1

Your screen should look similar to Fig. 25-12, depending on how long the centerlines were sketched. If yours has two centerlines, the hole and four vertical lines it's right on target. If not, erase the screen and try again!

Figure 25-12 *Your drawing has four vertical lines, one horizontal line, and a circle of 0.375 diameter. Line length isn't important.*

*Data can be input using embedded math functions in Mastercam and most other CAD/CAM software. Use this convenience to save time and prevent errors.

Figure 25-13 *Horizontal line of any length are drawn.*

Figure 25-14 *Adding Fillet Radii and trimming.*

TRY IT!

Now draw the six horizontal lines that define the geometry. Use the master drawing and the techniques we've just learned. They will define the horizontal perimeter and the pocket plus the boss. The finished image will look similar to Fig. 25-13.

Trim and Add Fillet Radii

At conclusion of this segment your drawing will look exactly like Fig. 25-14. It will be an exact replica of the part profile, top view.

Main Menu
  Create
    Fillet *Trim* = N   *First, we fillet datum A to the bottom line. No trim since the square corner remains on the geometry after the fillet.*

      Radius = 1.5
        Mouse Click Datum A and Bottom Line

Backup
  Trim = Y *YES, to trim out pocket corner radius*
    Radius = 1.0 *pocket corner*
      Click any two intersection pocket corner lines

Trade Tip

You may need a little help learning to pick lines for filleting and trimming operations if it's your first time drawing on screen. *Click on the portion of the line you want to keep,* not its extension beyond the trim point, the part that will be trimmed away. Use the *undo* tool to fix a bad fillet or trim. However, once you've left a given menu tree, where the error occurred, undo can't help you.

TRY IT

Add remaining 0.5 fillet radii and trim out pocket Your drawing will then look something like Fig. 25-14.

Trim the pocket
Do not trim the top corner to the radius.
Do not radius the center boss yet.

Main Menu *to finish center boss radii*
  Create
    Arc
      Tangent
        3 Ent *arc will touch 3 entity lines*
          Click 3 Lines
            *The three lines that define the 1.00 by 1.35 boss on the **left** of the vertical centerline.*

Delete *On upper toolbar to avoid menu tree hopping*
  Click on the Vertical Construction Line left side of the Boss
  Left edge of the 1.35 wide boss (Fig. 25-15). Delete the vertical line

Figure 25-15 *Using the delete tool speeds erasure of the temp. construction line.*

• **Main Menu**
  Modify
    Trim
    *Choose either **1 Entity** or **2 Entities** then trim out the drawing to match (Fig. 25-16).*
      Click Datums A and B
      Be sure to trim the short horizontal lines on the boss, to the radii. Also trim datum A and the bottom horizontal line.

Figure 25-16 *Trimming the boss and all radii.*

Figure 25-17 *Window selecting all entities—except the centerline axis and hole.*

Mirror Copy to Complete the Shape

Main Menu
  Xform
    Mirror
      Mouse click all elements to be copied
      Do not select the 0.375 hole because it's complete now. Could also use a window to select the elements to be mirrored but that would select the hole thus duplicating it. (See the Trade Tip.)
        Done *(Fig. 25-17)*
          Line *Chooses line mirror axis*
            Click the Vertical Centerline
              Copy *(Fig. 25-18)*
                1 copy only

Trade Tip

When copying drawing elements as with the mirroring exercise, be cautious of creating duplicate elements—one line or arc superimposed over the other. They don't show on the drawing but when doing a toolpath can become a problem that's difficult to detect.

Deleting Duplicate Elements When you suspect duplicate elements exist they might be a problem, go to the DELETE menu then click DUPLICATES to find and remove them.

Unselecting Elements in a Pick For the mirror copy, you could have used a window to pick all elements then use UNSELECT to remove the centerline and 0.375 hole from the pick, before copying it across the line.

Clear the Operation Colors *(Figs. 25-18 to 25-19)*

Figure 25-18 *The result of the mirroring is color-coded purple. It could be selected for another transformation, by color pick, until we clear the colors.*

Figure 25-19 *To restore the construction colors use color clear.*

Creating Intersection Points Where There Are None

Now we must create break points in the straight lines at the upper corner tangent points (Fig. 25-20). If this step isn't done, the computer would not be able to profile mill along the radiused corners at the 0.355 depth as the straight lines would not join the radiuses. Without a break, the straight lines would be continuous past the tangent points. To see small details, click on zoom then window the area to be magnified, as shown in Fig. 25-20.

Main Menu
  Modify
    Break
      2 Pieces
        Click Horizontal Line
        Fig. 25-20 line to be broken
          Intersection *to break it where the radius is tangent*
          Click On the Radius Line
          Breaks horizontal line at tangent point

● **TRY IT**
Repeat the operation on the vertical tangent lines, breaking them into two pieces at the needed intersection point. Then break six more places, at the other three corners both vertical and horizontal tangent points. When complete, all straight tangent lines will break at the tangency points, around the outer perimeter.

Question? Would it have been more efficient to break the lines back before the mirror copy?

Answer: Yes. In this example, this isn't a big deal but in some cases, thinking out steps that would be eliminated by correctly sequencing operations can save lots of drawing time.

Step 1 is Complete

The flat pattern is now finished per the master drawing, and should be saved if working on fully enabled software. It is not ready for machining yet.

KEYPOINT

A copy of the completed top view profile at this stage is on your student CD (SECT25-TOP VIEW 1.DXF).

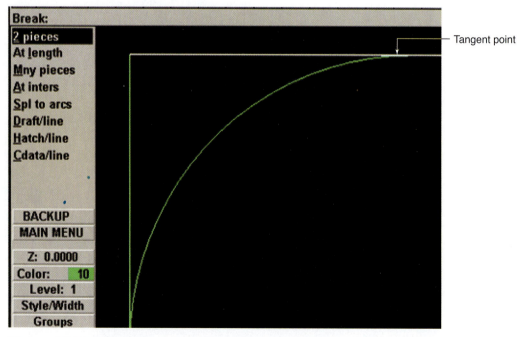

Figure 25-20 *Break the horizontal line at the radius tangent point in order for the upcoming toolpath to flow onto the curve at the tangent point.*

Step 2   Adding Z Depth

Objectives:

To change flat pattern elements to represent cut depths.

Refer back to Fig. 25-2 for dimensions.

Make pocket depth along with boss profile −0.500.

Make outer profile −0.375 deep from top surface.

We will *not* add depth to the hole to see how it can be done during the program parameter stage.

Gview

Isometric *to see Z depth as it is created.*

Main Menu—Xform

Translate *To move pocket rim 0.500 deeper using a Z distance move.*

> **Chain** (Fig. 25-21) *Selects entities that fasten together.*
>
>> **Click Any Entity on pocket rim—** White entities *(Fig. 25-21)*
>>
>>> **Done** *Ends this chain*
>>>
>>> **Done** *Ends selection function*
>>>
>>>> **Rectangular** *How to move elements.*
>>>>
>>>>> **X0, Y0, Z−0.5 [ENTER]**
>>>>> *Could enter just Z−.500*
>>>>>
>>>>>> ***Move***
>>>>>>
>>>>>> **OK**

Figure 25-21 *The pocket perimeter has been selected (white lines) by chaining.*

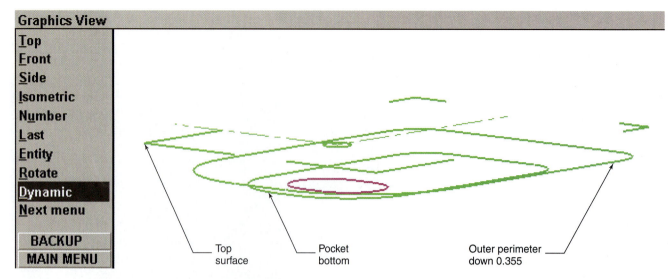

Graphics View

Top
Front
Side
Isometric
Number
Last
Entity
Rotate
Dynamic
Next menu

BACKUP
MAIN MENU

Top
surface

Pocket
bottom

Outer perimeter
down 0.355

Figure 25-22 *By "grasping" the geometry with a mouse pick, it can be moved dynamically.*

Now do a *Screen Fit* to display the translated lines.

- **TRY IT**
- Translate the oval boss perimeter −0.500 deeper into the geometry
- Translate the outer perimeter 0.355 deeper, including corner radii *but not the square corners.* See Fig. 25-22.

KEYPOINT

During the outer perimeter chaining, the cursor will come to **branch points** then wait for you to click which element to branch onto. Branch points are where three or more elements share a common coordinate. They are indicated by a red arrow in Mastercam.

Now change the view perspective to see if it's right (Fig. 25-22). If not, find and fix the problems.

Gview *Fig. 25-22*
  Dynamic
    Mouse click on geometry then move it to see depths

F9 *to display the axis orientation and the PRZ (brown lines Fig. 25-23)*

Gview—T *Return to Top view for step 3*

Remember, the square corners are at the top of the raw part. The pocket is deeper than the perimeter profile. Go to a front view to see the depths as straight lines. Return to the top view. Step 2 is now complete.

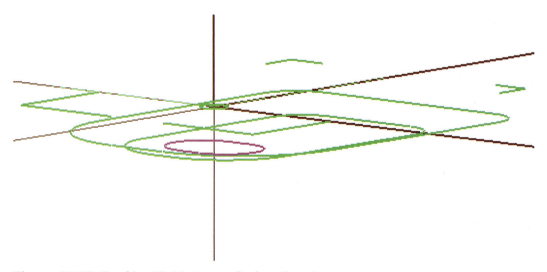

Figure 25-23 *Touching F9, Mastercam displays the axis system.*

Step 3   Setting Up to Machine— Axes and PRZ

Objectives We're going to perform two **transformations** (Xform) of the drawing (Fig. 25-24) to place the *X* axis as we want it and the left lower corner (LLC) at the PRZ. This, the more common PRZ location, is abbreviated to LLC in shop lingo.

1. Between points to bring the intersection of datums A and B ($A \times B$) to the origin ($X0$, $Y0$, $Z0$)

2. Rotate the part drawing to make datum A parallel to the *X* axis.

◆ TRY IT
- With your drawing or with the file Sect25
- Top View 1.dxf imported into Mastercam,

Begin in Top View
  Xform
    Translate *Moves entities Point-to-Point,*
    Window
    Rectangular window is automatic—can change if needed
      Window all elements using rectangle.
      Click window beyond upper corner.
      Click opposite corner.
      Elements selected turn white (Fig. 25-25)
      Done
        Between Points
        Endpoint
        Selects control point to use for move.
        Click on Datum A × B
        Corner
          Corner will be moved to next pick
          Origin
          Moves selected elements to X0, Y0, Z0
            Move
            Moves the elements rather than copies them.
              OK

As drawn　　　**Two transformations**

Origin

Origin

Rotate 90° about the origin

Translate from intersection A × B to origin

PRZ ready for program

Figure 25-24 *We'll be moving the drawing such that the intersection of Datums A and B will lie at the PRZ.*

Figure 25-25 *Window select the entire drawing for translation.*

Screen Fit Icon or Alt-F1

Backup

   Rotate *to rotate geometry with datum A parallel to X axis.*

     Window *Select all elements in geometry.*

      Done

       Origin *Point to rotate around*

        Move

         90°

          OK

F-3 or Clear Colors Icon

This step returns colors to their original drawing color. (See the Trade Tip.)

F-9

To be certain that the geometry lies with LLC at PRZ. (Fig. 25-26.)

Step 3 is now complete, this geometry is ready for machining.

This file has been saved on your disk as Sect25-Top View 2.DXF.

Figure 25-26 *A dynamic or isometric view will show the new location ready for machining.*

Importing Drawing Geometry

The final version of the drawing just completed can be used to create the program. But it's also saved on disk. So, before doing a program, let's stop and take run through importing a geometry file then getting it ready for machining, typical of readying a CAD drawing to be machined. The file you'll be importing from your student CD, is Sect25-Master.DXF.

Start Mastercam Mill *If it's not already running*

Files
  Converters
    Autodesk
      Read
        Select DXF format *looks for this extension.*
        Go to Student Disk—*Section25 files*
          Read Section25-Master.DXF
          OK

CLEAN THE DRAWING UP FOR MACHINING.

Levels
Remove the red check to make the drawing look like Fig. 25-28.

  Note, you can remove and restore layers repeatedly. Your levels list may not be the same as Fig. 25-27 due to revisions and file importation.
  OK

The remaining geometry is on layer 1 and must now be erased to cleanup the drawing for programming.

Delete
  Window *Defaults to rectangular shaped window.*
          Experiment with other window shapes later

Beyond this point, it's a matter of picking and deleting all entities other than the top view.

  We then translate *Z* depths as practiced earlier.

  Translate the geometry from *X*0, *Y*0 at the hole, to the PRZ lower corner, then rotate it so the *X* axis is per plan.

At this point, we have a program ready geometry again—we're back at the end of Step 3. Or are we?

Click to set control list remove level number 2

Figure 25-27 *Deselect CK-Level-2 to mask details.*

Figure 25-28 *The drawing needs a little more work, but the essential geometry is looking good—or is it?*

Critical Question

How can you know that a drawing acquired from an outside source is correctly sized?

Never assume a CAD drawing is exact. Before writing the program, check a few dimensions using the **Analyze** function in the main menu.

Analyze
  Between Points
    Endpoints *Click on the two widest endpoints (Fig. 25-29).*
    The data shows the correct width.
    Try a few more dimensions between known points.

Yes, it's a correct drawing. We're now ready to review then go to Unit 2, to write the program using this either this drawing or Sect25-Top View 2.DXF if time has run out for this session.

Figure 25-29 *Never assume a CAD drawing is dimensionally correct! Analyze it for size.*

UNIT 25-1 REVIEW

Review the Key Terms

Boss
A thickened feature on a machined or cast part, usually around a hole.

Branch points (Mastercam)
Points where more than two elements share a common *X-Y-Z* coordinate. Indicated by a red arrow in Mastercam.

Entities
Points, lines, arcs, and surfaces in a CAD drawing.

Graphic format
The file form in which drawing graphics are saved.

Layers (CAD/CAM drawing)
Placing drawing geometry on one level with details, notes, and dimension overlaid on one or more other clear levels.

Masking
Temporarily making invisible drawing details and entities on a given level.

Supported (file format)
A CAD geometry file format that the CAM system can import directly.

Transforming (Xform Mastercam menu)
A main menu pick that begins changing entities by moving, scaling, rotating, mirroring.

Translating (entities and/or geometry)
Moving selected entities from point to point.

Trimming
Making two or more elements have shared endpoints in three-space.

Vector (entities CAD/CAM)
CAD drawing geometry lines and arcs with a start, end, and direction, necessary for elements to be manipulated.

Replay the Key Points

◆ When it's to be used for CNC production, layers should always be used when creating the CAD drawing.

◆ It's no more time consuming to draw work accurately than to sketch lines and arcs.

◆ *Z* depth can be three dimensional, in which case the cut depth will be automatic, or if the drawing is a flat profile template, depth can be assigned during the CAM program creation.

◆ Learning the 10-point definitions is the essential skill for computer drawing.

◆ CAD/CAM geometry consists of vector entities, (defined lines and surfaces that can be manipulated) while everyday screen graphics consist of raster screens (all dots in a bit map). The dots can be changed one by one, but there is nothing that links them together other than color.

◆ You may need a little help learning to pick lines where they will trim correctly. There is a bit of practice required to pick the line at the correct location such that the trim occurs where you want it.

◆ During chaining, the cursor will come to branch points then wait for you to indicate which element to branch onto, indicated by a red arrow in Mastercam.

◆ Never assume a CAD drawing is dimensionally accurate, always check key dimensions using the analyze function.

◆ Duplicate lines may cause problems when performing toolpath operations. Find and eliminate them before using the drawing for programming.

Respond

1. Identify and describe at least five of the seven advantages CAM programming brings to the shop.

2. What is the only down-side to CAM programming? How is it overcome? (Name two ways.)

3. We wish to add depth to a flat top view, which CAD/CAM drawing function should be used?

4. When exporting a CAD drawing to a CAM what must be done to the file? What should be done to the drawing?

5. CAD/CAM drawings require different kinds of line and arc definitions compared to everyday screen graphics. Explain.

UNIT 25-2

Planning and Creating Tool Paths

Introduction: OK, task 1 is complete, the geometry is in place either by importation of Sect25-Top View 2 or by drawing it. Either way, we're ready for programming. At this stage, different CAM software diverges in sequence slightly but they all accomplish the four CAM tasks of

Geometry acquisition

Toolpath and machine parameters

Toolpath evaluation

Program management

In Mastercam, after acquiring the part drawing, the programmer first selects and links lines or surfaces, in a process called *toolpath element selection*. With the elements selected on screen, the user applies the operation parameters (tools, number of rough passes etc.) to that segment of the program. At this stage, Mastercam has created a transparent internal file. It's attached to the drawing file and will become named with an extension of *.NCI* (numerical control intermediate). The NCI file is not readable by a specific controller, only by Mastercam. But the next and final step is to select the specific *postprocessor* (specific translation filter) to change it into a set of commands for the machine control.

KEYPOINT

The Mastercam programming process is

A. Select Toolpath from the main menu.

B. Pick the method for element selection, then click entities to be machined in this segment.

C. Apply machining conditions and parameters. Machining conditions for each set of elements creates one program segment; for example, the pocket, or the drilled hole in our example.

D. Create remaining segments.

E. Postprocess the intermediate file to .NC. Creates both the .NCI and .NC files.

F. Evaluate the toolpath with the backplot function. Can be seen at any time in the process. No need to wait until it's all complete.

Postprocessing—Customizing the .NC File Similar to using a printer driver in your PC, postprocessing makes the generic commands just finished, specific for a given controller—called a *machine readable program*. If, due to shop loads, the job must be

transferred to another machine with a different controller, the generic NCI file can be quickly reposted to serve any other control in the shop.

Software such as Mastercam allows user customization of the **postprocessor** to suit shop preferences or special situations. For example: using all absolute or all incremental values in the program or always parking at a given position in the work envelope upon reading the program end command.

KEY TERMS

Dynamic rotation
The CAD/CAM ability to "grasp" the geometry and rotate it to any view perspective.

Handshake
The process that establishes two-way communication between PCs and controllers.

Island
In CNC work, this is a closed shape within another shape, around which the cutter must go.

Island avoidance
The ability of a CAM software to machine around an island.

Postprocessor
A machine-specific filter that readies a program for a given machine control.

CAM Task 2   Toolpath Element Selection

Job Plan

The program you are about to write has three segments:

1. Outside profile cut 0.355 in. deep from the top surface

    T01 = 0.50 dia. end mill

    Climb mill (cutter left)

    3 rough and 1 finish pass

2. Pocket cut 0.500 deep, avoiding the central boss

    T02 = 0.375 dia. end mill cutter

3. Drill 0.375-in.-dia hole—using peck drilling.

    T03 = 0.375 dia. drill

We'll be vise holding a rectangular block, already machined to 3.850 × 5.250 in., with

Y axis
X
Segment 2
cut pocket
avoid island
PRZ top
left corner
Solid jaw-datum A
Segment 3
drill hole
Segment 1
cut profile

Figure 25-30 *The setup sketch.*

the solid jaw against datum A and a stop against datum B (Fig. 25-30). The part will protrude above the vise jaws, a bit over $\frac{1}{2}$ in. to allow a 100 percent profile cut.

The PRZ is LLC—top surface per Fig. 25-30 and shown shortly in Fig. 25-32 as a red arrow.

TRY IT

After loading the geometry file, we'll be choosing elements on the part geometry that are linked together at common coordinate points using the chaining

process. Once a group of elements are selected, we'll then choose machining parameters to determine how they will be machined. We'll create three program segments that will be linked together to form the complete program.

Start Mastercam Mill

Files
  Converters
    Autodesk *(Autocad parent)*
      Read File
        Go to your completed geometry file or:
        D:\Sect25-Files\Sect25-Top View 2.DXF
          Note, your CD drive may not be D so substitute drive letter or C:\ if you have copied the CD to the hard drive.
        Set extension to list **.DXF files.**
          Open
            OK

Gview
  Top *(should be already)*
  Erase or mask unwanted notes, dimensions and other details if they aren't gone already.

Program Segment 1— Outside Profile Mill 0.500 Deep

Main Menu
  Toolpath
    Contour

      Chain *(Fig. 25-31)*

      Click on the horizontal line, near the left end, that causes the chaining arrow to go around the profile, in a clockwise path, thus climb milling. Note, beginning the chain at a tangent point is a good place to lock onto the geometry.

        Done—*Closes this chain*

        A machining parameters page appears Figs. 25-32 and 25-33. See Entering Parameters next.

Figure 25-31 *The way an element is selected to start the chain will determine which way it goes around the linked lines.*

Entering Parameters

After the chain closes and you've clicked **DONE,** parameters page appears (upper left in Fig. 25-32). You'll need to fill in highlighted data fields on this page and in five more pages linked to it. Check marks (✓) indicate boxes to be checked. Letters refer to the required data fields in Fig. 25-32. (Also, see the Shop Talk and Trade Tip.)

Data Field A) Cutting Tool Parameters

Right-click inside the large **tool setup window.**

Then click **Define** three cutting tools per plan (Figs. 25-32 and 25-33).

T01

Left-click on 0.5 endmill to select cutter for this contour cut.

T02

Select bull nose endmill to add 0.03 corner radius on T02.

T03

Drill 0.375 diameter.

Trade Tip

Instead of defining cutters as we did at Parameter Step A, you can also pick them from the cutting tool library. Right-click tool window, click **? Job Setup,**

        Tools,
                Get from Library,

Then pick the required cutting tools in the order in which you'll call them in the program

        T01 = 0.5 End Mill,
        T02 = 0.375 End Mill
        T03 = 0.375 Drill

Note that during parameter selection, you have the choice of cutter compensation in the computer or in the control (upper right corner of screen). That means Mastercam will generate either a cutter centerline (comp in computer) or a part path (comp in control, G41 or G42). The choice will be based on shop policy. However, if you use comp in control, then ramp on and ramp off lines must be drawn.

Figure 25-32 *The menu tree needed to set the parameters for the profile cut.*

Click to
auto-calc RPM

T01 = 0.5 Bull with 0.03 rad.
T02 = 0.375 Bull with 0.03 rad.

Tool
diam.

0.03 corner
radius

Figure 25-33 *Tool definition is important—use these parameters.*

Data Field B) Job Setup Per Fig. 25-32

To Define Blank Workpiece Boundaries and PRZ

Click **Job Setup**—right-click on large window, scroll down to

Define Workpiece = **3.85 × 5.25 × .5** (depth of deepest cut)

Feed Calculations based on Material

Select **2024 Aluminum** Per Drawing

Select **Origin**—to position the RPZ

Click on lower left corner of blank

Note, Job Setup must be done one time for T01, it then applies to all other tools unless conditions change.

OK

Data Field C) Contour Parameters

Click **Multi Passes**—*to make roughing then finish passes*

Click on the **Multi Pass Box** to open Parameters Box

Enter 2 Roughing passes @ .1 in.
1 Finish @ .050 in.

At Final Depth
Keep Tool Down
OK

Toolpath Segment 1 Is Complete

Blue Lines Show Selected Tool Path

If all parameter data has been supplied, then close all parameter windows with OK, then the toolpath automatically appears as blue lines (Fig. 25-34).

Figure 25-34 *Closing the parameters, the toolpath appears as a blue line. The roughing and finish passes show.*

SEGMENT 2—.355 in. Deep Pocket

Backup to Toolpath—*to start second segment of program*
  Pocket
    Chain (*Fig. 25-35*)
      Select Elements
      Done
        Enter Parameters (*Fig. 25-36*)
        OK
      Your pocket toolpath should look like Fig. 25-37.

KEYPOINT

By selecting both the pocket rim and then the boss (called an **Island** in CAM) *before choosing DONE,* Mastercam links them together and recognizes the boss to be a closed figure within the pocket, it is an **island to be avoided** when cutting the pocket. Be sure to select the elements such that the cutter is climb milling on the pocket rim and also around the island (Fig. 25-35).

Figure 25-35 *This time we'll chain both the pocket perimeter and the closed island together before closing the chain selection.*

Entering Pocket Parameters (*Fig. 25-36*)
  Tool Parameters
    Click on T02 = .375 Endmill
    Note that after setup was completed during Segment 1, the RPM is now autocalculated.

A) Pocketing Parameters
  Note that the cut depth is already set since we translated the pocket rim down to −0.355 in. deep on the geometry template. If we had not done so, you would need to tell MC how deep the pocket was to be or if you wanted it to go to a different depth, to rough the pocket, for example.
      Depth Cuts
      Max Steps down = .25
      1 Finish Cut @ .05
      Keep Tool Down
    Roughing/Finishing Parameters
      Choose Constant Overlap Spiral.
        Note—you could change the amount of overlap between one pass to another, but for now leave it at 75 percent.
      .025 Finish pass spacing
      OK

Segment 3—Drilling the Hole

Backup to Toolpath
  Drill
    Entities *to select the circle as the drilling target*
      Click inside or on the .375 circle
        Done *to end this group of selections*
          Done *to end drill location selection*
            Enter Parameters (*Fig. 25-38*)
              Note that you will need to enter the drill depth at 0.900 in.
            OK

Select to open

Figure 25-36 *The pocket parameter tree requires two pages—pocket and roughing/finishing. Use T02 tool definition, 0.375 in. bull nose end mill with 0.030 in. corner radius.*

Note that the depth had to be supplied during parameter completion, since the hole was left on the geometry top surface. Also during drill parameter selection, options for the drilling could be made—simple or peck drilling.

You have now completed task 2—the tool path is complete.

Figure 25-37 *The pocket toolpath.*

Figure 25-38 *The drill parameters for the 0.375-in.-diameter hole.*

Cam Task 3   Evaluating the Toolpath

At any point in the process after the first segment has been created, toolpath results can be played back showing various perspectives, front, top, and oblique, for example. **Dynamic rotation** is a very useful feature when trying to visualize the result. Mastercam allows the user to "mouse-grasp" the geometry on-screen and move it in space to show any perspective.

KEYPOINT

Our goal here is to introduce the safe habit of evaluating programs at several levels before trying them on a machine. There are many useful evaluation views and screen arrangements left to explore on your own within Mastercam and within the Metacut Utilities, the second program on your student CD. Metacut can actually display the finished part surface along with any machining errors that the program might create.

Try It
View Perspectives and Display Options

Depending on sophistication level, various CAM software programs feature different options for raw material display, cutting tools, machined features, and the most important item, the path the tool takes, including rapid travel movements. After displaying split-screen images of the path with several perspectives playing back simultaneously, explore the display options and single step evaluation. To change to a split screen (Fig. 25-39).

Main Menu
  Screen
    Next Menu
      Viewports
      Choose the four square screen.
      Note that by highlighting a particular subwindow, each view may be changed to suit the needed perspective.

Backplot Your Program

Change the backplot display to see some of the ways a toolpath might be evaluated better (see Fig. 25-40).

Main Menu
  NC Utils *(Utilities for CNC preparation and evaluate functions)*
    Backplot
    Display

• TRY IT

Change to single-step then click your way through the program—a very useful tool to see each movement. Try various levels of cutting tool display.

Figure 25-39 *A split-screen backplot shows many things about the toolpath.*

Task 4 Readying the Program for a Specific Machine

The toolpath seems to check out OK. Almost done, it's time to postprocess it then save the NC and NCI files.

The objective is to process the file into an .NCI then into .NC (CNC, EIA codes) that will operate a Fanuc mill control.

For your convenience, the resulting machine readable code file is saved and may be loaded and edited (Sect25 FINAL.NC).

Main Menu
  NC Utils *(Utilities)*
    Postprocess
      Change
      MPFan—Open *Select Fanuc mill post.*
        RUN
        Mastercam asks if you wish to create an NCI file. Specify the file name and location. Note,
          YES
          Name and save NCI/ NC file

Figure 25-40 *Changing the display parameters can help analyze the toolpath.*

```
%
O0000
(PROGRAM NAME - SECT25-M-FINAL2)
(DATE=DD-MM-YY - 29-04-02 TIME=HH:MM - 20:15)
N100G20
N102G0G17G40G49G80G90
(T01=.5 EM .03    T02=.3765 EM .03    T03=.375 drill)
N104T1M6
N106G0G90G54X-.4Y.5A0.S3056M3
N108G43H12.25
N110Z.1
N112G12-.355F24.45
N114Y1.925F48.9
N116Y3.35
N118G2X.5Y4.25R.9
N120G1X3.75
N122G2X5.65Y2.35R1.9
N124G1Y1.925
N126Y1.5
N128G2X3.75Y-.4R1.9
N130G1X.5
N132G2X-.4Y.5R.9
N134G1X-.3
N136Y1.925
N138Y3.35
N140G2X.5Y4.15R.8
N142G1X3.75
N144G2X5.55Y2.35R1.8
N146G1Y1.925
N148Y1.5
N150G2X3.75Y-.3R1.8
```

Figure 25-41 *Only the enabled version of the software will produce the CNC codes after selecting a postprocessor then applying it to the toolpath file. Note this code file is on your disk as Sect25-M-Final2.NC.*

Final Editing

After posting, the program may be edited in two ways (Fig. 25-41):

In NC, edit where the toolpath may be graphically altered or machining parameters can be changed.

In Text, edit as any text file.

Program sequence, tooling/ fixture notes and setup instructions might be added in brackets to separate them from machine commands. For example, special operator situations might be inserted where some action is required:

(NOTE—CLAMP CHANGE)

(Danger—Large Excess Chunk Will Disconnect)

Or another reason for editing might be to add quality intervention points during the operation where optional stops are needed.

Maximizing Cycle Times, Improving Quality and Safety

Extralong "air cuts" (the cutter is moving at feed rate but not touching metal) could be upgraded (maximized by longer rapid travel moves) at this point but they are more likely to be seen in production and edited at the controller.

Downloading is Not Direct Numerical Control (DNC)

There is some confusion about the terms downloading and DNC:

Downloading If the file is small enough to completely fit the control storage capacity, the program is downloaded, PC to controller, in one shot.

DNC If it does not fit, the file must be broken into packets by the host computer, then sent as needed; that action is DNC.

Uploading When data is transferred in reverse, from controller to host PC, it is uploaded.

Storing and Downloading the Program

File Storing This phase is similar to other aspects of data management on PCs. The file is given a name and stored in the hard drive and usually sent to some backup media for large, critical programs.

Downloading—Data Transfer Files can be hand carried, via disk, to the controller, but this is less common than hard-wired, serial connection between the CAM system and the CNC controller.

The Handshake Download/upload and DNC communications are sent over a serial communication protocol much the same as logging onto the Internet. To create a connection, between two micro-processors, they must first negotiate how they are to proceed, called a **handshake.** The handshake may require a first action by the operator at the listening device, the CNC controller, or at the host PC, depending on the direction the data is moving. Then a second action at the talking device triggers the handshake and transfer. On systems set up by a single supplier of both CAM and the controller, the handshake can be automatic with no walking between.

UNIT 25-2 REVIEW

Review the Key Terms

Dynamic rotation
The CAD/CAM ability to "grasp" the geometry and rotate it to any view perspective.

Handshake
The process that establishes two-way communication between PCs and controllers.

Island
In CNC work, this is a closed shape within another shape, around which the cutter must go.

Island avoidance
The ability of a CAM software to machine around an island.

Postprocessor
A machine-specific filter that readies a program for a given machine control.

Replay the Key Points

- ◆ The CAM process performs these four tasks: geometry management, toolpath creation, graphic evaluation of programs, and program data management.

Figure 25-42 *After importing the NC file, set up the tool display parameters as shown.*

◆ The Mastercam process is: (1) select toolpath from main menu; (2) pick the method for element selection, then click entities to be cut for one segment; (3) apply machining conditions and parameters machining conditions for that set of elements—creates one program segment—for example, the pocket, or the drilled hole in our example; (4) create all remaining machining toolpath segments; (5) postprocess the intermediate file; and (6) evaluate the toolpath with backplot function, this can be done here or at any time after step 3 linked together.

Respond

1. Try It Open your Metacut Utilities software on your Machining Technology student disk then:

> Load Sect25-toolpath.NC file
>
> Define tools as per Fig. 25-42
>
> Right-click on Tool in Tool Window
>
> Fill in:
>
> > Diameter D
> >
> > Flute length F
> >
> > Overall length L
>
> Leave tool holder parameters blank—upper parameters.
>
> Generate the solid per Fig. 25-43.

Figure 25-43 *Set up a solids view window to see the cyber-part clearly displayed.*

Click Icon on upper tool bar—under cursor in Fig. 25-43.

2. Briefly identify and describe the four CAM tasks.

3. At what time may a program be post-processed using the Mastercam methods?

4. Why must a software have machine-specific templates (postprocessors)?

5. What kinds of items might be edited into a program?

UNIT 25-3

Understanding Part Models (Introductory)

Introduction: There are actually three forms in which part geometry might be drawn: *wireframes, surfaces, and solid models.* Unit 25-3 is designed to give a sense of when each is right and when it's not suited to the job. It's here as an introduction to solid model part geometry. But this subject requires in-depth study of CAD/CAM, surfaces, then solids. It's beyond the scope of this textbook. However, it's a subject we could not end our CAM section without discussing.

KEY TERMS

Normal vector
A line perpendicular to a tangent plane on a curved surface. Cutting complex surfaces, CAM software maintains the cutter tip at that normal angle to the surface.

Solid models
A three-dimensional representation of the entire part including all features internal and external.

Surfaces
A two-dimensional irregular surface used to define warped or shaped planes.

Watertight geometry
A CAM term that symbolizes no gaps in solid models or in surfaces.

Comparing Part Geometry

Wireframe Geometry

This is the kind we've been using in this section. Wireframes are well suited to making parts with basic feature characteristics or

Figure 25-44 *This geometry surface was created by rotating the yellow curved lines 180° around the green axis.*

where the existing CAD drawing already is in wireframe form and it produces good parts. Unless a radical design revision is looming ahead, it's doubtful that it will be redrawn as a solid model, which means wireframes are going to remain popular for a long time to come. However, if they are converted to solids, several big advantages are gained even for simple shapes. We'll explore them in just a bit.

Surfaces are two-dimensional representations of a warped, shaped, or bent plane. They define the walls of shapes like a Coke® bottle and many other shapes too (Fig. 25-44). Even though they are warped, they are truly 2-D—without thickness and thus they limit how much intelligence the CAM software can apply to the programming process. Surfaces do have a definite place in CNC programming, however.

Some supercomplex shapes, such as a coons patch (a warped or wavy surface created by blending three projection lines into one complex surface—see Mastercam Help). You'll learn about them in a formal CAD or CAM course (Fig. 25-45).

Figure 25-45 *A coons patch is generated by blending three lines into one surface.*

Solid Model Part Geometry

Today **solid model** part geometry is becoming very popular. Correctly drawn with CNC machining in mind, a solid part definition brings several advantages that must be understood.

Rapid Evolution Today's consumers demand designs that are visually pleasing, ergonomic, and innovative, all of which challenge us to machine ever-more complex shapes. For example, think of the evolution of the cell phone. Today, it's more than features and function that markets them, it's also very much the way they look! The new model can't just be better, it must also look new!

Using wireframes and surfaces to define and produce rapidly evolving products like stylish phones is far too slow to compete in the global arena. To thrive, or to even survive, manufacturers must turn to methods and software that produce the wild shapes in the least time with the highest guaranteed quality. Using solid model geometry meets those challenges.

While solids are on the edge of technology, the ability to design with them isn't new. Previously, however, this power tool was available only to large companies with top-end design software running on giant computers. Today, however, the ability to define, create, and manage solid geometry has come to the small shop with a good desktop PC. The hardware is affordable, as is the software.

What Is a Solid Model? A solid model is a part that is fully defined from every view perspective. One begins the drawing as though carving pieces away from a defined block or bar. Other features may be developed and seamlessly added to the block as well as one would with modeling clay. When finished,

Figure 25-46 *The normal vectors (small violet arrows) indicate in which direction a cutter must contact the surface to be tangent at this point.*

Figure 25-47 *This connecting rod began as a rectangular solid billet.*

it's virtually the whole part in data form. The whole part can be manipulated because it's defined with formulae.

The trend toward solid models will continue to expand with new CAD drawings (Fig. 25-46). Today, once a solid has been drawn, the orthographic views can be generated from it with little extra work for the drafter.

True Solids Bring Flexibility and Control The formulae defining a solid are graphically friendly without gaps in data. It's almost like stretching a math skin over the wireframe of the part. The designer has full control of the part definition, thus lessening the possibility of a misinterpretation as compared to a wireframe. Solid data can be rendered to produce a photo-like realistic image of the part before production of a prototype model.

The solid can be used in conjunction with a coordinate measuring machine to perform a quality control check of the manufactured part. CMMs can import the part definition file if they have compatible translators. Importing the whole part becomes critical for scanning technology where the CMM sweeps its probe over the entire surface. Until solid models came on the scene, we had no way to truly inspect complex designs other than by comparing them to a master model shape we carved out. Scanning concepts will be explained in Section 28, CMMs.

Toolpaths from Solids It's important to understand in creating a 3-D toolpath that the key to success is **watertight geometry** (no gaps in the data). Creating a 3-D toolpath could be compared to pouring water over the part to be machined. The path the tool takes needs to flow evenly over the part without leaking into any small gaps. At times, you need a geometry container to keep the tool where you want it, and if a hole exists in the part, you may need to plug it temporarily.

Solid Model Advantages When constructed correctly, a solid is usually simpler to machine, especially for five-axis milling work where every micron of the part surface is known to the computer, thus it's easy to keep a cutter tangent to the surface, known as the **normal vector** to the surface (Fig. 25-47).

Mentioned previously, solids are graphic-friendly, which helps if the file itself is incomplete or there are gaps in data (Fig. 25-48). The toolpath can flow over the surface as long as it has programmer approval to do so. To accommodate changing contours, the programmer can enter a cutter path deviation tolerance to the model and a maximum step over value, which allows the software to compute the best step-over for every curve, as opposed to a constant step-over value in both directions for wire work. This makes a very efficient and smooth shape.

Conclusion

Solid model part geometry is the direction CAD/CAM/CNC work is taking. It brings many advantages and new capabilities to the

Figure 25-48 *Created using solid models, these two mold halves fit together perfectly.*

workplace. However, it will take time for it to dominate industry. As it gains momentum, as a student, you would be wise to become fluent in this technology trend.

UNIT 25-3 REVIEW

Review the Key Terms

Normal vector
A line perpendicular to a tangent plane on a curved surface. Cutting complex surfaces, CAM software maintains the cutter tip at that normal angle to the surface.

Solid models
A three-dimensional representation of the entire part including all features internal and external.

Surfaces
A two-dimensional irregular surface used to define warped or shaped planes.

Watertight geometry
A CAM term that symbolizes no gaps in solid models or in surfaces.

Replay the Key Points

- Solid model part geometry automates the machining process.
- It results in shorter programs and better finishes.
- Solid models are made by removing material from a defined billet and by adding to it.
- Solid models are mathematical representations of the entire part.

Respond

1. When is a solid model *not* the right choice? With what kind of part definition?
2. If no wireframe geometry exists, and the part is simple, would it be better to draw the part in solids or wireframe? Explain.
3. Describe and compare a solid model part geometry to wireframe.
4. Describe how a solid is drawn.
5. List the three kinds of part geometry models.

Section 25 Review

Unit 25-1

Learning your first CAD and CAM software is fun and easy. Starting off that way was the goal of Section 25. Using familiar point-click procedures combined with built-in help menus and tutorials, shorten and demystify the rest of the learning process. With more training and adequate practice, you'll be up and running in no time at all.

The real challenge happens a few years (or months) from now, when it's time to adapt to the next generation of software, and especially newer generations after that. We're not talking about upgrades but rather the whole new products that will compete for and take the market. Time to switch software, what a pain that can be! By then, you'll be totally at ease with "your" program. It's like learning to speak a new language.

Thus far in my career, I've adapted to five completely different word processing programs, three totally new CAM software programs, two different computer operating systems, and as I write this, I'm trying to force myself to learn yet a new CAD program. (But not until I finish this book!) You see, the problem isn't that the new program is more difficult. On the contrary, they're written to be ever-more friendly. The real problem is a combination of letting go and the anxiety of lost productivity during the transition.

Why mention this here? Because readapting over and again is part of our technical world.

Expect it from the start. It's easy to hold onto the old, saying "it still does the job, there's really no need to change," but we all know that the inability to change is what happened to the dinosaurs.

Unit 25-2

OK, back to more earthly subjects. In Units 25-1 and 25-2, we didn't teach you to use Mastercam, what we did was introduce the four tasks: geometry acquisition, toolpath selection, path evaluation, and program management.

Now, it's time to make a full study of your chosen CAM software. Also I wouldn't consider my education complete without a class in CAD as well. Today, in manufacturing, having the ability to use CAD is similar to keyboarding skills in the everyday world—it's basic.

Unit 25-3

At the Westech Productivity Expo, in the last 2 years, I've seen new CAM software demonstrating near miracles in how they can take a solid model part definition from design to program. The process requires nearly no programmer intervention from cut sequences, cutting tool selection, feed rates, and cutting speeds. No kidding, this technology is approaching the point where we show the software the part model, then stand back! Stay tuned to this technology.

Questions and Problems

Problems 1 through 11 are Critical Thinking Challenges.
Assume we will be making 25 parts using the finished program of Unit 25-2.

1. As promised, the program we created in Unit 25-2, made the part, proven with Metacut Utilities (Sect25-Final2.NC). But it had a serious inefficiency problem. What is it? *Compare your answer to mine, before proceeding to Question 2.*

2. How might you make toolpath selections to solve the air cuts?

3. There are potential edits that might fix the inefficient cuts. Discuss the pros and cons of editing the air cuts out of the program.

4. What is program associativity in a CAM software?

5. Using the associativity property of Mastercam, could you edit or change the geometry to solve these inefficient moves?

6. When we put the program into Metacut View, the part looks to be wrong (Fig. 25-43). Is it? What can be done?

7. Is machining the blocks to their final size of 3.85 by 5.25 in., the best job plan before placing them in the vise? *Again, compare your answer to this question to mine before proceeding.*

8. Is there another way we could plan this job to avoid the problem associated with using finished dimension material blocks?

9. What might be done to lock tool comp onto the part geometry better?

10. When machining the pocket, how does the cutter enter the work? Is there a better way? Name two improvements that could be made to the pocket segment. (*Hint:* See Fig. 25-36, look for an unchecked box [v] that might improve the program.)

11. In view of Question 10, what characteristic must the 0.375 endmill have to do the job?

12. Identify and describe the kind of elements we use in CAD/CAM work as compared to everyday screen graphics.

13. A. When is it best to use a universal graphic format translator?

    B. When is it necessary?

14. Due to production problems, there is a jam-up at a given CNC mill. You need to move the job to another machine. Using the Mastercam process, what must you do to rewrite a program originally compiled for a Fanuc CNC mill and put it on a Haas?

15. After we improved and evaluated the Section 25 part program in Metacut View, what additional steps would be necessary to ensure it would make a perfect part on the machine?

16. Which part geometry model automates much of the CAD/CAM/CNC process?

17. Describe a solid model part geometry.

18. True or false? All parts can be drawn in solids but not all need to be.

19. Surfaces are 2-D or 3-D part geometry?

20. Upon what kind of computer equipment can solid model part geometry be drawn today?

Section 25 Answers

ANSWERS 25-1

1. *Universal system* One work station can program any number of different CNC controllers. *Speedy* A program may be written in a fraction of the time for writing codes. *Writes shapes impossible otherwise* Using line and arc approximations, Cam can create shapes that cannot be defined using standard written commands. *Effortlessly solve math eliminating potential errors* If the line intersections, angles, and lengths exist on the screen, they are known to the computer. *Verification tool* The ability to see the toolpath sequence during its creation and during final evaluation. *Existing CAD drawings as geometry templates* The geometry template already exists, a real time saver for complex parts! *PC environment* The program is created away from the noise and interruptions of the work area using familiar operating systems.

2. Bulk data. Add more ram to the controller to hold the large program or send it DNC—as needed from the PC host to the CNC control.

3. Transform
   Translate
   Point to Point

4. Put into a graphic format supported by the CAM software or into a universal translation form. Details, notes, and dimensions should be removed from the drawing before exporting it.

5. The must be vector entities, having a start point in three-space, a direction, and an end point, so they can be manipulated.

ANSWERS 25-2

1. Per views in Figs. 25-42 and 25-43. If you need help, first rescan the tutorial, then if no luck, ask your instructor.

2. Geometry management, toolpath creation, graphic evaluation of programs, program data management

3. After element selection in the toolpath and generation of an NCI file.

4. To be able to customize any program to any machine control

5. Tool types and operator notes such as clamp changes

ANSWERS 25-3

1. When part geometry is simple and the wireframe already exists. Or for very complex shapes such as coons patches.

2. Solids because it automates the process even for simple shapes

3. A wireframe has data gaps—cutter step-overs must be interpolated between data lines. A solid is a completely sealed definition of the entire part.

4. Starting with a defined billet, features are hewn away or added to the shape.

5. Wireframe, surfaces, and solids

Answers to Section Review Questions

1. On the outside contour, the straight lines between corner radii, are theoretically air cuts, at feed rate but not touching metal. At least the roughing passes are, the finish pass might just graze the premachined blocks.

2. Best idea—do each corner individually as a contour without straight-line links between. That way only one roughing pass would be required on the small 0.5-in.-radius corners. As it is now, the cutter doesn't even touch metal anywhere other than the 1.5-in. radius corner until the final rough pass.

3. Code editing isn't a terribly good idea due to many changes and the chance for an error, but it would work. Edit rapid travel moves into the straight-line segments to change G01 to G00. No need to change back to feed rate, as the G02 codes on the corners will bring the movement back to the programmed rate.

4. It allows the user to change an element in the geometry and the NCI and NC program then automatically updates.

5. No. This is not a geometry problem. It's a planning and toolpath selection problem.

6. The problem is my setup of Metacut, not the program. Enter the correct raw material size. Metacut assumed the block was larger than the part. It isn't and needs to be changed.

7. The problem is if any variation in the block size will become a mismatch on the far side from datums A and B. The datum sides will be fine as they locate reliably.

8. Perhaps it would be better to square up datums A and B and then finish the 3.85 × 5.25 dims

after machining the profile. This would require new planning with a soft jaw location on the finished profile upside down.

9. A ramp-on line that brings the cutter toward the material close to the intended path.

10. The cutter plunges straight in. That's no big deal as it's 2024 aluminum, however, a spiral or ramp plunge would do a better job, see the check boxes [v]. Also spiraling from the pocket center outward, would mean better finished cuts, climbing off the final excess.

11. Center cutting

12. Vectors with descriptions that can be manipulated

13. A. When there's any doubt that all transferred details and entities might not be understood by the importing software.
    B. When the importing software doesn't support the native file format of the exporting software.

14. Change postprocessors to the Haas mill, then repost the NCI file.

15. Various dry run tests on the machine—we've done all we can on screen.

16. Solid models because they are complete math definitions of the part.

17. A full math representation of the entire part with all features.

18. False

19. Surfaces are 2-D geometry.

20. A typical PC

Section 26

Advancing Technology

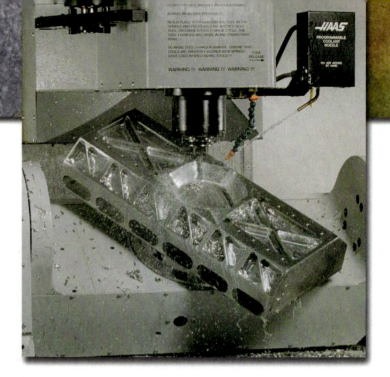

Introduction

Section 26 explores technologies that have changed our trade and continue to do so. Two units discuss advanced approaches to standard chip making and four explore some of the unusual ways metal is shaped. But first, we need to discuss two issues confronting modern machinists. The first is *how a craftsperson stays current in an ever-accelerating technical world.*

More than computerization, every aspect of engineering and manufacturing is evolving and it's accelerating as technologies multiply on each other. We must apply these new ideas when they improve quality and competitive edge.

Staying informed is easy here in training where information surrounds you, but it will not be so simple once your career begins. So how does one keep knowledge current? There are several ways.

Trade Publications

A time-tested solution comes in trade magazines and newsletters delivered to your company or home. Some are free, others are by subscription. Most employers subscribe because it's more than smart for them to stay ahead. It's not possible to experiment with all the new ideas ourselves, so we rely on articles contributed by other professionals to help us sort out the good ones.

Trade Shows

At the local, regional, and the national levels, trade shows are also excellent resources for the latest products and applications. On the inner leaf of this book there's a panorama of the giant Westech Manufacturing Exposition in Los Angeles, sponsored by the SME (Society of Manufacturing Engineers). Every 2 years, the IMTS, an even bigger manufacturing expo, comes to Chicago (International Machine Tool Show). It can't be over-stated how much there is to learn there! Here are a few *Trade Tips* on how to get the most from the experience.

Plan Your Experience Before stepping into the first pavilion, sit down with the program and plan out your day.

Plan It To Be a Two-Day Experience, If You Can Do it like a machining job, on the first day do the roughing then return to finish with those that show the most interest for your future.

Subscribe to Trade Publications Trade magazines don't want just anybody getting free subscriptions, but they assume being at the big shows is a sign of involvement and seriousness. Subscribe to publications and bulletins there.

Plan to Visit the Little Booths Here's my best tip: don't ignore the "little guys." While many of the main floor exhibits are not to be missed, many budding technologies and ideas are being shown in the small exhibits.

Ask Questions Don't just look, ask questions and dig deeper into their technology. You'll be amazed how much there is to learn and how many promotional items they have for visitors showing interest.

Technical Application Data and Seminars

Supplier technical bulletins and catalogs are two more ways to stay tuned in. They are loaded with free information, even though they are usually biased toward their product.

Stay in Touch with Your Tech School

Returning to the local tech college for upgrades is another way to keep knowledge current. They often sponsor short courses and seminars. Some are fast-paced training put on by suppliers, others are specific to an individual company's needs. Encourage your employer to sponsor or even host them.

Use the World Wide Web

It's fitting that this technology is a solution for staying informed about technology. As you already know, the Internet is a huge resource. Not only can one search out and share new ideas, it's possible at many technosites to be placed on mailing lists and receive periodical e-news.

The companion question in Section 26 is, how does one know when a new technology has moved from the experimental stage to the enabling arena, and that it's the right *time to adopt it?*

Timing based on facts is everything for purchasing and implementing a new technology! Jumping in too soon can be a costly waste of time, as the bugs are worked out or another takes its place almost immediately. Every new idea begins as an experiment and occasionally one will be launched a bit too soon. It's not a good idea to take the bait too early.

But waiting until competition forces the adoption is an even bigger mistake! In today's fast-paced world, catching up is often impossible.

To illustrate the point, all but one of the technologies in Section 26 were once in the wait-and-see category. First, we balked at the thought of cutting metal with pulses of electricity. Then came water and light as metal cutters? But then, a few progressive shops gave them a go, while most waited. In time, they became accepted practices.

What's next? Well, that's up to you to find out before the competition does. Here's the goals of Section 26, to get things going.

Objectives

Beyond Chip Making—Three Different Ways to Remove Metal

Introduction: In making a product, designers and planners can choose from a long and growing list of possible methods to cut, shape, and form their product. An informed machinist should build a working vocabulary of these processes, and know where they fit within the manufacturing loop. While there are many more methods, the selected three in Unit 26-1 are the ones our planning group felt might enhance career possibilities. They cut material using three very different media—plain tap water, high-energy light, and electric current. Each brings a unique advantage to the shop and a right and wrong application:

◆ Waterjet cutting
◆ Laser cutting and surface marking/finishing.
◆ Electrical discharge machining (EDM)

KEY TERMS

WATERJET CUTTING

Closed-loop recovery
Reusing water where contamination or scarcity are issues.

Heat-affected zone (HAZ)
Disturbed and modified material immediately surrounding many cutting processes, not present with waterjet cutting.

Intensifier Pump
The hydraulic piston that creates ultrahigh pressure.

Orifice
The tiny hole in the nozzle that focuses the energy.

Ultrahigh pressure
Pressure above 35,000 PSI up to 60,000 PSI.

LASER CUTTING

Emitter
Atoms, molecules, ions, or semiconducting crystals that produce light as they return down from an excited state to a normal energy level.

Exciter
Triggers the process by pumping energy into the medium thus raising the energy state of the emitter.

Gain medium/mirrors
A material that either supports or forms the emitter and will not release the beam until it is perfectly directed.

EDM

Current
The amount of energy flowing in the arc, measured in amperes.

Detritus
The EDM waste composed of mostly cooled work particles but also of electrode and destroyed fluid especially with oils.

Dielectric (fluid)
The insulator-path between the electrode and work. The absolutely essential component of EDM—water or oil.

Duty cycle
The ratio of on time compared to overall cycle time.

Heat-affected zone (HAZ)
The thin, undesirable, modified surface that is a by-product of EDM cutting.

Ion/ion path
An energy state whereby electrons are shifted and become electrically biased, resulting in a greater tendency to conduct current.

Overcut
The result of the arc gap beyond the electrode. Overcut can be regulated by varying voltage to control size. Similar to kerf in sawing operations.

Recast layer
The thin surface modification caused by metal refreezing to the parent.

Volts
In EDM, voltages are the potential to bridge a gap. The higher the voltage, the larger the gap it will jump. EDM voltages range from around 300 at open gap to 35 at the maximum cutting action during the on time.

White layer or white metal
The thin upper layer, surface modification of carbonaceous (oil dielectric) or oxidized (water dielectric) metal that was metallic vapor within the plasma bubble when it collapsed. In collapsing, it solidifies and refastens to the parent above the recast layer.

Waterjet Cutting*

The name is waterjet—no hyphen. Originally invented for the lumber industry, waterjet cutting has grown from a whizbang technology to become the fastest expanding

*Material and photographs courtesy of Flow International Corporation, Kent, Washington.

Figure 26-1 *A CNC, vertical waterjet with water absorption tank below.*

specialized machine tool cutting application in the industry (per Flow International, see Fig. 26-1). An *X-Y* CNC directed, vertical waterjet cutting machine can cleanly cut, without burrs or burn marks, nearly anything from soft paper, foam, and baby diapers to extremely hard materials such as titanium, marble, glass, and even ceramics! Cutting is the main use but there are several other ways a waterjet can be used. Using rotating multijets, surface preparation and paint stripping are accomplished too. Every day, the list of applications expands. The jet can also pierce through or to a given depth in some applications. Especially note that waterjets cut without burning the material (author).

The Basics

The first part of the process is to filter plain water in stages. Then it's pressurized from 35,000 to 60,000 pounds per square inch in a set of **intensifier pumps** (Fig. 26-2). The **ultrahigh pressure** water is then delivered through highly specialized plumbing, to be jetted through a tiny **orifice** from 0.003 in. up to 0.018 in. The tiny stream looks like a

Figure 26-2 *The intensifier pump creates the ultrahigh pressure required to cut material.*

Figure 26-3 *A double-head CNC waterjet making large, decorative, metal wagon wheels.*

silver thread when light reflects from it, but make no mistake, it packs a big wallop!

The jet is usually directed by *X-Y* CNC axis drives as seen in Fig. 26-3, a vertical cutting machine, the more common form of waterjet. Or it can be mounted as the terminal implement on a robotic, five-axis arm (*X-Y-Z-A-B*) to cut and pierce nearly anything from flat stock to complex 3-D shapes like auto headliners.

For safety and efficiency, the cutting head contains an instantaneous, off-on valve. Rather than stopping and starting the pump, this valve action is faster and safer. Opened with shop air pressure, should the supply fail, it quickly stops the stream and remains closed.

Adding Abrasives to Increase Cutting Power for Hard Material

When cutting hard material such as metal or glass, an abrasive is injected into the stream only *after* it exits the nozzle, for if we added it before passing even the hardest of nozzles, the precision orifice would be instantly blown out. Injecting the abrasive requires a merging and directing process to not diffuse the water stream's tight focus. This is accomplished using an acceleration tube, acting like a gun barrel, to join the grains to the water without disruption to the stream's focus. Typically, about 1 pound of 80-grit abrasive is consumed per minute in this application (Fig. 26-4).

Pressure Pumps—The Process Heart

There are two ways water is brought up to the level that it can cut material: intensifiers and gear displacement. The piston intensifier is the better in terms of peak pressure and overall life span, but they are more complex and costly as well.

Figure 26-4 *A close-up shows the nozzle, valve, jet stream, and the abrasive induction equipment.*

passing it through a set of precisely fitted gears. Gear pumps have a shorter life span, compared to intensifiers, but they are quick and economical to replace.

Waterjet Advantages

Using either pump method, due to the velocity of the jet, amazingly, nearly zero water is left behind on the product. It stays dry even when cutting absorbent diapers! Another bonus is that no heat is produced, a great advantage over other jet cutting methods. The lack of a **heat-affected zone (HAZ)** means no mechanical stress around the cut, no surface hardening, and no crack propagation.

ShopTalk

Remember the term HAZ, it's a significant comparison item we'll see again in other articles and processes coming up.

Intensifier Pumps

With two or more double-action pumps working in tandem to eliminate flat spots in pressure, water is delivered, surge free, at pressures and volumes varying by the application. A double-action pump puts out pressure on both the outward and inward strokes of the piston, but working by itself, there would be a turnaround stall as the piston reverses. So the second pump is added to pump 90 degrees out of phase from the first. That creates a continuous stream. The pumps work on the intensifier principle:

Primary pressure × Piston area = Secondary pressure × Piston area

So, large hydraulic pistons are powered with moderately high-pressure hydraulic fluid, which in turn pushes a much smaller piston against the water. The pressure gain, or intensification ratio is 20 to 1.

Positive Displacement Gear Pumps

Developing enough pressure to do many jobs (but not all), liquid can also be pressurized by

Multidirectional cutting is possible since the waterjet is simply a superfast stream, it can be directed in any axis as long as an energy-absorbing collector safety barrier is placed beyond the cut, to catch and stop the stream as it exits the cut material. Since the cutting head never actually touches the work, cutter dulling is never an issue. Due to the fast cutting action on softer materials such as sheet rock or paper, a waterjet can be used to slice a moving slab of material at a perfect 90 degrees, without stopping the slab. This is accomplished by moving the cutting head at an angle relative to the material speed, resulting in a square cut "on-the-fly" (Fig. 26-5). The final advantage is the availability of the water and its nonpolluting nature; a bonus, it can even be used over and over if necessary.

Environmental Issues for Waterjets

Since the jet is so tiny, the narrow kerf carries away very little target material *swarf* so cleanup of the tiny particles suspended in the water is simplified and economical before releasing it back to the environment. Further, since an amazingly small volume of

Figure 26-5 *A multijet, paper cutting machine uses waterjets without wetting the paper!*

top to be released. However, for toxic materials or where water is scarce, the stream may be efficiently recycled through the system again and again—called a **closed-loop recovery** system.

Safety Issues

Depending on the type and thickness of material being cut, and pressure used, as much as three-fourths of the original energy may still be in the stream after exiting the cut material. Stream catchers must be placed beyond the work. There are two types: for straight down cutting, a water pool with from 24 to 30 in. depth is required to absorb and dissipate the energy. The second method is a moving container that features a small hole for the jet to enter. It's filled with steel balls that deflect, dissipate, and absorb the excess energy with only 6 in. penetration. Obviously, operator protection is an absolute and the plumbing must be of the highest quality to withstand the pressure and movement of the nozzle head.

Industrial Applications of Waterjets

Due to their versatility and efficiency, waterjets can and have replaced routers, saws, cutting torches, mills drills, and many other processes within the factory. Here's an amazing partial list of places you might see waterjets used to cut material. Flow International Corporation, a world leader in the technology, states that growth has far exceeded even their expectations as new and exciting applications for waterjets arise.

water is actually used in waterjet cutting, it may often be harmlessly returned to the environment as-is as long as the small amount of suspended material is nonpolluting. For light contamination or where abrasive cutting must be used, the water is directed to a settling pool where clean water flows over the

Aerospace: Cutting of expensive metals or metal-plastic fiber and graphite-epoxy composites.

Automotive: Gaskets, plastic upholstery, floor mats, headlines, or any soft interior item.

Building and construction: Building tiles, insulation and wood. Waterjets can also be used to pierce (drill into) and remove concrete or other hard materials for shaping or for rework.

Food Industry: Chicken, frozen pizza, fish fresh, and frozen candy and vegetables (Fig. 26-6).

Job shops: Metal, glass, stone, and plastic.

Figure 26-6 *The tiny jet neatly slices this candy bar without disturbing its structure at all!*

Cutting with Laser Light[†]

Today, beams of high-energy laser light are doing an expanding array of tasks. Within the manufacturing world, lasers are used to heat in exact locations, to etch surfaces, and weld (Figs. 26-7 and 26-8). They also mark parts and cut material, the two applications you might be called on to use as a machinist. Lightweight lasers can be the terminus (end implement) on a five-axis robot arm, where in the auto industry they trim sheet metal after it is formed, and neatly cut other complex shapes.

But, beyond that, there's also a rapidly expanding market for lower-powered lasers. They're used for wildly diverse applications such as surveying, measuring, communications, tech-art displays called laser shows, delicate eye surgery, and marking extremely delicate or valuable objects such as serial numbers on diamonds and many more but they are out of our arena.

Part Marking and Vertical Cutting

These are the two applications we'll investigate here. For both, CNC programs either direct the laser beam across the work surface, or axis drives move the work similar to the vertical waterjet, through the beam.

The highly controllable nature of laser cutting makes it usable in unique applications not possible with other cutting methods. For example, as the tightly focused scalpel in delicate surgery or in the "flying optic" version where mirrors move in three axes to direct the beam thus expanding work envelopes without making the machine bigger. Here, both laser and work do not move, but rather, computer-positioned mirrors direct the energy to the work. Another redirection method is used in part marking where the mirror pivots to project the image to be etched onto nearly any 3-D work shape with a close control of penetration.

The controllability of laser means that different beam generation types with varying outputs can remove from only a fraction of a thousandth of the surface (Fig. 26-9) down to great depths. Varying kerf widths from a few

[†]Text and photos courtesy of Trumpf Incorporated of Farmington, Connecticut, and Synrad Corporation, Washington State.

Figure 26-7 *Computer direction of the moderate power laser, these machines can precisely and permanently mark a wide variety of items.*

thousandths and upward are possible with laser. Different from other thermal processes, due to the high density and small focused spot size, extremely intricate contours may be achieved with close tolerances and exact repeatabilities (Fig. 26-10).

Laser Advantages

Due to the narrowness and intensity of the beam, laser cutting can be quickly accomplished with little heat-affected zone (HAZ) as compared with *conventional* thermal cutting methods (conventional thermal cutting is either oxy-fuel flame or plasma cutting—author). Combining laser cutting and surface etching, designers can produce work not possible with any other method.

No tooling required

Easy to program

High accuracy

Can be projected to any surface image

Excellent sheet utilization (due to narrow swarf)

Quick part delivery

Quiet operation

Fast, permanent marking

KEYPOINT

Since a laser cut is made with a light beam, clamping/holding, work thickness, and turnaround time become nonissues, therefore this process is ideal for just-in-time manufacturing.

What Is a Laser?

An acronym for light amplification by stimulated emission of radiation, a laser beam is nothing more than light but it's created in a special form. In 1960, the first laser shot out from a mirrored artificial ruby, as a pale pink light. The inventor, Theodore Maiman, explained why laser light can tightly pack more useful energy into the beam compared to other sources. There are two reasons:

1. **Uniform Frequency—Purity**
   Lasers emit in one single frequency. Various **emitters** produce different frequencies but the uniformity of each means far more usable energy in one place compared to everyday light. Depending on the emitter, they range from infrared through visible light up to the ultraviolet regions of the

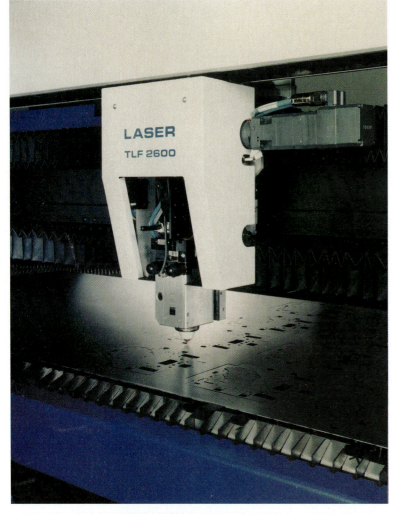

Figure 26-8 *A high-power, industrial CNC vertical laser cutting machine making intricate details in sheet metal components.*

Figure 26-9 *Note the tiny, precision bar code and part number etched upon the surface of this electronic component.*

Figure 26-10 *A five-axis robot moves a laser cutter to trim this complex sheet metal stamping.*

spectrum. In comparison, everyday light, like that from the sun or a flashlight, contains many rainbow-like frequencies. Thus it's difficult to focus its energy into a supertight beam.

2. Uniform Direction

Due to the generating process, all elements of the beam emit along an exact parallel axis. This means far less beam separation as it gets farther from the source—very different from other light sources.

Coherent Light Both properties combine to make laser beams usable as a process. The result is called *coherent* light, it stays closely focused over a long distance.

Laser Power Ratings Manufacturing outputs are rated in watts. Cutting fabric or paper requires 200 to 250 watts while part marking occurs down around 10 to 25 watts and there are literally hundreds of other uses below that, with outputs as low as a fraction of one watt. Cutting metal may require outputs from one or two thousand up to 30,000 watts or more depending on material and thickness.

Three Basic Components to Any Laser (Fig. 26-11)

1. The Exciter

Triggers the process by pumping energy into the gain medium thus raising the energy state of the emitter.

2. The Emitting Entity

Embedded within the gain medium, atoms, molecules, ions or semiconducting crystals first become excited by the pumped in energy, then they produce light to return down from the excited state to a normal energy level.

3. The Gain Medium/Mirrors

A material that either supports or forms the emitter and will not release the beam until it is perfectly directed.

The Four Gain Media Types

- Gas (CO_2 or argon, for example)
- Solid (glass, ruby, semiconducting diode)
- Liquid (with suspended dyes)
- Vapor (He-Cd)

Figure 26-11 *The three components of a laser.*

Two Differences within the Laser Process

Similar to everyday lights that are either iridescent (glowing element) or florescent (glowing gas) due to electrical energy put into the emitting element, the laser process also begins by raising the emitter to a higher energy state (Fig. 26-12). In returning to the normal state, light photons are emitted by the excited element, but the big difference for a laser is they give forth their radiation at a uniform single natural frequency. Everyday lights emit in a range of color frequencies.

The second difference is that the photons cannot escape until they are all pointed in the exact same direction. To ensure uniformly focused beams, photons can pass through a semireflective mirror, but only if they are at exact perpendicularity. Otherwise they bounce back and forth within the medium,

thus colliding with other charged molecules, which in turn begins a cascading effect, emitting further photons. Thus, they collide and bounce within the medium, until reaching exact perpendicularity to escape in a near-perfect beam. This bouncing action is known as a *laser oscillator*.

Light cannot leave the gain medium until it reaches the required angle with the mirror, therefore the parallel nature of the beam.

Environmental and Human Issues

It's vital that users understand that while laser is a form of radiation, it presents no special danger above other high-energy cutting

Laser Output (Radiation)

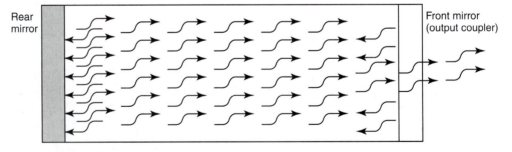

Laser beam exits resonator through an optical coupler
The output optic is 60% reflective 40% transmissive

Figure 26-12 *A laser oscillator features one fully reflective mirror and one at 60 percent reflectivity.*

methods. Other than direct contact with flesh or eyes as concentrated beam heat or ultraviolet light, it has no effect on people or equipment nearby. Laser is just light but it can and will cut and burn.

Since the beam is invisible in most cases, training and operator safeguards are an absolute must. Different types of protection are appropriate based on the frequency, cutting action, and output power. Also, when cutting certain materials, heated vapors and fumes must be recaptured, filtered, or recycled plus latent heat in the product can burn hands, but otherwise laser has no other danger or impact to people or environment.

Electrical Discharge Machining‡

Since its invention, over 50 years ago, this metal cutting technology has expanded outward from hand-fed, broken tap removers into tool and die applications where it has become a central utility. But modern CNC controls and updated power supplies for the cutting action have now made EDM a practical solution in nearly all aspects of production manufacturing from aerospace, electronics, and automotive through medical equipment manufacturing.

Pacing machine tool evolution in general, CNC-driven electrical discharge machines now feature tool changers, palletized part loading and auto-tool-set changing, and auto-slug removers. Some even have adaptive controls that adjust to varying cut conditions and autothreading of wire electrodes. Today, developments are coming fast, wire machines (explained next) have increased their cutting speeds by an amazing 30-fold in the last 10 years alone. Manufacturers see greater removal rates with better surface integrity and improved finishes, coming in the near future.

Why Use EDM?

This process is chosen when the work is too hard to machine conventionally or where the target shape is impossible to produce by other means. Finish is another reason. Modern equipment can produce mirror finishes with repeatabilities well below a thousandth of an inch.

‡Photographs and text courtesy of MC Machinery Division of Mitsubishi America.

Two EDM Processes—Wire Feed and Sinking Electrode

There are actually two distinct ways EDM is used to shape metal (Fig. 26-13). The wire machine, which performs similar to a continuous jewelers band saw (cutting a narrow kerf in any X-Y direction) but using a tightly stretched, slow-moving brass or alloy wire as a cutting electrode. The second is the vertical ram or electrode sinker, which makes pockets and holes by burning downward with a shaped electrode into the material. The sinker machines make use of shaped electrodes that can be easily machined into many complex shapes. The electrode is made to mirror the final shape (reverse of the desired finished shape).

Additionally, by moving laterally with a CNC axis drive and control, many other operations can be achieved with both types: wire or sinker. Since both machine types cut with spark erosion, we'll explore the EDM cutting action before looking at the machines themselves. Turn to Fig. 26-15 to see examples of industrial CNC EDM sinker and Fig. 26-16 for a CNC wire feed EDM.

Two types of electrical discharge machining

Continuous wire electrode | Sinker or solid electrode

Figure 26-13 *The two types of CNC EDM—wire feed and vertical ram/sinker.*

Pulse on

① Electrode is charged

Continuous fluid flush and/or total immersion — Ion path forms
Work opposite charge

② Arc forms on path

Current flows
Melting crater and gas bubble

Off

③ Bubble collapse starts flush

④ Final mechanical flush

Figure 26-14 *The four phases of one EDM arc.*

Figure 26-15 *A CNC-controlled sinker EDM showing the oil tank, CNC control, and vertical ram. Note the tool changer on the left.*

The EDM Process

The "cutting teeth" in this process are millions of tiny, electric arcs leaping between electrode and work, each melting and vaporizing metal to become a microcrater in the parent metal. The arc is an on-off pulse.

A fluid component between the work and electrode is necessary to not only cool but to control the action and flush away particles removed. This fluid, called the **dielectric,** performs several duties beyond cooling. Each is a separate but critical task. Truly, without the fluid, no controlled EDM could occur. Early tap burners used plain air as the dielectric material supporting the arc between the electrode and workpiece. But due to air's inconsistent control of the arc, they were incapable of accuracy.

One Arc Through One Cycle   There are four distinct phases to a single on-off cycle. Let's examine each, Fig. 26-14 shows them in sequence. This cycle occurs thousands of times per second, over an area that looks to be a spark curtain to the eye. For simplicity, we'll envision it as a single arc.

KEYPOINT

Pay close attention to the duties performed by the fluid.

The On—Cut Time

Phase 1—An *Ion Path* Forms   During the on time **voltage** potential builds between the electrode and work. Quickly, a charge builds between the electrode and grounded work, and atoms in the dielectric fluid within the gap between "line up electrically" in what is called an *ion path* or *flux tube*. Each atom, in that shortest path, shifts an electron, which causes a bias to conduct electricity.

No arc has started yet—just the path in the fluid, but within a nanosecond, the arc will follow the path. The fluid is supporting the path but it's also the insulator preventing premature arc formation. Without it, the arc would be unstable and jump uncontrollably as it does in the air supported tap burner.

The definition of a *dielectric* is a material that resists electrical **current** flow to a given point but through which an electric field may pass—thus holding off the arc until at a useful high-energy state.

Now, we've covered two of the three fluid duties, it supports the path and it acts as an

Figure 26-16 *A wire EDM machine showing the water tank and CNC control.*

insulator, stabilizing the process to avoid premature arc formation.

Phase 2—The Arc Then, the arc forms and current flows, doing the real work of melting one tiny crater. Remember, only a fraction of a second has elapsed since the on cycle began. During the spark time, the target metal first vaporizes into a gas bubble. Then as the crater forms, due to the expanding spherical crater surface absorbing the energy of the arc, the metal melts rather than vaporizing. Remember that aspect, we'll discuss it next. At this stage, both molten metal lines the crater and some metal exists as vapor in a gas bubble surrounding the arc/crater.

Now two things occur that begin to progressively slow the action. First, particles of not only the work, but unfortunately, destroyed electrode and burned fluid, build up in the gap. As the crater expands outward from the original point of contact, the arc become less focused and stable. Second, the energy is being spread over an expanding area as the crater becomes larger, therefore temperature drops to a minimum useful level. So, due to both factors, at a given portion of the cycle, the current must stop to flush the area of particles to ready it for the next arc within fresh fluid.

Off Time

Phase 3—Bubble Collapse Now, at a controlled time based on removal rates, desired finish, depth of cut and other factors, the controller stops the arc for a microsecond. Here, the plasma-gas bubble (highly charged gas) formed from the intense heat, collapses, which in turn forcefully pulls fresh fluid into the crater area. At this phase, the molten metal that wasn't flushed away, freezes back to the parent. Its chemical makeup is not changed but its structure has been modified. We'll discuss this again as the **recast layer.**

Phase 4—Mechanical Flush The bubble collapse starts the flush but a forced continuation is needed to completely sweep away particles and spent fluid. So, during off time flushing continues with fluid being pumped mechanically through the gap. The removed swarf is sometimes called **detritus** (total waste particles) in the EDM process. Contaminated dielectric fluid will be filtered for particles larger than 1 to 5 microns depending on the application then, sent back to be used again.

Phase 4 is complete. Now, with the gap swept clean, a single cycle is over, the control switches on again and a new ion path forms, many times per second.

Control Options and Factors

Since removal only occurs during the ON time, the objective is to extend the burn period relative to the overall cycle time, known as the **duty cycle.** But faster removal means big craters and rougher finishes. Other factors, such as depth of cut with decreased flushing ability come into play too.

KEYPOINT

As with all cutting operations, there are trade-offs when adjusting EDM cutting: operators must choose from varying balances of removal rate, accuracy, finish, and tool wear. Sounds familiar doesn't it?

Overcut and Electrode Wear

Because there must exist a gap between work and electrode both for flushing and to create voltage potential, the electrode must be smaller than the expected sinker pocket or wire kerf. A zero gap for either process would become a dead short! One prime function of the control, is to sense and regulate the advancement of the electrode to maintain the correct gap.

Overcut is the EDM term applied to the difference between the electrode size and the cut area. It is a factor to be calculated before machining an electrode or selecting a wire. It must also be regulated during operation.

KEYPOINT

Due to the EDM arc length, the resulting cut is larger than the electrode (Fig. 26-17). By varying voltage, amperage, or duty cycle, fine adjustments of the finished size and surface finish can be achieved.

So, in addition to using the original electrode size to create final work size, overcut can be adjusted to fine tune dimensions by varying voltage, amperage, and duty cycle times.

Roughing and Finishing on Sinker EDM Due to the unavoidable erosion of the electrode as well as parent metal, it is often necessary on sinker processes, to use two or more electrodes to rough then finish the work. Since the wire machine continuously spools new electrode wire through the cut, electrode wear becomes a minor factor.

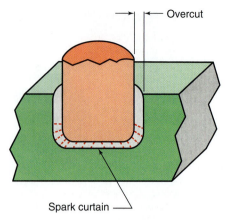

Figure 26-17 *Cavity size is a function of overcut due to the arc gap between the electrode and work.*

However, by misregulation of the wire speed or of the cut speed/power, the wire can be burned through, which brings the operation to a halt until it can be re-threaded again. During wire work, where extreme accuracy of finish is a requirement, a second skim cut is often taken.

Surface Modifications to the Work

Heat-Affected and Recast Zones Due to the nature of the process, two thin surface layer modifications are created and must be expected to varying degrees. The outer layer, known as the **white metal,** is actually material that was removed in the vapor but refastened to the parent as the bubble collapsed. Carbon from the oil dielectric, or oxygen from the water, combines with the metallic vapor. Due to intimate contact with the dielectric fluid, the white metal tends to be hard and carbonaceous on sinker machines and soft and oxidized on wire machines.

Then directly beneath is a second layer known as the recast metal. It was molten as the arc ceased but since it had not been removed, it quickly froze to the solid parent beneath, during the flush. Its physical structure has been affected but not to the degree of the white layer.

Combined, these undesirable surface modifications are called the **HAZ—heat-affected zone.** Both can cause cracking and other mechanical surface problems. But the good news is they can be controlled to varying degrees, depending on the operation. The HAZ can be almost eliminated completely in some cases with skilled machine settings. It can also be removed with secondary operations

Heat-affected zones for EDM

White layer, highly modified (oxidized or carbonaceous)

Recast metal structure modified

Figure 26-18 *The two thin surface modifications result from EDM removal of material.*

such as grinding or honing, where surface modification cannot be tolerated. Research is being conducted into methods of reducing and eliminating these effects. They are more prevalent on roughing cuts, shown in Fig. 26-18.

Two EDM Machine Types

Now, with an understanding of the cut process, let's take a second look at the two machine types: sinkers and wire feed.

Sinker Machines The vertical ram holds and advances the electrode (Fig. 26-19). On many verticals, a CNC control can move it in any number of axes of motion (X, Y common and A, B on fewer machines) to produce shapes beyond vertical pocket sinking.

Special Oil Dielectric is used as the fluid on these machines. It is formulated to be flash retardant and stable but it does smoke to some degree when heated. Its vapors can flash in extreme situations. EPA approved ventilation is a must when using oil flushed EDM equipment.

Carbon and Copper Alloy Electrodes on sinker machines break down into three groups: carbon or tungsten/copper alloys and composites for high-precision die work. The most common electrode material for sinker machines is fine-quality carbon (graphite), which withstands wear, is easily machined, and is relatively low cost. Carbon grain size is an issue, with the finer grain producing better results. Larger grain material is used for roughing electrodes. However, when the electrode must be machined into some thin or complex shape, technicians often switch to pure copper or the stronger copper graphite composite or copper tungsten materials.

Wire Feed Machines Using plain water that has been processed to be electrically neutral (deionized), as the dielectric fluid, and a brass wire as an electrode, these machines also feature advanced CNC controls. With extra *A-B* axis control beyond *X-Y* motion, the wire angle can tilt with respect to the work, as it profile cuts. This results in a continuous draft angle effect shown in Fig. 26-20. The tilting axis machines can cut a cone from flat bar-stock, for example.

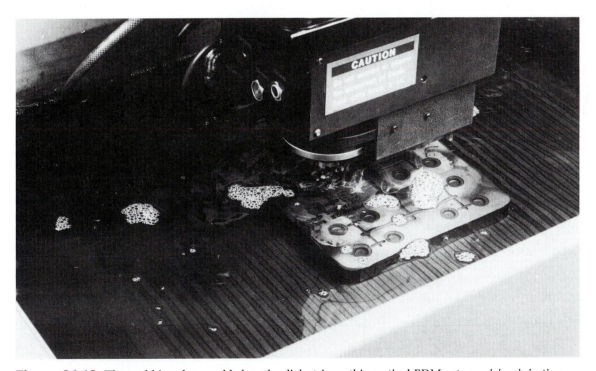

Figure 26-19 *The mold is submersed below the dielectric as this vertical EDM cuts precision injection mold cavities and plastic distribution runners.*

Draft angle with CNC four-axis wire EDM

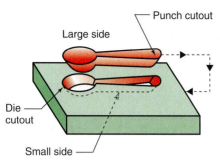

- Punch cutout
- Large side
- Die cutout
- Small side

Figure 26-20 *Continuously tilting the wire to create a draft angle, two parts can be made at the same time and can be sheared together for a perfect fit between punch and die!*

In the example, a die set is being made to blank (cut out by punching) a flat metal spoon. Both the die and punch are EDM hewn from the same piece of prehardened metal. The final operation will be to reverse their order, such that the wide part of the punch faces the narrow part of the die, then forcefully shear them together for a near perfect fit. The draft is carefully precalculated such that there is a minimum amount of interference between the components thus they shear easily. The advantage; the punch and die are made from a single piece of metal, in a single operation. The shearing removes the HAZ. Since EDM can cut hard metal as well as soft, EDM was the perfect technology to make these tiny detail punches. These blades (Fig. 26-21) were wire EDM cut from prehardened metal. They would have been very difficult to produce (or taken an impossibly long time) using any other process.

Wire Electrodes Thin brass or brass alloy wire is used as the conductive electrode. Similar to a band saw, the cut must start at the edge of the work or have a predrilled starter hole through which the wire is threaded for internal cuts. Modern EDM machines have now been developed that will thread their own wire, unattended, thus putting them into the production line making use of robotic part changing and cell technology.

To control wire tension and exact size, the brass wire is stretched tightly between two tension rolls and pulled through a gem quality, sizing die before entering the cut. This ensures electrode uniformity. The wire is a one-time-only cutter. Therefore electrode wear becomes a near nonissue. After exiting below the cut area, it is either wound up on a take-up spool or cut into short sections and put into a

Figure 26-21 *Nearly impossible to make by other means, this complex automotive gasket punch set was made with a wire feed EDM machine.*

bin, for recycling. Stratified wire with longer lasting outer shells, and special alloys within to withstand tensioning, are available to improve finish, accuracy, and performance.

UNIT 26-1 REVIEW

Review the Key Terms

WATERJET CUTTING

Closed-loop recovery
Reusing water where contamination or scarcity are issues.

Heat-affected zone (HAZ)
Disturbed and modified material immediately surrounding many cutting processes, not present with waterjet cutting.

Intensifier pump
The hydraulic piston that creates ultrahigh pressure.

Orifice
The tiny hole in the nozzle that focuses the energy.

Ultrahigh pressure
Pressure above 35,000 PSI up to 60,000 PSI.

LASER CUTTING

Current
The amount of energy flowing in the arc, measured in amperes.

Detritus
The EDM waste composed of mostly cooled work particles but also of electrode and destroyed fluid especially with oils.

Dielectric (fluid)
The insulator-path between the electrode and work. The absolutely essential component of EDM—water or oil.

Duty cycle
The ratio of on time compared to overall cycle time.

Electrical distance machining (EDM)
Removing conductive metal by pulses of electrical energy.

Emitter
Atoms, molecules, ions, or semiconducting crystals that produce light as they return down from an excited state to a normal energy level.

Exciter
Triggers the process by pumping energy into the medium thus raising the energy state of the emitter.

Gain medium/mirrors
A material that either supports or forms the emitter and will not release the beam until it is perfectly directed.

Heat-affected zone (HAZ)
The thin, undesirable, modified surface that is a by-product of EDM cutting.

Ion/ion path
An energy state whereby electrons are shifted and become electrically biased, resulting in a greater tendency to conduct current.

Overcut
The result of the arc gap beyond the electrode. Overcut can be regulated by varying voltage to control size. Similar to kerf in sawing operations.

Recast layer
The thin surface modification caused by metal refreezing to the parent.

Volts
In EDM, voltages are the potential to bridge a gap. The higher the voltage, the larger the gap it will jump. EDM voltages range from around 300 at open gap to 35 at the maximum cutting action during the on time.

White layer or white metal
The thin upper layer, surface modification of carbonaceous (oil dielectric) or oxidized (water dielectric) metal that was metallic vapor within the plasma bubble when it collapsed. In collapsing, it solidifies and refastens to the parent above the recast layer.

Replay the Key Points

- There are six major components in a waterjet system: water filtration, pump, special plumbing, nozzle head, the motion equipment, catchers and waste equipment.
- A waterjet can cut nearly anything soft to hard.
- There is no HAZ with waterjet cutting.
- Laser is just light so it poses no special danger over other high heat processes other than ultraviolet light for some and invisibility. For these, specialized safety training and protection is critical.
- Due to the narrowness and intensity of the beam, laser cutting can be quickly accomplished with little heat-affected zone (the HAZ) as compared with conventional thermal cutting methods.
- Some designs can only be produced by combined laser cutting and etching.
- EDM removes by melting then flushing small craters into the parent metal.
- Generally speaking, low frequencies are used for roughing and heavy metal removal.
- Any metal regardless of hardness may be cut by the EDM process. There are five major components in an EDM cutting

tool: *electrode*—brass wire or solid carbon/copper alloy; *power supply*—creates and controls the on-off, energy; *dielectric fluid*—water for wire machines and oil for sinkers—cools, paths the arc, and flushes away waste; *drive mechanism (machine tool)*—CNC drives the electrode causing vertical action and sideways cutting; *filter system*—cleaning the dielectric fluid in closed-loop recovery.

Respond

1. Of the three processes, laser, waterjet, and EDM, which affects the material the least—explain?

2. Define electrical discharge machining.

3. Why must an EDM pulse the arc rather than make them continuous?

4. True or false? Because it uses only plain water as the cutting medium, waterjet cutting is far less dangerous compared to laser. If it's false, what makes it true?

5. There are two kinds of EDM; describe them briefly.

UNIT 26-2

New Information About Chip Making

Introduction: In times past, the ceiling for metal removal was set by axis drives and spindle RPM, both being limited due to central

Figure 26-22 *Cutters capable of HSM must be highly engineered.*

processor speeds. But today, those challenges are over, at least until leapfrog technology enables the machines to once again go faster than the best cutters can withstand. Today machining frontiers lie in advanced cutter technology. These are the cutters that can withstand the heat and force, to take full advantage of lightning quick machines.

Advancing this aspect of machining, here too computers come into play. Today, we're not bound to experiment with cutter shapes to see how they perform. We can do early research before making the cutter, using engineering models in powerful fast simulation programs, to analyze how the metal flows over a given tool geometry. We can "see" without trying every possible cutter rake shape and chip flow control and what a slight modification to the tool's geometry might make in chip flow, finish, efficiency, and life expectancy (Fig. 26-22).

KEYPOINT

After 200 years of metal removal by cutting, we are learning more about removing chips than at any time in the past other than when carbide was introduced.

Combining fast machines and the new cutters, just how much metal can be removed accurately, reliably, and safely within the shortest possible time? Let's see what the experts have to say. First, Iscar Metals Cutting Tools will show us just how much—you won't believe what Iscar's research has determined! Then another major tool supplier, Kennametal Inc., will show us the tools that are performing these near-miracles today and what's new for tomorrow.

KEY TERMS

Aluminum oxide (Al$_2$O$_3$)
A coating applied to carbide insert tooling to improve surface hardness and resistance to friction.

Chip thinning
The effect of not taking a cutter swarf as wide as the cutter diameter. The chip is thinner than that for which the feed rate was calculated to produce. The feed may be increased by a factor to compensate. Also known as the tangential effect.

Coating
The thin layer or layers applied by CVD (chemical vapor deposition) or PVD (physical vapor

deposition) of materials harder than the base carbide.

Dynamic balance
Rotating cutter, holder combinations at or above 7,000 RPM then adjusting balance to within two grams.

High-speed machining
Mostly on milling machines, at spindle speeds above 10,000 RPM and at table feeds above 1,000 in. per minute.

Plastic flow
The desirable threshold whereby, during chip formation, work material loses resistance to deformation flow due to internal friction.

Polycrystaline diamond (PCD)
A modern material used either in chip or coating form to improve cutter hardness and friction resistance.

Radial engagement
The amount of side engagement of a cutter compared to its full-diameter swarf.

Substrate
The chemically or physically deposited layers applied to cutting tools to achieve a certain effect. See coatings.

Titanium nitride (TiN)
A coating commonly applied to carbide inserts to improve friction resistance and hardness.

High-Speed Machining[§]

High-speed machining (HSM) can be defined in a couple of ways but neither fully reveals the magic that you are about to learn:

◆ Using the greatest RPM/IPM (feed) available within the machine tool's capability.

So, HSM occurs at speeds and feeds somewhere above "normal" CNC machining. Older machine tools don't have the speed, but new machines or upgrades are available with the speed muscle needed to do HSM. But that means buying the high-tech cutters required to stand up to the job and take advantage of the machine's ability (Fig. 26-23A). But "above normal" is a pretty general definition.

◆ It is generally accepted that HSM occurs above 10,000 RPM and at feed rates above 300 in. per minute—today, but tomorrow it's sure to rise. At these rates, something happens quite different from what has been called normal machining.

Due to two issues, safety and a greater application range, HSM is more common on mills where the cutter spins rather than the

[§]Text and photographs courtesy of Iscar Metals, USA.

workpiece. We'll concentrate the remainder of this discussion on that side of the shop but it can be performed on lathes too if they have the spindle speed. So how fast are the new mills going? Today, 20 to 25,000 RPM is usual and axis drives are feeding parts accurately at 1,000 in. per minute and above!

HSM Is More Than a Metal-Moving Issue

While extreme productivity is the central goal, there's a lot more advantage to the subject than simply cutting chips quickly. To understand it we must investigate the phenomenon that's called the *threshold*. It's the center of this discussion. To grasp it, you'll need to unravel another chip-making mystery:

If during a cut, you turn up the RPM and feed rates proportionately to maintain a constant chip load, an unexpected plateau occurs! At first, as the removal goes up, so does the horsepower (hp) requirement. That makes sense. But at the threshold (which varies for each alloy being machined), the increase in needed hp levels off then, while still increasing speed and feed, it begins to descend. That doesn't compute, energy consumed goes down as removal rate increases?

Here's a clue; it involves surface speed and deformation. When machining aluminum below the threshold, it's accepted that 1 horsepower is required to remove 3 cubic in. of metal. But as the speed and feeds go above 5,000 ft per minute, the removal rate becomes 4 cubic in. per hp. A similar ratio is true for other metals as well, in different proportions.

Machine Design Issues That Had to be Solved

An HSM machine is born for extreme service. For example, spindle bearings are made with ceramic materials and their oils are cooled by refrigeration. Cutter balance is another challenge here at the limits.

Figure 26-23A *Advanced geometry and engineered cutters make HSM a reality.*

High-speed milling taper (HSK type)

Standard vee
groove for tool
change gripper

Flange surface
mates with mill
spindle

Reverse internal bore
cavity for expanding
internal clamp

Figure 26-23B *The HSK tool holder taper stays perfectly located within the mill spindle even at extreme RPMs.*

KEYPOINT

In HSM milling, exceeding an imbalance of two just two grams, three problems can occur: bearing wear, poor finishes, and short cutter insert life.

Iscar recommends that HSM holder/cutters be **dynamically balanced** at a minimum of 7,000 RPM. We'll see more on this issue in the next article, too. Advanced tool holders specifically designed for HSM, incorporate full contact, holder to spindle, qualified surfaces and provision for dynamic (moving) balancing. Dynamic balancing means that the cutter-holder pair are spun and balanced while in motion, not unlike a car tire on a wheel balancing machine.

The method of retaining the tool in the holder also had to be addressed. The old set screw or collet holder no longer work in HSM. Using hydraulic lock or heat, shrink fits, solves this issue. All these HSM issues are standardized in ISO-1947.

A Shop Example

Compare the data to see why HSM outperforms standard cutting. First, we'll set up the job using what would be a fast but normal operation. We'll then compare it to an HSM.

The Job	
Work Material	6061-T6 Aluminum
Operation	Reduce an area 9.0 × 30.0 in. down 3.50—a total volume of 945 cubic in.
Machine	25-horsepower vertical spindle motor RPM 20,000 max feed rate 1,200 max
Cutter	3-in.-diameter, three-flute, carbide face mill 0.875 depth of cut limit
Standard Setup Facts	
Speeds and feeds	3,900 Surface Speed 5,000 RPM @ 150 IPM
Depth of cut	0.350 in.
Width of cut	1.50 in.
Passes taken	60
Time required	**12 minutes**
High-Speed Setup	
Speeds and feeds	15,700 Surface Speed 20,000 RPM @ 1,200 IPM
Depth of cut	0.200 in.
Width of cut	0.350 in.
Passes	260
Time required	**6.5 minutes**

Even though four times the number of passes were performed using HSM, the operation required just over half the time. But time isn't the only advantage to HSM. Reducing the volume per pass but taking many fast fed passes lessens pressure to both cutter and bearings, so both tool life and machine wear are improved. Using HSM forces, thin wall items can be machined with less distortion and surface finishes are also improved greatly at these cutting rates. To perform these near miracles, new tooling is often needed (Fig. 26-24).

Chip thinning due to engagement amount

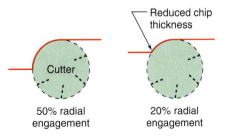

Figure 26-24 *For HSM, feed rates should be increased by a factor when reducing radial engagement due to reduction in chip thickness.*

Chip Thinning a Critical Factor in HSM

Below threshold machining, speeds and feeds are always important, but above the difference between success and catastrophic failure is very small and requires careful setup computation. Machine horsepower can easily be exceeded with stalling and breakage resulting. Or, if **chip thinout** isn't considered, work hardening, cutter overheating and mainly, the machine will be working below optimum.

KEYPOINT

Chip thinout is usually a minor problem below the threshold, but above it's a central issue and must be considered in the cutter path, chosen by the programmer.

Recall that feed calculations begin with a recommended chip load provided in bite per tooth, per revolution. For the previous example, 0.010 in. chip load, times a three-flute cutter, times its calculated RPM, yields the feed rate of 1,200 IPM.

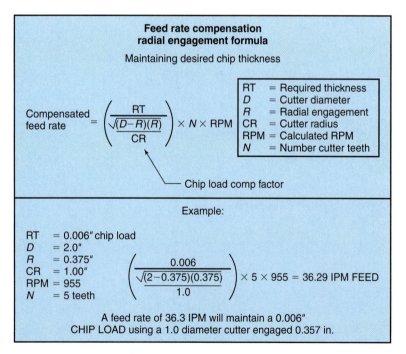

Feed rate compensation radial engagement formula

Maintaining desired chip thickness

$$\text{Compensated feed rate} = \left(\frac{\dfrac{RT}{\sqrt{(D-R)(R)}}}{CR} \right) \times N \times RPM$$

RT	= Required thickness
D	= Cutter diameter
R	= Radial engagement
CR	= Cutter radius
RPM	= Calculated RPM
N	= Number cutter teeth

Chip load comp factor

Example:

RT = 0.006" chip load
D = 2.0"
R = 0.375"
CR = 1.00"
RPM = 955
N = 5 teeth

$$\left(\frac{0.006}{\dfrac{\sqrt{(2-0.375)(0.375)}}{1.0}} \right) \times 5 \times 955 = 36.29 \text{ IPM FEED}$$

A feed rate of 36.3 IPM will maintain a 0.006" CHIP LOAD using a 1.0 diameter cutter engaged 0.357 in.

Figure 26-25 *Compensating formula for chip thinning.*

But in the previous example, as we arranged lighter cuts the cutter was backed out of the cut sideways, away from the work. It was deliberately set up with a tool path width that was less of the cutter's diameter. That's called **radial engagement** (Fig. 26-24). Compare the width of cut relative to the cutter's diameter. From the drawing note that with less radial engagement, the chip becomes thinner than that for which the load was originally calculated.

As a result, it's cutting a thinner chip than the feed rate predicts. At HSM rates, heat builds up in the thin chip and edge rubbing from the thin entry can cause work hardening and cutter wear. Additionally, the operation isn't maximized—the feed rate can be increased to achieve the recommended chip load.

KEYPOINT

The feed rate must be increased to account for the narrower radial engagement, otherwise the potential chip load will not be reached.

Calculating HSM Radial Engagement (RE)

If the width of cut relative to the cutter diameter is 50 percent or greater, then this effect need not be considered. But when RE is below 50 percent an upward adjustment to the IPM feed rate is necessary to maximize the setup.

This was the case in the example comparison, the 1.5 in. wide cut was reduced to 0.20 RE. Machinists often use this *tangential effect* (moving sideways until the cutter force is against the long dimension of the work rather than across its short axis) when cutter chatter or work deflection is a problem. It works similar to increasing the lead angle on a lathe bit, thinning the chip over a greater distance.

Trade Tip

Feed Rate Compensation Factors for Tangential Cuts Iscar recommends the following upward adjustments for feed rates based on RE, percentage of cut width compared to cutter diameter (Fig. 26-25).

Cut Width Percentage of Cutter Diameter	Divide IPM by Feed Rate to Get New Feed Rate
5%	0.45
10%	0.60
25%	0.85
30%	0.93
50%	1.00

For example, if the RE is 25 percent, divide the calculated feed rate by 0.85. Suppose initial feed is 1,000 IPM, then adjust:

$$1,000 \text{ IPM}/0.85 = 1,176 \text{ IPM}$$

You can extrapolate between the lines above or use the following formula for an exact calculation.

HSM Today and Tomorrow

The definition is sure to evolve as faster cutters, holders, spindle, and axis drives become available. But, for now, it's well above 10,000 RPM and 1,000 IPM, where chip deformation resistance lessens. What's next? Cutter geometry and coatings along with dry versus coolant machining seem to be the areas of most interest.

The Tools of Today and Tomorrow**

Today, cutter selection is a lot more complex than it once was. Progressive shops must fold far more factors into tooling selection in comparison. Over the last half-century, since tungsten carbide appeared, cutting tools and inserts have evolved from simple flat pieces of carbide brazed onto steel shanks, to become composite insert teeth with complex systems of substrates, coatings, and special geometries and holders. The new work materials being cut contribute to the need for wise cutter choice too, for example the new superhigh silicone aluminum. The old cutters just won't do the job! CAD/CAM/CNC evolution bringing us the complex shapes of today's products, also present unique cutter requirements. Then there's a new emphasis on responsibility to our environment with energy conservation and recycling challenges. Another factor we didn't have previously, is strong international competition requiring rock bottom manufacturing costs but yielding the highest quality possible.

**Text and photos courtesy of the Kennametal Inc., USA.

Economic factors affect development of tool related products too, such as quick-change tools and application-specific engineered cutters. They all add up to a new look at cutting tools *as a system* within any factory. It's no longer simple! Even the smallest shop can invest thousands of dollars in cutters—choices must be done wisely to remain profitable.

In developing today's cutting tool materials, we manufacturers must "mix and match" performance characteristics to meet the demands of widely varying tasks. There's no perfect solution.

KEYPOINT

Tool composition always represents trade-offs among attributes including hardness versus toughness and ultimate performance versus ultimate reliability.

For example, a cutter engineered for general purpose work on a variety of applications might combine a **substrate** (combinations of layers) engineered for toughness. They feature multilayered coatings featuring a thick, shard alumina layer to reduce friction and heat while producing good surface finish then other materials below for bond and wear (Fig. 26-26).

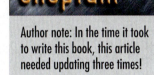

ShopTalk

Author note: In the time it took to write this book, this article needed updating three times!

On the other hand, a grade intended for HSM cuts on cast iron, for example, would feature harder substrates paired with a coating designed to maximize resistance to the free carbon abrasion and heat associated with CI machining.

The Coatings

Today, the majority of all inserts sold feature some form of insert coating such as the standard golden-colored coating, **titanium nitride (TiN)** or dull gray **aluminum oxide (Al_2O_3)**—a related material to grinder abrasive. But there are several others, and new specialized coatings appear regularly. This is an area where advancements occur every day!

For some years, **coatings** have been applied to the carbide base using a process called *chemical vapor deposition (CVD)*—a high-heat method of depositing the coating on the insert. However, coatings applied by *physical*

Figure 26-26 *Microphotograph shows layers of titanium nitride, aluminum oxide, and a final bond of titanium carbide nitride.*

vapor deposition (PVD) featuring lower heat, have also grown in popularity because it involves lower temperatures and causes no embrittlement of the parent material as with CVD thus sharper cutting edges can be produced that last longer.

KEYPOINT

The low-impact CVD process permits coating inserts with sharp edges without changing its physical characteristics.

This offers significant advantages in coating cutters for applications such as threading, cutoff, and other special milling operations. Today, research continues to improve the PVD process because it provides stronger bonds and deeper deposits. Soon it will effectively use medium temperatures to expand coating capabilities even further.

Making Future-Minded Decisions

Kennametal recommends that shops consider all the factors before choosing new tech cutters over standard tools. In the lean workplace of today, cutter decisions can be a make-or-break decision.

The Future Driven by Industrial Needs and Supplier Developments— A Partnership

A good example of a marketplace–material-driven development is the diamond-coated carbides. They were born in response to the newer very high silicon aluminum alloys (above 12 percent) showing up in the auto and aerospace industries. While aluminum below this threshold can be successfully machined with sharp edge, positive rake, uncoated tools, the abrasive qualities of the higher silicon material, demands a coating able to withstand the greater wear challenge.

There are two developments that might meet the need. Which is right?

Diamond Tips One features a **polycrystalline diamond (PCD)** chip, brazed into the cutting edge (artificial or natural diamonds). This tool lasts an exceptionally long time. However, this insert features only one cutting edge and fewer possible geometries. On the other hand, a diamond-coated carbide insert can solve these problems too but still be used elsewhere in the shop (Fig. 26-27).

Chip control geometry is another area of some development. Some shapes have been created for general application, whereas other inserts are engineered for specific extreme conditions or where buildup is a problem with low carbon steels for example. Note in Fig. 26-28 that many insert grade applications overlap. So before purchasing a specific insert for a job, be certain that a specialized tool is needed. Otherwise choose the better grade of standard insert to get more overall service from it.

Survey Your Needs

Kennametal recommends a thorough evaluation of your specific applications before moving into specialized tools. While they do offer advantages over standard tools, sometimes a standard insert that lasts moderately long, will be the more cost-effective tool over a longer period of application. Be certain your application justifies the extra inventory and costs. Specialized tools have their place but only when used with rational study. Staying informed of new developments is vital in this

Figure 26-28 *The decision of which insert is right, is made by balancing job requirements against manufacturing costs.*

Figure 26-29 *Is quick change tooling what your shop needs? Consider time allocation based on production.*

respect. As with all new products, they will become more available in greater varieties as product application and research continues. Read tool catalogs, brochures, and trade magazines and attend trade shows to stay up on the latest developments (Fig. 26-29).

Future Tool Development

What's next? There are two new factors driving the future of cutting tool technology.

Environmental Factors

A great deal of research is being conducted in "dry machining" to address the stricter regulations concerning disposal of spent fluids

Figure 26-27 *Diamonds form the surface of this special insert.*

(Author's) Higher cutting speeds and feeds from using specialized cutters do indeed increase production. But studies show that, in low- and medium-volume operations, cutting time represents just over 10 percent of overall available production time. Buying expensive tools with a single purpose, to upgrade 10 percent of the overall manufacturing cost may be the wrong place to invest! Tool changing, set up, and gaging consume around 20 percent—about twice as much (Fig. 26-30). Buying better holders, cutter preset-ting equipment, and fast turnaround of setups might be a wiser investment, on a case-by-case basis.

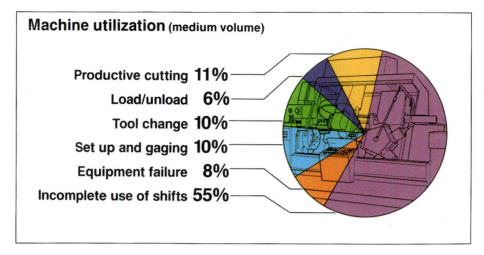

Machine utilization (medium volume)

Productive cutting **11%**
Load/unload **6%**
Tool change **10%**
Set up and gaging **10%**
Equipment failure **8%**
Incomplete use of shifts **55%**

Figure 26-30 *For medium production runs, cutting time consumes only 10% of the total time, whereas turnaround and gaging represents twice that.*

and for issues concerning employee exposure to these chemicals. Tool manufacturers are responding with thicker coatings and high-speed setups to reduce tool contact time thereby reducing tool heat buildup. These are among the options being explored toward the goal of less coolant needed in machining.

Overall Management of Cutting Tools

The second key to the future is partnerships between tooling manufacturers and users. On the shop floor, three elements must be combined for maximum profit, quality, and rock bottom cost:

Tools must be approached as a system—selecting compositions, coatings, and geometries based on genuine need studies

Quick-change, balanced modular tool systems well planned for the application

Managed inventories using lean manufacturing principles

UNIT 26-2 REVIEW

Review the Key Terms

Aluminum oxide (Al_2O_3)
A coating applied to carbide insert tooling to improve surface hardness and resistance to friction.

Chip thinning
The effect of not taking a cutter swarf as wide as the cutter diameter. The chip is thinner than that for which the feed rate was calculated to produce. The feed may be increased by a factor to compensate. Also known as the tangential effect.

Coating
The thin layer or layers applied by CVD (chemical vapor deposition) or PVD (physical vapor deposition) of materials harder than the base carbide.

Dynamic balance
Rotating cutter, holder combinations at or above 7,000 RPM then adjusting balance to within two grams.

High-speed machining
Mostly on milling machines, at spindle speeds above 10,000 RPM and at table feeds above 1,000 in. per minute.

Plastic flow
The desirable threshold whereby, during chip formation, work material loses resistance to deformation flow due to internal friction.

Polycrystalline diamond (PCD)
A modern material used either in chip or coating form to improve cutter hardness and friction resistance.

Radial engagement
The amount of side engagement of a cutter compared to its full-diameter swarf.

Substrate
The chemically or physically deposited layers applied to cutting tools to achieve a certain effect. See coatings.

Titanium nitride (TiN)
A coating commonly applied to carbide inserts to improve friction resistance and hardness.

Replay the Key Points

◆ HSM is chosen not only to increase productivity but to also improve finishes, to lower impact on tools, spindles and work and to machine delicate features not possible to cut otherwise.

◆ The feed rate must be increased to account for the narrow pass relative to the cutter diameter otherwise the true chip load is not achieved.

◆ Studies show that, in medium production, cutting time represents just over 10 percent of overall available production time. Buying expensive custom tools may be the wrong place to invest.

◆ Tool compositions always represent trade-offs among attributes including hardness versus toughness and ultimate performance versus ultimate reliability.

Respond

1. Describe at least two reasons to choose high-speed machining.

2. What is the threshold phenomenon in HSM?

3. Why must chip thinning be accounted for in a setup, when performing HSM?

4. What advantage does PVD (physical vapor deposition) have over CVD for coating carbide or other cutting tools?

5. What is the max limit for cutter balance when it is to be used for HSM?

UNIT 26-3

Putting Metal Together— Direct Deposition

Introduction: From the first gut-wrenching moment when an ancestral machinist realized he'd made his very first undersized part to this day, we've all wished for some way to put metal back on the part! Today, that wish has come true, granting us several processes that add rather than take away material— known as **additive manufacturing.** Search the Internet for the key words **direct deposition** *of metal* to read about the variety of ways it can be done. But as you'll discover, most add crude repair layers to existing parts, they do not create usable parts with complex shapes. One special combination of technologies goes beyond repair work to make new parts that are impossible to produce by any other method.

It's a magic brought to us through advancements in four different disciplines, all coming together to create solid or hollow, fully functioning metal parts starting with powder! The product can even be made of different metals combined into one part! These technologies are

Moderately high powered lasers

Powdered metals

Exact CNC computer direction of the laser beam

CAD/CAM tool paths

KEY TERMS

Additive manufacturing
Putting material together to make a part rather than removing it as in machining (subtractive).

Direct deposition
Adding or building small deposits of material together to create a model or real part. See additive manufacturing.

LENS (laser-engineered net shaping)
Registered trademark technology of Sandia National Labs that deposits powdered metal, laser melted into liquid then solid form, to create complex metal parts.

Assembling Metal— Here's How

While the results are like magic, the process is not difficult to envision. First, the user must plan an efficient application of the technology. Often it's not necessary to build the entire part by addition. Many objects can be started from a machined parent material, then features impossible to machine otherwise are added. Figure 26-31 shows how the colored cylinders could be built upon the gold starting block. To begin the addition, the solid part model is sliced into thin, stacked layers by the CAM software (Fig. 26-32). Each layer becomes a surface to be built upon the previous layer or the parent metal.

As the laser focal point scans across a layer, four nozzles inject highly controlled volumes of powdered metal into the 0.020-in.-wide beam at the surface of the receiving material. Precisely controlling the powder flow to a few grams per minute is a major key to this process.

Parent Base Metal

Pins added by lens process

Figure 26-31 *Use the process only when the object cannot be made otherwise. The orange cylinders should be created on the gold machined parent.*

Figure 26-32 *This simple mold has been sliced into layers. Each will be made into overlapping tool paths, then scanned by the laser.*

As the laser heats the part surface, the molten pool is increased to around 0.040 in. diameter by the incoming powder. Surface tension in the liquid metal builds height as well as width, up to around 0.015 in., dependent on alloy. The process creates a small bead string. Coming to the bead's end, the laser switches off, to reposition then make the next. The program ensures each successive bead overlaps the last to create a continuous surface with 100 percent density and penetration. Finishing all the connected beads in one layer, the laser moves up and a new slice is added on until the part is complete (Fig. 26-33).

To ensure metal purity without chemical reactions to room air, the process must take place in a contained environment charged with inert argon gas. The shielding gas excludes atmospheric oxygen and nitrogen because, de-

Design   Build   Finished Part

Figure 26-33 *The LENS process is not difficult to understand.*

pending on the alloy, one or both gases can seriously degrade various metals when they are elevated above their melting point

Advantages of the LENS® Process‡

The metal produced by this process is fully functional with remarkable properties in some cases. Called **laser-engineered net shaping (LENS),** it creates metal of equal or superior properties, compared to all other forms of the same alloy. Some exhibit finer, more uniform grain structure than every other form. It can be heat-treated, machined and ground, or processed in every way the alloy normally would be. An ideal example is 316 stainless made by LENS then processed for best strength, which exhibits nearly twice the tensile strength of conventional 316! See the chart in Fig. 26-34 for a few other examples.

Mechanical Property Comparison Chart	Tensile Yield Strength (ksi)	Ultimate Tensile Strength (ksi)	Elongation (%)
316SS typical	42	84	50
316 SS minimum	30	75	40
316SS LENS	40.3	96.5	66.5
Ti-6-4 typical	128	138	14
Ti-6-4 minimum	120	130	10
Ti-6-4 LENS	122.7	138.5	15
IN625 typical	63	136	51.5
IN625 minimum	60	120	40
IN625 LENS	68	139	45

Figure 26-34 *Properties of a few metals made by LENS®.*

‡LENS is a trademark of Sandia National Laboratories and Sandia Corporation.

Figure 26-35 *This titanium valve is made of two different alloys but it is one continuous metal!*

Figure 26-36 *This stainless mold could feature a copper core all fused together as one.*

Functionally Selected Metals

Using the LENS process, parts can be engineered with different alloys in a single part, to suit function. Figure 26-35 shows an engine valve that could be produced by fusion welding, but it would have a problem—a disturbed lattice structure at the joint where it would probably fail with use. The LENS-made valve resolves the weakness. It's made of two different alloys of titanium. The harder alloy is used for the valve, while the more flexible alloy forms the stem.

Figure 26-37 *The LENS-built injection mold can have complex internal cooling channels that follow the outer contour!*

Different metals are loaded into different feeders, and each is dropped into the pool when required (Fig. 26-36).

Impossible Geometry Made Possible

A second giant advantage suggests this question. What if you wanted to make an injection mold with internal cooling channels that follow the outside contour? A great idea! They would control the plastic flow more uniformly, and cool the part more quickly. But no machining process could do the trick nor could cast the mold then machine it. Additive production makes that mold. Plus no matter how wild the shape, the core could be contoured copper to help distribute heat more evenly, while the surface was hard steel, for example! (See Fig. 26-37.)

For another example, consider how you would make or machine a continuous surface of steel but with an engineered framework of internal braces to keep it light be strong (Fig. 26-38). Objects that overhang (Figs. 26-39 and 26-40) can be created with a draft angle (taper) of up to around 30° maximum due to the deposit metal's surface tension. Various metals and shapes dictate the results.

Figure 26-38 *A cutaway shows how a steel block could be made with an internal lattice of braces—impossible by any other method.*

Figure 26-39 *These metal bottles all connect together!*

Figure 26-40 *This hollow turbine blade was made by direct deposition.*

Repairing with LENS Technology

Flame spraying has been used to add metal to worn or broken parts for many years. It's an affordable, portable solution for general repair of metal parts. By spraying a powder, hopper is added to an oxy-fuel welding torch. Spraying does the job, but requires heating the entire part, plus it creates a coarse layer of metal that's mechanically fastened to the parent—it isn't fused. It's porous and it can peel away under load.

Repairing with the LENS process, however, lays down a nearly perfect bead of 100 percent density and penetration into the parent metal. To use LENS as a repair tool, the part is fixtured to be moved relative to the laser and feeders (Figs. 26-41).

Advantage Application

The best return on investment using this process happens when one is able to see clearly where it applies (and doesn't), and then build LENS into the job plan accordingly. Engineers for Optomec, the company commercializing the process, admit that it's not a

KEYPOINT

Today LENS is relatively slow at a max rate of 1 cubic in. of growth per hour. But keep in mind that's 1 cubic in. of high-quality, manufactured product impossible to produce by other methods!

Figure 26-41 *This expensive gear can be rebuilt using LENS technology.*

panacea. At this time, it's aimed at low-volume production and repair. There are a great many times when subtractive manufacturing makes sense over additive manufacturing.

Future Applications

It's predictable that accumulation speed will increase as the technology advances. Also, like other space-age ideas, as it emerges into the enabling arena, end users will discover an amazing range of applications undiscovered today. Will that be you, a subtractive machinist, or will it be the competition? See the review about shifting paradigms and the dangers of being too entrenched in a way of doing things. As a chip removing kind of person, it would be easy to be blind-sided to the potential of this technology!

UNIT 26-3 REVIEW

Review the Key Terms

Additive manufacturing
Putting material together to make a part rather than removing it as in machining (subtractive).

Direct deposition
Adding or building small deposits of material together to create a model or real part. See additive manufacturing.

LENS (laser-engineered net shaping)
Registered trademark technology of Sandia National Labs that deposits powdered metal, laser melted into liquid then solid form, to create complex metal parts.

Replay the Key Points

- Joining different metals together using LENS technology produces joints as fine as a single metal.
- Direct deposition creates metal parts from powder directly, with no mold.
- The LENS process creates metal that is as functional as any other form.

Respond

1. What three advantages does direct deposition offer?
2. List the four technologies required to perform laser-engineered net shaping of metal.
3. Define additive manufacturing.
4. Using LENS technology, what is the maximum overhang that can be produced.

Section 26 Review

Units 26-1, 26-2, and 26-3 Paradigm Shifts

A paradigm is a perceived model of how we do things—a set of mutually, widely accepted rules. We machinists live under a set that tells us how we cut and shape metal.

A shift in paradigm isn't an advance in the way things are done. It's a giant, and usually sudden, leap. It changes the rules. The problem is when any group or organization is entrenched in a paradigm (technology and procedures in this case), the new way will be made invisible by those it challenges. The Swiss example shows this.

In the past, the phrase "Fine Swiss watch" was a part of the world's vocabulary. They refined and kept secret the manufacture of microgears, escapements, springs, shafts, and other mechanical parts. The Swiss were unarguably the best at the craft. The industry was a large part of their nation's economy and of their image as the finest craftsmen worldwide.

But then the electronic digital timepiece appeared. Simple, with few moving parts, it was seen as an inexpensive novelty by the watchmakers but nothing more. Other nations saw the potential and began producing the new watch. In less than 5 years, the Swiss watchmaking industry was devastated. A final twist to this illustration: the digital watch was a Swiss invention in the first place! I'm sure you see the point, direct deposition is a quantum difference in the way we do things.

In Unit 26-1, we started with technologies that had to prove themselves to us but one-by-one they did. Then in Unit 26-2, we delved into the edge of cutter technology, a field that is changing so fast that it's nearly impossible to publish the facts before they change! Finally in Unit 26-3, a technology was presented that has the potential to be a paradigm shift.

Questions and Problems

1. At what pressure range does a waterjet cut?
2. What dangers are there to waterjets?
3. Is this statement true or false? The gear drive, intensifier pump is the better but more costly means of creating ultrahigh-pressure water for waterjets? If it is false, what makes it true?
4. What is the size of a waterjet stream?
5. What must be done to cut harder materials with a waterjet?
6. In your opinion, what are the main advantages of waterjet cutting?
7. Define EDM?
8. Define the phases of one EDM arc making one crater.
9. Describe briefly the two different kinds of EDM machines.
10. What materials can an EDM cut and why?
11. Each type of EDM uses a fluid. Describe each and list five different purposes for the fluid.
12. Which fluid is used on a wire feed EDM? Which is used on a sinker?
13. What does the acronym LASER mean?
14. What are the special dangers of laser cutting? Is there a radiation danger?
15. Which of the three processes, EDM, lasers, and waterjet cutting leaves behind a HAZ of two distinct layers? Describe each layer.
16. During HSM, a magic-like phenomenon occurs, the *threshold.* Explain it.
17. Compare HSM to regular milling.
18. What is the name and purpose of the familiar gold coating on cutting tools?

Critical Thinking

19. Which process is required to coat HSS cutting tools and why?
20. Describe how four technologies make the LENS® process possible?

Section 26 Answers

ANSWERS 26-1

1. Waterjet does not heat the material
2. Cuts by melting tiny craters of metal then flushing it away
3. Because the arc's energy is spread over a larger area as the crater expands and because the dielectric breaks down, which defocuses the ion path.
4. False. Waterjets can cut anything including flesh and bone!
5. Sinker driving a shaped electrode downward into the workpiece
   Wire—moving a thin wire in an *X-Y* motion similar to a jeweler's saw

ANSWERS 26-2

1. Faster metal removal; lower impact on machine spindle bearings; lower impact on material; longer tool life where breakage is a problem; better finishes
2. At a given elevated surface cutting speed, the chip begins to soften due to internal heat, thus it flows freely during the rake phase.
3. Chip load is supercritical in HSM with regard to tool life. When reducing radial engagement to perform HSM, the effective chip thickness is also reduced.
4. Lower temperatures; less impact on parent cutter; can create sharp edge, whereas CVD cannot
5. 2 grams

ANSWERS 26-3

1. High-quality material; functionally selected alloys in one part; shapes impossible to machine are made possible
2. Lasers, CNC directions, powdered metal, CAM software
3. Putting metal onto the part rather than taking it off
4. 30°

Answers to Section Review Questions

1. 18,000 to 60,000 PSI
2. Contact with flesh
3. False. There is no gear intensifier. There are two pump varieties. The *gear drive* or the piston intensifier. The intensifier is the better of the two.
4. From 0.004 in. to 0.018 inch
5. Add abrasive to the jet stream
6. No heat-affected zone; can cut anything hard or soft; no cutters; uses plain tap water as the media
7. Electrical discharge machining—removing metal by melting away tiny craters then flushing the suspended material away.
8. Ion path forms; arc on—crater forms but has limit as crater grows; arc off—bubble collapses; mechanical flush
9. *Wire feed*—a continuous thread of charged brass wire is moved through the material by CNC; *sinker electrode*—pushing shaped, charged electrodes into the workpiece and laterally by CNC.
10. Metals because they conduct electricity
11. It's the *dielectric fluid*. It cools the work; insulates (holds back the arc until voltage is at the right point); creates ion path for a stable focused arc; creates a bubble that collapses for the first stage of the flush; and mechanically flushs the remaining detritus away.
12. Water on wire feeders; oil (special) on sinkers
13. Light amplification by stimulated emission of radiation
14. No special dangers other than it cutting flesh, breathing the vapors, or eye damage from the intense light.
15. Laser leaves only one layer while EDM leaves behind two. The inner recast layer is slightly modified as it was liquid but not removed, when the arc went off. The outer white layer is highly modified as it had been removed from the parent but was frozen back during the bubble collapse off phase.
16. Deformation heat rises to the point that the chip flows as a semiplastic.
17. For HSM, RPM and feed are increased to above 10,000 RPM and 300 IPM. Cut depth and radial engagement are decreased to take smaller passes but many more are taken at a much faster feed rate.
18. Titanium-nitride (TiN) creates an extrahard antifriction surface.
19. Physical vapor deposition because it is a relatively low heat. Two reasons: hardness won't be drawn from the underlying cutter and PVD leaves sharp edges—HSS tools need them while carbide tools have slight radii on the cutting edge.
20. CAM software slices the object into thin (0.015-in.) stacked layers, then creates a tool path to cover each layer with passes of the CNC-directed laser. As it scans, powdered metal is injected into the molten puddle to form a bead. Successive beads overlap to form a continuous surface.

Section 27

Statistical Process Control (SPC)

Introduction

SPC (Statistical Process Control) helps machinists do their work better, consistently achieving the best quality at the lowest cost. SPC equips users with an early warning radar that detects variation *before* it can cause problems. It works on any process with a quality target, but especially well on machining processes. With the data SPC provides, we can learn how close to the target we should be able to machine a given feature, a difficult thing to deduce otherwise. Then, putting that powerful knowledge to work, we monitor, detect, and reduce variation when it exceeds norms. The ultimate goal is improvement and control of our overall process, to the point that we make all excellent parts!

We know that within any batch of parts, features are going to vary in size, position, and other characteristics. Although most variation is normal, it's our constant adversary. Accepting that we can't eliminate it, we take the right actions to identify the source and then control it as close to the target as economically practical, and sort out and eliminate special sources altogether. That requires three related abilities provided by SPC:

An understanding of the *nature of variation*. Arguably this is the most empowering aspect of SPC and a major goal of Section 27. Learning these concepts will help you make better decisions, even if you don't apply SPC to your work.

A clear picture of what we should *expect to happen*. After producing a number of parts (or operations), the capability of the process becomes known. That benchmark shows us how much variation is normal when everything is going right. That becomes the benchmark process capability.

A clear picture of *exactly what is happening during real-time machining*. Using the process capability as our guide, rather than job tolerance, we measure parts as they are made.

Special editorial thanks go to Mr. Joel Bisset, Quality Manager, Coastal Manufacturing of Mukilteo Washington, and Mr. Jeff Whitehead, Math Department Marysville School District, Marysville, Washington.

A Visual Data Tool

While SPC can be performed with paper and pencil, it's a lot more efficient on PCs or CNC controllers, where the software crunches measurements of critical part features into graphs that show capability and control charts (run charts) that are used as real-time guides during production.

Global Oversight

The same data are used in many other ways throughout the shop by management. With hard facts showing capability, they then bid or plan work that can be delivered with existing machines and tools. It also clarifies where equipment and tooling dollars should be spent to improve capabilities, or pursue new business.

Final Training

In Unit 27-1, we'll discuss the *what and why* of SPC for machinists. Then Unit 27-2 will present a lab activity. A great deal of the *how* depends heavily on the shop's application methods, the size of an average part run, general tolerances typical of the shop, and, to a lesser extent, on the specific software chosen. An employer survey showed they wanted graduates with an understanding of the benefits and a grasp of basic concepts and terms of SPC. Beyond that, they wanted to finish specific training themselves. Section 27 is written to provide that platform.

Objectives

Unit 27-1 What and Why SPC? (Pages 1002–1008)
List and describe the two objectives of SPC
Define normal variation and real-time control
Define the Cp ratio
Compare a histogram to a control chart
Define the two types of variation—special and common cause

Unit 27-2 An SPC Experiment (Pages 1008–1013)
Create a control chart and derive a Cpk ratio
Provided a number of manufactured items, select a critical dimension
Measure variation and produce your own histogram and control chart

What and Why SPC?

Introduction: All machinists, whether or not they use SPC, must understand the difference between the two kinds of variation: *assignable cause* and *common cause*. SPC helps us differentiate the two more clearly and early, so we can take the right action. If we can assign a reason for feature differences, then we can reduce or control them. But some variation is built into the process, so the process must be improved or we must accept the results as they are.

KEYPOINT

It's important to know the differences in variation because each is dealt with differently.

Common cause variation is that which results from the process. It can be quantified (given a numeric range) by SPC and then can be reduced but only by process improvement. It cannot be eliminated altogether. Examples include cutting tool wear, work-holding flex or runout, and tool holder or machine rigidity.

Assignable cause variation originates from special events that do change results but are not integral parts of the process. They often show as bumps or waves on production graphs. Examples might be a forklift driving past during a delicate operation, every 15 minutes, or the daily heat buildup in the shop, in a machine, or in the coolant. Using SPC, variation of this type can be identified as outside normal occurrences, then with a bit of detective work, the source *can be* eliminated.

KEY TERMS

Bell curve
The natural graphical shape of variation, tending to centralize about a target result with diminishing results to each side.

Centering (a process)
Adjusting such that results distribute equally around the target dimension.

Common cause (normal) variation
The unwanted but ever-present spread of results in a process.

Control limits, upper and lower (UCL and LCL)
The boundaries of normal variation in a process—they are not the drawing tolerance.

Key dimension
A dimension assigned for SPC tracking. May be noted on the drawing with brackets or notes. May be assigned in the work order.

Real-time control
Tracking production of key dimensions, in sequence, as they are made.

SPC (statistical process control)
A math/graphical method of seeing normal variation in any process with a quality outcome.

Spread (variation)
The natural bell-shaped tendency of variation to be domed with tapering to each side.

Trend
A process that is starting to deviate out of the normal control limits.

Setting Control Limits—The Boundary of Normal

Let's begin by clarifying an important key point:

> SPC detects and helps sort variation, but it does not show the reason—identification of the source and what to do about it, remains the machinist's job and is based on skill and experience.

OK—here's how it's done. The first step is to compile how much normal variation is part of the process, the **common cause.** That data is converted by the software into **upper** and **lower control limits (UCL** and **LCL).** Control limits set the bounds for production. Then as we produce ongoing parts using those natural control limits, we can see and distinguish variation.

KEYPOINT

The UCL and LCL boundaries are not related to drawing tolerance, they are the *natural limits* of normal variation within the process when everything is going right.

For example, a drawing shows a tolerance of ±0.001 in., applied to the diameter of a 1.000-in. hole. After drilling and measuring say 100 parts, then entering the size, the SPC software shows us how well we can control its diameter—consistently around 1.0002 in. diameter, with the variation spread at ±0.0003 in. Through statistical formulae, that data is converted to the control limits for diameter around 0.0006 in. total range. The UCL/LCL spread won't come out at exactly 0.0006 in. We'll see why in just a bit.

OK, we have the first ability—we know how well the process should go. For now, let's suppose there are no special causes—all variation is due to the nature of drilling this hole. That's the usual, most variation is common cause.

Provides Real-Time Comparison to Normal

Then with control limits set at 1.0005 UCL max diameter (1.0002 + 0.0003) and 0.9999 in. UCL (1.0002 − 0.0003) naturally occurring, the machinist continues to make parts and the **key dimension** feature is compared to the normal control limits. To be under control, the job must stay within the UCL/LCL limits.

To further improve, we could center the process on the design target of 1.0000-in. diameter using the 1.0003-in. UCL and 0.9997-in. LCL. But that step would be possible only once we have a clear understanding of the natural variation in this drilling process—hard data that are provided by the SPC software. To close the range we would need to change the process—use more coolant, implement a reamer following a smaller drill, or change the cutting speed, for example.

Figure 27-1 *You'll be working on charts similar to this at your workstation.*

occur, and indicate when the job is within control limits or show that intervention is needed. In Fig. 27-1, a key measurement is being tracked as parts are made. The key feature's target size was 0.6257 in.

You can see in Fig. 27-1, that things were going along pretty well until around 4:30 P.M. Until then, the diameter was coming out just a little big, but well within bounds, then a trend started. The measured key feature began deviating toward the upper control limit.

Critical Question If that trend continued across the UCL, would the parts be scrap?

Answer Not yet, remember, the UCL is the boundary of the normal. The job tolerance still lies beyond that. But something needs to be done now!

The cause of the problem might be cutting tool wear, or the setup may need to be tightened up, or heat might be building in the machine itself, but it might also be that it's at shift change and a new machinist just took over and is doing something different! The graph provides clues but not solutions.

A Mind Experiment—Finding Normal Variation

All the principles are illustrated by shooting two different rifles at two targets of the same size. The objective is to hit the bull's-eye (a nominal target dimension), but there will be some variation. In this example, the rifles will be bench-mounted and the ammunition standardized, so only the rifle's ability to shoot straight is in question.

KEYPOINT

SPC isn't a voluntary tightening of tolerances, but rather functions like a dipstick to show us how much variation should be occurring if everything is going right.

Properly executed, SPC empowers the machinist to predict future process outcomes, thus reducing the risk of poor quality.

Because SPC makes the process more predictable, budgets can be wisely spent on prevention, as opposed to detection and reworking.

Real-Time Control Charts

With some of the specific details left out, we now have a grasp on variation and we come to the graphic tool used at the machine. By tracking progress graphically, they focus attention on trends in the operation as they

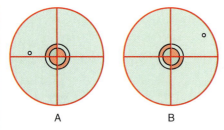

Figure 27-2 *These targets illustrate two different machines performing the same process.*

Each rifle could also represent two different CNC machines drilling a hole at a given geometric location. Both are attempting to locate the hole's axis at the same position, within the same nominal dimension. The bull's-eye represents the hole's true geometric position. The diameter of the target is the drawing tolerance for location. As long as the axis hits somewhere on the target, the work is acceptable.

OK, we fire each rifle one time and see what the results show (Fig. 27-2). Based on this initial information, it appears rifle A is a bit more accurate than rifle B, although that isn't much to go on. If you were to take corrective action at this time, what would you do? Move A to the right and B to the left and down. But do we have a complete enough picture?

Not yet. So we shoot a few more times (Fig. 27-3), and something very different begins to appear. Process A is not very well-suited to the job. It seems that normal variation for A consumes more than the job tolerance—it's predictably going to make a lot of scrap unless it's improved. The small crosses indicate that some shots missed the target completely.

— Missed shots

Figure 27-3 *After more data is collected, a clear picture forms.*

The two graphs shows us three pieces of valuable information:

1. Rifle A is not a good process for this job and rifle B has the capability to be very good.
2. B isn't centered—it's not aimed at the bull's-eye.
3. B has also missed the target one time as well.

Capability Shows on Histograms

Note that at this stage, it truly did not matter in which order the shots were fired. Given enough shots, the histograms show us the capability of both processes—rifles A and B. Those total capability charts are called histograms and they show overall capability in a process.

Control Limits

Next, a small circle can be drawn around B's spread (Fig. 27-4). That's representative of its capability, and it becomes the control limits for shooting B. Still we'd need to shoot and record many more times to determine an

Process B's capability for location accuracy

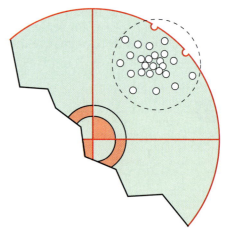

Figure 27-4 *B's record shows it has the capability to hit very near the bull's-eye once the process is centered.*

KEYPOINT

The capability plot indicates we can correct the process such that it will not make scrap unless new variation is introduced! Rifle B shows a good process that isn't centered on the target.

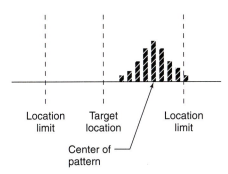

Figure 27-5 *A histogram in bar form shows the spread of B's variation.*

exact capability, but it gives us useful data. In fact, that's what the SPC software does: it continuously updates as more data are collected.

There are several kinds of histograms that we'll leave for factory training. But no matter what the form, they all compare process variation to the drawing feature tolerance. In Fig. 27-5, I've put the data into SPC software and it has created a bar graph histogram of B's hits.

Bell Curve Distribution

Notice the way rifle B's holes were distributed—most were near the center of a tight pattern and a diminishing few hit farther out. Notice that they concentrate at a central location then thin out from their natural center. That's called the variation or **spread.** The spread of variation normally assumes the same shape, concentrating at a point then thinning in a flattening curve to each side of the process center—this is called a bell-shaped curve or a **bell curve.**

Figure 27-6 is a training aid that shows the tendency of any process with a quality outcome to form a similarly shaped curve. If we dropped balls into the top, they would fall at random, bouncing off pegs. The drop point can be moved, to simulate a process that is not centered, as our example rifle B is. If enough balls are dropped, they naturally form a bell-shaped curve around a central location.

When enough data is present, the SPC software can create a well-defined definition of the variation in this process, to the point that it can predict future events. Fig. 27-7 graphs the spread, and notice that the curve goes out beyond the collected data. That's why control limits are a bit wider than early data. SPC

Figure 27-6 *The bell curve is starting to form. If all balls are dropped down, a full curve would appear.*

software can predict how wide the spread will be, based on the bell curve it produces.

Looking at Fig. 27-7, notice that the plotted curve extends out beyond the data we've collected. It shows that a significant number of shots will inevitably hit beyond the samples we've taken. That's why the circle drawn around the first few shots of process B wasn't exactly equal to its control limits. The extended bell curve shows that when a much greater number of parts are made, a predictable few will fall a bit farther from the pattern center. But they need not make scrap parts if we center the process.

Based on this distribution graph, the software can predict once the rifle is aimed correctly (**centering** the process) how many parts will exceed the job tolerance (become scrap or require a rework) within a batch of 100, 1,000, or 10,000 parts. That's powerful information!

Actions Based on Information

The target examples illustrate an important concept in SPC—*complete knowledge equals better control.*

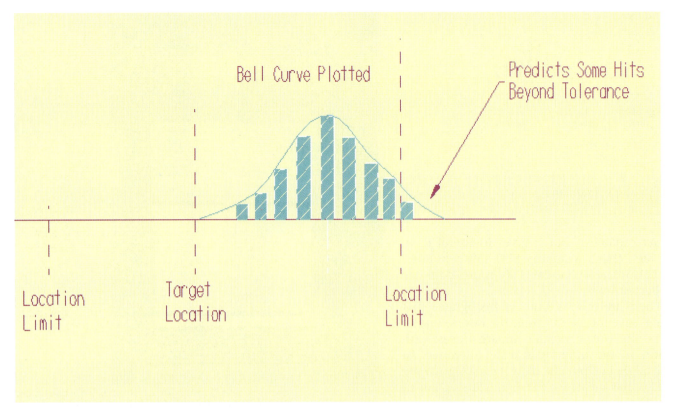

Figure 27-7 *The SPC software can display a histogram of the variation.*

Critical Question Going back to Fig. 27-2, right after the first shot, what would have happened had you corrected each setup before shooting a second time? After the early correction, where would the bullet hit on the second shot?

The answer, of course, is that you would have mistakenly aimed A to the right. But that misinformed correction would have then gone far off the target on the plus side of the tolerance, to the right—a scrap part. Truthfully, once a clear picture came to be, A's histogram showed us that it's actually aimed right—its process is centered on the target, but it's a crummy gun!

That questionable action of adjusting without knowing parallels a machinist adjusting the machine before having a clear understanding of the capability of the process. He might be "chasing normal" rather than refining the process. However, in reality, based on skills and knowledge, when it can be anticipated that the process will be within reasonable control, the machinist would still make early adjustments because no scrap is acceptable under any circumstances, even to get the data picture.

A Is Out! The problem isn't that the adjustment was wrong, it's that Rifle A is just not up to the task. So, from here on, we consider A as invalid. The histogram proves it. We'll continue with B only. Rifle B is more like the usual situation—a process able to deliver but requiring some refinement.

Critical Question OK, what correction would you have applied to rifle B after the first shot?

Making the early correction (Fig. 27-8), to the left and down, brought the process closer to the target, but that preliminary information would not perfectly center it. If the big picture was known (a complete histogram), then the best correction would have been from the *center* of B's pattern, to the bull's-eye

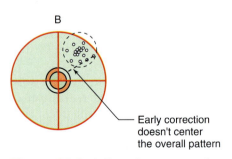

Early correction doesn't center the overall pattern

Figure 27-8 *A Cp ratio compares the normal spread to the target tolerance.*

center. But at that point, with only one shot, the pattern center was unknown.

That perfect centering data is available only if you can leave the process alone until the normal variation comes into focus on the histogram. However, it's impossible for a skilled machinist to make a bunch of questionable parts at the onset, to collect data. For if he did, one piece of data would be his final paycheck!

In real machining, that first correction would be made if experience told the operator that the process was known to exhibit tight repeatability as on Rifle B. But in making early corrections, the data collected would no longer be valid in the histogram database. After the initial correction to B, the parts would be reproducing very close but not perfectly centered on the target. At that time, the operator might make one more adjustment, then let the machine run for a while, producing high-quality parts but not the absolute best possible, with little (or no) further corrections for a time. That will then result in an accurate histogram with perhaps one last correction.

So you see, just having the SPC understanding of variation helps with decisions about corrections—even without the graphs. But with clear data, corrections can be spot-on!

The Cp Ratio Describes Capability

To rank the potential of a process, a ratio is formed by comparing the overall feature tolerance on the drawing, to the normal variation found by the software. Process capability *Cp,* equals:

$$\frac{\text{Feature tolerance}}{\text{Variation normal}}$$

In this example, it's determined by dividing B's spread diameter into the target's rim diameter, yielding the Cp. It is the capability of the process expressed as a decimal ratio.

Rifle A had a very poor Cp below 1.0, since, predictably, many of its hits were going to be off the target. Its normal variation consumed more than the tolerance. Even before centering B, the histogram showed it had a great Cp.

Left alone, aimed as it is, rifle B will make a bad part or two, but the potential to make 100 percent all good parts was evident. Centered or not, Rifle B shows a good Cp well above 1.

The Second Step—Control During Production

We now return to Rifle B, starting with a clean target but one whose outer rim is not the location tolerance. This time, it represents the control limits. During this phase, we're producing parts. We track each shot in sequence—that's known as **real-time control.** Since we have the data needed, we center on the target.

As Fig. 27-9 shows, all goes well for a while, the process is behaving normally. Then in

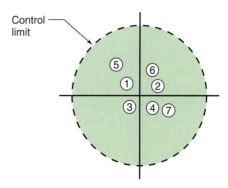

Figure 27-9 *Using the histogram, we've now centered the process.*

New? SPC is far from a new science. It's accepted that it was invented in the United States in 1920 at Bell Laboratories and was widely in use by 1950. But sadly, with little pressure from international competition, it faded from popularity here. And even worse, it rapidly found acceptance in Japan, then in the world.

Today, global competition has forced North American companies to readopt SPC. Today customers expect quality and they will buy from any supplier or nation delivering excellence at an affordable price.

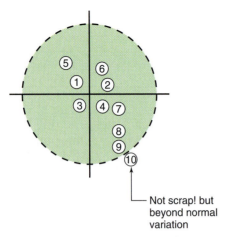

A trend developing

Not scrap! but beyond normal variation

Figure 27-10 *A trend is starting but no bad parts have been made yet.*

Fig. 27-10 we see that a problem begins developing, called a **trend.** The process is wandering off center. No scrap has been made as yet. However, left uncorrected, it soon will be. The process isn't performing normally. Now the machinist earns his pay. OK, those are the basics. Let's review.

UNIT 27-1 REVIEW

Review the Key Terms

Bell curve
The natural graphical shape of variation, tending to centralize about a target result with diminishing results to each side.

Centering (a process)
Adjusting such that results distribute equally around the target dimension.

Common cause (normal) variation
The unwanted but ever-present spread of results in a process.

Control limits, upper and lower (UCL and LCL)
The boundaries of normal variation in a process—they are not the drawing tolerance.

Key dimension
A dimension assigned for SPC tracking. May be noted on the drawing with brackets or notes. May be assigned in the work order.

Real-time control
Tracking production of key dimensions, in sequence, as they are made.

SPC (Statistical Process Control)
A math/graphical method of seeing normal variation in any process with a quality outcome.

Spread (variation)
The natural bell-shaped tendency of variation to be domed with tapering to each side.

Trend
A process that is starting to deviate out of the normal control limits.

Replay the Key Points

◆ SPC can be applied to any process with a measurable quality target but it works particularly well for precision machining.

◆ SPC signals clearly and sooner that a trend is occurring away from normal.

◆ SPC doesn't identify the root cause of a trend or excess variation. It isn't a voluntary tightening of tolerances—it highlights natural capability.

◆ A Cp above 1.0 is better—the larger the Cp, the better the chances of making all good parts.

◆ A histogram detects capability and normal variation.

Respond

1. What does SPC tell the machinist in real time?

2. What is the graph that shows the Cp of a process?

3. True or false? Because a histogram analyzes the spread of many measurements of key features, those measurements need not be in any order. If the statement is false, what makes it true?

4. True or false? A Cp of 1.2 is closer to 1.0, therefore better than a Cp of 1.3. If the statement is false, what makes it true?

5. Using 10 words or less, explain why SPC is useful in machining.

UNIT 27-2

An SPC Experiment

Introduction: In Unit 27-2, we proceed in two stages: Based on a batch of sample parts, you set up a histogram to determine UCL and LCL. Then you use that data to simulate production using a control chart of your own making.

Assignable cause
Variation specific to a factor outside the process.

Control chart
A real-time evaluation of ongoing production sometimes called a run chart that shows the unavoidable differences from one part to another.

Cpk ratio
The ratio comparison of the present production variation to the control limits.

Using Software

If your training lab has SPC software for student use, then by all means use it instead of paper and pencil as suggested here. Alternately, see the Shop Talk in Unit 27-1 about Internet resources for SPC downloads.

Making a Histogram

After the walk-through explanation, find 30 or 40 items that *should be* exactly the same size on a key dimension that can be easily measured with a micrometer. The number sampled doesn't matter in this exercise, but there should be enough to create a database. I'll use the inside diameter of steel washers as the key feature based on function—you could use any key dimension in a group of parts (Fig. 27-11). The nominal hole size should be a bit oversize, let's say it's 0.510 in. with a ±0.005-in. feature tolerance.

In the graph in Fig. 27-12, height represents the number of parts that measure a certain size. The horizontal graduations are 0.001-in. size difference. You may need to change the scale on yours to see the variation more clearly if the range of variation is tight. Remember,

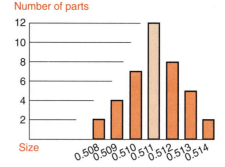

Example histogram

Number of parts

Size

Figure 27-12 *Measure and record the key feature on all parts, then create a bar graph similar to this.*

for a histogram, the order in which the washers are measured doesn't matter. We're only interested in the spread. Here are my results:

Number	Size
12	0.511 in.
8	0.512 in.
7	0.510 in.
5	0.513 in.
4	0.509 in.
2	0.508 in.
2	0.514 in.

Based on this small sample of data, the software can draw the bell curve (Fig. 27-13). From it we're reminded that the Cp for a given process isn't exactly the final spread until a large database (DB) is created. The curve extends beyond my data to predict that a few parts will vary more than those we've detected thus far.

Sample Range—The Dipstick

The larger the number of overall parts represented in the DB, the more accurate will be

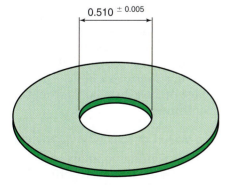

Selecting a critical dimension based on function

0.510 ± 0.005

Figure 27-11 *We'll use washers for the example parts.*

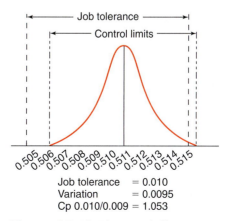

Job tolerance

Control limits

Job tolerance	= 0.010
Variation	= 0.0095
Cp 0.010/0.009	= 1.053

Figure 27-13 *Plot your bell curve.*

the histogram. If for example on a run of 4,000 washers, 5 out of every 100 are removed and measured, then the histogram represents a great deal more data. Taking 5 to represent every 100 parts is known as a *sample range*.

What Does the Graph Reveal?

The bell curve shows us that the process isn't centered. Diameters center around a slightly large 0.512 in. Looking at the diameter tolerances (vertical lines), it also shows it will predictably make a few parts with the hole too big at 0.5155 in. and left off center. There's not much chance of making an undersized part as the curve intersects the horizontal axis just before the lower size limit.

The Cp for my process isn't very good! It should be improved before going into production, but we'll continue with it as-is for the example. The print tolerance range adds up to 0.010-in. total diameter difference. The observed spread is ~0.0095, thus the Cp is 1.053:

$$\frac{0.010}{0.0095} = 1.053$$

Centering the process will slightly reduce the number of predictably bad ones, but still, this Cp shows the process will make reject parts!

Wired and Wireless Measuring Tools
Another SPC question I often hear is "doesn't all this measuring and data entering take time away from production?" The answer is yes it can, unless the machinists uses the right tools for data input, then plans the part turnaround to efficiently check while the next part is being made. Part of the solution is to use electronic measuring tools made for SPC data collection. They are equipped with data ports to automatically download measurements to the PC. Most use a data wire, but there are tools that transmit results to the PC with no wire involved.

KEYPOINT

If the process has a Cp of 1.5, then it can be statistically shown to be capable of making only 3 parts beyond tolerance, in 10,000! A Cp of 1.3 is the minimum acceptable for production; 2.0 shows no scrap will be made ever, as long as the machinist pays attention to trends and keeps the process centered.

Try It

Now, make your own histogram based on a batch of sample parts of your choosing that are easily measured with a micrometer.

Instructions

1. First, draw a horizontal line near the bottom of the paper to represent the scale.

2. Then divide that line into equal segments representing the resolution of the measuring tool. For this experiment, label them in 0.0001 increments.

3. Make sure to spread the horizontal graduations apart enough to discriminate between data points—the total range expected should consume about two-thirds the width of the paper.

4. It does not matter what scale you use for the vertical spacing for this exercise as long as they are consistent. Perhaps each tick should be 0.2 in. higher on the stack.

Measure, then mark all 30 or 40 parts on the chart. The bell curve may be a bit rough using a small sample set. Then, if not working on SPC software, roughly sketch the bell curve. The width of the curve at the baseline can represent the control limits for this experiment for now.

Find the Process Capability Now calculate the Cp for your samples. Remember, the Cp is derived by dividing the total tolerance available by the range of variation found under the bell curve.

Based on the feature you've chosen, what would be a reasonable feature tolerance? Choose 0.030 for this experiment. Use the measured spread on your chart as the variation range for this experiment.

Simulate a Control Chart

Instructions This time, we'll measure the same parts but in a sequence to simulate real-time production.

The most common **control chart** is the paired horizontal lines introduced back in Fig. 27-1, representing the upper and lower control limits UCL and LCL (Fig. 27-14). The space between represents normal variation. Just like driving a highway, the idea is to keep it centered between the lines!

Draw your control chart using the two derived control limits from your bell curve. Let's assume this time that each washer represents 5 measured from a range of 100 parts coming off the machine. So each dot represents the average of 5 washers taken from the run, every 15 minutes.

The Cpk Ratio—Capability Applied

Now the need for a numeric index arises to summarize how well the process is performing in real time. This ratio is called the **Cpk index.** It's the capability of the process—applied.

Control chart

(Normal production)

Upper control limit

Lower control limit

10.00 AM 10.15 AM 10.30 AM 10.45 AM 11.00 AM 11.15 AM 11.30 AM 11.45 AM 12 AM

Figure 27-14 *Measure each part as though it was coming off in real time, then make a control chart.*

It compares process capability to present results. It's derived by dividing the Cp by actual variation results. And again, a Cpk of 1.3 is good but 1.5 is better.

$$\frac{Cp}{\text{Present variation}}$$

Cpk is a moving value. It changes as we go along. If the operation is working well, the Cpk will remain above 1.3. If variation leaves the control limits, the Cpk falls below 1.00, and that's bad. The tighter the variation can be controlled in the process, the higher the Cpk will be. On the control graph (Fig. 27-15), the Cpk is dropping off toward a trend.

> **KEYPOINT**
> Differing from a Cp ratio, a good Cpk depends on keeping the process centered as well as controlling variables in the setup and outside it.

Detection, How We Earn Our Pay!

There's a possible problem developing. I need to solve a diminishing Cpk by investigating

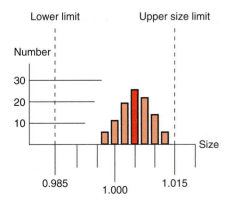

Range = 5 × 100

Lower limit   Upper size limit

Number

30
20
10

Size

0.985   1.000   1.015

Figure 27-15 *What does this histogram show?*

for the **assignable cause.** There are four general categories:

1. **Common**
   Not a cause for adjustment—key features can and will vary in repeating patterns or irregular patterns but they do not show any trend toward the control limits.

2. **Trend**
   A steady up or down change toward the control limits.

3. **Exceptional Repeating**
   An odd bump in production on the control chart—that will happen again if not corrected. It is probably assignable but will require some thought to sort out.

4. **Exceptional Nonrepeating**
   An odd bump in the control chart that investigation proves will not happen again. In this case, the bad parts are not entered into the database—they become nonevents.

> **KEYPOINT**
> Trends account for 95 percent of all our challenges. It's the daily things like tool wear, chatter, and heat. Generally, finding and fixing these things is easy.

Exceptional problems add a little mystery to the challenge. They fall into two categories. First are events that follow a pattern and can be assigned and corrected, but they don't happen very often (see the Shop Talk on the following page). For example, everyday at 4:00 P.M., a freight train goes by the factory. When it does, I make five washers that aren't the same as the rest. Or each Monday, I make several that are too big, before the

ShopTalk

Doesn't the Time Spent Doing SPC, Drive Up Manufacturing Costs? No, in fact, due to improved control and the lower rework costs, expenses actually go down. Quality-based manufacturing doesn't cost, it pays. It's always less costly to prevent than to detect errors.

Here's my best example of SPC detecting regular-exceptional variation. We were operating an ultraprecision jig boring lab making aircraft assembly fixtures. The lab was on the side of the shop near a large river. To our frustration, we found inexplicable variation on a day-to-day basis, yet the jig boring machines were the most precise in the factory and the machinists all top guns. We set up control charts and watched the trends. Soon after, we saw that hole locations were moving apart then back by about 0.0002 in. or so, in about a 6-hour swing. It took us a while to realize we were about 10 miles upstream from Puget Sound and we were looking at a tide chart! The river backed up enough to slightly change the floor shape! Knowing that weird fact, our machinists consulted the tide tables each shift, then took corrective action. No kidding, tide charts!

machine warms up. The second cause can be from needed improvements. Perhaps the carbide grade isn't right and breaks down or chatters too often. This is not routine tool wear, but, for example, we experience catastrophic failure at around 250 parts. That requires a bit more investigation to find and fix.

Nonrepeating exceptions are once-only events. They could be a sand pocket in a casting, or my machine goes offline due to a power failure so it looses data. These events are assignable, but won't happen again. Since they would affect your Cpk, yet you can't fix them, it's not cheating to eliminate their key data from the DB.

KEYPOINT

Summary
If the cause is common, it must be corrected.

If it's regular, exceptional, and controllable (I press slide hold, while the train goes by), then I take action.

If the exceptional cause is nonrepeating, but assignable, then I can eliminate its data from my collection of Cpk data.

OK—that's all you need to be ready for specific training on how to apply this super tool to your new career. Review to be sure you have the shop-ready concepts.

UNIT 27-2 REVIEW

Review the Key Terms

Assignable cause
Variation specific to a factor outside the process.

Control chart
A real-time evaluation of ongoing production sometimes called a run chart that shows the unavoidable differences from one part to another.

Cpk ratio
The ratio comparison of the present production variation to the control limits.

Replay the Key Points

◆ A Cpk of 1.5 has the ability to limit bad parts to only 3 within 10,000.

◆ An initial Cp can change over time even without refining the process.

◆ Differing from a Cp ratio, a good Cpk depends on keeping the process centered as well as controlling variables in the setup.

◆ Sample order does matter when computing a Cp.

Respond

1. Suppose a trend develops that takes a critical size just over the lower control limit line. This part must be thrown away as it is scrap. Is this statement true or false? If it is false, what will make it true?

2. What must be done assuming the trend continues in Question 1?

3. Is the compiled comparison of the overall results of a group of parts to the job tolerance the Cp or Cpk ratio?

4. The Cp ratio compares a group of machined results to _____.

5. Three companies are seeking your business. They each present their normal Cpk ratios under which they historically perform production. That of company A is 1.0, company B's is 1.33, and company C's is 1.65. If the price is similar, which company would get your purchase order?

Critical Thinking

6. A production run starts up with the goal of 2,000 parts. The early period shows a Cp of 1.5 and a Cpk of 1.3. After 1,000 parts, the Cpk is higher at 1.55. What might have happened?

Section 27 Review

Unit 27-1

When I was researching this section, a local shop manager told me, "We'd do SPC training for our people even if we didn't actually use it on our shop floor. Just knowing the concepts, our skilled people become better decision makers." The concepts of SPC create powerful control oversight. Job experience will prove this but take it from the experts, the smart machinist can prevent scrap, improve quality, and be in better control using SPC.

Unit 27-2

We've left some details out of this discussion. But most focus around statistical methods and terms. If you now understand and can compare Cp and Cpk ratios and can define histograms, normal variation, control limits, and charts, then you are ready for the workplace. Now from this end of the lesson, you can see that SPC is really a matter of common sense.

Questions and Problems

1. A *histogram* is compiled as shown Fig. 27-15, for a hole that's nominally 1.000-in. diameter with a ±0.015-in. tolerance.

    A. What's the approximate Cp? If in doubt how to proceed, see the hint in the answers.

    B. Based on the histogram, what should be done to this process?

    C. Can this process make all good parts? Is it an OK Cp?

    D. This histogram represents how many parts?

2. What does the control chart in Fig. 27-16 tell us?

3. Does the variation shown in Fig. 27-16 need to be corrected?

4. Identify some possible causes for the variation in Fig. 27-16.

Control chart

Figure 27-16 *Control chart.*

Critical Thinking

5. I have started a job and am recording on a control chart *starting with part number one off the machine*. How could I do that? Can my control chart limits vary over time?

6. The feature tolerance is ±0.001 in. After running 1,000 parts, I've found no more variation than 0.0005-in. total spread. What is the Cp of the process and what does it tell you?

7. Is the result of Problem 6 strictly correct—why or why not?

8. Identify the three kinds of causes of variation and your actions for each.

9. Define the four terms employers want you to know:
   A. Cp
   B. Cpk
   C. Histogram
   D. Control chart

10. Sum up what you believe SPC brings to the shop, the machinist, and factory.

ANSWERS 27-1

1. The process is normal or trending beyond control limit.
2. A histogram
3. True
4. False. 1.3 is better than 1.2.
5. SPC helps the machinist see variation clearly and sooner. (Nine words.)

ANSWERS 27-2

1. False. The control limits are the edges of normal variation, not the job tolerances. This part is probably still well within the job tolerance if the Cpk is high.
2. A correlation must be found to explain the trend and corrective action must then be taken.
3. That is the Cp ratio
4. The job tolerance
5. The Cpk ratio of 1.33 is better than 1, but I'd do business with the 1.65 company any day!
6. Experience showed improvements that could be made. The process improves with time. This is called "settling down" by machinists. On the next production run of the same parts, the Cp will be higher due to added data from the first run, thus the control limits might be tighter due to this new information.

Answers to Section Review Questions

1. To determine the normal spread, rough in the bell curve.
   (*Hint:* Sketch the approximate curve as in Fig. 27-17 with a spread = 0.0215.)
   A. 0.030/0.0215 = 1.395 Cp
   B. Center it.
   C. No. This Cp is just slightly better than the minimum acceptable for production and will make a few reject parts within 10,000.
   D. 25 sample parts measured with 5 for every 100 in the range $\frac{25}{5}$ = 5 samples × 100 parts = 500 parts total
2. Regular variation. The process is centered, within control.
3. There are two answers to this critical question. First, none of the samples exceeds the control

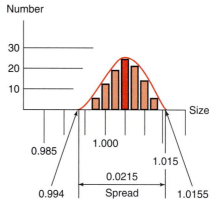

Figure 27-17 *To solve Problem 1, sketch this bell curve to determine the variation spread in the process.*

limits therefore the production is within control—do nothing. But if the reason is assignable, that is, can be found and eliminated, then the overall quality goes up—always our goal. A better Cpk.

4. Since the swing cycle seems to be every 2 hours (we don't know the number of parts made), the best guess is tool wear and replacement with a sharp cutter. A second guess might be that the operator takes a work break and stops and restarts the machine.

5. Previous data from running this or a similar job. Yes, the spread of normal variation often improves as the process becomes more mature. The machinist may learn ways to contain variation closer to the target.

6. 4 (Calc 0.002/0.0005). This process has the capability to never make a scrap part.

7. It's close as 1,000 parts represents a good DB. However, it doesn't account for variation outside, beyond the spread.

8. A. If the cause is common, it must be corrected.
   B. If it's regular, exceptional, and controllable, take action when it's called for.
   C. If it's exceptional, then you can eliminate its data from the Cpk.

9. Four terms:
   A. *Cp*—A ratio comparison of feature tolerance divided by spread—disregards centering of the process. Shows capability of the process. Above 1.3 is acceptable, with 1.5 desirable.
   B. *Cpk*—A ratio comparison of normal variation divided by actual variation in real time. A good Cpk depends on keeping the process centered.
   C. *Histogram*—A data chart showing normal variation in a process. Order doesn't matter. The upper and lower control limits are derived from these data and this chart.
   D. *Control chart*—Part features are measured and tracked on this real-time graph, hopefully keeping the key features between the derived UCL and LCL.

10. SPC provides a clear picture of a process's capability and it helps to see problem trends before scrap is made. Knowing the concepts improves decision making for all involved. The bottom line is that SPC saves money and time by reducing scrap and rework.

Section 28

Computer Coordinate Measuring

Introduction

Used skillfully, CMMs solve questions no other method can. Section 28 has two overall goals: to impart enough background so you can learn to use any generation CMM when you encounter it on the job and to show how this technology fits and functions within today's manufacturing plan.

In Unit 28-2, we'll be setting up and using a manually positioned CMM, since they are likely to be in a training lab due to their affordability. We'll also investigate the higher-level machines and how they collect and process inspection data automatically.

As much as anywhere in our trade, geometric understanding plays a big role in CMM success. Often there's as much creativity in planning and writing a CMM program as there was in the machining program that made the part. Without the insight of geometric dimensioning and tolerancing principles, writing inspection programs to evaluate complex features is nearly impossible. Just like cut sequences, there are usually several ways to solve a given geometric question, but only one or two yield best results or the functionality being sought.

CMM skills are a must-have competency for today's machinist. But that wasn't always so. In the shops of the past, the CMM was considered a specialist's job even among inspectors. It was programmed and run by a white-lab-coat person, not to be touched by shop people!

Today, however, in progressive shops, the quality philosophy is moving toward access. Three factors combine to make it so.

> The person most responsible for quality, the machinist, should also be armed with the best measuring tools to achieve it.

The equipment is evolving toward user-friendliness, thus less training and handling are required and it's also better suited to the shop environment. It does not need to be in a perfectly clean and temperature-controlled lab. Not all but some CMMs now deal with workplace dirt and temperature swings. We see these *shop hardened* versions conveniently placed within cells of machines.

Inspection has become a part of the machinist's job. In these progressive shops, inspectors turn much of their attention toward auditing the process and systems to ensure they are working right and they help the machinist assure quality. There is still a need for an inspection department and parts still must be sent there, but far less often than in the past. Once we get this new attitude right, costs go down, since effort (and money) is spent on perfecting the process not on finding the "bad" parts already made.

So, if I've made my case, you can see that while CMMs are the last chapter of this book, they are by no means an add-on subject for the modern tradesperson, they are a central competency. Here are the learning units designed to impart this vital background and skill.

Objectives

Unit 28-1 What Does a CMM Do? How Does It Work? (Pages 1020–1029)
Define the floating, work axis system
Describe the three different types of CMMs
List the advantages of computer inspection compared to standard methods
Compare data collection methods and know which is right for the task

Unit 28-2 Set Up a Manual CMM for Inspection (Pages 1029–1035)
Plan a complex CMM inspection routine
Describe and perform the three phases of setup: qualifying the machine—
   before inspection, aligning the workpiece—first part of inspection, and
   menu selections for the inspection process

What Does a CMM Do? How Does It Work?

Introduction: Without experience, most assume any computer measuring equipment does what the manual instruments and processes do, but they do it faster and more accurately. Yes, that's true in many cases (but not all), but it's only a part of the reason they improve quality and profitability. In Section 28, we're focusing on computer-coordinate machines; however, similar arguments could be presented for other computer-directed measuring equipment as well.

CMMs verify geometric requirements in ways no other process can. Proof coming up.

KEY TERMS

Conformal axis system
Axes are assigned to the work independent of machine axes also called a floating axis set.

Digitizing
Programming by hand probe motion—see teach-learn.

Hit (CMM)
Shop lingo for one recorded touch of a data collecting probe.

Nontactile measurement
Using optical methods to measure surface elements.

Scanning CMMs
Two types both of which collect entire surface element data. The tactile type maintains its probe in constant contact with a surface. The nontactile type collects entire surface element without touching.

Shop hardened
An instrument built to withstand the temperature swings and dirt of a shop environment.

Tactile measurement
Measuring by touching the object with a probe.

Teach-learn (read-learn)
Programming by recording hand probe movements and menu picks.

Computer Inspection Advantages

Cures Choke Points

High-speed mills and lathes often machine parts faster than they can be measured with mics, electronic height gages, and indicators. Without a CMM, there are three options.

In the least desirable, the operator must measure the sample while the machine cranks out more questionable parts! Or, the halt button must be used until the sample part is proven right. Either way, time is lost.

The third choke point solution is to load the cutting tool drum with a touch trigger probe (information coming) and then write inspection routines into the machining program itself (Fig. 28-1). In previous sections, we've seen probes testing cutting tools to autoload offsets. But, not all feature inspection lends itself to *in-process inspection,* nor can all CNC machines perform it. There are many times when the part must still go to the stand-alone CMM.

Figure 28-1 *In-process inspection helps check key features as they are made and keeps real-time SPC up to date.*

CMMs Evaluate Complete Shape Envelopes

Using CAM-generated programs and closed-loop CNC machining equipment, there's almost no shape that cannot be cut accurately. But, without a CMM, we are often put in the odd position of having confidence the shape is probably OK. But there's no definite way to prove it using standard measuring equipment! A tool path evaluation showed the program was right and no errors occurred when we cut the part, but we are unable to conclusively prove the part is right, to ourselves or more importantly to the customer!

KEYPOINT

A CMM can compare a CAD database or digital shape definition to the actual result.

Geometric Reports of True Values

The main reason these reports are so vital to part evaluation is that other methods of measurement don't always report data in a way that's useful, in two related ways:

1. **Element Analysis (Fig. 28-2)**

   A simple question is how far the hole location is from the datum edge. There are three possible answers depending on the function being sought.

   Edge to Datum The answer on the left in the figure is easy to get with a caliper, but by simply measuring the hole diameter, then adding the radius to the edge measurement, doesn't account for the datum and due to hole irregularities, can yield a low-reliability answer.

   Reserved Circle from Datum The middle test in Fig. 28-2 is accurate if the question is how far the reserved round space is from the datum.

To find it, the height of a snug fitting test pin is measured above the datum table. Here, a fairly accurate answer is obtained using standard tools. A typical example of this functionality would be that this hole is to receive a bolt in an assembly to another plate.

Mean Hole to Datum But one functionality can only be determined by computer analysis. Suppose the hole is a precision laser aperture or a spray nozzle that must exactly center its total output on a target location. The functional question then becomes how far the mean hole is from the datum. The mean hole accounts for all form irregularities. Only the CMM can sum the irregularities and then determine the location of its mean round diameter relative to the datum. Think about it, that distance will be slightly different from the pin test if the hole is not perfectly round. This is a simple example of the concept. But the conclusion is that when features and shapes become complex, only computer analysis can solve their total shape envelope before location or orientation questions can truly be answered.

2. **Geometric Value Opposed to Acceptance**

   The second justification for CMMs, with their ability to analyze elements, is that they report the value of the measurement rather than an acceptance within a tolerance range, which is often the result when using standard testing. Acceptance occurs when we measure the part to be within tolerance, but do not know the exact value of the variation.

A simple example is using a micrometer to measure a diameter and assuming the roundness is OK as well. We already know what's wrong with that assumption. Another acceptance example is illustrated in Fig. 28-3. Using shop methods, the test on the left determines that the TIR is 0.003 in. It's flat enough, and within the 0.005-in. tolerance, but truthfully the exact flatness is not known.

Different answers

Radius

Edge measure

Pin to datum

Mean diameter to datum

Datum feature irregularities

Datum

Figure 28-2 *Three answers to the question of how far the hole is from Datum A.*

Requirement ⊐ 0.005

Flat = 0.0015

0.003-in. TIR part is acceptable

Datum table

Standard test

CMM element evaluation

Figure 28-3 *Only the CMM can detect the true flatness value filtered from tilt.*

In fact, this part could be perfectly flat or slightly unflat, but also tilted to cause the 0.003-in. TIR. Part of the problem is that the height gage/indicator test is actually a parallelism test. Flatness is embedded, so we can say the part is good enough but not exactly how flat it is. When requested to report flatness, the CMM filters out tilt and finds true flatness to be 0.0015 in. The part in Fig. 28-4 illustrates this concept.

KEYPOINT

The CMM can filter out and report the value of the required characteristic from other variations; flatness from tilt, for this example.

If the true variation of any process is not known, then the capability of the process (Cp ratio, Section 27) will be wrong!

Reliable Accuracy

An obvious advantage of computer inspection is that it eliminates all the classic inaccuracies of standard measuring. There is no need to do it, but we could go back to the list in Section 6 and show how each one—heat, pressure, alignment, and so on—are eliminated altogether or reduced beyond any significant effect by CMMs. The repeatability of today's CMM is in

Figure 28-4 *Near perfect inspection of size, shape, and orientation of this tilted hole requires collecting much surface data. A scanner is the right CMM.*

the microvariation zone below 0.0001 in. or 0.003 mm.

Data Management

Another advantage of all electronic and computer testing equipment is direct input of data to the SPC database. All math is instant and to the tenth decimal place! It also reports in metric or imperial values with equal ease. Finally, a CMM provides printed inspection reports that take zero operator time other than setting them up and requesting them, for company records or if required by the customer.

An Assignable Axis System

The first difference one must understand about a CMM compared to any other CNC equipment is the **conformal axis system** also called a *floating axis*. Even though the CMM has *X*, *Y*, and *Z* axes, they are machine axes only (Fig. 28-5). The part to be inspected needn't be aligned to them. It can be set up in any convenient orientation—clamped or lightly held or even unrestrained in any way. In some cases, the work is not fastened if it's heavy enough to withstand 3 to 6 grams contact force from an electronic touch trigger probe or when the data will be collected optically.

Then at the beginning of the inspection, the operator or program determines in what orientation the part lies with respect to the machine axes. Through a routine, it assigns an *X-Y* flat plane and an axis direction and origin to the part—very convenient!

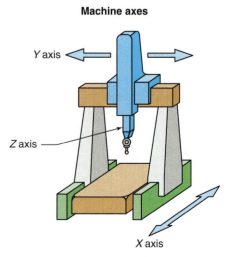

Figure 28-5 *CMMs have axes but the part needn't be aligned to them for inspection.*

Figure 28-6 *The part may be set conveniently on the table.*

Hard probe
Limited usefulness

Touch trigger probe

Hand position
only

Multipositional

Contact gives range
plus feedback

Horizontal
probes

Figure 28-7 *Hard probes versus touch trigger probes.*

KEYPOINT

Objects to be measured do not have to be aligned to the machine's axis system. The machine conforms axes to the part's orientation.

KEYPOINT

In most shops the touch trigger probe type (**tactile measurement**—*collecting data by touching*) is by far the more common method.

Notice in the illustration (Fig. 28-6) that the *X* and *Y* axes were nearly reversed relative to the machine axes. This is usually done for convenience. Part axes can be set up in any way useful. The orthogonal set may be chosen in any orientation within the envelope as long as it follows the right-hand rule used to identify axis orientation (Section 17). Can you see why the *Z* axis is minus in the illustration?

Data Collection—Probes

Feature data is collected in one of four ways. Understanding the difference is a big part of CMM skills.

Hard probe—limited usefulness and accuracy

Touch trigger probes—most common

3-D scanning probes—complex geometry or exceptional accuracy

Nontactile methods—not covered in this unit, specialized

Using a touch trigger probe is by far the more common solution. However, **nontactile** (not touching) **measurement** methods consisting of bounced laser or video camera are gaining popularity in certain specialized applications. Touching the probe to the feature surface remains the mainstay of machine shop part evaluation.

Hard Probe

A ground sphere of known diameter on a solid steel shank is shown in Fig. 28-7. The precision ground probe sphere, is brought into tangency with the object, then the screen position is read or the data is downloaded to a database. The diameter of the sphere is then accounted within the result.

Hard probes are used on manually operated machines only. Because they do not have any give range (information coming) nor do they provide feedback to CNC controls, they cannot be used when the machine is run automatically. As such, hard probes are not often used on modern CMMs, but they can be useful for quickly locating over holes as shown in the figure.

KEYPOINT

Hard probe, approach speed, bounce back, and hand pressure against the work are all controlled by operator skill and experience, thus they inject human inaccuracy factors into manual CMMs.

Touch Trigger Probes

A touch trigger probe (TTP) features a ruby or other jewel sphere on an omnidirectional joystick that allows a deflection range in every

Figure 28-8 *Only the CMM can determine if these ears are accurately machined relative to the part's datum axis.*

Silicon strain gages mounted on webs (1 out of 4 shown)

Electronics

Kinematics remain seated at low F_C

F_C

Figure 28-9 *A touch trigger probe is not a switch, it's a strain gage that changes values as the prove force F_C changes.*

How Does the Probe Signal at Zero Pressure Contact? The secret is that the probe does not signal the computer instantly as it touches the work. It's not a simple switch, it's a strain gage that reports displacement values—something like a test indicator. On contact, it begins to be displaced (moved away from neutral) by a touch in any direction against the tip sphere.

The probe produces a changing value database as it is deflected. Its position is continuously recorded with time at each position, which produces an approach displacement vector. The computer then computes backward in time to the instant that the probe touched the work. The probe controller calculates the exact position where contact was made. The switch needn't be extremely precise, or delicate. Computers do the work!

Critical Question?

How can a touch switch be sensitive enough that the instant it con-

direction including inward along the probe's shank (Figs. 28-8 and 28-9). Electronic touch probes are commonly used for either manual or programmed inspection. They automatically collect data by making single touches against the part surface, called **hits** in shop lingo.

Overshooting or overpressure are not a problem with touch triggers as the probe features a *give range*. After initial contact the probe deflects.

In CNC mode, the TTP signals the axis drives that the probe sphere is in contact so the control won't overdrive toward the work surface. They can contact the workpiece from any direction—*omnidirectional contact.*

KEYPOINT

Similar to a hard probe, the touch trigger probe takes individual hits. It doesn't sum the entire surface unless an infinite number of hits are taken—an impossibility.

tacts, at zero pressure, it changes conditions from off to on? Think this one through, the solution is a brilliant example of computer technology engineering. (See the Shop Talk on this page.)

This problem could be solved by reading the electrical contact between the probe and work. Many dedicated tool setting probes use that solution. But that would mean only conductive work could be inspected. So, how would you solve it to inspect all work, whether or not it conducts electricity? Switches can be only so sensitive! If it was sensitive enough to register at zero pressure, you couldn't even move it without it taking a false hit?

Scanning Probes

Depending on the accuracy required or the functionality being measured, individual touches on the part surface don't always detect the whole answer, as we've already demonstrated on that roundness element with a bump in it. Both hard and touch trigger probes do well in many inspection situations, but not all. Bumps in the surface can be missed. Of course, the inspection routine could include a larger number of hits, but you can see that the data density is incomplete no matter how many hits are taken. Additionally, it's going to take lots of time to collect dense data.

For the best accuracy, an entire element must be collected to determine its absolute shape. Third-generation CMMs use scanning

Seven touch trigger hits still miss the bump

Scanning probe traces and computes entire element

Figure 28-10 *Sometimes even a large number of hits can miss an unusual deviation in shape.*

probe technology to solve data density questions. Scanning probes look and work the same as touch triggers. How the machine collects the data makes the difference. The **scanning CMM** automatically traces and records the entire element Fig. 28-10. It does so by approaching then staying tangent to a given surface by keeping probe pressure within a range—while still knowing when the probe sphere was at zero pressure! Now that's engineering!

If features are complex or highly irregular or where absolute shape values are critical, scanning technology should be used to detect the truth.

CMM Differences

The planner, supervisor, and machinist must know which machine is right for a given inspection. Third-generation scanners can do the tasks of all three kinds of CMMs but they are a large investment if a lower-level machine is right for the job. There are three types of CMMs and each has its place in the shop:

Manually operated

Programmed CNC touch probe

Programmed surface scanning (third-generation equipment)

Manual CMM

These are the basic generation of CMMs with no axis drives (Fig. 28-11). Their probe must be led to the part by hand, as shown. While less sophisticated than the next two types, they are affordable and reliably robust and have a definite place in machine shops. Often overlooked because of their simplicity,

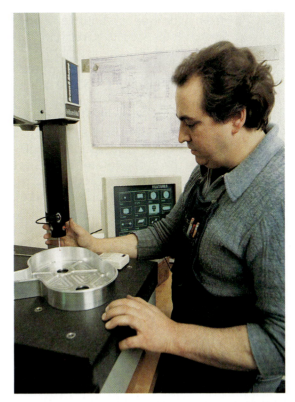

Figure 28-11 *This machinist is hand leading the probe to the machined part.*

manual CMMs are well suited to one-off and tooling work and where moderate inspection turnaround is acceptable. They are also at home where an affordable, **shop-hardened** CMM is required within the chip-making environment (see the upcoming Shop Talk).

With manual CMMs, parts can be quickly checked (without a program) for basic feature compliance, such as angularity, flatness, location, and the size of basic features or roundness. If the shop's general inspection load involves constantly changing parts with simple features or low batch numbers, then the best choice might be a manual machine.

For basic inspection on medium- to low-volume production, or for tooling work, the manual CMM is right for the job. It can also be placed near the mill and lathe for immediate quality control.

As mentioned previously, manuals are the only machines that use hard probes, which are very useful for location over holes. When the ball is in gentle contact and has settled

Beyond the Stone Age

CMMs require extremely accurate axis beams—their equivalent to ways on cutting machinery. Most are made of stable black granite, the same material from which a layout table is made. Most first-generation machines are in use today as their beams have a long life. But natural granite can be produced only in moderately long lengths. An alternative to nature-made granite is the composite beam made with resin and granite particles. But no matter what the source, nature- or man-made, these machines must be kept in a controlled environment to remain stable. Where ultimate inspection is required, inspection labs are the solution.

But when we put these CMMs out in the shop, accuracy problems arise when heat fluctuates. So, working toward absolute accuracy, ever-more accurate and stable beams weren't the answer. Brown and Sharpe, a world leader in metrology, began a new family of CMM based on a new evolutionary path.

Their solution was to apply computer power to the task. Shop-hardened CMMs (Fig. 28-12) are made using metal beams that do change with temperature. But these changes are predictable. Each machine then has an internal blueprint of its unique tendency for thermal expansion. The computer adjusts to compensate.

This not only reduces the cost of construction and improves shop access, but inspection turnaround time is also improved. To inspect work **(continued)**

from vibration or bounce, a foot switch is pressed to save that contact hit data.

More often, a touch trigger probe is used on manual CMMs to collect surface data. These increase speed and repeatability and eliminate human error. To use a TTP on a manual CMM, the operator gently moves the probe toward the object until the controller interface beeps, signaling that the probe sphere's position was known at zero pressure contact and one data point has been collected.

Manual Mode on CNC Machines

Often programmed CMMs can be operated in manual lead mode with the drives disconnected. They can also be operated in jog mode with a joystick control to perform the start of a program, until the control recognizes the *X-Y* plane and axis conformality.

Manual mode is also used for simple inspection or for low-volume work where there isn't a need to write a program (Fig. 28-13).

Manually Digitizing a Program

Second- and third-generation CMMs can be put into manual lead mode to compile programs where axis motion or approach vectors are too complex to write with commands. (See programming methods on page 1022.) As when you compile a macro routine on your PC, the CMM remembers all axis moves and screen picks, to be played back as a program. This may also be called lead-learn mode.

Programmed CMM Inspection

Touch Probe Machines— Second Generation

Besides programmed movements, the touch trigger probe can be moved similar to manual machining equipment with a joystick or another jog device on an operator control pendant (Fig. 28-13).

Figure 28-12 *A micro-val is hardened to be used by everyone in the shop.*

The programmed machine (Fig. 28-14) is best applied to medium- to high-volume production and complex but normal characteristics (not irregular surfaces).

Figure 28-13 *Similar to other CNC equipment, the automatic CMM can be driven by manual jog mode.*

Figure 28-14 *This CNC CMM can completely inspect the engine block in minutes.*

Scanning Third Generation

When geometry is complex, location and orientation tolerances are very tight, or where the entire part envelope must be quantified before testing a feature for close tolerance location and/or orientation, a scanning CMM is right for the job. The scanning CMM shown in Fig. 28-15 is testing the overall shape of a turbine blade, then it can verify its functionally critical position relative to the central axis. Only the scanner can do that.

Although a touch probe CMM could test the surface at many locations and compare them to a database describing the blade, it would never complete a full element analysis. It would still require connecting the dots between hits and could easily miss a bump!

Figure 28-15 *To determine shape, size, and location of each blade, a scanning probe is required.*

ShopTalk

Beyond the Stone Age (continued) on temperature-controlled CMMs meant large work had to be brought to the CMM room as much as 24 hours before it could be inspected in order for its temperature to equalize. Today, on shop-hardened CMMs, the part temperature and the CMM start at a similar point, since they are both within the shop environment. While machining heat remains a factor, environmental heat is dealt with by the computer.

Programming Methods

There are two ways a CNC inspection program can be compiled.

Manual Lead-Learn

Most CNC-driven machines can be moved by manual jog or hand motion. This is done to show the machine the part shape where commands cannot describe the movements. Programming by physical motion is called a **teach-learn** or **lead-learn** method. Sometimes it's called **digitizing** the program.

Robots are programmed in a similar way when axis movements are too complex to describe using programming languages; for example, a robot that paints a car or welds a fender to substructures. Movements are programmed by leading the machine through one cycle.

For a feature inspection example, consider writing a roundness test into a program. After moving the probe into the hole, the operator would select roundness, then lead the machine around the approximate circle or take several hits. The probe would be pulled up from the hole and the entire motion is remembered.

Proprietary Program Languages

Similar to CAM software, graphic/menu driven languages allow the operator to compile an inspection routine. They include commands that cause the machine to seek the part origin (see the work-alignment-for-inspection process in Unit 28-2) and commands to verify and report geometric characteristics.

UNIT 28-1 REVIEW

Review the Key Terms

Conformal axis system
Axes are assigned to the work independent of machine axes also called a floating axis set.

Digitizing
Programming by hand probe motion—see teach-learn.

Hit (CMM)
Shop lingo for one recorded touch of a data collecting probe.

Nontactile measurement
Using optical methods to measure surface elements.

Scanning CMMs
Two types both of which collect entire surface element data. The tactile type, maintains its probe in constant contact with a surface. The nontactile type collects entire surface element without touching.

Shop hardened
An instrument built to withstand the temperature swings and dirt of a shop environment.

Tactile measurement
Measuring by touching the object with a probe.

Teach-learn (lead-learn)
Programming by recording hand probe movements and menu picks.

Replay the Key Points

- Modern inspection helps pace measuring time to production time.
- A CMM can compare a CAD database or digital shape definition to the actual result.
- The CMM can filter out the value of the required characteristic from other variation; flatness from tilt, for example.
- If the true variation of any process is not known, then the capability of the process (Cp ratio), on which the control is based, is wrong.
- The CMM accepts part axes in whatever position they may be relative to the machine table. Objects to be measured do not have to be true to the machine's axis system.
- Probe approach speed, bounce back, and hand pressure against the work are controlled by operator skill and experience, thus they are an inaccuracy factor for manual CMMs.
- For basic inspection of medium- to low-volume production, or for tooling work where characteristics are not complex, the manual CMM is just right for the job.
- Touch trigger probes eliminate the inaccuracy of hand pressure and they eliminate the need for the foot switch.
- CNC touch probe CMMs are suited to inspection where the user needs to be freed from the task to concentrate on machining operations or where there are many features to be verified, thus a long routine can be taken over by the CPU.
- The scanning CMM tests the entire element then analyzes it, responding with a

near perfect value. They are best suited to complex shapes and where absolute values are required for data.

Respond

1. Identify and describe the three different kinds of CMM.
2. What are the three ways data is collected for CMM inspection?
3. You have three parts to check, with no complex surfaces. Of the three types, which is best suited to this task? Of the three, which could do the job and how?
4. Identify the two ways CNC CMMs are programmed.
5. Your CNC vertical mill makes one part every 55 seconds. You are using SPC to track progress by checking five key features—a distance, a square corner, and three hole diameters. The sample range is 3 parts in every 50. Which of the three CMMs would be ideal for your use?

UNIT 28-2

Set Up a Manual CMM for Inspection

Introduction: In Unit 28-2, we investigate setting up and using a manual CMM equipped with a touch trigger probe. These skills will transfer to higher-level CNC inspection equipment. There are four phases:

Planning the inspection Not all CMM inspection requires a plan. We'll withhold planning until after learning setup.

Machine qualification

Work qualification

Inspection of characteristics

KEY TERMS

Hit
One touch of the probe to the work registered in the computer.

Machine axes (CMM)
The rigid axes of the CMM but not necessarily of the work.

Machine zero
The constant position used to set the axis register to zero.

Qualifying
Bringing the machine up to operational level—the "wake up call."

Qualifying sphere
A sphere of known size used to determine the size and position of a probe tip.

Touch probe
An artificial ruby sphere that senses contact with the work electronically.

Work axes
The floating axis set that can be assigned to the work as it lies within the envelope.

Step 1   Machine Setup

Similar to homing a vertical mill then setting tool offsets, there are two tasks needed to get a CMM ready to measure: homing its axis system and setting up probe(s). After a bit of practice, both tasks combined shouldn't take over 5 minutes or so. Unless something unusual happens, this needn't be done more than once per day. If the machine remains powered up, then the start-up is valid until it is shut down.

Setting Machine Axes to Zero

While higher-level CMMs either home themselves or they are absolutely positioned and require no homining routine, the manual CMM must be at **machine zero** based at start-up (Fig. 28-16). Moving all three **machine axes,** the quill must be positioned at a predetermined location, at the edge of the envelope. This operation might be called **qualifying** the machine. Following a menu routine, or from a keypad, the position registers are set to zero.

Figure 28-16 *Some CMMs require a homing routine when first initialized.*

Figure 28-17 *Different probes may be qualified at setup time. Note the autochanger storage rack for other probes behind.*

Figure 28-18 *The probe is touched several times to the qualifying sphere.*

Probe Qualification

The second premeasure activity provides a sense of the size and position of its probe (Fig. 28-17). Each time a probe is changed or when first mounted at initial start-up, it must be registered in the control as to its

- Diameter
- Position relative to the quill
- Shank length—this determines its pivot distance within the probe body

These three facts can be set ahead of time for each probe or they can be found one at a time as the new probe is mounted.

All three probe factors are found by touching the probe sphere to a precise sphere located within the work envelope (Fig. 28-18).

The Qualifying Sphere A precision **qualifying sphere**, 19 mm in diameter, is mounted on the machine table (Fig. 28-18). Several touches of the probe around the sphere yield a data set. Since the CMM controller knows the size of the qualifying sphere, and is now in probe qualify mode, from five or more touches around the sphere, it can calculate the exact diameter of the **touch probe** and its position in three-space relative to machine home and the probe body. With these facts known, the CMM is ready to measure.

Multiprobe Qualification According to the inspection plan, all needed probes are usually qualified before the inspection begins. Each is assigned a number in the controller with the three parameters attached.

Step 2   Work Alignment for Inspection

OK the machine is ready to go. Inspection begins by conforming the axis system to the part and establishing the reference zero point upon the work. Every part setup is different according to its shape. We'll look at three basic ways in which this can be done.

Work conformation occurs in two steps:

1. Establishing the X-Y plane
2. Establishing the X and Y axes and their value direction; Z will be at right angles to the X-Y plane according to the right-hand rule, thus it's not necessary to establish it.

Since the controller conforms its axes to the **work,** it doesn't matter a great deal how the work is placed upon the table. However, to speed and simplify three-dimensional calculations within the CPU, it's good practice to place work roughly parallel with the machine axes.

X-Y Plane First

As shown in Fig. 28-19, part qualifying mode is selected, then with three (minimum) touches of the probe upon the surface that is bounded by the X and Y axes, the operator indicates to the computer which work surface represents the X-Y plane. At this stage, the control now recognizes how the X-Y plane is oriented within the machine.

KEYPOINT

To establish the best flat plane, hits must not be taken close to a straight line and should be spaced as far apart as possible on the surface. More than three touches are recommended if in doubt about flatness.

Note that in the setup sketches, individual hits are indicated by a star. The number shows the sequence in which they would be taken to perform a part alignment.

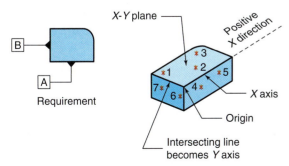

Using a line-line intersect routine

Figure 28-19 *Setting up part 1 for inspection.*

Trade Tip

When operating any CMM, help screens and icons will guide you through each step and show how many hits are required.

Minimum Touches and Part Geometry

In many cases, the number of **hits** required is just the minimum, more would yield better accuracy. For example, 10 hits on each edge of the example would have provided a more accurate picture of how straight or flat the edge was. However, two hits on an edge provide adequate information to establish a line, but not the line's geometric characteristics.

KEYPOINT

In some routines there are a required number of hits—no more, no less. In most it's a least number and more is OK, and usually yields better data. It's a judgment call based on geometric principles.

Establishing Axes and Program Reference Zero (PRZ)

X Axis First The operator has the flexibility to assign any direction on the work, parallel to the X axis. This is done by touching the X line in two locations—with the first being closer to the origin (*4 in Fig. 28-19) then the next being further from the origin (*5). That informs the CMM which direction is positive X.

KEYPOINT

Establishing the X axis is accomplished by first touching the feature closest to the origin then touching it farther away, which shows both the X axis line and which direction is positive.

X-Y Origin The next step depends on the part shape. If it's square as in Fig. 28-19, and it has a Y axis, then Y can be established in the same way as X. Here the Y axis is hit

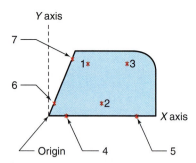

Figure 28-20 *On part 2, the intersecting lines determine the origin and the Y axis.*

twice, going away from zero. At that time, the CMM knows the intersection of each line forms the *X-Y* origin.

On part number two (Fig. 28-20), the *X* axis direction is first established by hits 4 and 5 then the origin is established at the intersection of the slanted line using hits 6 and 7. The *Y* axis then extends out from the origin according to the right-hand rule, even though it does not exist on the part.

In Fig. 28-12, the center of the hole is to be the origin. That's called a separate origin routine on the Brown & Sharpe Microval®, shown previously. After hits 1, 2, and 3 establish the plane, hits 4 and 5 establish the *X* angle and direction *but* see the Key Point.

KEYPOINT

*4 and *5 do not establish the *X* axis—only its angle and direction relative to the machine.

Using a separate origin menu with touches 6, 7, and 8 on the hole's rim determines its circular element, and thus its center to become the origin. For this part, the *X* and *Y*

Establishing a separate origin

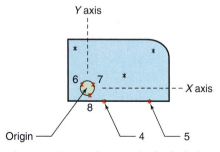

Figure 28-21 *On part 3, the hole becomes a separate origin for the X and Y axes.*

axes will then be automatically established according to the right-hand rule, at the hole's center.

Step 3   Work Inspection

OK, the work is qualified and ready to measure. Next a menu is chosen to inspect the characteristics in question, one at a time. For example, the size and position of a hole or the distance between two sides of a part and so on. Menus will guide you through the needed touches and movements (Fig. 28-22). Using screen icons and help questions, this is easy to do.

Saving Macros

When there are more than a few parts to measure, and inspection is complex with several menu selections, menu picks can be compiled into a macro, thus eliminating the need to repeat actions. That not only saves time, it frees the operator's hands from keyboard or touch screen duties.

Planning an Efficient Inspection

Unless it's a simple or one-off inspection, a little time spent planning can save lots of

Figure 28-22 *A typical screen selection for feature evaluation.*

Figure 28-23 *The machinist has planned this inspection using notes on the print.*

confusion and mistakes. Look at Fig. 28-23 to see how a CMM inspection for this test bolt project might be planned out. The handwritten notes identify features in their sequence. Notes keep the inspection process on track (and the inspector too):

Key features to be measured

The sequence in which they will be inspected

The menu picks (to become a program)

KEYPOINT

Without an inspection plan, it's easy to loose track of which dimension represents which feature on the CMM printout report. This is especially true for complex work or where many parts must be checked.

Notice the handwritten notes on the drawing. These are the routine sequences planned by the operator for this inspection. They will coincide with the actual report of the job.

Trade Tip

While writing on a drawing is taboo in many shops unless they use throwaway drawings, an inspection copy is often created that can be used for CMM planning and programming purposes only—but it must be clearly noted as such and never returned to the shop.

CMM Printout Reports

CMM reports vary widely based on controller manufacturer and the way the operator sets up the program. They can be a simple line-by-line recap of the menu picks and measured results. They can also display tolerances in both coordinate or geometric values. They can be set up with metric or imperial values as well and show them with graphics if needed. They can show compliance or beyond tolerance and they can be put into SPC Cpk format or even plotted on a control chart!

Figure 28-24 shows an inspection report such as this that would be easy to follow. Notice the SPC bar graph on the bottom. It looks as though a key feature is showing a trend.

Figure 28-24 A graphic inspection report helps digest the data.

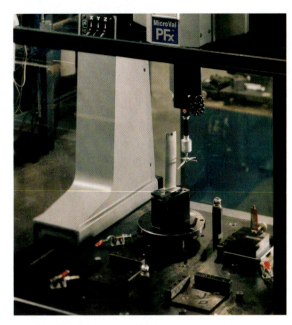

Figure 28-25 *Several part fixtures preset to speed inspection.*

Fixture Inspection (Fig. 28-25)

To speed inspection, part-holding devices such as precision vises and angle plates can be used and also fixtures can be made similar to those for mill work. They bring three advantages:

1. It stabilizes odd-shaped work that would not easily lie on the table.

2. It speeds repetitive inspection where the same part is checked many times during the day. That way, the initial part qualification can be automatic too, since it will always lie at the same place and orientation.

3. Two or more fixtures can be preset similar to fixure offsets in machining.

KEYPOINT

Making a part alignment each time can be time consuming for high-speed inspection. Seven to 10 probe touches can be eliminated from the plan (and program) if the part is reliably placed in a precision fixture.

UNIT 28-2 REVIEW

Review the Key Terms

Hit
One touch of the probe to the work registered in the computer.

Machine axes (CMM)
The rigid axes of the CMM but not necessarily of the work.

Machine zero
The constant position used to set the axis register to zero.

Qualifying
Bringing the machine up to operational level—the "wake up call." Setting probe diameter and length in the control.

Qualifying sphere
A sphere of known size used to determine the size and position of a probe tip.

Touch probe
An artificial ruby sphere that senses contact with the work electronically.

Work axes
The floating axis set that can be assigned to the work as it lies within the envelope.

Replay the Key Points

◆ The needed number of hits necessary to compile a complete geometric image of the object is an operator call based on experience. However the computer will always require a minimum number for the base function.

◆ There are four phases to setting up a CMM: planning the inspection (when complex), machine qualification (home and probe), work qualification (*X-Y* plane and axis set), and inspection of characteristics.

◆ To speed and simplify three-dimensional calculations within the CPU, it's good practice to place work roughly parallel with the machine axes.

◆ To establish a plane, touches must not be in a straight line. More than three touches are recommended if in doubt about flatness.

Respond

1. Describe two tasks that must be done to get a CMM ready to inspect.

2. What do you do when qualifying a probe sphere?

3. Why choose a touch trigger probe over a hard probe?

4. To begin an inspection of a part what must be done?

5. What advantage does a fixture offer for CMM inspection?

Unit 28-1

Because it came into industrial application later than CNC-programmed machine tools, CMM technology is now evolving at a faster pace than most other aspects of modern manufacturing. One advancement yet to be solved is the combining of surface finish testing with shape evaluation.

Looking closely, one sees that surface finish is actually feature variation at a microinch resolution. As the technology marches on toward absolute accuracy, elements at that level will need to be folded into the overall shape envelope of the feature.

Unit 28-2

With a background in geometrics, CNC machining, shop math, and metrology, you'll have no difficulty learning to operate any CMM. When first learning to use our new manual CMM in our tech school, I was able to decrate it, set it up, and do most of the basic functions all on a Sunday morning! Of course, the higher-level machines require more formal training but as with all computer equipment, they are becoming ever-more user-friendly.

Questions and Problems

1. A. Of the three CMM types, which would be ideally suited to evaluate the key features in Fig. 28-26 and why?

   B. Identify the hits by number and describe their purpose to inspect this drill gage. Follow this plan:
      (1) Set up line-to-line intersect origin A at B;
      (2) inspect 1.5000-in. key feature; and
      (3) inspect five holes for size, roundness, and position.

2. The CAM software cycle time is 4.5 minutes for the profile, pocket, and hole and you find that it takes you 2 minutes to change parts in the vise, which CMM is best for the task if the part count is to be 750 total?

3. Identify and describe the three probe types.

4. Which probe(s) take hits?

5. What ability must a probe have to be used with a CNC machine either a cutting machine or CMM? Explain.

Datum A

1.5000 key dim

Datum B

Figure 28-26 *Plan to inspect the key features of this drill gage.*

6. What inaccuracies exist when using a hard probe that do not occur with touch triggers and scanners?

Critical Thinking Questions

7. In your opinion, of the advantages of CMM technology, which is the most important? Explain.

8. Briefly describe how a touch trigger probe computes the zero pressure hit?

9. When during an inspection do you not need to qualify the probe?

10. Can a program be written for a manual CMM? If yes, describe it.

11. You need to know the mean distance and position of a complex contoured surface, from a datum. Which CMM can do the job? Explain.

12. When would a hard probe be a good application?

13. What are the two steps required to set up a manual CMM for inspection?

14. Describe qualifying a probe on a CMM.

15. List the three facts that probe qualifying tells the CMM control.

16. True or false? To inspect a hole for size, and location, I need not set up (qualify the part). If the statement is false, what makes it true?

17. Describe the number of hits required to determine the corner radii size on the drill gage drawing, Fig. 28-26.

18. Describe a typical part alignment on a CMM.

19. Define a conformal axis system.

Critical Thinking Problem

20. Add up the total, minimum number of hits that would be required to inspect the drill gage in Fig. 28-26.

Section 28 Answers

ANSWERS 28-1

1. Manual CMM—
   Set up and moved with hand action
   Can use a touch trigger probe or hard probe
   Can inspect with or without a program
   Best for tooling and simple work but quick to use
   CNC—Programmed CMM
   Moved by program or by jog or by hand motion on some when drives are disconnected
   Use touch trigger probes only to collect finite data
   Best for regular feature inspection and for large batch work
   CNC—Scanning CMM
   Moved by programs, manual jog, and some by hand lead with drives disconnected
   Use scanning probe to collect complete surface element(s)
   Best for complex shapes and very tight tolerances

2. Hard probe, touch trigger, scanning (touching or optical)

3. The manual CMM is best suited to this job. The touch probe or scanning CMMs could be used if a program was written or if the machine was operated in manual mode.

4. By teach-learn (digitizing) or by programming languages

5. The key facts here are that while there is a long time between testing sample ranges (55 seconds times 50) there is less than a minute between part turnaround cycle time. A manual CMM could do the job but a CNC touch probe machine would be best as long as it was programmed. Scanning could do the job too but it's not needed for this object.

ANSWERS 28-2

1. Home the axes; qualify the probe
2. Showing the machine where the sphere lies with respect to the machine axes (quill); determining the sphere diameter
3. Automatic downloading of data; eliminates bounce-back and touch pressure errors

4. Determine X-Y plane; determine X axis and positive direction; determine origin—from which the Y axis radiates per the right-hand rule
5. Quick placement of otherwise difficult shaped parts

Answers to Section Review Questions

1. A manual CMM.
   All three are able to evaluate this simple part but the CNC versions would be overkill!

Hit Numbers	Purpose
1, 2, 3	Establish X-Y plane
4, 5	Establish X, line-line intersect for origin
6, 7	Establish Y line = origin
8, 9	Measure 1.500 distance
10, 11, 12	Measure first hole—three hits sets size and location
13, 14, 15	Second hole
16, 17, 18	Third hole
19, 20, 21	Fourth hole
22, 23, 24	Fifth hole

Repeat three minimum hits for each hole—nine more hits

2. A CNC CMM with a program due to the time factor
3. *Hard probe*—ball on a rigid stick, *touch trigger probe*—ball on joystick, *scanning probe*—seeks entire surface keeping constant pressure
4. This was a trick question—both the hard and touch trigger probes take hits. The hard probe requires your touch of a foot switch, while the touch trigger probe does it automatically.
5. Feedback to the control to stop approaching the work
6. Bounce back, vibration, hand pressure, and to some extent hand heat transfer if you grasp the probe body to lead it (the usual method)
7. *Best answer:* Element analysis because no other method can do that. *Good answer:* Improved accuracy reduces cost and improves quality management.
8. The CPU stores changing position values as probe pressure builds on the strain gage. The computer calculates back in time to the position where zero pressure occurred.
9. Truthfully, the probe must always be qualified, no exception. However, if it was qualified at start-up or before the inspection, it need not be done again.
10. Yes. It's a string of menu picks glued together into a macro. It saves making menu choices but doesn't move the probe.

11. A scanning CMM because the entire surface must be summed up before its location and orientation can be determined.
12. Locating over drilled holes without considering their roundness
13. Home the machine; qualify one or more probes
14. Touching a precision sphere in several locations
15. The diameter of the probe; its position relative to the probe body; its pivot distance from the probe body
16. False. Hole size does not require part setup but location does. Since location requires a datum (place to start the location from) that surface or feature must be detected first.
17. A radius (fillet) is part of a circle. It can be measured as a circle, with the report as a radius value, or if the menu allows as a radius. It requires three hits minimum, just the same as a circle.
18. Determine X-Y plane; determine X axis and positive direction; determine Y axis or origin depending on part shape
19. One such that the machine accepts part axis orientation independent of the machine's axis system.
20. 24 hits

American Made Starrett®

Precision Quality Since 1880

The L.S. Starrett Company • Athol, MA 01331, U.S.A.

INCH / METRIC TAP DRILL SIZES & DECIMAL EQUIVALENTS

Inch Tap Drill Sizes (Part 1)

Fraction or Drill Size	Decimal Equivalent	Tap Size
NUMBER SIZE DRILLS		
80	.0135	
79	.0145	
1/64	.0156	
78	.0160	
77	.0180	
76	.0200	
75	.0210	
74	.0225	
73	.0240	
72	.0250	
71	.0260	
70	.0280	
69	.0292	
68	.0310	
1/32	.0312	
67	.0320	
66	.0330	
65	.0350	
64	.0360	
63	.0370	
62	.0380	
61	.0390	
60	.0400	
59	.0410	
58	.0420	0 - 80
57	.0430	
56	.0465	
3/64	.0469	
55	.0520	
54	.0550	
53	.0595	1 - 64,72
1/16	.0625	

Inch Tap Drill Sizes (Part 2)

Fraction or Drill Size	Decimal Equivalent	Tap Size
9/64	.1406	
27	.1440	
26	.1470	
25	.1495	
24	.1520	10 - 24
23	.1540	
5/32	.1562	
22	.1570	
21	.1590	10 - 32
20	.1610	
19	.1660	
18	.1695	
11/64	.1719	
17	.1730	
16	.1770	12 - 24
15	.1800	
14	.1820	12 - 28
13	.1850	
3/16	.1875	
12	.1890	
11	.1910	
10	.1935	
9	.1960	
8	.1990	
7	.2010	1/4 - 20
13/64	.2031	
6	.2040	
5	.2055	
4	.2090	
3	.2130	1/4 - 28
7/32	.2188	
2	.2210	

Inch Tap Drill Sizes (Part 3)

Fraction or Drill Size	Decimal Equivalent	Tap Size
23/64	.3594	
U	.3680	7/16 - 14
3/8	.3750	
V	.3770	
W	.3860	
25/64	.3906	7/16 - 20
X	.3970	
Y	.4040	
13/32	.4062	
Z	.4130	
27/64	.4219	1/2 - 13
7/16	.4375	
29/64	.4531	1/2 - 20
15/32	.4688	
31/64	.4844	9/16 - 12
1/2	.5000	
33/64	.5156	9/16 - 18
17/32	.5312	5/8 - 11
35/64	.5469	
9/16	.5625	
37/64	.5781	5/8 - 18
19/32	.5938	
39/64	.6094	
5/8	.6250	
41/64	.6406	
21/32	.6562	3/4 - 10
43/64	.6719	
11/16	.6875	3/4 - 16
45/64	.7031	
23/32	.7188	
47/64	.7344	
3/4	.7500	

Metric Tap Drill Sizes

Metric Tap	Tap Drill mm	Decimal Equiv. Inches
M1.6 x 0.35	1.25	.0492
M1.8 x 0.35	1.45	.0571
M2 x 0.4	1.60	.0630
M2.2 x 0.45	1.75	.0689
M2.5 x 0.45	2.05	.0807
M3 x 0.5	2.50	.0984
M3.5 x 0.6	2.90	.1142
M4 x 0.7	3.30	.1299
M4.5 x 0.75	3.70	.1457
M5 x 0.8	4.20	.1654
M6 x 1	5.00	.1968
M7 x 1	6.00	.2362
M8 x 1.25	6.70	.2638
M8 x 1	7.00	.2756
M10 x 1.5	8.50	.3346
M10 x 1.25	8.70	.3425
M12 x 1.75	10.20	.4016
M12 x 1.25	10.80	.4252
M14 x 2	12.00	.4724
M14 x 1.5	12.50	.4921
M16 x 2	14.00	.5512
M16 x 1.5	14.50	.5709
M18 x 2.5	15.50	.6102
M18 x 1.5	16.50	.6496
M20 x 2.5	17.50	.6890
M20 x 1.5	18.50	.7283
M22 x 2.5	19.50	.7677
M22 x 1.5	20.50	.8071

Tap and Drill Sizes

Number Size Drills

Drill	Decimal
52	.0635
51	.0670
50	.0700
49	.0730
48	.0760
5/64	.0781
47	.0785
46	.0810
45	.0820
44	.0860
43	.0890
42	.0935
3/32	.0938
41	.0960
40	.0980
39	.0995
38	.1015
37	.1040
36	.1065
7/64	.1094
35	.1100
34	.1110
33	.1130
32	.1160
31	.1200
1/8	.1250
30	.1285
29	.1360
28	.1405

Taps: 2-56,64 · 3-48 · 3-56 · 4-40 · 4-48 · 5-40 · 5-44 · 6-32 · 6-40 · 8-32,36

Letter Size Drills

Drill	Decimal
1	.2280
A	.2340
15/64	.2344
B	.2380
C	.2420
D	.2460
E — 1/4	.2500
F	.2570
G	.2610
17/64	.2656
H	.2660
I	.2720
J	.2770
K	.2810
9/32	.2812
L	.2900
M	.2950
19/64	.2969
N	.3020
5/16	.3125
O	.3160
P	.3230
21/64	.3281
Q	.3320
R	.3390
11/32	.3438
S	.3480
T	.3580

Fractional Drills

Drill	Decimal	Tap
49/64	.7656	
25/32	.7812	7/8-9
51/64	.7969	
13/16	.8125	7/8-14
53/64	.8281	
27/32	.8438	
55/64	.8594	
7/8	.8750	1-8
57/64	.8906	
29/32	.9062	
59/64	.9219	
15/16	.9375	1-14
61/64	.9531	
31/32	.9688	
63/64	.9844	1 1/8-7
1	1.0000	
1 3/64	1.0469	1 1/8-12
1 7/64	1.1094	1 1/4-7
1 1/8	1.1250	
1 11/64	1.1719	1 1/4-12
1 7/32	1.2188	1 3/8-6
1 1/4	1.2500	
1 19/64	1.2969	1 3/8-12
1 11/32	1.3438	1 1/2-6
1 3/8	1.3750	
1 27/64	1.4219	1 1/2-12
1 1/2	1.5000	

Metric

Thread	Drill (mm)	Decimal
M24 x 3	21.00	.8268
M24 x 2	22.00	.8661
M27 x 3	24.00	.9449
M27 x 2	25.00	.9843
M30 x 3.5	26.50	1.0433
M30 x 2	28.00	1.1024
M33 x 3.5	29.50	1.1614
M33 x 2	31.00	1.2205
M36 x 4	32.00	1.2598
M36 x 3	33.00	1.2992
M39 x 4	35.00	1.3780
M39 x 3	36.00	1.4173

PIPE THREAD SIZES (NPSC)

THREAD	DRILL
1/8-27	11/32
1/4-18	7/16
3/8-18	37/64
1/2-14	23/32
3/4-14	59/64
1-11 1/2	1 5/32
1 1/4-11 1/2	1 1/2
1 1/2-11 1/2	1 3/4
2-11 1/2	2 7/32
2 1/2-8	2 21/32
3-8	3 1/4
3 1/2-8	3 3/4
4-8	4 1/4

Drill Gage Drawing

Drill Speeds—Common Bit Sizes in Six Materials

Notes

◆ All speeds are based on high speed steel (HSS) drill bits.

◆ If your machine does not have the RPM listed, shift to next lower speed.

◆ Data is simplified for learning. For a complete listing of letter, number, fraction, and metric drill RPMs see *Machinery's Handbook*.

◆ Or compute your RPM using the following simplified formula:

	Mild Steel	Carbon Steel-Annealed	Aluminum	Soft Brass	Cast Iron	Annealed Stainless
Diam.						
1/8	3,200	2,880	8,000	5,600	3,200	2,880
3/16	2,133	1,920	5,333	3,733	2,133	1,920
1/4	1,600	1,440	4,000	2,800	1,600	1,440
5/16	1,280	1,152	3,200	2,240	1,280	1,152
3/8	1,066	960	2,667	1,867	1,067	960
7/16	914	823	2,285	1,600	914	823
1/2	800	720	2,000	1,400	800	800
9/16	711	640	1,778	1,244	711	711
5/8	640	576	1,600	1,120	640	576
11/16	581	523	1,454	1,018	582	523
3/4	533	480	1,333	933	533	480
13/16	492	443	1,230	862	492	443
7/8	457	411	1,143	800	457	411
15/16	426	384	1,066	747	426	384
1	400	360	1,000	700	400	360
1 1/16	376	339	941	658	376	339
1 1/8	355	320	888	622	355	320
1 3/16	336	303	842	589	336	303
1 1/4	320	288	800	560	320	288

Table title: $\dfrac{4 \times \text{Surface Speed}}{\text{Diameter of Rotating Object}}$

Appendix IV

Recommended Cutting Speeds for Six Materials: Surface Speeds in Feet Per Minute

Notes

◆ These numbers are biased toward safe/convenient learning.

◆ With experience, cutting speeds shown may be exceeded.

◆ Stronger setups, coolants and many factors combine to determine the final result

Cutting Tool	Mild Steel	Carbon Steel-Annealed	Aluminum	Soft Brass	Cast Iron	Annealed Stainless
HSS	100	80	250 to 350	175	100	80 to 100
Carbide	300	200	750 to 1000	500	250	200 to 250

Superabrasives

Superabrasives

Grinding Wheel Selection Guide
Surface, Cylindrical, Centerless and ID

Diamond and CBN Basics

USE DIAMOND FOR:	USE CBN FOR:
Cemented Carbide	High Speed Tool Steels
Glass	Hardened Carbon Steel
Ceramics	Alloy Steels
Fiberglass	Aerospace Alloys
Plastics	Abrasion-Resistant Ferrous Materials
Abrasives	

In general, CBN is used to grind ferrous materials, and diamond is used to grind nonferrous materials.

TYPES OF DIAMOND

RESIN BOND PRODUCTS

AMD: A blocky shaped armored diamond that prevents excessive wear when a high percentage (over 35%) of the area is steel or braze. 100-400 grit.

ASD: An armored diamond that is the most versatile of the diamond types. Used where 1/3 or less of the total area is steel. 60-800 grit.

ASDC: An armored diamond used mostly on dry applications where no steel or braze is contacted. A second choice to A2D when dry grinding 100% carbide, as it improves edge holding ability. 60-600 grit.

A2D: An armored diamond with the advantages of ASD, plus longer wheel life when no steel or braze is contacted. 60-400 grit.

A4D: An armored diamond, similar to ASD, except freer cutting and milder acting. 60-800 grit.

DED: A blocky shaped diamond similar to RMD and AMD. Used when grinding carbide and steel combination in excess of a 50/50 ratio.

SD: A synthetic diamond suitable for wet or dry grinding when freer cutting is desired. 60-400 grit.

VITRIFIED BOND PRODUCTS

D: A designation used for diamond micron sizes. Available in micron 40/60 through 2/4

RMD: A medium strength diamond specifically manufactured for vitrified bonds. Available in grits 80 through 320

METAL BOND PRODUCTS

M3D: Commonly used blend of strong, fragmented shaped diamond for general purpose applications on ceramics, glass and other non-metallics.

M4D: Most commonly used metal bonded diamond. Blend of strong blocky shaped diamond for performance measurable applications on glass, ceramics, refractories, and other non-metallics.

M5D: The strongest, blockiest, toughest, premium quality diamond for high performance applications on glass, refractories, and other non-metallics. M5D is premium priced.

TYPES OF CBN

RESIN BOND PRODUCTS

B: A strong uncoated CBN crystal. Used with resin bond.

CB: A coated CBN crystal having excellent abrasive retention. 60-400 grit.

C2B: A coated CBN crystal that is slightly freer cutting than CB.

C5B: The most durable CBN abrasive, C5B is a high performance, premium quality, coated CBN crystal used for production grinding. 60-320 grit.

VITRIFIED BOND PRODUCTS

1B: A strong uncoated CBN crystal commonly used for internal grinding. Available in grits 60 through 400.

VB: A premium quality CBN crystal designed to provide exceptional performance in high metal removal rate applications. Available in grits 80-320. Results show 10% to 25% less power draw compared to B types.

METAL BOND PRODUCTS

Superabrasive Specifications

		DIAMOND SPECIFICATION		CBN SPECIFICATION	
		WET		WET	DRY
Surface	Carbide	ASD150-R75B99			
	Ceramics, Composites	SD320-R100B69			
	Tool Steel (Rc 50+)			CB100-TB99	AZTEC III-100W
Cylindrical	Carbide	ASD180-R75B99			
	Ceramics, Composites	SD320-R100B80			
	Tool Steel (Rc 50+)				CB100-B99
Centerless	Carbide	ASD150-R75B99E			
	Ceramics, Composites	ASD150-R75B99E			
	Tool Steel (Rc 50+)			CB150-TBA	
ID	Carbide	SD100-R100B99			
	Ceramics, Composites	ASD320-R75B615			
	Tool Steel (Rc 50+)			B180-H150VI	

Examples of a Typical Specification:

DIAMOND TYPE	GRIT SIZE	GRADE	CONCENTRATION	BOND	BOND MODIFICATION	DIAMOND DEPTH
ASD	150	R	75	B	99	1/8

CBN TYPE	GRIT SIZE	GRADE	CONCENTRATION	BOND	BOND MODIFICATION	CBN DEPTH
B	240	H	150	V		1/8

Typical Superabrasive Wheel Shapes

6A2C WHEEL

D = Diameter
H = Hole
E = Back Thickness
T = Thickness
X = Abrasive Depth

TYPE 1A1 TYPE 1V1P TYPE 4A2P
TYPE 6A2 TYPE 11A2 TYPE 11V9

Troubleshooting Guide: Dry Grinding

PROBLEM	POSSIBLE CAUSES	SUGGESTED CORRECTIONS
Burning (excessive heat)	Wheel loaded or glazed	Dress wheel with a dressing stick
	Excessive feed rate	Reduce in-feed of wheel or work piece
	Wheel too durable	Use freer cutting specification or slow down wheel speed
Poor Finish	Grit size too coarse	Select a finer grit size
	Excessive feed rate	Reduce in-feed of wheel or work piece
Chatter	Wheel out of truth	True wheel, ensure its not slipping on mount

Troubleshooting Guide: Wet Grinding

PROBLEM	POSSIBLE CAUSES	SUGGESTED CORRECTIONS
Burning (excessive heat)	Wheel loaded or glazed	Re-dress wheel
	Poor coolant placement	Apply coolant directly to wheel/work piece interface
	Excessive material removal rate	Reduce down-feed and/or cross-feed
Poor Finish	Excessive dressing	Use lighter dressing pressure Stop dressing as soon as wheel starts to consume stick rapidly
	Grit size too coarse	Select a finer grit size
	Poor coolant flow or location	Apply heavy flood so it reaches wheel/work interface

Troubleshooting

Problem	Cause	Solution
Chatter		True wheel, ensure its not slipping on mount
Wheel will not cut	Glazed by truing / Wheel loaded	Dress lightly until wheel opens up / Dress lightly until wheel opens up / Increase coolant flow to keep wheel surface clean / Never run wheel with coolant turned off
Slow cutting	Low feeds and speeds	Increase feed rate, increase wheel speed (observe maximum wheel speed)
Short wheel life	Incorrect coolant flow / Low wheel speed / Excessive dressing / Wheel too soft or too hard	Apply coolant to flood wheel/work surface / Increase wheel speed / Use lighter dressing pressure / Change grit or grade, use higher concentration

Expected Surface Finish by Grit Size

Use these charts as guides only. Surface finish is affected by a number of variables i.e., machine type and condition, type of material ground, coolant, wheel speed, bond system, etc.

FOR CARBIDE

DIAMOND GRIT SIZE	EXPECTED FINISH MICRO INCH AA	MAXIMUM DEPTH OF CUT PER PASS FOR GRIT SIZE
100	24 TO 32	0.001" TO 0.002"
120	16 TO 18	0.001" TO 0.002"
150	14 TO 16	0.001" TO 0.002"
180	12 TO 14	0.0007" TO 0.001"
220	10 TO 12	0.0007" TO 0.001"
320	8	0.0004" TO 0.0006"
400	7 TO 8	0.0003" TO 0.0005"

FOR HIGH SPEED STEEL

CBN GRIT SIZE	EXPECTED FINISH WITH OSCILLATION	EXPECTED FINISH PLUNGE
100	35 TO 40	40 TO 45
120	30 TO 35	35 TO 40
150	25 TO 30	30 TO 35
180	20 TO 25	25 TO 30
220	15 TO 20	20 TO 25
320	10 TO 15	15 TO 20
400	4 TO 8	5 TO 10

Expected Surface Finish (RMS) for Sprayed Coatings

GRIT/CONCENTRATION	CERAMICS Cr Oxide 99%	Cr Oxide 96%	Al Oxide	AlfIt	METALS 420ss	NiCr	NiCrMo	CARBIDES 88/11	88/12	88/12 Wc/Co
150/75	19	25	21	22				22	25	26
220/75	15	19	17	18				17	19	17
280/75	13	13	13	15				14	15	12
500/75	7	10	7	8				8	8	8
B150/75					21	20	21			

Wheel types: TYPE 12A2, TYPE 12V9, TYPE 15V9

C3B: A strong coated CBN crystal used in all bond modifications. C3B is the most commonly used metal bond. Available in grits 80 through 400.

SUPERABRASIVE WHEEL GRIT SIZE
Superabrasive grit size rule of thumb is to go two to three grit sizes finer than a conventional abrasive grit size wheel, to achieve the same relative finish. A vitrified conventional wheel with 80 grit would require a 150 grit Diamond or CBN wheel to produce similar finishes.

DIAMOND WHEEL GRADES
The most common Diamond Resin Bond grades are H,N and R grade. H is least durable, N is midrange durability, R is most durable.

The most common Diamond Vitrified Bond grades are L (least durable), P (midrange durability), R (most durable).

The most common Diamond Metal Bond grades are L (least durable), N (most common). Unlike resin and vitrified bond the grade is not always designated for metal bonds.

CBN WHEEL GRADES
For Resin Bond CBN, the grade includes the relative amount of CBN: Q is least durable, T is midrange durability, W is very durable, Z is extremely durable (most CBN/highest concentration).

Vitrified Bond CBN is available in grades E (least durable/mildest acting) to K (most durable/hardest acting). Note vitrified CBN has lage and concentration markings.

Metal Bond CBN is most common grades are Q (freest cutting/least durable), T (freer cutting than W, but more durable than Q), W (most durable).

DIAMOND CONCENTRATION
Concentration is the relative amount of diamond by carat weight in a wheel. Concentrations can range from 25 to 200. Standard concentrations are equivalent in percentages of:

100 Concentration: 25% (diamond/CBN) volume of the abrasive section
75 Concentration: 18% (diamond/CBN) volume of the abrasive section
50 Concentration: 12.5 (diamond/CBN) volume of the abrasive section

The higher the number, the more superabrasive there is in the wheel thus more cutting teeth and the wheel would be harder acting.

BOND SYSTEMS
Resin Bonds (B): For most Precision Grinding operations, including cylindrical, surface, and internal grinding of Carbides and Ceramics. The exceptional fast and cool cutting action is the reason they are suited to sharpen multi-tooth cutters, reamers, etc. Easy to true and dress.

Metal Bonds (M): MSL available in various shapes for dry, offhand reconditioning of carbide tools and composite materials.

Vitrified Bonds (V): Popular in offhand grinding of carbide tools, suitable for Carbide Creep Feed and Internal grinding, and are very durable for holding form/shapes.

Rules of Thumb

Coarse Grits:	Best life and stock removal rate, good form holding
Fine Grits:	Low stock removal, best finish and good life
High Wheel Speed:	Better finish and form holding, hard acting
Low Wheel Speed:	Worse finish and form holding, softer acting, shorter life
High Concentration:	Longer life, better form holding, harder acting
Low Concentration:	Shorter life, soft acting
Diamond Wheels:	Wet = 5500 - 6500 SFPM; dry = 3500 - 4500 SFPM
CBN Wheels:	Wet grind = 6500 - 9500 SFPM
Water Coolant:	Highest stock removal rate, fair finish
Oil Coolant:	Lower stock removal rate, better finish

AFTER TRUING
Truing is defined as altering wheel geometry so that the wheel is precisely formed and running concentric with the center line of the machine spindle.

AFTER DRESSING
Dressing superabrasive wheels is a cleaning/sharpening process. For resin bond dressing sticks, the grain should be 1 or 2 grit sizes finer than the superabrasive grain in the wheel. A 120 grit wheel would require a 150 or 180 grit stick. For metal bonds, use same grit size or one size coarser than the wheel. Vitrified bonds generally do not require stick dressing.

SafetyTIPS

SAINT-GOBAIN ABRASIVES

HOW TO USE THIS GUIDE

1 Select contact area of part ground.
2 Select performance level required.
3 Select grinding application.
4 Select material type.
5 Wheel specification: first choice is preferred second choice is optional.
6 Adjust for grit size depending on finish and corner holding (refer to Grit Size Guide).

1 CONTACT AREA GUIDE
NARROW - area of grind less than 25% of wheel width or less than 1/4" wide.
MEDIUM - area of grind 25%-50% of wheel width or 1/4" - 1/2" wide.
WIDE - area of grind greater than 50% of wheel width or over 1/2" wide.

2 PERFORMANCE LEVEL GUIDE
Key variable in determining the abrasive type selected.
Total grinding cost and productivity will depend on the abrasive selection.

3 GRINDING APPLICATION
4 MATERIAL TYPE
5 WHEEL SPECIFICATION

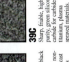

5SG60-MVFL	3SG60-MVFL, 32AA60-MVFL	53G60-MVFL
5TG120-KVFL, 5SG80-KVFL	3SG80-KVFL, 32AA80-KVFL	32A80-KVFL, 53A80-KVFL
	3TC46-JVK, 32AA46-JVFL	37C46-JVK, 32A46-JVFL

STRUCTURE

Select the most open structure specification that will hold the required form and tolerances for the application.

5, 6	Tight form holding, large amount of interrupted cuts.
8	General purpose, medium to wide contact area with good form holding.
10	Wide to medium contact area for good form holding, cooler cut & better chip clearance.
12	Wide to medium contact area for coolest cut & maximum chip clearance.

BONDS

Vitrified bonds ending with a iP" denote porous open structure type products offering a very cool cut and high chip clearance for heat sensitive materials in wide area of contact grinding applications.

BOND TYPE	DESCRIPTION
VBE, VBEP	Toolroom standard, exceptional versatility, Alundum® abrasives only
VS, VSP	High performance, SG abrasives only
VTR, VTRP	Enhanced performance & versatility for all toolroom grinding, 32AA abrasive only
VH	High performance form holding, for TG, SG, & Alundum abrasives
VFL	Pressed to size I.D. wheels less than 4-1/2" diameter, for TG, SG, & Alundum abrasives
V	Original Norton® vitrified bond, excellent form/corner holding, Alundum abrasives only
VK, VKP	For silicon carbide (Crystolon®) abrasives only

TYPES OF ABRASIVE GRAIN

TG	5SG	3SGP	32AA	38A	32A	23A	53A	37C	39C
Second generation ceramic abrasive. More durable than SG for extreme applications on the most difficult to grind materials offering maximum wheel life. Contact your Norton sales representative for specifications.	Maximum durability, for the most demanding applications & most difficult to grind materials offering exceptional wheel life.	High performance & durability for very demanding applications & very difficult to grind materials. Good value when maximum performance is not required.	A Norton exclusive, with Norton's TG Abrasive. More aggressive & durable, cooler cutting, higher stock removal rates with up to twice the life of 32A.	The most friable abrasive for maximum coolness of cut on very heat-sensitive tool steels. For light to medium cut rates.	The toolroom benchmark. A strong, sharp, very versatile premium abrasive for a wide range of materials & applications.	A tougher/ stronger intermediate abrasive for less heat sensitive hardened and soft steels.	A tough intermediate abrasive for less heat sensitive soft steels and cast irons.	Blocky shaped black silicon carbide abrasive, for all non-ferrous metals, cast irons, 300 series stainless steels, ceramics, plastics.	Sharp, friable, high purity, green silicon carbide for carbides, titanium, plasma sprayed materials.

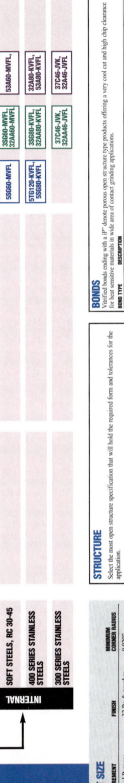

INTERNAL

SOFT STEELS, RC 30-45	
400 SERIES STAINLESS STEELS	
300 SERIES STAINLESS STEELS	

⑥ GRIT SIZE

	REQUIREMENT	FINISH
46	General Purpose	32 Ra & rougher
60	Commercial Finish	32 Ra & better
80	Fine Finish	20 Ra & better
120	Very Fine Finish	10 Ra & better
150	Corner-Form Holding	
180	Corner-Form Holding	
220	Corner-Form Holding	

MINIMUM CORNER RADIUS
0.020"
0.016"
0.0105"
0.006"
0.005"
0.0035"
0.0026"

RELATIVE ABRASIVE PERFORMANCE

700%	
600%	600%
500%	500%
400%	375%
300%	250%
200%	125% 100%
100%	
0%	3TGK 5SG 3SGP 32AA 32A 38A

© Norton Company 1999

GRINDING TROUBLESHOOTING

Check the obvious first. Before changing the grinding wheel specification, investigate the following most common causes for most grinding problems:

1. Diamond dressing tool condition (check if worn or dull, rotate tool or replace if necessary)
2. Coolant direction, volume & filtration
3. Wheel dressing procedures (dress more open to free up cut rate, dress more closed to improve finish)

PROBLEM	POSSIBLE CAUSE	CORRECTION
1. Burn	Poor coolant direction	Redirect coolant into grinding zone
	Restricted or low coolant volume	Increase coolant volume
	Too heavy cut rate	Reduce cut rate
	Wheel too hard	Use one grade softer wheel
	Wheel structure too closed	Use more open structure wheel
2. Loading & Glazing	Wheel too hard	Use one grade softer wheel
	Wheel structure too closed	Use a more open structure wheel
	Too graded abrasive	Use a sharper more friable abrasive
3. Chatter	Unsupported work	Increase work support
	Machine vibration	Check for worn bearings
	Too heavy cut rate	Reduce cut rate
	Wheel too hard	Use one grade softer wheel
	Wheel structure too closed	Use a more open structure wheel
	Wheel out of balance	Check wheel balance or try new wheel
4. Poor surface finish	Dirty coolant	Check coolant filter and quality
	Incorrect wheel dress	Dress wheel finer (slow down dressing tool traverse)
	Too coarse grit size	Use a finer grit size
5. Not holding form	Wheel too soft	Use one grade harder wheel
	Wheel structure too open	Use a more closed wheel structure
6. Not holding corner	Incorrect wheel dress	Dress wheel finer. Face and side true wheel
	Too large grit size	Use smaller grit size (maximum grit diameter less than 1.5 times corner radius)
	Wheel too soft	Use harder grade wheel
	Wheel structure too open	Use more closed structure wheel

TECH TIPS

GRINDING

1. Consider one grade harder starting spec for surface grinding applications with interrupted cut.
2. Use a grit size with grit diameter less than the corner radius required.
3. True the wheel face and sides to eliminate any wheel runout for the tightest corner holding control.
4. For ID grinding, recommend using a wheel diameter (after truing) no larger than 75% of the bore diameter.
5. Increase stock removal rate to minimize burn and chatter with too hard of a wheel.
6. Decrease stock removal rate to reduce wheel breakdown for too soft of a wheel.
7. Use Norton SG® diamond dressing tools for TG and SG wheels for the most consistent performance and maximum tool life.

DRESSING TOOLS

Single Point Diamond:

1. Infeed/pass should not exceed .0015" for aluminum oxide abrasives, .001" with Norton SG.
2. Dress traverse rate 10"-20" per minute for rough grinding & slower for finish grind.
3. Use a 10°-15° drag angle to the wheel centerline.
4. Rotate the diamond often to extend tool life.
5. Use coolant when possible to extend diamond life.

Multi-Point Diamond Nibs:

1. Infeed/pass less than .002" for aluminum oxide abrasives, .0015" with Norton SG.
2. Dress traverse rate 20"-40" per minute for rough grinding & slower for finish.
3. Use at 90° to wheel face.
4. With new tool, run 3-5 passes at .005" per pass to expose diamonds and to ensure full face contact between dressing tool and wheel face.
5. Use coolant when possible to extend dressing tool life.

NORTON
® Leading Technology, Leading Solutions™

Form #7505

Contact your Norton distributor or Norton sales representative for assistance, or check out the Norton Web Site: **www.nortonabrasives.com**

Credits

Chapter 7

7.1: © McGraw-Hill Higher Education, Inc./Photography by Prographics at Milwaukee Area Technical College; **7.3:** © McGraw-Hill Higher Education, Inc./Lake Washington Technical College, Kirkland, WA; **7.5:** © McGraw-Hill Higher Education, Inc./Photography by Prographics at Milwaukee Area Technical College; **7.9:** © McGraw-Hill Higher Education, Inc./Lake Washington Technical College, Kirkland, WA; **7.10, 7.11, 7.12, 7.16, 7.17, 7.18, 7.21, 7.23, 7.25, 7.26:** © McGraw-Hill Higher Education, Inc./Lake Washington Technical College, Kirkland, WA; **7.28, 7.30:** © McGraw-Hill Higher Education, Inc./Photography by Prographics at Milwaukee Area Technical College; **7.31, 7.34:** © McGraw-Hill Higher Education, Inc./Lake Washington Technical College, Kirkland, WA; **7.38:** Courtesy Mike Fitzpatrick; **7.39:** © McGraw-Hill Higher Education, Inc./Photography by Prographics at Milwaukee Area Technical College; **7.54:** Aerospace Manufacturing Technologies (AMT)/Arlington, WA; **7.55:** Courtesy Ambios Technology, Inc.; **7.56:** Courtesy Mike Fitzpatrick.

Chapter 8

8.3: Courtesy Mike Fitzpatrick; **8.4, 8.5:** © McGraw-Hill Higher Education, Inc./Lake Washington Technical College, Kirkland, WA; **8.8:** © McGraw-Hill Higher Education, Inc./Photography by Prographics at Milwaukee Area Technical College; **8.11, 8.13, 8.17, 8.20, 8.22, 8.23:** Courtesy Mike Fitzpatrick.

Chapter 9

9.3: Courtesy Mike Fitzpatrick; **9.12:** © McGraw-Hill Higher Education, Inc./Lake Washington Technical College, Kirkland, WA; **9.14, 9.15, 9.16:** Courtesy Mike Fitzpatrick; **9.19:** From "Modern Metal Cutting" 1996, © Sandvik Coromant; **9.20:** Courtesy Mike Fitzpatrick; **9.26:** © McGraw-Hill Higher Education, Inc./Lake Washington Technical College, Kirkland, WA.

Chapter 10

10.1: Courtesy L. S. Starrett Co.; **10.2:** © McGraw-Hill Higher Education, Inc./Photography by Prographics at Milwaukee Area Technical College; **10.3, 10.5:** Courtesy Mike Fitzpatrick; **10.7:** © McGraw-Hill Higher Education, Inc./Lake Washington Technical College, Kirkland, WA; **10.8:** Courtesy Mike Fitzpatrick; **10.10:** © McGraw-Hill Higher Education, Inc./Photography by Prographics at Milwaukee Area Technical College; **10.15, 10.16, 10.17, 10.18:** Courtesy Mike Fitzpatrick; **10.20:** © McGraw-Hill Higher Education, Inc./Photography by Prographics at Milwaukee Area Technical College; **10.23:** Courtesy Morgan Branch CNC/Arlington, WA; **10.24, 10.25, 10.26, 10.28, 10.29 (both):** Courtesy Mike Fitzpatrick; **10.30:** © McGraw-Hill Higher Education, Inc./Lake Washington Technical College, Kirkland, WA; **10.33, 10.34, 10.38:** Courtesy Mike Fitzpatrick; **10.42:** © McGraw-Hill Higher Education, Inc./Lake Washington Technical College, Kirkland, WA; **10.44:** Courtesy Bridgeport Machines, Inc./Bridgeport, CT; **10.57:** Courtesy Mike Fitzpatrick; **10.58:** © McGraw-Hill Higher Education, Inc./Photography by Prographics at Milwaukee Area Technical College; **10.62:** © McGraw-Hill Higher Education, Inc./Lake Washington Technical College, Kirkland, WA; **10.64, 10.66:** Courtesy Mike Fitzpatrick; **10.72:** Courtesy Smiths Aerospace-Actuation Systems/Yakima, WA; **10.75, 10.76:** © McGraw-Hill Higher Education, Inc./Lake Washington Technical College, Kirkland, WA; **10.77, 10.78, 10.91:** Courtesy Mike Fitzpatrick.

Chapter 11

11.10: Courtesy Mike Fitzpatrick; **11.23:** Courtesy Landis Threading Systems; **11.30, 11.31, 11.32:** © McGraw-Hill Higher Education, Inc./Lake Washington Technical College, Kirkland, WA; **11.37, 11.38, 11.40, 11.41, 11.42, 11.43, 11.44, 11.50, 11.56, 11.57:** © McGraw-Hill Higher Education, Inc./Photography by Prographics at Milwaukee Area Technical College; **11.38:** © McGraw-Hill Higher Education, Inc./Photography by Prographics at Milwaukee Area Technical College; **11.62, 11.63:** © McGraw-Hill Higher Education, Inc./Lake Washington Technical College, Kirkland, WA; **11.66, 11.69:** © McGraw-Hill Higher Education, Inc./Photography by Prographics at Milwaukee Area Technical College; **11.78:** © McGraw-Hill Higher Education, Inc./Lake Washington Technical College, Kirkland, WA; **11.85:** Courtesy Tyee Aircraft/Everett, WA; **11.91, 11.92, 11.94, 11.108:** © McGraw-Hill Higher Education, Inc./Lake Washington Technical College, Kirkland, WA;

11.110: Courtesy Okuma America Corp.; **11.111:** © McGraw-Hill Higher Education, Inc./Photography by Prographics at Milwaukee Area Technical College; **11.112:** Courtesy Morgan Branch CNC/Arlington, WA; **11.114:** From "Modern Metal Cutting, 1996, © Sandvik Coromant; **11.122:** © McGraw-Hill Higher Education, Inc./Photography by Prographics at Milwaukee Area Technical College; **11.125:** Courtesy Universal Aerospace/Arlington, WA; **11.127:** Alloris Tool Technology Co., Inc./Clifton, NJ; **11.133:** © McGraw-Hill Higher Education, Inc./Lake Washington Technical College, Kirkland, WA; **11.135, 11.136, 11.137:** © McGraw-Hill Higher Education, Inc./Photography by Prographics at Milwaukee Area Technical College; **11.138:** © McGraw-Hill Higher Education, Inc./Lake Washington Technical College, Kirkland, WA; **11.139:** Courtesy Kennametal Corporation; **11.140, 11.141, 11.143, 11.157:** © McGraw-Hill Higher Education, Inc./Lake Washington Technical College, Kirkland, WA; **11.164, 11.165:** Courtesy Mike Fitzpatrick.

Chapter 12

12.2: Courtesy Iscar Metals, Inc.; **12.4:** Courtesy Kennametal, Inc.; **12.7, 12.15:** © McGraw-Hill Higher Education, Inc./Lake Washington Technical College, Kirkland, WA; **12.22:** Courtesy Aloris USA; **12.23:** Courtesy Bridgeport Machines, Inc./ Bridgeport, CT; **12.28:** Courtesy Tyee Aircraft, Inc./Paine Field, WA; **12.31:** © McGraw-Hill Higher Education, Inc./Photography by Prographics at Milwaukee Area Technical College; **12.42, 12.43:** Courtesy Cincinnati Machine; **12.44:** Courtesy Okuma America Corp.; **12.54:** Courtesy Bridgeport Machines, Inc.; **12.56, 12.57:** © McGraw-Hill Higher Education, Inc./Photography by Prographics at Milwaukee Area Technical College; **12.58:** Courtesy Bridgeport Machines, Inc.; **12.59:** © McGraw-Hill Higher Education, Inc./Photography by Prographics at Milwaukee Area Technical College; **12.60:** Courtesy Bridgeport Machines, Inc.; **12.61, 12.62, 12.62a:** © McGraw-Hill Higher Education, Inc./Lake Washington Technical College, Kirkland, WA; **12.64:** Courtesy Tyee Aircraft/Everett, WA; **12.65:** © McGraw-Hill Higher Education, Inc./Photography by Prographics at Milwaukee Area Technical College; **12.74:** Courtesy Iscar Metals; **12.75, 12.80:** © McGraw-Hill Higher Education, Inc./Photography by Prographics at Milwaukee Area Technical College; **12.81:** © McGraw-Hill Higher Education, Inc./Lake Washington Technical College, Kirkland, WA; **12.82:** Courtesy Kurt Manufacturing Company; **12.86:** © McGraw-Hill Higher Education, Inc./Photography by Prographics at Milwaukee Area Technical College; **12.87:** © McGraw-Hill Higher Education, Inc./Lake Washington Technical College, Kirkland, WA; **12.96:** Courtesy Cincinnati Machine; **12.97:** Calculator by Iscar Metals (Blue) Kennametal, Inc. (Yellow); **12.101:** Courtesy Kurt Manufacturing/Industrial Products Division; **12.105, 12.106, 12.111:** © McGraw-Hill Higher Education, Inc./Lake Washington Technical College, Kirkland, WA; **12.125, 12.126, 12.127, 12.128:** Courtesy Kurt Manufacturing/Industrial Products Division; **12.133:** Courtesy Bridgeport Machines, Inc./Bridgeport, CT.

Chapter 13

13.1, 13.2: Courtesy Chevalier Machinery, Inc./Santa Fe Springs, CA; **13.3:** Courtesy Smiths Aerospace-Actuation Systems/Yakima, WA; **13.4, 13.6, 13.7:** Courtesy Chevalier Machinery, Inc.; **13.13:** Courtesy of Norton Company/Worcester, MA; **13.14, 13.18:** Courtesy Chevalier Machinery, Inc.; **13.21:** © McGraw-Hill Higher Education, Inc./Photography by Prographics at Milwaukee Area Technical College; **13.22, 13.23, 13.25:** Courtesy Chevalier Machinery, Inc.; **13.28:** Courtesy Landis Gardner/Waynesboro, PA; **13.29:** Courtesy Chevalier Machinery, Inc.; **13.30:** © McGraw-Hill Higher Education, Inc./Photography by Prographics at Milwaukee Area Technical College; **13.31:** Courtesy Chevalier Machinery, Inc.; **13.32, 13.34:** © McGraw-Hill Higher Education, Inc./Photography by Prographics at Milwaukee Area Technical College; **13.38, 13.41, 13.42, 13.43, 13.57:** Courtesy Chevalier Machinery, Inc.; **13.60:** © McGraw-Hill Higher Education, Inc./Photography by Prographics at Milwaukee Area Technical College; **13.63:** Courtesy Norton Company; **13.64:** Courtesy Okuma America; **13.67:** Courtesy Landis Gardner Company/Waynesboro, PA; **13.70:** Courtesy Maximum Advantage-Carolinas, Precision Grinding Machinery/Tega Cay, SC; **13.72: (Both)** Courtesy RSS Grinders & Automation, Inc.; **13.75:** Courtesy Get An Edge-Tool Grinding Service/Mukilteo, WA; **13.76:** Courtesy Get An Edge-Tool Grinding Service/Mukilteo, WA; **13.78:** Courtesy Smiths Aerospace-Actuation Systems/Yakima, WA; **13.79:** Courtesy

Mike Fitzpatrick; **13.81:** Courtesy Blanchard Division of DeVlieg Bullard II, Inc.

Chapter 14

14.1: Courtesy Universal Aerospace/Arlington, WA; **14.12:** © McGraw-Hill Higher Education, Inc./Photography by Prographics at Milwaukee Area Technical College; **14.13:** Courtesy RIGID Tools; **14.15, 14.16:** Courtesy Emhart Fastening Technologies; **14.30:** Courtesy Landis Threading Systems.

Chapter 15

15.1: Courtesy Mike Fitzpatrick; **15.2:** Courtesy Procedyne Corp.; **15.3, 15.7:** © McGraw-Hill Higher Education, Inc./Photography by Prographics at Milwaukee Area Technical College; **15.11, 15.12:** Courtesy Mike Fitzpatrick; **15.13:** Courtesy Engineered Production Systems/Div. of Engineered Product Sales Corp.; **15.14:** Courtesy Spectrodyne, Inc.; **15.15, 15.16, 15.17:** Courtesy Mike Fitzpatrick; **15.18:** © McGraw-Hill Higher Education, Inc./Photography by Prographics at Milwaukee Area Technical College; **15.19:** Courtesy Fluid-Therm Corp.; **15.20:** Courtesy George VanderVoort, Buehler, Ltd.; **15.21, 15.23:** Courtesy Mike Fitzpatrick; **15.24:** © McGraw-Hill Higher Education, Inc./Lake Washington Technical College, Kirkland, WA; **15.25:** Courtesy SECO/WARWICK Corp.; **15.26:** Courtesy Mike Fitzpatrick; **15.27:** Courtesy Tyee Aircraft/Paine Field, WA; **15.31:** Courtesy Mike Fitzpatrick; **PIII.1:** Courtesy Contour Aerospace/Everett, WA; **PIII.2:** © McGraw-Hill Higher Education, Inc./Lake Washington Technical College, Kirkland, WA.

Chapter 17

17.6: Courtesy Mike Fitzpatrick; **17.8:** Courtesy Tyee Aircraft/Everett, WA; **17.10:** Courtesy Cincinnati Machinery/Cincinnati, OH; **17.14:** Courtesy Smiths Aerospace-Actuation Systems, Yakima, WA; **17.30:** Left:Aerospace Manufacturing Technologies/AMT/Arlington, WA/Right:Contour Aerostructures, Corp/Everett, WA; **17.47:** Courtesy Bridgeport Machines/Bridgeport, CT.

Chapter 18

18.2: Courtesy Hass Automation, Inc.; **18.4:** Courtesy Wedin International, Inc.; **18.6:** Courtesy MIT Museum; **18.7:** Courtesy HAAS Automation, Inc./Oxnard, CA; **18.8:** Courtesy Mike's Machines/Arlington, WA; **18.9:** Courtesy Tyee Aircraft/Everett, WA; **18.10:** Courtesy Bridgeport Machines/Bridgeport, CT; **18.12:** Courtesy Cincinnati Machinery; **18.13:** Courtesy Smiths Aerospace-Actuation Systems/Yakima, WA; **18.14:** Courtesy Hass Automation; **18.15:** © McGraw-Hill Higher Education, Inc./Photography by Prographics at Milwaukee Area Technical College; **18.16, 18.17:** Courtesy Chip Blaster, Inc.; **18.18:** Courtesy Contour Aerostructures/Everett, WA; **18.19:** Courtesy Delta Tau Data Systems; **18.20, 18.21:** Courtesy Okuma America Corp.; **18.22:** Courtesy Obuljen & Stäubli Robotics; **18.23:** Courtesy JFA, Inc./Arlington, WA; **18.24:** Courtesy Aerospace Manufacturing Technologies (AMT), Arlington, WA; **18.25:** Courtesy Hass Automation, Inc.; **18.26:** Courtesy Tyee Aircraft/Everett, WA; **18.27:** Courtesy Hass Automation, Inc.; **18.28, 18.29:** Courtesy Bridgeport Machines/Bridgeport, CT; **18.30:** Courtesy Tyee Aircraft/Paine Field, WA; **18.31:** Courtesy Tyee Aircraft/Everett, WA; **18.32:** Courtesy Okuma America Corp.; **18.33:** Courtesy Mike Fitzpatrick; **18.34:** Courtesy Carboloy, Inc./Mesa, AZ; **18.35:** Courtesy Tyee Aircraft/ Everett, WA.

Chapter 19

19.2a: Courtesy Bridgeport Machines/Bridgeport, CT; **19.2b:** Courtesy Chavalier Machinery, Inc.; **19.2c:** Courtesy Haas Automation; **19.3:** Courtesy Mike Fitzpatrick; **19.4:** Courtesy Okuma America, Inc.; **19.5:** © McGraw-Hill Higher Education, Inc./Photography by Prographics at Milwaukee Area Technical College; **19.6:** Courtesy JFA Corporation; **19.7:** Courtesy Tyee Aircraft/Everett, WA; **19.15:** Courtesy Contour Aerostructures Corp./Everett, WA; **19.17, 19.18:** © McGraw-Hill Higher Education, Inc./Lake Washington Technical College, Kirkland, WA.

Chapter 20

20.9: © McGraw-Hill Higher Education, Inc./Photography by Prographics at Milwaukee Area Technical College; **20.16:** Courtesy Universal Machining/Arlington, WA; **20.19, 20.20:** © McGraw-Hill Higher Education, Inc./Photography by Prographics at Milwaukee Area Technical College; **20.22:** Courtesy Lake Washington Technical College; **20.23:** Courtesy Mike Fitzpatrick; **20.24:** Courtesy Tyee Aircraft/Everett, WA; **20.25:** Courtesy Tyee Aircraft/Paine Field, WA.

Chapter 21

21.9: © McGraw-Hill Higher Education, Inc./Lake Washington Technical College, Kirkland, WA; **21.13:** Courtesy Universal Aerospace/Arlington, WA; **21.15:** Courtesy Morgan Branch CNC/Arlington, WA; **21.17:** Courtesy Cincinnati Machine Company/Cincinnati, OH; **21.19:** Courtesy Contour Industries/Everett, WA; **21.20, 21.21:** Courtesy Tyee Aircraft/Everett, WA.

Chapter 22

22.1: Courtesy ANILAM/Miramar, FL; **22.21:** Courtesy Northwood Designs, MetaCut View® Software @ HTTP://www.MetaCutView.

Chapter 24

24.1: © McGraw-Hill Higher Education, Inc./Lake Washington Technical College, Kirkland, WA; **24.8:** Courtesy Mike Fitzpatrick; **24.9, 24.12:** Courtesy Haas Automation; **24.22, 24.23:** CNC Software/Mastercam Software; **24.24:** Courtesy Northwood Designs/Metacut Utilities®, Toolpath Evaluation Software; **24.25:** Courtesy Universal Aerospace/Arlington, WA; **24.26:** © McGraw-Hill Higher Education, Inc./Photography by Prographics at Milwaukee Area Technical College; **24.27:** Courtesy Tyee Aircraft/Everett, WA.

Chapter 26

26.1, 26.2, 26.3, 26.4, 26.5, 26.6: Courtesy Flow International, Corp.; **26.7:** Both courtesy Synrad of Mukilteo Washington State; **26.8:** Courtesy Trumpf Incorporated/Farmington, CT; **26.9:** Courtesy Synrad Corporation, Washington State; **26.10:** Courtesy Trumpf Incorporated/Farmington, CT; **26.15, 26.16:** Courtesy Mitsubishi Division/MC Machinery; **26.19:** Courtesy Jones Plastic and Mitsubishi Division/MC Machinery; **26.21:** Courtesy Thomas D. Neary, Inc.; **26.22:** Courtesy Iscar Metal, Inc.; **26.23, 26.26, 26.27, 26.29, 26.30:** Courtesy Kennametal, Inc.; **26.35, 26.36, 26.37, 26.38, 26.39, 26.40, 26.41:** Courtesy Optomec.

Chapter 27

27.6: Courtesy Tyee Aircraft/Paine Field, WA.

Chapter 28

28.1: Courtesy Haas Automation; **28.4:** Courtesy Renishaw, Inc.; **28.8:** Courtesy J.D. Ott Aerospace Machining/Seattle, WA; **28.11, 28.12:** Courtesy Brown and Sharpe Corporation; **28.13:** Courtesy J.D. Ott Aerospace Machining/Seattle, WA; **28.14:** Courtesy Brown and Sharpe Corporation; **28.15:** Courtesy Renishaw, Inc.; **28.17:** Courtesy Brown and Sharpe Corporation; **28.18:** Courtesy J.D. Ott Aerospace Machining/Seattle, WA; **28.25:** Courtesy Brown and Sharpe Corporation.

Index